HANDBOOK OF LIQUID CRYSTAL RESEARCH

Handbook of Liquid Crystal Research

PETER J. COLLINGS and JAY S. PATEL

Editors in Chief

OXFORD UNIVERSITY PRESS

New York Oxford

1997

To Diane and Susan

OXFORD UNIVERSITY PRESS

Oxford New York
Athens Auckland Bangkok Bogotá
Bombay Buenos Aires Calcutta Cape Town
Dar es Salaam Delhi Florence Hong Kong Istanbul
Karachi Kuala Lumpur Madras Madrid Melbourne
Mexico City Nairobi Paris Singapore
Taipei Tokyo Toronto

and associated companies in
Berlin Ibadan

Published by Oxford University Press, Inc.,
198 Madison Avenue, New York, New York 10016

Library of Congress Cataloging-in-Publication Data

Handbook of Liquid Crystal Research / Peter J. Collings and
Jay S. Patel, editors in chief.
p. cm.
Includes bibliographical references and index.
ISBN 0-19-508442-X
(hardcover : alk. paper)
1. Liquid Crystals. I. Collings, Peter J., 1947- .
II. Patel, Jay S.
QD923.H36 1997
530.4'29—dc20 96-31118 CIP

Printing (last digit): 9 8 7 6 5 4 3 2 1

Printed in the United States of America
on acid-free paper

CONTENTS

CONTRIBUTORS

R. Barberi
University of Reggio Calabria

L. M. Blinov
Institute of Crystallography, Moscow

Peter J. Collings
Swarthmore College

Gregory P. Crawford
Xerox PARC, Palo Alto

J. William Doane
Liquid Crystal Institute, Kent, Ohio

G. Durand
University of Paris—South, Orsay

J. W. Goodby
The University of Hull

A. W. Hall
The University of Hull

J. Hollingshurst
The University of Hull

Hiroo Hori
Toshiba Display Laboratory, Kawasaki, Japan

Shunsuke Kobayashi
Tokyo University of Agriculture and Technology

Michael R. Kuzma
Philadelphia, Pennsylvania

Y. Owechko
Hughes Research Laboratories, Malibu, California

Jay S. Patel
Pennsylvania State University

Robert A. Pelcovits
Brown University

Virgil Percec
Case Western Reserve University

E. Santamato
University of Naples

Alfred Saupe
Max Planck Working Group, Halle, Germany

Terry Scheffer
In Focus Systems Incorporated, Wilsonville, Oregon

Y. R. Shen
University of California, Berkeley

Yasuharu Tanaka
Toshiba Display Laboratory, Kawasaki, Japan

H. Yokoyama
Electrotechnical Laboratory, Ibaraki, Japan

Slobodan Žumer
University of Ljubljana, Slovenia

PREFACE

The last twenty-five years has seen enormous advances in both our understanding of liquid crystals and our ability to use them in applications. Early on during this period, it was possible for a person working in the field to be aware of most of the results, both pure and applied. Midway through this period, it was still feasible to compile a bibliography of all the scientific work done on liquid crystals. The situation is vastly different now: the sheer volume of what has been published in the field of liquid crystals makes any attempt to assemble an exhaustive summary impractical.

Yet it is extremely important that scientists and engineers working in one area of the liquid crystal field have ready access to what is known and what has been accomplished in other areas of the field. Likewise, people either considering working with liquid crystals or just starting to work with liquid crystals must be able to learn about the most important developments of this twenty-five-year period. For these two reasons, we have asked twenty-one scientists with extensive experience in different areas of liquid crystal research to write chapters detailing the most important and valuable advances in their areas. Since the progress made during this period has been both scientific and technological, we have collected an equal number of chapters emphasizing each.

The authors have not attempted to make their chapters truly comprehensive; instead, they have concentrated on those results that have been most important in either advancing our knowledge and opening up possibilities for applications. They have also emphasized work done in the last fifteen years and developments that have helped to both simplify and elucidate the phenomena. The result is a volume encompassing a huge amount of material with references to thousands of publications. Rather than being a summary, however, it brings together those developments that have had the most lasting impact on the field. It is therefore a critical progress report of the state of liquid crystal research as we enter the twenty-first century.

Peter J. Collings
Jay S. Patel
April 1996

ABBREVIATIONS OF JOURNALS

Acta Crystallogra. = *Acta Crystallographica*

Acta Phys. Pol. A = *Acta Physica Polonica, Series A*

Acta Polym. = *Acta Polymerica*

Adv. Colloid Interface Sci. = *Advances in Colloid and Interface Science*

Adv. Image Pickup Displ. = *Advances in Image Pickup and Display*

Adv. Mater. = *Advanced Materials*

Adv. Mater. Opt. Electr. = *Advanced Materials for Optics and Electronics*

Adv. Phys. = *Advances in Physics*

Adv. Polym. Sci. = *Advances in Polymer Science*

Am. Chem. Soc. Div. Polym. Chem., Polym. Prepr. = *American Chemical Society, Division of Polymer Chemistry: Polymer Preprints*

Am. Lab. = *American Laboratory*

Angew. Chem. Int. Ed. Eng. = *Angewandte Chemie, International Edition in English*

Ann. Inst. henri Poincaré = *Annales de l'Institut Henri Poincaré*

Ann. NY Acad. Sci. = *Annals of the New York Academy of Science*

Ann. Physik = *Annalen der Physik*

Ann. Physique = *Annales de Physique*

Ann. Rev. Phys. Chem. = *Annual Review of Physical Chemistry*

Appl. Opt. = *Applied Optics*

Appl. Phys. = *Applied Physics*

Appl. Phys. Lett. = *Applied Physics Letters*

Appl. Spec. = *Applied Spectroscopy*

Arch. Néerl. Sci. Exactes Natur. 2 = *Archives Néerlandaises des Sciences Exactes et Naturelles, Série 2*

Arch. Ration. Mech. Anal. = *Archive for Rational Mechanics and Analysis*

Archiv. Mikroskopische Anat. = *Archiv für Mikroskopische Anatomie*

Australian J. Chem. = *Australian Journal of Chemistry*

Ber. Bunsen-Ges. Phys. Chem. = *Berichte der Bunsen-Gesellschaft für Physikalische Chemie*

Br. Polym. J. = *British Polymer Journal*

Bull. Chem. Soc. Japan = *Bulletin of the Chemical Society of Japan*

Bull. Isr. Phys. Soc. = *Bulletin of the Israel Physical Society*

Bull. Soc. Chim. Fr. = *Bulletin de la Société Chimique de France*

Bull. Soc. Fr. Minéral. Cristallogr. = *Bulletin de la Société Française de Minéralogie et de Cristallographie*

C. R. Acad. Sci. Paris = *Comptes Rendus de l'Académie des Sciences, Paris*

Carbon = *Carbon*

Chem. Lett. = *Chemistry Letters*

Chem. Mater. = *Chemistry of Materials*

Chem. Phys. Lett. = *Chemical Physics Letters*

Chem. Rev. = *Chemical Reviews*

Chem.-Ztg. = *Chemiker-Zeitung*

Colloid Polym. Sci. = *Colloid and Polymer Science*

Colloids Surf. = *Colloids and Surfaces*

Color Res. Applic. = *Color Research and Application*

Commun. Pure Appl. Math. = *Communications on Pure and Applied Mathematics*

Complex Syst. = *Complex Systems*

Condens. Matt. News = *Condensed Matter News*

Coors. Chem. Rev. = *Coordination Chemistry Reviews*

Cryst. Res. Technol. = *Crystal Research and Technology*

Discuss. Faraday Soc. = *Discussions of the Faraday Society*

Displ. Imaging = *Display and Imaging*

Displays = *Displays: Technology and Applications*

Dokl. Akad. Nauk SSSR = *Doklady Akademii Nauk SSSR*

Electron. Lett. = *Electronics Letters*

Elektron. Tekh. = *Elektronnaya Tekhnika*

Eur. Polym. J. = *European Polymer Journal*

Europhys. Lett. = *Europhysics Letters*

Experientia = *Experientia*

Ferroelectrics = *Ferroelectrics*

Fiz. Tverd. Tela = *Fizika Tverdogo Tela*

Fortschr. Phys. = *Fortschritte der Physik (Berlin)*

Gazz. Chim. Ital. = *Gazzetta Chimica Italiana*

Helv. Chim. Acta = *Helvetica Chimica Acta*

Helv. Phys. Acta Suppl. = *Helvetica Physica Acta Supplementum*

High Perform. Polym. = *High Performance Polymers*

IEEE ASSP Mag. = *IEEE ASSP Magazine*

IEEE Commun. Mag. = *IEEE Communications Magazine*

IEEE J. Quantum Electron. = *IEEE Journal of Quantum Electronics*

IEEE Photon. Technol. Lett. = *IEEE Photonics Technology Letters*

IEEE Sig. Proc. Mag. = *IEEE Signal Processing Magazine*

IEEE Trans. Elec. Dev. = *IEEE Transactions on Electron Devices*

IEEE Trans. Inform. Theory = *IEEE Transactions on Information Theory*

Int. J. Nonlin. Opt. Phys. = *International Journal of Nonlinear Optical Physics*

Int. J. Optical Comput. = *International Journal of Optical Computing*

Int. J. Optoelectron. = *International Journal of Optoelectronics*

Itogi Nauki i Tekh. Fiz. Khim. = *Itogi Nauki i Tekhniki: Fizicheskaya Khimiya*

J. Am. Chem. Soc. = *Journal of the American Chemical Society*

J. Appl. Phys. = *Journal of Applied Physics*

J. Appl. Polym. Sci. = *Journal of Applied Polymer Science*

J. Appl. Polym. Sci., Appl. Polym. Symp. = *Jounal of Applied Polymer Science: Applied Polymer Symposium*

J. Chem. Phys. = *Journal of Chemical Physics*

J. Chem. Soc. = *Journal of the Cemical Society (London)*

J. Chem. Soc. Chem. Commun. = *Journal of the Chemical Society, Chemical Communications*

J. Chem. Soc. Faraday Trans. I = *Journal of the Chemical Society, Faraday Transactions I*

J. Chem. Soc. Faraday Trans. II = *Journal of the Chemical Society, Faraday Transactions II*

J. Chem. Soc. Perkin Trans. I = *Journal of the Chemical Society, Perkin Transactions I: Organic and Bio-Organic Chemistry*

J. Chem. Soc. Perkin Trans. II = *Journal of the Chemical Society, Perkin Transactions II: Physical Organic Chemistry*

J. Chim. Phys. = *Journal de Chimie Physique*

J. Coatings Technol. = *Journal of Coatings Technology*

J. Colloid Interface Sci. = *Journal of Colloid and Interface Science*

J. Elec. Imaging = *Journal of Electronic Imaging*

J. Inorg. Organometal. Polym. = *Journal of Inorganic and Organometallic Polymers*

J. Lightwave Technol. = *Journal of Lightwave Technology*

J. Macromol. Sci. A = *Journal of Macromolecular Science A: Chemistry*

J. Macromol. Sci. Pure Appl. Chem. A = *Journal of Macromolecular Science: Pure and Applied Chemistry A*

J. Mater. Chem. = *Journal of Materials Chemistry*

J. Math. Phys. = *Journal of Mathematical Physics*

J. Mod. Opt. = *Journal of Modern Optics*

J. Mol. Spectrosc. = *Journal of Molecular Spectroscopy*

J. Opt. Soc. Am. = *Journal of the Optical Society of America*

J. Opt. Soc. Am. A = *Journal of the Optical Society of America A: Optics, Image Science, and Vision*

J. Opt. Soc. Am. B = *Journal of the Optical Society of America B: Optical Physics*

J. Org. Chem. = *Journal of Organic Chemistry*

J. Phys. A: Math. Gen. = *Journal of Physics A: Mathematical and General*

J. Phys. Chem. = *Journal of Physical Chemistry*

J. Phys. Condens. Matt. = *Journal of Physics: Condensed Matter*

J. Phys. D: Appl. Phys. = *Journal of Physics D: Applied Physics*

J. Phys. Soc. Japan = *Journal of the Physical Society of Japan*

J. Physique = *Journal de Physique I: Physique général, physique statistique, matière condensée, domaines interdisciplinaires*

J. Physique II = *Journal de Physique II: Physique atomique et moléculaire, physico-chimie, mécanique et hydrodynamique*

J. Physique Coll. = *Journal de Physique, Colloque*

J. Physique Lett. = *Journal de Physique Lettres*

J. Polym. Sci. = *Journal of Polymer Science*

J. Polym. Sci. Polym. Chem. Ed. = *Journal of Polymer Science: Polymer Chemistry Edition*

J. Polym. Sci. Polym. Lett. Ed. = *Journal of Polymer Science: Polymer Letters Edition*

J. Polym. Sci. Polym. Phys. Ed. = *Journal of Polymer Science: Polymer Physics Edition*

J. Prakt. Chem. = *Journal für Praktische Chemie*

J. Reproduct. Med. = *Journal of Reproductive Medicine*

J. SID = *Journal of the Society for Information Display*

J. Stat. Phys. = *Journal of Statistical Physics*

J. Vac. Sci. Technol. B = *Journal of Vacuum Science and Technology B: Microelectronics and Nanometer Structures*

Japan. J. Appl. Phys. = *Japanese Journal of Applied Physics*

Japan. J. Appl. Phys. Lett. = *Japanese Journal of Applied Physics, Part 2, Letters*

JETP Lett. = *JETP Letters*

Kristallografiya = *Kristallografiya*

Langmuir = *Langmuir: The ACS Journal of Surfaces and Colloids*

Liq. Cryst. = *Liquid Crystals*

Macromol. Chem. Phys. = *Macromolecular Chemistry and Physics*

Macromol. Symp. = *Macromolecular Symposia*

Macromolecules = *Macromolecules*

Makromol. Chem. = *Makromolekulare Chemie*

Makromol. Chem. Macromol. Symp. = *Makromolekulare Chemie: Macromolecular Symposia*

Makromol. Chem. Rapid Commun. = *Makromolekulare Chemie: Rapid Communications*

Mater. Res. Bull. = *Materials Research Bulletin*

Math. Comput. Simul. = *Mathematics and Computers in Simulation*

Mikroelektronika = *Mikroelektronika*

Mod. Phys. Lett. B = *Modern Physics Letters B: Condensed Matter Physics, Statistical Physics, Applied Physics*

Mol. Cryst. = *Molecular Crystals*

Mol. Cryst. Liq. Cryst. = *Molecular Crystals and Liquid Crystals*

Mol. Cryst. Liq. Cryst. Lett. = *Molecular Crystals and Liquid Crystals, Letters*

Mol. Mater. = *Molecular Materials*

Mol. Phys. = *Molecular Physics*

Monatssch. Wiener Chem. Ges. = *Monatsschrift der Wiener Chemische Gesellschaft*

MRS Bull. = *MRS Bulletin/Materials Research Society*

MRS Symp. Proc. = *Materials Research Society Symposia Proceedings*

Nature = *Nature*

Naturwissensch. = *Naturwissenschaften*

Neural Networks = *Neural Networks*

New J. Chem. = *New Journal of Chemistry/Nouveau Journal de Chimie*

New Polym. Mater. = *New Polymeric Materials*

New Scientist = *New Scientist*

Nikkei Microdevices = *Nikkei Mikurodibaisu/Nikkei Microdevices*

Nucl. Phys. A = *Nuclear Physics A* (Amsterdam)

Nuovo Cimento = *Il Nuovo Cimento della Società Italiana di Fisica*

Opt. Commun. = *Optics Communications*

Opt. Eng. = *Optical Engineering*

Opt. Lett. = *Optics Letters*

Opt. Photon. News = *Optics and Photonics News*

Opt. Quantum Electron. = *Optical and Quantum Electronics*

Optica Acta = *Optica Acta*

Phase Trans. = *Phase Transitions*

Phil. Mag. = *Philosophical Magazine*

Phil. Mag. Lett. = *Philosophical Magazine Letters*

Phil. Trans. R. Soc. London A = *Philosophical Transactions of the Royal Society of London, Series A*

Philips J. Res. = *Philips Journal of Research*

Phys. Bull. = *Physics Bulletin*

Phys. Fluids = *Physics of Fluids*

Phys. Lett. = *Physics of Letters*

Phys. Rep. = *Physics Reports*

Phys. Rev. = *The Physical Review*

Phys. Rev. A = *Physical Review A: General Physics*

Phys. Rev. B = *Physical Review B: Condensed Matter*

Phys. Rev. E = *Physical Review E: Statistical Physics, Plasmas, Fluids, and Related Interdisciplinary Topics*

Phys. Rev. Lett. = *Physical Review Letters*

Phys. Scr. T = *Physica Scripta (Topical Issue)*

Phys. Status Solidi A = *Physica Status Solidi A: Applied Research*

Phys. Today = *Physics Today*

Phys. Z. = *Physikalische Zeitschrift*

Phys. Z. Sowjun. = *Physikalische Zeitschrift der Sowjetunion*

Physica A = *Physica A* (Amsterdam)

Physica B = *Physica B* (Amsterdam)

Pis'ma Zh. Eksp. Teor. Fiz. = *Pis'ma v Zhurnal Eksperimental'noi i Teoreticheskoi Fiziki*

Pis'ma Zh. Tekh. Fiz. = *Pis'ma v Zhurnal Tekhnicheskoi Fiziki*

Polym. Bull. = *Polymer Bulletin*

Polym. Comm. = *Polymer Communications*

Polym. J. = *Polymer Journal (Tokyo)*

Polymer = *Polymer*

Poverkhnost' = *Poverkhnost': Fizika, Khimiya, Mekhanika*

Prakt. Chem. = *Praktische Chemie*

Pramana = *Pramana: A Journal of Physics*

Pramana Suppl. = *Pramana Suppletment*

Proc. IEEE = *Proceedings of the IEEE* [Institute of Electrical and Electronics Engineers]

Proc. IRE = *Proceedings of the IRE* [Institute of Radio Engineers]

Proc. R. Soc. London A = *Proceedings of the Royal Society of London, Series A: Mathematical and Physical Sciences*

Proc. SID = *Proceedings of the Society for Information Display*

Proc. SPIE = *Proceedings SPIE* [Society for Photo-optical Instrumentation Engineers]

Prog. Colloid Polym. Sci. = *Progress in Colloid and Polymer Science*

Prog. Opt. = *Progress in Optics*

Pure Appl. Chem. = *Pure and Applied Chemistry*

Rep. Prog. Phys. = *Reports on Progress in Physics*

Rep. Res. Lab. Asahi Glass Co. = *Reports of the Research Laboratory, Asahi Glass Co.*

Rev. Mod. Phys. = *Reviews of Modern Physics*

Rev. Sci. Instrum. = *Review of Scientific Instruments*

Russ. Chem. Rev. = *Russian Chemical Reviews*

Sci. Am. = *Scientific American*

Science = *Science*

Sem. Lect. Notes SID = *Society for Informational Display, Seminar Lecture Notes*

SID Dig. Tech. Papers = *SID International Symposium Digest of Technical Papers*

Soc. Photo-Opt. Instru. Eng. = *Society of Photo-Optical Instrumentation Engineers*

Solid State Commun. = *Solid State Communications*

Solid State Phys. Suppl. = *Solid State Physics Supplement*

Sov. J. Quantum Electron. = *Soviet Journal of Quantum Electronics*

Sov. Phys. Dokl. = *Soviet Physics, Doklady*

Sov. Phys.-JETP = *Soviet Physics-JETP*

Sov. Phys. Solid State = *Soviet Physics, Solid State*

Sov. Phys. Usp. = *Soviet Physics, Uspekhi*

Sov. Tech. Phys. Lett. = *Soviet Technical Physics Letters*

Springer Ser. Chem. Phys. = *Springer Series in Chemical Physics*

Surf. Sci. = *Surface Science*

Symp. Faraday Soc. = *Symposia of the Faraday Society*

Synlett = *Synlett: Accounts and Rapid Communications in Synthetic Organic Chemistry*

Synth. Met. = *Synthetic Metals*

Tetrahedron = *Tetrahedron: International Journal of Organic Chemistry*

Tetrahedron Lett. = *Tetrahedron Letters*

Thin Solid Films = *Thin Solid Films*

Top. Curr. Chem. = *Topics in Current Chemistry*

Trans. Faraday Soc. = *Transactions of the Faraday Society*

Ukr. Fiz. Zh. = *Ukrainskyi Fizychnyi Zhurnal*

Usp. Fiz. Nauk = *Uspekhi Fizicheskikh Nauk*

Vysokomol. Soed. B = *Vysokomolekuliarnye Soedineniya, Series B*

Z. Chem. = *Zeitschrift für Chemie*

Z. Naturforsch. = *Zeitschrift für Naturforschung*

Z. Phys. B = *Zeitschrift für Physik B*

Z. Phys. Chem. = *Zeitschrift für Physikalische Chemie (Leipzig)*

Zh. Eksp. Teor. Fiz. = *Zhurnal Eksperimental'noi i Teoreticheskoi Fiziki*

Zh. Russ. Fiz. Khim. Obshch. = *Zhurnal Russkogo Fiziko-Kimicheskogo Obshchestva*

CHAPTER 1

Introduction to the Science and Technology of Liquid Crystals

PETER J. COLLINGS AND
JAY S. PATEL

1.1 LIQUID CRYSTALS AS A PHASE OF MATTER

1.1.1 Positional and Orientational Order

The molecules composing a solid generally possess both positional and orientational order, meaning that the centers of mass of the molecules occupy specific locations and the molecular axes point in certain directions. When the solid melts to a liquid, both the positional and orientational order vanish. In this state there are no preferred locations for the centers of mass or preferred directions for the molecular axes. The molecules in a solid may change their positions and orientations slightly due to thermal motion, but their positions are generally fixed at specific lattice points and the motion is with respect to the perfect geometrical lattice. In the liquid state, the molecules diffuse freely throughout the sample and the centers of mass move in random directions.

In addition to the solid and liquid phases, there are condensed phases that exhibit intermediate order. The simplest case is one in which the molecules are generally fixed at lattice points but in addition to vibration, may freely rotate. This type of material is referred to as a plastic crystal or "rotor phase."

If one axis of a molecule is much longer or shorter than the other two axes, then it is possible for additional phases to form in which there is some positional and orientational order, but much less than is found in solids or plastic crystals. These are the liquid crystal phases in which the molecules diffuse throughout the sample while maintaining some positional and orientational order. The simplest liquid crystal is the nematic phase, in which there is a higher probability that the molecular axes of neighboring molecules point in a certain direction. In other types of liquid crystals, there may be a higher probability that the centers of mass lie in layers, with the molecules moving freely from one lattice site to another. The ability of the molecules to move among the various lattice sites imparts fluidity to these structures, but since all directions within the phase are not identical, they are anisotropic rather than isotropic fluids.

1.1.2 Order Parameters

The amount of positional or orientational order can be described using parameters appropriately called order parameters. For example, the layering tendency of the molecules in smectic liquid crystal phases can be described by an average density of the centers of mass $\rho(z)$, which sinusoidally varies along the normal to the layers (here denoted by the z-axis). The amplitude of the sinusoidal part ψ describes the amount of positional order and can be used as an order parameter:

$$\rho(z) = \rho_0[1 + \psi\cos(2\pi z/d)].$$

Here ρ_0 is the average density, and d is the distance between layers.

Another example of an order parameter is the one used to describe orientational order. In a nematic liquid crystal phase with a preferred direction $\hat{n}$ for the long molecular axes (called the director), an average orientational distribution function $f(\theta)$ exits, where θ is the angle between the long axis of the molecule and the director. This distribution function can be expanded in a series of Legendre polynomials, but retaining only even powers since the director can be defined in either of two opposite directions, $\hat{n}$ or $-\hat{n}$:

$$f(\theta) = S_0P_0(\cos\theta) + S_2P_2(\cos\theta) + S_4P_4(\cos\theta) + S_6P_6(\cos\theta) + \cdots.$$

The Legendre polynomials are denoted by $P_n(\cos\theta)$ and the S_n are the coefficients describing the orientational distribution function. The dominant parameter is S_2 and when properly normalized can be used as an orientational order parameter.

Order parameters decrease as the temperature is increased and become zero at the transition to the phase that lacks the order represented by the order parameter. This vanishing of the order parameter can happen discontinuously, as is the case for S_2, or it can happen continuously, as is sometimes the case for ψ.

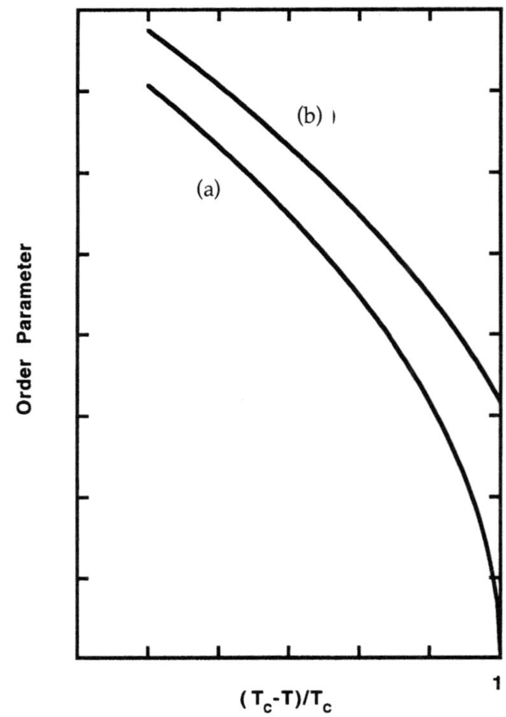

FIGURE 1.1. Temperature dependence of the order parameter at a continuous (a) and discontinuous (b) transition.

Examples of these two temperature dependencies are shown in FIGURE 1.1. Since the order can be altered by changing the temperature, one liquid crystal phase can be transformed to another.

One useful theoretical model for these phase transitions is the phenomenological theory first developed by Landau. The free energy is expressed in those powers of the order parameter that are allowed by the symmetry of the phase. By minimizing the free energy at various temperatures, the behavior of the order parameter at the transition can be calculated. With just second- and fourth-order terms in the free energy, the order parameter goes to zero continuously. If a third-order term is added to the free energy, the order parameter becomes zero at the transition discontinuously. Chapter 3 is directed at recent theoretical progress in understanding both liquid crystal phases and the transitions between them.

1.1.3 Types of Liquid Crystals

1.1.3.1 Calamitic Liquid Crystals

The most common type of liquid crystals, often called calamitic liquid crystals, are composed of rod-like molecules with one molecular axis much longer than the other two. In the nematic liquid crystal phase for rod-like molecules, the long axes of the molecules tend to point along a certain direction as they move from one location to another. The essential difference between this liquid crystal phase and the normal, isotropic liquid phase is shown in FIGURES 1.2 and 1.3, using snapshots of the molecules in time.

Both positional order and orientational order are necessary for a smectic liquid crystal phase. There are a number of such phases, but two important cases are when the director is parallel to the layer normal (smectic A phase) and when the director is at an angle to the layer normal (smectic C phase). These two liquid crystal phases are shown in FIGURE 1.4. In both of these phases, there is no positional order within the plane of the layers, so the centers of mass of the molecules form a two-dimensional liquid.

Other liquid crystal phases exist in which short-range positional order or bond orientational order develops within the plane of the layers. There are many examples, each with a different combination of in-plane order (of various types) and the angle between the layer normal and the director. These are called the smectic B, E, F, G, H, I, J, and K phases, with the letters denoting the chronological order of discovery of these phases. Some of these phases possess three-dimensional positional order, but with the molecules freely rotating about their long axes. These phases therefore should be called soft crystals or plastic crystals rather than liquid crystals. A full discussion of these phases is contained in Chapter 4.

FIGURE 1.2. Isotropic liquid.

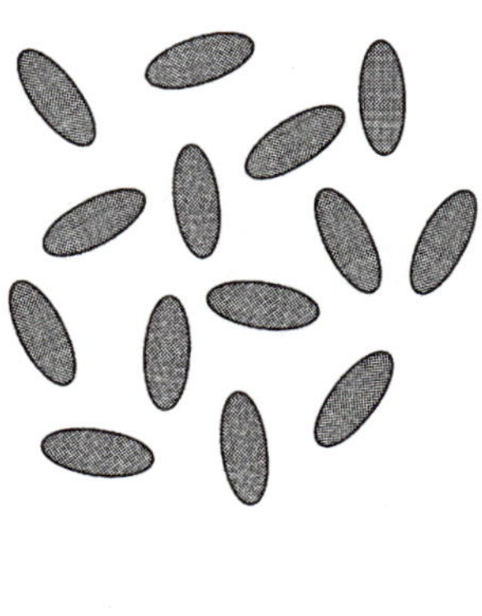

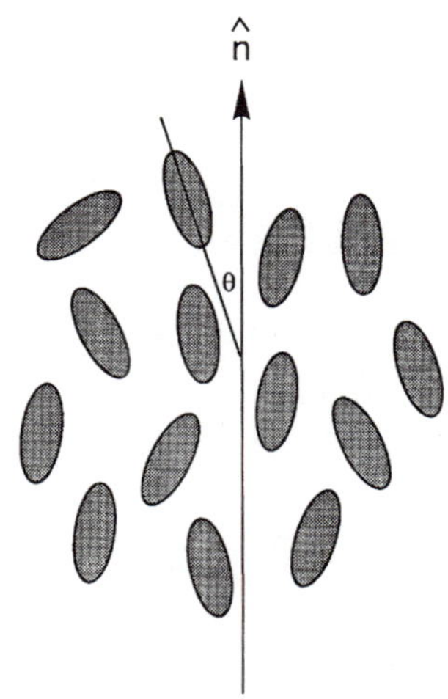

FIGURE 1.3. Nematic liquid crystal. The preferred orientation of the molecules, the director, is given by the unit vector $\hat{n}$. θ is the angle between the long axis of the molecule and the director.

If the molecules constituting these phases are chiral, which means that they do not possess inversion symmetry, than the nematic phase and some of the smectic phases do not occur. Instead, they are replaced by chiral versions of these phases with different physical structures. For example, in a chiral nematic phase (also called a cholesteric phase) the director rotates along a direction perpendicular to the director as shown in FIGURE 1.5. In principle, the helical axis could be in a direction other than perpendicular to the director, but no such phases are known. The distance over which the director rotates one revolution is called the pitch, although due to the symmetry between $\hat{n}$ and $-\hat{n}$ the

FIGURE 1.4. The smectic A and smectic C phases.

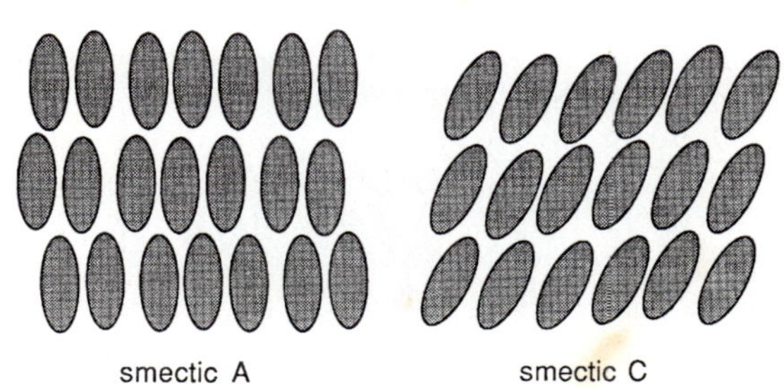

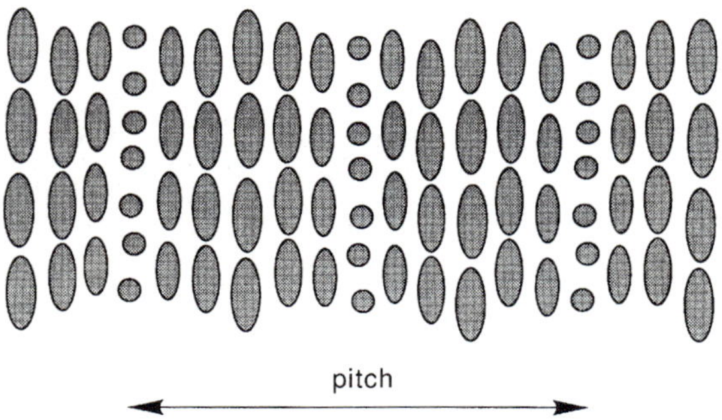

FIGURE 1.5. Chiral nematic phase. The preferred orientation rotates about a helical axis. The pitch is the distance for one full revolution of the director.

spatial period is actually half the pitch. The pitch can range from less than 100 nm to greater than meters. Similarly, in place of the smectic C phase, the chiral smectic C (sometimes called the smectic C* phase) results. In this phase the director maintains a constant tilt angle with respect to the layer normal and rotates about a cone in going from one layer to the next, as illustrated in FIGURE 1.6. In this case, the pitch is the distance for one full revolution around the cone and can be as short as 300 nm or as large as any laboratory sample. Chiral versions of all tilted smectic phases should exist and these are usually denoted with an asterisk after the letter. However, this nomenclature is not used exclusively. Some authors use an asterisk to denote a helical structure while others use it to indicate that the molecules are chiral. This confusion arises because chiral molecules do not always produce chiral phases, especially when subjected to surface forces and external fields.

FIGURE 1.6. Chiral smectic C phase. The preferred orientation rotates about a cone. The axis of the cone is perpendicular to the smectic layers. The pitch is the distance for one full revolution of the director around the cone.

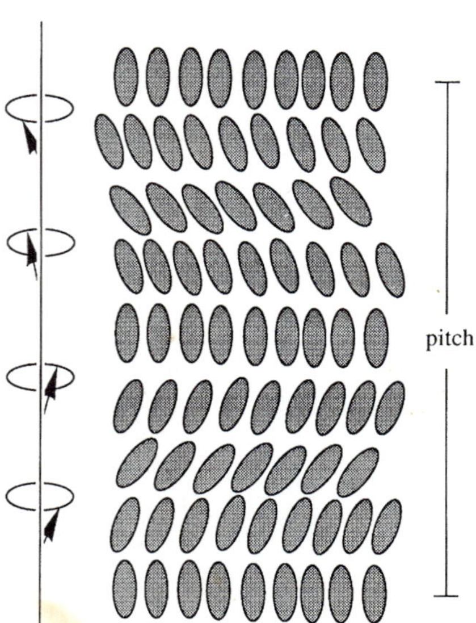

1.1.3.2 Discotic Liquid Crystals

In addition to rod-shaped molecules, disk-like molecules also form liquid crystals. The anisotropic phases formed by disk-like molecules are called discotic liquid crystals and the simplest type is the nematic phase in analogy with the nematic phase formed by rod-like molecules. A time snapshot is shown in FIGURE 1.7, where it is seen that the short axes of the molecules maintain a preferred orientation, again called the director. This simple phase of disk-like molecules has no positional order, just like nematic liquid crystals. Other discotic liquid crystal phases exist in analogy with smectic liquid crystals. For example, in the columnar phase the molecules are more likely to be found in columns rather than between the columns. This phase possesses the orientational order of the nematic phase and also positional order in directions perpendicular to the director. Like the simplest smectic liquid crystals, there are discotic liquid crystals with no positional order parallel to the director, making them one-dimensional liquids.

1.1.3.3 Lyotropic Liquid Crystals

In the above examples, all of the molecules are essentially identical. It is also possible to form liquid crystals by dissolving certain molecules in solution. These differ from the liquid crystal phases previously described because they are multi-component systems and the concentration of the various components is a new and important parameter. These liquid crystals are called lyotropic liquid crystals to differentiate them from the liquid crystal phases pure compounds form (also called thermotropic liquid crystals). Thus in the case of lyotropic liquid crystals, concentration plays the role that temperature does in thermotropic liquid crystals; the transformation from one phase to another can be accomplished by changing the concentration.

FIGURE 1.7. Two discotic liquid crystal phases. Nematic phase (left) and columnar phase (right).

A typical lyotropic liquid crystal molecule has two parts, one that is polar and therefore hydrophilic, and another that is nonpolar and therefore hydrophobic. Such molecules are called amphiphilic and form ordered phases when mixed with either a polar or nonpolar solvent. When dissolved in water, for example, these molecules form structures which present the hydrophilic part of the molecules toward the water and segregate the hydrophobic parts with each other. FIGURE 1.8 shows two of the structures that molecules with a polar "head" and hydrocarbon "tail" form in water. These micellar and lamellar (bilayer) structures usually form isolated structures at lower concentrations of the amphiphilic component and at higher concentrations can pack these structures together to form a much more ordered phase. FIGURE 1.9 illustrates a structure typically found for amphiphilic molecules with a large polar "head" and two hydrocarbon "tails." This structure is a combination of the micellar and lamellar structures and is called a vesicle. Shown in FIGURE 1.9 is a single lamellar vesicle; multilamellar vesicles have many single lamellar vesicles nested inside each other with water in between. These molecules also form a highly ordered lamellar phase at higher concentration. Similar structures form when amphiphilic molecules are mixed with a nonpolar solvent, the only difference being that the hydrocarbon "tails" are presented to the solvent instead of the polar "heads."

Lyotropic liquid crystals are extremely important to the study of surfactants, emulsions, and certain biological structures. They are discussed at length in Chapter 7.

FIGURE 1.8. Two of the structures formed by some amphiphilic molecules in a polar solvent.

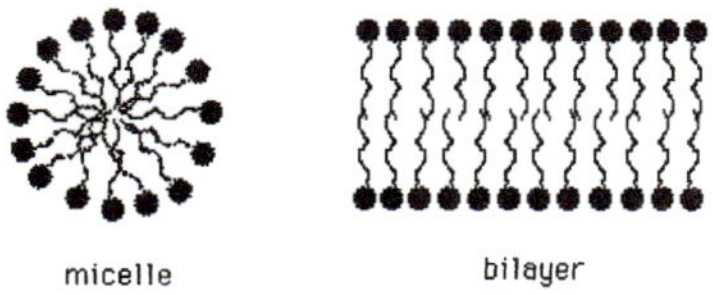

1.1.4 Defect Phases

One of the most important recent developments in the science of liquid crystals is the appreciation of liquid crystal phases only possible through the incorporation of periodic and nonperiodic defects. Defects in a liquid crystal phase are points, lines, and surfaces at which the orientational or positional order is not defined. These defects are energetically unfavorable, but become locked into a sample due to the influence of the sample container and the thermal history of the sample. Defects are normally present in liquid crystal phases, but they can be removed by specific experimental procedures. In all of the examples of liquid crystals so far, it is understood that the discussion and diagrams refer to regions in a sample free from these defects.

In a number of cases, liquid crystals form defects spontaneously throughout the sample as a new phase is formed. Unlike most defects which can be removed by certain procedures, these defects are part of the

FIGURE 1.9. One of the structures formed by some amphiphilic molecules in a polar solvent.

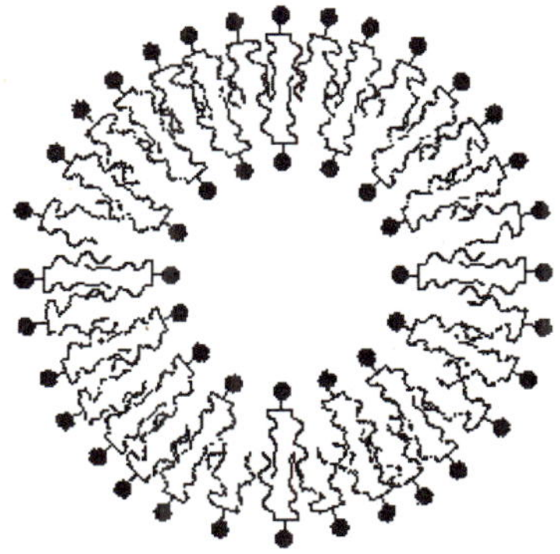

structure of the new phase. These phases can be thought of as superstructures in which individual regions form liquid crystals and the regions are separated by defects. It is possible, however, for these defects to have the order of an isotropic liquid in analogy with a melted lattice. These "melted" defect phases should transform continuously to an isotropic liquid as the liquid crystal order within the regions vanishes.

Two examples of such defect phases are the smectic twist grain boundary phase (TGB) and the blue phases (BPs) formed by some highly chiral molecules. In the case of the TGB phase, the tendency for these molecules to form a twisted phase is frustrated by an additional tendency to form layers. The result is a smectic A structure periodically interrupted by grain boundaries where the smectic A structure twists by a small angle. In the case of the BPs, regions with liquid crystalline order slightly different from the chiral nematic phase are separated by a lattice of line defects. Both of these phases are discussed in later chapters.

1.2 LIQUID CRYSTAL FORMING MOLECULES

1.2.1 Calamitic Liquid Crystals

1.2.1.1 Nematic Liquid Crystals

The typical nematic liquid crystal can be represented by the structure in FIGURE 1.10(a). The bridging group B together with the two ring structures O_1 and O_2 form the "rigid core" of the rod-like molecule. This rigidity is necessary to make the interactions with the other molecules anisotropic and therefore favorable for the formation of liquid crystal phases. Modifications to the "rigid core" which ruin its linear structure or increase its flexibility tend to reduce or eliminate liquid crystalline behavior. The two terminal groups R_1 and R_2 are usually necessary for the formation of liquid crystal phases, but may be quite short, a methoxy group for example, or nonexistent if the "rigid core" section is long enough. FIGURE 1.10(b) is an example of a molecule that forms a nematic liquid crystal phase.

FIGURE 1.10. Typical calamitic liquid crystal forming molecule. (a) Schematic and (b) example. R denotes an end group, O a ring structure, and B a bridging group.

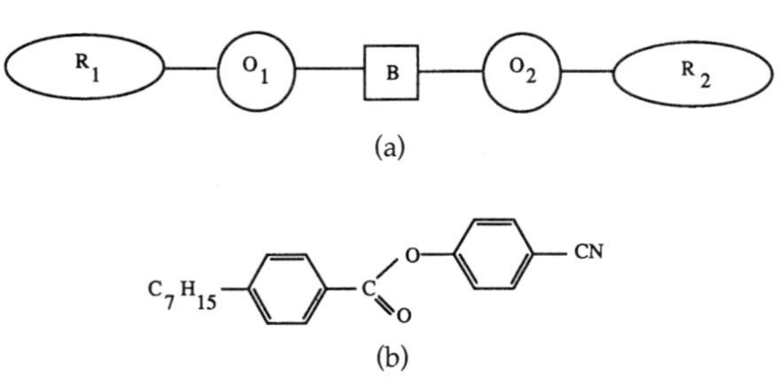

A number of different groups have been used as bridging groups. Some examples are the Schiff bases, diazo and azoxy compounds, nitrones, stilbenes, esters, and biphenyls. The ring structures are frequently aromatic rings, but an increasing amount of work is being done with cyclohexane and other ring structures. When synthesizing nematic liquid crystals with strong electrical properties, a polar terminal group is typically added to one end of the molecule. Typical examples are cyano and nitrous oxide groups.

1.2.1.2 Smectic Liquid Crystals

Typical molecules that form smectic phases are very similar to those forming nematic phases but tend to be slightly longer. This can be done by elongating the "rigid core" of the molecule to include three rings or by making the terminal groups longer. This is well illustrated by examining the nature of the phases present when molecules with the same "rigid core" but with longer terminal groups are investigated. When the terminal groups are very short, the compound melts from the solid to the nematic phase and ultimately to the isotropic liquid phase. As the terminal groups get longer, the compound melts first from the solid to a smectic phase, then to a nematic phase, and finally to the isotropic phase. With even longer terminal groups, the solid melts to a smectic phase and then directly to the isotropic liquid phase. This general behavior is shown in FIGURE 1.11, where the transition temperatures for the homologous series of alkyloxyazoxybenzenes are presented.

There is good evidence to suggest that long terminal groups, especially hydrocarbon chains, are necessary for smectic phases to exist. The behavior of many homologous series discussed above is just one indication. Long "rigid core" molecules without terminal groups may have nematic phases but they rarely have smectic phases. Likewise, structures containing strong dipoles perpendicular to the long axis of the molecule favor side-by-side organization and therefore tend to form smectic phases.

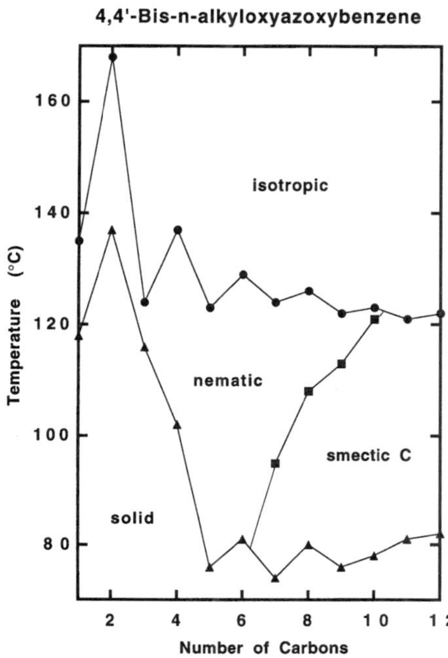

FIGURE 1.11. Transition temperatures for the homologous series of alkyloxyazoxybenzenes.

Due to the vast synthesis effort that has gone into liquid crystals due to their technological importance, a great deal is known about how to create liquid crystal materials with just the projected combination of properties. An important aspect of adjusting the physical properties is the ability of a given phase of one compound to be completely miscible with the same phase of another compound. Chapter 2 contains a detailed discussion of much of this synthesis work.

1.2.1.3 Chiral Liquid Crystals

If a rod-like molecule is chiral, then it forms the chiral nematic phase instead of the nematic phase and the chiral smectic C phase instead of the smectic C phase. Although it may form some nonchiral smectic phases, it cannot form the nematic or smectic C phases. Usually chirality is introduced to the molecule by incorporating a single chiral center, a good example being the inclusion of a methylbutyl group somewhere in one of the terminal groups. Naturally produced substances can form chiral liquid crystal phases, the best example being some of the esters of cholesterol, one of which incidentally was the first liquid crystal discovered in 1888 by Reinitzer (Reinitzer, 1888), an Austrian botanist.

The chirality of a phase can be gauged by the magnitude of the pitch of the chiral nematic or chiral smectic C phases. The shorter the pitch, the more chiral the phase. The chirality of the phases formed by molecules with two chiral centers of the same handedness is greater than for molecules with one chiral center. In addition, the closer the chiral center is to the "rigid core" of the molecule, the higher the chirality in general. One convenient way to vary the chirality of a phase is to mix one chiral compound with its optical isomer. The two pure isomers have the highest chirality (of opposite handedness, of course) and the mixture of equal amounts of the two isomers has no chirality (forms the nematic or smectic C phase).

1.2.2 Discotic Liquid Crystals

Molecules that form discotic liquid crystal phases usually are comprised of a core based on benzene, triphenylene, or truxene to which four or six groups have been added. These groups often resemble the molecules that form calamitic liquid crystals, with part of the "rigid core" toward the center of the molecule and a terminal group pointing away from the center. An example of such a liquid crystal forming molecule is shown in FIGURE 1.12.

There are many similarities between calamitic and discotic liquid crystals. Longer terminal groups favor

FIGURE 1.12. Typical discotic liquid crystal forming molecule.

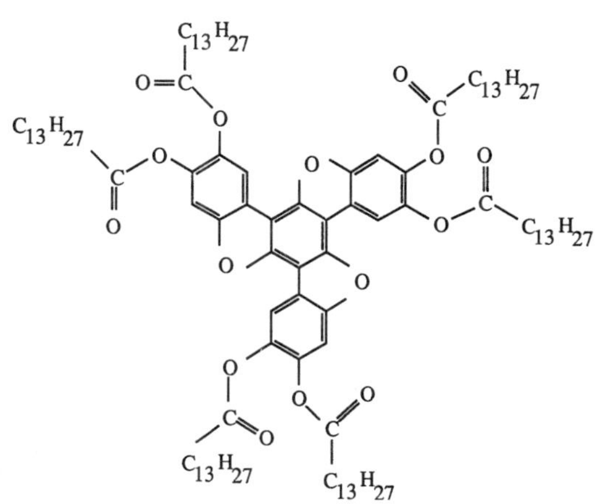

columnar discotic phases over nematic discotic phases just as they favor smectic phases over nematic phases in calamitic liquid crystals. Introduction of a chiral center to the molecule causes a chiral nematic discotic phase to form, and just as in a calamitic liquid crystal, the director is perpendicular to the helical axis. Liquid crystal blue phases can result when the chirality is high enough.

1.2.3 Lyotropic Liquid Crystals

1.2.3.1 Soaps

Conventional soap molecules are one of the best examples of amphiphilic molecules that form lyotropic liquid crystal phases. The polar head is formed by the bonding of a positive ion to a negatively charged ion which is linked to a hydrocarbon chain. Sodium laurate is the example shown in FIGURE 1.13(a). Although the polar part of the molecule does not have to be ionic, generally ionic structures are best for forming liquid crystal phases.

1.2.3.2 Phospholipids

Molecules that are the basis for the physical structure of various biological membranes are also fine examples of lyotropic liquid crystals. Compounds in this class are called phospholipids and they consist of a large polar group with a phosphorus atom plus two hydrocarbon chains. One such molecule is illustrated in FIGURE 1.13(b). Different polar groups containing phosphorus are present in liquid crystal forming molecules, and hydrocarbon chains of various lengths are possible. Double bonds in these chains are found in some lyotropic liquid crystal forming molecules.

It is the tendency of these molecules to form bilayers that causes vesicles to form rather than micelles and makes the lamellar phase the most prevalent phase at higher phospholipid concentrations.

1.3 POLYMER LIQUID CRYSTALS

1.3.1 Polymer Melts and Polymer Solutions

As the linear molecules increase in length, they can gradually crossover into the realm of polymers, which are formed by small monomeric units linked together. A wide variety of polymers is possible including block copolymers, in which the polymers are composed of flexible parts and rigid parts, with the rigid parts resembling the "rigid core" of calamitic liquid crystals. In these liquid crystal phases, the different parts of each polymer diffuse throughout the sample, with the "rigid core" parts maintaining some orientational and possibly positional order while the flexible parts show no long-range order.

Thermotropic polymer liquid crystals are polymers that possess at least one liquid crystal phase between the glass transition and the transition to the isotropic liquid. Lyotropic polymer liquid crystals are rigid polymers that form ordered phases when in solution. Many biological polymeric materials like DNA and tobacco mosaic virus (TMV) form lyotropic liquid crystals. These liquid crystal phases exist for some temperature range and for certain concentrations of the polymer. A complete account of the research into polymer liquid crystals is contained in Chapter 8.

1.3.2 Thermotropic Polymer Liquid Crystals

When the "rigid cores" and flexible parts are incorporated into single long polymer chains, the resultant polymer is a main chain polymer. Such a structure is depicted in FIGURE 1.14(a). When the small molecular weight mesogenic units are attached to long flexible polymer chains, a side chain polymer is the result. This is shown in FIGURE 1.14(b). Orientational and positional order parameters for the liquid crystal-like

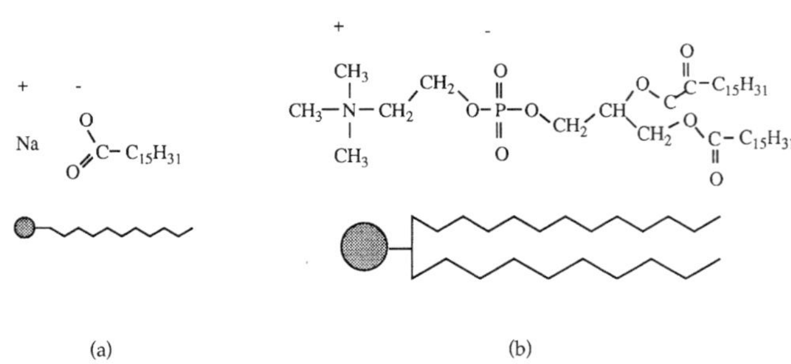

FIGURE 1.13. Two examples of amphiphilic molecules. (a) Sodium laurate and (b) dipamitoylphosphatidylcholine.

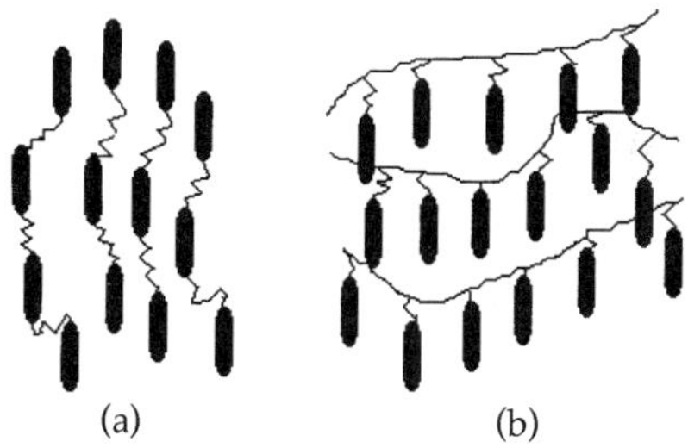

FIGURE 1.14. Two examples of liquid crystalline polymers. A main chain polymer is shown in (a) and a side chain polymer is shown in (b).

units can be defined in the same way as for thermotropic liquid crystals, and turn out to have similar temperature dependencies in polymer systems as in low molecular weight liquid crystals. Many of the liquid crystal phases previously discussed occur in both main chain and side chain polymer systems. These include nematic and smectic phases, chiral nematic and chiral smectic C phases, as well as the defect-containing blue phases. Examples of a main chain and a side chain liquid crystal are shown in FIGURE 1.15.

Several well known polymers form liquid crystal phases. These include the polyesters, polyethers, and polyamides. The tendency for certain polymers to form liquid crystal phases is crucial for the manufacturing of the new high-strength polymer fibers, as it is the liquid crystalline order that is "frozen in" the solid phase of the polymer producing its high strength.

FIGURE 1.15. Examples of a main chain (a) and a side chain (b) liquid crystal polymer.

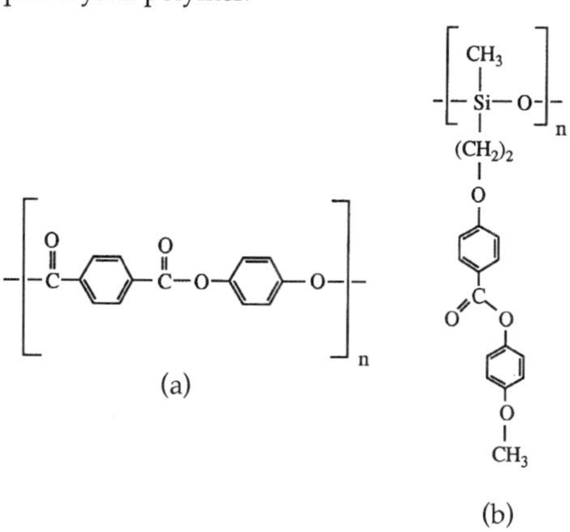

1.3.3 Lyotropic Polymer Liquid Crystals

Large macromolecules that form single- or double-stranded helical arrangements are good examples of polymers that form liquid crystal phases in solution. Specific examples include TMV, the synthetic polypeptide poly-γ-benzyl-L-glutamate, and some polyamides. In all cases, a critical volume fraction of the polymer is necessary to form an ordered arrangement of the macromolecules. The liquid crystal phase is readily distinguished from the isotropic liquid crystal phase by its optical birefringence. An abrupt change of viscosity also occurs at the critical volume fraction, but this may be an increase or decrease, depending on the component of the viscosity being measured and the nature of the macromolecule.

1.4 PROPERTIES OF LIQUID CRYSTALS

1.4.1 Mechanical Properties

Because liquid crystals are anisotropic fluids, various properties may not be the same when measured in different directions. Perhaps the most important mechanical property is viscosity. In the isotropic liquid phase, there is only one value for the viscosity. No matter how it is measured, the same value is obtained. Such is not the case for a liquid crystal. For example, if the viscosity is measured by sliding one plate over another, the magnitude of the viscous force depends on whether the director is perpendicular to the plates, parallel to the plates but perpendicular to the motion, or parallel to the plates and parallel to the motion. In each case a different viscosity is measured.

A similar phenomenon results when a liquid crystal is deformed. Starting with a region in which the director is constant, it turns out that there are three ways in which the director can be deformed from this undeformed condition. These are depicted in FIGURE 1.16, and the forces required to produce these distortions are generally different.

Surfaces strongly interact with liquid crystals, a property that produces a richness in behavior and allows for an enormous number of applications. Interfaces between liquid crystals and solid substrates

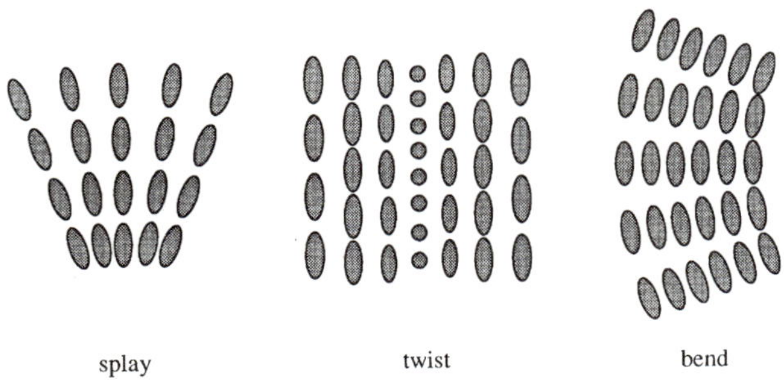

FIGURE 1.16. Types of deformation in liquid crystals.

along with the behavior of liquid crystal thin films are covered in Chapter 6.

1.4.2 Electromagnetic Properties

The anisotropy of a liquid crystal also affects its electromagnetic properties. The electric and magnetic susceptibilities are different along the director and perpendicular to the director. This anisotropy in the susceptibilities is very important, and is responsible for the director of a liquid crystal orienting parallel or perpendicular to an applied field, depending on which susceptibility is larger.

This anisotropy in the electric susceptibility can be made greater by the addition of a permanent electric dipole oriented either parallel or perpendicular to the long axis of the molecule, allowing liquid crystals for specific applications to be synthesized. The effect of electric fields on liquid crystals is discussed at length in Chapter 5.

1.4.3 Optical Properties

The anisotropy in the dielectric properties of a liquid crystal is responsible for the linear birefringence. Light polarized parallel to the director propagates with one index of refraction and light polarized perpendicular to the director propagates according to a different index of refraction. Thus light polarized parallel or perpendicular to the director propagates through the liquid crystal with no change to its polarization state. Linearly polarized light polarized at an angle other than 0° or 90° to the director suffers retardation of one component and thus emerges elliptically polarized in general. This accounts for liquid crystals appearing bright when viewed with the sample between crossed polarizers.

In the case of chiral nematic and chiral smectic C liquid crystal phases, the helical structure produces circular birefringence. This means that right circular polarized light and left circular polarized light travel at different velocities in these phases. The most important implication of this is that such phases are optically active, in that the direction of linearly polarized light is rotated as it passes through such phases. This optical activity is in addition to the inherent optical activity to be expected from the chiral molecules themselves.

The difference between the propagation characteristics of right and left circularly polarized light becomes extreme when the wavelength of the light in the liquid crystal is equal to the pitch. At this point, one of the two circular polarizations is completely reflected and the other is transmitted. Which one is reflected depends of the handedness of the liquid crystal phase. A right-handed phase reflects right circularly polarized light and vice versa. Since this occurs for the wavelength of light that equals the pitch, this phenomenon is often called selective reflection. It is responsible for the bright iridescent colors that sometimes can be seen when chiral liquid crystal phases are viewed in white light.

For most practical purposes, the passage of light through the liquid crystal does not affect the orientation of the liquid crystals molecules. However with increasing light intensity, the electric field of the light begins to play a role similar to that of an externally applied electric filed. Thus light itself modifies the refractive index of the liquid crystal medium as it propagates. Because liquid crystals are easily distorted by small forces, the nonlinear optical properties of light are very large although slow since the major contribution is from nuclear motion, rather then electronic. These properties make these materials quite unique. Furthermore, some exotic interaction of radiation with matter, such as the coupling of angular momentum of circularly polarized light with matter, can be examined using liquid crystals. These and other properties of liquid crystals are discussed in Chapter 14.

1.4.4 X-ray Diffraction

The order present in liquid crystals is apparent when one conducts x-ray diffraction experiments. Most pronounced is the Bragg-like scattering from the layers in smectic liquid crystals. Since the layer separation is on the order of a molecular length (tens of ångstroms), small-angle scattering must be employed when using x-rays with a wavelength on the order of an ångstrom or so.

X-ray scattering experiments have been extremely useful in elucidating the order present in liquid crystal phases. For example, a diffuse reflection at a larger angle is an indication that some short-range order exists between a liquid crystal molecule and its nearest neighbors. Likewise, diffraction from periodically repeating distances within the plane of a smectic liquid crystal indicates that positional order of the molecules in the smectic planes is present. Many such investigations have revealed that short-range, long-range, and quasi-long-range order (where the correlation function falls off algebraically) exist in various liquid crystal phases.

1.5 APPLICATIONS OF LIQUID CRYSTALS

1.5.1 Displays

The anisotropic electrical properties of liquid crystals that cause the director to respond to an applied electric field are the basis for most of the applications of liquid crystals. Because liquid crystals have fairly low electrical conductivity, little power is consumed when an electric field is applied. Thus liquid crystal displays (LCDs) have become the dominant display devices where power consumption is an issue.

The development of stable chemical compounds, reliable manufacturing procedures, and optimized optical characteristics have been crucial for the success of LCDs, not only in battery operated devices, but in almost all types of instrumentation. Switching speeds in most LCDs are in the millisecond range, which is fine for alphanumeric displays and for most video applications (where active circuits help switch the entire frame at the switching speed of a single element). Progress in new materials and device structures has steadily increased the switching speed of LCDs. It is only a matter of time before high-resolution video LCDs become commonplace.

1.5.1.1 Twisted Nematic and Supertwisted Nematic Displays

The "workhorse" LCD display is the twisted nematic display. It has been in production as a numeric display on wristwatches and calculators for many years, and is now becoming important for various machines (cash registers, gasoline pumps, electronic test equipment, etc.).

As shown in FIGURE 1.17, the nematic liquid crystal used in this display is contained between two pieces of glass. A transparent electrode of indium–tin oxide is deposited on the side of each piece of glass in contact with the liquid crystal, followed by a polymer coating for orienting the liquid crystal. The polymer coating is rubbed in one direction, which in turn causes the liquid crystal director adjacent to the coating to orient parallel to the rubbing direction. What makes a twisted nematic LCD work is that the rubbing directions for the two inside glass surfaces are perpendicular to each other. Therefore, the director rotates by 90° in the liquid crystal occupying the 10 μm space between the two pieces of glass.

Two polarizing films are added to the side of each piece of glass not in contact with the liquid crystal. These polarizing films on each piece of glass are oriented so that they allow light to pass if it is polarized parallel to the rubbing direction on the same piece of glass. Light striking the top polarizing film is polarized along the rubbing direction for the

FIGURE 1.17. Twisted nematic liquid crystal display.

top piece of glass, and if the thickness and optical properties of the liquid crystal are chosen appropriately, the polarization of the light rotates with the director as it passes through the liquid crystal. It therefore passes through the polarizing film on the bottom piece of glass. In short, the liquid crystal display is transparent when there is no applied voltage.

When a large enough voltage is applied to the electrodes, the director in the liquid crystal orients parallel to the electric field. This means that except for a thin boundary layer next to each piece of glass, the director is parallel to the light propagation direction. The light that is polarized by the top polarizing film remains polarized in this direction as it propagates along the director of the liquid crystal, so it is extinguished by the perpendicular polarizing film on the bottom piece of glass. Thus the LCD is opaque when a suitable voltage is applied.

This type of LCD can be used in either a transmissive or reflective mode. In the transmissive mode, a light behind the display is blocked for those electrodes carrying an applied voltage. By shaping the electrodes into patterns for numbers and letters, black characters appear on a bright background. In the reflective mode, a reflecting layer is placed below the bottom polarizing film. For electrodes with no voltage, light that has passed through the LCD is reflected and passes through again. These areas appear to be silver colored to the viewer. Electrodes with an applied voltage appear black, since the ambient light striking the LCD is extinguished by the crossed polarizing films.

A recent improvement to the contrast and viewing angle characteristics is to construct an LCD in which the director rotates 270° in going from one glass surface to the other, rather than 90°. This is called a supertwisted nematic display and it operates in essentially the same way as described above, except that the threshold voltage required to turn on the display element becomes very sharp and this allows better multiplexing as discussed in Chapter 11.

Some of these devices have rather unique properties such as bistability, and a variety of nematic bistable devices are discussed in Chapter 15. In these devices the bistability is either intrinsic, meaning that the liquid crystal structure itself provides for the bistability, or extrinsic, meaning that the external environment (such as the surface structure) provides for the bistability.

1.5.1.2 Surface Stabilized Ferroelectric Liquid Crystal Display

These LCDs have been the subject of a great deal of research because of their relatively faster switching speeds. This LCD relies on the spontaneous polarization that is characteristic of chiral smectic C liquid crystals. If the helical structure of the chiral smectic C liquid crystal is suppressed, then a spontaneous polarization is present parallel to the smectic layers and perpendicular to the director. Such liquid crystals are therefore ferroelectric. Since the polarization direction is coupled to the director, the orientation of the polarization causes the director to reorient by twice the tilt angle in the smectic C phase.

The helical structure can be suppressed by surface forces in a very thin cell, on the order of a micrometer thick, causing the director to lie parallel to the surface. The result is depicted in FIGURE 1.18, which is a top view of the display. The smectic layers are

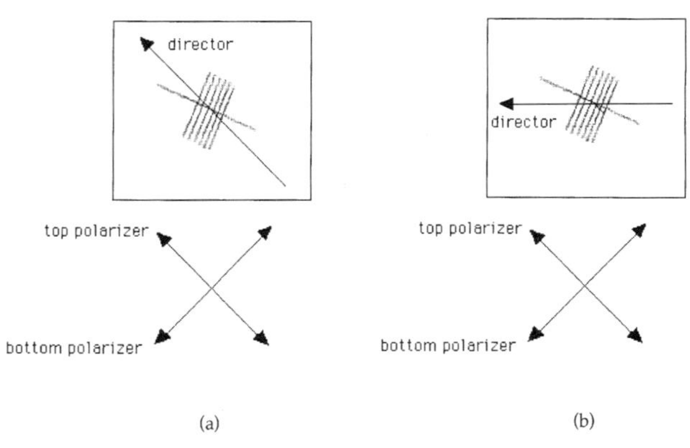

FIGURE 1.18. Top view of a surface stabilized ferroelectric liquid crystal display. The orientation of the smectic layers is shown in the center of each diagram. The angle between the director and normal to the smectic layers (the tilt angle) is 22.5°. (a) The orientation of the director in the off-state and (b) the orientation of the director in the on-state.

perpendicular to the glass surfaces and for best results the angle between the director and the normal to the smectic layers is 22.5°. In this configuration, the spontaneous polarization is parallel to the applied electric field, into the page in FIGURE 1.18(a) and out of the page in FIGURE 1.18(b)). Notice that the reversal of the direction of the electric field causes the director to rotate halfway around the normal to the smectic layers.

If the top and bottom polarizers are oriented as shown in FIGURE 1.18, then the light entering the LCD from above is polarized along the director in FIGURE 1.18(a). Since this polarization direction is not changed at all in propagating through the liquid crystal, the light is extinguished by the bottom polarizer. The LCD appears dark. In FIGURE 1.18(b), however, the light polarized by the top polarizer is at 45° to the director. If the thickness of the cell is matched to the birefringence of the liquid crystal in just the right way such that the cell acts as a half-wave plate, then the light emerging from the liquid crystal is polarized at 90° to its direction upon entering the LCD. Thus the light passes through the bottom polarizer and the LCD appears bright. Switching times in these surface stabilized ferroelectric LCDs can be on the order of microseconds.

1.5.1.3 Polymer Dispersed and Polymer Stabilized Liquid Crystal Displays

A more recent development in LCD technology involves incorporating liquid crystal droplets in a solid polymer matrix (polymer dispersed liquid crystal or PDLC) or incorporating a polymer network in a bulk liquid crystal (polymer stabilized cholesteric texture or PSCT). Both of these schemes possess important advantages over conventional liquid crystals and are receiving considerable attention for commercial applications, because these display elements do not require polarizers.

The PDLC display consists of a solid polymer containing droplets of nematic liquid crystalline material. The operating principle of this display is the refraction of light when it experiences a medium with a different refractive index. The higher contrast is obtained by the multiple scattering of light. When no electric field is applied to the PDLC film, the directors in the droplets point in all directions at random. This causes light passing through the film to encounter droplets with an index of refraction different from the polymer matrix and a large amount of scattering occurs. This makes the film cloudy white. However, when an electric field is applied, the directors of all the droplets align with the field so that light propagating through the PDLC is polarized perpendicular to the director in the droplets and the index of refraction of the polymer is chosen to minimize refraction. There is no mismatch in the indices, so no scattering occurs. This causes the film to appear clear. FIGURE 1.19 shows these two conditions for a PDLC display and a complete discussion is given in Chapter 9. Such films have already been commercialized as sunroof screens and window blinds. Chapter 9 covers both the scientific and technological issues connected with PDLCs.

A PSCT LCD is only slightly different from a display made entirely of a chiral nematic liquid crystal. If a chiral nematic material with a pitch in the visible is contained between two pieces of glass with the helical axis perpendicular to the glass, light with a wavelength in the vicinity of the pitch is rotated significantly and therefore passes through crossed polarizing films on the outer surfaces of the glass. A backlit display appears brightly colored. If an electric field is applied so that the director in the liquid crystal is everywhere perpendicular to the glass surfaces, no rotation of the light occurs so the display appears black due to the crossed polarizers. A small amount of crosslinking polymer is added to the chiral nematic material to overcome a problem of conventional chiral nematic displays, namely that it does not return to a defectless helical texture. The polymer network becomes the "memory" that allows the chiral nematic to return to its original zero-field texture each time. This also shortens the time for the display to switch when the electric field is removed (Yang and Doane, 1992).

FIGURE 1.19. Polymer dispersed liquid crystal display.

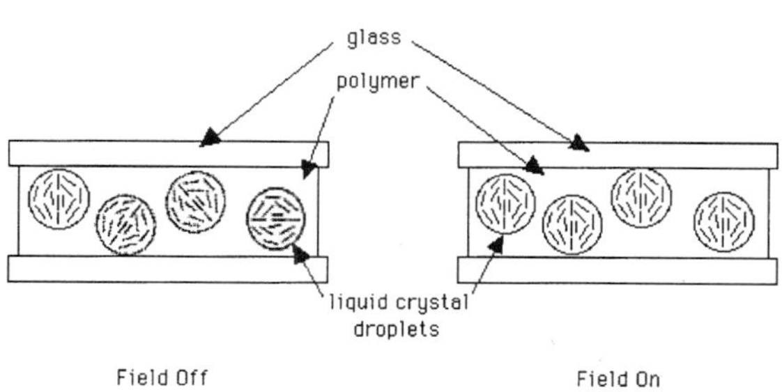

1.5.1.4 Multiplexing and Active Matrix Addressing

One of the technological problems that must be solved in any flat panel display device is how individual pixels are addressed. The magnitude of the problem becomes very apparent when one considers the number of pixels for a display on the order of several inches on a side with a resolution of 500 pixels per inch. Instead of making individual connections to each pixel, groups of electrodes are connected together and the voltages to these electrode groups are multiplexed.

In order to work effectively, the cycle time for refreshing individual pixels must be short enough so that the state of the pixel is maintained and that the display does not appear too sluggish. In addition, the threshold behavior of the switching from one state to the other must be extremely sharp, since each pixel is exposed to voltages below threshold between the times it is refreshed. Sharp thresholds are often achieved by transferring the threshold behavior from the display elements to active integrated devices placed at each pixel. For example, the use of a thin film transistor with a sharp threshold which in turn supplies the voltage to the pixel can allow the display to be multiplexed to a much higher degree. This scheme is known as active matrix addressing and is discussed in detail in Chapter 10.

1.5.2 Other Applications of Liquid Crystals

1.5.2.1 Sensors

Another area where liquid crystals have made a commercial impact is the field of temperature sensors. Chiral nematic liquid crystals provide an excellent opportunity for such sensors due to the fact that the wavelength of selective reflection depends on the pitch, which in turn depends on temperature. By mixing liquid crystal compounds with different pitches and chiral nematic ranges, the selective reflection can be made to vary throughout the visible for just about any range of temperatures.

For example, the pitch of a chiral nematic liquid crystal tends to diverge as the temperature is lowered toward a transition to a smectic phase. This divergence can be very sharp, producing a sensor in which color is extremely sensitive to temperature. Putting several of these materials side-by-side, with each sensitive to a slightly different temperature, is another way of making liquid crystal temperature sensors. These devices have found widespread use in fish tank thermometers, fever thermometers, diagnostic temperature sensors for the skin, and mood rings; even temperature-sensitive artwork has been realized using chiral nematic liquid crystals. It should be kept in mind that such sensors are usually not made by placing the liquid crystal between two pieces of glass like an LCD. Instead, chiral nematic material is either encapsulated in a polymer coating and placed in a solid matrix or simply dispersed as droplets in a solid polymer material much like a PDLC.

Because liquid crystals are sensitive to many applied influences in ways which change their optical properties significantly, there is the possibility they may be used in other sensors. For example, a device that can detect the presence of an electric or magnetic field can easily be fabricated using a liquid crystal. In addition, sensors for pressure, shearing stress, infrared and other radiation, and chemical vapors are all possible based on liquid crystal devices. Such devices are not expensive, so development will commence if a need for such a sensor arises (Sage, 1992).

1.5.2.2 Other Optical Applications

A light valve is a device that can change the intensity of light striking it, typically by modulating its polarization in conjunction with attenuation by a polarization-sensitive element. If the device has the capability of spatially modifying the beam of light, then it is called a spatial light modulator. By applying different electrical voltages to various pixels in a twisted nematic liquid crystal cell, the output can be made to vary in the cross-section of the beam in whatever way desired. While this is the simplest form of such a device, there are others that rely on a photoconducting layer to affect the voltage across the liquid crystal. These types of modulators are often used to convert incoherent light into coherent light, which is often necessary for image processing applications and other applications where phase coherence of light is necessary.

There are many other applications which use liquid crystal devices. In Chapter 12, a variety of devices suitable for telecommunication applications are discussed. These devices allow the passage of light to be switched from one optical fiber to another while others allow a particular wavelength to be separated from a group of wavelengths. The liquid crystal based devices are also suitable for a variety of optical information processors. In these applications, the liquid crystals can be used as reconfigurable two-dimensional interconnects or, using laser light, as two-dimensional signal processors using the coherence properties of light. In some of these applications the incoherent light must first be converted to coherent light, using devices such as the liquid crystal light valves. Many of these applications, including the fundamentals of information processing, are discussed in Chapter 13.

REFERENCES

Reinitzer, F. 1888. *Monatssch. Wiener Chem. Ges.* **9**: 421.

Sage, I., 1992. in *Liquid Crystals: Applications and Uses*, vol. 3, (ed. B. Bahadur), ch. 20. Singapore: World Scientific.

Yang, D. K., and Doane, J. W. 1992. *SID Dig. of Tech. Papers* **23**: 759.

CHAPTER 2

Chiral and Achiral Calamitic Liquid Crystals for Display Applications

A.W. HALL, J. HOLLINGSHURST, AND J.W. GOODBY

2.1 INTRODUCTION

Liquid-crystalline materials generally have molecular structures that are either rod-like, disk-like or spherulitic in shape. However, rod-like (calamitic) systems have been the most thoroughly investigated. This is because calamitic compounds possess properties that make them useful for applications in electrooptic devices. Thus, in this chapter we concentrate our attention on two calamitic systems: nematics and ferroelectrics. Both can be subdivided into achiral and chiral systems. In nematic systems: achiral materials form the basis of most devices whereas in ferroelectric systems chirality (or induced chirality) is essential to the operation of the display. In the following sections we describe the chemical characteristics of materials that exhibit nematic phases and are useful in twisted nematic and multiplexed twisted nematic devices, and their chiral analogs which are used in thermochromic applications. In contrast, the properties of chiral smectogens which exhibit ferroelectric behavior, and their achiral counterparts that are used as host systems for applications in fast-switching electrooptic shutters, are then discussed.

2.2 NEMATIC LIQUID CRYSTALS

In this section the general structure and properties of calamitic nematic molecules are described and discussed in terms of structure/property correlations.

2.2.1 Nematogens and Their Uses in Display Devices

2.2.1.1 Molecular Architecture

A schematic form of a typical calamitic nematogen is shown in FIGURE 2.1. As the term "calamitic" implies, the molecule is rod-like with a high length-to-breadth

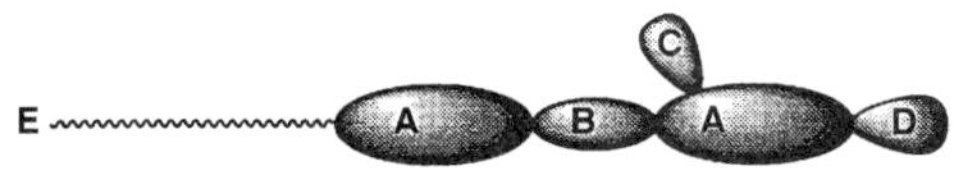

FIGURE 2.1. Schematic diagram of a nematogen.

ratio. The center, or core, of the molecule (A) is a fairly rigid section usually made up of one or more 1,4-disubstituted six-membered rings. These rings may be directly joined to one another or separated by a short linking group (B). Such groups include esters, Schiff bases and short alkyl or alkoxy chains. There may also be one or more lateral groups (C) appended to the core. The ends of the molecule may be flexible linear alkyl or alkoxy structures (E) or a combination of a small polar moiety (D) at one end with a chain at the other. These sections will be discussed individually in detail in section 2.2.2.

2.2.1.2 Nematogens in Electrooptic Display Devices
When the molecules are statistically aligned in the nematic phase, the physical properties of the bulk material, such as refractive index, viscosity and dielectric permittivity are different in directions perpendicular and parallel to the average direction of alignment or director. As this phase is by definition a fluid, it is thus possible to effect rapid molecular reorientation within a display device by the application of, for example, an external electric or magnetic field. The speed of this switching is, of course subject to the constraints of the bulk viscosity of the material in relation to surface forces: the higher the viscosity, the greater the resistance to molecular realignment. It is thus possible to conceive of a number of display devices in which the degree of transmission of light can be altered electrically between two contrasting states.

The dielectric anisotropy ($\Delta\epsilon$) governs the behavior of the molecules in an applied electric field (Maier and Meier, 1961). This may be either positive or negative depending on whether the dielectric permittivity along the director ($\epsilon_{\|}$) is greater than, or less than, the perpendicular component ($\epsilon_{\perp}$). The majority of electrooptic display devices are based upon strongly positive $\Delta\epsilon$ materials since this maximizes the interaction between the molecules and the applied field and thus minimizes the required driving voltage. Weakly negative $\Delta\epsilon$ materials are, however, needed in nematic mixtures for multiplexed devices in order to

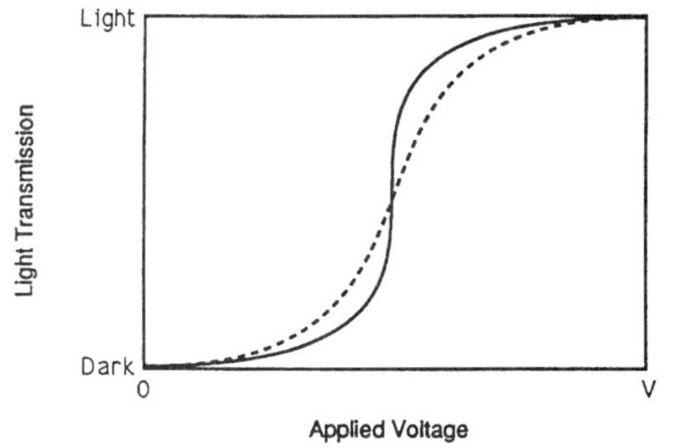

FIGURE 2.2. Typical optical switching characteristics of a nematic mixture with (full line) and without (dotted line) weakly negative $\Delta\epsilon$ components.

steepen the response/voltage curve and thereby reduce the potential between the 90% on and 90% off states (see FIGURE 2.2). This allows for better multiplexing in matrix-driven displays.

The use of strongly negative $\Delta\epsilon$ materials, whilst equally possible, is much less common: a lateral dipole moment is only half as effective at producing negative dielectric anisotropy as a terminal dipole is at producing positive dielectric anisotropy (Sage, 1987). Applications have, however been found in distorted alignment phase (DAP) cells which operate solely on a change of birefringence on switching from homeotropic to planar alignment. These materials are also used in positive contrast guest-host displays since the "on" state is opaque on a translucent background. Backlighting is therefore used to produce a black-on-white effect.

Refractive index anisotropy, or birefringence (Δn) is a consequence of the different speeds of light propagation parallel and perpendicular to the director. Highly birefringent nematics are primarily used for encapsulated nematic phase (NCAP) displays to maximize the contrast between the transmitting and scattering states. In the case of twisted nematic devices, however, low birefringence is preferable as a consequence of the Mauguin limit (Mauguin, 1911).

$$2d\Delta n/\lambda \gg 1$$

where d = cell thickness, and λ = wavelength of light.

The use of low birefringence materials permits the use of thicker, and hence more easily fabricated, cells. There is also a tendency for highly birefringent materials to appear colored between crossed polarizers, and in such cases there is a reduction in contrast between the "on" and "off" states.

The viscosity of nonisotropic fluids is complex. Nematic materials have three shear flow viscosities and a further two associated with the coupling between the director and the flow (de Gennes, 1974). Although these components are experimentally accessible, for most practical purposes viscosities are measured at several temperatures in the isotropic melt and then extrapolated back into the nematic phase. Alternatively, the bulk viscosity is measured in the phase using a suspension viscometer. The test material is either measured neat, or as a dopant in a suitable nematic host. In the latter case the viscosity is determined by extrapolation from several different concentrations of dopant. As stated earlier, a high viscosity opposes molecular realignment, and as a result reduces the speed of switching.

2.2.1.3 A Note on Comparisons

When comparing liquid crystals, the following factors must be borne in mind:

1. The term "phase stability" refers to the stability of the phase with respect to a higher temperature phase or the isotropic melt, and gives no indication of the phase range. Consequently, a nematic phase between 220 °C and 230 °C may be more stable than one between 10 °C and 120 °C, but the latter has a wider range, and is perhaps more likely to be of practical use in display devices.

2. Phase transitions may or may not be reversible. For example, in the former case a crystal may melt to a nematic liquid which then clears to an isotropic liquid at a higher temperature. On cooling, the phase sequence is reversed. However, in the case of monotropic phases, the crystal may melt straight to the isotropic liquid and only form a nematic phase on cooling from the isotropic. Such transitions are indicated, in the following text, by transition temperatures in parentheses. A further class of materials shows no phases on heating or cooling, but a transition can be determined by extrapolation in a phase diagram of a mixture of the compound and a compatible enantiotropic nematogen. These "virtual" transitions are denoted by square brackets.

3. The physical properties of nematic phases are generally temperature dependent, and account must be taken of the temperature at which measurements are made. From a pragmatic standpoint, values should be quoted at the likely operating temperature of any application (say 25 °C). In the case of materials which are not liquid crystalline at this temperature, the property may be quoted at a higher temperature or as an extrapolated measurement as a function of either temperature or, in the case of mixtures, composition. A more rigorous approach is to quote values at a given "degree of nematogenicity" or order parameter. This scales as the ratio of temperature to clearing temperature, and thus values are often quoted at a given "reduced temperature" (ΔT).

4. When comparing different homologous series of materials with different core lengths, it is often reasonable to compare molecules of similar overall length.

5. Nematic materials in display devices are almost always mixtures of three or more compounds. It is thus possible that a nematogen with one excellent characteristic (say, high $\Delta\epsilon$) but several poor characteristics (high Δn and high viscosity for example), may still be used in a display device. A number of theoretical and empirical expressions have been developed to account for, and to some extent predict, the physical properties of mixtures from the behavior of their constituent materials with varying degrees of success (Coates, 1991).

2.2.2 Structure/Property Correlations

2.2.2.1 Cores

The core of a nematogen is commonly made up of two or more 1,4-disubstituted six-membered rings. Although apparent one-ring systems have been produced (McDonnell, 1979), these molecules only become nematic upon dimerization as shown in FIGURE 2.3.

The *trans*-1,4-disubstituted ring system is comparatively rigid and the substituent ends are parallel or colinear, thus fulfilling the central criterion of a calamitic molecule, i.e., that its structure should be relatively linear. The other criterion that there should be anisotropy of intermolecular bonding in the fluid state, is met by differential dispersion forces along the

FIGURE 2.3. Dimerization of 4-alkyl cyclohexylcarboxylic acids to form a nematogen.

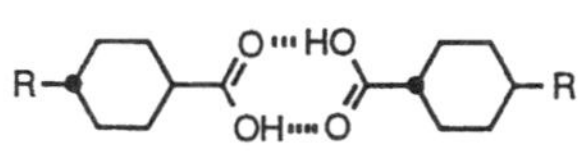

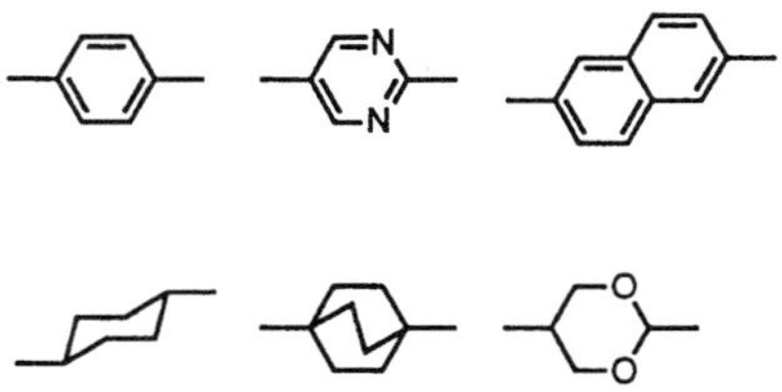

FIGURE 2.4. Some typical core components.

molecule, particularly in the case of polarizable aromatic rings; the "broadside" interaction between parallel molecules is greater than the end-to-end or end-to-side forces. Indeed, there is a wide-ranging if incomplete correlation between the anisotropy of polarizability and phase stability (Maier and Saupe, 1958, 1959, 1960). Thus, fused ring systems tend to have a smaller length-to-breadth ratio and are essentially less linear, but can still give nematic phases. This is exemplified by one of the earliest liquid crystals to be studied, cholesteryl benzoate (Reinitzer, 1888), which contains a number of fused rings (see FIGURE 2.30).

PHASE STABILITY Some of the more common core ring components are shown in FIGURE 2.4.

For the sake of comparison, the effects of these rings will be compared in two- and three-ring alkyl cyano mesogens shown schematically in FIGURE 2.5.

The thermal properties of some of these materials are summarized in TABLES 2.1 and 2.2.

From this it can be seen that the general order of phase stability is as shown in FIGURE 2.6 (Carr et al., 1981).

This sequence reflects somewhat the decreasing rigidity and increasing polarizability of the rings. It is also noteworthy that the presence of the cylindrical bicyclooctyl ring suppresses the formation of smectic phases (Gray and Kelly, 1981). This is particularly useful since "injected" smectic phases often occur in mixtures of nematogens which are not, in isolation,

FIGURE 2.5. Schematic two- and three-ring mesogens.

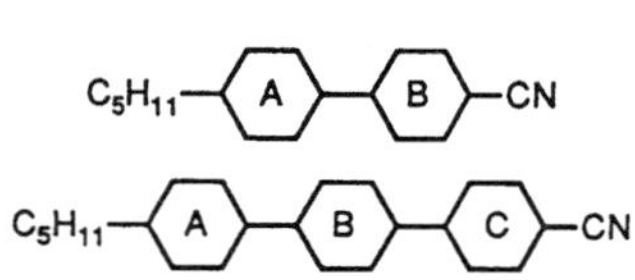

TABLE 2.1. The clearing points of some 4-pentyl-4′-cyano two-ring nematogens (Gray et al., 1973; Gray and Kelly, 1981; Gray, 1982; Pohl et al., 1977; Villiger et al., 1979; Eidenschink et al., 1978; and Inukai et al., 1978).

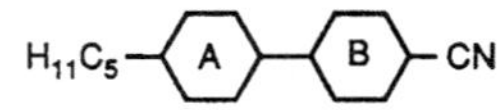

A	B	$T_{N\text{-}I}$ (°C)
		35
		55
		100
		(-25)
		(50)
		85
		129

smectic. Smectic injected phases are particularly common in mixed terminal cyano ($+\Delta\epsilon$) and noncyano ($-\Delta\epsilon$) systems used in multiplexed displays. The $-\Delta\epsilon$ material possibly breaks up the molecular pairs formed by the $+\Delta\epsilon$ material, resulting in the injection of a smectic phase into the system.

There are, however, a number of exceptions to this rule. Nematic phase stability is increased if the polarizable sections of the molecule are adjacent. Thus clearing points are drastically lowered when phenylene rings are not in conjugation (FIGURE 2.7), or when a phenylene ring is separated from another delocalized electron region or from linking or end groups (FIGURE 2.8).

TABLE 2.2. The clearing points of some 4-pentyl-4″-cyano three-ring nematogens (Gray et al., 1974; Gray, 1975; Gray and Kelly, 1981; Eidenschink et al., 1978a; and Villiger et al., 1979).

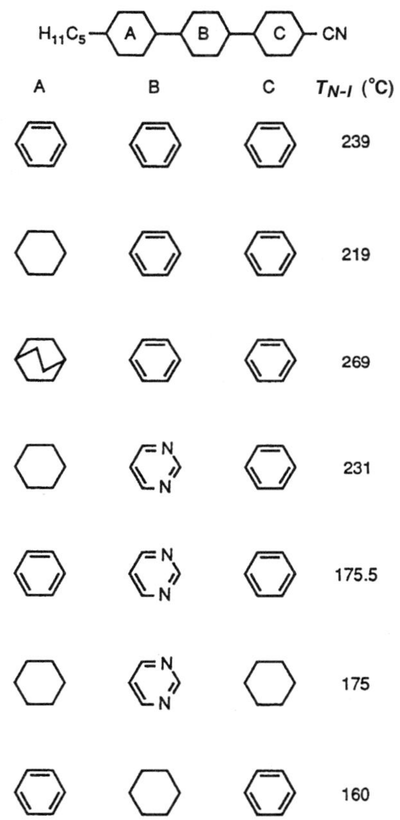

In three-ring systems it is generally observed that the relative order of the clearing points of the phenylene and cyclohexyl analog is reversed (Osman and Revesz, 1980; Osman, 1983; Osman and Huynh-Ba, 1984; and Osman, 1985). It has been proposed that at the generally higher clearing points of these nematogens, there is a significant population of molecules

FIGURE 2.6. The relative nematic stability conferred by different core components.

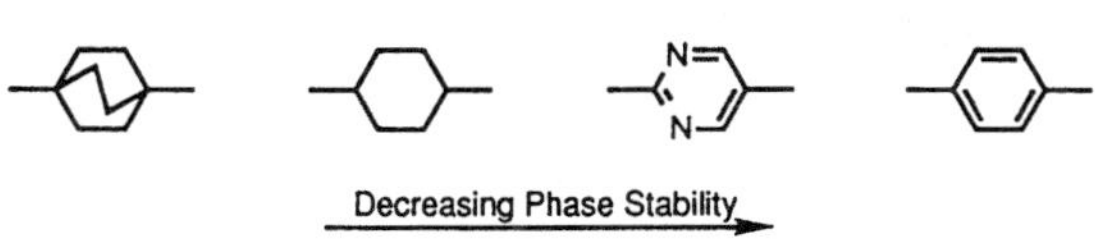

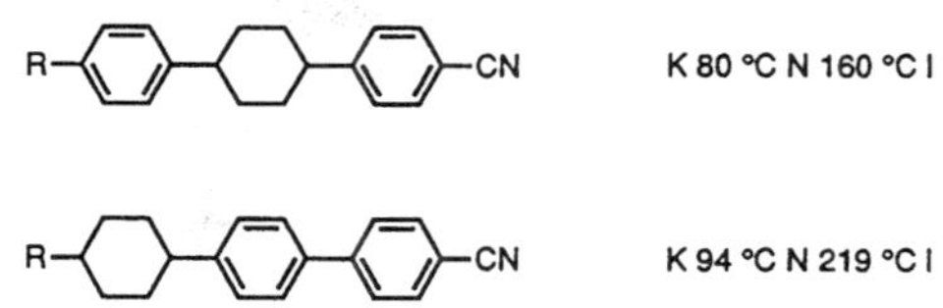

FIGURE 2.7. Illustration of the effect of inter-ring conjugation on nematic phase stability (Eidenschink et al., 1978a).

with a nonlinear energetically unfavorable diaxial conformation (FIGURE 2.7) (Gray and McDonnell, 1979).

The presence of even small amounts of this bent conformer disrupts, and thus destabilizes, the phase. There is no possibility of a comparable situation with either phenylene or bicyclooctane, and thus their relative rankings remain unchanged. Below the lower clearing temperatures of two-ring nematogens, the proportion of the diaxial conformer is negligible. An alternative reason for this reversal is that in the cases studied, a phenyl ring can increase the polarizability anisotropy by extended conjugation with the rest of the core whereas a cyclohexane unit cannot. The advantage of conjugation does not, apparently, outweigh that of the superior rigidity of the bicyclooctyl ring.

A number of oxygen and sulfur heterocyclic nematogens have also been reported and although none of these gives an enantiotropic phase it can be seen from FIGURE 2.10 that the monotropic phase stability correlates with the bulk of the heteroatom, and thus with the length-to-breadth ratio of the molecule.

It is also apparent that the heteroatoms adjacent to the phenylene ring extend the polarizable portion of the molecule, and thereby stabilize the phase with respect to the cyclohexyl analog.

Pyridine and pyrimidine rings have also been incorporated into nematogens, as shown in FIGURE 2.11.

FIGURE 2.8. The effect of end-group conjugation on nematic phase stability (Villiger et al., 1979; and Eidenschink et al., 1977).

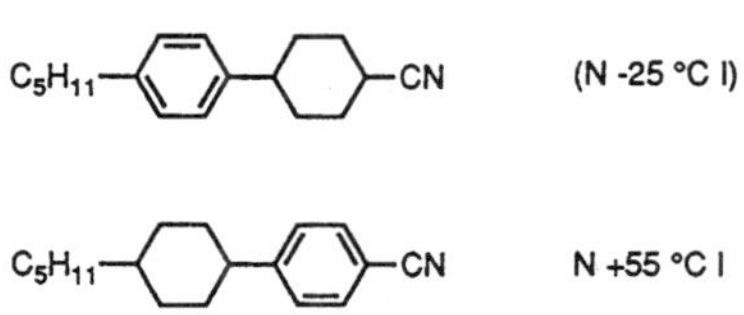

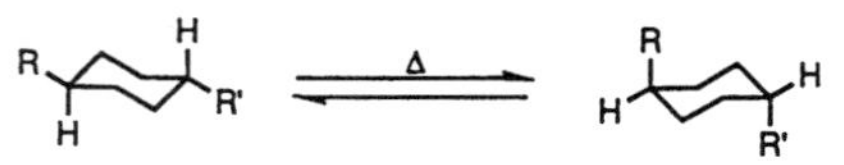

FIGURE 2.9. The temperature-dependant conformation of cyclohexyl core components in three-ring nematogens.

The latter, in particular, are highly polarizable and give more stable phases than their phenylene analogs. However, they also have the effect of raising melting points, thereby increasing the number of monotropic phases.

The comparison of fused ring systems with separate rings gives an interesting insight into the relative roles of molecular shape and polarizability. In the case of fused ring systems, the order of phase stability is as shown in FIGURE 2.12, whereas the order is altered in the nonfused analogs (FIGURE 2.13).

In the latter case, the fall in nematic phase stability mirrors the reduction in molecular polarizability, whereas in the former case the shape of the partially saturated systems dominates; that is, partial hydrogenation of one of the fused rings may increase the polarizability with respect to the fully unsaturated ring, but the resultant ring is neither flat, as in the aromatic parent compound, nor fully staggered. The result is an angular species which cannot pack efficiently in the nematic phase.

VISCOSITY The viscosities of a number of materials at 20 °C are shown in TABLE 2.3.

From this table it is apparent that short saturated cores with lower polarizabilities (and hence weaker

FIGURE 2.10. Saturated heterocyclic core components (Andrews et al., 1985; Haramoto and Kamogawa, 1985; and Eidenschink et al., 1977).

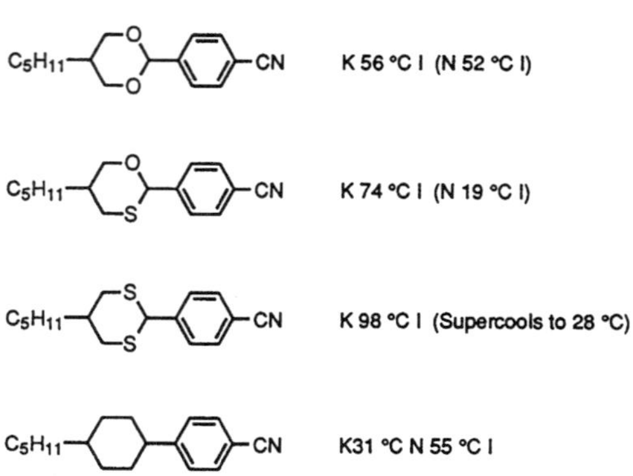

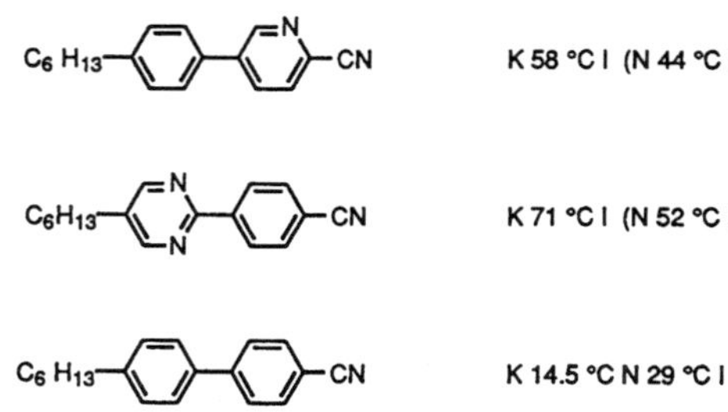

FIGURE 2.11. Aromatic heterocyclic core components (Karamysheva et al., 1981; Boller et al., 1977; and Gray, 1978).

intermolecular dispersion forces) give the lowest viscosity materials, whereas fused rings give more viscous systems. It should be noted that the bicyclooctane ring promotes high viscosity despite its saturation, presumably because of its three-dimensional steric bulk which causes steric hindrance to ease of flow.

DIELECTRIC ANISOTROPY TABLE 2.4 lists the dielectric anisotropies ($\Delta\epsilon$) of a number of nematogens. Detailed comparisons of $\Delta\epsilon$ are not possible from these data since the measurements were made at different reduced temperatures, and in some cases no temperature is quoted. It is nevertheless possible to conclude that the more highly polarizable molecules have the highest dielectric anisotropy; that is, the replacement of pyrimidyl or phenyl rings with saturated units drastically reduces $\Delta\epsilon$. The negligible contribution of saturated rings to dielectric anisotropy is further illustrated by the observation that the addition of an extra saturated ring to a biphenyl nematogen does not affect the dielectric anisotropy markedly. Within certain classes of materials it is also apparent that the locations and strengths of the local dipoles in the system play a part in the dielectric behavior. For

FIGURE 2.12. The relative phase stability of some nematogens with fused ring core components (Gray and Lacey, 1983).

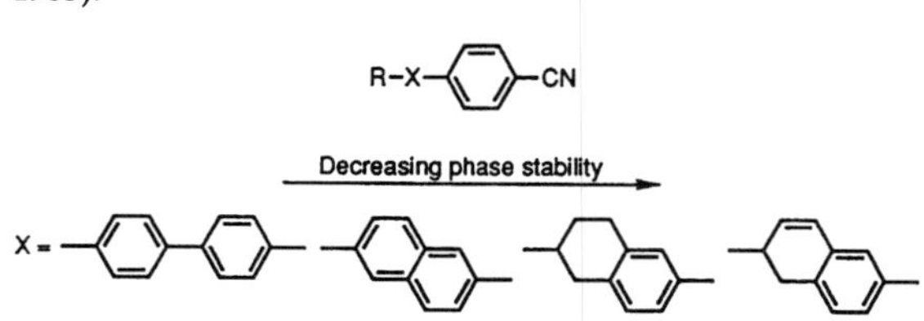

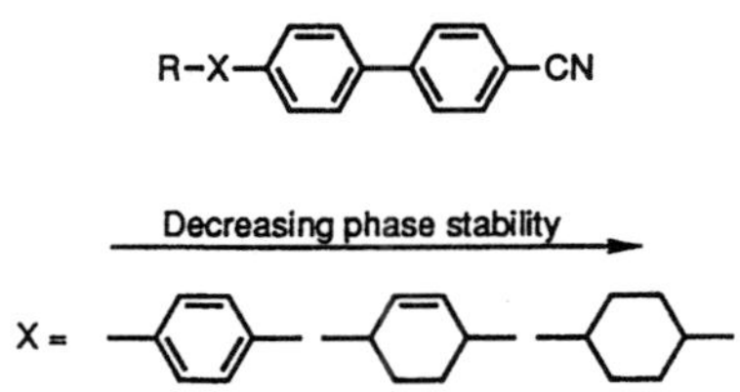

FIGURE 2.13. The relative phase stability of some nematogens with separate ring core components (Gray and Lacey, 1983).

instance, additive dipoles can increase $\Delta\epsilon$, whereas subtractive dipoles can reduce the relative value. This is exemplified in the case of the two pyrimidines in TABLE 2.4.

TABLE 2.3. The viscosities of some terminal cyano nematogens (Pohl et al., 1977; Gray and Kelly 1981; Gray and Lacey, 1983; and Boller et al., 1977).

	Viscosity (cP) 20°C
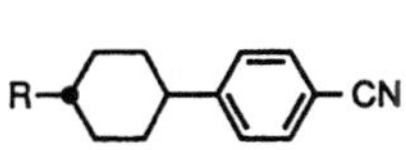	27
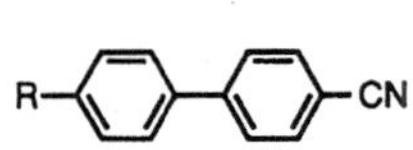	35
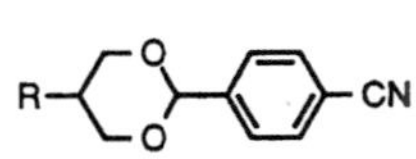	46
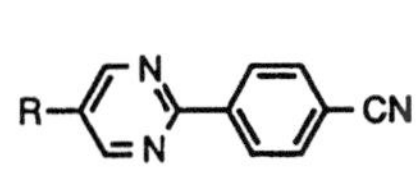	50
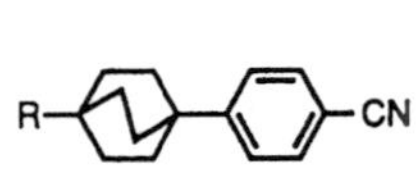	90
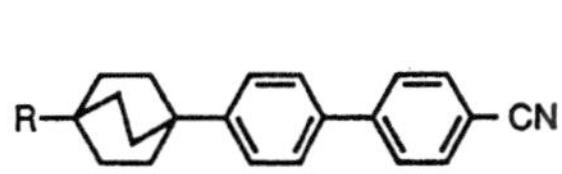	90
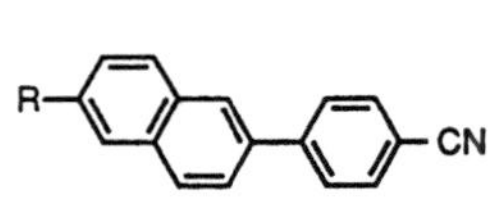	176
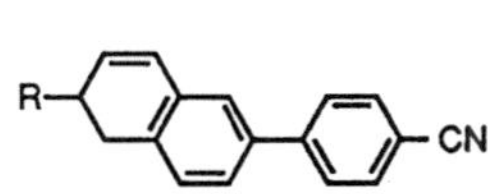	410

TABLE 2.4. The dielectric anisotropies of some terminal cyano nematogens (Pohl et al., 1978; Osman and Huynh-Ba, 1983a; Gray and Kelly, 1981, 1981a; Karamysheva et al., 1981; Haas, 1983; Boller et al., 1977; and Andrews et al., 1985).

	Typical dielectric anisotropy
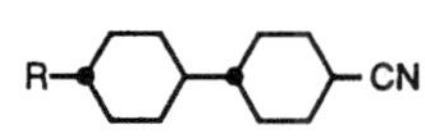	+3 to +4
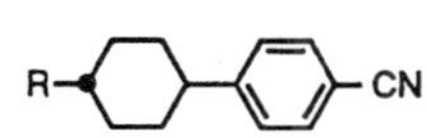	+8.3 at $\Delta T = 0.98$, $R = C_7H_{15}$
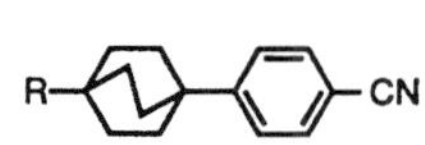	+10 at 25°C
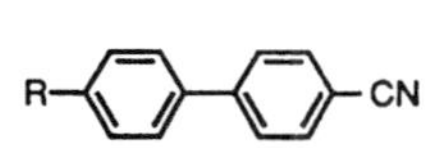	+8.5 at $\Delta T = 0.98$, $R = C_5H_{11}$
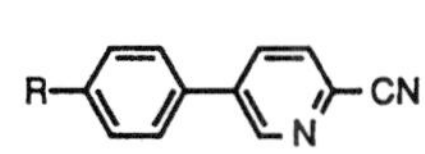	+9.4 at $\Delta T = 0.98$, $R = C_5H_{11}$
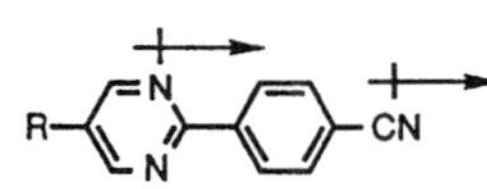	+18
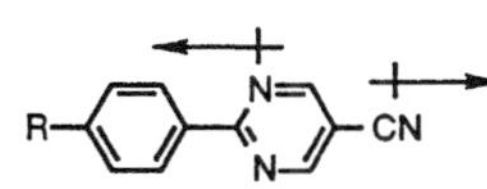	+3.4
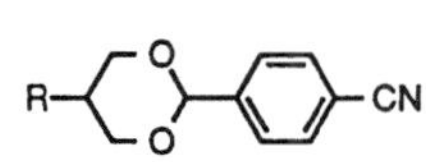	+11 at 42°C
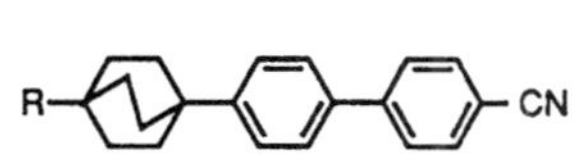	+13.3

BIREFRINGENCE From TABLE 2.5 it can be seen that, broadly speaking, birefringence correlates with aromaticity. It should be borne in mind that a 8 μm twisted nematic display requires that the nematic mixture must have a birefringence of at least 0.14 to give good contrast, whereas guest host displays need as high a birefringence as possible for good contrast.

2.2.2.2 Linking Groups

The introduction of a linking group into the middle of a nematic core increases the flexibility of the mesogen,

TABLE 2.5. The birefringence of some terminal cyano nematogens (Osman and Huynh-Ba, 1984; Andrews et al., 1985; Pohl et al., 1977; Gray, 1978; Gray and Kelly, 1981; and Boller et al., 1977).

	Typical birefringence for $C_8H_{17} > R > C_3H_7$
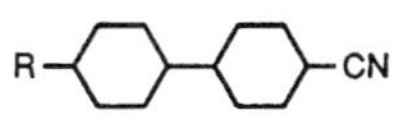	0.04 – 0.05
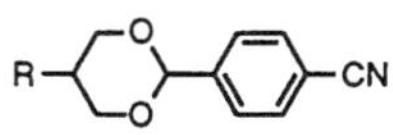	0.08 – 0.09
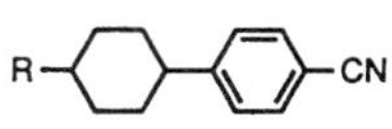	0.11 – 0.13
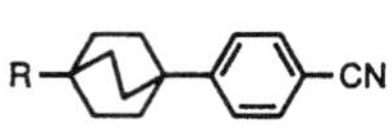	0.14
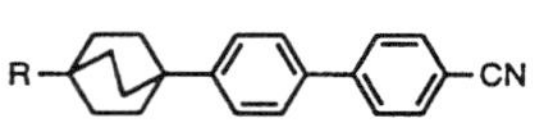	0.14
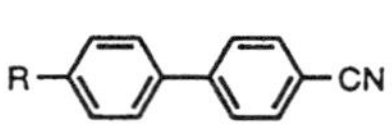	0.18
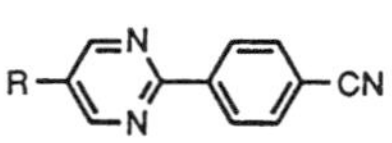	0.18

and is therefore a useful way of reducing transition temperatures and destabilizing the phase. In many cases though, this is at the cost of thermal, chemical or photochemical stability or coloration (Gray and Kelly, 1981a). Thus, although Schiff bases (Vaz et al., 1985; and Byron et al., 1980), diazides (Byron et al., 1985; and Bullock et al., 1984), and stilbenes (Young et al., 1971) have been successfully used in nematic mixtures, their practical uses are limited because of stability problems. The more stable ester function has been used in a large number of molecules (Deutscher et al., 1978; Steinstrasser, 1972; and Coates and Gray, 1978), particularly because of the relative ease of synthesis of these types of materials. Ether (Carr and Gray, 1985) and hydrocarbon linkages (Gray and McDonnell, 1979; and Carr et al., 1983) have also been employed, and these moieties clearly have the highest stability with respect to degradation. In particular, ethylene-bridged systems are very useful in display applications since they have beneficial properties in multiplexed displays.

PHASE STABILITY The three main factors governing the effect of a given linking group on a nematogen are the geometry and flexibility of the link, and compatibility of the link with the core in terms of polarizability and π conjugation. On the first count the link needs to have a favorable geometry in which both ends are at least parallel and preferably colinear. This generally rules out links with an odd number of chain atoms since they cannot meet this criterion without, for example, an atom with a bond angle of 180°. Examples of potentially useful mesogens with single atom links are rare and their phase ranges and stabilities are limited (Mitsuhiro et al., 1984). The number of atoms in the link also affects the flexibility: the more chain atoms there are in, for example, an alkyl link, the lower the population of the linear conformations. Thus examples of links with more than two chain linking atoms are uncommon except in cases in which the link is conformationally restricted by, for example, multiple bonding (Shigetsugu et al., 1984). A number of four-center linked mesogens have nonetheless been prepared and exhibit wide ranging and room-temperature nematic phases (Kelly, 1991, 1991a) (see FIGURE 2.14).

The third consideration, compatibility, is an extension of the arguments used for directly linked rings in section 2.2.2.1; linking groups which extend the conjugation and polarizability of the molecule stabilize the nematic phase whereas saturated nonpolar links pack more efficiently with cyclohexyl rings. Conversely, the separation of two aromatic sections of a core by a nonpolar link can destabilize the nematic phase.

FIGURE 2.14. The transition temperatures of some nematogens with four-center linking groups (Kelly, 1991b).

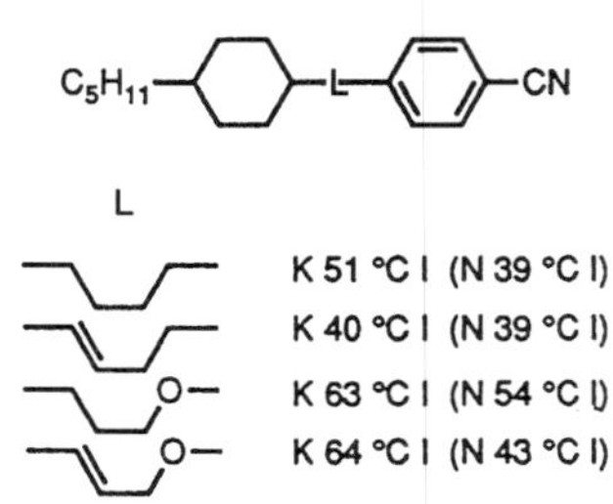

TABLE 2.6. Nematic phase stability as a function of linking group (Gray et al., 1971; and Carr and Gray, 1985).

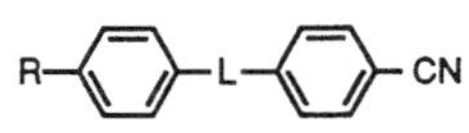

R	L	Transition Temperatures (°C)
C_7H_{15}	—	K 30 N 43 I
C_5H_{11}	CO_2	K 64.5 I (N 57.8 I)
C_5H_{11}	CH_2O	K 49 I [N-20 I]
C_5H_{11}	$(CH_2)_2$	K 62 I [N-24 I]

It should also be noted that the introduction of a linking group into a nematogen increases the overall molecular length, and this, in isolation, may improve the stability of the ensuing nematic phase. The following comparisons are therefore made with molecules of similar overall length.

In the wholly aromatic cases (TABLE 2.6) the introduction of an ester group between the phenylene rings stabilizes the nematic phase, with respect to the parent compound, by extending the polarizable region of the molecule. Replacement of the ester by the oxymethylene and dimethylene groups drastically reduces the phase stability as the two conjugated sections of the molecule are separated by a saturated unit.

In the phenyl cyclohexyl cases (TABLE 2.7) the ester has a similar effect as in the fully aromatic case, but the destabilization of the phase by the saturated linkages is not nearly so marked and may be accounted for by the increased flexibility of the molecule and the apparent increased size of the aliphatic part of the core.

The parent fully alicyclic compound (TABLE 2.8) has no nematic phase, but exhibits a more ordered smectic phase. The stability of this phase is reduced by the insertion of a dimethylene linkage, and the introduction of the more polar oxymethylene group promotes a

TABLE 2.7. Nematic phase stability as a function of linking group (Eidenschink et al., 1978; Gray and McDonnell, 1979; Deutscher et al., 1977; Carr et al., 1983; and Carr and Gray, 1985).

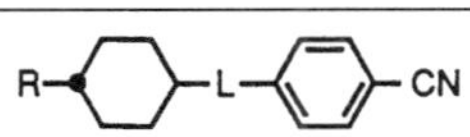

R	L	Transition Temperatures (°C)
C_7H_{15}	—	K 30 N (S_A 17 N) 59 I
C_5H_{11}	CO_2	K 48.3 N 79.2 I
C_5H_{11}	CH_2O	K 74.3 I (N 48.6 I)
C_5H_{11}	$(CH_2)_2$	K 30 N 51 I

TABLE 2.8. Nematic phase stability as a function of linking group (Osman, 1982, 1983b; and Sato et al., 1982).

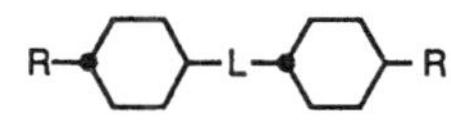

R	L	Transition Temperatures (°C)
C_3H_7	—	K 64.2 (S 58) S 81.8 I
C_3H_7	CO_2	K 22.8 N 36.6 I
C_3H_7	CH_2O	K 6.9 S 8 N 17.5 I
C_3H_7	$(CH_2)_2$	K 34.6 S 73.0 I

nematic phase. Although it might be predicted that the ester function should further destabilize the nematic phase, this is not the case: the isolated region of delocalized electrons enhances intermolecular attractions and stabilizes the phase by nearly 20 °C.

2.2.2.3 End Groups

The role of an end group on a nematogen is to act as either a flexible extension to the core, or as a dipolar moiety to impart dielectric and/or optical anisotropy. Since calamitic molecules have two ends, these may be different, as in 4-alkyl-4′ cyanobiphenyls, or similar, as in the case of dialkyl biphenyls; two similar dipolar end groups on a molecule would, of course, be self-defeating.

PHASE STABILITY It has been established that the general order of phase stability conferred by end groups is as shown in FIGURE 2.15 (Gray and Leadbetter, 1977).

This ranking can be broken down into several (interdigitated) subsets. Thus, for example, the general trend in the halides and pseudo halide (–CN) is for decreased phase stability as the size of the end group is reduced and its electronegativity is increased. The fluoride end group stands out of this ordering because although fluorine is the most electronegative atom, it also has a high electron density in its p orbitals and thus π back-donation of electrons competes with and partially negates the dipolar attraction of electron

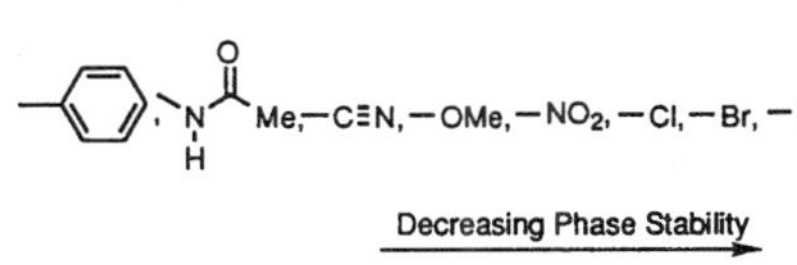

FIGURE 2.15. The relative phase stability conferred by some terminal core substituents.

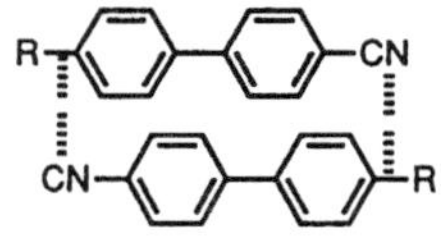

FIGURE 2.16. Antiparallel correlation in 4-alkyl,4″-cyanobiphenyl mesogens (Madhusudana and Chandrasekhar, 1973; Leadbetter et al., 1975; and Lydon and Coakley, 1975).

density through the σ bond. The position of the nitrile function (–CN) in the list is also anomalously high but can be accounted for by dispersion forces between the aromatic rings and nitrile functions of some, but not all, pairs of adjacent molecules. This is shown in FIGURE 2.16.

This antiparallel correlation stabilizes the phase but reduces the dipolar and dielectric properties of the phase. There is also a strong correlation between the viscosity of the nematic phase and the van der Waal radius of the end group (Takatsu et al., 1983). As can be seen in FIGURE 2.17, as the size of the end group is increased the system becomes more viscous.

It is also possible to increase phase stability by the incorporation of an end group which increases the effective length of the mesogen without altering its breadth significantly. Thus, alkyl and alkoxyl end groups can give good phase stability without the need for dipolar functions (Gray et al., 1973; and Gray, 1978). As the length of a terminal chain is increased the phase transition temperatures may either rise or fall, or even fall to a minimum before rising. However, it is to be expected that beyond a certain chain length, the phase stability should decrease as the weaker alkyl chain interactions effectively dilute the stronger intermolecular interactions (even to the extent of promoting smectic phases). There is often also a marked odd/even effect on phase stability as the chain length is increased (Gray

FIGURE 2.17. The phase transition temperatures and viscosities of some terminal halo-mesogens (Takatsu et al., 1983).

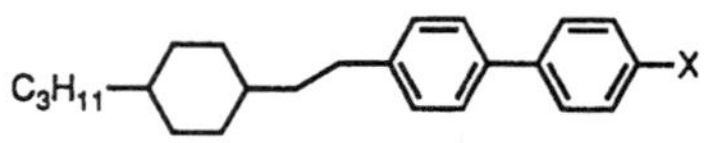

X	K - N	N - I	Viscosity (20 °C)
H	67 °C	82 °C	22.7 cPs
F	76 °C	125 °C	25.1 cPs
Cl	100 °C	158 °C	46.6 cPs
Br	125 °C	163 °C	63.0 cPs

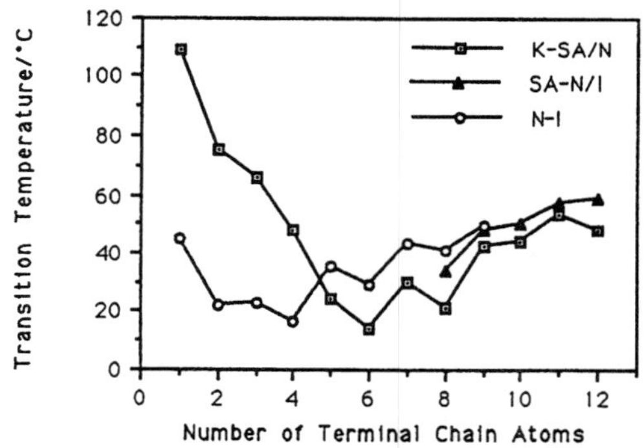

FIGURE 2.18. Illustration of the "odd–even" effect in 4-alkyl 4″-cyanobiphenyls (Gray, 1978).

et al., 1973; and Gray, 1978). Members of the series with an even number of chain atoms tend to lie on a lower curve than the odd members. This is illustrated in FIGURE 2.18.

This may be rationalized in terms of skeletal bond orientation, since in the preferred all-*trans* conformation members of the odd series have more skeletal bonds parallel to the average long molecular axis, and thus have a higher length-to-breadth ratio. This is not to imply that the all-*trans* state is the only conformation adopted by the chain, and indeed as the chain length is increased the probability of encountering *gauche* conformers increases, and the two curves converge.

The comparison of straight chain and branched alkyl end groups serves to illustrate the conformational effects upon phase stability. Methyl branched systems exhibit markedly lower clearing points than their linear analogs (Gray and Kelly, 1984); presumably because the methyl group sticks out at an angle from the molecular rotational cylinder, and thereby disrupts efficient packing by increasing the separation between neighbouring molecules. However, as the methyl branch is moved further away from the core of the mesogen, the depression of the clearing point is reduced. This effect is illustrated in FIGURE 2.19

This demonstrates the greater degree of conformational freedom further away from the core since the effect of branching is less marked as the methyl group nears the more mobile, and thus less linear, end of the alkyl chain.

Finally, one apparent contradiction to this effect is the behavior of terminal cyano and cyanoethyl groups on cyclohexyl cores (FIGURE 2.20). It might be expected

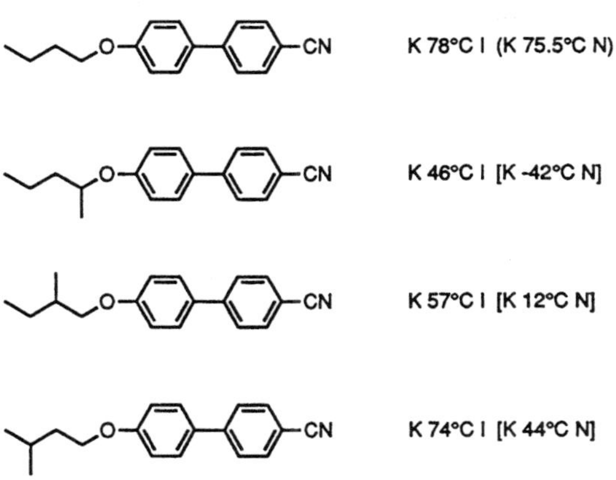

FIGURE 2.19. The effect of terminal chain branching on nematic phase stability (Gray and Kelly, 1984).

that decoupling the polar group from the core might destabilize the phase by increasing the orientational freedom of the nitrile. The converse is however true since the cyanoethyl group packs more efficiently than the nitrile with the staggered "staircase" architecture of the cores (Osman and Huynh-Ba, 1984). Thus the core apparently becomes longer because of this effect, thereby increasing the length-to-breadth ratio.

2.2.2.4 Lateral Groups

The effects of appending a lateral group to a nematic core may be divided into two subsections; steric and electronic. Dealing first with the steric properties, a lateral group effectively widens the core and increases the separation between neighbouring molecules resulting in reduced lateral interactions. Long alkyl substituents (Weissflog and Demus, 1983, 1984, 1985) may also dilute lateral interactions further. All of these effects can combine to destabilize the phase. Side groups may also be expected to increase the viscosity of the fluid as these angular groups inhibit the free flow of the rod-like molecules past one another. The electronic effects may, by definition, be resolved into

FIGURE 2.20. Comparison of the effects of cyano and cyanoethyl end groups on a bicyclohexyl core (Osman and Huynh-Ba, 1984).

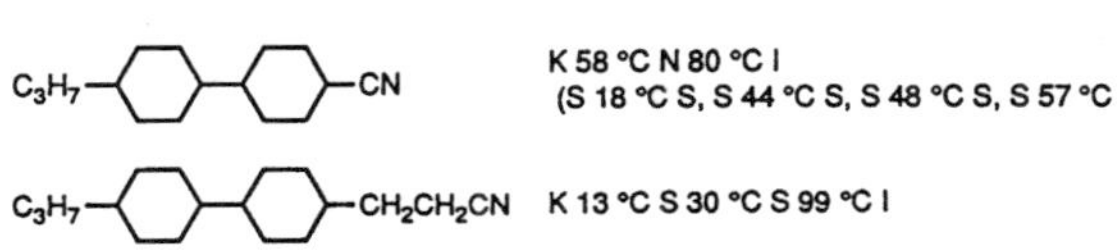

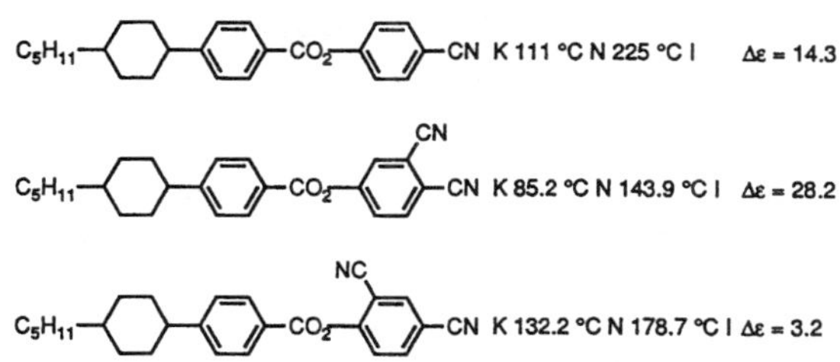

FIGURE 2.21. The effects of lateral cyano substitution on phase stability and dielectric anisotropy (Osman and Huynh-Ba, 1983a).

components along and perpendicular to the long molecular axis; the relative proportions being dependent upon the angle between the substituent and the core. Such groups may therefore enhance either $\epsilon_{\parallel}$ or $\epsilon_{\perp}$, or indeed both, to a greater or lesser extent.

In real examples the electronic and steric effects cannot be totally separated. Some of these considerations are illustrated in the following examples.

In this first example (FIGURE 2.21) there are two points to note. Firstly, the lateral substituent reduces the clearing point by broadening the molecular structure. This effect is most marked when the lateral and terminal nitriles are mutually *ortho*; the *meta* nitrile lies in the structural depression between the two rings, and thus its effect is somewhat shielded. Secondly, the dielectric anisotropy is reduced by the *meta* nitrile because these two dipolar groups are at least partially in opposition to one another. The converse is also true: the *ortho* nitrile adds to some degree to the effect of the terminal group. In this case the enhancement is particularly large, presumably because the lateral substituent sterically hinders some of the antiparallel correlations.

In the second example (see FIGURE 2.22), the parent compound (bottom of the list) shows no nematic phase, but forms a more highly ordered smectic phase. The addition of a nitrile group (top), albeit on the terminal carbon of the core, nonetheless provides a lateral substituent since it is in an axial ring position and therefore orthogonal to the molecular long axis. The dielectric anisotropy is negative and fairly large, especially when compared with the equatorial bicyclohexyl compound (middle). This material also has a particularly low birefringence.

The final example (see FIGURE 2.23) shows the effect of substituent size on the nematic phase.

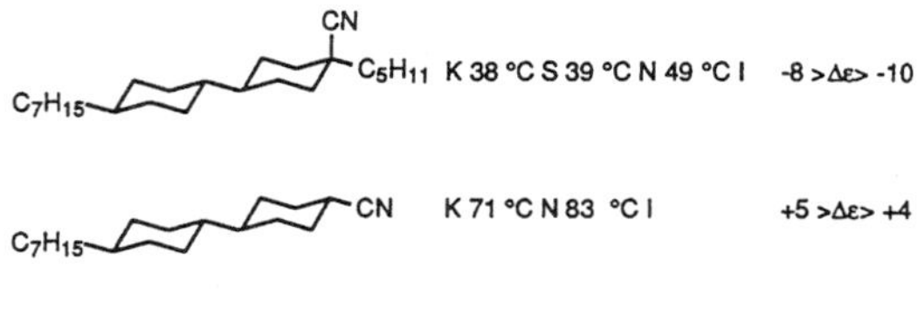

FIGURE 2.22. The effect of axial lateral cyano substitution on phase stability and dielectric anisotropy (Eidenschink et al., 1984; Pohl et al., 1978; Eidenschink et al., 1978; and Osman 1984a).

It can be seen that as the van der Waal radius of X increases, the clearing point is depressed, the viscosity increases and the birefringence falls. The heptyl homolog with an *ortho* fluoro substituent shows a particularly high dielectric anisotropy because it helps to inhibit the formation of antiparallel correlations and the cooperative effect of the dipoles of the terminal cyano and lateral fluoro groups. This material is, however highly viscous.

2.3 CHOLESTERIC LIQUID CRYSTALS

2.3.1 The Chiral Nematic Phase

The addition of a chiral center to a known nematogen forces the director; that is, the vector which describes the average direction of the long molecular axis, to undergo a helical distortion (Oseen, 1933). This may be envisaged as a succession of "layers" of nematic liquid crystals where the director of an individual sheet is slightly rotated with respect to the director in adjacent "layers." This helical distortion is constant from "layer" to "layer" and proceeds in a direction which is determined by the chirality of the constituent molecules. Eventually, the director turns through an angle of 360° and this represents the pitch length of the helix. This situation is illustrated diagrammatically in FIGURE 2.24.

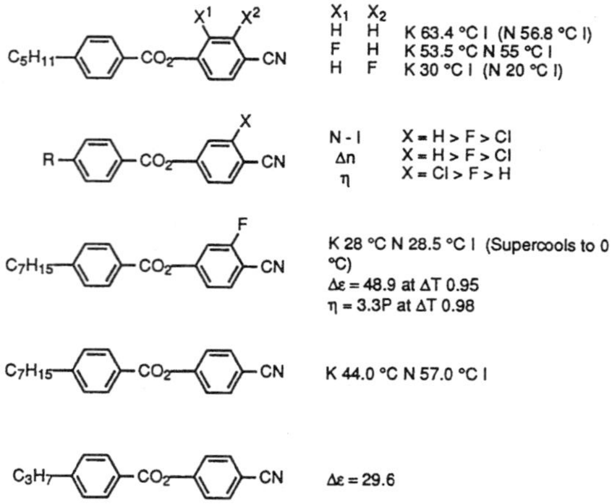

FIGURE 2.23. The effects of lateral halo-substitution on the physical properties of a cyanoester nematogen (Osman, 1985; Kelly, 1983, 1984; Sasaki et al., 1984; Schad and Kelly, 1984; and Gray and Lacey, 1983).

Compounds that possess liquid-crystalline phases with this structure are known as *cholesterogens* (because derivatives of cholesterol (Reinitzer, 1888) were the first compounds to exhibit this phase); however the phase is more accurately described as the *chiral nematic* (N*) phase.

Cholesteric or chiral nematic liquid crystal phases, therefore, have structures and optical properties that are due to the inherent chirality of the constituent molecules. Nonchiral nematic phases can also be made to be chiral by the addition of small amounts of chiral dopant molecules, which may not necessarily be liquid crystal materials at all (Stegemeyer and Mainvsch, 1971). The resultant materials are sometimes called *induced cholesterics*.

When a *racemate* (i.e., an equimolar mixure of both enantiomers of a compound) is resolved (Leclercq et al., 1969) into its constituent enantiomers, each enantiomer exhibits a chiral nematic phase. The helical twist sense of the resultant phases formed by the two enantiomers is of opposite sign, which indicates a direct relationship between the direction of the helical twist, either left- or right-handed, to the absolute spatial configuration of the molecules, *R* or *S* (section 2.3.2) (Gray and McDonnell, 1977a).

This relationship between spatial configuration and helicity can be illustrated by examining contact preparations of the enantiomers by light microscopy between crossed polarizers. If samples of each enantiomer are placed side by side on a microscope slide with a small contact boundary (FIGURE 2.25(a)), the contact region will show an ordinary nematic phase, indicating the formation of a racemic mixture. Away from the contact region, the pure enantiomers retain their cholesteric character. It is important to emphasize that in this case, only the twist sense of the helical structure is different. The pitch length of each enantiomer is identical.

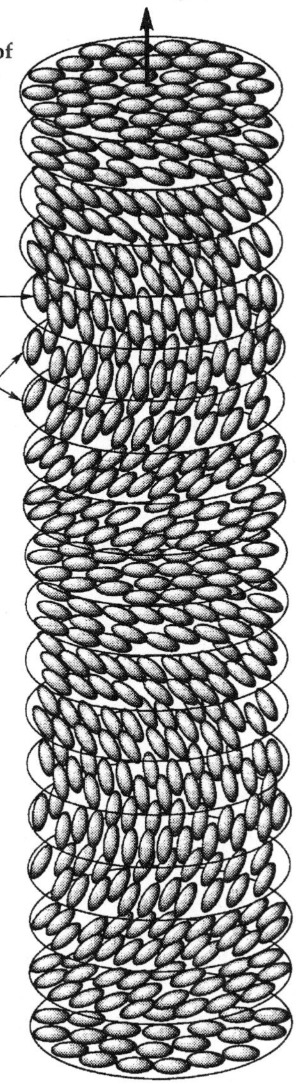

FIGURE 2.24. The structure of the chiral nematic (cholesteric) phase.

Alternatively, if two different compounds of the same handedness are allowed to form a contact in a similar manner, the preparation will show a continuous chiral nematic phase; that is, no discontinuity is found at the contact boundary (FIGURE 2.25(b)).

Examination by polarized light microscopy of a chiral nematic phase shows two distinctly different textures. The first of these (FIGURE 2.26(a)) is the focal–conic texture (Bouligard, 1973) which is analogous to (but not exactly the same as) the smectic focal–conic texture, and forms on cooling from the isotropic liquid. In this texture, there is no selective reflection since the structure of the phase in this state provides a regularly changing distribution of the optic axis leading, simply, to light scattering. If the focal–conic texture is mechanically disturbed by shearing, a second texture, the Grandjean plane texture (Bouligard, 1973) (FIGURE 2.26(b)) is formed where the helical axis of the phase is positioned perpendicular to the glass substrates, as shown in FIGURE 2.4. This texture is particularly important because of the selective reflection of light. When viewed in ambient light, in this texture the cholesteric phase appears free flowing with a highly iridescent color.

2.3.2 The Origin of Thermochromism in Chiral Nematic Liquid Crystals

Plane polarized light consists of left- and right-handed components. When it is passed through an isotropic liquid or solution containing optically active chiral molecules, the left- and right-handed components propagate at different velocities. This causes the plane of polarization of the emerging beam to be rotated to give an effect which is referred to as *circular birefringence*. In sugar solutions, for example, this rotation is of the order of $10^\circ\ mm^{-1}$.

In chiral nematic liquid crystals, the situation is more complex as a result of the two-fold activity caused by both molecular and macromolecular structures. The simple rotation effect caused by the chiral molecules is complicated by an additional interaction of the plane polarized light with the macromolecular helical structure. When plane polarized light propagates through this structure, the left- and right-handed components experience different *refractive indices* which result in a phase retardation between them. The resultant emerging plane of polarized light is rotated to a much greater extent than in simple solutions of optically active molecules and is of the order of $1000^\circ\ mm^{-1}$.

When the pitch length (p) of the helix (FIGURE 2.24) is of a similar order of magnitude to the wavelength of the incident white light, either the left- or right-handed component of the circularly polarized light is reflected depending on the handedness of the constituent

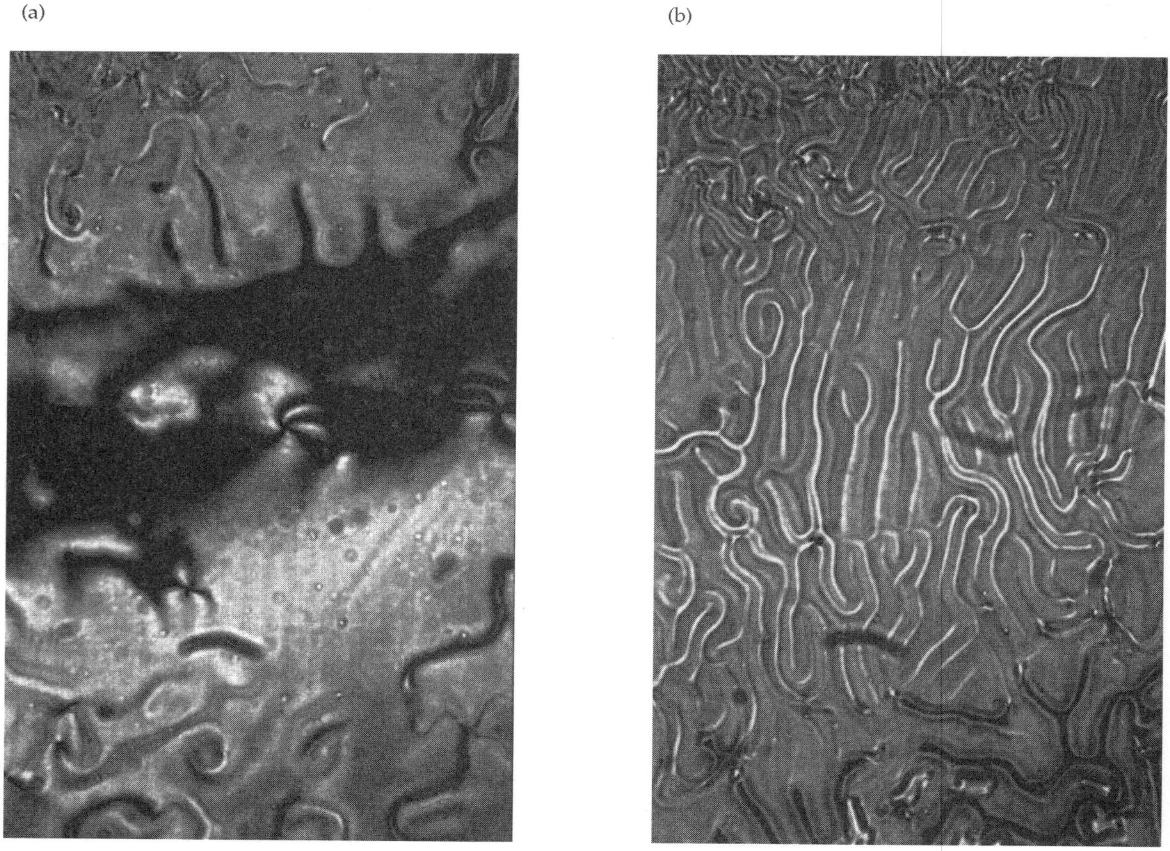

FIGURE 2.25. Contact preparations illustrate the effect of molecular structure on the helical twist sense in the cholesteric phase; (a) two compounds of opposite twist sense show a nematic phase at the contact boundary; (b) two compounds of the same twist sense show a cholesteric texture/phase across the contact boundary (×100).

molecules. The other component of the incident light is transmitted without attenuation.

If one considers the helical structure shown in FIGURE 2.24 as a series of nematic "layers," each layer has an associated director related to the long axes of the molecules. For each layer, therefore, two refractive indices may be defined, n_e parallel to the director, and n_o at right angles to the director in the same plane.

If a plane polarized beam of light of the same wavelength as the pitch length of the helix enters the helical structure, one circularly polarized component will be transmitted and the other circularly polarized component will be reflected (Ferguson, 1966, 1964). This is called *circular dichroism*, and the wavelength for this effect is expressed in the following formula

$$\lambda = \bar{n}p$$

where: $\bar{n}$ = average refractive index, and p = pitch length of helix.

Incident plane polarized light will have half of its intensity reflected. However, ordinary white light is made up of many wavelengths and each individual wavelength will experience a different refractive index, i.e., $\Delta\lambda = (\Delta n)p$ (FIGURE 2.28), thereby producing a reflective color band.

The most important feature of chiral nematic materials is the temperature–pitch relationship. The pitch length of a chiral nematic helical structure shows an

(a) (b)

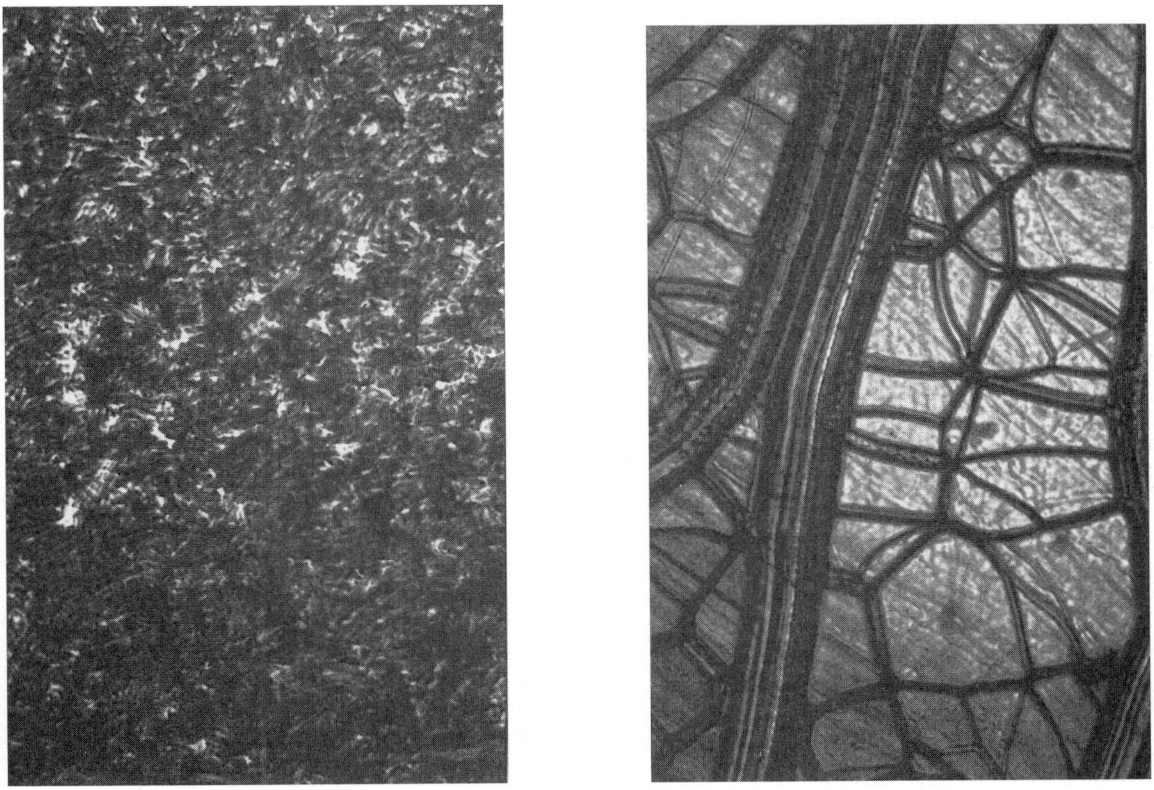

FIGURE 2.26. (a) The focal–conic texture and (b), the Grandjean plane texture of the chiral nematic phase (×100).

inverse temperature relationship, i.e., the pitch length decreases as the temperature rises and vice versa (McDonnell, 1987). In certain circumstances chiral nematics can have underlying smectic A* phases where the pitch is infinite. In the region near the $T_{S_{A^*}}$–N* phase transition, called the *pretransitional region*, very large changes in pitch length can occur over very small changes in temperature (FIGURE 2.29).

It is this relationship between temperature and pitch length, with the associated spectacular thermochromic color changes in ordinary white daylight, that make these materials interesting and useful as thermal sensors. They can therefore be used in thermometry, surface thermography, thermal imaging, shear-flow imaging and for decorative and novelty uses. Some applications of thermochromic chiral nematic liquid crystals will be described in section 2.3.5.

2.3.3 Materials that Exhibit Chiral Nematic Liquid–Crystalline Behavior

The first cholesteric liquid crystal was observed by the Austrian botanist Reinitzer in 1888 (Reinitzer, 1888) when he was performing melting experiments on cholesteryl benzoate (FIGURE 2.30).

Since this time, many other cholesterogens derived from cholesterol and other steroids have been synthesized. However, when considering these compounds for thermochromic applications, it is important to note that they possess some unfortunate disadvantages. Firstly, cholesterol and other steroids are natural

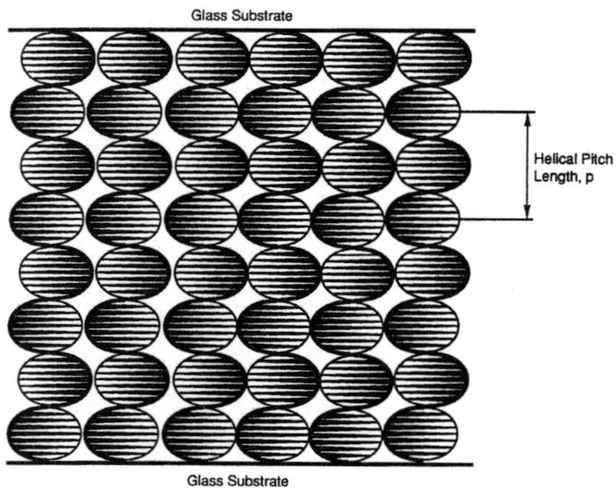

FIGURE 2.27. The structure of the Grandjean plane texture.

products and their enantiomeric and physical purities vary according to their source. The practical use of steroid esters is also limited by their chemical and photochemical stability as well as other physical limitations.

In recent years, nonsteroids have been synthesized in order to produce chiral nematogens. The work of Gray and McDonnell (Gray and McDonnell, 1976) has been particularly important in this field. Their work provided a new range of chiral nematic materials in which chiral alkyl groups have been incorporated into the structures of many compounds which are known

FIGURE 2.28. Reflectance spectrum of a chiral nematic liquid crystal in ambient white light.

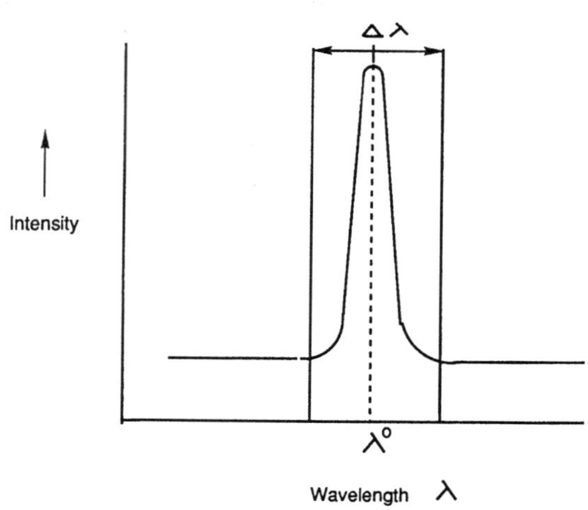

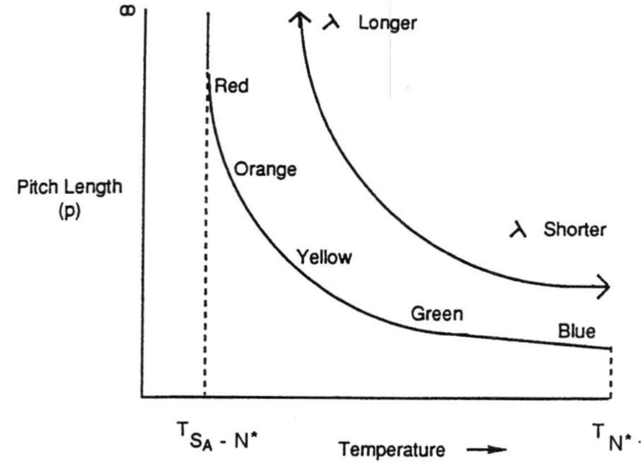

FIGURE 2.29. Changes in helical pitch length with associated color changes around smectic A*–N* pretransitional region.

to show nematic phases. This was principally achieved through the use of (*S*)-(−)-2-methylbutan-1-ol as a chiral alkyl substrate to be incorporated into the structures of standard nematogenic compounds.

The core structures of these nonsteroid chiral nematics are similar to those employed in the nematic liquid crystals which are discussed in section 2.2. When they are compared with the sterol cholesterogens, it can be seen that they are more adaptable for structural variations.

A systematic study of helical twist sense, pitch length and other physical characteristics was made by Gray and McDonnell (Gray and McDonnell, 1977). From this work a set of empirical rules was derived which indicates that the spatial configuration of the chiral center of the molecule is related to the twist direction of the helical structure in the following ways:

Sol	Sed
Rod	Rel

where **R** or **S** is the absolute spatial configuration of the chiral center, **o** or **e** refers to either an odd or even

FIGURE 2.30. Cholesteryl benzoate.

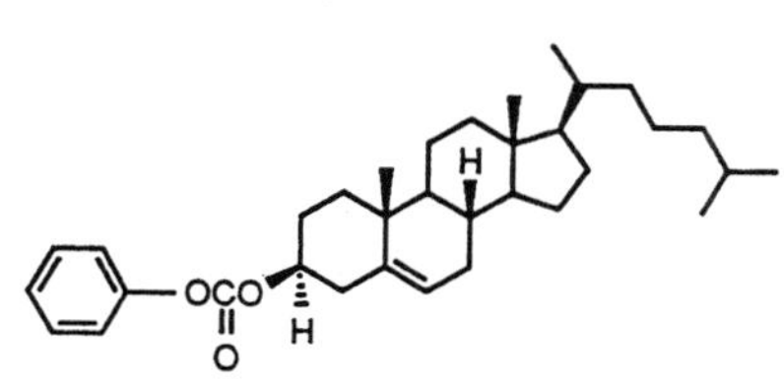

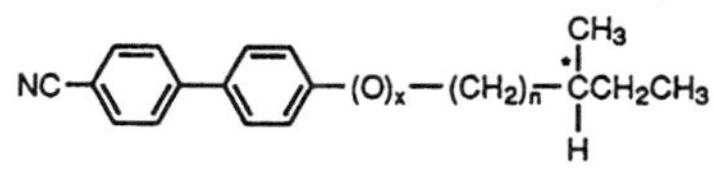

Compound	x	n	o or e	Pitch length (μm)	Optical twist sense
1	0	1	e	0.15	d
2	0	2	o	0.30	l
3	0	3	e	0.40	d
4	1	1	o	1.50	l
5	1	3	o	1.00	l

FIGURE 2.31. Determination of optical twist by Gray and McDonnell rules (Gray and McDonnell, 1977).

position of the chiral center relative to the rigid molecular core, (i.e., the parity), and **d** or **l** refer to either dextro (right) or laevo (left) optical twist senses of the helix.

These rules work well for simple molecules which exhibit chiral nematic phases, but are sometimes less appropriate for more complex systems. Their effectiveness in determining the twist senses of relatively simple systems is exemplified in the series of chiral (*S*)-4-alkyl-4′-cyanobiphenyls (Gray, 1975) shown in FIGURE 2.31.

This example, (FIGURE 2.31), also shows that the closer the chiral center is to the molecular core, the shorter is the pitch length of the helical structure. However, the pitch length of alkoxy compound **4**, where $n = 1$ is unexpectedly long compared with alkoxy compound **5**, where $n = 3$. In general, however, chiral alkyl groups give much shorter helical pitch lengths than chiral alkoxy groups (McDonnell, 1987).

Nonsterol chiral nematic liquid crystals have several other advantages over chiral nematic compounds derived from steroids (McDonnell, 1987). These include the following.

1. Greater in-built chemical and photochemical stability.
2. Higher enantiomeric purities as a result of the availability of enantiomerically pure chiral substrates.
3. Racemic modifications are often readily available for mixing specially compensated mixtures (cf. section 2.3.6).

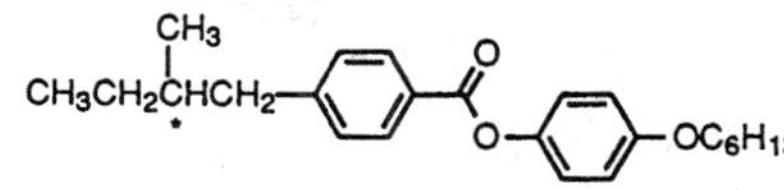

K 22.0 °C (N* 17.0 °C) I

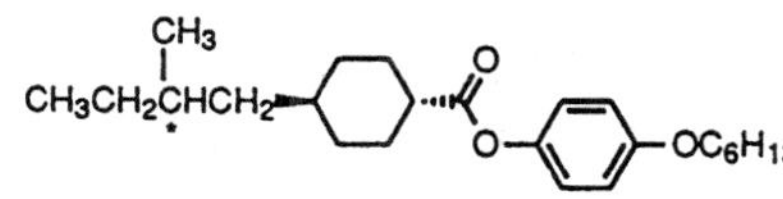

K 50.0 °C (N* 29.5 °C) I

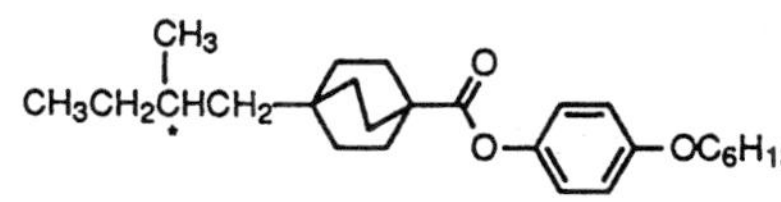

K 43.0 °C N* 50.0 °C I

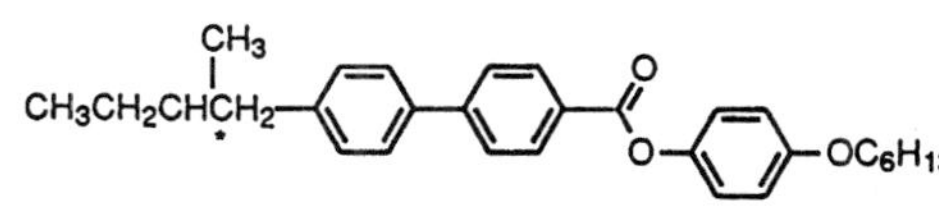

K 103.0 °C N* 115.5 °C I

FIGURE 2.32. Typical nonsterol chiral nematic liquid crystals.

4. Materials can be prepared that do not readily crystallize, which is important in some applications such as microencapsulation (cf. section 2.3.7).
5. The different chemical and physical properties of each component can be optimized/maximized in order to reduce the amount of chiral nematic mixture required to give a desirable response in an application, thus such mixtures can be incorporated into inks, paints and artificial fibers more cheaply (cf. section 2.3.6).

Some typical examples of nonsterol chiral nematic compounds with associated phase transition temperatures (°C) are shown in FIGURE 2.32 (Gray and McDonnell, 1978a, 1983a).

These compounds are chiral modifications of known nematogens. More recently, however, many chiral nematic compounds have been synthesized which have not been designed in this way.

For example, these include a series of (*R*)-1-alkoxy-2-[4-(2,3-difluoro-4′-pentylbiphenyl-4-ylcarbonyloxy)-phenoxy]propanes (Booth et al., 1993) (FIGURE 2.33)

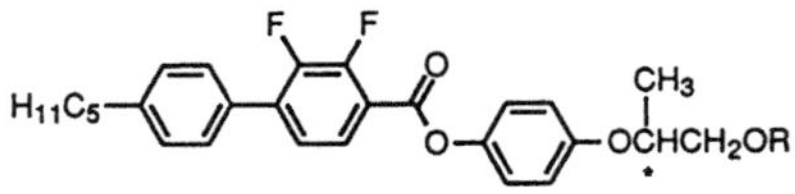

R	K to S_A*	S_A* to N*	N* to I
methyl	34.3	41.4	85.6
ethyl	17.5	45.6	74.5
propyl	18.2	25.1	54.1
butyl	1.4	9.5	32.8
pentyl	2.6	13.6	35.8
hexyl	0.6	15.8	42.1

FIGURE 2.33. Transition temperatures (°C) of the (*R*)-1-alkoxy-2-[4-(2,3-difluoro 4′-pentylbiphenyl-4-ylcarbonyloxy)-phenoxy]propanes.

synthesized from chiral substrates produced as a by-product of a pharmaceutical process.

The authors have also prepared a series of non-fluorinated homologues (Booth et al., 1993b) but only one compound (R = methyl) shows a chiral nematic phase, but this has only a very narrow temperature range ~ 4 °C. In these compounds, therefore, it is the lateral fluorination which appears to enhance the chiral nematic behavior. It can also be observed from the data that the melting point and phase transition temperatures fall as the terminal alkyl chain on the external side of the chiral center is extended. The phase transition temperatures, however, begin to rise again where R = hexyl.

Chirality can also be built into molecular structures by using chiral rings as adjuncts to the core structure, for instance. Scherowsky and Gay (Scherowsky and Gay, 1989) prepared some compounds which include either a chiral thiirane (a) or chiral oxirane (b) group (FIGURE 2.34).

Both of these systems exhibit chiral nematic phases as well as smectic C* phases. The compound which includes a chiral oxirane group also displays a blue phase.

Chiral oxirane units also feature in a series of laterally fluorinated terphenyl compounds (Styring et al., 1993) shown in FIGURE 2.35 which exhibit extra-ordinary phase properties and behavior.

For example, where R = *n*-propyl, the following phase sequence was found (FIGURE 2.36), including

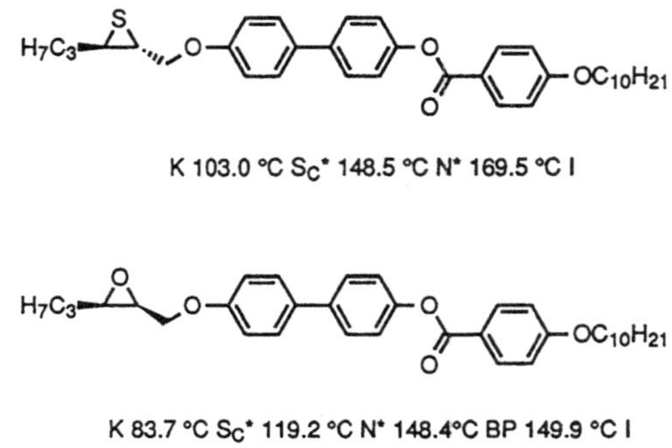

FIGURE 2.34. Chiral nematic compounds which include (a), a chiral thiirane group, and (b), a chiral oxirane group.

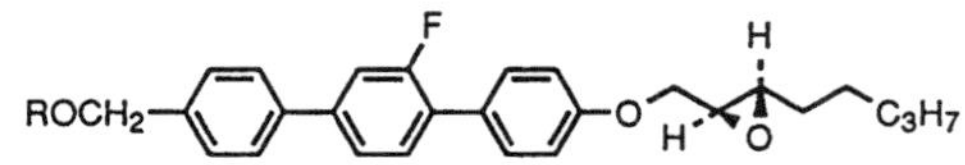

FIGURE 2.35. Laterally fluorinated terphenyl compound which includes a chiral oxirane group.

two separate chiral nematic phases of opposite twist sense which are separated by a nematic phase (or chiral nematic phase of infinite pitch), a sequence which arises as a result of an inversion in helical structure.

The underlying smectic C* phase in the propyl homologue also inverts via a smectic C phase that has an infinite pitch as a result of twist inversion. The twist inversion effect was explained in terms of a conformational model similar to that described in section 2.4.5. The three lowest energy conformations are shown in FIGURE 2.37.

The twist inversion is thought to arise because of a temperature-dependent competition between the

FIGURE 2.36. Transition data (°C) and phase sequence of the 4-[(2*S*,3*S*)-3-propyloxiran-2-ylmethoxy]-2′-fluoro-4″-propyloxymethyl [1,1:4,1″] terphenyl.

Iso	- BPII	164.6	
BPII	- BPI	162.9	BP = blue phase
BPI	- N^*_R	158.8	R and L are the handedness
N^*_R	- N*	112.1	of the helix
N*	- N^*_L	106.3	
N^*_R	- SmC_L*	103.3	
(SmCL*)	- (SmC)	46.8	
(SmC)	- (SmC_R*)	46.2	
M.pt.		59.0	

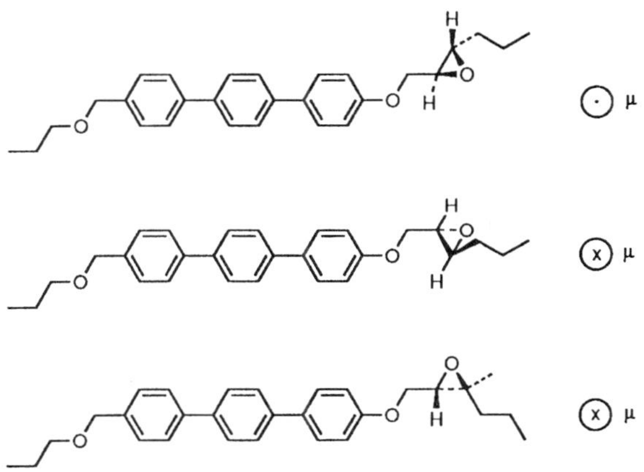

FIGURE 2.37. The three lowest energy conformations which are thought to contribute to the observed inversion of helical twist sense.

different conformers as previously described (Patel and Goodby, 1987). Interconversion between conformers (a) and (b) involves a thermodynamic energy barrier of 11.8 kJ mol^{-1} which is consistent with other studies (Patel and Goodby, 1987a).

The systems described so far place carbon at the center of the chiral moiety. These proliferate because they are more easily synthesized and are, in most cases, more stable. However, Nishide and coworkers (Nishide et al., 1990) produced compounds where the chirality is derived from a chiral 4-alkylsulfinyl group. The example shown in FIGURE 2.38 exhibits both chiral nematic and smectic C* phases.

Chirality can also be introduced into liquid crystal systems via dissymmetric structures that possess only axial symmetry. This contrasts with the systems described so far which have asymmetric point symmetry associated with a chiral atom. For example, Solladie and Zimmerman (Solladie and Zimmerman, 1985, 1985a) produced a series of compounds which

FIGURE 2.38. A chiral nematic compound which includes a chiral alkylsulfinyl group.

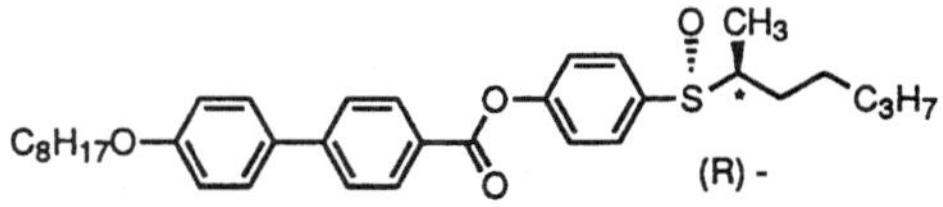

K 155 °C Sc* 158 °C N* 169 °C I

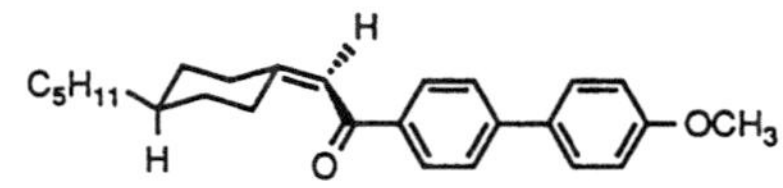

K 65 °C N* 124 °C I

FIGURE 2.39. A chiral nematic compound derived from 1-biphenylyl-2-cyclohexylidene ethanone.

uses a chiral cyclohexylidene ethanone unit to produce axial chirality. An example from this work is shown in FIGURE 2.39.

This compound shows poor light stability which makes it impractical in device applications. It does show, however, that molecules having axial chirality (in this case a 90° dihedral angle between the two major parts of the molecule), can exhibit chiral liquid crystalline properties.

Similarly, Geivandov and coworkers (Geivandov et al., 1989) have synthesized a remarkable series of compounds based on twistane; tricyclo (4,4,0,0) decane. This chiral unit has unique symmetry (D_2) and is chiral, being composed of five fused boat-form cyclohexane rings which are all twisted in the same sense. This chiral moiety exists in two enantiomeric forms (**a**) and (**b**), as shown in FIGURE 2.40. The chirality of twistane is therefore due to the asymmetry (or space asymmetry) of the molecule as a whole and does not depend on the presence of asymmetric atoms.

The series of compounds derived from twistane consist of 8-alkyltwistanol esters. Some of these compounds have been found to exhibit chiral nematic phases, as shown in FIGURE 2.41.

The compounds (FIGURE 2.41(b) and (c)) with the longer alkyl chains (*n*-pentyl) show lower melting points, which is more evident when a cyclohexyl ring (FIGURE 2.41(c)) forms part of the molecular core. These two compounds also show the widest chiral nematic temperature ranges.

Some side chain liquid crystal polymers also show chiral nematic phases. For example, Lewthewaite and

FIGURE 2.40. Enantiomeric forms of tricyclo (4,4,0,0) decane (twistane).

(a) (b)

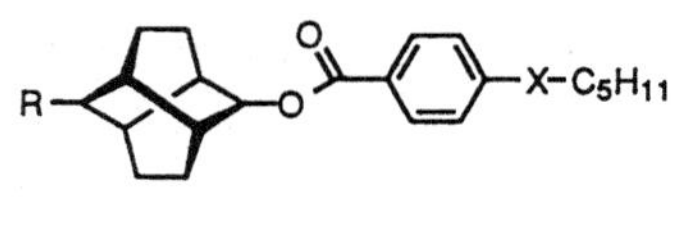

	R	X	K - N*	N* - I
(a)	C_3H_7	Phenyl	94.2	119.7
(b)	C_5H_{11}	Phenyl	71.8	126.4
(c)	C_5H_{11}	Cyclohexyl	65.0	124.7

FIGURE 2.41. Transition temperatures (°C) of compounds derived from 8-alkyltwistanols.

coworkers (Lewthewaite et al., 1993) have synthesized two laterally attached side chain polymers based on (a), a polyacrylate backbone and (b) a polymethacrylate backbone (FIGURE 2.42). Both of these polymers show wide temperature range chiral nematic phases. The presence of the methyl group on the polymethacrylate polymer (FIGURE 2.42(b)) has the effect of depressing the clearing point and increasing the melting point but, in each case, only to a small extent.

2.3.4 Induced Chiral Nematic Mesophases

The addition of small amounts of chiral dopants (solutes) to nematic host materials (solvents) can induce the formation of chiral nematic phases (Friedel, 1922). The twisting ability of chiral dopants to twist nematic phases can be expressed in the following way:

$$\beta = (pcr)^{-1}$$

where β is the twisting power of the chiral dopant, p is the pitch length (μm) of the twist induced in the nematic host, c is the molar concentration of dopant expressed as moles of dopant per mole of host, and r is the enantiomeric purity (ee) of the chiral dopant. The value of β may be either positive or negative, corresponding to right- or left-handed twists, respectively.

An understanding of the way in which the twisting power of the chiral dopant varies with molecular architecture, the structure of the nematic host, and with temperature, can give valuable information about the nature of the chiral nematic phase. In this respect, it appears that the transmission of chirality is from the chiral dopant molecule to the nearest host molecule which in turn influences the next host molecule and so on. This transmission effect continues until the chirally ordered domain attains a metastable dimension. It apparently then thermally disintegrates as it attempts to aggregate further. The value of β is therefore strongly dependent on the molecular structure of the chiral dopant and is related to the molecular structure of the host material. There are no clearly defined guidelines to determine the extent by which a chiral dopant will induce a twist in a nematic phase. A few observations (Gottarelli and Spadar, 1985; and Solladie and Zimmerman, 1984) have been recorded for some chiral dopants which have been devised for this purpose. FIGURE 2.43 shows some examples of chiral nematic dopants; the list illustrates how varied these structures can be, and includes (a) chiral α-bromoesters (Korte and Schrader, 1981), (b) binaphthyl compounds (Gottarelli et al., 1983); (c) epoxides (Gottarelli et al., 1983); and (d), spiranes (Korte and Schrader, 1981).

FIGURE 2.42. Transition temperatures (°C) of two laterally attached side chain liquid polymers.

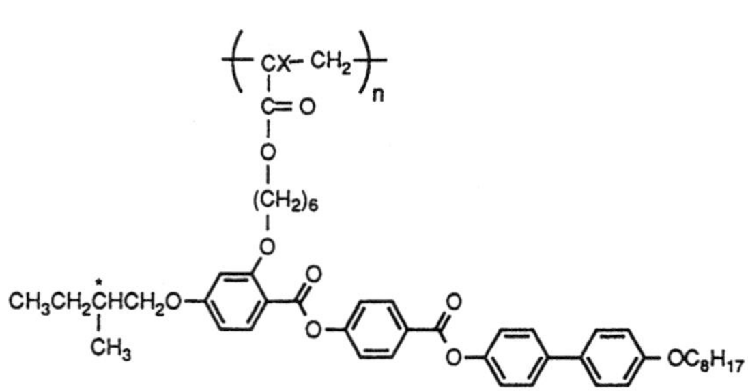

(a) X = H g 30.1 °C N* 110.8 °C I
(b) X = CH_3 g 37.8 °C N* 101.5 °C I

FIGURE 2.43. Some examples of chiral dopants used to induce chiral nematic phases in nematic host materials.

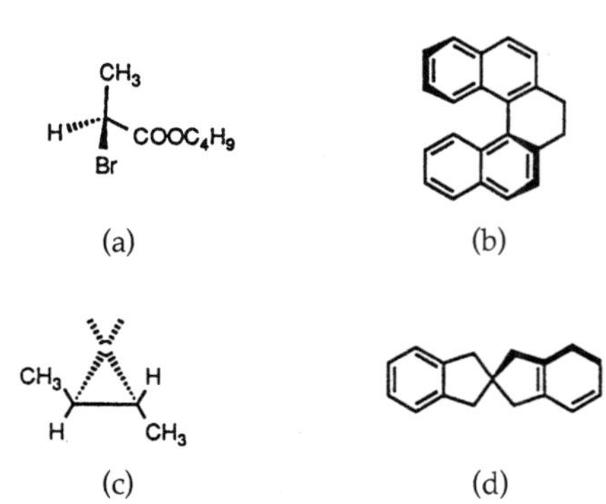

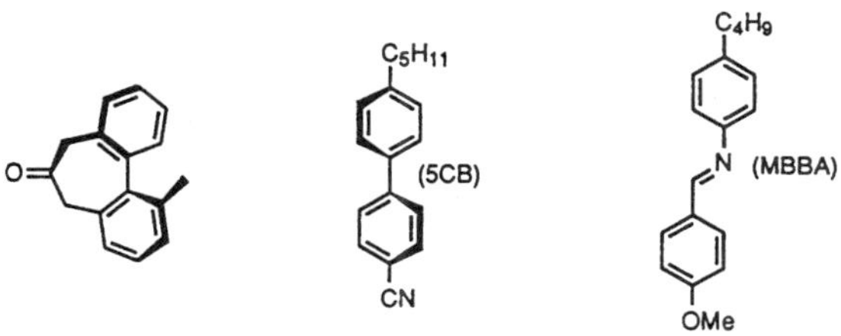

(a) (b) (c)

FIGURE 2.44. Twisting effect of (a) chiral dopant 4,5-dimethyl-9,11-dihydrodibenzo(a,c)cycloheptan-10-one, and two nematic hosts, (b) 5CB and (c) MBBA.

The value of β is very sensitive to the type of nematic host used in the mixture. The value is not only dependent on the solute–solvent interactions of the dopant and host, it is also affected by the nematic twist-elastic constant k_{22}. Therefore, whenever making comparisons between different nematic hosts, the scaling relationship $B_{corr} = Bk_{22}$ should be applied.

Generally, it has been observed (Gottarelli and Spada, 1985; and Solladie and Zimmerman, 1984) that the twisting power of a dopant is higher when the nematic host material has a structure which is similar to that of the chiral dopant. For example, 4,5-dimethyl-9,11-dihydrodibenzo(a,c)cycloheptan-10-one (FIGURE 2.44(a)) has a stronger twisting power in a biphenyl-like nematic phase such as 4-pentyl,-4′-cyanobiphenyl (5CB) (FIGURE 2.44(b)) where $\beta = 15.2$ when compared with *N*-(4-methoxybenzylidene)-4-*n*-butyl-aniline (MBBA) (FIGURE 2.44(c)), where $\beta = 10.2$ (Gottarelli and Spada, 1985).

2.3.5 Some Applications of Thermochromic Chiral Nematic Liquid Crystals.

2.3.5.1 Surface Thermometry

The thermochromic behavior of the chiral nematic phase has been successfully exploited in surface thermometry (McDonnell, 1987) and this field accounts for the most important products developed so far for cholesteric materials. One application (Parker, 1974) utilizes a series of temperature sensing labels with the temperature related to a digital, alphanumeric scale by means of certain selectively reflected colors.

Another application (Hilsum and McDonnell, 1985) utilizes the phase change N*–I–N* sequence. The liquid crystal, in this case, loses its Grandjean texture

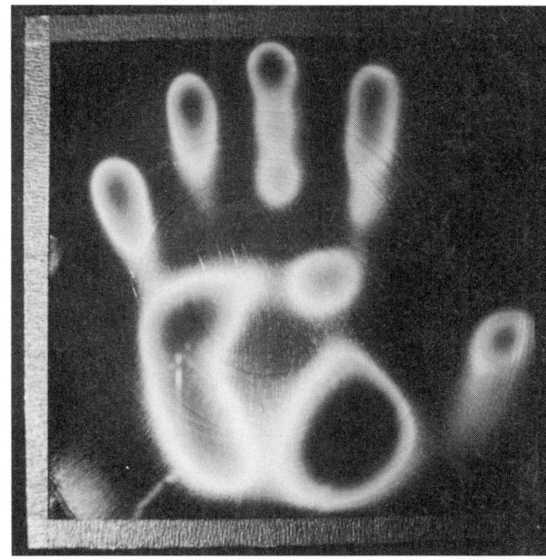

FIGURE 2.45. An example of chiral nematic liquid crystal surface thermometry.

as it becomes isotropic, but on cooling again, the N* phase regains its opaque, focal–conic texture and so no selective reflection is observed. Reorganizing to give the original Grandjean texture can be achieved by mechanically realigning the sample. This form of *maximum* thermometry is used in medical thermometers and in freezing point indicators.

The medical uses of chiral nematic thermography (Nyirjesy et al., 1977) include the use of these materials to acquire visual thermal maps (see FIGURE 2.45) of the body with diseases such as breast cancer, vascular disorders and skin grafting being shown as different temperature zones. However, in order to achieve good thermal contact, the materials must either be painted on top of black stain, which is inconvenient to patients, or stretched in a black elastomeric film over the affected area. Uneven contact in this case can lead to inaccuracies in the recorded data.

2.3.5.2 Electrooptic Displays

The most important use of chiral nematic liquid crystals in display technology is as dopants in twisted nematic displays. Their purpose is to suppress the formation of reverse twist defects (McDonnell, 1987; and Raynes, 1973) which form within TN devices. They are also used to enhance the contrast of the display (White and Taylor, 1974). When used to suppress twist defects, approximately 1% of dopant is required, but to enhance contrast about 10% of dopant is normally used.

2.3.5.3 Decorative Applications

The visual appeal of color changes with temperature has led to the development of several decorative applications of chiral nematic liquid crystals (Makow, 1979; Mace et al., 1981; Kurusawa, 1981; and James, 1974). Some of these exploit the temperature dependence of the selective region of the N*–S transition which can be raised by the application of pressure (Shashighar and Chandrasekha, 1975; Pollman and Stegemeyer, 1973; and Merran, 1970) or lowered by the addition of impurities. The effect of these influences is to produce color changes at a constant temperature.

Recently, interest has centered on the use of these materials for decorative purposes on large areas. Their use, in this case, is not primarily to obtain a thermochromic effect but to impart a characteristic optical quality such as a tint or iridescence.

These systems may be paint-based or polymeric films or coatings. The paint systems can be employed for a large number of applications. The polymeric films exhibit optical properties which allow them to function as filters and have mechanical properties similar to other oriented polymers (Hara et al., 1988; Chen and Jones, 1987, 1988; and Polk et al., 1988).

2.3.6 Thermochromic Liquid Crystal Mixtures

Some of the applications described in section 2.5 require thermochromic materials with low melting points, short pitch lengths and smectic A*–N* transitions which are just below the temperature-sensing region. Often, no single compound is capable of meeting these requirements, but mixtures of compounds can be formulated to provide the required properties. Commercially available mixtures include some which are sold under the trade name "Licritherm" by Merck UK (Liquid Crystal Product Data Sheet, Merck (UK)). These mixtures have colorplay ranges from −30 to +70 °C, and individuals can request mixtures to suit their own requirements.

An example of a typical chiral nematic mixture system (Merck UK) consists of combinations of the homolog of the phenyl- and biphenylesters shown in FIGURE 2.46.

(a)
(b)
(c)

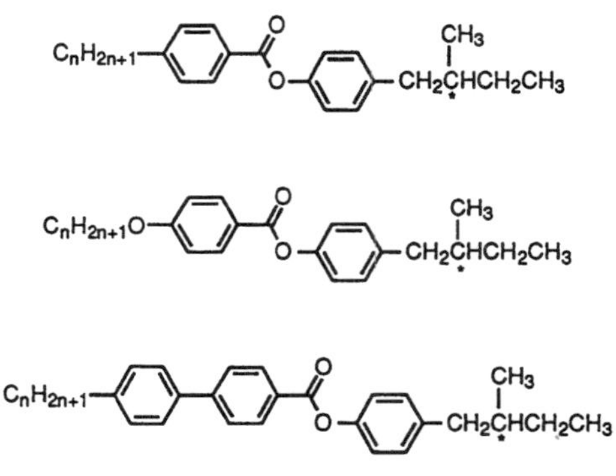

FIGURE 2.46. Phenyl- and biphenylesters used in thermochromic liquid crystal mixtures.

These esters incorporate the same optically active chiral 2-methylbutyl group which is appended to a phenyl ring and attached via an ester likage to either an alkyl- or alkoxyphenyl or alkylbiphenyl group. The mixtures are produced as high- (TM75A) (FIGURE 2.47) and low-temperature (TM74A) formulations which are mutually miscible in all proportions. On mixing these two formulations (TM75A and TM74A), the temperature at which a desired color play begins can be selected. This can be achieved at any point between the lower and higher S_A*–N* transition temperatures of TM74A and TM75A respectively, by mixing them in accordance with the phase diagram shown in FIGURE 2.47.

If the mixture contains 50% of the racemic modifications of the components of the mixture, the temperature range of the color play can be extended from approximately 0.5 to 5 °C (TM74B and TM75B). Other mixtures can also be obtained which have thermochromic properties to the higher temperature of 110 °C.

2.3.7 Microencapsulation of Thermochromic Liquid Crystals

Microencapsulation is a process in which solid particles or droplets of a liquid are coated with a resilient and tough protective wall (Vandegaer, 1974; Nixon, 1976; and Kondo and Valkenburg, 1979). There are numerous methods and processes for encapsulation including simple coacervation, complex coacervation, encapsulation by phase separation from aqueous solution, interfacial polymerization, *in situ* polymerization,

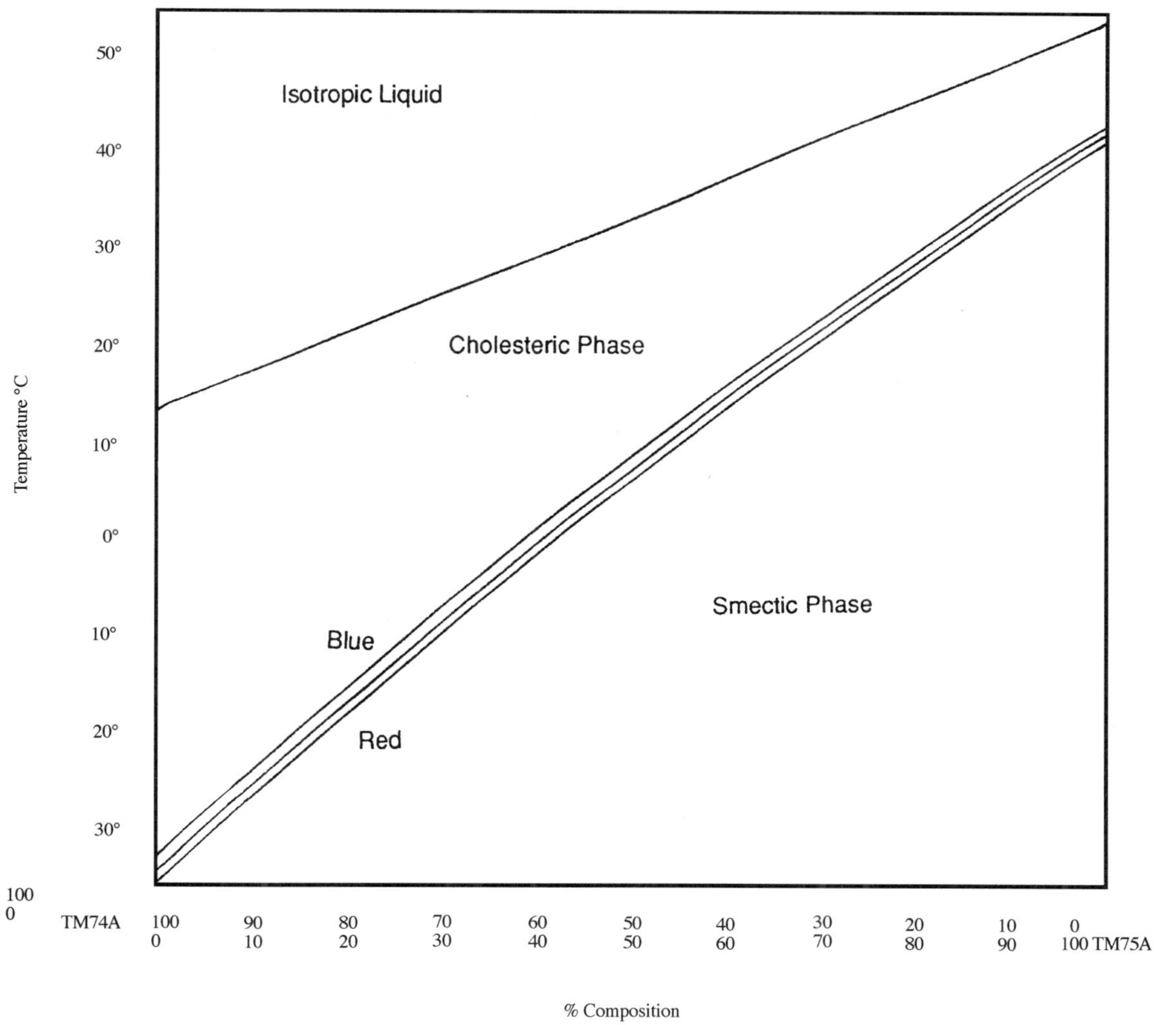

FIGURE 2.47. Phase diagram for mixtures of TM74A and TM75A (Merck UK).

and solidification in the liquid (orifice process). Detailed descriptions of these processes are beyond the scope of this article but can be obtained from the appropriate literature.

Encapsulation of thermochromic liquid crystals is necessary for the following reasons (Booth, 1992).

1. Chiral nematic liquid crystals are oily, viscous fluids which are not easily manipulated in this form.
2. The wall of the capsule protects the liquid crystal mixture from both atmospheric and other contaminants which may cause the colorplay to deteriorate.
3. For optimal reflectance properties the microcapsule must consist of about 20% of its weight of liquid crystal, the amount depending on application and use, and for this reason, microencapsulation provides an economic alternative to using a neat thermochromic mixture.
4. Undesirable crystallization of the liquid crystal is inhibited as the droplets of material are separated by the coacervate walls.
5. Microcapsules with different thermochromic ranges can be mixed to give independent or

overlapping regions of colorplay. This would not be possible when allowing neat mixtures of thermochromic materials to mix and has obvious attractions in the field of decorative applications.

2.4 FERROELECTRIC LIQUID CRYSTALS

2.4.1 The Origin of Ferroelectricity

The ferroelectric effect was first observed in 1921 by Valasek (Valasek, 1921) during experiments on crystals of Rochelle salt (potassium sodium tartrate). When a sample of nonferroelectric material is placed between two capacitor plates and an electric field is applied, the permanent or induced dipoles align with the field. This alignment causes a net imbalance of charge across the material and the sample becomes polarized in a direction opposite to that of the applied field. The polarization disappears when the electric field is removed. A marked departure from the normal dielectric behavior is observed when Rochelle salt is placed between the capacitor plates and subjected to a similar applied electric field. In this case, a residual polarization is observed after the electric field is removed. Valasek suggested that this behavior could be due to a hysteresis in the crystal which is analogous to ferromagnetic hysteresis, and his experiments confirmed this theory. The term ferroelectric was derived, therefore, from the analogous phenomenon of ferromagnetism, but ferroelectricity is not related to iron.

In order to understand the ferroelectric effect, it is necessary to examine the physical changes which occur in a material when it is subjected to mechanical stress. Rochelle salt is unusual in that it exhibits a *piezoelectric* effect, that is, an induced polarization resulting from a structural change in the crystal due to the application of mechanical stress. In this case, a structural change from an orthorhombic to a monoclinic crystal structure results in a lowering of symmetry in the stressed crystal. This causes an imbalance of electric charges within the material which induces the ferroelectric effect. In 1975, the American physicist Meyer suggested that liquid crystals in a tilted smectic phase could also exhibit ferroelectricity (Meyer et al., 1975). The physical basis of the ferroelectric effect in tilted liquid crystal phases is described in section 2.4.3.

Ferroelectricity in tilted smectic phases requires that the molecules are *chiral*. The concept of chirality, and how it influences the tilted smectic phase, is discussed in the following section. In this section the discussion will be confined to the chiral smectic C (S_C*) phase since materials exhibiting this phase are being investigated for device applications; however, the same arguments could equally apply to other tilted chiral smectic phases such as chiral smectic F* or I*, and crystal smectic phases H*, J*, K* and G*.

2.4.2 Chirality and the Smectic C (S_C*) Phase

The smectic C liquid crystal phase is a layered modification composed of molecules whose long molecular axes are inclined at a temperature-dependent tilt angle, θ, to the layer normal (FIGURE 2.48). Within each layer there is no positional order of the molecules with the correlation length extending only over a few molecular centers, but the tilt direction remains relatively uniform over large domains. The smectic C phase may be thought of as having been derived from the orthogonal smectic A phase by simple ordering of the random tilt in the A phase, a relationship which is reflected in a low enthalpy change in the transition from the smectic A to the smectic C phase. Alternatively the smectic C phase can be derived on cooling the nematic phase. Here the phase transition has been shown to be first order usually, although second-order transitions have been found for a few systems.

The chiral smectic C* phase is composed of *optically active* molecules and the optical activity arises as a result of molecular asymmetry. Generally, this takes the form of an *asymmetric* carbon atom, that is, a carbon atom which is bonded to four different groups which can be arranged in two different ways to produce two optical isomers or *enantiomers*. The two

FIGURE 2.48. Structure of the smectic C liquid crystal phase.

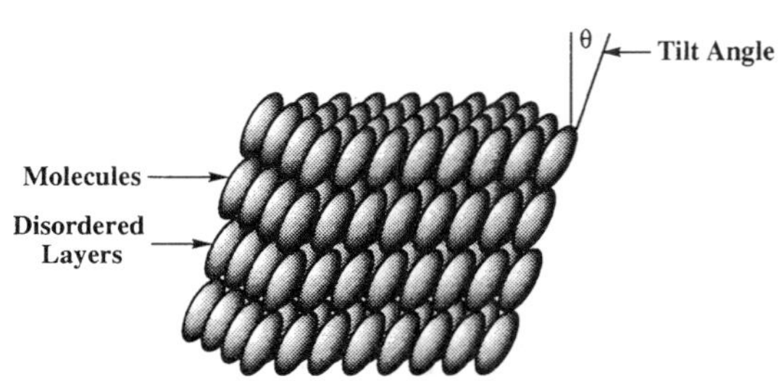

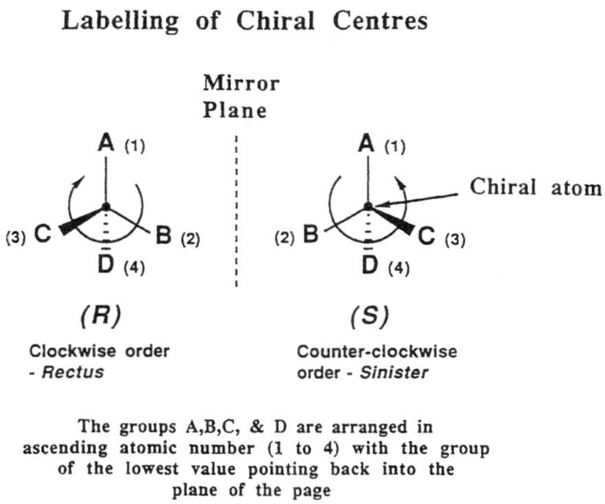

FIGURE 2.49. Spatial configuration and labelling of chiral centers.

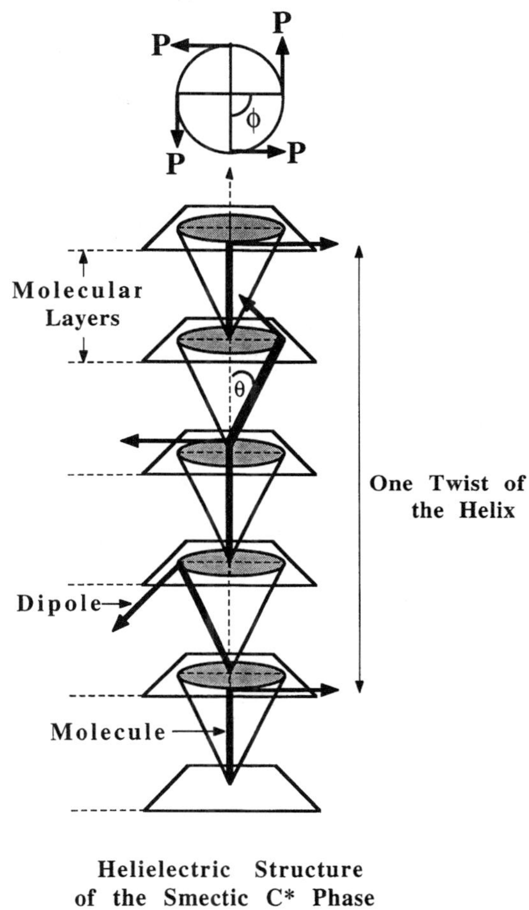

FIGURE 2.50. The structure of the chiral smectic C* phase.

optical isomers rotate plane polarized light in opposite directions to each other, a situation which is denoted by (+) when the optical rotation is to the right and (−) when the optical rotation is to the left.

The enantiomers are, however, more conveniently identified by determining their absolute spatial configuration using a system of priorities based on the atomic numbers of the atoms which are directly attached to the asymmetric carbon atom (see FIGURE 2.49) (Cahn and Ingold, 1951; and Cahn et al., 1956). The groups A, B, C, and D in FIGURE 2.49 are arranged in ascending atomic number, 1 to 4, with the group of the lowest priority pointing back into the plane of the page. When the atomic numbers ascend in a clockwise direction, the enantiomer is described as (*R*−) (*rectus*) and in the anticlockwise direction as (*S*−) (*sinister*). These directions, however, bear no relation to the optical rotation so enantiomers may be described as (*R*)(−), (*S*)(+), (*R*)(+) or (*S*)(−).

When the smectic C phase is composed of optically active molecules, a macroscopic helical arrangement of the molecules is formed. The helix occurs as a result of a precession of the molecular tilt about an axis perpendicular to the layer planes as shown in FIGURE 2.50. The tilt direction is rotated through an azimuthal angle (ϕ) on moving from one layer to the next. As the rotation is in a constant direction, a helix is formed which is either left- or right-handed. The helical twist sense is determined by the nature and position of the chiral center with respect to the central core of the mesogenic material. One 360° rotation of the helix for the smectic C* phase usually extends over hundreds of layers and the azimuthal angle is generally found to be of the order of 0.1–0.01°.

2.4.3 Ferroelectricity in Chiral, Tilted Smectic Liquid Crystal Phases

In section 2.4.1 it was noted that ferroelectricity in Rochelle salt arises as a result of the lowering of the symmetry elements as the crystal changes from orthorhombic to monoclinic under mechanical stress. The sample then becomes polarized as a result of dipole

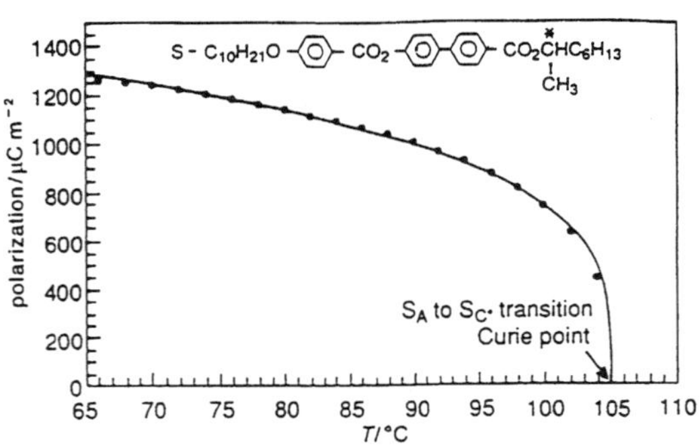

FIGURE 2.51. Increase in the spontaneous polarization as the temperature decreases.

alignment when subjected to an applied electric field. In the smectic C* phase, the *spontaneous polarization* (P_s) arises when the system is cooled from the preceding smectic A*, cholesteric or isotropic phase, initially rising very rapidly but thereafter rising more slowly, as shown in FIGURE 2.51. Typically, for a second-order phase transition, the spontaneous polarization is coupled to the change in tilt angle which occurs on cooling the mesophase. The tilt angle varies in the following way:

$$\theta_T = \theta_0 \, (T_C - T)^{\alpha}$$

where θ_T is the tilt angle at temperature T, θ_0 is a constant, T_C is the transition temperature for the A to C transition (Curie temperature), and α is an exponent theoretically equal to 0.5. The spontaneous polarization varies in a similar way, taking the form:

$$P_s = P_0(T_C - T)^{\beta}$$

where P_s is the spontaneous polarization at temperature T and β is an exponent theoretically equal to 0.5. The spontaneous polarization is usually measured in units of nanocoulombs per square centimeter (nC cm^{-2}).

With the lowering of temperature, the rotational freedom of the molecules, and the conformations of the molecules change as the tilt angle increases. These changes are analogous to the lowering of symmetry when Rochelle salt is transformed from an orthorhombic to a monoclinic structure.

For an achiral smectic C phase, the environment or local symmetry elements consist of a center of inversion, a mirror plane normal to the layers and a C_2 axis parallel to the layers and normal to the tilt direction

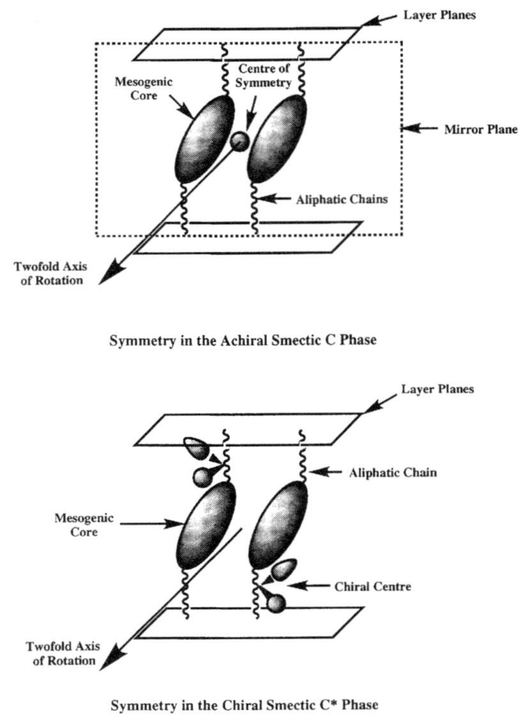

FIGURE 2.52. The symmetry in smectic C and smectic C* phases.

(see FIGURE 2.52(a)). This combination of elements gives C_{2h} symmetry.

When some of the molecules in the mesophase are made to be chiral, the center of symmetry and mirror plane are lost and only the C_2 axis remains, (see FIGURE 2.52(b)). As the molecules are polar, an imbalance arises with respect to the molecular dipoles along the C_2 axis (Meyer, 1976). This imbalance occurs even though the molecules are undergoing rapid reorientational motion about their long molecular axes. This time-dependent alignment of the dipoles along the C_2 axis causes the spontaneous polarization (P_s) to develop along this direction and parallel to the layer planes. Each individual layer, therefore, essentially has a spontaneous polarization associated with it, and since the layers are stacked one on top of the other in a helical arrangement (FIGURE 2.50), the layer polarizations are consequently averaged out to zero and the phase is therefore described as *helielectric* (Brand et al., 1985). However, if the helix is unwound, the layer polarizations point in the same direction and the phase then becomes ferroelectric.

The arguments for this reduction in symmetry are illustrated in FIGURE 2.52(b). In the layers there are as many molecules pointing up as there are down, and this general situation is illustrated with just two molecules to represent the whole layer. If we assume that each molecule has a lateral dipole pointing out of the plane of the paper then we see that the dipoles are pointing in the same direction and the diagram has a twofold axis pointing out of the page. The layer structure also has no center of inversion or mirror planes. A degenerate structure with the dipoles pointing back into the page cannot be produced, therefore, by simply rotating the molecules about their long molecular axes. An energetically equivalent layer structure can only be produced if the molecules are tilted in the opposite direction in the plane of the page with their dipoles pointing into the page.

This rotational degeneracy is used in the concept of fast switching surfacestabilized ferroelectric liquid crystal (SSFLC) displays (Clark and Lagerwall, 1980). In these displays, the smectic C* molecules are aligned parallel to the surfaces of a thin glass cell. The cells have to be very thin ($\simeq 2\,\mu$m) to allow the boundary forces to unwind the helix of the smectic C* phase. The direction of polarization is perpendicular to the glass plates of the cell, but by applying a small DC potential across the plates, the polarization direction can be changed from up to down and vice versa. This causes the molecules to rotate about a degenerate energy cone thereby changing direction through an angle of 2θ, as shown in FIGURES 2.53(a) and (b). If the cell is placed between crossed polarizers, this change can be observed optically and the whole cell operates as a half-wave plate. The optimum molecular tilt for this cell is 22.5° so that the change in angle described by the molecules as they undergo a directional change from one orientation to the other, is ideally 45°, (i.e., $2\theta = 45°$). In thin cells the two orientation states can be surface stabilized producing a bistable operating regime with an energy barrier between them. In this configuration the cell can be switched much more quickly (10–100 μs) than the conventional twisted nematic cell (10 ms). The good viewing angle and the bistable operation are the main advantages of SSFLC displays over twisted nematic (TN) displays. However, other factors, including the rotational viscosity of the liquid crystal material, can also have a profound effect on the switching times of SSFLC displays.

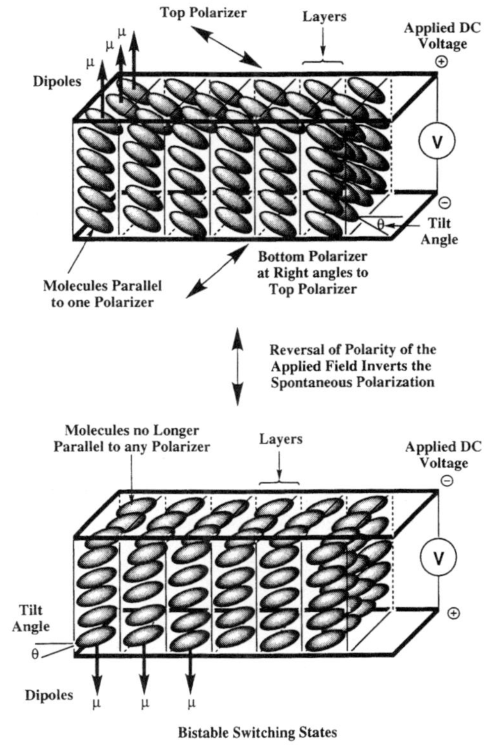

FIGURE 2.53. Optical transmission in a surface stabilized ferroelectric liquid crystal (SSFLC) display. (a) Light transmission and (b) light extinction.

Much of the synthetic work in this field has been directed towards finding ideal materials for the SSFLC device. Typically, for device applications a mixed system of liquid crystals is used; this is usually made up of a host material which is achiral into which a chiral dopant is added to create the ferroelectric properties. The host and dopant are often a mixture of individual materials, therefore, both chiral and achiral materials are required for applications. Many chiral compounds have been synthesized and some understanding of the property–structure relationships with respect to the spontaneous polarization (P_s) has been achieved. Some of this work will be reviewed in section 2.4.4. Achiral smectic C liquid crystals doped with chiral dopant molecules also display the ferroelectric effect and some examples of achiral systems will be reviewed in section 2.5.4.

2.4.4 Direction of the Spontaneous Polarization

Symmetry arguments show that the spontaneous polarization can point in one of two directions along the polar C_2 axis, so a material will also have one of two possible polarization directions, $P_s(+)$ or $P_s(-)$, associated with it as shown in FIGURE 2.54. (Clark and Lagerwall, 1984; Goodby, 1991). The definition of the orientation of the two directions was made arbitrarily and is now used as convention. The polarization direction is dependent on the absolute spatial configuration of the molecule (see FIGURE 2.49), and (*R*) and (*S*) enantiomers of the same molecule will have different polarization directions.

An attempt has been made (Goodby et al., 1986) to relate the direction of the spontaneous polarization to the nature, spatial configuration and location of the chiral center in the molecule in a manner similar to that used by Gray and McDonnell (Gray and McDonnell, 1977a) in order to characterize cholesteric helical structures. The following relationships were found to be of some value when characterizing smectic C* liquid crystal phases:

Sed Ps(−) +I, Sol Ps(+) +I, Sod Ps(−) −I, Sel Ps(+) −I

Red Ps(−) −I, Rol Ps(+) −I, Rod Ps(−) +I, Rel Ps(+) +I

where **R** or **S** is the absolute spatial configuration of the molecule, **o** or **e** refer to either an odd or even position (in terms of the number of connecting atoms) of the chiral center from the rigid molecular core, **l** or **d**, levo (left) or dextro (right), is the rotation of plane polarized light through the helix, $\mathbf{P_s(+)}$ or $\mathbf{P_s(-)}$ is the polarization direction, and **−I** or **+I** is the inductive effect of the off-axis constituent at the chiral center. This inductive effect essentially determines the direction of the dipole at the chiral center. These relationships are found to work well with simple molecules but they are less accurate with more complicated systems, a situation similar to that encountered with cholesteric helical structures (Goodby et al., 1986).

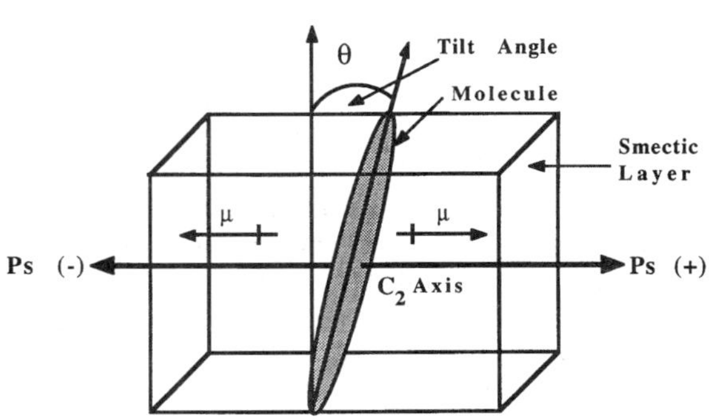

FIGURE 2.54. Polarization directions in the smectic C* phase.

2.4.5 Molecular Features that Influence the Magnitude of the Spontaneous Polarization in Smectic C* Phases

The magnitude of the spontaneous polarization in ferroelectric liquid crystals is relatively low when compared with inorganic ferroelectric materials. Low values are usually obtained because the molecules are undergoing rapid reorientational motion about their long molecular axes, and therefore it is only the time-dependent alignment of the lateral dipoles which contributes to the polarization. In addition, the molecules that form ferroelectric smectic phases are considerably less polar than inorganic ferroelectric materials. Furthermore, as the phases are fluid, the molecules also relax rapidly after being subjected to an applied electric field. As noted earlier, the magnitude of the spontaneous polarization in ferroelectric liquid crystals is mainly dependent on the tilt angle (θ) of the phase (Patel and Goodby, 1987b), the size of the dipole at the chiral center and the degree of freedom that the chiral center has to rotate about the long molecular axis. In this section, these factors will be discussed with reference to actual examples where the molecular architecture has been designed specifically to produce the maximum spontaneous polarization.

The first ferroelectric liquid crystal phase was discovered in (*S*)-4-n-decyloxybenzylideneamino-2-methylbutylcinnamate (DOBAMBC) (Meyer et al., 1975) (FIGURE 2.55(a)). Subsequently, it was found that the replacement of the 2-methylbutyl chiral unit with the isomeric 1-methylbutyl unit (DOBA-1-MBC) (Yoshino et al., 1984) caused a tenfold increase in the value of the spontaneous polarization (see FIGURE 2.55(b)). These two systems are directly comparable, except that the chiral center is positioned closer to the core, and hence the molecular dipole, in DOBA-1-MBC than it is in DOBAMBC.

Since the polarization arises as a result of a lowering of symmetry when a chiral center is introduced

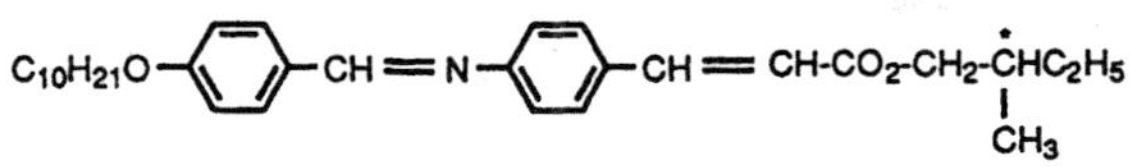

(a) 0.009 D/Molecule

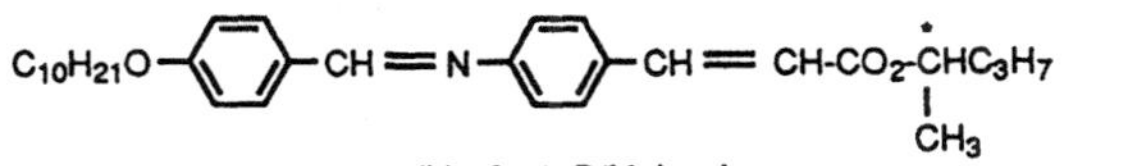

(b) 0.11 D/Molecule

FIGURE 2.55. Structures and spontaneous polarization values of (a) DOBAMBC and (b) DOBA-1-MBC.

into the molecule (FIGURE 2.52), the magnitude of the polarization will obviously be increased if the rotation of the *dipole* about the molecular long axis is restricted by an energy barrier. The magnitude of the molecular dipole, which is directly related to the spontaneous polarization, then depends on the height of this energy barrier (ΔE). Increasing this barrier to rotation is achieved for DOBA-1-MBC, for example, by bringing the lateral dipole into closer proximity with the chiral center resulting in an increase in the magnitude of the polarization.

Another example of how the polarization is affected by intramolecular rotational restriction is illustrated by a series of (*R*)- and (*S*)-1-methylalkyl 4-(*n*-octanoyloxy)biphenyl-4′-carboxylates (FIGURE 2.56) synthesized by Goodby and coworkers (Goodby et al., 1987). This work systematically investigated the effect

FIGURE 2.56. The effect of increasing the alkyl chain length on the absolute polarization values in a series of (*R*)- and (*S*)-1 methylalkyl 4-(*n*-octanoyloxy)biphenyl-4′-carboxylates.

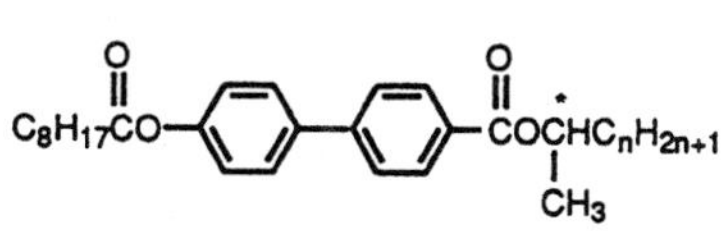

n	P (nC cm^{-2})
2	7.7
3	32.0
4	30.7
5	48.8
6	42.2

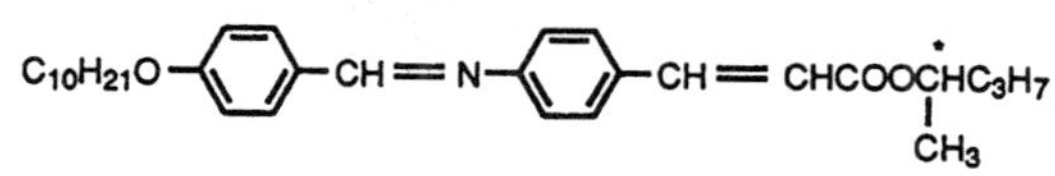

DOBA - 1 - MBC 42 nC cm^{-2}

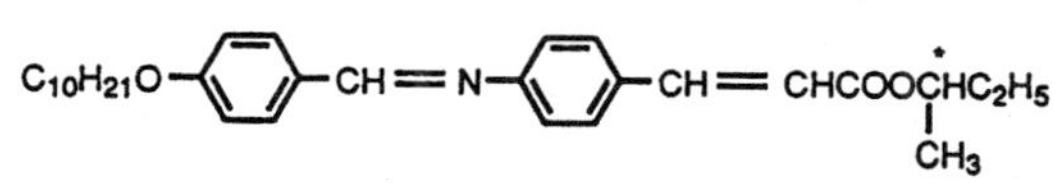

DOBA - 1 - MPC 17 nC cm^{-2}

FIGURE 2.57. The effect of the addition of one methylene group to DOBA-1-MPC on the value of the spontaneous polarization.

of the damping of the motion of the *chiral center* by extending the alkyl chain length on the external side of the chiral center, i.e., the motion of the chiral center was effectively reduced by positioning it more towards the center of the molecule rather than being in a terminal position.

The results show that there is a considerable increase in the value of the spontaneous polarization as the alkyl chain is lengthened, reaching a maximum where $n = 5$. The value declines after this point as the effect of the dipole is "diluted" when the overall size of the molecule increases. This is because the spontaneous polarization is proportional to the dipole density which in turn is affected by molecular size. Similar effects were found in other systems, for instance, the alkyl chain length on the external side of the chiral center was varied in DOBAMBC analogs, as shown in FIGURE 2.57. The spontaneous polarization was increased by a factor of 2.5 merely by extending the alkyl chain by one methylene unit.

In order to restrict the rotation of the chiral center even more efficiently, Kawada and coworkers (Kawada et al., 1990) have synthesized some compounds which possess two chiral centers linked together by an alkyl chain which forms part of an alicyclic ring as shown in FIGURE 2.58. These compounds show much higher spontaneous polarizations in comparison with those above.

The introduction of a dipole near (Yoshino et al., 1989) or into (Koden et al., 1989, 1990; and Sakaguchi et al., 1989) the central core was also found to increase the rotational hindrance. Ikemoto and coworkers (Ikemoto et al., 1989) have synthesized some compounds which include a chiral valerolactone ring and have

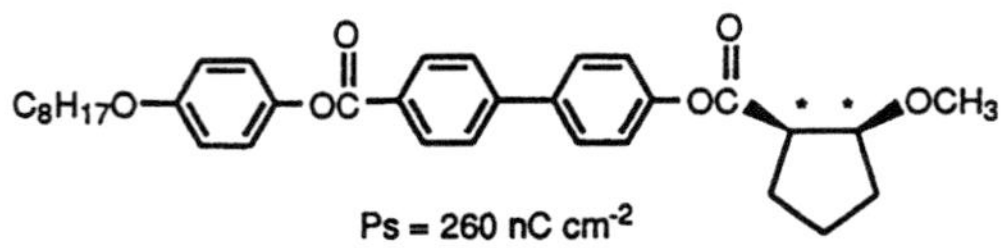

FIGURE 2.58. An example of a compound with two chiral centers linked together by an alkyl chain which forms part of an alicyclic ring.

reported that these compounds also display large polarization values (FIGURE 2.59). In this case, it can be seen that not only a dipole, but also the chiral center itself is introduced into the central core. This restricts the rotation of both the dipole moment and the chiral center.

Rotational freedom of the chiral center can also be restricted by intramolecular hydrogen bonding. For instance, Taniguchi and coworkers have reported (Taniguchi et al., 1988) that the introduction of a hydroxyl group into the molecular core near to the group carrying the dipole enhances the value of the polarization in (*R*)-4-(methoxycarbonyl-1-ethylcarbonyl)-3-hydroxyphenyl 4-(octyloxypentyl) benzoate (FIGURE 2.60(b)) when compared with its non-hydroxylated analog, (see FIGURE 2.60(a)). Hydrogen bonding between the hydroxyl group and the adjacent ester carbonyl group which restricts the free motion of the chiral center, as shown in FIGURE 2.60(c), is probably the reason for the increase in the value of the spontaneous polarization in this compound.

Another factor that affects the spontaneous polarization is conformational structure. The couplings between the lateral molecular dipole and various stable conformers formed by rotation about bonds adjacent to the chiral center have to be taken into account. Thus, the relationship between temperature and spontaneous polarization (cf. FIGURE 2.51) can sometimes produce some unusual results. For example, the polarization value can be seen to fall with decreasing temperature so that the sign of the spontaneous polarization inverts. Two examples (Goodby et al., 1987a; Mikami et al., 1986; Scherowsky et al.,

FIGURE 2.59. An example of a compound with two chiral centers contained within a valerolactone ring.

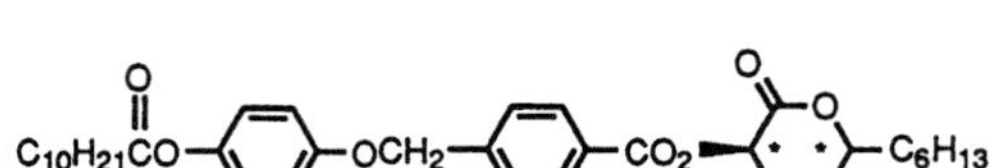

(a) (b) (c)

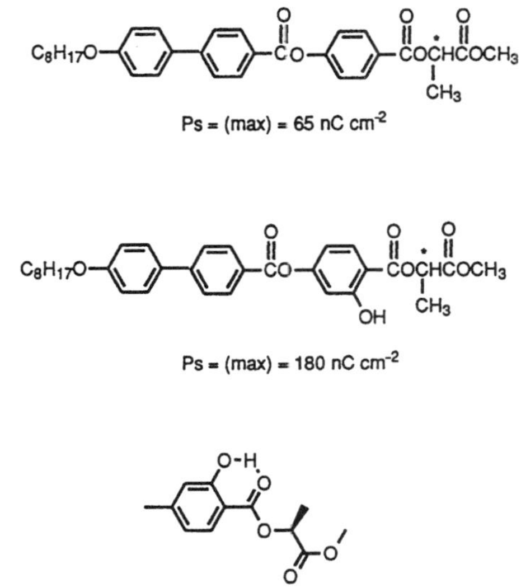

FIGURE 2.60. Increase in spontaneous polarization due to hydrogen bonding.

1990, 1992; and Koide et al., 1989) are shown in FIGURE 2.61, one of which is a side-chain liquid crystal polymer.

The first of these compounds, (*S*)-2-methylbutyl 4-*n*-nonoyloxybiphenyl-4-carboxylate (FIGURE 2.61(a)), has a negative spontaneous polarization at high temperatures, but as the temperature is reduced the polarization falls and its direction inverts and becomes positive at low temperatures, as shown in FIGURE 2.62.

Patel and Goodby (Patel and Goodby, 1987) have proposed a model for the polarization inversion in this compound based on the presence of at least two oppositely polarized species interconvertible via a small energy barrier. If a number of conformational species are present, the relative magnitude and direction of the dipole for each conformer and its rotational distribution about the director have to be taken into account in the generation of the polarization. For simplicity, it is assumed that these materials have two important conformations, one of which induces a positive polarization and the other a negative polarization, as shown in FIGURE 2.63.

The magnitude of the polarization also depends on the magnitude of the lateral dipole moment itself particularly when it is located near to the chiral center. Yoshizawa and coworkers (Yoshizawa et al., 1989), for example, have used a keto carbonyl group as

(a)

(b)

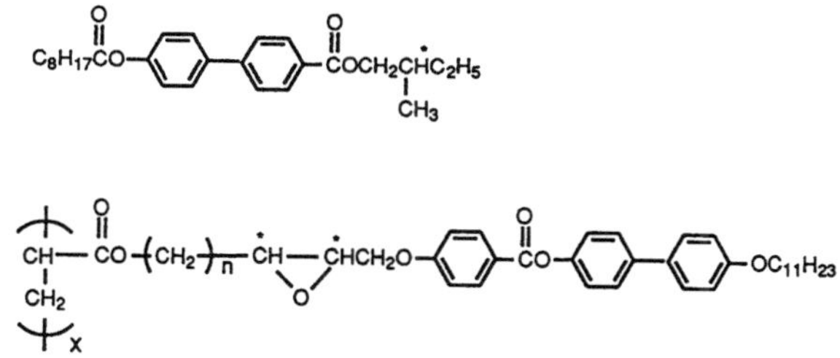

FIGURE 2.61. Compounds where the sign of the spontaneous polarization inverts.

a large dipole moment in the vicinity of the chiral center in order to produce a new ferroelectric liquid crystal which shows a large value for the spontaneous polarization (see FIGURE 2.64).

This compound has three dipolar groups, an ether, an ester carbonyl, and a keto carbonyl group. The keto carbonyl group has the largest dipole moment. The large spontaneous polarization arises because of the position of the keto carbonyl which is adjacent to the chiral center. The effect of the length of the alkyl chain on the external side of the chiral center on the magnitude of the polarization values is also significant (Hirai et al., 1991). The value of the polarization reaches a maximum at $n = 5–7$, but then decreases again for $n = 8$ as predicted. The data in FIGURE 2.65 show relatively high values for the polarization (due to the keto carbonyl group) which increase with increasing alkyl chain length on the external side of the chiral center until $n = 5–7$. The values diminish beyond that point as the increasing chain length "dilutes" the dipole by reducing the dipole density.

Combinations of the above molecular design features can be used to produce materials with very high spontaneous polarizations; Kobayashi and coworkers (Kobayashi and Ishibashi, 1990; and Kobayashi et al., 1990a, 1990b) have introduced a hydroxyl group at the *ortho* position of the phenyl ring which then hydrogen bonds to the adjacent keto carbonyl group as shown in FIGURE 2.66. This results in a very large increase in the value of the spontaneous polarization which is probably due to restricted dipole rotation as a result of the hydrogen bonding.

The effect of the strength of the dipole at the chiral center on the spontaneous polarization has been investigated through the introduction of fluorine atoms at the asymmetric center. Shiratori and coworkers (Shiratori et al., 1991) have positioned fluorine atoms around the chiral center to see how these affect the value of the spontaneous polarization. Several fluorinated compounds have been synthesized and the results compared with the nonfluorinated analog. The results are shown in FIGURE 2.67.

It can be seen that the 2-fluoro-2-methyl chiral structure, which has a donor methyl group and an acceptor fluorine atom attached to the chiral carbon atom, has the largest dipole and therefore produces the largest polarization value. In comparison, all of the fluorinated compounds give larger values of the spontaneous polarization than the nonfluorinated analog.

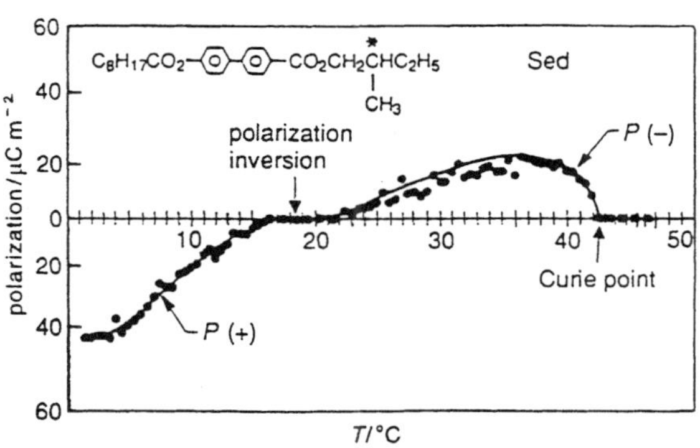

FIGURE 2.62. Temperature-related polarization inversion.

2.4.6 Antiferroelectric and Ferrielectric Phases

In section 2.4.3 we saw how ferroelectricity arises when the naturally helielectric smectic C* phase is

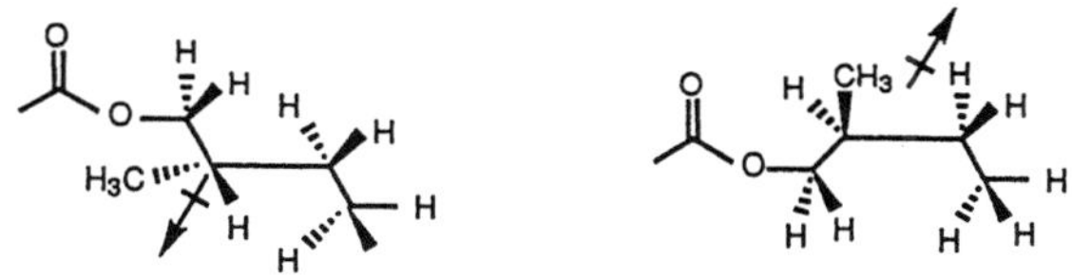

FIGURE 2.63. Conformations which show polarizations of opposite signs.

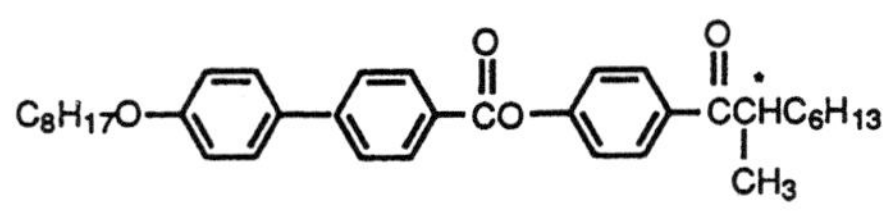

Ps(max) = 300 nC cm^{-2}

FIGURE 2.64. A ferroelectric liquid crystal which possesses a keto carbonyl group. This strong dipole in the vicinity of the chiral center induces a large value of the spontaneous polarization.

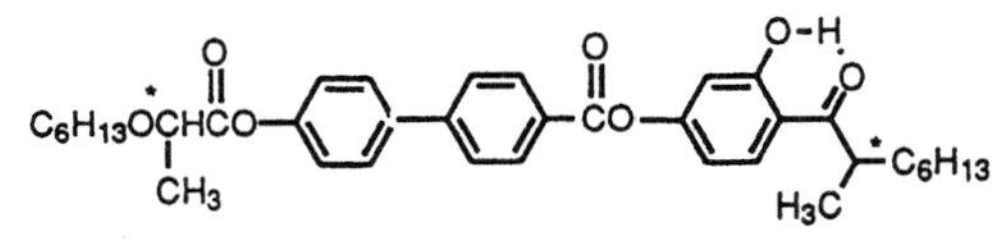

Ps = max =1130 nC cm^{-2}

FIGURE 2.66. Increased spontaneous polarization as a result of restricted rotation of the keto group by hydrogen bonding to the *ortho* hydroxyl group.

unwound to allow the layer polarizations to point in the same direction. This situation is shown in FIGURE 2.68. However, in recent years, two new associated phenomena (Chandani et al., 1989; Gorecka et al., 1990; and Inui et al., 1990) have been discovered which show fundamental differences to the situation shown in FIGURE 2.68. These discoveries led to two different arrangements of the molecules in their layer planes from that seen in the smectic C* ferroelectric phase with associated variations in polarization directions. These new phases are termed *antiferroelectric* and *ferrielectric* phases and their origins and structures are the subjects of this section.

The suggested structure of the antiferroelectric phase is shown in FIGURE 2.69. This shows that the molecular layers are arranged in such a way that the polarization directions in subsequent layers point in opposite directions which results in an averaging of the spontaneous polarization to zero. Evidence to support this hypothesis comes from the fact that when a strong electric field is applied to this phase the layer ordering is perturbed and the phase returns to a normal ferroelectric phase as shown by schematic switching of the molecules in FIGURE 2.70.

Three states are produced in the switching of antiferroelectric phases; one antiferroelectric and two ferroelectric. This *tristable* switching from the antiferroelectric to the ferroelectric phase occurs at a defined electric field, that is, there is a sharp switching threshold that differs from normal ferroelectric switching which appears to be continuous. In a normal liquid crystalline ferroelectric phase, switching occurs without a threshold field in cases where the switched states are not surface stabilized. The presence of a sharp switching threshold in antiferroelectrics, therefore, may be useful in display applications that require multiplexing with grey scales.

The switching process can be investigated by measuring the tilt angle as a function of applied voltage. No change in the tilt angle occurs at low voltages, but

FIGURE 2.65. Effect of the alkyl chain length on a homologous series that contains a keto carbonyl group.

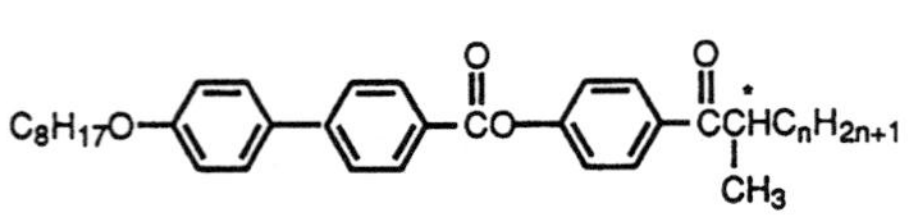

n	P_S (nC cm^{-2})
2	-49
5	-118
6	-112
7	-117
8	-87

FIGURE 2.67. The effect of the introduction of fluorine atoms at the chiral center.

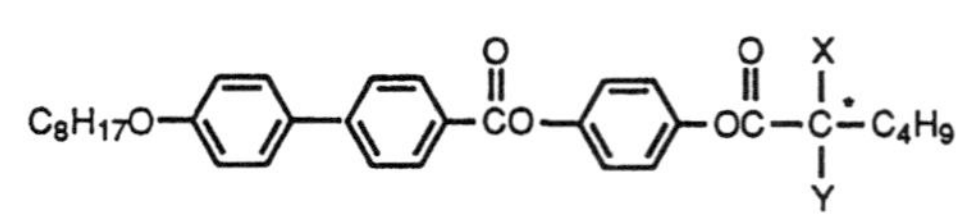

X	Y	Ps (nC cm^{-2})
CH_3	F	135
F	H	85
CF_3	H	95
CF_3	F	25
CH_3	H	15

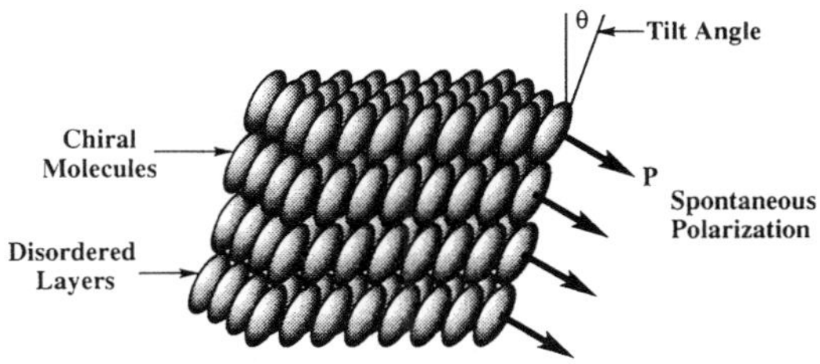

FIGURE 2.68. Structure of the ferroelectric smectic C* phase.

at a critical field the phase changes from antiferroelectric to ferroelectric. At this point, the tilt angle increases substantially and reaches the maximum of the normal smectic C* phase. There is a degree of hysteresis when the voltage is cycled up or down or when the polarity is reversed (see FIGURE 2.71) (Johno et al., 1989) leading to a double hysteresis loop in the field switching experiment.

Unlike the structure of the normal smectic C* phase which is repeated every 360° rotation of the helix, the helical structure of the antiferroelectric phase repeats every 180° rotation. The phase therefore appears to have a relatively short pitch, and also the pitch appears to change quite significantly as the temperature is altered.

In the *ferrielectric* phase, the layers are stacked in such a way that there is a net overall spontaneous polarization (see FIGURE 2.72). In this phase, the number of layers of opposite polarization is not equal. There may be, for example, twice as many layers where the polarization direction is opposite to that of the other layer type. It is also suggested that the stacking of the layers has two interpenetrating sublattices. Thus, there will be alternating layer structures, that is, two layers tilted to the right and one to the left, with this arrangement repeating itself through the

FIGURE 2.69. Structure of the antiferroelectric smectic C* phase.

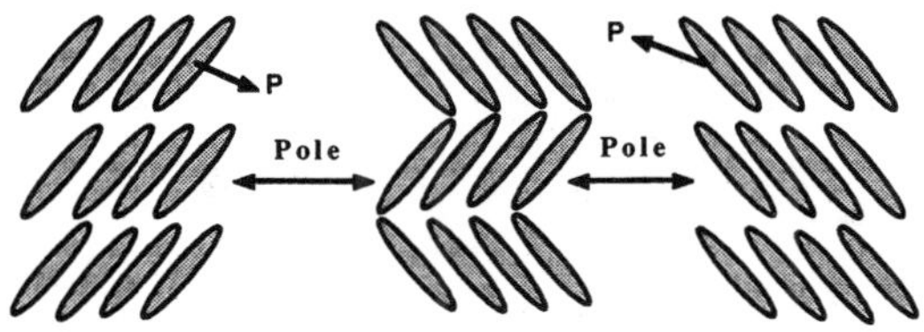

FIGURE 2.70. Antiferroelectric–ferroelectric switching.

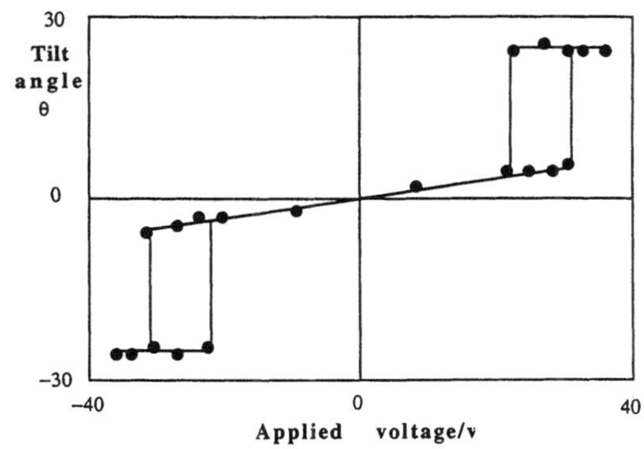

FIGURE 2.71 Switching from antiferroelectric to ferroelectric state in the smectic C* phase.

bulk of the phase. For this stacking arrangement, therefore, the ferrielectric phase will have a measureable polarization. Many ferrielectric layer combinations are of course possible.

Antiferroelectric behavior is usually found in materials where the chiral center is near to the rigid aromatic core of the molecule. Most of these materials have similar structures based around the chiral centers shown in FIGURES 2.73, (a) (Hiji et al., 1988) and (b) (Nishiyama et al., 1989).

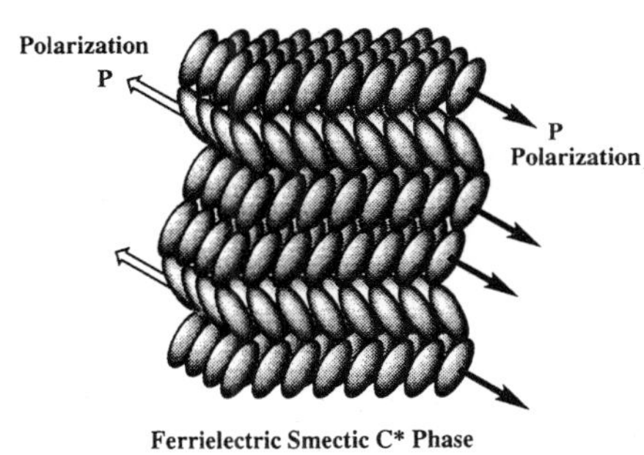

FIGURE 2.72. Structure of the ferrielectric smectic C* phase.

FIGURE 2.73. Moieties that promote the formation of antiferroelectric phases.

FIGURE 2.74. Moieties that do not promote the formation of antiferroelectric phases.

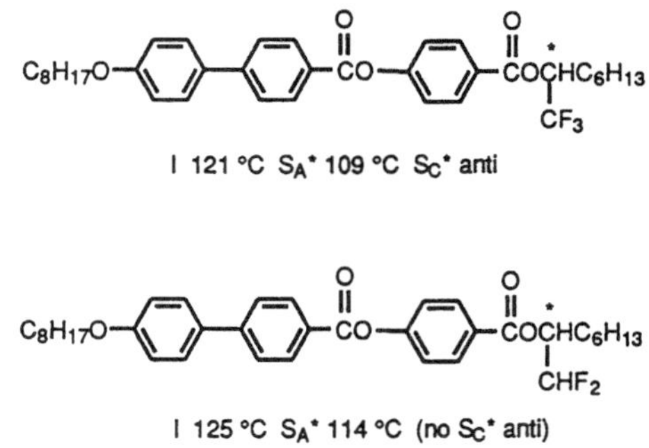

FIGURE 2.76. The effect of placing fluorine atoms at the chiral center.

On the other hand, certain structural arrangements around the chiral center, as shown in FIGURES 2.74(a) and (b), do not appear to favor antiferroelectric phases.

It has also been found that the replacement of the lateral methyl group with an ethyl group at the chiral center increases the stability of the smectic C* phase (see FIGURE 2.75) (Nishiyama et al., 1989).

Fluorine atoms at the chiral center also appear to increase the smectic C* antiferroelectric stability (Suzuki et al., 1989; and Hagiwara and Suzuki, 1989) but when hydrogen atoms are also present in the lateral chiral branch, the phase does not occur (see FIGURE 2.76).

The formation of smectic C* antiferroelectric phases is also influenced by the orientation of the ester groups in the molecule. FIGURE 2.77 shows how the reversal of an ester group can often inhibit the formation of this phase (Nishiyama et al., 1989; Suzuki et al., 1989; and Takeda et al., 1990). Thus, compound (a) in this figure shows antiferroelectric behavior, whereas compound (b) does not.

The usual molecular core structures for compounds that exhibit smectic C* antiferroelectric phases consist of three phenyl rings as shown in FIGURE 2.78(a) and (b) or two phenyl rings and one heterocyclic ring FIGURE 2.78(c) (Inui et al., 1990). All of the structures include at least one ester group.

In materials that display antiferroelectric behavior, the degree of "molecular chirality" is normally relatively high. In section 2.4.5, we saw how the length of the peripheral aliphatic chain influences the degree of chirality in compounds showing smectic ferroelectric C* phases. In these materials, the motion of the chiral center about the long molecular axis is damped, which increases the degree of the "molecular chirality". This is clearly seen for the examples of (*S*)- or (*R*)-1-methylalkyl 4-(*n*-decyloxybenzoyloxy)biphenyl-4′-carboxylates, shown in FIGURES 2.79(a) and (b) (Goodby, 1991; and Goodby et al., 1992).

When the alkyl chain is short in length, (e.g., propyl; see FIGURE 2.79(a)), normal phase behavior is observed. The material cools from the isotropic liquid

FIGURE 2.75. The effect of the replacement of a lateral methyl group with an ethyl group at the chiral center on antiferroelectric smectic C* stability.

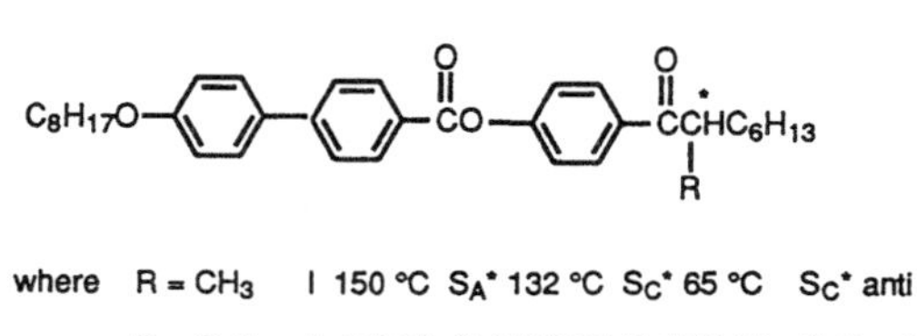

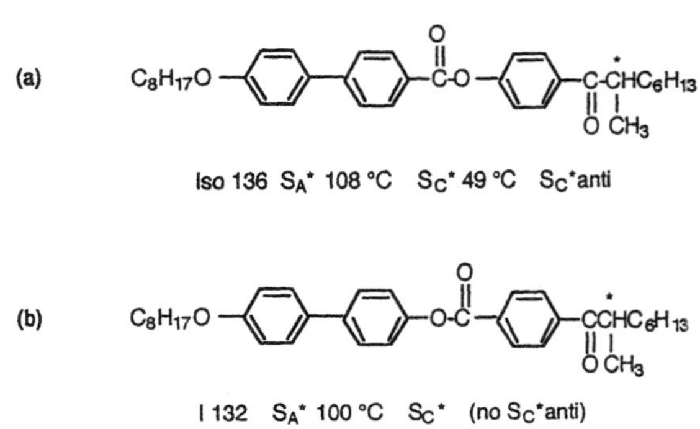

FIGURE 2.77. The effect of the orientation of ester groups on antiferroelectric behavior.

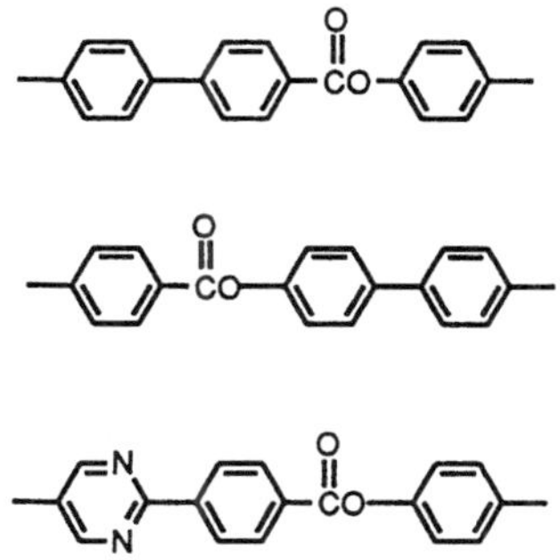

FIGURE 2.78. Molecular core structures of compounds that exhibit antiferroelectric phases.

into a smectic A phase and then undergoes a transition into the smectic C* phase, where the pitch of the helix is $\simeq 2$–$3\,\mu$m. However, when the alkyl chain is longer, (e.g., heptyl; see FIGURE 2.79(b)), the pitch of the smectic C* phase decreases to a value of $\simeq 0.5\,\mu$m. At the same time, the magnitude of the spontaneous polarization increases to a value six to seven times greater than that seen for the shorter (propyl) homolog. Further cooling of the longer chain compound results in the formation of antiferroelectric (and ferrielectric) phases. These phase changes are clearly discernible from the large changes in optical textures of the materials when observed by thermal polarized light microscopy.

Goodby and co-workers (Goodby et al., 1993) have produced a large series of phenylpropiolates, some of which show antiferroelectric and ferrielectric phases. This series includes a number of compounds where the terminal chiral moiety carries two aliphatic chains. An example of this type of compound is shown in FIGURE 2.80. This compound displays a narrow smectic C* phase and a wide antiferroelectric phase showing that, apparently, the chiral "swallow-tail" moiety stabilizes the *antiferroelectric* phase in preference to the ferro- or ferrielectric phases.

FIGURE 2.79. The effect of terminal chain length on the incidence of antiferroelectric phases. Compound (a) shows no antiferroelectric behavior, whereas (b) exhibits an antiferroelectric phase.

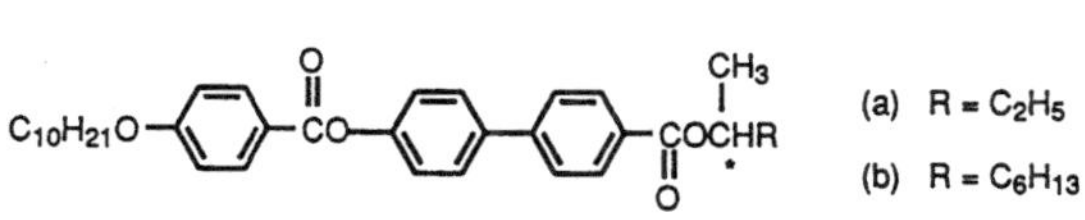

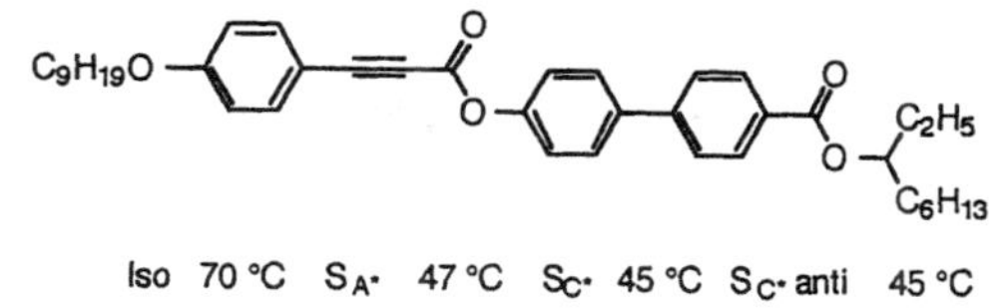

FIGURE 2.80. Chiral phenylpropiolate compound with two terminal aliphatic chains. This compound shows an antiferroelectric phase in preference to ferro- or ferrielectric phases.

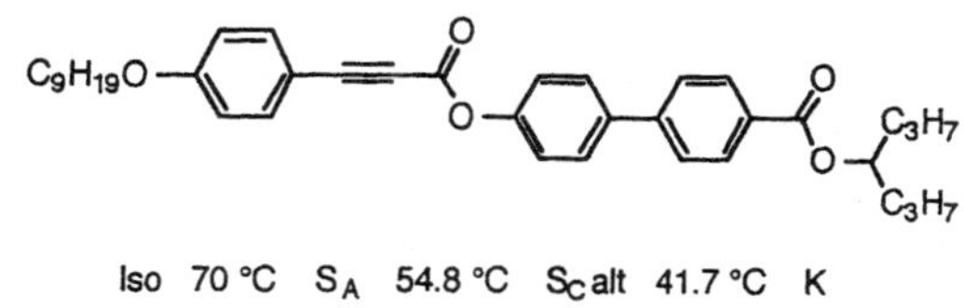

FIGURE 2.81. Nonchiral phenylpropiolate compound which exhibits an antiferroelectric-like phase.

Taking this structure to the extreme so that the two terminal alkyl chains are of the same length, the material has no chirality, but it still exhibits a phase that is structurally similar to an antiferroelectric phase on cooling from the smectic A phase (FIGURE 2.81). This indicates that chirality does not necessarily stabilize the formation of ferrielectric and antiferroelectric phases, but that the phases may be stabilized by the biaxial nature of the molecular structure.

2.4.7 Side-Chain Liquid Crystal Polymers that show Smectic C* Phases

The first side-chain liquid crystal polymer (SCLCP) to show a smectic C* phase was the polymethylacrylate P5*N which was synthesized by Shibaev and co-workers in 1984 (FIGURE 2.82) (Shibaev et al., 1984; and Shibaev and Plate, 1985).

In 1988 Ushida and colleagues (Ushida et al., 1988) synthesized a polymer with a spontaneous polarization of $3\,nC\,cm^{-2}$ and electrooptical switching times of

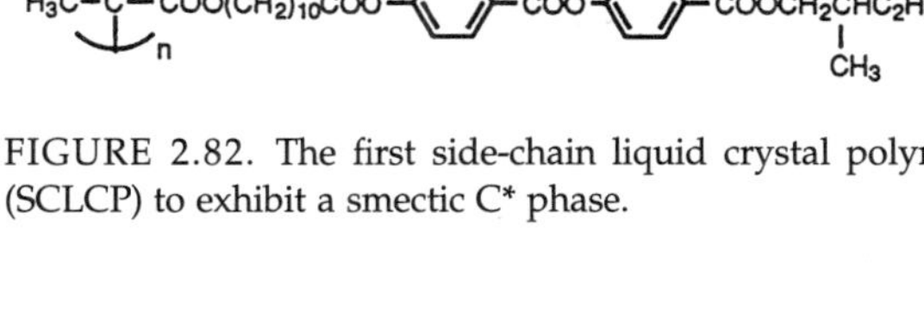

FIGURE 2.82. The first side-chain liquid crystal polymer (SCLCP) to exhibit a smectic C* phase.

(a)

(b)

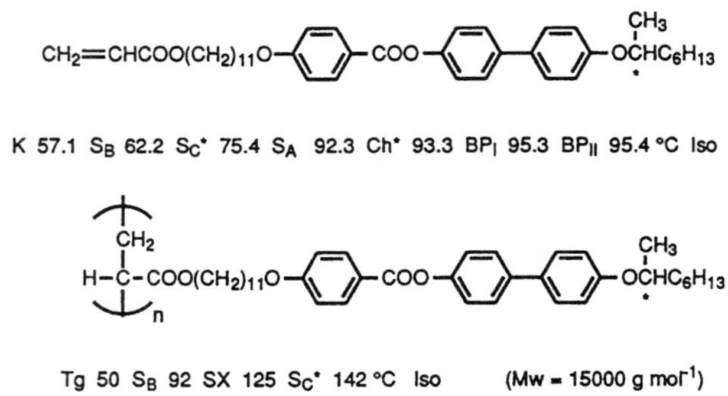

FIGURE 2.83. Polymer (b) derived from monomer (a)—both show wide-range smectic C* phases.

between 3 and 50 ms, the value being dependent on molecular weight and temperature. These switching times are considerably slower than for low molar mass ferroelectrics where speeds in the microsecond range are typically found.

Ferroelectric switching polymers are usually derived from chiral monomers which have a broad temperature range smectic C* phase (as would also be required for low molar mass ferroelectrics). Scherowsky and colleagues (Scherowsky, 1992a) used such a monomer (FIGURE 2.83(a)) for the synthesis of the polymer shown in FIGURE 2.83(b).

The monomer (FIGURE 2.83(a)) shows six mesophases including a smectic C* phase and two blue phases. After polymerization (FIGURE 2.82(b)), three smectic mesophases are observed; the cholesteric, blue phases and smectic A* phase of the monomer are no longer evident. The type of smectic phase, however, varies with the molecular weight of the polymer. In the case where the molecular weight is relatively low (mw = 15,000), the switching time in the smectic C* phase is between 200 and 400 μs which is appreciably faster than previously synthesized ferroelectric polymers.

Chiral side-chain liquid crystal polymers can also show antiferroelectric behavior; two examples (Kuehnpard and Springer, 1991; and Skarp et al., 1991) are shown in FIGURE 2.84(a) and (b).

In each case, an intermediate stable state exists when the polarity of an applied field (DC) is inverted on switching from one ferroelectric state to another, that is, tristate switching is observed (see FIGURE 2.70).

Decobert and colleagues (Decobert et al., 1986) have suggested that antiferroelectric SCLCPs have a gross ordering that is dependent on the conformational and configurational structure of the *backbone* of the polymer. Under certain circumstances, it is believed that the mesogenic groups are arranged in a zig-zag relationship with respect to the long axis of the backbone (see FIGURE 2.85). This zig-zag order forms the basis for the structure of the antiferroelectric phases.

In order to establish property/structure correlations, the chiral liquid-crystalline properties of a series of the low molar mass and polymeric compounds shown in FIGURE 2.86 have been investigated (Nishiyama and Goodby, 1993). These materials were found to exhibit ferrielectric and antiferroelectric phases in addition to ferroelectric phases.

(a)

(b)

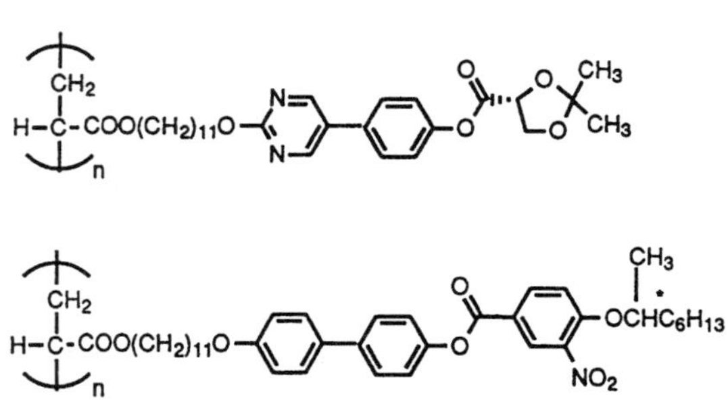

FIGURE 2.84. Two side-chain liquid crystal polymers that show antiferroelectric behavior.

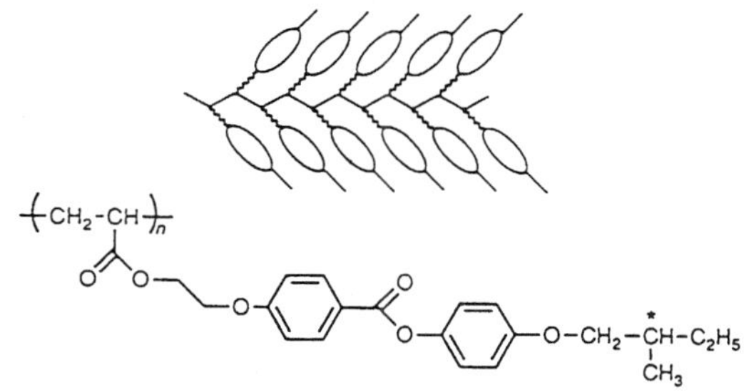

FIGURE 2.85. The zig-zag conformation of the polymer backbone which forms the basis of the structure of the antiferroelectric phase.

The following results were obtained where $q = 6$ and $r = 1$:

2.86(b) monomer

I–S_A*	S_A*–S_C*	S_C*–ferri	S_C*–ferri–anti	mp
85.7	73.5	45.2	38.5	54.2 °C

2.86(c) polymer

I–S_A*	S_A*–anti	anti–K
159.7–157.5	146.1	37.3 °C

Cooling of the isotropic liquid of the monomer produced a smectic A* phase which, on further cooling, underwent a transition to a ferroelectric smectic C* phase and, eventually, to an antiferroelectric phase.

For the polymer, cooling of the isotropic liquid produced a smectic A* phase. Further cooling produced a transition from the smectic A* phase to an antiferroelectric phase, but neither a ferro- nor a ferrielectric phase was evident in this compound. An examination of the molecular examples given in this section shows that the material requirements to produce ferroelectric (and ferri- and antiferroelectric) phases in polymers are largely similar to those for low molar mass smectic C* materials. Thus, molecular core structures, the orientation of ester groups, the position of the dipole relative to the chiral center etc., are also important features in polymeric smectic C* materials. The length of the spacer groups, that is, the alkyl chains which connect the mesogenic groups to the molecular backbones, normally approximate to the length of the alkyl chains in low molar mass compounds.

2.4.8 Chiral Dopants for Ferroelectric Liquid Crystal Mixtures

The switching time for a surface stabilized ferroelectric device is inversely proportional to the polarization (Clark and Lagerwall, 1980a) as expressed in the equation:

$$\tau \approx \eta / PE$$

where τ is the reorientational time, η is the viscosity, P is the polarization and E is the electric field. Increasing the polarization should, therefore, reduce the switching speed in electrooptic devices. It was found that increasing the polarization creates relatively strong internal fields that have to be overcome before the

(a)

(b)

(c)

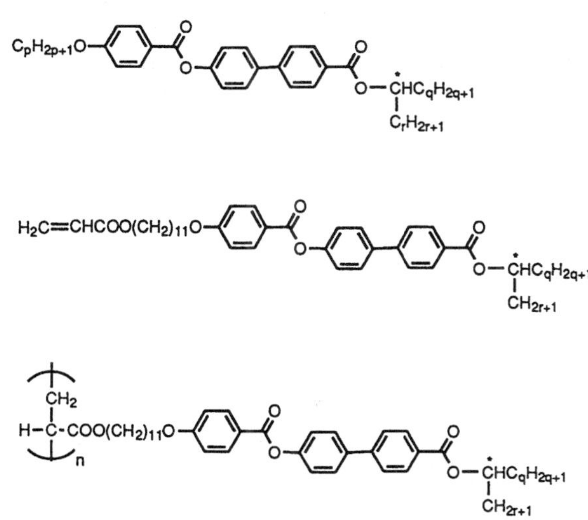

FIGURE 2.86. A comparison between low molar mass and polymeric antiferroelectric materials.

molecules will latch and switch. Conversely, it has been found that it is necessary to reduce the reorientational viscosity and the polarization in order to obtain switching times in the region of 1–10 μs which are necessary for video rate displays. Fast switching mixtures can be obtained by making mixtures of low-viscosity achiral materials that exhibit smectic C phases and doping with chiral materials that have large polarizations, but are not necessarily liquid crystals in their own right.

Chiral dopants are materials which possess a high spontaneous polarization (P_s) value and which can be used at low concentrations in a compatible achiral liquid-crystalline host material. The combination of host and dopant produces a mixture which gives a phase type with a temperature range which is appropriate to the desired application. In this way, an achiral host mixture can be blended with a high spontaneous polarization dopant to give an admixture with a moderate spontaneous polarization value because of the low concentration of dopant. The modification to the host system, therefore, is only slight in respect of phase behavior, transition temperatures and optical properties.

A high spontaneous polarization value in a chiral dopant is best achieved by placing a strongly dipolar group either near or directly attached to a chiral center. As in high spontaneous polarization smectic C* ferroelectric materials, the chiral center should be placed close to the molecular core. This is exemplified in materials prepared by Koden and colleagues (Koden et al., 1989) who have synthesized some compounds which include a chiral lactone ring (FIGURE 2.87). High polarization values have been achieved in this example by locating the chiral center adjacent to the molecular core.

From the commercial point of view, the optically active components should be readily available, chemically stable and should not require costly optical resolution. The compounds should be available in both enantiomeric forms and be stable to racemization.

FIGURE 2.87. High spontaneous polarization dopant with two chiral centers contained in a lactone ring.

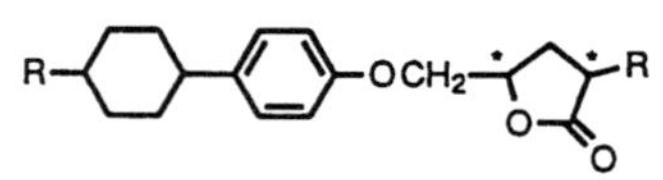

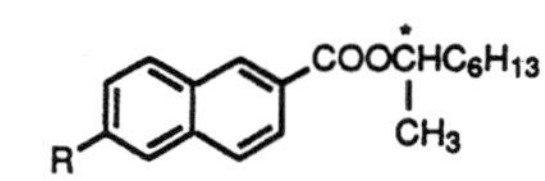

R = Primary Alkyl

FIGURE 2.88. Chiral dopant consisting of a 2-octyl group attached to a naphthyl core.

Readily available chiral precursors, in which the material is already 100% optically pure, are ideal provided that they can be used in reactions where there is no possibility of racemization. A simple example of a nonliquid-crystalline chiral dopant is shown in FIGURE 2.88 (Bishop et al, 1986). This compound possesses a chiral 2-octanol group attached to a naphthyl core. Again the spontaneous polarization arises as a result of the placing of the chiral center adjacent to the ester dipole.

Transversely positioned cyano groups (Chan et al., 1989) located near to the molecular core can also be used to produce high spontaneous polarization values in admixtures; a schematic representation is shown in FIGURE 2.89.

Y is the core system and X is an alkyl system or vice versa. In the next section, we will see how chiral materials can be used to produce ferroelectric mixtures for applications.

2.4.9 Mixtures of Ferroelectric Liquid Crystals

When considering the mixing of low molecular weight isotropic liquids, the axiom that "like mixes with like" is generally found to be true. Thus, two polar solvents will usually be miscible as will two nonpolar liquids, but a polar and a nonpolar fluid will not usually mix. This simple rationale can be extended to the smectic C mixtures, but in the case of an anisotropic phase the compatibility of the phase structures must also be considered.

Structurally, a smectic C phase can be characterized by three interrelated parameters: molecular length, layer spacing and tilt angle. If the phase is chiral,

FIGURE 2.89. Dopants with transversely positioned cyano groups.

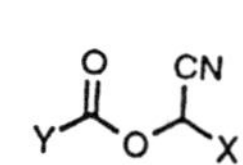

there are two further parameters: pitch length and twist sense. Ferroelectric materials have the added complication of the sign of the spontaneous polarization (P_S). It is thus not surprising that in many cases smectic C phases are immiscible, and when mixing does occur it is seldom thermodynamically ideal.

Ideal eutectic mixing is typified by the depression of the melting point, and consequently a widening of the temperature range of the phase. In nematic systems, the Schröder–van Laar equation (Schröder, 1893; and van Laar, 1903) can sometimes be used to predict the minimum melting point and composition of a mixture at the eutectic point from the melting points and enthalpies of melting of the pure components. This may be expressed as a series of equations of the form

$$\ln x_A = \frac{H_A}{R}\left(\frac{1}{T_A} - \frac{1}{T}\right)$$

where x_A = mole fraction of component A;
H_A = latent heat of melting of component A;
T_A = melting point of component A;
R = gas constant;
and T = melting point of the mixture.

There is one such equation for each component of the mixture. These equations can then be solved for the condition:

$$\sum_{i=A}^{N} x_i = 1$$

and thus the eutectic point may be found. It must be stressed that this approach is only applicable to ideal systems, and as such is only of use in a small proportion of cases. Smectic mixtures have the added complication that a compound may have several smectic phases. The Schröder–van Laar equation has therefore been modified to take account of the enthalpy of phase transition (Goodby and Leslie, 1984). Thus

$$\ln x_A = \frac{H_A}{R}\left(\frac{1}{T_A} - \frac{1}{T}\right) + \frac{H_A^{tr}}{R}\left(\frac{1}{T_A^{tr}} - \frac{1}{T}\right)$$

where x_A = mole fraction of component A;
T_A = melting point of component A;
T = melting point of the mixture;
H_A^{tr} = latent heat of phase transition of compound A;
R = molar gas constant;
and T_A^{tr} = latent heat of melting of compound A.

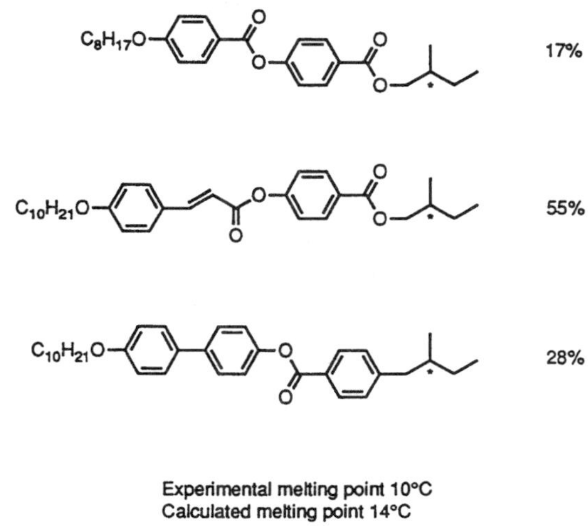

FIGURE 2.90. Comparison of experimental and theoretical eutectic melting points.

Once again, this approach is by no means universally applicable since it relies on ideal mixing. However, FIGURE 2.90 shows a particularly good example of a smectic mixture in which the experimental and predicted melting points differ by only 4 °C.

However, since nonideality and immiscibility are the norm in smectic C mixtures, the following examples illustrate a number of possible reasons for these effects, and demonstrate that even seemingly similar compounds can prove incompatible. FIGURE 2.91 shows the structures of two materials with smectic C* phases.

Compound A cools from the isotropic to the smectic C* phase via a cholesteric phase, whereas compound B has an orthogonal smectic A phase between the two chiral phases. Although this difference in phase sequences does not, of itself, prevent ideal mixing it gives an indication that compound A may have a considerably different tilt angle to compound B. This is in fact the case, and the tilt angles are 45° and 22° respectively. On diluting A with B, or vice versa, the stability of the smectic C* phase is dramatically reduced (FIGURE 2.92) and between 65 wt.% and 85 wt.% of B, the smectic A phase cools to an injected smectic B phase, which is foreign to both of the parent compounds.

This injection of an orthogonal phase is typical of nonideal mixtures and in this case indicates that the

(a)

(b)

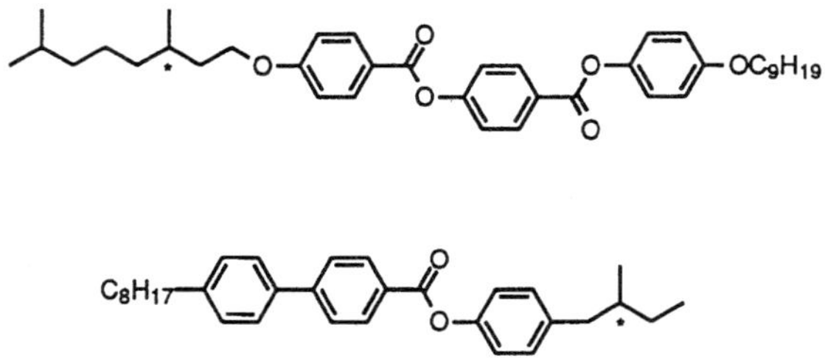

FIGURE 2.91. Two smectogens with smectic C* phases.

two materials are probably stucturally incompatible. On a simplistic level, a hypothetical smectic C mixture of these materials would have a prohibitively high free volume as the highly and less-highly tilted molecules would not be able to pack together efficiently.

A second example (see FIGURES 2.93 and 2.94) illustrates the general incompatibility of materials that contain Schiff bases and esters as linking groups in their molecular cores. Either the two groups have a geometric or electronic incompatibiliy, or conversely, they are mutually attracted into an orthogonal phase when compounds possessing these groups are mixed together. In either case, the stability of the two smectic C* phases is greatly reduced upon mixing, and an injected B* phase is formed.

FIGURE 2.92. The phase diagram for mixtures of the materials in FIGURE 3.44.

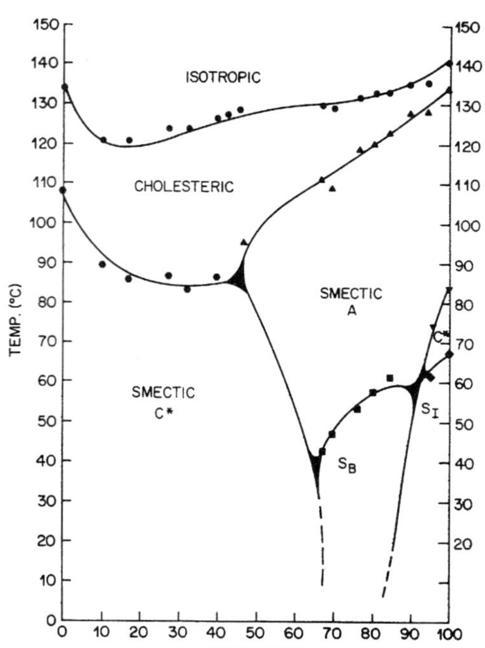

FIGURE 2.95 shows two compounds that may be expected to exhibit ideal mixing behavior.

The molecular structures of the compounds shown are similar and the phase sequences are identical. However, as FIGURE 2.96 shows, the two smectic A* to C* transition boundaries are not even convergent, and there is an injected B* phase once again.

It is perhaps unfortunate that although it is common for an orthogonal phase to be injected into a tilted mixture, the reverse is seldom true. The reason for the immiscibility in the above case may be due to the fact that the two components have the same polarization sense, whereas their helical twist senses are opposite. This mixture was prepared, in fact, in an attempt to produce a pitch-compensated mixture in which the two helical structures cancel whilst the spontaneous polarizations average to a nonzero value. Had this experiment worked, a composition point would have been obtained for which the mixture was not helielectric but was nevertheless ferroelectric.

Ferroelectric switching studies can also be used as a complementary diagnostic tool to optical microscopy for elucidating the phase behavior of ferroelectric mixtures. This is exemplified in the last case by monitoring the temperature-dependent switching of three

FIGURE 2.93. Two molecules with (respectively) ester and Schiff base core linkages.

(a)

(b)

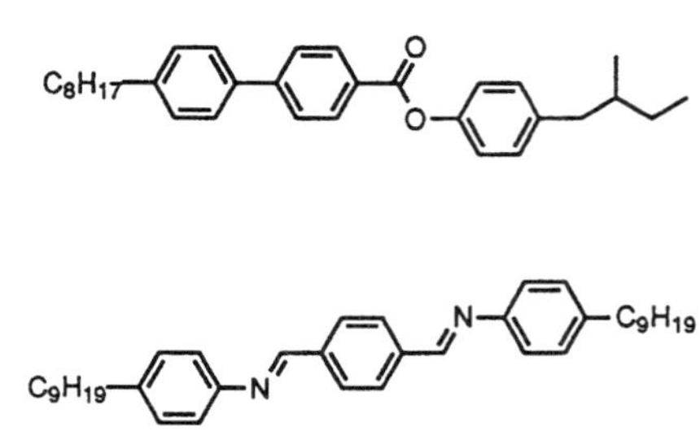

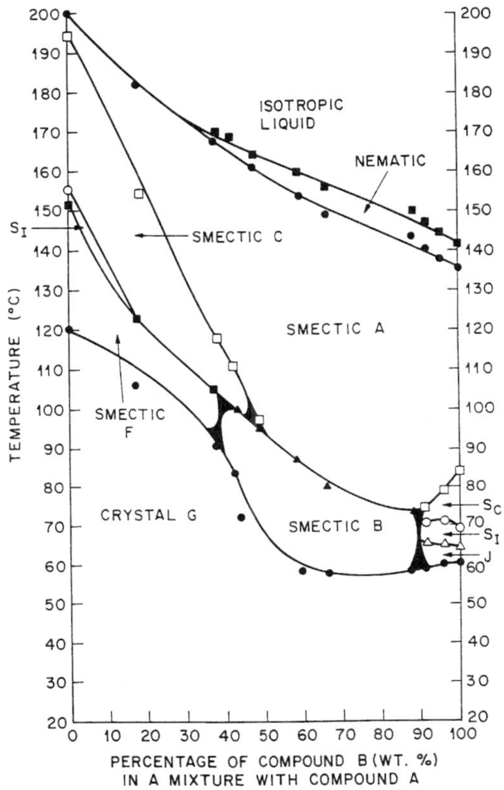

FIGURE 2.94. The phase diagram for binary mixtures of the ester and Schiff base materials in FIGURE 2.93.

mixtures denoted x, y, and z, in FIGURE 2.96. Each switching curve, shown in FIGURE 2.97, represents a vertical cross-section through the phase diagram and plots the temperatures at, and degree to which, optical switching occurs.

Experimentally, the mixture is placed in an aligned cell between two transparent electrodes. The cell is illuminated between crossed polarizers using an

FIGURE 2.95. Two structurally similar smectogens.

(a)

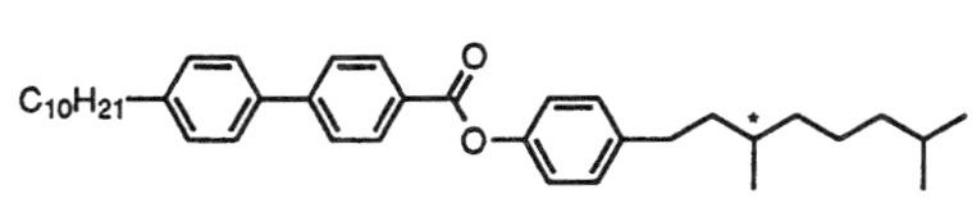

(b)

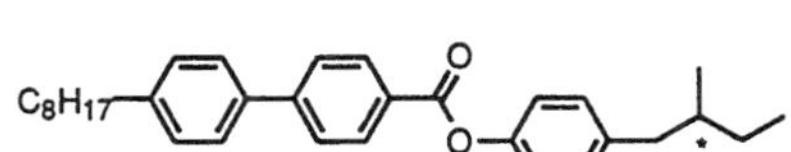

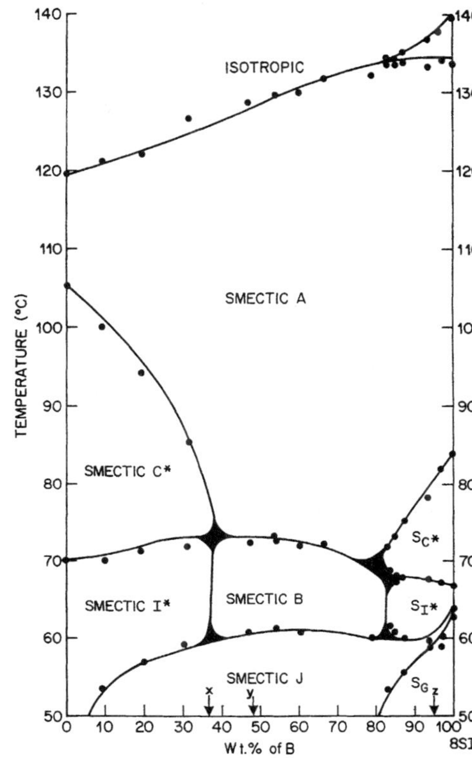

FIGURE 2.96. The phase diagram for the materials shown in FIGURE 2.95.

FIGURE 2.97. The temperature-dependent switching of three mixtures of the materials shown in FIGURE 2.95.

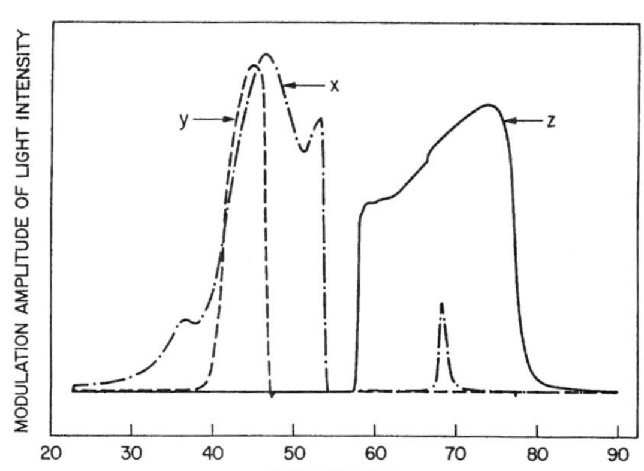

optical microscope. Switching pulses are then applied at a frequency of about 100 Hz and the intensity of light passing through the cell is measured using a photodiode. Clearly, switching can only occur in ferroelectric phases, and then only if the driving voltage is sufficient to re-align the molecules. Thus orthogonal phases and some viscous higher-order smectics do not effect modulation of the light intensity.

From FIGURE 2.97 it is apparent that mixture z commences switching (on cooling) at 80 °C in the smectic C* phase. There are then transitions, seen as discontinuities in the switching curve, to the smectic I* and crystal J* phases at 67 °C and 60 °C respectively. The final transition to the smectic G* phase results in the cessation of switching as the driving voltage is too low. Mixture x shows switching between 70 °C and 68 °C as it passes from smectic A* to smectic B* via a short smectic C* phase. Switching then recommences at 55 °C at the onset of the J* phase. Mixture y, which is an approximately equimolar blend of the two mesogens, does not begin switching until the crystal J* phase is reached at 48 °C. There is therefore no smectic C* phase above the smectic B* and the transition to the B* phase occurs straight from smectic A*. The discrepancies in phase transition temperatures between the optical and switching studies indicates that there may be another nonferroelectric phase below the smectic B* phase which was not observed by microscopy.

2.4.10 The Effects of Mixing upon Heli- and Ferroelectricity

A one-component ferroelectric phase may have either a left- or right-handed helical twist and the spontaneous polarization may either be positive or negative. On mixing two such materials, these properties are averaged, ideally with due weight being given to the signs and magnitudes of helical pitch and the spontaneous polarization with respect to the proportions of the two components. As with transition temperatures, however, the mixing behavior of the polarization and pitch are often nonideal. The following two examples serve to illustrate how helicity and ferroelectricity can be manipulated to optimize the performance of a mixture.

The ferroelectric effect is not dependent upon the helical twisting of the molecular tilt angle in the chiral smectic C phase. Indeed, in devices this twisting is disadvantageous and thus materials with large or

(a)

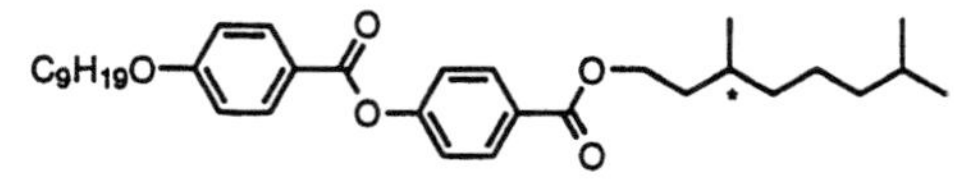

(b)

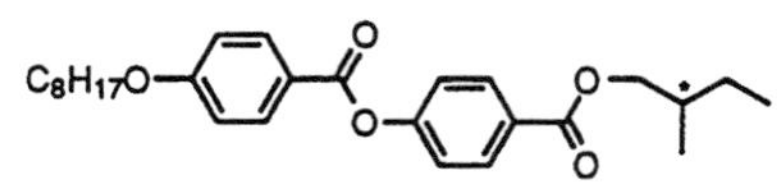

FIGURE 2.98. Two chiral smectic C* materials with mutually opposite twist senses.

infinite pitch are sought. In mixtures this may best be achieved by combining materials with opposite twist senses, since at some point in the phase diagram the left- and right-handed helices will effectively cancel.

This is the case in the mixture of (*S*-)3,7-dimethyloctyl 4-*n*-nonyloxybenzoyloxybenzoate with (*S*-)2-methylbutyl 4-*n*-octyloxybenzoyloxybenzoate (see FIGURE 2.98).

The phase digram (FIGURE 2.99) shows near-ideal mixing of the two components with only moderate depression of the smectic A* to C* transition.

This slight nonideality may well be caused by the difference in terminal chain lengths of the two

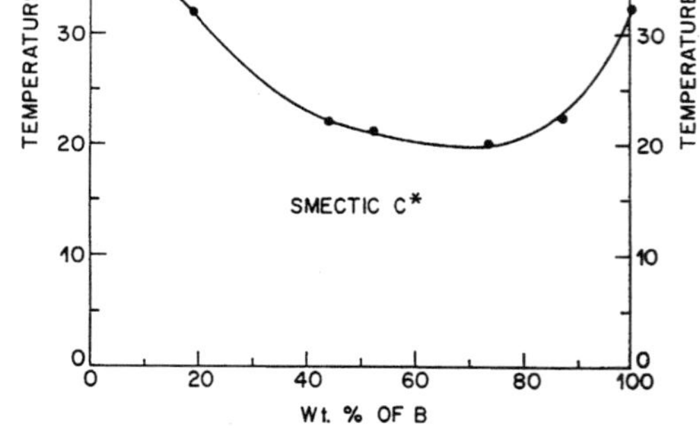

FIGURE 2.99. The phase diagram for the near-ideal mixing of the materials in FIGURE 2.98.

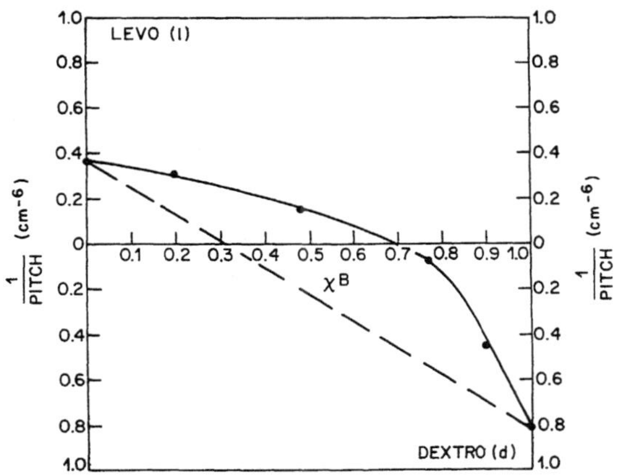

FIGURE 2.100. A plot of reciprocal pitch length versus composition for mixtures of the materials in FIGURE 2.98.

molecules since the core structure is common to both. Compound A has a left-handed helix whereas compound B has the opposite twist sense. Ideally, the reciprocal pitch length should vary linearly with the relative proportions of the two components. This is shown by the dotted line in FIGURE 2.100; the continuous line shows the experimental results.

Although the pitch should ideally become infinite at around 30 mol.% of B, the experimental result is nearer 70 mol.%. A similar graph plotting the magnitude of the polarization against mole fraction is shown in FIGURE 2.101. The data are corrected for the variation in smectic C* phase stability by taking the measurements at a given reduced temperature.

In this case the behavior is almost ideal, and since the two components have approximately equal and opposite polarization senses, the spontaneous polarization falls to zero at about 43 mol.% of compound B. Reference to FIGURE 2.100 shows that at this composition the mixture has a finite helical twist but it is, nonetheless, not ferroelectric. Since the points at which the polarization is zero and the pitch is infinite do not coincide, it is possible to formulate a mixture with a large pitch and a reasonable spontaneous polarization.

In this last example, the two materials have the same polarization direction but opposite twist senses. The molecular structures, shown in FIGURE 2.102, are similar, varying only in the terminal chain length.

This time the phase diagram (FIGURE 2.103) shows virtually ideal mixing. This is in contrast to one of the previous examples, despite the fact that the structural differences within the two pairs of molecules are almost the same. This further illustrates the fine balance between miscibility and smectic C* destabilization.

FIGURE 2.104 shows the near-ideal dependence of pitch upon composition for this mixture and that the infinite pitch point occurs at 80 mol.% compound B.

Similarly, the spontaneous polarization (FIGURE 2.105) behaves ideally except for a slight depression in the infinite pitch region.

It is therefore possible to combine the requirements of long pitch and non-zero value of the spontaneous polarization and thus produce a useful material for display devices.

FIGURE 2.101. A plot of spontaneous polarization versus composition for mixtures of the materials in FIGURE 2.98.

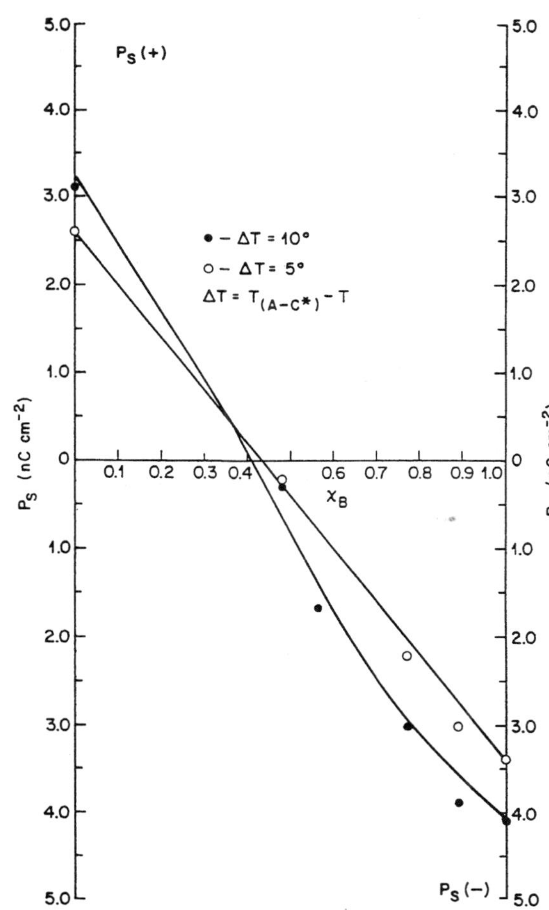

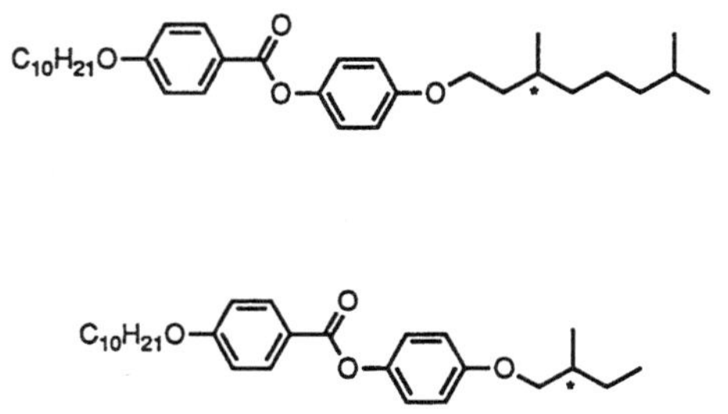

FIGURE 2.102. Two materials with smectic C* phases with the same polarization direction but opposite twist senses.

2.4.11 Use of Dopants and Hosts

Although the use of chiral mixtures is an elegant solution to the problem of producing useful ferroelectric materials, it suffers from one major disadvantage: all of the mesogens are chiral and thus have branched terminal chains. These branched chains cause the mixtures to be viscous, and thus their performance is impaired. To solve this problem, the proportion of branched, and hence chiral, material in the mixture must be kept to a minimum whilst maintaining an acceptable spontaneous polarization. This is achieved by preparing an achiral smectic C mixture

FIGURE 2.103. The phase diagram for mixtures of the materials in FIGURE 2.102.

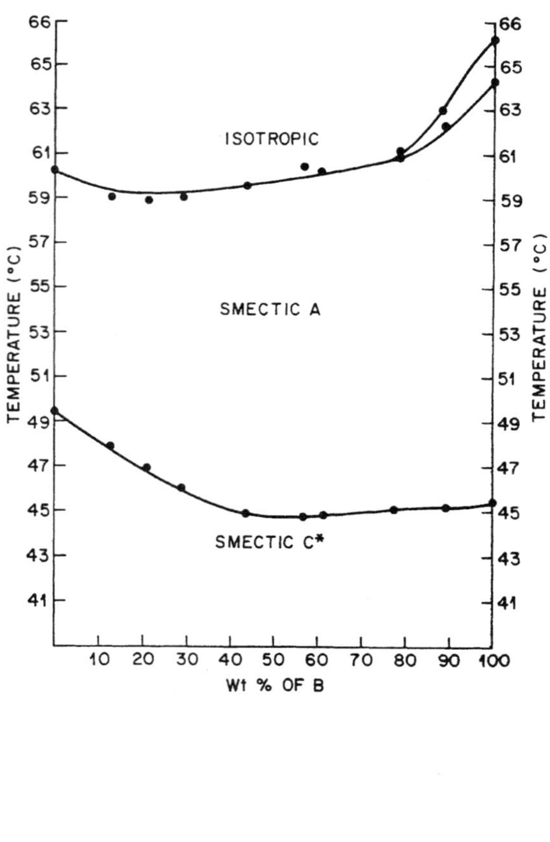

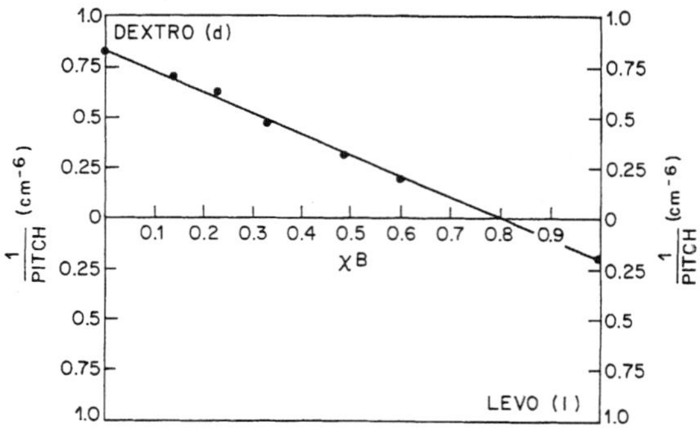

FIGURE 2.104. The near-ideal dependence of pitch upon composition for mixtures of the materials in FIGURE 2.102.

FIGURE 2.105. The near-ideal dependence of spontaneous polarization upon composition for mixtures of the materials in FIGURE 2.102.

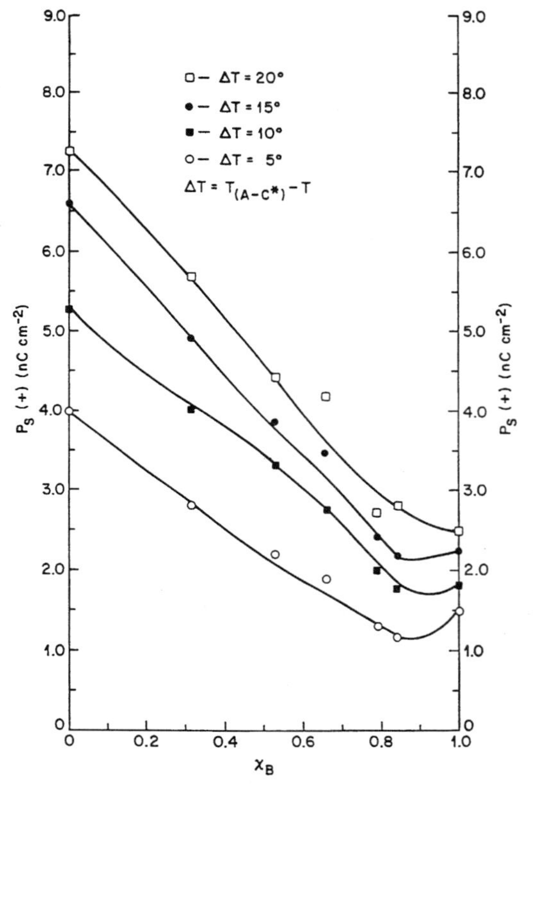

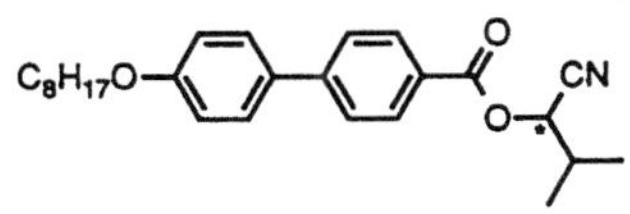

FIGURE 2.106. A cyanohydrin ester dopant.

with suitably low viscosity and negative dielectric anisotropy, and doping it with a small amount of a chiral material with a high spontaneous polarization. The mixture should then have the bulk mechanical properties of the pure host combined with the electro-optic properties of the diluted dopant. This is illustrated with reference to a cyanohydrin ester dopant (FIGURE 2.106) (Chan et al., 1989) in a terphenyl host mixture (FIGURE 2.107) (Hird, 1990).

The host is a eutectic mixture of *ortho*-difluoroterphenyls with both alkyl and alkoxy terminal chains. The latter two compounds in FIGURE 2.107 have the advantage of high negative dielectric anisotropies ($\simeq -4$) but are also quite viscous and have high clearing and phase transition temperatures. The two major constituents are used to lower the overall melting point and reduce the stability of the smectic C phase. This mixture does not have a smectic A phase and, at first sight, this would seem to be a disadvantage since good alignment in a cell is usually achieved by cooling into the smectic C phase from an smectic A phase. Fortuitously, the introduction of the cyanohydrin ester dopant injects the required phase into the sequence.

FIGURE 2.107. A difluoroterphenyl host mixture.

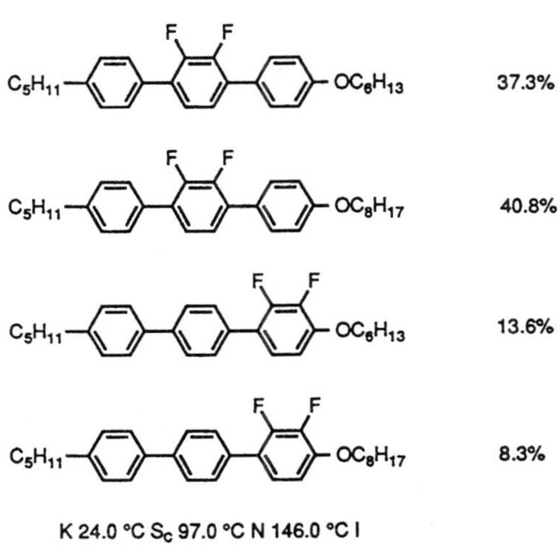

TABLE 2.9 shows the effect of increasing the proportion of chiral dopant in the mixture (Hird, 1990). Addition of the dopant depresses the smectic and nematic phase stabilities and thus the overall proportion of dopant is kept constant whilst the relative amounts of chiral and racemic cyanohydrin ester are varied. In this way the phase transition temperatures are kept constant throughout TABLE 2.9 and response time can be measured as a single function of spontaneous polarization.

As can be seen from TABLE 2.9, as the proportion of chiral material is increased, so the spontaneous polarization of the mixture rises and thus the ferroelectric response time falls.

Finally, TABLE 2.10 shows the effect of adding a second chiral dopant (FIGURE 2.108) to the mixture. The phase transition temperatures are favorably depressed and the response time is further reduced to 3 μs, a state-of-the-art figure. In addition, the helical twist senses of the two dopants are in opposition and thus the overall pitch is long.

2.5 ACHIRAL FERROELECTRIC HOST MATERIALS

The role of a host system is to provide a medium which can be switched quickly and efficiently by a low concentration of chiral dopant. The following material requirements are therefore of importance (Hird et al., 1992).

1. Low viscosity for fast switching.
2. High negative dielelectric anisotropy also for fast switching.
3. A suitable phase sequence for good smectic C alignment; typically S_C–S_A–N–I.
4. A tilt angle of 22.5° as a compromise for fast switching and good optical contrast.
5. An optical path length difference of around 0.28 μm for optimum contrast.

Real systems clearly have to be a compromise since, for example, a high dielectric anisotropy can best be produced by the addition of polar lateral groups to the core of the mesogen, however, such groups tend to promote higher viscosity.

There is considerable research interest in this area currently, however, commercial organizations are (understandably) reluctant to publish their most

TABLE 2.9. The spontaneous polarization, tilt angle and response time of a doped mixture as a function of the concentration of chiral dopant.

Mixture	Component	Proportion	Properties	
	Host	90.0%	P_S	3.3 nC cm^{-2}
A	Chiral dopant	1.0%	Tilt angle	23.5°
	Racemic dopant	9.0%	Response time	180 μs (10 V μm^{-1})
	Host	90.0%	P_S	9.0 nC cm^{-2}
B	Chiral dopant	2.5%	Tilt angle	23.5°
	Racemic dopant	7.5%	Response time	—
	Host	90.0%	P_S	21.0 nC cm^{-2}
C	Chiral dopant	7.5%	Tilt angle	21.0°
	Racemic dopant	2.5%	Response time	25 μs (10 V *m^{-1})
	Host	90.0%	P_S	38.0 nC cm^{-2}
D	Chiral dopant	10.0%	Tilt angle	27.0°
	Racemic dopant	0.0%	Response time	9 μs (10 V μm^{-1})

TABLE 2.10 The performance of a mixture doped with two chiral materials.

Mixture	Component	Proportion	Properties	
	Host	80.0%	P_S	39.0 nC cm^{-2}
E	Chiral dopant 1	7.6%	Tilt angle	18.5°
	Chiral dopant 2	12.4%	Response time	3 μs (10 V μm^{-1})

recent findings or techniques. There is therefore comparatively little literature on this topic. Similarly, the available results are seldom mutually compatible since the performance of a component in a ferroelectric host mixture is highly dependent upon the choice of mixture composition, alignment technique, driving scheme and dopant. It is therefore quite possible that materials classified as poor hosts by one experimenter may be considered excellent by another using a different test procedure.

2.5.1 Terphenyl Host Systems

A number of these materials have been prepared as ferroelectric hosts, for example the difluoro terphenyl systems (FIGURE 2.109) (Hird, 1990). The lateral fluorine substituents provide a high negative dielectric anisotropy, and although this is at the cost of increased viscosity, the substituents are relatively small, and the second fluorine *ortho* to the first actually reduces the viscosity slightly. It is also apparent that the viscosity is lowest when the lateral groups are on the middle ring. This suggests that difluorination, particularly in the 2′ and 3′ positions, merely broadens the molecule rather than acting as an angular projection at the side of the core. The choice of end group is also important. The electron-donating alkoxy chain feeds electron density into the aromatic π system to be withdrawn by the fluorines, especially when they are on the same ring. This has the overall effect of increasing the negative dielectric anisotropy. Conversely, if the end groups are not electron rich, as in the case of the dialkyl terphenyls, the dielectric anisotropy is considerably lower.

These materials also exhibit a marked tendency to produce smectic C phases, and the lateral groups help to lower the otherwise high melting and clearing temperatures of three-ring mesogens. The phase sequence is also ideal for good alignment.

FIGURE 2.108. A second chiral cyanohydrin ester dopant.

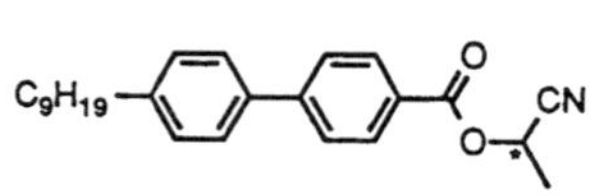

	Dielectric Anisotropy	Viscosity/cP
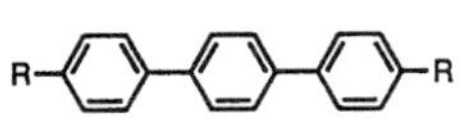	-	-
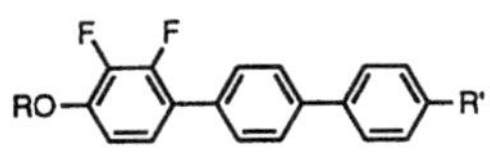	-4.2	130
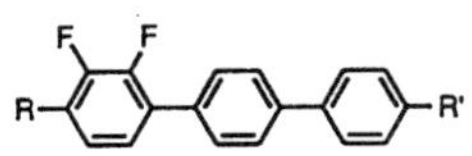	-2.0	80
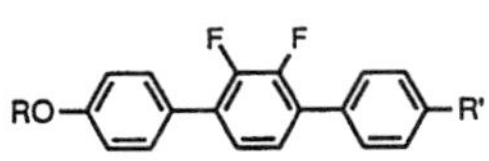	-2.1	60
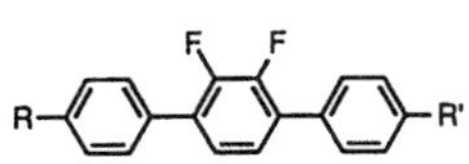	-2.0	35
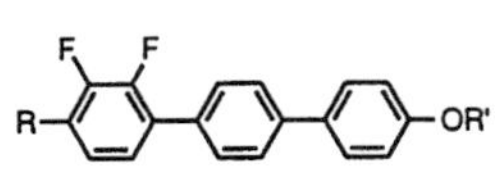	-2.1	130

R = C_5H_{11}: K 192 °C S_A 213 °C I

R = C_6H_{13}, R" = C_5H_{11}: K 97.5 °C S_C 145.5 °C N 166.0 °C I

R = C_5H_{11}, R" = C_5H_{11}: K 81.0 °C S_C 115.5 °C S_A 131.5 °C N 142.0 °C I

R = C_6H_{13}, R' = C_5H_{11}: K 54 °C S_C 67.0 °C N149.0 °C I

R = C_5H_{11}, R' = C_6H_{13}: K 54 °C S_C 67 °C N 149°C I

R = C_5H_{11}, R' = C_6H_{13}: K101.5 °C S_C 156.5 °C S_A 167.0 °C N 171.5 °C I

FIGURE 2.109. The phase transition temperatures, viscosities and dielectric anisotropies of some difluoroterphenyl smectogens (Hird, 1990).

2.5.2 Biphenyl Host Systems

Biphenyl systems (FIGURE 2.110) have also been studied (Hird, 1990), and these smectogens can prove more mobile. It has been noted, however, that the viscosity of these systems is generally limited by the viscosity of the chiral dopant, which is usually higher. There is also a general tendency for the dopant to depress the phase transition temperatures. A 10 °C depression in a terphenyl system is usually acceptable but in the case of two-ring systems, the smectic C phase stability can be reduced below the likely operating temperature of a device.

Similarly, biphenyl cyclohexyl systems have been investigated (FIGURE 2.111).

	Dielectric Anisotropy	Viscosity/cP
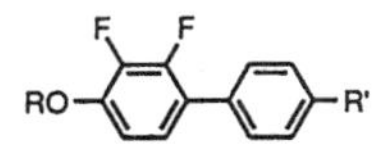	-4.7	57
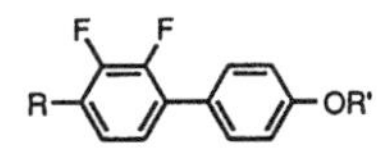	-2.6	80

R = C_8H_{17}, R' = C_7H_{15}: K 24 °C (S_C 6.0 °C S_A 13.0 °C N 23.0 °C) I

R = C_9H_{19}, R' = C_8H_{17} : K 25.0 (S_C 11.5 °C S_A 33.0 °C N 34.0 °C I

FIGURE 2.110. The transition temperatures, dielectric anisotropies and viscosities of some difluorobiphenyl smectogens (Hird, 1990).

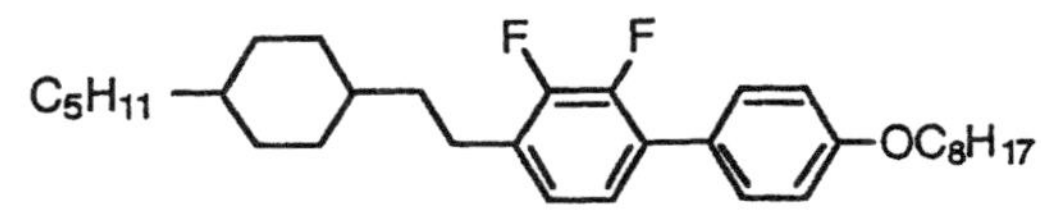

K 32.5 °C S_C 55 °C S_A 83 °C N 122 °C Iso

FIGURE 2.111. A biphenylcyclohexyl host material (Hird et al., 1992).

In this case, the saturated linking group is used to inhibit the formation of the higher-order smectic phases commonly associated with cyclohexyl mesogens. This also increases the flexibility of the mesogen and thereby reduces the transition temperatures. An ester group could be used to similar effect, but the higher polarity of the carbonyl moiety would increase the viscosity.

The next examples (FIGURE 2.112), based on 3-(4′-*n*-alkoxybiphenylyl)propanoic acid (Hall et al., 1992), illustrate the use of a four-center linking group and also show how the incorporation of fluorine substituents changes the smectic phase behavior.

Clearly, the difluorinated analog has the widest and lowest smectic C phase. It has, however, been

FIGURE 2.112. An illustration of the use of a four-center linking group in a host material.

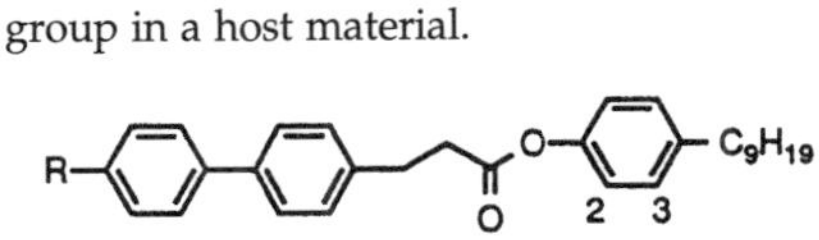

(a) R = C_8H_{17}
(b) R = $C_{10}H_{21}$

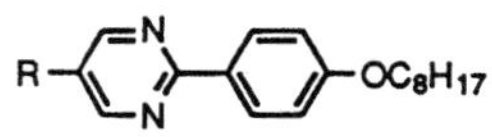

R	$T_{K\text{-}S_C}$/°C	$T_{S_C\text{-}S_A/N}$/°C	$T_{S_A\text{-}N/I}$	$T_{N\text{-}I}$/°C
C_7H_{15}	51	(45)	–	72
C_8H_{17}	34	56	64	70
C_9H_{19}	33	61	75	–

FIGURE 2.113. The phase transition temperatures of some phenylpyrimidine-based host materials.

found that four-center linking groups promote higher viscosity (Kelly et al., 1991c).

2.5.3 Heteroaromatic Core Systems

The 5-*n*-alkyl-2-(4-*n*-alkyloxyphenyl)pyrimidines (Zaschke, 1975) (FIGURE 2.113) are used in the majority of commercial ferroelectric host materials. Their widespread use is due to their low viscosities, wide-range smectic C phases, and low melting points.

A typical phenylpyrimidine host mixture is shown in FIGURE 2.114 (Kusamoto et al., 1990). When doped with a range of cyanohydrin ester dopants, this mixture has been switched at between 64 μs and 134 μs with a spontaneous polarization of between 3 and 9.1 nC cm^{-2}.

The smectic C phase range is quite wide and extends well above typical display device operating temperatures, although the crystal to smectic C transition is somewhat high. There are also nematic and smectic A phases, and thus good alignment should be possible. The fluorophenyl pyrimidine core is used in the minor component, presumably to increase the negative dielectric anisotropy, although this must be at the cost of increased viscosity. The viscosity of host mixtures is, however, seldom quoted since the most relevant estimate of viscosity is derived from switching speeds.

FIGURE 2.114. A typical phenylpyrimidine host mixture.

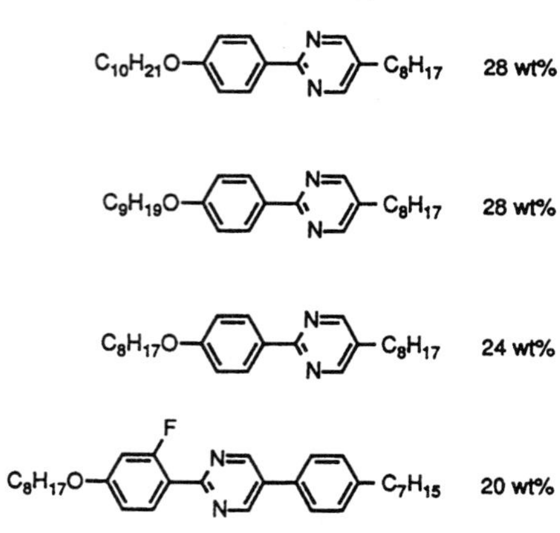

K 13 °C S_C 68.5 °C S_A 73.5 °C N 83.5 °C I

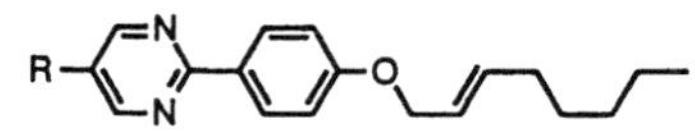

R	$T_{K\text{-}S_C}$/°C	$T_{S_C\text{-}S_A/N}$/°C	$T_{S_A\text{-}N/I}$	$T_{N\text{-}I}$/°C
C_7H_{15}	38	(35)	–	68
C_8H_{17}	24	56	–	68
C_9H_{19}	42	70	72	73

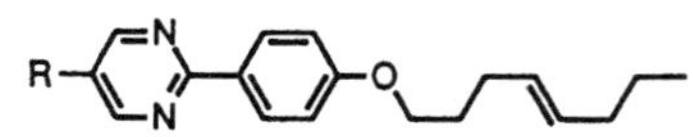

R	$T_{K\text{-}S_C}$/°C	$T_{S_C\text{-}S_A/N}$/°C	$T_{S_A\text{-}N/I}$	$T_{N\text{-}I}$/°C
C_7H_{15}	39	45	–	65
C_8H_{17}	32	56	59	63
C_9H_{19}	42	64	73	–

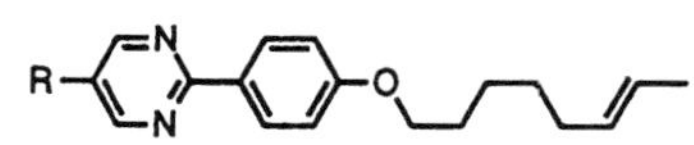

R	$T_{K\text{-}S_C}$/°C	$T_{S_C\text{-}S_A/N}$/°C	$T_{S_A\text{-}N/I}$	$T_{N\text{-}I}$/°C
C_7H_{15}	53	–	–	72
C_8H_{17}	40	52	57	67
C_9H_{19}	39	53	71	72

FIGURE 2.115. Phenylpyrimidine host materials with *trans* alkenyl terminal chains.

A recent modification to the phenylpyrimidine systems (Kelly, 1993) is the incorporation of double bonds into the terminal chains. The unsaturated units have the effect of fixing the geometric orientation of two chain atoms and thus alter the phase behavior. The *trans* alkenes (FIGURE 2.115) have, on average, the same transition temperatures as the unsaturated parent compounds. However, the variation is such that in the optimum case the smectic C phase range is broadened at the expense of the smectic A and a short nematic phase is induced.

In contrast, the *cis* alkenes (FIGURE 2.116) depress the phase transition temperatures, giving moderately wide smectic C phases at, and below, room temperature.

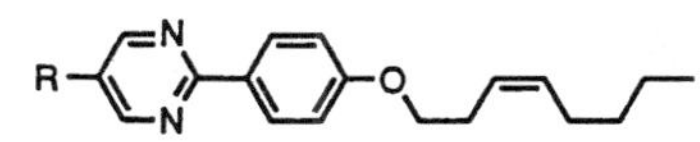

R	$T_{K\text{-}S_C}$/°C	$T_{S_C\text{-}S_A/N}$/°C	$T_{S_A\text{-}N/I}$	$T_{N\text{-}I}$/°C
C_7H_{15}	28	–	43	44
C_8H_{17}	25	38	45	–
C_9H_{19}	24	43	54	–

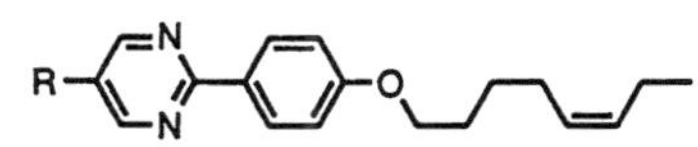

R	$T_{K\text{-}S_C}$/°C	$T_{S_C\text{-}S_A/N}$/°C	$T_{S_A\text{-}N/I}$	$T_{N\text{-}I}$/°C
C_7H_{15}	2	24	47	53
C_8H_{17}	10	38	53	–
C_9H_{19}	24	45	61	–

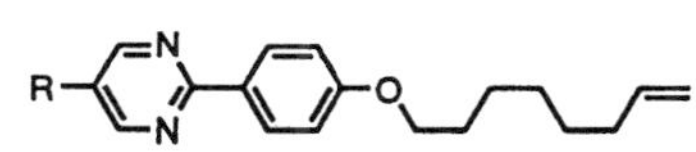

R	$T_{K\text{-}S_C}$/°C	$T_{S_C\text{-}S_A/N}$/°C	$T_{S_A\text{-}N/I}$	$T_{N\text{-}I}$/°C
C_7H_{15}	36	(33)	48	62
C_8H_{17}	22	34	56	60
C_9H_{19}	37	–	67	–

FIGURE 2.116. Phenylpyrimidine host materials with *cis* alkenyl terminal chains.

Both of these effects diminish as the unsaturated unit is moved further from the aromatic core until both *cis* and *trans* alkenes depress phase transition temperatures. This may be the result of induced dipole–dipole interactions between neighboring chains; such associations may be expected to disrupt the phase.

Some potentially useful pyridyl systems have also been reported (Kelly et al., 1993a) (FIGURE 2.117). In contrast to the analogous pyrimidyl alkanoates

FIGURE 2.117 Potential phenylpyridinyl alkanoate host materials.

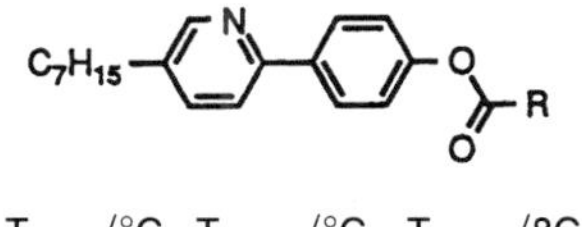

R	$T_{K\text{-}S_G}$/°C	$T_{S_G\text{-}S_F}$/°C	$T_{S_F\text{-}S_C}$/8C	T_{S_C}/°C
C_8H_{17}	50	68	76	77
C_9H_{19}	52	68	78	79
$C_{10}H_1$	56	69	79	81
$C_{11}H_{23}$	52	66	79	81
$C_{12}H_{25}$	54	68	79	82

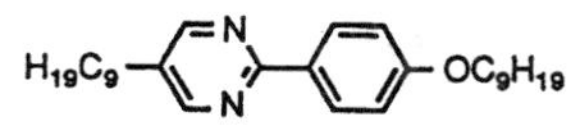

K 34 °C S_C 61°C S_A 75 °C I

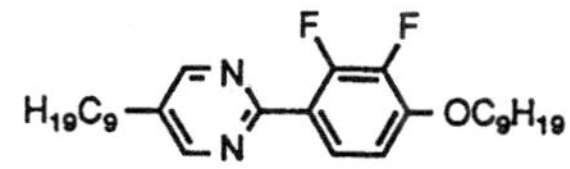

K 41 °C S_C 54 °C I

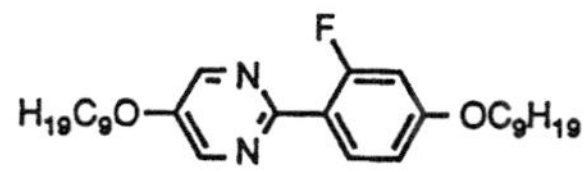

K 45 °C S_C 62 °C N 73 °C I

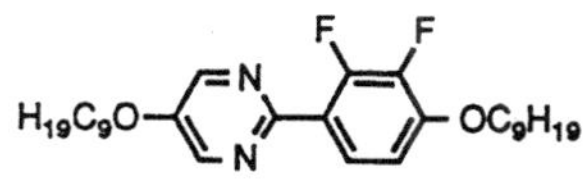

K 48 °C S_C 78 °C I

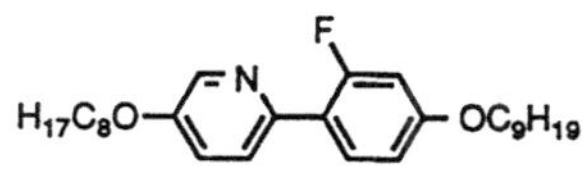

K 44 °C S_C (43 °C) N 70 °C I

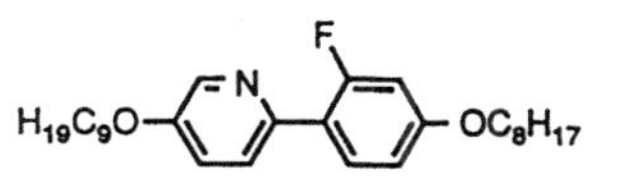

K 47 °C S_C 48 °C N 69 °C I

FIGURE 2.118. Fluorinated phenylpyrimidyl and phenylpyridyl host materials.

(Zaschke, 1975) these molecules exhibit smectic mesophases and individual examples have useful smectic C phases.

Mixtures incorporating these species have been switched at typical speeds of 100 μs. These systems do not, however, exhibit nematic phases and must therefore be used in conjunction with more easily aligned host materials.

Attempts to increase the negative dielectric anisotropy of heterocyclic hosts have led to the fluorination of phenyl core components (Pausch et al., 1992). Typical phase transition temperatures are shown in FIGURE 2.118. The phase ranges and sequences of these materials are not outstanding, but their dielectric properties may be useful in mixtures.

A further variation is the fluorination of one of the terminal chains. As can be seen from FIGURE 2.119, fluorination on the pyrimidine side of the molecule leads to a wider smectic C phase range than the reverse configuration. Furthermore, the fluorinated terminal chain gives rise to extremely wide smectic C phase ranges (FIGURE 2.120), and in the optimum case there is even a 4 °C smectic A range to aid alignment.

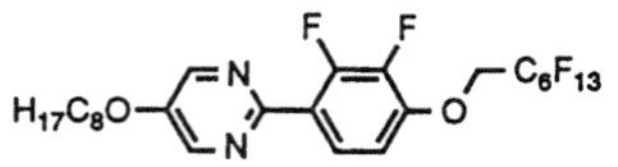

K 65 °C S_C (64 °C) S_A 138 °C

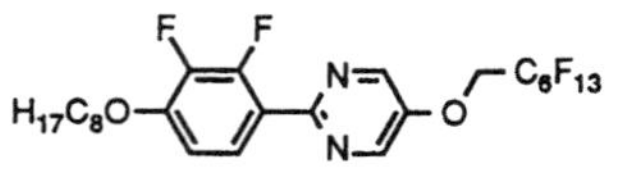

K 77 °C S_C 127 °C I

FIGURE 2.119. Difluorophenylpyrimidyl host materials with fluorinated terminal chains (Pausch et al., 1992).

2.5.4 Core Linking Groups

In common with nematic systems, the incorporation of a linking group in a ferroelectric host material increases the flexibility of the mesogen, effectively lengthens the core structure, and can introduce regions of high polarizability into the molecule. This is illustrated, in FIGURE 2.121, for a cyclohexyl phenylpyrimidine (Kelly et al., 1991c) as a 13 mol.% component of a chiral smectic C base mixture.

The parent compound, with the cyclohexyl ring directly bonded to the phenylpyrimidine, has only a 10 °C smectic C range above a more ordered smectic B phase. The introduction of an ethyl or oxymethylene linkage only slightly increases the onset temperature of the smectic C phase and there is no underlying smectic B phase. The smectic C phase range is increased significantly in the case of the oxymethylene linkage, but this is at the cost of the smectic A phase. The spontaneous polarization of the ethyl-linked material is almost identical to that of the parent compound, whereas that of the oxymethylene-containing mixture is some 20% higher.

The incorporation of an ester linkage significantly increases the spontaneous polarization of the mixture, but although the smectic C phase range is almost double that of the parent compound, there is no smectic A phase. This material also has the disadvantage of a higher-ordered smectic B phase.

The four-centered butylene linkage has a similar phase sequence to the parent compound, although the transition temperatures are depressed by as much as 40 °C. This material is also rather viscous and has a low spontaneous polarization. The use of small amounts of this class of compound in mixtures has, however, been reported to improve the surface alignment in surface-stabilized FLC devices.

Both the propyloxy and propenyloxy-linked materials have remarkably wide (41 °C) smectic C phases. They melt straight into the smectic C phase but do not, unfortunately, have a smectic A phase. The propyloxy-linked material has a reasonable spontaneous polarization although, in keeping with other findings (Kelly et al., 1991c) the viscosity associated with the four-centered linking group is somewhat high. In general,

FIGURE 2.120. Monofluorophenylpyrimidyl host materials with fluorinated terminal chains (Pausch et al., 1992).

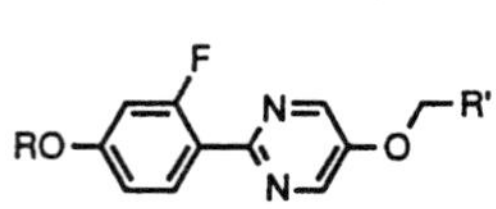

R	R′	$T_{K\text{-}S_C}$/°C	T_{S_CA}/°C	$T_{S_C/S_A\text{-}I}$/°C
C_8H_{17}	C_4F_9	67	–	104
C_8H_{17}	C_6F_{13}	65	–	127
C_8H_{17}	C_8F_{17}	89	141	145
$C_{12}8H_{25}$	C_4F_9	65	–	96
$C_{12}H_{25}$	C_6F_{13}	64	–	119

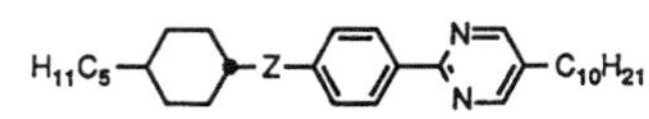

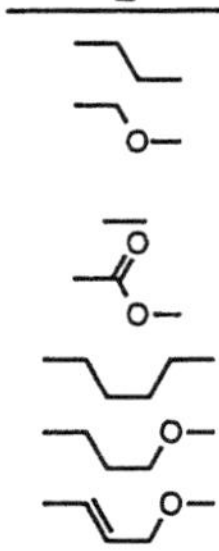

Z	$T_{K\text{-}S_B/S_C}$/°C	T_{S_B/S_C}/°C	$T_{S_C\text{-}S_A}$/°C	$T_{S_A\text{-}N}$/°C	$T_{N\text{-}I}$/°C
	84	-	98	118	132
	87	-	118	-	142
	60	83	93	131	152
	64	85	104	-	161
	66	71	86	102	109
	69	-	110	-	130
	69	-	110	-	134

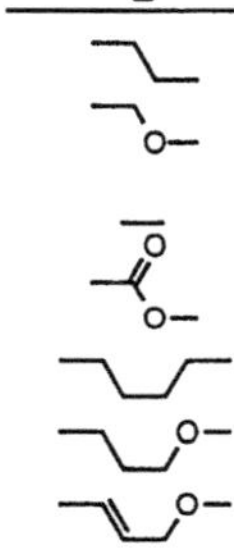

Z	Tilt Angle/°	Viscosity/mPa s	P_S/nC cm^{-2}
	20.0	30.0	5.66
	23.2	38.8	6.44
	18.7	26.6	5.64
	25.8	45.2	7.05
	19.3	31.9	4.97
	25.1	51.1	6.31
	-	-	-

FIGURE 2.121. The effect of different linking groups in a 13 mol.% component of a ferroelectric mixture on transition temperatures and performance.

the incorporation of a *trans* alkene or an oxygen atom into a four-centered linking group stabilizes the smectic C and nematic phases.

Acknowledgement
The authors would like to thank Dr. K. J. Toyne and Dr. M. Hird for useful discussions and the Science and Engineering Research Council (SERC) and Ministry of Defence (DRA), UK, for financial support. They also wish to thank Dr C. Booth for the use of photographs of the cholesteric liquid crystal phase.

REFERENCES

Andrews, B. M., Gray, G. W., and Bradshaw, M. J. 1985. *Mol. Cryst. Liq. Cryst.* **123**: 257.

Bishop, D., Jenner, J. A., and Sage, I. 1986. *Proceedings of the Eleventh International Liquid Crystal Conference [Berkeley 1986]*, FE **02**: 0852–0915. New York: Gordon and Breach.

Boller, A., Cereghetti, M., Schadt, M., and Scherrer, H. 1977. *Mol. Cryst. Liq. Cryst.* **42**: 215.

Booth, C. J. 1992. *PhD Thesis*, University of Hull.

Booth, C. J., Goodby, J. W., Hardy, J. P., Lettington, O. C., and Toyne, K. J. 1993a. *J. Mater. Chem.* **3**: 935.

Booth, C. J., Hardy, J. P., Lettington, O. C., and Toyne, K. J. 1993b. *J. Mater. Chem.* **3**: 821.

Bouligard, Y. 1973. *J. Physique* **34**: 603.

Brand, H. R., Cladis, P. E., and Finn, P. L. 1985. *Phys. Rev. A* **31**: 361.

Bullock, P. J., Byron, D. J., Harwood, D. J., Wilson, R. C., and Woodward, A. M. 1984. *J. Chem. Soc. Perkin Trans. II* **1984**: 2121.

Byron, D. J., Keating, D. A., O'Neill, M. T., Wilson, R. C., Goodby, J. W., and Gray, G. W. 1980. *Mol. Cryst. Liq. Cryst.* **58**: 179.

Byron, D. J., Wilson, R. C., Baker, P. A., Danilewicz, T., Hayer, G. S., Hewison, D. A., Taylor, J. M., and Wyer, J. M. 1985. *J. Chem. Soc. Perkin Trans. II* **1985**: 297.

Cahn, R. S., and Ingold, C. K. 1951. *J. Chem. Soc.* **1951**: 612.

Cahn, R. S., Ingold, C. K., and Prelog, V. 1956. *Experientia* **12**: 81.

Carr, N., Gray, G. W., and Kelly, S. M. 1981. *Mol. Cryst. Liq. Cryst.* **66**: 267.

Carr, N., Gray, G. W., and McDonnell, D. G. 1983. *Mol. Cryst. Liq. Cryst.* **97**: 13.

Carr, N., and Gray, G. W. 1985. *Mol. Cryst. Liq. Cryst.* **124**: 27.

Chan. L. K. M., Gray, G. W., Lacey, D., Scrowston, R. M., Shenouda, I. G., and Toyne, K. J. 1989. *Mol. Cryst. Liq. Cryst.* **172**: 125.

Chandani, A. D. L., Ouchi, Y., Takezoe, H., Fukuda, A., Terashima, K., Furukawa, K., and Kishi, A. 1989. *Japan. J. Appl. Phys. Lett.* **28**: 1261.

Chen, D. S., and Jones, F. N. 1987. *J. Polym. Sci.* **25**: 1109.

Chen, D. S., and Jones, F. N. 1988. *J. Coatings Technol.* **60**: 39.

Chin, E., Goodby, J. W., Gery, J. M., Patel, J. S., and Leslie, T. M. 1987. *Mol. Cryst. Liq. Cryst.* **146**: 325.

Clark, N. A., and Lagerwall, S. T. 1980a. *Liquid Crystals of One- and Two-Dimensional Order*, ed. W. Helfrich and G. Heppke, p. 222. Berlin: Springer.

Clark, N. A., and Lagerwall, S. T. 1980b. *Appl. Phys. Lett.* **36**: 899.

Clark, N. A., and Lagerwall, S. T., 1984. *Ferroelectrics* **59**: 25.

Coates, D., and Gray, G. W. 1978. *Mol. Cryst. Liq. Cryst.* **41**: 197.

Coates, D. 1990. *Liquid Crystals: Applications and Uses*, vol. 1, ed. B. Bahadur, p. 92. Singapore: World Scientific.

Decobert, G., Dubois, J. C., Esselin, S., and Noël, C. 1986. *Liq. Cryst.* **1**: 307.

de Gennes, P. G., and Prost, J. 1993. *The Physics of Liquid Crystals.* 2d ed. Oxford: Oxford University Press.

Deutscher, H.-J., Kuschel, F., König, S., Kresse, H., Pfeiffer, D., Wiegeleben, A., Wulf, J., and Demus, D. 1977. *Z. Chem.* **17**: 64.

Deutscher, H.-J., Laaser, B., Dolling, W., and Schubert, H. J. 1978. *J. Prakt. Chem.* **320**: 191.

Eidenschink, R., Erdmann, D., Krause, J., and Pohl, L. 1977. *Angew. Chem. Int. Ed. Engl.* **16**: 100.

Eidenschink, R., Krause, J., and Pohl, L. 1978a. *German Patent* 2, 701, 591.

Eidenschink, R., Erdmann, D., Krause, J., and Pohl, L. 1978b. *Angew. Chem. Int. Ed. Engl.* **17**: 133.

Eidenschink, R., Haas, G., Romer, M., and Scheuble, B. S. 1984. *Angew. Chem. Int. Ed. Engl.* **23**: 147.

Fergason, J. L. 1964. *Sci. Am.* **211**: 77.

Fergason, J. L. 1966. *Molecul. Cryst.* **1**: 293.

Friedel, G. 1922. *Ann. Physique* **18**: 373.

Geivandov, R. C., Goncharov, I. V., and Titov, V. V. 1989. *Mol. Cryst. Liq. Cryst.* **166**: 101.

Goodby, J. W., and Gray, G. W. 1979. *Mol. Cryst. Liq. Cryst.* **56**: 43.

Goodby, J. W. 1984. *Mol. Cryst. Liq. Cryst.* **110**: 205.

Goodby, J. W., Chin, E., Leslie, T. M., Geary, J. M., and Patel, J. S. 1986. *J. Am. Chem. Soc.* **108**: 4736.

Goodby, J. W., Chin, E., Geary, J. M., Patel, J. S., and Finn, P. L. 1987a. *J. Chem. Soc. Faraday Trans. I* **83**: 3429.

Goodby, J. W., Patel, J. S., and Chin, E. 1987b. *J. Phys. Chem.* **91**: 5151.

Goodby, J. W. 1991. *J. Mater. Chem.* **1**: 307.

Goodby, J. W., Patel, J. S., and Chin, E. 1992. *J. Mater. Chem.* **2**: 197.

Goodby, J. W., Nishiyama, I., Slaney, A. J., Booth, C. J., and Toyne, K. J. 1993. *Liq. Cryst.* **14**: 37.

Gorecka, E., Chandani, A. D. L., Ouchi, Y., Takezoe, H., and Fukuda, A. 1990. *Japan. J. Appl. Phys.* **29**: 131.

Gottarelli, G., Mariani, P., Spada, G. P., Samori, B., Forni, A., Solladie, G., and Hibert, M. 1983a. *Tetrahedron* **39**: 1337.

Gottarelli, G., Hibert, M., Samori, B., Solladie, G., Spada, G. P., and Zimmerman, R. G. 1983b. *J. Am. Chem. Soc.* **104**: 7318.

Gottarelli, G., and Spada, G. P. 1985. *Mol. Cryst. Liq. Cryst.* **123**: 277.

Gray, G. W., Harrison, K. J., and Nash, J. A. 1971. *Electron. Lett.* **18**: 127.

Gray, G. W., Harrison, K. J., and Nash, J. A. 1973. *Electron. Lett.* **9**: 130.

Gray, G. W., Harrison, K. J., and Nash, J. A. 1974. *J. Chem. Soc. Chem. Commun.* **1974**: 431.

Gray, G. W., and McDonnell, D. G. 1975a. *Electron. Lett.* **11**: 556.

Gray, G. W. 1975b. *J. Physique Coll.* **36**: C1–337.

Gray, G. W., and McDonnell, D. G. 1976. *Mol. Cryst. Liq. Cryst.* **37**: 189.

Gray, G. W., and McDonnell, D. G. 1977. *Mol. Cryst. Liq. Cryst.* **34**: 211.

Gray, G. W., and Leadbetter, A. J. 1977c. *Phys. Bull.* **28**: 28.

Gray, G. W., and McDonnell, D. G. 1978a. *Mol. Cryst. Liq. Cryst.* **48**: 37.

Gray, G. W. 1978b. *Advances in Liquid Crystal Materials for Applications.* Poole: BDH Ltd.

Gray, G. W., and McDonnell, D. G. 1979. *Mol. Cryst. Liq. Cryst.* **53**: 147.

Gray, G. W., and Kelly, S. M. 1981a. *J. Chem. Soc. Perkin Trans. II* **1981**: 26.

Gray, G. W., and Kelly, S. M. 1981b. *Angew. Chem. Int. Ed. Engl.* **20**: 393.

Gray, G. W. 1982. *Abstract 12 Freiburger Arbeitstagung.* 1983.

Gray, G. W., and McDonnell, D. G. 1983a. *Mol. Cryst. Liq. Cryst.* **95**: 101.

Gray, G. W., and Lacey D. 1983b. *Mol. Cryst. Liq. Cryst.* **99**: 123.

Gray, G. W., and Kelly, S. M. 1984. *Mol. Cryst. Liq. Cryst.* **104**: 335.

Haas, W. E. 1983. *Mol. Cryst. Liq. Cryst.* **94**: 1.

Hagiwara, T., Suzuki, Y., et al. 1989. *Proceedings of the Fifteenth Japanese Liquid Crystal Conference [Osaka]: Abstracts*, p. 304.

Hall, A. W., Lacey, D., Gray, G. W., and Benson, J. 1992. *Liq. Cryst.* **12**: 879.

Hara, H., Satoh, T., Toya, T., Iida, S., Orii, S., and Watanabe, J. 1988. *Macromolecules* **21**: 14.

Haramoto, Y., and Kamogawa, H. 1985. *Chem. Lett.* **1985**: 79.

Hiji, N., Chandani, A. D. L., Nishiyama, I., Ouchi, Y., Takezoe, H., and Fukuda, A. 1988. *Ferroelectrics* **85**: 99.

Hilsum, C., and McDonnell, D. G. 1985. *British Patent* 2,085,585B.

Hirai, T., Fukumasa, M., Nishiyama, I., Yoshizawa, A., Shiratori, N., Yokoyama, A., and Yamane, M. 1991. *Ferroelectrics* **114**: 251.

Hird, M. 1990. *PhD Thesis*, University of Hull.

Hird, M., Gray, G. W., and Toyne, K. J. 1992. *Liq. Cryst.* **11**: 531.

Ikemoto, T., Sakeshita, K., Hayashi, S., Kageyama, Y., Uematsu, M., Nakauchi, J., and Mori, K. 1989. *Proceedings of the Fifteenth Japanese Liquid Crystal Conference [Osaka]: Abstracts*, p. 18.

Inui, S., Kawano, S., Saito, M., Iwane, H., Takanishi, Y., Hiraoka, K., Ouchi, Y., Takezoe, H., and Fukuda, A. 1990. *Japan. J. Appl. Phys. Lett.* **29**: 987.

Inukai, T., Inoue, H., and Sato, H. 1978. *US Patent* 4,211,666.

James, B. G. 1974. *US Patent* 3,802,945.

Johno, M., Chandani, A. D. L., Lee, J., Ouchi, Y., Takezoe, H., Fukuda, A., Itoh, K., and Kitazuma, T. 1989. *Japan Display '89: Proceedings of the Ninth International Display Research Conference [Kyoto]*, p. 22. Playa del Rey, Calif.: Society for Information Display.

Karamysheva, L. A., Kovshev, E. I., Pavluchenko, A. I., Roitman, K. V., Titov, V. V., Torgova, S. I., and Grebenkin, M. F. 1981. *Mol. Cryst. Liq. Cryst.* **67**: 241.

Kawada, M., Uesugi, Y., Matsumura, K., Sudo, Y., Kondo, K., and Kitamura, T. 1990. *Proceedings of the Sixteenth Japanese Liquid Crystal Conference [Hiroshima]: Abstracts*, p. 10.

Kelly, S. M. 1983. *J. Chem. Soc. Chem. Commun.* **1983**: 366.

Kelly, S. M. 1984. *Helv. Chim. Acta* **67**: 1521.

Kelly, S. M. 1991a. *Liq. Cryst.* **10**: 261.

Kelly, S. M. 1991b. *Liq. Cryst.* **10**: 273.

Kelly, S. M., Fünfschilling, J., and Leenhouts, F. 1991c. *Liq. Cryst.* **10**: 243.

Kelly, S. M. 1991d. *Mol. Cryst. Liq. Cryst.* **204**: 27.

Kelly, S. M., Fünfschilling, J., and Villiger, A. 1993a. *Liq. Cryst.* **14**: 699.

Kelly, S. M. 1993b. *Liq. Cryst.* **14**: 675.

Kobayashi, S., Ishibashi, S., and Tsuru, S. 1990a. *Proceedings of the Sixteenth Japanese Liquid Crystal Conference [Hiroshima]: Abstracts*, p. 40.

Kobayashi, S., Ishibashi, S., Tsuru, S. 1990b. *Mol. Cryst. Liq. Cryst. Lett.* **7**: 105.

Kobayashi, S., and Ishibashi, S. 1990c. *Mol. Cryst. Liq. Cryst. Lett.* **7**: 53.

Koden, M., Kuratate, T., Funada, F., Sakaguchi, K., Takehira, Y., and Kitamura, T. 1989. *Japan Display '89: Proceedings of the Ninth International Display Research Conference [Kyoto]*, p. 34. Playa de Rey, Calif.: Society for Information Display.

Koden, M., Kuratate, T., Funada, F., Awane, K., Sakaguchi, K., Shiomi, Y., and Kitamura, T. 1990. *Japan. J. Appl. Phys. Lett.* **29**: 981.

Koide, N., Kaneko, T., and Aoyama, Y. 1989. *Proceedings of the Fifteenth Japanese Liquid Crystal Conference [Osaka]: Abstracts*, p. 88.

Kondo, A., and Valkenburg, J. 1979. *Microcapsule Processing and Technology.* New York: Dekker.

Korte, E. H., and Schrader, B. 1981. *Advances in Infrared and Raman Spectroscopy*, ed. R. J. Clark and R. E. Hester. London: Heyden.

Kuehnpard, K., and Springer, J. 1991. *Makromol. Chem. Rapid Commun.* **12**: 381.

Kurusawa, K. 1981. *Japanese Patent* 81118.859.

Kusumoto, T., Hanamoto, T., Hiyama, T., Takehara, S., Shoji, T., Osawa, M., Kuriyama, T., Nakamura, K., and Fujisawa, T. 1990. *Chem. Lett.* **1990**: 1615.

Leadbetter, A. J., Richardson, R. M., and Collings, C. N. 1975. *J. Physique* **36**, 37.

Lecleercq, M., Billard, J., and Jacques, J. 1969. *Mol. Cryst. Liq. Cryst.* **8**: 367.

Lewthewaite, R. A., Goodby, J. W., and Toyne, K. J. 1993. *J. Mater. Chem.* **3**: 241.

Liquid Crystal Product Data Sheets. Poole: Merck.

Lydon, J. E., and Coakley, C. J. 1975. *J. Physique* **36**: 45.

Mace, A., Bersan, J., and Luby, J. 1981. *French Patent* 2,461,008.

Madhusudana, N. V., and Chandrasekhar, S. 1975. *Liquid Crystals: Proceedings of the International Conference Held at the Raman Research Institute [Bangalore 1973]*, ed. S. Chandrasekhar, p. 57. Pramana Suppl., 1. Bangalore: Indian Academy of Science.

Maier, W., and Saupe, A. 1958. *Z. Naturforsch.* **13a**: 564.

Maier, W., and Saupe, A. 1959. *Z. Naturforsch.* **14a**: 882.

Maier, W., and Saupe, A. 1960. *Z. Naturforsch.* **15a**: 287.

Maier, W., and Meier, G. 1961. *Z. Naturforsch.* **16a**: 262.

Makow, D. 1976. *Color Res. Applic.* **1**: 25.

Mauguin, C. 1911. *Bull. Soc. Fr. Minéral. Cristallogr.* **34**: 71.

McDonnell, D. G. 1979. *PhD Thesis*, University of Hull.

McDonnell, D. G. 1987. *Thermotropic Liquid Crystals*, ed. G. W. Gray, p. 120. Chichester: Wiley.

Merran, R. 1970. *French Patent* 2,105,524.

Meyer, R. B., Liebert, L., Strzelecki, L., and Keller, P. 1975. *J. Physique Lett.* **36**: 69.

Meyer, R. B. 1976. *Mol. Cryst. Liq. Cryst.* **40**: 74.

Mikami, N., Higuchi, R., Sakurai, T., Ozaki, M., and Yoshimo, K. 1986. *Japan. J. Appl. Phys.* **25**: 1833.

Mitsuhiro, K., Hayashi, S., Nishimura, K., Kusabayashi, S., and Takenaka, S. 1984. *Mol. Cryst. Liq. Cryst.* **106**: 31.

Nishide, K., Nakayama, A., Kusumoto, T., Hiyama, T., Takehara, S., Shoji, T., Osawa, M., Kuriyama, T., Nakamura, K., and Fujisawa, T. 1990. *Chem. Lett.* **1990**: 623.

Nishiyama, I., Yoshizawa, A., Fukumasa, M., and Hirai, T. 1989. *Japan. J. Appl. Phys. Lett.* **28**: 2288.

Nishiyama, I., and Goodby, J. W. 1993. *J. Mater. Chem.* **3**: 169.

Nixon, J. R., ed. 1976. *Microencapsulation.* New York: Dekker.

Nyirjesy, I., Abernathy, M. R., Billingsley, F. S., and Bruns, P. 1977. *J. Reproduct. Med.* **18**: 165.

Oseen, C. W. 1933. *Trans. Faraday Soc.* **29**: 883.

Osman, M. A., and Revesz, L. 1980. *Mol. Cryst. Liq. Cryst.* **56**: 133.

Osman, M. A. 1982. *Mol. Cryst. Liq. Cryst.* **72**: 291.

Osman, M. A., and Huynh-Ba, T. 1983a. *Mol. Cryst. Liq. Cryst.* **82**: 331.

Osman, M. A. 1983b. *Z. Naturforsch.* **38a**: 693.

Osman, M. A. 1983c. *Z. Naturforsch.* **38a**: 779.

Osman, M. A. 1984a. *Mol. Cryst. Liq. Cryst.* **116**: 141.

Osman, M. A., and Huynh-Ba, T. 1984b. *Helv. Chim. Acta.* **67**: 959.

Osman, M. A. 1985. *Mol. Cryst. Liq. Cryst.* **128**: 45.

Parker, R. 1974. *US Patent* 3,822,594.

Patel, J. S., and Goodby, J. W. 1987a. *J. Phys. Chem.* **91**: 5838.

Patel, J. S., and Goodby, J. W. 1987b. *Mol. Cryst. Liq. Cryst.* **144**: 117.

Patel, J. S., and Goodby, J. W. 1987c. *Opt. Eng.* **26**: 373.

Patel, J. S., and Goodby, J. W. 1987b. *Phil. Mag. Lett.* **55**: 283.

Pausch, A. E., Geelhaar, T., Poetsch, E., and Finkenzeller, U. 1992. *Liquid Crystal Materials, Devices, and Applications I*, ed. P. Drzaic and U. Efron, p. 128. Proc. SPIE, vol. 1665. Bellingham, Wash.: SPIE.

Pohl, L., Eidenschink, R., Krause, J., and Erdmann, D. 1977. *Phys. Lett.* **60A**: 421.

Pohl, L., Eidenschink, R., Krause, J., and Weber, G. 1978. *Phys. Lett.* **65A**: 169.

Polk, M. B, Banks, H. D., Onwumere, F., Venkatasubramian, N., Nandu, N., Phingbodhipakkiya, M., and Samulski, E. T. 1988. *J. Polym. Sci.* **26**: 2405.

Pollman, P., and Stegemeyer, H. 1973. *Chem. Phys. Lett.* **20**: 87.

Raynes, E. P. 1973. *Electron. Lett.* **9**: 118.

Reinitzer, F. 1888. *Monatssch. Wiener Chem. Ges.* **9**: 421.

Sage, I. 1987. *Thermotropic Liquid Crystals*, ed. G. W. Gray, p. 64. Chichester: Wiley.

Sakaguchi, K., Takehira, Y., Kitamura, T., Koden, M., Kuratate, T., and Funada, F. 1989. *Proceedings of the Fifteenth Japanese Liquid Crystal Conference [Osaka]: Abstracts*, p. 34.

Sasaki, M., Takeuchi, K., Sato, H., and Takatsu, H. 1984. *Mol. Cryst. Liq. Cryst.* **109**: 169.

Sato, H., Takatsu, H., Takeuchi, K., Fujiti, Y., Tazume, M., and Ohnishi, H. 1982. *Japanese Patent* 118, 526.

Schad, H., and Kelly, S. M. 1984. *J. Chem. Phys.* **81**: 1514.

Scherowsky, G., and Gay, J. M. 1989. *Liq. Cryst.* **5**: 1253.

Scherowsky, G., Beer, A., Gay, J. M., Muller, U., Schliwa, A., Rhode, E., and Kuhnpast, K. 1991. *Proceedings of the Thirteenth International Liquid Crystal Conference [Vancouver 1990]*, vol. 1, ed. J. Springer and P. Harnischfeger, p. 105. New York: Gordon and Breach.

Scherowsky, G. 1992a. *Polymers for Advanced Technologies*, vol. 3, p. 219. Chichester: Wiley.

Scherowsky, G., Brauer, B., Grunegerg, K., Muller, U., Komitov, L., Lagerwall, S. T., Skarp, K., and Stebler, B. 1992b. *Mol. Cryst. Liq. Cryst.* **215**: 257.

Schroder, I. 1893. *Z. Phys. Chem.* **11**: 339.

Shashidhar, R., and Chandrasekhar, S. 1975. *J. Physique Coll.* **36**: C1–49.

Shibaev, V. P., Kozlovsky, M. V., Beresnev, L. A., Blinov, L. M., and Platé, N. A. 1984. *Polym. Bull.* **12**: 299.

Shivaev, V. P., and Platé, N. A. 1985. *Pure Appl. Chem.* **57**: 1589.

Shigetsugu, H., Takenaka, S., and Kusabayashi, S. 1984. *Bull. Chem. Soc. Japan* **57**: 283.

Shiratori, N., Yoshizawa, A., Nishiyama, I., Fukumasa, M., Hirae, T., and Yamane, M. 1991. *Mol. Cryst. Liq. Cryst.* **199**: 129.

Skarp, K., Andersson, G., Lagerwall, S. T., Kapitza, H., Poths, H., and Zentel, R. 1991. *Ferroelectrics* **122**: 127.

Solladie, G., and Zimmerman, R. G. 1984. *Angew. Chem. Int. Ed. Engl.* **23**: 348.

Solladie, G., and Zimmerman, R. G. 1985a. *J. Org. Chem.* **50**: 4062.

Solladie, G., and Zimmerman, R. G. 1985b. *Angew. Chem. Int. Ed. Engl.* **24**: 64.

Stegemeyer, H., and Mainvsch, K. J. 1971. *Naturwissensch.* **58**: 599.

Steinstrasser, R. 1972. *Z. Naturforsch.* **27**: 774.

Styring, P., Vuijk, J. D., Nishiyama, I., Slaney, A. J., and Goodby, J. W. 1993. *J. Mater. Chem.* **3**: 399.

Suzuki, Y., Hagiwara, T., Aihara, Y., Sadamure, Y., and Kawamura, I. 1989. *Proceedings of the Fifteenth Japanese Liquid Crystal Conference [Osaka]: Abstracts*, p. 302.

Takatsu, H., Takeuchi, K., and Sato, H. 1983. *Mol. Cryst. Liq. Cryst.* **100**, 345.

Takeda, H., Suzuki, Y., et al. 1990. *Proceedings of the Sixteenth Japanese Liquid Crystal Conference [Hiroshima]: Abstracts*, p. 134.

Taniguchi, H., Ozaki, M., Nakano, K., Yoshino, K., Tamasaki, N., and Satoh, K. 1988. *Japan. J. Appl. Phys.* **27**: 452.

Uchida, S., Morita, K., Miyoshi, K., Hashimoto, K., and Kawasaki, K. 1988. *Mol. Cryst. Liq. Cryst.* **155**: 93.

Valasek, J. 1921. *Phys. Rev.* **17**: 475.

Vandegaer, J. E. 1974. *Microencapsulation: Process and Application.* New York: Plenum.

van Laar, J. J. 1903. *Arch. Néerl. Sci. Exactes Natur.* 2 **8**: 264.

Vaz, N. A., Arora, S. L., Doane, J. W., and de Vries, A. 1985. *Mol. Cryst. Liq. Cryst.* **128**: 23.

Villiger, A., Boller, A., and Schadt, M. 1979. *Z. Naturforsch.* **34**: 1535.

Weissflog, W., and Demus, D. 1983. *Cryst. Res. Technol.* **18**: K21.

Weissflog, W., and Demus, D. 1984. *Cryst. Res. Technol.* **19**: 55.

Weissflog, W., and Demus, D. 1985. *Mol. Cryst. Liq. Cryst.* **129**: 235.

White, D. L., and Taylor, G. N. 1974. *J. Appl. Phys.* **45**: 4718.

Yoshino, K., Ozaki, M., Sakurai, T., Honma, M., and Sakamoto, K. 1984. *Japan. J. Appl. Phys. Lett.* **23**: 175.

Yoshino, K., Taniguchi, H., and Ozaki, M. 1989. *Ferroelectrics* **91**: 267.

Yoshizawa, A., Nishiyama, I., Fukumasa, H., Hirai, T., and Yamane, M. 1989. *Japan. J. Appl. Phys.* **28**: 1269.

Young, W. R., Aviram, A., and Cox, R. J. 1971. *Angew. Chem. Int. Ed. Engl.* **10**: 410.

Zaschke, H. 1975. *Prakt. Chem.* **9**: 1173.

CHAPTER 3

Theory and Computation

ROBERT A. PELCOVITS

3.1 INTRODUCTION

The variety of mesophases formed by anisotropic molecules has stimulated the interest of theorists for many decades, beginning with the work of Onsager (Onsager, 1949) on the nematic–isotropic transition. Subsequently, de Gennes' application of Landau theory to a wide range of liquid crystal systems led to an enormous growth of theoretical activity, especially the application of renormalization group techniques to the analysis of Landau–de Gennes free energies (de Gennes and Prost, 1993). The construction and analysis of phenomenological free energies has proven to be remarkably successful in predicting the behavior of liquid crystals, including phase transition properties, the nature of the molecular order in various liquid crystalline phases, as well as elastic and hydrodynamic properties. Renormalization group theory has been used successfully to understand the profound effects of thermal fluctuations on many of these properties. This success has come in systems composed of short rod-like molecules, as well as systems composed of long, flexible molecules. Complex phases composed of regular arrays of topological defects have also been studied in detail and with great success using this approach.

While these phenomenological theories have been very successful, they do not confront the question of how molecular parameters influence observed macroscopic behavior (except in very general terms of molecular symmetry), nor can they predict the numerical values of transition temperatures, elastic moduli and the like. Only microscopic theories can accomplish these goals. Onsager's model of long, thin rods is one example of such a theory. More recent work in the microscopic area includes density functional theories for the nematic–isotropic and nematic–smectic A transitions, the spin-gas model of frustrated smectics, and molecular field theories for flexible molecules. These models still include a large degree of phenomenological construction, necessitated by the complexity of the constituent organic molecules, but do go a step further than Landau theory in addressing how molecular shape and local molecular packing determine observed properties.

In the next section of this chapter we review a number of problems solved theoretically in the past fifteen years, using Landau theory and some of the more microscopic approaches. Topological defects, that is, disclinations and dislocations, play an important role in many of these problems, and future work in theoretical liquid crystal physics will probably see a continuing emphasis on their behavior. A number of complex liquid crystal phases such as the blue phases and TGB phase have been understood in terms of the crystalline or quasicrystalline ordering of defects. Thermal fluctuations are another important theme, leading to the breakdown of conventional hydrodynamics in a variety of liquid crystal systems, as well as playing a dominant role in many liquid crystal phase transitions.

Computer simulations of liquid crystalline systems have proliferated in the past fifteen years as computers have become more powerful. These simulations offer a very good means of probing microscopic theories of liquid crystals. In section 3.3 we review much of the simulation work done since 1980. This work has shown how molecular eccentricity affects the phase diagram, as well as determining the relative importance of repulsive and attractive molecular interactions. This area should continue to show tremendous activity in the coming years as more complex liquid crystalline phases are studied in greater detail.

Finally, in the concluding section we discuss a number of problems that we believe will attract considerable attention in the coming years, both from theoretical and computational viewpoints.

3.2 THEORY

3.2.1 Blue Phases

Blue phases occur in a narrow temperature range between the cholesteric and isotropic phases. Thus they occur only when the constituent molecules are chiral, though not all chiral compounds exhibit blue phases. Up to three distinct blue phases appear in many chiral compounds. The two low-temperature phases, blue phase I (BPI) and blue phase II (BPII), have cubic symmetry, while the highest temperature phase, blue phase III (BPIII), appears to be amorphous. The crystalline phases have lattice spacings on the order of several thousand ångstroms and can therefore Bragg-scatter visible light. Experiments appear to indicate that a single-crystalline space group characterizes BPI and a second describes BPII.

Here we review the theoretical work that has been done in the past decade to describe the crystalline blue phases. This work involves the construction of a Landau free energy whose minimization (approximately done in this case) provides theoretical predictions for the structure and properties of these phases. It is a wonderful example of the power of the Landau approach in a highly nontrivial problem. Two groups of workers have approached this problem from different points of view, though their approaches share the common theme of a phenomenological free energy. The first approach, initiated by Brazovskii and coworkers and developed much more fully by Hornreich and Shtrikman (Brazovski and Dmitriev, 1975; Brazovski and Filev, 1978; Hornreich and Shtrikman, 1979, 1980a, 1980b, 1981a, 1981b; Alexander, 1981; Alexander et al., 1981; Kleinert and Maki, 1981; Hornreich et al., 1982; Grebel et al., 1983, 1984; and Wright and Mermin, 1985; for a review see Hornreich and Shtrikman, 1988) uses a Landau free energy constructed from the most general tensorial order parameter for a nematic. The second approach due to Meiboom et al. (Meiboom et al., 1981; Meiboom et al., 1983; Meiboom et al., 1983b; Sethna, 1985) uses a Frank free energy for a chiral nematic and constructs blue phases from regular arrays of line defects (disclinations). As emphasized in a very comprehensive review by Wright and Mermin (Wright and Mermin, 1989), and discussed in more detail below, these two approaches are complementary rather than competing. The first approach is appropriate when the gradient terms in Landau energy dominate, which will occur when the molecular chirality is strong. The second approach is appropriate to the other limit where the nongradient (bulk) terms in the Landau energy are dominant; this is true when the chirality is low (it is the molecular chirality which induces nonzero gradients in equilibrium). In this limit, what remains of the Landau free energy beside the constant bulk terms are the gradient terms which form the Frank free energy.

The most general Landau energy for a nematic (chiral or not) is constructed from a symmetric, traceless tensor order parameter $\overleftrightarrow{Q}$ which is defined in terms of the local dielectric constant $\overleftrightarrow{\varepsilon}$ as follows:

$$Q_{ij} \equiv \varepsilon_{ij} - \frac{1}{3}(\mathrm{Tr}\varepsilon)\delta_{ij}. \tag{3.1}$$

By construction $\overleftrightarrow{Q}$ is identically zero in the isotropic phase.

The Landau energy is an expansion in powers of $\overleftrightarrow{Q}$ (the bulk free-energy terms) and its gradients. Each term must be rotationally invariant. Keeping terms through fourth order in $\overleftrightarrow{Q}$ we have

$$\begin{aligned} F = \int \mathrm{d}^3x \Bigg[&\frac{1}{4}K_1[(\vec{\nabla} \times \overleftrightarrow{Q})_{ij} + 2q_0 Q_{ij}]^2 \\ &+ \frac{1}{4}K_0[(\vec{\nabla} \cdot \overleftrightarrow{Q})_i]^2 + c(\mathrm{Tr}(\overleftrightarrow{Q}^2)) \\ &- \sqrt{6}b(\mathrm{Tr}(\overleftrightarrow{Q}^3)) + a[\mathrm{Tr}(\overleftrightarrow{Q}^2)]^2 \Bigg] \end{aligned} \tag{3.2}$$

where we have used the notation of Wright and Mermin in labeling the phenomenological coupling constants. The chiral nature of the compound is inherent in the presence of a nonzero parameter q_0.

Ideally this Landau free energy in its full form should be minimized to search for the solutions that will describe the allowed thermodynamic phases. This procedure however is not tractable and limiting cases must be considered instead, corresponding to the two theoretical approaches to this problem mentioned above. It is useful to first rescale variables in (3.2) as follows:

$$\tau = (a/b^2)c \qquad \overleftrightarrow{\chi} = \frac{a}{b}\overleftrightarrow{Q} \qquad r' = 2q_0 r \tag{3.3}$$

in which case the free energy can be expressed as

$$\begin{aligned} F = \int \mathrm{d}^3 \left(\frac{b^4}{a^3}\right) \Bigg\{ &\kappa^2 \Big[[(\vec{\nabla} \times \overleftrightarrow{\chi})_{ij} + \chi_{ij}]^2 \\ &+ \eta[(\vec{\nabla} \cdot \overleftrightarrow{\chi})_i]^2 \Big] \\ &+ \tau \mathrm{Tr}(\chi^2) - \sqrt{6}\mathrm{Tr}(\chi^3) \\ &+ [\mathrm{Tr}(\chi^2)]^2 \Bigg\} \end{aligned} \tag{3.4}$$

where κ is the chirality parameter (Grebel et al., 1983),

$$\kappa = q_0 \left(\frac{aK_1}{b^2}\right)^{1/2} \tag{3.5}$$

and $\eta = K_0/K_1$.

This rescaling indicates that the strength of the chirality parameter κ determines whether it is the gradient (large κ) or the bulk (small κ) terms that dominate the free energy. The former situation is appropriate to the model developed by Hornreich and Shtrikman. Their trial solutions for the minimization of the free energy are constructed from the exact solution χ_{ij}^{bh} for the quadratic portion of the free energy:

$$\chi_{ij}^{bh} = \mathrm{Re}\left[(\vec{a} - \mathrm{i}\vec{b})_i(\vec{a} - \mathrm{i}\vec{b})_j \mathrm{e}^{\mathrm{i}\vec{c}\cdot\vec{r}}\right] \tag{3.6}$$

where $\vec{a}, \vec{b}, \vec{c}$ are any trio of orthonormal vectors obtained by a rotation of the Cartesian unit vectors. This solution has been termed the biaxial helix by Wright and Mermin. It corresponds to a pitch axis specified by the wavevector $\vec{c}$ and is constant in the $\vec{a}$–$\vec{b}$ plane. However, it is a biaxial tensor (eigenvalues 1, 0, −1), unlike the uniaxial tensor (two of the eigenvalues are equal) that describes the simple cholesteric helix.

The blue phases are separated from each other and the isotropic and helical phases by first-order transitions. First-order transitions in Landau theory require a nonzero cubic term in the free energy (the term proportional to $\sqrt{6}$ above). A superposition χ of terms of the form χ^{bh} for different choices of the orthonormal triads will yield a nonzero value of $\mathrm{Tr}(\chi^3)$ only if there is a trio of wavevectors $\vec{c}, \vec{c}\,'$, and $\vec{c}\,''$ in the superposition that add up to zero. Eight distinct trios satisfying this condition can be formed from linear combinations of the twelve $\langle 110 \rangle$ directions of a face-centered-cubic lattice (Baym et al., 1971; and Alexander and McTague, 1978). The real lattice structure is then body-centered-cubic. However, a simple cubic structure is also possible if nonunit wavevectors that are linear combinations of the unit vectors $\vec{c}$ are allowed. To determine which structure is stable, the higher-order terms in the free energy must be evaluated. The resulting phase diagram is shown in FIGURE 3.1. It is important to note that the free-energy differences between the different cubic structures is small and higher-order terms in the Landau free-energy could alter this picture. Note that these cubic phases will appear only if κ exceeds a critical value

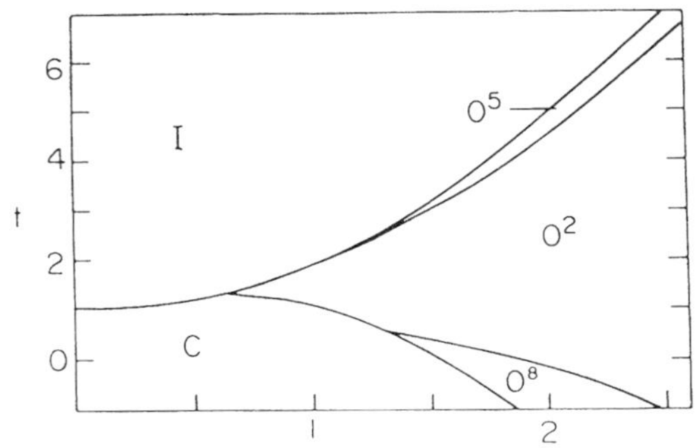

FIGURE 3.1. Phase diagram for cholesteric liquid crystals derived from Landau theory (Hornreich and Shtrikman, 1988) in terms of reduced temperature t and chirality parameter κ.

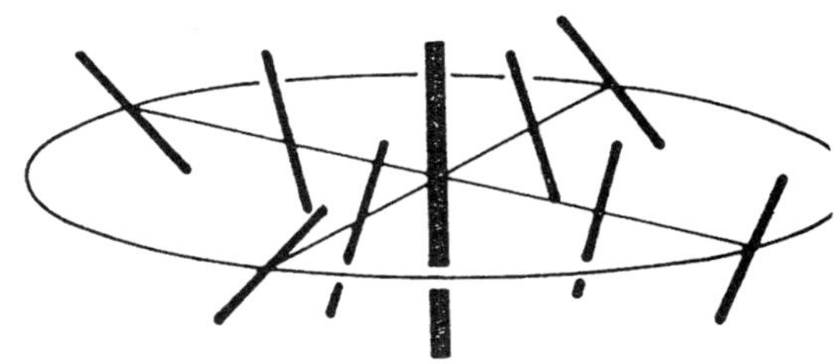

FIGURE 3.2. Illustration of double twist in the director field of a chiral liquid crystal. The dark lines show the orientation of the director field as given by (3.8). The point $r = 0$ is at the center of the figure (after Wright and Mermin, 1989).

(in qualitative agreement with experiment), which in the Hornreich–Shtrikman theory can be given an upper bound of 0.47.

The other theoretical approach to the blue phase problem, adopted by Meiboom et al., is appropriate to the low-chirality limit. As can be seen from the rescaled free energy (3.4), in this limit it is the bulk terms that dominate the energy. Thus, the starting point of this approach is to minimize the nongradient terms alone. It can be shown that the minimum of these terms corresponds to a uniaxial $\overleftrightarrow{Q}$ (due to the presence of the cubic term). With a uniaxial $\overleftrightarrow{Q}$, the free energy can be reexpressed in terms of a unit length director field $\vec{n}$ and an amplitude λ. In the one elastic constant approximation, the free energy can be written as

$$F = \frac{b^4}{a^3} \int d^3x \Big\{ \kappa^2 \Big[(\nabla\lambda)^2 - \frac{1}{2}\lambda^2 + 3\lambda^2 (\nabla_i n_j + \frac{1}{2}\varepsilon_{ijk} n_k)^2 \Big] + \tau\lambda^2 - \lambda^3 + \lambda^4 \Big\} \tag{3.7}$$

where ε_{ijk} is the antisymmetric tensor.

With the assumption $\lambda =$ constant, it can be shown that the free energy is minimized in the helical state. This will be the ground state when $\kappa = 0$. For $\kappa \neq 0$, however, the assumption of constant λ can be relaxed and nonhelical phases can assume lower energy. These phases (which are thus models for the experimentally observed blue phases) and their corresponding lower energies can be understood by visualizing chiral molecules as screw-like objects. A pair of neighboring molecules will pack well if they sit at a very small angle with respect to each other so their "threads" interleave. In the helical phase this twisting of the molecules occurs along one direction only. The twisting occurs along a direction which is perpendicular to the long axis of the molecule. This "one-dimensional" twisting, however, need not yield the lowest possible energy state of the system. In particular a double twist, illustrated in FIGURE 3.2, should be more favorable in a local scale. This can be seen mathematically from the gradient free energy. In the helical phase given by $\vec{n} = \hat{x}\cos q_0 z + \hat{y}\sin q_0 z$, the gradient term yields $\frac{1}{4}\kappa^2\lambda_0^2$ for a constant λ_0. However, the gradient term is bounded from below by the value $-\frac{1}{2}\kappa^2\lambda_0^2$. Locally this value can be obtained if $\nabla_i n_j + \frac{1}{2}\varepsilon_{ijk} n_k = 0$. This equation can be satisfied along lines. Using cylindrical coordinates centered on a line, the solution

$$\vec{n} = \hat{z}\cos\frac{1}{2}r - \hat{\phi}\sin\frac{1}{2}r \tag{3.8}$$

does indeed yield $\nabla_i n_j + \frac{1}{2}\varepsilon_{ijk} n_k = 0$ when $r \to 0$. This solution is precisely double twist and is illustrated in FIGURE 3.2.

Unfortunately, as indicated in FIGURE 3.3, a double twist introduces frustration, namely it cannot be continued through space indefinitely without either undoing one of the rotations or introducing defects. This is precisely what happens in the blue phases. However, the gain in energy achieved by introducing the double twist is offset by the necessity of having defects which themselves cost energy. Thus, at sufficiently low temperatures, the helical phase will once again be favorable.

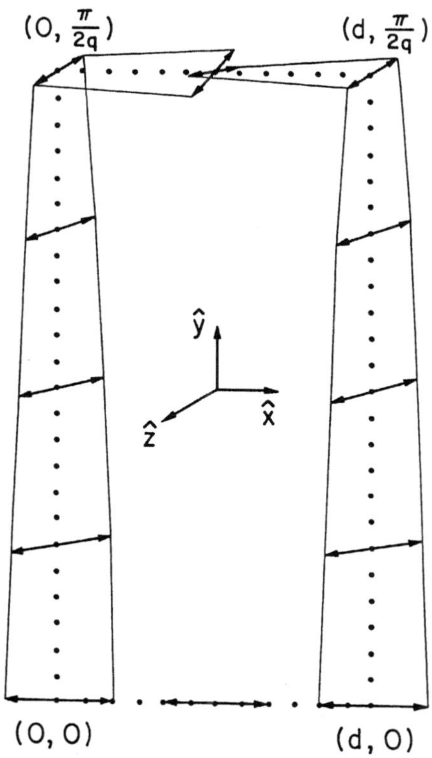

FIGURE 3.3. Frustration in the blue phase in the presence of double twist. With the director pointing in the x-direction at the origin, the double twist solution requires that the director rotate as we move up the y-axis. Along the x-axis the director does not rotate. Along the vertical line which intersects the x-axis at an arbitrary point d, the director again rotates as indicated, in the same fashion as along the y-axis. However, along the top dotted line the double twist solution cannot be satisfied with the director having the same orientation at the two ends, thus leading to frustration (after Sethna, 1985)

It is then a delicate and difficult problem to decide how to build up the blue phase structures from the double twist cylinders of FIGURE 3.2. Crystalline structures can be made from periodic arrays of these cylinders. Some of these proposals are shown in FIGURES 3.4 and 3.5. Detailed computer calculations (Sammon, 1982; Meiboom et al., 1983a; Meiboom et al., 1983b) are required in order to determine the free energies of these structures. In these calculations the director field $\hat{n}$ is assumed to have a fixed magnitude, dropping abruptly to zero at the cores of the defects.

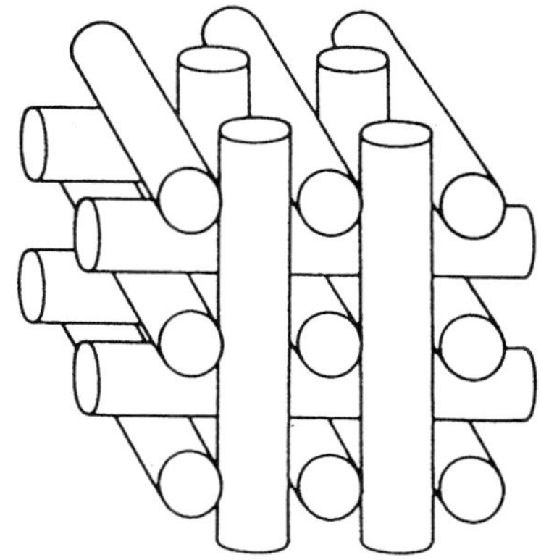

FIGURE 3.4. The arrangement of double twist cylinders in the O^2-structure (after Wright and Mermin, 1989).

The core energy of the defects is inferred from experimental data. Meiboom and coworkers have calculated the energies of these structures in the one elastic constant approximation as well as for the case $K_1 = K_3 = 2K_2$. In the former case the 0^2 structure appears to be most favorable, while in the latter 0^{8-} is preferred. These results are consistent with experiments that indicate that many materials exhibit two crystalline blue phases (BPI and BPII) with space groups 0^8 and 0^2 respectively. However, it is not clear how sensitive these calculations are to the choice of parameters, especially for the cost of defect cores.

FIGURE 3.5. The arrangement of double twist cylinders in the O^8-structure (after Wright and Mermin, 1989).

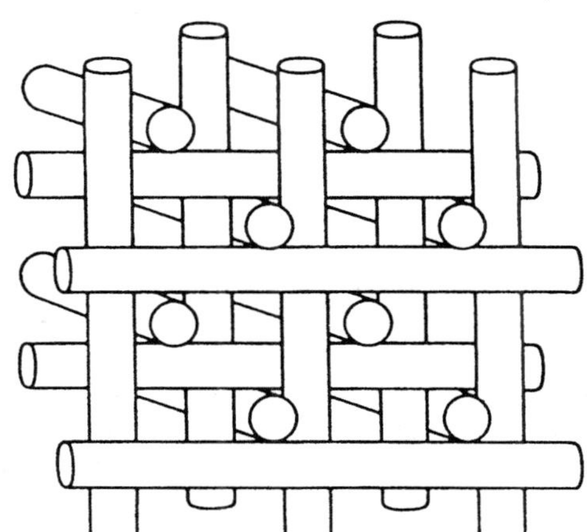

Two-dimensional versions of the low-chirality blue phase have been studied in thin ferroelectric smectic films both experimentally (Maclennan, 1990; and Bahr and Fliegner, 1993) and theoretically (Langer and Sethna, 1986; Hinshaw et al., 1988; and Hinshaw and Petschek, 1989). The structure of the phases is much simpler in this reduced dimensionality, but the physical ideas are very similar. The chiral interactions prefer a molecular ordering which leads to frustration and the presence of defects. These defects can be either lines or points leading to striped or hexagonal phases respectively. More recent work (Selinger et al., 1993) on smectic films composed of nonchiral molecules has shown that chiral symmetry can be spontaneously broken and modulated phases can again form.

3.2.2 TGB Phase

One of the great successes of the early application of Landau theory to liquid crystals was the prediction of an analogy between the nematic-to-smectic A transition and the normal-to-smectic transition in metals (de Gennes, 1972; and Halperin and Lubensky, 1974). The ordered phases of both systems are characterized by a complex order parameter ψ. In the smectic case, ψ is the complex amplitude of the density modulation, while in a metal it is the complex gap function. The direction $\vec{n}$ is the analog of the vector potential $\vec{A}$ in a metal. This analogy can be seen by writing down a very general de Gennes model (McMillan, 1972; Renn and Lubensky, 1988) for the free energy of a chiral material:

$$F = \int d^3x \left[|\psi|^2 + c|(\vec{\nabla} - iq_0\vec{n})\psi|^2 \right] + \frac{1}{2}g \int d^3x |\psi|^4 + F_n \tag{3.9}$$

where F_n is the usual nematic Frank free energy. This model is more general than the one originally written down by de Gennes which was restricted to small fluctuations $\delta\vec{n}$ of the director. The present model is more appropriate as we shall see for the TGB phase. This free energy can be compared with the Landau–Ginzburg Gibbs free energy for a superconductor in the presence of a magnetic field $\vec{H}$ (see e.g., de Gennes, 1966; and Tinkham, 1975):

$$F_{sc} = \int d^3x \left[r|\psi|^2 + \frac{1}{2m}\left|(i\hbar\vec{\nabla} - \frac{2e}{c}\vec{A})\psi\right|^2 \right] + \frac{1}{2}g \int d^3x |\psi|^4 + \int d^3x \left[\frac{1}{8\pi\mu}(\vec{\nabla}\times\vec{A})^2 - \frac{1}{4\pi}\vec{H}\cdot\vec{\nabla}\times\vec{A} \right]. \tag{3.10}$$

Comparing (3.9) and (3.10) we see that the elastic constant K_2 is the analog of $(4\pi\mu)^{-1}$, and the product $K_2 q_o$ is the analog of H. Thus a superconductor in a magnetic field is analogous to a liquid crystal with chirality. Chirality yields a nonzero average value for $\vec{\nabla}\times\vec{n}$, analogous to the nonzero magnetic induction, $\vec{B} = \vec{\nabla}\times\vec{A}$ in the superconductor.

Until the recent work of Renn and Lubensky which we will review here, the analogy between smectics and superconductors was incomplete because no analog of the Abrikosov flux lattice had been found, either theoretically or experimentally, in smectics. Abrikosov flux lattices occur in type-II superconductors in sufficiently strong magnetic fields. Superconductors are classified as either type I or type II depending on the value of the Ginzburg parameter κ which is the ratio of the penetration depth λ to the coherence length ξ. Type I have $\kappa < 1/\sqrt{2}$, while type II have $\kappa > 1/\sqrt{2}$. From the de Gennes model above we find the correlation length

$$\xi = (c/|r|)^{1/2} \tag{3.11}$$

and the twist penetration depth (recall $K_2 q_0$ is the analog of H),

$$\lambda_2 = q_0^{-1}\left(\frac{K_2 g}{2c|r|}\right)^{1/2}. \tag{3.12}$$

These lengths in turn yield the twist Ginzburg parameter,

$$\kappa_2 = \frac{1}{cq_0}\left(\frac{gK_2}{2}\right)^{1/2}. \tag{3.13}$$

As in superconductors, κ_2 determines what type of behavior will be observed in the smectic–cholesteric phase diagram, with the critical value of κ_2 being $1/\sqrt{2}$. We now focus on the type-II case, where $\kappa_2 > 1/\sqrt{2}$. In a type-II superconductor there are two critical magnetic fields, H_{c1} and H_{c2}. For magnetic fields H in the range $H_{c1} < H < H_{c2}$, an Abrikosov phase of vortices forms. The external magnetic field penetrates the superconductor through these vortices which are parallel to the field. The smectic analog of a magnetic vortex is a screw dislocation, so an obvious first guess of the nature of the smectic Abrikosov phase is a lattice of parallel screw dislocations. Unfortunately, this guess leads to a physically unreasonable phase as pointed out by Sethna (as reported in Renn and Lubensky, 1988). The difficulty with this phase is illustrated in FIGURE 3.6. A more creative suggestion is required and this was provided by Renn and Lubensky. Since in the superconductor the magnetic field at a vortex core parallels the external field, they suggested looking for a lattice of twist dislocations that produces, on average, the same twisted director field found in the cholesteric phase. The appropriate lattice structure is shown in FIGURE 3.7, and consists of planes of parallel twist dislocations whose axes rotate in a helix from plane to plane. Because the average director in the core of a twist dislocation is parallel to the dislocation axis, the director field is helical on average but the rate of twist is smaller than in the cholesteric phase. This latter statement has an analog in the superconductor, where the magnetic induction $\vec{B}$ in the Abrikosov phase is less than the corresponding value in the normal phase. Renn and Lubensky suggested the name twist grain boundary (TGB) state for their proposal for this lattice of dislocations.

The TGB phase is characterized by the distance ℓ_d between dislocations in a layer, the distance ℓ_b between grain boundaries, and the separation d between smectic layers (see FIGURES 3.7 and 3.8). From FIGURE 3.8 we see that a given smectic layer rises by a half layer spacing, $d/2$, in a half circuit around each dislocation. The difference $\Delta\theta$ in the angle of smectic plane normals on two sides of the grain boundary is given in the caption of FIGURE 3.8. The nature of the TGB phase depends on whether $\Delta\theta$ is a rational or irrational multiple of 2π. In the latter case, the lattice is incommensurate along the x-axis (the twist axis). In the former case, the lattice is commensurate along the x-axis and consists of unit cells containing q grain boundaries undergoing a total of P rotations of 2π, with $\Delta\theta = 2\pi P/q$. This lattice will have a q-fold screw axis. If $q = 1, 2, 3, 4$, or 6, it will be commensurate in the yz plane; otherwise it will be incommensurate and the lattice will have quasicrystalline symmetry.

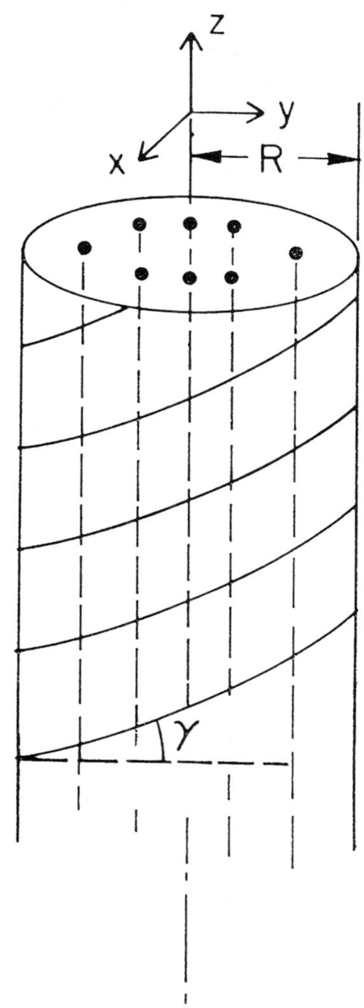

FIGURE 3.6. Parallel dislocations in a smectic liquid crystal. A circular area C of radius R contains an average areal density n_A of screw dislocations with Burgers vectors parallel to the z-axis and of magnitude d. The integral of u, the smectic layer displacement, around the boundary of C equals $\pi n_A R^2 d$, the number of dislocations enclosed in C. Thus $|\vec{\nabla} u| = \tan\gamma$ on the boundary of C is proportional to R, and the slope of the smectic layers relative to the xy-plane diverges as $R \to \infty$ (after Renn and Lubensky, 1988).

Many properties of the TGB phase can be calculated using the generalized de Gennes model (3.9). The calculations are simplified if we confine the director to the yz plane as indicated in FIGURE 3.7, and

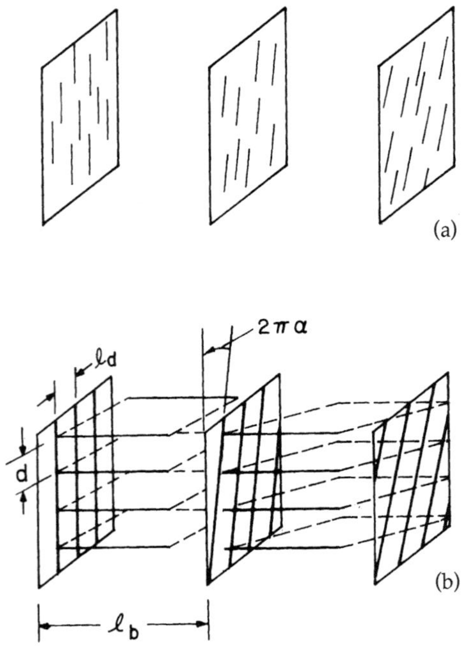

FIGURE 3.7. Schematic representation of the TGB state (after Renn and Lubensky, 1988). The twist grain boundaries are separated by a distance ℓ_b, while the twist dislocations within each grain boundary are separated by a distance ℓ_d. The axes of the dislocations in the nth grain boundary make an angle $2\pi\alpha n$ with the z-axis. Between the grain boundaries the molecular configuration is essentially identical to the smectic A phase with regularly spaced layers separated by a distance d.

$\vec{n} = \sin\theta\hat{y} + \cos\theta\hat{z}$. The lower critical twist field h_{c1} (the analog of the superconductor's H_{c1}, (Abrikosov, 1957)) can be found by calculating the energy $\varepsilon_{\mathrm{screw}}$ of a screw dislocation in the limit where the density of dislocations is low. In that case, the Gibbs free energy per unit volume is given by

$$G/V = \ell_d^{-1}\ell_b^{-1}(\varepsilon_{\mathrm{screw}} - hd_0). \tag{3.14}$$

The critical field h_{c1} is the value of h for which $G/V = 0$. Renn and Lubensky found

$$h_{c1} = \frac{h_c}{\sqrt{2}\kappa_2}\ln(\lambda_2/\xi). \tag{3.15}$$

In the extreme type-II limit, where h_c is the thermodynamic critical field (the value of h for which the cholesteric and smectic phases have equal free

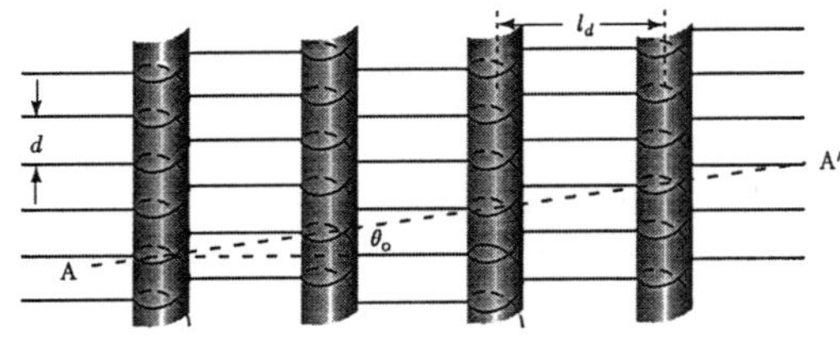

FIGURE 3.8. Structure of a single twist grain boundary consisting of parallel twist dislocations with darkened cores. The average slope of the smectic planes in front of the grain boundary is $d/2\ell_d$, while those behind the grain boundary have a slope opposite in sign and of the same magnitude. Thus $\Delta\theta$, the change in the angle of the normals to the smectic planes across the boundary, is approximately equal to d/ℓ_d (courtesy of T. C. Lubensky).

energies). This result is identical in form to that obtained for a superconductor in the extreme type-II limit.

The upper critical twist field h_{c2} is determined by the limit of stability of the cholestric phase. This can be determined again from the de Gennes model with the result

$$h_{c2} = \sqrt{2}\kappa_2 h_c \tag{3.16}$$

which again is of the same form as the Abrikosov case.

The TGB phase can be studied in detail near h_{c2}, in much the same way as Abrikosov constructed his famous solution (Abrikosov, 1957) for the superconductor in the corresponding limit. The smectic order parameter $\psi(\vec{x})$, is written as a linear combination of the functions which diagonalize the de Gennes model near h_{c2}. This *ansatz* is then inserted into the de Gennes free energy and the latter is minimized as a function of the coefficients of the linear combination.

A phase with the geometry of the TGB structure was also found theoretically in a model of polymers dissolved in a chiral nematic liquid crystal (Lubensky et al., 1991). In that case the polymers adopt configurations identical to those of the dislocations composing the twist grain boundaries.

The elasticity theory for the incommensurate TGB phase was studied by Toner (Toner, 1991) who found that thermal fluctuations destroy the long-range translational order, leading instead to algebraic singularities in the x-ray scattering. The hydrodynamics of the incommensurate phase was studied by Hatwalne and coworkers (Hatwalne, 1993), who found diverging viscosities, similar to the case of smectics (see the next section).

3.2.3 Breakdown of Conventional Smectic A Hydrodynamics

Smectic A liquid crystals are characterized by one-dimensional (1D) solid order, with the mass density modulated parallel to the long axes of the molecules. An elastic free energy can be constructed on the basis of this 1D order in the mass density. Since a 1D solid, or a stack of liquid layers, cannot support shear, the elastic energy is a function of the compressional and undulational (or bending) modes of the solid. These modes are illustrated in FIGURE 3.9, and can be described mathematically in terms of a function $u(\vec{x})$ which gives the local layer displacement when fluctuations are present. Compression corresponds to a nonzero value of the trace of the strain tensor $\overleftrightarrow{e}$. For a 1D solid we have

$$\mathrm{Tr}\,\overleftrightarrow{e} = \frac{\partial u}{\partial z} - \frac{1}{2}(\vec{\nabla} u^2) \tag{3.17}$$

where we have chosen a coordinate system with z perpendicular to the layers. The last term in the equation above is a nonlinear one which is often dropped. However, this term must be included to maintain rotational invariance for arbitrarily large rotations about an axis in the plane of the layers, and is ultimately the source of the breakdown of conventionalelasticity and hydrodynamics. Undulations correspond to a nonzero value of the two-dimensional Laplacian $\nabla_\perp^2 u$, where

$$\nabla_\perp^2 \equiv \frac{\partial^2}{\partial x^2} + \frac{\partial^2}{\partial y^2}$$

(x and y are in the in-layer coordinates). The elastic free energy of the smectic A is then written in terms of the two invariants, $\mathrm{Tr}\,\overleftrightarrow{e}$ and $\nabla_\perp^2 u$ as follows:

$$F = \frac{1}{2}\int \left[B(\mathrm{Tr}\,\overleftrightarrow{e})^2 + K(\nabla_\perp^2 u)^2\right] \mathrm{d}^3 r. \tag{3.18}$$

FIGURE 3.9. Normal modes of a smectic liquid crystal: (a) compressional, and (b) undulational (after Grinstein and Pelcovits, 1982).

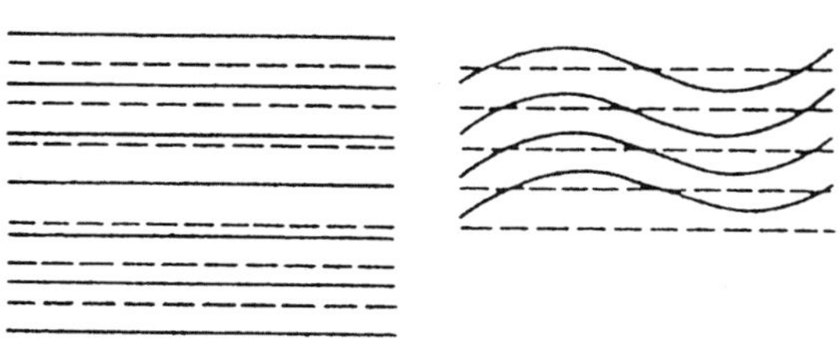

Equation (3.18) neglects higher-order terms in these invariants, which is reasonable if we restrict ourselves to long-wavelength fluctuations. The coupling B is the compression modulus, while K, which describes the energy cost of undulations, is a splay-like Frank modulus (if the molecules remain perpendicular to the layers, then undulations correspond to splay deformations of the nematic director).

The first analyses of (3.18) invoked the harmonic limit where the last term in (3.17) is ignored and (3.18) is thus quadratic in u. This limit is interesting in its own right. Many years ago, Landau and Peierls (Landau, 1937; and Peierls, 1935) demonstrated that thermal fluctuations within this harmonic limit destroy long-range density order and restore the uniform density of the nematic state. Subsequently, Caillé (Caillé, 1972) showed that density correlations decay algebraically and not exponentially as one would expect in a liquid. Thus, a smectic A exhibits quasi-long-range order similar to the ordering found in two-dimensional superfluids and magnets with continuous symmetry. This prediction has been confirmed experimentally (Als-Nielsen et al., 1977; for a review, see Als-Nielsen, 1981).

Because only quasi-long-range order is present, it should not be surprising that anharmonic terms which naturally account for large fluctuations should prove important, and should not be ignored. The elastic moduli B and K appearing in (3.18) are constants in the harmonic approximation, but a more general definition of the moduli is required in the anharmonic theory. Specifically, moduli are defined in terms of stress–strain relationships. For example, for a unilateral compression of wavevector $\vec{q}$ along the z-axis,

$$\langle \sigma_{zz}(q) \rangle = B^R(q) \left\langle \frac{\partial u(q)}{\partial z} \right\rangle \tag{3.19}$$

where σ_{zz} is a component of the stress tensor and B^R is the renormalized (or measurable) compression modulus. The brackets above refer to a thermal average over the ensemble specified by the anharmonic elastic energy. The renormalized elastic moduli could be calculated by straightforward perturbation theory in the anharmonic terms of the elastic energy, but that procedure will not work here because of logarithmic divergences. Instead, Grinstein and Pelcovits (Grinstein and Pelcovits 1981, 1982) carried out a renormalization group calculation directly in three dimensions. They found that $B^R(q)$ *vanishes* logarithmically as $q \to 0$. Specifically,

$$B^R(q) \simeq B\left(1 + \frac{5W}{128\pi} \ln \frac{\Lambda}{q_z^2 + \lambda^2 q_\perp^4}\right)^{-4/5} \tag{3.20}$$

where $W = k_B T B^{1/2} K^{-3/2}$, $\lambda = \sqrt{K/B}$, and Λ^{-1} is of the order of the layer spacing. The elastic modulus K^R which governs undulations behaves anomalously as well, diverging as $q \to 0$:

$$K^R(q) = K\left(1 + \frac{5W}{64\pi} \ln \frac{\Lambda}{q_z^2 + \lambda^2 q_\perp^4}\right)^{2/5}. \tag{3.21}$$

Application of a uniform ($q = 0$) stress also leads to anamolous behavior as B^R is found to depend non-analytically on the applied stress, implying the breakdown of Hooke's Law. A "poor man's" derivation of an approximate form of these results has been given by Prost (Prost, 1984).

Unfortunately, the logarithmic dependence of the moduli on the wavevector $\vec{q}$, makes experimental observation of this dispersion very difficult, if not impossible. However, more spectacular and observable anharmonic effects occur in the hydrodynamics of smectic A liquid crystals. The hydrodynamic variables of a smectic A include conserved quantities (mass, energy and momentum) and the additional variable associated with the breaking of one-dimensional translational symmetry (Martin et al., 1972). Ignoring the equation associated with energy, the hydrodynamic equations read:

$$\begin{aligned} &\frac{\partial \rho}{\partial t} + \vec{\nabla} \cdot \vec{g} = 0 \\ &\frac{\partial g_i}{\partial t} = -\nabla_j \sigma_{ij} + \xi_i \\ &\frac{\partial u}{\partial t} + \frac{\vec{g}}{\rho} \cdot \vec{\nabla} u = \frac{g_z}{\rho} - \Gamma \frac{\delta(F + F_\rho)}{\delta u} + \theta. \end{aligned} \tag{3.22}$$

Here $\vec{g}$ is the momentum density, σ_{ij} is the stress tensor, ρ is the mass density, and $\vec{\xi}$ and θ are noise sources associated with the thermal reservoir. The first of these equations expresses the conservation of mass; the second is Newton's second law with a random thermal force, and the third is a Langevin-type equation for the layer displacement (the first term on the right-hand side of this latter equation allows for permeation of the layer (Helfrich, 1978)). The free energy entering (3.22) is (3.18) supplemented by the two lowest-order terms allowed by symmetry that include fluctuations in the mass density heretofore neglected:

$$F_\rho = \frac{1}{2} \int d^3x \left[A \left(\frac{\delta\rho}{\rho_0}\right)^2 + C(\mathrm{Tr}\,\overleftrightarrow{e}) \frac{\delta\rho}{\rho_0} \right] \tag{3.23}$$

where $\delta\rho = \rho - \rho_0$, ρ_0 is the average mass density, A is the bulk modulus and C describes the coupling between layer compressions and density fluctuations.

The stress tensor σ_{ij} appearing in (3.22) has both a reactive and dissipative part, the former involving the various moduli and the latter involving five independent viscosities. These viscosities characterize the damping of the five hydrodynamic modes (there are six if thermal diffusion, i.e. the energy equation, is included). The modes are readily found if (3.22) is linearized, i.e. the harmonic approximation is used for the free energy as well as the reactive stress tensor. One finds first (or ordinary) sound with an anisotropic sound velocity, and when the wavevector is oblique with respect to the layers there is a "second sound" mode involving changes in the interlayer spacing at nearly constant density. If the wavevector lies in the plane of the layers, the propagating second sound mode is absent and is replaced by an over-damped undulation mode. When the wavevector is parallel to the layer normal, the second sound mode is replaced by a viscous permeation mode.

Mazenko and coworkers (Mazenko et al., 1982, 1983) studied the effects of including the anharmonic terms in (3.22), and found some striking behavior. Four of the five viscosities diverge as $1/\omega$, where ω is the frequency of the mode. The divergence of these four viscosities, conventionally labeled η_1, η_2, η_4 and η_5, leads to the breakdown of conventional hydrodynamics. In particular, the attenuations $\alpha_i(\omega)$ of the first ($i = 1$) and second ($i = 2$) sound no longer scale as ω^2 at low frequencies as they would in a conventional (i.e. harmonic) theory, but rather scale as ω. These effects have been observed experimentally for both first (Baumann et al., 1985; and Collin et al., 1986) and second (Battacharya and Ketterson, 1982) sound. The divergence of η_2 which characterizes shear flow within the layers leads to plastic-like behavior, namely a yield stress must be exceeded before flow can begin (Ramaswamy, 1984).

The breakdown of hydrodynamics described above occurs as well in hexatic B and cholesterics, which like smectics are characterized by 1D order. A similar breakdown occurs in discotics (Ramaswamy and Toner, 1983) which are characterized by two-dimensional order. In that case the viscosities diverge as $1/\sqrt{\omega}$. While short-rod nematic liquid crystals do not exhibit this anomalous elasticity, very long nematic polymers (essentially infinitely long flexible chains) will have elastic moduli that do not approach constants in the long-wavelength limit (Toner, 1992). An exact solution for the breakdown of hydrodynamics in a two-dimensional smectic was obtained by Golubovic and Wang (Golubovic and Wang, 1992).

3.2.4 Flexible Nematic Polymers

Significant theoretical effort in the past decade has been devoted to understanding the formation of liquid crystalline phases in systems of rigid chain macromolecules (for other reviews, see Odijk, 1986; and Semenov and Khokhlov, 1988). These molecules are characterized by a persistence length ℓ_P (the length over which an isolated chain is stiff) which is much larger than the characteristic thickness of the chain. Nematic as well as hexagonal (often called columnar) phases are present in the phase diagrams of these materials. The chains may have different degrees of flexibility depending on the molecular makeup of the material. If the total length L of the molecule is large compared with ℓ_P, as happens in DNA, then the flexibility can be neglected and the theories of Onsager (Onsager, 1949) and Flory (Flory, 1956; and Flory and Ronca, 1979) are applicable. On the other hand if $L >> \ell_p$, then the molecule is termed semiflexible, and there is much new physics to consider.

One of the first theoretical approaches to lyotropic liquid crystals composed of semiflexible molecules is due to Khokhlov and Semenov (Khokhlov, 1978; and Khokhlov and Semenov, 1981, 1982, 1984) who formulated an Onsager-like theory. They constructed a free energy consisting of conformational entropy for each individual chain plus a steric interaction term in the second virial approximation:

$$F = Nk_BT\left[\int \frac{[\vec{\nabla} f(\hat{u})]^2}{4f(\hat{u})}\frac{d\Omega}{4\pi} + \frac{1}{2}\left(\frac{N}{V}\right)\int f(\hat{u}_1)f(\hat{u}_2)B(\hat{u}_1\cdot\hat{u}_2)d\Omega_1 d\Omega_2\right]. \tag{3.24}$$

Here N is the number of chains, $f(\hat{u})$ is the orientational distribution function as a function of $\hat{u}$, the tangent to the chain, and B is the second virial coefficient of the interaction of two chains segments with orientations $\hat{u}_1$ and $\hat{u}_2$. The translational free energy can be neglected for long chains. Because the chains are locally rigid, the second virial coefficient is identical to Onsager's result for short, rigid rods.

The free energy (3.24) is then evaluated in a mean-field fashion by minimization with respect to all possible distribution functions $f(\hat{u})$. The resulting free energy exhibits a first-order phase transition at low polymer concentrations. The transition occurs when the ratio of densities in the nematic and isotropic phases is 1.23. The nematic order parameter at that point is 0.61 (using the more accurate solution of the Khokhlov–Semenov theory due to Odijk (Odijk, 1986)). These values can be compared to those of Onsager for rigid rods, who obtained a ratio of densities equal to 1.34 and an order parameter value of 0.84 at the transition. Khokhlov and Semenov (Khokhlov and Semenov, 1982) also developed interpolation formulas to explore the case where L is comparable in value to ℓ_P and found that even a small degree of flexibility suffices to make the system exhibit properties characteristic of the semiflexible limit.

The Khokhlov–Semenov theory was extended by Selinger and Bruinsma (Selinger and Bruinsma, 1991b) to includepositional ordering using density-functional theory. This hybrid theory produced a phase diagram (see FIGURE 3.10) that includes a hexagonal phase as well.

Khokhlov and Semenov (Khokhlov and Semenov, 1985) also analysed the situation of more concentrated solutions where attractive forces between the chains are important. The latter can be modeled in a mean-field fashion by including a Maier–Saupe potential in the free energy.

Nematic ordering in thermotropic systems has also received considerable attention using mean-field theory, by Khokhlov and Semenov (Khokhlov and Semenov, 1986) as well as other authors (Wang and Warner, 1986; Warner et al., 1985; and ten Bosch et al., 1983). The free energy of a worm-like chain model of a polymer chain in a nematic environment can be written

$$F = \frac{1}{2}\kappa\int_0^L ds\left|\frac{d\hat{u}(s)}{ds}\right|^2 - aS\int_0^L ds\left(\frac{3}{2}u_z^2(s) - \frac{1}{2}\right). \tag{3.25}$$

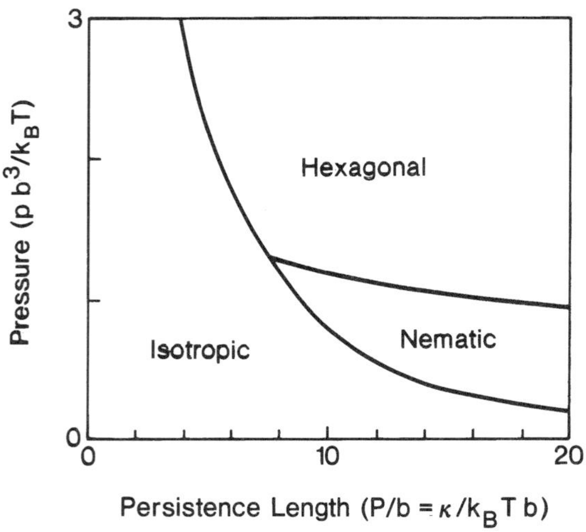

FIGURE 3.10. Theoretical phase diagram derived by Selinger and Bruinsma (1991b) for infinitely long, semiflexible chains interacting with steric, excluded volume repulsion. The chains have diameter b. All transitions are first-order.

The polymer bending rigidity is denoted by κ (in terms of which the persistence length is given by $\ell_P = \kappa/k_B T$). The nematic environment is modeled by a Maier–Saupe potential with order parameter S and coupling a. Associated with F is a partition function Z, which is a functional integral over all possible tangent fields $\widehat{u}(s)$. The partition function Z can be written in turn as an integral over a propagator for the "diffusion process" undergone by the tangent vector. Various thermodynamic quantities (entropy, latent heat, nematic order parameter) can be evaluated numerically for arbitrary values of L.

Using the free energy (3.25) it is also possible to define a "deflection length" $\lambda \equiv \sqrt{\kappa/aS}$ (Odijk, 1983, 1986), which is the distance a polymer wanders along the director before feeling the effects of neighboring chains, represented here by the nematic mean field. The correlation function $\langle \widehat{u}(s_1) \cdot \widehat{u}(s_2) \rangle$ between two tangent vectors at points s_1 and s_2 along the chain will decay exponentially $(\exp(-|s_1 - s_2|/\ell_p))$ for $|s_1 - s_2| << \lambda$, and will approach a nonzero value for $|s_1 - s_2| >> \lambda$.

Hydrodynamic theories of flexible nematic polymers have also been developed, starting with the work of de Gennes (de Gennes, 1975, 1982) and Meyer (Meyer, 1982). The simplest theories assume that the polymers are infinitely long. An elastic free energy can then be written in terms of the density fluctuations $\delta\rho$ and the tangent field $\widehat{u}$ (both variables are strictly speaking coarse-grained averages of microscopic fields). The free energy to quadratic order is given by

$$F = \frac{1}{2\rho_0^2} \int d^3 r \left[B(\delta\rho)^2 \right] + F_n \tag{3.26}$$

Here B is a bulk modulus for compressions and dilations perpendicular to the director, and F_n is the usual Frank elastic energy. However, fluctuations in $\delta\rho$ and the director field are not independent. The director field is the local average of the tangent field $\widehat{u}$. The assumption of infinite lines forces an equation of continuity to be satisfied (de Gennes, 1977; and Meyer, 1982), namely

$$\frac{\partial \delta\rho}{\partial z} + \vec{\nabla}_\perp \cdot \widehat{n} = 0 \tag{3.27}$$

where

$$\vec{\nabla}_\perp = \left(\frac{\partial}{\partial x}, \frac{\partial}{\partial y}, 0 \right).$$

Physically this statement says that a splay in the director must be accompanied by a change in density, and splay will not occur for infinitely long chains. This constraint suggests how a theory with finite-length chains can be formulated (Meyer, 1982). Namely, if the chains are long and entangled, their heads and tails can be treated as independent ideal gases. The free energy (3.26) is then replaced by:

$$F = F_\eta + \frac{1}{2} \int d^3 r \left[B \left(\frac{\partial \rho}{\rho_0} \right)^2 + G \left(\frac{\partial \delta\rho}{\partial z} + \vec{\nabla}_\perp \cdot \widehat{n} \right)^2 \right]. \tag{3.28}$$

The coupling G can be expressed in terms of $\langle \rho_H \rangle$ and $\langle \rho_T \rangle$, the average concentrations of heads and tails respectively, by,

$$G = \frac{k_B T}{\langle \rho_H \rangle + \langle \rho_T \rangle} \tag{3.29}$$

(the limit $\langle \rho_H \rangle = \langle \rho_T \rangle = 0$ then restores the constraint (3.27)). Because each chain has one head and one tail, it follows that

$$G = \frac{L k_B T}{2\rho_0} \tag{3.30}$$

for chains of average length L.

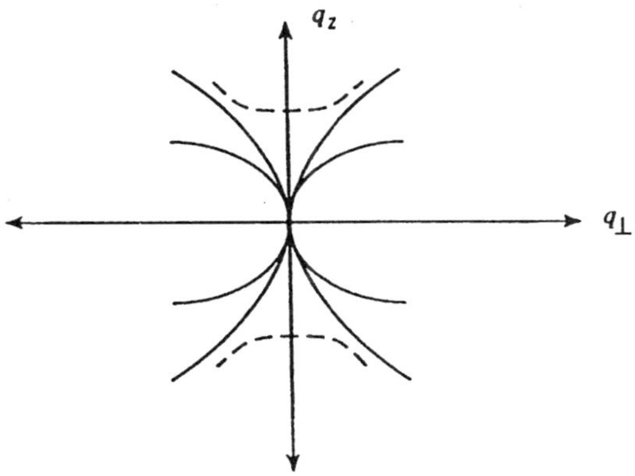

FIGURE 3.11. Contours of constant scattering intensity for nematic polymers from the theory of Kamien and coworkers (1992). The solid lines correspond to infinitely long chains (results first obtained by Selinger and Bruinsma, 1991a). Finite-length chains exhibit rounding-off of the square root contours as indicated by the dashed lines.

This harmonic free energy can be used to compute the structure function $S(\vec{q}_\perp, q_z) = \langle |\delta\rho(\vec{q}_\perp, q_z)|^2 \rangle$ with the result (Le Doussal and Nelson, 1991; Kamien et al., 1992):

$$S(q_\perp, q_z) = k_B T \frac{\rho_0^2 q_\perp^2 + (K_1 q_\perp^2 + K_3 q_z^2)/G}{B q_\perp^2 + \left(\dfrac{B}{G\rho_0^2} + q_z^2\right)(K_1 q_\perp^2 + K_3 q_z^2)}. \tag{3.31}$$

This result in principle will allow experimental determination of the Frank constants, B and L. FIGURE 3.11 shows contours of constant S for small $\vec{q}$ indicating pronounced anisotropy. For $L \to \infty$, S vanishes along the q_z axis and the contours scale as $q_z \propto q_\perp^{1/2}$ (a result first obtained by Selinger and Bruinsma, 1991a). Finite lengths introduce the rounding indicated by the dashed lines in the figure. The structure factor in the hexagonal phase was calculated by Selinger and Bruinsma (Selinger and Bruinsma, 1991a) who found enhanced diffuse thermal scattering compared with a conventional hexagonal solid due to the undulations of the chains.

Director correlation functions can also be computed using the free energy (3.28). Using these correlation functions renormalized Frank constants can then be defined which depend on wavevector as well as contour length L. In the nematic phase if the harmonic free energy (3.28) is used, the bend and twist constants do not renormalize, but the renormalized splay constant does depend on L as follows (Kamien et al., 1992):

$$K_1^R = K_1 + \frac{1}{2} k_B T L \rho_0 \tag{3.32}$$

in agreement with Meyer's prediction (Meyer, 1982). The effect of anharmonic terms in the elastic free energy was considered in the limit $L \to \infty$ by Toner (Toner, 1992) who found a breakdown of conventional elasticity similar to the one described above for smectics. The bend and twist elastic constants diverge at small q, while B vanishes in the same limit. The elastic constants in the hexagonal phase were calculated by Selinger and Bruinsma (Selinger and Bruinsma, 1991a) who found divergences in both splay and twist as $q \to 0$.

The hydrodynamics of nematic polymers is also unusual, as shown by Wen (Wen, 1991). He showed that the constraint (3.27), exactly true for infinitely long polymers, as well as the additional solvent component lead to the existence of a second sound mode. This mode is absent in low molecular weight nematics though it does occur in smectics. The finite length of real polymers is not expected to affect the experimental observation of this phenomenon.

A microscopic theory of nematic polymers was developed by Kamien et al. (Le Doussal and Nelson, 1991; and Kamien et al., 1992) using the following free energy for N polymer lines:

$$\begin{aligned} F_P = & \frac{1}{2}\kappa \sum_{j=1}^{N} \int \mathrm{d}z \left(\frac{\mathrm{d}^2\vec{r}_j}{\mathrm{d}z^2}\right)^2 \\ & + \frac{1}{2} g \sum_{j=1}^{N} \int \mathrm{d}z \left(\frac{\mathrm{d}\vec{r}_j}{\mathrm{d}z} - \delta\vec{n}(\vec{r}_j(z), z)\right)^2 \\ & + \frac{1}{2} \sum_{i \neq j}^{N} \mathrm{d}z V(|\vec{r}_i(z) - \vec{r}_j(z)|). \end{aligned} \tag{3.33}$$

The position of the ith polymer is described by a function $(\vec{r}_i(z), z)$, κ is the polymer bending rigidity and g controls the coupling between the local polymer direction and the nematic matrix. The potential $V(r)$ represents short-ranged excluded volume effects.

The free energy F_P models a polymer aligned by a nematic solvent and will also model dense nematic

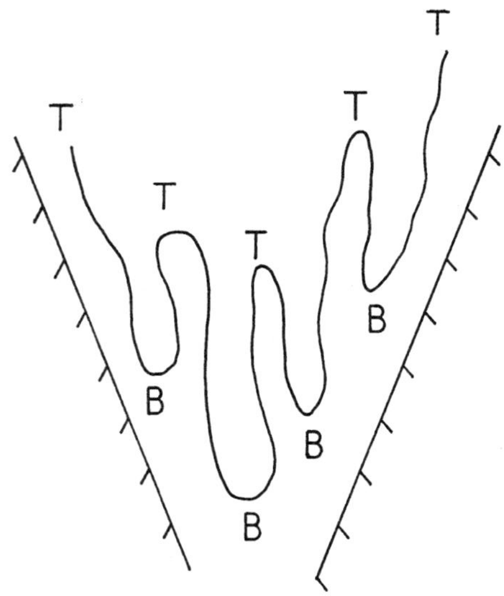

FIGURE 3.12. Illustration of top (T) and bottom (B) hairpin bends in a flexible chain undergoing splay (after Vroege and Odijk, 1988; reprinted with permission from the American Chemical Society).

polymers in an isotropic solvent if the director $\vec{n}$ is viewed as a coarse-grained field obtained by averaging over a volume of polymer tangents. Kamien et al. analysed their model using an analogy with two-dimensional quantum bosons at zero temperature. In the boson formalism it is possible to incorporate the finite length of the chains and compute the structure factor and renormalized elastic constants. These quantities reduce to the corresponding ones above in the limit of small q.

Defects in nematic polymers have also attracted considerable attention due to their unique nature. One important type of defect is the "hairpin" illustrated in FIGURE 3.12, whose importance was first emphasized by de Gennes (de Gennes, 1982). The presence of hairpins can have a significant effect on properties such as K_1. As illustrated in FIGURE 3.12, hairpins allow splay to occur more readily, as constant density can be maintained. Thus, for very flexible molecules, K_1 will *not* diverge as $L \to \infty$, except at very low temperatures. Various theories of this effect have been put forth (Selinger and Bruinsma, 1992; Kamien et al., 1992; Petschek and Terentjev, 1992; Warner et al., 1985; Williams and Warner, 1990; and Vroege and Odijk, 1988).

Finally, we note that molecular flexibility is also of interest in low-molecular weight liquid crystals, where the constituent molecules contain one or more flexible alkyl chains. This flexibility can have a profound effect on observed liquid crystalline properties. For example, the members of a homologous series with an odd number of atoms in the chain have higher entropies than those with an even number. Liquid crystal dimers where the alkyl chain joins two mesogenic groups can exhibit very strong odd–even effects.

A mean-field theory of flexible molecules was first developed by Marcelja (Marcelja, 1974), and subsequently corrected and extended by several groups (Counsell et al., 1985; Luckhurst, 1985; Ferrarini et al., 1992; and Photinos et al., 1990). In these theories the alkyl chain is assumed to exist in one of a discrete number of conformational states. The model is then analysed using mean-field theory. Application of these theories has yielded good qualitative agreement for the dependence of the nematic–isotropic transition temperature and the entropy of transition on the length of the flexible spacer (see, e.g., Ferrarini et al., 1993, 1994).

3.2.5 Defect-Mediated Phase Transitions

Theorists continue to be challenged by the problem of understanding and characterizing the phase transitions between liquid crystalline phases. In this section we discuss two examples where topological defects appear to play a dominant role in the transition.

Significant progress has been made in understanding the nematic–smectic A (NA) transition. The nature of this transition was puzzling to theorists who initially showed that the transition would be weakly first-order, whereas experimentally it is continuous. Theoretical studies have been based on the Landau–de Gennes free energy displayed in (3.9). Halperin and Lubensky (Halperin and Lubensky, 1974) analysed this free energy using the renormalization group ε expansion near four dimensions, and concluded that the transition would be weakly first-order. However, more recently, Dasgupta and Halperin (Dasgupta and Halperin, 1981; and Dasgupta, 1983) analysed this free energy in three dimensions (beyond the range of validity of the ε expansion) and showed that a continuous phase transition would occur, belonging to the inverted XY universality class. The exponents are those of the XY model (which describes superfluid helium), however, the high- and low-temperature sides of the transition are interchanged, for example, in the behavior of the specific heat. The latter quantity diverges at the transition according to

$$C_P \sim A^{\pm} t^{-\alpha} \tag{3.34}$$

with $t = (T - T_C)$, and the amplitudes A^+ and A^- referring to the smectic A and nematic phases respectively, whereas in helium A^+ and A^- refer to the high- and low-temperature phases. The universal ratio (A^+/A^-) is according to this theory then the inverse at the NA transition of what is found in helium.

Dasgupta and Halperin also demonstrated that the transition is mediated by topological defects which in this case are dislocation loops. The nematic phase can be viewed as a "gas" of many free dislocation loops. The NA transition can be viewed as the unbinding of these loops which are bound tightly together in the smectic phase. A continuum version of this model was analysed by Toner (Toner, 1982) using renormalization group methods. The correlation length ξ associated with the transition is identified as the average distance between dislocations in the smectic phase. The dislocation loop theory was further applied to the nematic–smectic A–smectic C multicritical point (NAC) by incorporating a tilt order parameter (Grinstein and Toner, 1984). Renormalization group arguments suggest that the density and tilt order parameter fluctuations are not coupled yielding a decoupled tetracritical point for the NAC transition. More tantalizing is the prediction of a biaxial nematic phase where the tilt order parameter is nonzero, but the density order parameter is zero. This phase has not yet been observed experimentally.

Topological defects may also play an important role at the NI transition as argued recently by Lammert and coworkers (Lammert et al., 1993). Using analytic and numerical techniques they analysed a model where the disclination line energy K can be varied. As K increases the NI transition becomes more weakly first-order and eventually splits into two continuous transitions which are defect mediated (see FIGURE 3.13). These continuous transition lines bound a new isotropic phase where disclinations do not have a long-range interaction but do cost energy to create with long lengths. The standard isotropic phase is characterized by a proliferation of defects including those of infinite length. The observed weakly first-order nature of the NI transition may be due to the close proximity

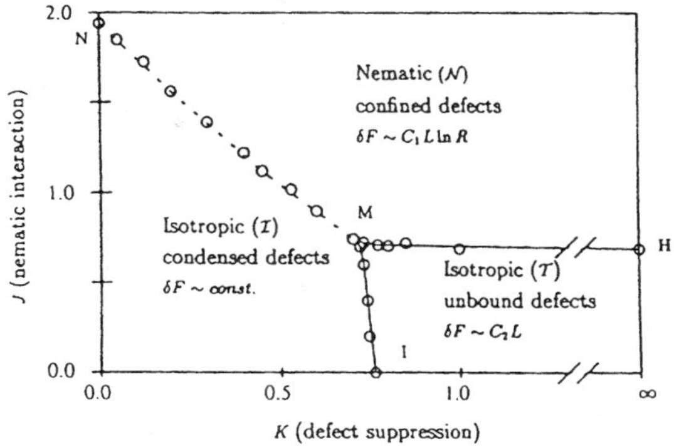

FIGURE 3.13. Theoretical phase diagram obtained by Lammert and coworkers (1993) for a lattice model of a nematic with an adjustable defect core energy K. The dotted line denotes a first-order transition and the unbroken lines denote continuous transitions. The open circles represent data from Monte Carlo simulations of the model.

of experimentally studied materials to the split in the phase transition lines.

3.2.6 Hexatic Phase

Renormalization group techniques have proven to be remarkably successful in understanding x-ray scattering from the hexatic smectic B phase (Aharony et al., 1986). In this phase, the molecules in each liquid layer are distributed *locally* on a triangular lattice. However, while the positional order is not correlated over large distances, the bond orientational order of the local triangular arrangements are correlated. This sixfold symmetry implies that the scattering intensity $I(\chi)$, a function of the angle χ between the in-plane component of the wavevector transfer $\vec{q}$ and a laboratory reference axis, should have the form

$$I(\chi) = I_0 + \frac{1}{2}\sum_{n=1}^{\infty} C_{6\eta} \cos(6\eta\chi) \tag{3.35}$$

including all harmonics. Perfect bond orientational order would require $C_{6\eta} = 1$. Aharony and coworkers calculated $C_{6\eta}$ near the smectic A–hexatic B transition using a Ginzburg–Landau free energy as the starting point. They found the following scaling behavior:

$$C_{6\eta} \sim C_6^{\sigma_\eta} \tag{3.36}$$

where in d dimensions, σ_η is given by

$$\sigma_\eta = \eta + x_\eta \eta(\eta - 1)/(d - 2 + \eta) \tag{3.37}$$

and $x_\eta \simeq 0.3 - 0.008\eta$. The exponent η is associated with the scaling of the correlation function and is nearly zero. This prediction agrees remarkably well with experiment ($d = 3$) on the material 8OSI where the data (Brock et al., 1986) can be fit to

$$\sigma_\eta = \eta + \lambda\eta(n - 1) \tag{3.38}$$

with $\lambda = 0.295$.

3.2.7 Density Functional Theories

Onsager's theory of the isotropic–nematic transition (Onsager, 1949) requires a system of long, thin rods in the dilute limit. While this is a realistic limit for rigid-rod nematic polymers, it is unreasonable for short rods. Attempts have been made to develop approximate theories of hard-rod systems that do not assume either low density or very large aspect ratio. Recent efforts have utilized the density functional theory (DFT) following its successful application to hard spheres (for a review see Baus, 1987). In this approach the Helmholtz free energy F is viewed as a functional, $F[\rho]$, of the local density $\rho(\vec{x})$. The true equilibrium state $\rho(\vec{x})$ should minimize $F[\rho]$. This requires that

$$\beta\frac{\delta^2 F[\rho]}{\delta\rho(\vec{x})\delta\rho(\vec{x}\,')} \equiv \frac{\delta(\vec{x} - \vec{x}\,')}{\rho(\vec{x})} - C(\vec{x}, \vec{x}\,'; [\rho]) \tag{3.39}$$

be a positive definite functional. This equation defines the two-body direction correlation function $C(\vec{x}, \vec{x}\,'; [\rho])$. Integrating both sides of (3.39) twice along a linear path in density space starting from a reference state ρ_0 yields the *exact* relation:

$$\beta F[\rho] = \beta F[\rho_0] + \int d\vec{x}\rho(\vec{x})\ln\left[\frac{\rho(\vec{x})}{\rho_0}\right] - \int d\vec{x}\int d\vec{x}^{\,1}\int_0^1 d\lambda(1-\lambda)C(\vec{x}, \vec{x}\,'; [\rho_0 + \lambda\Delta\rho]) \times \Delta\rho(\vec{x})\Delta\rho(\vec{x}\,') \tag{3.40}$$

where $\Delta\rho(\vec{x}) = \rho(\vec{x}) - \rho_0$. To utilize this equation an approximate expression for $C[\rho]$ must be chosen for each phase in terms of $C[\rho_0]$. Then $F[\rho]$ must be minimized in terms of a suitable parametrization of $\rho(\vec{x})$. The expressions for $F[\rho]$ and $\rho(\vec{x})$ at the minimum of F can then be used to calculate thermodynamic quantities. Several different approaches (Mulder and Frenkel, 1985; Singh and Singh, 1986; Marko, 1988; and Colot et al., 1988) have been taken in choosing the reference state ρ_0 and the approximate $C[\rho]$ for the isotropic and nematic phases. We refer the reader to the relevant literature for the details which are quite technical. Each of the theories produces a first-order I–N transition, however, at different densities and order parameter values (see Table I of Colot et al.). As the aspect ratio decreases the densities of the coexisting I and N phases shift to larger values and the order parameter at the transition shifts to lower values (qualitative features predicted by Landau theory as well). Colot and coworkers (Colot et al., 1988) have compared their DFT results with those of Landau and Onsager theories. They find that Landau theory yields a fair location of the transition but underestimates the strength of the order parameter jump as well as the width (i.e. the difference in I and N densities at the transitions). In the limit of large aspect ratio, the theory of Colot et al. reproduces Onsager's results but the approach to that limit is very slow, requiring aspect ratios of several hundred (such values however are found in some polymeric systems).

Density functional theories have also been developed for the nematic–smectic A transition (Mulder, 1987; Poniewierski and Holyst, 1988; and Holyst and Poniewierski, 1989) and provide reasonable agreement with computer simulations of hard sphero-cylinders (see section 3.3.2).

3.2.8 Frustrated Smectics

A number of experiments performed since the late 1970s (Sigaud et al., 1979; Hardouin et al., 1983; Shashidar and Ratna, 1989; Ratna et al., 1985; and Chan et al., 1985, 1986) found a variety of smectic A phases including: S_{A_1}, a monolayer smectic A with a Bragg peak at reciprocal wavevector $2\pi/\ell$ ($\ell =$ molecular length); S_{A_2} , a bilayer smectic with a peak reciprocal wavevector π/ℓ, and S_{A_d}, a partial smectic with a peak at wavevector $2\pi/\ell'$, where $\ell' =$ length of molecular pairs. In general $\ell' \neq 2\ell$. Some of these phases occur in reentrant phase diagrams, for example, as the temperature is lowered the system goes through the following phases:

$$\text{I--N--}S_{A_d}\text{--N--}S_{A_d}\text{--N--}S_{A_1}.$$

Prost (Prost, 1979) developed a Landau–de Gennes model to describe these phases and phase diagrams which utilizes two order parameters: the mass density $\rho(\vec{r})$ (used to describe conventional smectic A phases), and a polarization wave $P_z(\vec{r})$ which describes long-range head-to-tail correlations of asymmetric molecules. The z-axis is chosen normal to the smectic layers. If these two order parameters are decoupled, then, P_z would develop modulations at $q_1 = 2\pi/\ell'$, and $\rho(\vec{r})$ would exhibit a Bragg peak at $q_2 = 2\pi/\ell$, where we identify ℓ' as the length of a pair of anti-parallel partially overlapping molecules. In general, ℓ' and ℓ are incommensurate and it is impossible to satisfy both tendencies simultaneously; hence these systems are called frustrated smectics.

As in the case of conventional smectics, it is convenient to work with complex order parameters, whose real parts yield $P_z(\vec{r})$ and $\rho(\vec{r})$ as follows:

$$\begin{aligned} P_z(\vec{r}) &= \operatorname{Re}\psi_1(\vec{r}) \\ \rho(\vec{r}) &= \operatorname{Re}\psi_2(\vec{r}). \end{aligned} \tag{3.41}$$

Because P_z and ρ develop spatial modulations we write ψ_1 and ψ_2 as follows:

$$\begin{aligned} \psi_1(\vec{r}) &= |\psi_1|\, e^{i(\vec{Q}_1\cdot\vec{r}+\phi_1)} \\ \psi_2(\vec{r}) &= |\psi_2|\, e^{i(\vec{Q}_2\cdot\vec{r}+\phi_2)}. \end{aligned} \tag{3.42}$$

The Landau free energy can then be written as an expansion in ψ_1, ψ_2 and their derivatives:

$$\begin{aligned} F = \int d\vec{r} \Big\{ & \frac{a_1}{2}|\psi_1|^2 + \frac{D_1}{2}|(\nabla^2 + q_1^2)\psi_1|^2 \\ & + \frac{C_1}{2}|\vec{\nabla}_\perp \psi_1|^2 + \frac{u_1}{2}|\psi_1|^4 \\ & + \frac{a_2}{2}|\psi_2|^2 + \frac{D_2}{2}|(\nabla^2 + q_2^2)\psi_2|^2 \\ & + \frac{C_2}{2}|\vec{\nabla}_\perp \psi_2|^2 + \frac{u_2}{4}|\psi_2|^4 \\ & + \frac{u_{12}}{4}|\psi_1|^2|\psi_2|^2 - w\operatorname{Re}(\psi_1^2\psi_2^*) \Big\} \\ & + w'\operatorname{Re}(\psi_1\psi_2^*) \end{aligned} \tag{3.43}$$

where $\vec{\nabla}_\perp$ is the in-layer gradient operator. The terms proportional to D_1 and D_2 favor $Q_1^2 = q_1^2$ and $Q_2^2 = q_2^2$ respectively. The next-to-last term favors $2\vec{Q}_1 = \vec{Q}_2$ which is appropriate to the case $\ell' = 2\ell$, while the last term dominates when $\ell' \approx \ell$.

Minimizing (3.43) yields a mean field theory for this model with a very rich phase structure, depending strongly on the relative values of ℓ and ℓ'. When $\ell' \approx \ell$, ψ_1 and ψ_2 will adopt a common wavevector q if w is sufficiently large. The layer spacing is given by $a_0 = 2\pi/q$. The value of a_0 ranges between ℓ and ℓ' depending on the actual parameters of the free energy. When $a_0 \approx \ell$ we have an S_{A_1} phase, while for $a_0 \approx \ell'$, the S_{A_d} phase exists. For smaller values of w, incommensurate wavevectors Q_1 and Q_2 can occur in a phase labeled $S_{A_{i1}}$. When $\ell' \approx 2\ell$, six different phases can occur including "antiphases" $S_{\tilde{A}}$ and $S_{\tilde{C}}$ where a periodic modulation of the antiferroelectric order in one in-layer direction occurs.

Mean-field theory can also provide an explanation for single reentrance in the phase diagrams. Multiple reentrance however requires consideration of fluctuations or a suitable microscopic model (see below). Experiments suggest that double reentrance is associated with multicritical points where, for example, the N, S_{A_1} and S_{A_d} phases meet. At such points, renormalization group arguments (Prost, 1984) predict that the phase boundaries obey scaling relations of the form:

$$\widetilde{P} = W^{\pm}\tilde{t}^{x} \tag{3.44}$$

for some exponent x, where $\widetilde{P}$ and $\tilde{t}$ are linear combinations of the physically reduced temperature and pressure variables t and p. Suitable choices of these linear combinations lead to double reentrance. Consideration of fluctuations is also important near the critical point which terminates the first-order S_{A_2}--S_{A_d} line. Near this point the free energy can be mapped at the mean-field level onto the liquid–gas free energy. At the critical point it can be shown that the compressional elastic constant vanishes in analogy with the vanishing of the inverse compressibility in the liquid vapor problem. However, when fluctuations are included, it is found that the analogy with the liquid–vapor problem does not hold quantitatively and different critical exponents are obtained. These exponents are anisotropic, that is, the scaling of the correlation length is different parallel and perpendicular to the layers.

Because the compressional elastic constant vanishes at the critical point, the system is nematic-like there. It is reasonable to suppose then that this nematic point might be replaced by a "bubble." Answering this question requires consideration of dislocations which was done by Prost and Toner (Prost and Toner, 1987). They found that depending on the parameters of the system either a bubble or critical point may be observed.

A microscopic theory has been pursued by Berker and coworkers (Berker and Walker, 1981; Indeku and Berker, 1986; Indeku et al., 1987; Indeku and Berker, 1988; and Netz and Berker, 1992a) over the past decade to explain this reentrance phenomenon and more recently smectic C ordering. Their model, dubbed the frustrated spin-gas model, includes dipolar, van der Waals and benzene-ring steric interactions. It was analysed via approximated theoretical techniques, and more recently by Monte Carlo sampling.

The model is constructed by first noting that close-packed molecules with dipole moments, such as liquid crystal molecules, are necessarily in a frustrated situation where there is no unique energy minimum; different local orientational configurations equally minimize the energy. Choosing the z-axis for the director or smectic layer normal, the model assumes that the molecules occupy n consecutive sites along the z-dimension in a three-dimensional stacked triangular lattice. The triangular structure arises from the close-packing in the xy plane, and the segmentation into n units arises from molecular corrugation: two neighboring molecules have preferred positions relative to each other. The model does include, however, librational permeation which represents fluctuations about the corrugation minima. Lateral positional fluctuations within the xy plane do not appear to be important.

FIGURE 3.14. Sample theoretical phase diagram due to Netz and Berker for a moderate strength dipolar system. (From R. R. Netz and A. N. Berker, *Phase Transitions in Liquid Crystals*, New York, 1992, pp. 109–124).

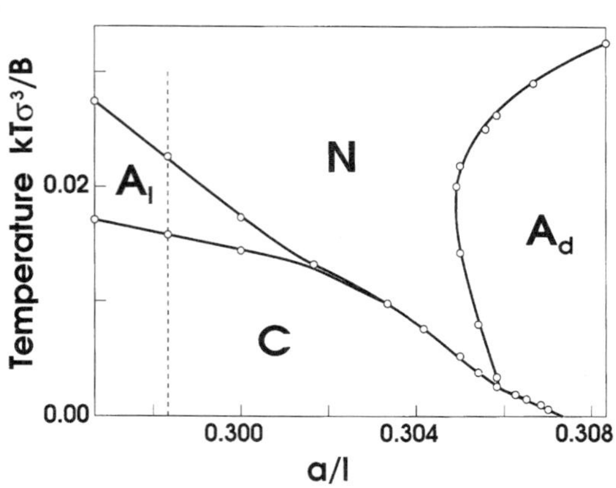

The inclusion of dipolar interactions and molecular corrugation and the neglect of librational permeation are sufficient to produce a phase diagram showing a reentrant nematic phase at temperatures below the smectic A_d phase. More complicated phase diagrams with multiple reentrance are predicted when van der Waals interactions are included. Including rotational steric hindrance yields a model for smectic C ordering. The full phase diagram for a moderate strength of dipolar interactions is shown in FIGURE 3.14.

3.3 COMPUTER SIMULATIONS

Ever increasing computer power has made simulations of liquid crystal systems a feasible enterprise. Larger, more complex systems can now be simulated for longer times and many interesting phenomena can be studied for the first time, albeit in some cases at a crude level. With the aid of computers it is possible to study the effects of molecular shapes, sizes and interactions on macroscopic behavior. While present day computers are not powerful enough to model a liquid crystal molecule in precise detail, it is possible to study simple models which capture the essence of the system. If we restrict our attention to long-wavelength, low-frequency phenomena, then these simple, phenomenological models should suffice in the same way that Landau free energies truncated at low order, can also describe these phenomena. The models simulated on a computer take us one step closer than Landau theory in understanding the molecular origin of observed behavior. They also allow us to study dynamical phenomena which in many instances are analytically intractable in Landau theory.

Three broad classes of liquid crystal models have been used in simulation work. The first is the Lebwohl–Lasher model (Lebwohl and Lasher, 1972) which is a lattice model for rotators. This model is appropriate for studying the N–I transition when viewed as a rotational order–disorder transition in an effective crystalline solid and all density fluctuations are thus ignored (however, a recent theory due to Tao and coworkers (Tao et al., 1993) suggests that this may be a poor assumption to make). Another class of models which has received much attention and models the translational as well as orientational degrees of freedom of a liquid crystal uses hard particles of various shapes which interact solely by excluded volume effects (for reviews, see Frenkel,

1992; and Allen, 1993). These models exhibit very rich phase diagrams including smectic, columnar and cubatic phases. Dynamical properties, including those involving translational motion, can of course be studied as well. Finally, in recent years there has been considerable numerical study of the Gay–Berne system (Gay and Berne, 1981) which is a fluid of point-like particles, each carrying a unit vector $\widehat{u}$ which acts as the long molecular axis. The particles interact via an anisotropic Lennard-Jones potential which depends on the relative orientation and location of a pair of molecules. This system displays rich behavior like the hard body models and allows one to study the effects of attractive forces absent for hard bodies.

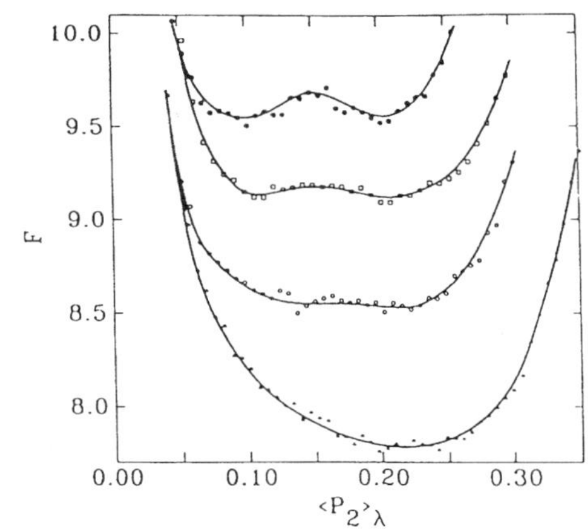

FIGURE 3.15. Free energy as a function of orientational order from Monte Carlo simulations of the Lebwohl–Lasher model (Zhang et al., 1992). The points on the curves correspond to the following lattice sizes from bottom to top: $L^3 = 16^3$, 20^3, 24^3, and 28^3. The solid lines are guides to the eye. The different sets of data are displaced vertically for the sake of clarity.

3.3.1 Lebwohl–Lasher Model

The Lebwohl–Lasher model is the lattice version of the Maier–Saupe model of a nematic. The Hamiltonian for the model is defined to be:

$$H = -J \sum_{\langle i,j \rangle} P_2(\widehat{s}_i \cdot \widehat{s}_j). \tag{3.45}$$

P_2 is the second Legendre polynomial, $\widehat{s}_i$ is a unit vector parallel to the long axis of the molecular at site i and J is the strength of the nematic coupling. Because the molecules are fixed on the sites of a lattice, translation motion is absent. Nevertheless, this model can be useful in studying the I–N transition and it has been intensively investigated using Monte Carlo techniques (Luckhurst and Simpson, 1982; Fabbri and Zannoni, 1986; Zannoni, 1986; and Biscarini et al., 1991).

The definitive work in this area has been done by Zhang and coworkers (Zhang et al., 1992; and Zhang et al., 1993) who simulated a simple cubic lattice with up to 28^3 sites. Their work is noteworthy for its use of the Lee–Kosterlitz finite-size scaling method (Lee and Kosterlitz, 1991). This method is very efficient in numerically locating first-order phase transitions and has been used extensively in simulations of magnetic systems. An order parameter distribution function $P(\lambda, T, L)$ is computed where λ is the nematic order parameter (the largest eigenvalue of the tensor order parameter $\overleftrightarrow{Q}$), T is the temperature and L the system size. This computation is done using the Ferrenberg–Swendsen reweighting techniques (Ferrenberg and Swendsen, 1988). The method of Lee and Kosterlitz defines a free-energy-like quantity $F(\lambda, T, L)$ via

$$F(\lambda, T, L) \sim -\ln P(\lambda, T, L) \tag{3.46}$$

which differs from the true bulk free energy by a T and L dependent additive quantity. At a first-order transition F will have a pronounced double minimum structure corresponding to phase coexistence. The height of the barrier $\Delta F(L)$ separating these minima is a measure of the interfacial free energy and scales like L^2 in three dimensions. A numerical measurement of $\Delta F(L)$ can then determine the nature of the transition. For a first-order transition ΔF will increase with L while it will approach a constant for a continuous transition. FIGURE 3.15 shows F for four different lattice sizes and the appearance of a double well structure at large L is very apparent. However, the absence of a detectable double well structure at smaller L indicates that the transition is very weakly first-order. The transition temperature T_c in the thermodynamic limit can be obtained by extrapolating three different measures of $T_c(L)$, the finite-size transition temperature. These measures correspond to the peaks in the specific heat and the order susceptibility, and the temperature where the double minima are of equal depth. All three measures of $T_c(L)$ extrapolate to the same value, $1.1232 \pm 0.0001 J/k_B$ as $L \rightarrow \infty$. The transition enthalpy ΔH can be estimated from the finite-size scaling of the specific heat with the result $\Delta H = (0.20 \pm 0.04)J$.

The free-energy function F also allows the determination of the limits of stability of the nematic and isotropic phases. Zhang et al. (1992) estimate that these temperatures are within 0.5×10^{-3} reduced temperature of T_c which is in reasonable agreement with experimental data.

The Lebwohl–Lasher model has also been extended to model nematic liquid crystals in porous media such as aerogels (Cleaver et al., 1995). This extended model is the analog of the random anisotropy model of magnetism (for a review, see Goldschmidt, 1992) and includes an interaction with a random easy axis quenched at each site i. The Hamiltonian is given by:

$$H = -J\sum_{i,j} P_2(\hat{s}_i \cdot \hat{s}_j) - D\sum_{i} P_2(\hat{s}_i \cdot \hat{n}_i) \qquad (3.47)$$

where D is the strength of the coupling to the random axis. The random axis mimics the local pore environment which selects a preferred direction for the nematic order within the pore. This model has been studied for the case $D/J = 1$ by Cleaver and coworkers using Monte Carlo techniques on lattices up to 64^3 in size. They find no evidence for a first-order transition and their data are consistent with the absence of nematic long-range order at low temperatures. This is precisely what occurs in the ferromagnetic version of this model. Furthermore, the two-particle orientational correlation $g_2(r)$ function decays as a power-law:

$$g_2(r) = \frac{B}{r^{\eta}} \qquad (3.48)$$

with an exponent $\eta \approx 0.15$ at $T/J = 1$. This result indicates that correlations fall off slowly and exist over length scales longer than a pore size.

3.3.2 Hard Particles

The seminal work of Alder and Wainwright (Alder and Wainwright, 1957) led to the conclusion that the freezing of atomic liquids is primarily an excluded volume effect. Attractive forces do affect the location of the freezing temperature, but the essential physics of the phase transition is contained in a model based purely on repulsive forces. This conclusion has motivated most of the work done on simulations of

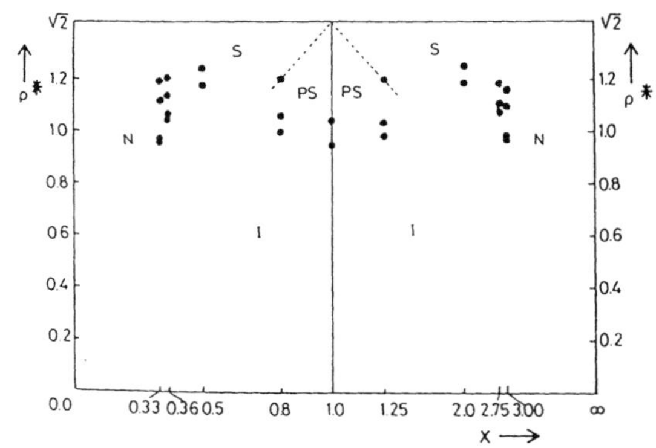

FIGURE 3.16. Phase diagram for hard ellipsoids of revolution obtained by Monte Carlo simulation (Frenkel and Mulder, 1985). The reduced density ρ^* is defined such that the density of regular close packing is equal to $\sqrt{2}$ for all x. The shaded areas indicate two-phase coexistence. The phases shown are: I, isotropic, S, orientationally ordered crystal; PS, plastic crystal (orientationally disordered); and N, nematic.

anisotropic hard particles to map out mesophase diagrams. This work was also motivated by Onsager's theory which showed that excluded volume effects will produce the I–N transition in a system of long, thin rods.

The first computer simulations of anisotropic hard particles were done by Vieillard-Baron in two dimensions over twenty years ago (Vieillard-Baron, 1974). The first systematic study of a three-dimensional system was done by Frenkel and Mulder (Frenkel and Mulder, 1985) using ellipsoids of revolution characterized by x, the ratio of the length of the major axis to the minor axis. The phase diagram they obtained for system sizes of order 100 particles is illustrated in FIGURE 3.16. The diagram is plotted in such a way as to illustrate the near-perfect symmetry between oblate and prolate ellipsoids, i.e. $x \to 1/x$, for $x \approx 1$. This symmetry, however, appears to weaken for more extremely shaped objects (i.e. $x \to 0$ or $x \to \infty$). Note that a nematic phase appears only in the range $x > 2.5$ or $x < 0.4$, however, these limits have not been firmly established (the values found by Frenkel and Mulder appear to change as larger systems have been studied (Zarragoicoechea et al., 1992)). The I–N transition for ellipsoids with $x = 3$ appears to be weakly first-order. Pretransitional nematic fluctuations in the isotropic phase have also been observed for $x = 3$ by Allen and Frenkel (Allen and Frenkel, 1987).

A biaxial phase can be produced if the molecules are no longer modeled as ellipsoids of revolution, but rather spheroids with unequal semi-axes a, b, c. A rough survey of the phase diagram of such objects has been carried out by Allen (Allen 1990). The biaxial phase is stable only for a narrow range of values of the semi-axes near the critical value $b/a \approx \sqrt{10}$. Again an approximate symmetry exists under the transformation $(a, b, c) \leftrightarrow (c, ca/b, a)$.

Unfortunately, hard core systems composed of ellipsoids cannot exhibit smectic phases. This is easily seen using an argument due to Frenkel (Frenkel, 1988). Consider a perfectly aligned nematic phase of hard ellipsoids with major axes $2a$ and minor axes $2b$ as a possible candidate for a smectic phase. However, if we transform the z-coordinate (assuming the director is along the z-axis) according to $z \rightarrow z(a/b)$ and transform momenta in the z-direction according to $p_z \rightarrow p_z(b/a)$ then the partition function is unchanged. However, this transformation changes ellipsoids into spheres which will certainly not exhibit a smectic phase.

Fortunately, it is possible to find a simple hard core model which exhibits a smectic phase, namely a system of parallel spherocylinders. A spherocylinder is a cylinder of length L, diameter D, capped at each end by hemispheres of diameter D. A stable smectic phase appears for $L/D > 0.5$ (Stroobants et al., 1986, 1987). The phase diagram is shown in FIGURE 3.17. At high densities and large values of L/D a columnar phase appears to form (Stroobants et al., 1987) but this may be an artifact of the small sample sizes used (on the order of 1000 particles).

FIGURE 3.17. Phase diagram for parallel hard spherocylinders obtained by Monte Carlo simulations (Frenkel, 1992). ABC denotes a solid phase consisting of stacked triangular planes, while AAA denotes a hexagonal structure.

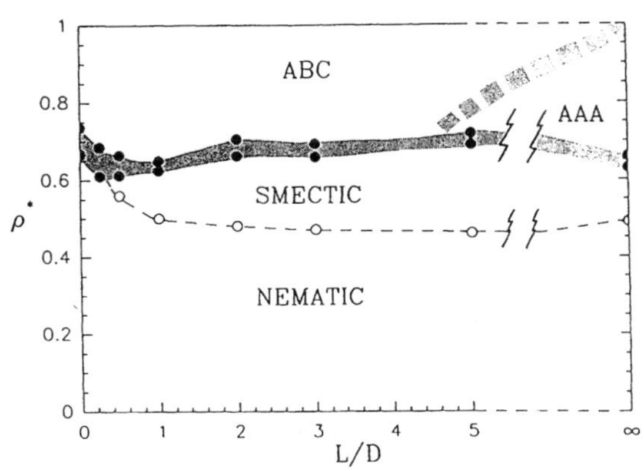

FIGURE 3.17 was obtained for perfectly parallel spherocylinders. When the particles are allowed to rotate a new phase diagram, is obtained (Frenkel, 1988; and Frenkel et al., 1988). Here the smectic phase disappears for $L/D \leq 3$.

Columnar phases can be found in systems of "oblate spherocylinders". A good choice for the latter object is the cut or truncated sphere, which is a sphere cut off at the top and bottom by two parallel cuts. It is characterized by a ratio L/D where L is the distance between the cuts. Veerman and Frenkel (Veerman and Frenkel, 1992) have studied the cases $L/D = 0/1, 0.2$ and 0.3. For $L/D = 0.1$, the system forms a nematic phase at reduced density $\rho^* \approx 0.33$ ($\rho^* = \rho/\rho_{cp}$, where ρ_{cp} is the density for close packing). At $\rho^* \approx 0.5$ the nematic undergoes a strong first-order transition to a columnar phase with $\rho^* \approx 0.53$. At yet higher density, $\rho^* = 0.8$, there is a columnar–crystalline transition. The case of $L/D = 0.2$ is even more interesting. At $\rho^* = 0.5$ there is a transition to a cubatic phase having cubic symmetry but no translational order. The nematic order parameter in this phase is zero, and the orientational correlation function $g_2 = \langle P_2(\hat{u}(0) \cdot \hat{u}(\vec{r}))\rangle$, ($\hat{u}$ is along the polar axis of the cut sphere) decays to zero quickly. However, $g_4 = \langle P_4(\hat{u}(0) \cdot \hat{u}(\vec{r}))\rangle$ is long-ranged. This cubatic order does not appear to be an artifact of boundary conditions, however it is stable only over a narrow range density. For $L/D = 0.3$ both the nematic and cubatic phases are absent and the isotropic fluid transforms directly to a solid.

3.3.3 Gay–Berne Model

The Gay–Berne (GB) potential is a phenomenological model which includes both attractive and repulsive forces. In that sense it shares some features with both the Lebwohl–Lasher and hard body models. Unlike the Lebwohl–Lasher model, the molecules in the GB system have translational degrees of freedom. Likewise, the GB model includes attractive forces absent in the hard body simulations, but models the repulsive hard core in a continuous, "softer" fashion.

The GB potential is based on the gaussian overlap model, suitably modified to remedy some unrealistic features found in the original Berne–Pechukas–Kushick potential (Berne and Pechukas, 1972; and Kushick and Berne, 1976). The latter was the first in a series of potentials based on gaussian overlap (for other examples, see Walmsley, 1977; Stone, 1979; Sediawan et al., 1988, 1989; and Kabadi, 1986). The GB potential was modeled to give the best fit to the pair potential for a molecule consisting of a linear array of four equidistant Lennard-Jones centers with separation of $2\sigma_o$ between the first and fourth sites (subsequent work by Luckhurst and Simmonds (Luckhurst and Simmonds, 1993) examined different fits; see below) The GB potential is given by

$$U(\hat{u}_1,\hat{u}_2,\vec{r}) = 4\varepsilon(\hat{u}_1,\hat{u}_2,\vec{r}) \times \left[\left\{\frac{\sigma_o}{r-\sigma(\hat{u}_1,\hat{u}_2,\vec{r})+\sigma_o}\right\}^{12} - \left\{\frac{\sigma_0}{r-\sigma(\hat{u}_1,\hat{u}_2,\vec{r})+\sigma_o}\right\}^{6}\right] \tag{3.49}$$

where $\hat{u}_1, \hat{u}_2$ are unit vectors giving the orientations of the two molecules separated by the position vector $\vec{r}$. The parameters $\varepsilon(\hat{u}_1,\hat{u}_2,\vec{r})$ and $\sigma(\hat{u}_1,\hat{u}_2,\vec{r})$ are orientation dependent and give the well depth and intermolecular separation where $U = 0$ respectively. The well depth is written as

$$\varepsilon(\hat{u}_1,\hat{u}_2,\vec{r}) = \varepsilon_o\varepsilon^{\nu}(\hat{u}_1,\hat{u}_2)\varepsilon'^{\mu}(\hat{u}_1,\hat{u}_2,\vec{r}) \tag{3.50}$$

where

$$\varepsilon(\hat{u}_1,\hat{u}_2) = (1-\chi^2(\hat{u}_1\cdot\hat{u}_2)^2)^{-1/2} \tag{3.51}$$

and

$$\varepsilon'(\hat{u}_1,\hat{u}_2,\hat{r}) = 1-\frac{\chi'}{2} \times \left\{\frac{(\hat{r}\cdot\hat{u}_1+\hat{r}\cdot\hat{u}_2)^2}{1+\chi'(\hat{u}_1\cdot\hat{u}_2)} + \frac{(\hat{r}\cdot\hat{u}_1-\hat{r}\cdot\hat{u}_2)^2}{1-\chi'(\hat{u}_1\cdot\hat{u}_2)}\right\}. \tag{3.52}$$

The shape anisotropy parameter χ is given by

$$\chi = \{(\sigma_e/\sigma_s)^2 - 1\}/\{(\sigma_e/\sigma_s)^2+1\} \tag{3.53}$$

where σ_e and σ_s are the separation of end-to-end and side-by-side molecules respectively. The parameter χ' is given by

$$\chi' = \{1-(\varepsilon_e/\varepsilon_s)^{1/\mu}\}/\{1+(\varepsilon_e/\varepsilon_s)^{1/\mu}\}. \tag{3.54}$$

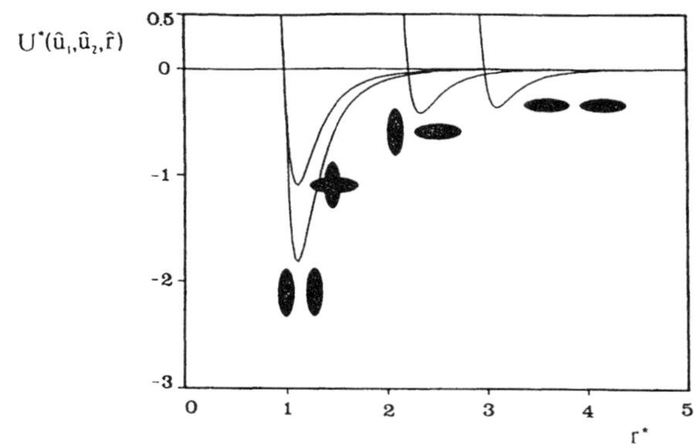

FIGURE 3.18. Illustration of the Gay–Berne potential for several choices of orientation of a pair of molecules (after Luckhurst et al., 1990). The quantity U^* is the scaled potential, U/ε_0.

The ratio of the well depths for end-to-end and side-by-side configurations is $\varepsilon_e/\varepsilon_s$.

The exponents ν and μ appearing in (3.50) and (3.54) are adjustable parameters. The best fit to the linear array of four Lennard-Jones centers yielded $\nu = 1, \mu = 2, \sigma_e/\sigma_s = 3$ and $\varepsilon_s/\varepsilon_e = 5$. A visualization of this complex potential is shown in FIGURE 3.18 where U/ε_o is plotted for four different molecular orientations as a function of r/σ_o. The FIGURE indicates that the GB potential captures the essential features of interaction of two mesogenic molecules, for example, the end-to-end and side-by-side configurations possess different well depths even though the molecules are parallel in both instances. Likewise the "X" and "T" configurations where the molecules are perpendicular have different potentials.

DeMiguel and coworkers (DeMiguel et al., 1991a, b) have carried out the most complete simulations to date of the GB potential with its original parametrization. They simulated 256 molecules using molecular dynamics in the NVT ensemble. The approximate phase diagram they obtained is shown in FIGURE 3.19. The I–N transition is found to be weakly first-order. The small system size studied did not allow the authors to ascertain whether the Sm B phase is crystalline or hexatic. The biaxial smectic phase B(t) is a tilted version of the Sm B. DeMiguel and coworkers also studied a purely repulsive GB potential, defined by subtracting off the attractive portion of the potential at a given relative molecular orientation. The I–N transition in this case occurs (for fixed T) at a slightly higher

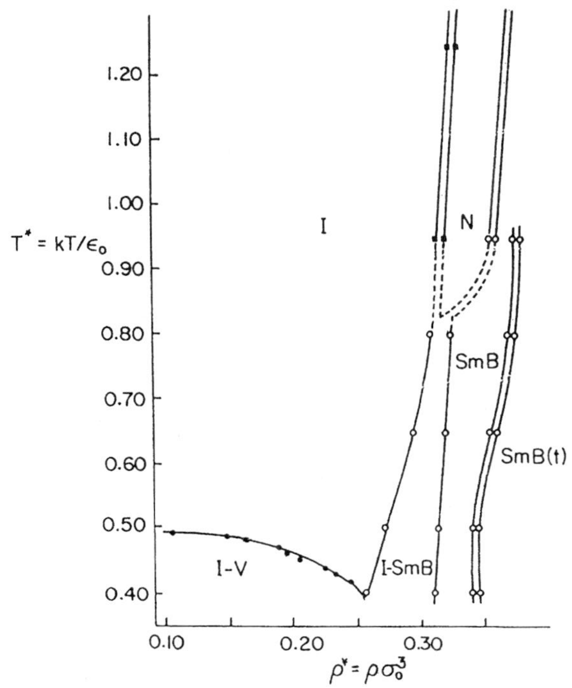

FIGURE 3.19. Phase diagram for the Gay–Berne fluid from the molecular dynamics simulations of DeMiguel and co-workers (DeMiguel et al., 1991a). The solid lines are guides to the eye, and the dashed lines are extrapolations of the data.

value of reduced density. Furthermore, the Sm B and Sm B(t) phases do not appear, and the system remains nematic even at high density, indicating the importance of the attractive tail of the potential in obtaining smectic phases.

Subsequently, DeMiguel and coworkers (DeMiguel et al., 1992) studied the rotational and translational dynamics of the GB fluid in the isotropic and nematic phases. Various autocorrelation functions (ACF) were studied including translational velocity, force, angular velocity and torque. The behavior of the translational velocity ACF suggests that in the nematic phase the molecules are diffusing along cylindrical cages whose long axes are parallel to the director. Rotational relaxation times were deduced from the self-particle reorientational correlation function, $\phi_\ell(t) \equiv \langle P_\ell(\widehat{u}(t_o) \cdot \widehat{u}(t_o + t)\rangle$, where P_ℓ is the ℓth-order Legendre polynomial. The Debye diffusion model where the reorientational process is a consequence of many small angular displacements appears to explain the data in the nematic phase, but no theoretical model used by the authors could explain the data in the isotropic phase.

Luckhurst's group (Luckhurst et al., 1990) carried out extensive simulations of a reparametrized GB potential, beginning with a simple change of the values of μ and ν. Rather than the original choices, $\nu = 1,\ \mu = 2$, the values $\nu = 2,\ \mu = 1$ were used. This latter choice makes the side-by-side configuration more stable relative to the cross and tee configurations which should help stabilize the smectic phase. Like DeMiguel and coworkers 256 particles were simulated using molecular dynamics (MD). In addition to the I, N and Sm B phases, a smectic A also appeared. A more systematic reparametrization was carried out by Luckhurst and Simmonds (Luckhurst and Simmonds, 1993) who constructed a site–site potential for p-terphenyl. This potential is not uniaxial, but the authors projected out the biaxiality and then mapped this result onto the GB potential by examining various configurations of molecular pairs. This procedure produced the following values: $\nu = 0.74$, $\mu = 0.8$, $\sigma_o = 3.65\text{Å}$, $\varepsilon_o/k_B = 4302°\text{ K}$ $\sigma_e/\sigma_s = 4.4$ and $\varepsilon_s/\varepsilon_e = 39.6$. These values are considerably different than those used in the original GB parametrization. Luckhurst and Simmonds then simulated 256 particles interacting with this new potential, and isotropic, nematic and smectic A phases were found. As in the study by DeMiguel's group, the nematic phase disappears if the reduced density is too low. Luckhurst and Simmonds also compared the results of a number of GB simulations with those of the hard ellipsoid system (see FIGURE 3.20). The good agreement indicated in the FIGURE shows that the I–N transition is dominated by excluded volume effects. On the other hand, attractive forces dominate the formation of the smectic phases which have been found in the GB system for $\sigma_e/\sigma_s = 3$ and 4.4. Hard ellipsoids do not form smectics and hard spherocylinders do so only if their length-to-breadth ratios exceed 4. In the GB system, Luckhurst and Simmonds found that $\varepsilon_s/\varepsilon_e$ must exceed a critical value of approximately 5 in order for the smectic phases to form. They also found a strong correlation between the presence of translational and orientational order, with the smectic phases appearing when the nematic order parameter is at least 0.73.

The GB potential can also be reparametrized to study discotics as shown by Emerson and coworkers

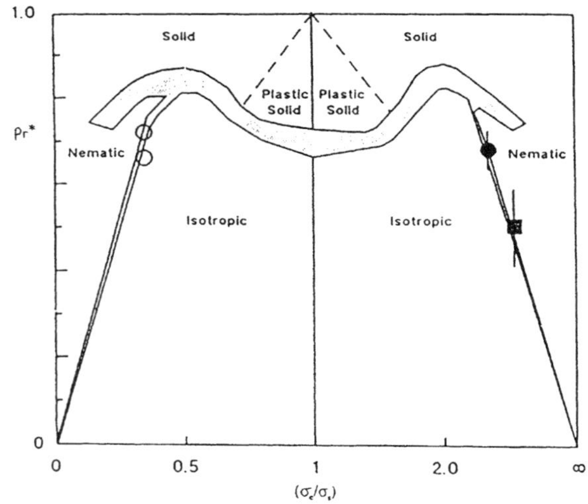

FIGURE 3.20. The phase diagram of FIGURE 3.16 (hard ellipsoids) along with data points resulting from simulations of the Gay–Berne potential due to Luckhurst and Simmonds (Luckhurst and Simmonds, 1993). Vertical lines indicate the approximate density range where a nematic–isotropic transition was observed at constant density. The symbols correspond to the following length-to-breadth ratios in the GB system: ■, $\sigma_e/\sigma_s = 4.4$, ●, $\sigma_e/\sigma_s = 3.0$, ○, $\sigma_e/\sigma_s = 0.345$.

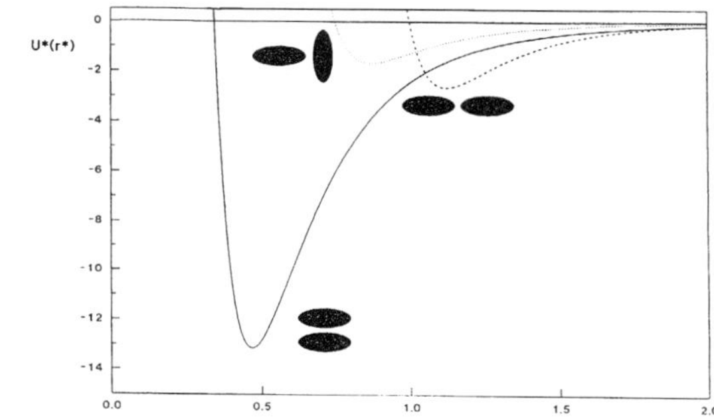

FIGURE 3.21. Illustration of the Gay–Berne discogen potential due to Emerson and coworkers (Emerson et al., 1994) for three molecular arrangements: face-to-face (solid line), edge-to-edge (dashed line) and edge-to-face (dotted line). U^* denotes the scaled potential U/ε_0 and r^* denotes r/σ_0.

(Emerson et al., 1994). They fit the GB potential to a model of interacting triphenylene molecules yielding the values $\sigma_f/\sigma_e = 0.345$ and $\varepsilon_e/\varepsilon_f \simeq 1/9$, where the subscripts e and f refer to edge-to-edge and face-to-face configurations respectively (the oblate counterparts of side-by-side and end-to-end). This value of $\varepsilon_e/\varepsilon_f$ resulted in the formation of cavities in the MD cell, so $\varepsilon_e/\varepsilon_f$ was set equal to 1/5, the inverse of the value used in the original GB model. The exponents ν and μ were set equal to the values used by Luckhurst's group (Luckhurst et al., 1990) for prolate systems, namely $\nu = 2$, $\mu = 1$. A plot of this potential for three molecular configurations is shown in FIGURE 3.21. A system of 256 molecules was simulated, yielding a phase diagram consisting of isotropic, discotic–nematic, and columnar phases. The latter phase is rectangular at high densities, but has D_{ho} symmetry at low densities.

The GB potential can also be modified to include chirality. This was first done by Tsykalo (Tsykalo, 1985) for the Berne–Pechukas potential, and more recently by Memmer and coworkers (Memmer et al., 1993a, b) for the GB potential. The total potential consists of the GB term plus an additional chiral interaction:

$$U_{\text{chiral}} = 4g\varepsilon(\widehat{u}_i, \widehat{u}_j, \widehat{r}) \times \left(\frac{\sigma_o}{r - \sigma(\widehat{u}_i, \widehat{u}_j, \widehat{r}) + \sigma_o}\right)^7 (\widehat{u}_i \cdot \widehat{u}_j)(\widehat{u}_i \times \widehat{u}_j \cdot \widehat{r}). \tag{3.55}$$

The parameter g is the strength of the chiral coupling, and the remaining parameters have the same definitions as the original GB model. Memmer and coworkers studied 256 particles interacting with this new potential (the GB parameters were chosen to have their original values). They explored the phase diagram by adjusting the chiral coupling g. For $0.6 < g < 0.7$ they found a cholesteric phase, while for $1.0 < g < 1.1$ they found a phase which appears to have the geometric structure of blue phase II. The nature of the phase at higher values of g is not yet clear. Furthermore, it is not yet clear whether the appropriate boundary conditions have been used to study this system. Additional study would be very interesting.

3.4 FUTURE DIRECTIONS

In this final section we discuss some problems in liquid crystal physics that we expect will receive considerable theoretical and computational attention in the coming years.

A considerable portion of the theoretical work profiled in section 3.2 above concerned the behavior of topological defects. One aspect of defect behavior that we did not discuss there was their role in coarsening dynamics. The latter refer to the dynamical behavior of a physical system after a quench from a high-temperature to a low-temperature phase, such as from the isotropic to the nematic phase. The accepted belief among statistical physicists is that the approach of the system to thermal equilibrium in the low-temperature phase is governed at long times by topological defects which are created in large numbers immediately after the quench and must eventually all but disappear as equilibrium is attained. The coarsening dynamics of nematic liquid crystals have received considerable attention from Turok, Yurke and collaborators (Chuang et al., 1991; Pargellis et al., 1991; and Chuang et al., 1991, 1993) as well as others (Blundell and Bray, 1992). Some of this attention was motivated by the mathematical equivalence of nematodynamics and the dynamics of cosmic strings. The striking feature of these studies is the existence of scaling laws for the density of disclination lines, the radius of disclination loops, and the structure factor. However, the numerical values of the exponents in these laws remain a subject of dispute and will certainly warrant further study. There has also been some work done on the coarsening dynamics of smectic C films (Pargellis et al., 1992). An even more interesting problem that has yet to be tackled either experimentally or theoretically is the nature of the coarsening dynamics of phases such as the blue and TGB. We expect that these systems might exhibit interesting coarsening phenomena because the defects must disentangle and form regular arrays rather than completely disappearing.

In the realm of computation we expect the most effort to be devoted to the Gay–Berne model and its chiral version. The structure of disclinations in these models has yet to be probed in detail, nor have systematic attempts been made to study possible blue phases or TGB phases in the chiral model. Coarsening dynamics in the Gay–Berne model will be an extremely fruitful study, though more powerful computers than presently available may be necessary in order to study sufficiently large systems for sufficiently long times.

A problem that has begun to attract attention recently and should do so for quite some time is the nature of the nematic–isotropic transition in porous media. This transition has been studied experimentally using light scattering (Wu et al., 1992) and calorimetry (Bellini et al., 1992). Some theoretical work has also been done (Maritan et al., 1994) using a simple model. In section 3.3.1 above we described a modification of the Lewohl–Lasher model that has been used to simulate this system. There is qualitative agreement between theory and experiment, but more detailed, realistic work remains to be done. The N–I transition in confined geometries is not a sharp one but rather a smooth evolution to a glassy state. Glassy nematic phases can also be produced by rapid quenches in conventional geometries. This problem has received little theoretical attention to date, and is another example where computer simulations might play an important role in elucidating the physics.

REFERENCES

Abrikosov, A. A. 1957. *Zh. Eksp. Teor. Fiz.* **32**: 1442 (1957 *Sov. Phys.–JETP* **5**, 1174).

Aharony, A., Birgeneau, R. J., Brock, J. D., and Litster, J. D. 1986. *Phys. Rev. Lett.* **57**: 1012.

Alder, B. J., and Wainwright, T. E. 1957. *J. Chem. Phys.* **27**: 1208.

Alexander, S., and McTague, J. P. 1978. *Phys. Rev. Lett.* **41**: 702.

Alexander, S. 1981a. in *Symmetry and Broken Symmetry in Condensed Matter Physics*, ed. N. Boccara, p. 141. Paris: IDSET.

Alexander, S., Hornreich, R. M., and Shtrikman, S., 1981b. *Symmetry and Broken Symmetry in Condensed Matter Physics*, ed. N. Boccara, p. 379. Paris: IDSET.

Allen, M. P., and Frenkel, D. 1987. *Phys. Rev. Lett.* **58**: 1748.

Allen, M. P., 1990. *Liq. Cryst.* **8**: 499.

Allen, M. P. 1993. *Phil. Trans. R. Soc. London A* **344**: 323.

Als-Nielsen, J., Birgeneau, R. J., Kaplan, M., Litster, J. D., and Safinya, C. R. 1977. *Phys. Rev. Lett.* **39**: 352.

Als-Nielsen, J. 1981. *Symmetry and Broken Symmetry in Condensed Matter Physics*, ed. N. Boccara, p. 107. Paris: IDSET.

Bahr, C., and Fliegner, D. 1993. *Liq. Cryst.* **14**: 573.

Battacharya, S., and Ketterson, J. B. 1982. *Phys. Rev. Lett.* **49**: 997.

Baumann, C., Marcerou, J. P., Prost, J., and Rouillon, J. C. 1985. *Phys. Rev. Lett.* **54**: 1268.

Baus, M. 1987. *J. Stat. Phys.* **48**: 1129.

Baym, G. A., Bethe, H. A., and Pethick, C., 1971. *Nucl. Phys. A* **175**: 225.

Bellini, T., Clark, N. A., Muzny, C. D., Wu, L., Garland, C. W., Schaefer, D. W., and Oliver, B. J. 1992. *Phys. Rev. Lett.* **69**: 788.

Berker, A. N., and Walker, J. S. 1981. *Phys. Rev. Lett.* **47**: 1469.

Berne, B. J., and Pechukas, P. 1972. *J. Chem. Phys.* **56**: 4213.

Biscarini, F., Zannoni, C., Chicccoli, C., and Pasini, P., 1991. *Mol. Phys.* **73**: 439.

Blundell, R. E., and Bray, A. J. 1992. *Phys. Rev. A.* **46**: 6154.

Brazovskii, S. A., and Dmitriev, S. G. 1979. *Zh. Eksp. Teor. Fiz.* **69:** 979 (1976. *Sov. Phys.–JETP* **42**: 497).

Brazovskii, S. A., and Filev, V. M. 1978. *Zh. Eksp. Teor. Fiz.* **75**: 1140 *Sov. Phys.–JETP* **48**: 573).

Brock, J. D., Aharony, A., Birgeneau, R. J., Evans-Luttrodt, K. W., Litster, J. D., Horn, P. M., Stephenson, G. B., and Tajbakhsh, A. R. 1986. *Phys. Rev. Lett.* **57**: 98.

Caillé, A. 1972. *C.R. Acad. Sci.*, Paris B274, 891.

Chan, K. K., Pershan, P. S., Sorensen, L. B., and Hardouin, F. 1985. *Phys. Rev. Lett.* **54**: 1694.

Chan, K. K., Pershan, P. S., Sorensen, L. B., and Hardouin, F. 1986. *Phys. Rev. A* **34**: 1420.

Chuang, I., Durrer, R., Turok, N., and Yurke, B. 1991a. *Science* **251**: 1336.

Chuang, I., Turok, N., and Yurke, B. 1991b. *Phys. Rev. Lett.* **66**: 2472.

Chuang, I., Yurke, B., Pargellis, A. N., and Turok, N., 1993. *Phys. Rev. E* **47**: 3343.

Cleaver, D. J., Kralj, D., Sluckin, T. J., and Allen, M. P. 1995. *Liquid Crystals in Complex Geometries Formed by Polymer and Porous Networks*, ed. G. P. Crawford and S. Žumer. London: Taylor and Francis.

Collin, D., Gallani, J. L., and Martinoty, P. 1986. *Phys. Rev. A* **34**: 2255.

Colot, J.-L., Wu, X.-G., Xu, H., and Baus, M. 1988. *Phys. Rev. A* **38**: 2022.

Counsell, C. J. R., Emsley, J. W., Heaton, N. J., and Luckhurst, G. R. 1985. *Mol. Phys.* **54**: 847.

Dasgupta, C., and Halperin, B. I. 1981. *Phys. Rev. Lett.* **47**: 1556.

Dasgupta, C. 1983. *Phys. Rev. A* **27**: 1262.

de Gennes, P. G. 1966. *Superconductivity of Metals and Alloys.* New York: Benjamin.

de Gennes, P. G. 1972. *Solid State Commun.* **10**: 753.

de Gennes, P. G. 1975. *J. Physique Lett.* **36**: 55.

de Gennes, P. G. 1977. *Mol. Cryst. Liq. Cryst. Lett.* **34**: 177.

de Gennes, P. G. 1982. *Polymer Liquid Crystals*, ed A. Ciferri, et al., ch. 5. New York: Academic.

de Gennes, P. G., and Prost, J., 1993. *The Physics of Liquid Crystals*, Oxford.

DeMiguel, E., Rull, L. F., Chalam, M. K., and Gubbins, K. E. 1991a. *Mol. Phys.* **74**: 405.

DeMiguel, E., Rull, L. F., Chalam, M. K., Gubbins, K. E., and Van Swol, F. 1991b. *Mol. Phys.* **72**: 593.

DeMiguel, E., Rull, L. F., and Gubbins, K. E. 1992. *Phys. Rev. A* **45**: 3813.

Emerson, A. P. J., Luckhurst, G. R., and Whatling, S. G. 1994. *Mol. Phys.* **82**: 113.

Fabbri, U., and Zannoni, C. 1986, *Mol. Phys.* **58**: 763.

Ferrarini, A., Moro, G. J., Nordio, P. L., and Luckhurst, G. R., 1992. *Mol. Phys.* **77**: 1.

Ferrarini, A., Luckhurst, G. R., Nordio, P. L., and Roskilly, S. J., 1993. *Chem. Phys. Lett.* **214**: 409.

Ferrarini, A., Luckhurst, G. R., Nordio, P. L., and Roskilly, S. J., 1994. *J. Chem. Phys.* **100**: 1460.

Ferrenberg, A. M., and Swendsen, R. H. 1988. *Phys. Rev. Lett.* **61**: 2635.

Flory, P. J. 1956. *Proc. R. Soc. London A* **234**: 73.

Flory, P. J., and Ronca, G. 1979. *Mol. Cryst. Liq. Cryst.* **54**: 289, 311.

Frenkel, D., and Mulder, B. 1985. *Mol. Phys.* **55**: 1171.

Frenkel, D. 1987. *Mol. Phys.* **60**: 1.

Frenkel, D. 1988a. *J. Phys. Chem.* **92**: 3280.

Frenkel, D., Lekkerkerker, H. N. W., and Stroobants, A. 1988b. *Nature* **332**: 822.

Frenkel, D. 1992. *Phase Transitions in Liquid Crystals*, ed. S. Martellucci and A. N. Chester, p. 67. New York: Plenum.

Gay, J. G., and Berne, B. J. 1981. *J. Chem. Phys.* **74**: 3316.

Goldschmidt, Y. Y. 1992. *Recent Progress in Random Magnets*, ed. D. H. Ryan, p. 151. Singapore: World Scientific.

Golubovic, L., and Wang, Z.-G. 1992. *Phys. Rev. Lett.* **69**: 2535.

Grebel, H., Hornreich, R. M., and Shtrikman, S. 1983. *Phys. Rev. A* **28**: 1114; 3669 (Errata).

Grebel, H., Hornreich, R. M., and Shtrikman, S. 1984. *Phys. Rev. A* **30**: 3264.

Grinstein, G., and Pelcovits, R. A. 1981. *Phys. Rev. Lett.* **47**: 856.

Grinstein, G., and Pelcovits, R. A. 1982. *Phys. Rev. A* **26**: 915.

Grinstein, G., and Toner, J. 1984. *Phys. Rev. Lett.* **51**: 2386.

Halperin, B. I., and Lubensky, T. C., 1974. *Solid State Commun.* **14**: 997.

Hardouin, F., Levelut, A. M., Achard, M. F., and Sigaud, G. 1983. *J. Chim. Phys.* **80**: 53.

Hatwalne, Y., Ramaswamy, S., and Toner, J. 1993. *Phys. Rev. Lett.* **70**: 2090.

Helfrich, W. 1978. *J. Physique* **39**: 1199.

Hinshaw, G. A., Petschek, R. G., and Pelcovits, R. A. 1988. *Phys. Rev. Lett.* **60**: 1864.

Hinshaw, G. A., and Petschek, R. G. 1989. *Phys. Rev. A* **39**: 5914.

Holyst, R., and Poniewierski, A. 1989. *Phys. Rev. A* **39**: 2742.

Hornreich, R. M., and Shtrikman, S. 1979. *Bull. Isr. Phys. Soc.* **25**: 46.

Hornreich, R. M., and Shtrikman, S. 1980a. *Liquid Crystals of One- and Two-Dimensional Order*, ed. W. Helfrich and G. Heppke, p. 185. Berlin: Springer.

Hornreich, R. M., and Shtrikman, S. 1980b. *J. Physique* **41**: 335.

Hornreich, R. M., and Shtrikman, S. 1981a. *Phys. Lett.* **82A**: 345.

Hornreich, R. M., and Shtrikman, S. 1981b. *Phys. Lett.* **84A**: 20.

Hornreich, R. M., and Shtrikman, S. 1981c. *J. Physique* **42**: 367 (Errata).

Hornreich, R. M., Kugler, M., and Shtrikman, S. 1982. *Phys. Rev. Lett.* **498**: 1404.

Hornreich, R. M., and Shtrikman, S. 1988. *Mol. Cryst. Liq. Cryst.* **165**: 183.

Indeku, J. O., and Berker, A. N. 1986. *Phys. Rev. A* **33**: 1158.
Indeku, J. O., Berker, A. N., Chiang, C., and Garland, C. W. 1987. *Phys. Rev. A* **35**: 1371.
Indeku, J. O., and Berker, A. N. 1988. *J. Physique* **49**: 353.
Kabadi, V. N. 1986. *Ber. Bunsen-Ges. Phys. Chem.* **90**: 327.
Kamien, R. D., Le Doussal, P., and Nelson, D. R. 1992. *Phys. Rev. A* **45**: 8727.
Khokhlov, A. R. 1978. *Phys. Lett.* **68A**: 135.
Khokhlov, A. R., and Semenov, A. N. 1981. *Physica A* **108**: 546.
Khokhlov, A. R., and Semenov, A. N. 1982. *Physica A* **112**: 605.
Khokhlov, A. R., and Semenov, A. N. 1984. *Macromolecules* **17**: 2678.
Khokhlov, A. R., and Semenov, A. N. 1985. *J. Stat. Phys.* **38**: 161.
Khokhlov, A. R., and Semenov, A. N. 1986. *Macromolecules* **19**: 373.
Kleinert, H., and Maki, K., 1981, *Fortschr. Phys.* **29**: 219.
Kushick, J., and Berne, B. J. 1976. *J. Chem. Phys.* **64**: 1362.
Lammert, P. E., Rokhsar, D. S. and Toner, J. 1993. **70**: 1650.
Landau, L. D. 1937. *Phys. Z. Sowjun.*, **2**: 3207.
Langer, S. A., and Sethna, J. P. 1986. *Phys. Rev. A* **34**: 5035.
Lebwohl, P. A., and Lasher, G. 1972. *Phys. Rev. A* **6**: 426.
Le Doussal, P., and Nelson, D. R. 1991. *Europhys. Lett.* **15**: 161.
Lee, J., and Kosterlitz, J. M. 1991. *Phys. Rev. B* **43**: 1268.
Lubensky, T. C., Tokihiro, T., and Renn, S. R. 1991. *Phys. Rev. A* **43**: 5449.
Luckhurst, G. R., and Simpson, P. 1982. *Mol. Phys.* **47**: 251.
Luckhurst, G. R. 1985. *Recent Advances in Liquid Crystalline Polymers*: ed. p. 105. Amsterdam: Elsevier.
Luckhurst, G. R., Stephens, R. A., and Phippen, R. W. 1990. *Liq. Cryst.* **8**: 451.
Luckhurst, G. R., and Simmonds, P. S. J. 1993. *Mol. Phys.* **80**: 233.
Maclennan, J. 1990. *Europhys. Lett.*, **13**: 435.
McMillan, W. L. 1972. *Phys. Rev. A* **6**: 936.
Marcelja, S. 1974. *J. Chem. Phys.* **60**: 3599.
Maritan, A., Cieplak, M., Bellini, T., and Banavar, J. R. 1994. *Phys. Rev. Lett.* **72**: 4113.
Marko, J. 1988. *Phys. Rev. Lett.* **60**: 325.
Martin, P., Parodi, O., and Pershan, P. S. 1972. *Phys. Rev. A* **6**: 2401.
Mazenko, G. F., Ramaswamy, S., and Toner, J. 1982. *Phys. Rev. Lett.* **49**: 51.
Mazenko, G. F., Ramaswamy, S., and Toner, J. 1983. *Phys. Rev. A* **28**: 1618.
Meiboom, S., Sethna, J. P., Anderson, P. W., and Brinkman, W. F. 1981. *Phys. Rev. Lett.* **46**: 1216.
Meiboom, S., Sammon, M., and Berreman, D. W. 1983a. *Phys. Rev. A* **28**: 3553.
Meiboom, S., Sammon, M., and Brinkman, W. F. 1983b. *Phys. Rev. A* **27**: 438.
Memmer, R., Kuball, H.-G., and Schonhofer, A. 1993a. *Liq. Cryst.* **15**: 345.
Memmer, R., Kuball, H.-G., and Schonhofer, A. 1993b. *Ber. Bunsen-Ges. Phys. Chem.* **97**: 1193.
Meyer, R. B. 1982. *Polymer Liquid Crystals*, ed. by A. Ciferri, et al., ch. 6. New York: Academic.
Mulder, B., and Frenkel, D. 1985. *Mol. Phys.* **55**: 1193.
Mulder, B. 1987. *Phys. Rev. A* **35**: 3095.
Netz, R. R., and Berker, A. N. 1992a. *Phys. Rev. Lett.* **68**: 333.
Netz, R. R., and Berker, A. N. 1992b. *Phase Transitions in Liquid Crystals*: ed S. Martellucci and A. N. Chester, p. 109. New York: Plenum.
Odijk, T. 1983. *Macromolecules* **16**: 1340.
Odijk, T. 1986. *Macromolecules* **19**: 2313.
Onsager, L. 1949. *Ann. N. Y. Acad. Sci.* **51**: 627.
Pargellis, A. N., Turok, N., and Yurke, B. 1991. *Phys. Rev. Lett.* **67**: 1570.
Pargellis, A. N., Finn, P. L., Goodby, J. W., Panizza, P., Yurke, B., and Cladis, P. E. 1992. *Phys. Rev. A* **46**: 7765.
Peierls, R. E. 1935. *Ann. Inst. Henri Poincaré* **5**: 177.
Petschek, R. G., and Terentjev, E. M., 1992. *Phys. Rev. A* **45**: 5775.
Photinos, D. J., Samulski, E. T., and Toriumi, H. 1990. *J. Phys. Chem.* **94**: 4688, 4694.
Poniewierski, A., and Holyst, R. 1988. *Phys. Rev. Lett.* **61**: 2461.
Prost, J. 1979. *J. Physique* **40**: 581.
Prost, J. 1984. *Adv. in Physics* **33**: 1.
Prost, J., and Toner, J. 1987. *Phys. Rev. A* **36**: 5008.
Ramaswamy, S., and Toner, J. 1983. *Phys. Rev. A* **28**: 3159.
Ramaswamy, S. 1984. *Phys. Rev. A* **29**: 1506.
Ratna, B. R., Shashidar, R., and Raja, V. N. 1985. *Phys. Rev. Lett.* **55**: 1476.
Renn, S. R., and Lubensky, T. C. 1988. *Phys. Rev. A* **38**: 2132.
Sammon, M. 1982. *Mol. Cryst. Liq. Cryst.* **7**: 59.
Sediawan, W. B., Gupta, S., and McLaughlin, E. 1988. *Mol. Phys.* **63**: 691.
Sediawan, W. B., Gupta, S., and McLaughlin, E. 1989. *J. Chem. Phys.* **90**: 1888.
Selinger, J. V., and Bruinsma, R. F. 1991a. *Phys. Rev. A* **43**: 2910.
Selinger, J. V., and Bruinsma, R. F. 1991b. *Phys. Rev. A* **43**: 2922.
Selinger, J. V., and Bruinsma, R. F. 1992. *J. Physique II* **2**: 1215.
Selinger, J. V., Wang, Z.-G., Bruinsma, R. F., and Knobler, C. M. 1993. *Phys. Rev. Lett.* **70**: 1139.
Semenov, A. N., and Khokhlov, A. R. 1988. *Usp. Fiz. Nauk* **156**: 427 (1988 *Sov. Phys. Usp.* **32**: 988).
Sethna, J. P. 1985. *Phys. Rev. B* **31**: 6278.
Shashidar, R., and Ratna, B. R. 1989. *Liq. Cryst.* **5**: 421.
Sigaud, G., Hardouin, F., and Achard, M. F. 1979. *Phys. Lett.* **72A**: 24.
Singh, U. P., and Singh, Y. 1986. *Phys. Rev. A* **33**: 2725.
Stone, A. J. 1979. *The Molecular Physics of Liquid Crystals*: ed. G. R. Luckhurst and G. W. Gray, p. 31. New York: Academic.

Stroobants, A., Lekkerkerker, H. N. W., and Frenkel, D. 1986. *Phys. Rev. Lett.* **57**: 1452.
Stroobants, A., Lekkerkerker, H. N. W., and Frenkel, D. 1987. *Phys. Rev. A* **36**: 2929.
Tao, R., Sheng, P., and Lin, Z. F. 1993. *Phys. Rev. Lett.* **70**: 1271.
ten Bosch, A., Maissa, P., and Sixou, P. 1983. *Phys. Lett.* **94A**: 298.
Tinkham, M. 1975. *Introduction to Superconductivity* New York: McGraw-Hill.
Toner, J. 1982. *Phys. Rev. B* **26**: 462.
Toner, J. 1991. *Phys. Rev. B* **43**: 8289.
Toner, J. 1992. *Phys. Rev. Lett.* **68**: 1331.
Tsykalo, A. L. 1985. *Mol. Cryst. Liq. Cryst.* **129**: 409.
Veerman, J. A. C., and Frenkel, D. 1992. *Phys. Rev. A* **45**: 5632.
Vieillard-Baron, J. 1972. *J. Chem. Phys.* **56**: 4729.
Vroege, G. J., and Odijk, T. 1988. *Macromolecules* **21**: 2848.
Walmsley, S. H. 1977. *Chem. Phys. Lett.* **49**: 320.
Wang, X. J., and Warner, M. 1986. *J. Phys. A* **19**: 2215.
Warner, M., Gunn, J. M., and Baumgartner, A. 1985. *J. Phys. A:* **18**: 3007.
Wen, X. 1991. *Phys. Rev. Lett.* **66**: 2887.
Williams, D., and Warner, M. 1990. *J. Physique* **51**: 317.
Wright, D. C., and Mermin, N. D. 1985. *Phys. Rev. A* **31**: 3498.
Wright, D. C., and Mermin, N. D. 1989. *Rev. Mod. Phys.*, **61**: 385.
Wu, X., Goldburg, W. I., Liu, M. X., and Xue, J. Z. 1992. *Phys. Rev. Lett.* **69**: 470.
Zannoni, C. 1986. *J. Chem. Phys.* **84**: 424.
Zarragoicocoechea, G. J., Levesque, D., and Weis, J. J. 1992. *Mol. Phys.* **75**: 989.
Zhang, Z., Mouritsen, O. G., and Zuckermann, M. J. 1992. *Phys. Rev. Lett.* **69**: 1803.
Zhang, Z., Zuckermann, M. J., and Mouritsen, O. G. 1993. *Mol. Phys.* **80**: 1195.

CHAPTER 4

Phase Structures and Transitions in Thermotropic Liquid Crystals

PETER J. COLLINGS

4.1 PHASE STRUCTURES

4.1.1 Nematic Liquid Crystals

4.1.1.1 Orientational Order

Since there is no long-range positional order in nematic liquid crystals, one is concerned exclusively with long-range orientational order. This is best described by an orientational distribution function, $f(\Omega)$, which is proportional to the probability of finding the long axis of a molecule oriented at solid angle Ω relative to a coordinate frame in which the z-axis points along the director, $\hat{n}$. Since the nematic phase is almost always uniaxial, $f(\Omega)$ does not depend on the azimuthal angle ϕ, but only on the polar angle θ. Therefore, complete knowledge of the long-range orientational order in a nematic liquid crystal comes from knowledge of the orientational distribution function, $f(\theta)$.

In order to generate an order parameter, averages of certain functions must be used. Since the director in a nematic liquid crystal can point in either of two directions, a logical choice for these functions is the even members of the series of Legendre polynomials:

$$\begin{aligned} P_2(\cos\theta) &= \tfrac{3}{2}\cos^2\theta - \tfrac{1}{2} \\ P_4(\cos\theta) &= \tfrac{1}{8}(35\cos^4\theta - 30\cos^2\theta + 3). \end{aligned} \tag{4.1}$$

The average of $P_2(\cos\theta)$ is called the orientation order parameter and given the symbol S.

$$S = \int_0^\pi \left(\frac{3}{2}\cos^2\theta - \frac{1}{2}\right)\sin\theta\,\mathrm{d}\theta \Big/ \int_0^\pi \sin\theta\,\mathrm{d}\theta. \tag{4.2}$$

If the molecules are perfectly ordered, then $S = 1$. If the molecules are randomly ordered as is true for the isotropic liquid phase, $S = 0$. The orientational order parameter can be determined using just about any method which measures the macroscopic anisotropy of the liquid crystal. This is described in the next paragraph. The average of $P_4(\cos\theta)$ can be measured by Raman scattering, elastic x-ray and neutron scattering, and by electron spin resonance. These methods give somewhat conflicting results, but seem to indicate that $P_4(\cos\theta)$ starts out small or even negative near the isotropic transition and grows more positive as the temperature is decreased (Jen et al., 1977; Miyano, 1978; Kohli et al., 1976; and Luckhurst and Yeates, 1976).

The relationship between the order parameter and the anisotropy of a macroscopic property can be illustrated quite easily. Let T_{xx}, T_{yy}, and T_{zz} be the three components of a molecular tensor property, such as the diamagnetic susceptibility. Let $T'_{x'x'}$, $T'_{y'y'}$, and $T'_{z'z'}$ be the components of the corresponding macroscopic tensor property, with the z'-axis along the director. If the direction cosines giving the orientation of molecular axis i relative to macroscopic direction i' are denoted by $A_{i'i}$ then the transformations relating T_{ij} to $T_{i'j'}$ are as follows.

$$\begin{aligned}\langle T'_{x'x'}\rangle &= \langle A^2_{x'x}\rangle\langle T_{xx}\rangle + \langle A^2_{x'y}\rangle\langle T_{yy}\rangle + \langle A^2_{x'z}\rangle\langle T_{zz}\rangle\\ \langle T'_{y'y'}\rangle &= \langle A^2_{y'x}\rangle\langle T_{xx}\rangle + \langle A^2_{y'y}\rangle\langle T_{yy}\rangle + \langle A^2_{y'z}\rangle\langle T_{zz}\rangle\\ \langle T'_{z'z'}\rangle &= \langle A^2_{z'x}\rangle\langle T_{xx}\rangle + \langle A^2_{z'y}\rangle\langle T_{yy}\rangle + \langle A^2_{z'z}\rangle\langle T_{zz}\rangle.\end{aligned} \quad (4.3)$$

The brackets denote an average over time and an assumption has been made that the intramolecular and intermolecular motions are uncorrelated. Equation (4.3) allows the relationship between the anisotropies in the microscopic and macroscopic properties to be determined.

$$\begin{aligned}\langle \Delta T'\rangle &= \langle T_{z'z'}\rangle - \tfrac{1}{2}(\langle T'_{x'x'}\rangle + \langle T'_{y'y'}\rangle)\\ \langle \Delta T'\rangle &= \tfrac{3}{2}(\langle A^2_{z'x}\rangle\langle T_{xx}\rangle + \langle A^2_{z'y}\rangle\langle T_{yy}\rangle\\ &\quad + \langle A^2_{z'z}\rangle\langle T_{zz}\rangle) - \tfrac{1}{2}\mathrm{Tr}\langle \overleftrightarrow{T}\rangle.\end{aligned} \quad (4.4)$$

If three order parameters are defined as follows,

$$S_{xx} = \tfrac{3}{2}\langle A^2_{z'x}\rangle - \tfrac{1}{2} \qquad S_{yy} = \tfrac{3}{2}\langle A^2_{z'y}\rangle - \tfrac{1}{2} \qquad S_{zz} = \tfrac{3}{2}\langle A^2_{z'z}\rangle - \tfrac{1}{2} \quad (4.5)$$

then the anisotropy can be written in terms of these order parameters. In fact, the sum of these three order parameters is zero, so that only two are necessary to describe the relationship between macroscopic and microscopic tensor properties. The usual choice for the two independent order parameters is $S = S_{zz}$, the orientational order parameter described previously, and $D = S_{yy} - S_{xx}$. With these choices, the macroscopic anisotropy has a simple form:

$$\begin{aligned}\langle \Delta T'\rangle &= (\langle T_{zz}\rangle - \tfrac{1}{2}[\langle T_{xx}\rangle + \langle T_{yy}\rangle])S\\ &\quad + (\langle T_{yy}\rangle - \langle T_{xx}\rangle)D.\end{aligned} \quad (4.6)$$

As noted before, S measures the tendency for the longest axis of the molecule to project along $\hat{n}$, while D measures the difference in tendencies for the two transverse molecular axes to project along $\hat{n}$.

Macroscopic properties for which this approach has been used include magnetic susceptibility, electric polarizability, dipolar and quadrupolar NMR, linear dichroism, Raman scattering, birefringence, and EPR. Since the molecular quantities in the parentheses of (4.6) must be known in order to determine the order parameter, some methods are more useful than others (de Gennes and Prost, 1993). Quadrupolar NMR also allows the order parameter for different parts of the molecule to be measured. For example, order parameters for parts of the hydrocarbon chain on the end of liquid crystal molecules tend to decrease the further away they are from the molecular core (Emsley, 1985).

Theoretically, noncylindrically symmetric molecules can form a biaxial nematic phase, that is, one in which all directions perpendicular to the director are not equivalent (Chandrasekhar, 1992). Such a phase has been found in both lyotropic (Yu and Saupe, 1980) and thermotropic systems (Praefcke et al., 1991).

4.1.1.2 Short-Range Positional Order

Positional order is best described through the density–density correlation function (ρ is the density):

$$G(\vec{r} - \vec{r}') = \langle \rho(\vec{r})\rho(\vec{r}')\rangle - \bar{\rho}^2. \quad (4.7)$$

When this correlation function does not decay to zero as $\vec{r} - \vec{r}' \to \infty$, then true long-range order is present. If the correlation function decays to zero as $|\vec{r} - \vec{r}'|^{-\eta}$, where η may be temperature-dependent, then the positional order is quasi-long-range. Finally, when the decay of the correlation function is exponential, $e^{-|\vec{r} - \vec{r}'|/\xi}$, where ξ is a correlation length, then the order is short-range only. Long-range order is typical of solids, short-range order is the norm for liquids, and all three forms of positional order are found in liquid crystals (Pershan, 1988).

X-ray scattering is the most valuable tool in measuring the positional order in liquid crystals. High resolution, intense sources, and the ability to produce monodomain samples have combined to produce detailed information on the structure of phases. Since the x-ray scattering cross-section can be expressed in terms of the Fourier transform of the density–density correlation function, the location of the scattering peak in reciprocal space determines the spatial periodicity present, while the structure of the peak indicates the spatial extent of this periodicity.

The positional order of nematic liquid crystals is short-range, but anisotropic. That is, the x-ray scattering cross-section of an oriented monodomain sample shows two diffuse spots. One represents a wavevector $q_{\|} = 2\pi/L$ along the director, where L is on the order of a molecular length, and the other represents a wavevector $q_{\perp} = 2\pi/d$ perpendicular to the director, where d is on the order of a molecular width. This indicates that there are two correlation lengths for short-range positional order in nematics, $\xi_{\|}$ along the director, and $\xi_{\perp}$ perpendicular to the director. The diffuse spot along the director sometimes grows in intensity and sharpens as the temperature is lowered in the nematic phase and a smectic A phase is approached. This is a precursor to the ordering present in smectic layers, and exactly how this order builds up has been the focus of a great deal of both theoretical and experimental research. This will be discussed at length in a later section.

4.1.2 Chiral Nematic Liquid Crystals

4.1.2.1 Intrinsic Twist

Although the nematic phase is characterized by a director which does not vary spatially, there are many liquid crystals with spatially varying directors. Most of these involve twist, so a general description of twist in liquid crystals is required. Liquid crystal phases with intrinsic twist are possible whenever the molecules of the liquid crystal are chiral, that is, lack a center of inversion symmetry. While chiral liquid crystal molecules possess both chiral and achiral liquid crystal phases, no achiral molecules form chiral liquid crystal phases. Perhaps the most convenient formulation to describe the twist is through the variation of a macroscopic tensor property, such as the anisotropic part of the electric susceptibility, ε_{ij}.

Imagine a chiral nematic liquid crystal with a pitch P and a helical axis directed along the z-axis. The components of the director $\hat{n}$ can be expressed quite easily:

$$n_x = \cos(2\pi z/P) \qquad n_y = \sin(2\pi z/P) \qquad n_z = 0. \tag{4.8}$$

The anisotropic part of the electric susceptibility is given by $\varepsilon_{ij} = \varepsilon_a(n_i n_j - \delta_{ij}/3)$ where ε_a is the anisotropy or difference between the electric susceptibility parallel and perpendicular to the director, $\varepsilon_{\|} - \varepsilon_{\perp}$. This means that ε_{ij} takes the following form:

$$\varepsilon_{ij} = \frac{\varepsilon_a}{3}\begin{pmatrix} \frac{1}{2}+\frac{3}{2}\cos\left(\frac{4\pi z}{P}\right) & \frac{3}{2}\sin\left(\frac{4\pi z}{P}\right) & 0 \\ \frac{3}{2}\sin\left(\frac{4\pi z}{P}\right) & \frac{1}{2}-\frac{3}{2}\cos\left(\frac{4\pi z}{P}\right) & 0 \\ 0 & 0 & -1 \end{pmatrix}. \tag{4.9}$$

It is important to see this expression for the anisotropic part of a tensor property for a chiral nematic liquid crystal as a specific example for a more general treatment of the problem. Since ε_{ij} is a symmetric tensor with zero trace, five independent quantities are necessary to specify it in general. For chiral systems, the most convenient way to do this is to define five basis tensors, each isomorphic to the spherical harmonics of order two (Brazovskii and Dmitriev, 1975; and Brazovskii and Filev, 1978):

$$T_{ij}^{\pm 2} = \frac{1}{2}\begin{pmatrix} 1 & \pm i & 0 \\ \pm i & -1 & 0 \\ 0 & 0 & 0 \end{pmatrix}$$

$$T_{ij}^{\pm 1} = \frac{1}{2}\begin{pmatrix} 0 & 0 & 1 \\ 0 & 0 & \pm i \\ 1 & \pm i & 0 \end{pmatrix} \tag{4.10}$$

$$T_{ij}^{0} = \frac{1}{\sqrt{6}}\begin{pmatrix} -1 & 0 & 0 \\ 0 & -1 & 0 \\ 0 & 0 & 2 \end{pmatrix}.$$

The anisotropic part of the electric susceptibility can then be written in terms of the basis tensors,

$$\varepsilon_{ij} = \frac{1}{2}\sum_{m=-2}^{+2}\left(\varepsilon_m T_{ij}^m \, e^{i\vec{q}\cdot\vec{r}} + \varepsilon_m^* T_{ij}^{m^*} \, e^{-i\vec{q}\cdot\vec{r}}\right) \tag{4.11}$$

where ε_m is the amplitude of the "structural mode" described by each basis tensor and $m = -2, -1, 0, +1, +2$ (Grebel et al., 1983, 1984). The $m = \pm 2$ modes describe a planar spiral variation of the director, with the -2 mode being right-handed and the $+2$ mode being left-handed. The $m = \pm 1$ modes describe a conical spiral variation of the director, with the principal

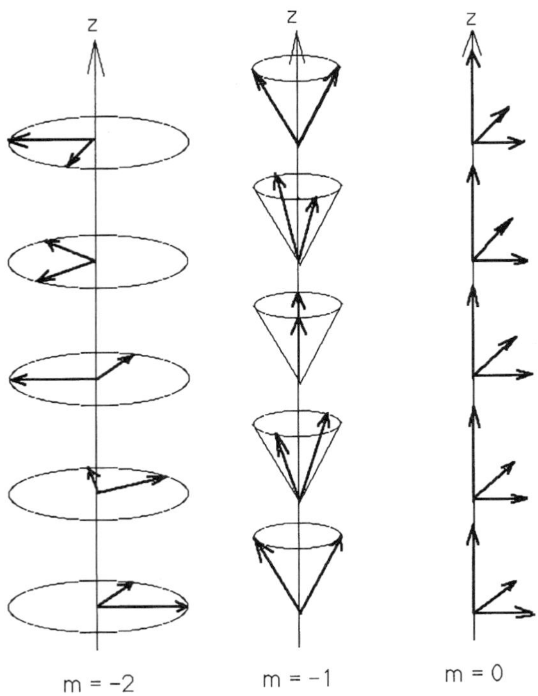

FIGURE 4.1. Three of the five structural "modes" useful in describing the orientational order in a chiral liquid crystal. The two additional "modes" are the $m = +2$ and $m = +1$ "modes," which are the left-handed analogs of the $m = -2$ and $m = -1$ "modes," respectively.

axes of the director making a constant angle of 45° to the helical axis. The $m = -1$ mode is right-handed and the $m = +1$ mode is left-handed. The $m = 0$ mode describes the achiral mode appropriate for a nematic liquid crystal. Using this formulation, ε_{ij} for the chiral nematic liquid crystal described by (4.9) is just a combination of the $m = -2$ and $m = 0$ modes with ε_m equal to $-\varepsilon_a/\sqrt{6}$ and $\varepsilon_a/2$, respectively, and with $q = 4\pi/P$. Figure 4.1 shows the principal axes for the $m = -2$, -1, and 0 structural modes.

While the smallest pitch encountered is about 100 nm (Yang and Crooker, 1987), there is no bound as to how long the pitch can be. In fact, mixing equal amounts of the stereoisomers produces a nematic phase, that is, a chiral nematic phase with an infinite pitch. The uniform variation in the refractive index due to the helical structure produces spectacular optical effects for light with a wavelength in the liquid crystal equal to the pitch. These include the reflection of circularly polarized light and anomalous optical activity (Chandrasekhar, 1992; and de Gennes and Prost, 1993). The fact that at normal incidence there are no higher order reflections indicates that the components of the director are well described by a sinusoidal function.

4.1.2.2 Local Nematic Order

Since the pitch is at least ten times larger than a typical molecular width, a thin slice perpendicular to the helical axis several molecular widths wide possesses orientational order which is very similar to a nematic. For this reason, chiral nematic liquid crystals are described as twisted nematic phases. It should be kept in mind, however, that the uniaxial symmetry of the nematic phase is broken by the twist, that is, one direction perpendicular to the local director is parallel to the helical axis and the other orthogonal direction is perpendicular to the helical axis. Since the twist is usually small on a molecular scale, this is a very small effect in most cases.

There is at least one situation where the fact that a chiral nematic is not locally uniaxial does become important. When the coherence length for orientational fluctuations and the pitch become comparable, the biaxial character of the local orientational order is crucial and new phases become stable. These are the so-called blue phases and are discussed later in the chapter.

4.1.3 Smectic Liquid Crystals

4.1.3.1 Positional Order

Short-range, quasi-long-range, and long-range positional order all occur in smectic liquid crystals. While the dominant positional order is a tendency to form a layered structure, positional order of various kinds can occur in the two directions parallel to the layers. Of course, when long-range order in all three directions is present, the phase is crystalline, but in many liquid crystalline materials a number of crystalline phases form in which the amount of disorder is quite high. These phases are characterized by only a small number of independent x-ray reflections, and large entropy and volume changes at the transition to the highly ordered crystal phase at lower temperatures.

The smectic A phase possesses positional order in one dimension, that is, there is a tendency for the molecules to be in layers which have normals parallel to the director. De Gennes first represented this order as a sinusoidal variation in the density:

$$\rho(\vec{r}) = \bar{\rho} + \mathrm{Re}\{\Psi \cdot e^{i(2\pi/L)z}\}. \tag{4.12}$$

Here L is the spacing between the layers and z is along the director. Ψ is a complex order parameter: $|\Psi|$ is a measure of the change in density in going from layer to layer, and the phase of Ψ describes where the layers are along the z-axis (de Gennes, 1972).

It has long been known that true long-range order in only one dimension is not possible (Peierls, 1934; and Landau, 1965). These arguments can be extended to liquid crystalline systems (Caille, 1972), where it is shown that the density–density correlation function decays as some power of the distance (quasi-long-range order). Experimental confirmation of this has been achieved in both thermotropic and lyotropic systems (Als-Nielsen et al., 1980; and Safinya et al., 1986). The positional order within the layers is short-range in smectic A liquid crystals, with correlation lengths of only a few molecular widths.

4.1.3.2 Non-polar Smectics

Positional order in the plane perpendicular to the director can take many forms. In the hexatic smectic B phase, this positional order is quasi-long-range in the direction normal to the layers and short-range, but with much longer correlation lengths than in the smectic A phase, within the plane of the layers. The interesting additional feature is the presence of long-range bond orientational order within the layers. Bond orientational order implies that in spite of the fact that positional order within the layers is only short-range and is weakly correlated from layer to layer, the in-plane crystallographic axes are maintained over large distances both parallel and perpendicular to the layers. Locally the molecules are packed with hexagonal symmetry, and the orientation of the sixfold symmteric axes is maintained with true long-range order. The fact that a crystal with two-dimensional order can melt to a phase with hexatic order was first proposed by a number of workers (Halperin and Nelson, 1978; Nelson and Halperin, 1979; and Young, 1979). The reason for this is an instability against the spontaneous creation of dislocations, which produces short-range positional order but maintains long-range order in the orientation of the crystallographic axes.

With the presence of long-range positional order in the plane perpendicular to the director, disordered crystalline phases are formed. The crystal B phase is similar to the hexatic B phase, but now long-range positional order exists both within the layers and from layer to layer. While the molecular centers are packed with hexagonal symmetry within the layers, the molecules are free to rotate about the normal to the layers. This rotational freedom is frozen out in the crystal E phase, where the transverse axes of the four nearest neighbors are rotated by 90° relative to the transverse axes of the center molecule. All molecules undergo rapid rotations of 180° about their long axes. This results is what is called a herringbone pattern within the layers, and a lattice that has rectangular symmetry. Both the crystal B and E phases show sharp Bragg-like x-ray reflections, indicative of structures with long-range order (Leadbetter, 1987; and Pershan, 1988).

There are also versions of these phases in which the director is not normal to the layers, but tilted at an angle of up to 45°. The tilted analog of the smectic A phase is called the smectic C phase. In many cases the tilt angle increases from a very small value just below the smectic A to smectic C transition to a value typically around 30°. There is still short-range positional order within the planes. The hexatic smectic F and I phases are tilted versions of the hexatic smectic B phase. In the hexatic smectic F phase, the molecular tilt is toward the midpoint of the line connecting two nearest neighbors; in the hexatic smectic I phase, the molecular tilt is toward a nearest neighbor. Since these phases have long-range bond orientational order, the direction of tilt is maintained over large distances both within and perpendicular to the planes. The crystal G and K phases are the tilted analogs to the crystal B phase, with the tilt being toward the side of the hexagon in the crystal G phase, and toward the apex of the hexagon in the crystal J phase. The molecules in both phases rotate freely about their long molecular axes. Finally, the crystal H and K phases correspond to tilted versions of the crystal E phase. In the crystal H phase the tilt is toward a next nearest neighbor, like the crystal G phase. In the crystal K phase, the tilt is toward a nearest neighbor, like the crystal J phase. Rotation motion about the long molecular axis is restricted as in the crystal E phase (Leadbetter, 1987; and Pershan, 1988).

Finally, there are chiral versions of these tilted smectic and crystal phases. This occurs if the molecule itself is chiral, that is, it lacks inversion symmetry. Such molecules can form smectic phases which are nonchiral, such as the smectic A phase, but tilted smectic phases of these molecules have a helical structure. If the normal to the smectic layers is the z-axis, then the director rotates about a cone from layer to layer. The generating angle of the cone is just the tilt angle of the smectic phase. The smallest pitch for a chiral smectic C phase (denoted smectic C^*) is around 300 nm.

A chiral smectic C liquid crystal with a pitch of P and a helical axis in the z-direction has components of the director in all directions at some point in the structure:

$$n_x = \sin\theta\cos(2\pi z/P) \qquad n_y = \sin\theta\sin(2\pi z/P) \qquad n_z = \cos\theta \tag{4.13}$$

Therefore, ε_{ij} takes the following form:

$$\varepsilon_a \begin{pmatrix} \sin^2\theta\cos^2\left(\frac{2\pi z}{P}\right) - \left(\frac{1}{3}\right) & \sin^2\theta\sin\left(\frac{2\pi z}{P}\right)\cos\left(\frac{2\pi z}{P}\right) & \sin\theta\cos\theta\cos\left(\frac{2\pi z}{P}\right) \\ \sin^2\theta\sin\left(\frac{2\pi z}{P}\right)\cos\left(\frac{2\pi z}{P}\right) & -\sin^2\theta\sin^2\left(\frac{2\pi z}{P}\right) - \left(\frac{1}{3}\right) & \sin\theta\cos\theta\sin\left(\frac{2\pi z}{P}\right) \\ \sin\theta\cos\theta\cos\left(\frac{2\pi z}{P}\right) & \sin\theta\cos\theta\sin\left(\frac{2\pi z}{P}\right) & \cos^2\theta - \left(\frac{1}{3}\right) \end{pmatrix}. \tag{4.14}$$

As in the case of chiral nematics, ε_{ij} can be expressed as a linear combination of the structural modes described by the tensor in (4.10). Three modes contribute and the amount of each depends on the tilt angle. For a right-handed helicoidal phase, the amount of $m = 0$ contribution is proportional to $(3\cos^2\theta - 1)/2$, the amount of $m = -1$ contribution is proportional to $\sin\theta\cos\theta$, and the amount of $m = -2$ contribution is proportional to $\sin^2\theta$, where θ is the tilt angle.

Phases of Non-polar Molecules with Positional Order

Non-tilted	Tilted	Chiral	Positional Order: Between Planes	Positional Order: Within Planes	Bond Orientational Order
Smectic A	Smectic C	Smectic C*	S	S	S
Hexatic B	Smectic F,I	Sm. F*,I*	Q	S	L
Crystal B	Crys. G,J	Crys. G*,J*	L	L	L
Crystal E	Crys. H,K	Crys. H*,K*	L	L	L

S = short-range order Q = quasi-long-range order L = long-range order

FIGURE 4.2. Chart of layered liquid crystal and "soft crystal" phases.

There are also antiferroelectric and ferrielectric forms of the chiral smectic C phase (Chandari et al., 1989; Gorecka et al., 1990; and Inui et al., 1990). In the antiferroelectric phase, even though the director rotates in helical fashion along the direction perpendicular to the layers, neighboring layers have in-layer projections of the director which are approximately in opposite directions. The electric polarization is in the plane of the layer, perpendicular to the director, and antiparallel in neighboring layers, producing an antiferroelectric phase. The ferrielectric phase is thought to be similar to the chiral smectic C antiferroelectric phase, but with unequal numbers of layers with polarization in opposite directions. This creates a phase with net electric polarization. These helicoidal phases are both scientifically and technologically important, and are discussed in Chapter 5.

Figure 4.2 summarizes the phases found in nonpolar liquid crystals along with the most important properties of each.

4.1.3.3 Polar Smectics

The smectic phases formed by highly polar molecules or mixtures of highly polar molecules include some of the phases described for nonpolar molecules, but also include a number of additional phases. When a smectic A or smectic C phase is present with complete head-to-tail disorder within the layers, then these phases are called smectic A_1 or smectic C_1 if the molecules are polar. However, polar molecules also form smectic A and C phases in which there is antiferroelectric ordering of pairs of layers. Since the spatial periodicity is two layers in these phases, they are called smectic A_2 and smectic C_2. In some systems, there is a good deal of molecular association of pairs of molecules (parallel long axes, partial overlap). A smectic phase with a periodicity of between one and two molecular lengths sometimes occurs in these compounds and is called the smectic A_d or smectic C_d phase, depending on the orientation of the director relative to the layer normal. Finally, a modulated phase occurs in some highly polar systems, in which antiferroelectric ordering of neighboring layers is modulated in two directions, one parallel to the director and the other perpendicular to the director. These are called the smectic $\tilde{A}$ and smectic $\tilde{C}$ phases or smectic A and C fluid antiphases. All of the smectic phases of polar molecules are sometimes referred to as frustrated smectic phases. This term stems from the fact that one periodicity is locked in for each phase; the other periodicities are frustrated and are absent.

These phases are usually identified from x-ray scattering experiments. The smectic A_1 phase produces an on-axis quasi-Bragg peak at a reciprocal wavevector of $2\pi/L$, where L is on the order of a molecular length. The x-ray pattern for the smectic A_2 phase shows a reciprocal wavevector at π/L, while the smectic A_d phase has a peak at $2\pi/L'$, where L' is between L and $2L$. All these peaks are on-axis. The smectic $\tilde{A}$ phase shows weak, diffuse off-axis peaks. In fact, these spots can be seen in the smectic A_1 phase, near the transition to the smectic $\tilde{A}$ phase (Young et al., 1994). The smectic C phases corresponding to these A phases have similar x-ray reflections, but the tilt of the director relative to the layer normal shifts the peaks off-axis.

X-ray measurements on these highly polar smectic systems have also revealed that phases with more than one periodicity may occur. In the smectic A_{inc} phase (inc = incommensurate), peaks indicating both A_d and A_2 ordering are present (Ratna et al., 1985). Likewise, the smectic A_{cren} phase (cren = crenelated) shows peaks indicating a modulation of the antiferroelectric order as in the smectic $\tilde{A}$ phase, but with regions of opposite polarization not equal in size (Levelut, 1984). The existence of both of these phases is doubtful, as revealed by recent experiments which showed that these phases are probably broad two-phase regions due to extremely slow conversion of one phase to the other (Kumar et al., 1991; Patel et al., 1992; and Young et al., 1994). For example, following a change in temperature across a phase transition in a mixture of a nonpolar and polar liquid crystal, x-ray peak intensities changed with a time constant of about half a day.

A phenomenological theory is useful in understanding the existence of these phases in polar smectic liquid crystals (Prost, 1984; Wang and Lubensky, 1984; and Lubensky et al., 1988). In this theory, two order parameters are used to describe the tendency of the molecules to form layers. The first is the normal smectic order parameter describing the amplitude of the mass density wave. The second is the amplitude of a wave describing the antiparallel association of either molecules or layers. While the first order parameter is associated with a wavevector equal to $2\pi/L$, where L is a molecular length, the second order parameter is associated with a wavevector equal to $2\pi/L'$, where L' is between one and two molecular lengths. Terms in the free energy which include both order parameters compete with the elastic energy and produce several phases: the A_1 phase with only the first order parameter nonzero; the A_2 phase with both order parameters about equal; and the A_d phase with the second order parameter significantly greater than the first. This theory also predicts that smectic phases with incommensurate wavevectors are possible when the mismatch between the two wavevectors is great.

One other property of polar smectics is the existence of reentrant phases. First discovered in a mixture of two polar compounds, cooling the sample resulted in the phase sequence isotropic–nematic–smectic A_d–nematic–solid (Cladis, 1975). The most extreme example of reentrant behavior is probably the compound 4-nonyloxyphenyl-4′-nitrobenzoyloxybenzoate, which has the following phase sequence on cooling: iso–N–A_d–N–A_d–N–A_1–$\tilde{C}$–A_2–C_2–solid (Tinh et al., 1982). Such behavior can be understood both in terms of mean-field theory if certain coefficients have appropriate temperature and pressure dependences (Prost, 1979) and by packing considerations for molecules with a tendency for antiparallel pairing (Cladis, 1980).

One final phase which is worthy of note is the "smectic D" phase, first discovered in a laterally substituted carboxylic acid compound. This phase is optically isotropic and occurs between the smectic C and smectic A phases or between the smectic C and isotropic phases. Recent x-ray measurements have shown that it is a three-dimensional cubic crystal, with a unit cell of about 800 molecules. A second type of "cubic smectic" phase has also been found in nonsubstituted, nonpolar liquid crystals. This phase exists between the crystal and smectic C phase and seems to have a unit cell about half the size of the first "smectic D" phase described. Clearly these phases possess three-dimensional order and should be referred to as crystals rather than smectic phases (Leadbetter, 1987; and Pershan, 1988).

4.1.3.4 Mixture and Pressure Studies

The study of mixtures has been an important component of research into smectic phases. Typically, various concentrations of two compounds are prepared and the phases present as a function of temperature are determined. This results in a phase diagram for the mixture. If a phase of each pure compound is connected across the phase diagram, then that is good evidence that the phases in each pure compound are in fact the same. Much more important, however, is the fact that firstly, "induced phases" are sometimes present for some concentration range of the mixture when that phase does not exist in either of the pure compounds, and secondly, phase boundaries in the mixtures show interesting topologies when either the two pure compounds possess different phase sequences, or induced phases are present. Much of the work on reentrant phases has been done in mixtures, and mixtures of polar and nonpolar smectic liquid crystals have provided a wealth of information on the polymorphism in smectic liquid crystals. A good example of the polymorphism in a binary mixture is shown in FIGURE 4.3, where four different smectic phases are present, with one of them occuring in two regions (Levelut et al., 1981).

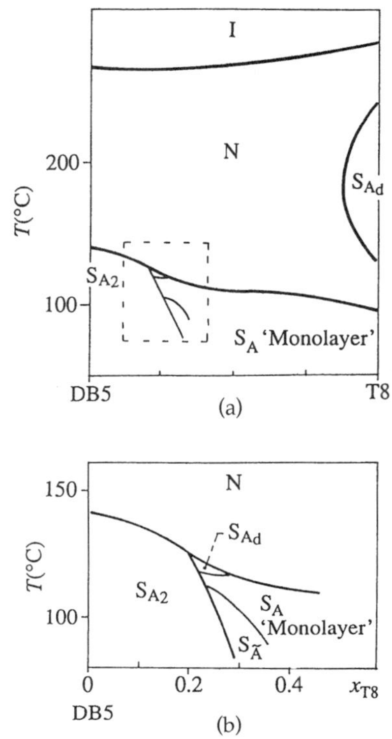

FIGURE 4.3. Polymorphism in a binary mixture of two liquid crystals. The entire phase diagram is shown in (a) and an enlarged inset is shown in (b). (From Levelut et al., 1981.)

Pressure studies have also been important in understanding smectic phases. Because high-pressure experiments are much more difficult than mixture experiments, many fewer of them have been performed. One of the important consequences of these pressure experiments has been to confirm many of the findings coming out of experiments on mixtures. While pressure is a true thermodynamic variable which can be changed for a system of a single composition, the concentration in a mixture experiment suffers from the fact that the actual system being studied changes as the concentration is varied. When the behavior of a single compound under presssure corresponds to the behavior of a mixture as the concentration is changed, there is reason to have some faith that the behavior can be explained in simple terms. A good example of this is the pressure experiment on the reentrant behavior of a single polar compound that compared favorably with the experiments on mixtures of two similar polar compounds (Cladis et al., 1977).

4.1.4 Defect Phases

4.1.4.1 Blues Phases

While chiral nematics can be viewed as slightly twisted nematics in systems where the chirality is low (and the pitch is long), such is not the case for highly chiral systems. The presence of a large additional term in the free energy promotes the formation

(a) (b)

FIGURE 4.4. Double twist cylinder, in which the director rotates by 45° in all directions radial to the central axis. (a) Top view; (b) side view.

of a double twist structure rather than the single twist structure typical of the chiral nematic phase. In the double twist structure, the director rotates about every axis in a plane as shown in FIGURE 4.4(a). The free energy of such a structure is lower than the free energy of the chiral nematic structure only near the center. This means that macroscopic phases with double twist must be comprised of many of these structures. For example, if the double twist structure ends when the director is rotated by 45°, then a macroscopic phase can be made from many of these double twist cylinders (see FIGURE 4.4(b)). Although such double twist cylinders which meet at right angles can have a continuous rotation of the director in going from one cylinder to the next, the points in the region between three orthogonal double twist cylinders possess a defect where the orientational order goes to zero and the director is undefined. Examples of a simple cubic structure and a body centered structure composed of double twist cylinders are shown in FIGURES 4.5(a) and (b), respectively. The defects in both structures are line defects in which the director rotates by 180° in going around the defect. The lattice constant for these structures is roughly one half the pitch. One can also imagine a random assortment of double twist cylinders, with an accompanying arrangement of line defects typical of an amorphous phase. As has been pointed out, chiral structures are inherently biaxial, although the biaxiality in chiral nematics is very small and of little importance. Such is not the case for the

(a)

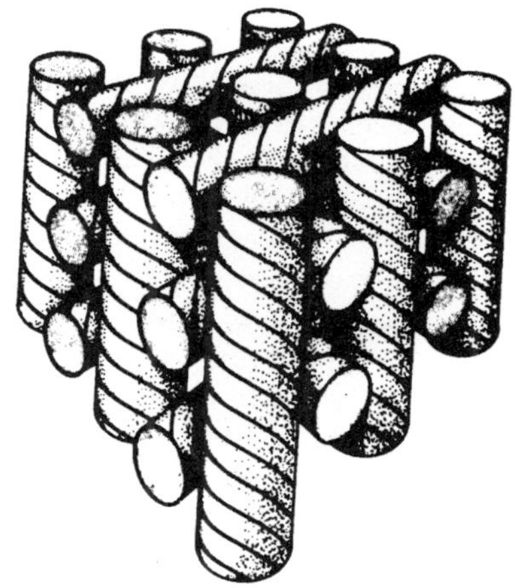

(b)

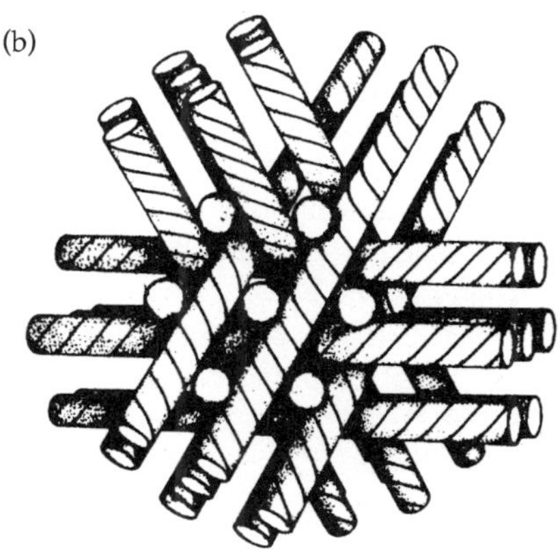

FIGURE 4.5. Two possible defect phases composed of double twist cylinders. (a) a simple cubic structure; (b) a body centered cubic structure. (From Grebel et al., 1984.)

phases of highly chiral substances made from double twist cylinders.

These three structures (simple cubic, body centered cubic, and amorphous) exist in chiral liquid crystals if the pitch is small. Usually they are stable in a very narrow temperature region (on the order of 1 K) between the chiral nematic and isotropic phases. These phases are called the blue phases, because they were first discovered in materials which reflected ultraviolet light in the chiral nematic phase and blue light in the phases with cubic symmetry. The color of reflected light is determined by the pitch of the structure in exact analogy to x-ray diffraction from crystals, so blue phases which reflect at wavelengths throughout the visible spectrum have been observed. Several recent reviews excellently summarize the theoretical and experimental work on blue phases (Wright and Mermin, 1989; Crooker, 1989; Hornreich and Shtrikman, 1989; Seideman, 1990; and Keyes, 1991).

A useful way to view the stability of the blue phases is through experiments on mixtures of the two optical isomers of the same chiral compound. In some senses these mixtures should be thermodynamically identical except for the pitch, which is a minimum for either of the pure isomers and infinite for the racemic mixture. FIGURE 4.6 shows the phases which are stable as the chiral fraction (amount of pure isomer divided by the sum of the amounts of pure isomer and racemic mixture) is increased from zero to one for one particular liquid crystal. Notice (1) that at a certain chiral fraction, the body centered cubic phase (BPI) forms; (2) that at a higher chiral fraction, the simple cubic phase (BPII) forms; (3) that at a higher chiral fraction, the amphorous phase (BPIII) forms; and (4) that at an even higher chiral fraction, BPII becomes unstable, leaving only BPI and BPIII. In compounds with a longer pitch, the pure isomer may possess only BPI, or BPI and BPII, or BPI, BPII, and BPIII. These compounds therefore represent a region of the phase diagram shown in FIGURE 4.6 extending from the left-hand axis to some point in the middle of the diagram. All compounds which have been investigated to date, however, follow this scheme of BPI first, BPII second, BPIII third, and loss of BPII as the chiral fraction is increased (Bowling et al., 1993).

Single crystals of a blue phase with dimensions on the order of millimeters have been grown, and thus there have been numerous experiments performed on both powder samples and single crystals in addition to optical diffraction. Facets on these single crystals are even visible in some cases. Optical microscopy, including the observation of Kossel diagrams, optical activity measurements, and nuclear magnetic resonance experiments have been used to probe the blue phases.

4.1.4.2 Twist Grain Boundary Phases

Chirality is responsible for a slightly different phenomenon in smectic liquid crystals. Because of the layered structure, smectic liquid crystals do not allow for twist, so in many cases chiral molecules possess the nonchiral smectic A phase. However, in compounds with a very short pitch, the free energy of the phase is lowered by introducing twist grain boundaries at regular intervals. The phase is therefore a succession of undistorted smectic blocks, separated at regular intervals by defect or fluid-like regions where the layer structure and director rotate by a small angle. This structure is shown in FIGURE 4.7, where it can be seen that the twist axis is in the plane of the layers with the director perpendicular to this twist axis. In one compound, a rotation of the smectic blocks by $\Delta\Psi = 17^\circ$ occurs at twist grain boundaries spaced $l_b = 24\,\text{nm}$ apart. This results in a pitch of about

FIGURE 4.6. Temperature–chirality phase diagram. (From Bowling et al., 1993.)

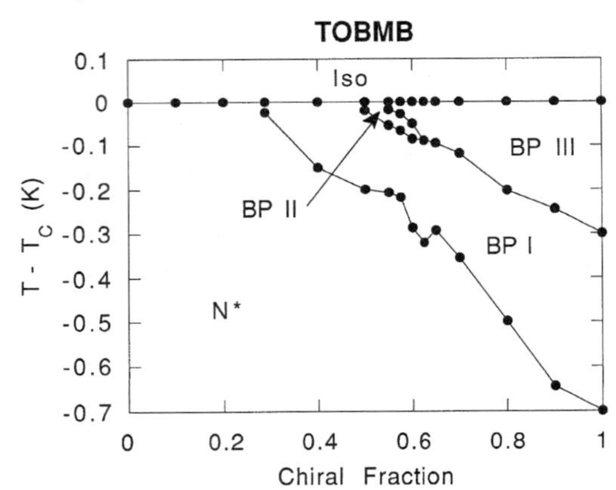

FIGURE 4.7. Twist grain boundary (TGB) phase. (Reprinted with permission from Ihn et al., *Science* 258, 1992. Copyright American Association for the Advancement of Science.)

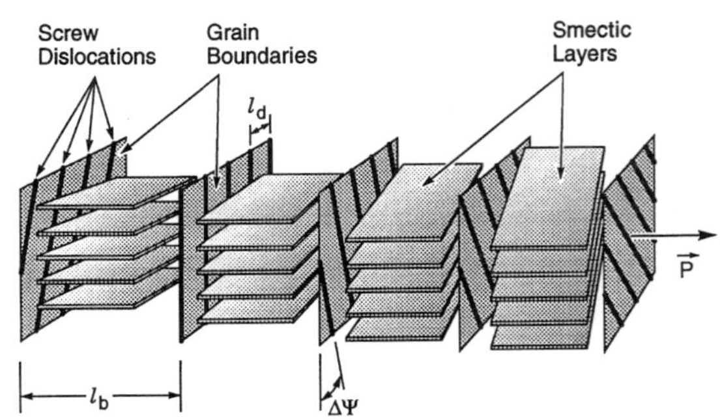

500 nm and a resulting optical reflection of about 800 nm (Ihn et al., 1992). Such a twist grain boundary A phase (TGB A phase) was theoretically predicted (Renn and Lubensky, 1988) and experimentally observed (Goodby et al., 1989) at about the same time. Since that time there has been both additional theoretical (Renn, 1992) and experimental (Navailles et al., 1993) work on the twist grain boundary C phase (TGB C) in addition to the TGB A phase.

This work on the twist grain boundary phases has extended the analogy between superconductors and smectic A liquid crystals first pointed out by de Gennes. Both possess a complex order parameter and a free energy of the same form (de Gennes, 1972). Twist distortion in smectic A liquid crystals plays the same role as the magnetic field for superconductors. Type-I superconductors expel the magnetic field just as smectic A liquid crystals possess no twist distortion, even when the molecules are chiral. Type-II superconductors allow the magnetic field to penetrate in flux vortices just as the TGB A phase incorporates twist in grain boundaries. The TGB A phase, with its regular spaced array of twist grain boundaries, is analogous to the lattice of flux vortices of the Abrikosov phase in superconductors.

4.1.5 Discotic Liquid Crystals

4.1.5.1 Nematic Phase

One of the most important findings of the last fifteen years is that disk-shaped molecules can form liquid crystal phases (Chandrasekhar and Ranganath, 1990). These molecules are called discotic liquid crystals and usually possess a flat core connected to four, six, or eight long chains via ester or ether linkages. FIGURE 4.8 shows a molecule with two discotic phases. X-ray scattering and NMR spectroscopy experiments show that the cores are well ordered and the chains are not; yet the chains seem to be crucial to the stability of these phases.

The most simple discotic phase is the nematic phase, sometimes given the notiation N_D, so as not to confuse it with the nematic phase of rod-like molecules. As shown in FIGURE 4.9(a), it is the short axes of the molecules which preferentially orient along the director. There is no long-range positional order. While the symmetry of the nematic phases for both rod-like and disk-like molecules is the same, the nematic phase of rod-like molecules is diamagnetically and optically positive while the nematic phase of disk-like molecules is negative. The dielectric anisotropy of both nematic phases can be either positive or negative.

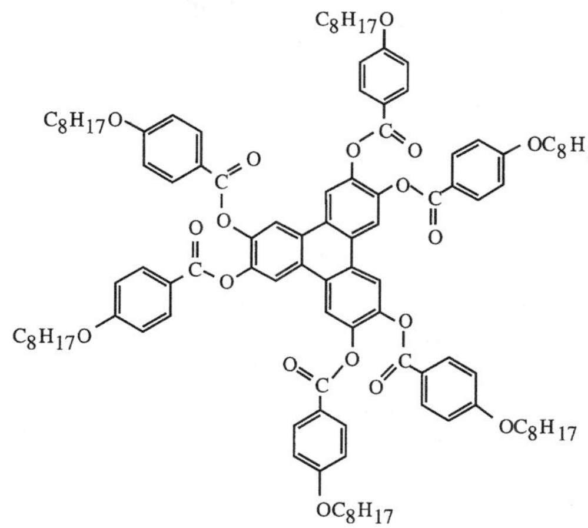

FIGURE 4.8. Discotic liquid crystal possessing both columnar and nematic phases.

Nematic phases occur in compounds with shorter chains. In fact, lengthening the chains in a homologous series narrows the nematic phase and eventually eliminates it, just as for the nematic phase of rod-like molecules. There has not been a great deal of work done on the N_D phase, but what has been done reveals that the elastic constants for disk-like and rod-like systems are similar, and the viscosity of the disk-like systems is much higher (Chandrasekhar, 1992). A chiral nematic phase has been observed in discotic systems, with the director perpendicular to the helical axis, just as in the chiral nematic phase of rod-like

FIGURE 4.9. Structure of the nematic (a) and columnar (b) phases.

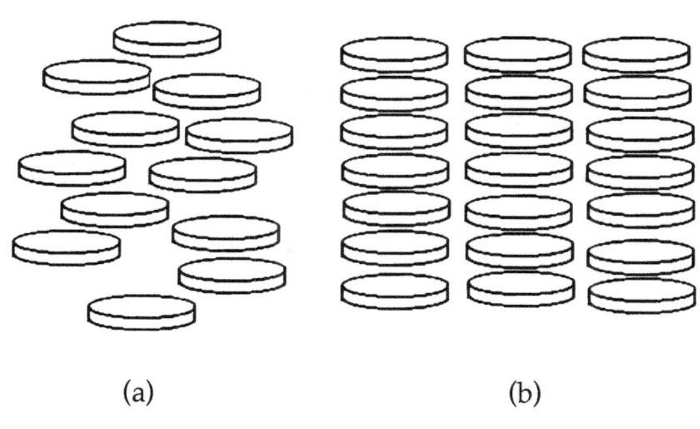

molecules (Destrade et al., 1980). Finally, a reentrant N_D phase exists between two columnar phases of certain truxene derivatives (Frank and Chandrasekhar, 1980).

4.1.5.2 Columnar Phase

FIGURE 4.9(b) shows the type of positional order present in the columnar phases of discotic liquid crystals. There is still no long-range positional order along the columns, but there is two-dimensional positional order of the columns themselves. The columns can either form a rectangular lattice or a hexagonal lattice. Along the columns, long-range orientational order of the molecular cores is present, but the flexible chains are again highly disordered.

In the rectangular columnar phase D_r, the columns occupy a rectangular lattice but the short axis of the molecules in each column is tilted away from the column axis. The tilt direction of the columns is a herringbone arrangement as shown in FIGURE 4.10(a). The higher-temperature hexagonal columnar phase D_h possesses a hexagonal lattice of columns with the molecules tilted within the columns, but there is no order in the azimuthal direction of the tilt. This arrangement is shown in FIGURE 4.10(b).

There is also evidence for phases in which the positional order of the molecules along the column is long-range in addition to the two-dimensional order of the columns themselves. Since true long-range positional order in one dimension is forbidden by the Landau–Peierls instability, there must be ordering between the molecules in neighboring columns.

FIGURE 4.10. The rectangular (a) and hexagonal (b) columnar phases. The molecules are tiled with respect to the column axis in both phases.

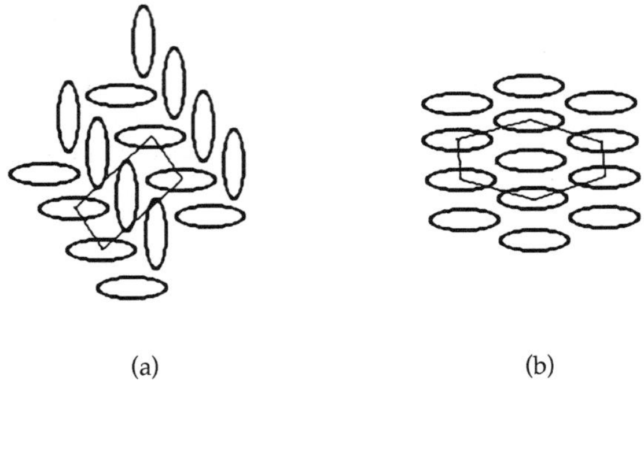

(a) (b)

Experiments with freely suspended strands of these ordered phases reveal a helicoidal stacking of the molecular cores and triangular superlattice (Fontes et al., 1988). These "ordered" discotic phases obviously have much in common with the three-dimensionally ordered smectic phases, and therefore should be referred to as soft crystals rather than liquid crystals.

4.1.5.3 Biaxial Nematic Phase

The existence of a biaxial nematic phase has been the subject of discussion for a long time. In such a phase, the molecular axes perpendicular to the long axis of a rod-shaped molecule possess long-range orientational order just as the long molecular axis does. The most common example is a fluid of orthorhombic molecules, in which each molecular axis defines three orthogonal directors. Such a phase was first discovered in a lyotropic system, where for certain concentration and temperature conditions, the micelles did not possess cylindrical symmetry (Yu and Saupe, 1980).

The first thermotropic biaxial nematic phases were seen in molecules containing both rod-like and disk-like features. In one case, the molecular core was a metal complex with four attached groups: two long rod-like groups at opposite ends and two small side groups perpendicular to the long axis of the molecule (Chandrasekhar et al., 1986). In another case, two disk-like groups with a benzene core were tethered together to form a long molecule (Praefcke et al., 1990). Both optical and x-ray experiments were used to ascertain that the molecular order was indeed biaxial.

4.1.6 Free Standing Films

4.1.6.1 Smectic Films

A technique whereby free-standing films of smectic liquid crystal material can be made has been recently revived in an effort to study the transition from three-dimensional systems to two-dimensional systems. In this technique, some smectic material is drawn across a hole in a thin piece of glass or metal using the edge of another piece of glass or metal. The size of the hole is on the order of a few millimeters. By adjusting the amount of material, the speed at which it is drawn over the hole, and the temperature, free-standing smectic films of varying thicknesses (from a few layers up to thousands of layers) can be fashioned.

The smectic film is quite stable with the smectic layers in the plane of the film, and experiments can be performed on it for hours. The angular spread of the normal to the layers is extremely low in these films (on the order of a tenth to a hundredth of a degree compared with about one degree for a bulk liquid crystal aligned by a magnetic field). The majority of experiments on smectic films have been x-ray diffraction experiments probing the positional and orientation order within the film. Polarized light microscopy, light reflection spectroscopy, and heat capacity experiments have also been performed.

The stability of the smectic film is due to the energy barrier necessary to create a hole. The energy required to create a hole has two contributions: (1) the positive energy required to create an air–liquid crystal interface line, and (2) the negative energy from the decrease in the air–liquid crystal interface area. Since the first is proportional to the radius of the hole, while the second is proportional to the radius squared, the positive contribution wins for small radii. However, once a hole is created, it grows and the film pops (de Gennes and Prost, 1993).

In many cases, free-standing smectic films provide the ordered material which allows the various types of positional and orientational order in smectic liquid crystals to be understood (Moncton and Pindak, 1979; and Pindak et al., 1981). But thin films can have smectic phases not found in the bulk. For example, in one material in which the bulk possesses only crystal G and B phases, thin films show two additional crystal B phases and two tilted hexatic phases (Sirota et al., 1987).

In tilted smectic films of only a few layers, line defects where the orientation changes by multiples of $\pi/2$ have recently been studied. These lines form closed loops or terminate in fractional vortices and can be found in smectic C, hexatic smectic F, hexatic smectic I, and antiferroelectric phases (Pang et al., 1992).

4.1.6.2 Surface Ordering

Free-standing films have also provided the means for a good deal of work probing the ordering that takes place near surfaces (in this case, the air–liquid crystal surface). For example, the liquid crystal 750BC has the following phase sequence upon cooling in the bulk: smectic A (65 °C)–hexatic B (59 °C)–crystal E. Yet in a ten-layer free-standing film of this material, the two outermost layers undergo a transition to the hexatic B phase at 71 °C, the next two outermost layers transform at 65.2 °C, followed by the bulk at 64.6 °C. Likewise, the two outermost layers transform to the crystal E phase at a higher temperature than the internal layers. This crystal E ordering of the two outermost layers also dramatically affects the hexatic B–smectic A transition for the inner layers (Geer et al., 1992).

Surface ordering has also been studied in bulk materials with an air–liquid crystal interface. An example is the smectic A ordering at the air–liquid crystal interface that occurs in nematic liquid crystals (Chen et al., 1989). Thin nematic films on an isotropic liquid substrate have also been used recently to study both the defects in these films (Lavrentovich and Nastishin, 1990) and the so-called "surface" elastic energy terms (Lavrentovich, 1991).

4.1.6.3 Columnar Strands

When discotic material in a columar phase is spread across a hole, freely suspended strands form. The strands are typically tens of microns thick, meaning that they contain on the order of a thousand molecular columns. High-resolution x-ray experiments have revealed that the column axes are very well ordered (less than 0.3° variation on average) even though the strand contains several domains. The number of domains can be decreased by proper annealing of the strands (de Gennes and Prost, 1993).

4.2 PHASE TRANSITIONS

4.2.1 Universality Classes

One reason for the continued interest in liquid crystal phase transitions is that these transitions provide numerous examples for much of the recent theoretical work on critical phenomena. Because liquid crystal transitions are either weakly discontinuous or continuous, they should display the behavior associated with critical points, including strong fluctuations and diverging susceptibilities. One of the most important results of this theoretical work is that near such a transition, the microscopic details of the system become unimportant in describing how the transition takes place. Instead, the range of the interactions, the physical dimension of the system, and symmetry of the order parameter determine how the system behaves very close to the transition.

Model	Order Parameter Symmetry	α	β	γ	ν
2D Ising	scalar	0	0.125	1.75	1
3D Ising	scalar	0.1	0.33	1.24	0.63
3D XY	2D vector	0.01	0.34	1.30	0.66
3D Heisenberg	3D vector	-0.12	0.36	1.39	0.71
2D Potts	scalar	0.33	0.111	1.44	0.83

FIGURE 4.11. Some models for phase transitions.

The striking implication of this is that phase transitions in very different systems, superconductors and liquid crystals for example, may behave identically if the interactions, physical dimension, and symmetry of the order parameter are the same. To provide predictions, theorists have developed models of systems in which they can calculate the important thermodynamic properties. By convention these models are for magnetic systems, but they really correspond to a general interacting system with a particular type of interaction, dimension, and order-parameter symmetry. Some of these models, along with their dimensions and order-parameter symmetries, are described in FIGURE 4.11.

One parameter which is extremely important in describing the physical behavior near a phase transition is a critical exponent. This parameter gives the functional form of the divergence of a particular property. For example, the critical exponent α describes how the specific heat at constant magnetic field, C_H, diverges as the transition is approached by changing the temperature

$$C_H \propto |t|^{-\alpha} \tag{4.15}$$

where $t = (T - T_C)/T_C$ is the reduced temperature and T_C is the transition temperature. This critical exponent is determined experimentally from the slope of a log–log plot of C_H versus t,

$$\alpha = -\lim_{t \to 0} \left[\frac{\ln(C_H)}{\ln(t)} \right]. \tag{4.16}$$

Other critical exponents describe the temperature variation of the order parameter (β), the susceptibility (γ), correlation length (ν), etc. Values for the critical exponents from experiments are then compared with those predicted by the various models, to see how well the transition is described by the model. Some of these exponents for the various models are also given in FIGURE 4.11.

4.2.2 Liquid Crystal–Isotropic Transition

4.2.2.1 Nematic–Isotropic Transition

The transition from the nematic phase to the isotropic phase is always a discontinuous one, with a latent heat typically about $0.5\,\text{kJ}\,\text{mol}^{-1}$, which is nearly two orders of magnitude smaller than for the melting of a solid to either the liquid crystal or the isotropic liquid phase. Due to the smallness of the latent heat, there are pretransitional effects on either side of the transtion. Values for the critical exponent α range from 0.3 and 0.4, depending on the range of temperatures used for the fit, while the critical exponent β is around 0.25 (Thoen, 1992). These exponents are closer to those of a classical tricritical point ($\alpha = 0.5$ and $\beta = 0.25$), than those of mean-field theory ($\alpha = 0$ and $\beta = 0.5$). In general, the latent heat of the transition increases in a homologous series as the smectic–nematic transition gets closer to the nematic–isotropic transition.

The nematic–isotropic transition is usually described by the mean-field theory of Maier and Saupe or the phenomenological theory of Landau and de Gennes. In the latter, the free energy is expanded in powers of the order parameter

$$F = F_0 + \tfrac{1}{2} A S_{\alpha\beta} S_{\beta\alpha} + \tfrac{1}{3} B S_{\alpha\beta} S_{\beta\gamma} S_{\gamma\alpha} + \tfrac{1}{4} C (S_{\alpha\beta} S_{\beta\alpha})^2. \tag{4.17}$$

The constants B and C are greater than zero and $A = A_0(T - T^*)$, where A_0 is a constant. F_0 is the free energy of the isotropic phase. Since the transition occurs near where $A = 0$, T^* sets the temperature of the transition. At temperatures well above T^*, the isotropic phase with $S_{\alpha\beta} = 0$ is stable. Below a temperature slightly above T^*, the nematic phase with $S_{\alpha\beta} \neq 0$ is stable. The presence of the cubic term insures that the transition will be discontinuous with both critical exponents α and β equal to 0.5. As shown in FIGURE 4.12, enthalpy data show the discontinuous nature of the transition. Due to the smallness of the discontinuity, critical-like pretransition effects are present, but the values for critical exponents tend to depend on the range of temperature used for the fit.

A similar phenomenological theory is used to describe the short-range nematic order that is present in the isotropic phase just above the transition. The free

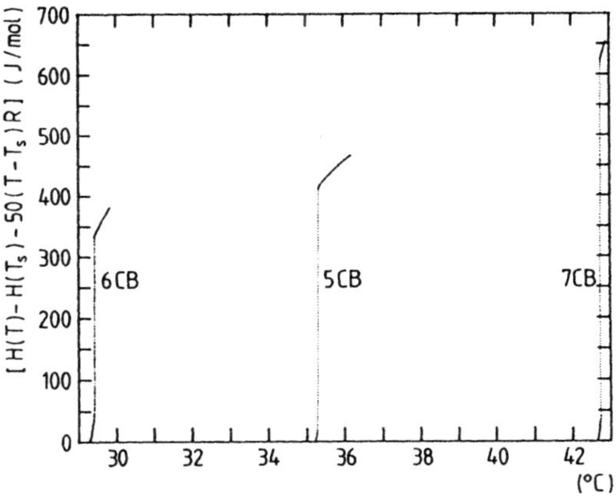

FIGURE 4.12. Enthalpy in the vicinity of the nematic–isotropic transition for three alkylcyanobiphenyls (nCB). (From Thoen et al., 1992.)

energy is expanded in terms of the order parameter and the spatial variation of the order parameter

$$F = F_0 + \tfrac{1}{2}AS_{\alpha\beta}S_{\beta\alpha} + \tfrac{1}{2}L_1\left(\frac{\partial S_{\beta\gamma}}{\partial x_\alpha}\right)\left(\frac{\partial S_{\beta\gamma}}{\partial x_\alpha}\right) + \tfrac{1}{2}L_2\left(\frac{\partial S_{\alpha\gamma}}{\partial x_\alpha}\right)\left(\frac{\partial S_{\beta\gamma}}{\partial x_\beta}\right). \tag{4.18}$$

$A = A_0(T - T^*)$ as before and L_1 and L_2 are constants. By expressing this free energy in terms of the Fourier transform of $S_{\alpha\beta}$ and then applying the equipartition theorem, the mean-square order parameter value in the isotropic phase varies with temperature as $(T - T^*)^{-1}$. This temperature dependence has been verified by light scattering experiments (Litster and Stinson, 1972), although it may break down close to the transition if a smectic phase exists below a narrow nematic phase (Gohin et al., 1983 and Anisimov et al., 1987).

4.2.2.2 Blue Phase–Isotropic Transition

For chiral substances, the transition from the liquid crystal to isotropic phase can be from the chiral nematic, BPI, BPII, or BPIII phase to the isotropic liquid. The phenomenological Landau–de Gennes theory predicts that these transitions may be weaker than for nonchiral phases. The free energy for a chiral system has the following additional gradient term allowed by symmetry

$$\varepsilon_{\alpha\gamma\beta}S_{\alpha\beta}\left(\frac{\partial S_{\delta\beta}}{\partial x_\gamma}\right) \tag{4.19}$$

where $\varepsilon_{\alpha\gamma\delta}$ is the totally antisymmetric third rank tensor. It is convenient to express the free energy in terms of the Fourier transform of the order parameter and utilize the five basis tensors described previously. Then the free energy becomes

$$F = F_0 + \frac{1}{2}\sum_m \left\{A - mBq_0q + \left[B + \left(\frac{C}{6}\right)(4 - m^2)q^2\right]\right\}|S_m(q)|^2 \tag{4.20}$$

where $q_0 = 4\pi/P$ and P is the pitch. The parameter A has the same temperature dependence as before, while B and C are constants (Grebel et al., 1983, 1984). The effect of the new term is to shift the temperature dependence of the orientational order fluctuations in the isotropic phase for four of the structural modes. For the two modes with the same handedness as the liquid crystal phases, T^* is shifted to higher temperature, and thus closer to the actual transition temperature. This has the effect of making the transition less discontinuous. This shift is proportional to q_0^2, so this effect should be strongest for the most chiral substances.

The blue phase to isotropic transition has been studied using heat capacity, optical activity, and light scattering measurements (Thoen, 1992; and Collings, 1992). For all but one of the compounds studied, all transitions between the chiral nematic and blue phases, between the blue phases themselves, and between the blue phases and the isotropic phase are first-order, although some have latent heats which are almost two orders of magnitude smaller than is typical for transitions from liquid crystals to the isotropic phase. The effect of increasing chirality on the transition to the isotropic phase is quite dramatic, as evidenced from the adiabatic scanning calorimetry on two systems in which the concentration of optical isomers was varied to change the chirality without affecting other physical parameters of the system (Voets, 1992; and Voets and Van Dael, 1993). For one compound, the latent heat of the BPIII–isotropic transition decreased from $1627\,\mathrm{mJ\,g^{-1}}$ in the racemic mixture to $785\,\mathrm{mJ\,g^{-1}}$ for the pure optical isomer. For another compound, this decrease was from $957\,\mathrm{mJ\,g^{-1}}$ to $147\,\mathrm{mJ\,g^{-1}}$. Very recent heat capacity and dynamic light scattering measurements on one of the most chiral compounds available indicates that the BPIII–isotropic transition is actually a supercritical

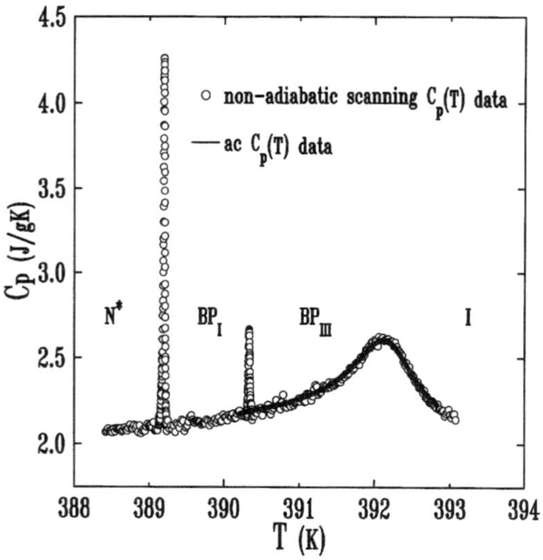

FIGURE 4.13. Specific heat in the chiral nematic phase (N*), first blue phase (BPI), third blue phase (BPIII), and isotropic phase (I) for a highly chiral liquid crystal. The N*–BPI and BPI–BPIII transitions are weakly first-order, while BPIII converts to I without a phase transition. (From Kutnjak et al., 1995.)

conversion and not a phase transition at all (Kutnjak et al., 1995). This is shown in FIGURE 4.13 where the latent heats for the chiral nematic–blue phase I and blue phase I–blue phase III transitions are clearly evident, but the blue III–isotropic transition region shows only a rounded conversion without a phase transition. The most important implication of this finding is that the macroscopic symmetry of the BPIII and isotropic phases must be the same.

4.2.2.3 Smectic–Isotropic Transition

The latent heat for a direct transition from the smectic phase to the isotropic phase is considerably larger than for the nematic–isotropic transition (Thoen, 1992). Pretransitional effects, which include both orientational order and positional order fluctuations are less pronounced.

4.2.3 Smectic A–Nematic Transition

4.2.3.1 Heat Capacity and X-ray Experiments

The transition from the smectic A phase to the nematic phase involves the melting of one-dimensional translational order. For the past two decades, this transition has received more theoretical and experimental attention than any other, partially due to the analogy with the superconducting–normal transition. Most of the focus has been on the critical exponents, in the hope of determining the universality class of this transition. In spite of all this research, the situation remains very complicated with numerous questions unresolved. A list of references to much of the theoretical and experimenal work is contained in a recent article (Garland and Nounesis, 1994).

The critical exponents that have been measured most extensively are α, γ, $\nu_{\|}$, and $\nu_{\perp}$, where the last two correspond to the divergence of the correlation lengths parallel and perpendicular to the director, respectively. In summarizing the data, it is convenient to specify the McMillan ratio, T_{NA}/T_{NI}, where T_{NA} is the nematic–smectic A transition temperature and T_{NI} is the nematic–isotropic transition temperature (McMillan, 1973). At first, differential scanning calorimetry (DSC) seemed to indicate that most of the observed smectic A–nematic transitions were discontinuous, with only those below the tricritical value of 0.87 being continuous, as suggested by McMillan. Later work utilizing adiabatic scanning calorimetry and AC calorimetry showed that pretransitional effects had been interpreted as latent heats in the DSC experiments, and that continuous behavior occurred up to much higher values of the McMillan ratio (Kasting et al., 1980; and Thoen et al., 1984). Experimental advances continue to be made in detecting small latent heats; a good example is the dynamical method using propagation characteristics of the interface between two phases (Cladis et al., 1989; and Anisimov et al., 1990).

The critical exponents for the susceptibility and correlation lengths are measured from diffuse x-ray scattering experiments in the nematic phase (Ocko et al., 1984). First, the x-ray structure factor is fit to the following form

$$\frac{\chi}{1+\xi_{\|}^2 q_z^2+\xi_{\perp}^2 q_{\perp}^2(1+C\xi_{\perp}^2 q_{\perp}^2)} \tag{4.21}$$

where C is a constant. Then the temperature dependences of χ, $\xi_{\|}$, and $\xi_{\perp}$ are fit to power laws to determine the critical exponents γ, $\nu_{\|}$, and $\nu_{\perp}$, respectively. The power laws typically hold over three decades in reduced temperature (de Gennes and Prost, 1993).

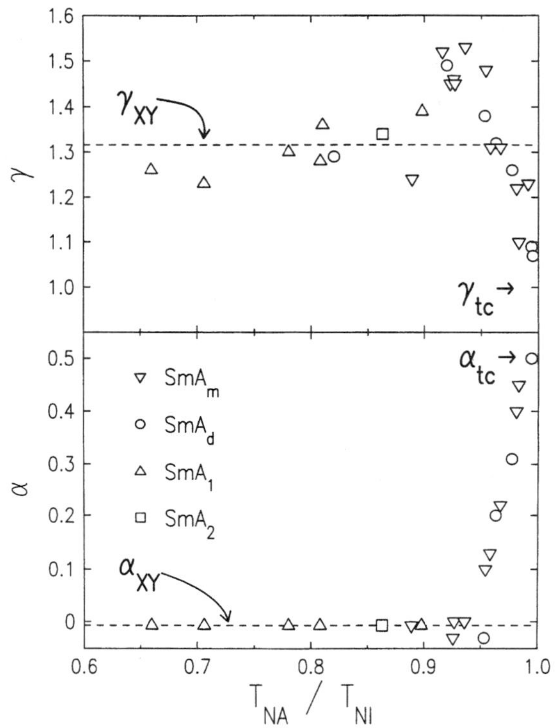

FIGURE 4.14. Specific heat exponent (α) and susceptibility exponent (γ) as a function of the ratio of the N–A transition temperature (T_{NA}) to the N–I transition temperature (T_{NI}). (From Garland and Nounesis, 1994.)

If one looks at the measurements of continuous smectic A–nematic transitions, the critical exponents are clearly dependent on the McMillan ratio. For all ratios below about 0.93, they are close to those of the 3D XY universality class ($\alpha = -0.007$, $\gamma = 1.316$, $\nu_{\|} = \nu_{\perp} = 0.669$). Above a ratio of roughly 0.93, they are no longer fairly constant, but tend to approach their tricritical values ($\alpha = 0.5, \gamma = 1.0, \nu_{\|} = \nu_{\perp} = 0.5$) as the McMillan ratio approaches unity. This trend for α and γ is shown in FIGURE 4.14. In addition, the correlation is no longer isotropic at these high McMillan ratios, with $\nu_{\|}$ falling above and $\nu_{\perp}$ falling below their tricritical values. It should be pointed out that the value of the McMillan ratio at the tricritical point is not universal and that the crossover from 3D XY values to tricritical values is not the same for different homologous series, but the trend in the critical exponents is very similar for all compounds studied. A clear demonstration

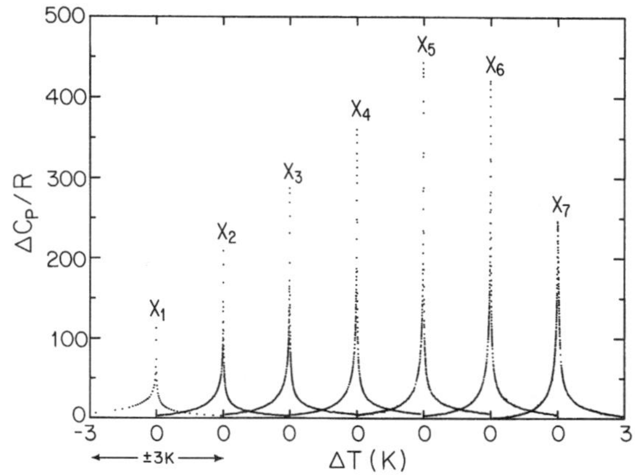

FIGURE 4.15. Excess specific heat near the nematic–smectic A transition for seven different binary mixtures. The mole fraction for one of the compounds increases from 0 (X_1) to 0.5 (X_7). (From Stine and Garland, 1989b.)

showing the change in α in approaching the tricritical point along with the increase in the size of the specific heat anomaly is given in FIGURE 4.15. For this binary mixture, the tricritical concentration is close to the one labeled X_5 in the figure.

4.2.3.2 Theory

The mean-field description for the smectic A–nematic transition begins with an expansion for the free energy in terms of the amplitude of the smectic order parameter $|\psi|$

$$F = F_0 + \tfrac{1}{2}a|\psi|^2 + \tfrac{1}{4}b_0|\psi|^4 + \cdots \tag{4.22}$$

where $a = a_0(T - T_{NA})$ and $b_0 > 0$. Only even powers are allowed since changing the sign of $|\psi|$ is the same as changing the reference point for the sinusoidal density variation and therefore cannot affect the free energy. This free energy predicts a continuous transition at T_{NA}.

Coupling between the amount of nematic order (given by the order parameter S) and the smectic order parameter $|\psi|$ complicates the theory. This means that the value of S which minimizes the nematic free energy, S_0, may not be the value of S which minimizes the total free energy if a term coupling S and $|\psi|$ is present. Letting $\Delta S = S - S_0$, the free energy becomes

$$F = F_0 + \tfrac{1}{2}a|\psi|^2 + \tfrac{1}{4}b_0|\psi|^4 + \frac{(\delta S)^2}{2\chi} - c|\psi^2|(\delta S) \tag{4.23}$$

where χ is the susceptibility and $c > 0$. Minimizing F with respect to ΔS, one obtains

$$\begin{aligned} F &= F_0 + \tfrac{1}{2}a|\psi|^2 + \tfrac{1}{4}(b_0 - 2c^2\chi)|\psi|^4 \\ &= F_0 + \tfrac{1}{2}a|\psi|^2 + \tfrac{1}{4}b|\psi|^4. \end{aligned} \tag{4.24}$$

For a small value of the McMillan ratio, χ is small and $b > 0$, resulting in a continuous transition. For high values of the McMillan ratio, χ is large and $b < 0$, resulting in a discontinous transition (assuming that a sixth-order term in $|\psi|$ is present for stability). A tricritical point thus exists at the McMillan ratio where $b = 0$. The critical exponents have their classical values.

Near the transition, fluctuations in both the smectic order parameter and nematic director are important. If one introduces gradient terms of the smectic order parameter into the free-energy expression but with isotropic coefficients, the situation is analogous to superfluid helium and thus 3D XY behavior is expected (Prost, 1984). If fluctuations of the director are also included, the situation resembles the normal superconducting transition. Some of the analogies to superconductivity are quite interesting. For example, just as a superconductor expells magnetic fields (the Meissner effect), smectic A liquid crystals expell twist and bend distortions. Accordingly, the twist and bend elastic constants in the nematic phase diverge as the transition to the smectic A phase is approached.

A variety of conclusions have been reached through theoretical work on this problem. The transition seems always to be first-order in four dimensions due to the coupling between the smectic order parameter and the director fluctuations (Halperin et al., 1974). In three dimensions, the behavior on the low and high sides of the transition can be reversed from 3D XY behavior, i.e., an inverted 3D XY model (Dasgupta and Halperin, 1981; and Lubensky, 1983). A dislocation-loop melting theory, in which a divergence in the density of dislocation loops destroys the smectic order, yields anisotropic critical behavior in the correlation lengths (Nelson and Toner, 1981; and Toner, 1982). Noninverted behavior is seen in Monte Carlo simulations (Dasgupta, 1987). Finally, self-consistent one-loop theory employing intrinsically anisoptropic coupling between the director fluctuations and the smectic order parameter predicts a gradual crossover from isotropic behavior to strongly anisotropic behavior (Patton and Andereck, 1992; and Andereck and Patton, 1994).

Clearly the nematic–smectic A transition is very complicated, with many factors affecting the behavior near the transition. Coupling between the smectic order parameter and the nematic order drives the transition toward tricritical behavior, while coupling between the smectic order and nematic director fluctuations drives it into an anisotropic regime. While the McMillan ratio is a convenient indicator of the strength of both of these couplings, it is quite imprecise. Still, general trends with the McMillan ratio are plainly evident (Garland and Nounesis, 1994).

4.2.3.3 Polar Smectics

The smectic A phase of polar compounds can be A_1, A2, or A_d, depending on whether the layer spacing is roughly one molecular length, two molecular lengths, or between one and two molecular lengths, respectively. The smectic A_d–N transition falls into the category of crossover behavior from 3D XY to tricritical as the McMillan ratio is increased as described previously, and in fact this transition has been used for quite a few experiments. When a reentrant nematic phase, N_r, is present, the smectic A_d–N_r transition is similar. The heat capacity peak at this transition is often extremely small (Wu et al., 1992).

A large number of experiments on the smectic A_1–N transition show critical exponents consistent with the 3D XY universality class (Garland, 1990). Typical specific heat data are contained in FIGURE 4.16. One of the reasons for this may be the large

FIGURE 4.16. Specific heat associated with the smectic A_1–nematic transition. (From Garland et al., 1989.)

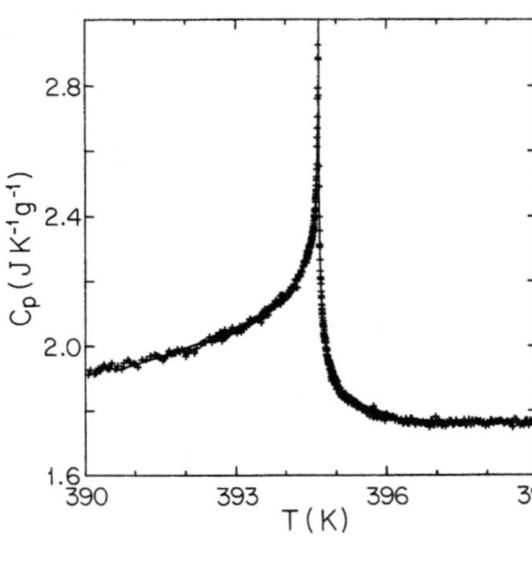

nematic range that occurs in many of these systems. More recent heat capacity measurements have also shown a tendency for this system to crossover to tricritical-like behavior (Wu et al., 1992). The reason for this may be the same as for the smectic A–N transition in nonpolar compounds (or the smectic A_d–N transition for polar compounds), namely a trend to a larger nematic susceptibility that drives the transition toward first order through the coupling of the nematic and smectic order parameters. For the smectic A_1–N transition, however, the larger nematic susceptibility may be due to a nearby nematic (dimers)–nematic (monomers) transition rather than a nearby nematic–isotropic transition.

Although most smectic A_2–N transitions are first-order with high values of the McMillan ratio, at least one continuous transition has been observed (Wen et al., 1991). In this case 3D XY critical exponents were obtained from heat capacity measurements. As with the other smectic–nematic transitions, it is expected that crossover from 3D XY to tricritical behavior occurs for the smectic A_2–N transition also. So far, experimental results showing this have not been reported.

It must be pointed out that the reentrant phenomenon in the case of polar molecules creates some very special circumstances that have only been hinted at in the previous discussion. An isolated nematic region replacing a critical point at the end of a smectic A_d–smectic A_1 coexistence line in binary mixtures has been theoretically described by a dislocation loop melting model (Prost and Toner, 1987) and experimentally observed (Hardouin et al., 1986; and Wu et al., 1992). A frustrated spin-gas model also predicts such a "nematic lake," but completely surrounded by the smectic A_d phase (Marko et al., 1989). The dislocation loop melting theory also predicts that a transition between a nematic phase with strong A_1 fluctuations, N_1, to a nematic phase with strong A_d fluctuations, N_d, occurs in this region, but such a transition has not been observed.

In the case of mixtures, especially ones involving dissimilar molecules, Fischer renormalization is sometimes important (Fischer, 1968; and Anisimov, 1988). This is a renormalization of the critical exponents in systems with positive α due to large changes in the smectic A–nematic transition temperature as the concentration is varied. The critical exponents are divided by $1-\alpha$, except for α itself, which is divided by $\alpha-1$.

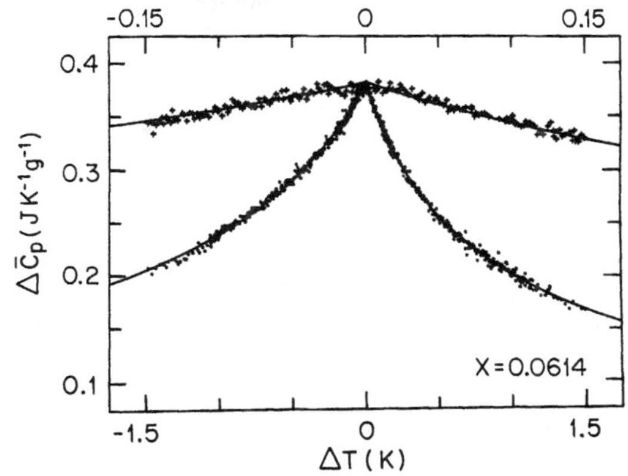

FIGURE 4.17. Excess specific heat at the smectic A_d–nematic transition for a tricritical mixture. The solid line is a fit with a fully Fischer-renormalized exponent. (From Huster et al., *Phys. Rev. A* 36, 2364 (1987)).

Good examples of this occur in mixtures of polar and nonpolar molecules (Huster et al., 1987; Hill et al., 1989; and Stine and Garland, 1989). FIGURE 4.17 shows heat capacity data for the tricritical mixture of two dissimilar molecules. The data fit a curve with a renormalized value for α of -1.

4.2.4 Smectic A–Smectic C Transitions

The order parameter that must be specified in the smectic A–smectic C transition is the magnitude and azimuthal direction of the tilt of the director relative to the layer normal. Thus a two-component (complex) order parameter and free energy are involved, both of which are formally identical to the condensation of superfluid helium (de Gennes, 1972). This leads to the expectation that the transition is continuous and of the 3D XY universality class. The heat capacity critical exponent should be about -0.007 and below the transition the tilt angle should vary with temperature with a power of about 0.35.

The regime of critical fluctuations, however, may lie outside of experimental observation. The Ginzburg criterion indicates that for this transition a reduced temperature as low as 10^{-5} may be necessary to observe critical phenomena. The experimental data show mean-field Landau behavior with a large sixth-order term (Huang and Viner, 1982; and Birgeneau et al., 1983). The heat capacity exponent α is too small to measure and the tilt angle dependence below the transition is in the neighborhood of 0.35.

In smectic A–smectic C* transitions, the lower temperature phase possesses both a helicoidal twist of the director about the layer normal and a spontaneous polarization in the plane of the layers. Both first- and second-order smectic A–smectic C* transitions have been observed; the continuous transitions and tricritical point are again explained well by a Landau free energy with a large sixth-order term (Garland, 1992).

In polar compounds a smectic A–smectic C transition is possible, but in both cases involving bilayers (smectic A_2–smectic C_2). Heat capacity studies on this transition for two compounds show a step in the heat capacity at the transition with a constant value below the transition. A mean-field model for the data in the smectic A_2 phase requires an extremely small sixth-order term (Jeong et al., 1988; and Wen et al., 1991).

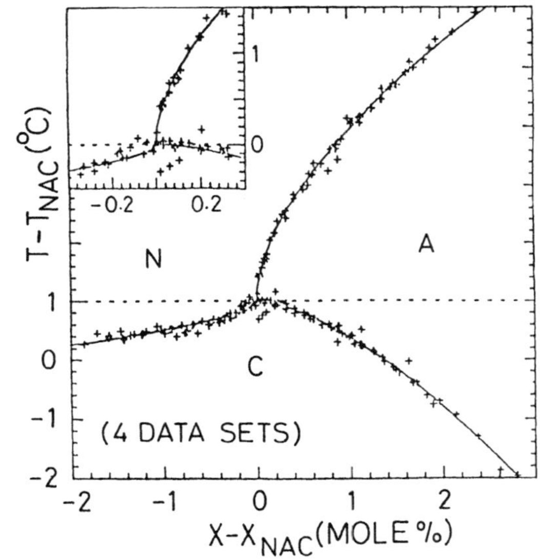

FIGURE 4.18. Temperature–concentration phase diagram in the vicinity of the NAC point for four binary mixtures. (From Shashidhar et al., *Phys. Rev. Lett.* 53, 2141 (1984)).

4.2.5 Nematic–Smectic A–Smectic C Point

By varying the concentration in a binary mixture or by varying the pressure in certain single-component systems, it is possible to obtain a point where the smectic A–N, smectic A–smectic C, and smectic C–N phase boundaries meet (NAC point). This region of the phase diagram has been studied extensively, with very interesting results.

Perhaps the most important finding is that there is universal behavior near the NAC point. FIGURE 4.18 shows that the phase diagrams for four different binary systems fall right on top of each other (Brisbin et al., 1983; Shashidhar et al., 1984). FIGURE 4.19 shows that a single-component system under pressure has the same topology in the neighborhood of the NAC point (Shashidhar et al., 1984). The boundaries between the phases in a temperature (T)–concentration (X) phase diagram obey simple power laws of the form

$$T - T_{NAC} = A|X - X_{NAC}|^{\phi} + B(X - X_{NAC}), \qquad (4.25)$$

where A and B are constants and ϕ is equal to about 0.6 for the smectic A–N and smectic C–N transitions, and about 1.5 for the smectic A–smectic C transition. An NAC point also exists for some reentrant nematic phases. In many cases the critical regions near this NAC point seem too small to be realized, but in the proper system singularities in the phase boundaries and the same universal behavior have been observed (Somasekhara et al., 1986).

4.2.5.1 Theory

Any theoretical description of the NAC point must utilize an order parameter which can describe a mass density wave not just normal to the layers, but in rings

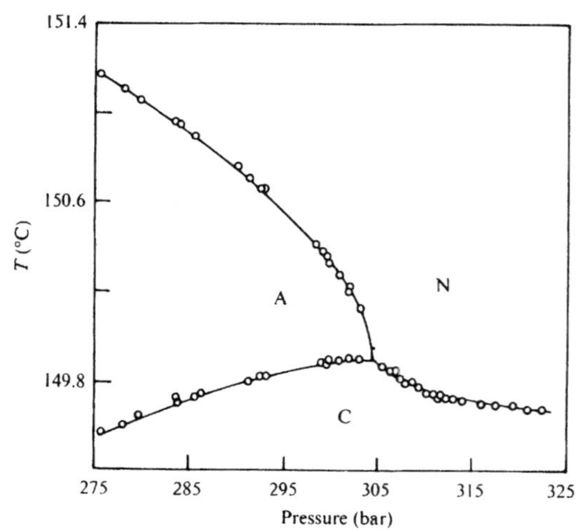

FIGURE 4.19. Temperature–pressure phase diagram in the vicinity of the NAC point for a single compound. (From Shashidhar et al., 1984.)

around the layer normal. A single order parameter involving an integration in wavevector space can be used in a free energy involving this order parameter and gradients of this order parameter (Chen and Lubensky, 1976). Depending on the sign of one of the coefficients, the fluctuations in the nematic phase can be either smectic A-like or smectic C-like. The point where both this coefficient and the coefficient in front of the term quadratic in the order parameter are zero is the NAC point and implies that the NAC point is a Lifshitz point (Hornreich et al., 1975). The amplitude of the mass density wave increases in moving away from the Lifshitz point and the x-ray scattering profile should change from being Lorentzian away from the NAC point to quartic at the NAC point. Other theoretical approaches utilize two order parameters, one for the mass density wave and the other for the tilt angle.

A dislocation loop model can be developed by using the renormalization group technique (Grinstein and Toner, 1984). The interesting addition in this treatment is the prediction of a biaxial nematic phase between the nematic and smectic C phases. This means that the NAC point is really a tetracritical point. The theory of Chen and Lubensky can also be modified to predict such a biaxial nematic phase if fluctuations are included (Grinstein et al., 1986).

4.2.5.2 Experiment

The NAC point has been investigated most thoroughly using calorimetry but there have been some high-resolution x-ray scattering and light scattering experiments also. The first detailed study was done on a binary mixture of two thioesters, revealing second-order smectic A–N and smectic C–smectic A transitions and a first-order smectic C–N transition, although the latent heat for this last transition decreased to zero at or near the NAC point (De Hoff et al., 1982). Experiments on other systems confirm these first results and the nature of the heat capacity anomalies indicate that a classical tricritical point may exist at or near the NAC point (Anisimov et al., 1985; Garland and Huster, 1987; and Thoen and Parret, 1989). Alternatively, the x-ray scattering (Safinya et al., 1981, 1983; and Martinez-Miranda et al., 1986) and light scattering experiments (Witanachi et al., 1983; Solomon and Litster, 1987; and Huang and Ho, 1987, 1990) are in full agreement with the NAC point being a Lifshitz point.

Although the biaxial nematic phase between the nematic and smectic C phases has not been observed, a broad peak in the heat capacity in the nematic phase has been observed in two systems. The first was a binary mixture of two nonchiral compounds (Anisimov et al., 1985) and the second was a binary mixture of two chiral substances with chiral nematic and chiral smectic C phases (Wen et al., 1990). These broad peaks may be the manifestation of biaxial or chiral fluctuations.

4.2.6 Smectic A–Hexatic Smectic B Transition

The order that sets in at the smectic A–hexatic smectic B phase transition is bond orientational order with sixfold symmetry in the plane of the layers. The appropriate order parameter is much like the smectic A order parameter, namely a bond angle variation of the form

$$\psi_6 = I_6\, e^{6i\phi} \tag{4.26}$$

where ϕ is the azimuth angle and I_6 is the complex order parameter. This complex order parameter should again put the transition into the 3D XY universality class. However, instead of a critical exponent α near -0.007, heat capacity and optical studies yield values for α of about 0.6 (Pitchford et al., 1985; and Nounesis et al., 1986). The heat capacity data can be seen in FIGURE 4.20. While these results are clearly inconsistent with the 3D XY model, they are close to the tricritical value of 0.5, raising the question of whether the transition occurs in the neighborhood of a tricritical point. In order to predict such a tricritical

FIGURE 4.20. Specific heat in the vicinity of a hexatic B–smectic A transition. (From Nounesis et al., 1986.)

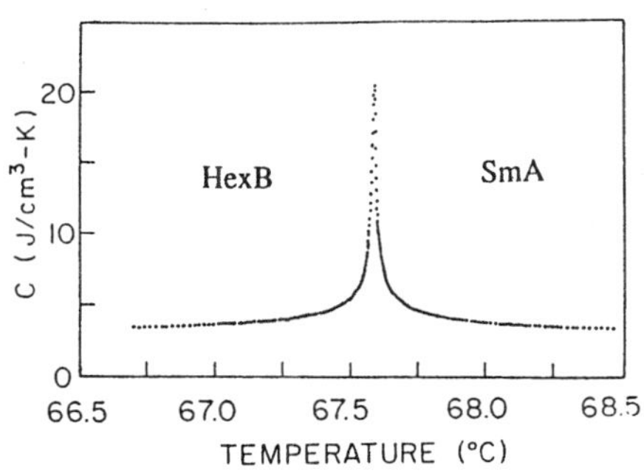

point, theoretical work includes a coupling of I_6 to some other type of local order, but more recent heat capacity results show that this high value of α is maintained over wide composition ranges and in many compounds, making the explanation of a tricritical point less likely (Huang et al., 1989; and Nounesis et al., 1989).

The transition to hexatic order has also been studied in tilted phases. In this case the transition is from the smectic C phase to either the smectic F or smectic I phases. The molecular tilt acts as a temperature-independent field, which promotes some bond orientational order in the smectic C phase. In at least one compound the heat capacity peak at the smectic C–smectic I transition is clearly rounded (Garland et al., 1989), while in other compounds the smectic C–smectic F transition is strongly first-order (Stine and Garland, 1990).

Some recent work on free-standing liquid crystal films has been very revealing. For very thin smectic A films of three or four layers, a single heat capacity peak appears at the smectic A–hexatic B transition. For thicker films (five to ten layers) two heat capacity peaks appear, one at a higher temperature representing the establishment of bond orientational order in the outer two layers and another at lower temperature when the rest of the film undergoes the phase transition (Geer et al., 1989). These thinnest films with only one heat capacity peak have the chance of representing a truly two-dimensional system. The record for the number of individual layer transitions observed is four. Using a different compound in the same homologous series, heat capacity studies have revealed the smectic A–hexatic B transition at 74 °C for the outermost layers, 67 °C for the second outermost layers, 66.3 °C for the third outermost layers, and the fourth outermost layers transforming just above the bulk transition. These layer transitions are evident in the data shown in Figure 4.21 (Geer et al., 1993). All these layer transitions are continuous (Stoebe et al., 1992).

4.2.7 Smectic A–Smectic A Transitions

The large number of smectic phases exhibited by polar molecules creates many interesting phase diagrams. Many involve transitions between different smectic A phases as well as interesting critical points, reentrant phases, and isolated nematic regions.

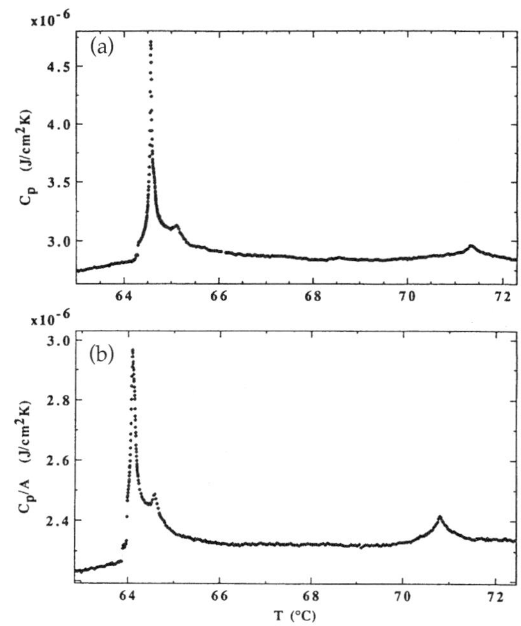

FIGURE 4.21. Specific heat near the hexatic B–smectic A transition in thin films. (a) Ten layers and (b) eight layers. (From Geer et al., *Phys. Rev. E* 48, 408 (1993)).

4.2.7.1 Smectic A_1–Smectic A_2 Transition

The transition from the smectic A_1 phase to the smectic A_2 phase involves the doubling of the spatial period due to dipolar ordering, so it can be first- or second-order. Theoretical work places this transition in the 3D Ising universality class with critical exponents $\gamma = 1.24$, $\nu = 0.63$, and $\alpha = 0.11$ (Prost and Barois, 1983; and Wang and Lubensky, 1984). Heat capacity and high-resolution x-ray studies in binary mixtures of a nonpolar and a polar compound have observed a tricritical point where the transition changes from first- to second-order, but the measured exponents have significantly different values, namely $\gamma = 1.46 \pm 0.05$, $\nu = 0.74 \pm 0.03$, and $\alpha \simeq -0.16$ (Chan et al., 1986; and Das et al., 1990). These different results can be nicely explained by Fischer renormalization, since the slope of the temperature–concentration phase boundary is large (Fischer, 1968; Anisimov, 1988). When the renormalized values of the critical exponents are calculated, one obtains $\gamma = 1.39$, $\nu = 0.71$, and $\alpha = -0.124$, which are all close to the measured values. Figure 4.22 shows the heat capacity data.

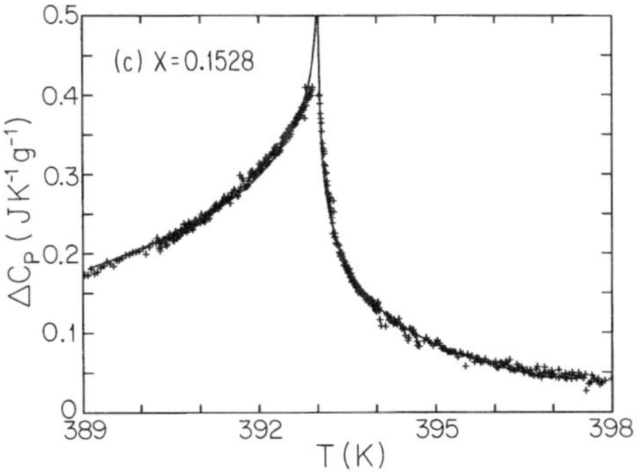

FIGURE 4.22. Excess specific heat associated with the smectic A_2–smectic A_1 transition. (From Das et al., 1990.)

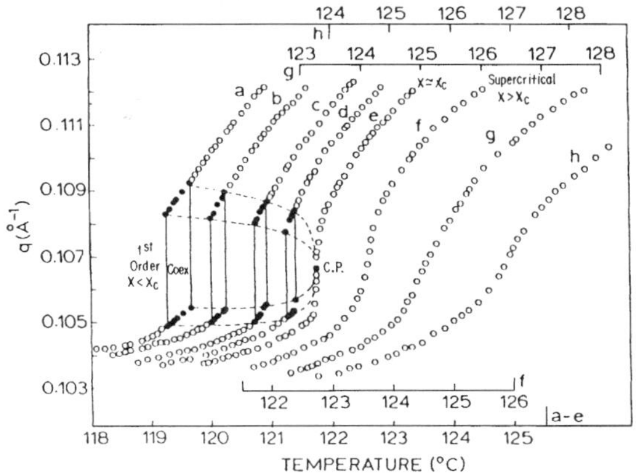

FIGURE 4.23. Smectic wavevector in the vicinity of the smectic A_d–smectic A_2 transition for eight binary mixtures. (From Shashidhar et al., 1987.)

4.2.7.2 Smectic A_d–Smectic A_2 Transition

The smectic A_d and smectic A_2 phases both involve dipolar ordering of pairs of molecules in layers, so they possess the same macroscopic symmetry. The phase boundary between them must represent a first-order transition and the boundary can end at a critical point. This has been confirmed by x-ray measurements that measured the layer periodicities in the neighborhood of the transition. In a binary mixture of two polar compounds, changing the temperature for certain mixtures caused the transition to take place. As this was repeated for differing concentrations, what was a discontinuous jump in the layer spacing at the transition (with a two-phase region) became less and less of a jump, finally reaching a critical concentration for which one phase continuously evolved into the other phase (Shashidhar et al., 1987). This behavior, which is shown in FIGURE 4.23, is strikingly similar to what happens to the density at the gas–liquid critical point. Theoretical arguments involving fluctuations and order parameter coupling to the elastic degrees of freedom, however, suggest that this critical point may belong to a new universality class with upper critical dimension six and anisotropic scaling (Park et al., 1988). Experiments seem to confirm this (Prost et al., 1990; and Wen et al., 1992).

4.2.7.3 Smectic A_d–Smectic A_1 Transition

In a similar fashion, it is predicted theoretically that the smectic A_d–smectic A_1 transition line can end at a critical point (Park et al., 1988). A disclination loop model, however, predicts a number of other interesting possibilities (Prost and Toner, 1987). One of them involves a "nematic lake" at the end of the smectic A_d–smectic A_1 transition line, as has been mentioned previously. This "nematic lake" is separated from the smectic phases around it by a second-order transition boundary, but the two ends of this boundary connect to the first-order smectic A_d–smectic A_1 phase transition line at two different points. This creates a first-order smectic A_d–N phase transition and finally a first-order N–N transition. This N–N transition line ends in the nematic lake at a critical point. Inside the nematic lake and on opposite sides of the N–N transition line, there are two nematic phases. One is characterized by strong A_d-type smectic fluctuations and the other by strong A_1-type smectic fluctuations. As is clear from FIGURE 4.24, such a "nematic lake" has been observed, but high-resolution experiments capable of detecting some of these predicted transitions have not been performed (Hardouin et al., 1986; and Wu et al., 1992).

4.2.7.4 Other Transitions

Transitions are possible between both the smectic A_1 and smectic A_2 phases and the modulated antiferroelectric phases (antiphase $\tilde{A}$ or crenelated phase A_{cren}). The smectic A_1–smectic $\tilde{A}$ transition is expected to be first-order due to Brazovskii fluctuations (Prost and Barois, 1983). While two-phase coexistence has been observed for this transition, there are strong pretransitional effects, especially on the smectic A_1 side of the transition. The transition to the smectic A_{cren} phase is

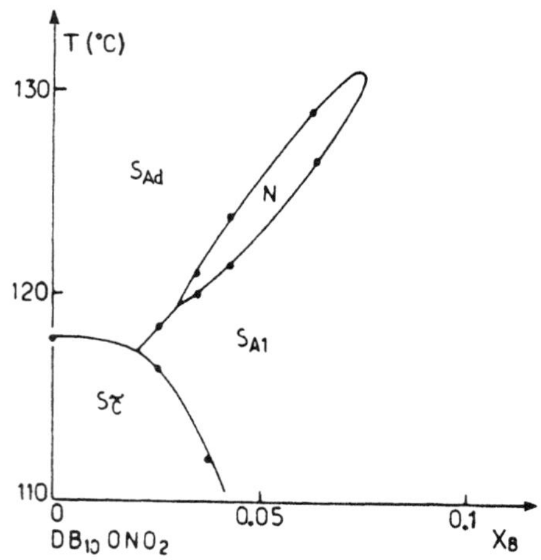

FIGURE 4.24. Partial temperature–concentration phase diagram for a binary mixture. (From Hardouin et al., 1986.)

very different, with a heat capacity peak which is truncated. The interpretation is that the smectic A_{cren} phase is more like a periodic repeating of smectic $\tilde{A}$ and smectic A_2 slabs (Garland, 1990).

It should be noted that all smectic A phases possess the same macroscopic symmetry. One way to realize this comes from the fact that the smectic A_1 phase can convert to the smectic A_d phase without a phase transition and that the smectic A_d phase can convert without a phase transition to the smectic A_2 phase. Therefore, it is possible for the smectic A_1 phase to convert to the smectic A_2 phase without a phase transition. This does not rule out, however, the possibility of a first- or second-order transition from the smectic A_1 to the smectic A_2 phase (de Gennes and Prost, 1993).

REFERENCES

Als-Nielsen, J., Litster, J. D., Birgeneau, R. J., Kaplan, M., Safinya, C. R., Lindegaard, A., and Mathiesen, S. 1980. *Phys. Rev. B* **22**: 312.

Andereck, B. S., and Patton, B. R. 1994. *Phys. Rev. E* **49**: 1393.

Anisimov, M. A., Voronov, V. P., Kulkov, A. O., and Kholmurodov, F. 1985. *J. Physique* **46**: 2137.

Anisimov, M. A., Labko, V. I., Nikolaenko, G. L., and Yudin, I. K. 1987. *Sov. Phys.–JETP* **45**: 111.

Anisimov, M. A. 1988. *Mol. Cryst. Liq. Cryst.* **162A**: 1.

Anisimov, M. A., Cladis, P. E., Gorodetskii, E. E., Huse, D. A., Podneks, V. E., Taratuta, V. G., van Saarloos, W., and Voronov, V. P. 1990. *Phys. Rev. A* **41**: 6749.

Birgeneau, R. J., Garland, C. W., Kortan, A. R., Lister, J. D., Meichle, M., Ocko, B. M., Rosenblatt, C., Yu, L. J., and Goodby, J. W. 1983. *Phys. Rev. A* **27**: 1251.

Birgeneau, R. J., Garland, C. W., Kortan, A. R., Litster, J. D., Meichle, M., Ocko, B. M., Rosenblatt, C., Yu, L. J., and Goodby, J. W. 1984. *Phys. Rev. A* **29**: 1371.

Bowling, M. B., Collings, P. J., Booth, C. J., and Goodby, J. W. 1993. *Phys. Rev. E* **48**: 4113.

Brazovskii, S. A., and Dmitriev, S. G. 1975. *Zh. Eksp. Teor. Fiz.* **69**: 979 (1976. *Sov. Phys.–JETP* **42**: 497).

Brazovskii, S. A., and Filev, V. M. 1978. *Zh. Eksp. Teor Fiz.* **75**: 1140 (1978. *Sov. Phys.–JETP* **48**: 573).

Brisbin, D., Johnson, D. L., Fellner, H., and Neubert, M. E. 1983. *Phys. Rev. Lett.* **50**: 178.

Caillé, A. 1972. *C. R. Acad. Sci. Paris B* **274**: 891.

Chan, K. K., Pershan, P. S., Sorensen, L. B., and Hardouin, F. 1986. *Phys. Rev. A* **34**: 1420.

Chandari, A. D. L., Ouchi, Y., Takezoe, H., Fukuda, A., Terashima, K., Furukawa, K., and Kishi, A. 1989. *Japan. J. Appl. Phys. Lett.* **28**: 1261.

Chandrasekhar, S., Sadashiva, B. K., Ramesha, S., and Srikanta, B. S. 1986. *Pramana* **27**: L713.

Chandrasekhar, S., and Ranganath, G. S. 1990. *Rep. Prog. Phys.* **53**: 57.

Chandrasekhar, S. 1992. *Liquid Crystals.* 2d ed. Cambridge: Cambridge University Press.

Chen, J. H., and Lubensky, T. C. 1976. *Phys. Rev. A* **14**: 1202.

Chen, W., Martinez-Miranda, L. J., Hsiung, H., and Shen, Y. R. 1989. *Phys. Rev. Lett.* **62**: 1860.

Cladis, P. E. 1975. *Phys. Rev. Lett.* **35**: 48.

Cladis, P. E., Bogardus, R. K., Daniels, W. B., and Taylor, G. N. 1977. *Phys. Rev. Lett.* **39**: 720.

Cladis, P. E. 1980. *Liquid Crystals: Proceedings of an International Conference Held at the Raman Research Institute [Bangalore 1979]*, ed. S. Chandrasekhar, p. 105. London: Heyden.

Cladis, P. E., van Saarloos, W., Huse, D. A., Patel, J. S., Goodby, J. W., and Finn, P. L. 1989. *Phys. Rev. Lett.* **62**: 1764.

Collings, P. J. 1992. *Mod. Phys. Lett. B* **6**: 425.

Crooker, P. P. 1989. *Liq. Cryst.* **5**: 751.

Das, P., Nounesis, G., Garland, C. W., Sigaud, G., and Tinh, N. H. 1990. *Liq. Cryst.* **7**: 883.

Dasgupta, C. 1987. *J. Phys.* **48**.

Dasgupta, C., and Halperin, B. I. 1981. *Phys. Rev. Lett.* **47**: 1556.

de Gennes, P. G. 1972a. *C. R. Acad. Sci. Paris B* **274**: 758.

de Gennes, P. G. 1972b. *Solid State Commun.* **10**: 753.

de Gennes, P. G., and Prost, J. 1993. *The Physics of Liquid Crystals.* 2d ed. Oxford: Oxford University Press.

De Hoff, R., Biggers, R., Brisbin, D., and Johnson, D. L. 1982. *Phys. Rev. A* **25**: 472.

Destrade, C., Tinh, N. H., Malthete, J., and Jacques, J. 1980. *Phys. Lett.* **79A**: 189.

Emsley, J. W., ed. 1985. *NMR of Liquid Crystals.* Darmstadt: Reidel.

Fischer, M. E. 1968. *Phys. Rev.* **176**: 257.

Fontes, E., Heiney, P. A., and de Jeu, W. H. 1988. *Phys. Rev. Lett.* **61**: 1202.

Frank, F. C., and Chandrasekhar, S. 1980. *J. Physique* **41**: 1285.

Garland, C. W. 1992. *Phase Transitions in Liquid Crystals,* eds. S. Martellucci and A. N. Chester, p. 175. New York, Plenum.

Garland, C. W., and Huster, M. E. 1987. *Phys. Rev. A* **35**: 2365.

Garland, C. W., Litster, J. D., and Stine, K. J. 1989. *Mol. Cryst. Liq. Cryst.* **170**: 71.

Garland, C. W. 1990. *Geometry and Thermodynamics,* ed. J.-C. Tolédano, p. 221. New York: Plenum.

Garland, C. W., and Nounesis, G. 1994. *Phys. Rev. E* **49**: 2964.

Geer, R., Huang, C. C., Pindak, R., and Goodby, J. W. 1989. *Phys. Rev. Lett.* **63**: 540.

Geer, R., Stoebe, T., Huang, C. C., and Goodby, J. W. 1992. *Phys. Rev. A* **46**: 6162.

Geer, R., Stoebe, T., and Huang, C. C. 1993. *Phys. Rev. E* **48**: 408.

Gohin, A., Destrade, C., Gasparoux, H., and Prost, J. 1983. *J. Physique* **44**: 427.

Goodby, J. W., Waugh, M. A., Stein, S. M., Chin, E., Pindak, R., and Patel, J. S. 1989. *J. Am. Chem. Soc.* **111**: 8119.

Gorecka, E., Chandani, A. D. L., Ouchi, Y., Takezoe, H., and Fukuda, A. 1990. *Japan. J. Appl. Phys.* **29**: 131.

Grebel, H., Hornreich, R. M., and Shtrikman, S. 1983. *Phys. Rev. A* **28**: 1114.

Grebel, H., Hornreich, R. M., and Shtrikman, S. 1984. *Phys. Rev. A* **30**: 3264.

Grinstein, G., and Toner, J. 1984. *Phys. Rev. Lett.* **51**: 2386.

Grinstein, G., Lubensky, T. C., and Toner, J. 1986. *Phys. Rev. B* **33**: 3306.

Halperin, B. I., Lubensky, T. C., and Ma, S.-K. 1974. *Phys. Rev. Lett.* **32**: 292.

Halperin, B. I., and Nelson, D. R. 1978. *Phys. Rev. Lett.* **41**: 121.

Hardouin, F., Achard, M. F., Nguyen, H. T., and Sigaud, G. 1986. *Mol. Cryst. Liq. Cryst. Lett.* **3**: 7.

Hill, J. P., Keimer, B., Evans-Lutterodt, K. W., Birgeneau, R. J., and Garland, C. W. 1989. *Phys. Rev. A* **40**: 4625.

Hornreich, R. M., Luban, M., and Shtrikman, S. 1975. *Phys. Rev. Lett.* **35**: 1678.

Hornreich, R. M., and Shtrikman, S. 1989. *Mol. Cryst. Liq. Cryst.* **165**: 183.

Huang, C. C., and Viner, J. M. 1982. *Phys. Rev. A* **25**: 3385.

Huang, C. C., Nounesis, G., Geer, R., Goodby, J. W., and Guillon, D. 1989. *Phys. Rev. A* **39**: 3741.

Huang, J., and Ho, J. T. 1987. *Phys. Rev. Lett.* **58**: 2239.

Huang, J., and Ho, J. T. 1990. *Phys. Rev. A* **42**: 2449.

Huster, M. E., Stine, K. J., and Garland, C. W. 1987. *Phys. Rev. A* **36**: 2364.

Ihn, K. J., Zasadzinski, J. A. N., Pindak, R., Slaney, A. J., and Goodby, J. W. 1992. *Science* **258**: 275.

Inui, S., Kawano, S., Saito, M., Iwane, H., Takanishi, Y., Hiraoka, K., Ouchi, Y., Takezoe, H., and Fukuda, A. 1990. *Japan. J. Appl. Phys. Lett.* **29**: 987.

Jen, S., Clark, N. A., Pershan, P. S., and Priestley, E. B. 1977. *J. Chem. Phys.* **66**: 4635.

Jeong, Y. H., Stine, K. J., Garland, C. W., and Tinh, N. H. 1988. *Phys. Rev. A* **37**: 3465.

Kasting, G. B., Garland, C. W., and Lushington, K. J. 1980. *J. Physique* **41**: 879.

Keyes, P. H. 1991. *Mater. Res. Bull.* **16**: 32.

Kohli, M., Otnes, K., Pynn, R., and Riste, T. 1976. *Z. Phys. B* **24**: 147.

Kumar, S., Chen, L., and Surendranath, V. 1991. *Phys. Rev. Lett.* **67**: 322.

Kutnjak, Z., Garland, C. W., Passmore, J. L., and Collings, P. J. 1995. *Phys. Rev. Lett.* **74**: 4859.

Landau, L. D. 1965. *Collected Papers of L. D. Landau.* Ed. D. ter Haar. New York: Gordon and Breach.

Lavrentovich, O. D., and Nastishin, Y. A. 1990. *Europhys. Lett.* **12**: 135.

Lavrentovich, O. D. 1991. *Phys. Scr. T* **39**: 394.

Leadbetter, A. J. 1987. *Thermotropic Liquid Crystals,* ed. G. W. Gray, p. 1. New York: Wiley.

Levelut, A. M., Tarento, R. J., Hardouin, F., Achard, M. F., and Sigaud, G. 1981. *Phys. Rev. A* **24**: 2180.

Levelut, A. M. 1984. *J. Physique Lett.* **45**: 603.

Litster, J. D., and Stinson, T. 1972. *Phys. Rev. Lett.* **28**: 688.

Lubensky, T. C. 1983. *J. Chem. Phys.* **80**: 31.

Lubensky, T. C., Ramaswamy, S., and Toner, J. 1988. *Phys. Rev. A* **38**: 4284.

Luckhurst, G. R., and Yeates, R. N. 1976. *J. Chem. Soc. Faraday Trans. II* **72**: 996.

Marko, J. F., Indekeu, J. O., and Berker, A. N. 1989. *Phys. Rev. A* **39**: 4201.

Martinez-Miranda, L. J., Kortan, A. R., and Birgeneau, R. J. 1986. *Phys. Rev. Lett.* **56**: 2264.

McMillan, W. L. 1973. *Phys. Rev. A* **7**: 1419.

Miyano, K. 1978. *J. Chem. Phys.* **69**: 4807.

Moncton, D. E., and Pindak, R. 1979. *Phys. Rev. Lett.* **43**: 701.

Navailles, L., Barois, P., and Nguyen, H. T. 1993. *Phys. Rev. Lett.* **71**: 545.

Nelson, D. R., and Halperin, B. I. 1979. *Phys. Rev. B* **19**: 2457.

Nelson, D. R., and Toner, J. 1981. *Phys. Rev. B* **24**: 363.

Nounesis, G., Huang, C. C., and Goodby, J. W. 1986. *Phys. Rev. Lett.* **56**: 1712.

Nounesis, G., Geer, R., Lin, H. Y., Huang, C. C., and Goodby, J. W. 1989. *Phys. Rev. A* **40**: 5468.

Ocko, B. M., Birgeneau, R. J., Litster, J. D., and Neubert, M. E. 1984. *Phys. Rev. Lett.* **52**: 208.

Pang, J., Muzny, C. D., and Clark, N. A. 1992. *Phys. Rev. Lett.* **69**: 2783.

Park, Y., Lubensky, T. C., Barois, P., and Prost, J. 1988. *Phys. Rev. A* **37**: 2197.

Patel, P., Keast, S. S., Neubert, M. E., and Kumar, S. 1992. *Phys. Rev. Lett.* **69**: 301.

Patton, B. R., and Andereck, B. S. 1992. *Phys. Rev. Lett.* **69**: 1556.

Peierls, R. E. 1934. *Helv. Phys. Acta Suppl.* **2**: 81.

Pershan, P. S. 1988. *Structure of Liquid Crystal Phases.* Singapore: World Scientific.

Pindak, R., Moncton, D. E., Davey, S. C., and Goodby, J. W. 1981. *Phys. Rev. Lett.* **46**: 1135.

Pitchford, T., Nounesis, G., Dumrongrattana, S., Viner, J. M., Huang, C. C., and Goodby, J. W. 1985. *Phys. Rev. A* **32**: 1938.

Praefcke, K., Krohne, B., Singer, D., Demus, D., Pelzl, G., and Diele, S. 1990. *Liq. Cryst.* **7**: 589.

Praefcke, K., Krohne, B., Gundogan, B., Singer, D., Demus, D., Diele, S., Pelzl, G., and Bakowsky, U. 1991. *Mol. Cryst. Liq. Cryst.* **198**: 393.

Prost, J. 1979. *J. Physique* **40**: 581.

Prost, J., and Barois, P. 1983. *J. Chim. Phys.* **80**: 65.

Prost, J. 1984. *Adv. Phys.* **33**: 1.

Prost, J., and Toner, J. 1987. *Phys. Rev. A* **36**: 5008.

Prost, J., Pommier, J., Rouillon, J. C., Marcerou, J. P., Barois, P., Benzekri, M., Babeau, A., and Nguyen, H. T. 1990. *Phys. Rev. B* **42**: 2521.

Ratna, B. R., Shashidhar, R., and Raja, V. N. 1985. *Phys. Rev. Lett.* **55**: 1476.

Renn, S. R., and Lubensky, T. C. 1988. *Phys. Rev. A* **38**: 2132.

Renn, S. R. 1992. *Phys. Rev. A* **45**: 953.

Safinya, C. R., Birgeneau, R. J., Litster, J. D., and Neubert, M. E. 1981. *Phys. Rev. Lett.* **47**: 668.

Safinya, C. R., Martinez-Miranda, L. J., Kaplan, M., Litster, J. D., and Birgeneau, R. J. 1983. *Phys. Rev. Lett.* **50**: 56.

Safinya, C. R., Roux, D., Smith, G. S., Sinha, S. K., Dimon, P., Clark, N. A., and Bellocq, A. M. 1986. *Phys. Rev. Lett.* **57**: 2718.

Seideman, T. 1990. *Rep. Prog. Phys.* **53**: 659.

Shashidhar, R., Ratna, B. R., and Prasad, S. K. 1984. *Phys. Rev. Lett.* **53**: 2141.

Shashidhar, R., Ratna, B. R., Prasad, S. K., Somasekhara, S., and Heppke, G. 1987. *Phys. Rev. Lett.* **59**: 1209.

Sirota, E. B., Pershan, P. S., Sorensen, L. B., and Collett, J. 1987. *Phys. Rev. A* **36**: 2890.

Solomon, L., and Litster, J. D. 1987. *Phys. Rev. Lett.* **56**: 2268.

Somasekhara, S., Shashidhar, R., and Ratna, B. R. 1986. *Phys. Rev. A* **34**: 2561.

Stine, K. J., and Garland, C. W. 1989a. *Phys. Rev. A* **39**: 1482.

Stine, K. J., and Garland, C. W. 1989b. *Phys. Rev. A* **39**: 4919.

Stine, K. J., and Garland, C. W. 1990. *Mol. Cryst. Liq. Cryst.* **188**: 91.

Stoebe, T., Geer, R., Huang, C. C., and Goodby, J. W. 1992. *Phys. Rev. Lett.* **69**: 2090.

Thoen, J., Marynissen, H., and Van Dael, W. 1984. *Phys. Rev. Lett.* **52**: 204.

Thoen, J., and Parret, R. 1989. *Liq. Cryst.* **5**: 479.

Thoen, J. 1992. *Phase Transitions in Liquid Crystals*, ed. S. Martellucci and A. N. Chester, p. 155. New York: Plenum.

Tinh, N. H., Hardouin, F., Destrade, C., and Levelut, A. M. 1982. *J. Physique Lett.* **43**: 33.

Toner, J. 1982. *Phys. Rev. B* **26**: 462.

Voets, G. 1992. *PhD Thesis*, University of Louvain.

Voets, G., and Van Dael, W. 1993. *Liq. Cryst.* **14**: 617.

Wang, J., and Lubensky, T. C. 1984. *Phys. Rev. A* **29**: 2210.

Wen, X., Garland, C. W., and Wand, M. D. 1990. *Phys. Rev. A* **42**: 6087.

Wen, X., Garland, C. W., and Heppke, G. 1991. *Phys. Rev. A* **44**: 5064.

Wen, X., Garland, C. W., Shashidhar, R., and Barois, P. 1992. *Phys. Rev. B* **45**: 5131.

Witanachi, W., Huang, J., and Ho, J. T. 1983. *Phys. Rev. Lett.* **50**: 594.

Wright, D. C., and Mermin, N. D. 1989. *Rev. Mod. Phys.* **61**: 385.

Wu, L., Garland, C. W., and Pfeiffer, S. 1992. *Phys. Rev. A* **46**: 973.

Yang, D. K., and Crooker, P. P. 1987. *Phys. Rev. A* **35**: 4419.

Young, A. P. 1979. *Phys. Rev. B* **19**: 1855.

Young, M. J., Wu, L., Nounesis, G., Garland, C. W., and Birgeneau, R. J. 1994. *Phys. Rev. E* **50**: 368.

Yu, L. J., and Saupe, A. 1980. *Phys. Rev. Lett.* **45**: 1000.

CHAPTER 5

Electric Field Effects in Liquid Crystals

L. M. BLINOV

5.1 INTRODUCTION

As it is well known, liquid crystals are very sensitive to an electric field [1] and it is this property which allows their application to display and other optical device technology. Since the beginning of the century, the electrooptical properties of liquid crystals have been studied very actively and many important phenomena such as the Frederiks transition [2], electrohydrodynamic instability [3,4], field-induced domain patterns [5,6], flexoelectric distortion [7] and spatially periodic structures [8–10], ferroelectric switching [11,12], etc., have been discovered.

Today, the electrooptical properties of liquid crystals form well developed branches in both the physics and the technology of liquid crystals. In addition, electrooptical measurements are the basis for a number of precise methods of determining the physical parameters of a material, such as its elastic and viscous coefficients, optical anisotropy, spontaneous polarization, flexoelectric coefficients, anchoring energies at interfaces, etc.

The earlier theoretical and experimental results on the physical properties of liquid crystals, including certain electric field effects, were reviewed in books written by de Gennes [1], Chandrasekhar [13], de Jeu [14], Belyakov and Sonin [15], Vertogen and de Jeu [16] and others. Electrooptical effects were discussed in detail by Kapustin [17], Pikin [18], and Blinov and Chigrinov[19]; many results related to the properties of liquid crystalline materials, and their applications in devices can be found in [20–22].

Recently, a great deal of novel field-induced effects were observed in various liquid crystal phases. They may be divided into three groups: (i) direct influence of the electric field on the structure (phase transition temperatures, order parameters, etc.) of liquid crystals, when the process of the director reorientation is excluded; (ii) the director reorientation, both uniform and spatially periodic, also induced by the direct interaction of the electric field with the dielectrically anisotropic or ferroelectric medium; and (iii) the director reorientation (indirect) by a flow due to the electric field interaction with space charge in a liquid crystal.

Examples of the novel effects are the field-induced biaxiality and the periodic structures in nematics caused by the quadratic coupling of an electric field with the dielectric anisotropy of the substance, the linear electrooptical effects in nematics and cholesterics caused by the flexoelectric coupling with the external field, field distortion of the blue phases, ferro- and antiferro- behavior of chiral smectic structures and so on.

The present paper is devoted to a discussion of electric field effects in various liquid crystal phases. The emphasis is on recent achievements in the area, although important classical results are cited. Most attention is paid to the physical aspects of the phenomena. Only the direct influence of the field on the structure and the director distribution is described. Some important but special problems such as electrohydrodynamic instabilities are only mentioned; some others, such as the behavior of liquid crystals in a spatially nonuniform field or electromechanical effects, are completely excluded from consideration. The electrooptical effects in polymer dispersed liquid crystals and nonlinear effects of a strong optical field are discussed in other chapters.

The Frederiks transition resulting both in uniform and spatially periodic structures, as well as the flexoelectric effect are considered for nematics, both low molecular mass and polymer. A detailed discussion of the anchoring problem is provided in the chapter on surface phenomena. For cholesteric liquid crystals, dielectrically and flexoelectrically induced instabilities and the field behavior of the blue phases are discussed. Electrooptical and dielectric properties of polymer liquid crystals were recently discussed in detail [23]. Thus, here, only specific features of polymer mesophases and novel electric field effects are considered. The field-induced distortions in both ferro- and antiferroelectric chiral smectics are reviewed and specific features of novel amphiphilic (chiral) and polyphilic (achiral) polar mesophases are considered in more detail.

5.2 DIRECT INFLUENCE OF AN ELECTRIC FIELD ON THE STRUCTURE OF LIQUID CRYSTALS

When a field interacts with a dielectrically anisotropic medium, it can shift phase transition points, change the order parameters, or induce a new symmetry, that is, influence the structure and thermodynamic properties of the medium [24]. Such phenomena must be distinguished from the field effects related to the reorientation of the director, which is assisted by strong changes in optical properties of liquid crystals (the latter are discussed in subsequent sections).

5.2.1 Shift of Phase Transitions

When an electric field is applied to a nonpolar uniaxial liquid crystal, a quadratic-in-the-field term must be added to the expression for the free energy of the medium

$$g_{el} = -\frac{1}{8\pi}(\varepsilon_\perp E^2 - \Delta\varepsilon(\mathbf{n}\cdot\mathbf{E})^2) \tag{5.1}$$

where $\varepsilon_\parallel$, $\varepsilon_\perp$ are principal dielectric permittivities referred to the director $\mathbf{n}$ and $\Delta\varepsilon = \varepsilon_\parallel - \varepsilon_\perp$. For $\Delta\varepsilon > 0$ and $\mathbf{E}\,||\,\mathbf{n}$, the second term in (5.1) reduces the free energy and stabilizes the anisotropic phase. Strictly speaking, even for a weak field, the symmetry of the nematic and the isotropic phase is broken and the N–I phase transition disappears (chapter 4 of [18]). However, the apparent N–I phase transition temperature increases by the value

$$\Delta T_{NI} \simeq \frac{(\varepsilon_\parallel - \varepsilon_{is})E^2}{8\pi\Delta H} \tag{5.2}$$

where ΔH is the enthalpy of the N–I phase transition and ε_{is} is the dielectric permittivity of the isotropic phase [25]. The same approach can be used for any other first-order phase transition, even for a weak one like the smectic A–nematic transition.

In the case of second-order transitions, a gain in the electric energy must be compared with a loss of the soft-mode elastic energy. For example, for the smectic A–smectic C phase transition, the shift of the T_{CA} point is [18]:

$$\Delta T_{CA} \simeq \frac{\Delta\varepsilon(\mathbf{E}\cdot\mathbf{n})^2}{8\pi\cdot a'} \tag{5.3}$$

where a' is a parameter of the zero-field Landau expansion for the A–C transition

$$F = a\theta^2 + c\theta^4 \\ a = a'(T - T_{AC}) \tag{5.4}$$

and θ is the tilt angle of the smectic C phase. The sign of the shift depends on the field direction. If **E** is parallel to the normal, to the smectic layers **h** and $\Delta\varepsilon > 0$, the field stabilizes the smectic A phase and T decreases. An oblique field ($\Delta\varepsilon > 0$) induces a tilt, thus stabilizing the smectic C phase, and T_{CA} increases.

For the chiral smectic C* phase, instead of the quadratic term $\Delta\varepsilon E^2/8\pi$, a linear term proportional to the spontaneous polarization (P_sE) must be taken into consideration.

5.2.2 Influence of the Field on Order Parameters

Since the process of the field-induced director reorientation occurs at rather weak fields, the geometries which are being discussed are: (i) the field parallel to the director and $\Delta\varepsilon > 0$; (ii) the field perpendicular to the director and $\Delta\varepsilon < 0$. In the first case, the uniaxial symmetry is conserved and the field stabilizes thermal fluctuations and increases the orientational order parameter, both in the nematic and smectic A phases. A qualitative picture is shown in FIGURE 5.1, taken from [24]. With increasing field the orientational order parameter S grows and the T_{NA} temperature increases. The translational order parameter, being coupled with S, also increases.

FIGURE 5.1. Field-induced changes in the order parameters for a smectic A liquid crystal. S is the orientational order parameter, R is an order parameter describing the coupling of translational and orientational order. The field increases from curve 1 to curve 3 [24]. (Reprinted with permission from the *Journal of Physical Chemistry* 92, p. 1406. Copyright 1988 American Chemical Society.)

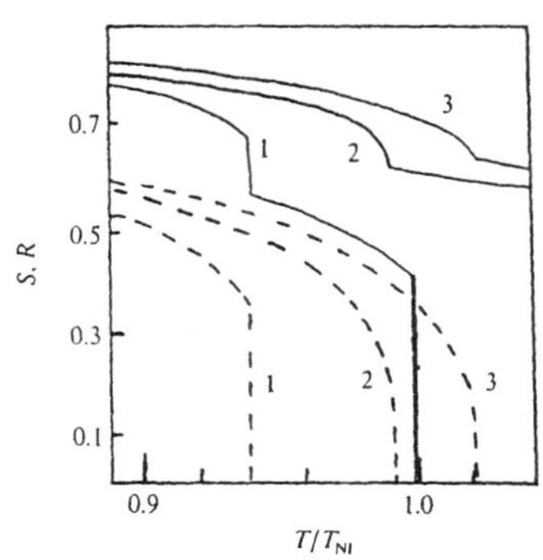

The additional optical anisotropy induced by the field is proportional either to E^2 or to $|E|$, for weak and strong fields respectively [26].

5.2.3 Changes in Symmetry

In fact, an example of a change in symmetry has already been mentioned: a field-induced tilt in the smectic A phase, and thus a change in symmetry from $D_{\infty h}$ to C_{2h}. The induced tilt angle is proportional to the field squared and the factor $(T_{CA}/T - T_{CA})^\gamma$ where $\gamma \simeq 1.3$ [1]. For small $T - T_{CA}$ the field-induced biaxiality can be measured by optical methods.

5.2.3.1 Induced Biaxiality in Nematics

Another example of the symmetry change is the field-induced biaxiality in nematic liquid crystals with negative dielectric anisotropy [27]. In this case the field is applied perpendicular to the director. The orientational order at a point **r** in a nematic liquid crystal is defined by the ordering matrix

$$S_{\alpha\beta}(\mathbf{r}) = \tfrac{1}{2}S_o[3n_\alpha(\mathbf{r})n_\beta(\mathbf{r}) - \delta_{\alpha\beta}] \tag{5.5}$$

where S_o is the order parameter defined with respect to a local director $\mathbf{n}(\mathbf{r})$. Averaging over time and spatial fluctuations of **n** gives the ordering matrix in a principal axis system as:

$$\langle S_{\alpha\beta}\rangle \begin{bmatrix} -1/2(Q-B) & 0 & 0 \\ 0 & -1/2(Q+B) & 0 \\ 0 & 0 & Q \end{bmatrix} \tag{5.6}$$

where Q and B are uniaxial and biaxial order parameters

$$Q = S_o(1 - \tfrac{1}{2}\langle n_x^2 + n_y^2\rangle) \tag{5.7}$$

$$B = (3/2)S_o\langle n_x^2 - n_y^2\rangle. \tag{5.8}$$

Now Q and B are presented in terms of the director fluctuations which are not assumed to be isotropic in the xy-plane. For $n_x^2 = n_y^2$, the biaxial parameter vanishes and the uniaxial one becomes equal to the well known parameter:

$$S = \langle 3\cos^2\vartheta - 1\rangle/2 \tag{5.9}$$

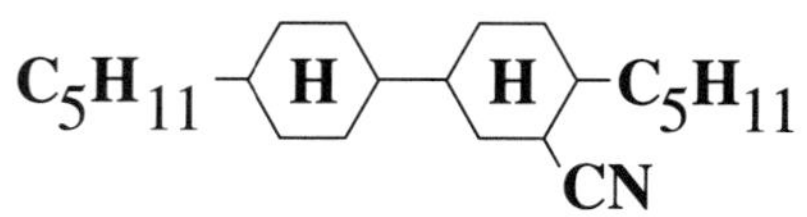

FIGURE 5.2. Compound demonstrating field-induced biaxiality.

where ϑ is an angle between the longitudinal axis of a molecule and the director of the nematic mesophase.

When the field is applied, for example, along the x-direction and $\varepsilon_{xx} = \varepsilon_{yy} = \varepsilon_{\perp} > \varepsilon_{zz} = \varepsilon_{\|}$, the n_x fluctuations are quenched to some extent and B increases. The theory [28] considers the quenching of fluctuations with various wavevectors and results in the following field dependence of the biaxial order parameter

$$B = \frac{3S_o k_B T}{8\pi K}\left(\frac{\Delta\varepsilon E^2}{4\pi K}\right)^{1/2} \tag{5.10}$$

where K is an average elastic constant.

The field-induced biaxiality has been observed in a planarly oriented layer of the compound shown in FIGURE 5.2, with $\Delta\varepsilon = -5.4$ by measuring the ellipticity of laser light transmitted through a sandwich cell ($E \,||$ light wavevector) [27].

5.2.3.2 Kerr Effect

The Kerr effect in the isotropic phase may also be discussed in terms of the field-induced symmetry change of the medium. A birefringence is induced in an ordinary liquid if a uniform electric field is applied at right-angles to the direction of a beam passing through the cell; its value is related to the field strength E

$$\Delta n(E) = \lambda B E^2 \tag{5.11}$$

where B is the Kerr constant and λ is the wavelength of light. The intensity of the light at the exit of the cell placed between an analyser and a polarizer is determined by the phase lag:

$$\Delta\Phi = 2\pi \Delta n l/\lambda \tag{5.12}$$

where l is the path length of the beam in the field.

At temperatures significantly above the transition point from the nematic to the isotropic liquid, materials forming liquid crystals behave like normal liquids. In an electric field, they display the Kerr effect, with an order of magnitude equal to or less than that of nitrobenzene ($+4.1 \times 10^{-12}\,\mathrm{m\,V^{-2}}$). However, as the temperature decreases toward T_{NI}, the Kerr constant of a nematogen changes considerably. For some substances it diverges when $T \to T_{NI}$, being either positive or negative. In general there is a correlation between signs of the Kerr effect in the isotropic phase and the dielectric anisotropy of the nematic phase.

The growth of the Kerr constant is accounted for by considerable contribution of fluctuations of the orientational order parameter to dielectric properties of the isotropic phase. This contribution can be calculated in the framework of the Landau theory [29]. A field E induces the orientational order [30]

$$Q(E) = \rho\Delta\varepsilon E^2/12\pi a_0(T - T^*) \tag{5.13}$$

where ρ is the number density of molecules and a_0 is the parameter of the Landau expansion.

The field-induced order can be related to the Kerr constant

$$B = \Delta n^0 Q(E)/\lambda E^2 \tag{5.14}$$

where Δn^0 is the birefringence of the perfectly aligned liquid crystal [31]. Experiments carried out on 5CB showed that the law $B \propto (T - T_{NI})^{-1}$ is fulfilled within the range $T - T_{NI}$ from several K down to 0.5 K (the Kerr constant is equal to $6 \times 10^{-11}\,\mathrm{m\,V^{-2}}$ at $T - T_{NI} = 1\,\mathrm{K}$ and $\lambda = 633\,\mathrm{nm}$). Closer to T_{NI}, the law is violated and the discrepancy can be accounted for by higher-order terms in the Landau expansion [31].

The Kerr effect in the pretransition temperature regions has been investigated actively [32–37]. In [37], values of $B > 0$ (phenylbenzoates) and $B < 0$ (cyanostilbenes) were obtained close to T_{NI}. These values exceed the absolute value of the Kerr constant for nitrobenzene by factors of 200 and 6.5, respectively. Such high values of B enable the control voltage of the Kerr cell to be reduced several-fold, and permit a decrease of an order of magnitude or more in power consumption [38].

Relaxation times τ_1 associated with order parameter fluctuations increase when approaching T_{NI}, but still remain short enough for light modulation (within the range 0.1–1 μs for 5CB at $T - T_{NI} = 5 - 0.5\,\mathrm{K}$, FIGURE 5.3) [39]. The other time (τ_2) related to molecular relaxation is independent of temperature (FIGURE 5.3).

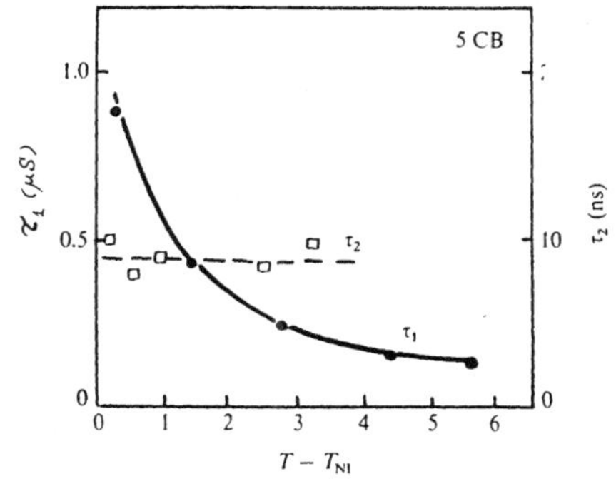

FIGURE 5.3. Relaxation times associated with nematic order fluctuations (τ_1) and with molecular relaxation (τ_2) as functions of temperature (Kerr effect in 5CB) [39].

5.3 NEMATICS

5.3.1 Classical Frederiks Transition

The process of the field-induced reorientation of the director, called the Frederiks transition [2], is usually discussed for the two typical geometries shown in FIGURE 5.4. They correspond to the splay (with subsequent development of the bend) and bend (combined with a splay) distortions.

5.3.1.1 The Simplest Model

In the simplest model (no magnetic field, no conductivity and no flexoelectric coupling), the director **n** reorients in the electric field under the action of the dielectric torque, which is proportional to the dielectric anisotropy $\Delta\varepsilon$. The corresponding contribution g_E to the density of the nematic free energy is written (D is dielectric displacement vector):

$$g_E = -DE/8\pi = -\varepsilon_\perp E^2/8\pi - \Delta\varepsilon(\mathbf{E}\cdot\mathbf{n})^2/8\pi \quad (5.15)$$

that is, director **n** tends to align itself along the field ($\mathbf{n} \,||\, E$) if $\Delta\varepsilon > 0$, and perpendicular to it, ($\mathbf{n} \perp E$), provided that $\Delta\varepsilon < 0$.

The elastic torque supports the initial director orientation, fixed by the boundary conditions on the surface (S). In the case of the strong (infinite) anchoring, the condition $\mathbf{n}|_S = \mathbf{n}_0$ holds during the reorientation process and, as a result, a compromised director profile is produced which satisfies the condition of the minimum free energy:

$$F_V = \int_V (g_K + g_E)\, d\tau \quad (5.16)$$

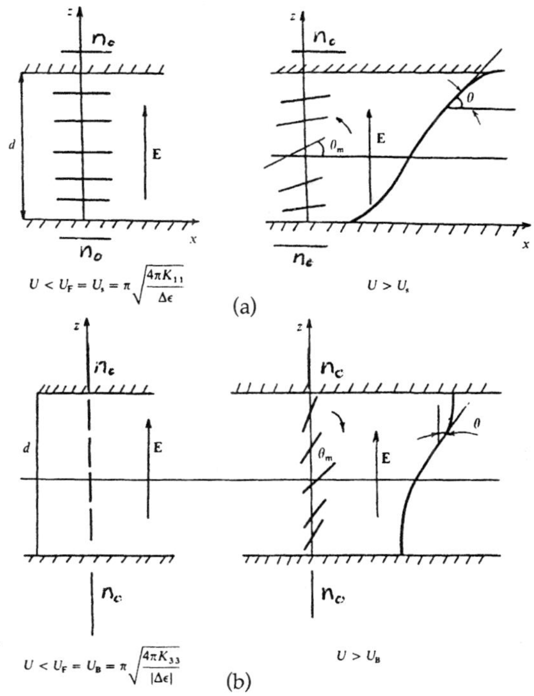

FIGURE 5.4. $S(a)$ and $B(b)$ Frederiks transitions in homogeneous (a) and homeotropic (b) nematic cells with positive (a) and negative (b) values of the dielectric anisotropy. The transitions take place when the applied voltage U exceeds a certain threshold value U_S (a) or U_B (b).

where g_k is the Frank energy:

$$g_k = 1/2[K_{11}(\text{div}\,\mathbf{n})^2 + K_{22}(\mathbf{n}\,\text{curl}\,\mathbf{n})^2 + K_{33}(\mathbf{n}\times\text{curl}\,\mathbf{n})^2]. \quad (5.17)$$

In a more general case of a finite director anchoring at the boundaries, the total energy F of the liquid crystal is written as follows:

$$F = F_V + F_S \quad (5.18)$$

where F_S is the surface energy.

Let us consider the splay Frederiks transition (all expressions are also valid for the bend effect if the following exchange of parameters is made: $K_{11} \leftrightarrow K_{33}$, $\varepsilon_{||} \leftrightarrow \varepsilon_\perp$, $n_{||} \leftrightarrow n_\perp$). In the case of the uniform orientation of the director in the plane of the sample (the xy-plane, wavevectors of distortion q_x, $q_y = \infty$), the director profile **n** takes a two-dimensional form:

$$\mathbf{n}(z) = (\cos\theta(z), 0, \sin\theta(z)) \tag{5.19}$$

which can be described by elliptic functions [19].

The distortion starts at a well-defined threshold voltage independent of the cell thickness:

$$U_S = \pi(4\pi K_{11}/|\Delta\varepsilon|)^{1/2}. \tag{5.20}$$

With increasing thickness d and the voltage kept constant, the decrease in the field strength is compensated for by a weaker elastic energy necessary to maintain a distortion with a larger wavevector $q_z \cong \pi/d$.

$$\Delta\varepsilon E_S^2/4\pi \cong K_{11}q_z^2. \tag{5.21}$$

For a weak surface anchoring (surface energy $W < \infty$), the voltage threshold becomes thickness-dependent decreasing according to the equation:

$$U_S(W) = U_S d/(d+b) \tag{5.22}$$

where $b = K_{11}/W$ is the so-called surface extrapolation length. However, any deviation of the director from the surface in the field-off regime (a pretilt angle) results in a loss of the threshold character of the effect (a continuous distortion instead of the Frederiks transition).

The dynamics of the splay and bend distortions inevitably involve the flow processes coupled with the director rotation. Such a backflow effect usually renormalizes viscosity coefficients. The same is also true for the combined twist distortion. Only the pure torsional distortion is not accompanied by the flow. In the latter case, and for the infinite anchoring energy, the equation of motion of the director expresses the balance between the torques due to the elastic and viscous forces and the external field (and does not contain the fluid velocity):

$$K_{22}\partial^2\varphi/\partial z^2 + (4\pi)^{-1}\Delta\varepsilon E^2 \sin\varphi\cos\varphi = \gamma_1\partial\varphi/\partial t \tag{5.23}$$

This equation describes the director rotation in the electric field with the inertia term $I\partial^2\varphi/\partial t^2$ being disregarded; $\gamma_1 = \alpha_3 - \alpha_2$ is the rotational viscosity, where α_i are Leslie's coefficients. In the limit of small φ angles $|\varphi| \ll 1$, equation (5.23) reduces to:

$$K_{22}\partial^2\varphi/\partial z^2 + (4\pi)^{-1}\Delta\varepsilon E^2\varphi = \gamma_1\partial\varphi/\partial t \tag{5.24}$$

with the solution:

$$\varphi = \varphi_m \exp(t/\tau_r)\sin(\pi z/d) \tag{5.25}$$

where

$$\tau_r = \gamma_1/[(4\pi)^{-1}\Delta\varepsilon E^2 - K_{22}\pi^2/d^2] \tag{5.26}$$

is the reaction or switching-on time.

The corresponding relaxation or decay times are found from (5.24) for $E = 0$ in a similar way:

$$\tau_d = \gamma_1(K_{22}\pi^2/d^2). \tag{5.27}$$

Disregarding the backflow, the same equation can be used for the splay (with K_{11}) and bend (K_{33}) distortions.

Weak boundary anchoring increases both the τ_r and τ_d times. This can be easily understood if it is kept in mind that a finite anchoring energy W results in an increase in the apparent thickness of a cell according to equation (5.22).

The field reorientation of the director results in the electrically controlled birefringence effect. Let, in its initial position, the director be oriented along the x-axis and the nematic liquid crystal layer birefringence be $\Delta n = n_e - n_o = n_{\parallel} - n_{\perp}$. If the field exceeds the threshold value, the director deviates from its orientation along the x-axis while remaining perpendicular to the y-axis. Therefore, the refractive index for the ordinary ray remains unchanged, $n_o = n_{\perp}$. At the same time, the refractive index for the extraordinary ray (n_e) decreases, tending towards n_o. It is simple to relate the magnitude of n to the angle of orientation of the director $\theta(z)$:

$$n_e(z) = \frac{n_{\parallel}n_{\perp}}{[n_{\parallel}^2\sin^2\theta(z) + n_{\perp}^2\cos^2\theta(z)]^{1/2}}. \tag{5.28}$$

The phase difference between the extraordinary and the ordinary ray for monochromatic light of wavelength λ is found by integrating over the layer depth (d):

$$\Delta\Phi = \frac{2\pi}{\lambda}\int_0^d (n_e(z) - n_o)\,dz = \frac{2\pi d\langle\Delta n(z)\rangle}{\lambda}. \tag{5.29}$$

The intensity of light passing through the cell depends on the angle φ_0 between the polarization vector of the incident beam and the initial orientation of the director of the nematic liquid crystal:

$$I = I_0\sin^2(2\varphi_0)\sin^2(\Delta\Phi/2) \tag{5.30}$$

where I_0 is the intensity of the plane polarized light incident on the cell. Thus, the external electric field changes the orientation of the director, $\theta = \theta(E, z)$ and consequently the values of $\Delta n(E, z)$ and $\Delta\Phi(E)$.

A change in the phase difference $\Delta\Phi$, in turn, results in an oscillatory dependence of the optical signal at the exit of the analyser. The maximum amplitude of these oscillations corresponds to an angle $\varphi_0 = 45^\circ$ and the maximum possible number of oscillations (e.g., the number of maxima during a complete reorientation of the director) is approximately $(n_{\|} - n_{\perp})d/\lambda$.

5.3.1.2 Experimental data on conventional nematics

Up to now, the classical results just discussed have been applied successfully for the determination of the elastic moduli K_{ii} and Leslie's viscosity coefficients α_i of conventional nematics. The experimental data may be found in books [14,19,20]. In fact, the Frederiks transition is the most reliable technique for measuring those coefficients which are of great importance in the development of new materials for displays [40].

The measurement of the director profile throughout the cell including surface layers (e.g., by a birefringence technique) allows the anchoring energy to be calculated. The profile of the director distortion above the threshold of the Frederiks transition is shown in FIGURE 5.5. The difference between curves 1 ($W^s = \infty$) and 2 ($0 < W^s < \infty$) is a measure of the anchoring energy. The simplest estimate comes from the comparison of the electric field ξ_E^c coherence length at the threshold field E_c [41–43]:

$$\xi_E^c = \frac{1}{E_c}\left(\frac{4\pi K_i}{\Delta\varepsilon}\right)^{1/2} \tag{5.31}$$

with the sum $d + 2b$, (equation (5.22)). Such a technique requires the use of thin cells ($d \simeq 2b$) and high fields to have ξ^c and b of the same order of magnitude. If a strong field is available one can measure even higher-order terms in the expansion of the anchoring energy. When the applied field is much higher than the threshold of the Frederiks transition, the field coherence length becomes comparable to the surface extrapolation length itself, $\xi_E^* \simeq b$. This condition corresponds to the second threshold of the complete reorientation of a liquid crystal including surface layers, curve 3 in FIGURE 5.5. Thus, the second threshold field allows "b" (and W^s) to be determined.

FIGURE 5.5. The geometry of the bend Frederiks transition (a) and the profile of the director deviation induced by an external field. Curve 1: $H > H_c$, $W^S = \infty$; curve 2: $H^* > H > H_c$, $0 < W^S < \infty$; curve 3: $H > H^*$, $0 < W^S < \infty$.

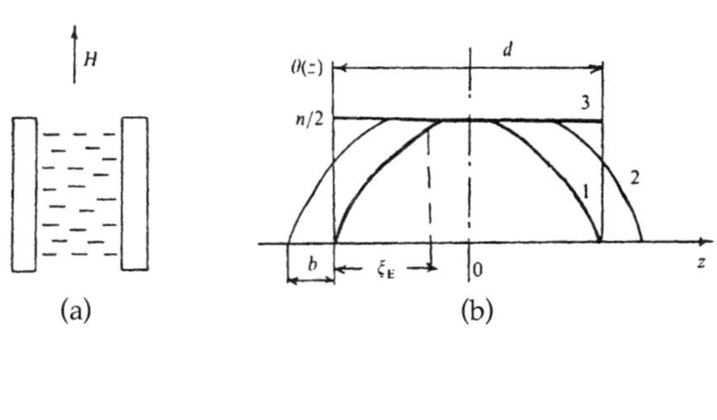

The anchoring energy was shown to depend on cell thickness [44,45]. The thinner the cell, the higher the anchoring energy. It is one of the nonlocality effects discussed in [41]. One of the possible microscopic reasons for the effect is the field of the space charge of the ions adsorbed at opposite interfaces. Such a field can stabilize or destabilize the director in the surface layers in accordance with the sign of the dielectric anisotropy and initial orientation of the nematic [46]. In any case, it results in an increase of W_s in thin cells.

There are dozens of papers in which anchoring energies were measured for various substances [47,48]. It should be noted that the anchoring energy for homeotropically oriented samples is generally of the order of 10^{-3}–10^{-2} erg cm^{-2}. For planar oriented samples, the anchoring energy is generally higher, of the order of 10^{-2}–1 erg cm^{-2}. For a more detailed discussion of the anchoring problems see chapter 6.

Some special cases of the Frederiks transition were recently studied in conventional thermotropic nematics. A discussion of these follows.

The electrooptical behavior of quasihomeotropic oppositely pretilted nematic layers has been investigated [49]. In these layers, due to the boundary conditions

$$\theta|_{z=-(d/2)} = \theta_0, \theta|_{z=+(d/2)} = -\theta_0 \qquad \theta_0 \neq 0, \pi/2 \tag{5.32}$$

the director profile is initially bent, that is, the director angle varies linearly with the coordinate

$$\theta = 2\theta_0 z/d, \tag{5.33}$$

coming to zero at the center of the layer. The Frederiks transition in this case assumes a quasithreshold behavior for negative values of $\Delta\varepsilon$, that is, the angle at the center of the layer remains zero for voltages U lower than the threshold value

$$U_B = \pi(4\pi K_{33}/|\Delta\varepsilon|)^{1/2}.$$

For large enough pretilt angles ($\theta \cong 45^\circ$), $\Delta\varepsilon > 0$, and the splayed initial configuration of the director, an increase in the external voltage results in a jump-like orientational transition from the splayed configuration to a bent structure (a first-order transition instead of the second-order one) [50,51]. At a certain voltage there are two stable director distributions over the cell thickness, evidently separated by a potential barrier (a bistability effect). In [52] even three stationary states (two stable and one metastable) were observed.

The Frederiks transition in a hybrid (or homeoplanar) cell was considered both for the fields in the plane of the director distortion [53] and perpendicular to it [54]. It was shown that in the latter case, the effect has a threshold character and the threshold is lower than that for the homogeneous orientation [53].

Specific features of the Frederiks transition are observed in the case of a simultaneous action of electric and magnetic fields. If the magnetic field is applied parallel to the nematic director $\mathbf{H} \| \mathbf{n}_0$, it increases the apparent elastic coefficient by an additional quantity of $\Delta\chi H^2 d^2/\pi^2$, which results in higher values of the corresponding threshold voltages [55].

However, this is not the only result of a magnetic field [56]. If a stabilizing magnetic field is applied along the homogeneously (FIGURE 5.6(a)), or homeotropically (FIGURE 5.6(b)) oriented director and a destabilizing electric field is perpendicular or parallel to the substrate plane, the first-order Frederiks transition with a discontinuous jump of the director orientation at the threshold voltage may take place for sufficiently high H values.

In the case when $\mathbf{E} \perp \mathbf{H} \perp n_0$ and both the $\mathbf{E}$ and $\mathbf{H}$ fields are destabilizing, several distortion regions occur [57]. For sufficiently high magnetic fields and small electric fields, twist distortion takes place in the direction of $\mathbf{H}$ (FIGURE 5.6(c)). If the magnetic field is low and the electric field is strong, the splay–bend distortion or S-effect appears (FIGURE 5.6(d)). In the region where the electric and magnetic field actions are comparable, a mixed splay–bend–twist deformation arises (FIGURE 5.6(e)). Phase transitions of both second and first order occur between the differently distorted regions. First-order transitions are accompanied by bistability and hysteresis [58].

A consideration of the influence of the anisotropy of the electrical conductivity on the Frederiks transition showed that the anisotropy has no influence on the threshold voltage, but it does affect the above-threshold variation of the director throughout the depth of the cell [59-61].

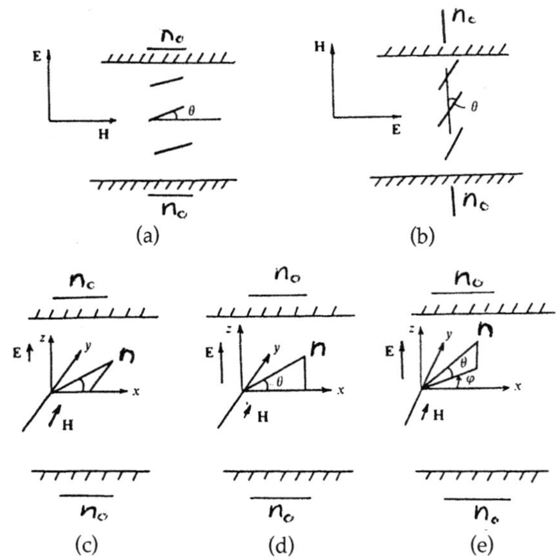

FIGURE 5.6. Crossed electric and magnetic fields. (a), (b) magnetic field stabilizes director orientation, the first-order Frederiks transitions in an electric field are possible; (c), (d), (e) both magnetic and electric fields are destabilizing; twist (c), splay–bend (d) and twist–splay–bend (e) distortions are possible, dependent on the value of the electric and magnetic fields.

The effect of the flow that accompanies the reorientation process in the cases of the S (splay) and B (bend) distortions (the backflow effect mentioned above) was shown to renormalize the rotational viscosity of a nematic [62–64]:

$$\gamma_1^* = \gamma_1 - \frac{2\alpha_3^2}{\alpha_3 + \alpha_4 + \alpha_6} \quad \text{for the S effect}$$
$$\gamma_1^* = \gamma_1 - \frac{2\alpha_2^2}{\alpha_4 + \alpha_5 - \alpha_2} \quad \text{for the B effect.} \tag{5.34}$$

These relationships show that backflow considerably alters the response times and should be taken into account. The other consequence of the backflow is a nonmonotonic relaxation of the director after switching the field off.

Weak boundary anchoring increases the τ_r time and slows down τ_d [65,66], as expected from equations (5.26) and (5.27).

A fast electrooptical response may be achieved with the Frederiks transition when the electrically controlled birefringence (ECB) of monochromatic light is used.

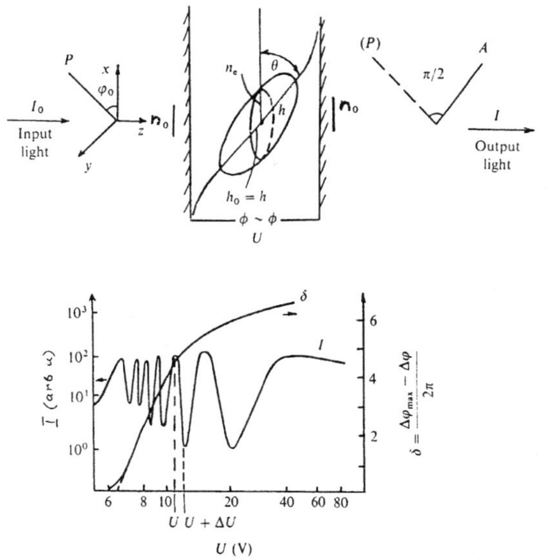

FIGURE 5.7. Electrically controlled birefringence in a homogeneous cell (S effect). Polarizer (P) is oriented at an angle φ_0 with respect to the initial director $\mathbf{n}_0$ parallel to the substrate plane; the analyzer is crossed with the polarizer. Below: the intensity I of the output light and birefringence S are shown versus applied voltage U [69]. When the voltage varies from U to $U + \Delta U$, the transmitted intensity decreases from $I = I_{max}$ to $I = I_{min}$.

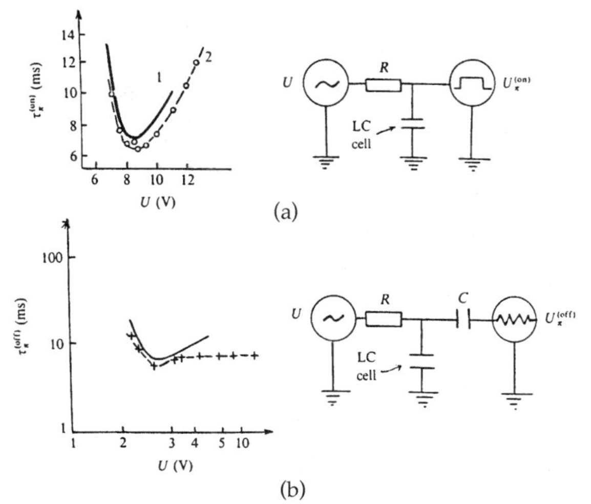

FIGURE 5.8. Fast switching in the S effect by $\Delta\Phi$ equal to π [69,70]. (a) τ_π^{on} times reveal a minimum at intermediate voltages (1 is calculated curve, backflow allowed for; 2 is experiment). (b) Dependence of τ_π^{off} times on bias voltage qualitatively resembles that of τ_π^{on} (solid curve–calculations, (+)—experiment). The principal schemes of driving circuits for low- (a) and high- (b) frequency addressing are also given.

The characteristic curves of the transmitted intensity in the ECB effect obey equation (5.30) and are shown in FIGURE 5.7. The contrast ratio I_{max}/I_{min} is not in principle limited, but determined in the experiment by the quality of initial director orientation. The output light, in general, becomes elliptically polarized [55].

The problem is to provide full switching from I_{max} to I_{min} with minimal response times [67,68]. Such switching is realized, when the phase difference $\Delta\Phi$ is changed by exactly π. If a nematic cell is subjected to voltage U, corresponding to a maximum intensity, I_{max}, then to attain another state with I_{min}, one must supply additional voltage $U_\pi > \Delta U$ (FIGURE 5.7). Time τ_π, which corresponds to the total intensity variation $I_{max} \leftrightarrow I_{min}$ with increasing applied (bias) voltage U, passes through a minimum at the intermediate voltages $U \simeq 1.3U_S$ [69,70], where U_S is the S effect threshold voltage (FIGURE 5.8). For large amplitudes of a switching pulse $U_\pi \gg \Delta U$, the reaction times for a phase lag change by π can reach values of 20 μs at room temperatures [68].

For the biased cells discussed, the off times are usually three to four orders of magnitude larger. Using a special form of the applied voltage (relaxation in the field-off regime) and operation on the last (highest voltage) peak of the optical transmission, the decay time may be reduced considerably. A frame time of about 50 μs has been demonstrated for a room temperature nematic [71]. A decrease of the decay time down to the value of τ_r^π may also be achieved if, as a result of the low-frequency relaxation of $\varepsilon_{\|}$, the dielectric anisotropy of the liquid crystal becomes negative at a certain frequency. It is then possible to address the cell with a high-frequency signal and to force the director to rotate rapidly into the homogeneous "off" configuation [72]. Minimal times of the total switching cycle $I_{max} \rightarrow I_{min} \rightarrow I_{max}$, obtained in the S effect with the double frequency addressing technique, achieved 40–50 μs at room temperatures [73].

5.3.1.3 Polymer Systems

Field behavior of linear-chain and comb-like polymers was investigated recently. In general, in analogy with low-molecular mass compounds, an electric field can reorient polymer nematics according to the sign of the dielectric anisotropy (Frederiks transition).

On the other hand, an electric field causes some specific transformations in the structure of polymeric mesophases. As was pointed out by Shibayev [74], the field considerably changes the correlation length for the short-range smectic order in the nematic phase of comb-like polymers. The conformation and the orientational order parameter for flexible spacers separating mesogenic units from the backbone are also changed [75]. The anisotropy in the orientation of polymeric backbones is induced by the electric field due to a torque exerted by the mesogenic units. In some cases the field induces smectic phases in thermodynamically stable nematic phases [76].

Thus, in general, electric and viscoelastic properties of liquid crystal polymers are field-dependent, and the response of the materials to an external field is essentially nonlinear. Unfortunately, in most of these electrooptical experiments this nonlinearity is not taken into account and the results are interpreted in terms of the conventional nematodynamics with constant material parameters.

Field-induced reorientation of nematic polymers, caused by the coupling of the electric field with the dielectric anisotropy, has been studied in a variety of papers [74,76–83]. The optical techniques and measurements of the birefringence are reported in detail [77,79]. Unfortunately, only in a few papers, for example [82], is a certain preliminary orientation of a polymer specified and one can speak of the true Frederiks transition with a well defined threshold voltage. Nevertheless, qualitative estimates of the

FIGURE 5.9. Frequency dependence of the threshold voltage for the Frederiks transition in a nematic polymer with sign inversion of the dielectric anisotropy [78].

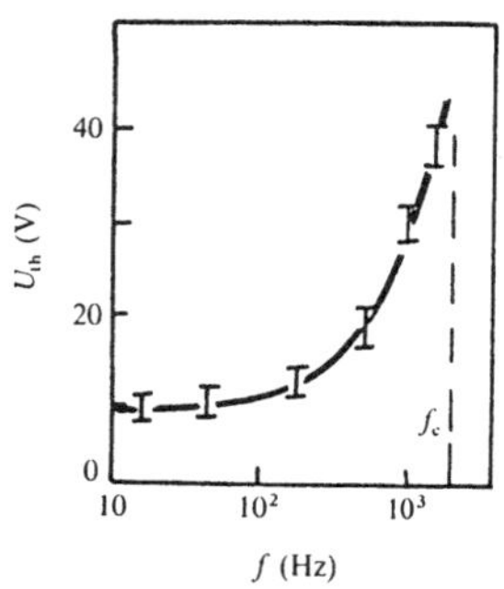

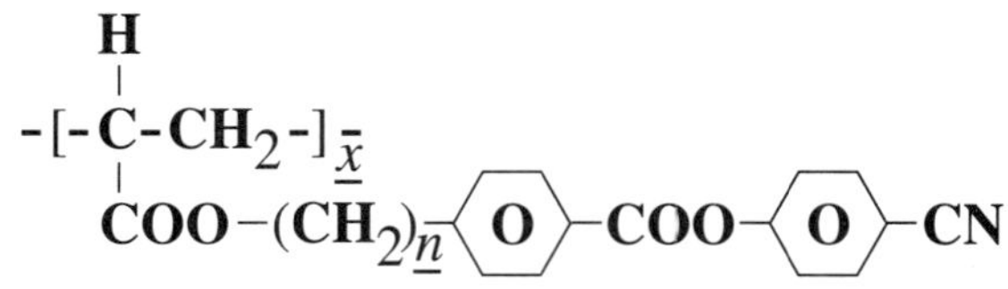

FIGURE 5.10. Comb-like nematic polymer.

elastic moduli in both comb-like [78–80] and linear-chain [81,82] nematic polymers have been carried out.

FIGURE 5.9 shows the role of the dielectric anisotropy in the process of the field reorientation of a comb-like nematic polymer (see FIGURE 5.10) [78a]. Due to the chemical coupling of the cyanobiphenyl moieties to the backbone, there is an inversion of the sign of $\Delta\varepsilon$ at a rather low frequency, which results in a divergence of the threshold voltage–frequency curve for the Frederiks transition in planarly oriented cells.

The experimental results show that the Frank elastic moduli are determined mainly by the structure of the mesogenic units which are similar for conventional nematics and thermotropic polymers (the situation changes considerably for lyotropic solutions of long rod-like polymer molecules). On the other hand, the dynamics of reorientation are strongly influenced by the backbone. Field response and relaxation times depend dramatically on the molecular mass of the polymer, though in the first approximation they obey the same equations (5.26, 5.27). This may be demonstrated in FIGURE 5.11, where the inverse of the field response time is proportional to $U^2 - U_S^2$ as predicted by (5.26). The elastic modulus $K_{11} = 3 \times 10^{-7}$ dyn and the twist viscosity coefficient $\gamma = 10$–$100\,\mathrm{P}$ were estimated from the threshold voltage (offset of the line) and the slope of the curve, respectively [77].

FIGURE 5.11. Reciprocal field-response times as a function of voltage squared for a siloxane polymer [77].

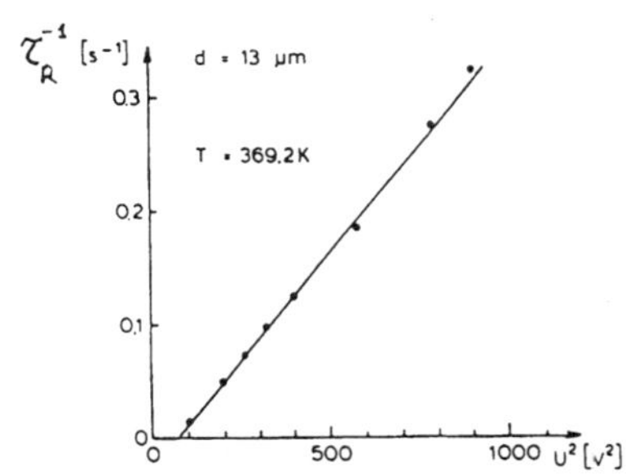

Polymer dispersed liquid crystals are out of the scope of this chapter and are discussed in chapter 9. Here only the case when polymer threads are introduced into a nematic solvent and are chemically cross-linked to form a network for the liquid crystal molecules is mentioned [84]. This is a novel form of the guest–host system and many new physical effects are still to be studied. A novel type of display (randomly oriented nematic displays) has already been proposed [84].

5.3.2 Dielectrically Induced Periodic Structures

Under certain conditions, the field-induced reorientation caused by the dielectric torque (energy term (5.15)) may result not in the uniform director distribution in the plane of a liquid crystal layer, but in spatially periodic distortions, either steady-state or transient.

5.3.2.1 Steady-State Patterns

At present, at least three types of the steady-state dielectrically driven patterns are known for nematics.

The electric-field-induced periodic bend distortion in the form of parallel stripes has been observed in a homeotropically oriented layer of 5CB ($\Delta\varepsilon \cong 13$) in the presence of a stabilizing magnetic field H_Z [85]. The stripes with wavevector q_y were parallel to the electric field E_X and stationary at low fields. It was shown that a stable periodic pattern of the director minimizes the free energy of the cell when elastic moduli K_{11} and K_{33} do not differ very much from each other [85]. In these experiments, the Frederiks transition is first-order; the nondeformed and deformed areas coexist at a given voltage and the interface between them may propagate along the y direction perpendicular to both fields [86].

The other stationary periodic pattern has been observed in polymer lyotropic liquid crystals [87]. The smallness of the ratio K_{22}/K_{11} of a poly-γ-benzyl-glutamate solution results in a structure with a period of the longitudinal domains on the order of the cell thickness. Due to the high viscosity of the material ($\gamma_1 \cong 34\,\mathrm{P}$), the time of the domain formation exceeds two hours. If the director was initially oriented along the x-axis (n_x) and the field (magnetic in [87]) acts along the cell normal z, the structure of the distortion

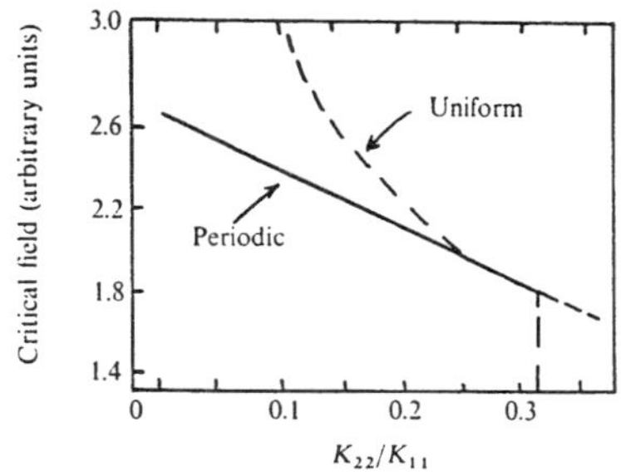

FIGURE 5.12. Calculated values of the critical field in units $(\pi/d)(K_{22}/\Delta\chi)^{1/2}$ for the periodic splay–twist and uniform splay distortions [87].

may be found in the form

$$n_z = f(z)\cos qy \qquad n_y = g(z)\sin qy \qquad n_x \cong 1 \tag{5.35}$$

(the uniform Frederiks distortion corresponds to $q = 0$).

Solution of the Euler equation with the splay and twist distortions results in a bifurcation point for the threshold voltage when plotted as a function of K_{22}/K_{11} (FIGURE 5.12) [88]. The curves shown in the figure were calculated numerically for zero anchoring energy ($W = 0$) at both boundaries. It is seen that for $K_{22}/K_{11} < 0.303$, the periodic splay–twist distortion is more favorable than a uniform splay.

The same result was confirmed in [89,90] for a finite anchoring energy. The critical ratio K_{22}/K_{33} increases with increasing W and reaches 0.5 for infinitely strong anchoring. The periodic distortion has been observed in the magnetic field, but there must be an electric analog. Different geometries for studying the effect were reviewed by Kini [91].

The steady-state modulated structure is also observed in homogeneously oriented nematic layers at frequencies of the applied AC voltage close to the frequency of the sign inversion of the dielectric anisotropy [92,93]. According to the Debye model, the frequency dependence of the dielectric constant $\varepsilon_{\|}$ is:

$$\begin{aligned} \varepsilon_{\|} &= \varepsilon' - \mathrm{i}\varepsilon'' \\ \varepsilon' &= \varepsilon_\infty + \frac{\varepsilon_0 - \varepsilon_\infty}{1 + (\omega/\omega_D)^2} \\ \varepsilon'' &= \frac{(\varepsilon_0 - \varepsilon_\infty)\omega/\omega_D}{1 + (\omega/\omega_D)^2} \end{aligned} \tag{5.36}$$

where ω_D is the Debye dipole relaxation frequency, $\varepsilon_0 = \varepsilon_{\|}\,(\omega = 0)$, $\varepsilon_\infty = \varepsilon_{\|}\,(\omega \to \infty)$, and ε' and ε'' are the real and imaginary parts of the dielectric constant $\varepsilon_{\parallel}$, respectively. The dielectric anisotropy sign inversion point $\omega = \omega_i$ means that $\varepsilon' = \varepsilon_{\perp}$. The calculations show that the free energy is a minimum when the periodic splay distortion and periodic electric potential distribution in the plane of a layer occur. The threshold and period of the distortion at the threshold are [92,93]:

$$U_{th} = \pi\varepsilon'(\varepsilon'')^{-3/2}(54\pi K)^{1/2}(1 + 2\varepsilon''^2/9\varepsilon'^2)$$
$$S_{th} = d(\tfrac{1}{2} + 2\varepsilon''^2/9\varepsilon'^2)^{-1/2} \quad (5.37)$$

where for simplicity it is assumed $K_{11} = K_{33} = K$. As seen from (5.37), threshold conditions include only elastic and dielectric parameters of a nematic liquid crystal, as in the ordinary Frederiks transition. The threshold voltage U_{th} becomes infinite for $\varepsilon_0 \to \varepsilon_{\perp}$ and $\varepsilon_{\perp} \to \varepsilon_{\infty}$.

The frequency region of the modulated Frederiks structure is very narrow near $\omega = \omega_i$. For $\omega < \omega_i$, the uniform Frederiks transition becomes more favorable, while for $\omega > \omega_i$, the threshold, calculated according to (5.37) rapidly diverges (FIGURE 5.13). If an initial director orientation is homeotropic, the threshold curves for U_{th} and S_{th} reverse their slope with respect to the axis $\omega = \omega_i$ [93].

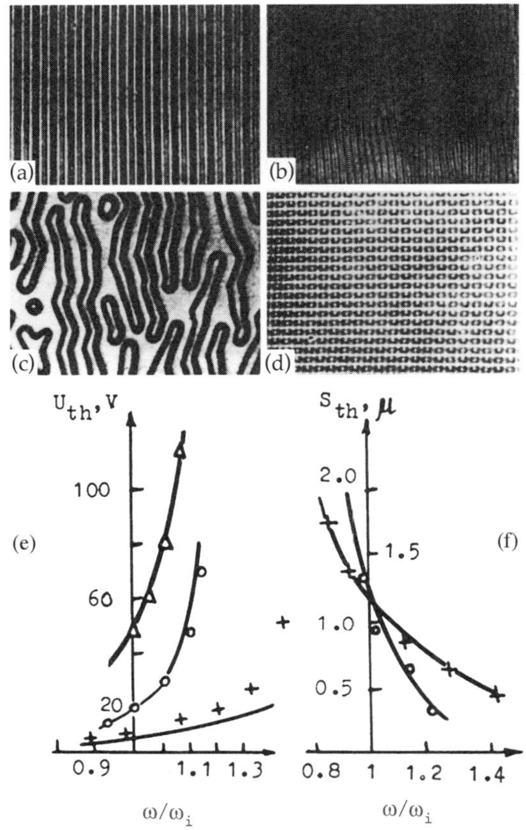

FIGURE 5.13. Frederiks domains near the dielectric anisotropy sign inversion frequency [92,93]. (a)–(d) are experimental pictures. (a) $\omega < \omega_i$, $U = U_{th}$; (b) $\omega > \omega_i$, $U = U_{th}$; (c) $\omega < \omega_i$, $U > U_{th}$; (d) $\omega > \omega_i$, $U > U_{th}$. Below: dependences of the threshold voltage (e) and domain period (f) on reduced frequency ω/ω_i (solid line—calculations; experimental values—crosses: $\varepsilon_0 - \varepsilon_{\perp} = +4.75$, circles: $\varepsilon_0 - \varepsilon_{\perp} = +0.35$, triangles: $\varepsilon_0 - \varepsilon_{\perp} = +0.05$.

5.3.2.2 Transient Patterns

The formation of transient domain patterns aligned perpendicular to the initial director during the relaxation process of the magnetically driven Frederiks transition has been known since the earliest observations [94,95]. The observed parallel stripes are, in fact, walls separating splayed regions with the clockwise and counterclockwise rotations of the director, arising due to a coupling of the director orientation with a flow of a liquid (the backflow effect). The same mechanism is responsible for the appearance of oblique structures [96].

In some cases, magnetically induced transient twist distortions were observed both in thermotropic (MBBA [97]) and lyotropic (PBG [98]) systems, which cannot be explained by the backflow effect since the latter does not accompany twist distortions. The physical origin of this phenomenon could be explained by a faster response time for the modulated structures compared with the uniform ones. When the equilibrium director distribution is approached, that is, a relaxation process is over, the transient structures disappear. The emergence and subsequent evolution of the spatial periodicity of the transient structures have been considered theoretically [97,98].

Recently, another transient pattern was observed when a rather strong electric field was applied to a homogeneously oriented layer of 5CB having a large dielectric anisotropy [99,100]. Domain stripes observed were oriented parallel to the initial orientation of the director. Thus the mechanism of the pattern formation should be quite different from that corresponding to the case of small dielectric anisotropy (or to the magnetically driven transitions since the magnetic anisotropy is always small). A nonlinear theory of the transient patterns seems to be necessary to account for the appearance of the longitudinal domains [101,102].

5.3.3 Flexoelectric Phenomena

When a liquid crystal possesses an electric polarization P (spontaneous or induced by external factors), in addition to the quadratic-in-the-field coupling of an external electric field with the dielectrically anisotropic medium (free-energy term (5.15)), a new, linear-in-the-field interaction appears (free-energy density $\mathbf{P}\cdot\mathbf{E}$). One of the sources of the electric polarization is an orientational distortion of a liquid crystal.

5.3.3.1 Flexoelectric Effect: Classical Results

A macroscopic dipole moment per unit volume, that is, the electric polarization, may result from both the nonuniform director orientation [7]

$$\mathbf{P}_f = e_1 \mathbf{n}\,\mathrm{div} - e_3 \mathbf{n} \times \mathrm{curl}\,\mathbf{n} \qquad (5.38)$$

and spatially nonuniform orientational order parameter [103]

$$\mathbf{P}_o = r_1(\mathbf{n}\,\mathrm{grad}\,S)\mathbf{n} + r_2\,\mathrm{grad}\,S. \qquad (5.39)$$

The flexoelectric polarization $\mathbf{P}_f$ depends on the curvature of the director field ($\mathrm{div}\,\mathbf{n}$ and $\mathrm{curl}\,\mathbf{n}$) at constant modulus of the order parameter S. For uniform director orientation ($\mathbf{n} = \mathrm{const}$) $\mathbf{P}_f \equiv 0$. The molecular picture of the phenomenon is shown in FIGURE 5.14(a) and (b). The second term is the so-called order polarization. It depends on the gradient of the nematic quadrupolar order parameter (5.5) and does not vanish for the uniform director distribution. For instance, near a solid wall the nematic order parameter is a function of the z-coordinate. The order electricity coefficients $r_{1,2}$ are of the order of the flexoelectric coefficients, $e_{1,3} \approx 10^{-4}$ CGS units. The order electricity is not further discussed as it is mostly related to the surface properties reviewed in chapter 6.

The flexoelectric effect results in a distortion of the dielectrically stable configuration (homeotropic nematic with $\Delta\varepsilon < 0$) provided, that the director anchoring at the boundaries is weak (FIGURE 5.14(c)) [19]. The corresponding distortion is shown in FIGURE 5.14(d). It differs from the distortion in a Frederiks transition (see, for example, FIGURE 5.4) in that the maximum angle of deviation of the director occurs at the restricting surface, and the angle of deviation in the middle is zero. This is due to a particular geometry of the distortion: the polarization in a field occurs uniformly throughout the volume, and as a result the torque capable of rotating the director through an angle θ occurs only at the surfaces. Because of this, weak anchoring of the nematic liquid crystal at the surface is a necessary requirement for the occurrence of such a flexoelectric distortion.

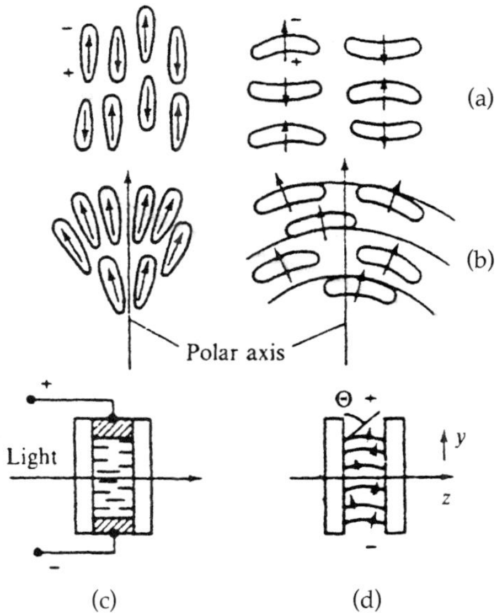

FIGURE 5.14. Flexoelectric effect. (a) The structure of an undeformed nematic liquid crystal with pear- and banana-shaped molecules; (b) the same nematic liquid crystal subjected to splay and bend deformations, respectively; (c) experimental configuration; (d) distribution of the director along the z-axis due to the flexoelectric effect.

The anchoring energy may be determined from the flexoelectric experiment (see, for example, [45]). It should be noted, however, that there is currently some discrepancy between the values of the anchoring energy and flexoelectric coefficients derived by various methods, even for the same material, in particular, MBBA. This may be caused by not allowing for the surface polarization of the nematic liquid crystal, caused by the polar nature of its molecules [104].

A special case of the flexoelectric distortion in a hybrid aligned nematic (HAN) configuration was studied in [105,106]. In HAN cells the director is homeotropic on one of the substrates and homogeneous on the other. Measurement of transmitted intensity in HAN samples allows for the evaluation of both the anchoring energy for the homeotropic orientation $W \cong 6 \times 10^{-3}\,\mathrm{erg\,cm^{-2}}$ (MBBA) and the sum of flexoelectric moduli: $e_{11} + e_{33} \cong -4.5 \times 10^{-4}\,\mathrm{dyn}^{1/2}$. In an electric field normal to the initial director plane, the flexoelectric effect results in the appearance of the twist deformation. From the measurements of the pyroelectric effect in a HAN cell, the absolute value of the flexoelectric polarization has been found [107].

An interesting effect of the surface bistability in nematics has been proposed in [108]. Two stable degenerate states are realized when SiO is obliquely evaporated onto the glass substrate, provided that a certain evaporation angle and the thickness of evaporated SiO are chosen [109]. The bistable switching between two surface states takes place due to the linear flexoelectric interaction between the electric field E in the plane of the substrate and the nematic director.

5.3.3.2 Dynamics of the Flexoelectric Distortion

The dynamics of the flexoelectric effect must be studied in more detail. In an earlier paper [110], the frequency spectrum was studied of a helium–neon laser beam scattered by a phase diffraction lattice occurring in a liquid crystal as a result of the flexoelectric effect. A characteristic relaxation frequency was found, which agreed in order of magnitude with the calculated value $K_{11}\pi^2/d^{*2}\gamma_1$, where d^* is the separation between adjacent interdigital electrodes.

Reference [111] examined theoretically the dynamic behavior of a nematic layer (with a geometry similar to that of FIGURE 5.14(c) in an AC field with allowance for the flexoelectric effect. Euler's equation for the volume of a sample written in linearized form ($\sin\theta \cong \theta$, $\cos\theta \cong 1$):

$$K_{33}\partial^2\theta/\partial z^2 - \Delta\varepsilon E^2\theta/4\pi = \gamma_1\partial\theta/\partial t \tag{5.40}$$

does not contain a destabilizing flexoelectic torque, as it is linear in the derivatives of the director. However the flexoelectric term appears in the boundary conditions:

$$K_{33}\partial\theta/\partial z\big|_{z=\pm d/2} \pm W_S^i\theta\big|_{z=\pm d/2} - e_{33}E = 0. \tag{5.41}$$

Here W_S^i corresponds to different anchoring energies at the two opposing surfaces ($i = 1, 2$), which have coordinates $\pm d/2$ (d is the thickness of the layer). The solution of the linear equations (5.40, 5.41) is given for various values of $\Delta\varepsilon$ and W_S^i.

According to the above-mentioned results, the flexoelectric response is limited to a narrow frequency region below the structural relaxation frequency $f_s \simeq K\pi^2/\gamma_1 d^2$, where K is an average elastic modulus and d is a layer thickness.

In a special case studied in [112], the flexoelectric torque acting on a nematic liquid crystal (doped MBBA and 5CB) at the interface with an electrode of a sandwich-type cell results in a viscoelastic wave of the director curvature decaying into the bulk of the substance. In experiments with MBBA the dielectric torque was excluded by a special adjustment of $\Delta\varepsilon$ to zero, and the tilted orientation was used to increase the flexoelectric torque ($\theta_0 = 72°$, insert to FIGURE 5.15). The wavevector of the curvature oscillations increases with increasing field frequency, $q \cong (\omega\gamma_1^*/2K)^{1/2}$, where γ_1^* is effective rotational viscosity and K is an average elastic modulus.

The linear-in-the-field flexoelectric response in the boundary region was observed by modulation ellipsometry. In such a case, the response is observed up to frequencies on the order of 100 kHz. The characteristic frequency, $f_c \cong K/\pi\gamma_1^*\lambda^2$, is determined by a crossover of two distances, the spatial period of the curvature wave and the penetration length of the optical evanescent wave (a few tenths of the light wavelength λ) (FIGURE 5.15). The comparison of the theoretical and experimental data allows for the evaluation of the

FIGURE 5.15. Profile of the curvature wave $\delta\theta(z)$ in a nematic. The solid surface is at $z = 0$. A normalized evanescent optical wave (a) for $q_\omega\lambda \ll 1$; (b) for $q_\omega\lambda \cong 1$. Inset: definition of angles.

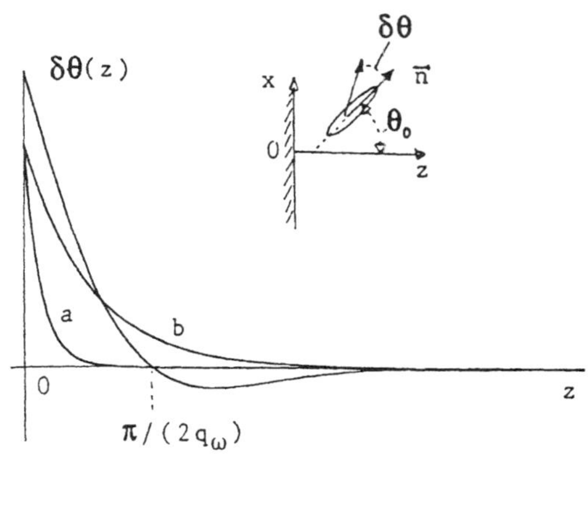

anchoring energy for the oblique orientation, $W \cong 6 \times 10^{-2}\,\mathrm{erg\,cm^{-2}}$, and the sum of the flexoelectric moduli, $e_{11} + e_{33} = 3 \times 10^{-4}\,\mathrm{dyn}^{1/2}$.

Recently, a ferroelectric-type switching and optical second-harmonic generation have been observed in a classical nematic substance, 5CB, obliquely oriented by polyimide films [113]. The interpretation was given in terms of polar, near-surface layers, possessing macroscopic polarization; however, another interpretation based on consideration of flexoelectric phenomena cannot be ruled out either.

5.3.3.3 Flexoelectric Domains

The flexoelectric term δF in the nematic free energy

$$\delta F_f = -\int_v [e_{11}(\mathbf{E}\cdot\mathbf{n})\,\mathrm{div}\,\mathbf{n} + e_{33}\mathbf{E}(\mathrm{curl}\,\mathbf{n}\times\mathbf{n})]\,d\tau \quad (5.42)$$

at least, in one particular case, may result in a periodic (modulated) structure. When an external DC electric field is applied to a homogeneously oriented nonconductive nematic layer, a pattern of domains arises parallel to the initial director orientation [8]. The period of the stripes w_{th} and the threshold voltage U_{th} of their appearance has been found [9] by minimizing the free energy of the nematic in an electric field, taking into account the term (5.42). The equations were linearized with respect to the splay and twist angles θ, φ of the director deviation from the initial orientation and the threshold values of w_{th} and U_{th} were found as solutions of the corresponding dispersion relation (assuming $K_{ii} = K$) [9]:

$$U_{th} = \frac{2\pi K}{|e|(1+\mu)} \qquad w_{th} = \frac{d}{\pi}\left(\frac{1+\mu}{1-\mu}\right) \quad (5.43)$$

where $e = e_{11} - e_{33}$ and $\mu = \Delta\varepsilon K / 4\pi e^2$. In [114] the values of U_{th} and w_{th} have been calculated more precisely for the case of $K_{11} \neq K_{22}$.

An experimental study of the dependences of the threshold and period of the longitudinal domains on the dielectric anisotropy and other parameters of nematic liquid crystals [10] has shown a very good

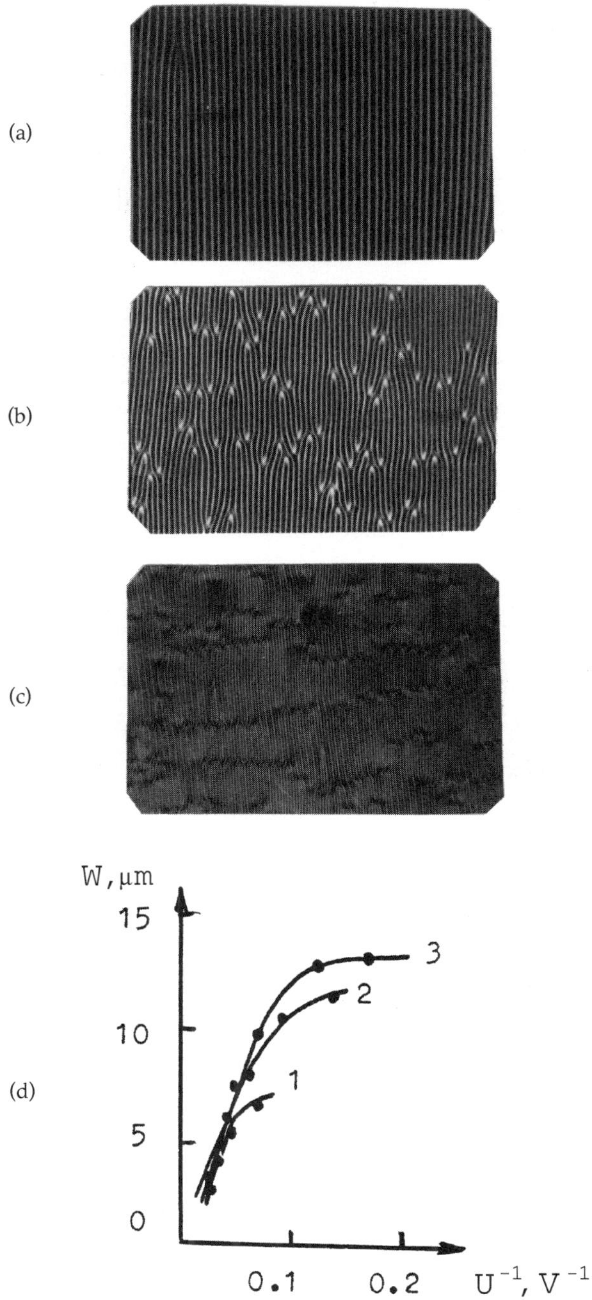

FIGURE 5.16. Flexoelectric domains in the nematic liquid crystal 4-butyl-4′-methoxyazoxybenzene [10]: (a) $U = 16\,\mathrm{V}$, (b) $U = 25\,\mathrm{V}$, (c) $U = 50\,\mathrm{V}$ ($\Delta\varepsilon = -0.25$, $d = 11.7\,\mu\mathrm{m}$); (d) variation of the domain period for different $\Delta\varepsilon$ values: 1, $\Delta\varepsilon = -0.25$; 2, $\Delta\varepsilon = 0$; 3, $\Delta\varepsilon = 0.15$.

agreement between theory and experiment. Thus, it was possible to calculate the flexoelectric coefficients and moduli of elasticity involved.

The flexoelectric modulated structure in nematic liquid crystals is also known as the "variable grating mode" [115–120], since for $U > U_{th}$, the period of domains w is varied according to the law $w \sim U^{-1}$ (FIGURE 5.16). The possibility of using flexoelectric domains in the regime of "controlled diffraction grating" in optical high-resolution data processing systems has been shown [118–120]. However, for sufficiently high voltages, dislocations are produced which destroy the regular domain pattern (FIGURE 5.16(b) and (c)). From a theoretical point of view [121], the formation of dislocations cannot be avoided above a certain critical voltage.

5.4 TWISTED NEMATICS AND CHOLESTERICS

In this chapter we will first consider electric field effects in pure nematics twisted by boundary conditions, then dielectric instabilities of nemato–cholesteric (supertwisted) mixtures, and helix unwinding of cholesterics. After the discussion of flexoelectric phenomena we shall briefly consider the electrooptical properties of blue phases.

5.4.1 Twist and Supertwist Structures

5.4.1.1 Twist Effect

If the directions x and y of the planar orientation of nematic liquid crystal molecules on opposite electrodes are perpendicular to each other, and the material has a positive dielectric anisotropy, then, when an electric field is applied along the z axis (FIGURE 5.17), a director reorientation occurs which is a combination of the splay, bend, and pure torsional distortions [122]. In the absence of the field (FIGURE 5.17(a)), the light polarization vector follows the director and, consequently, the structure rotates the light polarization vector through the angle $\pi/2$ [123]. This specific waveguide regime (Mauguin's regime) takes place, when the phase delay satisfies the condition:

$$\Delta n d/\lambda \gg 1. \tag{5.44}$$

When the applied voltage exceeds a certain threshold value:

$$U_{tw} = \pi[\pi \Delta\varepsilon^{-1}(4K_{11} + K_{33} - 2K_{22})]^{1/2} \tag{5.45}$$

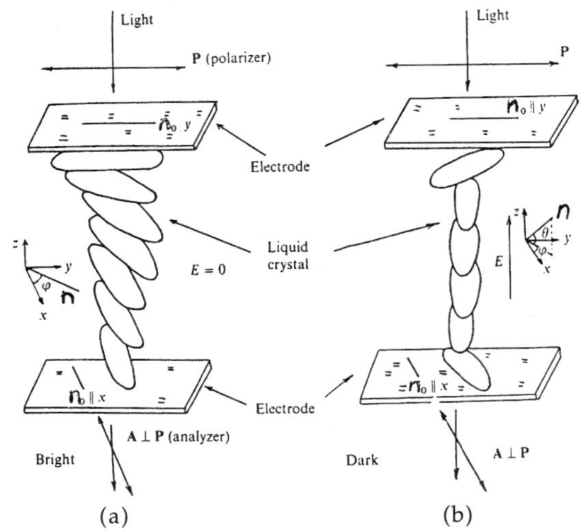

FIGURE 5.17. Twist effect. Field off (a) and on (b) states.

the director **n** deviates from the initial orientation so that the linear dependence of the azimuthal angle $\varphi(z)$ disappears and the tilt angle $\theta(z)$ becomes nonzero (FIGURE 5.17(b)). The qualitative character of the functions $\theta(z)$ and $\varphi(z)$ for different voltages is shown in (FIGURE 5.18). Since the director tends to orient perpendicular to the substrates, the average value of $\langle \Delta n \rangle$ in (5.44) decreases and for a certain voltage (optical threshold of the twist effect U_{opt}), the waveguide regime vanishes.

At zero field, the waveguide regime is violated for the small values of $\Delta n d/\lambda$. FIGURE 5.19 shows the dependences of the transmitted light intensity (I), ellipticity (e), and the rotation angle (ψ) of the long ellipse axis with respect to the polarizer P axis on the

FIGURE 5.18. Dependences of director angles $\theta(z)$ and $\varphi(z)$ in the twist effect for different voltages. (a) $U \leq U_{tw}$; (b) $U_{tw} < U_1 < U_2 < U_3$; U_{tw} is the threshold voltage, $\theta(U_{tw}) \equiv 0$, $\varphi(U_{tw}) = \pi z/2d$. The distributions θ_i, φ_i are induced by the voltage $U = U_i$, $i = 1, 2, 3$.

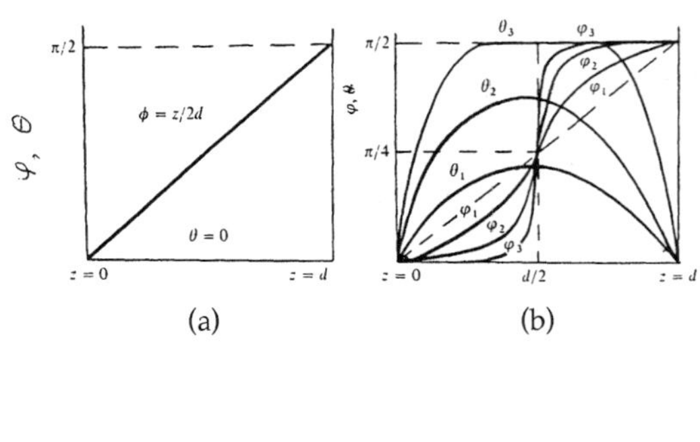

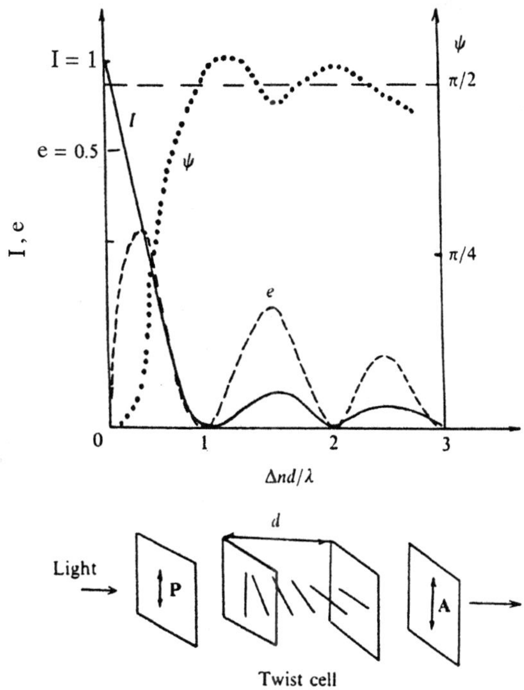

FIGURE 5.19. Optical characteristics of the twist cell in the absence of the field. I: intensity (solid line), e: ellipticity (dashed line), ψ: rotation angle (dotted line) versus the parameter $\Delta nd/\lambda$. Without the twist cell ($d = 0$) the intensity between the parallel polarizer (P) and the analyzer (A) is taken equal to $I = 1$. The director at the first substrate of the twist cell is parallel to the polarizer (P) [124].

parameter $\Delta nd/\lambda$ [124] (the director at the front substrate is parallel (or perpendicular) to the polarizer (P) and $\mathbf{P} \parallel \mathbf{A}$).

As can be seen from FIGURE 5.19, the exact Mauguin conditions $I = 0$, $\psi = \pi/2$, and $e = 0$ are satisfied not only for infinitely large $\Delta nd/\lambda$ values, but also at certain discrete points [124,125]

$$\Delta nd/\lambda = (4m^2 - 1)^{1/2}/2 \qquad m = 1, 2, 3, \ldots \quad (5.46)$$

usually called Mauguin minima.

The twist effect is very important for applications, as it may provide rather fast electrooptical response at any wavelength in the visible spectrum. It is widely used in the displays of watches, clocks, calculators, etc. For other displays (e.g., in scientific calculators), the number of the addressed rows (multiplexing ratio) must be enhanced and, to this end, the optical contrast–voltage curve (CVC) has to be as steep as possible.

Recently a number of efforts were made to improve the steepness of the CVC. It should be optimized for any specific case. There is no need to perform numerous experiments, as a computer simulation can solve the problem with a high degree of accuracy. At present, many authors have considered the problem of the calculation of the electrooptical characteristics of twist cells, using both analytical estimates and computer programs [126–131]. Modeling of the twist cell electrooptical behavior includes two steps. First the distributions of the director are found for various voltages (FIGURE 5.18). The second step consists of solving Maxwell's equations for light, propagating in anisotropic liquid crystal media.

Certain new possibilities of twist cells with weak boundary anchoring were predicted and observed [132–142].

1. The effect of saturation, that is, the total reorientation of the director to the homeotropic structure at a finite voltage. This effect may be used for the creation of memory states [133].
2. A considerable increase in the multiplexing capability of a twist cell for low anchoring energies, [132,133]. The number of the lines addressed is very sensitive to the form of the anchoring potential. For sufficiently small anchoring energies, the twist cell response reveals bistability and hysteresis [134]. The same effect occurs when a special form of the anchoring potential is assumed, for example, described by elliptical functions [136]. It should be pointed out that the above mentioned is true only for so-called polar anchoring (for deviation of the director through a polar angle θ). The azimuthal anchoring (deviation through angle φ) should be kept strong in all cases, otherwise the twist alignment disappears for a certain critical thickness [137].
3. If the directors in a twist cell are tilted at the boundaries then, for sufficiently high voltages the tilt tends to increase, and, at a certain critical angle, the twist state disappears and transforms to a quasihomeotropic configuration. Transformations of this kind take place without moving disclination lines since the twist and quasihomeotropic alignment are topologically equivalent [138,139].
4. If the electrodes of a cell which have been pretreated by rubbing are further coated with a layer of

surfactant (e.g., lecithin), the resulting orientation of the molecules of the nematic liquid crystal will be homeotropic. A cell formed in such a way that the directions of initial rubbing are perpendicular to each other transmits light under parallel polarizers. In an electric field, a nematic liquid crystal with negative dielectric anisotropy (e.g., MBBA) reorients into a twist structure, and transmission through the cell decreases to zero (the converse twist effect observed in [140]). If the two opposite electrodes of a twist cell are not treated at all but the pitch of a liquid crystal and the cell thickness are chosen so that $d = P/4$, a 90° twist structure again occurs with better viewing angle characteristics [141].

5. A combination of the twist and electrically controlled birefringence modes, called the TN–ECB effect, has been proposed [142]. The effect is based on the strong violation of the Mauguin regime, when linearly polarized light is transformed by a twist cell to circularly polarized light (e.g., the twist angle $\varphi_m = 63°$, $\Delta nd = 0.2\,\mu$m or $\varphi_m = 200°$, $\Delta nd = 0.56\,\mu$m). In the off-state, the circularly polarized light at the output of a twist cell is reflected from a mirror, changes the direction of circular polarization, and, when passing the twist cell for the second time, transforms back to linearly polarized light, this time of perpendicular polarization. Thus the off-state appears dark, as the input polarizer absorbs the light component perpendicular to its axis. In the homeotropic on-state, the twist cell becomes "transparent" and does not affect the polarization characteristics of light. This TN–ECB mode seems effective for twist displays operating in the reflection regime.

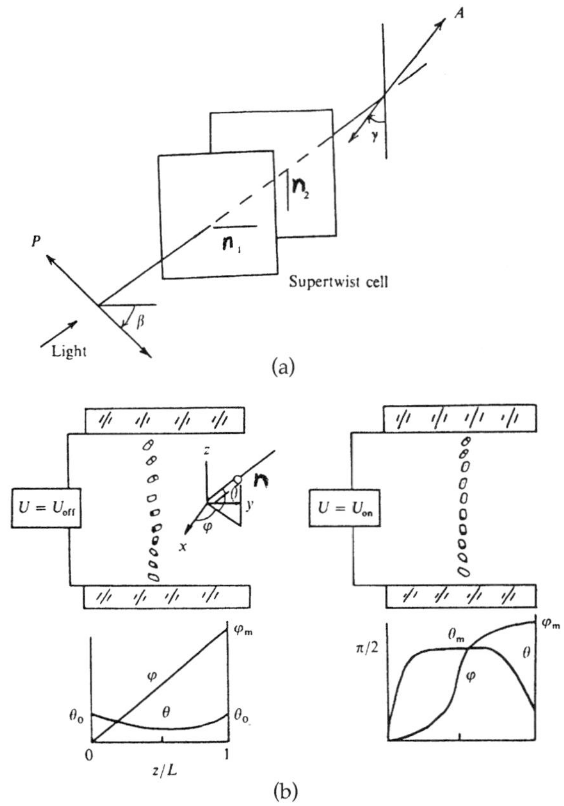

FIGURE 5.20. The geometry of the SBE display. (a) SBE cell with the directors twisted at an angle of 270° ($\mathbf{n}_1$, $\mathbf{n}_2$ show the projections of the directors at the two substrates) is placed between polarizer (P) and analyzer (A) oriented at the angles β and γ with respect to $\mathbf{n}_1$ and $\mathbf{n}_2$ respectively. (b) Distribution of the director angles inside the supertwist cell in the "off" and "on" states; θ_0 is the director pretilt angle, φ_m is the maximum rotation angle.

5.4.1.2 Supertwist Structures

Supertwisted, that is, twisted through an angle exceeding 90°, liquid crystal cells were known long ago [134], however, their importance for display applications was not clear. The effects of bistability and hysteresis in supertwisted nematic layers were first investigated in [143]. To obtain twist angles larger than 90°, nematics were doped with a small amount of an optically active material. Thus a cholesteric or chiral nematics with a large pitch P was created, so that the helical pitch could be fitted to the boundary conditions for the directors at the substrates. A display based on the supertwist birefringent effect (SBE) was proposed [144].

The geometry of an SBE display is typical of supertwisted displays (FIGURE 5.20). A 270°-supertwisted nematic layer is oriented with a 28° director tilt at the boundaries to prevent the appearance of light-scattering domain structures. In an SBE cell the ratio $d/P = 0.75$, that is, 3/4 of the helix pitch is fitted within the layer thickness d. When the field is switched on, the director reorients to a nearly homeotropic configuration (dielectric anisotropy $\Delta\varepsilon > 0$). The two polarizers used in the SBE display are located at angles β and γ with respect to the director projection on the input and output substrates.

The position of polaroids (β, γ) and optical path difference Δnd define the color and contrast characteristics of the SBE device [145]. A considerable wavelength dependence of the optical transmission of SBE displays imposes certain limitations on their applications.

The director deformation for the zero director pretilt at the boundaries starts at a certain threshold voltage [126,146]:

$$U_{\varphi_m} = U_S\{1 + [(K_{33}/K_{11} - 2K_{22}/K_{11})\varphi_m\pi^{-1} + 4K_{22}d/K_{11}P]\varphi_m\pi^{-1}\}^{1/2} \tag{5.47}$$

where for the SBE geometry $\varphi_m = 3\pi/2$, $d/P = 0.75$ and U_S is the S effect threshold (5.20).

For a finite director pretilt at the boundaries $\theta = \theta_0 \neq 0$, the deformation threshold (5.47) disappears; however, it is possible to find the voltage when the director angle at the center of the layer becomes equal to the boundary pretilt angle θ_0 [147]. This voltage is very close to the optical threshold for the supertwist transition. A significant advantage of the SBE mode is a steep transmission voltage characteristic, that correlates with a sharp growth in the director deformation with voltage.

The high contrast and uniformity of the transmission characteristics at oblique incidence are also useful features of the SBE mode. Better viewing angles than those of the 90° twist structure seem to be a common feature of highly twisted chiral nematics. FIGURE 5.21 demonstrates this for a 200° supertwist cell in comparison with the usual 90° twist cell [148].

The optical mode interference (OMI) effect provides a weak wavelength dependence of the transmission in the visible region [149,150]. The effect requires a low optical path difference, leading to a strong interference of two polarization modes propagating through an OMI cell (here, Mauguin's waveguide regime (5.46) is not valid). The electrooptical characteristics of OMI cells are much more tolerant to the cell gap inhomogeneity than in the STN case, and are less temperature dependent [151]. However, one of the main disadvantages of the OMI display is low brightness in the off-state, which makes it impossible to use the reflection mode. The brightness of OMI displays can be improved by increasing the twist angles up to 210° or 240°, making an appropriate choice of Δnd and angular positions of the polarizers.) Thus the OMI cells are very attractive for high information content devices.

FIGURE 5.21. Transmission–voltage characteristics. (a) 90° twist nematic structure; (b) 200° supertwisted chiral nematic. i_θ are angles of light incidence, $i_\varphi = 180°$ [148].

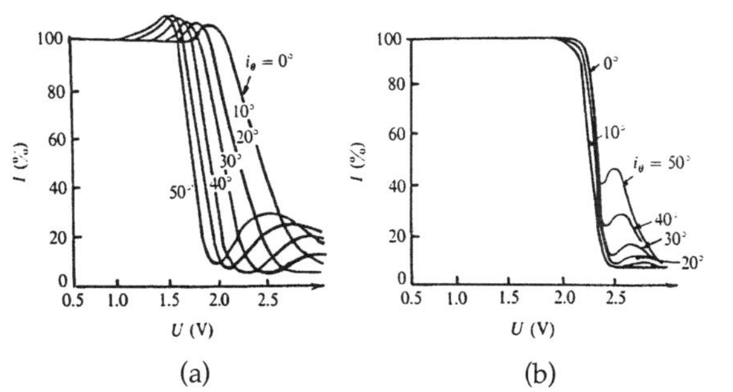

5.4.1.3 Electrical Bistability

The bistable states are often observed in nemato–cholesteric mixtures [142,152]. As chapter 15 is devoted to consideration of bistability effects, here only one example is considered [142,153], the cholesteric Grandjean texture with a tilted director orientation on the boundaries (FIGURE 5.22(a)). In the absence of tilt, the free (elastic) energy g is minimum at the following thicknesses d:

$$d/P_o = n/2 \qquad n = 1, 2, 3, \ldots \tag{5.48}$$

corresponding to $n/2$ turns of the helix, located along the layer normal. If the directors at the boundaries are tilted, then the corresponding free-energy minima are no longer equidistant, as it follows from FIGURE 5.22(b).

The bistability appears for switching between the $0/2$ and $2/2$ cholesteric states (FIGURE 5.22(a)). If the thickness-to-pitch ratio $(d/P_o) = 0.89$, then, at zero field, the most energetically favorable state is $2/2$ (FIGURE 5.22(b)), while at a sufficiently high field it is $0/2$ (FIGURE 5.22(c)). If the electric field is turned off slowly, the director remains in the $0/2$ state since it corresponds to a local minimum of the free energy. In contrast, a strong field turned off suddenly results in the $2/2$ state. The tilt angle at the boundaries and the layer thickness must be chosen to be optimum in order to avoid unnecessary switching to the "parasitic" state, $1/2$.

The optical transmission in the $0/2$ and $2/2$ states can be made considerably different by doping a liquid crystal with dichroic dyes or locating the cell between crossed polarizers.

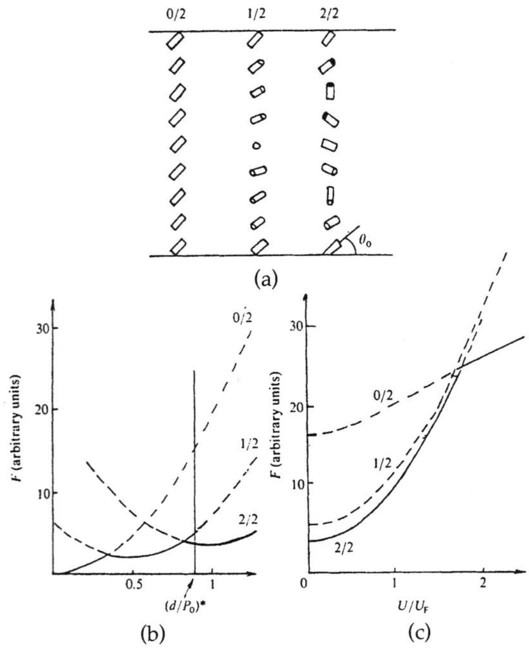

FIGURE 5.22. Bistable switching in long pitch cholesterics with a tilt of the director θ_0. (a) Tilted states with $n/2$ turns in zero field, $\theta_0 = 55°$. Free energies F as functions (b) of the thickness-to-pitch ratio d/P_0 at zero field, $(d/P_0)^* = 0.89$ is the operating point; (c) of reduced voltage U/U_F, where U_F is the Frederiks transition threshold.

5.4.2 Instability of the planar cholesteric texture

5.4.2.1 Texture Transitions

Let us assume that the frequency of the electric field is sufficiently high, or the electrical conductivity of a cholesteric is sufficiently low. Then the alignment of the molecules by the electric current can be neglected and the interaction between a cholesteric and the electric field can be regarded as purely dielectric (in this section the flexoelectric terms are also disregarded). The alignment of the cholesteric material in an electric field occurs so as to minimize the free energy of the material in the field. The term $(-\Delta\varepsilon(\mathbf{E}\cdot\mathbf{n})^2/4\pi)$ is added to the free energy per unit volume in a static electric field **E**. Thus, in an electric field for a material with $\Delta\varepsilon < 0$, a texture with the direction of the helical axis parallel to the field is energetically favored. In contrast, when $\Delta\varepsilon > 0$, the favored orientation of the helical axis is perpendicular to the field. Thus when an electric field is applied to a liquid crystal cell, a texture transition occurs to minimize the free energy of the system. These texture changes in cholesteric liquid crystals are physically similar to the Frederiks transition in a nematic liquid crystal and result in a significant change in the optical properties of the layer. The texture transitions have been reviewed before [19] with allowance for the sign of $\Delta\varepsilon$, the initial texture, and the direction of the applied electric field. Here, the instability of the planar cholesteric texture, which has been widely discussed in recent literature, will be considered.

5.4.2.2 Optical Properties of the Planar Cholesteric Textures

The following are the most characteristic features of the planar textures of cholesteric liquid crystals:

1. There is a strong rotation (10–100 revolutions per mm) of the plane of polarization of light with a change of the sign of the rotation at a particular wavelength λ_o.
2. The regions of rotation with different signs are not separated by an absorption band as in typical gyrotropic materials, but by a selective reflection band with a particular circular polarization. The opposite circular polarization is transmitted without variation.
3. The electric vector of the incident and reflected light is rotated in the same direction when both cases are viewed along the path of the beam. In typical gyrotropic materials there is a change in the sign of rotation.
4. The wavelength λ_o depends on the angle of incidence ϕ of the beam by the approximate relation $\lambda_o = P_0\langle n\rangle\cos\phi$, where P_0 is the pitch of the helix.

5.4.2.3 Field Behavior of the Planar Texture

The physical cause of the purely dielectric instability is well understood. When an electric field is not sufficiently high to untwist a helix with a short pitch, the dielectric torque tends to reorientate the local directors into an arrangement corresponding to the fingerprint texture. In contrast, the elastic forces tend to preserve the distribution of the cholesteric layers, which is shown in FIGURE 5.23(a). With fields that are not too large, a pattern which undulates in the x and y directions appears as a compromise distribution of

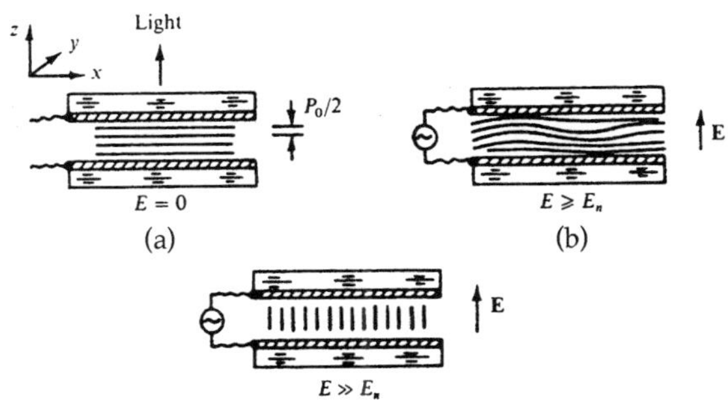

FIGURE 5.23. Illustration of a dielectric instability in a planar cholesteric texture when $\Delta\varepsilon > 0$.

the director (FIGURE 5.23(b)). The wavevector of this deformation pattern along the z-axis is approximately π/d. Related to the two-dimensional director pattern is a two-dimensional periodicity in the distribution of the refractive index and a phase lag between the ordinary and extraordinary rays, and hence a two-dimensional optical grating is formed. With an increase in voltage, the angle of deviation of the director increases, tending towards the limit $\theta = \pi/2$ for the whole layer (helix untwisting).

The threshold field for the formation of the periodic distortion has been calculated [154,155] based on the expression for the free energy of a cholesteric liquid crystal:

$$g = \tfrac{3}{16}K_{33}\left(\frac{\pi}{w}\right)^2\theta_m^2 + \tfrac{1}{2}K_{22}\left(\frac{wq_0}{d}\right)^2\theta_m^2 - \tfrac{1}{16}\Delta\varepsilon E^2\theta_m^2. \tag{5.49}$$

The conditions of large layer thickness ($d \gg P_o$) and small $\Delta\varepsilon \ll \langle\varepsilon\rangle$ were assumed. Also, the assumptions are made of small deviations of the axis of the cholesteric helix from the normal to the glass surfaces or, equivalently, small deviations of the director from the xy plane, that is, $\sin\theta \simeq \theta$, $\cos\theta \simeq 1$. Further, the deformation is assumed to be sinusoidal along both the x-axis (the half-period equals w) and the z-axis (the half-period equals d) and is described, in contrast to the nematic case, by two variables. One is the angle of deviation of the helical axis from the normal to the surfaces:

$$\theta = -\theta\sin(\pi z/d)\sin(\pi x/w) \tag{5.50}$$

and the other is the difference in the wavevectors for the distorted and equilibrium helices, $\Delta q = q - q_o$ ($q_o = 2\pi/P_o$):

$$\Delta q = \Delta q_m \cos(\pi z/d)\sin(\pi x/w). \tag{5..51}$$

The maximum compression Δq_m of the cholesteric planes and the maximum angle of deviation θ_m of the helical axis from the normal can be related using purely geometric considerations:

$$\Delta q_m = \theta_m(q_o w/d)$$

and this value is substituted into the term with the K_{22} coefficient in the equation for the free energy, equation (5.49). The numerical factors in (5.49) appear as a result of averaging of the energy over the period of the deformation [155].

The minimization of the elastic part of the free energy without a term containing the field by a choice of the proper value of w gives the period of the deformation

$$(2w)^2 = (3K_{33}/2K_{22})^{1/2}P_o d \tag{5.52}$$

and, using the condition $g = 0$ (with allowance for the field term), the threshold voltage for the deformation is

$$U_{th}^2 = \frac{8\pi^3(6K_{22}K_{33})^{1/2}}{\Delta\varepsilon P_o}d. \tag{5.53}$$

When the dielectric anisotropy is not small, the term $\Delta\varepsilon/4\pi$ in (5.53) must be replaced by $\varepsilon_\perp(\varepsilon_\parallel - \varepsilon_\perp)/[2\pi(\varepsilon_\perp + \varepsilon_\parallel)]$. The threshold field is independent of frequency up to the dispersion region ω_D of the dielectric permittivity.

The action of the field on a planar texture of a cholesteric liquid crystal for cell thicknesses comparable to the pitch and with a rigid anchoring of the directors to the surfaces does not cause the two-dimensional deformations. In this case, one-dimensional periodic patterns are observed [19] with the orientation of the domains depending on the number of half-turns of the helix contained within the cell thickness. For instance, when $d = P_o/2$, the domains are parallel to the direction of the rubbing of the glass surfaces, and when $d = P_o$ they are perpendicular to it, that is, the domains are always perpendicular to the director in the middle of the layer.

A theoretical consideration of the case of a pitch comparable to the layer thickness for a purely dielectric destabilization ($\Delta\varepsilon > 0$) of a planar texture in a

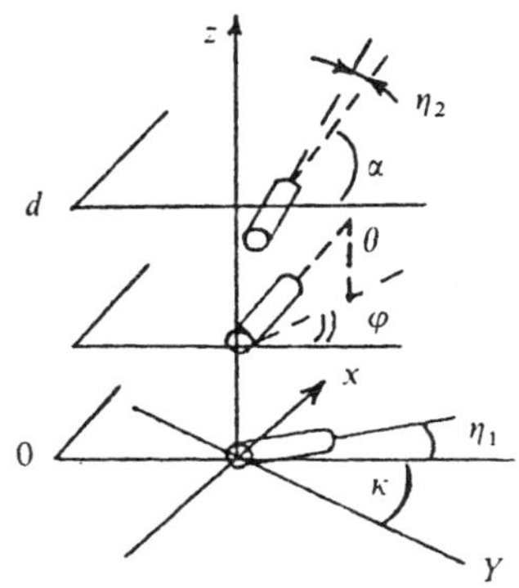

FIGURE 5.24. Geometry of a twisted cholesteric film in the planar texture. κ is the angle between the wavevector of the distortion and the director projection onto the lower plate.

field $\mathbf{E} \perp \mathbf{h}$ has been given both numerically [156] and analytically [157,158]. In the latter case, perturbation theory was used to search for the structure of the director field just above the threshold of the instability. The geometry of the problem is shown in FIGURE 5.24. Two variables, the polar angle θ and the azimuthal angle φ, were considered with an orientation of the director at opposite walls differing by twist angle α (pretilt angles η_1 and η_2 are also taken into account). It has been shown that two types of instability can be observed depending on the elastic moduli of the material ($k_2 = K_{22}/K_{11}$, $k_3 = K_{33}/K_{11}$), the total twist of the structure between two boundaries α (or reduced twist $\omega = \alpha/\pi$), relationship between the thickness of the cell d and the cholesteric pitch P_o ($\beta = 2\pi d/P_o\alpha$), and dielectric anisotropy $\Delta\varepsilon$ ($\gamma = \Delta\varepsilon/\varepsilon_\perp$). The first type is a homogeneous distortion analogous to the Frederiks transition in nematic liquid crystals discussed in sections 5.3.1 and 5.4.1. The second type is a periodic distortion (one-dimensional in the layer plane) with a wavevector $\mathbf{q}$ directed at certain angles with respect to the director at the boundaries.

The reduced threshold field of distortion

$$U_F = U_S\sqrt{R} \tag{5.54}$$

includes all the parameters mentioned above, namely, the threshold for the Frederiks transition U_S for the untwisted nematic (equation (5.20)), and

$$R = 1 + \omega^2(k_3 - 2k_2 + 2k_2\beta). \tag{5.55}$$

With increasing voltage one of the regimes (uniform or periodic distortion) is entered, separated by the Lifshitz point where the wavevector of the distortion is zero. The corresponding phase diagram has already been discussed for nematic polymers (FIGURE 5.12).

The periodic instability is favored when the original twist (α) and the normalized cell thickness (d/P_o) increase and dielectric anisotropy ($\Delta\varepsilon/\varepsilon_\perp$) decreases. This is illustrated by FIGURE 5.25, where lines of the Lifshitz points are presented for the case of zero pretilt angles η_1 and η_2. The lines separate regimes of the homogeneous (higher K_2 values) and periodic (below the line) distortions. The principal predictions of the theory, namely, the disappearance of the periodic instability in thin cells for small twist angles agree with experiments performed on the Grandjean planar texture with varying d/P_o [156].

Qualitatively, the same results were obtained in [159]. Additionally, it was claimed that the presense of a pretilt in the planar cholesteric texture decreases the tendency of the system to adopt the periodically modulated (ripple) phase.

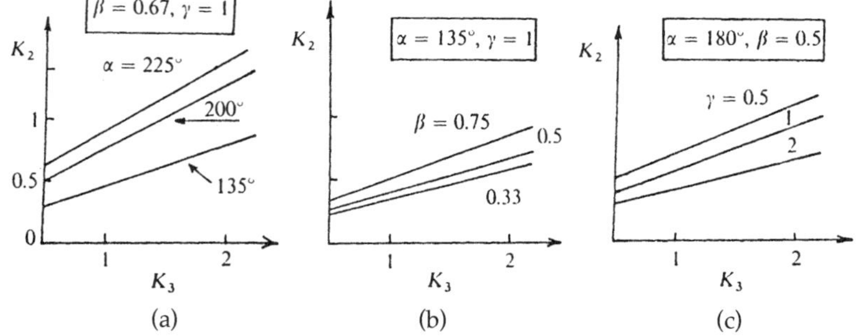

FIGURE 5.25. Lines of Lifshitz points separating regimes of uniform (upper) and periodic (lower) distortions as functions of material parameters α (a), $\beta = 2\pi d/P_0\alpha$ (b) and $\gamma = \Delta\varepsilon/\varepsilon_\perp$ (c) [158].

5.4.3 Field Untwisting of the Cholesteric Helix

Untwisting of a helix is observed only with positive dielectric anisotropy and only in a field perpendicular to the helical axis. Thus, in some cases, the director reorientation (or a texture transition) has to occur first, and then the untwisting is possible.

5.4.3.1 The Static Case

The model investigated by Meyer and de Gennes [160,161] is shown in FIGURE 5.26. Let us assume that we have a helix with pitch P_o in the absence of a field, and that the thickness of the sample is sufficiently large (relative to P_o) that the boundary conditions can be neglected. If $\mathbf{E} \perp \mathbf{h}$ and $\varepsilon \gg \Delta\varepsilon > 0$, then, in the initial state, in certain parts of the helix (A) the molecules are arranged favorably relative to the field, but in other parts (B) they are arranged unfavorably and therefore tend to reorient themselves with their axes along the field. As a result, the A regions will be increased in size not so much through a sharp decrease in the dimensions of the B regions (this would require too large an elastic energy $K_{22}\,(\partial\phi/\partial z)^2$), but through a decrease in their number. Therefore, the helical pitch increases and the helix itself becomes nonideal, that is, the sinusoidal dependence of the azimuth of the director φ upon the z-coordinate is destroyed. When the last B region is eliminated from the sample, the liquid crystal becomes a nematic, oriented uniformly along the field. Consequently, a cholesteric-to-nematic liquid crystal phase transition has occurred in the field.

The process can be described within the framework of the elasticity theory of cholesterics. The free energy

FIGURE 5.26. Schematic representation of the untwisiting of a cholesteric helix by a magnetic field [1]. In the text we discuss the electric analog of this model.

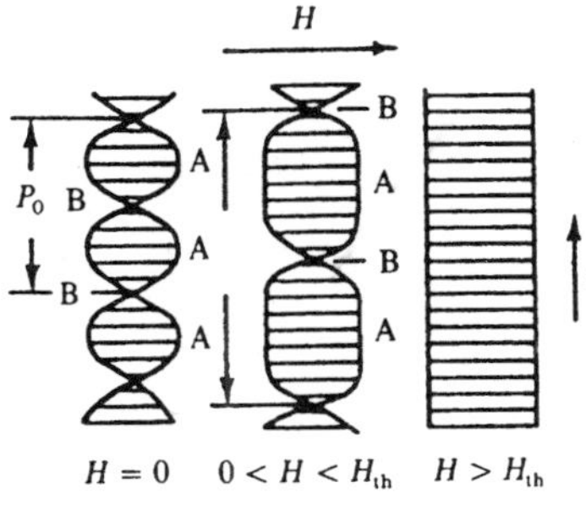

of cholesterics in an electric field is

$$F = \frac{1}{2}\int \left[K_{22}\left(\frac{\partial\varphi}{\partial z} - q_o\right)^2 - \frac{\Delta\varepsilon E^2}{4\pi}\sin^2\varphi\right] dz. \quad (5.56)$$

The function $\varphi(z)$ should satisfy the condition for minimum g, or Euler's equation ("nematic" equation (5.23) written for $\partial\varphi/\partial t = 0$). Its solution may be expressed in terms of the elliptic function

$$\sin\varphi(z) = \mathrm{sn}[(z/\xi k), k]. \quad (5.57)$$

The values $u = z/\xi k$ and k are the argument and modulus of the elliptic function, respectively, and

$$\xi^2 = 4\pi K_{22}/\Delta\varepsilon E^2 \quad (5.58)$$

is the electric coherence length mentioned earlier (see a more general equation (5.31)).

The helical pitch does not enter into equation (5.23), but it appears again when the constant of integration (or the modulus k) is adjusted in order to minimize the free energy (5.56). The condition for the minimum is

$$q_o\xi = 2E(k)/\pi k \quad (5.59)$$

and the new pitch of the helix satisfying the condition of a minimum in g is

$$P(H) = P_o(2/\pi)^2 F(k)E(k) = 4\xi k F(k). \quad (5.60)$$

Here $F(k)$ and $E(k)$ are complete elliptic integrals of the first and second kind. When $k \to 1$, the elliptic integral $F(k)$ diverges logarithmically, that is the helix pitch $P(H) \to \infty$. Simultaneously, when $k \to 1$, the integral $E(k) = 1$, and equation (5.59) gives the critical field for the untwisting of the helix:

$$E_C = \tfrac{1}{2}\pi q_o\left(\frac{4\pi K_{22}}{\Delta\varepsilon}\right)^{1/2} = \frac{\pi^2}{P_o}\left(\frac{4\pi K_{22}}{\Delta\varepsilon}\right)^{1/2}. \quad (5.61)$$

The pitch of the helix increases gradually, beginning with infinitesimally small fields. The expansion of equation (5.60) gives:

$$P(H) = P_o\left(1 + \frac{(\Delta\varepsilon)^2 P_o^4 E^4}{128(2\pi)^6 K_{22}^2} + \cdots\right). \quad (5.62)$$

The increase of the pitch with increasing electric (or magnetic) field has been confirmed in numerous experiments carried out with thick samples [19].

Complete untwisting of the helix results in a conversion of the optically negative and optically active planar texture to the optically positive and inactive texture. However, the spectrum of reflection from the cholesteric medium is changed even with partial

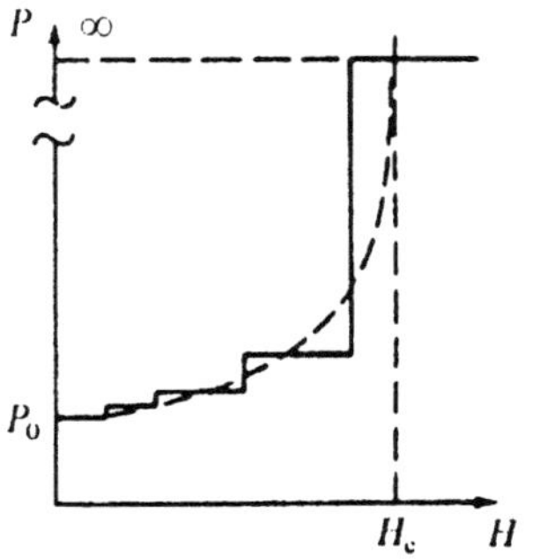

FIGURE 5.27. Illustration of the field-induced untwisting of a cholesteric helix in a thin cell with a planar orientation by the walls.

untwisting of the helix. First, the wavelength of the reflection maximum, $\lambda_o = P\langle n\rangle$, is increased, since the pitch P is increased: second, for normal incidence, higher orders of reflection should appear [162] due to the conversion of the ideal helix into the soliton-like structure shown in FIGURE 5.26. In experiment, the field-induced second- and third-order reflections from the planar texture have been observed [163].

Field untwisting of a helix for thin cells with planar boundary conditions occurs differently from the case of an infinite cholesteric medium. For a field perpendicular to the axis of a cholesteric helix in a planar texture, a stepwise change in the pitch with increase in the field is predicted [164] (FIGURE 5.27). The size of the step increases with a decrease in the ratio $2d/P_o$.

Stepwise untwisting of the helix by an electrical field perpendicular to its axis has been observed for the case of a planar cholesteric texture with $\Delta\varepsilon > 0$ and with strong anchoring of the molecules to the limiting surfaces [165]. It is interesting that the relaxation of the field-induced, or untwisted, state under these conditions is accompanied by the formation of spatially modulated structures in the form of strips or grids.

The role of the boundary conditions was discussed theoretically in [166]. When the helical axis is parallel to the limiting walls and the field $\mathbf{E} \perp \mathbf{h}$, the threshold for the helix untwisting decreases in going from planar to homeotropic anchoring. For cell thickness $\alpha < \sqrt{2}P_o/\pi$, homeotropic boundaries unwind the helical structure even in the field-off state. For thicker cells, the threshold in the one-constant approximation is:

$$E_C = \tfrac{1}{2}\pi q_o \left[\frac{4\pi K}{\Delta\varepsilon}\left(1 - \frac{8}{d^2 q_o^2}\right)\right]^{1/2}. \qquad (5.63)$$

The lowering of the threshold field with decreasing cell thickness and increasing surface tilt angle was observed in [167] and [168], respectively.

The linear-in-the-field flexoelectric coupling discussed in section 5.3.3 for nematics can modify the threshold field for helix untwisting in cholesterics. The flexoelectric effect stabilizes the helical structure, thus, it results in a decrease of an "apparent dielectric anisotropy" [169]:

$$E_C = \tfrac{1}{2}\pi q_o \left(\frac{4\pi K_{22}}{\Delta\varepsilon - \pi^3 e^2 K^2}\right)^{1/2}. \qquad (5.64)$$

Here $e = e_1 = e_3$ is the average flexoelectric coefficient and $K = K_{11} = K_{33}$ is the average splay–bend elastic modulus.

The discussion above did not allow for the conical deformation of the helical structure, that is, the polar angle θ was considered to be equal to zero. In a more general case, the components of the director are taken in the form

$$L = [\cos\theta(z)\cos qz, \cos\theta(z)\sin qz, \sin\theta(z)]. \qquad (5.65)$$

Even in thermodynamic equilibrium, $\theta(z)$ can form a periodic soliton-like structure with a period incommensurate with the helical pitch $q_o = 2\pi/P_o$ [170]. When a field is applied, the conical distortion can appear, especially for certain values of the twist elastic constants, $K_{22} > K_{33}$ [160,171]. Such a conical distortion can be accompanied by a compression of the helical pitch. The results of some earlier controversial observations have been discussed in [19]. Probably, the conical deformation has been observed in a ternary cholesteric mixture by measuring field-induced changes in the optical activity of the planar texture [172]. The results with polymer lyotropic systems are still unclear [173] because of controversial information on the elastic moduli K_{33} and K_{22}.

5.4.3.2 Dynamics of the Untwisting

In the simplest case, when the conical distortion is disregarded, the dynamics of the helix unwinding is described by the same equation (5.23), ($\mathbf{E} \perp \mathbf{h}$), where $\varphi(z)$ varies from 0 to 2π and cannot be considered small enough to linearize the equation. The approximate solution of (5.23) results in the following expression for the response τ_r and decay τ_d times for the field-induced helix distortion [174]:

$$\tau_r = \frac{\gamma_1}{K_{22}q^2 \pm \Delta\varepsilon E^2/8\pi} \tag{5.66}$$

$$\tau_d = \frac{\gamma_1}{K_{22}q^2}. \tag{5.67}$$

Here γ_1 and K_{22} are the rotational viscosity and twist elastic constant, q is a wavevector of the distortion mode under investigation (it depends on the geometry of the experiment, in particular, on the light scattering angle), and signs (+) and (−) in (5.66) are related to the cases $q \geq q_o$ and $q \leq q_o$, where $q_o = 2\pi/P_o$ and P_o is the equilibrium pitch.

Untwisting of the helix is usually observed under conditions where the bounding surfaces play a significant role. However, the role of the surface anchoring in the dynamics of the effect has not been analyzed carefully. Most experiments have been performed on the usual sandwich cells, with an electric field applied perpendicular to the walls of the cell. Here the initial turbid focal–conic texture changes to a transparent homeotropic nematic texture. This opens up the possibility of using this effect for practical purposes as a field-controllable cholesteric-to-nematic phase transition [19]. The dynamics of this process, as a rule, also includes texture transformations, for example, the transition from a focal–conic to a fingerprint texture with subsequent untwisting of the cholesteric helix [175]. When the initial cholesteric texture is planar, the process passes through several stages: first, a periodic instability in the form of a grid appears; then rotation of the axis of the cholesteric helix occurs so that the axis assumes an orientation perpendicular to the field, after which untwisting of the helix occurs. Relaxation can occur in the reverse order, in which case the texture changes are the slowest processes.

Generally, the dynamics of untwisting of a helix has not been investigated sufficiently, and only the following fundamental facts have so far been established. The relaxation time is determined by the helical pitch P_o when the thickness $d \gg P_o$, and by the thickness of the cell when $d \ll P_o$. The state of complete untwisting of the helix is metastable, and relaxation begins at defects in the structure. The reaction times in the field are inversely proportional to E^2. These conclusions conform qualitatively with the theory presented above.

5.4.3.3 Bistable Structures

Bistability is observed, if one of the bistable states is disordered, resulting in strong light scattering [176–180]. Consider the unwinding of the cholesteric structure with homeotropically oriented directors at the boundaries (FIGURES 5.28(a) and (b)). In this case an electric field induces the cholesteric–nematic transition which is of a hysteretic type, so that the transparent homeotropic texture is metastable and has a long lifetime. FIGURE 5.28(c) shows the intensity of the light passed through the corresponding liquid crystal cell, placed between two crossed polarizers. An increase of the field up to $E = E_\uparrow$ results in the homeotropic texture (FIGURE 5.28(b), $I = 0$), which returns back to the scattering focal-conic texture with a lowering of the field down to $E = E_\downarrow$ (FIGURE 5.28(a), $I = 1$).

As seen from FIGURE 5.28(c), within the field region $E_\downarrow < E < E_\uparrow$, two bistable states exist, leading to the

FIGURE 5.28. Thermodynamic bistability in cholesteric liquid crystals with homeotropically oriented directors at opposite boundaries. (a) scattering state; (b) transparent homeotropic state; (c) intensity curve for a cell, placed between crossed polarizers, $\Delta\varepsilon > 0$.

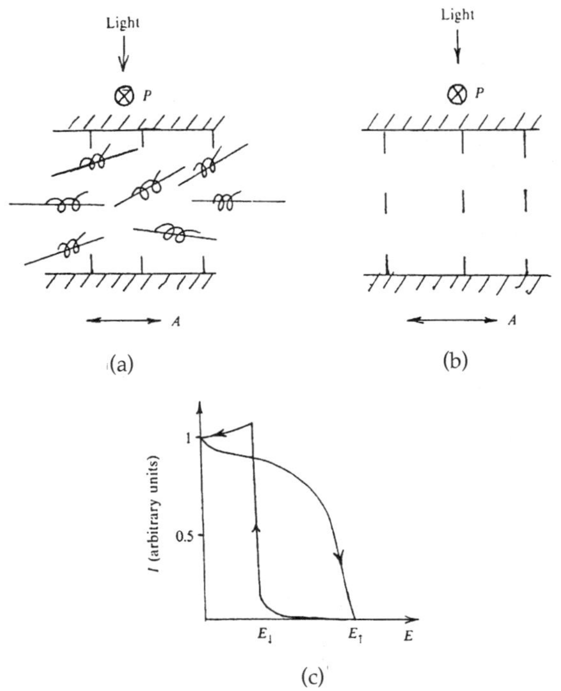

homeotropic and focal–conic configurations, dependent on the prehistory. The switching between these states takes place via the formation of the so-called quasi-Grandjean structure at $E = E_{\downarrow}$, with the helix pitch P larger than its equilibrium value $P > P_o$ [179]. The initial focal–conic texture is formed at $E = 0$. The realization of this "thermodynamic" bistability depends on the value of d/P_o as well as the liquid crystal elastic anisotropy K_{33}/K_{11}, K_{33}/K_{22} [176,177]. Let us note, that the d/P_o ratio must be sufficiently large, because, for $d/P_o \lesssim 1$, the cholesteric helix is unwound due to the strong homeotropic anchoring at the surfaces [177,180].

The planar texture formed by short-pitch cholesterics does not show any field-induced periodic instability [181]. The field drives a cholesteric structure directly to the unwound homeotropic state. The back relaxation may lead to either a scattering focal–conic state or the transparent planar texture (the bistability effect). The choice between the two final states can be made by using a proper voltage applied during the relaxation process [181].

5.4.4 Flexoelectric Effects

Flexoelectric effects, which were discussed for nematics in section 5.3.3, are caused by the linear coupling of an external field with the flexoelectric polarization. In cholesterics, they manifest themselves in two specific phenomena to be considered below.

5.4.4.1 Fast Linear-in-the-Field Rotation of the Cholesteric Helix

This effect is observed in a geometry where the cholesteric axis **h** is homogeneously oriented in the plane of the cell (along x) and an electric field is applied to the electrodes of a sandwich cell along the z-axis [166,182,183]. In this case, the helical structure, even the ideal one, is incompatible with the planar boundary conditions and splayed and bent regions form near the boundaries. Thus, according to equation (5.38), the flexoelectric polarization arises in those regions that can interact with the electric field.

In experiment, a deviation of the optical axis in the plane of a cell is observed. The sign and the magnitude of the deviation angle depend on the polarity and strength of the applied field, respectively. In the field-off state, the helix is undistorted and the cell behaves as a uniaxial optical plate with the optical

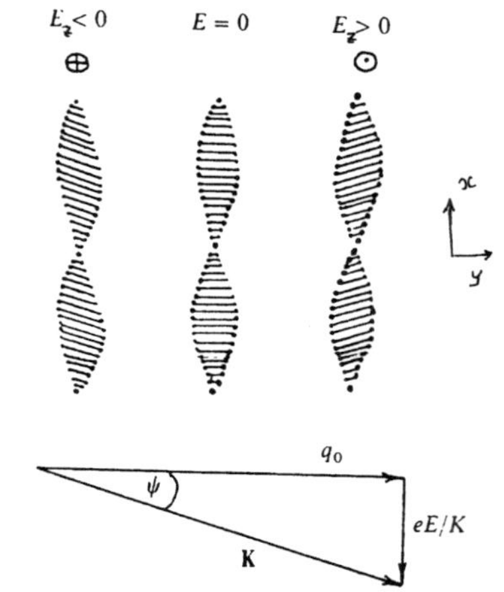

FIGURE 5.29. The pattern of the director rotation induced by an electric field applied perpendicular to the plane of the drawing [182]. Below: components of the wavevector **k** of the distorted helix.

axis coincident with **h**. When the field is applied, the director leaves the y,z-plane due to the flexoelectric deformation coming from the surface regions where the flexoelectric torque $M_f = e_f E$ $(e_f = e_1 = e_3)$ is developed. Now the optical axis does not coincide with the initial orientation of the helix **h** but forms an angle Ψ with respect to it, proportional to E. The field-induced distortion of the helix is shown in FIGURE 5.29 [182].

For director components parallel to the z,y-plane, $n_z = \cos\varphi$, $n_y = \sin\varphi$, the free-energy density of the system in an electric field is written

$$g = \tfrac{1}{2}K\left(\frac{\partial\varphi}{\partial y}\right)\tfrac{1}{2}K_2\left(q_o - \frac{\partial\varphi}{\partial x}\right)^2 - e_f E\frac{\partial\varphi}{\partial y} + \frac{1}{8\pi}\Delta\varepsilon E^2 \sin^2\varphi \tag{5.68}$$

where $K = K_{11} = K_{33}$.

For negligible dielectric anisotropy, $\Delta\varepsilon \to 0$, the minimization of (5.68) resuts in $\partial\varphi/\partial x = q_o$ and $\partial\varphi/\partial y = eE_f/K$ (the anchoring energy is considered to be small). If the rotated axis of the helix is represented by a wavevector **k**, then $|\mathbf{k}|\cos\Psi = q_o$ and $|\mathbf{k}|\sin\Psi = e_f E/K$ (FIGURE 5.29) and the rotation angle is given by

$$\tan\Psi = e_f E/q_o K \tag{5.69}$$

which is linear in E for small rotations.

Experiments show that the effect of the reorientation of the helical axis is fairly fast, in the range of 10–100 μs, and that the speed of the response is independent of field strength. This can be accounted for under the assumption that the pitch of the helix does not change in the field-on state $|\mathbf{k}| = q_o$. Then, the free energy becomes a function of angle Ψ only [182]. For small Ψ:

$$\eta \frac{\partial \Psi}{\partial t} = -Kq_o^2 \Psi + e_f E q_o \quad (5.70)$$

where η is an effective viscosity coefficient for the rotation of the helix. From (5.70) it follows that the relaxation time is independent of field

$$\tau = \eta / K q_o^2. \quad (5.71)$$

Thus, the flexoelectric effect in cholesterics is as fast as the helix untwisting effect. By placing the sample between crossed polarizers and applying an AC voltage, the optical modulation is observed. The transmitted intensity is:

$$I = I_o \sin^2(2\beta) \sin^2(\Phi/2) \quad (5.72)$$

where I_o is the intensity of incident light, $\phi = 2\pi \Delta n d/\lambda$ is the phase delay in the sample, and β is the angle between the zero-field optical axis and the polarizer axis. When an electric field is applied, it changes the angle β due to rotation of the optical axis by angle $\Delta\beta = \Psi$. Consequently, the relative change in the transmitted intensity is given by

$$\Delta I = \frac{4 \arctan(l_f E / I_o K)}{\tan 2\beta} \quad (5.73)$$

which is proportional to E for small fields.

It should be noted that phenomenologically, this effect is analogous to the "deformed helix electro-optical mode" observed in ferroelectric liquid crystals where coupling of an electric field with the spontaneous (instead of flexoelectric) polarization is used [184,185].

When the helical structure of the chiral nematic phase is unwound by the influence of limiting walls, one can observe a linear-in-the-field light modulation which is caused by a small molecular tilt [186]. The effect is analogous to the electroclinic effect observed in the smectic A* phase as a pretransitional phenomenon in the vicinity of the A*–C* transition (section 5.5.2.4).

5.4.4.2 Flexoelectric Domains in Cholesterics

In section 5.3.3.3 we discussed the longitudinal flexoelectric domains in nematics whose period depends on the inducing field, $w \sim E^{-1}$. Upon changing the sign of the field, the phases of the distortion angles $\theta(y)$ and $\varphi(y)$ are reversed, but the direction of the wavevector of the instability (perpendicular to the initial director orientation $\mathbf{n}_o$) remains unchanged. In cholesterics, the mirror symmetry is broken and the same instability has specific features [187,188].

First of all, orientation of the domains depends on the number of half-turns of the helix between cell boundaries. In the Grandjean texture, the direction of the domains is different in neighbouring Grandjean zones [189], for the same polarity of the field. When the polarity of an applied voltage is changed, the directions of domain lines are also changed. With an oscillating field (at very low frequencies, of the order of 1 Hz), the direction of domains oscillates between two positions, characterized by an angle $\pm\alpha$ with respect to the director in the middle of the cell (in the first two Grandjean zones or in a twist nematic cell). The threshold voltage for the domain formation, as well as the domain period and angle α, have been calculated in [188].

Figure 5.30 shows the experimental data (obtained on 4-butyl-4′-methoxy-azoxybenzene, BMAOB, doped with 1% of cholesteryl caprinate) and the results of numerical calculations of the threshold voltage and the reduced wavevector of the flexoelectric instability for the first three Grandjean zones (0,1,2) for various dielectric anisotropies.

For the zero zone with directors at opposite boundaries parallel to each other (no twist) and zero dielectric anisotropy, the threshold voltage and the wavevector can be calculated analytically [188]:

$$U_{th} \simeq \frac{2\pi (K_{11} K_{22})^{1/2}}{|e^*|} \times \left[1 - \frac{K_{22}^2}{K_{33}(K_{11} + K_{22})} \left(\frac{2d}{P}\right)_o^2\right] \quad (5.74)$$

$$q_{th} \simeq \frac{\pi}{d} \left[1 - \left(\frac{d}{P_o}\right)^2 \frac{K_{22}^2}{K_{33}^2} \left(\frac{K_{33}}{K_{11}} + \frac{K_{33}}{K_{22}} - 2\right)\right]. \quad (5.75)$$

Both magnitudes depend on the cholesteric pitch despite the completely unwound state of the helix. The sign and the amplitude of the angle of the field-induced domain rotation (at threshold) depends on

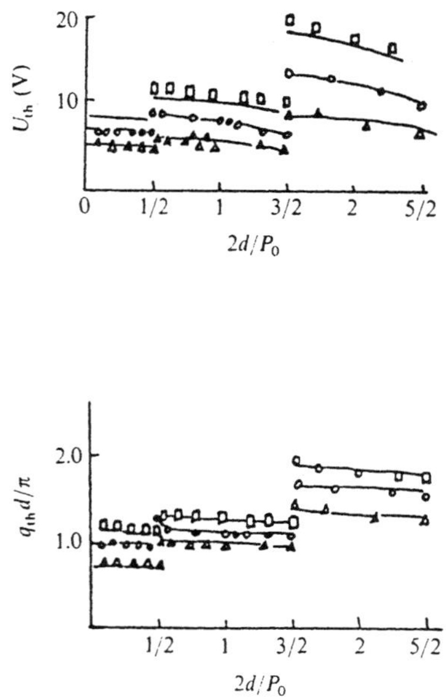

FIGURE 5.30. Threshold voltage (upper) and the reduced wavevector (lower) for flexoelectric domains in the lowest three Grandjean zones for various dielectric anisotropies. $\Delta\varepsilon = -0.1$ (rectangles), 0 (circles) and $+0.25$ (triangles). Solid lines are computer calculations [188].

the handedness and pitch of the helix

$$\alpha_{th} \simeq \frac{2d}{P_o}\frac{K_{22}}{K_{33}}\operatorname{sign}(e^*U). \tag{5.76}$$

The sign of α (at a given polarity of the voltage) depends also on the sign of the difference in flexoelectric coefficients $e^* = e_{11} - e_{33}$.

With increasing frequency of a pulsed electric field, $f > \tau^{-1}$, where $\tau \cong \gamma_1 d^2/K\pi^2$ is the relaxation time for the director, the deviation angle decreases and the threshold voltage diverges [189].

The other flexoelectric structure can arise in a very thin planar cholesteric layer at a certain set of material parameters. The instability occurs in the form of spiral domains [190] whose handedness depends on the sign of an electric field. While the threshold voltage for the linear domains discussed above is almost independent of cell thickness, $U_{th} \simeq 2\pi K/|e_{11} - e_{33}|$, the threshold for spiral domains is proportional to thickness

$$U_{th} \simeq \frac{5Kq_o d}{e_{33} + 0.08e_{11}} \tag{5.77}$$

and number of turns of the helix seen under the microscope in the plane of the layer is

$$m_o = 0.2\left(\frac{E}{E_{th}} - 1\right). \tag{5.78}$$

It should be noted that with increasing field, the spiral domains appear first if $d \lesssim q_o^{-1}$ and $(e_{11} - e_{33}) < e_{33}$. Spiral domains have been observed in experiments [191].

5.4.5 Blue Phases in an Electric Field

5.4.5.1 Optical Features

Nowadays, three thermodynamically stable blue phases are known: BPI, BPII and BPIII (or foggy phase). The structures of the first two are already established; BPI is body centered cubic phase (symmetry group 0^8 or $I4_132$), and BPII is simple cubic (symmetry group 0^2 or $P4_232$) [192]. The foggy phase may be described using an amorphous or quasicrystal model [193].

The three-dimensional crystalline structure of the blue phases with lattice periods comparable to the wavelength of visible light results in optical diffraction dependent on the orientation of the light wavevector with respect to crystalline planes [194]. Thus, selective reflection of light occurs in different spectral ranges for different experimental geometries. This is well illustrated by FIGURE 5.31 taken from paper [195]

FIGURE 5.31. Selective reflections versus temperature in a cholesteric, BPI and BPII phases of CB15-E9 mixture (42–58%). $\beta = \pi/4$ for the (110) curve, $\beta = \pi/2$ for all other curves. Wavelengths of reflection maxima are normalized to $1.6/n$ (BPI supercools below 42.1 °C) [195].

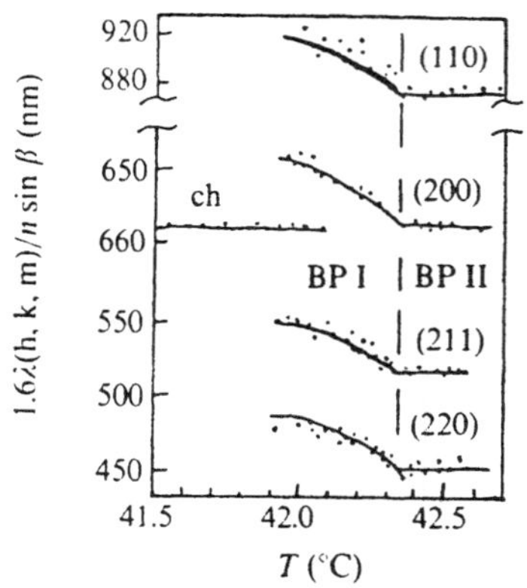

where selective reflection maxima (normalized to 1.6n) of a nemato–cholesteric mixture

$$\lambda(h,k,m) = \frac{2nl \sin\beta}{(h^2+k^2+m^2)} \tag{5.79}$$

are plotted as functions of temperature for different directions of incident light, characterized by the Miller indices h, k, m (in (5.79), n is the average refractive index, l is the lattice constant, and β is the Bragg angle). Four different Bragg reflections (from the crystalline planes marked on the curves by Miller's indices) are observed for BPI and BPII in FIGURE 5.31, and only one reflection for the cholesteric phase.

The most pronounced features of the optics of blue phases BPI and BPII are [194]: 1) selective reflection in the visible range which provides blue color of the phases (in contrast to conventional cholesterics, blue phases manifest several reflection orders); only one circular polarization, as in cholesterics, is scattered backward; 2) blue phases are optically active and the sense of the light polarization rotation changes sign at the wavelengths of selective scattering maxima; 3) linear birefringence is absent, that is, the blue phases are optically isotropic, though multiple scattering in perfect samples can result in a small amount of optical anisotropy.

Optical properties of the blue phases are defined completely by the tensor of the high-frequency dielectric permittivity [194]:

$$\hat{\varepsilon}(\mathbf{r}) = \hat{\varepsilon}_o + \Delta\hat{\varepsilon}_{ik}(x,y,z) \tag{5.80}$$

where $\hat{\varepsilon}_o$ is the average dielectric permittivity and $\Delta\hat{\varepsilon}_{ik}$ is its three-dimensional periodic part. The tensor $\hat{\varepsilon}(\mathbf{r})$ can be expanded in the Fourier series

$$\hat{\varepsilon}(\mathbf{r}) = \sum_{\mathbf{H}} \hat{\varepsilon}_{\mathbf{H}} \exp(\mathbf{iHr}) \tag{5.81}$$

where $\mathbf{H}$ is the vector of the reciprocal lattice of a blue phase, which can be presented in the form

$$\mathbf{H} = \frac{2\pi}{l}(h\mathbf{x} + k\mathbf{y} + m\mathbf{z}) \tag{5.82}$$

where $\mathbf{x}, \mathbf{y}, \mathbf{z}$ are coordinate axes, and h, k, m are the Miller's indices mentioned above, and l is the lattice constant.

Fourier harmonics of tensor (5.80):

$$\hat{\varepsilon}_{\mathbf{H}} = \frac{1}{d^3}\int \hat{\varepsilon}(\mathbf{r}) \exp(-\mathbf{iHr})\, d\mathbf{r} \tag{5.83}$$

play the role of the order parameters for blue phases. Their symmetry, different for BPI and BPII, is responsible for a number of observed diffraction maxima and polarization properties of the scattered and transmitted light, as well as for the electrooptical behavior of the blue phases. BPIII has very specific diffraction properties which are not discussed here.

5.4.5.2 Field Behavior

An external electric field interacts with the local dielectric anisotropy of a blue phase and contributes to the energy of the liquid crystal with a term $\Delta\varepsilon E^2/4\pi$ [196]. Flexoeffects in the blue phases seem to have not been discussed yet. The field distorts the cubic lattice and results in a change in the angular (or spectral) positions of the Bragg reflections. Moreover, field-induced phase transitions into novel phases have been observed [192,197]. The field can also induce birefringence parallel to the field direction due to the optical biaxiality of the distorted cubic lattice [198].

In a weak field, no phase transition occurs and a cubic lattice is distorted according to the sign of the local dielectric anisotropy of the medium. This is illustrated in FIGURE 5.32 taken from [199]. The effect

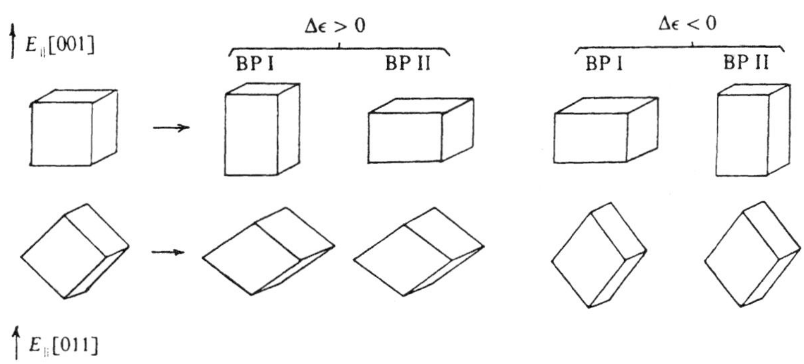

FIGURE 5.32. Deformation of the unit cell due to electrostriction for a blue phase with positive and negative dielectric anisotropy when the field is parallel either along a fourfold or along a twofold axis [199a].

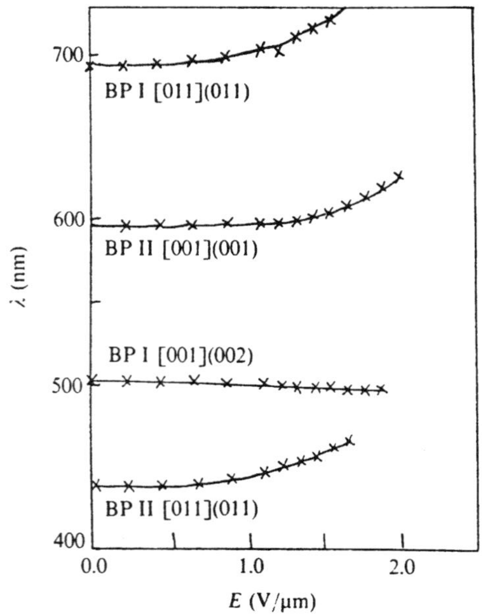

FIGURE 5.33. Wavelengths of the Bragg peaks corresponding to planes (hkm) versus applied field strength for BPI (34.1 °C) and BPII (34.3 °C) oriented with either a fourfold [001] or a twofold [011] axis parallel to the field (CB15-E9 mixture, 49.2–50.8%) [199a].

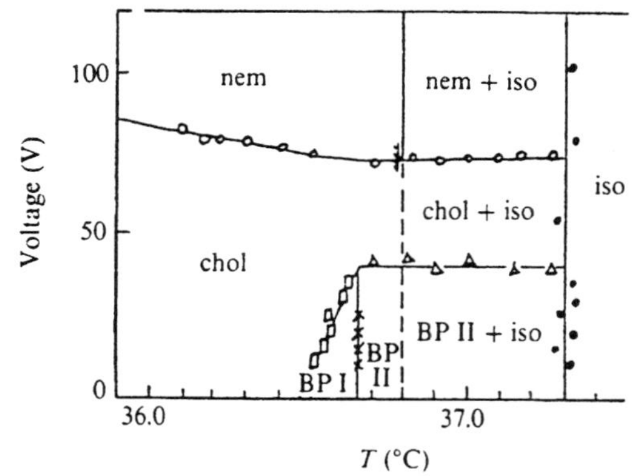

FIGURE 5.34. Voltage–temperature phase diagram for 47–53 mol.% mixture of CB15 and 6OCB [192].

is described by the strain tensor

$$U_{ij} = \sum_{k=1} \sum_{e=1} \gamma_{ijke} E_k E_e \tag{5.84}$$

where γ_{ijkl} is the fourth-rank electrostriction tensor [196,199].

The distortion of the lattice is accompanied by a change in the Fourier harmonics (5.83) of the dielectric tensor (5.80) and the corresponding changes in optical properties. For example, wavelengths of the Bragg peaks for a mixture (49.2–50.8%) of CB15 and E9 with $\Delta\varepsilon > 0$ (the same mixture as in FIGURE 5.31) increase or decrease with increasing field, depending on the field direction with respect to the crystallographic axes (FIGURE 5.33) [199a]. As a consequence, the spectra of the optical activity and circular dichroism also change. When the dielectric anisotropy is negative, the signs of the components of the electrostriction tensor γ_{ijkl} are inverted and the field-induced red shifts of the Bragg maxima are replaced by blue shifts and vice versa [199b]. Red and blue shifts may reach values of 26 and 35 nm, respectively, in the rather weak field of about 0.2–0.4 V μm^{-1} and negative $\Delta\varepsilon$ [200].

With increasing external field, a series of field-induced phase transitions BPI → cholesteric, BPII → cholesteric and then cholesteric → nematic are observed. This is illustrated by FIGURE 5.34 [192], where a voltage–temperature phase diagram is presented for a mixture (47–53 mol.%) of chiral CB15 with 4-n-hexyloxycyanobiphenyl (60CB).

BPI loses its stability first, transforming into the cholesteric phase, because the transition enthalpy ΔH is extremely small ($\simeq$50 J mol^{-1}) for the BPI–Ch transition. This enthalpy, normalized to a unit volume and compared with the difference in electrostatic energy density for the two phases $\delta\varepsilon \cdot E^2$ ($\delta\varepsilon = \varepsilon_{\mathrm{BPI}} - \varepsilon_{\mathrm{ch}} \simeq 0.2$), explains the observed results. For other materials, novel phases induced by an external field have been observed. Among them, a tetragonal (BPX) and two hexagonal (a three-dimensional BPH^{3d}, and a two-dimensional BPH^{2d}) phases were distinguished [197]. In the identification of new phases, the optical technique of Kossel diagrams is especially informative [197]. For the CB15/E9 mixtures mentioned above, the voltage–concentration phase diagram is shown in FIGURE 5.35. Only those phases that appear first with cooling of the substance from the isotropic phases are indicated. In [200], the field-induced transition from BPII to the isotropic phase was observed in a mixture with $\Delta\varepsilon > 0$.

When an electric field is applied to the BPIII (foggy) phase with $\Delta\varepsilon > 0$, the broad selective reflection peak typical of this less-ordered phase decreases in intensity

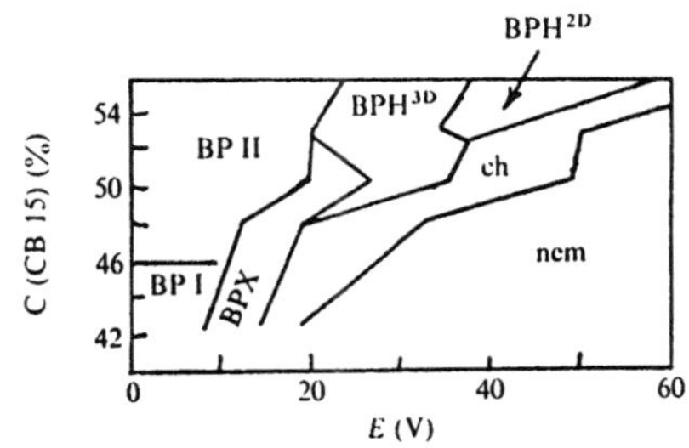

FIGURE 5.35. Voltage–composition phase diagram for CB15-E9 mixtures [197].

and, at some threshold voltage, is replaced by a sharp peak at longer wavelength [193]. Thus, the phase transition into a new phase occurs. The symmetry of this latter phase has not been established yet. For a system with $\Delta\varepsilon < 0$ [201], an increase in the field results in considerable growth of the selective reflection peak. An explanation of this behavior as well as many other examples of the field effects in the blue phase may be found in a very comprehensive review paper by Kitzerow [202].

The field-induced distortions of blue phases decay rather fast. In the geometry of a Kerr cell (the field is perpendicular to the light wavevector), the field induces birefringence δn which results in electrooptical modulation ΔI of the transmitted light. For nearly crossed polarizers, the amplitude of modulation is

$$\frac{\Delta I}{I_m} = \sin^2\left(\pi \frac{d}{\lambda} \delta n\right) \tag{5.85}$$

where d is the light path and I_m is the maximum transmitted intensity for parallel polarizers [203]. Modulation characteristics are shown in FIGURE 5.36 for a blue phase with negative dielectric anisotropy, $\Delta\varepsilon = -0.75$.

The relative field-induced optical anisotropy $\delta n/\Delta n$ can be estimated by comparing the electric $\Delta\varepsilon E^2$ and elastic Kq^2 energy densities:

$$\frac{\delta n}{\Delta n} \approx \left(\frac{\Delta\varepsilon E^2}{4\pi}\right) \Big/ \frac{4\pi^2 K}{P_o} \tag{5.86}$$

where P_o is the equilibrium pitch and K is a Frank modulus. The time constant is well approximated by the simplest expression

$$\tau = \frac{\gamma_1}{Kq^2} = \frac{\gamma_1 P_o^2}{4\pi^2 K}. \tag{5.87}$$

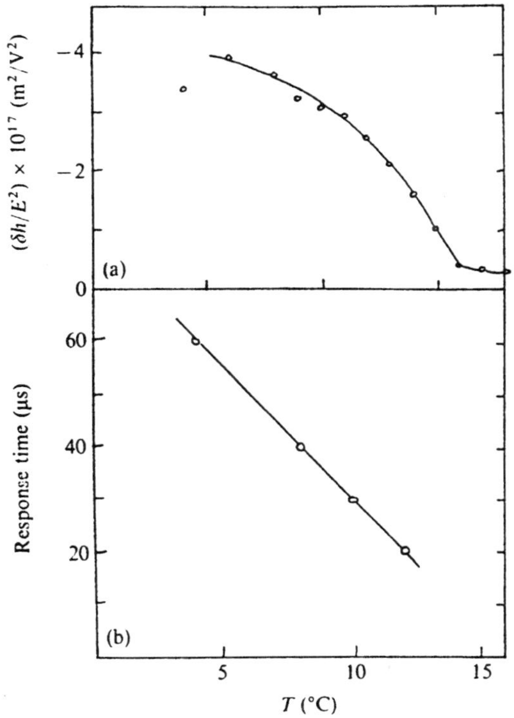

FIGURE 5.36. Electrooptical modulation by a blue phase. (a) normalized field-induced birefrigence at a voltage of 170 V and a frequency of 50 Hz (the position of the analyzer was adjusted at each temperature); (b) response time versus temperature [203].

With parameters $\Delta n \approx 0.07$, $\Delta\varepsilon \approx -0.7$, $K = 4 \times 10^{-7}$ dyn, $P_o \approx 0.2\,\mu$m and $\gamma_1 \approx 2.7$ P, expressions (5.86) and (5.87) give the right orders of magnitude for δn and τ.

From the point of view of applications, the electrooptical effects in blue phases may become of interest for relatively low-voltage and relatively fast light modulators [203]. Color switching in the submillisecond range has been demonstrated using blue phases in [204].

5.5 NONFERRO-, FERRO-, AND ANTIFERROELECTRIC SMECTICS

5.5.1 Electric Field Behavior of Non-Ferroelectric Smectics

5.5.1.1 Frederiks Transition in a Smectic A

In the smectic A phase, the director is always perpendicular to the plane of the smectic layers. Thus, only the splay distortion leaves the interlayer distance unchanged and only the elastic modulus K_{11} is finite while K_{22} and K_{33} diverge when approaching the smectic A phase from the nematic phase. On the other hand, the compressibility of the layered structure and the corresponding elastic modulus $\bar{B}$ must be taken into account when discussing the elastic properties of smectic phases. The free energy density for the smectic A phase subjected to the action of an external electric field is [1]:

$$g = \frac{1}{2}\left[\bar{B}\left(\frac{\partial u}{\partial z}\right)^2 + K_{11}(\mathrm{div}\,\mathbf{n})^2 - \frac{\Delta\varepsilon}{4\pi}(\mathbf{nE})^2\right] \qquad (5.88)$$

where $u(\mathbf{r})$ is the displacement of the smectic layers from their equilibrium position, and the z-direction coincides with the normal to the layers.

Thus, it makes sense to investigate the Frederiks transition in the geometry corresponding to the splay distortion only. Let us assume a so-called bookshelf geometry where smectic layers are perpendicular to substrates made of conductive glasses, forming a sandwich-type cell of thickness d with the director oriented in the plane of the cell, for example, along the x-axis. If the dielectric anisotropy is positive ($\Delta\varepsilon > 0$) and a voltage is applied to electrodes, the Frederiks transition should take place. The threshold of the transition coincides formally with that for a nematic (equation (5.20)), however, the amplitude of the distortion θ_m for fields exceeding the threshold is very small, of the order of λ_s/d [205] where a smectic characteristic length

$$\lambda_s = (K_{11}/\bar{B})^{1/2} \qquad (5.89)$$

is on the order of a few layer thicknesses. Thus, even in this favorable case of allowed splay distortion, the Frederiks transition is in fact unobservable (ghost transition). Instead, one observes a texture transition accompanied by the appearance of a number of defects.

5.5.1.2 Dielectrically Induced Texture Transitions

The texture transitions are especially pronounced in materials with large dielectric anisotropies. A high-quality planar texture undergoes a transition to a quasihomeotropic optically transparent texture via intermediate structural defects [206,207]. A homeotropic texture can also be obtained by applying the field when erasing the defect structure induced by thermal action of a laser or by an electrohydrodynamic instability [208]. The threshold voltage observed experimentally for a transition from a planar to a homeotropic texture depends on the layer thickness according to $U_o \propto d^{1/2}$ (for the Frederiks transition the threshold voltage would be independent of thickness). A model accounting for the experimental data (at least partly) has been developed by Parodi [209], who assumed the formation of transition layers between the surface and bulk of the sample.

The discrepancy between the calculated (about 0.2 μm [209]) and observed (few μm) periods of the texture instability may be caused by a nonuniform, "chevron" structure of the original "bookshelf" geometry. Chevron structures with smectic layers broken by a certain angle are often observed in smectic C liquid crystal cells (see below). However, recently, they have been discovered in a certain temperature range for smectic A layers [210].

The kinetics of the relaxation of the homeotropic field-induced texture to a focal–conic one depends on the surface treatment [211–213]. A rough surface which provides strong anchoring facilitates relaxation [213].

A texture transition can also be observed for smectic A liquid crystals with negative dielectric anisotropy [214]. In this case, the transition from a homeotropic to a planar texture occurs. The threshold for this dielectric transition can be modified (lowered) at low frequencies of the applied field by the anisotropy of the electrical conductivity of a substance.

The field-induced wave-like instability discussed theoretically in [215] is still to be observed in experiment.

5.5.1.3 Frederiks Transition in a Smectic C

Depending on the arrangement of the smectic layers and the director of the liquid crystal relative to the limiting surfaces, four different configurations can be identified. For each of them, there are at least three alternatives for the direction of the field. Thus, twelve variations in all can be obtained, and these have been studied theoretically by Rapini [205]. In most of the configurations, the Frederiks transition is a ghost one, since it requires the collapse of the smectic layers. However, in three instances [216], the field only induces rotation of the director around the normal to the layers, leaving the layer structure unchanged.

Let us identify the three main values of the dielectric susceptibility ε_i ($i = 1, 2, 3$), where ε_3 corresponds to the direction along the director, ε_1 corresponds to the direction perpendicular to the the plane of the tilt, and ε_2 corresponds to the direction perpendicular to the preceding two [19].

When the field is oriented along the director, a Frederiks transition is possible when $\varepsilon_2 > \varepsilon_3$. The director rotates around the normal to the smectic layers not changing its angle Ω. The corresponding threshold field is proportional to $(\varepsilon_2 - \varepsilon_3)^{-1/2}$. A field perpendicular to the director and lying in the plane of the tilt induces the same distortion when $\varepsilon_2 > \varepsilon_1$. Now the threshold for reorientation of the director is proportional to $(\varepsilon_2 - \varepsilon_1)^{-1/2}$. Finally, when the field is perpendicular to the tilt plane, the same rotation takes place at a threshold proportional to $(\varepsilon_1 \cos^2 \Omega + \varepsilon_3 \sin^2 \Omega - \varepsilon_2)^{-1/2}$.

Extensive measurements of the Frederiks transition in various smectic C liquid crystals have been carried out by the Halle group [217–219]. In experiments, the conventional sandwich cells with optically transparent electrodes were used. The director was oriented uniaxially by rubbing the electrodes, and the smectic layers were tilted with respect to the cell plane (a "bookshelf" geometry). An electric field applied along the cell normal (z-direction) (FIGURE 5.37), induces a director rotation around the normal to the smectic layers due to positive dielectric anisotropy. Two optical effects are observed: firstly, the deviation of the optical axis of a cell from the x-direction due to the deviation of the director from the xz-plane and secondly, a change in the optical path for the light beam due to the deviation of the director from the xy-plane.

The threshold field for the director deviation can be calculated neglecting distortion of the smectic layers [218] and assuming $\varepsilon_2 \cong \varepsilon_3$. It depends on the angle μ

FIGURE 5.37. Geometry of the Frederiks transition for a planar oriented smectic C liquid crystal.

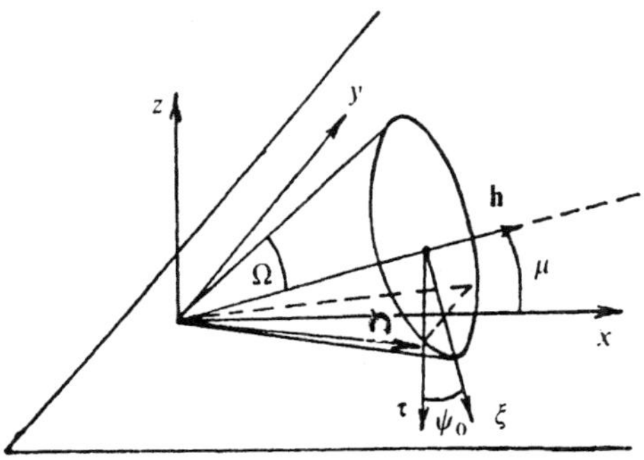

between the normal and the xy-plane. For $\mu = 0$, when the xy-plane cuts a half of the director cone (with angle Ω at the apex), the threshold field is

$$E_C = \frac{\pi}{d \sin \Omega} \sqrt{\frac{4\pi K^*}{\Delta\varepsilon}} \tag{5.90}$$

that is, $(\sin \Omega)^{-1}$ times higher than for nematics with the same physical parameters.

When the cone just touches the xz-plane ($\mu = \Omega$) (FIGURE 5.37), the threshold field is approximately

$$E_C' = \frac{4.5}{d \sin \Omega \cos \Omega} \sqrt{\frac{4\pi K^*}{\Delta\varepsilon}}. \tag{5.91}$$

Both cases were observed experimentally using various substances [218]. For a certain set of material parameters, the theory predicts a bistable behavior of the smectic C phase in a strong field. Such bistability has been observed in the form of the field-induced motion of domain walls and a hysteresis in the process of the director reorientation [219].

The dynamic behavior may be described by adding a viscous torque to the elastic and electric ones. For the field-off state, $E_z = 0$, the equation (see FIGURE 5.37 for a definition of ψ)

$$K^* \frac{\partial^2 \psi}{\partial z^2} - \gamma_1 \frac{\partial \psi}{\partial t} = 0 \tag{5.92}$$

results in a distortion decaying with the time constant

$$\tau_d = \gamma_1 d^2 / K^* \pi^2 \tag{5.93}$$

analogous to the case of a nematic. The time of response to an electric field τ_E can only be calculated for the simplest case when $\mu = 0$

$$\tau_E = \frac{\gamma_1}{\Delta\varepsilon \sin^2 \Omega (E^2 - E_c^2)} \tag{5.94}$$

that is, the response of the smectic C phase is $\sin^2 \Omega$ times slower than that of the nematic phase. However, in experiments the same substance often responds faster in the smectic C phase than in nematic phase [218]. This may be accounted for by a smaller value of γ_1 when the motion of the director is confined by the cone surface. The same phenomenon has been observed for the ferroelectric smectic C* phase [220].

5.5.2 Ferroelectric Liquid Crystals

Since their discovery in 1975 [11], ferroelectric liquid crystals (FLCs) remain the object of intensive investigation. A little later [221] a model was suggested for the thermodynamical description of FLC physical properties, including their response to external fields. A contemporary version of the general phenomenological theory, dynamic behavior included, is presented in the book by Pikin [18]. Many experimental results and the discussion of the FLC parameters may be found in review papers [222,223]. Here, basic ideas and recent results are discussed.

5.5.2.1 Total Free Energy with Allowance for Anchoring

The symmetry of the ferroelectric smectic C* phase corresponds to the polar point group C_2. When proceeding along the z-coordinate perpendicular to the smectic layers, the director $\mathbf{n}$ and the polarization vector $\mathbf{P}$, directed along the C_2, axis rotate

$$\mathbf{n}(z+h) = \mathbf{n}(z) \qquad P(z+h) = P(z) \tag{5.95}$$

that is, a helix with pitch h forms (FIGURE 5.38).

In the absence of external fields, the equilibrium helix pitch h_0 corresponds to the minimum of the FLC free energy, that is

$$\begin{gathered} \mathbf{n} = (\sin\theta\cos\varphi, \sin\theta\sin\varphi, \cos\theta) \\ \varphi = q_o z = 2\pi z/h_0 \end{gathered} \tag{5.96}$$

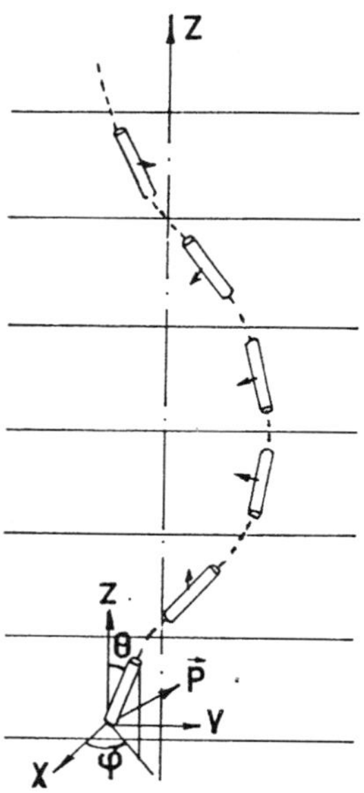

FIGURE 5.38. Macroscopic structure of the ferroelectric C* phase.

where q_o is the wavevector of the structure. Often $\sin\theta \sim \theta$ is assumed since the tilt angle θ usually does not exceed 30°. In equilibrium, the average polarization of the bulk is zero.

The elastic properties of FLCs may be discussed using the density of the elastic energy f in the form:

$$\begin{aligned} f_{el} = (1/2)[K_1(\operatorname{div}\mathbf{n})^2 + K_2(\mathbf{n}\operatorname{curl}\mathbf{n} - t)^2 \\ + K_3([\mathbf{n}x\operatorname{curl}\mathbf{n}] - \mathbf{b})^2] \end{aligned} \tag{5.97}$$

where K_i are elastic moduli for different deformations, as in nematic liquid crystals, and the linear terms $\mathbf{n}\operatorname{curl}\mathbf{n}$ and $[\mathbf{n}\times\operatorname{curl}\mathbf{n}]$ indicate the lack of mirror symmetry in FLCs. This form of energy implies that the system is far from the smectic A–C* transition and the tilt angle is not affected by the field. In the vicinity of the transition, Landow type terms (5.4) must be added to the right part of (5.97) [18,221].

The FLC free energy should also include surface terms, with the polar $\tilde{W}_p = -W_p(\mathbf{Pv})$ and dispersion $\tilde{W}_d = -W_d(\mathbf{Pv})^2$ contributions, where W_p and W_d are the corresponding anchoring strength coefficients, $\mathbf{v}$ is the layer normal, and $\mathbf{P}$ is the polarization. Thus the total free energy F_d

$$F_d = \int_V f_{el}\,d\tau + \int_S (\tilde{W}_p + \tilde{W}_d)\,d\sigma \tag{5.98}$$

consists of a sum of the volume (V) and surface (S) integrals. The helical state occurs [224] when the surface energy is small compared with the energy of the helix, that is

$$Kdq_0^2/2 \gg W_d, W_p, \tag{5.99}$$

which takes place for sufficiently large values of the cell thickness d. If the polar anchoring coefficient W_p is much higher than the dispersion anchoring coefficient W_d and helix energies, that is

$$W_p \gg W_d, Kdq_0^2/2, \tag{5.100}$$

then a specific twisted state appears (only a half of the helical pitch throughout a cell) [224].

In the presence of the external field, the total free energy includes the energy of director deformation F_d and the energy F_E of the interaction of the ferroelectric phase with the field $\mathbf{E}$:

$$F = F_d + F_E = F_d + \int [-(\mathbf{PE}) - \mathbf{DE}/8\pi]\,d\tau. \tag{5.101}$$

The anchoring energies for an FLC may be determined [225]: (i) by measuring the width of the coercivity loop ΔV in the voltage dependence of the polarization

$$W_d = (1/8)\Delta V P_S \tag{5.102}$$

and (ii) by measuring the free relaxation times τ_d to the bistable states for the FLC director:

$$W_d = \gamma_\varphi d/4\tau_d \tag{5.103}$$

where γ_φ is the viscosity for the φ relaxation of the director.

5.5.2.2 The Clark–Lagerwall Effect

Let us consider the best known electrooptical phenomenon in FLCs, the Clark–Lagerwall effect, which results in the director reorientation from one bistable state to another, when the external electric field changes its sign [12] (FIGURE 5.39). In this case, smectic layers are perpendicular to the substrates and the director moves along the surface of a cone, whose axis is normal to the layers and parallel to the cell substrates. In each final position of its deviation, the director remains parallel to the substrates, thus transforming a cell into a uniaxial phase plate. The origin of switching the director is an interaction of the polarization **P** perpendicular to the director with the electric field **E**. The maximum variation of the transmitted intensity is achieved when an FLC cell is placed between crossed polarizers, so that an axis of the input polarizer coincides with one of the final director positions. The total angle of switching equals twice the tilt angle θ. The Clark–Lagerwall effect is observed in the so-called surface stabilized FLC structures (SSFLC) [12,226,227]. In SSFLC cells, $d \ll h$ and the helix is unwound by the walls.

FIGURE 5.39. An FLC cell with the smectic layers (3) perpendicular to the substrates (1) and current conducting layers (2). **E** – electric field, **n** – director.

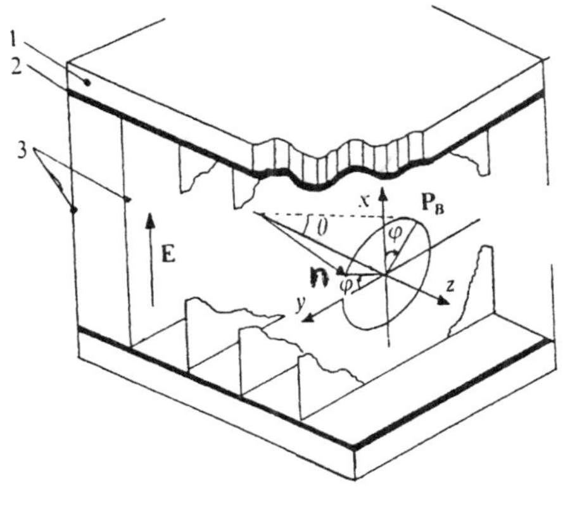

The variation of the azimuthal director angle φ in the Clark–Lagerwall effect is described by the equation for the torque equilibrium, which follows from minimization of the free energy (5.101):

$$\gamma_\varphi \frac{\partial \varphi}{\partial t} + K \frac{\partial^2 \varphi}{\partial x^2} = P_S E \sin\varphi + \frac{\Delta\varepsilon E^2}{4\pi} \sin\varphi \cos\varphi \tag{5.104}$$

assuming the FLC to be uniaxial, $\Delta\varepsilon = (\varepsilon_\parallel - \varepsilon_\perp)\sin^2\theta$, and K is a combination of K_1, K_2, and K_3 from (5.97). When the helix is unwound by the cell walls, the second term in the left part vanishes.

The boundary conditions are:

$$K\frac{\partial \varphi}{\partial x} + W_p \sin\varphi \pm W_d \sin 2\varphi|_{x=d,0}. \tag{5.105}$$

For polarizations $P_S > 10\,\mathrm{nC\,cm^{-2}}$, driving fields $E < 10\,\mathrm{V\,\mu m^{-1}}$ and dielectric anisotropy $|\Delta\varepsilon| < 1$, we have

$$|\Delta\varepsilon E/4\pi| \ll P \tag{5.106}$$

and, consequently, the second term in the right-hand part of (5.104) may be omitted. In this case the response times in the Clark–Lagerwall effect are given by

$$\tau_\varphi = \gamma_\varphi / P_S E. \tag{5.107}$$

If inequality (5.106) is invalid, which occurs for sufficiently high fields, $|\Delta\varepsilon E/4\pi| \simeq P_S$, the response times of the Clark–Lagerwall effect sharply increase for positive $\Delta\varepsilon$ values. In contrast, for negative values of $\Delta\varepsilon$, the corresponding switching times become shorter [228,229]. This is especially important for practical applications because it promotes an increase in the information capacity of FLC displays. For $|\Delta\varepsilon E/4\pi| \gg P_S$, the FLC switching times τ are approximately governed by the field squared, $\tau \simeq 4\pi\gamma_\varphi/\Delta\varepsilon E^2$, as in the Frederiks effect in nematics.

Reference [230] shows that two regimes of switching exist in the Clark–Lagerwall effect, separated by threshold field E_{th}:

$$E_{th} \cong 4W_d/P_S d. \tag{5.108}$$

For $E < E_{th}$, one observes the motion of domain walls, separating the regions of differently oriented polarization **P** and $-$**P**. The switching time is defined by the motion of the walls. If $E > E_{th}$ (the Clark–

Lagerwall regime), the switching time is determined by the τ_φ value (5.107). Electrooptical response times of FLCs strongly depend on their real structure (alignment technique, presence of defects, etc.).

The optical transmittance I in the Clark–Lagerwall effect is calculated as follows [231,232].

$$I = I_0 \sin^2 4\theta \sin^2 \Delta\Phi/2 \tag{5.109}$$

where $\Delta\Phi = 2\pi\Delta n d/\lambda$ is the phase difference, $\Delta n = n_{\|} - n_{\perp}$. As follows from (5.109), the maximum contrast is obtained for $\theta = \pi/8$ (22.5°) and $\Delta n d/\lambda = 1/2$. It means that very thin films, about $2\,\mu$m, have to be used.The variation of the cell thickness of $\Delta d = \lambda/8\Delta n$ from the optimum value $d = \lambda/2n$ results in a considerable difference in the FLC electrooptical response [233,234]. Display performance however, requires even more precise thickness, $\Delta d \cong \pm 0.2\,\mu$m.

The space charge, which accumulates at the interface between the orienting layer and the FLC medium [235,236] due to the spontaneous polarization, influences considerably the switching properties of the Clark–Lagerwall effect. In fact, the field caused by the space charge screens the external field, thus the threshold for the switching and the response times increase.

In order to avoid screening, thin orienting layers with high dielectric permittivity and sufficiently thick FLC cells with low spontaneous polarization have to be used. A decrease of the charge influence is also observed when charge transfer complexes are introduced either into the bulk of an FLC [237] or into an orienting layer [238]. These conductive complexes neutralize accumulated surface charges induced by spontaneous polarization and ionic impurities. On the other hand, charge controlled switching was suggested [239,240] to obtain the grey scale in SSFLC devices. A theoretical description of the charge transport phenomena in FLCs was given in [241,242].

The switching characteristics, especially purely bistable switching, depend critically on the homogeneity of an SSFLC structure. One of the standard techniques often used to produce the homogeneous smectic C* layer is a preliminary preparation of a perfectly aligned smectic A sample. If an FLC material does not possess the nematic phase, the smectic A phase can be oriented directly from the isotropic state. In thick (>100 μm) cells and even in thin ones with weak anchoring of the director to boundaries, it is possible to grow well oriented smectic A layers using either a magnetic field or a temperature gradient. In both cases, the preferred alignment direction is predetermined by a rubbed polymer layer [243] or by an edge of the spacer fixing the cell thickness and playing the role of a crystallization center [244].

FLCs can be aligned using a parallel shift of one cell substrate with respect to the other. The smectic layers tend to orient perpendicular to the substrates ("bookshelf geometry"), when mechanical shear is applied perpendicular to the rubbing direction.

FLC mixtures having no smectic A phase can be oriented by cooling down the nematic phase in an AC electric field which defines the direction of smectic layers via an electrohydrodynamic process, while the director is oriented by rubbing the polymer films on the substrates [245].

One of the best aligning techniques, which allows one to avoid zig-zag defects between two adjacent smectic layers bent in opposite directions (chevrons) (FIGURE 5.40(a)), remains oblique evaporation of silicon monoxide [246-248]. Zig-zag defects are avoided

FIGURE 5.40. Schematic picture of zig-zag defects. Zig-zag defects (a) are avoided by preparing (b) the preferred orientation of chevrons (left) or uniform tilt (right) of FLC layers. Short arrows in (b) indicate the directions of oblique evaporation of the orienting layers.

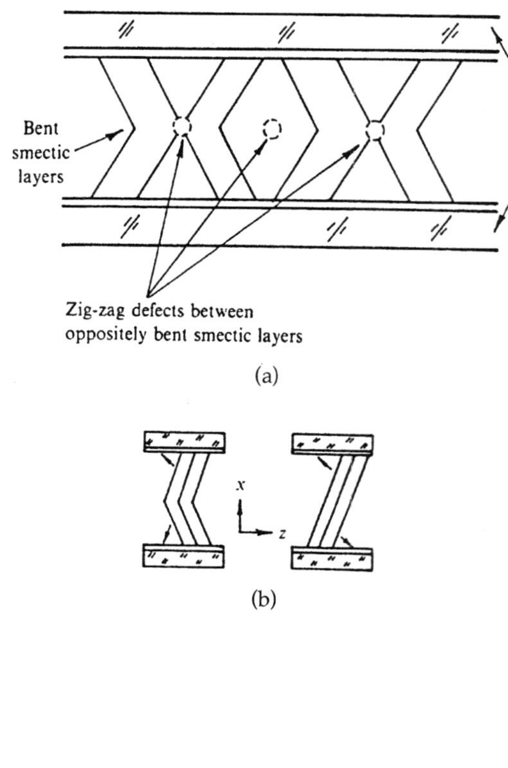

by promotion [249] of only one possible tilt of the smectic layers due to the specific oblique director orientation at the boundaries (FIGURE 5.40(b)). Samples with the antiparallel evaporation direction contain uniformly tilted layers. Those with parallel directions exhibit the chevron structure of tilted layers free of zig-zag defects. Optical properties [250] and x-ray diffraction [251] of chevron patterns have been studied.

The chevron structure and zig-zag defects were shown to disappear if a low-frequency electric field of sufficient amplitude is applied to an FLC cell [252,253]. The voltage required for reorientation decreases with increasing spontaneous polarization. The texture that appeared resembles the "bookshelf" geometry with smectic layers perpendicular to the substrates. This "quasi-bookshelf" structure has minimum response time and maximum transmission difference between the two stable states mentioned.

Reliable reproduction of bistability conditions is crucial for technological applications. The chevron geometry, even in zig-zag-free samples, hinders the occurrence of bistability. Chevrons and, more generally, the tilt of smectic layers, should be eliminated from the cell by any means, for example, using a low-frequency AC field to arrange "quasi-bookshelf" structure or developing FLCs with very short helical pitch preventing chevron appearance [254]. However, even in the most favorable case of smectic layers perpendicular to the substrates, it is important to optimize other FLC parameters such as polarization, helix pitch, elastic moduli and anchoring energies.

Qualitative analysis of bistability based on the catastrophe theory [255] shows that bistability appearance depends crucially on polar anchoring energy. For large W_p, a third stable state with a splayed director profile was shown to occur. If the polar energy is small compared with the dispersion one (in agreement with experimental data [225]), the following phenomenological relationships between bistability threshold and FLC parameters were obtained [255]:

$$\begin{aligned} &\text{(a)} \quad E_{th} \propto W_d && \text{for small } W_d \\ &\text{(b)} \quad E_{th} \propto 1/P_S && \text{for high } P_S \\ &\text{(c)} \quad E_{th} \propto d^{-\alpha} && \alpha \simeq 1\text{--}1.5. \end{aligned} \qquad (5.110)$$

A review of the development of liquid crystal devices based on the Clark–Lagerwall effect may be found in [256].

5.5.2.3 Deformed Helix Ferroelectric effect

The deformed helix ferroelectric (DHF) effect, observed recently in a short-pitch FLC [257], is a particular case of a more general phenomenon of field-induced helix distortion observed in the very first investigations on FLCs [11,258]. Its theoretical explanation was given in [259–261]. The geometry of the FLC cell with the DHF effect is presented in FIGURE 5.41. The polarizer at the first substrate makes an angle β with the helix axis and the analyzer is crossed with the polarizer. The FLC layers are perpendicular to the substrates and the layer thickness d is much larger than the value of the helical pitch h_0. A light beam with aperture $a \gg h_0$ passes parallel to the FLC layers through the cell.

In an electric field $\pm E$, the FLC helical structure becomes deformed so that the corresponding

FIGURE 5.41. Deformed helix ferroelectric effect.

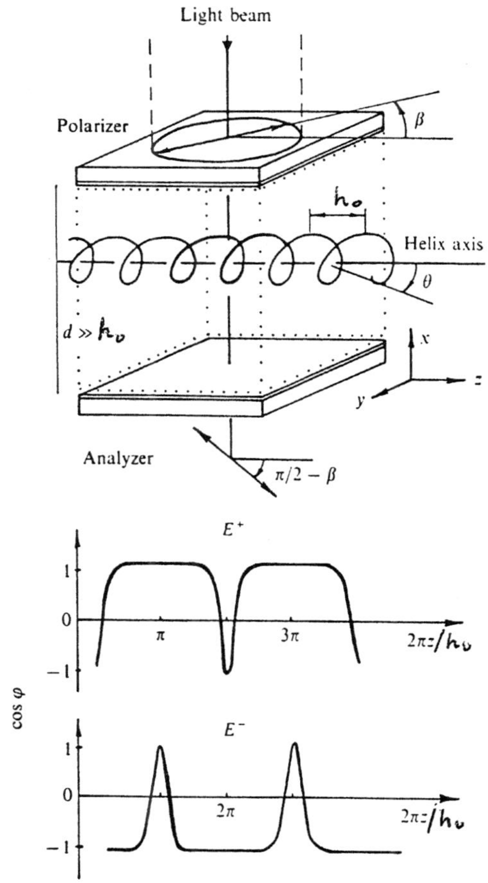

dependence of the director distribution $\cos\varphi$ as a function of coordinate $2\pi z/h_0$ oscillates between the two pictures shown in FIGURE 5.41. These oscillations result in variation of the effective refractive index, that is, electrically controlled birefringence appears. The effect takes place up to the fields for the FLC helix unwinding (here K_2 is defined so that it includes the θ^2 term, compare, e.g., with [18]):

$$E_u = \frac{\pi^2}{16}\frac{K_2 q_0^2}{P_S}. \tag{5.111}$$

The characteristic response times τ_c of the effect in low fields $E/E_u \ll 1$ are independent of FLC polarization P_S and field E, and defined only by the rotational viscosity γ_φ and helix pitch h_0:

$$\tau_C = \frac{\gamma_\varphi}{K_2 q_0^2} = \frac{\gamma_\varphi h_0^2}{K_2 \pi^2} \tag{5.112}$$

where K_2 is the elastic coefficient for helix deformation. For E close to the unwinding field E_u, the pitch h sharply increases $h \gg h_0$. Consequently, the times for the helix relaxation τ_d to the initial state also rise:

$$\tau_d/\tau_C \propto h^2/h_0^2 \tag{5.113}$$

that is for $E \simeq E_u$, it is possible to establish the memory state of the FLC structure [262]. In this regime, the electrooptical switching with the DHF effect reveals a pronounced hysteresis [257,263,264]. However, if the FLC helix is not deformed too much, fast and reversible switching in the DHF mode could be obtained [264].

The optical transmission of a DHF cell can be calculated as follows:

$$I = I_0 \sin^2\frac{\pi\Delta n(z)d}{\lambda}\sin^2 2[\beta-\alpha](z) \tag{5.114}$$

where β is the angle between the z-axis and the polarizer,

$$\alpha(z) = \arctan(\tan\theta\cos\varphi(z)) \tag{5.115}$$

is the angle between the projection of the optical axis on the y,z-plane and the z-axis, and $\Delta n(z) = \tilde{n}_{||}(z) - n_\perp$ is the effective birefringence,

$$\tilde{n}_{||}(z) = \frac{n_{||}n_\perp}{[n_\perp^2 + (n_{||}^2 - n_\perp^2)\sin^2\theta\sin^2\varphi(z)]^{1/2}}. \tag{5.116}$$

The diffraction of light by the helical structure is avoided for short helical pitch FLC mixtures, $h_0 \cong$–0.3 μm, and the helical structure behaves like a uniaxial plate with the direction of the optical axis controlled by the field strength [262].

In the case of small angles θ, $|\theta| \ll 1$, the transmission can be expanded in θ:

$$I \propto (\sin^2 2\beta - 2\theta\sin 4\beta\cos\varphi + 4\theta^2\cos 4\beta\cos^2\varphi) \times \sin^2\frac{\pi\Delta nd}{\lambda} \tag{5.117}$$

and, a variation in β may result in either pure quadratic or pure linear light modulation (or mixed).

Let us point to certain advantages of the DHF electrooptical effect for applications as compared with the Clark–Lagerwall mode.

1. High operation speed is achieved for low driving voltages. This takes place because a slight distortion of the helix near the equilibrium state results in considerable change in the transmission. Consequently, the instantaneous response of the FLC cell is provided without the so-called delay time, inherent to the Clark–Lagerwall effect.
2. The DHF effect is also less sensitive to the surface treatment and more tolerant to the cell gap inhomogeneity. Thus the same aligning technique which has been developed for nematic liquid crystals may be used.
3. The DHF effect allows for the implementation of a "natural," that is, dependent on voltage amplitude, gray scale both linear and quadratic in the voltage. Moreover, at $E \simeq E_u$, long-term optical memory states are possible.

FLC structures with very short helical pitch show bistable properties [264]. These types of FLC materials are called short pitch bistable ferroelectrics (SBF). A specific feature of SBF materials is a sufficiently large value of the spontaneous polarization.

5.5.2.4 Electroclinic Effect near the Smectic A → C* Phase Transition

The electroclinic effect or the field-induced tilt angle in the smectic A phase near the smectic A–C* phase transition was, for the first time, reported in [265]. The electrooptical characteristics of the effect were studied in detail [266–269] and novel materials based on the electroclinic effect were suggested [270,271].

The electroclinic effect may be understood in the framework of the Landau phase transition theory [18]. If the helical structure near the phase transition is disregarded the free energy density f of smectic A may be expanded in terms of the tilt angle θ as follows:

$$f = f_o + A'\theta^2/2 + P^2/2\chi_\perp - PE - \mu_p P\theta - \varepsilon_\infty E^2/8\pi \tag{5.118}$$

where f_o represents all contributions to f from the undistorted smectic A, $A' = a(T - T_c)$, $a > 0$, $\varepsilon_\infty = 1 + 4\pi\chi_\infty$ is the high-frequency background contribution to $\varepsilon_\perp$. Minimization of f with respect to θ and polarization P results in renormalization of the A' coefficient

$$A^* = A' - \mu_p^2\chi_\perp \tag{5.119}$$

and field-dependent tilt angle

$$\theta \cong \frac{\mu_p\chi_\perp E}{A'} = e_c E. \tag{5.120}$$

The coefficient $e_c = \mu_p\chi_\perp/A'$ is known as the electroclinic coefficient. Because $A' = a(T - T_c)$, the electroclinic coefficient becomes infinite at $T = T_c + \mu_p^2\chi_\perp$ (in fact, it is limited by the term Kq_c^2 in the denominator when the helical structure is taken into account [18]). This effect is also called the soft mode effect and occurs at $T < T_c$ also, but in the smectic C* phase the Goldstone mode related to the variation of the azimuthal director angle φ is much more pronounced.

The switching time of the director tilt angle in the electroclinic effect is independent of the electric field and is defined only by the rotational viscosity γ_θ and the elastic modulus A'. The corresponding switching times are derived from the Landau–Khalatnikov equation for the balance of the viscous and elastic torques:

$$\gamma_\theta \frac{\partial\theta}{\partial t} + A^*\theta = 0 \tag{5.121}$$

with

$$\tau_\theta = \gamma_\theta / A^*.$$

The switching time τ_θ may be calculated more precisely if the fourth-order term in θ is allowed for in expansion (5.118). In this case, a decrease in τ_θ for larger electric fields E is observed [268]. The experimental data confirm the linear increase in the tilt angle with increasing E for the electroclinic effect and the independence of τ_θ on E, FIGURE 5.42 [266].

The geometry of the cell in which the electroclinic effect is observed is also shown in FIGURE 5.42. In an electric field the optical axis of the cell rotates by an angle θ, measured as a difference between two

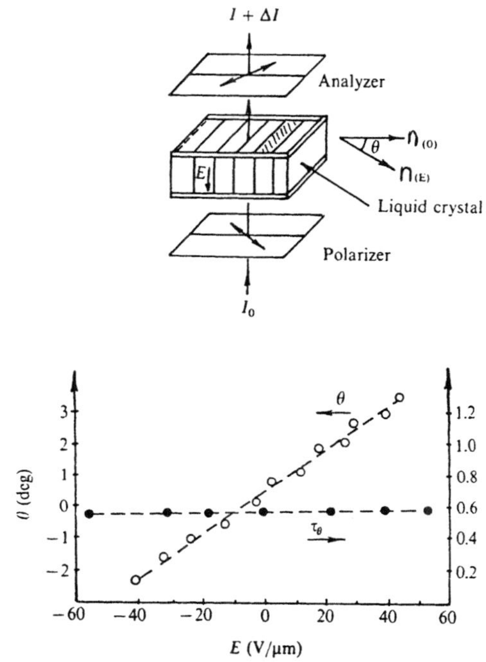

FIGURE 5.42. Electroclinic effect in an FLC cell. Typical dependences of the tilt angle θ and the response time τ_θ are given below, $T = 25\,°C$ [266].

directors with subscripts (E) and (0). The transmission is calculated according to the formula [266]:

$$I = \sin^2 2(\psi_o + \theta(E)) \tag{5.122}$$

where $\psi_o + \theta(E)$ is the angle between the polarizer and director projection on the substrate plane. In this case, the phase factor $\sin^2 \pi d\Delta n/\lambda$ is virtually independent of the modulation of θ and is chosen to be equal to 1, $d\Delta n/\lambda = 1/2$ and the optimum linear response to field $E = E_o \sin\omega t$ corresponds to $\psi_o = \pi/8$ [266,267]:

$$\Delta I \propto 2e_c E_o \sin\omega t. \tag{5.123}$$

For substances with a very high electroclinic coefficient, a considerable field-induced optical biaxiality has been measured in the smectic A phase by the transient dielectric spectrosopy technique [272].

The tilt angle in the smectic C* phase is also field dependent and a corresponding "electroclinic" coefficient of the same order of magnitude may be measured in both the smectic A and C* phases [273].

At present, the electroclinic effect is the fastest one among the electrooptical effects in liquid crystals. The response times for the effect measured at room temperature is on the order of 1 μs for voltages of about 10 V and layer thicknesses of about 1 μm. The electroclinic effect can be successfully used in high-speed electrooptical modulators (up to frequencies of a few MHz).

5.5.2.5 Domain Mode

A remarkable domain structure has recently been observed in FLCs possessing high spontaneous polarization, higher than 100 nC cm^{-2} [274]. A very stable spatially periodic optical pattern arises in layers of FLCs with molecular orientation parallel to substrates after a DC field treatment of the cell. The stripes are oriented perpendicular to the director orientation and their period D is inversely proportional to the polarization magnitude squared (FIGURE 5.43) [275], and independent of cell thickness. The optical pattern may be observed for several months after switching off the DC field.

The appearance of domains is accompanied by an additional (to the Goldstone and soft modes) relaxation process observed by dielectric spectroscopy [276]. Two relaxation times were resolved; the slower was related to bulk phenomena and the faster was related to surface effects.

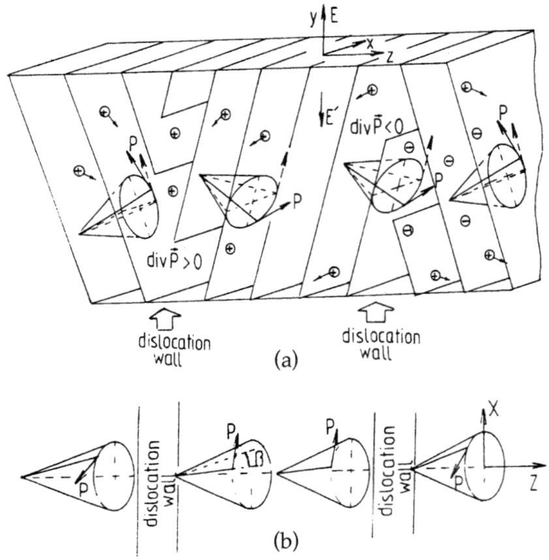

FIGURE 5.44. A model for charged dislocation walls between domains with oppositely tilted smectic layers. (a) A side view of dislocations and polarization vectors rotated over the cone surfaces under action of the internal electric field $\mathbf{E}'$; (b) a top view of the structure, angle β being the divergence angle between the director projections onto the substrate plane responsible for the optical contrast of the pattern of neighbor domains [275].

FIGURE 5.43. The period D and wavevector $q_D = 2\pi/D$ of the domain structure versus spontaneous polarization ($T = 20\,°$C, $\theta = 31°$, cell thickness =17mm) [275].

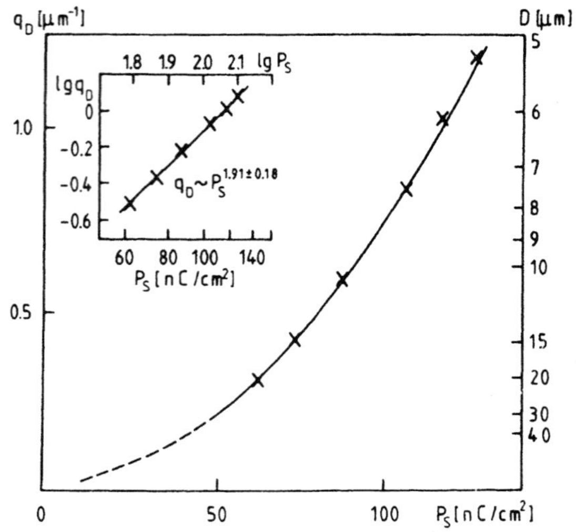

The theoretical model worked out in [275] is based on consideration of the stabilizing role of structural defects in an FLC layer. Domains may arise only in the structure where smectic layers are alternately tilted, even at a small angle θ, with respect to substrates. Areas with the opposite tilt direction are separated by thin walls (of thickness Δz) formed by edge dislocations (FIGURE 5.44). The z-projections of the spontaneous polarization perpendicular to the walls, P_z have opposite signs, thus a high density of polarization charge $\rho_p = -\Delta P_z/\Delta z$ is created in the wall area. The polarization charge controls the motion of the free carriers after switching off the DC field, which determines the initial direction of the vector of spontaneous polarization (the y-projection (directed up) in FIGURE 5.44). The process of compensation of the polarization charge by free carriers goes hand in hand with the reorientation of the polarization (and the director), which is governed by the internal field of the ions separated earlier by the external field.

The whole process is described by the energy balance per unit volume

$$F = -D^{-1}(\operatorname{div}\mathbf{P})^2 + (\theta/D)^2 - \mathbf{PE} \qquad (5.124)$$

where the first term is the energy of the Coulomb attraction between charges accumulated near the dislocation walls, which includes the neutralizing effect of the polarization charges by the free carriers mentioned; the second term describes the elastic energy of wall repulsion from each other, and the third term determines the orientation of the total polarization **P** by the external field.

In the field-off regime, the minimum of the free energy determines the optimum value of the domain width

$$D \propto (\theta/\operatorname{div}\mathbf{P})^2 \propto (\Delta z/\mu\theta)^2 \qquad (5.125)$$

where for small θ angles it was assumed $P_z = \mathbf{P}_s \sin\theta \propto \mu\theta^2$ (μ is the piezoelectric coefficient determining the magnitude of the spontaneous polarization [18]). Thus the most distinguished feature of the domain pattern, namely $D \propto P^{-2}$, can be accounted for. In fact the model developed in [275] agrees with all of the most important experimental data available at present. Moreover, the tilted layer structure shown in FIGURE 5.44 has been confirmed recently by direct x-ray investigation [277].

It should be noted that recently [278], the field-off domain structure has also been observed after the application of an electric field to the smectic A phase of a ferroelectric substance with very high spontaneous polarization, $P_S \cong 500\,\mathrm{nC\,cm^{-2}}$. This phenomenon is assumed to arise due to a break of the smectic layers (very similar to that discussed above) induced by a strong electroclinic effect.

5.5.3 Antiferroelectrics

5.5.3.1 Structural Features

Tristable switching typical of solid antiferroelectrics has recently been reported [279] for the chiral liquid crystalline phases of 4-(1-methylheptyloxycarbonyl)phenyl - 4′ - octyloxybiphenyl - 4 - carboxylate (MHPOBC) (see FIGURE 5.45). The third state, in

FIGURE 5.45. MHPOBC compound structure.

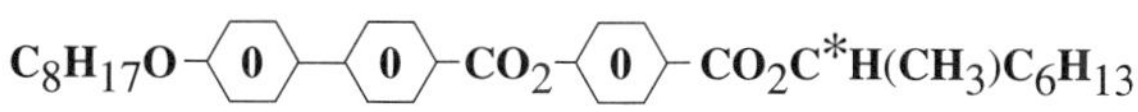

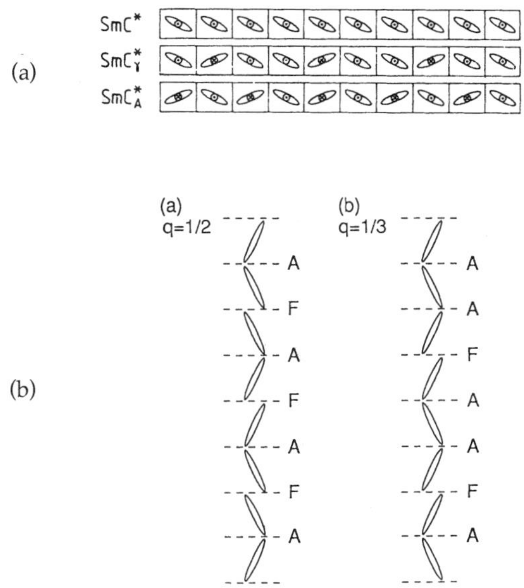

FIGURE 5.46. (a) The models of the ferro (SmC*), ferri (SmC^*_γ) and antiferro (Sm^*_A)-electric phases of MHPOBC (the normal to the smectic layers is horizontal) [283]. (b) Ferrielectric configurations with $q = 1/2$ and $1/3$ [287].

addition to the two stable states known for the Clark–Lagerwall effect, was shown to be stable [280] and corresponded to a layered structure with alternating direction of the molecular tilt [281]. Thus the direction of the in-plane polarization also alternates and the whole structure is antiferroelectric.

In fact, MHPOBC shows a very rich polymorphism [282];

$$SmI^*_A\text{–}64C\text{–}SmC^*_A\text{–}118.4C\text{–}SmC^*_\gamma\text{–}119.2C\text{–}SmC^*\text{–}120.9C\text{–}SmC^*_\alpha\text{–}122C\text{–}SmA\text{–}148C\text{–}I$$

where SmC^*_A and SmC^*_α are antiferroelectric phases, SmC* is a typical ferroelectric phase discussed in the previous section, and SmC^*_γ is a ferrielectric phase. The structures of the ferro-, ferri- and antiferroelectric phases are shown in FIGURE 5.46 [283]. On a macroscopic scale the phases are twisted with a characteristic pitch depending on temperature.

The antiferroelectric phases were observed not only in MHPOBC but in many other compounds [282]. Various techniques were used to prove the antiferroelectric packing of molecules, among them selective reflection and conoscopy [281], dielectric studies [284], observation of characteristic defects [285], and scanning tunneling microscopy [286]. The latter allowed a direct observation of the alternating tilt of molecules in neighboring layers, which is especially important in the absence of x-ray diffraction data.

The C_α^* phase observed in MHPOBC was carefully investigated both in the same substance [287] and another compound (MHPOCBC) [288] which manifests a much wider range of the phase (6 °C). With decreasing temperature a gradual change in its characteristics was observed, showing the gradual structural transformation from the antiferroelectric to the ferrielectric packing displayed in FIGURE 5.46(a). It was understood that the ferrielectric phase SmC_γ^* (shown once more on the right-hand side of FIGURE 5.46(b)) is not the only one. In fact, many other types of ferrielectric packing may exist, like that shown on the left-hand side of FIGURE 5.46(b). Each of these structures may be defined with a fraction q of the "ferroelectric states" F (the same sign of the molecular tilt in neighboring smectic layers) against a background of "antiferroelectric states" A (opposite tilt in neighboring layers). Thus the antiferroelectric phase AAAAA... corresponds to $q = 0/1$. In the left configuration, AFAFAF... of FIGURE 5.46(b) where $q = 1/2$, the position of each F state is $2, 4, 6 \ldots i/q$. For the right-hand configuration we have $q = 1/3$ and each F state can be found in a $3, 6, 9 \ldots i/q$ position. In general, a variety of ferrielectric phases with $q = m/n$ (m, n are integers) may be found [288] and each time the position of the F state is determined by the simple formula

$$z_i = \text{integer}(i/q). \tag{5.126}$$

Therefore, the so-called Devil's staircase is formed in a deep analogy with the one-dimensional Ising model where the F and A states correspond to the *up* and *down* states of interacting spins on a plane square lattice. Such a case also arises when discussing incommensurate crystalline phases [289].

5.5.3.2 Electric Field Behavior

An electric field affects considerably the phase transitions in MHPOBC [290]. The electric field–temperature phase diagram is shown in FIGURE 5.47. A very narrow low-temperature ferrielectric phase is seen (FIL) between the SmC_A^* and SmC_γ^* phases mentioned earlier. In addition, even a weak field strongly influences the boundary between the SmC_α^* and SmC^* phases. Further investigation has shown [291] that the field-induced transition between the two phases

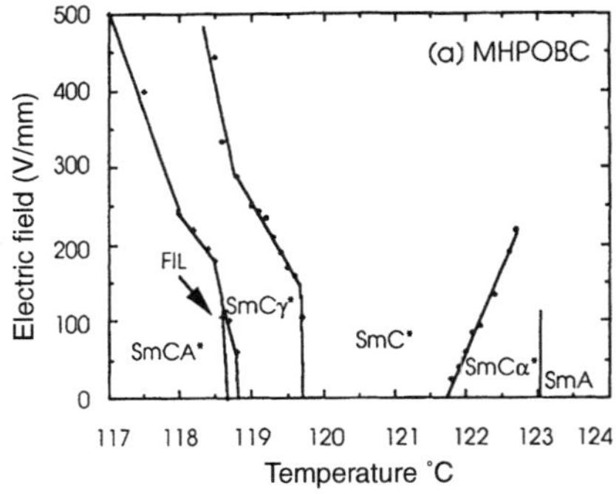

FIGURE 5.47. Electric field–temperature phase diagram of MHPOBC [290].

in MHPBOC involves the formation of a variety of ferrielectric configurations. In fact, the field-induced Devil's staircase is formed. This is especially apparent in FIGURE 5.48, where the field-induced change $\vartheta(E)$ of the apparent tilt angle in the SmC_α^* phase of the other substance (MHPOCBC) is shown together with its field derivative [292]. Thus both the temperature dependence and the field-induced Devil's staircases have been observed. An even richer field-induced polymorphism has been observed in binary mixtures of MHPOBC and MHPOCBC [290].

A typical double hysteresis loop for electrooptical switching of the antiferroelectric SmC_A^* phase is shown

FIGURE 5.48. Apparent tilt angle ϑ and the derivative $d\vartheta/dE$ as functions of the applied field in the SmC_A^* phase of MHPOCBC. Note the stepwise increase of ϑ (Devil's staircase) [292].

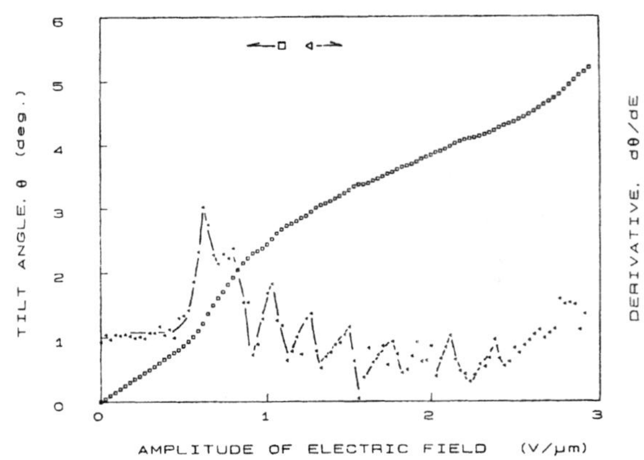

(a)

(b)

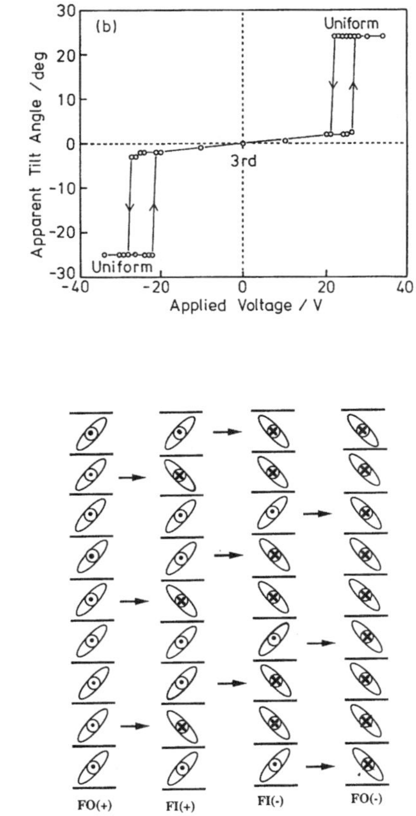

FIGURE 5.49. A double hysteresis loop (a) typical of the electrooptical switching of the antiferroelectric phase [293] and ferrielectric structures (b) appearing during the switching process [282].

in FIGURE 5.49(a). There are two field-induced uniform states and the zero-field third state which may be either uniform or twisted [293]. As was discussed above, the transition from the antiferroelectric to the ferroelectric state proceeds through intermediate ferrielectric states. An example is shown in FIGURE 5.49(b). With increasing frequency in the range of 10–100 Hz, the double hysteresis loop gradually transforms into the conventional ferroelectric loop [294]. The characteristic time corresponds to the time of the formation of spatial domains of the antiferroelectric phase. The low-frequency switching (0.2 Hz) of the SmC_{γ}^{*} phase reveals a triple hysteresis loop with two intermediate (+ and −) ferrielectric states and two final ferroelectric states. At higher frequency, the conventional ferroelectric switching is observed.

It should be noted that the antiferroelectric phase with zero macroscopic polarization even in the untwisted state has a great advantage over the ferroelectric phase as there is no accumulated charge preventing true bistable switching of the latter. This opens up good prospects for its application in display technology [295].

5.5.4 Polymers, Lyotropics, Polyphilics, Discotics

5.5.4.1 Ferroelectric Liquid Crystal Polymers

The symmetry requirements necessary for ferroelectricity in low molecular mass compounds are also valid for polymer mesophases. If a tilted chiral smectic phase is stable after a polymerization process it must be ferroelectric. Following this idea, the first polymer liquid crystalline ferroelectric has been synthesized by Shibayev and coworkers [296]. It was a comb-like homopolymer (FIGURE 5.50) with a polymethacrylate chain as a backbone, flexible spacers $COO(CH_2)_{10}$–COO–, separating phenyl benzoate mesogenic units from the backbone. Transverse (with respect to the mesogenic units) dipole moments provided by several COO– groups are weakly coupled to the chiral terminal fragment, thus, spontaneous polarization of the substance is not very high (about $3\,nC\,cm^{-2}$) and does not differ very much from the precursor monomer [297]. Later, a variety of comb-line polymers manifesting chiral smectic C*, F* and other phases were synthesized (see comprehensive reviews [298,299]).

There is also a wide class of ferroelectric polymers that contain mesogenic units both in the main chain and as lateral substitutents. A very high value of the

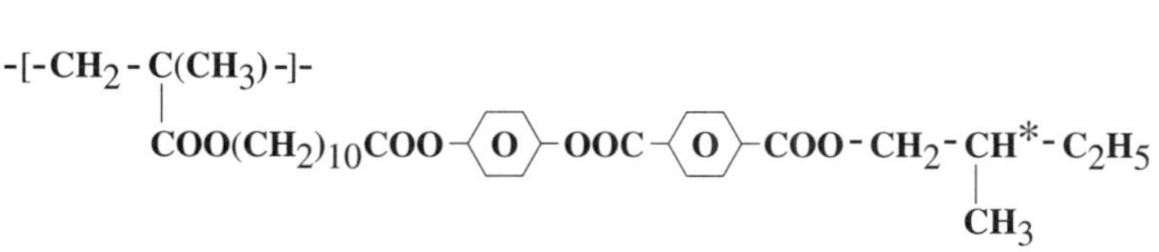

FIGURE 5.50. The first polymer liquid crystalline ferroelectric compound structure.

spontaneous polarization (500 nC cm^{-2}) has been reported recently for a poly(tetramethylene tartrate) with two chiral centers in the backbone and achiral side mesogenic groups [300]. In all cases, the thermodynamic and other properties of polymer ferroelectrics strongly depend on the molecular mass of the substance.

The dielectric and physical properties of ferroelectric liquid crystals were recently discussed in detail [23]. The temperature dependence of the spontaneous polarization is usually smoother than the corresponding curves for low molecular compounds but the general behavior is the same.

The process of the electric field-induced director reorientation in polymer ferroelectrics, like in their low molecular mass counterparts, involves changes in the tilt (θ) and azimuthal (φ) angles. These two modes are characterized by quite different rates. The θ process corresponds to the soft-mode distortion, and the corresponding time τ_θ is generally short, but almost diverges at the C*–A phase transition point. The τ_φ process means the motion of the director is on the conical surface centered on the normal to the smectic layer (the Goldstone mode). The latter involves the twisting–untwisting mode in the helical structure. The two times differ considerably from each other. This can be seen in both the dielectric spectra and in the pyroelectric and electrooptical response.

θ-SWITCHING. The simplest way to measure the rate of the relaxation of the θ angle in the C* phase is to apply a heat pulse to a sample and then follow the relaxation of the pyroelectric effect which reflects the relaxation of the spontaneous polarization proportional to the tilt angle. FIGURE 5.51(a) shows the divergence of τ_θ times in the smectic C* phase near the C*–A phase transition. The times were measured in studying the dynamics of the pyroelectric response and lie within the range 100 μs–1 μs [299]. They do not differ very much from the corresponding times for the low molecular mass analog of the polymer investigated. This means that the polymer backbone does not contribute very much to the dynamics of the soft mode of the material, that is, to the viscosity coefficient γ_θ.

The tilt angle may also be induced by an electric field in the smectic A phase near the A–C* phase transition temperature (electroclinic effect). The amplitude and sign of the θ angle depend linearly on the field (equation (5.120)). At the A–C* transition, the field-induced tilt can be as high as 10°. For a siloxane ferroelectric polymer [301], the temperature divergence of the field-induced tilt in the smectic A phase is shown in FIGURE 5.52. The field-independent relaxation time of the electroclinic effect is shown in FIGURE 5.51(b). We can see that the switching times (10–400 μs) for the θ process are only slightly longer

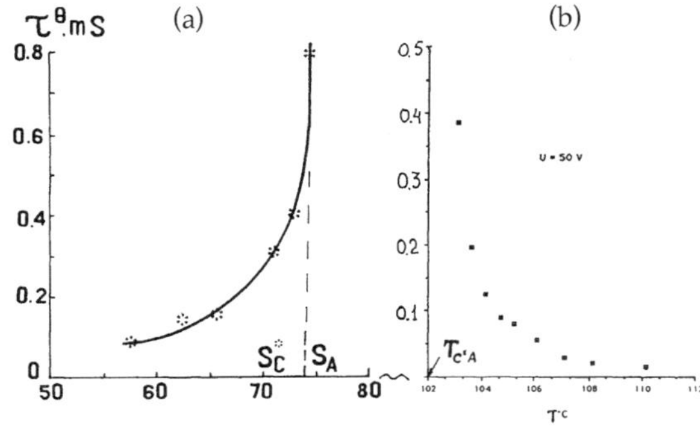

FIGURE 5.51. Tilt angle relaxation times as functions of temperature. (a) Relaxation of the pyroelectric response in the smectic C* phase of the polymer studied in [299]; (b) relaxation of the electroclinic effect in the smectic A phase of the siloxane polymer studied in [301].

FIGURE 5.52. Temperature behavior of the field-induced tilt angle in the smectic A phase of the same siloxane polymer as in FIGURE 5.51(a) [301].

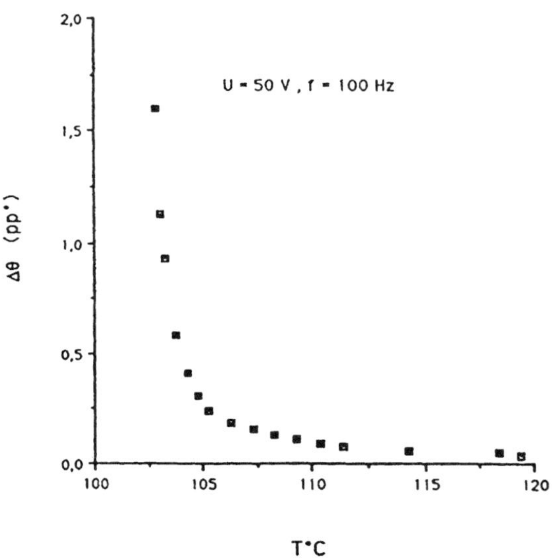

than the values typical of low molecular mass compounds.

A nonlinear dependence of the induced tilt angle on an external electric field with a very specific frequency dependence of the tilt has been observed in an acrylate side chain polymer with strongly polar mesogenic groups [302]. The features observed were attributed to the field-induced antiferroelectric–ferroelectric transition. The antiferroelectric switching of a polymer liquid crystal has also been reported [303].

φ-**SWITCHING.** For a field much lower than the threshold for the ferroelectric helix untwisting (equation (5.111)), the time of the response to an electric field is given by equation (5.112). For higher fields the reaction time is approximately given by equation (5.107). When the field is switched off, the relaxation time (or the off-time) is governed either by the helix twisting process, $\tau_R = \gamma_\varphi / Kq^2$ or by surface conditions.

τ_φ times measured from electrooptical switching are much longer than those in low molecular mass liquid crystals. They directly depend on the viscosity of the liquid crystal, that is, indirectly on its molecular mass. The field-on (reaction) times τ_r are shown in FIGURE 5.53 for the polymer liquid crystal studied in [304]. They are three to five orders of magnitute longer than typical τ_r values for conventional liquid crystalline ferroelectrics taken at the same field strength. They are governed by viscosity γ_φ which, in contrast to γ_θ, depends strongly on the motion of the backbones. The shortest τ_r times obtained in a polymer substance with very high polarization ($P_s \simeq 1\,\mu\text{C cm}^{-2}$) are on the order of 1 mS at an enhanced temperature (about 100 °C) [305]. Generally, the order of magnitude for γ_φ ranges from 10 to 10^4 P, depending on the temperature and molecular weight of the substance.

FIGURE 5.53. Field-on switching times τ_r as functions of temperature for a polymer with various molecular mass M_n [304].

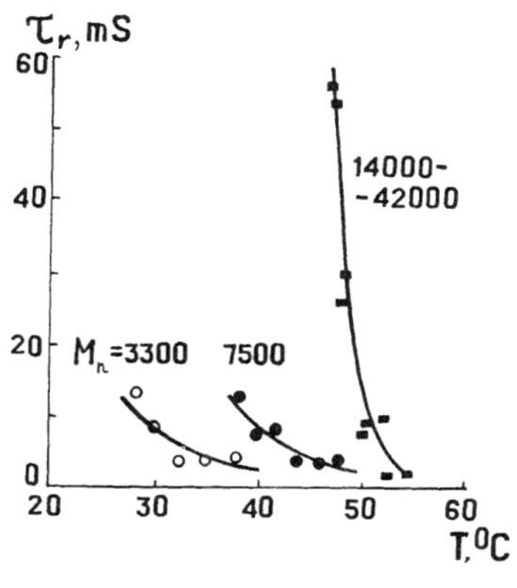

5.5.4.2 Lyotropic Systems

Water solutions of certain amphiphilic molecules, such as lipids, form lamellar lyotropic phases with a tilt of the molecules with respect to the layer normal [306]. When the same molecules are chiral or other chiral molecules (guests) are introduced into the lamellae (host), the lamellar phase satisfies all of the symmetry requirements for a ferroelectric phase (in a formal analogy to SmC*). One of the best chiral additives to be introduced into lyotropic phases is cholesterol, which induces high spontaneous polarization in thermotropic smectic C matrices [307]. To study the ferroelectric properties of lyotropic systems, ethylene glycol [308] and water [309] solutions of the synthetic dipalmitoyl phosphatidyl choline (DPPC or lecithin) were chosen. The ethylene glycol solutions are less conductive and more suitable for electric measurements. On the other hand, water solutions of lipids represent a close analogy to lipid membranes, whose ferroelectric properties might be of great importance for biological processes.

Thermotropic ferroelectric mesophases were shown to manifest a very high piezoelectric response when a low-frequency acoustic excitation was applied to a sample [310]. The same technique was chosen to demonstrate ferroelectric behavior in lyotropic systems. FIGURE 5.54 shows a piezoelectric response of pure ethylene glycol (30 wt.%)–DPPC (70 wt.%) mixture (curve 1) and the same mixture doped with 15 wt.% of chiral cholesterol (curve 2). In the L'_β phase, where matrix molecules are tilted, the chiral mixture manifests an order of magnitude higher response than the undoped mixture. An even more striking difference between the undoped and doped samples (two orders of magnitude) is observed when a DC poling field is applied to the samples [308]. The piezoresponse grows with increasing concentration of the dopant [309].

The temperature dependence of the piezoresponse of an aqueous lyotropic DPPC mixture (10% of water) is shown in FIGURE 5.55 [309]. Again, a dramatic

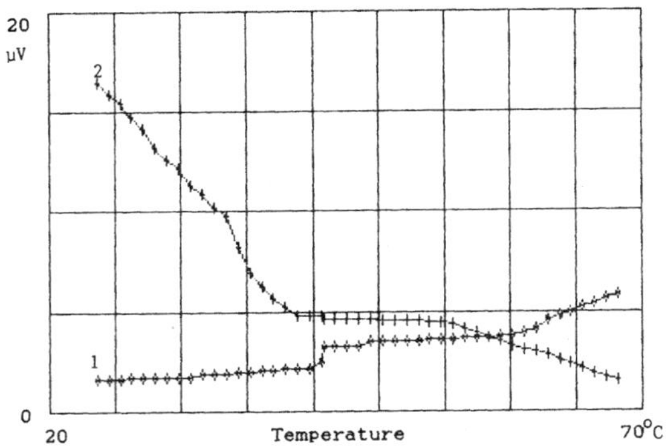

FIGURE 5.54. Piezoelectric response of pure ethylene glycol (30 wt.%)–DPPC (70 wt.%) mixture (curve 1) and the same mixture doped with 15 wt.% of chiral cholesterol (curve 2). Heating cycle [308].

difference between the undoped and doped (15% of cholesterol) samples is seen. Thus, the electric behavior of chiral analogs of the lyotropic phases is certainly much more pronounced than that of their achiral counterparts, and we conclude that cholesterol induces ferroelectricity into both systems studied (based on ethylene glycol and water).

5.5.4.3. Polyphilic Compounds

According to their symmetry all polar crystalline and mesomorphic phases are divided into two groups. One of them, which is composed of chiral molecules, has a point group symmetry ranging from C_1 to C_∞. Examples of this type are liquid crystalline chiral phases (smectic C* discussed above, F*, I* etc.) differing in character by the in-plane packing of their molecules within a smectic layer. These ferroelectric phases manifest reversible switching of their spontaneous polarization.

FIGURE 5.55. Temperature dependence of the piezoresponse in an aqueous lyotropic DPPC mixture (10% water) undoped (crosses) and doped with 15% of cholesterol (diamonds). Cooling cycle [309].

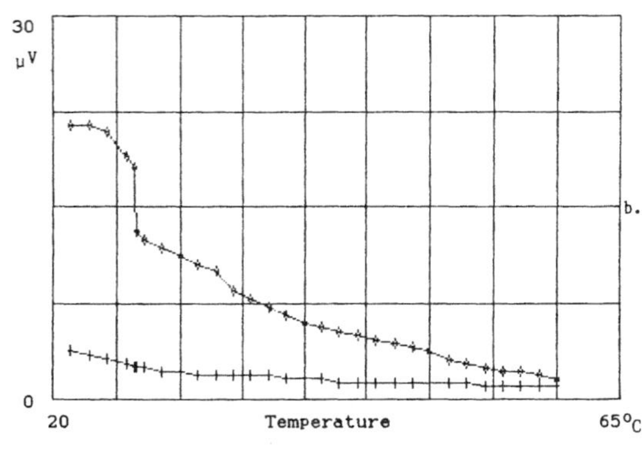

The other group of polar materials comprises mirror symmetric (achiral) phases with point symmetry groups from C_{2v} to $C_{\infty v}$, and among them one finds many classical solid crystalline ferroelectrics (like $BaTiO_3$, triglycinsulphate, thiourea, etc.). Recently, the ferroelectric properties have also been observed in noncrystalline (mesomorphic) achiral molecular material [311–313]. A chemical approach to construct new polar materials has been suggested in [314]. The basic idea was to use the polyphilic effect to form "building elements" of a polar phase. According to this concept, chemically different moieties of a molecule tend to segregate to form polar aggregates or lamellas which result in the formation of polar phases. The direction of the spontaneous polarization is allowed to be in the plane of the lamellas or layers (transverse ferro- or antiferroelectrics) or perpendicular to the layers (longitudinal ferro- or antiferroelectrics discussed theoretically earlier [315,316]).

The structure of the polyphilic compound (PC1) is shown in FIGURE 5.56. It consists of three distinct parts: perfluoroalkyl and alkyl chains, and a biphenyl rigid core. X-ray structure investigations showed the following phase sequences: on heating–crystal (98 °C) smectic A (115 °C) isotropic; on cooling–isotropic (115 °C) smectic A (82 °C) smectic X. The latter is stable down to room temperature and crystallized on the next heating cycle.

Two fundamental properties characteristic of ferroelectric substances were studied, namely, the field-induced switching of the piezoelectric response (discussed above) and the pyroelectric effect which is

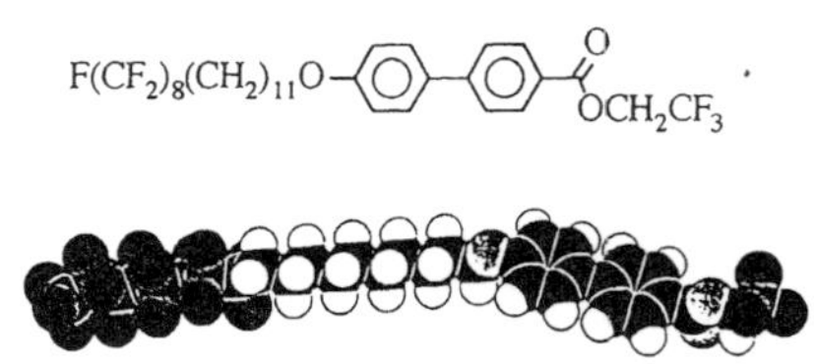

FIGURE 5.56. Polyphilic compound PC1. The different moieties have a tendency to segregate.

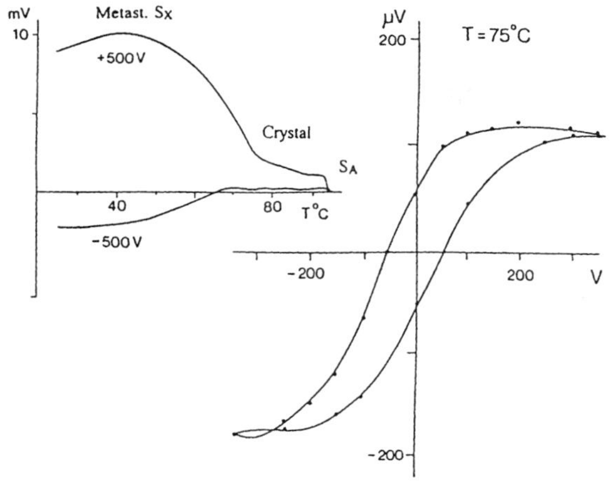

FIGURE 5.57. Left: temperature dependences of the piezoelectric response for PC1 poled either with +500 V or −500 V on cooling down from the isotropic phase. Right: the hysteresis of the piezovoltage taken at 75 °C) [313].

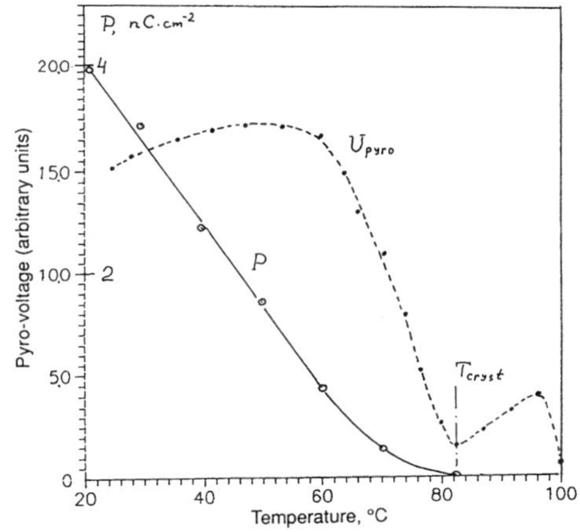

FIGURE 5.58. Pyroelectric voltage and the field-induced polarization of PC1 as functions of temperature (poling DC voltage +30 V, cell thickness 11 μmm) [313].

proportional to the temperature derivative of the macroscopic polarization (the pyrocoefficient $\gamma = \partial P/\partial T$). Thus the polarization is calculated by integrating $\gamma(T)$ over the temperature starting from the phase transition to the smectic A (or other paraelectric) phase.

In experiments with PC1, the polar state can be induced using a poling voltage. If the sample is poled by a DC field during its cooling from the isotropic phase (e.g., for 1 hour), the piezoelectric response stable for days is observed in the metastable smectic X phase. Its amplitude is comparable to the signal observed for conventional (chiral) ferroelectrics, that is, three orders of magnitude higher than the background level typical of nonpolar compounds.

On heating, the piezovoltage of PC1 decreases even in the case when no crystallization is observed (FIGURE 5.57, the curve for +500 V). If, by chance, crystallization occurs, the signal drops dramatically and can even change its sign (the curve for −500 V). It is impossible to repolarize the sample at room temperature, but repolarization is easily attained with increasing temperature. The higher the temperature, the faster the process of repolarization. The corresponding hysteresis loop (in coordinates applied DC voltage–piezovoltage) is shown in the same figure. The characteristic time of the repoling process at this temperature (75 °C) is on the order of a few minutes.

The temperature behavior of the pulsed pyroelectric response of a layer of PC1 is shown in FIGURE 5.58. The sample was poled with voltage of +30 V. The calculated value of the field-induced macroscopic polarization is also shown in the same figure. The magnitude of P at room temperature is about 10 nC cm^{-2} which is one order of magnitude lower than a value typical of chiral liquid crystalline ferroelectrics.

Thus, the X phase of the polyphilic compound PC1 manifests some ferroelectric properties. The direction of the macroscopic polarization can be switched by an external field with hysteresis loops typical of ferroelectrics. No electrooptical effect has been observed, which means that the polarization and the director are decoupled. Recent x-ray studies showed that the X phase is spatially modulated and strongly disordered [317]. The latter may be related to a certain domain structure at the microscopic level. Certainly this class of compounds requires further investigation.

5.5.4.4 Discotics–Ferroelectrics

Chiral discotic mesophases also manifest ferroelectric switching [318]. The switching was demonstrated in the tilted columnar phase consisting of amphiphilic disc-shaped molecules. By applying positive and negative voltages, the direction of the minimum refractive index could be rotated between −20° and +20° from the direction of columns (parallel to substrates). The switching times were on the order of 1 s (field $6 \times 10^4\,\mathrm{V\,cm^{-1}}$, temperature 85 °C). Another example of a ferroelectric discotic has been studied in [319].

5.6 CONCLUSION

In conclusion, electric field effects in liquid crystals are a very well developed branch of condensed matter physics. The field behavior of nematic liquid crystals in the bulk is quite well understood and the main focus of investigation has shifted to studying surface phenomena (anchoring effects), nematic polymers and heterogeneous systems (e.g., polymer dispersed nematics). To a certain extent, the same is true for the cholesteric mesophase, although the discovery of bistability phenomena and field effects in blue phases has opened up a block of new fundamental problems to be solved. Ferroelectric mesophases are the subject of current controversy, the physics of surface phenomena, the problems of bistability, and the high-speed response of conventional smectic C* materials are still a challenge to researchers. The other ferroelectric mesophases, such as those observed in lyotropic, polyphilic and discotic mesogens, should attract more attention in the near future. The same is true for a variety of polymer ferroelectric substances, including elastomers.

REFERENCES

1a. de Gennes, P. G. 1974. *The Physics of Liquid Crystals.* Clarendon Press, Oxford.

1b. de Gennes, P. G., and Prost, J. 1993. *The Physics of Liquid Crystals.* 2d ed. Oxford: Oxford University Press.

2a. Frederiks, V., and Zolina, V. 1930. *Zh. Russ. Fiz. Khim. Obshch.* **62**: 457.

2b. Frederiks, V., and Zolina, V. 1933. *Trans. Faraday Soc.* **29**: 919.

3. Frederiks, V., and Tsvetkov, V. N. 1935. *Dokl. Akad. Nauk SSSR* **4**: 123.

4a. Heilmeier, G. H., Zanoni, L. A., and Barton, L. A. 1968. *Proc. IEEE* **56**: 1162.

4b. Heilmeier, G. H., Zanoni, L. A., and Barton, L. A. 1970. *IEEE Trans. Elec. Dev.* **ED-17**: 22.

5. Zvereva, G. E., and Kapustin, A. P. 1961. *Primeneniye Ultraakustiki k Issledovaniyu Veshchestva* (Application of Ultraacoustics to a Study of Matter), vol. 15, p. 69. Moscow: MOPI.

6. Williams, R. 1963. *J. Chem. Phys.* **39: 384.**

7. Meyer, R. B. 1969. *Phys. Rev. Lett.* **22**: 319.

8. Vistin', L. K. 1970. *Dokl. Akad. Nauk SSSR* **194**: 1318.

9. Bobylev, Y. P., and Pikin, S. A. 1977. *Zh. Eksp. Teor. Fiz.* **72**: 369.

10. Barnik, M. I., Blinov, L. M., Trufanov, A. N., and Umansky, B. A. 1977. *Zh. Eksp. Teor. Fiz.* **73**: 1936.

11. Meyer, R. B., Liebert, L., Strzelecki, L., and Keller, P. 1975. *J. Physique Lett.* **36**: 69.

12. Clark, N. A., and Lagerwall, S. T. 1980. *Appl. Phys. Lett.* **36**: 899.

13. Chandrasekhar, S., 1992. *Liquid Crystals.* 2d ed. Cambridge: Cambridge University Press.

14. de Jeu, W. H. 1980. *Physical Properties of Liquid Crystalline Materials.* New York: Gordon and Breach.

15. Belyakov, V. A., and Sonin, A. S. 1982. *Optika Kholestericheskikh Zhidkikh Kristallov* (Optics of Cholesteric Liquid Crystals). Moscow: Nauka.

16. Vertogen, G., and de Jeu, W. H. 1988. *Thermotropic Liquid Crystals: Fundamentals.* Berlin: Springer.

17. Kapustin, A. P. 1973. *Electroopticheskiye i Akusticheskiye Svoistva Zhidkikh Kristallov* (Electrooptical and Acoustical Properties of Liquid Crystals). Moscow: Nauka.

18. Pikin, S. A. 1991. *Structural Transformations in Liquid Crystals.* New York: Gordon and Breach.

19a. Blinov, L. M. 1983. *Electro-Optical and Magneto-Optical Properties of Liquid Crystals.* Chichester: Wiley.

19b. Blinov, L. M., and Chigrinov, V. G. 1994. *Electrooptic Effects in Liquid Crystal Materials.* New York: Springer.

20. Grebenkin, M. F., and Ivashchenko, A. V. 1989. *Zhidkokristallicheskiye Materialy* (Liquid Crystalline Materials). Moscow: Khimiya.

21. Loseva, M. V., Pozhidaev, E. P., Rabinovich, A. Z., Chernova, N. I., and Ivashchenko, A. V. 1990. *Itogi Nauki i Tekh. Fiz. Khim.* **3**: 3.

22. Bahadur, B. 1984. *Mol. Cryst. Liq. Cryst.* **109**: 1.

23. Blinov, L. M., and Haase, W. 1993. *Mol. Mater.* **2**: 145.

24. Dunmur, D. A., and Palffy-Muhoray, P. 1988. *J. Phys. Chem.* **92**: 1406.

25. Helfrich, W. 1970. *Phys. Rev. Lett.* **24**: 201.

26a. Dunmur, D. A., Waterworth, T. F., and Palffy-Muhoray, P. 1985. *Mol. Cryst. Liq. Cryst.* **124**: 73.

26b. Dunmur, D. A., and Waterworth, T. F. 1990. Preprint. *American Chemical Society, Division of Polymer Chemistry, Polymer Preprints.*

27. Dunmur, D. A., Szumulin, K., and Waterworth, T. F. 1987. *Mol. Cryst. Liq. Cryst.* **149**: 385.

28. Malraison, B., Poggi, Y., and Guyon, E. 1980. *Phys. Rev. A* **21**: 1012.

29. Madhusudana, N. V., and Chandrasekhar, S. 1974. *Liquid Crystals and Ordered Fluids,* vol. 2, ed. J. F. Johnson and R. S. Porter, p. 657. New York: Plenum.

30. Dunmur, D. A., and Palffy-Muhoray, P. 1988. *J. Phys. Chem.* **92**: 1406.

31. Dunmur, D. A., and Tomes, A. E. 1981. *Mol. Cryst. Liq. Cryst.* **76**: 231.

32. Schadt, M., and Helfrich, W. 1972. *Mol. Cryst. Liq. Cryst.* **17**: 355.
33. Coles, H. J., and Jennings, R. R. 1976. *Mol. Phys.* **31**: 571.
34. Filippini, J. C., and Poggi, Y. 1975. *J. Phys. D: Appl. Phys.* **8**: 201.
35. Ratna, B. R., Vijaya, M. S., Shashidhar, R., and Sadashiva, B. K. 1975. *Pramana Suppl.* **1**: 69.
36. Yamamoto, R., Ishihara, S., and Morimoto, K. 1978. *Phys. Lett.* **69A**: 276.
37. Tsvetkov, V. N., Ryumtsev, E. I., Kovshik, A. P., Denis, G. I., and Daugvila, Y. 1974. *Dokl. Akad. Nauk SSSR* **216**: 1105.
38. Filippini, J. C. 1975. *J. Phys. D: Appl. Phys.* **8**: 201.
39. Tsvetkov, V. N., and Ryumtsev, E. I. 1986. *Mol. Cryst. Liq. Cryst.* **133**: 125.
40. Schadt, M., and Gerber, P. R. 1982. *Z. Naturforsch.* **37a**: 165.
41. Blinov, L. M., Kats, E. I., and Sonin, A. A. 1987. *Usp. Fiz. Nauk* **152**: 449.
42. Yokoyama, H. 1988. *Mol. Cryst. Liq. Cryst.* **165**: 265.
43. Jérôme, B. 1991. *Rep. Prog. Phys.* **54**: 391.
44. Blinov, L. M., and Kabayenkov, A. Y. 1987. *Zh. Eksp. Teor. Fiz.* **93**: 1757.
45. Blinov, L. M., Kabayenkov, A. Y., and Sonin, A. A. 1989. *Liq. Cryst.* **5**: 645.
46. Barbero, G., and Durand, G. 1990. *J. Physique* **51**: 281.
47. Cognard, J. 1982. *Mol. Cryst. Liq. Cryst. Suppl.* **A5**: 1.
48. Marusii, T. Y., Reznikov, Y. A., Reshetnyak, V. Y., and Chigrinov, V. G. 1990. *Poverkhnost'* **9**: 5.
49a. Komitov, L., and Petrov, A. G. 1983. *Phys. Status Solidi A* **79**: 623.
49b. Grover, C. P. 1985. *Mol. Cryst. Liq. Cryst.* **127**: 331.
50. Boyd, G. D., Cheng, J., and Ngo, P. D. T. 1980. *Appl. Phys. Lett.* **36**: 556.
51. Cheng, J. 1981. *J. Appl. Phys.* **52**: 724.
52. Sukhenko, E. P. 1983. *Zh. Eksp. Teor. Fiz.* **84**: 2099.
53. Simoni, F., Bartolino, R., Barbero, G., and Aiello, P. 1984. *Mol. Cryst. Liq. Cryst.* **113**: 303.
54. Barbero, G. 1984. *Z. Naturforsch.* **39a**: 575.
55. Chigrinov, V. G. 1982. *Kristallografiya* **27**: 404.
56. Frisken, B. J., and Palffy-Muhoray, P. 1989. *Phys. Rev. A* **40**: 6099.
57. Barbero, G., Miraldi, E., Oldano, C., and Taverna Valabrega, P. 1988. *Z. Naturforsch.* **43a**: 547.
58. Barbero, G., Miraldi, E., and Oldano, C. 1988. *Phys. Rev. A* **38**: 3027.
59. Bata, L., Buka, A., and Jánossy, I. 1974. *Solid State Commun.* **15**: 647.
60. Deuling, H. J., and Helfrich, W. 1979. *Appl. Phys. Lett.* **25**: 129.
61. Gruler, G., and Cheung, L. 1975. *J. Appl. Phys.* **46**: 5097.
62. Brochard, F., Pieranski, P., and Guyon, E. 1972. *Phys. Rev. Lett.* **26**: 1681.
63. Pieranski, P., Brochard, F., and Guyon, E. 1973. *J. Physique* **34**: 35.
64. Schiekel, M., Fahrenschon, K., and Gruler, H. 1975. *Appl. Phys.* **7**: 99.
65. Nehring, J., Kmetz, A. R., and Scheffer, T. J. 1976. *J. Appl. Phys.* **47**: 850.
66. Yamada, H., and Ishii, S. 1981. *Mol. Cryst. Liq. Cryst.* **76**: 113.
67. Basov, N. G., Berezin, P. D., Blinov, L. M., Kompanets, I. N., Morozov, V. N., and Nikitin, V. V. 1972. *Pis'ma Zh. Eksp. Teor. Fiz.* **15**: 200.
68. Belyaev, V. V., Blinov, L. M., and Rumyantsev, V. G. 1976. *Mikroelektronika* **5**: 552.
69. Chigrinov, V. V., and Belyaev, V. V. 1977. *Kristallografiya* **22**: 603.
70. Belyaev, V. V., and Chigrinov, V. G. 1978. *Kristallografiya* **23**: 811.
71a. Wo, S.-T. 1990. *Appl. Phys. Lett.* **57**: 986.
71b. Wo, S.-T., and Wo, C.-S. 1991. *Mol. Cryst. Liq. Cryst.* **207**: 1.
72a. Labrunie, G., and Robert, J. 1973. *J. Appl. Phys.* **44**: 4869.
72b. Labrunie, G., and Robert, J. 1974. *J. Appl. Phys.* **45**: 3222.
73. Belyaev, V. V. 1980. *PhD Thesis*, Phys. Tech. Inst., Moscow.
74. Shibaev, V. P. 1988. *Mol. Cryst. Liq. Cryst.* **155**: 189.
75. Smolyansky, A. L., Roganova, Z. A., Tal'rose, R. V., Kazarin, L. A., and Kostromin, S. G. 1987. *Kristallografiya* **32**: 265.
76. Finkelmann, H., Naegele, D., and Ringsdorf, H. 1979. *Makromol. Chem.* **180**: 803.
77. Pranoto, H., and Haase, W. 1983. *Mol. Cryst. Liq. Cryst.* **98**: 299.
78a. Shibaev, V. P., Kostromin, S. G., Tal'rose, R. V., and Platé, N. A. 1981. *Dokl. Akad. Nauk SSSR* **259**: 1147.
78b. Tal'rose, R. V., Sinitsyn, V. V., Korobeinikova, I. A., Shibaev, V. P., and Platé, N. A. 1984. *Dokl. Akad. Nauk SSSR* **274**: 1149.
79. Haase, W., and Pranoto, H. 1985. *Polymeric Liquid Crystals*, ed. A. Blumstein, p. 313. New York: Plenum.
80. Finkelmann, H., Kiechle, U., and Rehage, G. 1983. *Mol. Cryst. Liq. Cryst.* **94**: 343.
81. Ujiie, S., Koide, N., and Iimura, K. 1987. *Mol. Cryst. Liq. Cryst.* **153**: 191.
82. Tsvetkov, V. N., Andreeva, L. N., and Filippov, A. P. 1987. *Mol. Cryst. Liq. Cryst.* **153**: 217.
83. Kohjiya, S., Ono, A., Kishimoto, T., Yamashita, S., Yanase, H., and Asada, T. 1990. *Mol. Cryst. Liq. Cryst.* **185**: 183.
84. Takatsu, H., Takeuchi, K., and Umezu, Y. 1993. *Mol. Cryst. Liq. Cryst.* **225**: 81.
85a. Frisken, B. J., and Palffy-Muhoray, P. 1989. *Phys. Rev. A* **39**: 1513.
85b. Allender, D. W., Frisken, B. J., and Palffy-Muhoray, P. 1989. *Liq. Cryst.* **5**: 735.

86. Palffy-Muhoray, P., Yuan, H. J., Frisken, B. J., and van Saarloos, W. 1990. *Nonlinear Evolution of Spatio-Temporal Structures in Dissipative Continuous Systems*, ed. F. H. Busse and L. Kramer, p. 313. NATO ASI Series B, vol. 225. New York: Plenum.
87. Longberg, F., and Meyer, R. B. 1985. *Phys. Rev. Lett.* **55**: 718.
88. Taratuta, V. G., Hurd, A. J., and Meyer, R. B. 1985. *Phys. Rev. Lett.* **55**: 246.
89. Oldano, C. 1986. *Phys. Rev. Lett.* **56**: 1098.
90. Zimmermann, W., and Kramer, L. 1986. *Phys. Rev. Lett.* **56**: 2655.
91. Kini, V. D. 1987. *Mol. Cryst. Liq. Cryst.* **153**: 1.
92. Barnik, M. I., Blinov, L. M., Trufanov, A. N., Chigrinov, V. G., and Korkishko, T. V. 1984. *Zh. Eksp. Teor. Fiz.* **87**: 196.
93. Chigrinov, V. G., Korkishko, T. V., Barnik, M. I., and Trufanov, A. N. 1985. *Mol. Cryst. Liq. Cryst.* **129**: 283.
94. Carr, E. F. 1977. *Mol. Cryst. Liq. Cryst. Lett.* **34**: 159.
95. Guyon, E., Meyer, R., and Salan, J. 1979. *Mol. Cryst. Liq. Cryst.* **54**: 261.
96. Hurd, A. J., Fraden, S., Longberg, F., and Meyer, R. B. 1985. *J. Physique* **46**: 905.
97. Longberg, F., Fraden, S., Hurd, A. J., and Meyer, R. B. 1984. *Phys. Rev. Lett.* **52**: 1903.
98. Rey, A. D., and Denn, M. M. 1989. *Liq. Cryst.* **4**: 409.
99. Buka, A., Juarez, M., Kramer, L., and Rehberg, I. 1989. *Phys. Rev. A* **40**: 7427.
100. Winkler, B. L., Richter, H., Rehberg, I., Zimmermann, W., Kramer, L., and Buka, A. 1991. *Phys. Rev. A* **43**: 1940.
101. Buka, A., and Kramer, L. 1992. *Phys. Rev. A* **45**: 5624.
102. Buka, A., and Kramer, L. 1992. *J. Physique II* **2**: 315.
103. Barbero, G., Dozov, I., Palierne, I., and Durand, G. 1986. *Phys. Rev. Lett.* **56**: 2056.
104. Derzhanski, A. I., Petrov, A. G., and Mitov, M. D. 1978. *J. Physique* **39**: 273.
105a. Dozov, I., Martinot-Lagarde, P., and Durand, G. 1983. *J. Physique Lett.* **44**: 817.
105b. Barbero, G., and Strigazzi, A. 1985. *J. Appl. Phys.* **58**: 2544.
106a. Barbero, G., Taverna Valabrega, P., Bartolino, R., and Valenti, B. 1986. *Liq. Cryst.* **1**: 483.
106b. Valenti, B., Bertoni, C., Barbero, G., Taverna Valabrega, P., and Bartolino, R. 1987. *Mol. Cryst. Liq. Cryst.* **146**: 307.
107. Blinov, L. M., Radzhabov, D. Z., Yablonsky, S. V., and Yakovenko, S. S. 1990. *Nuovo Cimento* **12D**: 1353.
108a. Barberi, R., Boix, M., and Durand, G. 1989. *Appl. Phys. Lett.* **55**: 2506.
108b. Barberi, R., and Durand, G. 1990. *Eurodisplay '90: Proceedings of the Tenth International Display Research Conference [Amsterdam]*, p. 142. Berlin: VDE.
109. Monkade, M., Boix, M., and Durand, G. 1988. *Europhys. Lett.* **5**: 697.
110. Prost, J., and Pershan, P. S. 1976. *J. Appl. Phys.* **47**: 2298.
111. Derzhanski, A. I., and Petrov, A. G. 1979. *Acta Phys. Pol. A* **55**: 747.
112. Blinov, L. M., Durand, G., and Yablonsky, S. V. 1992. *J. Physique II* **2**: 1287.
113a. Mochizuki, A., Motoyoshi, K., and Kobayashi, S. 1990. *Japan. J. Appl. Phys. Lett.* **29**: 1898.
113b. Mochizuki, A., Motoyoshi, K., and Kobayashi, S. 1991. *Japan. J. Appl. Phys. Lett.* **30**: 1959.
113c. Mochizuki, A., Sotoyama, W., Tatsuura, S., Ishitsuka, T., Motoyoshi, K., and Kobayashi, S. 1991. *Japan. J. Appl. Phys. Lett.* **30**: 504.
114a. Bobylev, Y. P., Chigrinov, V. G., and Pikin, S. A. 1979. *J. Physique Coll.* **40**: C3–331.
114b. Schiller, P., Pelzl, G., and Demus, D. 1990. *Cryst. Res. Technol.* **25**: 111.
115. Khinov, K., and Vistin', L. K. 1979. *J. Physique* **40**: 269.
116. Vistin', L. K., Yakovenko, S. S., Chickeren, A. E., and Yarosheva, A. I. 1981. *Kristallografiya* **26**: 871.
117. Khinov, K. 1982. *Z. Naturforsch.* **37a**: 334.
118. Soffer, B. H., Margerum, J. D., Lackner, A. M., Boswell, D., Tanguay, A. R., Strand, T. C., Sawchuk, A. A., and Chavel, P. 1981. *Mol. Cryst. Liq. Cryst.* **70**: 145.
119. Tanguay, A. R., Wu, C. S., Chavel, P., Strand, T. C., and Sawchuk, A. A. 1983. *Opt. Eng.* **22**: 687.
120. Jeon, H., and Sawchuk, A. A. 1987. *Appl. Opt.* **26**: 261.
121. Pikin, S. A. 1986. *Incommensurate Phases in Dielectrics 2*, ed. R. Blinc and A. Levanyuk, p. 319. Amsterdam, Elsevier.
122. Schadt, M., and Helfrich, W. 1971. *Appl. Phys. Lett.* **18**: 127.
123. Mauguin, C. 1911. *Bull. Soc. Fr. Minéral. Cristallogr.* **34**: 71.
124a. Gooch, C. H., and Tarry, H. A. 1974. *Electron. Lett.* **10**: 2.
124b. Gooch, C. H., and Tarry, H. A. 1975. *J. Phys. D: Appl. Phys.* **8**: 1575.
125. Lien, A., and Suzuki, S. 1989. *Japan Display '89: Proceedings of the Ninth International Display Research Conference [Kyoto]*, p. 286. Playa del Rey, Calif.: Society for Information Display.
126a. Berreman, D. W. 1973. *J. Opt. Soc. Am.* **63**: 1374.
126b. Berreman, D. W. 1983. *Phil. Trans. R. Soc. London A* **309**: 203.
127. Ide, M., Oh-Ide, T., and Kobayashi, S. 1987. *Mol. Cryst. Liq. Cryst.* **144**: 225.
128. Allia, P., Oldano, C., and Trossi, L. 1987. *Mol. Cryst. Liq. Cryst.* **143**: 17.
129. Allia, P., Oldano, C., and Trossi, L. 1986. *J. Opt. Soc. Am. B* **3**: 424.
130. Burkhanov, A. I., Lipatova, T. A., Simonenko, G. V., Finkel, A. G., and Tsoy, V. I. 1985. *Elektron. Tekh.* **3(108)**: 50.
131. Wohler, H. F., Haas, G., Fritsch, M. G., and Mlynski, D. A. 1988. *J. Opt. Soc. Am. A* **5**: 1554.

132. Nehring, J., Kmetz, A. R., and Scheffer, T. J. 1976. *J. Appl. Phys.* **47**: 850.
133a. Yang, K. H. 1983. *J. Appl. Phys.* **54**: 6864.
133b. Yang, K. H. 1983. *J. Physique* **44**: 1051.
134. Becker, M. E., Nehring, J., and Scheffer, T. J. 1985. *J. Appl. Phys.* **57**: 4539.
135. Yang, K. H. 1983. *Appl. Phys. Lett.* **43**: 171.
136. Tsoy, V. I. 1993. *Mol. Mater.* **2**: 121.
137. Strigazzi, A. 1985. *J. Physique* **46**: 1507.
138. Porte, G., and Jadot, J. P. 1978. *J. Physique* **39**: 213.
139a. Cheng, J., Thurston, R. N., and Berreman, D. W. 1981. *J. Appl. Phys.* **52**: 2756.
139b. Thurston, R. N. 1981. *J. Appl. Phys.* **52**: 3040.
140a. Oyana, M., Tani, C., and Saito, F. 1976. *Electron. Lett.* **12**: 70.
140b. de Rossi, D., and Robert, J. 1978. *J. Appl. Phys.* **49**: 1139.
141. Toko, Y., Sugiyama, T., Katoh, K., Iimura, Y., and Kobayashi, S. 1993. *J. Appl. Phys.* **74**: 2021.
142. Sonehara, T., and Okumura, O. 1989. *Japan Display '89: Proceedings of the Ninth International Display Research Conference [Kyoto]*, p. 192. Playa del Rey, Calif.: Society for Information Display.
143a. Berreman, D. W., and Heffner, W. R. 1980. *Appl. Phys. Lett.* **37**: 109.
143b. Berreman, D. W., and Heffner, W. R. 1981. *J. Appl. Phys.* **52**: 3032.
144. Scheffer, T. J., and Nehring, J. 1984, *Appl. Phys. Lett.* **45**: 1021.
145. Scheffer, T. J., and Nehring, J. 1985. *J. Appl. Phys.* **58**: 3022.
146. Raynes, E. P. 1986. *Mol. Cryst. Liq. Cryst. Lett.* **4**: 1.
147. Breddels, P. A., and van Sprang, H. A. 1985. *J. Appl. Phys.* **58**: 2162.
148. Akatsuka, M., Katoh, K., Sawada, K., and Nakayama, M. 1987. *Proc. SID* **28**: 159.
149. Schadt, M. 1988. *Mol. Cryst. Liq. Cryst.* **165**: 405.
150. Schadt, M., and Leenhouts, F. 1987. *Appl. Phys. Lett.* **50**: 236.
151. Katoh, K., Akatsuka, M., and Sawada, K. 1987. *Rep. Res. Lab. Asahi Glass Co.* **37**: 285.
152. Thurston, R. N., Cheng, J., and Boyd, G. D. 1980. *IEEE Trans. Elec. Dev.* **ED-27**: 2969.
153a. Heffner, W. R., and Berreman, D. W. 1982. *J. Appl. Phys.* **53**: 8599.
153b. Berreman, D. W. 1984. *J. Appl. Phys.* **55**: 806.
154. Helfrich, W. 1970. *Appl. Phys. Lett.* **17**: 531.
155. Hurault, J. P. 1973. *J. Chem. Phys.* **59**: 2068.
156. Chigrinov, V. G., Belyaev, V. V., Belyaev, S. V., and Grebenkin, M. F. 1979. *Zh. Eksp. Teor. Fiz.* **77**: 2081.
157. Schiller, P., and Schiller, K. 1990. *Liq. Cryst.* **8**: 553.
158. Schiller, P. 1990. *Phase Trans.* **29**: 59.
159. Cohen, G., and Hornreich, R. M. 1990. *Phys. Rev. A* **41**: 4402.
160. Meyer, R. B. 1968. *Appl. Phys. Lett.* **12**: 281.
161. de Gennes, P. G. 1968. *Solid State Commun.* **6**: 163.
162. Belyakov, V. A., and Dmitrienko, V. E. 1978. *Fiz. Tverd. Tela* **65A**: 33.
163a. Blinov, L. M., Belyaev, S. V., and Kizel', V. A. 1978. *Phys. Lett.* **65A**: 33.
163b. Belyaev, S. V., Blinov, L. M., and Kizel', V. A. 1979. *Pis'ma Zh. Eksp. Teor. Fiz.* **29**: 344.
164. Dreher, R. 1973. *Solid State Commun.* **13**: 1571.
165. Belyaev, S. V., and Blinov, L. M. 1979. *Pis'ma Zh. Eksp. Teor. Fiz.* **30**: 111.
166a. Goossens, W. J. A. 1982. *J. Physique* **43**: 1469.
166b. Brokx, J., Vertogen, G., and van Groesen, E. W. C. 1983. *Z. Naturforsch.* **38a**: 1.
167. Mochizuki, A., and Kobayashi, S. 1993. *Mol. Cryst. Liq. Cryst.* **225**: 89.
168. Hauk, G., and Koswig, H. D. 1987. *Cryst. Res. Technol.* **22**: 1333.
169. Patel, J. S., and Meyer, R. B. 1987. *Phys. Rev. Lett.* **58**: 1538.
170. Belotsky, E. D., Il'chishin, I. P., Lev, B. I., Tolmachev, A. V., Tomchuk, P., and Shpak, M. T. 1990. *Pis'ma Zh. Eksp. Teor. Fiz.* **51**: 216.
171. Leslie, F. M. 1970. *Mol. Cryst. Liq. Cryst.* **12**: 57.
172. Bartolino, R., and Simoni, F. 1980. *Optica Acta* **27**: 1179.
173. Ginsburg, B. M., Shepelevsky, A. A., Alumyan, J. A., and Frenkel', S. Y. 1986. *Vysokomol. Soed. B* **28**: 723.
174. Parson, J. D., and Hayes, C. F. 1974. *Phys. Rev. A* **9**: 2652.
175. Hinov, H. P., Mitov, M. D., and Kukleva, E. 1986. *Mol. Cryst. Liq. Cryst.* **136**: 281.
176. Greubel, W. 1974. *Appl. Phys. Lett.* **25**: 5.
177. Belyaev, S. V., Rumyantsev, V. G., and Belyaev, V. V. 1977. *Zh. Eksp. Teor. Fiz.* **73**: 644.
178. Kawachi, M., and Kogure, O. 1977. *Japan. J. Appl. Phys.* **16**: 1673.
179. Tani, C. 1980. *Japan. J. Appl. Phys.* **19**: 211.
180. Lin-Hendel, C. G. 1981. *Appl. Phys. Lett.* **38**: 615.
181. Gerasimov, A. A. 1993. *Mol. Mater.* **2**: 255.
182. Patel, J. S., and Lee, S.-D. 1989. *J. Appl. Phys.* **66**: 1879.
183. Lee, S.-D., Patel, J. S., and Meyer, R. B. 1990. *J. Appl. Phys.* **67**: 1293.
184. Beresnev, L. A., Blinov, L. M., Dergachev, D. I., and Kondrat'yev, S. B. 1987. *Pis'ma Zh. Eksp. Teor. Fiz.* **46**: 328.
185. Beresnev, L. A., Chigrinov, V. G., Dergachev, D. I., Pozhidaev, E. P., Fünfschilling, J., and Schadt, M. 1989. *Liq. Cryst.* **5**: 1171.
186. Komitov, L., Lagerwall, S. T., Stebler, B., Andersson, G., and Flatischler, K. 1991. *Ferroelectrics* **114**: 167.
187. Umansky, B. A., Blinov, L. M., and Barnik, M. I. 1980. *Pis'ma Zh. Tekh. Fiz.* **6**: 200.
188. Umansky, B. A., Chigrinov, V. G., Blinov, L. M., and Pod'yachev, Y. B. 1981. *Zh. Eksp. Teor. Fiz.* **81**: 1307.
189. Chigrinov, V. G. 1983. *Kristallografiya* **28**: 825.
190. Terentjev, E. M., and Pikin, S. A. 1985. *Kristallografiya* **30**: 227.

191. Bourdon, L., Sommeria, M., and Kléman, M. 1982. *J. Physique* **43**: 77.
192. Stegemeyer, H., Blümel, T., Hiltrop, K., Onusseit, H., and Porsch, F. 1986. *Liq. Cryst.* **1**: 3.
193. Yang, D. K., and Crooker, P. P. 1988. *Phys. Rev. A* **37**: 4001.
194. Belyakov, V. A. 1988. *Difraktsionnaya Optika Periodicheskikh Sred Slozhnoi Struktury* (Diffraction Optics of Periodical Media with Complex Structure). Moscow: Nauka.
195. Johnson, D. L., Flack, J. H., and Crooker, P. P. 1980. *Phys. Rev. Lett.* **45**: 641.
196. Dmitrienko, V. E. 1989. *Liq. Cryst.* **5**: 847.
197a. Jérôme, B., and Pieranski, P. 1989. *Liq. Cryst.* **5**: 799.
197b. Heppke, G., Jérôme, B., Kitzerow, H.-S., and Pieranski, P. 1989. *Liq. Cryst.* **5**: 813.
198. Porsch, F., Stegemeyer, H., and Hiltrop, K. 1984. *Z. Naturforsch.* **39a**: 475.
199a. Heppke, G., Jérôme, B., Kitzerow, H.-S., and Pieranski, P. 1991. *J. Physique* **52**: 2991.
199b. Heppke, G., Jérôme, B., Kitzerow, H.-S., and Pieranski, P. 1989. *J. Physique* **50**: 549.
200. Chilaya, G., and Petriashvili, G. 1993. *Mol. Mater.* **2**: 239.
201. Kitzerow, H.-S., Crooker, P. P., Kwok, S. L., and Heppke, G. 1990. *J. Physique* **51**: 1303.
202. Kitzerow, H.-S. 1991. *Mol. Cryst. Liq. Cryst.* **202**: 51.
203. Gerber, P. R. 1985. *Mol. Cryst. Liq. Cryst.* **116**: 197.
204. Coles, H. J., and Gleeson, H. F. 1989. *Mol. Cryst. Liq. Cryst.* **167**: 213.
205. Rapini, A. 1972. *J. Physique* **33**: 237.
206. Hareng, M., Le Berre, S., and Metzger, J. J. 1975. *Appl. Phys. Lett.* **27**: 575.
207. Pelzl, G., Deutscher, H.-J., and Demus, D. 1981. *Cryst. Res. Technol.* **16**: 603.
208. Chirkov, V. N., Aliyev, D. F., and Zeinally, A. K. 1977. *Pis'ma Zh. Tekh. Fiz.* **3**: 1016.
209. Parodi, O. 1972. *Solid State Commun.* **11**: 1503.
210. Takanishi, Y., Ouchi, Y., Takezoe, H., and Fukuda, A. 1989. *Japan. J. Appl. Phys. Lett.* **28**: 487.
211. Hinov, H. P. 1988. *Liq. Cryst.* **3**: 1481.
212. Hinov, H. P., and Avramova, K. 1988. *Liq. Cryst.* **3**: 1505.
213. Jákli, A., and Saupe, A. 1992. *Mol. Cryst. Liq. Cryst.* **222**: 101.
214. Aliyev, D. F., Akhundov, C. G., and Zeinally, A. K. 1982. *Kristallografiya* **27**: 156.
215. Kamenskii, V. G. 1987. *Zh. Eksp. Teor. Fiz.* **92**: 97.
216. Stephen, M. J., and Straley, J. P. 1974. *Rev. Mod. Phys.* **46**: 617.
217. Pelzl, G., Kolbe, P., Preukschas, V., Diele, S., and Demus, D. 1979. *Mol. Cryst. Liq. Cryst.* **53**: 167.
218. Pelzl, G., Schiller, P., and Demus, D. 1987. *Liq. Cryst.* **2**: 131.
219. Schiller, P., Pelzl, G., and Demus, D. 1987. *Liq. Cryst.* **2**: 21.
220. Pozhidaev, E. P., Osipov, M. A., Chigrinov, V. G., Baikalov, V. A., Beresnev, L. A., and Blinov, L. M. 1988. *Zh. Eksp. Teor. Fiz.* **94**: 125.
221. Pikin, S. A., and Indenbom, V. L. 1978. *Usp. Fiz. Nauk* **125**: 251.
222. Beresnev, L. A., Blinov, L. M., Osipov, M. A., and Pikin, S. A. 1988. *Mol. Cryst. Liq. Cryst.* **158A**: 1.
223. Skarp, K., and Handschy, M. A. 1988. *Mol. Cryst. Liq. Cryst.* **165**: 405.
224. Isogai, M., Kondo, K., Nonaka, S., Odamura, M., and Mukoh, A. 1985. *Proceedings of the Fifth International Display Research Conference [Orlando, Fl.]*, p. 225. Playa del Rey, Calif.: Society for Information Display.
225. Pozhidaev, E. P., Chigrinov, V. G., Panarin, Y. P., and Vorflusev, V. P. 1993. *Mol. Mater.* **2**: 225.
226. Handschy, M. A., Clark, N. A., and Lagerwall, S. T. 1983. *Phys. Rev. Lett.* **51**: 471.
227. Clark, N. A., and Lagerwall, S. T. 1984. *Ferroelectrics* **59**: 25.
228. Xue, J. Z., Handschy, M. A., and Clark, N. A. 1987. *Ferroelectrics* **73**: 305.
229. Orihara, H., Nakamura, K., Ishibashi, Y., Yamada, Y., Yamamoto, N., and Yamanaki, M. 1986. *Japan. J. Appl. Phys. Lett.* **25**: 839.
230. Schiller, P. 1986. *Cryst. Res. Technol.* **21**: 167, 301.
231. Handschy, M. A., and Clark, N. A. 1984. *Ferroelectrics* **59**: 69.
232. Xue, J. Z., Handschy, M. A., and Clark, N. A. 1987. *Liq. Cryst.* **2**: 707.
233. Barnik, M. I., Baikalov, V. A., Chigrinov, V. G., and Pozhidaev, E. P. 1987. *Mol. Cryst. Liq. Cryst.* **143**: 111.
234. Chigrinov, V. G., Pozhidaev, E. P., Baikalov, V. A., and Barnik, M. I. 1989. *Kristallografiya* **34**: 406.
235. Yang, K. H., and Chien, T. C. 1989. *Japan. J. Appl. Phys. Lett.* **28**: 1599.
236. Saxena, K., Bawa, S. S., Biradar, A. M., Chandra, S., and Rup, R. 1990. *Japan. J. Appl. Phys.* **29**: 2041.
237. Nitta, M., Ozaki, N., Suenaga, H., Nakaya, K., and Kobayashi, S. 1988. *Japan. J. Appl. Phys. Lett.* **27**: 477.
238. Nakaya, K., Zhang, B., Yoshida, M., Isa, I., Shindoh, S., and Kobayashi, S. 1989. *Japan. J. Appl. Phys. Lett.* **28**: 116.
239. Hartmann, W. 1989. *J. Appl. Phys.* **66**: 1132.
240. Ikeno, H., Maeda, H., Yoshida, M., Zhang, B., Kimura, M., and Kobayashi, S. 1989. *Proc. SID* **30**: 329.
241. Nakagawa, N., and Akahane, T. 1986. *J. Phys. Soc. Japan* **55**: 1516.
242. Pikin, S. A., and Terentjev, E. M. 1988. *Liq. Cryst.* **3**: 259.
243. Williams, D., and Davis, L. 1986. *J. Phys. D: Appl. Phys.* **19**: 1.
244. Ishikawa, K., Ouchi, Y., Uemura, T., Tsuchiya, T., Takezoe, H., and Fukuda, A. 1990. *Mol. Cryst. Liq. Cryst.* **122**: 175.
245a. Patel, J. S., and Goodby, J. W. 1987. *Opt. Eng.* **26**: 373.

245b. Patel, J. S., and Goodby, J. W. 1987. *J. Appl. Phys.* **59**: 2355.
246. Yamamoto, N., Yamada, Y., Mori, K., Orihara, H., and Ishibashi, Y. 1989. *Japan. J. Appl. Phys.* **28**: 524.
247. Rieker, T., Clark, N. A., Smith, G., Safinya, C. R. 1989. *Liq. Cryst.* **6**: 565.
248. Bawa, S. S., Biradar, A. M., Saxena, K., and Chandra, S. 1990. *Appl. Phys. Lett.* **57**: 1398.
249. Bartolino, R., Doucet, J., and Durand, G. 1978. *Ann. Physique* **3**: 389.
250. Macgregor, A. 1990. *J. Mod. Opt.* **37**: 919.
251. Ouchi, Y., Lee, J., Takezoe, H., Fukuda, A., Kondo, K., Kitamura, T., and Mukoh, A. 1988. *Japan. J. Appl. Phys. Lett.* **27**: 725.
252. Sato, Y., Tanaka, T., Kobayashi, H., Aoki, K., Watanabe, H., Takeshita, H., Ouchi, Y., Takezoe, H., and Fukuda, A. 1989. *Japan. J. Appl. Phys. Lett.* **28**: 483.
253. Hartmann, W., and Luyekx-Smolders, A. 1990. *J. Appl. Phys.* **67**: 1253.
254. Fünfschilling, J., and Schadt, M. 1990. *SID Dig. Tech. Papers* **21**: 106.
255. Derfel, G. 1990. *Liq. Cryst.* **8**: 331.
256. Lagerwall, S. T., Clark, N. A., Dijon, J., and Clerc, J. F. 1989. *Ferroelectrics* **94**: 3.
257. Beresnev, L. A., Blinov, L. M., Dergachev, D. I., and Kondrat'yev, S. B. 1987. *Pis'ma Zh. Eksp. Teor. Fiz.* **46**: 328.
258. Ostrovsky, B. I., Rabinovich, A. Z., Sonin, A. S., and Strukov, B. A. 1978. *Zh. Eksp. Teor. Fiz.* **74**: 1748.
259. Ostrovsky, B. I., and Chigrinov, V. G. 1980. *Kristallografiya* **25**: 560.
260. Ostrovsky, B. I., Rabinovich, A. Z., and Chigrinov, V. G. 1980. *Advances in Liquid Crystal Research and Applications*, ed. L. Bata, p. 469. Oxford: Pergamon.
261. Ostrovsky, B. I., Pikin, S. A., and Chigrinov, V. G. 1979. *Zh. Eksp. Teor. Fiz.* **77**: 1631.
262. Beresnev, L. A., Blinov, L. M., and Dergachev, D. I. 1988. *Ferroelectrics* **85**: 173.
263. Beresnev, L. A., Blinov, L. M., Dergachev, D. I., Loseva, M. V., and Chernova, N. I. 1988. *Pis'ma Zh. Tekh. Fiz.* **14**: 260.
264. Fünfschilling, J., and Schadt, M. 1989. *J. Appl. Phys.* **66**: 3877.
265. Garoff, S., and Meyer, R. B. 1979. *Phys. Rev. A* **19**: 338.
266. Andersson, G., Dahl, I., Komitov, L., Lagerwall, S. T., Skarp, K., and Stebler, B. 1989. *J. Appl. Phys.* **66**: 4983.
267. Andersson, G., Dahl, I., Kuczynski, W., Lagerwall, S. T., Skarp, K., and Stebler, B. 1988. *Ferroelectrics* **84**: 285.
268. Lee, S.-D., and Patel, J. S. 1989. *Appl. Phys. Lett.* **55**: 122.
269. Dupont, L., Glogarova, M., Marcerou, J. P., Nguyen, H. T., Destrade, C., and Lejcek, L. 1991. *J. Physique II* **1**: 831.
270a. Bahr, C., and Heppke, G. 1987. *Liq. Cryst.* **2**: 825.
270b. Geelhaar, T. 1988. *Ferroelectrics* **88**: 329.
271. Nishiyama, S., Ouchi, Y., Takezoe, H., and Fukuda, A. 1987. *Japan. J. Appl. Phys. Lett.* **26**: 1787.
272. Kimura, M., Yang, Y. B., Iimura, Y., and Kobayashi, S. 1993. *Liq. Cryst.* **14**: 1785.
273. Yang, Y. B., Mochizuki, A., Nakamura, N., and Kobayashi, S. 1991. *Japan. J. Appl. Phys. Lett.* **30**: 612.
274a. Beresnev, L. A., Loseva, M. V., Chernova, N. I., Kononov, S. G., Adomenas, P. V., and Pozhidaev, E. P. 1990. *Pis'ma Zh. Tekh. Fiz.* **51**: 457.
274b. Beresnev, L. A., Pfeiffer, M., Haase, W., Loseva, M. V., Cherova, N. I., and Adomenas, P. V. 1991. *Pis'ma Zh. Tekh. Fiz.* **53**: 170.
275. Pikin, S. A., Beresnev, L. A., Hiller, S., Pfeiffer, M., and Haase, W. 1993. *Mol. Mater.* **3**: 1.
276. Haase, W., Hiller, S., Pfeiffer, M., and Beresnev, L. A. 1993. *Ferroelectrics* **140**: 37.
277. Beresnev, L. A., Schumacher, E., Pickin, S. A., Fan, X., Ostrovski, B. I., Hiller, S., Onoklov, A. P., and Haase, W. 1995. *Japan J. Appl. Phys.* **31**: 1409.
278. Skarp, K., Andersson, G., Hirai, T., Yoshizawa, A., Hiraoka, K., Takezoe, H., and Fukuda, A. 1992. *Japan. J. Appl. Phys.* **31**: 1409.
279. Chandani, A. D. L., Hagiwara, T., Suzuki, Y., Ouchi, Y., Takezoe, H., and Fukuda, A. 1988. *Japan. J. Appl. Phys. Lett.* **27**: 729.
280. Chandani, A. D. L., Ouchi, Y., Takezoe, H., Fukuda, A., Terashima, K., Furukawa, K., and Kishi, A. 1989. *Japan. J. Appl. Phys. Lett.* **28**: 1261.
281. Chandani, A. D. L., Gorecka, E., Ouchi, Y., Takezoe, H., and Fukuda, A. 1989. *Japan. J. Appl. Phys. Lett.* **28**: 1265.
282. Li, J., Takezoe, H., and Fukuda, A. 1991. *Japan. J. Appl. Phys.* **30**: 532.
283. Li, J., Ouchi, Y., Takezoe, H., Fukuda, A., and Watanabe, J. 1990. *J. Phys. Condens. Matt. Suppl. A* **2**: 271.
284. Hiraoka, K., Ouchi, Y., Takezoe, H., Fukuda, A., Inui, S., Kawano, S., Saito, M., and Iwane, H. 1991. *Mol. Cryst. Liq. Cryst.* **199**: 197.
285. Takanishi, Y., Takezoe, H., Fukuda, A., Komura, H., and Watanabe, J. 1992. *J. Mater. Chem.* **2**: 71.
286. Hara, M., Umemoto, T., Takezoe, H., Garito, A. F., and Sasabe, H. 1991. *Japan. J. Appl. Phys. Lett.* **30**: 2052.
287. Takanishi, Y., Hiraoka, K., Agrawal, V. K., Takezoe, H., Fukuda, A., and Matsushita, M. 1991. *Japan. J. Appl. Phys.* **30**: 2023.
288. Isozaki, T., Hiraoka, K., Takanishi, Y., Takezoe, H., Fukuda, A., Suzuki, Y., and Kawamura, I. 1992. *Liq. Cryst.* **12**: 59.
289. Bak, P. 1986. *Phys. Today* **39**: 38.
290. Isozaki, T., Fujikawa, T., Takezoe, H., Fukuda, A., Hagiwara, T., Suzuki, Y., and Kawamura, I. 1992. *Japan. J. Appl. Phys. Lett.* **31**: 1435.
291. Hiraoka, K., Takanishi, Y., Skarp, K., Takezoe, H., and Fukuda, A. 1991. *Japan. J. Appl. Phys. Lett.* **30**: 1819.

292. Hiraoka, K., Takanishi, Y., Takezoe, H., Fukuda, A., Isozaki, T., Suzuki, Y., and Kawamura, I. 1992. *Japan. J. Appl. Phys.* **31**: 3394.
293. Johno, M., Chandani, A. D. L., Lee, J., Ouchi, Y., Takezoe, H., Fukuda, A., Itoh, K. 1990. *Proc. SID* **31**: 129.
294. Lee, J., Chandani, A. D. L., Itoh, K., Ouchi, Y., Takezoe, H., and Fukuda, A. 1990. *Japan. J. Appl. Phys.* **29**: 1122.
295. Yamada, Y., Yamamoto, N., Mori, K., Nakamura, K., Hagiwara, T., Suzuki, Y., Kawamura, I., Orihara, H., and Ishibashi, Y. 1990. *Japan. J. Appl. Phys. Lett.* **29**: 1757.
296. Shibaev, V. P., Kozlovsky, M. V., Beresnev, L. A., Blinov, L. M., and Platé, N. A. 1984. *Polym. Bull.* **12**: 299.
297. Kozlovsky, M. V., Beresnev, L. A., Kononov, S. G., Shibaev, V. P., and Blinov, L. M. 1987. *Fiz. Tverd. Tela* **29**: 98.
298. Le Barny, P., and Dubois, J. C. 1989. *Side Chain Liquid Crystalline Polymers*, ed. C. B. McArdle, p. 130. New York: Chapman and Hall.
299. Kozlovsky, M. V., and Beresnev, L. A. 1992. *Phase Trans.* **40**: 129.
300. Ujiie, S., and Iimura, K. 1990. *Chem. Lett.* **1990**: 1031.
301. Flatischler, K., Komitov, L., Skarp, K., and Keller, P. 1991. *Mol. Cryst. Liq. Cryst.* **209**: 105.
302. Skarp, K., Andersson, G., Lagerwall, S. T., Kapitza, H., Poths, H., and Zentel, R. 1991. *Ferroelectrics* **122**: 127.
303. Skarp, K., Andersson, G., Gouda, F., Lagerwall, S. T., Poths, H., and Zentel, R. 1992. *Polym. Adv. Technol.* **3**: 241.
304. Uchida, S., Morita, K., Miyoshi, K., Hashimoto, K., and Kawasaki, K. 1988. *Mol. Cryst. Liq. Cryst.* **155**: 93.
305. Coles, H. J., Gleeson, H. F., Scherowsky, G., Schliwa, A. 1990. *Mol. Cryst. Liq. Cryst. Lett.* **7**: 117.
306. Ivkov, V. G., and Berestovsky, G. N. 1981. *Dinamicheskaya Struktura Lipidnogo Bisloya* (Dynamic Structure of a Lipid Bilayer). Moscow: Nauka.
307. Beresnev, L. A., Blinov, L. M., and Kovshev, E. I. 1982. *Dokl. Akad. Nauk SSSR* **265**: 210.
308. Blinov, L. M., Davidyan, S. A., Petrov, A. G., Todorov, A. T., and Yablonsky, S. V. 1988. *Pis'ma Zh. Eksp. Teor. Fiz.* **48**: 259.
309. Petrov, A. G., Todorov, A. T., Bonev, P., Blinov, L. M., Yablonsky, S. V., Subachyus, D. B., and Tvetkova, N. 1991. *Ferroelectrics* **114**: 415.
310. Blinov, L. M., Baikalov, V. A., Barnik, M. I., Beresnev, L. A., Pozhidaev, E. P., and Yablonsky, S. V. 1987. *Liq. Cryst.* **2**: 121.
311. Tournilhac, F., Blinov, L. M., Simon, J., and Yablonsky, S. V. 1992. *Nature* **359**: 621.
312. Tournilhac, F., Bosio, L., Simon, J., Blinov, L. M., and Yablonsky, S. V. 1993. *Liq. Cryst.* **14**: 405.
313. Tournilhac, F., Blinov, L. M., Simon, J., Subachyus, D. B., and Yablonsky, S. V. 1993. *Synth. Met.* **54**: 253.
314. Tournilhac, F., Bosio, L., Nicoud, J. F., and Simon, J. 1988. *Chem. Phys. Lett.* **145**: 452.
315. Prost, J., and Barois, P. 1983. *J. Chim. Phys.* **80**: 65.
316. Petschek, R. G., and Wiefling, K. M. 1987. *Phys. Rev. Lett.* **59**: 343.
317. Blinov, L. M., Lobko, T. A., Ostrovsky, B. I., Sulianov, S. N., and Tournilhac, F. 1993. *J. Physique II* **3**: 1121.
318. Bock, H., and Helfrich, W. 1992. *Liq. Cryst.* **12**: 697.
319. Bock, H., and Helfrich, W. 1994. *Liq. Cryst.* **14**: 345.

CHAPTER 6

Interfaces and Thin Films

H. YOKOYAMA

6.1 HISTORICAL SKETCH

Liquid crystals respond to their boundary conditions with an extreme sensitivity owing to the flexible orientational and positional ordering in their constituent molecules. Ever since the discovery of liquid crystals at the end of last century, the boundary effects of liquid crystals have attracted the attention of such pioneering scientists as Lehmann (1922), Mauguin (1911), Grandjean (1916), Zocher and Coper (1928), Chatelain (1937, 1943) and many others, offering not only an intriguing subject of scientific inquiry but at the same time a trick to prepare tractable experimental specimens.

Although reports on the interfacial properties of liquid crystals remained more or less sporadic and empirical till the end of 1960s, these early studies described many of the major boundary effects in liquid crystals, that is, epitaxy on crystal surfaces (Grandjean, 1916), influence of unidirectional rubbing (Mauguin, 1911; Chatelain, 1937, 1943), prevalence of homeotropic alignment (molecules aligned perpendicular to the surface) on a surfactant-coated glass surface (Lehmann, 1922), to name but a few, which constitute the foundation of many of today's surface-alignment techniques. During the subsequent two decades, however, interest in this field has soared and broadened a great deal, reflecting, on the one hand, the importance of liquid crystal displays (LCDs) necessitating an elaborate control of alignment, and, on the other, the rise of general interest in the physics of surface and interfacial phenomena, and in lower-dimensional and confined systems, *per se.*

Among others, the impetus brought about by LCDs has stimulated continued efforts to develop industry-quality alignment techniques. This has resulted in the invention of novel methodologies such as the oblique vacuum-deposition technique (Janning, 1972; Urbach et al., 1974; Crossland, 1975; Raynes et al., 1976; Goodman et al., 1977; Johnson and Penz, 1977; Pollack et al., 1977; Cheng et al., 1980; Heffner et al., 1980; Uchida et al., 1980b; Hauck, 1982; Levy et al., 1985; Monkade et al., 1988, Jérôme and Pieransky 1988a) and, more lately, the so-called "optical buffing," an alignment technique based on the optically induced surface anisotropies (Ichimura et al., 1988; Ichimura et al., 1989; Gibbons et al., 1991; Seki et al., 1992a, 1992b; Aoki et al., 1992a, 1992b; Ichimura et al., 1992; Yamamura et al., 1992; Seki et al., 1993; Iimura et al., 1993; Shannon et al., 1994), hand-in-hand with the massive development of and sophistication in a wide variety of surface-aligning agents (Cognard, 1982).

At the same time, a large number of theoretical and experimental efforts have begun to be directed to elucidating the physico-chemical mechanisms underlying the preferential orientation of liquid crystals thus produced. The theoretical approaches range from the statistical mechanical (Parsons, 1976, 1980; Murakami, 1977; Croxton, 1980; Rosenblatt and Ronis, 1981; Okano, 1983; Kimura and Nakano, 1985, 1986; Kimura, 1993; Poniewierski and Sluckin, 1985; Luckhurst et al., 1986; Telo da Gama, 1984a, b, 1987; Tjipto-Margo and Sullivan, 1988; Poniewierski and Holyst, 1988; Holyst and Poniewierski, 1988a, b; Somoza and Tarazona, 1991; Sharlow and Gelbart, 1992; Osipov and Sluckin, 1993; Teixeira et al., 1993b), molecular mechanical (van Aerle, 1994), quasithermodynamic or phenomenological (Parsons, 1975; Dubois-Violette and de Gennes, 1975; Parsons, 1978; Mada, 1979b; Marcus, 1983; Barbero and Oldano, 1985; Oldano and Barbero, 1985a, b; Sen and Sullivan, 1987; Teixeira and Sluckin, 1992a, b; Teixera et al., 1993a), macroscopic (Kats, 1971; Okano and Murakami, 1979; Mada, 1979c; Bernasoni et al., 1980; Okano et al., 1982), elastic continuum (Berreman, 1972, 1973; Hinov, 1977, 1987; Faetti, 1987b), mechanical (Lin and Taylor, 1994), interfacial scientific (Porte, 1976; Perez et al., 1977, 1988; Perez and Proust, 1979; Naemura 1980, 1981; Köhler, 1982; Mada, 1981a), even to empirical ones (Creagh and Kmetz, 1973). Those efforts proved to be quite fruitful, not necessarily in the sense that they fully accounted for the phenomenon, rather in that they led to new concepts, highlighted by the surface ordering (Sheng, 1976; Schröder, 1977; Akahane and Tako, 1979; Allender et al., 1981; Sheng, 1982; Telo da Gama, 1984b, 1987; Poniewierski and Sluckin, 1985; Sluckin and Poniewierski, 1985, 1986; Sen and Sullivan, 1987; Barberi and Durand, 1990; Mirantsev, 1992; Teixeira et al., 1993a), with which we become able to view the subject more sharply, and which suggest new phenomena that would illuminate the hitherto unknown aspects of the surface-alignment (Sluckin and Poniewierski, 1986). It is, nevertheless, obvious that all sorts of intermolecular interactions existing at the interface play a role in one way or another in the formation of the surface-aligned states, possibly extending far into the bulk due to the inherent order in the liquid crystal. Unfortunately, we are still far from the position where a clear distinction can be made among these working molecular interactions for a given real liquid crystal interface on firm theoretical ground; though we are certainly approaching the level of sophistication that our understanding of the bulk liquid-crystallinity has by now reached. Based on this conviction, we feel it is somewhat unfair to see that "the mechanism of surface alignment is not well understood" is still a cliche at the beginning of papers on surface alignment.

From an experimental point of view, on the other hand, characterization of liquid crystal interfaces has always been and still is a tantalizing challenge. A simple polarizing microscope is a powerful starting instrument, yet too qualitative to allow access to detailed quantitative information. It is in fact during the last two decades that detailed, especially microscopic, information on the liquid crystal interface has become progressively available by an introduction of advanced surface analysis techniques such as the surface-sensitive X-ray diffraction/reflection (Moncton and Pindak, 1979; Als-Nielsen et al., 1982; Pershan and Als-Nielsen, 1984; Ocko et al., 1986; Gramsbergen et al., 1986; Amador et al., 1989; Shi et al., 1993), optical harmonic generation (Guyot-Sionnest et al., 1986; Hsiung and Shen, 1986a; Hsiung et al., 1986b; Hsiung and Beckerbauer, 1992; Shen, 1989a, b; Zhao et al., 1991; Xue et al., 1992; Feller et al., 1991; Chen et al., 1992; Ouchi et al., 1992), scanning tunelling microscopy (Foster and Frommer, 1988; Smith, 1991; Smith et al., 1989, 1992; Spong et al., 1989a, b; Iwakabe et al., 1990, 1991; Shigeno, 1990; Hara et al., 1990, 1991; Mizutani et al., 1990; Bandow et al., 1991; Parks et al., 1993; Matsushige, 1993; Patrick and Beebe, 1994; Walba et al., 1995), and atomic force microscopy (Schwartz et al., 1992; Eckhardt et al., 1993; Seo et al., 1993a), surface-force measurements (Horn et al., 1981; Idziak et al., 1994), along with the furthering of the conventional approaches like the surface-induced birefringence measurement (Miyano, 1979a, b; Tarczon and Miyano, 1980, 1981; Beaglehole, 1982; Yokoyama et al., 1982, 1983, 1984a, 1987; van Sprang 1983a, c; Hsiung et al., 1986b; Moses and Shen, 1991; Moses et al., 1993; Chen et al., 1992; Ouchi et al., 1992) as well as the anchoring energy measurement (Rapini and Papoular, 1969; Kléman and Williams, 1973; Williams et al., 1973; Vitek and Kléman, 1975; Sicart, 1976; Ryschenkow and Kléman, 1976; Naemura, 1978, 1979, 1981; Riviere et al., 1979; Yang, 1982; Yang and Rosenblatt, 1983d; Barbero and Barberi, 1983; Barbero et al., 1984a; Komitov, 1983; Barnik et al., 1983; Rosenblatt, 1984a; Chiarelli et al., 1984; van Sprang and Aartsen, 1984b; Faetti, 1990; Faetti and Palleschi, 1984a, b, 1987a; Faetti et al., 1985a, 1991; Yokoyama and van Sprang, 1985a; Yokoyama et al., 1987; Yokoyama, 1988b; Yokoyama and Patel, 1994; Perez et al., 1988; Eidner et al., 1988; Oh-ide et al., 1988; Glogarova and Durand, 1988; Blinov et al., 1989; Valenti et al., 1990; Di Lisi et al., 1990; Blinov et al., 1991; Nobili et al., 1992b; Di Lisi et al., 1992; Seo et al., 1992b, d, 1993a; Cagnon and Durand, 1993; Durand, 1993; Murthy et al., 1993; C. S. Parks et al., 1994; Iimura et al., 1994.).

In the light of these experimental developments, the interfaces and thin films of liquid crystals have become a subject of active fundamental physical studies, as a model system to study exotic phase transitions predicted in low dimensional and confined systems (Poniewierski and Sluckin, 1987; Yokoyama, 1988a). In particular, the ease of preparing a few molecular layer-thick free standing films of smectic liquid crystals has opened a realistic way to test these predictions (Meyer and Pershan, 1973; Moncton and Pindak, 1979; Rosenblatt et al., 1979, 1980a; Rosenblatt and Amer, 1980b; Pindak et al., 1980a, b; Tarczon and Miyano, 1981; Pindak et al., 1982; Bishop et al., 1982; Cladis et al., 1985; Dierker et al., 1986; Langer and Sethna, 1986; Hsiung and Shen, 1986a; Cheng et al., 1988; van Winkle and Clark, 1988; Maclennan, 1990; Maclennan and Seul, 1992; Pang et al., 1992; Stoebe and Huang, 1992a; Stoebe et al., 1992b; Pettey and Lubensky, 1993; Bahr and Fliegner, 1993a, b; Schwartz et al., 1993b; Hauck et al., 1993; Maclennan et al., 1994). The finding of the hexatic phase and the accompanying defect-unbinding transition is a particularly notable example.

In recent years, considerable theoretical and experimental attention in this area has been focused on spontaneous formation of interfacial patterns, caused essentially by the broken symmetry at the liquid crystal-air interface (Cladis et al., 1985; Dierker et al., 1986; Maclennan, 1990; Maclennan and Seul, 1992; Pang et al., 1992; Schwartz, 1993a; Maclennan et al., 1994). Physically, this shares common ground (Seul and Andelman, 1995) with similar pattern formation phenomena, which have been independently observed at the surface of a hybrid aligned-nematic liquid crystal layer (Sparavigna et al., 1991, 1992a, b, 1994; Pergamenshchik, 1993a; Lavrentovich and Pergamenshchik, 1994; 1995), and also in two-dimensional liquid crystals formed at the air–water interface (Seul and Chen, 1993; Ruiz-Garcia et al., 1993; Tabe and Yokoyama, 1994).

The interface physics of liquid crystals involves almost as large an area of research as liquid crystal physics itself. The history and the present state seem rather chaotic, jumping from one point of interest to another without unified understanding. Nevertheless, some of the recent advances, as noted in the last paragraph, appear to offer illuminating examples that herald the emerging coherence of the field with the common thread of symmetry, that gives the liquid crystal its beauty.

This chapter is devoted to summarizing, on a critical basis from the author's viewpoint, the development and current status of our understanding of the interfaces and thin films (including confined systems) of liquid crystals. Strong emphasis will be placed on expounding the unifying physical principles, rather than on the technical or theoretical details, insofar as possible, in view of the infancy of the field. For this reason, we will not go into the chemical aspects of the surface alignment phenomena. Those who are interested in these aspects are referred to the review articles by Cognard (1982) and by Uchida and Seki (1992).

6.2 GENERAL VIEW OF LIQUID INTERFACES

The interface of a liquid is by no means a mathematical plane where the structures and properties of the bulk phase are suddenly terminated; rather, it is a physical entity with its own unique characteristics. The goal of studying the interfaces of molecular liquids in general is to know how these structures and properties emerge and behave themselves, to explain how these follow

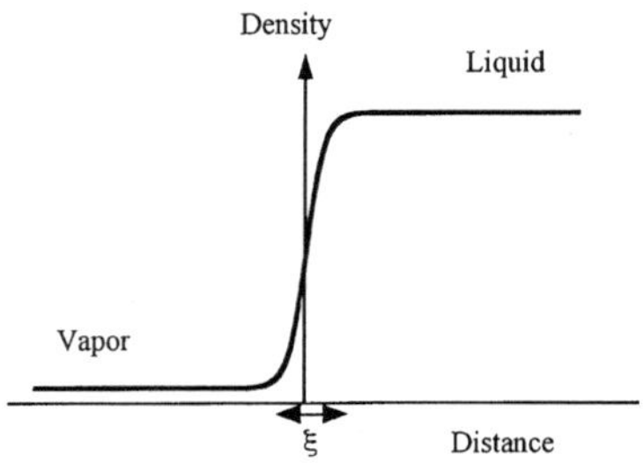

FIGURE 6.1. Density profile at an interface (ξ denotes the thickness of the interface).

from the properties of their constituent molecules, and to elucidate how these are related to their bulk counterparts.

The density profile at the interface of an ordinary isotropic liquid with its vapor is known to exhibit, as schematically depicted in FIGURE 6.1, a monotonic gradual change from the density of the bulk liquid ρ_l to that of the vapor ρ_v. At temperatures and pressures far from the liquid–vapor critical point, the thickness ξ of the interface stays on the order of a molecular dimension, but as one approaches the critical point, the interface grows thicker and, finally at the critical point, it becomes completely diffuse and disappears.

When this liquid is put into contact with a hard wall, the density profile at the interface undergoes a profound change compared with that of the liquid–vapor interface. Indeed, the interface often develops an oscillating density profile, reminiscent of the crystal structure, which decays rapidly into the bulk phase (see FIGURE 6.2). The period of oscillation is comparable to the hard-core diameter of the molecule, indicating the occurrence of interfacial layering induced by the hard wall. The decay length, in general, grows as the crystallization condition is approached.

FIGURE 6.2. Oscillating density profile at a solid–liquid interface. ξ denotes the thickness of the oscillating region.

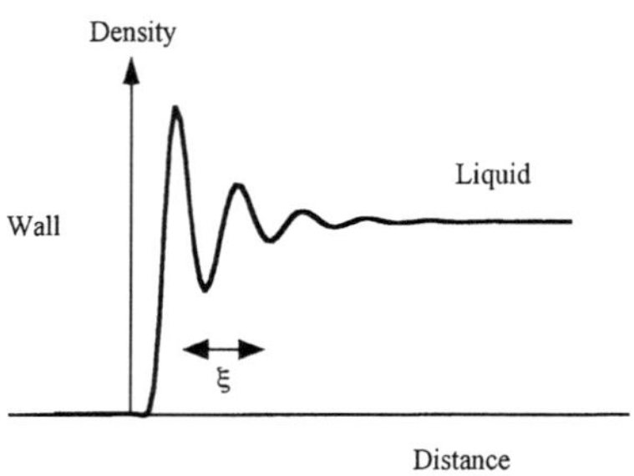

In statistical mechanical terms, these density profiles are basically a manifestation of the correlation between molecular positions and orientations existing in the liquid in question (Rowlinson and Widom, 1982). More precisely, when the interface is subjected to an external field depicted by the potential $\delta V_{ext}(\mathbf{r})$, the density at the point $\mathbf{r}'$ suffers a change:

$$\delta\rho(\mathbf{r}') = \frac{1}{k_B T}\int \mathrm{d}r\, \delta V_{ext}(\mathbf{r}) H(\mathbf{r}, \mathbf{r}') \tag{6.1}$$

where $H(\mathbf{r}, \mathbf{r}')$ denotes the density–density correlation function; for brevity of notation, only the position variable is explicitly indicated. When the potential couples with a particular mode of the liquid structure, say the crystalline order, the above formula determines how a local crystalline structure develops and decays as one moves away from the interface.

The density–density correlation function is known to obey roughly an exponential dependence on the intermolecular separation:

$$H(\mathbf{r}, \mathbf{r}') \sim \exp\left(-\frac{|\mathbf{r}-\mathbf{r}'|}{\xi}\right) \tag{6.2}$$

where ξ is a quantity called the correlation or coherence length of the mode. When ξ is finite, the correlation is said to be of short range, and when $\xi = \infty$, the correlation is of long range as in the orientational order in nematics; an intermediate case in which the correlation function decays in a power of $|\mathbf{r}-\mathbf{r}'|$ is the case of "quasi-long range order," which actually appears in the free-standing films of certain types of smectic liquid crystals. We, therefore, understand that the structure of an interface, be it either liquid–vapor or liquid–solid, may basically be determined by a couple of factors: first, the type (or symmetry) of the interfacial interaction; and second, the correlation length of the corresponding structural mode. In particular, the thickness of an interface is on the order of the positional correlation length.

Equation (6.1) reveals that any structural mode can be induced in the liquid to such an extent that the symmetry of the interaction potential and the liquid crystal allows. This argument immediately leads to the concept of "surface order," which may be identical to or different from the one prevailing in the bulk (see FIGURE 6.3) (Sluckin and Poniewierski, 1986). If it is identical, it may be more appropriate to refer to it as

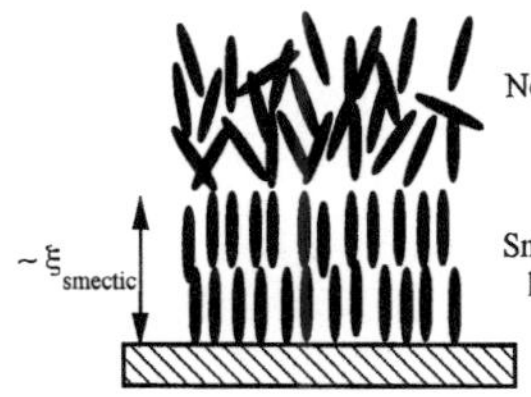

FIGURE 6.3. Surface-induced smectic order at a nematic–wall interface. The length over which the smectic order extends from the wall is given roughly by the correlation length of smectic order in the bulk $\xi_{smectic}$.

the surface "excess" order. The theoretical analysis of the surface order was initiated in the mid-1970s by Sheng (1976) and Schröder (1977), and it has grown into one of the leading concepts in the surface physics of liquid crystals. The existence of surface order now has firm experimental support for surface-induced isotropic, nematic, smectic and polar structures.

In this respect, the surface-induced alignment of liquid crystals may be stated as a phenomenon in which such interface structures merge into a bulk mode having an infinite range of correlation, for example, the director in nematics. The merging zone may be termed as the "transition region," which would have a mesoscopic size of 1–100 nm in ordinary nematics. Consequently, the ultimate understanding of the interface should correctly address the microscopic, mesoscopic and macroscopic structures and properties. We begin our review with the macroscopics.

6.3 THERMODYNAMICS

The thermodynamics of liquid interfaces was established by Gibbs in the late 19th century (Gibbs, 1961) and is now serving as the most fundamental basis for understanding the interfacial phenomena of liquids (Adamson, 1976). Although the interfaces of liquid crystals are no less appropriate a system to apply the Gibbs theory, only very few attempts have so far been made to rigorously extend the Gibbs theory to liquid crystals (Yokoyama et al., 1987; Yokoyama, 1988b). It must, however, be noted that as long as the variables associated with the liquid crystalline order, for example, the director, are not explicitly taken into account (being tacitly fixed at their equilibrium values), the thermodynamic formalism for ordinary liquids still applies to liquid crystals, though it of course fails to provide direct information on the thermodynamic implications of interfacial liquid crystalline variables.

Based on the latter property, it was Creagh and Kmetz (1973) who first employed a heuristic thermodynamic argument to give a practical criterion to decide the direction of surface alignment, homeotropic or planar, according to the relative magnitude of the surface tensions of the liquid crystal and of the substrate. This criterion, called the Creagh–Kmetz condition, has been subjected to a substantial number of experimental tests, elaborated to some extent, and has by now acquired a certain validity, yet with quite a few exceptions. In this section, we will first review the Gibbsian thermodynamics of nematic interfaces with a special emphasis on the anchoring transition, and return to the Creagh–Kmetz condition in the light of general thermodynamics. In passing, we will also briefly consider the thermodynamic treatments of thin films and confined systems.

6.3.1 Gibbs Thermodynamics of the Nematic Liquid Crystal Interface

6.3.1.1 Gibbs–Duhem Equation for Spherical Nematic Interface

We consider an interface between a nematic liquid crystal and a distinct solid, liquid or gaseous medium, hereafter referred to as the substrate. According to the hydrostatics of nematics (de Gennes, 1974), the pressure in a nematic liquid crystal increases as it is orientationally deformed following the formula,

$$P = P_0 + f_d \tag{6.3}$$

where f_d is the equilibrium density of Frank elastic energy and P_0 is the pressure in the undeformed nematic held at the constant temperature and chemical potential. Therefore, even when the interface is initially planar in the absence of orientational deformations, the interface tends to be curved so as to achieve a mechanical equilibrium as the nematic is orientationally deformed. The resultant curvature of the interface is related to the pressure difference across the interface by the well-known Laplace equation (Adamson, 1976):

$$\Delta P = f_d = \gamma\left(\frac{1}{r_1} + \frac{1}{r_2}\right) \tag{6.4}$$

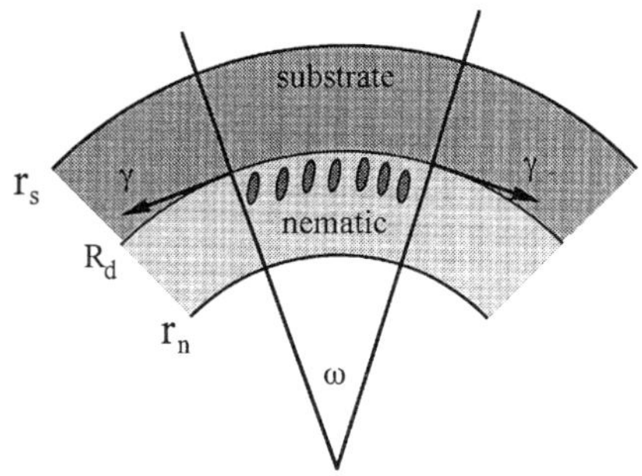

FIGURE 6.4. Spherical interface between an orientationally deformed nematic liquid and a substrate.

where r_1 and r_2 are the principal radii of curvature measured from the nematic side. We have assumed here that the interface is planar when the nematic is undeformed. As this equation indicates, the curvature is linear in f_d; thus, its influence on the director configuration in the nematic should be higher than first order in f_d. Consequently, if we restrict our attention to the small deformation regime, we are allowed to treat the interface as if it is planar as far as the orientational properties are concerned.

In what follows, we consider for simplicity a spherical interface as shown in FIGURE 6.4 circumvented by a conical side wall subtending a solid angle ω and a pair of spherical boundaries, concentric with the nematic–substrate interface. We assume that the orientation is uniform in the plane parallel to the interface, resulting in one-dimensional director fields depending only on r. Based on the Frank theory of curvature elasticity (de Gennes, 1974), the deformation free energy F_d in the bulk region for this quasi-one-dimensional configuration is approximately given by

$$F_d(\mathbf{n}(r)) = \int_{r_1}^{r_2} f_d(\mathbf{n}, \mathrm{d}\mathbf{n}/\mathrm{d}r)\omega r^2 \mathrm{d}r \tag{6.5}$$

where f_d denotes the Frank elastic energy density. Hence, for fixed boundary positions, the variation of the free energy associated with an infinitesimal change in the director profile is written as

$$\begin{aligned} \delta F_d = & \int_{r_1}^{r_2} \left(\frac{\partial f_d}{\partial n_i} - \frac{1}{r^2}\frac{\mathrm{d}}{\mathrm{d}r} r^2 \pi_i \right) \delta n_i \omega r^2 \mathrm{d}r \\ & + \omega r_2^2 \pi_i(r_2)\delta n_i(r_2) - \omega r_1^2 \pi_i(r_1)\delta n_i(r_1) \end{aligned} \tag{6.6}$$

where the summation convention over repeated indices has been employed with π_i representing the component of the orientational torque defined by

$$\pi_i = \frac{\partial f_d(\mathbf{n}, \mathrm{d}\mathbf{n}/\mathrm{d}r)}{\partial(\mathrm{d}n_i/\mathrm{d}r)}. \tag{6.7}$$

In terms of the polar and the azimuthal angles θ and ϕ, with which the director is $\mathbf{n} = (\sin\theta\cos\phi, \sin\theta\sin\phi, \cos\theta)$, the Frank elastic energy density becomes

$$\begin{aligned} f_d = & \frac{1}{2}(K_1 \sin^2\theta + K_3\cos^2\theta)\left(\frac{\mathrm{d}\theta}{\mathrm{d}r}\right)^2 \\ & + \frac{1}{2}\sin^2\theta(K_2\sin^2\theta + K_3\cos^2\theta)\left(\frac{\mathrm{d}\phi}{\mathrm{d}r}\right)^2 \end{aligned} \tag{6.8}$$

which leads to the expression for the orientational torque as

$$\begin{aligned} \boldsymbol{\pi} = & (K_1\sin^2\theta + K_3\cos^2\theta)\frac{\mathrm{d}\theta}{\mathrm{d}r}\frac{\mathrm{d}\mathbf{n}}{\mathrm{d}\theta} \\ & + \sin^2\theta(K_2\sin^2\theta + K_3\cos^2\theta)\frac{\mathrm{d}\phi}{\mathrm{d}r}\left(\frac{1}{\sin\theta}\frac{\mathrm{d}\mathbf{n}}{\mathrm{d}\phi}\right) \end{aligned} \tag{6.9}$$

$$\equiv \pi_p \frac{\mathrm{d}\mathbf{n}}{\mathrm{d}\theta} + \pi_a \frac{1}{\sin\theta}\frac{\mathrm{d}\mathbf{n}}{\mathrm{d}\phi} \tag{6.10}$$

with π_p and π_a being the polar and the azimuthal torque, respectively. The equilibrium director profile is the one that satisfies the Euler–Lagrange equation,

$$\frac{\partial f_d}{\partial n_i} - \frac{1}{r^2}\frac{\mathrm{d}}{\mathrm{d}r} r^2 \pi_i = \lambda n_i \tag{6.11}$$

where λ is a Lagrange multiplier. Then, for any equilibrium variation of the director profile, equation (6.6) reduces to

$$\delta F_d = \omega r_2^2 \pi_i(r_2)\delta n_i(r_2) - \omega r_1^2 \pi_i(r_1)\delta n_i(r_1). \tag{6.12}$$

It directly follows from equation (6.11) that the quantity

$$H \equiv r^4 f_d \tag{6.13}$$

is one of the first integrals of the Euler–Lagrange equation that stays constant through any equilibrium director field.

Based on these formulas for the orientational work at the boundary, the energy conservation of the entire volume embracing the nematic–substrate interface is

$$\begin{aligned} \mathrm{d}U = & T\,\mathrm{d}S + \sum_{j=1}^{m} \mu_j \,\mathrm{d}N^{(j)} + A_n P\,\mathrm{d}r_n - A_s P_0\,\mathrm{d}r_s \\ & - A_n \boldsymbol{\pi}\cdot \mathrm{d}\mathbf{n} + \eta\,\mathrm{d}\omega \end{aligned} \tag{6.14}$$

where T is the absolute temperature, S is the entropy,

r_n and r_s are the radii of the boundaries taken in the nematic and the substrate phases, $A_n = \omega r_n^2$ and $A_s = \omega r_s^2$ are the areas of these boundaries, μ_j is the chemical potential of the jth molecule, $N^{(j)}$ is the number of the jth molecule, $\mathbf{n}$ is the director at the lower boundary, $\boldsymbol{\pi}$ is the orientational torque per unit area, and η is the work needed for a unit extension of ω. In particular, integrating the above equation with respect to ω at fixed temperature, chemical potential and orientation, we find

$$\eta\omega = U - TS - \sum_{j=1}^{m} \mu_j N^{(j)} \equiv \Omega \tag{6.15}$$

where Ω is the grand thermodynamic potential.

In the spirit of Gibbs thermodynamics, we set a hypothetical interface, known as the dividing surface, at the point $r = R$, up to which each of the bulk phases are to be extrapolated. For the resultant hypothetical bulk phases, energy conservation can be written as

$$\begin{aligned} dU_n = {} & T\,dS_n + \sum_{j=1}^{m} \mu_j\,dN_n^{(j)} + A_n P\,dr_n - A_d P\,dR \\ & - A_n \boldsymbol{\pi} \cdot d\mathbf{n} + A_d \boldsymbol{\pi}_d \cdot d\mathbf{n}_d \\ & - \left[\int_{r_n}^{R} (P_0 - f_d) r^2\,dr\right] d\omega \end{aligned} \tag{6.16}$$

$$\begin{aligned} dU_s = {} & T\,dS_s + \sum_{j=1}^{m} \mu_j\,dN_s^{(j)} - A_s P_0\,dr_s + A_d P_0\,dR \\ & - \frac{r_s^3 - R^3}{3} P_0\,d\omega \end{aligned} \tag{6.17}$$

where f_d denotes the equilibrium Frank elastic energy density. It should be noted here that the boundary directors, $\mathbf{n}$ and $\mathbf{n}_d$, are not mutually independent as a thermodynamic variable, but either of them is sufficient to specify the director profile in the volume, because of the presence of the nematic–substrate interface which is regarded as an agent to uniquely determine the equilibrium director profile for a given boundary condition at the other interface.

Subtracting these equations from equation (6.14), we obtain

$$\begin{aligned} d\Delta U = {} & T\,d\Delta S + \sum_{j=1}^{m} \mu_j\,d\Delta N^{(j)} - A_d \boldsymbol{\pi}_d \cdot d\mathbf{n}_d \\ & + \gamma\,dA_d + \left(f_d - 2\frac{\gamma}{R}\right) A_d\,dR \end{aligned} \tag{6.18}$$

where

$$\Delta U = U - U_n - U_s = A_d u^s \tag{6.19}$$

$$\Delta S = S - S_n - S_s = A_d s^s \tag{6.20}$$

$$\Delta N^{(j)} = N^{(j)} - N_n^{(j)} - N_s^{(j)} \equiv A_d \Gamma_j \tag{6.21}$$

$$\gamma = (\Omega - \Omega_n - \Omega_s)/A_d \tag{6.22}$$

$$\begin{aligned} \Omega_n = {} & -\int_{r_n}^{R} (P_0 - f_d)\omega r^2\,dr = -P_0 V_n \\ & + \omega H\left(\frac{1}{r_n} - \frac{1}{R}\right) \end{aligned} \tag{6.23}$$

$$\Omega_s = -P_0 V_s \tag{6.24}$$

$$V_n = \omega(R^3 - r_n^3)/3 \tag{6.25}$$

$$V_s = \omega(r_s^3 - R^3)/3 \tag{6.26}$$

$$A_d = \omega R^2 \tag{6.27}$$

$$V = V_n + V_s \tag{6.28}$$

where u^s, s^s, and Γ_j are the surface energy, surface entropy and adsorption, respectively, and γ is the surface tension, which is, as in ordinary liquids, identical to the surface excess of the grand thermodynamic potential. The last term in equation (6.18) stands for the effect of a notional (unphysical) change in the position of the dividing surface (Rowlinson and Widom, 1982).

Integration of (6.18) with respect to the interface area A_d and substitution of the resultant equation in (6.18) yield the Gibbs–Duhem equation for the nematic interface:

$$d\gamma = -s^s\,dT - \sum_{j=1}^{m} \Gamma_j\,d\mu_j - \boldsymbol{\pi}_d \cdot d\mathbf{n}_d + \left(f_d - 2\frac{\gamma}{R}\right) dR. \tag{6.29}$$

It may be worthwhile to remind the reader of the fact that the director $\mathbf{n}_d$ at the dividing surface is a variable to specify the bulk orientational state of the nematic, being extrapolated from the bulk profile based on the Frank formalism, and hence has, in principle, nothing directly to do with the real director, if any, at this point.

6.3.1.2 Surface of Tension and Surface of Extrapolation: Curvature Dependence of Surface Tension

Unlike planar interfaces, the surface tension of a curved interface is no longer independent of the choice of the dividing surface. We find from equation

(6.29) that the variation of the surface tension associated with a notional change in R satisfies

$$\left[\frac{d\gamma}{dR}\right] = -\boldsymbol{\pi}_d \cdot \frac{d\mathbf{n}_d}{dR} + f_d - 2\frac{\gamma}{R} = -f_d - 2\frac{\gamma}{R}. \quad (6.30)$$

As for ordinary liquid interfaces, the surface of tension can be defined as such a choice of a dividing surface whereby the Laplace equation is satisfied:

$$f_d - 2\frac{\gamma_s}{R_s} = 0 \quad (6.31)$$

where γ_s denotes the corresponding surface tension. By integrating (6.30) from R_s to R, we find that the surface tension for an arbitrary dividing surface can be written as

$$\begin{aligned}\gamma &= \frac{R_s^2(2R_s - R)}{(R_s + R)^3}\gamma_s \\ &\approx \left(1 - 4\frac{R - R_s}{R_s} + 9\frac{(R - R_s)^2}{R_s^2}\right)\gamma_s.\end{aligned} \quad (6.32)$$

In contrast to the case of ordinary liquid interfaces (Rowlinson and Widom, 1982), the surface of tension for a nematic interface does not correspond to a choice of a dividing surface where the surface tension hits the minimum. This happens because the notional change in the dividing surface involves not only the translational work (as in ordinary liquid cases), but also an extra work associated with the orientation changes in the boundary director. The latter, in general, does not vanish even at the surface of tension and gives rise to a strong position dependence of surface tension as manifested in equation (6.32). This position dependence is inherited even in the planar limit $R \to \infty$. So, rigorously speaking, the surface tension ceases to be a universal thermodynamic quantity in a nematic as long as it is orientationally deformed, although it is always meaningful to specify the surface of tension. This implies that the value of the surface tension in a deformed nematic would change depending on the method of measurement one uses; however, the difference is on the order of $(R - R_s)K/l^2$ with K and l being the Frank constant and the characteristic length of deformation, and hence it remains to be a negligibly small correction unless l does not approach the molecular dimension.

In a similar fashion to the way we defined the surface of tension, we can conceive of another special surface for a nematic interface which is characterized by vanishing orientational work:

$$-\boldsymbol{\pi}_{ex} \cdot \frac{d\mathbf{n}(R_{ex})}{dH} = 0. \quad (6.33)$$

We shall refer to this surface, located at R_{ex}, as the surface of extrapolation after the definition of the extrapolation length due to de Gennes (de Gennes, 1974; Dubois-Violette and de Gennes, 1976). It should indeed be clear that the point of extrapolation in de Gennes' sense, whereby the director is fixed even when the nematic is subjected to orientational torque, satisfies equation (6.33).

The physical (not notional) change in the surface tension in response to the orientational torque can be expressed with the help of the surface of extrapolation. With the dividing surface taken at the surface of tension, the Gibbs–Duhem equation, (6.29), is

$$d\gamma_s = -s^s\, dT - \sum_{j=1}^{m} \Gamma_j\, d\mu_j - \boldsymbol{\pi}_s \cdot d\mathbf{n}_s. \quad (6.34)$$

Since

$$\frac{d}{dr} r^2 \pi_i \left(\frac{\partial n_i(r)}{\partial H}\right) = \frac{1}{r^2},$$

there exists such R_c that satisfies $r^2 \pi_i (\partial n_i(r)/\partial H) = 1/R_c - 1/r$. By substituting $r = R_{ex}$ and combining it with (6.33), we obtain

$$\pi_i \left(\frac{\partial n_i(r)}{\partial H}\right) = \frac{1}{V^2}\left(\frac{1}{R_{ex}} - \frac{1}{r} + 2H\frac{\partial R_{ex}^{-1}}{\partial H}\right).$$

Using this equation in equation (6.33) for orientational deformations at fixed temperature and chemical potential, the Gibbs–Duhem equation is rewritten as:

$$\begin{aligned}\frac{d\gamma_s}{dH} &= \frac{1}{R_s^2}\left[\frac{1}{R_s} - \frac{1}{R_{ex}} + 2H\frac{d}{dH}\left(\frac{1}{R_s} - \frac{1}{R_{ex}}\right)\right] \\ &= \frac{1}{R_s^2}\left[\frac{d}{dH}H\left(\frac{1}{R_s} - \frac{1}{R_{ex}}\right) + H\frac{d}{dH}\left(\frac{1}{R_s} - \frac{1}{R_{ex}}\right)\right]\end{aligned} \quad (6.35)$$

where the first and the second terms on the right-hand side represent the orientational and translational work, respectively. A formal integration of this equation leads to

$$R_{ex} - R_s = \frac{R_{ex}}{R_s\sqrt{f_d}} \int_0^{\sqrt{f_d}} \frac{d\gamma_s}{df_d}\, d\sqrt{f_d} \quad (6.36)$$

where f_d denotes the Frank energy density at the surface of tension.

Alternatively, (6.35) can be integrated in powers of $\delta = R_{ex} - R_s$ to give

$$\gamma_s = \gamma_0 \left(1 + 2\frac{\delta_0}{R_s} + \frac{\delta_0^2 + 3\delta_1}{R_s^2} \right) \tag{6.37}$$

to second order in δ / R_s, where γ_0 is the surface tension when the nematic is undeformed, and hence the interface is planar, and δ_0 and δ_1 are constants defined by

$$\delta = \delta_0 + \frac{\delta_1}{R_s} + \ldots . \tag{6.38}$$

This is the nematic version of the Gibbs–Tolman–König–Buff equation (Ono and Kondo, 1960), describing the curvature effect on the surface tension.

To lowest order in f_d, equations (6.36) and (6.37) are rewritten as

$$\gamma_s = \gamma_0 + \delta_0 f_d \tag{6.39}$$

which shows that the surface of extrapolation specifies the point to which the bulk nematic must be extrapolated when the interfacial contribution to the orientational free energy is regarded as being completely bulk-like. The quantity,

$$d_e \equiv R_{ex} - R_s \tag{6.40}$$

serves, therefore, as a measure of the anchoring strength, referred to as the extrapolation length, which was first introduced by de Gennes (de Gennes, 1974; Dubois-Violette and de Gennes, 1976) from a geometrical argument as a length by which the linear profile of director field in the bulk nematic should be extrapolated beyond the interface to reach the point of fixed orientation; a more detailed account of the anchoring phenomena will be given in a later section. The extrapolation length, as defined above, is not necessarily limited to infinitesimal deviations from the easy axis, but remains meaningful for any degree of deformation; so d_e is in general a function of f_d and hence H.

6.3.1.3 Orientation Equilibria and Anchoring Transition

We now proceed to derive a variety of thermodynamic relations involving the surface orientational variables. With the explicit use of the polar and azimuthal angles, the Gibbs–Duhem equation is rewritten as

$$d\gamma_s = -s^s\, dT - \sum_{j=1}^{m} \Gamma_j\, d\mu_j - \pi_{p_s}\, d\theta_s - \pi_{a_s}\, d\phi_s \tag{6.41}$$

or equivalently

$$d\gamma_s^* \equiv d(\gamma_s - \pi_{p_s}\theta_s - \pi_{a_s}\phi_s) = -s^s\, dT - \sum_{j=1}^{m} \Gamma_j\, d\mu_j + \theta_s\, d\pi_{p_s} + \phi_s\, d\pi_{a_s}. \tag{6.42}$$

The corresponding equation for undeformed states is:

$$d\gamma_0 = -s_0^s\, dT - \sum_{j=1}^{m} \Gamma_{0j}\, d\mu_j. \tag{6.43}$$

From (6.41), we have in equilibrium the following relations expressing the balance of orientational torque at the interface:

$$\left(\frac{\partial \gamma_s}{\partial \theta_s} \right)_{T,\mu,\phi_s} = -\pi_{p_s} \qquad \left(\frac{\partial \gamma_s}{\partial \phi_s} \right)_{T,\mu,\theta_s} = -\pi_{a_s}. \tag{6.44}$$

In order for this balance to be locally stable, the free energy change associated with any infinitesimal orientational deviation must be positive:

$$0 \leq \delta\pi_{p_s}\delta\theta_s + \delta\pi_{a_s}\delta\phi_s + \delta^2\gamma_s. \tag{6.45}$$

As illustrated in FIGURE 6.5, there could in general be a multiple number of interface orientations that satisfy

FIGURE 6.5. Balance of the bulk and the interfacial torques. Only the polar-angle dependence of the surface tension $\gamma_s(\theta_s)$ is considered, with the three local minima at $\pi/2$, $\sim\pi/4$, and $\sim 3\pi/40$. The solid straight line in the upper part shows the torque $-\pi$ transmitted through the bulk, and the broken line shows the interfacial torque $d\gamma_s/d\theta_s$. There are six points, numbered 1 through 6, where the torque balance equation is satisfied; but only 1, 3, and 5 are locally stable, with point 1 giving the global minimum.

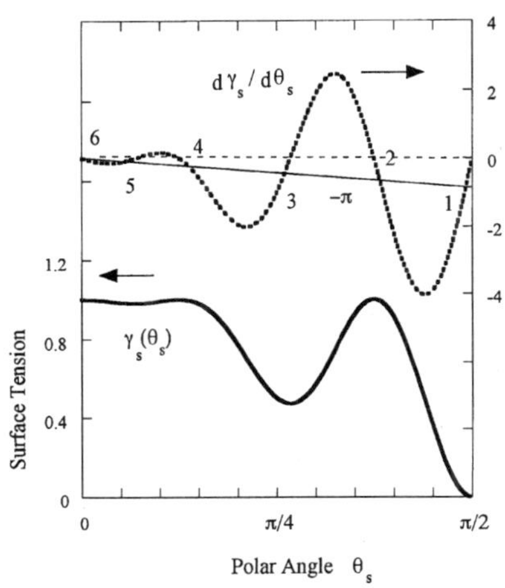

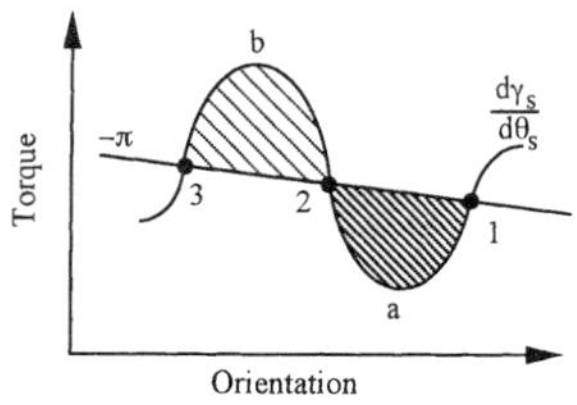

FIGURE 6.6. Maxwell's equal areas construction for the interface orientation equilibrium. The orientations at the points 1 and 3 are in equilibrium when the hatched areas above and below the $-\pi$ line are equal, that is, $S_a = S_b$.

the stability as well as the torque balance conditions (Sluckin and Poniewierski, 1986; Jérôme and Pieranski, 1988a, b; Monkade et al., 1988; Jérôme et al., 1990). The equilibrium orientation is the one with the minimum free energy for fixed boundary conditions. According to (6.22), the free-energy difference between the local minima, say the points 1 and 3, is written as

$$\Delta\Omega = \Omega_n(1) - \Omega_n(3) + A_s[\gamma_s(1) - \gamma_s(3)] \\ = A_s \int_{n_3}^{n_1} \left(\frac{d\gamma_s}{d\mathbf{n}} + \pi \right) d\mathbf{n}. \tag{6.46}$$

As depicted in FIGURE 6.6, $\Delta\Omega$ is identical to the difference of the hatched areas, that is, $S_a - S_b$. Hence, when $S_a > S_b$, point 1 gives the orientation of global stability, and when $S_a < S_b$, point 3 does. At $S_a = S_b$, the two orientations are equally stable, and a transition from one orientation to the other can occur. The procedure to locate the transition point is exactly the same as the so-called Maxwell's theorem of equal areas for the liquid–vapor equilibria (Hill, 1956).

This orientational transition, referred to as the anchoring transition (Jérôme, 1991), can occur due to changes in temperature (Ryschenkow and Kleman, 1976; Chiarelli et al., 1983, 1984; Hiltrop and Stegemeyer, 1984; Di Lisi et al., 1990; Flatischler et al., 1991; Petrov et al., 1992; Patel and Yokayama, 1993; Basturk, 1993), composition (Proust et al., 1972; Ohgawara and Itchida, 1981a, b; Pieranski and Jérôme, 1989; Bechhoefer et al., 1990a, b; Parmar et al., 1990; Jérôme et al., 1993) and the orientational torque induced by external electric or magnetic fields or by the boundary condition (Proust and Parez, 1977; Faetti and Fronzoni, 1978; Proust et al., 1979). In line with the liquid–vapor transition, the anchoring transition is fundamentally governed by the phase rule. We imagine for generality that p interfacial orientational states are in equilibrium; then, there arise $3p - 1$ equality relations from the torque and the free energy balance equations. For an m-component liquid crystal in the absence of external fields, the number of independent variables is $m + 2p + 4$, including $2p + 2$ orientational variables, that is, the interface and the boundary orientation angles, and the nematic layer thickness. Consequently, the number of degrees of freedom left in equilibrium amounts to

$$f = m - p + 5. \tag{6.47}$$

When, in particular, the orientation at the boundary and the thickness of the nematic layer are fixed (so as to be comparable to the normal cases of temperature- or composition-driven transitions, see FIGURE 6.7), the above relation is reduced to

$$f = m - p + 2. \tag{6.48}$$

Under fixed composition and pressure, therefore, $f = -p + 2$, and hence only two orientational states can coexist at a single temperature.

The easy axis corresponds to a special case of equilibrium that is characterized by the absence of orientational torque, that is, $\pi = 0$. When an anchoring transition takes place between two easy axes $\mathbf{n}_e(I)$ and $\mathbf{n}_e(II)$, the equilibrium line separating these two states satisfies

$$\gamma_s(\mathbf{n}_e(I)) = \gamma_s(\mathbf{n}_e(II)). \tag{6.49}$$

Taking the total derivative of both sides with the use of the Gibbs–Duhem equation, we obtain

$$-s^s(I)\, dT_t - \sum_{j=1}^{m} \Gamma_j(I)\, d\mu_j = -s^s(II)\, dT_t - \sum_{j=1}^{m} \Gamma_j(II)\, d\mu_j \tag{6.50}$$

where T_t is the temperature of the anchoring transition for given chemical potentials. Restricting our attention to a two-component system and taking the dividing surface at the zero-adsorption surface for the major component ($j = 1$) so that $\Gamma_1(I) = \Gamma_1(II) = 0$, we obtain the following Clausius–Clapeyron formula for the anchoring transition:

$$\frac{dT_t}{d\mu_2} = -\frac{\Gamma_2(II) - \Gamma_2(I)}{s^s(II) - s^s(I)}. \tag{6.51}$$

This formula indicates, first of all, that when the transition temperature shifts with composition, as in

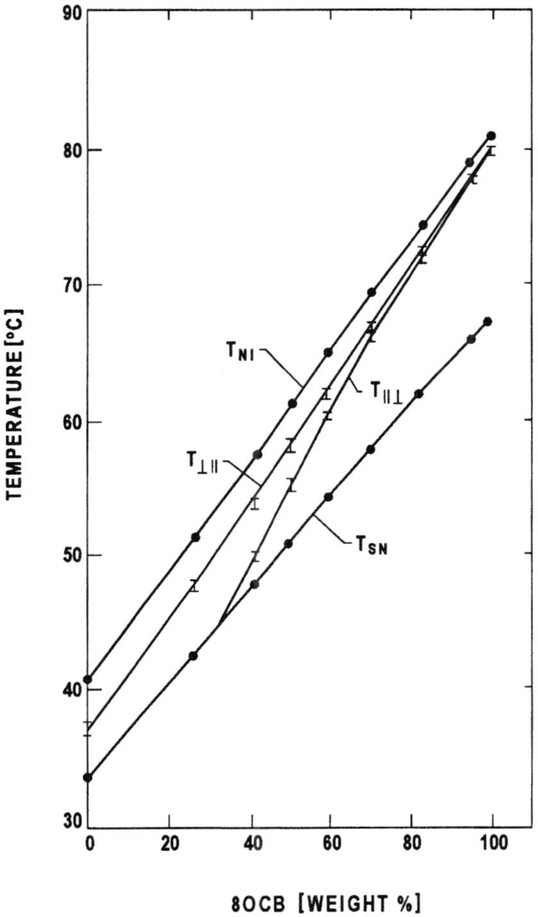

FIGURE 6.7. Temperature–composition phase diagram for anchoring transitions in the mixture of octyl- and octyloxycyanobiphenyls on substrates coated with fluorocarbon polymer. Reproduced by permission from (Birecki, 1984).

FIGURE 6.7, the adsorption as well as the surface entropy should undergo a discontinuous change at the transition. Jérôme and coworkers (Jérôme et al., 1993) have indeed observed such an abrupt jump in adsorption by measuring optical second harmonic generation from a nematic–substrate interface. Another feature to be noted is that, when $dT_t/d\mu_2 > 0$, $\Gamma_2(II) - \Gamma_2(I)$ and $s^s(II) - s^s(I)$ have opposite signs; so, if the adsorption is enhanced as the orientation is transformed from state I to state II, then the surface entropy should be lower in state II than in state I.

6.3.1.4 Anchoring Energy Coefficients and Continuous Anchoring Transition

Stability of the easy axis requires that the second-order deviation of γ_s around (θ_e, ϕ_e) is always positive. In conventional liquid crystal parlance (Yokoyama, 1988b), the second-order derivatives of the surface tension at its minima are referred to as the anchoring energy coefficients:

$$W_p = \left(\frac{\partial^2 \gamma_s}{\partial \theta_s^2}\right)_{T,\mu} \tag{6.52}$$

$$W_a = \left(\frac{\partial^2 \gamma_s}{\partial \phi_s^2}\right)_{T,\mu} \tag{6.53}$$

$$W_{pa} = \left(\frac{\partial^2 \gamma_s}{\partial \phi_s \partial \theta_s}\right)_{T,\mu} = \left(\frac{\partial^2 \gamma_s}{\partial \theta_s \partial \phi_s}\right)_{T,\mu} = W_{ap}. \tag{6.54}$$

Accordingly, for small deviations of the director from the easy axis, the surface tension is given by a quadratic form of the angular deviations as follows:

$$\gamma_s = \gamma_0 + \frac{1}{2} W_p \delta\theta_s^2 + W_{pa} \delta\theta_s \delta\phi_s + \frac{1}{2} W_a \delta\phi_s^2 \tag{6.55}$$

with $\delta\theta_s = \theta_s - \theta_e$ and $\delta\phi_s = \phi_s - \phi_e$. Then, the stability condition of the easy axis simply is

$$W_p > 0 \qquad W_a > 0 \qquad W_p W_a - W_{ap}^2 > 0. \tag{6.56}$$

As in liquid–vapor equilibria, the line of first-order anchoring transitions described by (6.51), if any, could be terminated by a critical point where the two easy axes merge. Along with the torque balance equations, that is,

$$\frac{\partial \gamma_s}{\partial \theta_s} = \frac{\partial \gamma_s}{\partial \phi_s} = 0 \tag{6.57}$$

the condition of a critical point can be stated in terms of the anchoring energy coefficients as

$$W_p W_a - W_{ap}^2 = 0. \tag{6.58}$$

With this condition, (6.55) is rewritten as

$$\gamma_s = \gamma_0 + (\sqrt{W_p}\delta\theta_s + \sqrt{W_a}\delta\phi_s)^2/2.$$

Therefore, when γ_s is expanded in powers of ψ, which is connected to $\delta\theta_s$ and $\delta\phi_s$ by

$$\delta\theta_s = \sqrt{W_a/(W_a + W_p)}\delta\psi$$

and

$$\delta\phi_s = -\sqrt{W_p/(W_a + W_p)}\delta\psi$$

the above condition means that the leading term is cubic in $\delta\psi$. Hence, in order for the critical point itself to be stable, we need further to require that this term also disappears (Landau and Lifshitz, 1959):

$$\frac{d^3\gamma_s}{d\delta\psi^3} = 0. \tag{6.59}$$

As a consequence of these conditions, we find, for the m-component system under the same fixed boundary conditions as for (6.48), that the number of degrees of freedom is given by $f = m - 1$, including the pressure. For example, a two-component system held at a fixed pressure can exhibit an isolated critical point on the temperature–composition plane, as observed experimentally by Birecki (1984) (see FIGURE 6.7).

The above conditions apply not only to the critical point for first-order anchoring transitions, but also to continuous (second-order) transitions in general. The well-controlled experiments by Bechhoefer and coworkers (Bechhoefer et al., 1990b) indicate the existence of a line of second-order anchoring transitions in a four-component system under fixed temperature and pressure (see FIGURE 6.8) in agreement with the phase rule. In apparent contradiction with the phase rule, however, there have also been several reports of a critical anchoring transition in single-component systems (Chiarelli et al., 1983, 1984; Di Lisi et al., 1990; Patel and Yokayama, 1993). According to the phase rule, we have $f = 0$ including the pressure; so the critical transition, if any, should take place at an isolated point on the temperature–pressure plane. Therefore, it seems quite unlikely that the critical pressure was coincident with atmospheric pressure, thereby making the transition observable in all these different experiments. A particularly notable example in this respect is the observation by Chiarelli and coworkers (Chiarelli et al., 1983, 1984) for a free surface of nematic liquid crystal; at a temperature slightly below the nematic–isotropic transition, the easy axis, which is tilted at lower temperatures, becomes oriented perpendicular to the free surface, accompanied by a vanishing anchoring energy coefficient. The key to resolving this discrepancy lies in the symmetry of the system under consideration (Pieranski and Jérôme, 1989). The critical anchoring transitions reported so far on single-component systems occurred at the point of the tilted-to-homeotropic transformation ($\theta_e \to 0$) on substrates having a complete rotation symmetry around the surface normal. Due to this symmetry property, the interfacial tension satisfies $\gamma_s(\theta) = \gamma_s(-\theta)$. Then, it automatically assures the absence of the cubic term as required by (6.59), which results in the desired increase of the degrees of freedom to $f = 1$. This argument holds for liquid crystal orientations on crystal surfaces, as well, having an appropriate symmetry (Jérôme et al., 1990).

FIGURE 6.8. Anchoring phase diagram in a four-component system. Reproduced by permission (Bechhoefer et al., 1990b).

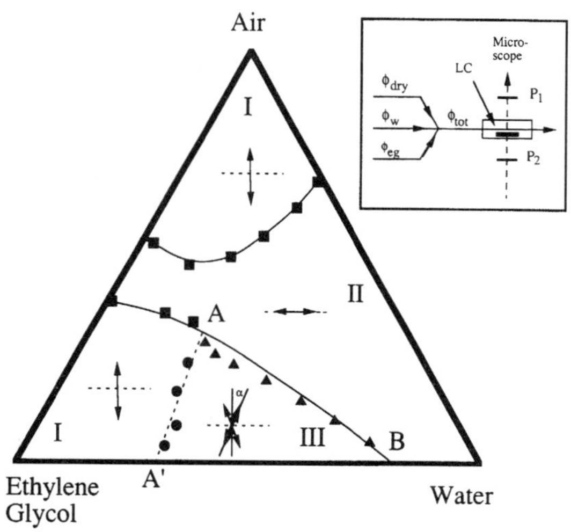

The anchoring energy coefficients are, in effect, surface-specific curvature elastic constants, so they could play a significant role in any situation involving the mechanical equilibrium between the bulk and the surface as in field-effect LCDs. The parameter that is in this respect completely equivalent to the anchoring energy coefficients is the extrapolation length which we have already defined. Directly differentiating the surface tension with respect to f_d using (6.39), we obtain a defining equation for the extrapolation length:

$$d_e = \lim_{f_d \to 0} \left(\frac{\partial \gamma_s}{\partial f_d} \right)_{T,\mu,\beta}. \tag{6.60}$$

Here, β is a variable, independent of f_d, specifying the mode of deformation defined by

$$\beta = \left(\frac{K_a}{K_p} \frac{\pi_p^2}{\pi_a^2} + 1 \right)^{-1} \quad 0 \le \beta \le 1 \tag{6.61}$$

where

$$K_p \equiv K_1 \sin^2\theta_e + K_3 \cos^2\theta_e \tag{6.62}$$

$$K_a \equiv \sin^2\theta_e (K_2 \sin^2\theta_e + K_3 \cos^2\theta_e). \tag{6.63}$$

Obviously, $\beta = 0$ and $\beta = 1$ correspond, respectively, to the polar and the azimuthal deformations, and a β inbetween corresponds to a mixed state. Then, substitution of (6.55) into equation (6.60) finally yields

$$d_e = \frac{W_p W_a}{W_p W_a - W_{pa}^2}\left[\frac{K_p}{W_p}(1-\beta) + \frac{K_a}{W_a}\beta - 2\frac{W_{pa}\sqrt{K_p K_a}}{W_p W_a}\sqrt{(1-\beta)\beta}\right]. \tag{6.64}$$

We now derive thermodynamic relations bridging between the orientational and other conventional variables. It must first be remembered that the temperature and the chemical potentials are not totally independent, but are coupled with each other by the Gibbs–Duhem equations governing the bulk phases:

$$0 = -\left[s_{0n} + \sum_{l=1}^{3} f_d^{(l)}\left(\frac{\partial \ln K_l}{\partial T}\right)_{\mu}\right] dT + dP_0 - \sum_{j=1}^{m}\left[\rho_{0n}^{(j)} + \sum_{l=1}^{3} f_d^{(l)}\left(\frac{\partial \ln K_l}{\partial \mu_j}\right)_{T,\mu'}\right] d\mu_j \tag{6.65}$$

$$0 = -s_s\, dT + dP_0 - \sum_{j=1}^{m} \rho_s^{(j)}\, d\mu_j \tag{6.66}$$

where the subscript 0 implies that the variable corresponds to undeformed states, s and ρ are the specific entropy and the number density of molecules, and $f_d^{(1)}$, $f_d^{(2)}$, and $f_d^{(3)}$ are the splay, twist and bend components of the Frank elastic-energy density, each of which is approximately uniform in the nematic in the weak deformation regime. By eliminating P_0, we obtain

$$\Delta s\, dT + \sum_{j=1}^{m} \Delta\rho^{(j)}\, d\mu_j = 0 \tag{6.67}$$

where Δs and $\Delta\rho$ denote the differences between the specific entropy and density between the nematic and substrate phases including orientational contributions. To be specific, let us imagine that the first species is the major component of the nematic phase. Then, defining a length L_1 by

$$L_1 = \Gamma_1 / \Delta\rho^{(1)} \tag{6.68}$$

and eliminating μ_1 in (6.41) and (6.42), we obtain

$$d\gamma_s = -(s^s - L_1\Delta s)\, dT - \sum_{j=2}^{m}(\Gamma_j - L_1\Delta\rho^{(j)}) \times d\mu_j - \pi_{p_s}\, d\theta_s - \pi_{a_s}\, d\phi_s \tag{6.69}$$

$$d\gamma_s^* = -(s^s - L_1\Delta s)\, dT - \sum_{j=2}^{m}(\Gamma_j - L_1\Delta\rho^{(j)}) \times d\mu_j + \theta_s\, d\pi_{p_s} + \phi_s\, d\pi_{a_s}. \tag{6.70}$$

Here, $\Gamma_j - L_1\Delta\rho^{(j)}$, for example, signifies the adsorption of the jth species with respect to the zero-adsorption plane of the major component of the nematic.

From these equations, we obtain the following Maxwell relations:

$$-\left(\frac{\partial \Delta s^s}{\partial \pi_{p_s}}\right)_{T,\boldsymbol{\mu},\pi_{a_s}} = \left(\frac{\partial \theta_s}{\partial T}\right)_{\boldsymbol{\mu},\pi_{p_s},\pi_{a_s}} \tag{6.71}$$

$$-\left(\frac{\partial \Delta\Gamma_j \rho^{(j)}}{\partial \pi_{p_s}}\right)_{T,\boldsymbol{\mu},\pi_{a_s}} = \left(\frac{\partial \theta_s}{\partial \mu_j}\right)_{T,\boldsymbol{\mu}',\pi_{p_s},\pi_{a_s}} \tag{6.72}$$

where $\Delta s^s \equiv s^s - L_1\Delta s$ and $\Delta\Gamma_j \equiv \Gamma_j - L_1\Delta\rho^{(j)}$. Replacing π_{p_s} and θ_s by π_{a_s} and ϕ_s, we get analogous formulas for the azimuthal mode. In particular, if we take $\pi_p = \pi_a = 0$, these equations relate the temperature and concentration dependences of the easy axis to the variations in the surface entropy and the adsorption associated with the orientational deformation. For example, (6.72) indicates that when the increase in the concentration of a surface-active impurity (say $j = 2$) induces the change of surface alignment toward a homeotropic alignment, that is, $\partial\theta_e/\partial\mu_2 < 0$, then the adsorption of the impurity should decrease as an orientational torque is introduced in such a way as to oppose the move toward homeotropic alignment. Similarly, (6.71) connects the temperature variation of the easy axis to a torque-induced change in the surface entropy.

Using (6.69), we reach another form of the Maxwell relation:

$$\left(\frac{\partial \Delta s^s}{\partial \theta_s}\right)_{T,\boldsymbol{\mu},\phi_s} = \left(\frac{\partial \pi_{p_s}}{\partial T}\right)_{\boldsymbol{\mu},\theta_s,\phi_s} \tag{6.73}$$

$$= -\left(\frac{\partial \pi_{p_s}}{\partial \theta_s}\right)_{T,\boldsymbol{\mu},\phi_s}\left(\frac{\partial \theta_s}{\partial T}\right)_{\boldsymbol{\mu},\pi_{p_s},\pi_{a_s}} - \left(\frac{\partial \pi_{p_s}}{\partial \phi_s}\right)_{T,\boldsymbol{\mu},\theta_s}\left(\frac{\partial \phi_s}{\partial T}\right)_{\boldsymbol{\mu},\pi_{p_s},\pi_{a_s}}. \tag{6.74}$$

Under the zero-deformation limit, $\pi_p, \pi_a \to 0$, the above equation reduces to

$$\left(\frac{\partial \Delta s^s}{\partial \theta_s}\right)_{T,\mu,\phi_s=\phi_e} = W_p\left(\frac{\partial \theta_e}{\partial T}\right)_\mu + W_{pa}\left(\frac{\partial \phi_e}{\partial T}\right)_\mu. \tag{6.75}$$

The corresponding equation for adsorption is:

$$\left(\frac{\partial \Delta \Gamma_j}{\partial \theta_s}\right)_{T,\mu',\phi_s=\phi_e} = W_p\left(\frac{\partial \theta_e}{\partial \mu_j}\right)_{T,\mu'} + W_{pa}\left(\frac{\partial \phi_e}{\partial \mu_j}\right)_{T,\mu'} \tag{6.76}$$

Finally, using (6.73), we also obtain the second derivatives of the entropy at $\pi = 0$ as

$$\left(\frac{\partial^2 \Delta s^s}{\partial \theta_s^2}\right)_{T,\mu,\pi=0} = -\left(\frac{\partial W_p}{\partial T}\right)_\mu$$
$$\left(\frac{\partial^2 \Delta s^s}{\partial \phi_s \partial \theta_s}\right)_{T,\mu,\pi=0} = -\left(\frac{\partial W_{pa}}{\partial T}\right)_\mu \quad \text{etc.} \tag{6.77}$$

and similar equations for the adsorption.

Differentiating, at fixed temperature and chemical potentials, the expression for the surface energy,

$$\Delta u^s = T\Delta s^s + \sum_{j=2}^{m} \Delta\Gamma_j \mu_j + \gamma \tag{6.78}$$

and taking the limit $\pi \to 0$, we reach for example

$$\left(\frac{\partial \Delta u^s}{\partial \theta_s}\right)_{T,\mu,\pi=0} = -W_p\left[T\left(\frac{\partial \theta_e}{\partial T}\right)_\mu + \sum_{j=2}^{m}\left(\frac{\partial \theta_e}{\partial \mu_j}\right)_{T,\mu'}\right] - W_{pa}\left[T\left(\frac{\partial \phi_e}{\partial T}\right)_\mu + \sum_{j=2}^{m}\left(\frac{\partial \phi_e}{\partial \mu_j}\right)_{T,\mu'}\right]. \tag{6.79}$$

Similarly, we obtain

$$\left(\frac{\partial^2 \Delta u^s}{\partial \theta_s^2}\right)_{T,\mu,\pi=0} = W_p - T\left(\frac{\partial W_p}{\partial T}\right)_\mu - \sum_{j=2}^{m}\left(\frac{\partial W_p}{\partial \mu_j}\right)_{T,\mu'}. \tag{6.80}$$

This formula reveals that, unless the easy axis is fixed with regard to changes in temperature and chemical potential, the surface energy should be linearly dependent on the deviation of the boundary director

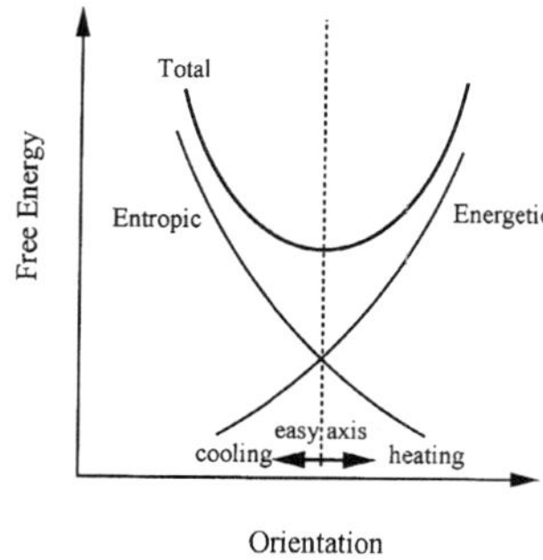

FIGURE 6.9. Thermodynamic origin of the stability of the easy axis. When the easy axis undergoes a temperature- or composition-induced variation, the surface energy and the surface entropy do not in themselves possess minima at the easy axis, but the competition between energy and entropy is responsible for achieving the thermodynamical stability.

(showing no minimum at the easy axis), which is exactly cancelled by the linear dependence of surface entropy and adsorption, resulting in the stability of the easy axis (see FIGURE 6.9). Put differently, in such systems that undergo a variation of the easy axis, neither the energy nor the entropy contribution alone can be the source of interfacial orientation; but the competition between these two entities is essential and gives rise to the variation itself.

Temperature-driven variations of the easy axis and anchoring energy, not necessarily associated with well-defined anchoring transitions, have been observed at a wide variety of nematic interfaces such as the nematic–air interface (Bouchiat and Langevin-Cruchon, 1971; Chiarelli et al., 1983, 1984), the solid–nematic interface (Yamashita and Amemiya, 1976; Toda et al., 1978; van Sprang and Aartsen, 1983b, 1984a; Faetti et al., 1985a; Yokoyama et al., 1987; Di Lisi et al., 1990). A typical example of the varying easy axis, shown in FIGURE 6.10, is in this respect a manifestation of the antagonistic origin of the alignment.

Concerning the second-order anchoring transitions, the above equations yield a set of inequalities among the relevant critical exponents. We can follow the procedure taken for the liquid–vapor transitions (Stanley, 1971; Baker, 1990), employing the following translation of variables: $\theta_s, \phi_s \to V$ and $\pi_{ps}, \pi_{as} \to P$. Let T_c and $\mathbf{n}_c = (\theta_c, \phi_c)$ be the temperature and the easy axis at the critical anchoring

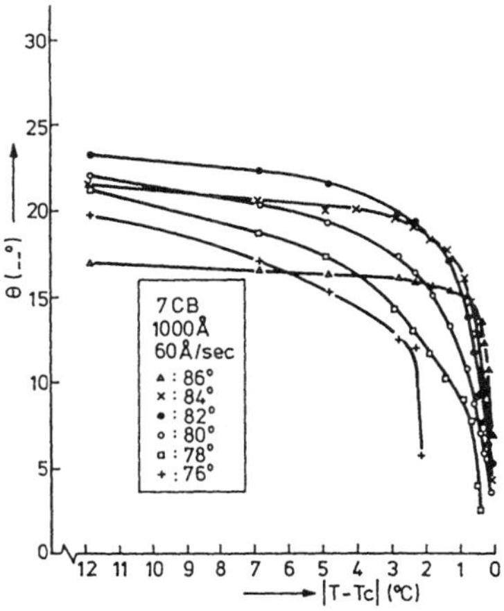

FIGURE 6.10. Temperature-driven variation of the easy axis at the obliquely vacuum-deposited SiO–nematic liquid crystal interface near the nematic–isotropic transition temperature for various deposition angles. Reproduced by permission from (van Sprang and Aartsen, 1984a).

transition, respectively. Then, we define the critical exponents by

$$C_\pi^s = T\left(\frac{\partial \Delta s^s}{\partial T}\right)_\pi \propto (T - T_c)^{-\alpha} \qquad T > T_c$$
$$\propto (T_c - T)^{-\alpha'} \qquad T < T_c \tag{6.81}$$

$$\mathbf{n}_e - \mathbf{n}_c \propto (T_c - T)^\beta \qquad T < T_c \tag{6.82}$$

$$W_p W_a - W_{pa}^2 \propto (T - T_c)^\gamma \qquad T > T_c$$
$$\propto (T_c - T)^{\gamma'} \qquad T < T_c \tag{6.83}$$

$$|\boldsymbol{\pi}_s| \propto |\mathbf{n}_e - \mathbf{n}_c|^\delta \qquad T = T_c \tag{6.84}$$

where C_π^s denotes the surface excess specific heat at a fixed orientational torque.

Combining the following identity

$$\left(\frac{\partial \Delta s^s}{\partial T}\right)_{\mu,\pi_s} - \left(\frac{\partial \Delta s^s}{\partial T}\right)_{\mu,n_s}$$
$$= \left(\frac{\partial \Delta s^s}{\partial \theta_s}\right)_{T,\mu,\phi_s} \left(\frac{\partial \theta_s}{\partial T}\right)_{\mu,\pi_s} \tag{6.85}$$
$$+ \left(\frac{\partial \Delta s^s}{\partial \phi_s}\right)_{T,\mu,\theta_s} \left(\frac{\partial \phi_s}{\partial T}\right)_{\mu,\pi_s}$$

with (6.74), we obtain in the limit $\pi_s \to 0$,

$$\left(\frac{\Delta C_\pi^s - \Delta C_{\mathbf{n}_s}^s}{T}\right)$$
$$= W_p \left(\frac{\partial \theta_e}{\partial T}\right)^2_{\mu,\pi_{ps},\pi_{as}} + 2W_{pa}\left(\frac{\partial \phi_e}{\partial T}\right)_{\mu,\pi_{ps},\pi_{as}} \tag{6.86}$$
$$\times \left(\frac{\partial \theta_e}{\partial T}\right)_{\mu,\pi_{ps},\pi_{as}} + W_a\left(\frac{\partial \phi_e}{\partial T}\right)^2_{\mu,\pi_{ps},\pi_{as}}.$$

Although the surface excess specific heat is not necessarily a positive quantity, the critical part of the surface excess must be so in order to assure the stability of the entire system, provided the bulk does not show a critical behavior at T_c. Taking account only of the critical part, therefore, we have $\Delta C_\pi^s \geq \Delta C_\pi^s - \Delta C_{n_s}^s$. Use of this inequality in equation (6.86) in combination with the definitions of the critical exponents leads to

$$\alpha' + 2\beta + \gamma' \geq 2 \tag{6.87}$$

which is the anchoring analog of the Rushbrooke inequality (Stanley, 1971; Baker, 1990). Similarly, we can derive the analogs of the Griffiths inequalities (Griffiths, 1965):

$$\beta(\delta + 1) \geq 2 - \alpha' \tag{6.88}$$
$$\gamma'(\delta + 1) \geq (2 - \alpha')(\delta - 1) \tag{6.89}$$

the latter of which rests on an assumption that $(\partial\pi/\partial\delta\psi)_{T,\mu}$ is a nondecreasing function of $\delta\psi$.

The exponents β and γ' have so far been experimentally determined only in two instances of a free surface (Chiarelli et al., 1983, 1984) and of an interface between a nematic and fluoropolymer layers (Patel and Yokoyama, 1993; Yokoyama and Patel, 1994). In both cases, the tilted orientation at low temperatures was continuously transformed into a homeotropic alignment; the exponents were found to agree with the mean-field predictions, $\beta = 1/2$ and $\gamma' = 1$. These are consistent with the Rushbrooke inequality for $\alpha' \geq 0$. Although the measurement of α' has not yet been carried out for any liquid crystal interfaces, the Griffiths inequalities demand for the above cases that $\delta \geq 3 - 2\alpha'$.

6.3.1.5 Remarks on Solid Substrates

So far in this section, we have not explicitly considered the solid substrates, despite their paramount practical significance. From the thermodynamic viewpoint, the solid surfaces do not allow unambiguous formulation due to their rigidity, long relaxation times, anisotropy, and inhomogeneity in structure and chemistry. Nevertheless, as far as the orientational states of flexible liquid crystal phases in contact with a solid are concerned, it is believed to be a good approximation to treat the bulk solid as perfectly rigid, and to ignore the change in thermodynamic states of the bulk solid even when orientational deformation occurs in the liquid crystalline phase. The rigidity keeps the interface planar.

The surface tension in which there is no contribution from the bulk of the solid substrate was designated as the boundary tension by Navascués and Berry (1977). For the boundary tension, we can repeat the same thermodynamic manipulations as presented here; the only formal difference is that the interface is now kept planar, so that by taking the limit $R \to \infty$, we can translate all the formulas to the case of the solid substrate. The sole reason we have adopted the surface of tension as the dividing surface through the preceding formulation is that the translational work on the interface vanishes with this choice. With a rigid substrate, however, the translational work is always absent regardless of the position of the dividing surface. Consequently, we are allowed to replace the surface of tension by an arbitrary dividing surface located at z_d, giving the Gibbs–Duhem equation for the boundary tension:

$$\mathrm{d}\gamma_B = -s^s\,\mathrm{d}T - \sum_{j=1}^{m} \Gamma_j\,\mathrm{d}\mu_j - \boldsymbol{\pi}_d \cdot \mathrm{d}\mathbf{n}_d + f_d\,\mathrm{d}z_d. \quad (6.90)$$

6.3.2 Thermodynamics of Thin Films

6.3.2.1 Kelvin Equation

Thermodynamic treatment of phase transitions in thin films dates back to the pioneering work of Lord Kelvin on capillary condensation at the end of last century. The liquid crystalline phase transitions in thin films can be treated in completely the same way (Poniewierski and Sluckin, 1987). Let us here imagine a thin film of single-component nematic liquid crystal of thickness d confined between a pair of identical substrates inducing no orientational deformations in the nematic phase (see FIGURE 6.11); the entire system is furthermore immersed in a large reservoir of molecules to fix the chemical potential. This type of film is often referred to as an open film (Casalnuovo et al., 1984) in that transfer of molecules between the film and the reservoir is allowed.

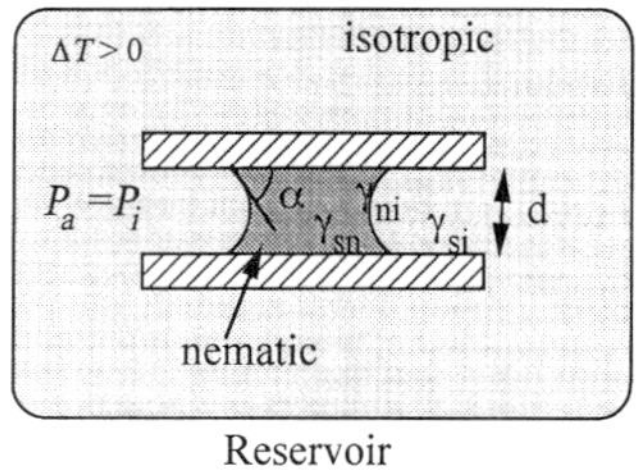

FIGURE 6.11. Nematic–isotropic coexistence in a thin film configuration. The thickness of the film is d and the contact angle of the nematic phase at the line of the nematic–isotropic-substrate three-phase contact is α.

At the nematic–isotropic transition of such an open film, the grand thermodynamic potentials of both phases should be equal to each other, that is, $\Omega_n = \Omega_i$. Including the interfacial tension, the condition of equilibrium is

$$-P_n V + \gamma_{sn} A = -P_i V + \gamma_{si} A \quad (6.91)$$

where γ_{sn} and γ_{si} denote the substrate–nematic and the substrate–isotropic interfacial tensions, respectively, and V and A are the volume and the surface area. This equation immediately shows that,

$$\Delta P = P_i - P_n = \frac{A}{V}(\gamma_{si} - \gamma_{sn}) \quad (6.92)$$

which indicates that the pressures in the coexisting nematic and isotropic phases are no longer generally equal in confined systems; this equation holds for any geometry, not only thin films but also cylinders and spheres, which have been the target of many recent studies. Introducing the contact angle α based on Young's equation (see FIGURE 6.11),

$$\gamma_{si} = \gamma_{sn} + \gamma_{ni} \cos\alpha \quad (6.93)$$

with γ_{ni} being the nematic–isotropic interfacial tension, we can rewrite (6.92) for a thin film with the thickness d as

$$\Delta P = 2\frac{\gamma_{ni} \cos\alpha}{d}. \quad (6.94)$$

Since $\mathrm{d}/(2\cos\alpha)$ is just the radius of curvature of the nematic–isotropic interface R_{ni}, this equation turns out to be of exactly the same form as the Laplace equation, (6.4), assuring the mechanical equilibrium of the interface; similar equations can be obtained for cylinders and spheres by replacing $2/d$ by $2/R$ and $3/R$, respectively, with R being the radius.

The nematic–isotropic transition in the bulk ($d \to \infty$) occurs at the temperature satisfying $\mu_n(T_{ni}, P_a) = \mu_i(T_{ni}, P_a)$, where P_a is the common ambient pressure. At a finite film thickness, the transition temperature, if it still exists, should shift from T_{ni} in such a way as to conform with the pressure equilibrium in (6.92). Taking the derivative of the condition of equilibria $\mu_n(T, P_n) = \mu_i(T, P_i)$ with the help of the Gibbs–Duhem equation for the bulk phase (6.65), we generally find

$$\left(\frac{s_i}{\rho_i} - \frac{s_n}{\rho_n}\right) \mathrm{d}T = \frac{1}{\rho_i}\,\mathrm{d}P_i - \frac{1}{\rho_n}\,\mathrm{d}P_n. \tag{6.95}$$

If the density change at the transition is negligibly small as in the nematic–isotropic transition, this equation can be reduced simply to

$$\mathrm{d}\Delta P = (s_i - s_n)\,\mathrm{d}T \tag{6.96}$$

where $\rho \approx \rho_i \approx \rho_n$. As $(s_i - s_n)T_{ni} \equiv L$ is the latent heat of the nematic–isotropic transition per unit volume, we can integrate the above equation to the temperature shift in question as

$$\Delta T_{ni} = 2T_{ni}\frac{\gamma_{si} - \gamma_{sn}}{dL} \tag{6.97}$$

$$= 2T_{ni}\frac{\gamma_{ni}\cos\alpha}{dL}. \tag{6.98}$$

This is the nematic version of the Kelvin equation (Poniewierski and Sluckin, 1987), and indicates that the temperature shift is either upward or downward depending on whether the contact angle of the nematic phase is smaller or larger than $\pi/2$ at the isotropic–substrate interface (cf. FIGURE 6.12). Using representative values for the latent heat and the interfacial tension, $L \sim 2 \times 10^6\,\mathrm{J\,m^{-3}}$ and $\gamma_{ni} \approx 2 \times 10^{-2}\,\mathrm{mN\,m^{-1}}$, in (6.98), the temperature shift is estimated to stay at most $10^{-2}\,\mathrm{K}$ even for a film of $1\,\mu$m thickness.

Experimental verifications of this and related effects in a confined geometry have been attempted by a number of authors (Kuzma and Labes, 1983; Iimura et al., 1984; Yokoyama, 1988a; Iannacchione and Finotello, 1992; Crawford et al., 1992; Bellini et al., 1992; Iannacchione et al., 1993a; Iannacchione and Finotello, 1993b; Iannacchione et al., 1993c; Tripathi et al., 1994). It was in fact demonstrated by Yokoyama (1988a) that the temperature shifts are of the correct order of magnitude as (6.98) predicts, and are correlated with the value of the contact angle, which is known to range from $\alpha = 0$ (perfect wetting by the nematic phase) to $\alpha = \pi$ (perfect wetting by the isotropic phase) in real systems (Yokoyama et al., 1983) for films of nematics with submicron thicknesses.

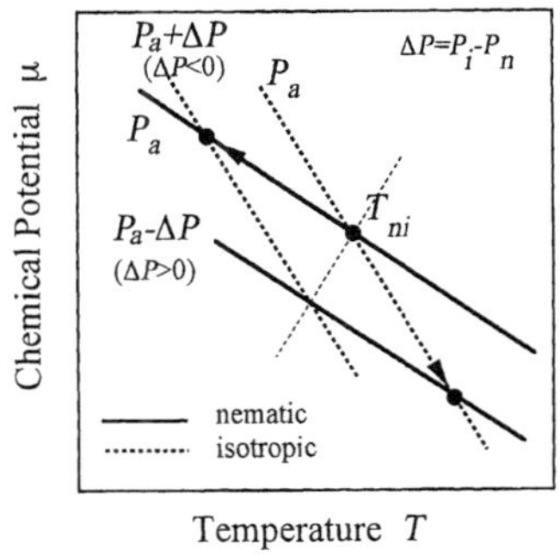

FIGURE 6.12. Shift of transition temperature in open thin films. When the pressure difference between the nematic and the isotropic phases in the thin film is negative, that is, $\Delta P = P_i - P_n \le 0$, the transition temperature is depressed along the bulk line of the nematic phase held at the ambient pressure P_a, while it is enhanced along the line of isotropic phase at P_a, for $\Delta P > 0$.

Detailed calorimetric studies by Iannacchione and Finotello (Iannacchione and Finotello, 1992; Iannacchione et al., 1993a) on nematics and smectics, confined in submicron-diameter cylindrical pores, have indicated a downward shift of the transition temperature and also a marked suppression of the specific heat (see FIGURE 6.13). Within the Kelvin regime, the shift of the transition temperature is essentially driven by the pressure difference caused by the surface tension. Therefore, the thermodynamics of the system is naturally expected to be very close to that of the bulk state. In accordance with the decomposition of the entropy, the heat capacity of the thin film consists of bulk and surface contributions,

$$C_t = c_b V + c_s A \tag{6.99}$$

where c_s is the surface specific heat defined by $c_s \equiv T\,\partial s^s/\partial T = -T\,\partial^2\gamma/\partial T^2$. In an open film configuration, a depression of the transition temperature occurs along the bulk equilibrium line of the nematic as shown in FIGURE 6.12. Consequently, the difference of specific heat in bulk and thin films should be regarded as coming only from the surface specific

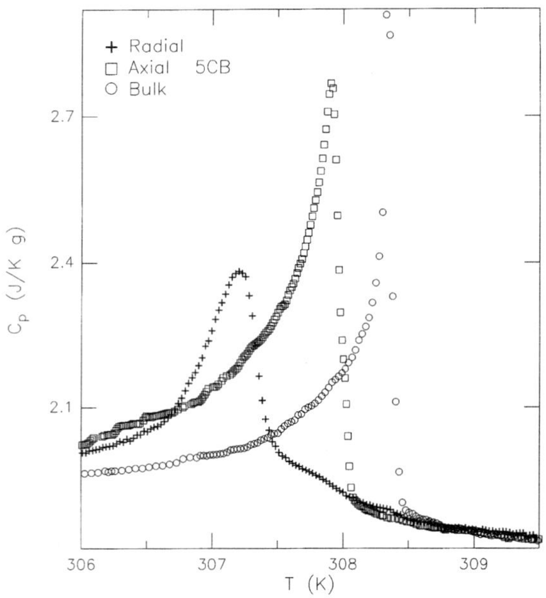

FIGURE 6.13. Specific heat as a function of temperature at the nematic–isotropic transition for 4-pentyl-4′-cyanobiphenyl (5CB) confined in 0.2 μm-diameter cylindrical pores in alumina matrix. The effect of director configurations, radial and axial, in the pore on the transition behavior has been examined. Reproduced by permission from (Iannacchione and Finotello, 1992).

heat with the dividing surface set at the zero-adsorption plane.

In a closed film configuration, which seems to apply in most of the calorimetric studies so far conducted on confined liquid crystal systems, the pressure in the mother phase is not necessarily fixed at the ambient pressure due to the absence of the reservoir. If we are allowed to imagine a situation as depicted in FIGURE 6.14, the pressure in the nematic mother phase is given by

$$P_n = P_a - 2\frac{\gamma_n \cos\alpha_n}{d} = P_a - 2\frac{\gamma_{sa} - \gamma_{sn}}{d} \tag{6.100}$$

where γ_n and γ_{sa} are the tension of the nematic–air interface and the substrate–air interface, respectively, and α_n is the contact angle of the nematic phase at the nematic–substrate–air three phase contact. When the nematic phase wets the substrate in air ($\alpha_n < \pi/2$), which is practically a condition that the cell can be impregnated with the liquid crystal, the pressure in the nematic becomes less than the ambient pressure. After

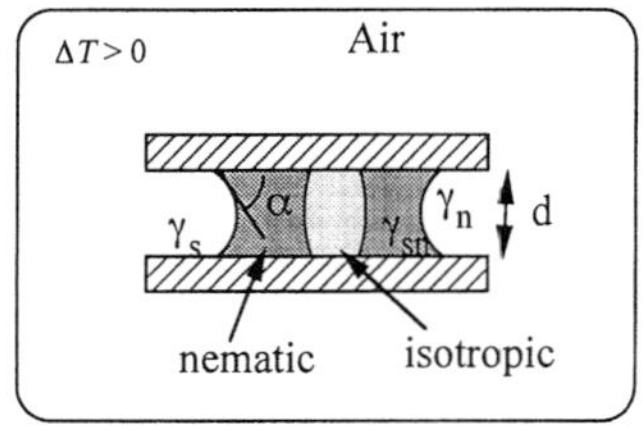

FIGURE 6.14. Nematic–isotropic transition in a closed film of thickness d.

transformation to the isotropic phase, the film exhibits a pressure given by

$$P_i = P_a - 2\frac{\gamma_i \cos\alpha_i}{d} = P_a - 2\frac{\gamma_{sa} - \gamma_{si}}{d}. \tag{6.101}$$

Naturally, the pressure *difference* between the two phases is the same as in the open film (cf. equation (6.94)). This time, however, the variation of the pressure from the ambient value causes a shift of the bulk transition point itself according to the Clausius–Clapeyron equation:

$$\Delta T_{ni} = -2T_{ni}\frac{\rho_n - \rho_i}{\rho_n}\frac{\gamma_n \cos\alpha_n}{dL}. \tag{6.102}$$

This gives, for the situation in FIGURE 6.14, a negative shift quantitatively comparable to that in an open film. Since the pressure in a closed film is difficult to control in reality, the interpretation of the temperature shift in a closed film entails a careful consideration.

6.3.2.2 Disjoining Pressure and Phase Transitions in Thin Films

In writing down the grand thermodynamic potential in equation (6.91), we have tacitly assumed that the film thickness is still sufficiently large that the interaction between the two opposing interfaces may be neglected. A convenient thermodynamic concept to account generally for this interaction is the disjoining pressure (Derjaguin et al., 1954; Perez et al., 1988; Poniewierski and Sluckin, 1987) defined by

$$\Pi = -\frac{1}{A}\left(\frac{\partial \Omega_t}{\partial d}\right)_{T,\mu} = -\frac{1}{A}\left(\frac{\partial \Omega_f}{\partial d}\right)_{T,\mu} - P_0 \tag{6.103}$$

where Ω_t is the grand thermodynamic potential of the total system comprised of the film and the reservoir, both of which are in an identical phase, and P_0 represents the pressure in the reservoir. Then, it follows that the disjoining pressure takes on the meaning of an extra pressure acting on the substrate from inside in the open film configuration.

Including the disjoining pressure, Poniewierski and Sluckin (1987) introduced an effective surface tension for thin films held between identical interfaces by

$$2\tilde{\gamma} = \frac{1}{A}\left(\Omega_f - d\frac{\partial \Omega_f}{\partial d}\right) = \frac{\Omega_f}{A} + d(\Pi + P_0). \qquad (6.104)$$

When the interfaces are not identical, the left-hand side of this definition should be replaced by the sum of the surface tensions $\tilde{\gamma}_1 + \tilde{\gamma}_2$. Since $\Pi \to 0$ as $d \to \infty$, it should be obvious here that $\tilde{\gamma}$ converges to the ordinary surface tension in the infinite thickness limit.

Combining (6.103) and (6.104), we obtain

$$\frac{\Omega_f}{A} = -(P_0 + \Pi)d + 2\tilde{\gamma}. \qquad (6.105)$$

The condition of nematic–isotropic coexistence is now written as

$$P_i - P_n = -\Pi_i + \Pi_n + 2\frac{\tilde{\gamma}_{si} - \tilde{\gamma}_{sn}}{d} \equiv \Delta P \qquad (6.106)$$

where P_n and P_i are the bulk pressures of the nematic and isotropic phases at $T_{ni}(d)$ and $\mu_{ni}(d)$. Taking the total differential of this equation along the transition line on the T–d plane with a fixed ambient pressure P_a, we get

$$\left(\frac{\partial T_{ni}}{\partial d}\right)_{P_a} = -2T_{ni}(d)\frac{\tilde{\gamma}_{si} - \tilde{\gamma}_{sn}}{d^2\tilde{L}} \qquad (6.107)$$

where $\tilde{L}$ is the latent heat per unit volume of the thin film at constant thickness and ambient pressure given by

$$\begin{aligned}\frac{\tilde{L}}{T_{ni}(d)} &= \rho_{res}(s_i - s_n) + \frac{\partial(\Pi_i - \Pi_n)}{\partial T} \\ &\quad - \frac{2}{d}\frac{\partial(\tilde{\gamma}_{si} - \tilde{\gamma}_{sn})}{\partial T} \qquad (6.108)\\ &= -\frac{1}{dA}\left(\frac{\partial \Omega_{f_i}}{\partial T}\right)_{d,P_a} + \frac{1}{dA}\left(\frac{\partial \Omega_{f_n}}{\partial T}\right)_{d,P_a}. \qquad (6.109)\end{aligned}$$

This is a generalized Kelvin equation, expressed in the rigorous Clausius–Clapeyron form.

Unlike the original Kelvin equation, this generalized equation applies to orientationally deformed as well as undeformed nematic films via the use of the disjoining pressure. As Perez and coworkers (Perez et al., 1978, 1988) pointed out, even at a macroscopic thickness before the effects of van der Waals and other intermolecular forces become important, the curvature elastic deformation yields a substantial disjoining pressure. On account of the hydrostatic equation of state, equation (6.3), the elastic disjoining pressure, namely the excess pressure, is simply given by

$$\Pi_d = f_d. \qquad (6.110)$$

Based on the identity

$$f_d = \frac{1}{d}\int_{-d/2}^{d/2} f_d \, \mathrm{d}z$$

the thickness derivative of the disjoining pressure is shown to satisfy

$$\begin{aligned}\left(\frac{\partial \ln \Pi_d}{\partial d}\right)_{T,\mu} &= -2\frac{1}{d+d_e}\left[1 + \frac{\partial d_e}{\partial d}\right] \\ &\quad + \frac{1}{d+d_e}\frac{\partial \beta}{\partial d}\left[\frac{1}{f_d}\int_0^{f_d}\frac{\partial d_e}{\partial \beta}\,\mathrm{d}f_d\right]\end{aligned} \qquad (6.111)$$

where d_e is the sum of the extrapolation lengths associated with the upper and the lower interfaces, considered to be a function of the deformation, and β is the parameter, previously defined by (6.61), to specify the mode of deformation in the nematic layer. In the absence of the last term on the right hand side, which corresponds to the cases in which the thickness changes do not modify the deformation mode, or the extrapolation length is isotropic, this equation can be integrated to give

$$\Pi_d = \frac{C}{(d + d_e)^2} \qquad (6.112)$$

where C is the constant of integration, readily determined by taking the $d \to \infty$ limit in the expression of the Frank elastic energy. The disjoining pressure increases as $1/d^2$ at large film thicknesses, yet it saturates at $d \approx d_e$. This behavior was experimentally verified by Horn and coworkers (Horn et al., 1981) as shown in FIGURE 6.15. However, in a special case as illustrated in FIGURE 6.16 in which the optimum orientation for the interface with higher anchoring strength coincides with the orientation giving the local maximum of the interfacial tension of the other, it can be shown that $d_e \to \infty$ as $d \to 0$. Hence, the disjoining pressure begins to decrease when the thickness becomes less than a certain threshold value, which is roughly equal to the extrapolation length of the interface with weaker anchoring.

Noting further that $\Omega_f/A = -(P_0 - f_d)d + \gamma_1 + \gamma_2$ from (6.22) and (6.23), we can express the pressure

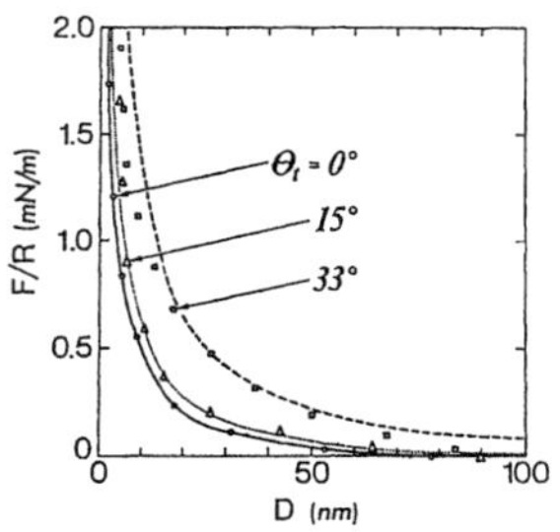

FIGURE 6.15. Curvature–elastic contribution to the disjoining pressure for 4-pentyl-4′-cyanobiphenyl (5CB) held between two mica sheets as a function of the twist angle. The deviation from zero-twist line (solid line) gives the elastic part, showing good agreement with the theoretical curve (broken lines). Reproduced by permission from (Horn, 1981).

difference between the nematic and isotropic reservoirs, respectively in equilibrium with the coexisting nematic and isotropic phases in the thin film, as

$$P_{0_i} - P_{0_n} = -\Pi_d + \frac{\gamma_{i_1} - \gamma_{n_1}}{d} + \frac{\gamma_{i_2} - \gamma_{n_2}}{d}. \tag{6.113}$$

Following the argument leading to the generalized Kelvin equation, we find the elastic shift in the transition temperature as

$$\Delta T_{ni} = \frac{T_{ni}}{L}\left(-\Pi_d + \frac{\gamma_{i_1} - \gamma_{n_1}}{d} + \frac{\gamma_{i_2} - \gamma_{n_2}}{d}\right). \tag{6.114}$$

Clearly, the elastic disjoining pressure leads to a negative shift in transition temperature. For a hybrid-aligned nematic film of $d = 1\,\mu$m, we see that $\Pi_d \sim K/d^2$ is on the order of $1\,\mathrm{N/m^{-2}}$, well below the interfacial part; but, due to its $1/d^2$ dependence, it could overwhelm the latter for thinner samples.

FIGURE 6.16. Balance of torque in a sandwich cell with interfaces of different anchoring strength. Interface I has a larger anchoring strength than Interface II. The easy axis is at $\theta_{e1} = \pi/2$ at Interface I and at $\theta_{e2} = 0$ at Interface II, yet $\theta_{s2} = \pi/2$ is an unstable stationary orientation for Interface II. As the thickness of the layer is decreased, θ_{s2} moves from 0 to $\pi/2$, whereas θ_{s2} starts from $\pi/2$, decreases, and returns finally to $\pi/2$ where no deformation persists.

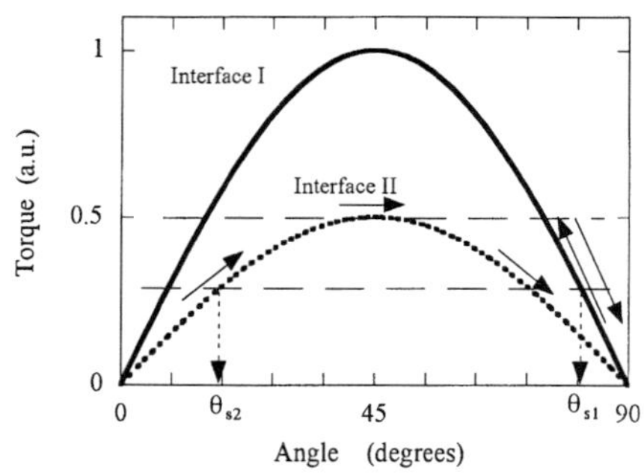

6.3.3 Creagh–Kmetz Condition: Thermodynamics of Surface Alignment

Creagh and Kmetz (1973) argued, in the course of a study aimed at comparing the roles of topography and surface contamination, that if the surface tension of a nematic liquid crystal–vapor interface, γ_n, is larger than the critical surface tension (Adamson, 1976) of the substrate, γ_c, then a homeotropic alignment prevails when the nematic is put into contact with the substrate; whereas, if γ_n is less than γ_c, then a planar alignment should occur. Their rationale for this hypothesis derives from the following: first, many nematics are known to assume homeotropic alignment at the free surface; and second, the surface tension may be regarded as a rough measure of the strength of intermolecular interaction (the anisotropic part as well), so that a substrate with a larger surface tension is expected to exert a stronger (possibly anisotropic) force on the neighboring liquid crystal molecules, thereby making them rotate toward the planar orientation.

Later, Porte (1976) put the argument on an analytical basis, taking account of the work of adhesion, $W(\theta_s)$, given as a function of the surface tilt angle:

$$\gamma_{sn}(\theta_s) = \gamma_s + \gamma_n(\theta_s) - W(\theta_s) \tag{6.115}$$

where γ_s and γ_{sn} are the interfacial tensions of the substrate–air and the substrate–nematic interfaces, respectively. If we employ the Girifalco–Good–Fowkes relation, which is known to yield a rather good estimate of the work of adhesion in terms of the surface tensions of the individual surfaces to be put into contact (Adamson, 1976), we can rewrite the above equation as

$$\begin{aligned}\gamma_{sn}(\theta_s) &= \gamma_s + \gamma_n(\theta_s) - 2\sqrt{\gamma_s\gamma_n(\theta_s)} \\ &= (\sqrt{\gamma_s} - \sqrt{\gamma_n(\theta_s)})^2.\end{aligned} \tag{6.116}$$

The fact that γ_{sn} vanishes when $\gamma_s = \gamma_n$ clearly indicates the nature of the Girifalco–Good–Fowkes relation as resulting from an assumption to conceptually equate the *surface energy* with the surface tension. Naturally, therefore, the formula is expected to work well when the entropy contribution to the surface tension may be neglected or at least is invariant before and after the contacting process, which unfortunately would not necessarily be the case for liquid crystal interfaces. We will return to this point after the experimental situation of the condition has been reviewed.

On the basis of equation (6.116), the easy axis at the substrate–nematic interface satisfies

$$\frac{\partial \gamma_{sn}}{\partial \theta_s} = \left(1 - \sqrt{\frac{\gamma_s}{\gamma_n(\theta_e)}}\right) \frac{\partial \gamma_n}{\partial \theta_s} = 0 \tag{6.117}$$

and, the stability of the easy axis (c.f. equation (6.56)), reads

$$\frac{\partial^2 \gamma_{sn}}{\partial \theta_s^2} = \left(1 - \sqrt{\frac{\gamma_s}{\gamma_n(\theta_e)}}\right) \frac{\partial^2 \gamma_n}{\partial \theta_s^2} + \frac{1}{2\gamma_n(\theta_e)} \sqrt{\frac{\gamma_s}{\gamma_n(\theta_e)}} \left(\frac{\partial \gamma_n}{\partial \theta_s}\right)^2 > 0. \tag{6.118}$$

Combining (6.117) and (6.118), we find

(I) $\mathrm{Min}(\gamma_n) \geq \gamma_s$
$\longrightarrow \theta_e = \theta_{min}$ where $(\gamma_n(\theta_{min}) = \mathrm{Min}(\gamma_n))$

(II) $\mathrm{Min}(\gamma_n) \geq \gamma_s \geq \mathrm{Max}(\gamma_n)$
$\longrightarrow \theta_e = \theta_{int}$ where $(\gamma_n(\theta_{int}) = \gamma_s)$

(III) $\mathrm{Max}(\gamma_n) \leq \gamma_s$
$\longrightarrow \theta_e = \theta_{max}$ where $(\gamma_n(\theta_{min}) = \mathrm{Max}(\gamma_n))$,

where $\mathrm{Min}(\gamma_n)$ and $\mathrm{Max}(\gamma_n)$ indicate the minimum and the maximum values of the nematic–air interfacial tension for all possible orientations. Although the content of Porte's treatment is essentially more general (Warenghem, 1982), it agrees with the Creagh–Kmetz condition, provided that the easy axis at the nematic–air interface is homeotropic and that $\gamma_s = \gamma_c$.

Given in familiar surface-scientific terms, the Creagh–Kmetz condition immediately became the subject of extensive experimental tests (Kahn et al., 1973; Haller, 1974; Dubois et al., 1976; Hiltrop and Stegemeyer, 1978; Uchida et al., 1980a; Ohgawara and Uchida, 1981b; Mada, 1981a; Naemura, 1980, 1981; Uchida and Seki, 1992). Despite the early experiments supporting the Creagh–Kmetz condition (Kahn et al., 1973; Dubois et al., 1976), subsequent scrupulous experiments (Haller, 1974; Hiltrop and Stegemeyer, 1978; Uchida et al., 1980a) revealed quite a few cases in which the condition is clearly violated.

As is immediately clear in (6.116), Porte's treatment derives the orientational anisotropy of the substrate–liquid crystal interface from that of the surface tension of the liquid crystal alone, neglecting the direct contribution from the substrate. Specifically, therefore, it follows from Porte's theory that, when γ_n does not depend on the orientation (i.e. vanishing anchoring energy), nor does γ_{sn}, in apparent contradiction with our intuition. It is possible to look into this point a little further by defining the orientation-dependent disjoining pressure $\Pi_{sn}(h, \theta_s)$ on the nematic liquid separated by a distance h from the substrate (see FIGURE 6.17). Clearly, the work of adhesion, used in Porte's argument, is connected to $\Pi_{sn}(h)$ by

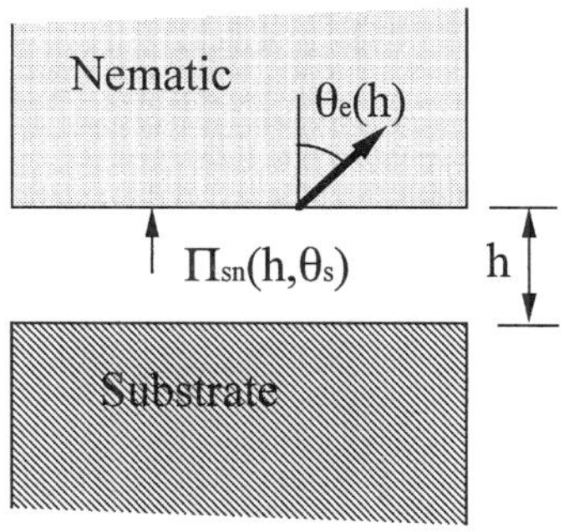

FIGURE 6.17. Definition of disjoining pressure between a substrate and a nematic liquid crystal when they are separated by a distance h. The equilibrium surface orientation is given by $\theta_e(h)$.

$$W(\theta_s) = -\int_0^\infty \Pi_{sn}(h, \theta_s)\, dh. \tag{6.119}$$

By differentiating the orientational equilibrium condition, (6.57), by h, in view of the definitions of the anchoring energy and the disjoining pressure, we obtain the following thermodynamic relation,

$$\frac{d\theta_e(h)}{dh} = -\frac{1}{W_p(h)} \left(\frac{\partial \Pi_{sn}(h, \theta_s)}{\partial \theta_s}\right)_{\theta_s = \theta_e(h)}. \tag{6.120}$$

Integration of this equation from infinite separation to direct contact ($h = 0$) yields

$$\theta_e|_{solid-nematic} = \theta_e|_{nematic} + \int_0^\infty \frac{1}{W_p(h)} \times \left(\frac{\partial \Pi_{sn}(h, \theta_s)}{\partial \theta_s}\right)_{\theta_s = \theta_e(h)} dh. \tag{6.121}$$

This formula shows that the anisotropic part of the disjoining pressure at *all* separations should be responsible for the final orientation, and the weaker the anchoring, the more pronounced the effect should be. In particular, therefore, when interfacial structures develop in the transition region such as those probed by the surface force measurements (Horn et al., 1981; Idziak et al., 1994), the surface orientation is expected to be more or less influenced by the anisotropic nature of these structures (see FIGURES 6.1 and 6.2). Indifference to this role of the transition region is the most serious shortcoming of the Creagh–Kmetz condition, and leads to the failure of its prediction.

A slight generalization of the Creagh–Kmetz condition has been attempted by extending Porte's argument to give a separate account of the contributions from different interaction sources such as polar (P), hydrogen bonding (HB), etc., along with the van der Waals (VdW) interactions (Perez et al., 1977; Perez and Proust, 1979; Mada, 1981a; Naemura, 1980, 1981). They replace, in one way or another, the work of adhesion by a sum of separate contributions as

$$W = 2\sqrt{\gamma_s(\mathrm{VdW})\gamma_n(\mathrm{VdW})} + 2\sqrt{\gamma_s(\mathrm{P})\gamma_n(\mathrm{P})} + 2\sqrt{\gamma_s(\mathrm{HB})\gamma_n(\mathrm{HB})} + \ldots \tag{6.122}$$

on an assumption of the additivity of each contribution in surface tension, for example, $\gamma_s = \gamma_s(VdW) + \gamma_s(P) + \gamma_s(HB) + \ldots$. Although these treatments, involving more parameters, have, as expected, been more successful in explaining the surface alignment, they still inherit the fundamental limitations of the original Creagh–Kmetz condition due to the neglect of the anisotropic contribution from the substrate and of the interfacial transition region, which cannot be properly taken into account through the individual surface tensions of the substrate and the nematic liquid crystal alone.

6.4 STATISTICAL MECHANICS AND PHENOMENOLOGICAL APPROACHES

Statistical mechanics allows us to explore those subtleties of the substrate–liquid crystal interface from a microscopic standpoint, and has been the target of continued investigations in the last two decades. Just as in their bulk counterparts, they rarely go beyond the mean-field level (exceptions are Chakravarty and Woo, 1975a, b); nevertheless, they proved to be quite useful in providing qualitative and semiquantitative understanding of the phenomena, as far as the orientational properties are concerned. They can be classified basically into two categories, molecular statistical mechanical and phenomenological, depending on whether it takes a direct account of the molecular nature or begins with a more coarse-grained view of the system. Although the phenomenological approach must be ultimately justified on statistical mechanical grounds, it can be particularly powerful to study relationships among mutually related phenomena based on a small number of phenomenological parameters. As thoroughly demonstrated by the marvellous success of the Landau–de Gennes theory, based on a properly defined order parameter, the phenomenological approach has provided us with a universal basis on which a variety of orientational interfaces can be tackled. It is, however, only on the molecular statistical mechanical basis that we can address such issues as the relative importance of repulsive and attractive intermolecular interactions in surface phenomena.

6.4.1 General Framework: Density Functional Theory

According to the density functional theory (Evans, 1979), the equilibrium density profile of a nonuniform liquid can be determined as the one that minimizes the grand thermodynamic potential Ω, based on the theorem that, as a functional of the single-particle distribution function $\rho(\mathbf{r}, \boldsymbol{\omega})$ at the point $\mathbf{r}$ and the orientation $\boldsymbol{\omega}$, the grand thermodynamic potential satisfies a variational inequality, $\Omega[\rho_{eq}] \leq \Omega[\rho]$, under fixed temperature and chemical potential with ρ_{eq} being the equilibrium density. Specifically, the grand thermodynamic potential functional has the following general form:

$$\Omega(\rho) = F_{id}(\rho) + F_{ex}(\rho) - \mu \int \rho(\mathbf{r}, \boldsymbol{\omega})\, d\mathbf{r}\, d\boldsymbol{\omega} + \int \rho(\mathbf{r}, \boldsymbol{\omega}) V_{ext}(\mathbf{r}, \boldsymbol{\omega})\, d\mathbf{r}\, d\boldsymbol{\omega} \tag{6.123}$$

where $V_{ext}(\mathbf{r}, \boldsymbol{\omega})$ is the single-particle external potential like the one that acts between a liquid crystal molecule and a neighboring substrate, $F_{id}(\rho)$ is the Helmholtz free energy of an ideal (noninteracting) gas given by

$$F_{id}(\rho) = k_B T \int \rho(\mathbf{r}, \boldsymbol{\omega})(\ln(\lambda^3 \rho(\mathbf{r}, \boldsymbol{\omega})) - 1)\, d\mathbf{r}\, d\boldsymbol{\omega} \tag{6.124}$$

with λ being the thermal wavelength, and $F_{ex}(\rho)$ is the excess part of the free energy due to the interparticle interactions.

Once $F_{ex}(\rho)$ is given, the equilibrium density is found from the condition $\delta\Omega/\delta\rho = 0$, which leads to the formal solution for the density:

$$\rho_0(\mathbf{r},\omega) = z\exp\left(-\frac{1}{k_BT}V_{ext}(\mathbf{r},\omega) + c(\rho_0 : \mathbf{r},\omega)\right) \quad (6.125)$$

where $z \equiv \lambda^{-3}\exp(\mu/k_BT)$ is the fugacity, and $c(\rho_0 : \mathbf{r},\omega) \equiv (k_BT)^{-1}\delta F_{ex}(\rho)/\delta\rho(\mathbf{r},\omega)$ denotes the effective single-particle potential.

In the mean-field approximation, the excess free energy is given by

$$F_{ex} = -\frac{k_BT}{2}\int f_2(\mathbf{r}_{12},\omega_1,\omega_2)\rho(\mathbf{r}_1,\omega_1)\rho(\mathbf{r}_2,\omega_2) \times d\mathbf{r}_1\, d\omega_1\, d\mathbf{r}_2\, d\omega_2 \quad (6.126)$$

where

$$f_2(\mathbf{r}_{12},\omega_1,\omega_2) = \exp\left[-\frac{1}{k_BT}\phi(\mathbf{r}_{12},\omega_1,\omega_2)\right] - 1 \quad (6.127)$$

is the Mayer function with ϕ being the intermolecular potential (Poniewierski and Holyst, 1988). Using this expression in (6.125), we obtain

$$\rho_0(\mathbf{r}_1,\omega_1) = z\exp\left[-\frac{1}{k_BT}V_{ext}(\mathbf{r}_1,\omega_1) + \int f_2(\mathbf{r}_{12},\omega_1,\omega_2) \times \rho(\mathbf{r}_1,\omega_1)\rho(\mathbf{r}_2,\omega_2)\, d\mathbf{r}_2\, d\omega_2\right]. \quad (6.128)$$

6.4.2 Perturbation Methods

6.4.2.1 The Kirkwood–Buff Formula

As we are mainly interested in orientational properties of the liquid crystal interfaces, it is in most cases fruitful to consider the reality as resulting from a small anisotropic perturbation imposed on a suitably chosen reference system (Evans, 1979; Rowlinson and Widom, 1982; Gray and Gubbins, 1984). Such an anisotropic perturbation may be invoked in the intermolecular interaction as well as in the single-particle external potential via

$$\phi(\mathbf{r}_{12},\omega_1,\omega_2) = \phi_r(\mathbf{r}_{12},\omega_1,\omega_2) + \delta\phi(\mathbf{r}_{12},\omega_1,\omega_2)$$
$$V_{ext}(\mathbf{r}_1,\omega_1,\mathbf{r}_2,\omega_2) = V_{ext}(\mathbf{r}_1,\mathbf{r}_2) + \delta V_{ext}(\mathbf{r}_1,\omega_1,\mathbf{r}_2,\omega_2).$$

In response to these perturbations, the grand thermodynamic potential Ω, given as a functional of the single-particle density $\rho(\mathbf{r},\omega)$, undergoes a variation written as

$$\delta\Omega(\rho) = \int \rho(\mathbf{r},\omega)\delta V_{ext}(\mathbf{r},\omega)\, d\mathbf{r}\, d\omega + \frac{1}{2}\int \rho_2(\mathbf{r}_1,\omega_1,\mathbf{r}_2,\omega_2) \times \delta\phi(\mathbf{r}_1,\omega_1,\mathbf{r}_2,\omega_2)\, d\mathbf{r}_1\, d\omega_1\, d\mathbf{r}_2\, d\omega \quad (6.129)$$

where $\rho_2(\mathbf{r}_1,\omega_1,\mathbf{r}_2,\omega_2)$ denotes the two-particle distribution function.

An immediate application of this result yields the Kirkwood–Buff formula for the surface tension. For this purpose, we consider a rectangular volume circumventing the interface, and then normalize the coordinates by the length of each side of the volume in question, for example $\bar{x} = x/L_x$, in the original expression of the grand potential. We now imagine an infinitesimal change in the length as $L_x \to L_x(1+\xi)$, $L_y \to L_y$ and $L_z \to L_z(1+\xi)^{-1}$, which changes the surface area by $A(= L_xL_y) \to A(1+\xi)$, with the volume, $V = L_xL_yL_z$, unchanged. With this notional perturbation, the external and the intermolecular potentials undergo variations as

$$\delta V_{ext} = \xi\left(x\frac{\partial V_{ext}}{\partial x} - z\frac{\partial V_{ext}}{\partial z}\right) \quad (6.130)$$

$$\delta\phi = \xi\left(x_{12}\frac{\partial\phi}{\partial x_{12}} - z_{12}\frac{\partial\phi}{\partial z_{12}}\right) \quad (6.131)$$

to first order in ξ. Substituting these equations into (6.129) and comparing with $\delta\Omega = \gamma\delta A$, we obtain

$$\gamma = \int \rho(\mathbf{r},\omega)\left(x\frac{\partial V_{ext}}{\partial x} - z\frac{\partial V_{ext}}{\partial z}\right)A^{-1}\, d\mathbf{r}\, d\omega + \frac{1}{2}\int \rho_2(\mathbf{r}_1,\omega_1,\mathbf{r}_2,\omega_2) \times \left(x_{12}\frac{\partial\phi}{\partial x_{12}} - z_{12}\frac{\partial\phi}{\partial z_{12}}\right)A^{-1}\, d\mathbf{r}_1\, d\omega_1\, d\mathbf{r}_2\, d\omega. \quad (6.132)$$

This is the extended Kirkwood–Buff formula (Navascués and Berry, 1977), incorporating the effect of an external single-particle potential. This result remains valid for an interface with a rigid wall, provided the surface tension on the left-hand side is understood as the boundary tension defined in section 6.3.1.5.

6.4.2.2 Expansion in Terms of Density

We have already seen that a perturbation in the external potential results in a change in the single-particle density as indicated in equation (6.1). The concomitant variation of the free energy, given by the first term of (6.129), is equal to the work done by the external potential. After subtracting the free energy stored in the external potential, we are left with the intrinsic (positive) increase in the grand thermodynamic potential of the system, which can be written as follows (Evans, 1979):

$$\delta\Omega[\rho] = \frac{k_B T}{2} \int \left(\frac{\delta(\mathbf{r}_1 - \mathbf{r}_2)\delta(\omega_1 - \omega_2)}{\rho(\mathbf{r}_1, \omega_1)} - c[\rho : \mathbf{r}_1, \omega_1, \mathbf{r}_2, \omega_2] \right) \times \delta\rho(\mathbf{r}_1, \omega_1)\delta\rho(\mathbf{r}_2, \omega_2)\, d\mathbf{r}_1\, d\omega_1\, d\mathbf{r}_2\, d\omega_2 \tag{6.133}$$

where $c[\rho : \mathbf{r}_1, \omega_1, \mathbf{r}_2, \omega_2]$ is the Ornstein–Zernike direct correlation function defined by

$$c[\rho : \mathbf{r}_1, \omega_1, \mathbf{r}_2, \omega_2] = -\frac{1}{k_B T} \frac{\delta^2 F_{ex}(\rho)}{\delta\rho(\mathbf{r}_1, \omega_1)\delta\rho(\mathbf{r}_2, \omega_2)}. \tag{6.134}$$

In the mean field approximation, it follows from (6.126) and (6.127) that the direct correlation function is equal to the Mayer function, which is always of short range.

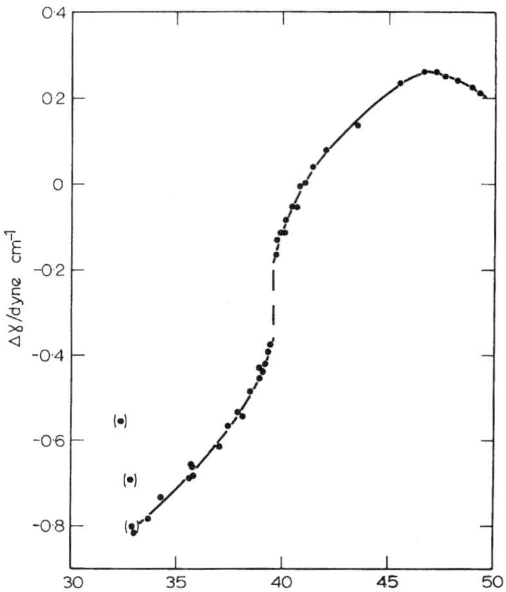

FIGURE 6.18. Surface tension of 8CB relative to its value (26.2 mN m^{-1}) at 41 °C. (Reproduced by permission from M. G. J. Gannon and T. E. Faber, "The Surface Tension of Nematic Liquid Crystals," *Philosophical Magazine A*, vol. 37, 1978).

6.4.2.3 Application to the Free Surface

The free surface of nematic liquids is the earliest case of nematic interfaces that was subjected to quantitative experimental studies (Gray, 1962). As shown in FIGURE 6.18, the orientational aspects of the interface are manifested in surface tension as an anomalous increase with temperature and a jump near and at the nematic–isotropic transition point, though the absolute value of the surface tension is comparable to that of ordinary liquids. This behavior was first observed by Ferguson and Kennedy (1938), yet soon followed by a conflicting observation (Naggiar, 1943). Owing to more recent careful experiments (Krishnaswamy and Shashidhar, 1976; Gannon, 1978), however, it is now believed that, as Ferguson and Kennedy first indicated, the surface tension of nematics does show, in many cases, an increasing behavior near the transition point, in contrast to experience with ordinary liquids.

For a single-component nematic liquid crystal, it follows from the Gibbs–Duhem equation (6.29) that, in the absence of orientational deformations, the slope of the surface tension γ_0 is connected to the surface entropy s_0^s by

$$\frac{d\gamma_0}{dT} = -s_0^s \tag{6.135}$$

where the dividing surface has been taken at the surface of zero adsorption. When the surface is subject to an orientational torque, however, this well-known formula must be generalized to

$$\frac{d\gamma^{(0)}}{dT} = -s_0^s + \frac{d}{dT}(d_e^{(0)} + \Delta)f_d \tag{6.136}$$

up to the lowest order in the Frank energy density f_d, by combining (6.32) and (6.39) with (6.135). Here, $\gamma^{(0)}$ is the surface tension with respect to the surface of zero adsorption, and $d_e^{(0)}$ and Δ denote the extrapolation length and the distance of the surface of tension measured from the dividing surface. Only when the experiment has been performed so that the deformation effect is negligible, can the positive slope of the surface tension be equivalent to a negative surface entropy, which is indicative of the fact that the surface region is more *ordered*, in one way or another, compared with the bulk.

The jump of the surface tension at the nematic–isotropic transition is also subject to a fundamental thermodynamic condition that

$$-\gamma_{ni} \leq \gamma_n - \gamma_i \leq \gamma_{ni}. \tag{6.137}$$

This condition, due basically to Gibbs, indicates that the absolute value of the jump should never exceed the nematic–isotropic interfacial tension γ_{ni}. Let us imagine a situation at the transition temperature wherein the air–nematic interface is separated by a macroscopically thick layer of isotropic phase; then we obtain $\gamma_n = \gamma_i + \gamma_{ni}$, the equality referred to as Antonow's rule. On the other hand, from the air–isotropic interface intervened by a nematic layer, we have $\gamma_i = \gamma_n + \gamma_{ni}$. The real interface should always have a lower surface tension from those physically possible yet hypothetical interfaces; this leads to the above inequality. The jump of surface tension observed by Gannon and Faber is on the order of $0.1\,\mathrm{mN\,m^{-1}}$ for cyanobiphenyls. However, the nematic–isotropic interfacial tensions for the same materials (Faetti and Palleschi, 1984b; Yokoyama et al., 1985b) are experimentally known to be of the order of $0.01\,\mathrm{mN\,m^{-1}}$, clearly violating the above inequality. Such an overevaluation of the jump in surface tension is seen in some cases (Krishnaswamy and Shashidhar, 1976), indicating experimental error caused by, say, insufficient equilibration and/or temperature control, etc., in these experiments.

The statistical mechanical treatment of a nematic free surface has been attempted via three different routes, based either on the Kirkwood–Buff formula of surface tension (Parsons, 1976; Murakami, 1977; Croxton, 1980; Parsons, 1980), or on the van der Waals picture of the interface (Parsons, 1975; Gannon, 1978); for the former approach, see the next section. Under both of these approaches lies the fundamental conviction that the density is of primary importance in determining the structure of the interface, and the orientational degree of freedom may be treated as a perturbation.

In applying the Kirkwood–Buff formula, one needs to know the pair distribution function in the interfacial region, about which little is known either theoretically or experimentally. To reach a tractable model, these authors resorted to the Fowler approximation which assumes a sharp boundary at the interface such that

$$\rho(\mathbf{r}, \boldsymbol{\omega}) = \begin{cases} \rho_b(\boldsymbol{\omega}) & z > 0 \\ 0 & z < 0 \end{cases} \tag{6.138}$$

where $\rho_b(\boldsymbol{\omega})$ is the single-particle density in the bulk phase. The pair distribution function is then decomposed into an approximate form as

$$\rho_2(\mathbf{r}_1, \boldsymbol{\omega}_1, \mathbf{r}_2, \boldsymbol{\omega}_2) = \rho(\mathbf{r}_1, \boldsymbol{\omega}_1)\rho(\mathbf{r}_2, \boldsymbol{\omega}_2)g_2(r_{12}/\sigma) \tag{6.139}$$

where $g_2(r_{12}/\sigma)$ is the pair correlation function assumed to be a function only of the separation between the molecules, r_{12}, divided by a scale factor σ which was taken to be a constant (Parsons, 1976; Murakami, 1977; Croxton, 1980), yet dependent on orientation in (Parsons, 1976) to account for an anisotropic repulsive interaction.

For the Maier–Saupe-type anisotropic attractive interaction, Parsons (1976) and Croxton (1980) derived the surface tension as a function of order parameter for both perpendicular and planar alignment in the following form:

$$\gamma_\perp = \gamma_0\left(1 + \frac{2Q}{3} + \frac{Q^2}{6}\right) \tag{6.140}$$

$$\gamma_\parallel = \gamma_0\left(1 - \frac{4Q}{9} + \frac{8Q^2}{27}\right) \tag{6.141}$$

where γ_0 is given by

$$\gamma_0 = \frac{3\pi\rho^2 v}{4}\int_0^\infty g_2(r)\frac{1}{r^3}z,\mathrm{d}r \tag{6.142}$$

with v being the parameter expressing the magnitude of intermolecular interaction. The absolute value of surface tension that this formula gives, being on the correct order of $10\,\mathrm{mN\,m^{-1}}$, agrees with the observations. However, the above result indicates that $\gamma_\parallel$ is always smaller than $\gamma_\perp$, so the alignment should be planar at the free surface. This contradicts the majority of previous experimental findings (Gannon, 1978; Yokoyama, 1983; Chiarelli et al., 1983, 1984) that the alignment is perpendicular to the air–nematic interface. Furthermore, the jump of surface tension resulting from the above formula is

$$\gamma_\parallel(I) - \gamma_\parallel(N) = \gamma_0\frac{4}{9}\Delta Q > 0 \tag{6.143}$$

which, although having the proper sign, greatly overestimates the jump to be $1\,\mathrm{mN\,m^{-1}}$, if $\Delta Q \sim 0.3$ is used. This discrepancy results first of all from the negligence of the spatial relaxation of the orientational order parameter, as anticipated from the thought experiment we invoked in connection with Antonow's rule. Since this relaxation occurs over a length scale much larger than that of the number density at the free surface (occurring probably over a few molecular distances), the order parameter appearing in the above formulas must be regarded as the surface order parameter Q_s rather than the bulk order parameter (Parsons, 1976). Hence, the above result must be understood to show instead that the jump of the surface order parameter should be extremely small even when the bulk undergoes a conspicuous order–disorder transition. As Parsons (1980) noted, this implies that a considerable degree of excess order should persist at the free surface of an isotropic phase; this has indeed been experimentally verified (Beaglehole, 1982; Yokoyama et al., 1983).

In order to investigate the role of anisotropic hard-core repulsion, which is unsurprisingly a predominant factor making up the liquid crystal structure in the bulk, Kimura and Nakano (1985) and Holyst and Poniewierski (Holyst, 1988a, b) developed a low-density Onsager theory of surface tension. These theories are based on the mean-field version of the free energy functional equations (6.123), (6.126), and (6.127). Again, adopting the Fowler approximation for the system of spherocylinders, both of these theories have shown that the anisotropic hard-core repulsion favors a perpendicular orientation at the free surface. It has furthermore been shown (Kimura and Nakano, 1985) that the competition between the attractive van der Waals interaction and the hard-core repulsion can explain the critical anchoring transition from tilted to perpendicular observed at the surface of MBBA and EBBA (Bouchiat and Langerin-Cruchon, 1971; Chiarelli et al., 1983, 1984), following

$$\theta \propto \begin{cases} (T_0 - T)^{1/2} & T < T_c < T_{ni} \\ 0 & T_c < T < T_{ni}. \end{cases} \tag{6.144}$$

The physical rationale for the preferred perpendicular orientation is intuitively quite simple, as shown in FIGURE 6.19.

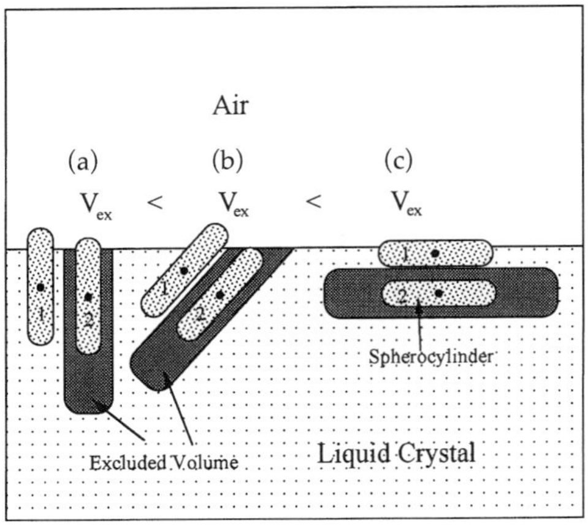

FIGURE 6.19. Excluded volume at the free surface of perfectly aligned spherocylindrical molecules as a function of tilt angle. (a) Smallest excluded volume for perpendicular alignment; (b) medium for tilted; (c) largest for planar.

6.4.3 Van der Waals Theory

The above formula for the free-energy increase in terms of density deviations has a significant utility in formulating a variety of tractable approximation schemes based on the expansion of the free energy with respect to the spatial gradients of relevant variables. The best known example of this gradient-expansion approach is the van der Waals theory of the liquid–vapor interface of a simple fluid, in which the local excess free energy density is expanded in powers of the density gradient up to second order as

$$\delta\Omega/A = \int_{-\infty}^{+\infty} W[\rho(z)] + \frac{1}{2}m\left(\frac{d\rho}{dz}\right)^2 dz \tag{6.145}$$

with the assumption that the density is uniform in the plane parallel to the interface (xy-plane), where $W[\rho(z)]$ is the excess part of the free energy density over that of coexisting bulk liquid and vapor phases, and m is a kind of elastic constant expressing the resistance of the system to becoming inhomogeneous.

The van der Waals theory follows naturally by employing the following Taylor expansion in equation (6.133),

$$\delta\rho(z_2) = \delta\rho(z_1) + z_{12}\frac{d\rho(z_1)}{dz} + \frac{1}{2}z_{12}^2\frac{d^2\rho(z_1)}{dz^2} \tag{6.146}$$

and by performing a by-parts integration of the integrand to obtain

$$\begin{aligned} \delta\Omega(\rho)/A &= \frac{k_B T}{2}\int\left[\int\left(\frac{\delta(\mathbf{r}_{12})}{\rho(z)} - c(\rho:\mathbf{r}_{12})\right)d\mathbf{r}_{12}\right]\delta\rho(z)^2\,dz \\ &\quad + \frac{k_B T}{4}\int\left(\int z_{12}^2 c(\rho:\mathbf{r}_{12})\,d\mathbf{r}_{12}\right)\left(\frac{d\rho(z)}{dz}\right)^2 dz. \end{aligned} \tag{6.147}$$

This result offers the molecular meaning of the constant m as

$$m = \frac{k_B T}{2} \int z^2 c(\rho : \mathbf{r})\, d\mathbf{r} = \frac{k_B T}{6} \int r^2 c(\rho : \mathbf{r})\, d\mathbf{r} \quad (6.148)$$

which reduces, in the mean-field approximation, to the original form due to van der Waals,

$$m = -\frac{1}{6} \int r^2 \phi(r)\, d\mathbf{r}. \quad (6.149)$$

The first term on the right-hand side of (6.147) represents the extra work needed for uniform compression of the fluid by an amount $\delta\rho(z)$. Since the extra work of compression is given thermodynamically by

$$\delta W = \frac{V}{2\rho(z)} \delta P \delta\rho(z) \quad (6.150)$$

with δP being the pressure increase, we find by comparison of these equations the well-known compressibility relation,

$$\chi^{-1} = \rho\frac{\partial P}{\partial \rho} = k_B T \rho \left[1 - \rho \int c(\rho : \mathbf{r})\, d\mathbf{r}\right] \quad (6.151)$$

where χ denotes the compressibility.

The equilibrium density profile can then be obtained by minimizing (6.145) by the variational calculus. Consequently, the surface tension is formally expressed as

$$\gamma = \frac{1}{A} \int \delta\Omega[\rho_{eq}]\, d\mathbf{r} = \int_{-\infty}^{+\infty} m \left(\frac{d\rho}{dz}\right)^2 dz \quad (6.152)$$

$$= 2 \int_{-\infty}^{+\infty} W[\rho(z)]\, dz \quad (6.153)$$

$$= \int_{\rho_v}^{\rho_l} [2mW(\rho)]^{1/2}\, d\rho. \quad (6.154)$$

Combined with (6.148), the first of the above expressions is reduced to

$$\gamma = \int_{-\infty}^{+\infty} \left[\frac{k_B T}{6} \int r^2 c(\rho : \mathbf{r})\, d\mathbf{r}\right] \left(\frac{d\rho}{dz}\right)^2 dz \quad (6.155)$$

which is an approximation to the exact formula derived by Triezenberg and Zwanzig (1972), known as an equivalent alternative to the Kirkwood–Buff formula.

6.4.4 Curvature Elasticity

When applied to nematics, the gradient expansion approach yields, first of all, the Oseen–Frank theory of curvature elasticity, as repeatedly shown by a number of authors (Poniewierski and Stecki, 1979; Stecki and Poniewierski, 1980; Singh, 1984; Lipkin, 1985; Singh and Singh, 1986a, b, 1987). Poniewierski and Stecki (1979) assumed, in their pioneering work, that the single-particle density depends on the position, only through the director, employing a functional form as $\rho(\mathbf{r}, \boldsymbol{\omega}) = \rho(\mathbf{n}(\mathbf{r}) \cdot \boldsymbol{\omega})$. The validity of this quasi-homogeneous approximation was questioned by Somoza and Tarazona (1991), pointing out the possibility that the density could relax further from this form to give a lower free energy in the distorted state. However, the density correction, which is closely related to the flexoelectricity, disappears if the molecule has an inversion symmetry. For simplicity, we shall restrict our attention to the quasi-homogeneous regime in what follows. The density fluctuations to be used in (6.133) then read:

$$\begin{aligned}\delta\rho(\mathbf{r}_2, \boldsymbol{\omega}_2) &= \delta\rho(\mathbf{r}_1, \boldsymbol{\omega}_2) + [\mathbf{r}_{12} \cdot \nabla(\mathbf{n}(\mathbf{r}_1) \cdot \boldsymbol{\omega}_2)]\rho_0'(\mathbf{n}_u \cdot \boldsymbol{\omega}_2) \\ &\quad + \frac{1}{2}[\mathbf{r}_{12} \cdot \nabla(\mathbf{n}(\mathbf{r}_1) \cdot \boldsymbol{\omega}_2)]^2 \rho_0''(\mathbf{n}_u \cdot \boldsymbol{\omega}_2) \quad (6.156) \\ &\quad + \frac{1}{2}[(\mathbf{r}_{12} \cdot \nabla)^2(\mathbf{n}(\mathbf{r}_1) \cdot \boldsymbol{\omega}_2)]\rho_0'(\mathbf{n}_u \cdot \boldsymbol{\omega}_2)\end{aligned}$$

where $\rho_0(\mathbf{n}_u \cdot \boldsymbol{\omega})$ denotes the equilibrium single-particle density in a nematic that is uniformly oriented along $\mathbf{n}_u$. Because of the Nanbu–Goldstone degeneracy of nematics (no restoring force for homogeneous rotation of a uniform director field), we have the following identity, valid in the bulk:

$$\begin{aligned}0 = \frac{k_B T}{2} \int \left(\frac{\delta(\mathbf{r}_1 - \mathbf{r}_2)\delta(\boldsymbol{\omega}_1 - \boldsymbol{\omega}_2)}{\rho(\mathbf{r}_1, \boldsymbol{\omega}_1)} - c[\rho : \mathbf{r}_1, \boldsymbol{\omega}_1, \mathbf{r}_2, \boldsymbol{\omega}_2] \right) \\ \times \delta\rho(\mathbf{r}_1, \boldsymbol{\omega}_1)\delta\rho(\mathbf{r}_1, \boldsymbol{\omega}_1)\, d\mathbf{r}_1\, d\boldsymbol{\omega}_1\, d\mathbf{r}_2\, d\boldsymbol{\omega}_2. \quad (6.157)\end{aligned}$$

Then, using (6.157) in (6.133), we can rearrange the free-energy increase into the Oseen–Frank elastic energy functional:

$$\begin{aligned}\delta\Omega = \int \frac{1}{2}[K_1(\nabla \cdot \mathbf{n})^2 + K_2(\mathbf{n} \cdot \nabla \times \mathbf{n})^2 \\ + K_3(\mathbf{n} \times \nabla \times \mathbf{n})^2] \\ + K_{13}\nabla \cdot [\mathbf{n}(\nabla \cdot \mathbf{n})] - \frac{1}{2}(K_2 + K_{24})\nabla \\ \cdot [\mathbf{n}(\nabla \cdot \mathbf{n}) + \mathbf{n} \times \nabla \times \mathbf{n}]\, d\mathbf{r}\end{aligned} \quad (6.158)$$

with the molecular expressions of the elastic constants as

$$K_1 = \frac{k_B T}{2} \int x_{12}^2 \omega_{1_x} \omega_{2_x} c[\rho_0 : \mathbf{r}_1, \boldsymbol{\omega}_1, \mathbf{r}_2, \boldsymbol{\omega}_2] \times \rho_0'(\mathbf{n} \cdot \boldsymbol{\omega}_1) \rho_0'(\mathbf{n} \cdot \boldsymbol{\omega}_2)\, \mathrm{d}\mathbf{r}_{12}\, \mathrm{d}\omega_1\, \mathrm{d}\omega_2 \tag{6.159}$$

$$K_2 = \frac{k_B T}{2} \int x_{12}^2 \omega_{1_y} \omega_{2_y} c[\rho_0 : \mathbf{r}_1, \boldsymbol{\omega}_1, \mathbf{r}_2, \boldsymbol{\omega}_2] \times \rho_0'(\mathbf{n} \cdot \boldsymbol{\omega}_1) \rho_0'(\mathbf{n} \cdot \boldsymbol{\omega}_2)\, \mathrm{d}\mathbf{r}_{12}\, \mathrm{d}\omega_1\, \mathrm{d}\omega_2 \tag{6.160}$$

$$K_3 = \frac{k_B T}{2} \int z_{12}^2 \omega_{1_x} \omega_{2_x} c[\rho_0 : \mathbf{r}_1, \boldsymbol{\omega}_1, \mathbf{r}_2, \boldsymbol{\omega}_2] \times \rho_0'(\mathbf{n} \cdot \boldsymbol{\omega}_1) \rho_0'(\mathbf{n} \cdot \boldsymbol{\omega}_2)\, \mathrm{d}\mathbf{r}_{12}\, \mathrm{d}\omega_1\, \mathrm{d}\omega_2 \tag{6.161}$$

with the z-axis taken along the initial director $\mathbf{n}_u$. In the bulk phase, the direct correlation function depends only on $\mathbf{r}_{12} = \mathbf{r}_2 - \mathbf{r}_1$, so the elastic constants become spatially invariant. The above expressions are in fact more general and embrace the interfacial transition zone, provided the form of single-particle density as assumed above remains valid.

It was demonstrated by Yokoyama and coworkers (Yokoyama et al., 1987) that such an interfacial deviation of the elastic constants from their bulk values could have a significant influence on the macroscopic anchoring strength. The fundamental point is very simple: the orientational torque transmitted through the liquid crystal is constant whatever inhomogeneity exists there. The continuity condition of the torque in the framework of the curvature elasticity yields the additional extrapolation length as

$$\delta d_e = \int_0^\infty \frac{K_b - K(z)}{K(z)}\, \mathrm{d}z \tag{6.162}$$

where K_b is the elastic constant in the bulk (Yokoyama et al., 1987; Barbero and Durand, 1991a; Alexe-Ionescu et al., 1994). Due to the presence of $K(z)$ in the denominator, this contribution becomes important particularly when the surface disorder develops, as found in some weak anchoring systems (Yokoyama, 1987; Seo et al., 1992b, d).

6.4.5 Surface-like Elastic Constants

The surface-like elastic constants, K_{24} and K_{13}, referred to as the saddle-splay and the second-order splay–bend constants, respectively, were first introduced by Oseen (1933) from a phenomenological viewpoint, and later by Nehring and Saupe (1971) from a molecular standpoint (equation (6.158)). Since these constants do not enter the Euler–Lagrange equation for bulk equilibria, they tend to be omitted in ordinary curvature elastic treatments with a fixed boundary condition.

In dealing with the interfacial problems, however, these *bulk* contributions are converted to a surface free energy by Gauss's theorem, apparently involving the first-order gradient of the director:

$$f_s = -\tfrac{1}{2}(K_2 + K_{24})\mathbf{k}[\mathbf{n}(\nabla \cdot \mathbf{n}) + \mathbf{n} \times \nabla \times \mathbf{n}] + K_{13}[\mathbf{k} \cdot \mathbf{n}(\nabla \cdot \mathbf{n})] \tag{6.163}$$

where $\mathbf{k}$ denotes the unit vector along the surface normal. Straightforward application of variational calculus to the total system with this surface free energy leads to a pathological behavior of the director field near the interface when $K_{13} \neq 0$ (Oldano and Barbaro, 1985a, b; Barbero and Oldano, 1985; Barbero and Strigazzi, 1989a; Barbero et al., 1989b; Barbero, 1991; Pergamenshchik, 1993b). Indeed, in the presence of the K_{13} term, the total free energy can be made arbitrarily small by introducing a large enough director normal distortion at the interface; the K_{24} term, however, does not contain a normal derivative $\mathbf{k} \cdot \nabla \mathbf{n}$ at a flat interface that causes this anomaly.

Nevertheless, there is no *a priori* reason, in the realm of phenomenology at least, to disregard the derivative-dependent surface free energy of bulk origin, over the ordinary anchoring energies we have considered so far. Furthermore, a few authors (Dubois-Violette and Parodi, 1969; Mada, 1979b) have envisaged directly a free-energy term proportional to the normal derivative of the director without mentioning the accompanying mathematical difficulties. In view of the exotic nature of these terms, there naturally arises a conceptual problem of how to deal with and understand them. As a consequence, the reality of such surface-like bulk elastic constants, including their experimental determinations, has become the subject of a large number of recent publications (Ericksen, 1966; Hinov, 1977; Derzhanski and Hinov, 1977; Hinov and Derzhanski, 1979; Barbero and Oldano, 1985; Oldano and Barbero, 1985a, b; Hinov, 1987; Yokoyama, 1988b; Barbero and Strigazzi, 1989a; Barbero et al., 1989b; Faetti, 1993a, b; Somoza and Tavazona, 1991; Allender et al., 1991; Sparavigna et al., 1991, 1992a, b, 1994; Pergamenshchik, 1993a, b, c; Teixeira et al., 1993b; Alexe-Ionescu et al., 1993; Faetti, 1994a, b, c; Polak et al., 1994; Laventovich and Pergamenshchik, 1994, 1995; Crawford and Žumar, 1995).

Experimental determination of the surface-like elastic constants is undoubtedly a difficult task, since their effects, if any, should always be interfacial and hence are hard to discriminate from those of ordinary surface anchoring (Faetti, 1993a). The first measurement of K_{24} was made by Allender and coworkers (Allender et al., 1991), by confining a liquid crystal in a fine cylindrical cavity of submicron radius and carefully examining the resultant director configurations, which are presumed to be dependent on the magnitude of K_{24}, based on their characteristic NMR spectra. The value they provided for K_{24} in small-molecule nematics is nearly of the same order of magnitude as the ordinary Frank elastic constant K. The subsequent experiments by the same group (Polak et al., 1994), using larger cavity sizes combined with optical microscopy observations, confirmed this estimate. The quantity they actually determined is $\sigma = RW_0/K + K_{24}/K - 1$ with R and W_0 being the radius of the cavity and the anchoring energy at the nematic–cavity wall interface, respectively, as a function of R. They took the anchoring energy into account in the simple Rapini–Papoular form, that is,

$$f_{RP} = \frac{1}{2} W_0 \sin^2 \phi \tag{6.164}$$

while ignoring the curvature dependence of the interfacial free energy as given by (6.37). Only considering the orientational work across a fixed cylindrical interface (equation (6.37) contains the translational work as well), the curvature correction amounts to replacing their anchoring energy by the following planar-equivalent effective anchoring energy:

$$W_{eff} = W_0 \frac{d_e}{R \ln(1 + d_e/R)} \tag{6.165}$$

where $d_e = K/W_0$ is the extrapolation length. For $d_e \ll R$, the above equation simply reduces to $W_{eff} = W_0 + K/2R$; so that K_{24}/K is subject to a correction of $-1/2$, which is far from negligible.

A more direct and reliable way to infer the reality of the surface-like elastic constants is, as pointed out by Pergamenshchik (1993a, b) and exploited by Sparavigna and coworkers (Sparavigna et al., 1992b, 1994) and Lavrentovich and Pergamenschik (1994, 1995), to seek an orientational behavior that is qualitatively different from anything possible in the absence of the surface-like elastic constants. The most fundamental and characteristic consequence of the existence of free-energy terms, linear in the gradient, may be that the uniformly oriented state could be energetically destabilized, and taken over by a spontaneously modulated structure under an appropriate condition, as is well known in the occurrence of spontaneous twist in cholesteric phases (de Gennes, 1974). Such an instability has indeed been observed in a thin nematic layer floating on an isotropic liquid substrate, as shown in FIGURE 6.20, and has been quantitatively explained in terms of the action of either K_{24} (Sparavigna et al., 1992b, 1994; Pergamenshchik, 1993a) or K_{13} (Lavrentovich and Pergamenshchik, 1994; Pergamenshchik, 1993b). Apart from the fact that K_{13} involves a normal derivative of the director, which as mentioned above entails special care for curing its disease (Barbero and Strigazzi, 1989a; Pergamenshchik, 1993b), the root of the observed spontaneous modulation lies in the linearity of these terms to the lateral derivative of director, and hence is common to

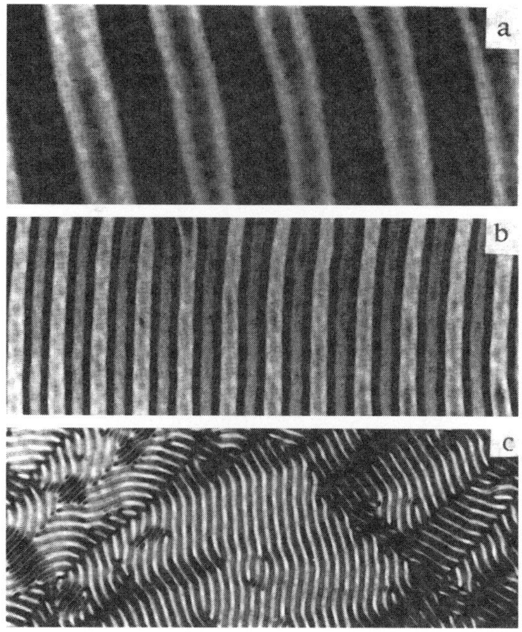

FIGURE 6.20. Stripe patterns observed in a nematic (4-pentyl-4′-cyanobiphenyl) thin layer at the air–isotropic substrate (glycerin) interface with different layer thicknesses: (a) 0.44 μm; (b) 0.32 μm; (c) 0.16 μm. Reproduced by permission from (Lavrentovich and Pergamenshchik, 1994.)

both K_{24} and K_{13}. For one-dimensional lateral distortions (assumed along the x-axis) as in FIGURE 6.20, the surface-like elastic free energy is rewritten per unit area as

$$f_s \sim -\frac{1}{2}(K_2 + K_{24})\left(n_z \frac{dn_x}{dx} - n_x \frac{dn_z}{dx}\right) + K_{13} n_z \frac{dn_x}{dx} \tag{6.166}$$

where the normal derivative has been omitted. Noting further that a complete differential, say $d(n_x n_z)/dx$, does not contribute to the free-energy gain in the thermodynamic limit as long as the director profile is continuous, the above equation can be reduced to the following equivalent form:

$$f_s \sim [K_{13} - (K_2 + K_{24})] n_z \frac{dn_x}{dx}. \tag{6.167}$$

It should be noted here that, if both surfaces of the nematic layer are identical, the surface-like elastic free energy has an opposite sign with the same magnitude so that the contributions from both surfaces cancel out. In order for this term to be operable, therefore, the layer should be asymmetric with respect to a reflection about the surface plane.

If we introduce here a sinusoidal perturbation (wavenumber $= q$) in the director as

$$n_x = n_{x_0} + \delta n_x \sin qx \qquad n_z = n_{z_0} + \delta n_z \cos qx \tag{6.168}$$

then the surface free energy experiences an average gain

$$\delta f_s \sim [K_{13} - (K_2 + K_{24})] \delta n_z \delta n_x q. \tag{6.169}$$

Since the bulk Frank elasticity and the ordinary surface anchoring give rise to a positive definite quadratic form for δn_x and δn_z (involving the layer thickness and the wavenumber in a rather complicated manner), there occurs a second-order transition from the unperturbed state to a modulated state as $|K_{13} - (K_2 + K_{24})|$ is increased. For fixed surface-like elastic constants, the same instability takes place as a function of the layer thickness. Since the bulk contribution, which tends to prevent the instability, diminishes with the decrease in thickness, the spontaneous modulation is expected to take place in a sufficiently thin layer as long as the surface anchoring is weak. These predicted behaviors for a thin nematic layer have indeed been experimentally confirmed by Sparavigna and coworkers (Sparavigna et al., 1991, 1992a, b, 1994) and by Lavrentovich and Pergamenshchik (1994, 1995).

Although the presence of surface free-energy terms involving the lateral derivatives of the director is now unquestionable, this does not necessarily mean that it should be of purely bulk origin, as implied by the concept of surface-like elastic constants. Stripe textures very similar to the above have long been known to develop in free-standing smectic films, and have been understood in a rather different context that ascribes their origin to the surface-specific interactions associated with the broken mirror symmetry at the interface (Meyer and Pershan, 1973; Maclennan, 1990; Maclennan and Seul, 1992; Maclennan et al., 1994; Seul and Andelman, 1995). From a statistical mechanical point of view, indeed, the unique existence of those surface-like bulk elastic constants, as resulting only from the properties of the liquid crystal (independent of the substrate), is highly disputable, because of the ambiguity associated with the construction of the local free-energy density from the essentially nonlocal intermolecular interactions, as indicated by equation (6.133) (Rowlinson and Widom, 1982; Yokoyama, 1988b). Namely, the gradient expansion rests on the following transformation,

$$\int \Phi(\mathbf{r}_1, \mathbf{r}_2)\, d\mathbf{r}_1\, d\mathbf{r}_2 = \int \phi(\mathbf{r})\, d\mathbf{r}. \tag{6.170}$$

If ϕ can be uniquely defined, then we are allowed to call it a local free energy density. In going from (6.133) to the Oseen–Frank elastic energy, we have expanded the density about one of the two interacting points $\mathbf{r}_1$ and $\mathbf{r}_2$, and allocated the entire free energy to the free-energy density at this very point with $\mathbf{r}_1$ being identified with $\mathbf{r}$ in the above. This is nonetheless a unique choice of constructing a local free-energy density in the above sense; for example, we could have taken an arbitrary point $\mathbf{r}_m = \alpha\mathbf{r}_1 + (1 - \alpha)\mathbf{r}_2$ on the line connecting these two points equally well, and constructed the local free-energy density at $\mathbf{r}_m$ by expanding the density about this point. Since there is no particular reason to take a certain value of α, there arises a criterion which a physically consistent definition of local densities must satisfy, that is, that it must at least be invariant to the choice of α. This point has been explicitly articulated in (Rowlinson and Widom, 1982) drawing upon the ambiguity in defining the local energy density, entropy density, etc.

As demonstrated by Poniewierski and Stecki (1979), the gradient expansion of the nonlocal excess free energy (6.133) yields microscopic expressions not only for the bulk Frank constants as given by (6.161), but for the surface-like elastic constants. Yet only the bulk Frank constants and K_{24} given by

$$K_{24} = \frac{k_B T}{2} \int x_{12} y_{12} \omega_{1_x} \omega_{2_y} c[\rho_0 : r_1, \omega_1, r_2, \omega_2] \times \rho_0'(n \cdot \omega_1) \rho_0'(n \cdot \omega_2) \, dr_{12} \, d\omega_1 \, d\omega_2$$

are shown to be physically consistent in the above sense, being independent of the choice of α against the aforementioned shift of the point of reference. In addition, these constants satisfy within the quasi-homogeneous approximation the following Nehring–Saupe relation (Nehring and Saupe, 1971; Teixeira et al., 1993b),

$$K_{24} = (K_1 - K_2)/2,$$

which illustrates the bulk origin of K_{24}. Mathematically, the K_{13} surface-like elastic contribution arises from the by-parts integration of the term of the form $\delta\rho(\mathbf{r}_1, \boldsymbol{\omega}_1)(\mathbf{r}_{12} \cdot \nabla)^2 \rho(\mathbf{r}_1, \boldsymbol{\omega}_2)$ in (6.133). So, it is readily shown that this part becomes proportional to $\alpha^2 + (1-\alpha)^2$; this result means that the surface-like elastic energy depends strongly on the unphysical choice of the reference point, which is not to be physically tolerated. Retaining higher-order terms in the Taylor expansion of density, as proposed by Barbero and Strigazzi (1989a), does not essentially alleviate this difficulty, though it apparently helps eliminate the mathematical anomaly at the interface at this order; it does, however, introduce another anomaly at a higher order.

More fundamental improvement is brought about when the expansion in the gradient of the director is summed up to infinite order, as suggested qualitatively by Pergamenshchik (1993b) and by Faetti (1994b, c). In view of the density functional formula (6.133), however, this procedure is nothing more than a recovery of the original nonlocal free energy. Since there is a wealth of evidence to believe that the Frank elastic energy functional is at least asymptotically correct for $\nabla\mathbf{n} \to 0$, it seems quite natural to guess that the correction from the infinite-order summation is important only in the interfacial transition region (Yokoyama, 1988b). This is essentially the point that Pergamenshchik advocated in a heuristic manner (Pergamenshchik, 1993b), although this was criticized by Faetti (1994a, c) as being physically inconsistent. The advantage of the density functional approach we have taken so far is that this desirable feature of the nematic elasticity can be naturally reached as follows. For this end, it is only necessary to employ the identity,

$$\begin{aligned} \delta\rho(\mathbf{r}_1, \boldsymbol{\omega}_1)\delta\rho(\mathbf{r}_2, \boldsymbol{\omega}_2) = {} & \frac{1}{2}[\delta\rho(\mathbf{r}_1, \boldsymbol{\omega}_1)\delta\rho(\mathbf{r}_1, \boldsymbol{\omega}_2) \\ & + \delta\rho(\mathbf{r}_2, \boldsymbol{\omega}_1)\delta\rho(\mathbf{r}_2, \boldsymbol{\omega}_2)] \\ & + \frac{1}{2}[\delta\rho(\mathbf{r}_1, \boldsymbol{\omega}_1)\delta\rho(\mathbf{r}_2, \boldsymbol{\omega}_2) \\ & - \delta\rho(\mathbf{r}_1, \boldsymbol{\omega}_2)\delta\rho(\mathbf{r}_2, \boldsymbol{\omega}_1)] \\ & + \frac{1}{2}[\delta\rho(\mathbf{r}_1, \boldsymbol{\omega}_1) - \delta\rho(\mathbf{r}_2, \boldsymbol{\omega}_1)] \\ & \times [\delta\rho(\mathbf{r}_2, \boldsymbol{\omega}_2) - \delta\rho(\mathbf{r}_1, \boldsymbol{\omega}_2)]. \end{aligned} \quad (6.171)$$

The first and the second terms, when substituted into (6.133), disappear in the bulk nematic due to the Nanbu–Goldstone degeneracy and the inversion symmetry of bulk nematics; these terms can therefore make a finite contribution only in the interfacial region. The last term yields the Frank elastic energy including K_{24} upon gradient expansion, but due to its difference form, it does not generate a K_{13} surface-like elastic term.

Inspection of each term shows that the surface-like elastic term results from a formal expansion of the first two terms in the above identity. This clearly reveals the interfacial origin of the surface-like elasticity (Oldano and Barbero, 1985b; Yokoyama, 1988b; Pergamenshchik, 1993b; Faetti, 1994b, c). The lowest-order term of the expansion of the first two terms in (6.172) with respect to the lateral gradient of the director gives the ordinary anchoring energy, and the rest is to be identified with Pergamenshchik's infinite sum R_∞.

6.4.6 Landau–de Gennes Model of Nematic Interface

6.4.6.1 Formulation

The Landau–de Gennes model and its variations have found extensive applications to important interfacial phenomena, such as the surface excess ordering and wetting transition (Sheng, 1976, 1982; Miyano, 1979a, b; Tarczon and Miyano, 1980; Allender et al., 1981; Mauger et al., 1984; Sen and Sullivan, 1987; Poniewierski and Sluckin, 1984, 1987; Sluckin and Poniewierski, 1985, 1986; Yokoyama, 1988a), surface tension and anchoring (de Gennes, 1971a; Parsons,

1975; Gannon, 1978; Marcus, 1983; Goossens, 1985; Yokoyama et al., 1987; Barbero, 1991a), contact angles (Yokoyama et al., 1987), anchoring transition (Parsons, 1978; Barbero, 1991a; Teixeira et al., 1993a), defect core structures (Schopohl and Sluckin, 1987, 1988), etc. If the single-particle density is considered to be a function of the tensor order parameter instead of the director, the gradient expansion of the free energy reduces to the Landau–de Gennes model of a nematic interface (de Gennes, 1971a). Although care should be taken as regards the treatment of surface-like terms as mentioned above, the gradient expansion yields all the symmetry-allowed terms in the free energy and enables us to express the phenomenological parameters appearing in the Landau–de Gennes model in molecular terminology (Tjipto-Margo and Sullivan, 1988).

In the principal axis frame, the tensor order parameter is

$$Q = \frac{1}{2}\begin{pmatrix} -Q-P & 0 & 0 \\ 0 & -Q+P & 0 \\ 0 & 0 & 2Q \end{pmatrix} \qquad (6.172)$$

where Q and P are the uniaxial and biaxial scalar order parameters, respectively. The density perturbation is written as $\delta\rho(\mathbf{r},\boldsymbol{\omega}) = \rho(\omega_i\omega_j Q_{ij}(\mathbf{r})) - \rho_0$, where ρ_0 stands for the density of the isotropic phase, and the summation convention is implied. When the nematic is uniaxial ($P=0$), the tensor order parameter is related to the director by

$$Q_{ij} = \frac{1}{2}Q(3n_in_j - \delta_{ij}) \qquad (6.173)$$

where $Q = \langle 3(\mathbf{n}\cdot\boldsymbol{\omega})^2 - 1\rangle/2$ with $\langle\cdot\rangle$ denoting the ensemble average.

Unlike the Oseen–Frank elasticity, a homogeneous perturbation of the tensor order parameter is in general accompanied by an increase in the free energy due to the variation in the scalar order parameter. Hence, the first term of (6.171) does not vanish even in the bulk. The Landau–de Gennes free energy takes this homogeneous part of the free energy increase up to the fourth order, while truncating the gradient expansion at the second. Further adding up the interfacial free-energy contributions, now coming from the first two terms in (6.171), the most general form of the Landau–de Gennes free energy for a nematic–substrate interface can be written as follows (Sluckin and Poniewierski, 1986):

$$\Omega = \int_V f_{LD}(Q(\mathbf{r})) + f_g(\nabla Q(\mathbf{r}))\,\mathrm{d}\mathbf{r} + \int_A f_s(Q(\mathbf{r}_s))\,\mathrm{d}\mathbf{r}_s \qquad (6.174)$$

where $\mathbf{r}_s$ denotes a point at the interface, f_{LD} is the Landau–de Gennes free-energy density (de Gennes, 1971), which is, for a uniaxial system, given by

$$f_{LD}(Q(\mathbf{r})) = f_0 + \frac{1}{2}AQ^2 - \frac{1}{3}BQ^3 + \frac{1}{4}CQ^4 \qquad (6.175)$$

where the coefficients B and C are constants and A is assumed to be of the form $A = a(T - T^*)$ in the Landau–de Gennes theory with T^* being the temperature below which the isotropic phase becomes absolutely unstable. f_g is the gradient part which reads

$$f_g = \frac{1}{2}L_1\frac{\partial Q_{ij}}{\partial x_k}\frac{\partial Q_{ij}}{\partial x_k} + \frac{1}{2}L_2\frac{\partial Q_{ij}}{\partial x_j}\frac{\partial Q_{ik}}{\partial x_k}. \qquad (6.176)$$

and f_s is the surface free energy, given generally by

$$f_s = f_{s0} - GQ_{ij}k_ik_j + 4a_sQ_{ij}Q_{ij} + 4b_sQ_{ij}k_jQ_{ik}k_k + 4c_s(Q_{ij}k_ik_j)^2 \qquad (6.177)$$

where $\mathbf{k}$ denotes, as before, the unit vector along the surface normal. Here, the substrate has been assumed to be isotropic in the plane of the interface, which appears to be a good first approximation, since the out-of-plane anchoring is known in most cases to be at least an order of magnitude larger than the in-plane anchoring. In (6.177), f_{s0} is the surface free energy of the isotropic–substrate interface; the second term together with the last two takes care of the liquid crystal–substrate interaction, while the third term represents part of liquid crystal–liquid crystal interaction, corresponding respectively to the first and second terms in the Kirkwood–Buff–Navascués formula, equation (6.132). When the liquid crystal is assumed to be uniaxial, the surface free energy reduces to:

$$f_s = f_{s0} + w_0(Q_s) + \frac{1}{2}w_2(Q_s)[1 - (\mathbf{n}_s\cdot\mathbf{k})^2)] + \frac{1}{4}w_4(Q_s)[1 - (\mathbf{n}_s\cdot\mathbf{k})^2]^2 \qquad (6.178)$$

where Q_s and $\mathbf{n}_s$ are the surface order parameter and the director at the interface, not in the thermodynamic sense but the ensemble average of molecular orientation at this very point, and

$$\begin{aligned} w_0(Q_s) &= -GQ_s + (a_s + 4b_s + 4c_s)Q_s^2 \\ &\equiv -GQ_s + \frac{1}{2}UQ_s^2 \\ w_2(Q_s) &= 3GQ_s - 6(b_s + 4c_s)Q_s^2 \\ &\equiv G_2Q_s + \frac{1}{2}U_2Q_s^2 \end{aligned} \tag{6.179}$$

$$w_4(Q_s) = 36c_sQ_s^2 \equiv \frac{1}{2}U_4Q_s^2. \tag{6.180}$$

This van der Waals-type model, though rather simple, embraces almost all the phenomenological theories so far proposed in the literature, and has been found to offer a powerful theoretical framework for investigating a wide range of orientational properties of the liquid crystal interface, as mentioned at the beginning of this section; more general treatments incorporating biaxiality were conducted by Sluckin and Poniewierski (1985) and by Sen and Sullivan (1987).

In an assumption of uniaxial symmetry, the nematic–isotropic transition temperature T_{ni}, the bulk order parameter at the transition Q_{ni}, and the latent heat of transition L, are derived from the Landau–de Gennes theory as follows:

$$T_{ni} = T^* + \frac{2B^2}{9aC} \tag{6.181}$$

$$Q_{ni} = \frac{2B}{3C} \tag{6.182}$$

$$L = \frac{1}{2}aQ_{ni}^2T_{ni}. \tag{6.183}$$

6.4.6.2 Surface Tension and Alignment at the Nematic–Isotropic Interface

As one of the first applications of the theory, de Gennes took up the alignment and tension at a planar nematic–isotropic interface (de Gennes, 1971a). Let us assume that the liquid crystal remains uniaxial and the director is uniformly oriented throughout the interfacial region. The gradient part of the free energy, equation (6.176), is then rewritten as

$$f_g = \frac{1}{2}m(\mathbf{n})\left(\frac{\mathrm{d}Q}{\mathrm{d}z}\right)^2 \tag{6.184}$$

where

$$m(\mathbf{n}) = \frac{6L_1 + L_2}{4} + \frac{3L_2}{4}(\mathbf{n}\cdot\mathbf{k})^2. \tag{6.185}$$

Further, at $T = T_{ni}$ where the nematic and isotropic phases can coexist, we have

$$f_{LD}(Q) = \frac{c}{4}Q^2(Q_{ni} - Q)^2. \tag{6.186}$$

Here, the surface potential f_s is of course absent. Hence, in view of (6.154), we obtain the nematic–isotropic interfacial tension as

$$\begin{aligned} \gamma_{ni}(\mathbf{n}) &= \int_0^{Q_{ni}} [2m(\mathbf{n})f_{LD}(Q)]^{1/2}\,\mathrm{d}Q \\ &= \frac{c^{1/2}[2m(\mathbf{n})]^{1/2}}{12}Q_{ni}^3. \end{aligned} \tag{6.187}$$

The profile of the order parameter across the interface is described by

$$Q(z) = \frac{Q_{ni}}{1 + \exp(-z/\xi)} \tag{6.188}$$

where

$$\xi = \frac{1}{Q_{ni}}\left(\frac{2m(\mathbf{n})}{c}\right)^{1/2} \tag{6.189}$$

denotes the coherence length of the orientational ordering, and gives a measure of the thickness of the interface.

The Landau–de Gennes theory gives a remarkably good estimate for small interfacial tension on the order of $10^{-2}\,\mathrm{mN\,m^{-1}}$, measured experimentally by light scattering (Lengevin and Bauchiat, 1973), light reflectometry (Faetti and Polleschi, 1984a, b), sessile drop (Williams, 1976), droplet coalescence dynamics (Kahlweit and Ostner, 1973; Smith, 1984; Smith et al., 1985), and forced interface deformation (Yokoyama et al., 1985b) techniques. It also successfully reproduces from these observations the coherence length of 10–20 nm, in excellent agreement with the result of direct measurement.

As regards the orientation of the director at the interface, however, the experiments (Vilanove, 1974; Yokoyama et al., 1984b; Faetti and Palleschi, 1984a, b) have revealed an oblique angle ranging universally around 60° from the interface normal for cyanobiphenyls, etc., whereas the Landau–de Gennes theory predicts, as (6.185) and (6.187) indicate, only the perpendicular ($L_2 < 0$) or planar ($L_2 > 0$) alignment. Marcus (1983) removed the assumption of uniaxiality in the original analysis, and studied the energy and structure of the nematic–isotropic interface by numerical analysis of the Landau–de Gennes model. He found that, although a strong biaxiality emerges at the isotropic side of the interface when $L_1 \neq L_2$, the equilibrium macroscopic alignment is little affected.

To account for this discrepancy, Kimura and Nakano (1986) developed a molecular mean-field theory taking into account the anisotropic intermolecular hard-core repulsion as well as long-range attraction, based on the mean-field version of (6.133), involving the Mayer function. They showed, for the system of spherocylinders, that the repulsive interaction favors an orientation normal to the interface, while the attraction prefers a planar alignment, and that the final alignment could be oblique by the balance between these antagonistic contributions. Their theory, however, may suffer from the approximate form of the hard-core repulsion and the negligence of the diffuse interfacial transition zone (i.e., orientational Fowler approximation), which is an important attribute of the nematic–isotropic interface. More recently, Holyst and Poniewierski (1988a) studied the same model without the attractive interaction and showed that the hard-core interaction alone can explain the oblique orientation observed at the nematic–isotropic interface, when the repulsive interaction is treated more accurately. Both theories adopted the Fowler approximation, however, the validity of which remains open.

6.4.6.3 Orientational Wetting Transition

When the orientational surface potential is such as to induce a sufficiently large orientational ordering (or disordering) at the interface compared with the bulk phase, a macroscopic layer of nematic (or isotropic) phase intervenes at the substrate–isotropic (nematic) phase interface as the transition temperature T_{ni} is approached from above (below). This phenomenon, referred to as the orientational wetting transition, is one of the subjects which has been most actively explored in the past two decades in the area of surface physics of liquid crystals (Sluckin and Poniewierski, 1986). It was initiated by the seminal theoretical paper by Sheng (1976), followed by a series of experimental demonstrations by Miyano (Miyano, 1979a, b) and Tarczon and Miyano (1980). Through subsequent theoretical (Allender et al., 1981; Sheng, 1982; Mauger et al., 1984; Poniewierski and Sluckin, 1984, 1985, 1987; Telo da Gama, 1984b, 1987 and Tarazona, 1990; Sluckin and Poniewierski, 1985; Sen and Sullivan, 1987; Pawlowska et al., 1987) and experimental (Yokoyama et al., 1982, 1984a; Yokoyama, 1988a; Beaglehole, 1982; van Sprang, 1983a, c; Rosenblatt, 1984b; Hsiung et al., 1986b; Ocko et al., 1986; Moses and Shen, 1991, Moses et al., 1993; Ouchi et al., 1992; Crawford et al., 1993a, b) investigations, orientational wetting is now phenomenally well established and is recognized as one of the key observations through which the nature of the liquid crystal interface can be inferred. Although Jérôme placed in her excellent review article (Jérôme, 1991) somewhat reduced credit on the significance of the phenomenon in view of the inconsistent results in the existing literature, careful examination of these results indicates that the inconsistency is primarily due to careless experimental designs in some reports. The orientational wetting transition offers one of the most sensitive and reproducible clues presently available for the study of liquid crystal interfaces.

The term "wetting transition" stems from the preferential wetting phenomenon in quasicritical liquid mixtures, which was predicted by Cahn in (1977), soon after Sheng's work. Cahn's wetting transition has received, for its very generic nature, an explosion of interest in a wider context including liquid mixtures, magnetic systems, superfluid He, as well as liquid crystals (Sullivan and Telo de Gama, 1986).

We imagine a semi-infinite liquid crystal ($z > 0$) bounded by a planar boundary exerting a surface potential of (6.180) on the basis of the Landau–de Gennes theory. For brevity of notation, the variables are normalized to a dimensionless form as follows:

$$q = Q/Q_{ni} \tag{6.190}$$

$$\tilde{z} = z/\xi \tag{6.191}$$

$$g = GQ_{ni}/6\gamma_{ni} \tag{6.192}$$

$$u = UQ_{ni}^2/6\gamma_{ni} \tag{6.193}$$

$$t = (T - T^*)/(T_{ni} - T^*). \tag{6.194}$$

Then, the Landau–de Gennes free energy density is reduced to

$$\tilde{f}_{LD}(q, t) \equiv f_{LD}\xi/3\gamma_{ni} = tq^2 - 2q^3 + q^4 \tag{6.195}$$

giving rise to the total normalized free energy per unit

area as

$$\tilde{\Omega} \equiv \Omega/3\gamma_{ni}$$
$$= \int_0^\infty \tilde{f}_{LD}(q,t) + \left(\frac{dq}{d\tilde{z}}\right)^2 d\tilde{z} + 2\left(-gq_s + \frac{1}{2}uq_s^2\right). \tag{6.196}$$

The equilibrium profile of the order parameter satisfies the Euler–Lagrange equation,

$$2\frac{d^2q}{d\tilde{z}^2} = \frac{\partial \tilde{f}_{LD}(q,t)}{\partial q} \tag{6.197}$$

along with the boundary condition at $z = 0$,

$$-\left(\frac{dq}{d\tilde{z}}\right)_{\tilde{z}=0} = -g + uq_s. \tag{6.198}$$

The first integral of (6.197) is given by

$$\left(\frac{dq}{d\tilde{z}}\right)^2 = \tilde{f}_{LD}(q,t) - \tilde{f}_{LD}(q_b,t) \tag{6.199}$$

where $q_b = 0$ for $T > T_{ni}$ or $q_b = [3 + (9 - 8t)^{1/2}]/4$ for $T < T_{ni}$ denotes the point of absolute minimum of $\tilde{f}_{LD}$. The surface free energy for a given q_s is, therefore, written as

$$\gamma = \int_{\min(q_s,q_b)}^{\max(q_s,q_b)} 2\sqrt{\tilde{f}_{LD}(q,t) - \tilde{f}_{LD}(q_b,t)}\, dq + 2\left(-gq_s + \frac{1}{2}uq_s^2\right) \tag{6.200}$$

and the boundary condition as

$$\sqrt{\tilde{f}_{LD}(q,t) - \tilde{f}_{LD}(q_b,t)} = g - uq_s. \tag{6.201}$$

As the graphical solution of this equation in FIGURE 6.21 indicates, when g and/or u are sufficiently large (so the interface is highly ordered or disordered), there exists only a unique solution for q_s. However, when the surface potential is weak, there are multiple solutions, between which a transition can take place as the temperature is changed. This boundary transition, noted first by Sheng (1982), is identical to the prewetting transition in the liquid mixture systems. Although the prewetting was actually observed in a binary liquid system (Schmidt and Moldovea, 1986), no conclusive evidence has yet been reported for liquid crystal systems except for a dubious claim of observation by Iimura and coworkers (Iimura et al., 1984), due probably to the weakly first-order nature of the nematic–isotropic transition that makes the range of possible surface potential parameters extremely narrow (Sheng, 1982; Sluckin and Poniewierski, 1986).

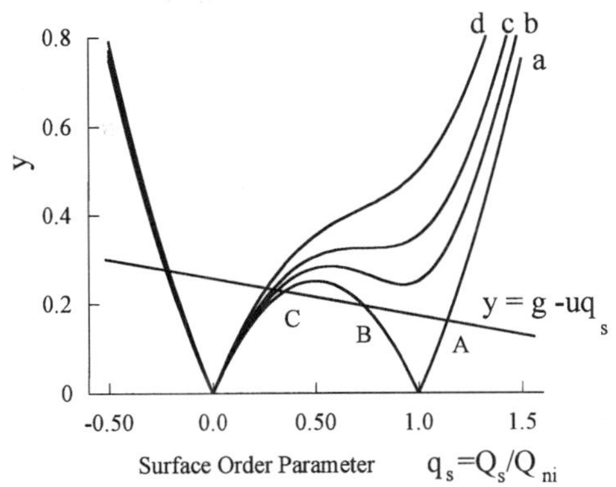

FIGURE 6.21. Graphical representation of the boundary equilibrium equation for the surface order parameter q_s: (a) $t = (T - T^*)/(T_{ni} - T^*) = 1$; (b) $t = 17/16$; (c) $t = 9/8$; (d) $t = 5/4$. The intercepts A and C correspond to locally stable solutions and B to an unstable solution. The absolute minimum can be determined by Maxwell's theorem of equal areas.

FIGURE 6.22 shows the profile of the order parameter near the interface as a function of temperature for fixed surface order parameters. A remarkable feature is that when the surface order parameter exceeds the bulk order parameter at the nematic–isotropic transition, that is, $q_s \geq 1$, the ordered interfacial layer grows to infinity as the temperature approaches T_{ni} from above, following a logarithmic dependence on the temperature difference. However, when $q_s < 1$, the ordered layer remains localized near the interface down to the transition temperature, whose extension is basically governed by the coherence length ξ of the orientational order in the bulk phase. Quite the opposite situation emerges in the bulk nematic range when the surface order parameter is lower than 0 in the bulk nematic regime; this is the case of surface-induced disordering, and a macroscopic thick layer of isotropic phase grows as $\ln(T_{ni} - T)$ as the temperature is raised.

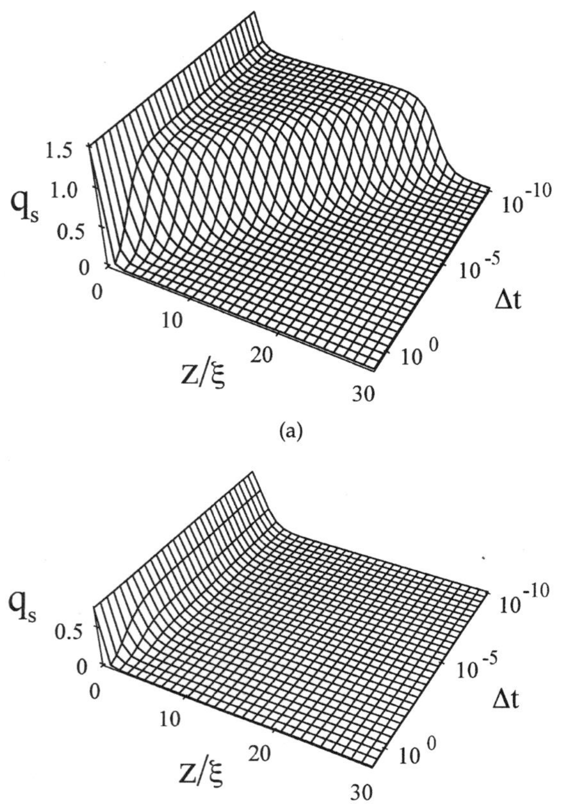

FIGURE 6.22. Temperature dependence of the order parameter profile for (a) strongly ordered ($q_s = 1.5$) and (b) moderately ordered ($q_s = 0.75$) interfaces.

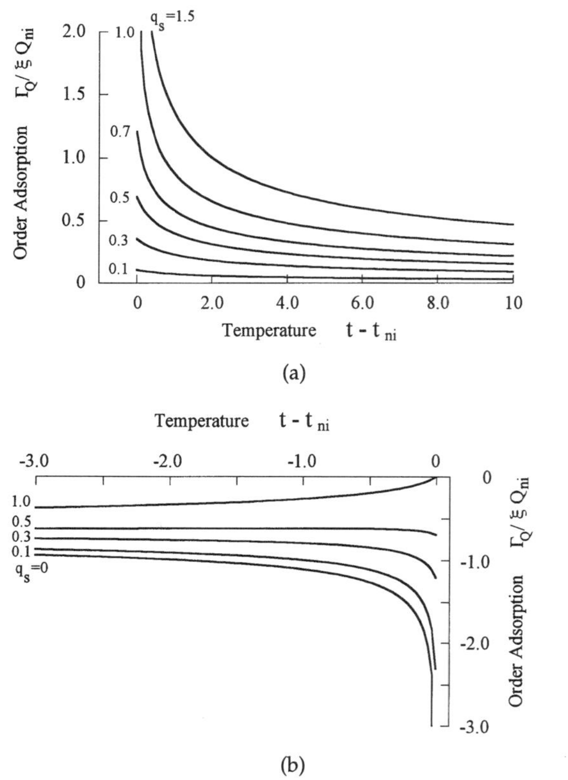

FIGURE 6.23. Temperature dependence of the orientational order adsorption for (a) the surface ordering ($T > T_{ni}$) and (b) the surface disordering ($T > T_{ni}$) with fixed surface order parameters.

An experimentally amenable quantity is the adsorption of orientational order defined by

$$\begin{aligned}\Gamma_Q &\equiv \int_0^\infty Q(z) - Q_b \, \mathrm{d}z \\ &= Q_{ni}\xi \int_0^\infty q(z) - q_b \, \mathrm{d}\tilde{z}.\end{aligned} \tag{6.202}$$

From the above equations, we readily find

$$\frac{\Gamma_Q}{\xi Q_{ni}} = \ln \frac{t^{1/2} + 1}{(q_s^2 - 2q_s + t)^{1/2} - q_s + 1} \qquad T > T_{ni} \tag{6.203}$$

$$= \ln \frac{q_s + q_b - 1 + (q_s^2 + 2(q_b - 1)q_s + q_b(q_b - 1))^{1/2}}{2q_b - 1 + (4q_b^2 - 3q_b)^{1/2}} \qquad T < T_{ni}. \tag{6.204}$$

FIGURE 6.23 shows the temperature dependence of the adsorption for fixed surface order parameters. As already mentioned, the adsorption logarithmically diverges as $T \to T_{ni}$ for $q_s \geq 1$ in the bulk isotropic regime or $q_s \leq 0$ for the nematic regime. There is, however, a fundamental limitation in confirming this divergence in real systems, because the Kelvin shift of the transition temperature associated with the finite sample size interrupts the growth of the interfacial wetting layer before the divergence takes place. In view of the Kelvin equation, (6.98), we can estimate the maximum observable wetting layer thickness as

$$l_{max} \sim \xi \ln d. \tag{6.205}$$

Experimental methods so far used to observe the order adsorption include the detection of surface-induced birefringence by transmission (Miyano, 1979a, b;

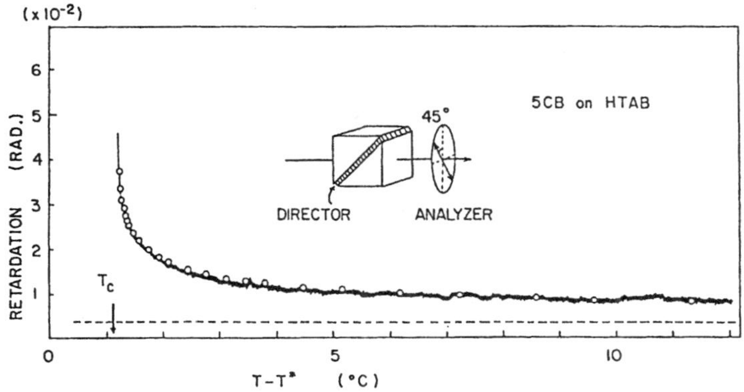

FIGURE 6.24. Orientational wetting transition observed through surface-induced birefringence. The sample of 5CB was held between homeotropically aligning substrates treated with hexadecyltrimethylammonium bromide (HTAB). Reproduced by permission from (Miyano 1979a).

Tarczon and Miyano, 1980; Yokoyama et al., 1982, 1984a; van Sprang, 1983a; van Sprang and Aartsen, 1983b; Ouchi et al., 1992; Moses et al., 1993) and reflection (Beaglehole, 1982; Hsiung, 1986b; Moses and Shen, 1991) ellipsometry, interferometry (Mada, 1979a; Sugimura et al., 1981), optical dilatometry (Wu and Efron, 1986), light scattering (Iimura et al., 1984) and surface fluorimetry (Salamon and Skibinski, 1983). But, due to the limited level of meaningful signals, only the ellipsometry seems to be sensitive enough to give sufficiently reliable results (see FIGURE 6.24). In particular, the surface disordering is difficult to observe, since it occurs over a large background from the bulk phase, and only evanescent wave ellipsometry (Hsiung, 1986b) has so far been able successfully to yield direct evidence (Moses and Shen, 1991). In this respect, it must be noted that Barberi and Durand (1990) attempted a direct determination of the surface order parameter q_s, not the adsorption, by utilizing the flexoelectric effect in the nematic phase, and showed that the disordering is enhanced as the roughness on the SiO_x substrate surface is increased; however, the role played by the interfacial inhomogeneity was not considered in their treatment, thus leaving room for discretion.

In TABLE 6.1, the surface order parameters determined from the observation of the order adsorption are compiled, together with a description of the type of interface and method of measurement. Roughly speaking, the rubbed surfaces of stiff polymers are likely to induce enhanced surface order, while high surface energy and/or rough surfaces, such as obliquely evaporated SiO induce an orientational disorder, though both of these surfaces uniformly align nematic liquid crystals apparently equally well. The enhanced surface order can naturally be understood as resulting from the anisotropic forces exerted by the substrate on adjacent liquid crystal molecules. The fact that rubbed polymer surfaces tend to induce higher order, while the unrubbed polymer barely has this ability, is in line with the conjecture that the uniform alignment of liquid crystals on rubbed polymer surfaces results from the anisotropic van der Waals interaction between the liquid crystal and a preferentially aligned polymer chain through the rubbing treatment (Geary et al., 1987). On the other hand, the mechanism underlying the surface disorder is less straightforward to imagine. A plausible explanation is, as suggested by many, to conceive the frustration in the orientational arrangement of molecules caused either by the randomly oriented adsorbed layer (Yokoyama, 1984a) or more or less equivalently by the roughness of the substrate surface (Monkade et al., 1988; Barberi and Durand, 1991).

In terms of the surface free energy, a complete wetting of the substrate–isotropic interface by a macroscopic nematic layer implies the following Antonow's rule:

$$\gamma_{si} - \gamma_{sn} = \gamma_{ni}. \tag{6.206}$$

According to the Landau–de Gennes theory of orientational wetting, this occurs if and only if the surface order parameter exceeds the bulk order parameter at the transition, and if $q_s < 1$, the interface is only partially wetted by the nematic phase. Such a correspondence between the wetting behavior and the growth of the ordered surface layer was directly and favorably confirmed, as shown in FIGURE 6.25 (Yokoyama et al., 1982, 1983) and was utilized by Clark (1985) in an analysis of the surface memory effect.

Table 6.1 Surface order parameter at the nematic-substrate interface.

Substrate	Treatment	Liquid crystal	Surface order	Method	Source
SiO	Oblique evaporation	5CB	0	Trans. ellipsometry	Miyano, 1979a
HTAB	Physisorption	5CB	0.20, 0.32	Trans. ellipsometry	Miyano, 1979a,b
DMOAP	Chemisorption	5CB	0.17	Trans. ellipsometry	Miyano, 1979b
MAP	Chemisorption	5CB	0	Trans. ellipsometry	Miyano, 1979b
MAP	Chemisorption + rubbing	5CB	0.18, 0.17	Trans. ellipsometry	Miyano, 1979b
Glass	Rubbing	5CB	0.22	Trans. ellipsometry	Miyano, 1979b
Glass	None	MBBA	0	Trans. ellipsometry	Tarczon and Miyano, 1980
Glass	Rubbing	MBBA	0.05, 0.14	Trans. ellipsometry	Tarczon and Miyano, 1980
HTAB	Physisorption	MBBA	0.32	Trans. ellipsometry	Tarczon and Miyano, 1980
DMOAP	Chemisorption	MBBA	0.05	Trans. ellipsometry	Tarczon and Miyano, 1980
MAP	Chemisorption	MBBA	−0.02	Trans. ellipsometry	Tarczon and Miyano, 1980
MAP	Chemisorption + rubbing	MBBA	0.06,0.09	Trans. ellipsometry	Tarczon and Miyano, 1980
PVA	Dip coating	MBBA	0.01	Trans. ellipsometry	Tarczon and Miyano, 1980
PVA	Dip coating + rubbing	MBBA	0.36	Trans. ellipsometry	Tarczon and Miyano, 1980
PVA	Dip coating	5CB	0	Trans. ellipsometry	Tarczon and Miyano, 1980
PVA	Dip coating + rubbing	5CB	0.17	Trans. ellipsometry	Tarczon and Miyano, 1980
SiO	Oblique evaporation	5CB	0.005	Trans. ellipsometry	Yokoyama et al., 1982
PVA	Rubbing	5CB	0.27	Trans. ellipsometry	Yokoyama et al., 1982
Air	None	5CB	0.44	Refl. ellipsometry	Beaglehole, 1982
Parylene	Rubbing	5CB, 7CB	0.34	Trans. ellipsometry	van Sprang, 1983a
Polyimide	Rubbing	5CB, 7CB	0.36	Trans. ellipsometry	van Sprang, 1983a
Glass	Rubbing	7CB	0.12	Trans. ellipsometry	van Sprang, 1983a
SiO	Rubbing	5CB, 7CB	0	Trans. ellipsometry	van Sprang, 1983a
DMOAP	Chemisorption	5CB	0.08	Refl. ellipsometry	Hsiung, 1986
SiO	Oblique evaporation	5CB	−0.05	Refl. ellipsometry	Moses and Shen, 1991

HTAB: hexadecyltrimethylammonium bromide.
DMOAP: dimethyl-n-octadecyl-3-aminopropyltrimethoxysilyl-chloride.
MAP: methyl-3-aminopropyltrimethoxysilane.
PVA: polyvinylalcohol.

Hsiung and coworkers (1986b), however, reported an observation that is in apparent contradiction with the prediction of the Landau–de Gennes theory. By using evanescent wave ellipsometry, they determined separately the surface order parameter and the thickness of the ordered surface layer, and found that, although the surface order parameter is only a quarter of the bulk order parameter, the surface layer diverges as $T \to T_{ni}$, according to $(T - T_{ni})^{-1/2}$. To resolve this discrepancy, Telo da Gama (1987) pointed out that, if the density is also allowed to change along with the orientational order parameter, the lower surface order parameter and the diverging ordered layer can be simultaneously accommodated in the theory. However, no experimental evidence has been given yet.

6.4.6.4 Surface-induced Non-nematic Order

SMECTIC A ORDER. The phenomenon of surface-induced order and disorder is very general in nature and is not restricted to the nematic-like ordering, as specifically mentioned above. As emphasized in section 6.2, surfaces and interfaces can support many different structures that are nonexistent in the bulk phase, as long as their symmetry allows. What is remarkable about the wetting transition, then, is the fact that a local structure connected with the daughter phase that the system is going to be transformed into, undergoes a pretransitional enhancement owing to the energetic constraint and/or the increased coherence length of the relevant structure toward the transition point.

The surface-induced nematic ordering is primarily driven by the energetics of the system due to the weak, yet conspicuous first-order phase transition from the isotropic phase. On the other hand, we can expect that the latter mechanism will operate as well, when the phase transition is of second-order, as for many nematic–smectic A transitions. The occurrence of smectic layer ordering at the free surface of a nematic liquid crystal was first demonstrated by Als-Nielsen

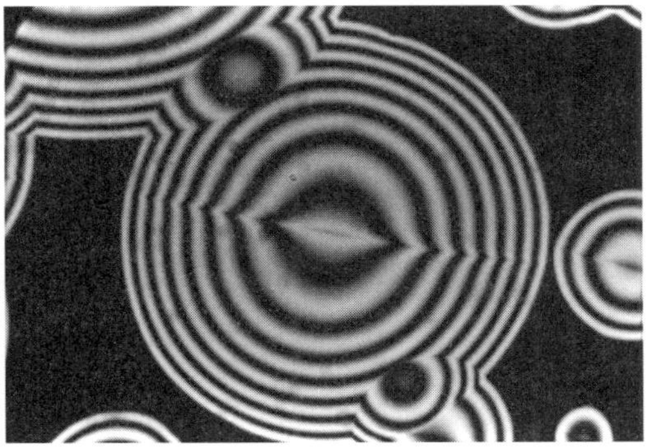

100μm
(a)

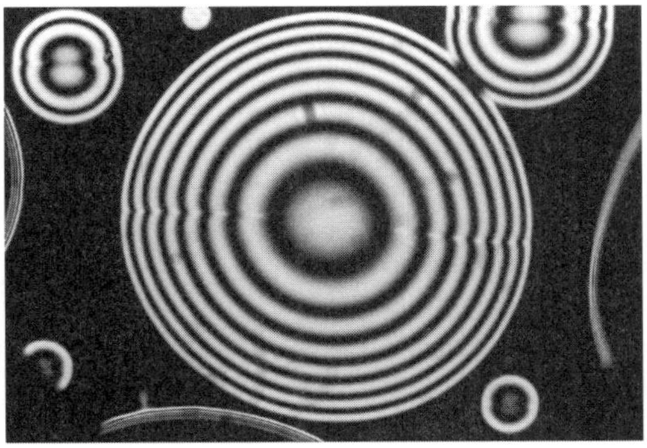

100μm
(b)

FIGURE 6.25. Sessile drops of (a) a nematic 5CB at the rubbed PVA–isotropic phase interface and (b) an isotropic 5CB at the SiO–nematic interface in equilibrium at $T_{ni} = 35.3\,^{\circ}\mathrm{C}$. Observed with monochromatic light under a crossed polarizers. The contact angle of the nematic liquid is about (a) 20° and (b) 160°, indicating situations close to complete wetting by the nematic phase and the isotropic phase, respectively. The interfacial line defect running across the circular drops reflects the oblique anchoring condition at the nematic-isotropic interface.

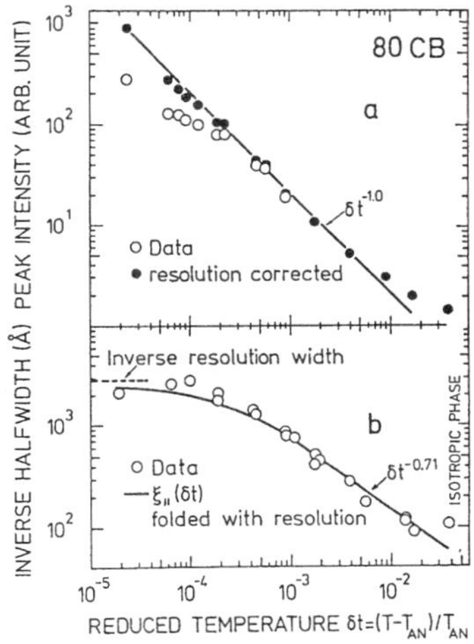

FIGURE 6.26. Temperature dependence of the x-ray diffraction peak intensity (a) and the inverse half-width (b) from the surface-induced smectic layer of 8OCB. Reproduced by permission from (Als-Nielsen et al., 1982).

and coworkers (Als-Nielsen et al., 1982) and Pershan and Als-Nielsen (1984) using the synchrotron x-ray diffraction and reflection techniques, respectively. They showed that the intensity and width of the diffraction peak at the wavenumber corresponding to the smectic layer underwent a power-law increase toward the nematic–smectic A transition temperature T_{AN}, as shown in FIGURE 6.26. In particular, the inverse of the peak width, which is thought to be proportional to the thickness of the ordered surface layer, was indeed found to have the same critical exponent as that of the longitudinal coherence length in the bulk phase.

Similar surface-induced smectic order was found at the free surface of 12CB, which undergoes a first-order isotropic–smectic A transition (Ocko et al., 1986). Unlike the second-order case, the growth of the smectic layer proceeded on a layer-by-layer basis, just like the multilayer adsorption phenomenon. This analogy suggests the energetic mechanism for the surface ordering. Pawlowska and coworkers (1987) employed the Lebwohl–Lasher lattice model, as a succession of the previous works by Rosenblatt and Ronis (Ronis and Rosenblatt, 1980; Rosenblatt and Ronis, 1981), and showed that the surface-induced layering could qualitatively be accounted for by a proper combination of the surface-induced nematic order and the surface forces which couple to the density field.

The surface-induced smectic order near T_{AN} was also found at a solid–nematic interface, via magnetic Fréedericksz transition experiments (Rosenblatt, 1984b). He observed that as the temperature approached T_{AN} from above, an anomalous increase of the threshold field appeared in a cell of a few microns thickness with a critical divergence toward T_{AN}. Although being indirect evidence, this observation strongly suggests that the homeotropically aligning substrate brought about an interfacial layer having smectic ordering, whose thickness was again governed by the bulk coherence length.

POLAR ORDER. Another important class of surface ordering is the one that is not connected with any of the possible bulk structures. A typical example is the surface-induced polar ordering of molecules in systems which assume only apolar bulk phases. The existence of such a polar ordering can be probed by means of the optical second-harmonic generation (SHG) technique, which has been extensively explored by Shen (Shen, 1989a, b). The power of the SHG technique derives from the fact that it is sensitive only to surfaces when dealing with apolar bulk phases such as nematics and nonchiral smectics. At the surface, the mirror symmetry of the bulk phase is broken so that it becomes SHG active.

The SHG method has been applied to surfaces and interfaces of nematics (Guyot-Sionnest et al., 1986; Feller et al., 1991; Ouchi et al., 1992; van Aerle, 1994), smectics (Hsiung and Shen, 1986a; Chen et al., 1992), and Langmuir monolayers (Zhao et al., 1991, Hsiung and Beckerbauer, 1992; Xue et al., 1992). Based on the analysis of nonlinear polarizability, the angular and polarization dependences of SHG data can be converted (on a few assumptions) to the orientational distribution function of the molecules. The correspondence between the SHG data and molecular orientation can be extremely simplified, when the frequency ω_l of the excitation light is so chosen that $2\omega_l$ falls in the absorption band of the molecule.

Using the SHG method, Shen showed that the liquid crystal molecules exhibit a temperature-independent polar order on treated substrates with the principal axis inclined by a quasi-universal tilt angle of 60–80°, regardless of the method of surface treatment and the type of liquid crystal (Shen, 1989a, b; Guyot-Sionnest et al., 1986). From the thickness dependence of the SHG intensity, they concluded that polar ordering is confined to the first monolayer, and by comparison with the SHG from the same monolayer spread at the air–water interface, they estimated that almost all the molecules in the first monolayer on the solid substrate point their polar heads toward the substrate, tilted by the universal angle, as schematically illustrated in

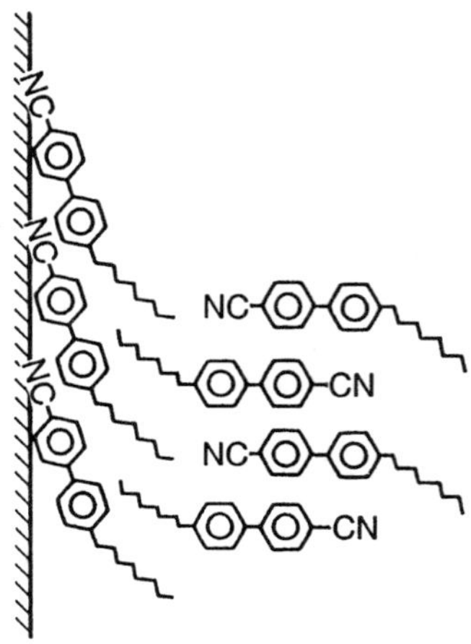

FIGURE 6.27. Polar ordering of 8CB molecules at a solid–nematic interface, as determined by optical second-harmonic generation. Reproduced by permission from (Shen, 1989a).

FIGURE 6.27. At the free surface of 8CB, however, only a negligible SHG was detected, indicating the absence of polar ordering.

Since polar molecules like cyanobiphenyls are known to have a strong tendency to form a quadrapolar pair in the bulk (Chandrasekhar, 1977), the coherence length of the polar ordering ξ_p in such a system should be of molecular dimensions. So, the surface-induced polar ordering is expected to rapidly decay into the bulk, in agreement with the results of SHG experiments.

CRYSTALLINE ORDER. The development of scanning probe microscopes (SPM), such as the scanning tunnelling microscope (STM) and the atomic force microscope (AFM), have made it possible to observe individual molecules sitting on a substrate surface. These techniques have revolutionized our understanding of the surface phenomena in general, and have been applied to liquid crystal samples bound on a single crystal surface as being one of the few testable subjects with SPM in the whole area of organic molecular systems (Foster and Frommer, 1988; Smith et al., 1989; Smith, 1991; Smith et al., 1992; Spong et al., 1989a, b; Iwakabe et al., 1990, 1991; Shigeno, 1990; Hara et al., 1990, 1991; Mizutani et al., 1990; Brandow et al., 1991; Schwartz et al., 1992; Eckhardt et al., 1993; Seo et al., 1993a; Parks et al., 1993; Matsushige et al., 1993; Patrick and Beebe, 1994; Walba et al., 1995).

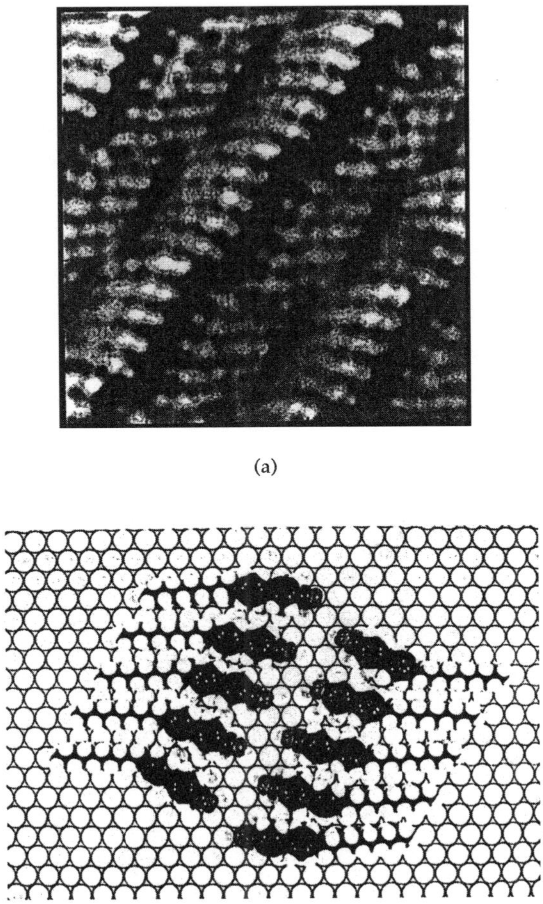

(a)

(b)

FIGURE 6.28. STM image (a) and the molecular model (b) for 10CB adsorbed on a single-crystal surface of MoS_2. Reproduced by permission from (Iwakabe et al., 1990).

As shown in FIGURE 6.28, a clear two-dimensional crystalline structure is usually observable by STM with or without an epitaxial relation with the substrate. Here, the following two remarks should be made: first, these observations of two-dimensional crystalline order at solid–nematic interfaces *never* imply that the first monolayer of liquid crystal is *always* in the crystalline phase. There is a strong self-selection of subjects in STM observations in the sense that only those with a sufficient mechanical stability and a near-crystalline order can be observed with STM. Although a defect structure, such as a grain boundary, has recently been observed (Parks et al., 1993), an amorphous or liquid layer does not usually allow observation of molecular details because of the high mobility of constituent molecules; second, the liquid crystal being observed by STM is in a state of strong confinement between the substrate and the STM tip, whose radius is small but normally a few tens of nanometers and thus much bigger than the molecular dimensions. Hence, the size effect must always be taken into account in interpreting the resultant STM images, which may induce a liquid to solidify (Bhushan et al., 1995). Although there are some controversies on the nature of the polymorphism in the adsorbed layer of liquid crystal molecules and also on their relationship with the bulk phase structures, the above points have more or less been overlooked.

6.5. MACROSCOPIC SURFACE ANCHORING

Surface anchoring has been one of the highlights in the surface physics of liquid crystals (Yokoyama, 1988b). During the past two decades, quite a large number of papers have been published on this subject (Rapini and Papoular, 1969; Kleman and Williams, 1973; Williams et al., 1973; Vitek and Kleman, 1975; Sicart, 1976; Ryschenkow and Klemen, 1976; Naemura, 1978; Riviere et al., 1979; Naemura, 1979, 1981; Yang, 1982, 1983d; Barbero and Barberi, 1983; Barbero et al., 1984a; Komitov, 1983; Barnik et al., 1983a, b; Rosenblatt, 1984a; Chiarelli et al., 1984; van Sprang and Aartsen, 1984b; Faetti and Palleschi, 1984a, b; Faetti et al., 1985a; Faetti and Palleschi, 1987a; Faetti, 1990; Faetti et al., 1991; Yokoyama and Van Sprang, 1985a; Yokoyama et al., 1987; Yokoyama, 1988b; Yokoyama and Patel, 1994; Perez et al., 1988; Eidner et al., 1988; Oh-ide et al., 1988; Glogarova and Durand, 1988; Blinov et al., 1989, 1991; Valenti et al., 1990; Di Lisi et al., 1990, 1992; Nobili et al., 1992; Seo et al., 1992b, d, 1993a; Cagnon and Durand, 1993; Durand, 1993; Murthy et al., 1993; Park et al., 1994; Iimura et al., 1994; Ruipeng et al., 1994). However, despite its straightforward phenomenological meaning, the lack of reliable experimental techniques to measure the anchoring strength coefficient has seriously hampered rapid progress in the field.

The measurement of the anchoring strength, except for an indirect physico-chemical approach, consists in creating an orientational torque at the interface and monitoring the degree of deviation of the director from the easy axis. The fundamental criterion to make a reliable measurement is that the characteristic length associated with the applied torque should be at least of the same order of magnitude as the extrapolation length. The recent development of reliable experimental techniques takes explicit account of this condition: Yang and Rosenblatt (1983d) prepared a cell a few microns thick for performing the Fréedericksz transition, and Yokoyama and van Sprang (1985a) applied a high electric field to meet this criterion even with the use of a cell of ordinary thickness. As it is the most widely used method for measuring the out-of-plane anchoring strength, we describe here the high electric field technique (Yokoyama and van Sprang, 1985a; Yokoyama et al., 1987; Yokoyama, 1988b). A detailed account of other techniques may be found elsewhere (Yokoyama, 1988b).

The high electric field technique is based on the scaling property of the Oseen–Frank theory of curvature elasticity. To expound the principle, it is only necessary to note that the free energy of a semi-infinite nematic liquid crystal in an electric field applied perpendicular to the substrate–nematic interface at $z = 0$ is written as:

$$\Omega = \int_0^\infty f_d(\mathbf{n}(z), \mathrm{d}\mathbf{n}/\mathrm{d}z) + \frac{D^2}{2\varepsilon_{eff}(n_z)}\,\mathrm{d}z + f_s(\mathbf{n}_s) \quad (6.207)$$

where D denotes the electrical displacement, which is constant throughout the volume, and $\varepsilon_{eff}(n_z)$ is the effective dielectric constant, which is assumed to take the maximum (i.e., positive dielectric anisotropy) when the director is oriented along the z-axis. By using $\zeta = Dz$, the above equation is rewritten as

$$\Omega = D\int_0^\infty f_d(\mathbf{n}(\zeta), \mathrm{d}\mathbf{n}/\mathrm{d}\zeta) + \frac{1}{2\varepsilon_{eff}(n_z)}\,\mathrm{d}\zeta + f_s(\mathbf{n}_s). \quad (6.208)$$

Hence, for a given $\mathbf{n}_s$, the director is generally written in the scaled form,

$$\mathbf{n}(z) = \mathbf{n}_g[\mathbf{n}_e, D(z + d_e)] \quad (6.209)$$

where $\mathbf{n}_e$ and d_e are the easy axis and the extrapolation length, respectively.

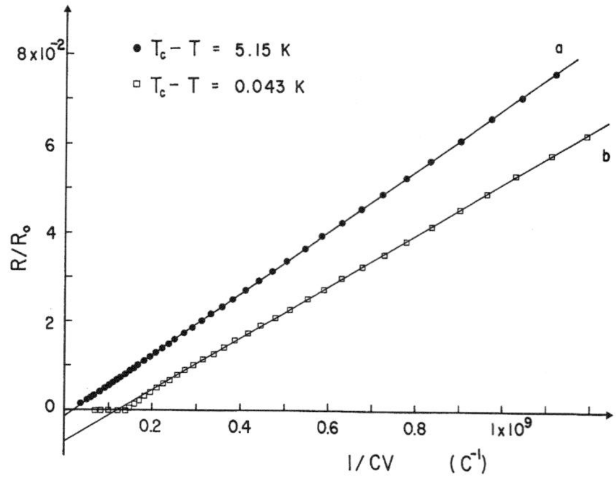

FIGURE 6.29. Retardation (R) versus reciprocal electrical displacement (1/CV) plot for anchoring strength measurements by the high electric field technique. 5CB on obliquely evaporated SiO. Reproduced by permission from (Yokoyama et al., 1987).

If we measure the optical retardation R to infer the interfacial orientation, we obtain

$$R = \frac{2\pi}{\lambda}\int_0^\infty \Delta n_{eff}(\mathbf{n})\,\mathrm{d}z \quad (6.210)$$

$$\approx \left[\frac{2\pi}{\lambda}\int_0^\infty \Delta n_{eff}[\mathbf{n}_g(\mathbf{n}_e, \zeta)]\mathrm{d}\zeta\right]\frac{1}{D} - \frac{2\pi\Delta n_{eff}(\mathbf{n}_e)}{\lambda}d_e \quad (6.211)$$

where $\Delta n_{eff}(\mathbf{n})$ is the effective birefringence for the director $\mathbf{n}$. This is the fundamental equation of the high electric field technique and indicates that, when the retardation is plotted against $1/D$, a straight line should result, whose intercept with the R-axis directly gives the extrapolation length. In order for the semi-infinite approximation to hold in a real finite-thickness cell, the applied electric field should be taken sufficiently high that the inhomogeneity is localized near the interface. FIGURE 6.29 shows an example of the R versus $1/D$ plot performed for a 5CB obliquely evaporated SiO interface. As this example shows, the linearity of the plot is usually extremely good, even up to the point of saturation ($\mathbf{n}$ is everywhere along the field) (Yang, 1983a, b; Sluckin and Poniewierski, 1986).

6.6 CONCLUDING REMARKS

The surface studies of liquid crystals have a vast spectrum of research and development. I have tried here to review the current status of the field with a special emphasis on expounding the unifying physical principles, rather than on compiling individual experimental and theoretical findings. As a consequence, it was inevitable that a strong bias was placed on a relatively mature area, and many potentially important results, that would have otherwise deserved much closer attention, were left out. This is particularly true as regards the molecular mechanism of surface alignment of nematics, as well as the surface properties of non-nematic liquid crystals, on which detailed molecular-level information is gradually becoming available with the help of advanced surface analysis techniques such as synchrotron x-ray diffraction, scanning probe microscopy, nonlinear optical spectroscopy, etc. These techniques yield, to a larger extent, information on the chemical aspects of the interfacial orientation of molecules, with unprecedented detail and precision (Toney et al., 1995). As yet, however, we have not come up with any language with which we can properly weave the new knowledge into the established phenomenology of liquid crystals. The role of a working hypotheses should not be underestimated in such circumstances, but it should also be stressed that once they begin to seek a dichotomous answer, such as the groove versus intermolecular-force question, they tend to become harmful in moving the field in the right direction. The truth should always be somewhere in between, and we should seek a way of coherent understanding, which I hope, in view of the present rise of interest, comes before very long.

REFERENCES

Abraham, F. F. 1980. Melting in two dimensions is first order: an isotherma–isobaric Monte Carlo study. *Phys. Rev. Lett.* **44**: 463.

Adamson, A. W. 1976. *Physical Chemistry of Surfaces.* New York: Wiley.

Akahane, T., and Tako, T. 1979. Orientational order of surface-aligned nematic liquid crystals. *Japan. J. Appl. Phys.* **18**: 19.

Akamatsu, S., and Rondelez, F. 1991. Fluorescence microscopy evidence for two different LE–LC phase transitions in Langmuir monolayers of fatty acids. *J. Physique II* **1**: 1309.

Alexe-Ionescue, A. L., Barberi, R., Barbero, G., Beica, T., and Moldovan, R. 1992. Surface energy for nematic liquid crystals: a new point of view. *Z. Naturforsch.* **47a**: 1235.

Alexe-Ionescu, A. L., Barbero, G., and Durand, G. 1993. Temperature dependence of surface orientation of nematic liquid crystals. *J. Physique II* **3**: 1247.

Alexe-Ionescu, A. L., Barberi, R., Barbero, G., and Giocondo, M. 1994. Anchoring energy for nematic liquid crystals: contributions from the spatial variation of the elastic constants. *Phys. Rev. E* **49**: 5378.

Allender, D. W., Henderson, G. L., and Johnson, D. L. 1981. Landau theory of wall-induced phase nucleation and pretransition birefringence at the isotropic–nematic transition. *Phys. Rev. A* **24**: 1086.

Allender, D. W., Crawford, G. P., and Doane, J. W. 1991. Determination of the liquid-crystal surface elastic constant K_{24}. *Phys. Rev. Lett.* **67**: 1442.

Als-Nielsen, J., Christensen, F., and Pershan, P. S. 1982. Smectic-A order at the surface of a nematic liquid crystal: synchrotron x-ray diffraction. *Phys. Rev. Lett.* **48**: 1107.

Amador, S., Pershan, P. S., Stragier, H., Swanson, B. D., Tweet, D. J., Sorensen, L. B., Sirota, E. B., Ice, G. E., and Habenschuss, A. 1989. Synchrotron studies of the first-order melting transition of hexatic monolayers and multilayers in freely suspended liquid-crystal films. *Phys. Rev. A* **39**: 2703.

Andelman, D., Brochard, F., and Joanny, J. F. 1987. Phase transitions in Langmuir monolayers of polar molecules. *J. Chem. Phys.* **86**: 3673.

Aoki, K., Seki, T., Suzuki, Y., Tamaki, T., Hosoki, A., and Ichimura, K. 1992a. Factors affecting photo-induced alignment regulation of cyclohexancarboxylate-type nematic liquid crystals of azobenzene molecular films. *Langmuir* **8**: 1007.

Aoki, K., Tamaki, T., Seki, T., Kawanishi, Y., and Ichimura, K. 1992b. Regulation of alignment of cyanobiphenyl liquid crystals by azobenzene molecular films. *Langmuir* **8**: 1014.

Aoyama, H., Yamazaki, Y., Matsuura, N., Mada, H., and Kobayashi, S. 1981. Alignment of liquid crystals on the stretched polymer films. *Mol. Cryst. Liq. Cryst.* **72**: 127.

Bahr, C., and Fliegner, D. 1993a. Selective reflection and textural studies of free-standing films of a chiral liquid crystal. *Liq. Cryst.* **14**: 573.

Bahr, C., and Fliegner, D. 1993b. Ferroelectric–antiferroelectric phase transition in a two-molecular-layer free-standing liquid crystal film. *Phys. Rev. Lett.* **70**: 1842.

Baker, G. A., Jr. 1990. *Quantitative Theory of Critical Phenomena.* Boston: Academic.

Barberi, R., and Durand, G. 1990. Order parameter of a nematic liquid crystal on a rough surface. *Phys. Rev. A* **41**: 2207.

Barberi, R., and Durand, G. 1991. Electrochirally controlled bistable surface switching in nematic liquid crystals. *Appl. Phys. Lett.* **58**: 2907.

Barberi, R., Barbero, G., Gabbasova, Z., and Zvezdin, A. 1993. Flexoelectricity and alignment phase transition in nematic liquid crystals. *J. Physique II* **3**: 147.

Barbero, G., and Barberi, R. 1983. Critical thickness of a hybrid aligned nematic liquid crystal cell. *J. Physique* **44**: 609.

Barbero, G., Madhusudana, N. V., and Durand, G. 1984a. Weak anchoring energy and pretilt of a nematic liquid crystal. *J. Physique Lett.* **45**: 613.

Barbero, G., Madhusudana, N. V., and Durand, G. 1984b. Anchoring energy for nematic liquid crystals: an analysis of the proposed forms. *Z. Naturforsch.* **39a**: 1066.

Barbero, G., and Oldano, C. 1985. Derivative-dependent surface-energy terms in nematic liquid crystals. *Nuovo Cimento* **6D**: 479.

Barbero, G., Taverna Valabrega, P., Bartolino, R., and Valenti, B. 1986. Evidence for flexoelectric effect in a nematic liquid crystal cell. *Liq. Cryst.* **1**: 483.

Barbero, G., and Strigazzi, A. 1989a. Second-order elasticity in nematics: a new anchoring source. *Liq. Cryst.* **5**: 693.

Barbero, G., Madhusudana, N. V., and Oldana, C. 1989b. Possibility of a deformed ground state in free-standing nematic films. *J. Physique* **50**: 2263.

Barbero, G., and Durand, G. 1990a. Selective ion adsorption and nonlocal anchoring energy in nematic liquid crystals. *J. Appl. Phys.* **67**: 2678.

Barbero, G., and Madhusudana, N. V. 1990b. On a theoretical analysis on the influence of nonuniformity of the order parameter on the surface energy in nematics. *Liq. Cryst.* **7**: 299.

Barbero, G., and Durand, G. 1991a. Order parameter spatial variation and anchoring energy for nematic liquid crystals. *J. Appl. Phys.* **69**: 6968.

Barbero, G., Gabbasova, Z., and Kosevich, Y. A. 1991b. On a generalization of the Rapini–Papoular expression of the surface free energy for nematic liquid crystals. *J. Physique II* **1**: 1505.

Barbero, G., Gabbasova, Z., and Osipov, M. A. 1991c. Surface order transition in nematic liquid crystals. *J. Physique II* **1**: 691.

Barbero, G., Beica, T., Alexe-Ionescu, A. L., and Moldovan, R. 1992. Anchoring energy and easy direction of nonuniform surfaces. *J. Physique II* **2**: 2011.

Barbero, G., Chuvyrov, A. N., Kaniadakis, G., Miraldi, E., and Rastello, M. L. 1993. Surface-polarization effect in the alignment of nematic liquid crystals. *J. Physique II* **3**: 165.

Barbero, G., Beica, T., Alexe-Ionescu, A. L., and Moldovan, R. 1993. Influence of the surface topography on the surface transitions in nematics. *Liq. Cryst.* **14**: 1125.

Barnik, M. I., Blinov, L. M., Korkishko, T. V., Umansky, B. A., and Chigrinov, V. G. 1983a. Investigation of NLC director orientational deformations in electric field for different boundary conditions. *Mol. Cryst. Liq. Cryst.* **99**: 53.

Barnik, M. I., Blinov, L. M., Korkishko, T. V., Umansky, B. A., and Chigrinov, V. G. 1983b. New type of boundary conditions for orientation deformations in homeotropic layers of nematic liquid crystals. *Sov. Phys.–JETP* **58**: 102.

Basturk, N. 1993. Thermal surface switching and thermo electrooptical properties of nematic liquid crystals. *Liq. Cryst.* **14**: 525.

Beaglehole, D. 1982. Pretransition order on the surface of a nematic liquid crystal. *Mol. Cryst. Liq. Cryst.* **89**: 319.

Bechhoefer, J., Jérôme, B., and Pieranski, P. 1990a. Systematic studies of the anchoring transition in nematic liquid crystals. *Phys. Rev. A* **41**: 3187.

Bechhoefer, J., Duvail, J.-L., Masson, L., Jérôme, B., Hornreich, R. M., and Pieranski, P. 1990b. Critical behavior in anchoring transitions in nematic liquid crystals. *Phys. Rev. Lett.* **64**: 1911.

Becker, M. E., Nehring, J., and Scheffer, T. J. 1985. Theory of twisted nematic layers with weak boundary coupling. *J. Appl. Phys.* **57**: 4539.

Bellini, T., Clark, N. A., Muzny, C. D., Wu, L., and Garland, C. W. 1992. Phase behavior of the liquid crystal 8CB in a silica aerogel. *Phys. Rev. Lett.* **69**: 788.

Bernasconi, J., Strassler, S., and Zeller, H. R. 1980. Van der Waals contribution to the surface and anchoring energies of nematic liquid crystals. *Phys. Rev. A* **22**: 276.

Berreman, D. W. 1972. Solid surface shape and the alignment of an adjacent nematic liquid crystal. *Phys. Rev. Lett.* **28**: 1683.

Berreman, D. W. 1973. Alignment of liquid crystals by grooved surfaces. *Mol. Cryst. Liq. Cryst.* **23**: 215.

Bhushan, B., Israelachvili, J. N., and Landman, U. 1995. Nanotribology: friction, wear, and lubrication at the atomic scale. *Nature* **374**: 607.

Biradar, A. M., Bawa, S. S., Sanena, K., and Chandra, S. 1993. Surface electroclinic effect in ferroelectric liquid crystal. *J. Physique II* **3**: 1787.

Birecki, H. 1984. Liquid crystal alignment on substrate surfaces: a collective phenomenon. *Liquid Crystals and Ordered Fluids*, vol. 4, ed. A. C. Griffin and J. F. Johnson, p. 852. New York: Plenum.

Biscari, P. G. Capriz, and Virga, E. G. 1994. On surface biaxiality. *Liq. Cryst.* **16**: 479.

Bishop, D. J., Sprenger, W. O., Pindak, R., and Neubert, M. E. 1982. Mechanical measurements of the melting transition in thin liquid-crystal films. *Phys. Rev. Lett.* **49**: 1861.

Blinov, L. M., Kabayenkov, A. Y., and Sonin, A. A. 1989. Experimental studies of the anchoring energy of nematic liquid crystals. *Liq. Cryst.* **5**: 645.

Blinov, L. M., Subachyus, D. B., and Yablonsky, S. V. 1991. The hydrodynamics of surface layers of nematic liquid crystals studied by modulation ellipsometry. *J. Physique II* **1**: 459.

Böttger, A., Frenkel, D., Joosten, J. G. H., and Krooshof, G. 1988. Smectic ordering in nematic and smectic liquid-crystalline films probed by means of surface light scattering. *Phys. Rev. A* **38**: 6316.

Bouchiat, M. A., and Langevin-Cruchon, D. 1971. Molecular order at the free surface of a nematic liquid crystal from light reflectivity measurements. *Phys. Lett.* **34A**: 331.

Brandow, S. L., DiLella, D. P., Colton, R. J., and Shashidhar, R. 1991. Evidence of pronounced positional order at the graphite–liquid crystal interface of a bulk nematic material. *J. Vac. Sci. Technol. B* **9**: 1115.

Brézin, E., Halperin, B. I., and Leibler, S. 1983. Critical wetting in three dimensions. *Phys. Rev. Lett.* **50**: 1387.

Cagnon, M., Ceccaldi, D., and Durand, G. 1990. Coherent molecular slip under shear in a plastic crystal. *Europhys. Lett.* **12**: 357.

Cagnon, M., and Durand, G. 1993. Positional anchoring of smectic liquid crystals. *Phys. Rev. Lett.* **70**: 2742.

Cahn, J. W. 1977. Critical point wetting. *J. Chem. Phys.* **66**: 3667.

Casalnuovo, S. A., Mockler, R. C., and O'Sullivan, W. J. 1984. Rayleigh-linewidth measurements on thin critical fluid films. *Phys. Rev. A* **29**: 257.

Chakravarty, S., and Woo, C.-W. 1975a. Short-range correlations in two-dimensional liquid crystals. *Phys. Rev. A* **11**: 713.

Chakravarty, S., and Woo, C.-W. 1975b. Short-range correlations in two-dimensional liquid crystals. II. Smectic and crystalline phases. *Phys. Rev. A* **12**: 245.

Chandrasekhar, S. 1966. Surface tension of liquid crystals. *Liquid Crystals: Proceedings of the International Conference on Liquid Crystals Held at Kent State University, August 16–20, 1965*, ed. G. H. Brown et al., p. 331. New York: Gordon and Breach.

Chandrasekhar, S. 1992. *Liquid Crystals.* 2d ed. Cambridge: Cambridge University Press.

Chatelain, P. 1943. Sur l'orientation des cristaux liquides par les surfaces frottées. *Bull. Soc. Fr. Minéral. Cristallogr.* **66**: 105.

Chen., W., Ouchi, Y., Moses, T., and Shen, Y. R. 1992. Surface electroclinic effect on the layer structure of a ferroelectric liquid crystal. *Phys. Rev. Lett.* **68**: 1547.

Cheng, J., Boyd, G. D., and Storz, F. G. 1980. A scanning electron microscope study of columnar topography and liquid-crystal alignment on obliquely deposited oxide surfaces at low rates. *Appl. Phys. Lett.* **37**: 716.

Cheng, M., Ho, J. T., Hui, S. W., and Pindak, R. 1988. Observation of two-dimensional hexatic behavior in free-standing liquid-crystal thin films. *Phys. Rev. Lett.* **61**: 550.

Chiarelli, P., Faetti, S., and Fronzoni, L. 1983. Structural transition at the free surface of the nematic liquid crystals MBBA and EBBA. *J. Physique* **44**: 1061.

Chiarelli, P., Faetti, S., and Fronzoni, L. 1984. Critical behaviour of the anchoring energy of the director at the free surface of a nematic liquid crystal. *Phys. Lett.* **101A**: 31.

Chuvyrov, A. N., Lachinov, A. N., Avzyanov, V. S., and Sonin, A. S. 1977. Investigation of the interface between nematic liquid-crystal and semiconductor phases. *Sov. Phys. Solid State* **19**: 695.

Cladis, P. E., Couder, Y., and Brand, H. R. 1985. Phase winding and flow alignment in freely suspended films of smectic-C liquid crystals. *Phys. Rev. Lett.* **55**: 2945.

Clark, N. A. 1985. Surface memory effects in liquid crystals: influence of surface composition. *Phys. Rev. Lett.* **55**: 292.

Cognard, J. 1981. The interfacial layer of inorganic solid surfaces and liquid crystals: implication on the anchoring energy. *Mol. Cryst. Liq. Cryst.* **64**: 331.

Cognard, J. 1982. Alignment of nematic liquid crystals and their mixtures. *Mol. Cryst. Mol. Liq. Suppl.* **1**: 1.

Crawford, G. P., Steele, L. M., Ondris-Crawford, R. J., Iannacchione, G. S., Yeager, C. J., Doane, J. W., and Finotello, D. 1992. Characterization of the cylindrical cavities of Anopore and Nuclepore membranes. *J. Chem. Phys.* **96**: 7788.

Crawford, G. P., Ondris-Crawford, R. J., Žumer, S., Keast, S. S., Neubert, M. E., and Doane, J. W. 1993a. Alignment and ordering mechanisms at a liquid crystal–solid interface. *Liq. Cryst.* **14**: 1573.

Crawford, G. P., Ondris-Crawford, R. J., Žumer, S., and Doane, J. W. 1993b. Anchoring and orientational wetting transitions of confined liquid crystals. *Phys. Rev. Lett.* **70**: 1838.

Crawford, G. P. and Žumer, S. 1995. Saddle-splay elasticity in nematic liquid crystals. *Int. J. Mod. Phys. B* **9**: 2469.

Creagh, L. T., and Kmetz, A. R. 1973. Mechanism of surface alignment in nematic liquid crystals. *Mol. Cryst. Liq. Cryst.* **24**: 59.

Crossland, W. A. 1975. Birefringence in silicon monoxide films used for aligning liquid crystal layers. *Appl. Phys. Lett.* **26**: 598.

Croxton, C. A. 1980. Statistical thermodynamics of the liquid crystal surface, smectic, nematic, and isotropic systems. *Mol. Cryst. Liq. Cryst.* **59**: 219.

Datta, P., Kaganowicz, G., and Levine, A. W. 1981. Alignment and wetting properties of nematic liquid crystals on plasma-polymerized films. *J. Colloid. Interface Sci.* **82**: 167.

de Gennes, P. G. 1971a. Short-range order effects in the isotropic phase of nematics and cholesterics. *Mol. Cryst. Liq. Cryst.* **12**: 193.

de Gennes, P. G. 1971b. Possible experiments on two-dimensional nematics. *Symp. Faraday Soc.* **no. 5**: 16.

de Gennes, P. G., and Prost, J. 1993. *The Physics of Liquid Crystals.* 2d ed. Oxford: Oxford University Press.

Derfel, G. 1991. Electric field effects in nematic layers with weak boundary anchoring. *Liq. Cryst.* **10**: 29.

Derjaguin, B. V., Titijevskaia, A. S., Abricossova, I. I., and Malkina, A. D. 1954. Investigation of the forces of interaction of surfaces in different media and their application to the problem of colloid stability. *Discuss. Faraday Soc.* **18**: 24.

Derzhanski, A. I., and Hinov, H. P. 1977. Polar flexoelectric deformations and second-order elasticity in nematic liquid crystals. *J. Physique* **38**: 1013.

Dierker, S. B., Pindak, R., and Meyer, R. B. 1986. Consequences of bond-orientational order on the macroscopic orientation patterns of thin tilted bexatic liquid-crystal films. *Phys. Rev. Lett.* **56**: 1819.

Di Lisi, G. A., Rosenblatt, C., Griffin, A. C., and Hari, U. 1990. Behaviour of the anchoring strength coefficient near a structural transition at a nematic-substrate interface. *Liq. Cryst.* **7**: 353.

Di Lisi, G. A., Rosenblatt, C., Akins, R. B., Griffin, A. C., and Hari, U. 1992. Anchoring strength coefficient of a monomer and its dimer at a polymer-coated interface. *Liq. Cryst.* **11**: 63.

Dubois, J. C., Gazard, M., and Zann, A. 1976. Liquid-crystal orientation induced by polymeric surfaces. *J. Appl. Phys.* **47**: 1270.

Dubois-Violette, E., and Parodi, O. 1969. Émulsions nématiques: effets de champ magnétiques et effets piézoélectriques. *J. Physique Coll.* **30**: C4–57.

Dubois-Violette, E., and de Gennes, P. G. 1975. Local Frederiks transitions near solid–nematic interface. *J. Physique Lett.* **36**, 255.

Dubois-Violette, E., and de Gennes, P. G. 1976. Effects of long-range van der Waals forces on the anchoring of a nematic fluid at an interface. *J. Colloid Interface Sci.* **57**: 403.

Dunmur, D. A., and Szumilin, K. 1989. Field quenching of director fluctuations in thin films of nematic liquid crystals. *Liq. Cryst.* **6**: 449.

Durand, G. 1993. Recent advances in nematic and smectic A anchoring on amorphous solid surfaces. *Liq. Cryst.* **14**: 159.

Ebner, C., and Saam, W. F. 1977. New phase-transition phenomena in thin argon films. *Phys. Rev. Lett.* **38**: 1486.

Ebner, C., Saam, W. F., and Sen, A. K. 1985. Wetting transitions in systems with van der Waals forces. *Phys. Rev. B* **32**: 1558.

Eckhardt, C. J., Peachey, N. M., Swanson, D. R., Takacs, J. M., Khan, M. A., Gong, X., Kim, J.-H., Wang, J., and Uphaus, R. A. 1993. Separation of chiral phases in monolayer crystals of racemic amphiphiles. *Nature* **362**: 614.

Eidner, K., Mayer, G., and Schuster, R. 1988. Determination of the surface anchoring energy of a nematic liquid crystal by polarization azimuth measurement. *Mol. Cryst. Liq. Cryst.* **159**: 27.

Elston, S. J. 1994. The alignment of a liquid crystal in a smectic A phase in a high surface tilt cell. *Liq. Cryst.* **16**: 151.

Ericksen, J. L. 1966. Inequalities in liquid crystal theory. *Phys. Fluids* **9**: 1205.

Evans, R. 1979. The nature of the liquid-vapour interface and other topics in the statistical mechanics of nonuniform, classical fluids. *Adv. Phys.* **28**: 143.

Evans, R., and Marini Bettolo Marconi, U. 1986. Fluids in narrow pores: adsorption, capillary condensation, and critical points. *J. Chem. Phys.* **84**: 2376.

Faetti, S., and Fronzoni, L. 1978. Molecular orientation in nematic liquid crystal films with two free surfaces. *Solid State Commun.* **25**: 1087.

Faetti, S., and Palleschi, V. 1984a. Molecular orientation and anchoring energy at the nematic–isotropic interface of 7CB. *J. Physique Lett.* **45**: 313.

Faetti, S., and Palleschi, V. 1984b. Nematic–isotropic interface of some members of the homologous series of 4-cyano-4′-(n-alkyl)biphenyl liquid crystals. *Phys. Rev. A* **30**: 3241.

Faetti, S., Gatti, M., Palleschi, V., and Sluckin, T. J. 1985a. Almost critical behavior of the anchoring energy at the interface between a nematic liquid crystal and an SiO substrate. *Phys. Rev. Lett.* **55**: 1681.

Faetti, S., and Palleschi, V. 1985b. Experimental investigation of surface deformations at the nematic–isotropic interface: a new method to measure the Nehring–Saupe elastic constant K_{13}. *J. Physique* **46**: 415.

Faetti, S., and Palleschi, V. 1987a. The twist elastic constant and anchoring energy of the nematic liquid crystal 4-n-octyl-4′-cyanobiphenyl. *Liq. Cryst.* **2**: 261.

Faetti, S. 1987b. Azimuthal anchoring energy of a nematic liquid crystal at a grooved interface. *Phys. Rev. A* **36**: 408.

Faetti, S. 1990. Anchoring at the interface between a nematic liquid crystal and an isotropic substrate. *Mol. Cryst. Liq. Cryst.* **179**: 217.

Faetti, S., Nobili, M., and Schirone, A. 1991. Experimental measurement of the azimuthal anchoring energy function at an SiO–nematic interface. *Liq. Cryst.* **10**: 95.

Faetti, S. 1993. On the K_{13} surfacelike elastic constant: an experimental method to test different theoretical models. *Liq. Cryst.* **15**: 807.

Faetti, S. 1994a. The K_{13} surfacelike elastic constrant: analysis of the predictions of different theoretical models. *Mol. Cryst. Liq. Cryst.* **241**: 131.

Faetti, S. 1994b. Theory of surface like elastic contributions in nematic liquid crystals. *Phys. Rev. E* **49**: 4192.

Faetti, S. 1994c. Resummation of higher-order terms in the free-energy of nematic liquid crystals. *Phys. Rev. E* **49**: 5332.

Feller, M. B., Chen, W., and Shen, Y. R. 1991. Investigation of surface-induced alignment of liquid-crystal molecules by optical second-harmonic generation. *Phys. Rev. A* **43**: 6778.

Ferguson, A., and Kennedy, S. J. 1938. On the surface tensions of liquid crystals. *Phil. Mag.* **26**: 41.

Filas, R. W., and Patel, J. S. 1987. Chemically induced high-tilt surfaces for liquid crystals. *Appl. Phys. Lett.* **50**: 1426.

Firpo, J. L., Dupin, J. J., Albinet, G., Bois, A., Casalta, L., and Baret, J. F. 1978. A theoretical study of phase transitions in bidimensional monolayers. *J. Chem. Phys.* **68**: 1369.

Firpo, J. L., Dupin, J. J., Albinet, G., and Baret, J. F. 1981. Model for the tricritical point of the chain-melting transition in a monomolecular layer of amphiphilic molecules. *J. Chem. Phys.* **74**: 2569.

Flatischler, K., Komitov, L., Lagerwall, S. T., and Stebler, B. 1991. Surface-induced alignment transition in a nematic layer with symmetrical boundary conditions. *Mol. Cryst. Liq. Cryst.* **198**: 119.

Foster, J. S., and Frommer, J. E. 1988. Imaging of liquid crystals using a tunnelling microscope. *Nature* **333**: 542.

Fournier, J. B. 1993. Curvature elasticity of smectic-A textures with virtual surface singularities. *Phys. Rev. Lett.* **70**: 1445.

Frank, F. C. 1958. Liquid crystals: on the theory of liquid crystals. *Discuss. Faraday Soc.* **25**: 19.

Frenkel, D., and McTague, J. P. 1979. Evidence for an orientationally ordered two-dimensional fluid phase from molecular-dynamics calculations. *Phys. Rev. Lett.* **42**: 1632.

Frenkel, D., and Eppenga, R. 1985. Evidence for algebraic orientational order in a two-dimensional hard-core nematic. *Phys. Rev. A* **31**: 1776.

Frenkel, D. 1992. Computer simulations of phase transition in liquid crystals. *Phase Transitions in Liquid Crystals*, ed. S. Martellucci and A. N. Chester, p. 67. New York: Plenum.

Frolova, E. K., Sarbey, O. G., and Sybashvily, A. S. 1984. Alignment of nematic liquid crystals on single crystal surfaces. *Mol. Cryst. Liq. Cryst.* **104**: 111.

Fukuro, H., and Kobayashi, S. 1988. Newly synthesized polyimide for aligning nematic liquid crystals accompanying high pretilt angles. *Mol. Cryst. Liq. Cryst.* **163**: 157.

Gannon, M. G. J. and Faber, T. E. 1978. The surface tension of nematic liquid crystals. *Philos. Mag. A* **37**: 117.

Garlerne, Y., and Liebert, L. 1990. Smectic O films. *Phys. Rev. Lett.* **64**: 906.

Garoff, S., Deckman, H. W., Dunsmuir, J. H., Alvarez, M. S., and Bloch, J. M. 1986. Bond-orientational order in Langmuir–Blodgett surfactant monolayers. *J. Physique* **47**: 701.

Gaub, H. E., Moy, V. T., and McConnell, H. M. 1986. Reversible formation of plastic two-dimensional lipid crystals. *J. Phys. Chem.* **90**: 1721.

Gay, J. M., Bienfait, M., and Suzanne, J. 1984. First-order wetting transition in CF_4 solid films adsorbed on the (0001) graphite surface. *J. Physique* **45**: 1497.

Geary, J. M., Goodby, J. W., Kmetz, A. R., and Patel, J. S. 1987. The mechanism of polymer alignment of liquid-crystal materials. *J. Appl. Phys.* **62**: 4100.

Gibbons, W. M., Shannon, P. J., Sun, S.-T., and Swetlin, B. J. 1991. Surface-mediated alignment of nematic liquid crystals with polarized laser light. *Nature* **351**: 49.

Gibbs, J. W. 1961. *The Scientific Papers of J. Willard Gibbs*. New York: Dover.

Gleeson, J. T., Palffy-Muhoray, P., and van Saarloos, W. 1991. Propagation of excitations induced by shear flow in nematic liquid crystals. *Phys. Rev. A* **44**: 2588.

Glogarova, M., and Durand, G. 1988. Anchoring strength of a nematic liquid crystal on a ferroelectric crystal interface. *J. Physique* **49**: 102.

Goodman, L. A., Mcginn, J. T., Anderson, C. H., and Digeronimo, F. 1977. Topography of obliquely evaporated silicon oxide films and its effect on liquid-crystal orientation. *IEEE Trans. Elec. Dev.* **ED-24**: 795.

Goossens, W. J. A. 1985. Bulk, interfacial, and anchoring energies of liquid crystals. *Mol. Cryst. Liq. Cryst.* **124**: 305.

Gramsbergen, E. F., de Jeu, W. H., and Als-Nielsen, J. 1986. Antiferroelectric surface layers in a liquid crystal as observed by synchrotron x-ray scattering. *J. Physique* **47**: 711.

Grandjean, F. 1916. L'orientation des liquides anisotropes sur les cristaux. *Bull. Soc. Fr. Minéral. Cristallogr.* **39**: 164.

Gray, C. G., and Gubbins, K. E. 1984. *Theory of Molecular Fluids*. New York: Oxford University Press.

Gray, G. W. 1962. *Molecular Structure and the Properties of Liquid Crystals*. London: Academic.

Griffiths, R. B. 1965. Ferromagnets and simple fluids near the critical point: some thermodynamic inequalities. *J. Chem. Phys.* **43**: 1958.

Gubbins, K. E. 1986. Molecular orientation at the free liquid surface. *Fluid Interfacial Phenomena*, ed. C. A. Croxton, p. 469. New York: Wiley.

Guttman, G. D., and Andelman, D. 1993. Electrostatic interactions in two-component membranes. *J. Physique II* **3**: 1411.

Guyot-Sionnest, P., Hsiung, H., and Shen, Y. R. 1986. Surface polar ordering in a liquid crystal observed by optical second-harmonic generation. *Phys. Rev. Lett.* **57**: 2963.

Hall, R., Miyachi, K., Newton, D., Takezoe, H., and Fukuda, A. 1992. Cell thickness effects in the determination of elastic constant ratios by observing Rayleigh light scattered intensity. *Japan. J. Appl. Phys. Lett.* **31**: 329.

Haller, I. 1974. Alignment and wetting properties of nematic liquids. *Phys. Rev. Lett.* **24**: 349.

Han, K. Y., Vetter, P., and Uchida, T. 1993. Determination of molecular inclination in rubbed polymer for liquid crystal alignment by measuring retardation. *Japan. J. Appl. Phys. Lett.* **32**: 1242.

Hara, M., Iwakabe, Y., Tochigi, K., Sasabe, H., Garito, A. F., and Yamada, A. 1990. Anchoring structure of smectic liquid-crystal layers on MoS_2 observed by scanning tunnelling microscopy. *Nature* **344**: 228.

Hara, M., Umemoto, T., Takezoe, H., Garito, A. F. and Sasabe, H. 1991. Novel quadruple zigzag structure of antiferroelectric liquid crystal molecules observed by scanning tunneling microscopy. *Jpn. J. Appl. Phys.* **30**: L2052.

Hatta, A. 1977. Application of infrared ATR spectroscopy to liquid crystals. I. Surface-induced orientation in thin films of nematic MBBA. *Bull. Chem. Soc. Japan* **50**: 2522.

Hatta, A. 1980. Application of infrared ATR spectroscopy to liquid crystals. IV. A comparison of molecular alignments of nematic EBCA in bulk and boundary layers under an external DC electric field. *Mol. Cryst. Liq. Cryst.* **62**: 251.

Hauck, G. 1982. Liquid crystal alignment on obliquely deposited Al_2O_3 surfaces. *Phys. Status Solidi A* **70**: K43.

Hauck, G., Koswig, H. D., and Labes, U. 1993. Field-induced reorientation in free-standing ferroelectric liquid crystal films. *Modern Topics in Liquid Crystals*, ed. A. Buka, p. 215. Singapore: World Scientific.

Heffner, W. R., Berreman, D. W., Sammon, M., and Meiboom, S. 1980. Liquid crystal alignment on surfactant treated obliquely evaporated surfaces. *Appl. Phys. Lett.* **36**: 144.

Hikmet, R. A. M., and Zwerver, B. H. 1991. Dielectric relaxation of liquid crystal molecules in anisotropic confinements. *Liq. Cryst.* **10**: 835.

Hill, T. L. 1956. *Statistical Mechanics*. New York: McGraw-Hill.

Hiltrop, K., and Stegemeyer, H. 1978. Contact angles and alignment of liquid crystals on lecithin monolayers. *Mol. Cryst. Liq. Cryst. Lett.* **49**: 61.

Hiltrop, K., and Stegemeyer, H. 1984. On the orientation of liquid crystals by monolayers of amphiphilic molecules. *Liquid Crystals and Ordered Fluids*, vol. 4, ed. A. C. Griffin and J. F. Johnson, p. 515. New York: Plenum.

Hinov, H. P. 1977. Influence of second-order elasticity on local Frederiks transitions near a nematic crystal–anisotropic crystal interface. *J. Physique Lett.* **38**: 215.

Hinov, H. P., and Derzhanski, A. I. 1979. Second-order elasticity and first-order Frederiks transitions in nematic liquid crystals. *J. Physique Coll.* **40**: C3–305.

Hinov, H. P. 1987. On the variation of the K_{13} nematic surfacelike volume energy and the nematic surface energy of Mada. *Mol. Cryst. Liq. Cryst.* **148**: 197.

Hinshaw, G. A., and Petschek, R. G. 1988a. Transitions and modulated phases in centrosymmetric ferroelectrics: Mean-field and renormalization-group predictions. *Phys. Rev. B* **37**: 2133.

Hinshaw, G. A., and Petschek, R. G. 1988b. Modulated phases in thin ferroelectric liquid-crystal films. *Phys. Rev. Lett.* **60**: 1864.

Hinshaw, G. A., and Petscheck, R. G. 1989. Transitions and modulated phases in chiral tilted smectic liquid crystals. *Phys. Rev. A* **39**: 5914.

Holyst, R., and Poniewierski, A. 1988a. Director orientation at the nematic-phase–isotropic-phase interface for the model of hard spherocylinders. *Phys. Rev. A* **38**: 1527.

Holyst, R., and Poniewierski, A. 1988b. Nematic-wall surface tension for perfectly aligned hard linear molecules. *Mol. Phys.* **65**: 1081.

Horn, R. G., Israelachvili, J. N., and Perez, E. 1981. Forces due to structure in a thin liquid crystal film. *J. Physique* **42**: 39.

Hsiung, H., and Shen, Y. R. 1986a. Probing the structure of freely suspended smectic-A films by optical second-harmonic generation. *Phys. Rev. A* **34**: 4303.

Hsiung, H., Rasing, T., and Shen, Y. R. 1986b. Wall-induced orientational order of a liquid crystal in the isotropic phase: an evanescent-wave-ellipsometry study. *Phys. Rev. Lett.* **57**: 3065.

Hsiung, H., and Beckerbauer, R. 1992. Orientational ordering of perfluoroalkyl-terminated amphiphiles at the air–water interface: a comparison between polymeric and monomeric monolayers. *Chem. Phys. Lett.* **193**: 123.

Iannacchione, G. S., and Finotello, D. 1992. Calorimetric study of phase transitions in confined liquid crystals. *Phys. Rev. Lett.* **69**: 2094.

Iannacchione, G. S., Strigazzi, A., and Finotello, D. 1993a. Interface influenced phase transitions in alkylcyanobiphenyl liquid crystals: a calorimetric study. *Liq. Cryst.* **14**: 1153.

Iannacchione, G. S., and Finotello, D. 1993b. Confinement and orientational study of liquid crystal phase transitions. *Liq. Cryst.* **14**: 1135.

Iannacchione, G. S., Crawford, G. P., Žumer, S., Doane, J. W., and Finotello, D. 1993c. Randomly constrained orientational order in porous glass. *Phys. Rev. Lett.* **71**: 2595.

Ichimura, K., Suzuki, Y., Seki, T., Hosoki, A., and Aoki, K. 1988. Reversible change in alignment mode of nematic liquid crystals regulated photochemically by command surface modified with an azobenzene monolayer. *Langmuir* **4**: 1214.

Ichimura, K., Suzuki, Y., Seki, T., and Kawanishi, Y. 1989. Reversible alignment change of liquid crystals induced by photochromic molecular films, 2. *Makromol. Chem. Rapid Commun.* **10**: 5.

Ichimura, K., Hayashi, Y., and Ishizuki, N. 1992. Photocontrol of in-plane alignment of a nematic liquid crystal by a photochromic spiropyran monolayer absorbing linearly polarized light. *Chem. Lett.* **1992**: 1063.

Idziak, S. H., Safinya, C. R., Hill, R. S., Kraiser, K. E., Ruths, M., Warriener, H. E., Steinber, S., Liang, K. S., and Israelachvili, J. N. 1994. The x-ray surface forces apparatus: structure of a thin smectic liquid crystal film under confinement. *Science* **264**: 1915.

Iimura, Y., Mada, H., and Kobayashi, S. 1984. Measurement of the phase transition of p-heptyl-p′-cyanobiphenyl in the bulk and boundary. *Phys. Lett.* **103**: 342.

Iimura, Y., Kusano, J., Kobayashi, S., Aoyagi, Y., and Sugano, T. 1993. Alignment control of a liquid crystal on a photosensitive polyvinylalcohol film. *Japan. J. Appl. Phys. Lett.* **32**: 93.

Iimura, Y., Kobayashi, N., and Kobayashi, S. 1994. A new method for measuring the azimuthal anchoring energy of a nematic liquid crystal. *Japan. J. Appl. Phys. Lett.* **33**: 434.

Imry, Y., and Ma, S.-K. 1975. Random-field instability of the ordered state of continuous symmetry. *Phys. Rev. Lett.* **35**: 1399.

Iwakabe, Y., Hara, M., Kondo, K., Tochigi, K., Mukoh, A., Garito, A. F., Sasabe, H., and Yamada, A. 1990. Two types of anchoring structure in smectic liquid crystal molecules. *Japan. J. Appl. Phys. Lett.* **29**: 2243.

Iwakabe, Y., Hara, M., Kondo, K., Tochigi, K., Mukoh, A., Yamada, A., Garito, A. F., and Sasabe, H. 1991. Correlation between bulk orderings and anchoring structures of liquid crystals studied by scanning tunneling microscopy. *Japan. J. Appl. Phys.* **30**: 2542.

Jákli, A., and Palffy-Muhoray, P. 1991. Observations of non-steady Poiseuille flow of the liquid crystal 8CB. *Mol. Cryst. Liq. Cryst.* **206**: 37.

Janning, J. L. 1972. Thin film surface orientation for liquid crystals. *Appl. Phys. Lett.* **21**: 173.

Jérôme, B., and Pieranski, P. 1988a. Bistable anchoring of nematics on SiO films. *Europhys. Lett.* **5**: 693.

Jérôme, B., and Pieranski, P. 1988b. Wetting and multistable anchorings in nematics. *J. Physique* **49**: 1601.

Jérôme, B., and Pieranski, P. 1989. Statics and dynamics of wetting: its contribution to the anchoring of nematics. *Liq. Cryst.* **5**: 683.

Jérôme, B., Bosseboeuf, A., and Pieranski, P. 1990. Anchoring of nematic liquid crystals on terraced silicon surfaces. *Phys. Rev. A* **42**: 6032.

Jérôme, B. 1991. Surface effects and anchoring in liquid crystals. *Rep. Prog. Phys.* **54**: 391.

Jérôme, B., and Boix, M. 1992. Wetting instability in nematic liquid crystals. *Europhys. Lett.* **17**: 169.

Jérôme, B., O'Brien, J., Ouchi, Y., Stanners, C., and Shen, Y. R. 1993. Bulk reorientation driven by orientational transition in a liquid crystal monolayer. *Phys. Rev. Lett.* **71**: 758.

Ji, Y., Kelly, J. R., and West, J. L. 1993. A study of the surface anchoring at a polymer/liquid crystal interface in the neighbourhood of the glass transition. *Liq. Cryst.* **14**: 1885.

Johnson, M. R., and Penz, P. A. 1977. Low-tilt-angle nematic alignment compatible with frit sealing. *IEEE Trans. Elec. Dev.* **ED-24**: 805.

Kahlweit, M., and Ostner, W. 1973. An estimation of the interfacial tension between the nematic and isotropic states of a liquid crystal. *Chem. Phys. Lett.* **18**: 589.

Kahn, F. J., Taylor, G. N., and Schonhorn, H. 1973. Surface-produced alignment of liquid crystals. *Proc. IEEE* **61**: 823.

Kalia, R. K., Vashishta, P., and Mahanti, S. D. 1982. Orientational order–disorder transition on a surface. *Phys. Rev. Lett.* **49**: 676.

Känel, H. V., Litster, J. D., Melngailis, J., and Smith, H. I. 1981. Alignment of nematic butoxybezilidene octylaniline by surface-relief gratings. *Phys. Rev. A* **24**: 2713.

Kats, E. I. 1971. Van der Waals forces in nonisotropic systems. *Sov. Phys.–JETP* **33**: 634.

Kimura, H., and Nakano, H. 1985. Statistical theory of surface tension and molecular orientations at the free surface in nematic liquid crystals. *J. Phys. Soc. Japan* **54**: 1730.

Kimura, H., and Nakano, H. 1986. Statistical theory of surface tension and molecular orientations in nematic liquid crystals. II. The nematic–isotropic interface. *J. Phys. Soc. Japan* **55**: 4186.

Kimura, H. 1993. Statistical theory of surface tension and molecular orientations in nematic liquid crystals. III. On hard flat walls. *J. Phys. Soc. Japan* **62**: 2725.

Kitzerow, H.-S. 1994. Polymer-dispersed liquid crystals from the nematic curvilinear aligned phase to ferroelectric films. *Liq. Cryst.* **16**: 1.

Kléman, M., and Williams, C. E. 1973. Anchoring energies and the nucleation of surface disclination lines in nematics. *Phil. Mag.* **28**: 725.

Kléman, M., and Ryschenkow, G. 1976. A note on two-dimensional nematic systems. *J. Chem. Phys.* **64**: 413.

Kobayashi, S., and Iimura, Y. 1994. Surface liquid crystal molecular alignment in LCDs and their electrooptical performance. *Liquid Crystal Materials, Devices, and Applications III*, ed. R. Shashidhar, p. 122. Proc. SPIE, vol. 2175. Bellingham, Wash.: SPIE.

Köhler, A. E. 1982. Die orientierungswirkung von substratoberflächen auf flüssigkristallmoleküle. I. Modellbetrachtungen. *Z. Phys. Chem.* **263**: 273.

Komitov, L. and Petrov, A. G. 1983. Optical method for determination of anchoring energy of tilted nematic layers. *Phys. Stat. Sol. (9)* **76**: 137.

Koshida, N., and Kikui, S. 1982. Magnetic field assisted alignment of nematic liquid crystal on a polymeric surface. *Appl. Phys. Lett.* **40**: 541.

Krishnaswamy, S., and Shashidhar, R. 1976. Experimental studies of the surface tension of nematic liquid crystals. *Mol. Cryst. Liq. Cryst.* **35**: 253.

Kutty, T. R. N., and Fischer, A. G. 1983. Planar orientation of nematic liquid crystals by chemisorbed polyvinyl alcohol surface layers. *Mol. Cryst. Liq. Cryst.* **99**: 301.

Kuzma, M., and Labes, M. M. 1983. Liquid crystals in cylindrical pores: effects on transition temperatures and singularities. *Mol. Cryst. Liq. Cryst.* **100**: 103.

Kwon, O'D., Beaglehole, D., Webb, W. W., Widom, B., Schmidt, J. W., Cahn, J. W., Moldover, M. R., and Stephenson, B. 1982. Thickness of liquid vapor wetting layer. *Phys. Rev. Lett.* **48**: 185.

Lachinov, A. N., Chuvyrov, A. N., and Sonin, A. S. 1982. Electroreflection in a system formed by a semiconductor and a nematic liquid crystal. *Sov. Phys. Solid State* **24**: 142.

Landau, L. D., and Lifshitz, E. M. 1959. *Statistical Physics*. London: Pergamon.

Langer, S. A., and Sethna, J. P. 1986. Textures in chiral smectic liquid-crystal film. *Phys. Rev. A* **34**: 5035.

Langevin, D., and Bouchita, M. A. 1973. Molecular order and surface tension for the nematic–isotropic interface of MBBA, deduced from light reflectivity and light scattering measurements. *Mol. Cryst. Liq. Cryst.* **22**: 317.

Lavrentovich, O. D., and Pergamenshchik, V. M. 1994. Stripe domain phase of a thin nematic film and the K_{13} divergence term. *Phys. Rev. Lett.* **73**: 979.

Lavrentovich, O. D., and Pergamenshchik, V. M. 1995. Patterns in thin liquid crystal films and the divergence ("surfacelike") elasticity. *Int. J. Mod. Phys. B* **9**: 2389.

Lee, S.-D., and Patel, J. S. 1990a. Surface effect on the field-induced structural change near the isotropic–cholesteric transition. *Phys. Rev. A* **42**: 997.

Lee, S.-D., Patel, J. S., and Meyer, R. B. 1990b. Effect of flexoelectric coupling on helix distortions in cholesteric liquid crystals. *J. Appl. Phys.* **67**: 1293.

Lehmann, O. 1922. *Handbuch der biologischen Arbeitsmeithoden*, ed. E. Abderhalden, p. 123. Berlin: Urban and Schwarzenberg.

Lejcek, L., Bechhoefer, J., and Oswald, P. 1992. Finite-size effects in smectic A droplets: steps, edge dislocations, and angular matching. *J. Physique II* **2**: 1511.

Levelut, A. M. 1975. Croissance épitaxique de phases smectieus ordonnées sur du mica. *J. Physique* **36**: 1029.

Levesque, D., and Wies, J.-J. 1985. Orientational structure of dipolar hard spheres near a hard neutral wall. *J. Stat. Phys.* **40**: 29.

Levy, Y., Jurich, M., and Swalen, J. D. 1985. Optical properties of thin layers of SiO_x. *J. Appl. Phys.* **57**: 2601.

Lin, B., and Taylor, P. L. 1994. Liquid crystal anchoring transition induced by thermal motion. *Liq. Cryst.* **16**: 831.

Lipkin, M. D., Rice, S. A., and Mohanty, U. 1985. The elastic constants of condensed matter: a direct-correlation function approach. *J. Chem. Phys.* **82**: 472.

Lipowsky, R., and Speth, W. 1983. Semi-infinite systems with first-order bulk transitions. *Phys. Rev. B* **28**: 3983.

Lipowsky, R. 1984. Upper critical dimension for wetting in systems with long-range forces. *Phys. Rev. Lett.* **52**: 1429.

Luckhurst, G. R., Sluckin, T. J., and Zewdie, H. B. 1986. Computer simulation studies of anisotropic systems. XV. The surface properties of the Lebwohl–Lasher model nematogen. *Mol. Phys.* **59**: 657.

Maclennan, J. 1990. Spontaneous director rotation in freely suspended ferroelectric liquid-crystal films. *Europhys. Lett.* **13**: 435.

Maclennan, J., and Seul, M. 1992. Novel stripe textures in nonchiral hexatic liquid-crystal films. *Phys. Rev. Lett.* **69**: 2082.

Maclennan, J., Sohling, U., Clark, N. A., and Seul, M. 1994. Textures in hexatic films of nonchiral liquid crystals: symmetry breaking and modulated phases. *Phys. Rev. E* **49**: 3207.

Mada, H., and Kobayashi, S. 1979a. Surface and bulk order parameters of a nematic liquid crystal. *Appl. Phys. Lett.* **35**: 4.

Mada, H. 1979b. Study on the surface alignment of nematic liquid crystals: temperature dependence of pretilt angles. *Mol. Cryst. Liq. Cryst.* **51**: 43.

Mada, H. 1979c. Study on the surface alignment of nematic liquid crystals: determination of the easy axis and temperature dependence of its field energy. *Mol. Cryst. Liq. Cryst.* **53**: 127.

Mada, H. 1981a. Surface stress anisotropy of treated glass and liquid crystal alignment. *J. Chem. Phys.* **75**: 372.

Mada, H., and Kobayashi, S. 1981b. Surface order parameter of 4-n-heptyl-4′-cyanobiphenyl. *Mol. Cryst. Liq. Cryst.* **66**: 57.

Mada, H. 1981c. Wall effect of director deformation in nematic liquid crystals. *Appl. Phys. Lett.* **39**: 701.

Mada, H. 1982. Wall effect of Frederiks transition in nematic liquid crystals. *Mol. Cryst. Liq. Cryst. Lett.* **82**: 53.

Mada, H., and Sonoda, T. 1993. A proposal on and verification of surface alignment of liquid crystals aligned by frictional rubbing. *Japan. J. Appl. Phys. Lett.* **32**: 1245.

Madhusudana, N. V., and Sumathy, K. R. 1983. Networks of point disclinations at the nematic–isotropic interface. *Mol. Cryst. Liq. Cryst. Lett.* **92**: 193.

Marcus, M. A. 1983. Structure of a nematic–isotropic interface in Landau theory. *Mol. Cryst. Liq. Cryst.* **100**: 253.

Maritan, A., Cieplak, M., Bellini, T., and Banavar, J. R. 1994. Nematic–isotropic transition in porous media. *Phys. Rev. Lett.* **72**: 4113.

Matsushige, K., Taki, S., Okabe, H., Takebayashi, Y., Hayashi, K., Yoshida, Y., Hoiuchi, T., Hara, K., Takehara, K., Isomura, K., and Taniguchi, H. 1993. Scanning tunneling microscope observation of a polar liquid crystal and its computer simulation. *Japan. J. Appl. Phys.* **32**: 1716.

Mauger, A., Zribi, G., Mills, L., and Toner, J. 1984. Substrate-induced orientational order in the isotropic phase of liquid crystals. *Phys. Rev. Lett.* **53**: 2485.

Mauguin, C. 1911. Les cristaux liquides de Lehmann. *Bull. Soc. Fr. Minéral. Cristallogr.* **34**: 71.

Mazenko, G. F., and Zannetti, M. 1984. Growth of order in a system with continuous symmetry. *Phys. Rev. Lett.* **53**: 2106.

McConnell, H. M., and Moy, V. T. 1988. Shapes of finite two-dimensional liquid domains. *J. Phys. Chem.* **92**: 4520.

McConnell, H. M. 1990. Harmonic shape transitions in lipid monolayer domains. *J. Phys. Chem.* **94**: 4728.

Meyer, R. B. 1972. Point declinations at a nematic–isotropic liquid interface. *Mol. Cryst. Liq. Cryst.* **16**: 355.

Meyer, R. B., and Pershan, P. S. 1973. Surface polarity induced domains in liquid crystals. *Solid State Commun.* **13**: 989.

Mirantsev, L. V. 1992. Oscillations in thin nematic layers. *Liq. Cryst.* **11**: 421.

Mirantsev, L. V. 1993. Oscillations of the Fréedericksz critical field in a superthin nematic layer in the vicinity of the second order nematic–smectic-A phase transition. *Liq. Cryst.* **15**: 799.

Miyano, K. 1979a. Wall-induced pretransitional birefringence: a new tool to study boundary aligning forces in liquid crystals. *Phys. Rev. Lett.* **43**: 51.

Miyano, K. 1979b. Surface-induced ordering of a liquid crystal in the isotropic phase. *J. Chem. Phys.* **71**: 4108.

Mizutani, W., Shigeno, M., Ono, M., and Kajimura, K. 1990. Voltage-dependent scanning tunneling microscopy images of liquid crystals on graphite. *Appl. Phys. Lett.* **56**: 1974.

Mochizuki, A., and Kobayashi, S. 1993. Surface effect on the threshold electric fields of cholesteric–nematic phase transition and its reverse process. *Mol. Cryst. Liq. Cryst.* **225**: 89.

Moldover, M. R., and Cahn, J. W. 1980. An interface phase transition: complete to partial wetting. *Science* **207**: 1073.

Moncton, D. E., and Pindak, R. 1979. Long-range order in two- and three-dimensional smectic-B liquid-crystal films. *Phys. Rev. Lett.* **43**: 701.

Monkade, M., Boix, M., and Durand, G. 1988. Order electricity and oblique nematic orientation on rough solid surfaces. *Europhys. Lett.* **5**: 697.

Moses, T., and Shen, Y. R. 1991. Pretransitional surface ordering and disordering of a liquid crystal. *Phys. Rev. Lett.* **67**: 2033.

Moses, T., Ouchi, Y., Chen, W., and Shen, Y. R. 1993. Pretransitional surface phenomena in ferroelectric liquid crystals. *Mol. Cryst. Liq. Cryst.* **225**: 55.

Murakami, J. 1977. Molecular theory of surface tension of liquid crystals. *J. Phys. Soc. Japan* **42**: 210.

Murthy, P. R. M., Raghunathan, V. A., and Madhusudana, N. V. 1993. An AC electrooptic technique of comparing the two anchoring energies of a hybrid aligned nematic cell. *Liq. Cryst.* **14**: 1107.

Muzny, C. D., and Clark, N. A. 1992. Direct observation of the Brownian motion of a liquid-crystal topological defect. *Phys. Rev. Lett.* **68**: 804.

Myrvold, B. O., Kondo, K., and Oh-hara, S. 1993. The relationship between liquid crystal alignment in the nematic and smectic C* phases. *Liq. Cryst.* **15**: 429.

Naemura, S. 1978. Measurement of anisotropic interfacial interactions between a nematic liquid crystal and various substrates. *Appl. Phys. Lett.* **33**: 1.

Naemura, S. 1979. Anisotropic interactions between MBBA and surface-treated substrates. *J. Physique Coll.* **40**: C3–514.

Naemura, S. 1980. Polar and nonpolar contributions to liquid-crystal orientations on substrates. *J. Appl. Phys.* **51**: 6149.

Naemura, S. 1981. Physicochemical study on liquid crystal–substrate interfacial interactions. *Mol. Cryst. Liq. Cryst.* **68**: 183.

Naggiar, P. V. 1943. Phénomènes d'orientation dan les liquides nématiques: une nouvelle méthode de mesure de la tension superficielle applicable à ces liquids. *Ann. Physique* **18**: 5.

Nakanishi, H., and Fisher, M. E. 1982. Multicriticality of wetting, prewetting, and surface transitions. *Phys. Rev. Lett.* **49**: 1565.

Nakatani, N., and Hirota, M. 1981. Alignment of nematic MBBA on the cleavage surface of ferroelectric TGS. *Japan. J. Appl. Phys.* **20**: 2281.

Napiórkowski, M., and Dietrich, S. 1986. Significance of the bulk correlation length for wetting transitions. *Phys. Rev. B* **34**: 6469.

Navascués, G. 1976. Adsorption of fluids: improved calculation of the density profile. *J. Chem. Soc. Faraday Trans. II* **72**: 2035.

Navascués, G., and Berry, M. V. 1977. The statistical mechanics of wetting. *Mol. Phys.* **34**: 649.

Nehring, J., and Saupe, A. 1971. On the elastic theory of uniaxial liquid crystals. *J. Chem. Phys.* **54**: 337.

Nehring, J., Kmetz, A. R., and Scheffer, T. J. 1976. Analysis of weak-boundary-coupling effects in liquid-crystal displays. *J. Appl. Phys.* **47**: 850.

Nelson, P., and Powers, T. 1992. Rigid chiral membranes. *Phys. Rev. Lett.* **69**: 3409.

Nicastro, A. J., and Keyes, P. H. 1984. Electric-field-induced critical phenomena at the nematic–isotropic transition and the nematic–isotropic critical point. *Phys. Rev.* **30**: 3156.

Nobili, M., and Durand, G. 1992a. Disorientation-induced disordering at a nematic-liquid-crystal–solid interface. *Phys. Rev. A* **46**: R6174.

Nobili, M., Lazzari, C., Schirone, A., and Faetti, S. 1992b. Azimuthal anchoring energy at an SiO_x–nematic interface. *Mol. Cryst. Liq. Cryst.* **212**: 97.

Ocko, B. M., Braslau, A., Pershan, P. S., Als-Nielsen, J., and Deutsch, M. 1986. Quantized layer growth at liquid-crystal surfaces. *Phys. Rev. Lett.* **57**: 94.

Ohgawara, M., and Uchida, T. 1981a. Effect of impurities on liquid crystal molecular orientation on inorganic surfaces. *Japan. J. Appl. Phys. Lett.* **20**: 75.

Ohgawara, M., and Uchida, T. 1981b. Relation between polarity of substrate surface and liquid crystal molecular orientation. *Japan. J. Appl. Phys. Lett.* **20**: 237.

Oh-ide, T., Kuniyasu, S., and Kobayashi, S. 1988. Surface coupling between nematic liquid crystals and rubbed polyimide substrates for pure twist deformation: dependence on rubbing strength. *Mol. Cryst. Liq. Cryst.* **164**: 91.

Okano, K., and Murakami, J. 1979. Van der Waals dispersion force contribution to the interfacial free energy of nematic liquid crystals. *J. Physique Coll.* **40**: C3–525.

Okano, K., Matsuura, N., and Kobayashi, S. 1982. Alignment of a liquid crystal on an anisotropic substrate. *Japan. J. Appl. Phys. Lett.* **21**: 109.

Okano, K. 1983. Anisotropic excluded volume effect and alignment of nematic liquid crystal in a sandwich cell. *Japan. J. Appl. Phys. Lett.* **22**: 343.

Oldano, C., and Barbero, G. 1985a. Possible boundary discontinuities of the tilt angle in nematic liquid crystals. *J. Physique* **46**: 451.

Oldano, C., and Barbero, G. 1985b. An *ab initio* analysis of the second order elasticity effect on nematic configurations. *Phys. Lett.* **100A**: 213.

Oliveira, E. A., Photinos, P. J., and Figueiredo Neto, A. M. 1993. Anchoring properties of a lyotropic and nematic discotic C_sPFO water mixture. *Liq. Cryst.* **14**: 837.

Ondris-Crawford, R. J., Crawford, G. P., Žumer, S., and Doane, J. W. 1993. Curvature-induced configuration transition in confined nematic liquid crystals. *Phys. Rev. Lett.* **70**: 194.

Ong, H. L., Meyer, R. B., and Hurd, A. J. 1984. Multistable orientation in a nematic liquid crystal cell induced by external field and interfacial interaction. *J. Appl. Phys.* **55**: 2809.

Ong, H. L., Hurd, A. J., and Meyer, R. B. 1985. Alignment of nematic liquid crystals by inhomogeneous surfaces. *J. Appl. Phys.* **57**: 186.

Ong, H. L. 1987. Interaction of liquid crystals with electromagnetic fields and with homogeneous and inhomogeneous surfaces. *Mol. Cryst. Liq. Cryst.* **144**: 17.

Ono, S. and Kondo, S. 1960. Molecular theory of surface tension of liquid. *Handbuch der Physik X*, ed. S. Flügge, p. 134. Berlin: Springer.

Oseen, C. W. 1933. The theory of liquid crystals. *Trans. Faraday Soc.* **29**: 883.

Osipov, M. A., and Sluckin, T. J. 1993. Molecular theory of order electricity. *J. Physique II* **3**: 793.

Oswald, P., Melo, F., and Germain, C. 1989. Smectic A–smectic B interface: faceting and surface free energy measurement. *J. Physique* **50**: 3527.

Oswald, P. 1991. Role of elastic effects in the secondary instabilities of the nematic–isotropic interface. *J. Physique II* **1**: 571.

Oswald, P., and Melo, F. 1992. Attachment kinetics at the smectic A–Smectic B interface. *J. Physique II* **2**: 1345.

Ouchi, Y., Feller, M. B., Moses, T., and Shen, Y. R. 1992. Surface memory effect at the liquid crystal–polymer interface. *Phys. Rev. Lett.* **68**: 3040.

Pandit, R., Schick, M., and Wortis, M. 1982. Systematics of multilayer adsorption phenomena on attractive substrates. *Phys. Rev. B* **26**: 5112.

Pang, J., Muzny, C. D., and Clark, N. A. 1992. String defects in freely suspended liquid-crystal films. *Phys. Rev. Lett.* **69**: 2783.

Papenfuss, C., and Muschik, W. 1992. Balance equations for liquid crystals in two dimensions. *Mol. Mater.* **2**: 1.

Parks, C. S., Copic, M., Mahmood, R., and Clark, N. A. 1994. Dynamic behaviour at a nematic liquid crystal–rubbed nylon interface using evanescent wave photon correlation spectroscopy. *Liq. Cryst.* **16**: 135.

Parks, D. C., Clark, N. A., Walba, D. M., and Beale, P. D. 1993. Scanning tunneling microscopy of coexisting 2D crystalline and 1D stacking-disordered phases at the chiral liquid crystal–graphite interface. *Phys. Rev. Lett.* **70**: 607.

Parmar, D. S., Clark, N. A., Keller, P., Walba, D. M., and Wand, M. D. 1990. Physical properties and alignment of a polymer–monomer ferroelectric liquid crystal mixture. *J. Physique* **51**: 355.

Parsons, J. D. 1975. Ordering and distortion at the nematic–isotropic fluid interface. *Mol. Cryst. Liq. Cryst.* **31**: 79.

Parsons, J. D. 1976. A molecular theory of surface tension in nematic liquid crystals. *J. Physique* **37**: 1187.

Parsons, J. D. 1978. Structural critical point at the free surface of a nematic liquid crystal. *Phys. Rev. Lett.* **41**: 877.

Parsons, J. D. 1980. A statistical theory of orientational ordering at the free surface of a liquid of ellipsoidal molecules. *Mol. Phys.* **42**: 951.

Patel, J. S., and Yokoyama, H. 1993. Continuous anchoring transition in liquid crystals. *Nature* **362**: 525.

Patrick, D. L., and Beebe, T. P. 1994. Substrate defects and variations in interfacial ordering of monolayer molecular films on graphite. *Langmuir* **10**: 298.

Pawlowska, Z., Kventsel, G. F., and Sluckin, T. J. 1987. Theory of smectic wetting and layering. *Phys. Rev. A* **36**: 992.

Perez, E., Proust, J. E., and Ter-Minassian-Saraga, L. 1977. Interfacial origin of liquid crystal anchorage. *Mol. Cryst. Liq. Cryst.* **42**: 167.

Perez, E., Proust, J. E., and Ter-Minassian-Saraga, L. 1978. Structural disjoining pressure in thin film of liquid crystals. I. Thermodynamics and Frederiksz transition with surface fields. *Colloid Polym. Sci.* **256**: 784.

Perez, E., and Proust, J. E. 1979. Anisotropic interactions across the interface of liquid crystals–glass with adsorbed monolayers of polysoaps. *J. Colloid Interface Sci.* **68**: 48.

Perez, E., Proust, J. E., and Ter-Minassian-Saraga, L., 1988. Thin films from liquid crystals. *Thin Liquid Films: Fundamentals and Applications*, ed. I. Ivanov, p. 891. New York: Dekker.

Pergamenshchik, V. M. 1993a. Surfacelike elasticity-induced spontaneous twist deformations and long-wavelength stripe domains in a hybrid nematic layer. *Phys. Rev. E* **47**: 1881.

Pergamenshchik, V. M. 1993b. Phenomenological approach to the problem of the K_{13} surfacelike elastic term in the free energy of a nematic liquid crystal. *Phys. Rev. E* **48**: 1254.

Pergamenshchik, V. M., Teixeira, P. I. C., and Sluckin, T. J. 1993c. Distortions induced by the K_{13} surfacelike elastic term in a thin nematic liquid-crystal film. *Phys. Rev. E* **48**: 1265.

Pershan, P. S., and Als-Nielsen, J. 1984. X-ray reflectivity from the surface of a liquid crystal: surface structure and absolute value of critical fluctuations. *Phys. Rev. Lett.* **52**: 759.

Petrov, M., Levelut, A. M., and Durand, G. 1982. Wall orientation of smectic C liquid single crystals. *Mol. Cryst. Liq. Cryst. Lett.* **82**: 221.

Petrov, M., Braslau, A., Levelut, A. M., and Durand, G. 1992. Surface-induced transitions in the nematic phase of 4-n octyloxybenzoic acid. *J. Physique II* **2**: 1159.

Pettey, D., and Lubensky, T. C. 1993. Star defects on flat and spherical surfaces. *J. Physique II* **3**: 1571.

Pieranski, P. 1980. Two-dimensional interfacial colloidal crystals. *Phys. Rev. Lett.* **45**: 569.

Pieranski, P., and Jérôme, B. 1989. Adsorption-induced anchoring transitions at nematic-liquid-crystal–crystal interfaces. *Phys. Rev. A* **40**: 317.

Pindak, R., Bishop, D. J., and Sprenger, W. O. 1980a. Mechanical measurements of the smectic B–A transition in liquid-crystal films. *Phys. Rev. Lett.* **44**: 1461.

Pindak, R., Young, C. Y., Meyers, R. B., and Clark, N. A. 1980b. Macroscopic orientation patterns in smectic-C films. *Phys. Rev. Lett.* **45**: 1193.

Pindak, R., Sprenger, W. O., Bishop, D. J., Osheroff, D. D., and Goodby, J. W. 1982. Mechanical measurements on free-standing films of smectic liquid-crystal phases. *Phys. Rev. Lett.* **48**: 173.

Pismen, L. M., and Rubinstein, B. Y. 1992. Motion of interacting point defects in nematics. *Phys. Rev. Lett.* **69**: 96.

Polak, R. D., Crawford, G. P., Kostival, B. C., Doane, J. W., and Žumer, S. 1994. Optical determination of the saddle-splay elastic constant K_{24} in nematic liquid crystals. *Phys. Rev. E* **49**: R978.

Pollack, J. M., Haas, W. E., and Adams, J. E. 1977. Topology of obliquely coated silicon monoxide layers. *J. Appl. Phys.* **48**: 831.

Poniewierski, A., and Stecki, J. 1979. Statistical theory of the elastic constants of nematic liquid crystals. *Mol. Phys.* **38**: 1931.

Poniewierski, A., and Sluckin, T. J. 1984. Statistical mechanics of a simple model of the nematic liquid crystal–wall interface. *Mol. Cryst. Liq. Cryst.* **111**: 373.

Poniewierski, A., and Sluckin, T. J. 1985. Statistical mechanics of a simple model of the nematic liquid crystal–wall interface II. *Mol. Cryst. Liq. Cryst.* **126**: 143.

Poniewierski, A., and Sluckin, T. J. 1987. Theory of the nematic–isotropic transition in a restricted geometry. *Liq. Cryst.* **2**: 281.

Poniewierski, A., and Holyst, R. 1988. Nematic alignment at a solid substrate: the model of hard spherocylinders near a hard wall. *Phys. Rev. A* **38**: 3721.

Porte, G. 1976. Tilted alignment of MBBA induced by short-chain surfactants. *J. Physique* **37**: 1245.

Povse, T., Musevic, I., Zeks, B., and Blinc, R. 1993. Phase transitions in ferroelectric liquid crystals in a restricted geometry. *Liq. Cryst.***14**: 1587.

Proust, J. E., Ter-Minassian-Saraga, L., and Guyon, E. 1972. Orientation of a nematic liquid crystal by suitable boundary surfaces. *Solid State Commun.* **11**: 1227.

Proust, J. E., and Perez, E. 1977. Films minces smectiques symétriques et asymétriques. *J. Physique Lett.* **38**: 91.

Proust, J. E., and Ter-Minassian-Saraga, L. 1979. Films minces de cristaux liquides. *J. Physique Coll.* **40**: C3–490.

Qiu, X., Ruiz-Garcia, J., Stine, K. J., Knobler, C. M., and Selinger, J. V. 1991. Direct observation of domain structure in condensed monolayer phases. *Phys. Rev. Lett.* **67**: 703.

Raghunathan, V. A. 1995. Undulatory instability of the nematic–isotropic interface. *Phys. Rev. E.* **51**: 896.

Rapini, A., and Papoular, M. 1969. Distorsion d'une lamelle nématique sous champ magnétique conditions d'ancrage aux parois. *J. Physique Coll.* **30**: C4–54.

Raynes, E. P., Rowell, D. K., and Shanks, I. A. 1976. Liquid crystal surface alignment treatment giving controlled low angle tilt. *Mol. Cryst. Liq. Cryst. Lett.* **34**: 105.

Riviere, D., Levy, Y., and Guyon, E. 1979. Determination of anchoring energies from surface tilt angle measurements in a nematic liquid crystal. *J. Physique Lett.* **40**: 215.

Robert, M. 1984. Critical behavior of fluid interfaces in two dimensions. *Phys. Rev. A* **30**: 2785.

Ronis, D., and Rosenblatt, C. 1980. Unified model of the smectic-A, nematic, and isotropic phases for bulk, interfaces, and thin films. I. Bulk. *Phys. Rev. A* **21**: 1687.

Rosenblatt, C., Pindak, R., Clark, N. A., and Meyer, R. B. 1979. Freely suspended ferroelectric liquid-crystal films: absolute measurements of polarization, elastic constant, and viscosities. *Phys. Rev. Lett.* **42**: 1220.

Rosenblatt, C., Meyer, R. B., Pindak, R., and Clark, N. A. 1980a. Temperature behavior of ferroelectric liquid-crystal thin films: a classical XY system. *Phys. Rev. A* **21**: 140.

Rosenblatt, C., and Amer, N. M. 1980b. Optical determination of smectic A layer spacing in freely suspended thin films. *Appl. Phys. Lett.* **36**: 432.

Rosenblatt, C., and Ronis, D. 1981. Unified model of the smectic-A, nematic, and isotropic phases for bulk, interfaces, and thin films. II. Interfaces and thin films. *Phys. Rev. A* **23**: 305.

Rosenblatt, C. 1984a. Temperature dependence of the anchoring strength coefficient at a nematic liquid crystal–wall interface. *J. Physique* **45**: 1087.

Rosenblatt, C. 1984b. Liquid-crystal Fréedericksz transitions and surface-induced smectic ordering. *Phys. Rev. Lett.* **53**: 791.

Rowlinson, J. S., and Widom, B. 1982. *Molecular Theory of Capillarity*. New York: Oxford University Press.

Ruipeng, S., Ximin, H., Kai, M., Zongkai, W., and Min, J. 1994. Alternative method of investigating surface torsional anchoring energy for pure nematic liquid crystals. *Phys. Rev. E* **50**: 1253.

Ruiz-Garcia, J., Qiu, X., Tsao, M.-W., Marshall, G., Knobler, C. M., Overbeck, G. A., and Möbius, D. 1993. Splay stripe textures in Langmuir monolayers. *J. Phys. Chem.* **97**: 6955.

Ryschenkow, G., and Kléman, M. 1976. Surface defects and structural transition in very low anchoring energy nematic thin films. *J. Chem. Phys.* **64**: 404.

Salamon, Z., and Skibinski, A. 1983. Fluorescence emission of guest molecules as a tool of surface alignment study in nematic liquid crystals. *Mol. Cryst. Liq. Cryst.* **90**: 205.

Sanda, P. N., Dove, D. B., Ong, H. L., Jansen, S. A., and Hoffmann, R. 1989. Role of surface bonding on liquid-crystal alignment at metal surfaces. *Phys. Rev. A* **39**: 2653.

Sano, M., and Kunitake, T. 1992a. Influence of phase and chain length of liquid crystals on wetting phenomena of graphite. *Langmuir* **8**: 320.

Sano, M., Sasaki, D. Y., Isayama, M., and Kunitake, T. 1992b. Interfacial and bulk structures of homologous series of azobenzene amphiphiles studied by scanning tunneling microscopy and x-ray diffraction. *Langmuir* **8**: 1893.

Schmidt, J. W., and Moldover, M. R. 1983. First-order wetting transition at a liquid–vapor interface. *J. Chem. Phys.* **79**: 379.

Schmidt, J. W., and Moldover, M. R. 1986. A search for the presetting line. *J. Chem. Phys.* **84**: 4563.

Schopohl, N., and Sluckin, T. J. 1987. Defect core structure in nematic liquid crystals. *Phys. Rev. Lett.* **59**: 2582.

Schopohl, N., and Sluckin, T. J. 1988. Hedgehog structure in nematic and magnetic systems. *J. Physique* **49**: 1097.

Schröder, H. 1977. Molecular statistical theory for inhomogeneous nematic liquid crystals with boundary conditions. *J. Chem. Phys.* **67**: 16.

Schwartz, D. K., Garnaes, J., Viswanathan, R., and Zasadzinski, J. A. N. 1992. Surface order and stability of Langmuir–Blodgett films. *Science* **257**: 508.

Schwartz, D. K. 1993a. Instant patterns in thin films. *Nature* **362**: 593.

Schwartz, D. K., Viswanathan, R., and Zasadzinski, J. A. N. 1993b. Commensurate defect superstructures in a Langmuir–Blodgett film. *Phys. Rev. Lett.* **70**: 1267.

Seki, T., Sakuragi, M., Kawanishi, Y., Suzuki, Y., Tamaki, T., Fukuda, R., Hiramatsu, H., Fujiwara, K., and Ichimura, K. 1992a. Structural defects of LB molecules on the command-surface-type photocontrols of liquid crystal alignment. *Journal of Photopolymer Science and Technology.* **5**: 569.

Seki, T., Sakuragi, M., Kawanishi, Y., Suzuki, Y., Tamaki, T., Ichimura, K., Fukuda, R., and Hiramatsu, H. 1992b. Modulated photoregulation of liquid crystal alignment by azobenzene Langmuir–Blodgett layers: reversible alignment changes of liquid crystals induced by photochromic molecular films, Part 11. *Thin Solid Films* **210–211**: 836.

Seki, T., Sakuragi, M., Kawanishi, Y., Suzuki, Y., Tamaki, T., Fukuda, R., and Ichimura, K. 1993. Command surface of Langmuir–Blodgett films: photoregulations of liquid crystal alignment by molecularly tailored surface azobenzene layers. *Langmuir* **9**: 211.

Selinger, J. V. 1991. Dynamics of tilted hexatic phases in liquid-crystal films. *J. Physique II* **1**: 1363.

Selinger, J. V., Wang, Z.-G., Bruinsma, R. F., and Knobler, C. M. 1993. Chiral symmetry breaking in Langmuir monolayers and smectic films. *Phys. Rev. Lett.* **70**: 1139.

Sen, A. K., and Sullivan, D. E. 1987. Landau–de Gennes theory of wetting and orientational transitions at a nematic-liquid–substrate interface. *Phys. Rev. A* **35**: 1391.

Seo, D.-S., Kobayashi, S., and Mochizuki, A. 1992a. Generation of the pretilt angles in nematic liquid crystal (5CB) aligned on the rubbed polypyrrole films. *Appl. Phys. Lett.* **60**: 1025.

Seo, D.-S., Iimura, Y., and Kobayashi, S. 1992b. Temperature dependence of the polar anchoring strength of weakly rubbed polyimide films for the nematic liquid crystal (5CB). *Appl. Phys. Lett.* **61**: 234.

Seo, D.-S., Kobayashi, S., and Nishikawa, M. 1992c. Study of the pretilt angle for 5CB on rubbed polyimide films containing trifluoromethyl moiety and analysis of the surface atomic concentration of F/C (%) with an electron spectroscope for chemical analysis. *Appl. Phys. Lett.* **61**: 2392.

Seo, D.-S., Muroi, K., Isogami, T., Matsuda, H., and Kobayashi, S. 1992d. Polar anchoring strength and the temperature dependence of nematic liquid crystals (5CB) aligned on rubbed polystyrene films. *Japan. J. Appl. Phys.* **31**: 2165.

Seo, D.-S., Matsuda, H., Oh-ide, T., and Kobayashi, S. 1993a. Alignment of nematic liquid crystal (5CD) on the treated substrates: characterization of orientation films, generation of pretilt angles, and surface anchoring strength. *Mol. Cryst. Liq. Cryst.* **224**: 13.

Seul, M., and Chen, V. S. 1993. Isotropic and aligned stripe phases in a monomolecular organic film. *Phys. Rev. Lett.* **70**: 1658.

Seul, M., and Andelman, D. 1995. Domain shapes and patterns: the phenomenology of modulated phases. *Science* **267**: 476.

Shannon, P. J., Gibbons, W. M., and Sun, S.-T. 1994. Patterned optical properties in photopolymerized surface-aligned liquid-crystal films. *Nature* **368**: 532.

Sharlow, M. F., and Gelbart, W. M. 1992. On the parallel–perpendicular transition for a nematic phase at a wall. *Liq. Cryst.* **11**: 25.

Shen, Y. R. 1989a. Studies of liquid crystal monolayers and films by optical second-harmonic generation. *Liq. Cryst.* **5**: 635.

Shen, Y. R. 1989b. Surface properties probed by second-harmonic and sum-frequency generation. *Nature* **337**: 519.

Sheng, P. 1976. Phase transition in surface-aligned nematic films. *Phys. Rev. Lett.* **37**: 1059.

Sheng, P. 1982. Boundary-layer phase transition in nematic liquid crystals. *Phys. Rev. A* **26**: 1610.

Shi, Y., Cull, B., and Kumar, S. 1993. Morphology and ordering in thin liquid-crystal films on flat glass substrates. *Phys. Rev. Lett.* **71**: 2773.

Shigeno, M., Mizutani, W., Suginoya, M., Ohmi, M., Kajimura, K., and Ono, M. 1990. Observation of liquid crystals on graphite by scanning tunneling microscopy. *Jpn. J. Appl. Phys.* **29**: L119.

Sicart, J. 1976. Méthode de mesure de l'énergie d'ancrage d'un nématique: application à l'ancrage sur une paroi non traitée. *J. Physique Lett.* **37**: 25.

Simons, B. D. 1992. Stability of dipolar fluid membranes. *J. Physique II* **2**: 1141.

Singh, Y. 1984. Molecular theory of liquid crystals: application to the nematic phase. *Phys. Rev. A* **30**: 583.

Singh, Y., and Singh, K. 1986a. Density-functional theory of curvature elasticity in nematic liquids. I. *Phys. Rev. A* **33**: 3481.

Singh, K., and Singh, Y. 1986b. Density-functional theory of curvature elasticity in nematic liquids. II. Effect of long-range dispersion interactions. *Phys. Rev. A* **34**: 548.

Singh, K., and Singh, Y. 1987. Density-functional theory of curvature elasticity in nematic liquids. III. Numerical results for the Berne–Pechukas pair-potential model. *Phys. Rev. A* **35**: 3535.

Sluckin, T. J., and Chalker, J. T. 1984. Re-entrant wetting transitions in molecular fluid mixtures. *J. Physique Lett.* **45**: 863.

Sluckin, T. J., and Poniewierski, A. 1985. Novel surface phase transition in nematic liquid crystals: wetting and the Kosterlitz–Thouless transition. *Phys. Rev. Lett.* **55**: 2907.

Sluckin, T. J., and Poniewierski, A. 1986. Orientational wetting transitions and related phenomena in nematics. *Fluid Interfacial Phenomena*, ed. C. A. Croxton, p. 215. New York: Wiley.

Smith, D. P. E., Hörber, H., Gerber, C., and Binning, G. 1989. Smectic liquid crystal monolayers on graphite observed by scanning tunneling microscopy. *Science* **245**: 43.

Smith, D. P. E. 1991. Defects in alkylcyanobiphenyl molecular crystals studied by scanning tunneling microscopy. *J. Vac. Sci. Technol. B* **9**: 1119.

Smith, D. P. E., Heckl, W. M., and Klagges, H. A. 1992. Ordering of alkylcyanobiphenyl molecules at MoS_2 and graphite surfaces studied by tunneling microscopy. *Surf. Sci.* **278**: 166.

Smith, G. W. 1984. The nematic/isotropic interfacial energy: determination from the Frenkel relation for droplet coalescence. *Mol. Cryst. Liq. Cryst. Lett.* **102**: 65.

Smith, G. W., White, J. L., and Buechler, M. 1985. Mesophase/isotropic phase interfacial energy: determination from coalescence kinetics of mesophase pitch. *Carbon* **23**: 117.

Somoza, A. M., and Tarazona, P. 1991. Density functional theory of the elastic constants of a nematic liquid crystal. *Mol. Phys.* **72**: 911.

Sparavigna, A., Komitov, L., Stebler, B., and Strigazzi, A. 1991. Static splay-stripes in a hybrid aligned nematic layer. *Mol. Cryst. Liq. Cryst.* **207**: 265.

Sparavigna, A., Komitov, L., and Strigazzi, A. 1992a. Influence of a magnetic field on the periodic splay-stripes in hybrid aligned nematics. *Mol. Cryst. Liq. Cryst.* **212**: 289.

Sparavigna, A., Komitov, L., Lavrentovich, O. D., and Strigazzi, A. 1992b. Saddle-splay and periodic instability in a hybrid aligned nematic layer subjected to a normal magnetic field. *J. Physique II* **2**: 1881.

Sparavigna, A., Lavrentovich, O. D., and Strigazzi, A. 1994. Periodic stripe domains and hybrid-alignment regime in nematic liquid crystals: threshold analysis. *Phys. Rev. E* **49**: 1344.

Spong, J. K., LaComb, L. J., Dovek, M. M., Frommer, J. E., and Foster, J. S. 1989a. Imaging of liquid crystals with tunneling microscopy. *J. Physique* **50**: 2139.

Spong, J. K., Mizes, H. A., LaComb, L. J., Dovek, M. M., Frommer, J. E., and Foster, J. S. 1989b. Contrast mechanism for resolving organic molecules with tunnelling microscopy. *Nature* **338**: 137.

Sprokel, G. J. 1977. Molecular order induced by cell walls. *Mol. Cryst. Liq. Cryst.* **42**: 233.

Stanley, H. E. 1971. *Introduction to Phase Transitions in Critical Phenomena*. Oxford: Oxford University Press.

Stecki, J., and Poniewierski, A. 1980. Application of the statistical theory of elastic constants of nematics to spherocylindrical molecules with attractive forces. *Mol. Phys.* **41**: 1451.

Stoebe, T., and Huang, C. C. 1992a. Simultaneous measurement of heat capacity and in-plane density of thin freestanding liquid-crystal films. *Phys. Rev. Lett.* **68**: 2944.

Stoebe, T., Geer, R., Huang, C. C., and Goodby, J. W. 1992b. Layer-by-layer surface ordering near a continuous transition in free-standing liquid-crystal films. *Phys. Rev. Lett.* **69**: 2090.

Sugimura, A., Yamamoto, N., and Kawamura, T. 1981. High surface ordering of nematic liquid crystal using periodicity grating. *Japan. J. Appl. Phys.* **20**: 1343.

Sullivan, D. E. 1981. Surface tension and contact angle of a liquid-solid interface. *J. Chem. Phys.* **74**: 2604.

Sullivan, D. E., and Telo de Gama, M. M. 1986. Wetting transition and multilayer adsorption of fluid interfaces. *Fluid Interfacial Phenomena*, ed. C. A. Croxton, p. 45. New York: Wiley.

Swanson, B. D., Stragier, H., Tweet, D. J., and Sorensen, L. B. 1989. Layer-by-layer surface freezing of freely suspended liquid-crystal films. *Phys. Rev. Lett.* **62**: 909.

Swendsen, R. H. 1982. First- and second-order phase transitions in the $d = 2XY$ model. *Phys. Rev. Lett.* **49**: 1302.

Tabe, Y., and Yokoyama, H. 1994. Novel liquid crystalline structure in Langmuir monolayers of amphiphillic azobenzene derivative. *J. Phys. Soc. Japan* **63**: 2472.

Tanaka, T., Honda, Y., Tani, T., Terahara, A., Tabe, Y., and Sugi, M. 1994. Alignment of surface-stabilized ferroelectric liquid crystal by the self-assembled monolayers of amphiphilic diacetylene derivatives. *Japan. J. Appl. Phys.* **33**: 295.

Tarazona, P., and Evans, R. 1983. Wetting transitions at models of a solid-gas interface. *Mol. Phys.* **48**: 799.

Tarczon, J. C., and Miyano, K. 1980. Surface-induced order of a liquid crystal in the isotropic phase. II. 4-methoxybenzylidene-4′-n-butylaniline (MBBA). *J. Chem. Phys.* **73**: 1994.

Tarczon, J. C., and Miyano, K. 1981. Shear mechanical properties of freely suspended liquid-crystal films in the smectic-A and smectic-B phases. *Phys. Rev. Lett.* **46**: 119.

Teixeira, P. I. C., Pergamenshchik, V. M., and Sluckin, T. J. 1993b. A model calculation of the surface elastic constants of a nematic liquid crystal. *Mol. Phys.* **80**: 1339.

Teixeira, P. I. C., and Sluckin, T. J. 1992a. Microscopic theory of anchoring transitions at the surfaces of pure liquid crystals and their mixtures. I. The Fowler approximation. *J. Chem. Phys.* **97**: 1498.

Teixeira, P. I. C., and Sluckin, T. J. 1992b. Microscopic theory of anchoring transitions at the surfaces of pure liquid crystals and their mixtures. II. The effect of surface adsorption. *J. Chem. Phys.* **97**: 1510.

Teixeira, P. I. C., Sluckin, T. J., and Sullivan, D. E. 1993a. Landau–de Gennes theory of anchoring transitions at a nematic liquid crystal–substrate interface. *Liq. Cryst.* **14**: 1243.

Teletzke, G. F., Scriven, L. E., and Davis, H. T. 1982a. Gradient theory of wetting transitions. *J. Colloid Interface Sci.* **87**: 550.

Teletzke, G. F., Scriven, L. E., and Davis, H. T. 1982b. Wetting transitions: first order or second order? *J. Chem. Phys.* **77**: 5794.

Teletzke, G. F., Scriven, L. E., and Davis, H. T. 1983a. Wetting transitions. II. First order or second order? *J. Chem. Phys.* **78**: 1431.

Telo de Gama, M. M., and Evans, R. 1983b. Adsorption and wetting transitions at a model of the interface between a solid and a binary fluid mixture. *Mol. Phys.* **48**: 687.

Telo da Gama, M. M. 1984a. The interfacial properties of a model of a nematic liquid crystal. I. The nematic–isotropic and the nematic–vapour interfaces. *Mol. Phys.* **52**: 585.

Telo da Gama, M. M. 1984b. The interfacial properties of a model of a nematic liquid crystal. II. Induced orientational order and wetting transitions at a solid–fluid interface. *Mol. Phys.* **52**: 611.

Telo da Gama, M. M. 1987. Wall-induced order of a liquid crystal. *Phys. Rev. Lett.* **59**: 154.

Telo da Gama, M. M., and Tarazona, P. 1990. First-order and continuous transitions in confined liquid crystals. *Phys. Rev. A* **41**: 1149.

Tikhomirova, N. A., and Guinsberg, A. V. 1980. Behavior of the nematic liquid crystal drop on the anisotropic surface. *Advances in Liquid Crystal Research and Applications*, ed. L. Bata, p. 651. Oxford: Pergamon.

Tjipto-Margo, B., and Sullivan, D. E. 1988. Molecular interactions and interface properties of nematic liquid crystals. *J. Chem. Phys.* **88**: 6620.

Toda, A., Mada, H., and Kobayashi, S. 1978. Temperature dependence of pretilt angle for nematic liquid crystals. *Japan. J. Appl. Phys.* **17**: 261.

Toney, M. F., Russell, T. P., Logan, J. A., Kikuchi, H., Sands, J. M., and Kumar, S. 1995. Near-surface alignment of polymers in rubbed films. *Nature* **374**: 709.

Toxvaerd, S. 1980. Phase transitions in a two-dimensional system. *Phys. Rev. Lett.* **44**: 1002.

Triezenberg, D. G., and Zwanzig, R. 1972. Fluctuation theory of surface tension. *Phys. Rev. Lett.* **28**: 1183.

Tripathi, S., Lu, M., Terentjev, E. M., Petschek, R. G., and Rosenblatt, C. 1991. Surface-induced polarization at chiral-nematic–substrate interface. *Phys. Rev. Lett.* **67**: 3400.

Tripathi, S., Rosenblatt, C., and Aliev, F. M. 1994. Orientational susceptibility in porous glass near a bulk nematic–isotropic transition. *Phys. Rev. Lett.* **72**: 2725.

Uchida, T., Ishikawa, K., and Wada, M. 1980a. Liquid crystal alignments and surface energy. *Mol. Cryst. Liq. Cryst.* **60**: 37.

Uchida, T., Ohgawara, M., and Wada, M. 1980b. Liquid crystal orientation on the surface of obliquely evaporated silicon monoxide with homeotropic surface treatment. *Japan. J. Appl. Phys.* **19**: 2127.

Uchida, T., and Seki, H. 1992. Surface alignment of liquid crystals. *Liquid Crystals: Applications and Uses*, vol. 3, ed. B. Bahadur, p. 1. Singapore: World Scientific.

Urbach, W., Boix, M., and Guyon, E. 1974. Alignment of nematics and smectics on evaporated films. *Appl. Phys. Lett.* **25**: 479.

Valenti, B., Grillo, M., Barbero, G., and Taverna Valabrega, P. 1990. Surface anchoring energy and ions adsorption: experimental analysis. *Europhys. Lett.* **12**: 407.

van Aerle, N. A. J. M. 1994. Anchoring of 4-n-alkyl-4′-cyanobiphenyl liquid crystal molecules on rubbed polyimides. *Liq. Cryst.* **17**: 585.

Vanderlick, T. K., and Möhwald, H. 1990. Mode selection and shape transitions of phospholipid monolayer domains. *J. Phys. Chem.* **94**: 886.

van Himbergen, J. E. 1984. From continuous to first-order transition in a simple XY model. *Phys. Rev. Lett.* **53**: 5.

van Sprang, H. A. 1983a. Surface-induced order in a layer of nematic liquid crystal. *J. Physique* **44**: 421.

van Sprang, H. A., and Aartsen, R. G. 1983b. Temperature dependence of liquid crystal tilt angles. *Appl. Phys. Lett.* **42**: 669.

van Sprang, H. A. 1983c. Surface order in nematics. *Mol. Cryst. Liq. Cryst.* **97**: 255.

van Sprang, H. A., and Aartsen, R. G. 1984a. The temperature dependence of liquid crystal tilt angles. *Appl. Phys. Lett.* **56**: 251.

van Sprang, H. A., and Aartsen, R. G. 1984b. Torsional anchoring of 5CB and 5PCH on various substrates. *Mol. Cryst. Liq. Cryst.* **123**: 355.

van Winkle, D. H., and Clark, N. A. 1988. Direct measurement of orientation correlations in a two-dimensional liquid-crystal system. *Phys. Rev. A* **38**: 1573.

Vetter, P., Ohmura, Y., and Uchida, T. 1993. Study of memory alignment of nematic liquid crystals on polyvinyl alcohol coatings. *Japan. J. Appl. Phys. Lett.* **32**: 1239.

Vilanove, R., Guyon, E., Mitescu, C., and Pieranski, P. 1974. Mesure de la conductivé thermique et determination de l'orientation des molécules a l'interface nématique isotrope de MBBA. *J. Phys. France* **35**: 153.

Vitek, V., and Kléman, M. 1975. Surface disclinations in nematic liquid crystals. *J. Physique* **36**: 59.

Walba, D. M., Stevens, F., Parks, D. C., Clark, N. A., and Wand, M. D. 1995. Near-atomic resolution imaging of ferroelectric liquid crystal molecules on graphite by STM. *Science* **267**: 1144.

Warenghem, M. 1982. A test for surface energy anisotropy sign determination. *Mol. Cryst. Liq. Cryst.* **89**: 15.

Williams, C. E., Pieranski, P., and Cladis, P. E. 1972. Nonsingular $S = +1$ screw disclination lines in nematics. *Phys. Rev. Lett.* **29**: 90.

Williams, C. E., Vitek, V., and Kléman, M. 1973. Surface disclination lines in 4-methoxy-benzylidene-4′-n-butylaniline (MBBA). *Solid State Commun.* **12**: 581.

Williams, R. 1976. Measurement of the interfacial tension between nematic and isotropic phases of MBBA. *Mol. Cryst. Liq. Cryst.* **35**: 349.

Wilson, T., Boyd, G. D., Westerwick, E. H., and Storz, F. G. 1983. Alignment of liquid crystals on surfaces with films deposited obliquely at low and high rates. *Mol. Cryst. Liq. Cryst.* **94**: 359.

Wu, S.-T., and Efron, U. 1986. Optical properties of thin nematic liquid crystal cells. *Appl. Phys. Lett.* **48**: 624.

Xue, J. Z., Jung, C. S., and Kim, M. W. 1992. Phase transitions of liquid-crystal films on an air–water interface. *Phys. Rev. Lett.* **69**: 474.

Yamamura, S., Tamaki, T., Seki, T., Sakuragi, M., Kawanishi, Y., and Ichimura, K. 1992. α-hydrazono-β-keto esters as command molecules. *Chem. Lett.* **1992**: 543.

Yamashita, M., and Amemiya, Y. 1976. Effect of substrate surface on alignment of liquid crystal molecules. *Japan. J. Appl. Phys.* **15**: 2087.

Yang, K. H. 1982. On the determination of liquid crystal-to-wall anchoring anisotropy by the surface-plasmon polariton technique. *J. Appl. Phys.* **53**: 6742.

Yang, K. H. 1983a. Weak boundary storage effect in homogeneous liquid crystal cells. *Japan. J. Appl. Phys.* **22**: 389.

Yang, K. H. 1983b. Freedericksz transition of twisted nematic cells. *Appl. Phys. Lett.* **43**: 171.

Yang, K. H. 1983c. Multiplexing of twisted nematic cells under nonrigid boundary conditions. *J. Appl. Phys.* **54**: 6864.

Yang, K. H., and Rosenblatt, C. 1983d. Determination of the anisotropic potential at the nematic liquid crystal-to-wall interface. *Appl. Phys. Lett.* **43**: 62.

Yang, Y. B., Bang, T., Mochizuki, A., and Kobayashi, S. 1991. The surface anchoring dependence of the dielectric constant of an FLC material showing electroclinic effect. *Ferroelectrics* **121**: 113.

Yokoyama, H., Kobayashi, S., and Kamei, H. 1982. Effect of substrate surfaces on the formation of nematic phase at the isotropic–nematic transition. *Appl. Phys. Lett.* **41**: 438.

Yokoyama, H., Kobayashi, S., and Kamei, H. 1983. Boundary dependence of the formation of new phase at the isotropic–nematic transition. *Mol. Cryst. Liq. Cryst.* **99**: 39.

Yokoyama, H., Kobayashi, S., and Kamei, H. 1984a. Role of surface adsorption in the surface-induced alignment of nematic liquid crystals on evaporated SiO. *J. Appl. Phys.* **56**: 2645.

Yokoyama, H., Kobayashi, S., and Kamei, H. 1984b. Measurement of director orientation at the nematic–isotropic interface using a substrate-nucleated nematic film. *Mol. Cryst. Liq. Cryst.* **107**: 311.

Yokoyama, H., and van Sprang, H. A. 1985a. A novel method for determining the anchoring energy function at a nematic liquid crystal–wall interface from director distortions at high fields. *J. Appl. Phys.* **57**: 4520.

Yokoyama, H., Kobayashi, S., and Kamei, H. 1985b. Deformation of a planar nematic–isotropic interface in uniform and nonuniform electric fields. *Mol. Cryst. Liq. Cryst.* **129**: 109.

Yokoyama, H., Kobayashi, S., and Kamei, H. 1987. Temperature dependence of the anchoring strength at a nematic liquid crystal–evaporated SiO interface. *J. Appl. Phys.* **61**: 4501.

Yokoyama, H. 1988a. Nematic–isotropic transition in bounded thin films. *J. Chem. Soc. Faraday Trans. II* **84**: 1023.

Yokoyama, H. 1988b. Surface anchoring of nematic liquid crystals. *Mol. Cryst. Liq. Cryst.* **165**: 265.

Yokoyama, H., and Patel, J. S. 1994. Critical anchoring transition at a solid–nematic interface. *Proceedings of the Fifteenth International Liquid Crystal Conference [Budapest]*, vol. 1, p. 468. New York: Gordon and Breach.

Young, C. Y., Pindak, R., Clark, N. A., and Meyer, R. B. 1978. Light-scattering study of two-dimensional molecular-orientation fluctuations in a freely suspended ferroelectric liquid-crystal film. *Phys. Rev. Lett.* **40**: 773.

Young, C. Y., and Clark, N. A. 1981. Dynamics of freely suspended lyotropic films. I. An inelastic light scattering study of thermal surface fluctuations. *J. Chem. Phys.* **74**: 4171.

Zel'dovich, B. Y., and Tabiryan, N. V. 1981. Fluctuations of the director of a nematic liquid crystal in a cell of finite thickness. *Sov. Phys.–JETP* **54**: 922.

Zhao, X., Subrahmanyan, S., and Eisenthal, K. B. 1991. Orientational fluctuations and phase transitions of long chain molecules at the air/water interface. *Phys. Rev. Lett.* **67**: 2025.

Zocher, H., and Coper, K. 1928. Erzeugung der Anisotropie von Oberflächen. *Z. Phys. Chem.* **132**: 295.

CHAPTER 7

Structure and Phase Transitions of Amphiphilic Lyotropic Liquid Crystals

MICHAEL R. KUZMA AND
ALFRED SAUPE

7.1 INTRODUCTION

Lyotropic liquid crystals owe their existence to the schizophrenic nature of amphiphilic molecules [1]. In dilute aqueous solution, the amphiphiles are driven by hydrophobic short-range forces [2] to assemble into clusters. Above a certain concentration (the critical micelle concentration or CMC), the hydrophilic head groups form a flexible crust which largely shields the hydrocarbon chains of the cluster from the water. The cluster, or micelle, is stable in a thermodynamic sense, and maintains its approximate shape and size under the constant assault of thermal motion and exchange with monomers present in the solvent.

When the concentration of amphiphile is small (but greater than the CMC), it is useful to view the micelle as an independent particle, at least over time scales of an experimental probe. The shape may be pictured as an ellipsoidal glob, rod-like with spherical end caps, or disk-like with hemitoroidal rims, etc. The smallest diameter, l, of the micelle is slightly less than twice the all-*trans* length of the amphiphile.

Many purified biological amphiphiles such as phospholipids have double chains with zwitterionic head groups and are not usually observed to aggregate into small micelles in equilibrium. Instead, they either form bilayer vesicles enclosing a given amount of solvent, or arrange into lamellar, hexagonal, inverted hexagonal, and bicontinuous phases [3] at higher concentrations.

Reviews of the detailed processes behind micelle, vesicle and membrane formation in dilute solutions are found in references [4,5,6].

At higher concentrations of amphiphile, interactions between aggregates allow the formation of liquid crystalline phases which, in many cases, may be classified as classical nematic and smectic lamellar or hexagonal phases. The picture of independent micelles continues to be useful as a model of the nematic phase, in spite of the strong interactions present in dense ionic surfactant/water systems. The reasonable anisometric shapes are not drastically different from the dilute case, and the micellar shape distribution therefore hovers near disk-like, rod-like, or ellipsoidal (see TABLE 7.1).

The generic classes of smectic (soap) phases [7,8,9] are the lamellar phase (neat soap) and the hexagonal phase (middle soap). In the simplest case, the lamellar phase is modeled as a stack of continuous bilayer membranes of thickness l, each separated by a given thickness of water. The hexagonal phase consists of water-separated tubes of the same diameter (l) and indefinite length, arranged on a two-dimensional hexagonal lattice.

In equilibrium, each phase may accommodate a certain density of curvature or structural defects [10]. Such defects provide channels for transitions between phases upon varying temperature or concentration, and are currently under intense experimental and theoretical investigation [11,12].

The hydrophobic tails interior to the micelle or bilayer are oriented on average perpendicular to the local micellar surface. Therefore, if the micelle has an anisotropic shape, it obtains both a polarizibility and diamagnetic anisotropy, both of which originate in the ordering of the chains imposed by the geometric shape of the micelle.

This anisotropy is exploited in many experiments on systems where the equilibrium physical observable is the chain order, for example, in birefringence and magnetic resonance measurements. Accordingly, an average micelle shape or membrane arrangement compatible with the data is inferred [13].

It is most often the case that several experimental methods are combined to deduce the microscopic structure. We will not review the beautiful techniques of neutron scattering, x-ray, or NMR here; instead we will refer to the appropriate literature that has proved useful to us in interpreting micellar lyotropics. Where applicable, we will emphasize the usefulness of a good polarizing microscope, heat stage, and camera.

7.2 CLASSIFICATION AND STRUCTURE OF LYOTROPIC PHASES

7.2.1 Nematic Phases

The macroscopic properties of lyotropic nematics are dramatically different from low molecular weight thermotropic nematics. They typically appear as a clear viscous liquid, whereas thermotropics are generally turbid with a much lower relative viscosity. The visual difference is due to the fact that the dielectric anisotropy (ε_a) at optical frequencies of a typical lyotropic nematic mixture ($DACl/NH_4Cl/H_2O$) is on the order 7.6×10^{-3}, while that of a thermotropic (MBBA) is about 0.7 [14]. The scattering intensity from director fluctuations of wavevector q is

$$I(q) \sim \left(\frac{\varepsilon_a \omega^2}{4\pi c^2}\right)^2 \frac{V k_B T}{K q^2} \tag{7.1}$$

where ω is the frequency; c is the speed of light; T is the temperature; V is the volume; and k_B is the Boltzmann's constant. The scattering intensity is thus reduced by a factor of 10^{-4} relative to low molecular weight thermotropics. In the above, we have used the fact that the Franck elastic constants (K) of lyotropics and thermotropics are of the same order of magnitude (see section 7.3.3). The theory of light scattering from director fluctuations [14] assumes $\varepsilon_a \ll \bar{\varepsilon}$ (mean dielectric constant), and therefore more properly applies to lyotropic liquid crystals than the aromatic thermotropic compounds.

The current picture of nematic phases is distilled from the work of Charvolin [15], Reeves [16], Saupe [17], and their coworkers. It is based on the model in which the micelles are finite in size, with a mean aggregation number stable enough for the micelle–micelle interactions to sustain long-range orientational and translational order.

Most of the examples of the micellar nematic and smectic phases considered here are formed by: (potassium laurate [KL]/decanol/water) [18,19], (decylammonium chloride [DACl]/ammonium chloride [ACl]/water) [20,21], (sodium decyl sulphate [SDS]/decanol/water) [19,22] and (cesium perfluorooctanoate [CsPFO]/water) [23].

A microscopic determination of the relative sizes of the micelles of (SDS/decanol/water), (DACl/ACl/water) and (KL/decanol/water) using x-ray and neutron scattering was carried out by the lyotropic group at Orsay [15,24]. The CsPFO system was extensively characterized using x-rays, along with other techniques, by the group originally centered at the University of Leeds [23].

A partial list of micelle sizes is given in TABLE 7.1. All dimensions given in TABLE 7.1 (except for the electron microscope images, where shadows of replicas are

TABLE 7.1. Approximate micellular dimensions of selected lyotropic nematics (||, ⊥ parallel and perpendicular to the director, respectively).

Phase	System wt.%	$l_{\parallel}A$	$l_{\perp}A$	Experimental method[a]	Ref.
N_L	DACl/ACl/D_2O				
	(a) 40.91/4.09/55.00 $T = 47.5\,°C$	39	60	X	[21]
	(b) 43.97/1.76/54/27 $T = 41\,°C$	40	105	TEM	[76]
	(c) 43.3/4.33/52.35 $T = 37\,°C$	40	60	TEM	[77]
	KL/decanol/D_2O				
	(a) 26.50/6.68/66.82 $T = 10\,°C$	47	75	X	[31]
	(b) 25.25/6.28/68.47 $T = 22.5\,°C$	59	70	N	[19]
	SDS/decanol/D_2O				
	36.32/6.63/57.05 $T = 23\,°C$	41.9	59.8	N	[19]
	CsPFO/D_2O				
	(a) 42·5/57.5 $T = 33\,°C$	54	68	N	[23a]
	(b) 50.0/50.0 $T = 40\,°C$	46	60	N	[23c]
N_c	KL/decanol/D_2O				
	(a) 26.50/6.68/66.82 $T = 25\,°C$	72	47	X	[31]
	(b) 26.34/6.28/67.38 $T = 22.5\,°C$	87.2	53	N	[19]
	SDS/decanol/D_2O				
	(a) 36.32/6.63/57.05 $T = 30\,°C$	—	42.3	N	[19]

[a] N, neutron scattering; X, x-ray; and TEM, transmission electron microscopy.

observed) correspond to observed peaks in the respective scattering intensities, not the actual micelle sizes.

The above observations, combined with polarizing microscopy, lead to three nematic phases with distinct physical signatures. The lamellar (N_L) and cylinder (N_C) nematics, formed from disk-like and rod-like aggregates, respectively, comprise the uniaxial phases. Each of these phases may undergo a transition to a biaxial nematic phase (N_B) as a function of concentration or temperature. Incorporation of a chiral solute into the nematic phases renders the corresponding cholesteric phases. We abstain from a discussion of cholesteric surfactant lyotropics [25] since many of their microscopic properties are similar to nematics.

Nematic phases are not confined to single-chain ionic surfactants. For example, double-chain surfactants [26] have been found to form cylinder (N_C) and lamellar (N_L) nematics, and work has been published on nonionic surfactants in aqueous solution which develop N_L and N_C phases [27].

7.2.1.1 Lamellar (or Discotic) Nematics (N_L)

These are formed from disk-like or oblate spheroid micelles. The N_L phases frequently, but not always, border lamellar phases.

When observed in flat (0.2 mm interior thickness) glass capillaries under a polarizing microscope, the typical N_L phase displays a schlieren texture (FIGURE 7.1(a)) [28]. Left for a few hours, this will relax to a uniform pseudoisotropic texture with the bilayer disks oriented parallel to the surfaces (optical axis

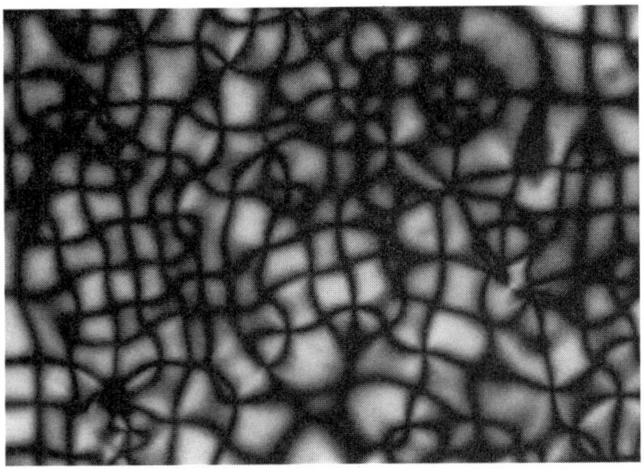

(a)

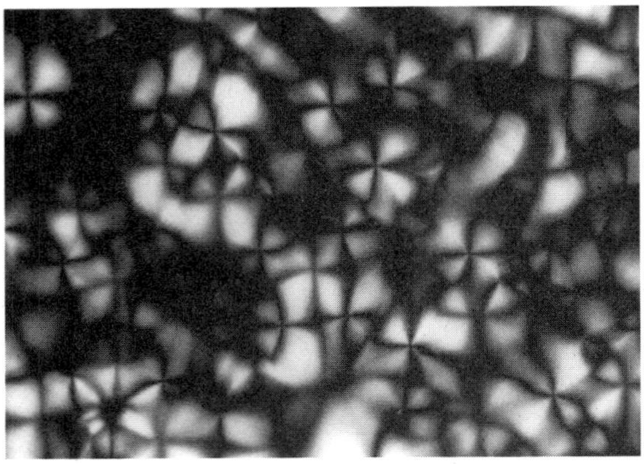

(b)

FIGURE 7.1. Photomicrograph of $(CsPFO/CsCl/H_2O) =$ (41.9/3.0/55.1) wt.% mixture; Mag. 200 ×, crossed polarizers, sample thickness 0.3 mm. (a) $T = 37.2\,°C$, nematic schlieren texture, (b) $T = 37.1\,°C$, beginning formation of lamellar fan texture at transition. Note the effect of the divergence of the Franck elastic constants on singularities.

perpendicular to surface). The sign of the birefringence (Δn) in the pseudoisotropic texture may be determined from the colors of the conoscopic interference figure when a λ plate is inserted between the crossed polarizers [29].

The apparent birefringence increases near the capillary edges and somewhat less near the meniscus, due to the orientation imposed by the surface. The texture includes a $\frac{1}{2}$ disclination running along the edge of the sample, and possibly a few other disclinations, but is otherwise smooth.

7.2.1.2 Cylindrical or Calamitic Nematics (N_C)

Formed from rod-like or prolate spheroid micelles of indefinite length, they are related in cross-section to the cylinders found in the hexagonal phases.

The N_C phases reveal a characteristic texture under a polarizing microscope. This texture relaxes to the partially oriented colorful planar alignment of the director field shown in FIGURE 7.2(a). When the polarizers are uncrossed the texture should appear smooth with no evidence of boundaries indicating a phase separation.

If a uniform planar alignment is realized (e.g., by a flow or magnetic field), the sign of the birefringence may be found using the conoscopic method of flash figures [29]. N_C phases do not require any special treatment of the glass surfaces to stay in a homogeneous planar alignment.

7.2.1.3 Biaxial Nematics (N_{Bx})

Usually found between the N_L and N_C phases, they require a certain amount of diligence to prepare and observe. Reviews of phase properties, textures, and earlier work are given in [28b,30].

In the N_L–N_{Bx}–N_C region of the phase diagram, experimental evidence points to an ellipsoidal micellar shape, which is a sensitive function of surfactant, cosurfactant, and temperature (see FIGURE 7.3) [31].

The above nematic textures are readily disturbed by flow or a magnetic field, and this is usually taken as a trademark of a nematic phase. When the disturbance is small, the relaxation of the optic axis is exponential in time. We take the exponential time dependence of the relaxation of an observable, for example, the birefringence, as a qualitative measure of the existence of the nematic phase. This test fails for certain special cases; such as for nematic gel phases [32], near the nematic–lamellar transition (section 7.3.3) or when polymeric solutes (section 7.4) are added. The system becomes "glassy," in the sense that an average response time of the observable becomes longer than a feasible dwell time between experimental points. It may be necessary to wait for hours (in principle, for years), for an equilibrium texture to establish itself.

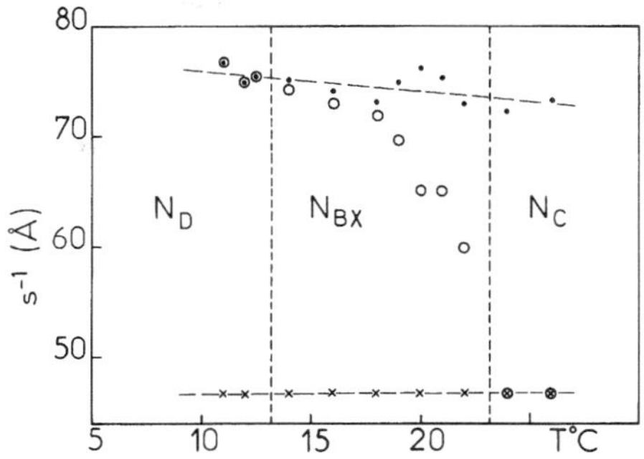

FIGURE 7.3. Synchrotron x-ray measurements of spacings s_1^{-1} (filled circles), s_2^{-1} (open circles), and s_3^{-1} (crosses) as functions of temperature in the three nematic phases. The expermental error is about 0.8 Å. Mixture (KL/decanol/D_2O) = (26.50/6.68/66.82) wt.%. From [31].

7.2.2 Lamellar Phases

The structure of a classic lamellar phase (or neat soap) consists of a stacked array of amphiphilic bilayer sheets separated by a solvent. A manifestation of this microscopic order is the formation of pseudoisotropic textures with oily streaks and focal conic textures [33]. Focal conics may bundle together filling the space between the glass slides, forming a fan-like texture. FIGURE 7.1(b) shows the beginning of the development of a lamellar fan texture just after the N_L–L transition.

Lyotropic lamellar phases have much in common with the smectic A phase of thermotropics, with one appealing exception: under certain conditions they can be greatly diluted while still retaining the layer structure [34].

Two striking examples of the degree to which the layers may be isolated were discovered by: (1) Larche et al. [35], by pushing the layers apart with oil; (2) Hoffmann et al. [36]; Satoh and Tsujii [37]; and Strey et al. [38], who diluted their system with water.

Larche et al. [35] began with a ternary mixture based on an anionic surfactant: (octylbenzene sulfonate [OBS^-Na^+]/water/pentanol) = (39.5/42.5/18.0) wt.%. The lamellar layer spacing was measured (for

(a)

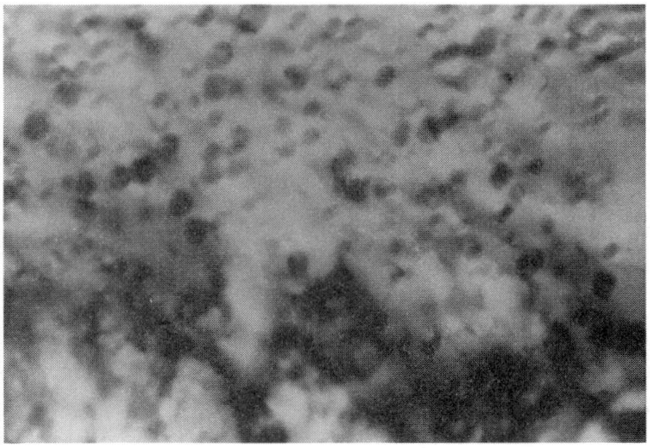

(b)

(c)

FIGURE 7.2. Photomicrograph of (KL/KCl/H_2O) = (35.32/2.35/62.33) wt.% mixture. Mag. 100 ×, crossed polarizers, sample thickness 0.3 mm. (a) $T = 41.3\,°C$, partially aligned N_c planar texture. (b) $T = 41.6\,°C$ N_c–H_I transition. (c) $T = 68.2\,°C$, H_I + isotropic.

the highest concentrations of surfactant) to be about twice the length of a surfactant molecule. The OBS mixture was then diluted with a solution of (decane/pentanol) = (92/8)wt.%, until it occupied a volume fraction of about 1%. The ordered phase displayed a birefringence which persisted over distances of 1 cm; presumably the size of the mixing tube. The layers formed parallel to the glass surfaces and light scattering was used to determine spacings of up to 6500 Å. The next higher-order quasi-Bragg peak was also observed, in contradiction to the existing theory of scattering from smectics A [39]. This discrepancy was briefly discussed in [35] and [40].

The stability of the OBS oil diluted lamellar phase was attributed to repulsive undulation forces [41]. Electrostatic forces between layers are negligible in this case because: (i) the ratio of the dielectric constants in the hydrocarbon region compared with the water is about 40, essentially confining the electric field lines to the aqueous region; and (ii) the headgroups of the ionic amphiphiles and counterions are restricted to the water layer of thickness l_w. Since in the swollen phases $d \gg l_w$, the layers appear as thin membranes of neutral charge.

In aqueous solutions of the nonionic surfactant tetradecyldimethylamine (C_{14}DMAO) with small additions of decane and hexanol, Hoffmann [36] observed phases which showed iridescent colors when illuminated with white light. The thickness of the layers may vary from 20–40 Å depending on the decane and hexanol content, but the layer spacings, d, are on the order of visible light. The layer spacing decreased over the selected range of concentrations of C_{14}DMAO, from about 2500 Å (20 mM) to 1300 Å (45 mM).

The stability of this iridescent lamellar phase was thought to be due to electrostatic forces. The line of thought given by Hoffmann proceeds from the observation that in water, the amine head groups of the C_{14}DMAO acquire protons, changing the pH of the solution to around 9, and imparting a small charge density to the surface. The concentration of $[OH^-]$ at 25 °C is (using the ionic product $[H^+] \times [OH] = 10^{-14}$) about 10^{-5} moles liter^{-1}. Assuming the concentration of other ions in solution is negligible and taking n_{OH} as the density of OH^- ions, we find the Debye length, $l_D = (8\pi n_{OH} l_B)^{-1/2}$ is of the order of 10^3 Å. The Bjerrum length, with ε as the dielectric

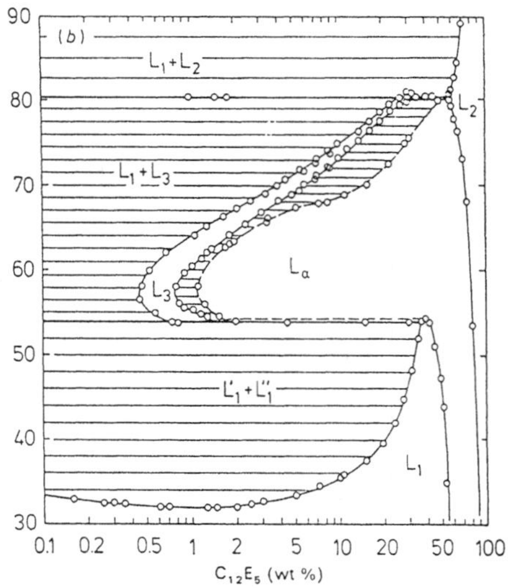

FIGURE 7.4. Semilog plot of ($C_{12}E_5$/water) system. L is the lamellar phase, L_3 is the sponge phase, $L_{1,2,3}$ are the isotropic phases. from [38].

constant of water is,

$$l_B = \frac{e^2}{\varepsilon k_B T} \simeq 7\text{Å for } T \simeq 300\ K.)$$

It is therefore possible for electrostatic forces to stabilize the observed layer structure. Therefore, addition of salt would be expected to destabilize the lamellar arrangement. This was found to be the case for the system Hoffmann studied. A possible instability to salt was not examined in [37].

Recent work of Strey and coworkers [38] presented a detailed analysis of the polyoxyethylene surfactant $C_{12}E_5$ $[CH_3(CH_2)_{10}CH_2(OCH_2CH_2)_5OH]$ in water (see FIGURE 7.4). They followed the layer spacing under dilution by water with neutron and light scattering. To see if the phase was stabilized by electrostatic interactions, enough salt was added to an iridescent green sample of volume fraction 1.7% to reduce the Debye length to 3 Å; colors apparently changed but still persisted in the sample. It was concluded that the stability of this lamellar phase was due to repulsive Helfrich interactions [41,42]. Remarkably, the lamellar phase was unstable to small additions of oil (for volume fractions of $C_{12}E_5$ in the range 1.26–3.03%). Traces of oil removed the iridescence.

It is possible that some oil may be intercalated between the monolayers altering the bilayer thickness.

FIGURE 7.5. Fan texture of H_I phase (KL/decanol/water), $T = 22°C$. Sample had slight water loss due to evaporation.

In this circumstance, modes of density fluctuation involving variations in bilayer thickness must be included [43]; these modes provide a path for break-up of the layers into cylinders, ribbons or globules.

7.2.3 Hexagonal Phases

Hexagonal phases frequently appear at temperatures above the N_C phase. FIGURE 7.2 displays a typical N_C–H_I transition under a polarizing microscope. The two-dimensional packing of the micelle cylinders allows a striated fan texture [33] to develop as shown in FIGURE 7.5. Equilibrium pseudoisotropic textures have not yet been observed. This is due mostly to the parallel alignment enforced on the rod-like micelles at the glass/liquid crystal interface of the sample cells. The rod-like micelles generally do not have a circular cross-section, and this induces a number of hexagonal phases of lower symmetry [10, 24]. (The hexagonal phase of the potassium laurate system in FIGURE 7.5 is possibly one of these modifications.)

Examination of the phase diagrams available with hexagonal phases [8,9] reveals they are unstable to dilution with water. On the other hand, oil dilution through swelling of the cylinder diameter is achievable in at least one case: the phase diagram [9] for the mixture (Na octanoate/methyloctanoate/water) demonstrates that up to 40% by weight of methyloctanoate may be solubilized inside the cylinders. With suitable additives or new polymeric surfactants, it may be possible to distend this phase even further. The additives must not favor a Rayleigh-like

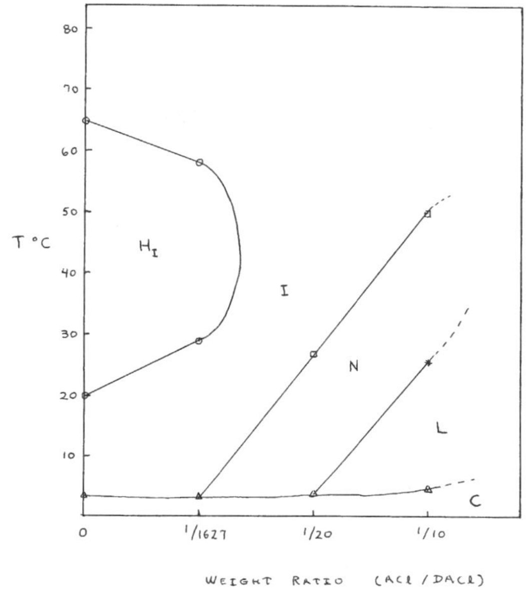

FIGURE 7.6. Transition temperatures versus approximate log of weight ratio NH_4Cl/DACl. Two-phase regions are not shown. Note the extreme sensitivity of the bubble of H_I phase to added salt. Points taken from [20].

instability that will lead to break-up of the cylinders into oil filled globules.

Many hexagonal phases are also unstable with respect to the addition of salt. An example is the DACL/ACl/water system [20] (FIGURE 7.6). The instability of the H_I phase is due to screening of the electrostatic interactions of the head groups, producing a decrease in curvature of the local monolayer.

Models of the I–N_C–H_I phases have been developed in recent years. The continuum theory of Selinger and Bruinsma (SB) [44] is based on infinite flexible chains of persistence length $P = K/kT$ and area density ρ_0. K is a bending elastic constant of the chain. As the chains are brought together they interact orientationally through excluded volume effects of neighboring persistence lengths, and translationally by a knocking together of thermally induced undulations (Helfrich interaction). The resultant phase diagram is given in FIGURE 7.7.

One difficulty with the SB theory, which was discussed by the authors, is the fact that it predicts an H_I lattice constant about 23% smaller than the hard-core diameter of the chains. While it is not clear how the SB

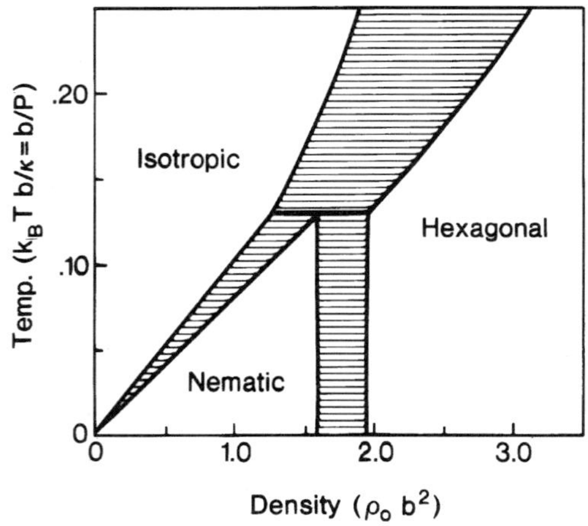

FIGURE 7.7. Diagram of I–N_C–H_I phases, temperature (ratio of molecular length, b to persistence length P($P = K/k_bT$)) versus area density of chains. After [44].

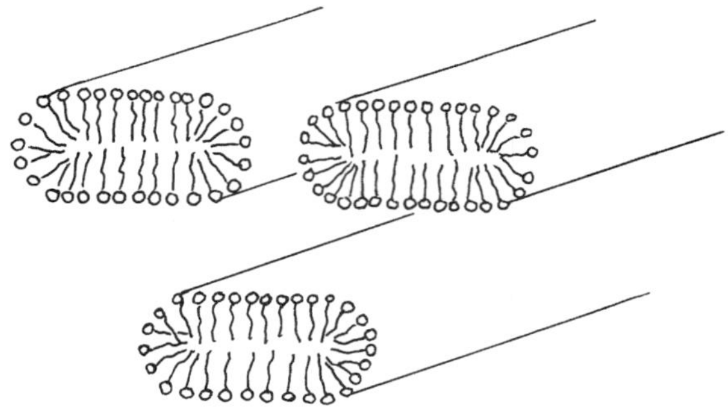

FIGURE 7.8. Schematic of local structure of ribbon phases.

theory may be modified to correct this inconsistency, the qualitative conclusions concerning the two-phase regions and temperature dependence of the transitions agree with experimental results [44]. The SB theory also omits the effects of "hairpins," or 180° kinks in the chains (see, e.g., [45]). A certain number of hairpins must be present in the nematic phase after the I–N transition impresses an alignment of an initially isotropic coil. A finite density of hairpins will also interfere with the establishment of an H_I phase. In a polymeric micellar nematic phase, it is possible that the high local curvature energy of a hairpin favors breakage of the chain and replacement by hemispherical end caps.

At about the same time, Taylor and Herzfeld (TH) [46] developed a theory of the I–N_C–H_I and I–N_L–L phases for a system of inflexible polydisperse finite rods and disks respectively. The I–N_C–H_I phase diagram deduced by TH is similar to that in FIGURE 7.7.

We note that the columnar phase model of TH bears some similarities to the so-called "tunnel model" of fluids proposed by Barker [47]. In Barker's model, M hard-sphere particles are confined in each tunnel. The system is then divided into K close-packed hexagonal tunnels such that $KM = N$, the total number of particles. The partition function factorizes into a one-dimensional integral over interactions parallel to the tunnel, and a two-dimensional integral giving the free cross-sectional area accessible to the center of a sphere in a tunnel. The particles need only be generalized to perfectly aligned hard polydisperse spherocylinders within the tunnels, corresponding to the TH model of the columnar phase.

Because of the similarity with the hexagonal phases and possible relevance to the N_L–lamellar transition, we briefly comment on the ribbon phases. They are modeled as flat ribbons of bilayer, with the edges approximated by half cylinders (FIGURE 7.8). The ribbons may arrange themselves into a two-dimensional rectangular lattice, and have been observed in (SDS/decanol/water) [24] and (potassium palmitate/potassium laurate/water) [48]. The phase diagram of [24] allows the possibility that the two-dimensional lattice of ribbons merges, by a first-order transition, into a lamellar phase.

The details of the interface order in the lamellar phase of (sodium dodecylsulphate [SDoS]/decanol/water) were recently examined by nuclear spin relaxation anisotropy (field strength 2.35 Tesla of the sodium counterions [49]. A contribution to the relaxation from the diffusion of the counterion spins over the curvature of the surfactant–water surface was observed. The findings of [49] are compatible with: (1) the bilayers of the lamellar phase being formed to some extent from a planar array of polymeric ribbons of undetermined length; and (2) correlations of the ribbon orientations perpendicular to the layers are short-ranged since the phase is uniaxial (the uniaxiality was determined from the line shape of a powder sample). If we take the ribbon model seriously, a magnetic field applied parallel to the layers should induce a small birefringence due to a partial alignment of the ribbons within the layers. The induced phase will be biaxial lamellar, of symmetry D_{2h}.

7.2.4 Ringing Gels [50]

Ringing gels are microemulsion gel phases that develop in ternary mixtures of water, hydrocarbon, and surfactant. They are formed when rod-like micelles undergo a rod-to-sphere transition under the addition of hydrocarbon. The volume fraction of water in the gels is slightly greater than $\frac{1}{2}$, the oil being confined to globules surrounded by surfactant. The resultant densely packed spherical globs form amorphous and cubic phases. The ringing gels were so-called because a macroscopic container filled with such a gel vibrates, after being tapped, at a natural audible resonant frequency (damping time about 1 second). Evidently, elastic waves in the kilohertz range propagate freely in the ringing gels. All other microemulsion gels mentioned in [50], are soundproof, that is, they efficiently damp sound vibrations in the kHz range.

At least two ternary systems form ringing gel phases. They were fabricated from the nonionic surfactants [50], alkyldimethylamine oxides and the alkyl polyglycol ethers (C_nE_m). (Ternary phase diagrams of a series of C_nE_m/oil/water systems have been discussed in detail in [51].)

As noted in section 7.2.2, some of the homologs of the above surfactants have demonstrated the ability to sustain lamellar phases and the sponge phases under extreme dilution. The addition of ringing gels to their repertoire expands the interest in understanding their influence on the physical properties of mixtures.

7.2.5 Sponge Phases

Sponge phases (L_3) are isotropic phases formed from a randomly connected distribution of amphiphilic bilayers. They usually occur between vesicle and lamellar phases. We refer the reader to a recent review of their exceptional properties [52]. (See also the last paragraph of section 7.3.3.)

7.3 PHYSICAL PROPERTIES OF NEMATIC AND LAMELLAR PHASES

7.3.1 Nematic–Isotropic Transitions

The simplest and perhaps most useful way to interpret and anticipate experimental results is through the Landau–de Gennes expansion of the free energy [14]. To set the notation, we write the free energy per unit volume due to the nematic order as

$$F = aS^2 + bS^3 + cS^4 + dS^5 + eS^6 - (1/3)\rho_s\, \beta_a\, SH^2. \tag{7.2}$$

The coefficient of the quadratic term takes the explicit temperature dependence, $a = a_0(T - T^*)$; b–e depend implicitly on temperature and concentration; ρ_s is the number density of amphiphile; β_a is the diamagnetic anisotropy of one amphiphile; T^* is the temperature of the lower limit of stability of the nematic phase; and H is the magnetic field strength in Gauss. One may choose to write $\rho_s\beta_a$ as $\phi(\beta_a/v_a)$ where ϕ is the volume fraction of amphiphile and v_a is the average volume of an amphiphile; typically $\rho_s\beta_a \sim 10^{-9}$ cgs [53]. For reference, we mention the N–I transition occurs when $a = b^2/4c$ or $T_0 = T^* + b^2/(4a_0c)$. The order parameter at $T = T_0$ is $S_0 = -b/2c$. A thorough discussion of (1), including terms allowing a biaxial nematic phase, is given in [54].

The order parameter is defined as $S = (1/2)\langle 3\cos^2\alpha - 1\rangle$ where α is the angle between the symmetry axis of an amphiphile (assumed for simplicity to be cylindrically symmetric) and the applied magnetic field. The angular brackets imply a thermal average. When $H = 0$, the system will spontaneously pick out an arbitrary axis, and in this instance, α is defined with respect to this direction. In the N_L phase, the average amphiphile aligns parallel to the symmetry axis, thus $S > 0$ and $b < 0$. In the N_C phase, the average amphiphile orders perpendicular to the symmetry axis; consequently $S < 0$ and $b > 0$.

Ideally, when $a > (9/16)b^2/c$, and there are no applied fields, the isotropic phase is the only allowed phase. Within the isotropic phase, some alignment may be produced by exposing the system to shear flow [55], surface effects [56], and electric and magnetic fields [54,57].

In the isotropic phase, above the N_C or N_L phases, the dependence of S on the coefficients and magnetic field is found by setting $\partial F/\partial S = 0$ giving

$$(1/3)\rho_s\beta_a H^2 = 2aS + 3bS^2 + 4cS^3 + 5dS^4 + 6eS^5 + \ldots. \tag{7.3}$$

Inverting this series

$$S(h) = A_1h + A_2h^2 + A_3h^3 + A_4h^4 + \ldots \tag{7.4}$$

with $h = (1/3)p_s\beta_a H^2$, and $A_1 = 1/2a$, $A_2 = 3b/8a^3$, $A_3 = (9b^2\text{-}4ac)/16a^5$etc.

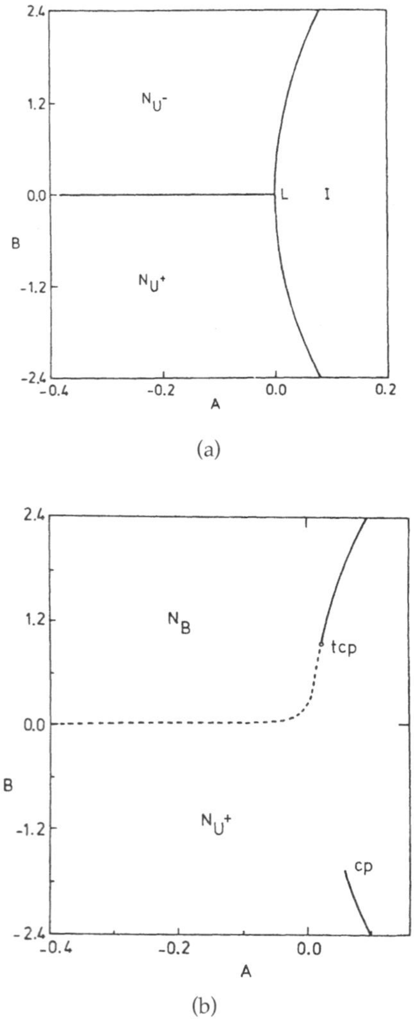

FIGURE 7.9. Phase diagram of vicinity of Landau point (L) in zero (a) and finite (b) magnetic field. N_{U+} corresponds to N_C, N_{U-} corresponds to N_L. N_B is the induced biaxial phase. TCP is the tricritical point, CP is the critical point. In our notation $B \sim -b, A \sim a$. After [54].

The magnetic field induces a small paranematic order in the isotropic phase

$$S = (1/6)\rho_s \beta_a H^2/a. \qquad (7.5)$$

Most hydrocarbon chain surfactants have $\beta_a < 0$, and the order induced in the isotropic phase is negative. Fluorocarbon chain surfactants like cesium perfluro-octanoate have $\beta_a > 0$ and consequently display a positive induced order. (It is often possible to predict the diamagnetic anisotropy of a molecule from the anisotropies of its constituent parts [58].)

The application of a magnetic field has a nontrivial effect on the structure of the phase diagram (see FIGURE 7.9). Close to the Landau point (L), the field smooths out the I–N_C transition. (The N_C paranematic alignment has the same symmetry axis as the spontaneous order.) As b increases, a critical point is reached beyond which there is a first-order transition between the para-N_C and the N_C phase. In contrast, the I–N_L transition does not disappear. The field-induced order within the N_L phase is biaxial. In FIGURE 7.9(b), N_B denotes the field-induced biaxial order. The symmetry of the paranematic N_L and contiguous N_C regions is uniaxial. Two regions with different symmetries must be separated by a line of at least second-order transitions. Considerations based on the Landau–de Gennes expansion [54], show that the second-order line terminates at a tricritical point (TCP) and changes into a line of first-order transitions.

In the next section, we study the birefringence induced in the isotropic phase by a magnetic field. In this case, the symmetry of the induced phase is always uniaxial and equation (1) holds.

For experimental studies within the nematic and lamellar phases, we refer the reader to the works of Haven and coworkers [59] on the refractive indices and birefringence of aligned N_L and lamellar phases of the (DACl/ACl/D_2O) system, and Boonbrahm and Saupe [60] for the N_L–N_{Bx}–N_C transitions in a (KL/decanol/water) mixture.

7.3.2 Magnetically Induced Birefringence in the Isotropic Phase

The phase shift induced by the magnetic field may be precisely determined by measuring the relative change in intensities between two orthogonal linearly polarized light waves emerging from the sample [61], say I_1 and I_2. The phase shift may be written as

$$\phi = 2\tan^{-1}(I_2/I_1)^{\frac{1}{2}}. \qquad (7.6)$$

The phase shift ϕ is related to the birefringence by

$$\phi = \frac{2\pi d}{\lambda}\Delta n \qquad (7.7)$$

where

$$\Delta n = \Gamma S(h) \qquad (7.8)$$

TABLE 7.2. Coefficients of Landau expansion in erg cm^{-3}.

	$T - T^*$ (mK)	a	$10^{-2}\,b$	$10^{-4}\,c$	$10^{-6}\,d$	$10^{-7}\,e'$
N_L	326	260				
	95.3	61.3	−22			
	30.2	17.7	−12	4.8		
	9.0	5.3	−5.3	7.3		
N_C	146	216				
	49.1	76	23			
	21.1	32.7	14	4.8		
	9.2	12	6	4.2	1.5	2.7

and

$$\Gamma = \frac{2\pi}{9}\rho_s \alpha_a \frac{(n_2^2 + 2)^2}{n_2}. \tag{7.9}$$

$S(h)$ is given by (7.4). λ is the wavelength of the light source, d is the sample thickness, α_a is the polarizibility anisotropy per amphiphile, n_2 is the ordinary refractive index. The magnitude of the induced birefringence for a typical lyotropic nematic is $\sim 10^{-5}$ (when $(T - T^*) \simeq 0.1°$ C and $H = 10$ kG).

The temperature and magnetic field behavior of the phase shift induced in a 0.5 mm film of a (KL/decanol/D_2O) mixture was reported by Melnik and coworkers [62]. By fitting the isotherms of ϕ/H^2 using (7.3), they were able to obtain the Landau coefficients up to sixth order (TABLE 7.2). They were also able to monitor the change in sign and magnitude of the coefficient b with temperature (see FIGURE 7.10). This was seen nicely by observing the variation of

$$\frac{\partial(\phi/H^2)}{\partial H^2} \sim \frac{\partial(|S/h|)}{\partial|h|} \sim \frac{b}{a^3}. \tag{7.10}$$

Combining the birefringence results in the isotropic phase and the x-ray results (FIGURE 7.3) within the N_L (oblate) and N_C (prolate) phases, we find reasonable support for the working hypothesis that the shape of the micelle is a smoothly changing function of concentration and temperature (except possibly in the vicinity of first-order transition lines).

The observation of the coexistence of the N_L and N_C phases [17] (droplets of N_L in N_C or vice versa), obliges one to consider the possibility that such a solution may be a mixture of rod and disk micelles. Under these circumstances, the cost in free energy per micelle to assemble either the rod or disk micelle should be nearly equal. If the two shapes are to coexist, a barrier somewhat larger than kT must separate the free-energy minima, locking the micelle into one shape or the other.

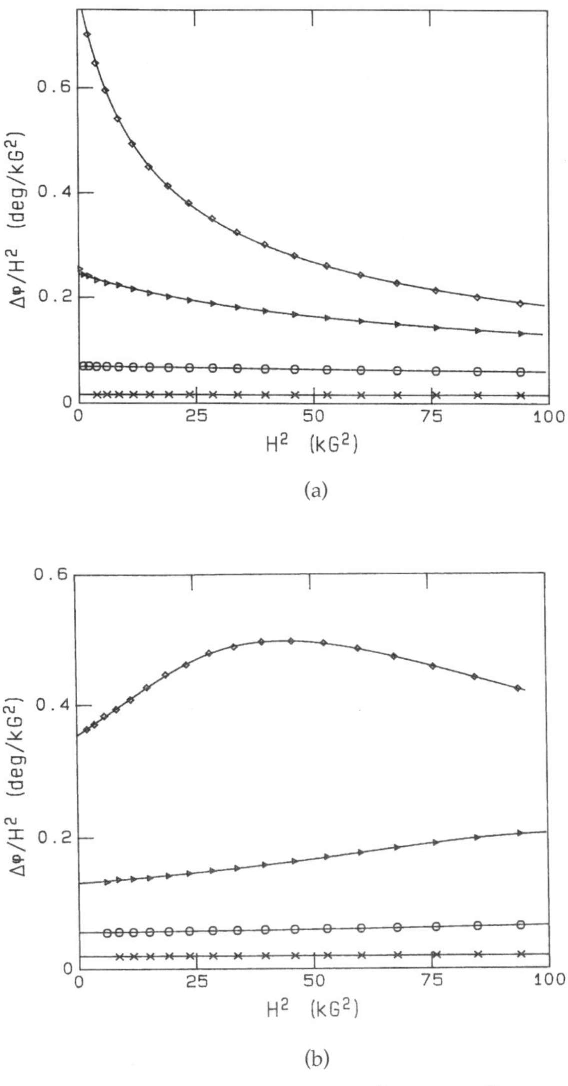

FIGURE 7.10. (a) Isotherms of ϕ/H^2 versus H^2 in the isotropic phase above N_L for $(T - T^*) = 326$ mK (crosses), 95.3 mK (circles), 30.2 mK (triangles), and 9 mK (diamonds). (KL/decanol/D_2O) = (25.25/6.35/68.4) wt.%. (b) Isotherms of ϕ/H^2 versus H^2 in the isotropic phase above N_C for $(T - T^*) = 141$ mK (crosses), 49.1 mK (circles), 21.1 mK (triangles), and 9.2 mK (diamonds). (KL/decanol/D_2O) = (25.32/6.35/68.33) wt.%. After [62].

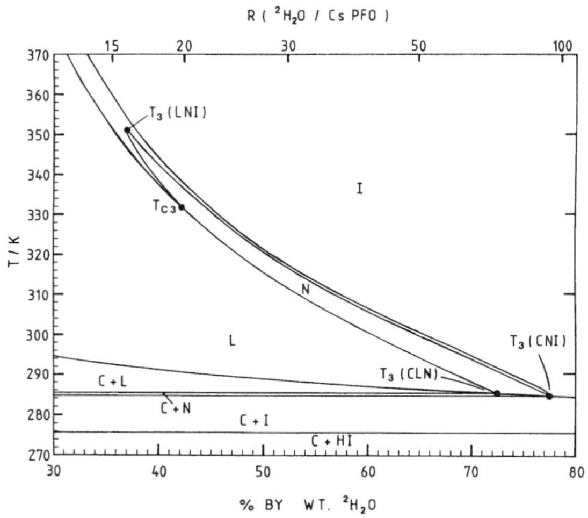

FIGURE 7.11. Phase diagram of CsPFO/water system. After [23a].

Discussions about the shapes of micelles are not purely academic. It has been shown that chemical reaction rates for a number of unimolecular and bimolecular reactions in lyotropic nematic solutions depend strongly on whether they take place in the N_C or N_L phase [63]. Presumably, the portion of the molecule sensitive to a reaction is screened for one micellar shape and exposed for the other.

The I–N_L transition in the CsPFO/water system has received considerable attention, because this phase change appears to become continuous, or approach a Landau point, as the weight fraction of surfactant is reduced (FIGURE 7.11) [64]. The approach to a Landau point is evidenced by: (1) the decrease of $\Delta T = (T_0 - T)^* = b^2/(4a_0c) = cS_0^2/a_0$; (2) the lowering of the latent heat of transition, $\Delta H = a_0 T_0 S_0^2$; and (3) a decrease in the jump of the order parameter at the transition, $S_0 = -b/2c$.

The behavior of ΔT versus weight fraction of surfactant w_s was deduced from the Cotton–Mouton constant [65] , $C \sim$ const. $(T - T^*)^{-1}$, the temperature T^* being obtained by extrapolation. It is found that $\Delta T \sim w_s$ for w_s less than about 0.4. For $w_s > 0.55$, ΔT increases roughly as (1-const. $w_s)^{-2}$. Similar trends in ΔT were reported from specific heat [66] and conductivity [67] measurements.

The specific heat measurements of Shin and coworkers [66] give an approximately linear decrease in the latent heat of the N–I transition (from about 1.3×10^{-2} to $4.25 \times 10^{-3}\,\mathrm{J\,g^{-1}}$) as the weight percent of surfactant is reduced from 55 to 39%.

Conclusions about the discontinuity of the order parameter S_0 versus w_s at the transition are more difficult. The birefringence measurements of Larson [68] show a decrease of $\Delta n(T_{ni}) = 1.2 \times 10^{-3}$ ($w_s = 0.524$) to 0.35×10^{-3} ($w_s = 0.438$). This is a larger reduction than can be accounted for by a decrease in surfactant density, indicating a decrease in the order of the surfactant chains at T_{ni}. This is consistent with the conductivity measurements of Boden and coworkers [67], who deduce a reduction in S_c from 0.35 ($w_s = 0.558$) to 0.16 ($w_s = 0.2$). In [67], S is taken as the order parameter of the micelles rather than the order of the surfactant chains. The latter is what is observed in a birefringence measurement; whereas the conductivity will depend on micellar shape and orientation.

Theoretically, the dependence of (ΔT, ΔH, and S_0) on w_s might be explained by: (1) coupling to biaxial order parameters [69] (controlled by the shape of the micelle); (2) reduction of the intermicellar interactions under dilution by water [70]; or (3) a combination of the above.

7.3.3 Nematic–Lamellar Transition

The nematic–lamellar transition impresses a strong signature on many observable properties. This is due to the continuous nature of the transition and its clothing of thermal fluctuations. The transition is identical in symmetry to the nematic–smectic *A* transition of thermotropic liquid crystals. This was emphasized by the work of the MIT and Kent groups on (DACl/ACl/water) [71,72], where it was shown that as the lamellar phase is approached from the nematic side, the bend and twist curvature elastic constants k_{33}, k_{22}, and viscosity γ_1 diverge with nearly the same exponents as thermotropic systems.

Within the lyotropic nematic phases, the Frank elastic constants are similar in magnitude to thermotropics [72]. This is likely to be a reflection of the flexibility of the micelles and points to the elasticity as originating mostly from the order of the surfactant chains. In contrast, the viscosity exhibits considerable differences. For example, consider the rotational viscosity associated with the heat dissipated per unit volume when the director field rotates relative to the stationary liquid. The rotational viscosity γ_1 of MBBA at room temperature is $\simeq 0.80$ poise. For a (CsPFO 42.6 wt.%/water) mixture, in the middle of the temperature range of the N_L phase (T = 28 °C), $\gamma_1 \simeq 10$ poise [73]. This difference reflects the size and connectedness of the micelles. It is sensible to expect the micelle size to fluctuate and diverge near the N–N_L transition. It is unclear how size fluctuations affect the temperature reversibility of the N–N_L transition [74]. As the N_L–lamellar transition is approached, the difference between lyotropic and thermotropic viscosities becomes smaller relative to the smectic fluctuation contribution [14]. In this temperature region, the smectic correlation length dominates any length associated with micelle size.

The first measurements of the conduction of ions at an N_L–L transition were performed by Photinos and coworkers [75]. The theoretical idea behind the experiment was to see how the formation of the lamellar phase affected the transport of charge. If the insulating layers are formed unbroken, the transport parallel to the optic axis should change from a continuous drift of charge around the disconnected lamellar micelles of the N_L phase to a hopping mechanism with a much longer relaxation time in the lamellar phase. Therefore, one should expect a large change in conductivity. This was not observed (see FIGURE 7.12), and was rather surprising at the time. Clearly there must exist an equilibrium density of defects which allows passage of ions, but does not affect the long-range order.

Recent measurements on an unaligned dilute lamellar phase of ($C_{12}E_5$(1.1% by volume)/water) show that the conductivity increases by as much as two orders of magnitude under a transition to the adjacent (in temperature) isotropic phases [38]. Ostensibly, this demonstrates that the layers are relatively intact in the $C_{12}E_5$ system.

Freeze fracture electron microscopy (FFEM) supports this view to explain the conductivity measurements. Near the N_L–L transition of the (DACl/ACl/water) system [76,77] FFEM reveals the layers of individual micelles that fuse to form the bilayers. (see FIGURE 7.13). This is consistent with the observation of a small change in conductivity around the transition. Upon completion of the transition, the lamellar phase is laced with a distribution of screw dislocations [76], which allows charge transport parallel to the layer normal.

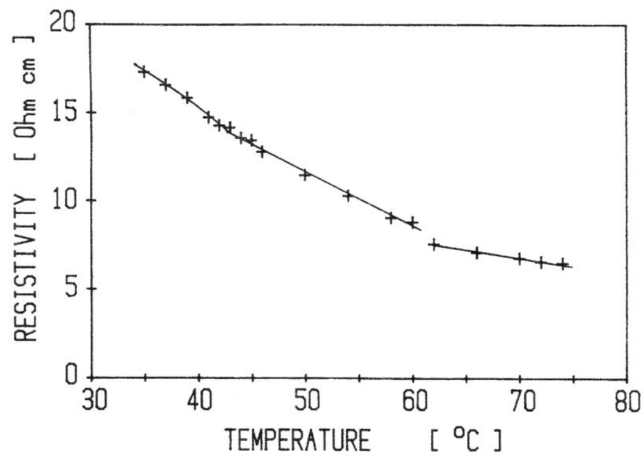

(a)

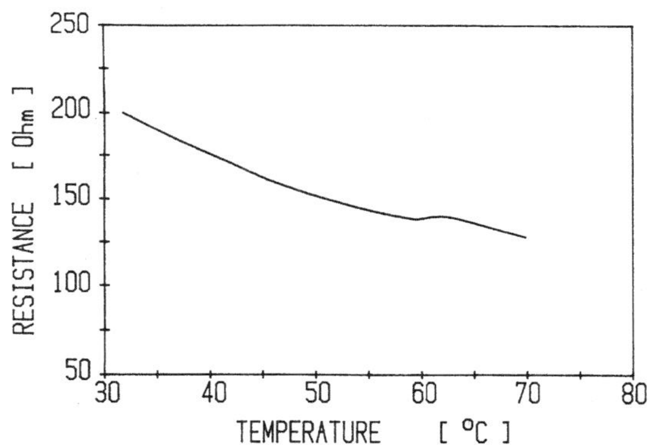

(b)

FIGURE 7.12. Resistivity versus temperature in isotropic (I), aligned nematic (N_L), and lamellar (L) phases of (DACl/ACl/water). (a) Current parallel to director. (b) Current perpendicular to director. After [75]. I–N_L–L transitions also determined optically.

The FFEM results of Strey and coworkers [78] show in a direct manner the substantial change in structure of the dilute lamellar phase relative to the isotropic L_3 (sponge phase) of the mixture (cetylpyridinium chloride/hexanol/NaCl/water). It is reasonable to assume that this structural change is representative of the discontinuity mentioned for the conductivity of the ($C_{12}E_5$/water) mixture. In this case there are no small micelles; the bilayers are largely intact and the change in conductivity is caused by a topological reorganization of the layers that control the conducting pathways.

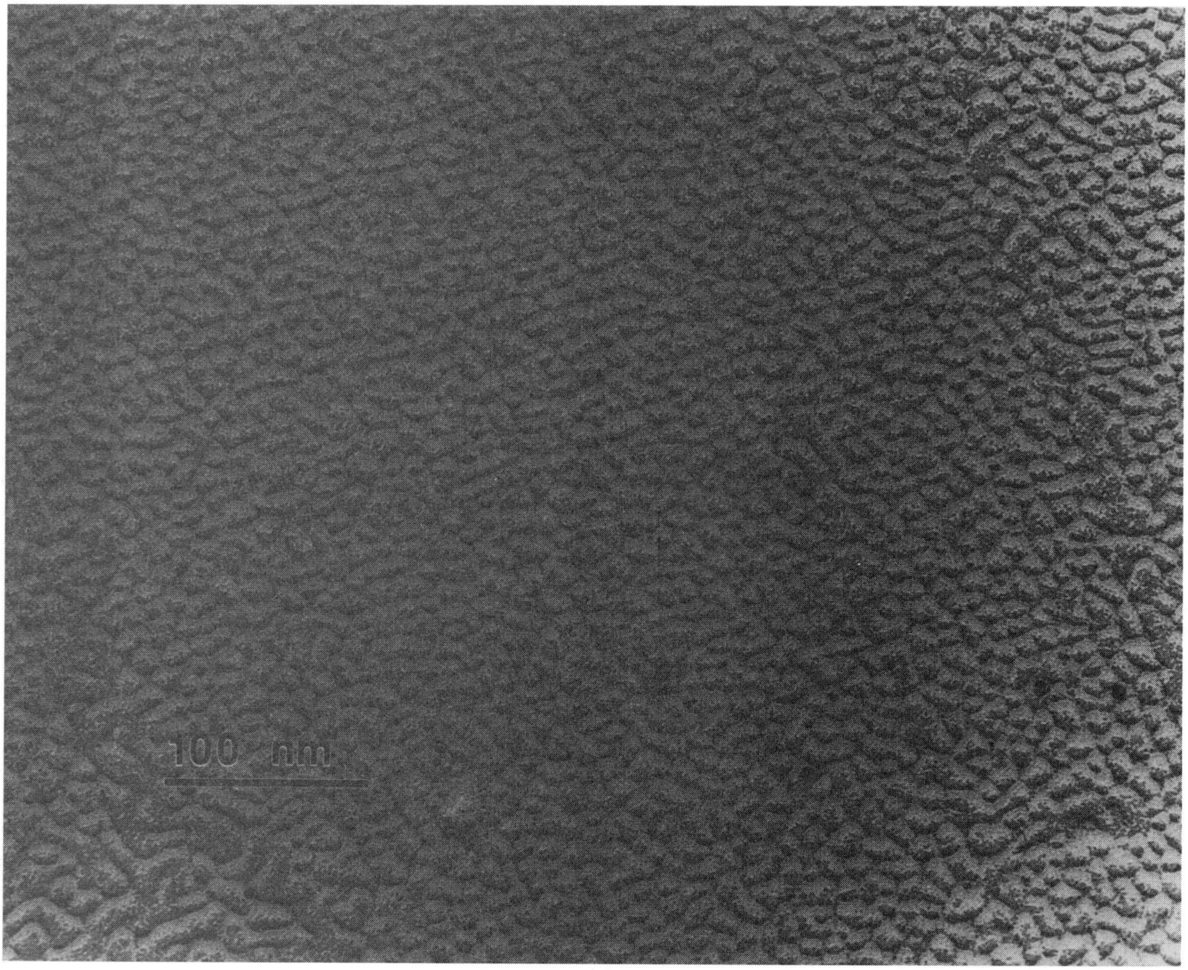

(a)

FIGURE 7.13. Transmission electron micrographs of ($DACl/ACl/D_2O$)=(43.97/1.76/54.27) wt.% [76]. (a) Disk micelles quenched from 41°C ($T_{ni} = 47.7\,°C$), optic axis normal to page. Average intermicellular spacing is 105 Å. (b) Quenched from 35 °C ($T_{nl} = 32.8\,°C$). Beginning of transition to lamellar phase.

7.4 EFFECTS OF POLYMERIC SOLUTES ON NEMATIC AND LAMELLAR PHASES

One of the most studied polymers is polyethylene oxide (PEO) [79]. It is known that PEO has strong attractive interactions with anionic surfactants and weak interactions with cationic surfactants. This trend is continued when the surfactants form micelles. (Recent work on a variety of polymers and surfactants may be found in [80].)

Using neutron scattering on sodium dodecylsulfate(SDoS)/water/PEO solutions, Cabane and Duplessix [81] confirmed the model by which a polymer of size R adsorbs to micelles of size $D(R \gg D)$ forming a necklace-like string of micelles. The adsorption energy was determined to be about $0.3kT$ per surfactant molecule.

The primary adsorption sites of PEO are the oxygen atoms spaced every 3.67 Å apart in the *trans* configuration [79]. Due to the electronegative character of the

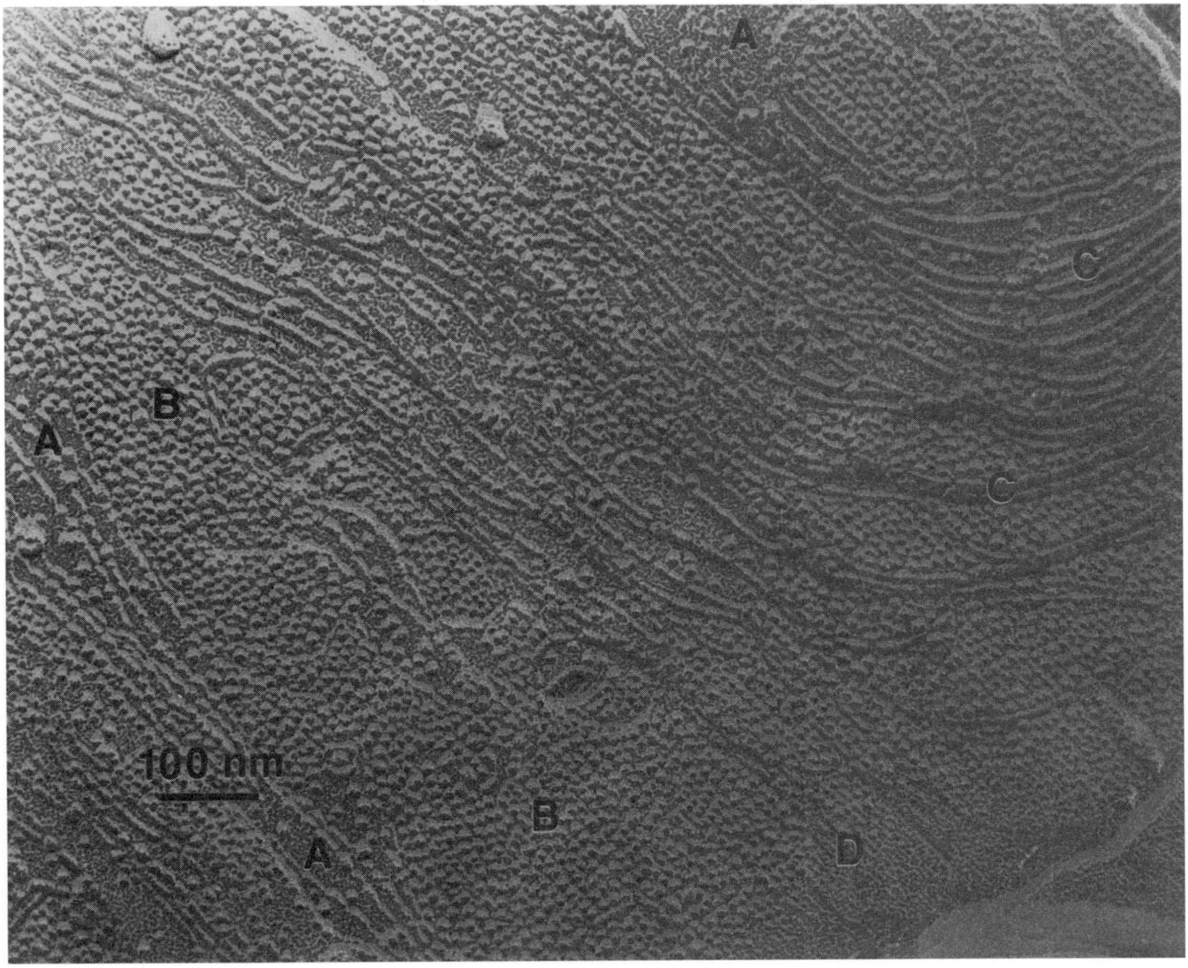

(b)

FIGURE 7.13. Continued.

oxygens, it might seem plausible to expect a strong interaction with cationic surfactants, however the opposite happens. This is likely to be due to the special properties of water. The oxygen atoms of PEO acquire protons from the solvent and gain an average net positive charge [82]; the micelles of anionic surfactant possess a net negative charge, hence the polymer binds to the micellar surface. The above scenario suggests that the polymer–micellar interaction should depend on pH or added salt. These effects might be difficult to pin down for ionic surfactants above the CMC, since the micellar shapes and sizes depend strongly on added ions. Some evidence of a pH dependence of PEO binding was given in [82]. It should be mentioned that the hydration shell, or orientation of the water dipoles around the polar parts of the polymer, may participate in the preferential binding of anionic surfactant to, for example, PEO [83].

The theoretical work on interactions between polymers and surfactants is rather sparse. Both Nagarajan [84] and Ruckenstein and coworkers [85] investigated the effect of added polymer on micelle size distribution and CMC.

Reference [84] assumes the main effects of the adsorbed segments are to: (1) increase the average area per surfactant in a micelle; and (2) help shield the interior of the micelle from contact with water. These effects depend on a parameter, a_{pol}, equal to the area per surfactant occupied by polymer segments on a given micelle surface. The difference between ionic and zwitterionic head groups is taken into account via an additive electrostatic term in the micellar chemical potential; for nonionic surfactants this term is set to zero. Interactions between the micelles are neglected.

We may attempt a connection to current theories of polymer adsorption on particles if we simply write $a_{pol} = N^* a_1/g$. Here N^* is the number of adsorbed segments on a micelle and is a function of the adsorption energy, geometry of the micelle, and quality of the solvent [86]; a_1 is the area of a segment adsorption site and g is the aggregation number. The parameter a_{pol} is treated in a mean-field sense. The micelles are thought to be immersed in a solvent with a uniform semidilute concentration of polymer of size $N \gg N^*$ [87]; a_{pol} thus represents the collective effect of the fraction of adsorbed sites on the micellar free energy.

Within this approximation, Nagarajan [84] finds: (1) the CMC of anionic surfactants is substantially reduced by the added polymer; (2) the CMC of nonionic surfactants is little affected by added polymer; and (3) rod-like micelles, bilayers and vesicles composed of both ionic and nonionic surfactants are unstable to adsorbed polymer and break up into smaller bound micelles.

The work of Ruckenstein and coworkers [85], allows polymer-bound micelles to coexist with polymer-free globular micelles, which is observed experimentally [80]. This permits a study of the effect of polymer concentration. The main parameter is $\Delta\sigma$, which gives the shift in oil/water interfacial tension when polymer is added to the aqueous phase. For a given concentration of polymer, increasing $\Delta\sigma$ reduces the CMC of surfactant which interacts with the polymer. The free surfactant begins to aggregate after the polymers have been saturated with bound micelles.

Current efforts at understanding the interactions of polymers with the dilute lamellar phases (section 7.2.2), vesicles, and surfactant monolayers, are documented in [86–89].

A lamellar phase formed in sodium decylsulfate (SDS)/water/PEO mixtures (up to 80 wt.% SDS), was used by Kekicheff and coworkers [92] to try to squeeze the polymer into a pancake shape. They found that rather than accepting confinement, the dissolved polymer pokes holes in the bilayers and retains an overall layered distribution of segments. The mixtures are homogeneous until about 5% of the water is replaced by PEO. At higher concentrations of PEO, the mixture separates into lamellar, isotropic, and other uncharacterized phases.

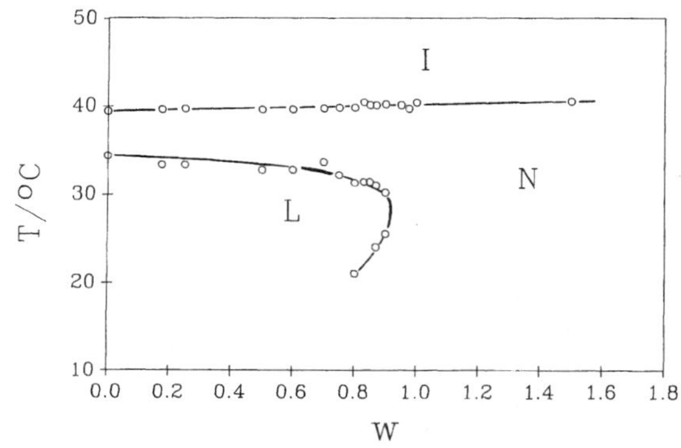

FIGURE 7.14 Transition temperatures versus weight percent of water replaced by PEO [MW = 10^5], CsPFO 49 wt.% (I, isotropic; N, nematic; L, lamellar).

From the general standpoint of stability of lyotropic liquid crystalline phases, it is of interest to see how polymers may affect the boundaries between different phases; in particular, the well studied isotropic–nematic and nematic–lamellar transitions. The binary system of (CsPFO/water) (see FIGURE 7.11), was well suited for a first study in this direction [93].

FIGURE 7.14 shows a cut of the phase diagram when PEO of molecular weight (MW) 10^5 was added to a 49 wt.% CsPFO solution. Water was replaced by PEO gram for gram, with the weight percentage of water replaced by PEO denoted by w. The phase diagram was determined by optical microscopy using textures characteristic of nematic and smectic phases. A similar phase diagram was obtained using a 42.6 wt.% CsPFO mixture.

The nematic transition temperature (including the width of the two-phase region) is hardly affected by the polymer at the considered concentrations. The formation of orientational order is apparently insensitive to the decoration of the micelles with the polymer. This is consistent with the irregular coil configuration assumed by the polymer necklace. It is probable the average configuration of the polymer changes at the N–I transition. It might be possible to observe this small effect by NMR using deuterated PEO.

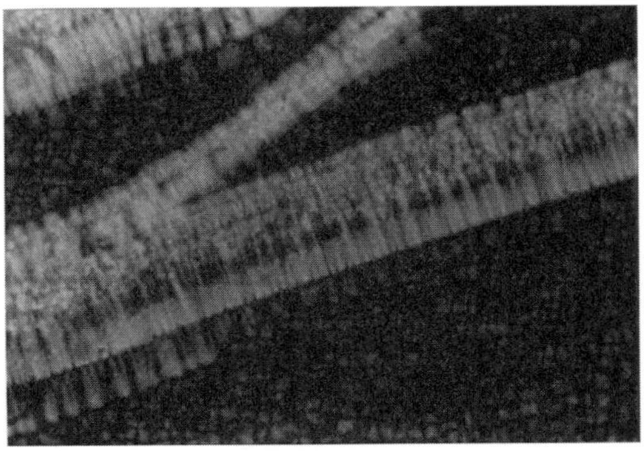

(a)

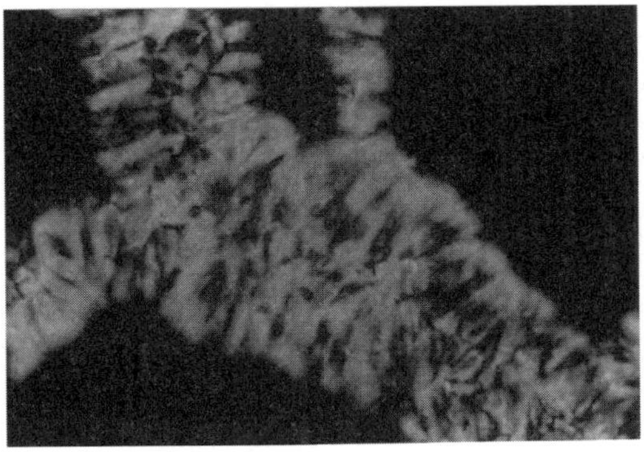

(b)

(c)

FIGURE 7.15. Photomicrographs of CsPFO/(water + PEO) system. 49 wt.% CsPFO, with given wt.% of PEO in water. T = 21 °C, Mag. 80×, crossed polarizers. Progression from lamellar to nematic with added PEO. (a) 0.5 wt.% PEO, (b) 0.65 wt.% PEO, (c) 1.0 wt.% PEO.

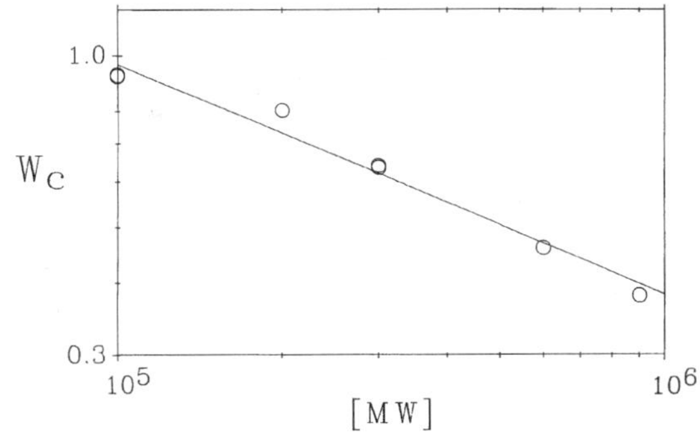

FIGURE 7.16. Critical concentration versus molecular weight of PEO.

Anisotropic polymer configurations have been observed in thermotropic nematics [94].

A PEO coil of MW = 10^5 has a radius of order 200–400Å; when the MW = 4×10^6, the size may be from 1200–3800 Å [79]; the actual dimensions depend on the quality of the solvent and the size of a segment. (A certain fraction of monomers of PEO retain a helical conformation in aqueous solution. This factor may be taken into account through a redefinition of the segment size [79].) In our case, the solvent is distilled water exposed to air, and has a pH of about 6. In addition, the micelles themselves should perturb the average size of the polymer through adsorption and excluded volume effects, but the average size of the polymer should always be large compared with the layer distance of 50 Å [23].

The PEO has little effect on the N–L transition temperature until a critical concentration of polymer, w_c, is reached. For $w > w_c$, the lamellar phase is unstable and is replaced by the nematic phase, see FIGURES 7.14 and 7.15. The critical concentration w_c depends on the molecular weight; the dependence for the available molecular weights is shown in FIGURE 7.16 for a 49 wt.% CsPFO solution. A linear fit yields a power law with an exponent 0.43 ± 0.07. Within the error, the same dependence was found for a 42.57 wt.% CsPFO solution.

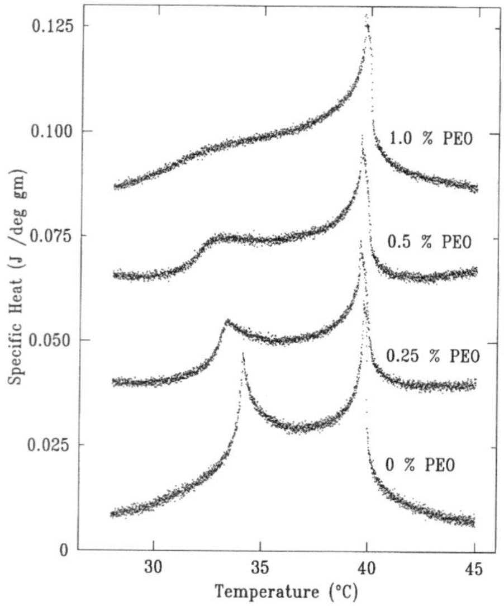

FIGURE 7.17. Specific heat of CsPFO/(water + PEO [MW = 10^5]) system, CsPFO 49 wt.% with indicated amount of PEO. A linear background has been subtracted from each curve. Curves with added PEO are shifted for clarity.

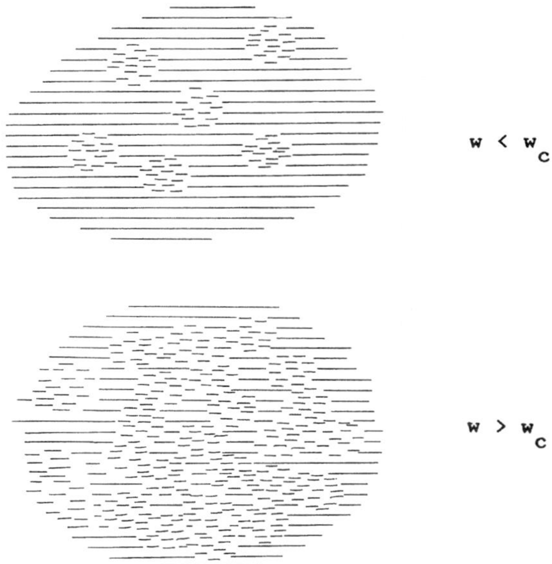

FIGURE 7.18. Two-dimensional representation of nematic droplet model.

FIGURE 7.17 displays calorimetric measurements made on a series of 49 wt.% CsPFO mixtures with PEO[MW] = 10^5. The difference between the two transitions is remarkable. The N_L–I peak is essentially unchanged, while the N_L–L peak broadens and shifts to a slightly lower temperature. Further experiments on samples near w_c did not reveal a second peak corresponding to the reentrant nematic transition. In the lamellar phase, when $w < w_c$, we have the picture of isolated droplets of polymer-induced nematic, (FIGURE 7.18). (The local break-up of the layers by the polymer is in accord with the theoretical result of Nagarajan [84].) When $w > w_c$, the smectic matrix becomes disconnected, the layers in different regions become decorrelated, and the macroscopic smectic order disappears. A distribution of short-range smectic order is indicated from the broad hump in C_p centered around T_{NL} when $w \gtrsim w_c$.

We note here a resemblance (in appearance) to the phase diagrams and heat capacities of Lushington and coworkers [95] for thermotropic octyloxycyanobiphenyl(8OCB)/hexyloxycyanobiphenyl (6OCB) mixtures. This mixture also shows a reduction in the N–SmA heat capacity peak with increasing wt.% of 6OCB, and a reentrant nematic shoulder near a critical concentration of 27 wt.% 6OCB. The behavior in the thermotropic system was attributed to: (1) a density dependence of the interactions which cause formation of the smectic phase; (2) frustration effects involved in arranging nearest and next-nearest dipoles of the molecular cyano groups; and (3) a combination of both of these.

FIGURE 7.19 demonstrates that the rotational viscosity increases in the expected manner with polymer

FIGURE 7.19. Rotational viscosity versus wt.% PEO, 42.57% CsPFO. After [73].

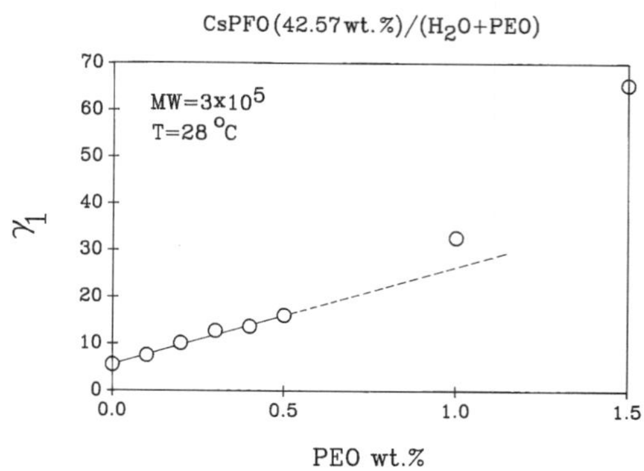

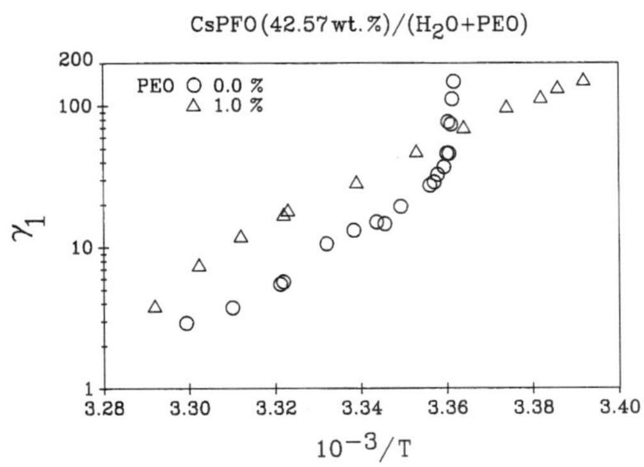

FIGURE 7.20. Rotational viscosity versus inverse temperature for 0% PEO and 1% PEO [MW = 10^5], 42.57 wt.% CsPFO. After [96].

concentration within the nematic phase. However, near T_{n1}, γ_1 is known to have a strong dependence on smectic fluctuations [72]; the data of FIGURE 7.20 exhibit the suppression of this divergence by the polymer [73]. The results of FIGURE 7.19 were extracted from the time dependence of the conductivity under small rotations of the director field [17,96]. The measurements of FIGURE 7.20 used the rotating field method of [53]. Careful independent measurements of the dependence of γ_1 on PEO concentration, especially near w_c and the reentrant nematic phase, have been made by Wedler [97].

We identify w_c with w^*, the weight fraction for which the polymer coils overlap if they are homogeneously distributed. Assuming the coils are nearly spherical, the MW dependence of w_c follows the simple power law, $w_c \sim [\mathrm{MW}]^{-p}$. The exponent may take on a value in the range 0.5 to 0.8, depending on the quality of the solvent. The value derived from FIGURE 7.16), is near the ideal solution value of $\frac{1}{2}$. This suggests that the dense ionic surfactant solution is a nearly ideal solvent for the PEO chains. It would be interesting to see if neutron scattering measurements on deuterated PEO reveal the same behavior.

Another point of interest is the self-diffusion coefficient of PEO in the nematic and lamellar phases. It would be reasonable to expect the diffusion of the polymer to become highly anisotropic in the lamellar phase. Diffusion perpendicular to the layer normal involves the viscosity of the two-dimensional liquid of the bilayers, with most of the hydrophilic chain translating in the aqueous environment. Conversely, motion parallel to the layer normal entails unfavorable transfer of the chain segments through the bilayers. In any case, the polymer must drag a cloud of induced disorder with it as it moves, and this alone should greatly slow the overall diffusion relative to the nematic phase.

Near the N–L transition, when the smectic correlation length, ξ, becomes comparable to the average distance between domains $1 \sim w_c^{-1/3}$, finite-size effects become important [98] and cause a rounding or loss of sharpness in the transition. We may write this condition as

$$\xi = \xi_0 \left(\frac{T - T_c}{T_c} \right)^{-\nu} \sim l \qquad (7.11)$$

so that

$$\frac{\Delta T}{T_e} \equiv \frac{T_c(l) - T_{c\ (bulk)}}{T_{c\ (bulk)}} \sim (\xi / l)^{1/\nu}. \qquad (7.12)$$

If we let $\nu = 1/2$, micellar size $\xi_0 = 50$ Å, and $1 \simeq 1000$ Å, corresponding to a MW $= 10^5$ homogeneous PEO solution of 0.25 wt.%, we find $\Delta T \sim 1\,^\circ\mathrm{C}$. This compares favorably with the observed broadening in FIGURE 7.17. The broadening is seen well before the critical concentration is reached ($W_c(MW = 10^5) = 0.91$ wt.%).

Following Grinstein and Luther [99], we may view the polymer-induced domains as impurity clusters which cause a measurable loss of sharpness in phase transitions occurring in the matrix. A new feature is that now "the size of the cluster can be systematically varied" by varying the molecular weight of the polymer.

We end this section with a remark on the well-known analogy of the nematic–smectic transition to the normal–superconducting transition [14]. Most high-T_c materials (e.g., $La_{2\text{-}n}(Sr_n)CuO_4$) at the end of preparation, appear to be mixtures of spatial domains; whether or not a given domain is superconducting depends on the doping concentration (n) within the domain, see for example, [100]. Macroscopic superconductivity occurs when superconducting domains percolate through the system. The consequent divergence of the diamagnetic susceptibility then parallels the divergence of the Franck elastic constants and rotational viscosity in the liquid crystal [14,71,72]. The polymer therefore plays the role of inducing a nematic (normal) domain of a given size in a lamellar (superconducting) matrix; the reverse situation being superconducting grains suspended in a normal matrix [101].

7.5 CONCLUSION

Lyotropic liquid crystals encompass a large field within the newly designated branch of physics known as complex fluids. Much of the experimental work in the field is guided by intuition, and a great portion of the theory is phenomenological in nature. The generality of intuition and phenomenology allow analogies with other somewhat exotic areas of physics. As an example, consider the frequently cited parallel between lattice gauge theories of particle physics and systems of self-avoiding membrane surfaces [12,102]. Some might view such analogies as pretentious, but it is undeniable that the audience of research workers is considerably broadened and hopefully inspired to produce theories with predictions which can be economically tested in independent, already-existing laboratories.

REFERENCES

1. Safran, S. 1994. *Statistical Thermodynamics of Surfaces, Interfaces, and Membranes.* Reading, Mass.: Addison-Wesley.

2a. Tanford, C. 1973. *The Hydrophobic Effect.* New York: Wiley.

2b. Tanford, C. 1974. *J. Phys. Chem.* **78**: 2469.

3. Gruner, S. 1989. *J. Phys. Chem.* **93**: 7562.

4. Puvvada, S., and Blankschtein, D. 1992. *J. Phys. Chem.* **96**: 5567.

5 Nagarajan, R., and Ruckenstein, E. 1979. *J. Colloid Interface Sci.* **71**: 580.

6. Israelachvili, J. N., Mitchell, D. J., and Ninham, B. W. 1976. *J. Chem. Soc. Faraday Trans. II* **72**: 1525.
   For dilute interacting systems see, e.g., W. M. Gelbart, A. Ben-Shaul, A. Masters, and W. McMullen, *Physics of Amphiphiles: Micelles, Vesicles, and Microemulsions*, ed. V. Degiorgio and M. Corti, p. 394 (Amsterdam: North Holland, 1985).

7. Tiddy, G. J. T. 1980. *Phys. Rep.* **57**: 1.

8. Ekwall, P. 1975. *Advances in Liquid Crystals*, vol. 1, ed. G. H. Brown, p. 1. New York: Academic.

9. Ekwall, P., Mandell, L., and Fontell, K. 1969. *Mol. Cryst. Liq. Cryst.* **8**: 157.

10. Helfrich, W. 1981. *Physics of Defects*, ed. R. Balian et al., p. 713. USMG/NATO ASI, Les Houches, Session 28. Amsterdam: North Holland.

11. Langevin, D., Meunier, J., and Boccara, N., eds. 1987. *Physics of Amphiphilic Layers.* Springer Proceedings in Physics, vol. 21. Heidelberg: Springer.

12. Nelson, D. R., Piran, T., and Weinberg, S., eds. 1989. *Statistical Mechanics of Membranes and Surfaces.* Singapore: World Scientific.

13. Kuzma, M. 1985. *J. Chem. Phys.* **89**: 4124.

14. de Gennes, P. G., and Prost, J. 1993. *The Physics of Liquid Crystals.* 2nd ed. Oxford: Oxford University Press.

15. Charvolin, J. 1983. *J. Chim. Phys.* **80**: 16.

16. Forrest, B. J., and Reeves, L. W. 1981. *Chem. Rev.* **18**: 1.

17. Saupe, A., Xu, S. Y., Plumley, S., Zhu, Y. K., and Photinos, P. J. 1991. *Physica A* **174**: 195.

18. Yu, L. J., and Saupe, A. 1980. *Phys. Rev. Lett.* **45**: 1000.

19. Hendrix, Y., Charvolin, J., Rawiso, M., Liebert, L., and Holmes, M. C. 1983. *J. Phys. Chem.* **87**: 3991.

20. Rizzatti, M. R., and Gault, J. D. 1986. *J. Colloid Interface Sci.* **110**: 258.

21. Holmes, M. C., and Charlovin, J. 1984. *J. Phys. Chem.* **88**: 810.

22. Yu, L. J., and Saupe, A. 1980. *J. Am. Chem. Soc.* **102**: 4897.

23a. Boden, N., and Holmes, M. C. 1984. *Chem. Phys. Lett.* **109**: 76.

23b. Boden, N., and Corne, S. A., and Jolley, K. W. 1987. *J. Phys. Chem.* **91**: 4092.

23c. Leaver, M. S., and Holmes, M. C. 1993. *J. Physique II* **3**: 105.

24. Hendrix, Y., and Charvolin, J. 1981. *J. Physique* **42**: 1427.

25. Radley, K., and Saupe, A. 1978. *Mol. Phys.* **35**: 1405.

26. Hertel, G., and Hoffmann, H. 1988. *Prog. Colloid Polym. Sci.* **76**: 123.

27a. Luhmann, B., and Finkelmann, H. 1986. *Colloid Polym. Sci.* **264**: 189.

27b. Boden, N., Bushby, R. J., and Hardy, C. 1985. *J. Physique Lett.* **46**: 325.

28a. Plumley S., and Kuzma, M. 1991. *Mol. Cryst. Liq. Cryst.* **200**: 33.

28b. General properties of lyotropic nematics are also reviewed by Sorir, A. S. 1987. *Sov. Phys. Vsp.* **30**: 875.

29. Wahlstrom, E. E. 1979. *Optical Crystallography*, p. 263. 5th ed. New York: Wiley.

30a. Saupe, A., Boonbrahm, P., and Yu, L. J. 1983. *J. Chim. Phys.* **80**: 7.

30b. For a discussion of biaxial cholesteric phases see Pleiner, H. and Brand, H. 1985. *Phys. Rev. Lett.* **54**: 1817.

31. Figueiredo Neto, A. M., Galerne, Y., Levelut, A. M., and Liebert, L. 1985. *J. Physique Lett.* **46**: 499.

32a. Campbell, J., Kuzma, M., and Labes, M. M. 1983. *Mol. Cryst. Liq. Cryst.* **95**: 45.

32b Casalnuovo, S., Heeger, A. J. 1984. *Phys. Rev. Lett.* **53**: 2254.

33a. Bouligand, Y. 1972. *J. Physique* **33**: 525.

33b. Saupe, A. 1976. *J. Colloid Interface Sci.* **58**: 549.

33c. Amaral, L. Q., Gulik, A., Itri, R. and Mariani. *Phys. Rev. A* **46**: 3548 (1992).
34. Roux, D., and Bellocq, A. M. 1985. *Physics of Amphiphiles*, ed. V. Degiorgio and M. Corti, p. 842. Amsterdam: North Holland.
35a. Larche, F. C., Appell, J., Porte, G., Bassereau, P. and Marignan, J. 1986. *Phys. Rev. Lett.* **56**: 1700.
35b. Larche, F. C., El Qebbaj, S. and Marignan, J. 1986. *J. Phys. Chem.* **90**: 707.
36a. Hoffmann, H. 1990. *Adv. Colloid Interface Sci.* **32**: 123.
36b. Thunig, C., Hoffmann, H., and Platz, G. 1989. *Prog. Colloid Polym. Sci.* **79**: 297.
37. Satoh, N., and Tsujii, K. 1987. *J. Phys. Chem.* **91**: 6629.
38. Strey, R., Schomacker, R., Roux, D., Nallet, F., and Olsson, U. 1990. *J. Chem. Soc. Faraday Trans.* **86**: 2253.
39. Gunther, L., Imry, Y., and Lajzerowicz, J. 1980. *Phys. Rev. A* **22**: 1733.
40. Nallet, F., Roux, D., and Milner, S. T. 1990. *J. Physique* **51**: 2333.
41. Helfrich, W. 1978. *Z. Naturforsch.* **33a**: 305.
42. Huse, D. A., and Leibler, S. 1988. *J. Physique* **49**: 605.
43. Nallet, F., Roux, D., and Proust, J. E. 1989. *J. Physique* **50**: 3147.
44. Selinger, J. W., and Bruinsma, R. F. 1991. *Phys. Rev. A* **43**: 2910, 2922.
45a. de Gennes, P. G. 1982. *Polymer Liquid Crystals*, ed. A. Ciferri et al., p. 115. New York: Academic.
45b. Halperin, A., and Williams, D. 1992. *Europhys. Lett.* **20**: 601.
46. Taylor, M. P., and Herzfeld, J. 1991. *Phys. Rev. A* **43**: 1892.
47a. Barker, J. A. 1960. *Australian J. Chem.* **13**: 187.
47b. Barker, J. A. 1961. *Proc. R. Soc. London A* **259**: 442.
48. Chidichimo, G., Golemme, A., Doane, J. W., and Westermann, P. 1984. *J. Chem. Phys.* **82**: 536.
49. Quist, P., and Halle, B. 1993. *Phys. Rev. E* **47**: 3374.
50. Hoffmann, H., and Ebert, G. 1988. *Angew. Chem. Int. Ed. Engl.* **27**: 902.
51. Kahlweit, M., Lessner, E., and Strey, R. 1983. *J. Phys. Chem.* **87**: 5032.
52. Roux, D., Coulon, C., and Cates, M. E. 1992. *J. Phys. Chem.* **96**: 4174.
53. Plumley, S., Zhu, Y. K., Hui, Y., and Saupe, A. 1990. *Mol. Cryst. Liq. Cryst.* **182B**: 215.
54. Gramsbergen, E. F., Longa, L., and de Jeu, W. H. 1986. *Phys. Rep.* **135**: 195.
55. Olmstead, P., and Goldbart, P. 1991. *Mol. Cryst. Liq. Cryst.* **198**: 265.
56. Sheng, P. 1982. *Phys. Rev. A* **26**: 1610.
57a. Hornreich, R. M. 1985. *Phys. Rev. A.* **109**: 232.
57b. Vause, C. 1986. *Phys. Lett.* **114A**: 485.
58. Bothner, A. A., and Pople, J. A. 1965. *Ann. Rev. Phys. Chem.* **16**: 43.
59. Haven, T., Radley, K., and Saupe, A. 1981. *Mol. Cryst. Liq. Cryst.* **75**: 87.
60. Boonbrahm, P. and Saupe, A. 1984. *J. Chem. Phys.* **81**: 2076.
61. Boonbrahm, P. 1983. *PhD Thesis*, Kent State University.
62. Melnik, G., Photinos, P. J., and Saupe, A. 1989. *Phys. Rev. A* **39**: 1597.
63. Ramesh, V., and Labes, M. M. 1987. *Mol. Cryst. Liq. Cryst.* **152**: 57.
64. Rosenblatt, C., Kumar, S., and Litster, J. D. 1984. *Phys. Rev. A* **29**: 1010.
65. Rosenblatt, C. 1985. *Phys. Rev. A* **32**: 1924.
66a. Shin, S. T., and Kumar, S. 1991. *Phys. Rev. Lett.* **66**: 1062.
66b. Shin, S. T., Kumar, S., Finotello, D., Keast, S. S., and Neubert, M. E. 1992. *Phys. Rev. A* **45**: 8683.
67. Boden, N., Clements, J., Dawson, K., Jolley, K. W., and Parker, D. 1991. *Phys. Rev. Lett.* **66**: 2883.
68. Larson, B. 1986. *PhD Thesis*, Massachusetts Institute of Technology.
69. Straley, J. P. 1974. *Phys. Rev. A* **10**: 1881.
70a. Kuzma, M., Gelbart, W. M., and Chen, Z. Y. 1986. *Phys. Rev. A* **34**: 2531.
70b. Kimura, H. 1974. *J. Phys. Soc. Japan* **36**: 1280.
71. Kumar, S., Yu, L. J., and Litster, J. D. 1983. *Phys. Rev. Lett.* **50**: 1672.
72. Zhou, E., Stefanov, M., and Saupe, A. 1988. *J. Chem. Phys.* **88**; 5137.
73a. Kim, D. R. 1994. *PhD Thesis*, Kent State University.
73b. Kim, D. R., and Kuzma, M. 1990. Unpublished research.
74. Photinos, P. J., and Saupe, A. 1990. *Phys. Rev. A* **41**: 954.
75. Photinos, P. J., Yu, L. J., and Saupe, A. 1981. *Mol. Cryst. Liq. Cryst.* **67**: 277.
76a. Sammon, M., Zasadzinski, J. A. N., and Kuzma, M. 1986. *Phys. Rev. Lett.* **57**: 2843.
76b. Zasadzinski, J. A. N., Sammon, M., and Kuzma, M. 1988. *Ordering and Organization in Ionic Solutions*, ed. N. Ise and I. Sogami, p. 5. Singapore: World Scientific.
77. Mihailovic, M. 1987. *C. R. Acad. Sci. Paris. Sér. II* **304**: 875.
78. Strey, R., Jahn, W., Porte, G., and Bassereau, P. 1990. *Langmuir* **6**: 1635.
79. Molyneux, P. 1983. *Water Soluble Synthetic Polymers: Properties and Behavior.* Vol. 1. Boca Raton, Fla.: Chemical Rubber Company.
80. Goddard, E. D., and Ananthapadmanabdan, K. P. eds. 1993. *Interactions of Surfactants with Polymers and Proteins.* Boca Raton, Fla.: Chemical Rubber Company.
81a. Cabane, B., and Duplessix, R. 1987. *J. Physique* **48**: 651.
81b. Cabane, B., and Duplessix, R. 1982. *J. Physique* **43**: 1529.
81c. Cabane, B., and Duplessix, R. 1985. *Colloids Surf.* **13**: 19.
82. Schwuger, M. J. 1973. *J. Colloid Interface Sci.* **43**: 491.
83. Lindman, B., and Thalberg, K. 1973. *J. Colloid Interface Sci.* **43**: 203.
84. Nagarajan, R. 1989. *J. Chem. Phys.* **90**: 1980.
85. Ruckenstein, E., Huber, G., and Hoffmann, H. 1987. *Langmuir* **3**: 382.

86. Pincus, P., Sandroff, C., and Witten, T. 1984. *J. Physique* **45**: 725.
87. Alexander, S. 1977. *J. Physique* **38**: 977.
88. de Gennes, P. G. 1990. *J. Phys. Chem.* **94**: 8407.
89. Andelman, D., and Joanny, J. F. 1993. *J. Physique II* **3**: 121.
90. Brooks, J. T., and Cates, M. E. 1993. *J. Chem. Phys.* **99**: 5467.
91. Ligoure, C., Bouglet, G., and Porte, G. 1993. *Phys. Rev. Lett.* **71**: 3600.
92a. Kekicheff, P., Cabane, B., and Rawiso, M. 1984. *J. Colloid Interface Sci.* **102**: 51.
92b. de Gennes, P. G. 1978. *Solid State Phys. Suppl.* **14**: 1.
93a. Kuzma, M., Wedler, W., Saupe, A., Shin, S. T., and Kumar, S. 1992. *Phys. Rev. Lett.* **68**: 3436.
93b. Smith, A. M., and Holmes, M. C. 1994. *Proceedings of the Fifteenth International Liquid Crystal Conference [Budapest]*, vol. 2, p. 616. New York: Gordon and Breach.
94. Dubalt, A., Ober, R., Veyssie, M., and Cabane, B. 1985. *J. Physique* **46**: 1227.
95. Lushington, K. J., Kasting, G. B., and Garland, C. W. 1980. *Phys. Rev. B* **22**: 2569.
96. Zhu, Y. K., and Kuzma, M. 1990. Unpublished research.
97. Wedler, W. 1992. Unpublished.
98. Barber, M. N. 1983. *Phase Transitions and Critical Phenomena*, vol. 8, ed. C. Domb and J. L. Lebowitz, p. 145. New York: Academic.
99. Grinstein, G., and Luther, A. 1976. *Phys. Rev. B* **13**: 1329.
100a. Klein, L. 1991. *Physica A* **179**: 62.
100b. Blumenfeld, R. 1991. *J. Physique II* **1**: 159.
101. John, S., Lubensky, T. C. 1986. *Phys. Rev. B* **34**: 4815.
102a. Huse, D. A., Leibler, S. 1991. *Phys. Rev. Lett.* **66**: 437.
102b. Karowski, M., and Thun, H. 1985. *Phys. Rev. Lett.* **54**: 2556.

CHAPTER 8

From Molecular to Macromolecular Liquid Crystals

VIRGIL PERCEC

8.1 HISTORICAL REVIEW

Onsager (1949) and Flory (1956) predicted that rigid rod-like macromolecules should display liquid crystallinity. The first main chain thermotropic liquid crystalline polymer was reported in the open literature in 1975 (Roviello and Sirigu, 1975). This publication was independent of the flexible spacer concept advanced by de Gennes (1975) in the same year. However, patents on thermotropic main chain liquid crystalline polymers were published starting with 1973 (Jackson, Jr. and Kuhfuss, 1976; Kuhfuss and Jackson, Jr., 1973, 1974). Some interesting publications dealing with the driving force behind the development of main chain thermotropic crystalline polymers and various historical variants of these developments are available (Ballauff, 1989; Blumstein and Blumstein, 1988; Economy, 1989, 1990a; Finkelmann, 1987; Gordon and Platé, 1984a, b; Jackson, Jr., 1989).

Although the field of thermotropic side chain liquid crystalline polymers is older than that of thermotropic main chain liquid crystalline polymers, systematic investigations in this field started only after Ringsdorf and coworkers introduced the spacer concept (Finkelmann et al., 1978a, b). Monographs and review articles on side chain and main chain liquid crystalline polymers are available (Blumstein, 1978a, b; 1983, 1985; Chapoy, 1985; Chiellini and Lenz, 1989; Ciferri et al., 1982; Ciferri, 1991; Collins, 1990; Donald and Windle, 1991; Gordon and Platé, 1984a, b; Gray, 1987; Griffin and Johnson, 1984; Luckhurst and Gray, 1979; McArdle, 1989; Noël and Navard, 1991; Percec and Pugh, 1989a; Percec and Tomazos, 1992q; Percec et al., 1992; Platé and Shibaev, 1987; Platé, 1993; Saeva, 1979; Weiss and Ober, 1990; Zentel, 1989b).

The goal of this chapter is to discuss the relation between various molecular parameters including structure, molecular weight, architecture, etc., and phase behavior. An attempt will be made to follow this trend on passing from a molecular to a macromolecular liquid crystal. The main goal is to provide the reader with the presently available tools which can be used to tailor-make liquid crystalline polymers. A brief and general introduction to low molar mass and polymeric liquid crystals which is required for this purpose will also be made.

8.2 INTRODUCTION TO LIQUID CRYSTALS

8.2.1 Molecular Liquid Crystals and Definitions

Mesogens or mesogenic groups are compounds that, under suitable conditions, give rise to liquid crystalline phases, that is, mesophases. They can be classified as nonamphiphilic (i.e., most frequently hydrogenated or hydrophobic compounds exhibiting a rod-like or disk-like shape) and amphiphilic (i.e., compounds that contain within the same molecule hydrophobic and hydrophilic groups, which have the ability to dissolve in organic solvents and water, respectively). Thermotropic mesophases are induced by a change in temperature, while lyotropic mesophases are induced by a solvent, which can be either water or organic. Some mesogens can generate amphotropic mesophases, that is, the liquid crystalline phase can be induced either by a change in temperature, by a solvent, or by both (Ringsdorf et al., 1988). Other highly immiscible groups, like perfluorinated and paraffinic, available within the structure of the same molecule, also give rise to amphotropic mesophases (Hopken et al., 1988; Mahler et al., 1985; Viney et al., 1989a, b).

Depending on their thermodynamic stabilities with respect to the crystalline phase, thermotropic mesophases can be virtual (unstable with respect to the crystalline phase), monotropic (metastable with respect to the crystalline phase), or enantiotropic (stable with respect to the crystalline phase) (Keller et al., 1990b; Percec and Keller, 1990e). Virtual mesophases exist only below the melting and crystallization temperatures and therefore cannot be observed. Monotropic mesophases can be observed only during cooling, due to the fact that the crystallization process is kinetically controlled and therefore is supercooled, while the liquid crystalline phase is thermodynamically controlled and is not supercooled. Enantiotropic mesophases can be observed both on heating and cooling. A detailed discussion on the conversion of a mesophase from virtual into monotropic and into enantiotropic will be presented later.

Traditionally, thermotropic mesophases exhibited by rod-like or calamitic mesogens can be classified as uniaxial, biaxial, and chiral nematic (cholesteric), and a number of different smectic phases that form either an untilted layer (smectic A, s_A etc.) or a tilted layer (smectic C, s_C etc.) structure. Characteristic for the nematic phase is a parallel orientation of the molecules along an axis that corresponds to the long axis of the mesogen. While uniaxial nematic phases (biaxial nematic (Galerne, 1988) phases will not be discussed here) exhibit a one-dimensional order, smectic phases can exhibit a two-dimensional (s_A, s_C, exhibiting unstructured layer structures) or a three-dimensional (s_B, s_G, s_E, s_H, exhibiting structured layer structures) order (Gray and Goodby, 1984). Therefore, s_B, s_G, s_E, and s_H liquid crystalline phases are in fact crystalline. Disk-like mesogens exhibit various columnar or discotic mesophases (Chandrasekhar, 1982, 1992; Chandrasekhar and Ranganath, 1990) that is, racemic and chiral nematic (n_D), columnar hexagonal or discotic (Φ_h or D), etc. Plastic crystals are a class of mesophases in which spherical shaped molecules, though orientationally completely disordered, reside (with minor fluctuations) at the points of a spatial lattice, that is, there is positional order but orientational disorder (e.g., methane, carbon tetrachloride, carbon tetrafluoride, camphor, cyclohexanol, etc.).

Upon increasing their concentration, amphiphilic mesogens generate isotropic, and then micellar solutions. At higher concentrations, rod-like micelles generate hexagonal mesophases, while spherical micelles generate cubic mesophases. Alternatively, an isotropic solution of an amphiphilic molecule can assemble directly into a lamellar liquid crystalline phase (DeGiorgio et al., 1985; Kelker and Hatz, 1980; Lehman, 1922; Pershan, 1979; Rinne, 1933; Seddon and Templer, 1991; Skoulios and Guillon, 1988; Winsor, 1968). Cubic mesophases do not exhibit an anisotropic texture since they are optically isotropic. The discotic hexagonal mesophase obtained from discotic mesogens, the hexagonal mesophase of amphiphilic molecules, and the columnar hexagonal mesophase displayed by conformationally disordered polymer chains (Ungar, 1993), are from a symmetry point of view identical.

8.2.2 Macromolecular Liquid Crystals

Based on the place of the polymer where the mesogen is inserted, that is, within the main chain, as side groups, or both within the main chain and as side groups, liquid crystalline polymers can be classified into main chain, side chain and combined (FIGURE 8.1). Polymers with more complex architectures are also possible (Demus, 1988, 1989; Percec and Pugh, 1989a; Ringsdorf et al., 1988). The mesogen used in the construction of the liquid crystalline polymer can be rod-like, disk-like, amphiphilic, etc. (Demus, 1988, 1989; Finkelmann et al., 1984; Hopken et al., 1988; Jahns and Finkelmann, 1987; Luhmann and Finkelmann, 1986, 1987; Mahler et al., 1985; Percec and Pugh, 1989a; Ringsdorf et al., 1988; Ringsdorf and Wustefeld, 1990; Viney et al., 1989a, b). In addition to linear polymer structures, cyclic (Percec and Hahn, 1989c; Richards et al., 1990), hyperbranched (Kim, 1992a, b; Percec et al., 1992i, 1994g; Percec and Kawasumi, 1992l), dendrimeric (Percec, 1995b; Percec and Chu, 1995c) and crosslinked (Gleim and Finkelmann, 1989; Zentel, 1989a) architectures have also been synthesized. Therefore, liquid crystalline polymers display thermotropic and lyotropic mesophases which are similar to those exhibited by thermotropic and lyotropic low molar mass liquid crystals. In addition, rigid-rod-like main chain nonamphiphilic polymers like poly(phenyleneterephthalamide) (Kevlar®) (Bair et al., 1977; Kwolek et al., 1977; Morgan, 1977; Panar and Beste, 1977), poly(p-phenylene-2,6-benzobisthiazole) and poly(p-phenylene-2,6-benzobisoxazole) (Papkov, 1984) also exhibit lyotropic nematic mesophases.

FIGURE 8.1. Schematic representation of main chain, side chain, and combined liquid crystal polymers.

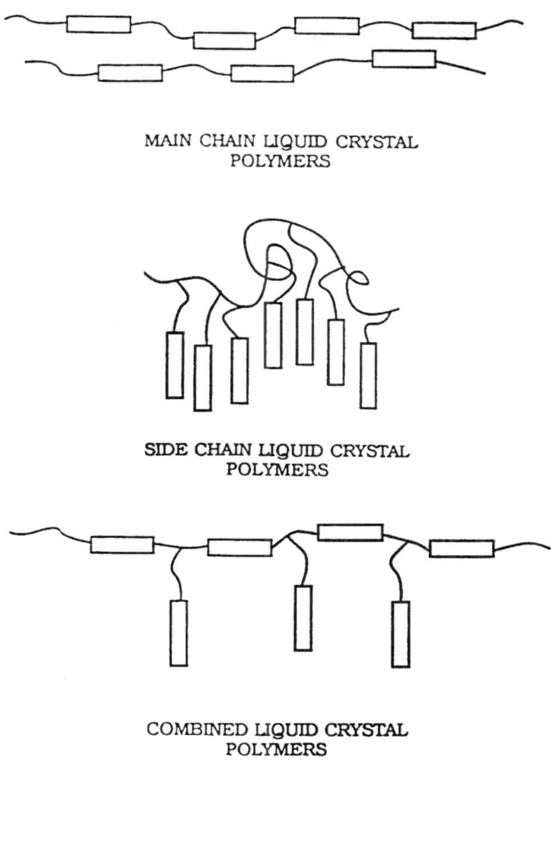

The backbone of side chain liquid crystalline polymers adopts a random-coil conformation only in ideal solutions and in isotropic melts. In the nematic phase the conformation of the polymer backbone is slightly distorted, while in a smectic phase it is highly distorted and can even be confined to the smectic layer (Davidson et al., 1991; Davidson and Levelut, 1992; Noël, 1988, 1989; Noël and Navard, 1991; Noirez et al., 1988, 1995; Pepy et al., 1988; Percec and Pugh, 1989a; Percec et al., 1990c; Percec and Tomazos, 1990i).

8.3 ISOMORPHISM OF LIQUID CRYSTALS

The miscibility of low molar mass liquid crystals and of polymers is important both for the identification of various liquid crystalline phases and for the preparation of mixtures with well defined phase transitions. Arnold, Sackmann, and Demus developed the miscibility rules (Arnold and Sackmann, 1960a, b; Demus et al., 1983; Krigbaum, 1985; Sackmann and Demus, 1973) which can be summarized as follows: if two liquid crystalline phases are miscible, they are isomorphic, and therefore belong to the same type of mesophase. However, the reverse is not true. The situation is more complicated for mixtures of liquid crystalline polymers with low molar mass liquid crystals and for mixtures of liquid crystalline polymers. Often, similar phases of liquid crystalline polymers and low molar mass liquid crystals are not miscible (Achard et al., 1988; Benthack-Thoms and Finkelmann, 1985; Casagrande et al., 1982; Krigbaum, 1985; Ringsdorf et al., 1982, Sigaud et al., 1987; Tinh et al., 1990). The same is the case for similar phases of two different liquid crystalline polymers (Percec and Tsuda, 1991q). A general discussion on this topic and a comparison of the isomorphism of crystalline and liquid crystalline polymers and of the isomorphism of structural units in copolymers is available (Percec and Tsuda, 1991q). When two compounds are isomorphic within a certain mesophase, both their thermal transition temperatures and corresponding thermodynamic parameters exhibit continuous dependences on composition. The same is the case for a copolymer. When its structural units are isomorphic within a certain phase, the corresponding transition temperatures and thermodynamic parameters exhibit continuous dependences on composition. This means that both the components of mixtures and the structural units of copolymers behave like an ideal solution. Therefore, by knowing the temperature transitions and thermodynamic parameters of the parent compounds, we can apply the equations of Schroeder and Van Laar (equation (8.1)) and predict the phase diagram of the mixture or of the copolymer (Achard et al., 1988; Schroeder, 1893; Van Hecke, 1979; Van Laar, 1908).

$$F_1 = \left(1 - \frac{\Delta H_1^0 T_2 (T - T_1)}{\Delta H_2^0 T_1 (T - T_2)}\right)^{-1} \tag{8.1}$$

where F_1 is the molar fraction of component 1; T_1 and ΔH_1 are the transition temperature and the enthalpy change of pure component 1; T_2 and ΔH_2 have the same meaning for component 2; while T is the transition temperature corresponding to molar fraction F_1. T should be lower than T_2 and higher than T_1. Thus, a phase diagram that exhibits a higher or lower temperature transition than the maximum or minimum T_2 or T_1 respectively, violates the temperature criteria discussed above and must therefore be discussed by the introduction of nonideality of solution of structural units of the copolymer. The negative, positive, or linear dependence of T versus F_1 is determined by the parameter A from (8.1)

$$A = \frac{\Delta H_1^0 T_2}{\Delta H_2^0 T_1} \tag{8.2}$$

8.4 ON THE "RIGIDITY" OF "ROD-LIKE" MESOGENS

The conventional procedure used to design low molar mass liquid crystals (Demus, 1989, 1988; Gray, 1962, 1974, 1979, 1982), main chain (Ballauff, 1989; Blumstein and Blumstein, 1988; Economy, 1989, 1990a; Finkelmann, 1987; Gordon and Platé, 1984a; Jackson, Jr., and Kuhfuss 1976; Jackson, Jr., 1989; Kuhfuss and Jackson, Jr., 1974) and side-chain (Finkelmann et al., 1978a, b, 1984; Percec and Pugh, 1989a) liquid crystalline polymers is based on the use of *rigid rod-like* mesogenic units. Recently, the concept of *flexible rod-like* mesogens or *rod-like* mesogens based on *conformational isomerism* was introduced and used in the synthesis of both side chain (Hsu and Percec, 1987, 1989; Percec and Pugh, 1989a) and main chain (Bilibin et al., 1989; Memeger, 1991; Percec and Yourd, 1988a, 1989d, 1989i; Percec and Tsuda, 1990b, 1990d; Ungar et al., 1990) liquid crystalline polymers. Low molar mass liquid crystals based on *flexible rod-like* mesogens were discussed elsewhere (Hsu and Percec, 1987, 1989; Percec and Pugh, 1989a; Percec and Yourd, 1988a, 1989d, 1989i; Percec and Tsuda, 1990b, 1990d; Ungar et al., 1990). Depending on their conformational or configurational rigidity, we have suggested the classification of *rod-like* mesogenic groups as *rigid rod-like, semirigid,* or *semiflexible rod-like* and *flexible rod-like* mesogens (Percec and Zuber, 1991o).

8.4.1 Rigid Rod-Like Mesogens

Rigid rod-like mesogens shown in FIGURE 8.2, such as diphenylacetylene (no. 1), oligo-p-phenylene (no. 2), and benzoxazole (no. 3) and other similar derivatives are based on linearly substituted aromatic or heterocyclic rings. They exhibit free rotation about certain C–C bonds as in the case of FIGURE 8.2, nos. 1–3, but this rotation does not perturb their elongated or extended shape, and therefore these molecules retain their rigid rod-like character. Consequently, in rigid rod-like mesogenic units, rigid refers to the rigidity of the linear shape.

FIGURE 8.2. Rigid rod-like mesogens based on diphenylocetylene (1), oligo(p-phenylene) (2), and benzoxatole (3).

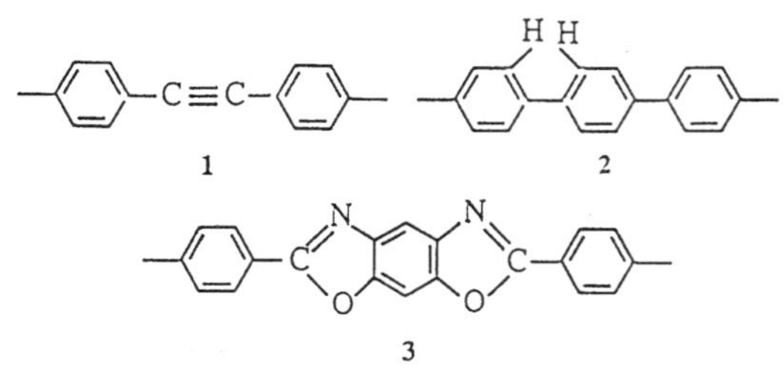

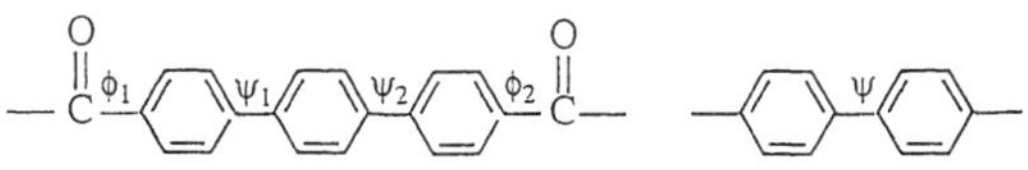

4 5

FIGURE 8.3. Rigid rod-like mesogens containing terphenyl (4) and biphenyl (5).

In oligo-p-phenylene (FIGURE 8.2, no. 2) the steric interaction of the *ortho* hydrogens impedes this internal rotation and the conjugation effect reinforces the rigidity due to the double bond character of the C–C bond between the phenylene rings. An energy barrier of 5 kcal mol^{-1} for the internal rotation of the outer rings of 4 was obtained experimentally (Tékély et al., 1983), while conformational calculations provided for this rotation a value of about 4.8 kcal mol^{-1} (Meurisse et al., 1984). The minimum energy is at $\phi_1 = \phi_2 = 0$ or $\phi_1 = 0$, $\phi_2 = 180°$.

In the case of FIGURE 8.3, no. 5, the energy barrier is 1.9 kcal mol^{-1} and the minimum energy is at $\phi = 40°$. This value is in agreement with the experimental data obtained in the vapor state (i.e., $\phi = 42°$) (Meurisse et al., 1984). If the aromatic rings are connected via a double bond as in the case of stilbene (FIGURE 8.4, no. 6), the *trans* isomer is energetically more favorable. It can be isomerized into the *cis* form, but the activation energy for this transformation is very high, making this molecule rigid. An activation energy of 42.8 kcal mol^{-1} was reported for the thermal uncatalyzed isomerization of stilbene (Kalinowski and Kessler, 1973).

8.4.2 Semirigid Rod-Like Mesogens

Semirigid rod-like mesogens such as amides (FIGURE 8.5, no. 7) and esters (FIGURE 8.5, no. 8) have both conformational and configurational character. For electronic reasons, the rotation about the C–O or C–N bonds of 7 and 8 is retarded and or even prohibited.

FIGURE 8.4. Rigid rod-like groups based on stilbene (6).

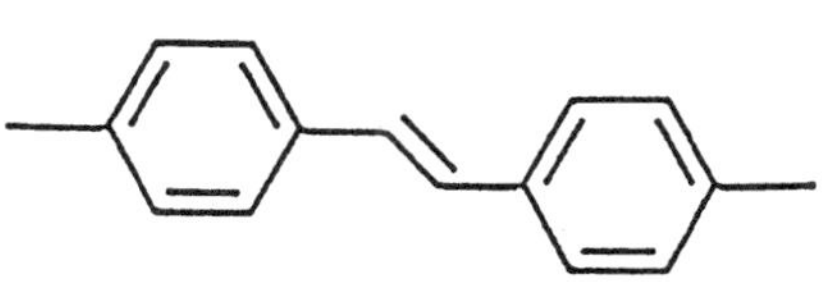

6

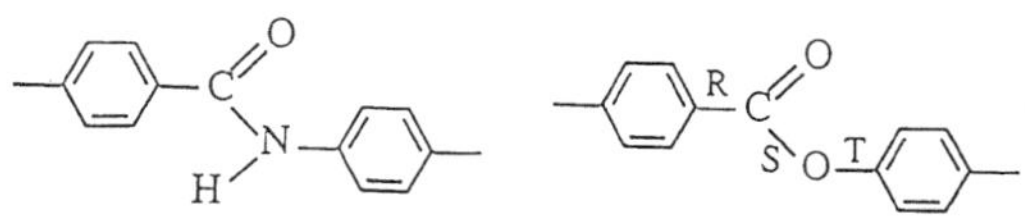

7 8

FIGURE 8.5. Semiflexible repeat units.

In esters like 8, there is rotation about the R, S, and T bonds.

The most significant rotation is about the S bond, since it determines the kinked (*cis*) or extended (*trans*) conformation of the molecule. Theoretical calculations (Coulter and Windle, 1989) suggest that the *trans* conformation is the most stable. The *trans* form of esters and acids has a better O_n–C_{σ^*} "hyperconjugation" which decreases their free energy (Coulter and Windle, 1989). The steric effect in the *cis* form makes it less stable than the *trans* form. The potential energy of the *cis* form is higher than that of the *trans* form with 7.9 kcal mol^{-1}. In the energy profile, there is a hump at 90° which is due to the O_p–C_{π^*} overlap that creates a double-bond character on the O–C bond. This barrier is smaller for aromatic esters than for aliphatic esters because the delocalization of the p-electrons of the oxygen on the aromatic ring attached to the oxygen gives the structure shown in FIGURE 8.6, no. 9, which competes with the contributor (FIGURE 8.6, no. 10). This conclusion is supported by the difference between the C–O bond length of aliphatic and aromatic esters. The C–O bond length is 1.33 Å in alkyl benzoate and 1.36 Å in aryl benzoate (Coulter and Windle, 1989). This value is the closest to the typical ether C–O bond length of 1.42–1.43 Å. This double-bond character creates a configurational character for the *cis* and *trans* conformers of esters and amides and subsequently, these compounds behave

FIGURE 8.6. Resonance structures of esters and amides.

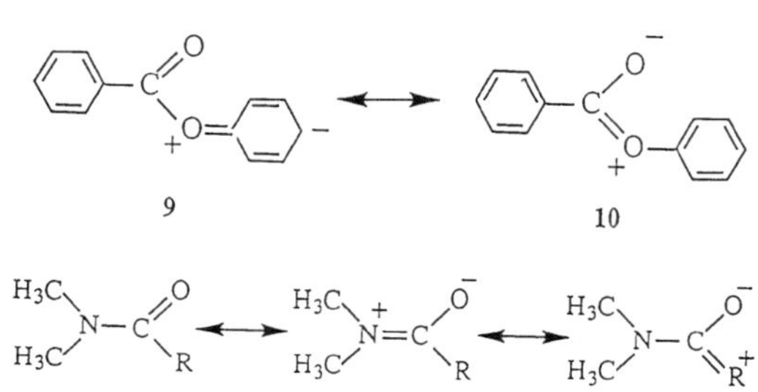

9 10

11 12 13

similarly to the class of rigid mesogens. Therefore, they should be classified as *semirigid* or *semiflexible rod-like mesogens*.

The structure and conformational behavior of amides (FIGURE 8.5, no. 7) is very close to that of esters (FIGURE 8.5, no. 8). The different resonance structures (FIGURE 8.6, nos. 11, 12, and 13) of a typical amide molecule (Kessler, 1970) are similar to those of esters. Due to its double-bond character, the contribution of no. 12 hinders the rotation about the C–N bond.

A rotational energy barrier of 21 kcal mol^{-1} which corresponds to about 40% double-bond character for the C–N bond of amides was predicted (Kessler, 1970). This double-bond character decreases due to the participation of no. 13, especially when R = aryl, and was demonstrated by NMR studies performed on $(CH_3)_2N$–C(O)–R. When R is CH_3–, the rotational energy barrier is 18.9 kcal mol^{-1} (Mannschreck, 1965) and decreases to 14.4 kcal mol^{-1} (Mannschreck et al., 1967) for R = phenyl. A rotational energy barrier of about 20 kcal mol^{-1} for the rotation about the C–N bond of phenyl benzamide was experimentally obtained (Tashiro et al., 1977).

8.4.3 Flexible Rod-Like Mesogens

1,2-Disubstituted ethane (FIGURE 8.7, no. 14) and methyleneoxy or benzylethers (FIGURE 8.7, no. 15) belong to the class of flexible rod-like mesogenic units. Let us briefly recapitulate on the conformational behavior of n-butane (FIGURE 8.7, no. 16). Out of its conformational isomers, the most stable are the *gauche* (FIGURE 8.7, nos. 17 and 18) and the *anti* (FIGURE 8.7, no. 19). Depending on the experimental technique used, an energy difference of 0.6 (Bartell and Kohl, 1963), 0.68 (Woller and Garbisch, Jr., 1972), 0.77 (Szasz et al., 1948), or 0.966 ± 0.054 (Verma et al., 1974) kcal mol^{-1} between the *gauche* and the *anti* conformers of n-butane was reported.

Theoretical calculations (Darsey and Rao, 1981) led to a rotational energy barrier at $\phi_2 = 60°$ (when the CH_3– group is eclipsed with a hydrogen atom) of 3.7 kcal mol^{-1} and to a potential energy difference of 1.19 kcal mol^{-1}. A rotational energy barrier of 2.8 kcal mol^{-1} and a potential energy difference of 0.53 kcal mol^{-1} were calculated (Abe et al., 1966). Consequently, the theoretical and experimental results on n-butane agree.

An x-ray study (Brown, 1954) of diphenyl ethane and IR and Raman investigations (Chiu et al., 1972) performed on 1,2-diphenyl ethane and 1,2-di(p-chlorophenyl)ethane (FIGURE 8.8, no. 20) demonstrated that all three molecules exist in the solid state as *anti* conformers. An optical anisotropy study (Unanue and Bothorel, 1965) performed by diffusion Rayleigh depolarization on solutions of 1,2-diphenylethane in CCl_4 and cyclohexane, shows that the *anti* conformer is preferred also in solution. Solution and Nujol mulls (Chiu et al., 1972) investigations of *para* substituted 1,2-diphenylethane derivatives showed at 25 °C a ratio of *anti* to *gauche* of 84 : 16. However, in case of 1,2-diphenylethane (FIGURE 8.7, no. 14) the concentration of the *gauche* conformer is too low to be detected by IR spectroscopy. A ^{13}C-NMR spectroscopy study (Reuvers et al., 1969) of 1,2-di(2,6-dimethylphenyl)-ethane performed in CCl_4 solution showed a dynamic equilibrium of 7 : 3 between the *anti* and *gauche* conformers.

The results of theoretical calculations performed on 1,2-diphenylethane are contradictory. It has been shown that the *gauche* conformer of no. 14 is more

FIGURE 8.7. Flexible repeat units and their various conformers.

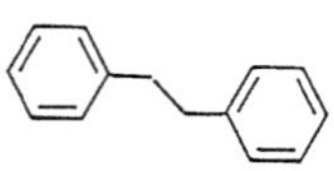

14

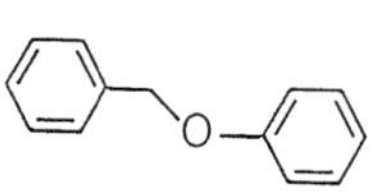

15

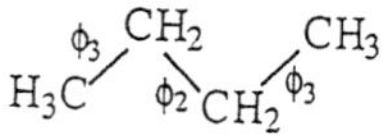

16

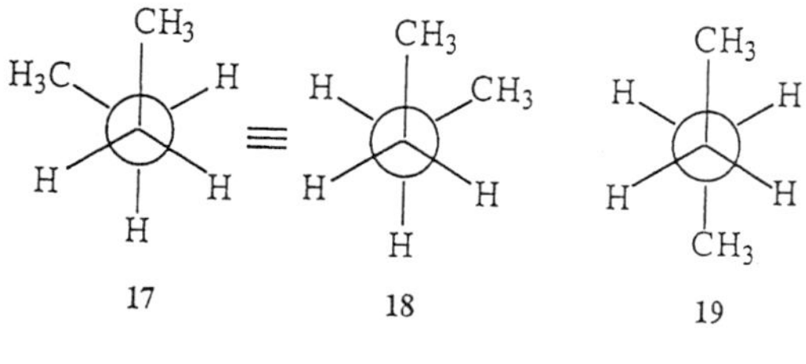

17 18 19

FIGURE 8.8. Flexible mesogens.

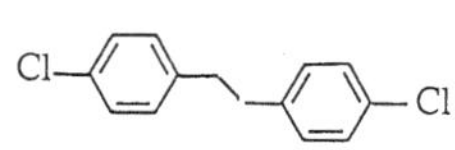

20

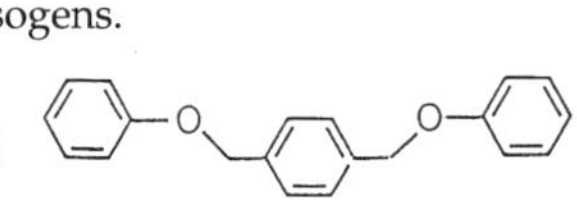

21

stable than the *anti* one by 0.57 kcal mol^{-1} (Ivanov et al., 1976). Jacobus (1976) reported that the *gauche* conformer is more stable by 1.15 kcal mol^{-1} than the *anti* conformer. A theoretical study (Petterson and Liljefors, 1987) provided data which are in agreement with the experimental results, suggesting that the *anti* form is more stable than the *gauche* form. The energy difference between these conformers is 0.95 kcal mol^{-1} and increases to 1.59 kcal mol^{-1} when point charge and electrostatic interactions are also considered. Jonsson and coworkers (Jonsson et al., 1989, 1990a, b) calculated the conformational energy barrier of α,α'-diphenyl-p-xylene (FIGURE 8.8, no. 21), which is a model compound for the polymers based on benzyl ether mesogenic units. The potential energy barrier to move one of the outer rings to the position where it becomes linear is 3.87 kcal mol^{-1}.

A theoretical calculation on main chain polyethers containing flexible benzyl ether mesogens showed that 50% of the mesogenic groups should be in their *gauche* conformation at a temperature in the liquid crystalline phase (Petraccone and Pirozzi, 1991). The mesogenic groups of the polymer are, however, at this temperature in their *anti* conformation. The formation of the liquid crystalline phase which favors a linear chain conformation provides the driving force for the change of the dynamic equilibrium between the *anti* and *gauche* conformers.

From this discussion it is clear that in the crystalline state only the *anti* conformer exists. However, in solution both the *anti* and the *gauche* conformers exist in a dynamic equilibrium. To our knowledge, the rotational energy barrier of 1,2-diphenylethane is not available. Ivanov and coworkers (Ivanov et al., 1976) calculated the potential energy at different values of the torsional angle (Ph–C–C–Ph), that is, when they are in the *anti*, *gauche*, and *eclipse* forms. In the *eclipse* form, the two rings are in the closest geometry (the steric and other nonbonded interactions are maximum). Depending on the method used, the calculated energy of the *eclipse* form is 3.61, 3.49, and 1.95 kcal mol^{-1}. This conformer has the maximum potential energy suggesting that the conformational energy barrier should be lower than the above mentioned values because during the *gauche* to *anti* or *anti* to *gauche* conformational change, the benzene rings eclipse with the ethylenic hydrogen atoms, and this steric interaction is less than it is for two benzene rings.

In conclusion, based on the difference between the rotational energy barrier of different configurational isomers or conformers of rod-like mesogens, their classification into three categories was suggested (Percec and Zuber, 1991o): *rigid*, *semirigid*, or *semiflexible*, and *flexible*. *Rigid rod-like* groups are rigid units whose shape is rigid, like for example, 1, 2, and 3 in FIGURE 8.2, and configurational isomeric units which require a high rotational energy barrier or activation energy, like for example, stilbene ($\Delta E_a = 42.8$ kcal mol^{-1}). *Semirigid* or *semiflexible rod-like* mesogens are conformationally flexible, but for electronic reasons they behave like configurational isomers with a medium rotational energy barrier. Classic examples are aromatic amides and esters ($\Delta E_a = 14.4$ kcal mol^{-1} and 7.1 kcal mol^{-1} respectively). The rotational energy barrier of these molecules is less than half that of rigid molecules, but is much higher than that of flexible molecules. *Flexible rod-like* mesogens are conformationally flexible groups whose rotational energy barrier is within the same range of values for butane ($\Delta E_a =$ 2.8–3.7 kcal mol^{-1}). Classic examples are phenylbenzylether ($\Delta E_a = 3.87$ kcal mol^{-1}) and 1,2-diphenylethane ($\Delta E_a = 3.61$ kcal mol^{-1}) based molecules.

8.5 MOLECULAR DESIGN OF PHASE TRANSITIONS THROUGH STRUCTURAL VARIATIONS. SOME THERMODYNAMIC CONSIDERATIONS

8.5.1 Equilibrium States

For the discussion which follows (Keller et al., 1990; Percec and Keller, 1990e) it will be sufficient to consider the basic thermodynamic relationship:

$$dG = V\,dP - S\,dT \qquad (8.3)$$

where G is the free energy, S is the entropy, V is the volume, P is the pressure, and T is the temperature. We will consider first the melting of a true crystal at constant pressure ($dP = 0$). The free energies of both the crystal (G_k) and isotropic liquid (G_i) decrease with increasing temperature, where the decrease in G_i is steeper due to $S_i > S_l > S_k$ (FIGURE 8.9, equation (8.3)). Where G_i crosses G_k, the crystal melts (i.e., at $T_{k-i} = T_m$). In this case, the free energy of any hypothetical liquid crystalline phase, G_{lc}, cannot fall below both G_k and G_i. While G_{lc} decreases faster with

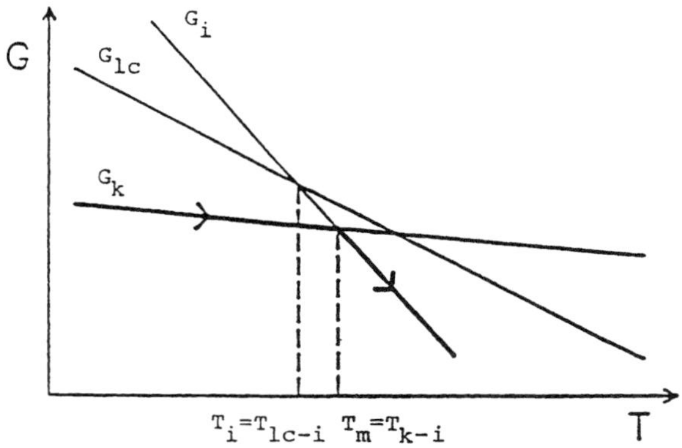

FIGURE 8.9. Schematic plot of free energies versus temperature for a system that does not show a mesophase. G_k, G_{lc}, and G_i are, respectively, the free energies of the crystalline, mesomorphic (virtual), and isotropic liquid states. $T_{\text{k-i}} = T_m$ is the crystalline melting point. Here, as in subsequent FIGURES 8.3 to 8.6, the heaviest lines correspond to the stablest state at a given temperature.

T than G_k, it will only cross G_k at a point which is above G_i, that is, where the isotropic liquid is already the most stable phase. Thus the mesophase is virtual and remains unrealizable as a stable phase.

In order to create a stable mesophase, a section of the G_{lc} versus T curve will need to be brought beneath both G_k and G_i. This can be achieved either by (a) raising G_i (FIGURE 8.10), or (b) by raising G_k (FIGURE

FIGURE 8.10. Schematic plot of free energies versus temperature for the system in FIGURE 8.2 but with G_i raised (to G_i') so as to "uncover" the mesophase. T_{k-lc} and T_{lc-i} are the crystal–mesophase transition and the isotropization transition temperatures.

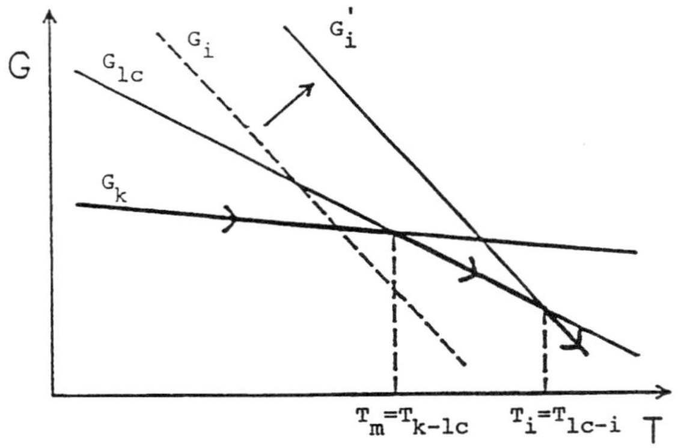

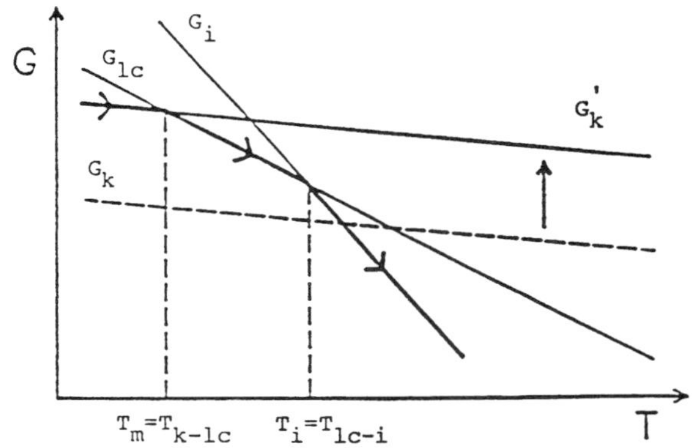

FIGURE 8.11. Schematic plot of free energies versus temperature for the system in FIGURE 8.2 but with G_k raised (to G_k') so as to "uncover" the mesophase.

8.11), or by a combination of both (a) and (b). As seen in FIGURES 8.10 and 8.11 the mesophase will be "uncovered" in a temperature range bounded by T_{k-lc} and $T_{lc-i} = T_i$, corresponding to temperatures of crystal melting and isotropization respectively. (The lowering of G_{lc} would have the same effect, but changes in G_{lc} are expected to be small compared with those in G_i and G_k, and will be disregarded in what follows.) In general, the raising of G_i (FIGURE 8.10) arises from the lowering of the melt entropy (i.e., by increasing the rigidity of the molecule via a permanent structural change or via a temporary change such as shear or any other kind of orientation, constraint, or pressure), while the raising of G_k (FIGURE 8.11) results from a reduction in the perfection of the crystal (i.e., by copolymerization or other kind of chemical and/or physical modification) so T_m decreases and therefore G_k increases. Note that in the case of FIGURE 8.10, the crystal melting point, T_m, is raised, while in FIGURE 8.11 it is lowered.

We can generalize further by considering the influence of a change in pressure, that is, the $V\,dP$ term in (8.3). Usually, the specific volume is larger for the liquid than for the crystal, with the mesophase expected to lie in between; hence $(dG_k/dP)_T < (dG_{lc}/dP)_T < (dG_i/dP)_T$. In principle, it could therefore happen that at some P, G_{lc} falls below G_k and thus a mesophase may become "uncovered." It has been found that in most experimental systems (Bassett, 1982) the effect of increased hydrostatic pressure is to promote the mesophase ("barophyllic" behavior).

However, there is no fundamental reason that would make this a general rule. This is clearly illustrated by the example of the sequence of alkanes → polyethylene (as it is described in detail elsewhere (Keller et al., 1990) where the "barophobic" behavior of short n-alkanes changes continuously with increasing chain length toward the "barophyllic" behavior of polyethylene (Ungar, 1986).

8.5.2 Metastable States

FIGURES 8.9–8.11 refer to states of thermodynamic equilibrium. However, systems may not respond immediately when passing from one stable regime to another within the phase diagram, hence metastable phases can often arise. The most commonly encountered metastability is that arising on crystallization. Usually, crystallization sets in after a certain degree of supercooling. In polymers in particular, crystallization temperature $T_{i \to k}$ can be appreciably below the equilibrium melting point $T_{k \to i} = T_m$. On the other hand, the formation of a mesophase generally requires less supercooling. Now, if the temperature from the isotropic liquid to a normally unstable mesophase is located between $T_{k \to i}$ and $T_{i \to k}$, such a "virtual" mesophase may materialize on cooling. As the temperature is lowered still further, crystallization will occur. At this stage, two extreme situations may be envisaged (FIGURE 8.12 and 8.13). On crystallization, the free energy either drops from G'_k to G_k, that is, the value for the perfect crystal (FIGURE 8.12, where the dotted line indicates a possible pathway), or else there is no continuous change in G, that is, a highly imperfect crystal is formed with its free energy remaining at G'_k (FIGURE 8.13). The realistic path would be somewhere in between these two extremes, where some decrease in G is expected, which may not quite reach the level of G_k.

We shall first consider the extreme situations of FIGURES 8.12 and 8.13. When the perfect crystal of FIGURE 8.12 is reheated, it melts directly into the isotropic liquid at $T_{k \to i}$; thus such a system displays the mesophase only on cooling, and is called "monotropic." On the other hand, the imperfect crystal of FIGURE 8.13 first changes back into the mesophase at $T_{k'-lc}$ and then into the isotropic liquid at T_{lc-i} on reheating. Subsequently, the mesophase occurs both on cooling and heating and is called "enantiotropic." The latter case clearly illustrates that an enantiotropic

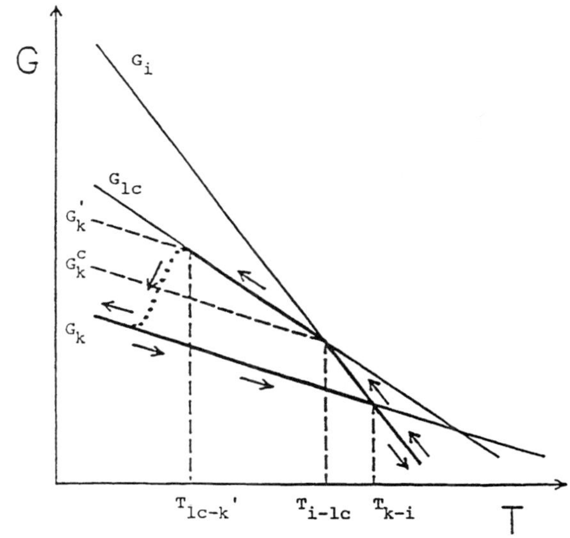

FIGURE 8.12. Schematic plot of the free-energy diagram illustrating the origin of a monotropic liquid crystal. Here, as in FIGURE 8.2, the arrows represent heating and cooling pathways.

FIGURE 8.13. Schematic free-energy diagram for an enantiotropic liquid crystal, where the mesophase is metastable.

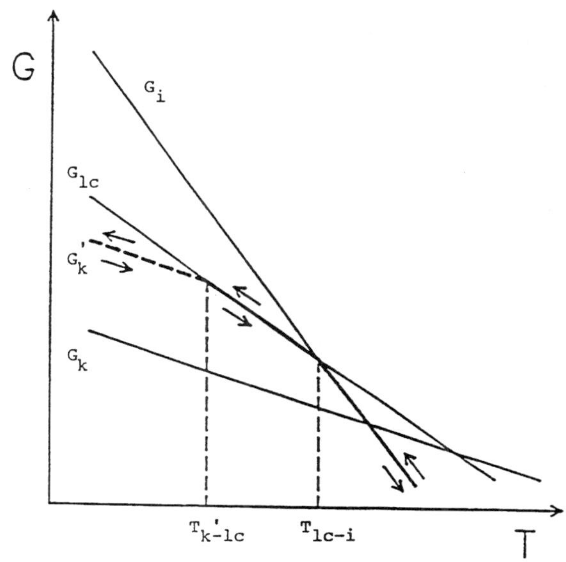

mesophase does not necessarily mean stability of the mesophase, as sometimes implied, although a stable mesophase, naturally, must be enantiotropic. As mentioned before, real systems are in between those described by FIGURES 8.12 and 8.13. Some decrease below G_k' will occur upon crystallization, the magnitude of the drop depending, among other things, on crystallization kinetics. Accordingly, neglecting possible perfectioning on subsequent heating, mono- and enantiotropic behavior are distinguished by the magnitude of the drop in G on crystallization; if G stays above a critical value (G_k^c), the system is enantiotropic, if it falls below, it is monotropic, the definition of G_k^c being apparent from FIGURE 8.12. It is easily seen how crystal perfectioning on annealing can lead to a "conversion" of an enantiotropic into a monotropic mesophase, an effect frequently observed in both polymeric and low molar mass liquid crystals.

It is worth noting further that, under certain conditions, polymers will also display superheating effects, in which case the mesophase may only appear on heating; this can be regarded as "monotropic" behavior in the reversed sense. An example of this in connection with polyethylene was discussed previously (Keller et al., 1990) and an example with a side chain liquid crystalline polymer was reported recently (Percec et al., 1991s).

Finally, a further kinetic factor, the glass transition (T_g), will be considered. On cooling, the system becomes immobilized at T_g, hence phase transformations will be arrested or altogether prevented. The inverse will apply when a previously immobilized system is heated above T_g when the system will again be able to follow its course towards the equilibrium state. In practice, this will lead to inaccessibility of certain portions of the phase diagram, or conversely, lead to the freezing in of the liquid or of the liquid crystalline state, enabling their attainments at temperatures where by thermodynamic criteria they would be unobtainable otherwise (isotropic or liquid crystalline glass, respectively). Even if T_g is not a thermodynamic quantity, the indication of its location in the phase diagram can therefore serve a useful purpose.

General examples from the literature following the patterns of the thermodynamic schemes described above are provided elsewhere (Keller et al., 1990).

8.5.3 Influence of Molecular Weight on Phase Transition Temperatures

We will now discuss the relationship between phase transition temperatures and polymer molecular weight for three different situations (Keller et al., 1990; Percec and Keller, 1990e). Upon increasing the molecular weight from monomer to polymer, the entropy of the liquid phase (S_i) decreases. The decrease of the entropies of mesomorphic and crystalline phases is lower than that of the isotropic phase. For simplicity, the decrease of the entropy of the crystalline phase will be neglected. The decrease in S_i and S_{lc} tends asymptotically to zero with increasing molecular weight (M). M_0 from FIGURES 8.14 to 8.16 refers to the molecular weight of the polymer structural unit. M_1 to M_∞ from the same figures are arbitrary molecular weights of the corresponding polymer. It follows that G_i and G_{lc} increase with the increase of the polymer molecular weight asymptotically; above a certain molecular weight, we may consider both parameters as remaining almost constant.

8.5.3.1 Both Monomeric Structural Unit and Polymer Display an Enantiotropic Mesophase

The first situation considered refers to the case in which the monomeric structural unit displays an enantiotropic mesophase. Upon increasing its molecular weight to dimer, trimer, etc., S_i decreases and, therefore, G_i increases. Beyond a certain molecular weight, G_i remains, for all practical considerations, constant. FIGURE 8.14 transforms the free energy (G) versus transition temperature (T) dependence, into a transition temperature (T) versus molecular weight (M) dependence. The T versus M plot in FIGURE 8.14 demonstrates that both the melting (T_{k-lc}) and isotropization (T_{lc-i}) temperatures increase with molecular weight up to a certain range of M values, beyond which T_{k-lc} and T_{lc-i} remain approximately constant, with the T_i line lying above the line for T_m for all molecular weights. However, the slope of the increase of T_{lc-i} is steeper than that of T_{k-lc}. The difference between these two slopes determines the relative thermodynamic stabilities of the mesomorphic versus crystalline phases at different polymer molecular weights.

For this particular case, the higher slope of the T_{lc-i}–M curve versus that of the T_{k-lc}–M dependence

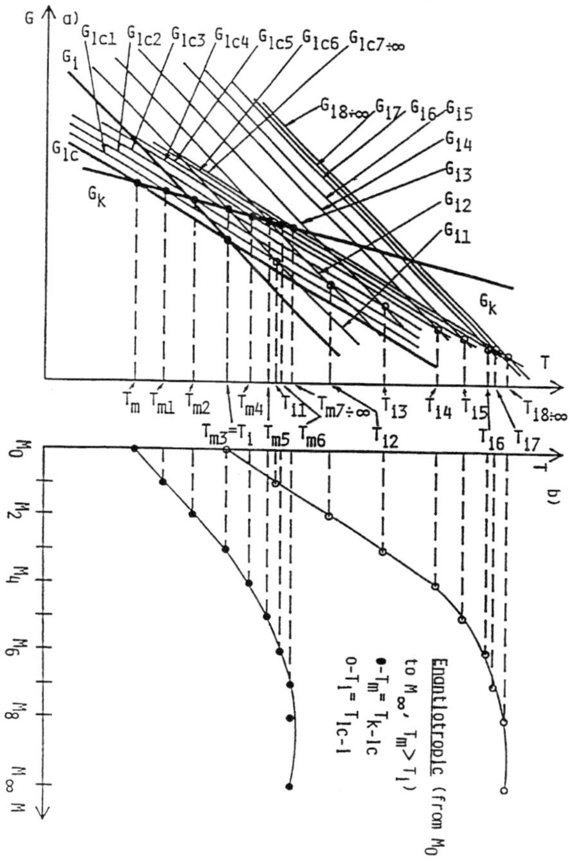

FIGURE 8.14. The broadening of the temperature range of an enantiotropic mesophase of the monomeric structural unit (M_0) by increasing the degree of polymerization. The upper part (a) describes the influence of molecular weight on the dependence of the free energies of the crystalline (G_k), liquid crystalline (G_{lc}), and isotropic (G_i) phases on transition temperatures. The translation of this dependence into the dependence of the phase transition temperature on molecular weight is presented in the lower part (b).

leads to a widening of the temperature range between the two curves with increasing molecular weights of the polymer. This widening of the liquid crystal temperature regime with molecular weight agrees with experimental data reported for the case of both main chain (Blumstein et al., 1982; Feijoo et al., 1988; Godovsky et al., 1985; Percec et al., 1987b) and side chain (Kostromin et al., 1982; Percec and Pugh, 1989a; Percec and Hahn, 1989f; Percec and Lee, 1991a; Perec et al., 1991c, g, h; Perec and Lee, 1991i, j, n; Percec and Tomazos, 1992a; Percec et al., 1989j, 1991b, e, f, p, t, 1992r) liquid crystalline polymers. This effect has been repeatedly labeled as the "polymer effect," especially in the case of side chain liquid crystal polymers (Percec and Pugh, 1989a).

8.5.3.2 The Monomeric Structural Unit Displays a Virtual or a Monotropic Mesophase; The Polymer Displays a Monotropic or an Enantiotropic Mesophase

The steeper slope of the $T_{lc-k(i)}$–M dependence versus that of the $T_{k-i(lc)}$–M dependence has even more important implications on the molecular weight–phase transition temperature dependence when the monomer structural unit displays only a monotropic or a virtual mesophase (FIGURE 8.15).

As seen, the two lines T_i versus M and T_m versus M intersect. This arises from the fact that $T_i < T_m$ for low molecular weights with a steeper slope for T_i. Specifically (for the illustration in FIGURE 8.15) the T_i (i.e., T_{lc-k}) values are below the corresponding T_m (i.e., T_{k-i}) values for M_0 to M_4, hence in the range M_0 to M_4, that is, the monomer together with its low oligomers up to four display only a virtual mesophase. Beyond M_4, the mesophase becomes stable, hence the system is enantiotropic. In addition to the thermodynamic criterion, kinetics also influence the phase transitions. A certain amount of supercooling of the isotropic–mesomorphic and especially of the mesomorphic–crystalline transitions is possible which can lead to monotropic behavior for molecular weights slightly below and at the intersection point (i.e., at M_4). In view of the fact that T_i and T_m are expected to be continuous functions of molecular weight, the intersection point can be arbitrarily closely approached (from below), hence realization of a metastable liquid crystal phase and subsequent monotropic behavior is to be expected for appropriate molecular weights in cases to which FIGURE 8.15 pertains. This effect has been observed experimentally both in the case of main chain (Feijoo et al., 1988; Majnusz et al., 1983; Percec and Nava, 1987a; Zhou et al., 1986) and side chain (Percec and Tomazos, 1992a; Stevens et al., 1984) liquid crystalline polymers, and was labeled "transformation of a monotropic mesophase into an enantiotropic mesophase by increasing the molecular weight of the polymer." A series of quantitative experiments along this line will be described later.

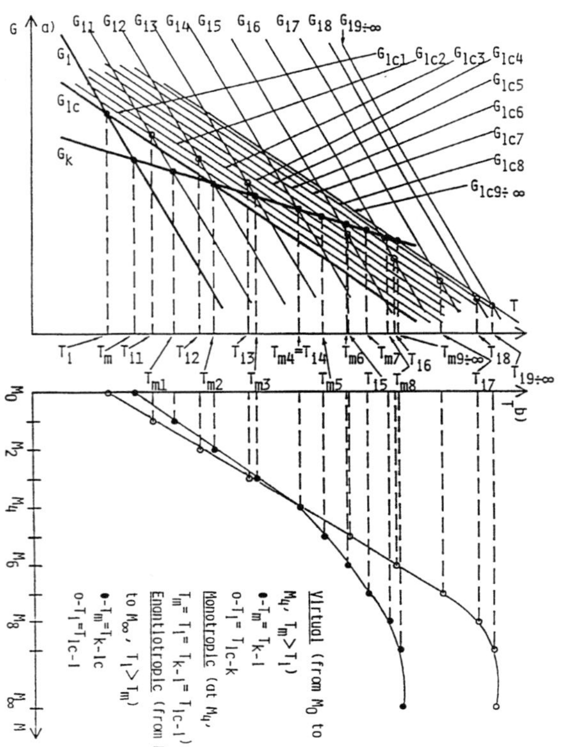

FIGURE 8.15. Transformation of a virtual or monotropic mesophase of the monomeric structural unit (M_0) into an enantiotropic mesophase by increasing the degree of polymerization. The upper part (a) describes the influence of molecular weight on the dependence of the free energies of the crystalline (G_k), liquid crystalline (G_{lc}), and isotropic (G_i) phases on transition temperatures. The translation of this dependence into the dependence of the phase transition temperature on molecular weight is presented in the lower part (b).

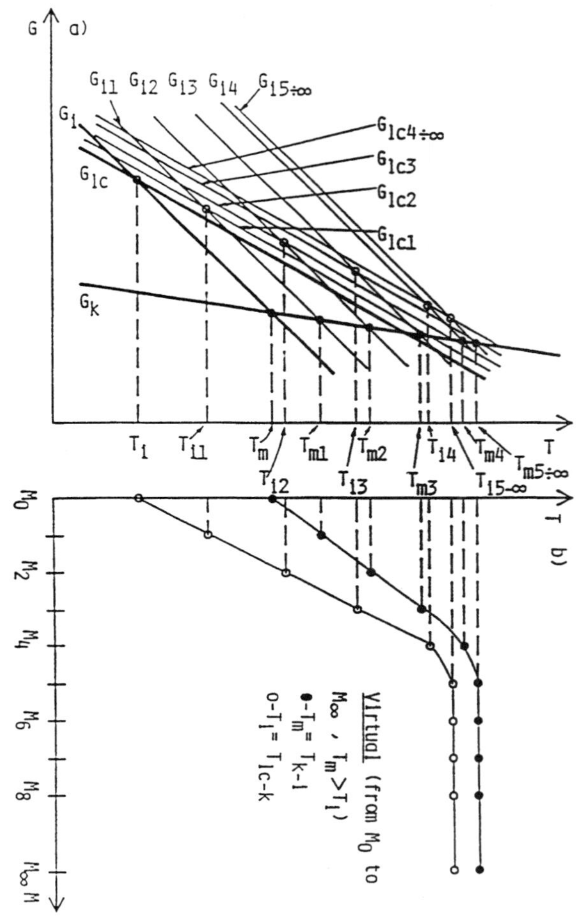

FIGURE 8.16. The narrowing of the temperature range of a virtual mesophase of the monomeric structural unit (M_0) by increasing the degree of polymerization. The upper part (a) describes the influence of molecular weight on the dependence of the free energies of the crystalline (G_k), liquid crystalline (G_{lc}), and isotropic (G_i) phases on transition temperatures. The translation of this dependence into the dependence of the phase transition temperature on molecular weight is presented in the lower part (b).

8.5.3.3 The Monomeric Structural Unit Displays a Virtual Mesophase; The Polymer Displays a Virtual Mesophase

The third situation is illustrated in FIGURE 8.16, and refers to a different case in which the monomeric unit displays only a virtual mesophase. Here as before, the slope of the T_i (i.e., T_{lc-k})–M dependence is higher than the T_m (i.e., T_{k-i})–M dependence; the latter lies above the former throughout, hence the two curves do not intercept each other. Therefore, the resulting polymer also displays a virtual mesophase. This thermodynamic situation was recently applied to the synthesis of liquid crystal polyethers containing both flexible mesogens and flexible spacers and exhibiting virtual liquid crystalline phases (Percec and Yourd, 1988a, 1989d, i; Percec and Tsuda, 1990b, d; Ungar et al., 1990) and will be detailed in a subsequent section.

8.5.3.4 Rigid Rod-Like Polymers

The discussion from cases 1 to 3 refers to semiflexible or semirigid and flexible polymer systems which exhibit first-order transitions (i.e., melting, isotropization) which are molecular weight dependent only up to a certain degree of polymerization. In these systems, the melting temperature is determined by the length of the chain fold. Rigid rod-like polymers such as poly(*p*-phenylene)s, poly(p-phenylenebenzobisthiazole) (PBT), poly(p-phenylenebenzobisoxazole) (PBO) etc. should exhibit a linear dependence of their melting transition over their entire range of molecular weights. Consequently, the dependences of their first-order transitions on molecular weight should follow the pattern in FIGURE 8.17. TABLE 8.1 summarizes the dependence of the first order transition temperatures of poly(*p*-phenylene)s H–(C_6H_4)–H as functions of their degree of polymerization and their axial ratio ($x = L/d$) (Flory and Ronca, 1979; Irvine et al., 1984; Sigaud, 1992) (FIGURE 8.18(a)). As predicted by theory (Flory, 1956; Flory and Ronca, 1979), when the axial ratio ($x = L/d$) of rod-like molecules reaches a value of about 6.2, the compound should exhibit an enantiotropic mesophase. This is indeed the case. For n of 6 or 7, poly(p-phenylene)s exhibit an enantiotropic

FIGURE 8.17. The dependence of various first-order transitions on molecular weight for a rigid rod-like polymer.

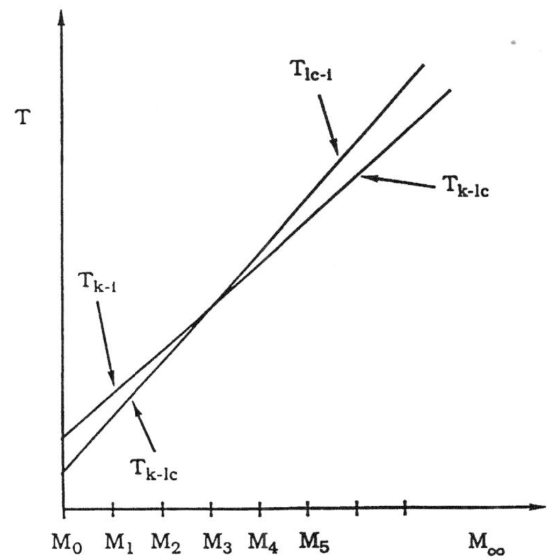

TABLE 8.1. Melting and isotropization transitions of poly(p-phenylene)s.

n	Axial ratio $x = L/d$	T_{n-i} (°C) Estimated[a]	Polymorphism
2	≈ 2	—	$T_{k-i} = 70\,°C$
3	≈ 3	—	$T_{k-i} = 210\,°C$
4	≈ 4	≈ 120	$T_{k-i} = 320\,°C$
5	≈ 5	≈ 450	$T_{n-i} = 430\,°C$
6	≈ 6	≈ 1900	$T_{n-i} = 560\,°C$
7	≈ 7	not isotropic	$T_{k-n} = 550\,°C$

[a] The estimated T_{n-1} has been normalized on quinquephenyl ($n = 5$).

nematic mesophase (as predicted by FIGURE 8.17). At lower n values the nematic mesophase is virtual. However, for oligomers with n larger than 7, the nematic mesophase cannot be observed since the decomposition temperature first overlaps the melting transition and then becomes lower than the melting temperature. Other rigid rod-like polymers like PBT and PBO, and even semirigid systems like fully aromatic and unsubstituted polyamids and polyesters decompose before melting.

While in rigid rod-like polymers, the axial ratio $x = L/d$ (FIGURE 8.17(a)) defines the chain dimension responsible for the generation of an enantiotropic mesophase, in the case of semirigid and flexible polymers the ratio l/d (FIGURE 8.17(b)) (where l is now a renormalized length i.e., the Kuhn's segment) defines

FIGURE 8.18. (a) Definition of the axial ratio ($x = L/d$) in the model of rigid rod-like mesogen and polymer; (b) model for flexible and semi-rigid main chain liquid crystalline polymer (l defines the Kuhn's segment).

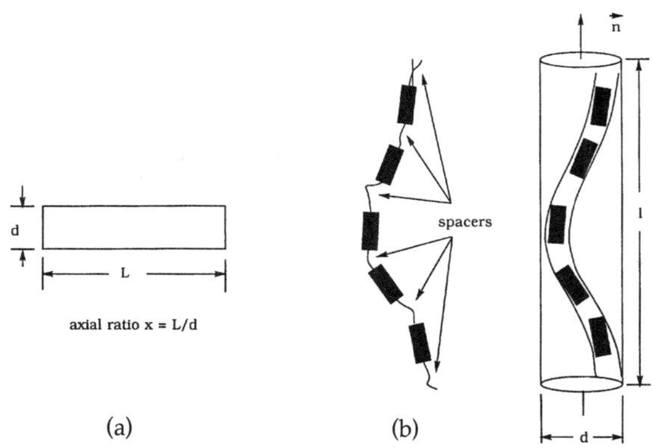

it. In this second case, the higher l is, the higher the stiffness of the chain (Sigaud, 1992).

In conclusion, the steeper slope of the isotropization temperature–molecular weight dependence versus that of the melting temperature–molecular weight dependence should have the following effects on transforming a monomer into a polymer. When the monomer structural unit displays an enantiotropic mesophase, the resulting polymer will display a broader enantiotropic mesophase. When the monomeric unit displays a monotropic mesophase, the resulting polymer will, probably, display an enantiotropic mesophase. When the monomeric unit displays a virtual mesophase, the resulting polymer may display either an enantiotropic, monotropic, or virtual mesophase. For the case of side chain (Percec et al., 1989j; Stevens et al., 1984) and main chain (Blumstein et al., 1982; Percec and Nava, 1987a; Zhou et al., 1986) liquid crystalline polymers containing flexible spacers and displaying a single mesophase, the nature of the mesophase displayed by the polymer is most frequently identical to that of its monomeric unit. Primarily, the molecular weight–phase transition dependences are determined by the relationship between the free energies of the crystalline, liquid crystalline, and isotropic phases of the monomeric structural unit, but influenced by the molecular weight of the main chain backbone through its effect on the melt entropy as in case 1 above. Combinations of more than one mesomorphic phase of different thermodynamic stabilities in a monomeric structural unit will follow the same molecular weight dependence trend as those described for the monomers displaying a single mesophase (Percec and Lee, 1991a, h; Percec et al., 1991b, e, f, p, 1992r). However, there are few examples where the nature of the mesophase displayed by the main chain and side chain liquid crystalline polymers is molecular-weight dependent. Both in the case of the main chain (Blumstein and Blumstein, 1988; Kumar et al., 1988;) and side chain (Percec et al., 1991e; Percec and Lee, 1991a, g, h; Percec et al., 1994f; Percec and Tomazos, 1992a; Percec and Oda, 1994h, i, 1995a) liquid crystalline polymers, it has been demonstrated that this change in the mesophase represents a continuous dependence of molecular weight. In the case of rigid rod-like polymers, both melting and isotropization temperatures reach higher values than the decomposition temperature within the range of low molecular weights.

8.6 MAIN CHAIN LIQUID CRYSTALLINE POLYMERS

8.6.1 Soluble and Fusible Main Chain Liquid Crystalline Polymers

Amorphous polymers are soluble in suitable solvents at low temperatures, while crystalline polymers are soluble only at temperatures which sometimes are close to their melting point. The solubility of any polymer increases with a decrease of its ability to crystallize. A suppression of the melting temperature below the glass transition temperature (i.e., a kinetically prohibited crystalline phase) makes the corresponding polymer noncrystallizable. Therefore, since most rigid polymers have much higher glass transition temperatures than flexible polymers, it is easier to decrease the crystallization tendency of the rigid polymers and therefore, to make them soluble. This is also the case for main chain liquid crystalline polymers. A decrease of the melting transition temperature can be accomplished by raising G_k, as shown in FIGURE 8.11. An increase in G_k can be accomplished by any synthetic technique that increases S, that is, introduces defects into the polymer structure. Most conveniently, this is realized by copolymerization of a few monomers, addition of lateral substituents that provide a high configurational entropy (alkyl groups, *t*-butyl groups), insertion of flexible spacers, increasing the overall flexibility of the molecule, and insertion of structural defects such as kinks, etc. The alternative solution is to increase both the glass transition temperature of the polymer and its entropy simultaneously. In order to maintain a high ability towards mesophase formation, the chemical reactions used to increase the solubility of the polymer should not disturb the ability of its backbone to generate a rod-like conformation. Therefore, the insertion of kinks or any other nonlinearities is not favorable since it increases the solubility and decreases the ability to generate the mesophase. The ability to decrease the crystallization tendency is not so much related to the size of the lateral substituent as it is to the number of different configurational and conformational structural units which can result from the monomer containing the side group. In fact, according to thermodynamic considerations, a high entropy can be realized either by a chain containing a large number of different structural units with different small substituents, or by a chain containing in each structural unit a single substituent that yields a large number of conformers. The second situation, however, reduces drastically the interchain interactions and therefore suppresses not only the ability towards crystallization, but also the ability to generate a conventional liquid crystalline phase. Therefore, the most profitable synthetic technique is to use the copolymerization of a few monomers containing a small substituent with different electronegativities and the ability to generate more than one constitutional isomeric structural unit from each monomer. Various research groups have used these principles, although in a nonsystematic way, and have obtained fusible and soluble polymers. In the following subsections, we will enumerate only selected examples of polymers whose melting temperatures were lowered and therefore, were solubilized by using these principles.

8.6.1.1 Poly(p-phenylene) and Its Derivatives

The first soluble oligo(p-phenylene)s were synthesized by placing methyl substituents on the phenyl ring (Kern et al., 1960). The thermal behavior of these oligomers was first reported by Heitz (Heitz, 1986, 1989a, 1991a). These oligomers and their phase transitions are outlined in FIGURE 8.19. After substitution, their solubility increases (Kern et al., 1960) and their melting points decrease. Methyl substituted oligo(p-phenylene)s have an axial ratio (x) equal to 0.8 times the number of benzene rings. According to theoretical predictions, hexamethylsexiphenyl which has $x = 4.8$, and lower oligomers, should not exhibit an enantiotropic mesophase. This is indeed the case. Octamethyloctaphenyl which has $x = 6.4$ and longer oligomers (i.e., those with x larger than the critical value $x = 6.2$) should and indeed do display an enantiotropic nematic phase. Insertion of *meta* linkages suppresses their ability to form a mesophase (FIGURE 8.19).

Soluble poly(p-2,5-di-n-alkylphenylene)s were prepared by Ni catalyzed coupling of 1,4-dibromo-2,5-di-n-alkylbenzene (Rehahn et al., 1989a) and Pd(0)-catalyzed coupling of 4-dibromo-2,5-di-n-alkylbenzeneboronic acid (Rehahn et al., 1989b, 1990a; Wahlenkamp and Wegner, 1994; Wilteler et al., 1993) (FIGURE 8.20). These high molecular weight polymers display liquid crystalline phases (Wahlenkamp and Wegner, 1994; Wilteler et al., 1993).

Soluble phenylated poly(p-phenylene)s were also synthesized by Ni(II) catalyzed polymerization of

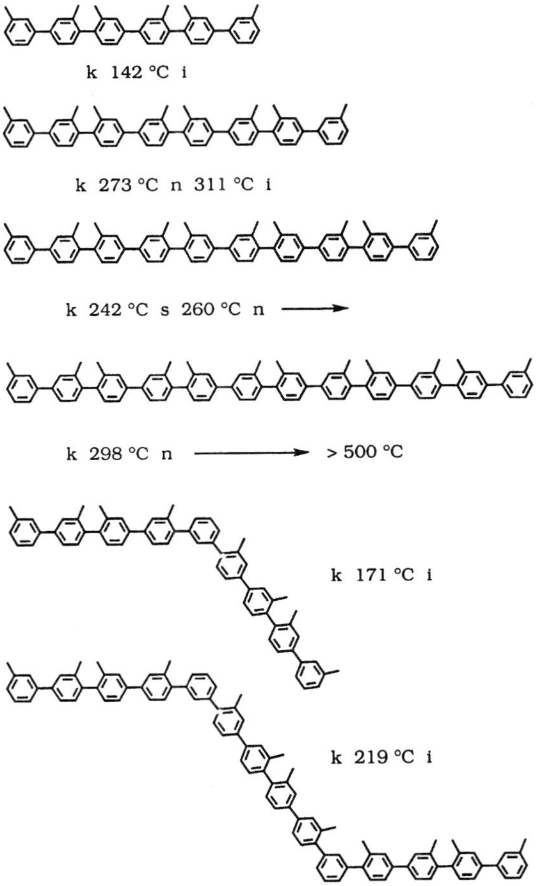

FIGURE 8.19. Effect of structure on phase transitions in oligophenylenes.

FIGURE 8.20. Direct synthesis of substituted poly(p-phenylene)s via Suzuki reaction

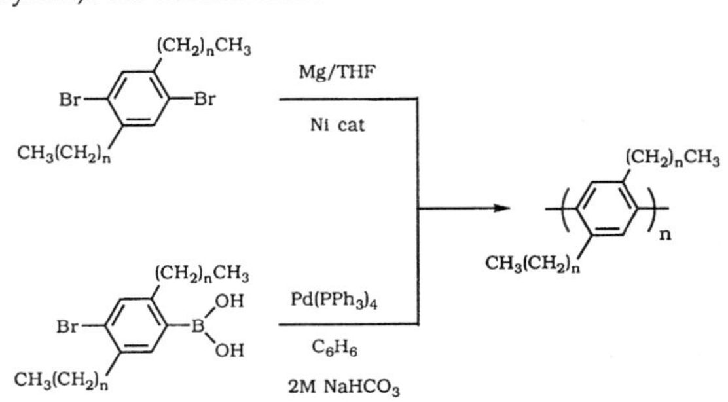

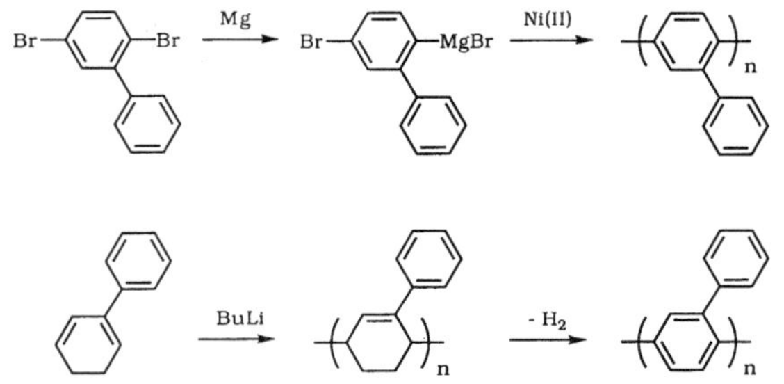

FIGURE 8.21. Direct synthesis of regioregular (top) and indirect synthesis of regioirregular (bottom) substituted poly(p-phenylene)s.

2,5-dibromobiphenyl, and by anionic polymerization of 2-phenyl-1,3-cyclohexadiene followed by aromatization (FIGURE 8.21) (Noll et al., 1990). Probably both synthetic methods lead to regioregular phenylated poly(p-phenylene)s. The high molecular weight polymer fractions exhibit an enantiotropic nematic mesophase.

Regioirregular substituted poly(p-phenylene)s were synthesized from substituted hydroquinones via Ni(0) catalyzed homocoupling of their bistriflates (FIGURE 8..22) (Percec et al., 1992j). All these polymers are soluble in conventional solvents, and many of them display liquid crystalline phases.

Other synthetic methods for the preparation of poly(p-phenylene)s were reviewed (Percec and Tomazos, 1992q; Schlter and Wegner, 1993; Tour, 1994). Systematic research is still required to design poly(p-phenylene)s with liquid crystalline properties.

8.6.1.2 Metal Containing Poly(yne)s

Metal containing poly(yne)s are a relatively new class of rigid rod-like polymers. They can be prepared by a copper chloride triethylamine catalyzed coupling of the appropriate metal halide with a dialkyne (FIGURE 8.23) (Takahashi et al., 1982a, 1984). A review on their synthesis is available (Hagihara et al., 1980). These

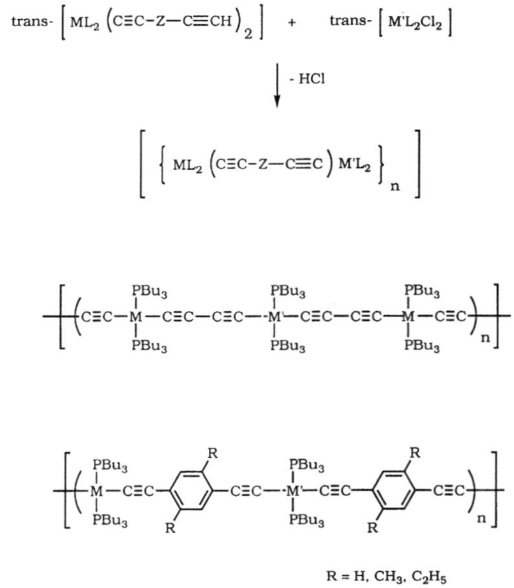

FIGURE 8.23. Synthesis of metal containing poly(yne)s.

polymers exhibit lyotropic liquid crystalline phases (Abe et al., 1991; Takahashi et al., 1982b). Recently, two novel methods for the synthesis of this class of polymers were reported. The first one consists of the reaction of *bis*-trimethylstannyl(acetylide)s with *trans*-[M$(PBu_3)_2Cl_2$] (Davis et al., 1991), while the second one consists of the reaction of [Rh$(PMe_3)_4$Me] with diacetylenes (Chisholm, 1991; Fyfe et al., 1991) (FIGURE 8.24). These metal containing poly(yne)s can be prepared with weight average molecular weights of up to 100,000 (Davis et al., 1991) and are therefore of interest both for their rigid rod-like character and for their nonlinear optical properties (Chisholm, 1991). Additional classes of metal containing liquid crystalline polymers were reviewed (Espinet et al., 1992; Giroud-Godquin and Maitlis, 1991; Hudson and Maitlis, 1993; Polishchuk and Timofeeva, 1993).

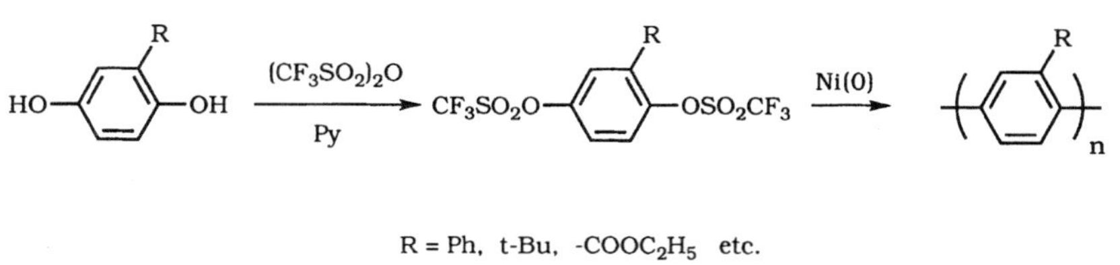

FIGURE 8.22. Synthesis of regioirregular substituted poly(p-phenylene)s from hydroquinones.

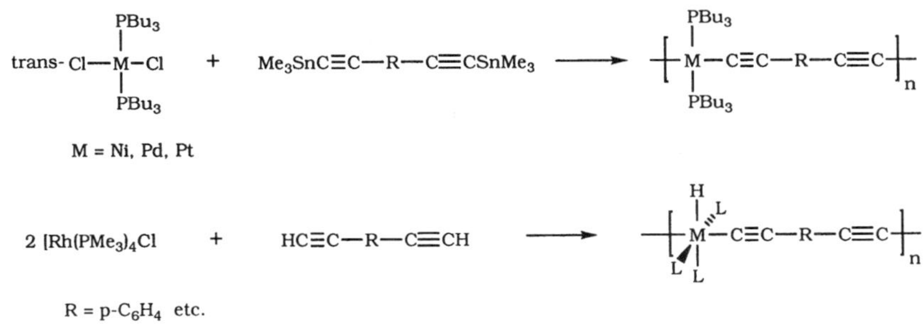

FIGURE 8.24. Synthesis of metal containing poly(yne)s.

8.6.1.3 Spinal Columnar Liquid Crystalline Polymers

Dihydroxysilicon (IV) phthalocyanine can be condensed in the solid state to yield insoluble phthalocyaninatopolysiloxanes (Orthmann et al., 1983; Orthmann and Wegner, 1986). However, octaalkyl-substituted derivatives of dihydroxytin (IV) phthalocyanine (Sirlin et al., 1987), dihydroxysilicon (IV) octakis-(dodecyloxymethyl)phthalocyanine (Sirlin et al., 1988a, b) and octaalkoxy-substituted silicondihydroxy phthalocyanines (Caseri et al., 1988; Sauer and Wegner, 1988) undergo polycondensation in a liquid crystalline phase and yield soluble spinal columnar liquid crystalline polymers (FIGURE 8.25). These polymers exhibit columnar hexagonal liquid crystalline phases which can range from −7 °C up to 300 °C (Sauer and Wegner, 1988; Sauer, 1993; Sirlin et al., 1987, 1988a, b). The diameter of the column is determined by the length of the alkyl side groups (Sauer and Wegner, 1988; Sauer, 1993).

FIGURE 8.25. Synthesis of spinal columnar liquid crystalline polymers.

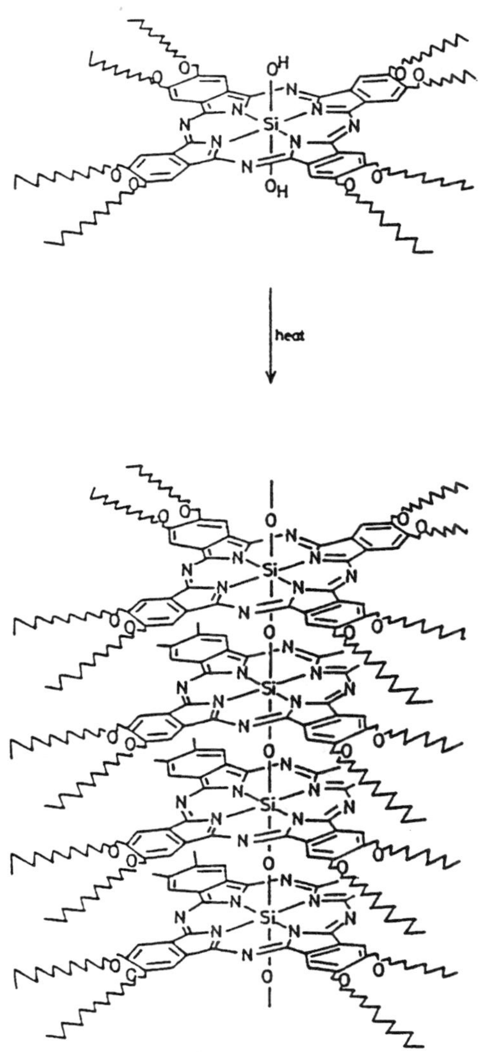

8.6.1.4 Aromatic Polyamides and Polyesters

Synthetic procedures for the preparation of soluble aromatic polyamides and polyesters were developed mainly by Gaudiana and coworkers (Gaudiana et al., 1987; Rogers et al., 1985a, b, 1986; Sinta et al., 1987) Their research was recently reviewed (Gaudiana et al., 1989). Depending on their structures, these polymers can be either amorphous or liquid crystalline (Gaudiana et al., 1989). A discussion of the influence of various molecular factors such as the position, polarizability and size of the substituents on the solubility of aromatic polyamides is available (Gaudiana et al., 1987, 1989). TABLE 8.2 (Gaudiana et al., 1987) summarizes some representative polymer structures. Their solubilities are described in TABLE 8.3. An inspection of the structures from TABLE 8.2 and of the

Table 8.2. Molecular structures of aromatic polyamides.

Head-to head phenyl and biphenyl polyamides

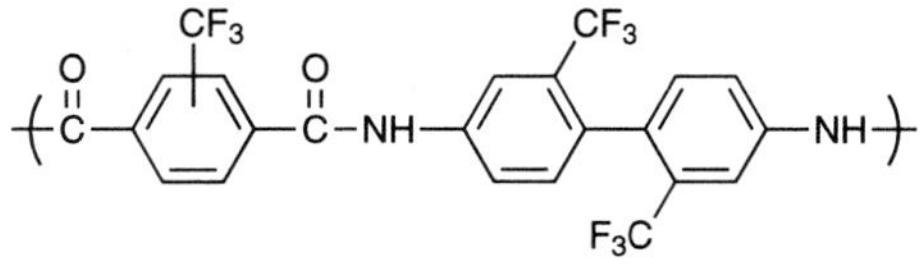	I
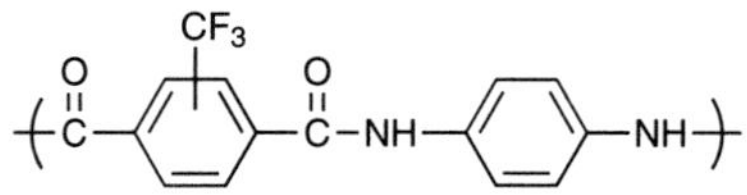	II
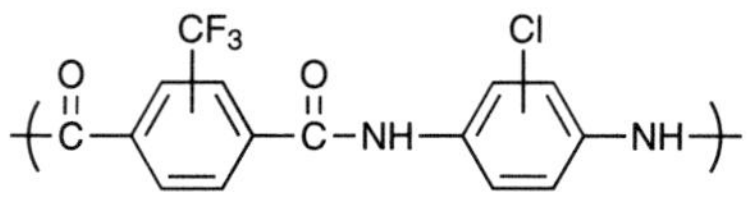	III
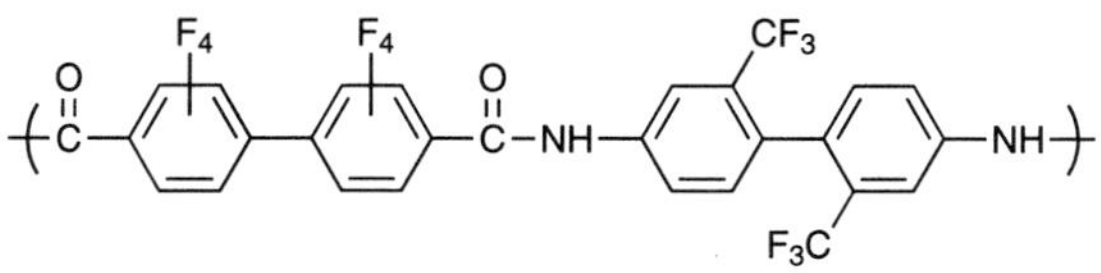	IV
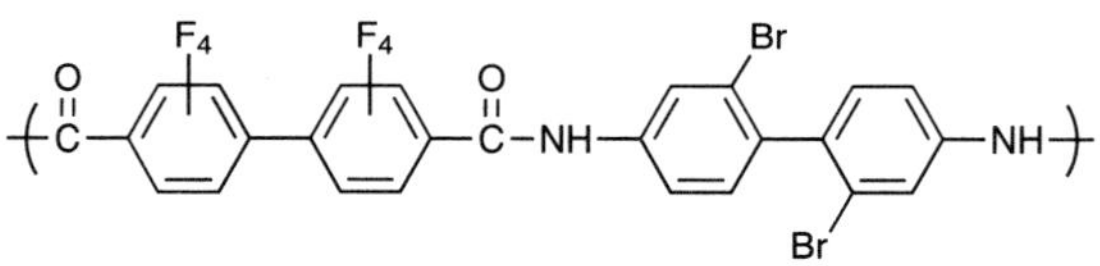	V
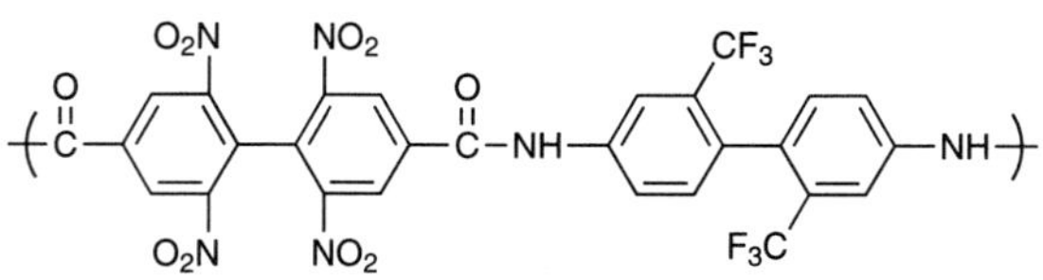	VI
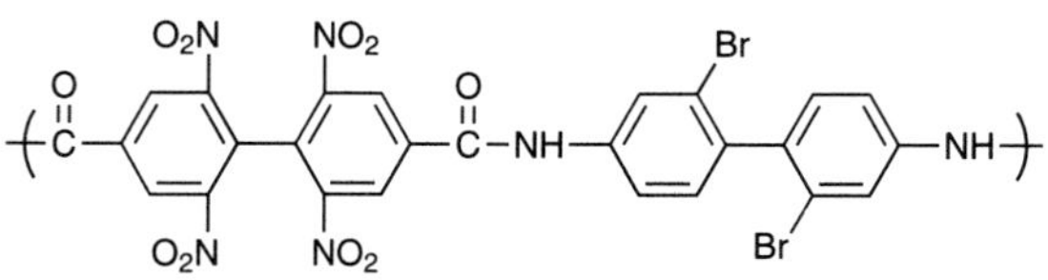	VII
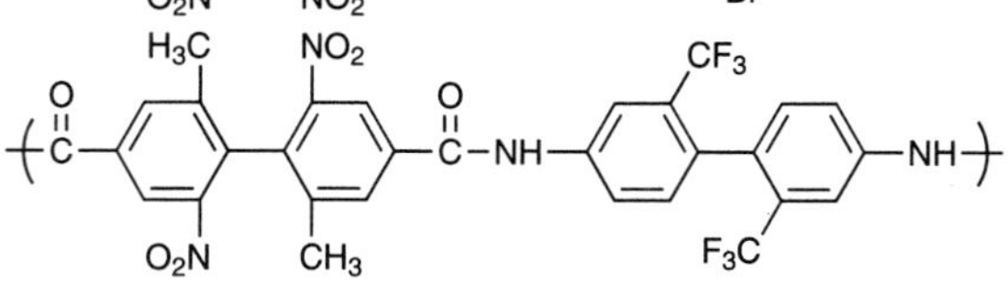	VIII
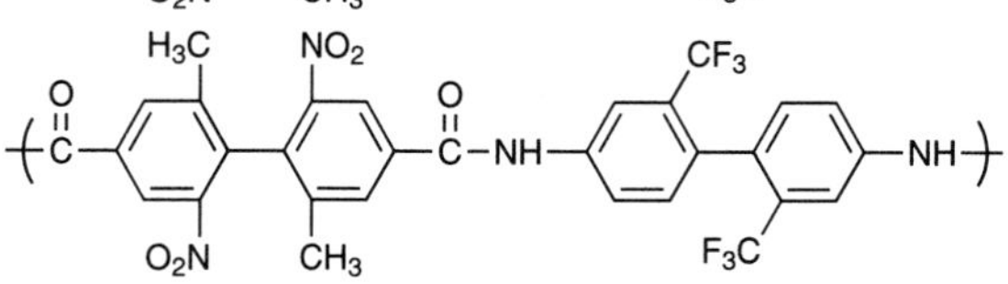	IX
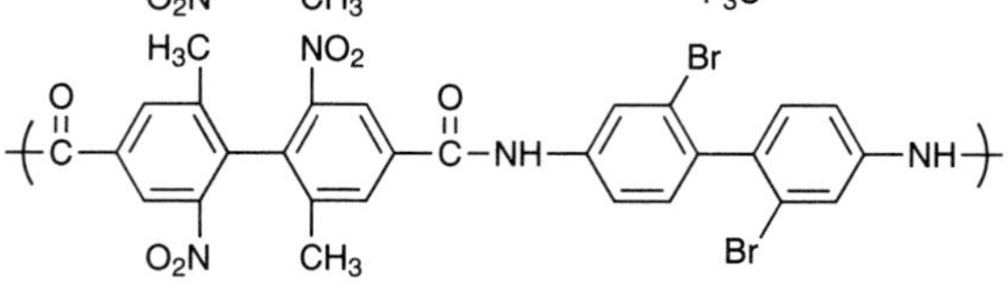	X

Table 8.2. (Continued)

Head-to head phenyl and biphenyl polyamides (Continued)	
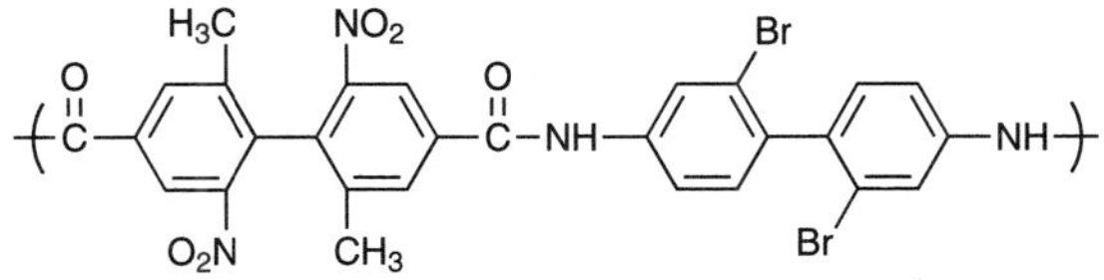	XI
Head-to-tail biphenyl polyamides	
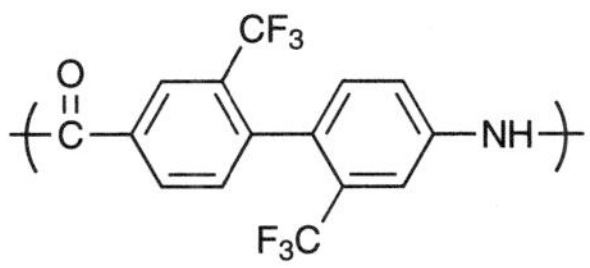	XII
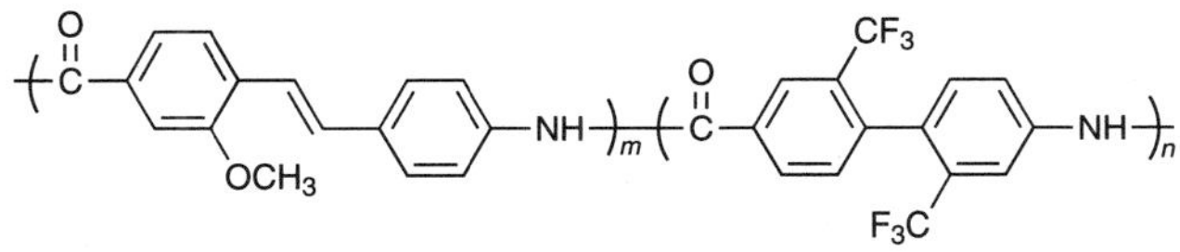	XIII
$m = 0.24$, $n = 0.76$	
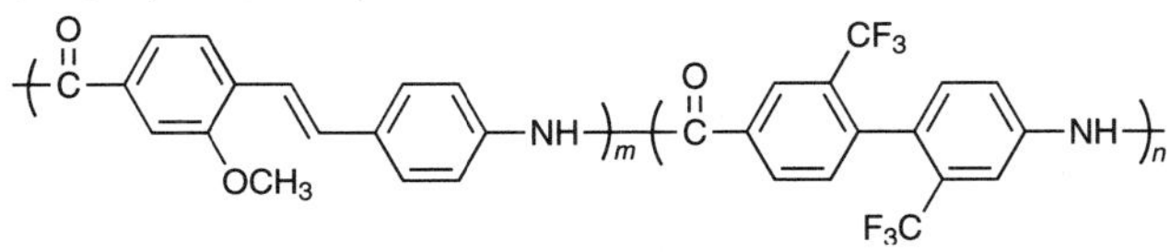	XIV
$m = 0.10$, $n = 0.76$	
Substituted terphenyl and quaterphenyl polyamides	
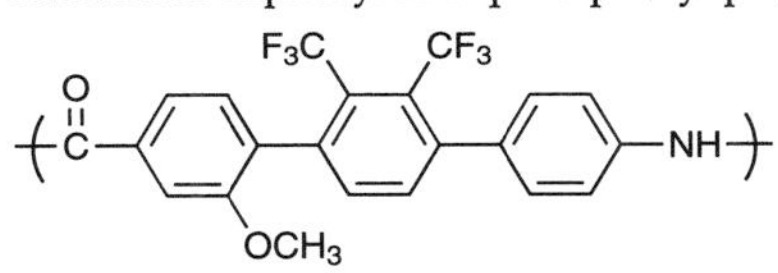	XV
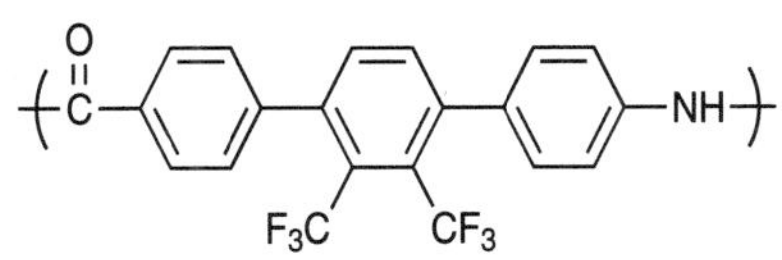	XVI
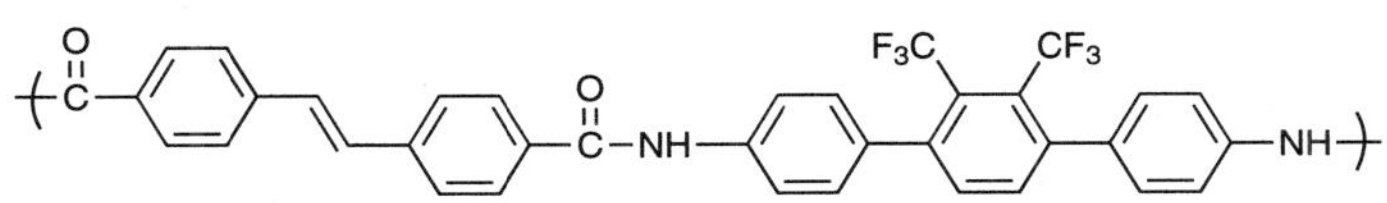	XVII
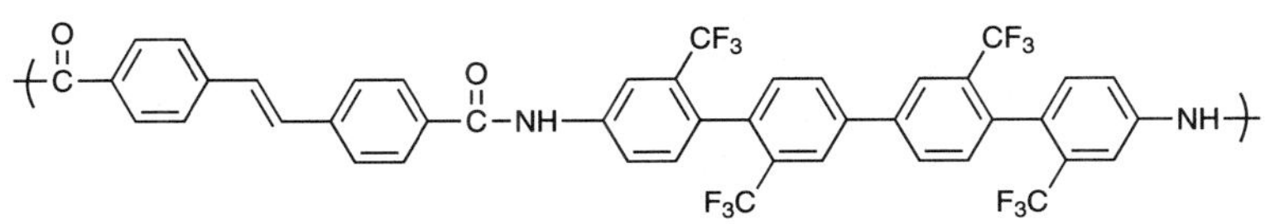	XVIII
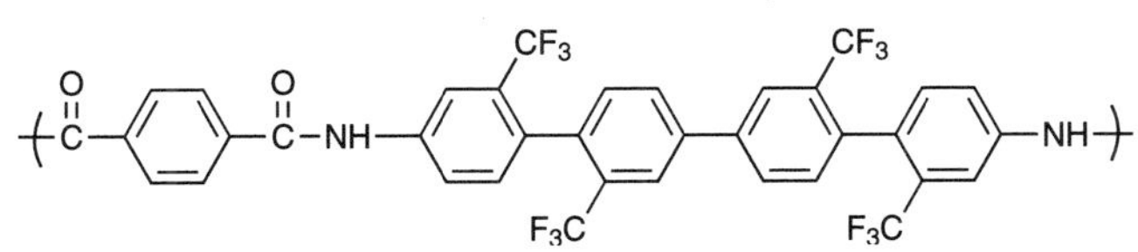	XIX

TABLE 8.3 Viscosity and solubility properties of biphenyl polyamides.

		Solubility			
Polymer	$\eta_{inh}(dl/g)$[a]	LiCl/amides[b]	Amides	Ethers[c]	Ketones[d]
I	3.23	S[e]	44–48[f]	Sw	Ins
II	1.84 (DMAc)	S	S	Ins	Ins
III	2.34 (DMAc)	S	S	Ins	Ins
IV	1.19S	S	Sls	Ins	Ins
V	1.68	S	Ins	Ins	Ins
VI	1.27	S	S	S[g]	Ins
VII	1.10	S	S	Ins	S
VIII	1.75	S	S	36 (THF)	S
IX	0.91	S	S	25 (THF)	S
X	2.46	S	S	17 (THF)	Ins
XI	1.68	S	S	14 (THF)	Ins
XII	5.0	S	S	S	S
XIII	4.74	S	S	Ins	Ins
XIV	4.94	S	S	S	S
XV	1.39	S	S	Ins	Ins
XVI	—	Ins	Ins	Ins	Ins
XVII	2.30	S	Ins	Ins	Ins
XVIII	1.31	S	S	S	S
XIX	6.04	S	S	Sls	S

[a] 0.5 polymer (w/v)/5% (w/v) LiCl/DMAc at 30 °C.
[b] Amide solvents: DMAc, TMU, and NMP.
[c] Ether solvents: THF, glymes, but not diethyl ether, dioxane.
[d] Ketone solvents: acetone and cyclohexanone.
[e] S, soluble; Sw, swells; Sls, slightly soluble; Ins, insoluble.
[f] w/v%.
[g] Methocy ethanol but THF.

solubilities of these polymers presented in TABLE 8.3 suggests that the noncoplanar conformation of the biphenyl moiety (introduced by the presence of substituents in their 2,2′-positions), the statistical distribution of the biphenyl enantiomers, and the presence of structurally dissimilar diacids and diamines, all contribute to an enhanced solubility. A combination of these factors diminishes or even eliminates the interchain correlations necessary for close packing in crystals, and may diminish van der Waals forces of attraction. These orientation-dependent intermolecular attractive forces are important in the formation of all classes of liquid crystals (Flory, 1984a; Flory and Matheson, 1984b).

Soluble aromatic polyamides with a much lower solubility were also realized by attaching phenyl substituents either on the p-phenylene diamine or on the terephthalic acid monomer (Jadhav et al., 1988, 1989; Krigbaum et al., 1991). Some representative structures together with their solubilities are presented in TABLE 8.4 (Jadhav et al., 1988). All these polymers exhibit lyotropic liquid crystalline phases or at least melt before decomposition.

A reversible solubilization of poly(p-phenylene terephthalamide) was accomplished by the polycondensation of the chromium tricarbonyl complex of p-phenylenediamine with terephthaloyl chloride in N,N-dimethylacetamide (Dembek et al., 1993; Jin and Kim, 1987). The polymer containing tricarbonyl chromium complexed structural units is soluble in dipolar aprotic solvents. Decomplexation can be achieved either by oxidation with I_2 or by heating above 150 °C.

Soluble poly(p-phenylene terephthalamide)s with phenyl and biphenyl units in the terephthalic acid were also synthesized (Hatke et al., 1991). The melting temperatures of these polymers were in the range 440–490 °C and overlap their decomposition temperatures.

Aromatic polyamides containing alkyl side chains were reported (Ballauff, 1986a, b; Ringsdorf et al., 1987; Schoenherr et al., 1986). When each structural unit contains four alkoxy groups, depending on their

TABLE 8.4 Phenyl-substituted aromatic polyamides.

	η_{inh} $(dL/g)^a$		Solubility in DMAc or NMP, $g/100g^d$
	$H_2SO_4{}^c$	DMAc[d]	
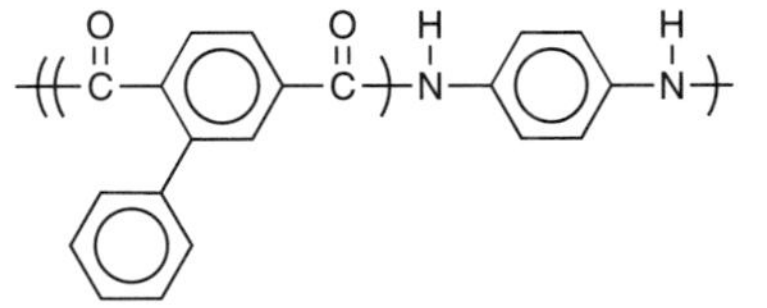	0.61	1.10	50
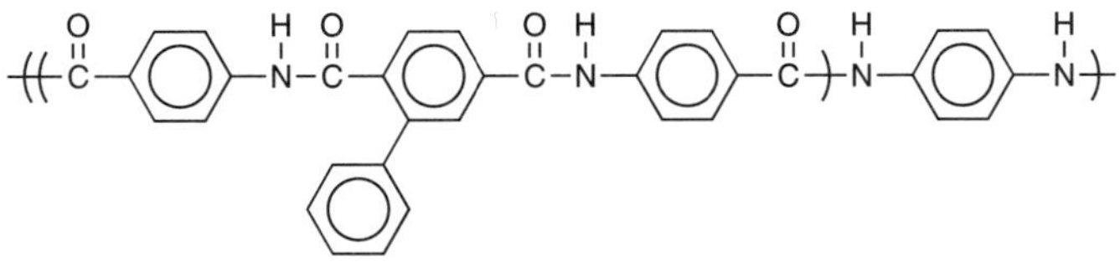	3.28	7.55	18
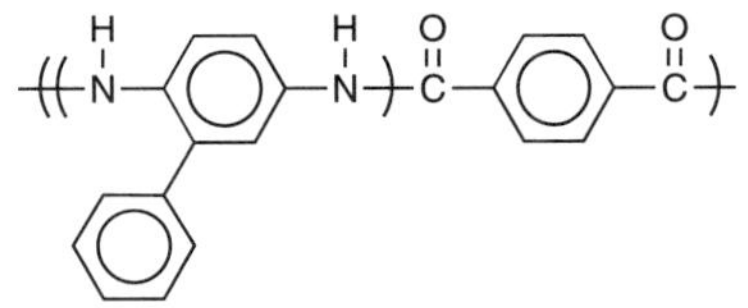	1.45	2.52	20
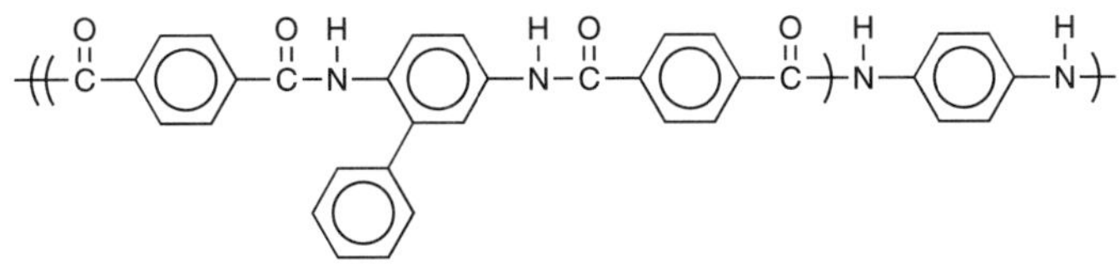	2.28	3.32	14–15

[a] All polymers prepared by the phosphorylation method.
[b] 0.5% concentration at 25 °C.
[c] 96% H_2SO_4.
[d] Containing 4% LiCl.

length, these polymers exhibit a board-like, "sanidic," mesophase (Ringsdorf et al., 1987; Schoenherr et al., 1986; Voight-Martin et al., 1995a, b) (FIGURE 8.26). However, when there are only two alkoxy groups per one of the two monomeric units, these polyamides display a new layered liquid crystalline mesophase (Ballauff, 1986b; Ballauff and Schmidt, 1987). The corresponding polyesters containing four alkoxy groups in each monomeric unit or two alkoxy groups in only one monomeric unit display mesophases similar to those of the polyamides (Ballauff, 1986b; Ballauff and Schmidt, 1987; Ringsdorf et al., 1987; Schoenherr et al., 1986; Voight-Martin et al., 1995a, b).

Soluble aromatic polyesters were synthesized by using similar synthetic procedures to those employed in the polymerization of soluble aromatic polyamides. The research groups from Mainz reported on the solubilization of aromatic polyesters by the addition of alkyl side groups (Ballauff, 1986b; Ballauff and Schmidt, 1987; Ringsdorf et al., 1987; Rodriguez-Parada et al., 1989; Schoenherr et al., 1986; Voight-Martin et al., 1995a, b). Solubility is easily accomplished for aromatic polyesters containing alkyl side groups when the alkyl groups derived from the terephthalic acid and hydroquinone monomers are highly dissimilar in length (Rodriguez-Parada et al., 1989) (FIGURE 8.27). For example, the polyester from the top of FIGURE 8.26, PTA16HQ-16, is highly crystalline and melts directly into an isotropic phase. However, PTA16HQ-6 exhibits much less crystallinity, is highly soluble, and melts into a liquid crystalline phase. It is also instructive to compare the phase behavior of polyesters derived from

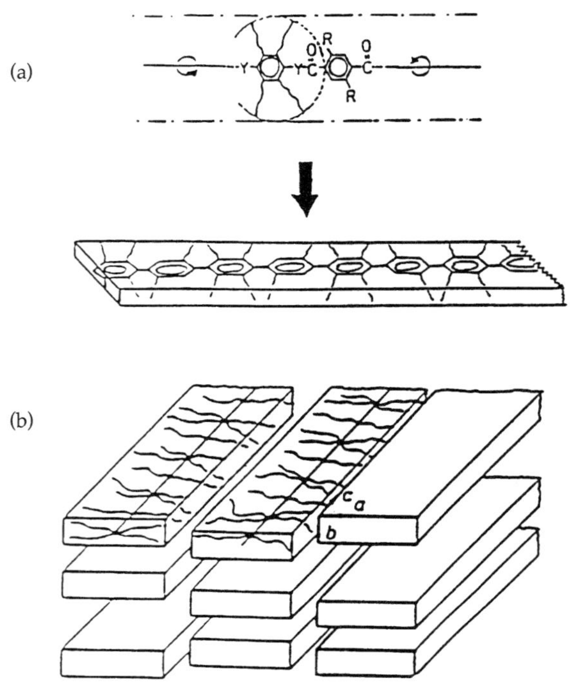

FIGURE 8.26. (a) Board formation of highly substituted rigid rod polymers, caused by hindered rotation of their mesogenic units (Ringsdorf et al., 1987); (b) schematic representation of a sanidic (from the Greek for board-like) mesophase (Schoenherr et al., 1986).

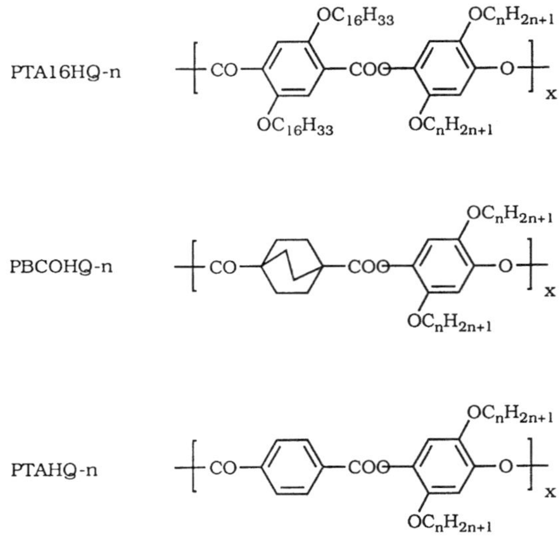

FIGURE 8.27. Examples of flexible aromatic polyesters.

bicyclo[2.2.2]octane-1,4-dicarboxylic acid with 2,5-dialkoxyhydroquinone diacetate (PBCOHQ-n) with those obtained from terephthalic acid and 2,5-dialkoxyhydroquinones (PTAHQ-n). Representative data on these polymers are available in TABLES 8.5 and 8.6. As we can observe from the data reported in these tables, all PTAHQ-n melt into a liquid crystalline phase while PBCOHQ-n melt directly into an isotropic phase. These results suggest that interchain interactions provided by the wholly aromatic backbone are necessary for the stabilization of a mesophase in this class of

TABLE 8.5. Characterization of polymers from bicyclo[2.2.2]octane-1,4-dicarboxylic acid and 2,5-dialkoxyhydroquinone diacetates (PBCOHQ-n).

			Thermal transitions[c]			
			Heating		Cooling	
n[a]	$[\eta]$[b]	Layer spacing d (Å)	T_s	T_m	T_s	T_m
6	1.10	10.1	63[d]	315	—	254
8	0.99	11.5	60[d]	255	—	210
10	1.21	12.8	55[d]	243	—	204
12	0.53	15.2	28	223	−2	186
14	0.87	16.5	46	206	30	166
16	0.59	17.8	60	193	49	160

[a] Number of carbon atoms in the side chain.
[b] In dL/g, measured in CH_2Cl_2/TFA (95/5 (v/v)) at 20 °C.
[c] In °C. Heating and cooling rates: 20 °C min^{-1}.
[d] Transition observed only in the first heating scan.

TABLE 8.6 Characterization of polymers from terephthalic acid and 2,5-dialkoxyhydroquinones (PTAHQ-n).

				Thermal transitions[e]					
		Layer spacing		Heating			Cooling		
n[a]	$[\eta]$[b]	d (Å)[c]	d (Å)[d]	T_s	T_m	T_i	T_s	T_m	T_i
6	1.20	10.3	11.7	79[f]	275	292	—	234	279
8	1.71	12.8	14.0	75[f]	255	282	—	217	256
10	1.21	13.3	—	78[f]	270	—	—	245	—
12	0.73	15.5	18.2	78	236	—	32	181	215
14	0.72[g]	17.6	20.5	91	215	230	62	160	207
16	0.70[g]	19.1	21.8	104	190	215	83	162	202

[a] Number of carbon atoms in the side chain.
[b] In dL/g, measured in CH_2Cl_2/TFA (95/5 (v/v)) at 20 °C.
[c] Measured at toom temperature.
[d] Measured at 10 °C below the isotropization temperature.
[e] In °C. Heating and cooling rates: 20 °C min^{-1}.
[f] Transition observed only in the first heating scan.
[g] In dL/g, measured in $CHCl_3$/TFA (95/5 (v/v)) at 25 °C.

TABLE 8.7 Properties of polyesters 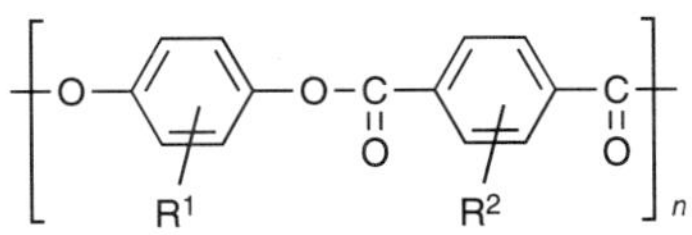.

						Solubility[b]		
R[1]	R[2]	η_{inh} (dl/g)	T_g (°C)	T_m (°C)	Phase $>T_m$	pCP	TCE	$CHCl_2$
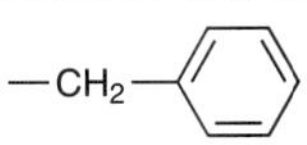	–H	1.0	—	380[a]	—	++	+	–
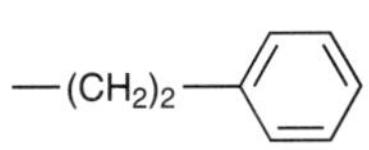	–H	0.7	—	340	aniso.	++	+	–
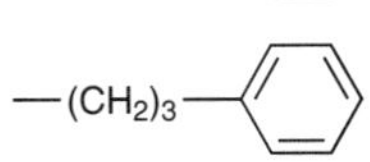	–H	4.2	—	325	aniso.	++	+	–
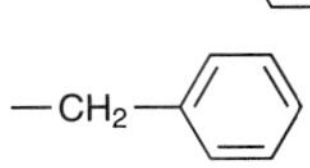	–Br	2.0	91	(207)[c]	nem.	++	+	–
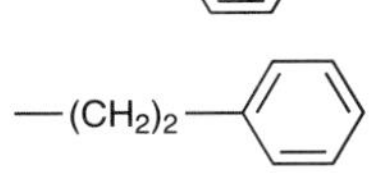	–Br	2.5	79	(195)[c]	nem.	++	+	–
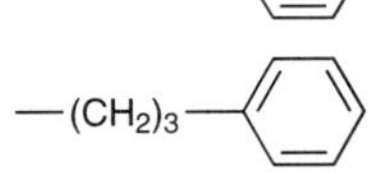	–Br	1.0	58	(247)[c]	nem.	++	+	++

[a] Melting in the decomposition range.
[b] ++: Stable solution (>0.5 g/dl) at room temperature; +: soluble on heating, precipitation on cooling; –: insoluble; pCP: p-chlorophenol; TCE: 1,1,2,2-tetrachloroethane.
[c] Weak endotherm on annealed samples.

TABLE 8.8 Properties of polyesters 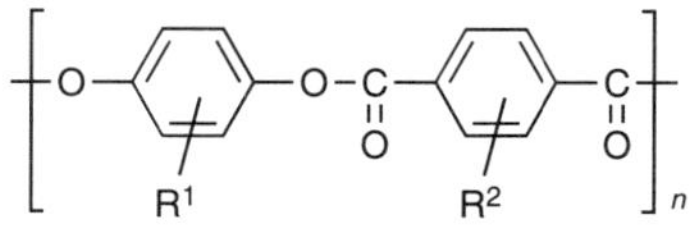

R^1	R^2	η_{inh} (dl/g)	T_g[a] (°C)	T_m[a] (°C)	Phase $>T_m$	Solubility[b] pCP	Solubility[b] TCE	Solubility[b] $CHCl_3$
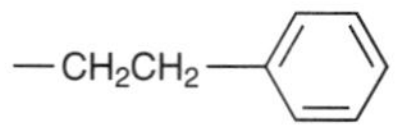	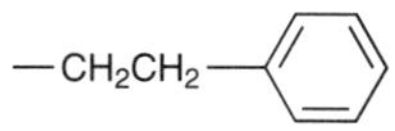	0.3–3.6	60	195	nem.	++	++	++
$-CH_3CH_3$	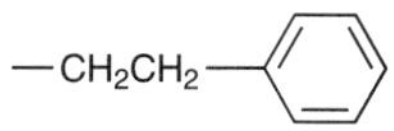	0.3–5.0	100	225	nem.	++	++	++
$-CH_3CH_3$	–Br	0.2–2.6	131	—	—	+	O	–
$-CH_3CH_3$	$-CF_3$	0.3–1.9	141	228	nem.	++	++	++
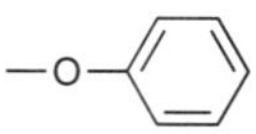	–Cl	1.8	85	241	nem.	++	++	O
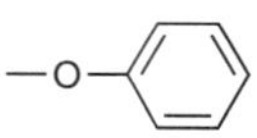	–Br	0.5–3.0	90	220	nem.	++	++	++
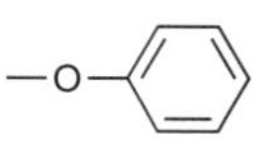	$-CF_3$	0.35	88	—	iso.	++	++	++
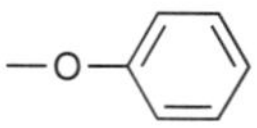	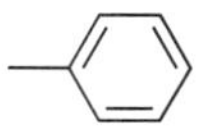	0.4–0.7	112	—	iso.	++	++	++

[a] Measured in p-chlorophenol at 45 °C.
[b] ++: Stable solution ($>0.5\,g/dl$) at room temperature; +: soluble on heating, precipitation on cooling; –: insoluble; pCP: p-chlorophenol; TCE: 1,1,2,2-tetrachloroethane.

polymers (Rodriguez-Parada et al., 1989). Soluble homologs of poly(p-hydroxybenzoate) were also synthesized by attaching pendant alkyl groups (Stern et al., 1989). Thus, while poly(p-hydroxybenzoate) and poly(3-n-alkyl-4-hydroxybenzoate)s with alkyl groups shorter than six are only crystalline, the corresponding polymers with alkyl groups from six to sixteen are soluble and exhibit a thermotropic mesophase (Stern et al., 1989). It is interesting to mention that the polyesters with alkyl side groups display a layered mesophase that is different from the classic nematic or smectic mesophases exhibited by conventional rod-like polymers.

A large variety of soluble polyesters were prepared (Gaudiana et al., 1989; Sinta et al., 1987) by using the same synthetic principles as those used in the preparation of soluble polyamides.

A large group of soluble aromatic polyesters was prepared by attaching various large substituents such as phenylalkyl (Brugging et al., 1988; Vogel and Heitz, 1990), perfluoroalkyl (Freund et al., 1989), phenoxy and tert-butyl (Heitz and Niessner, 1990a), phenyl and biphenyl (Heitz, 1991a, b), or twisted biphenylene units such as 2,2′-dimethylbiphenylene (Classen et al., 1990; Schmidt and Guo, 1988), and 1,1′-binaphthyl-4,4′-ylene (Hohlweg and Schmidt, 1989). The dependence of the solubility of the resulting polyesters on the type of various structural units was reviewed (Heitz, 1991b; Heitz and Schmidt, 1990b; Jin, 1994a; Lee, K. S. et al., 1994; Schmidt, 1989; Tiesler et al., 1994).

TABLE 8.9 Properties with twisted biphenyl units

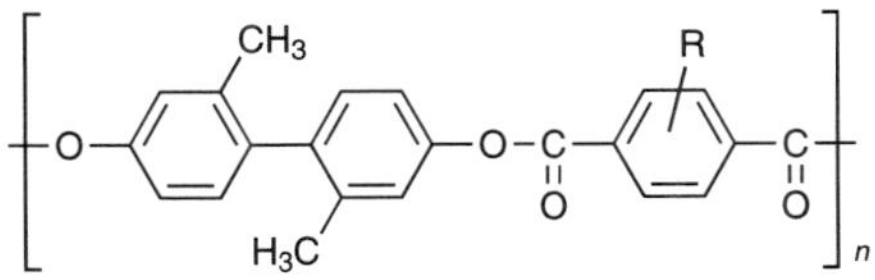

1

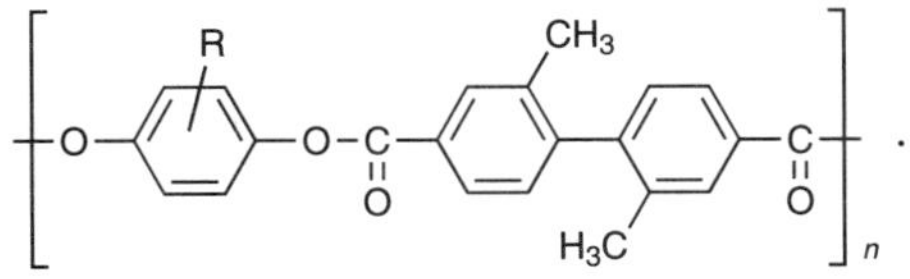

2

Polymer	R	η_{inh} (dl/g)	T_g[a] (°C)	T_m[a] (°C)	Phase	Solubility[b]		
						pCP	TCE	$CHCl_3$
1	H	1.43		nontractable		+	−	−
	Br	1.21	100	230	nem.	+	−	−
	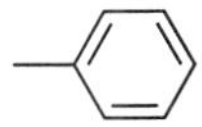	1.10	125	—	nem.	+	−	−
2	CH_3	0.81		nontractable		+	+	−
	tert-Butyl	1.43	150	—	nem.	+	+	+

[a] Measured in p-chlorophenol at 45 °C.
[b] ++: Stable solution (>0.5 g/dl) at room temperature; +: soluble on heating, precipitation on cooling; −: insoluble; pCP: p-chlorophenol; TCE: 1,1,2,2-tetrachloroethane.

TABLES 8.7 and 8.8 summarize the relation between structure, thermal properties, and solubilities of polyesters based on substituted hydroquinone and substituted terephthalic acid (Heitz and Schmidt, 1990b; Hsiao et al., 1991; Kromer et al., 1991). The presence of dissimilar substituents on both monomers generates highly soluble polymers. TABLES 8.9 and 8.10 show the relation between the properties of polyesters obtained with twisted biphenyl monomers and their structure (Heitz and Schmidt, 1990b; Hsiao et al., 1991; Kromer et al., 1991).

Nematic polyesters were obtained by the condensation of 2-phenylthioterephthalic acid, 2-(p-tolylthio)terephthalic acid, 2-(p-chlorophenylthio)terephthalic acid with the diacetates of hydroquinone, methylhydroquinone or 2,5-biphenyldiol (Kricheldorf et al., 1988e). 2-(2-naphthylthio)terephthalic acid was also used in conjunction with the same bisnucleophiles to obtain noncrystallizable nematic copolyesters (Kricheldorf et al., 1988f). The influence of various interconnecting groups present in the structure of bisphenols (i.e., O, CO, SO_2, S, $C(CH_3)_2$) on the phase behavior of polyesters with phenylthioterephthalic acid was also investigated (Kricheldorf and Erxleben, 1990e). Similar polyesters were synthesized from phenoxyterephthalic acid (Kricheldorf et al., 1989c), 4-fluoro-, 4-chloro, 4-bromo-, 4-phenyl, 4-cyclohexyl-, 4-phenoxy-, and 4-cumyl- phenol (Kricheldorf and Engelhardt, 1990a) with hydroquinone or 4,4′-biphenyldiol. The influence of the various polyesterification reaction conditions on the final structure and properties of polymers was also investigated. The synthetic methods used most frequently are outlined in FIGURE 8.28 (Kricheldorf et al., 1988a, 1989c; Kricheldorf and Engelhardt, 1990a).

For example, the acetate method results in partial debromination of 4-bromophenoxyterephthalic acid units during their condensation with hydroquinone diacetate, but not during the polymerization via the other two methods (Kricheldorf and Engelhardt, 1990a) (FIGURE 8.28). Model reactions have shown that both phenoxyterephthalic acid and its acid chloride yield anthrone-3-carboxylic acid above 180 °C and this product terminates the polymer chain end (Kricheldorf et al., 1988a) (FIGURE 8.29). This reaction is less probable when the acid chloride is reacted with the silylated hydroquinone. The interest in aryloxy- and arylthio-substituted terephthalic acid was mainly determined by their convenient synthesis from dimethyl nitroterephthalate via aromatic

TABLE 8.10. Low molar mass liquid crystals based on benzyl ether and methyleneoxy flexible units (Carr and Gray, 1985; Osman, 1982a,b).

No.	Structure	Phase transitions (°C)
1	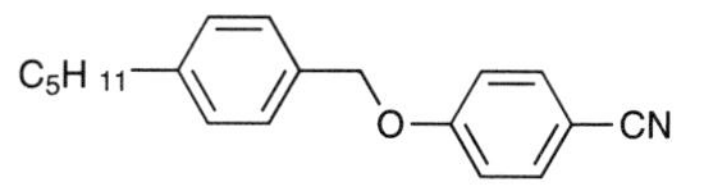	k 49 n [−20] i[a]
2	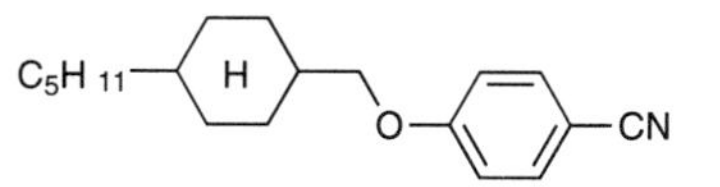	k 74.3 n (48.6) i[b]
3	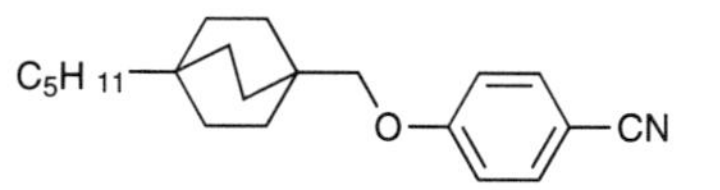	k 72 n 73 i
4	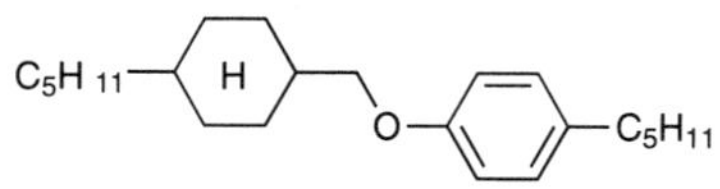	k 35 s_B (31) n [20] i
5	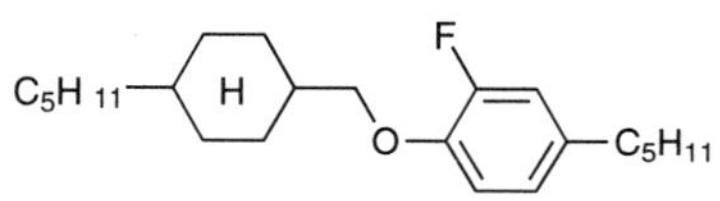	k 43 n [−10] i
6	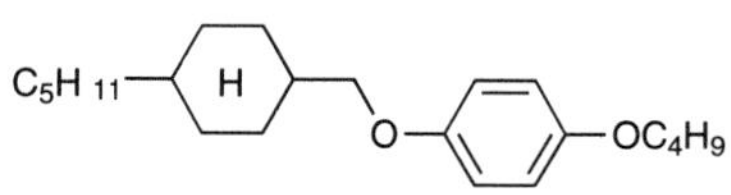	k 55 s_B (47) n (53) i
7	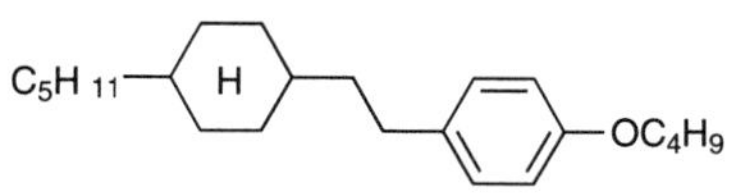	k 22 s_B 44.5 n 45.5 i

[a] [ ]: Virtual.
[b] ( ): Monotropic.

nucleophilic displacement with the corresponding phenolate or thiophenolate.

Recently, the influence of the regioregularity on the solubility and thermotropic properties of main chain polyesters was investigated (Jin et al., 1994b). Soluble liquid crystalline aromatic polyesters were also synthesized from terephthalic acid derivatives, *tert*-butylhydroquinone, and aromatic hydroxy-functional polystyrene macromonomers (Heitz et al., 1989b). The number average molecular weight of polystyrene side chains varied from 1000 to 2000 (FIGURE 8.30). The resulting polyesters contained polystyrene grafts and are of interest as emulsifiers for nematic aromatic polyesters with polystyrene (Heitz et al., 1989b). A graft copolymer containing a polyester main chain and polymethylmethacrylate graft units was synthesized from a hydroquinone substituted with polymethacrylate, *t*-butyl hydroquinone, and bromoterephthaloyl chloride (Heitz and Webster, 1991c). The hydroquinone containing the polymethylmethacrylate substituent was prepared by group transfer polymerization (FIGURE 8.31). Block copolymers containing thermotropic polyester and glassy polystyrene segments were also reported (Galli et al., 1994).

8.6.1.5 Rod-Like Soluble Polyimides

Rigid rod-like polyimides were synthesized by the sequence of reactions presented in FIGURES 8.32 and 8.33 (Metzmann et al., 1989; Wenzel et al., 1987). The series of polymers from FIGURE 8.32 contain alkyl side groups (Wenzel et al., 1987) while the one from FIGURE 8.33 contains oligooxyethylene side groups

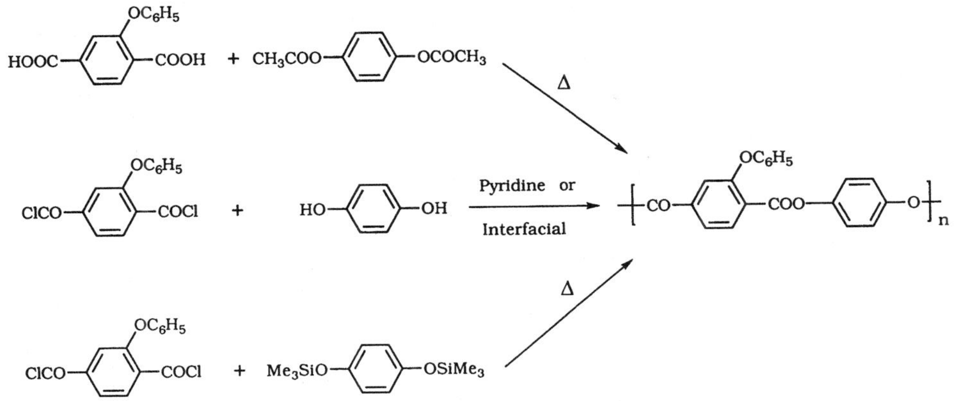

FIGURE 8.28. Various synthetic methods for the preperation of aromatic polyesters.

(Metzmann et al., 1989). The merit of these two synthetic schemes is that the resulting polymers contain only imide rings. Soluble polyimides were also reported by the conventional sequence of reactions from FIGURE 8.34 (Duran et al., 1988). All these rigid rod-like polyimides form layered mesophases (Duran et al., 1988; Metzmann et al., 1989; Wenzel et al., 1987). It is interesting to mention that polymers containing identical lengths of their alkyl side groups are isomorphic within their crystalline phase and therefore form solid solutions regardless of the nature of their polymer backbone (Duran et al., 1988). For example, the polyimide from the top part of FIGURE 8.34 is isomorphic with the polyester which has an identical number of carbons in its side groups (bottom part of FIGURE 8.34) (Duran et al., 1988).

Soluble polyimides were also synthesized from 3,6-diphenylpyromellitic dianhydride with various diamines (Harris and Hsu, 1989). A lyotropic polyimide was obtained from the twisted 2,2′-*bis*(trifluoromethyl)-4,4′-diaminobiphenyl and 3,6-diphenylpyromellitic dianhydride (FIGURE 8.35) (Harris and Hsu, 1989). A brief review of liquid crystalline polyimides is available (Kricheldorf, 1994).

FIGURE 8.29. Side reactions in aromatic polyesters.

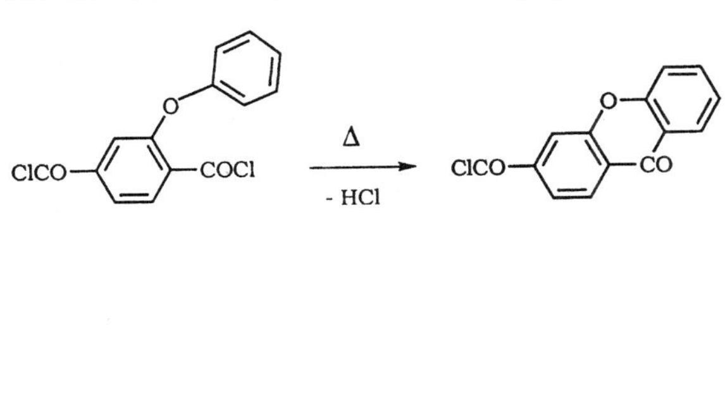

8.6.1.6 Thermotropic Poly(1,4-arylenevinylene)s

Poly(phenylenevinylene)s can be synthesized by a variety of reactions including the Wittig reaction (Horhold and Opfermann, 1970; Horhold et al., 1987; Kossmehl and Samandari, 1985), Hofmann elimination (Antoun et al., 1986), the McMurry reaction (Cooke and Wagener, 1991; Feast and Millichamp, 1983; Rehahn, 1990b), the Heck reaction (Heitz et al., 1988) and the Siegrist reaction (Kretzschmann and Meier, 1991; Siegrist, 1981).

Liquid crystalline oligo(p-phenylenevinylene)s were first reported by Campbell and McDonald

FIGURE 8.30. Synthesis of a graft copolymer containing a liquid crystalline polyester as backbone and amorphous polystyrene grafts.

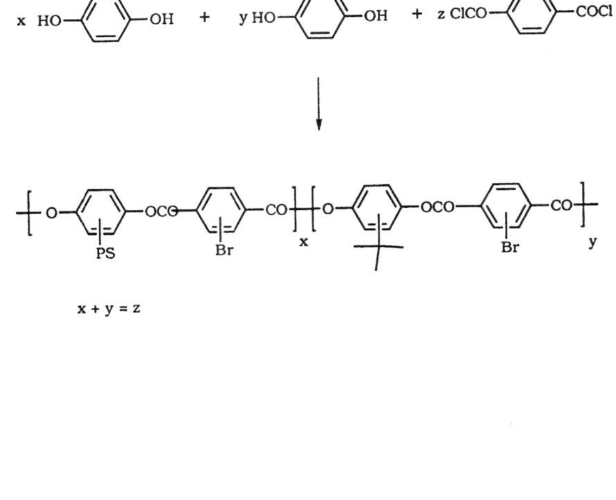

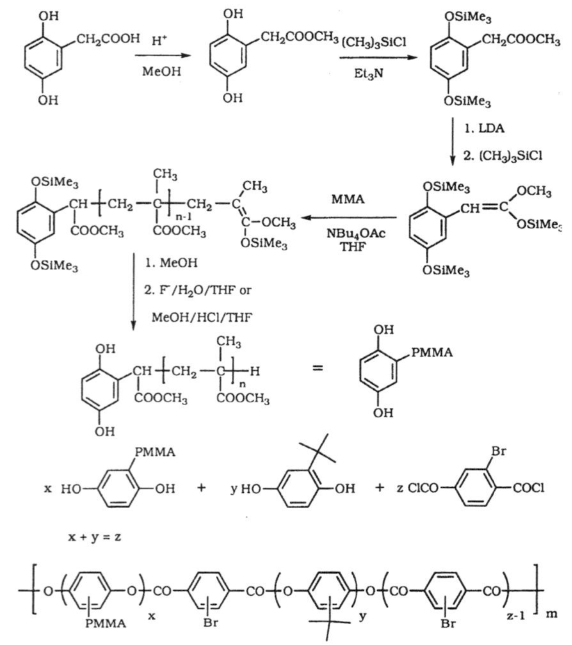

FIGURE 8.31. Synthesis of a graft copolymer containing polyester backbone and poly(methylmethacrylate) grafts.

(1959). High molecular weight poly(p-phenylenevinylene)s are insoluble and infusible. The first thermotropic poly(phenylenevinylene)s were recently reported (Memeger, Jr., 1989); they were synthesized by the Wittig reaction. The insertion of suitable substituents such as Cl on the phenyl ring, the replacement of phenyl rings with anthracene or naphthalene, and of vinylene with ethane, allowed the synthesis of thermotropic poly(arylenevinylene)s (Memeger, Jr., 1989).

Thermotropic poly(arylenevinylene)s were also synthesized by various synthetic methods based on the Heck reaction (Brenda et al., 1990; Greiner and Heitz, 1988; Heitz, 1991b; Heitz et al., 1988). Some of these procedures are outlined in FIGURE 8.36. The highest solubility is obtained for the polymer with R_1=H and R_2=C_6H_5. The polymers with R_1=H and R_2=C_6H_5 or CF_3 exhibit thermotropic mesophases (Greiner and Heitz, 1988). Additional examples of poly(p-arylene vinylene)s are presented in FIGURE 8.37 (Martelock et al., 1991). As expected, copolymers with highly dissimilar structural units which can also yield constitutional isomeric structural units yield soluble, thermotropic liquid crystalline polymers that are conjugated (Martelock et al., 1991). Only qualitative information is available on their liquid crystalline

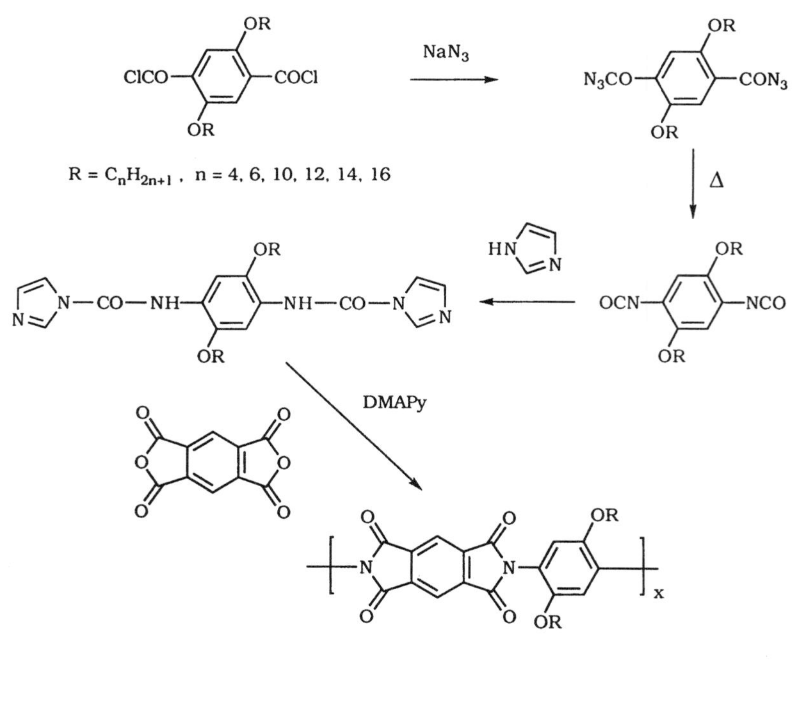

FIGURE 8.32. Synthesis of soluble poly(imide)s.

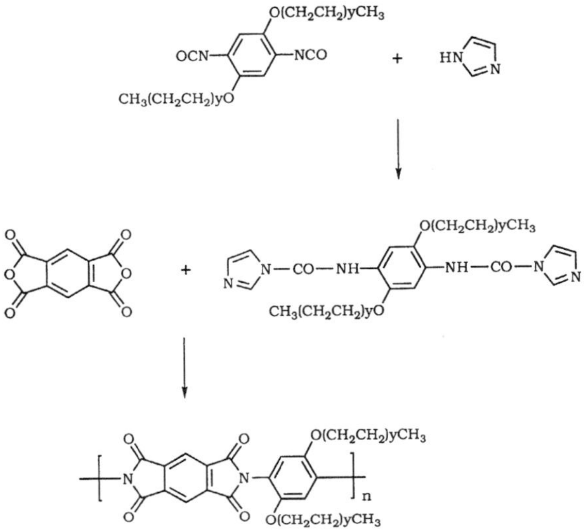

FIGURE 8.33. Soluble poly(imide)s.

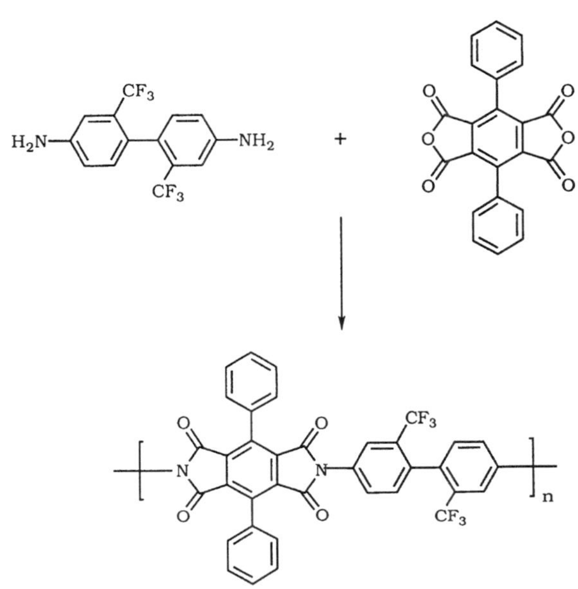

FIGURE 8.35. Rod-like polyimides.

behavior. Thermotropic poly(phenylenevinylene)s containing flexible spacers were also synthesized by the Heck reaction (Suzuki et al., 1990). More recently, liquid crystalline poly(p-phenylene vinylene)s were obtained by the Heck coupling of p-divinylbenzene with 1,4-dialkyloxy-2,5-diiodobenzene (Bao et al., 1993).

8.6.1.7 Other Conjugated Polymers

Alternating copolymers such as poly(1,4-phenylene-2,5-thiophene) were prepared by the Stille cross-coupling reaction. When the 1,4-phenylene repeat units contain 2,5-dialkyloxy substituents, these conjugated copolymers exhibit nematic mesophases (Yu et al., 1993). Both a-sexithienyl (Destri et al., 1993) and poly(α-hexathiophene) (Qian et al., 1994) were recently shown to display nematic mesophase.

8.6.1.8 Polyurethanes

Thermotropic polyurethanes were reported beginning in 1981 (Iimura et al., 1981; Tanaka and Nakaya, 1984, 1986, 1987, 1988). Most of them were thermally unstable in their liquid crystalline phases. The first examples of thermally stable polyurethanes were reported by Kricheldorf and coworkers (Kricheldorf and Awe, 1988d, 1989e; Kricheldorf and Jenssen, 1989a) Smectic polyurethanes were synthesized from the bischloroformates of 4,4′-alkylenedioxydiphenols

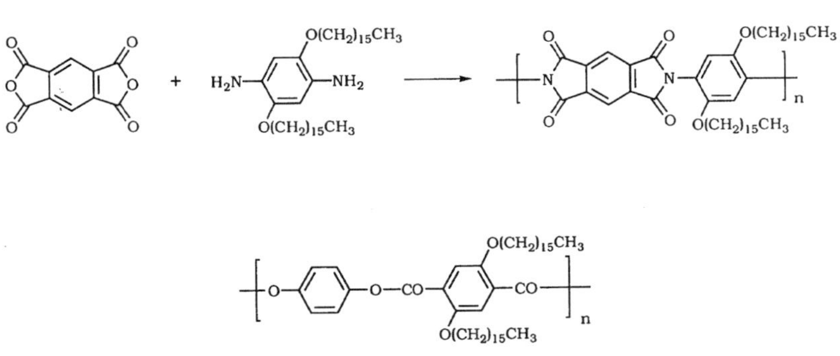

FIGURE 8.34. Polyimides and polyesters.

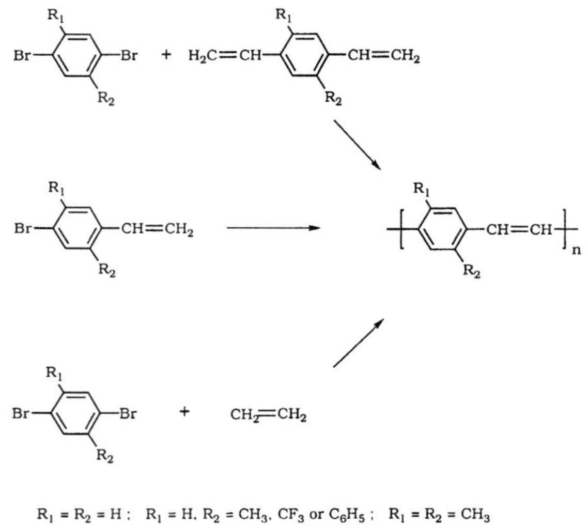

FIGURE 8.36. Synthesis of poly(1,4-arylenevinylene)s by Heck's reaction.

with piperazine, *trans*-2,5-dimethylpiperazine, and 4,4′-bipiperidine by interfacial polycondensation (FIGURE 8.38) (Kircheldorf and Awe, 1989e). Polyurethanes without flexible spacers were synthesized from 4,4′-bipiperidine, 1,2-*bis*(4-piperidinyl)ethane, and 1,3-*bis*(4-piperidinyl)propane with the bischloroformates of p-phenylene, 2-methyl-1,4-phenylene

FIGURE 8.37. Synthesis of various poly(arylenevinylene)s by Heck's reaction.

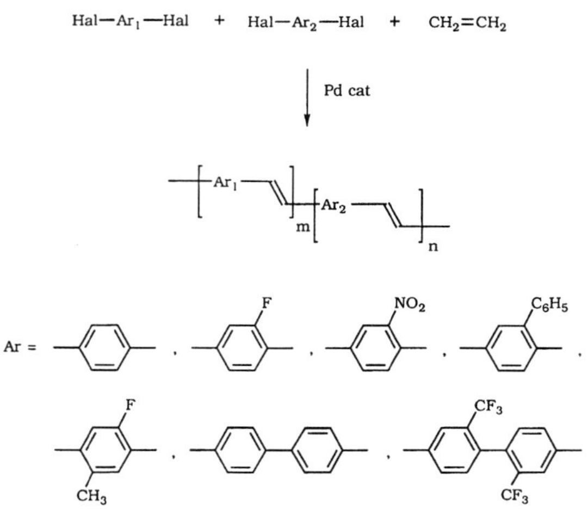

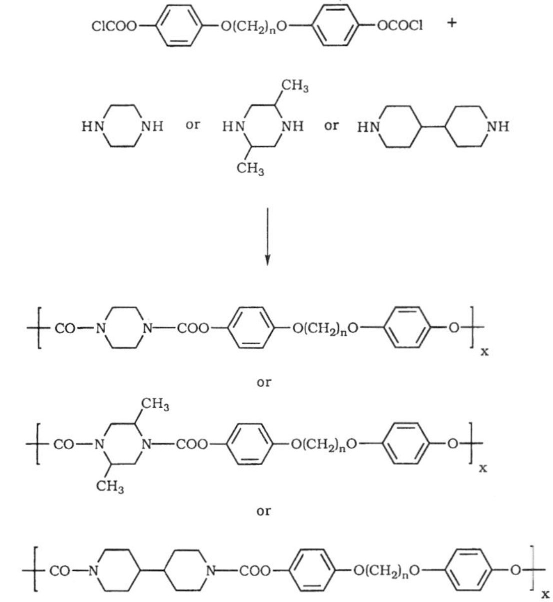

FIGURE 8.38. Synthesis of smectic polyurethanes.

or 2,5-*bis*phenylylene. Copolyurethanes were also synthesized (Kricheldorf and Awe, 1989f). The polymers derived from hydroquinone or methylhydroquinone are semicrystalline with short-term stability up to 310 °C. The polymers obtained from phenylhydroquinone are amorphous with thermal stability up to 360 °C. Polyurethanes and copolyurethanes prepared from 4,4′-biphenylylene units form a smectic layered structure and a nematic mesophase. While the research group of Kricheldorf decreased the crystallization ability of polyurethanes by suppressing the hydrogen bonding ability of the urethane groups, Mormann and coworkers have generated thermally stable polyurethanes from diisocyanates of substituted phenyl benzoates (Mormann and Hissmann, 1987; Mormann and Brahm, 1989a, b, 1991b; Mormann and Hohn, 1989b; Mormann and Baharifar, 1990a; Mormann and Benadda, 1991a; Mormann et al., 1995). The resulting polyester urethanes are more flexible and generate a broader range of the mesophase by decreasing the melting temperature (Mormann and Benadda, 1991a; Mormann and Brahm, 1991b). Additional examples of polyurethanes were reported by MacKnight and coworkers (Stenhouse et al., 1989). Recently, liquid crystalline

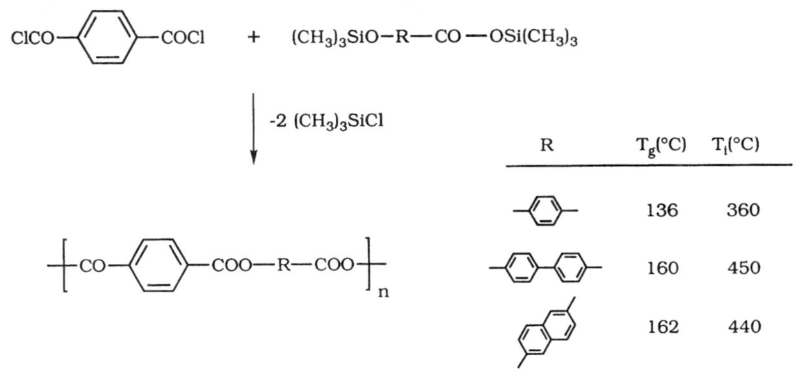

FIGURE 8.39. Synthesis of thermotropic poly(ester anhydride)s.

polyurethanes with mesogenic side groups were reported (Koide et al., 1994).

8.6.1.9 Polycarbonates

The first examples of polycarbonates containing flexible spacers were reported by Roviello and Sirigu (1979, 1980) and by Sato and coworkers (Sato et al., 1986, 1988). Poly(ester carbonate)s from *tert*-butylhydroquinone and terephthalic acid were first reported by Prevorsek and coworkers (Lai et al., 1988). Soluble aromatic copolycarbonates were synthesized by Kricheldorf and Lubbers (1989d, 1990b). They are conveniently prepared by copolymerization of three or four bisphenols with bis(trichloroacetyl)carbonate. For example, copolycarbonates based on 4,4′-dihydroxybiphenyl, methylhydroquinone, and 4,4′-dihydroxybiphenyl ether or 4,4′-dihydroxybenzophenone, are soluble in aprotic organic solvents such as methylene chloride (Kricheldorf and Lubbers, 1990b). Poly(ester carbonate)s were also synthesized by melt polycondensation of substituted hydroquinones and diphenylterephthalate with diphenylcarbonate (Lai et al., 1990) and from p-hydroxybenzoic acid, 4,4′-dihydroxybiphenyl, and diphenylcarbonate (Kawabe et al., 1990).

8.6.1.10 Poly(ester Anhydride)s

Thermotropic poly(ester anhydride)s were synthesized from terephthaloyl chloride with silylated hydroxy acids (Kricheldorf and Lubbers, 1990d). The polymers presented in FIGURE 8.39 are noncrystallizable and exhibit a nematic phase over a broad range of temperatures (Kricheldorf and Lubbers, 1990d).

8.6.1.11 Poly(ester Imide)s

Thermotropic poly(ester imide)s were developed by Kricheldorf's group. The first series of thermotropic poly(ester imide)s was synthesized starting from trimellitic anhydride and α,ω-diaminoalkanes containing from four to twelve methylene units. The resulting α,ω-diaminoalkane *bis*(trimellitimide)s were reacted with various 4,4′-diacetoxybisphenols (FIGURE 8.40) including those based on 4,4′-dihydroxybiphenyl (Kricheldorf and Pakull, 1988b), hydroquinone, methyl-, chloro-, phenyl-, and tetrachlorohydroquinone, 1,5-, 1,4-, 2,7- and 2,6-dihydroxynaphthalenes (Kricheldorf and Pakull, 1987b). All these polymers crystallize in a layered structure. The polymers with Ar = 4,4′-biphenylyl melt into a smectic mesophase (Kricheldorf and Pakull, 1988b). The melting temperature varies from 297 °C for the polymer with $n = 12$ to 393 °C for the polymer with $n = 4$. The liquid crystalline phase undergoes isotropization at 386 °C for the polymer with $n = 12$ and at 467 °C for the polymer with $n = 4$. The rate of crystallization of all these polymers is unusually fast. With the exception of the polymers based on Ar = 1,4-phenylene, 1,4-tetrachlorophenylene, and 2,6-naphthalene, which exhibit a smectic mesophase, the other polymers melt directly into an isotropic phase (Kricheldorf and Pakull, 1987b).

The second procedure used in the preparation of poly(ester imide)s is based on the reaction of pyromellitic dianhydride or benzophenone-3,3′,4,4′-tetracarboxylic dianhydride with aminoacids or lactams, followed by the condensation of the resulting diacids with diacetates of hydroquinone, 2,6-dihydroxynaphthalene or 4,4′-dihydroxybiphenyl (Kricheldorf

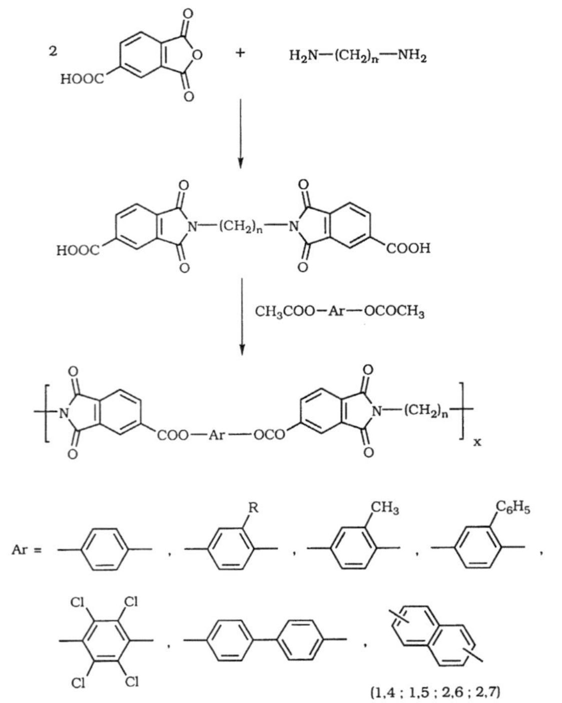

FIGURE 8.40. Synthesis of regioregular poly(ester imide)s.

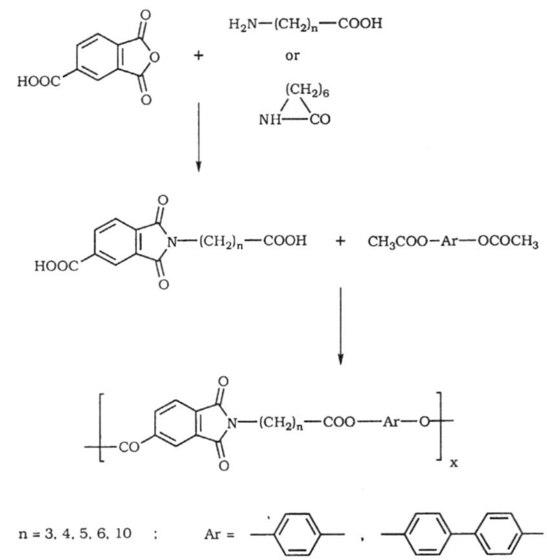

FIGURE 8.41. Synthesis of regioirregular aromatic poly(ester imide)s.

et al., 1988c). All these polymers crystallize in a layered structure. With the exception of the polymer with $n = 11$, all the polymers based on Ar = 4,4′-biphenylyl exhibit a smectic mesophase. The polymer with $n = 10$ and Ar = 1,4-phenylene exhibits a smectic mesophase. The other polymers based on Ar = 1,4-phenylene or 2,6-naphthylene melt directly into an isotropic phase (Kricheldorf et al., 1988c).

An additional method for the synthesis of thermotropic poly(ester imide)s is based on the reaction of trimellitic dianhydride with aminoacids or lactams, followed by condensation with acetates of bisphenols (Figure 8.41). This reaction can be performed as a "one-pot procedure" (Kricheldorf et al., 1989b). Again, all homopolymers exhibit a lamellar crystalline phase, which in all cases melts into an enantiotropic smectic phase. However, the corresponding copolymers exhibit a nematic mesophase. It is interesting to note that the homopoly(ester imide)s from Figure 8.41 are in fact copolymers since their structural units have two constitutional isomers. Finally, fully aromatic poly(ester imide)s were reported by the same research group by using several different synthetic procedures, which are outlined in Figure 8.42 (Kricheldorf and Pakull, 1985, 1989g; Kricheldorf et al., 1989h; Kricheldorf and Huner, 1990c). The imide monomers from Figure 8.42 were homopolymerized and copolymerized either between themselves or with other aromatic dicarboxylic acids, bisphenols, or hydroxyacids to give fully aromatic poly(ester imide)s with a complex thermal behavior (Kricheldorf and Pakull, 1989g; Kricheldorf et al., 1989h). However, the fully aromatic poly(imide ester)s from Figure 8.43 exhibit glass transition temperatures between 140 and 180 °C and nematic mesophases which undergo isotropization between 375 and 500 °C (Kricheldorf and Huner, 1990c).

8.6.1.12 Polyhydrocarbons

In addition to the class of poly(arylenevinylene)s described in section 8.6.1.6 and poly(phenylene)s and poly(arylene)s described in section 8.6.1.1, liquid crystalline polymers containing rigid rod-like mesogens and flexible spacers and based only on carbon–carbon bonds, were also synthesized by the

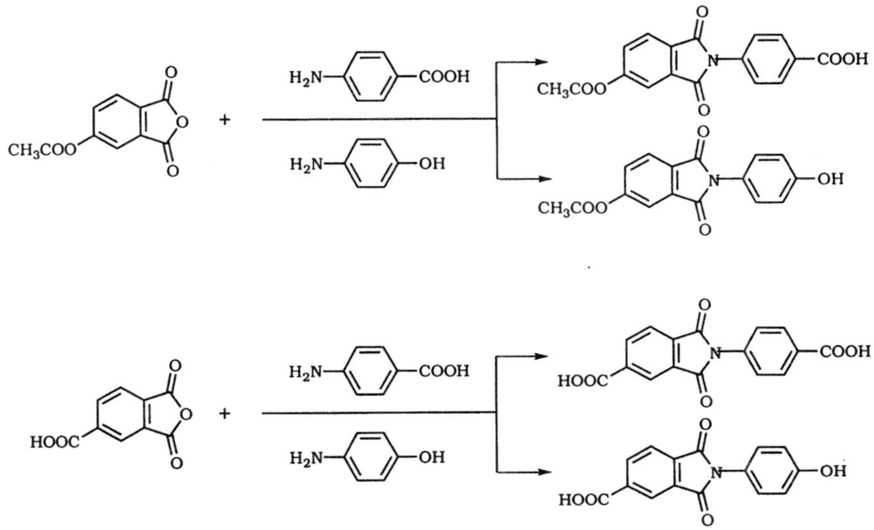

FIGURE 8.42. Synthesis of aromatic monomers for the preparation of poly(ester imide)s.

homocoupling of *bis*(4-bromophenyl)alkanes (Sung et al., 1990).

8.6.1.13 Polyethers

Thermotropic polyethers based on rigid rod-like mesogens and flexible spacers are of interest since they do not undergo transesterification or other rearrangement reactions during their characterization that requires extensive annealing at high temperatures. The effect of thermally induced rearrangement reactions will be discussed in a subsequent section. Polyethers can be conveniently prepared by the phase transfer catalyzed polyetherification of mesogenic bisphenols with α,ω-dibromoalkanes. The main criterion required for their

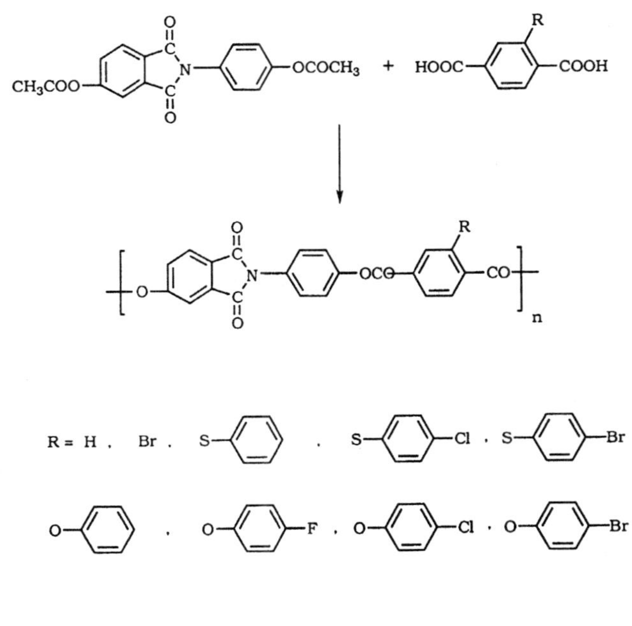

FIGURE 8.43. Aromatic poly(ester imide)s.

synthesis is good solubility of the resulting polymer in the reaction medium (Percec, 1988b, c; Percec et al., 1991l, m).

8.6.1.14 Liquid Crystalline Fluorocarbon–Hydrocarbon Copolymers

Main chain fluorocarbon–hydrocarbon alternating copolymers were synthesized by the radical initiated addition of α,ω-diiodoperfluoroalkanes to α,ω-alkanedienes, followed by the reduction of the secondary iodide groups with Bu_3SnH (Davidson et al., 1995; Wilson and Griffin, 1993). Since hydrocarbon and fluorocarbon segments are highly immiscible, these "microblock" copolymers form lamellar structures which display smectic B mesophases.

Semifluorinated main chain polyesters based on aliphatic dicarboxylic acids and aliphatic diols, one being perfluorinated, the other being perhydrogenated, also exhibit thermotropic smectic B phases (Wilson, 1994a, 1995).

Polyesters with hydrogenated backbones and fluorinated paraffinic side groups display smectic A and B phases (Wilson and Griffin, 1994c, d).

8.6.2 Persistence Lengths of Soluble Main Chain Liquid Crystal Polymers

The persistence length and the Mark–Houwink coefficient, a, were determined both for soluble polyesters and polyamides. For soluble polyesters, the persistence length data obtained by various authors are 95 Å ($a = 0.95$) (Heitz and Schmidt, 1990b), 120 ± 10 Å ($a = 0.85$, but estimated to be 1.0) (Kromer et al., 1991), and 110 Å ($a = 1.0$–1.1) (Hsiao et al., 1991). For substituted aromatic polyamides, the values of the persistence length reported are 85 ± 26 Å (Krigbaum et al., 1991) and 200 Å ($a = 1.1$–1.2) (Gaudiana et al., 1989). Both values are lower than those reported for poly(p-phenylenediamine terephthalic acid) (150–290 Å) and for poly(p-benzamide) (240–750 Å (Krigbaum et al., 1991). All these results demonstrate that both polyesters and polyamides do not resemble rigid rods, but are comparable to typical semiflexible chains. In addition, the values obtained for the Mark–Houwink coefficient, a, also support the semiflexible character of these chains (a is 0.8 for flexible chains and 2.0 for rigid chains). While aromatic polyamides, depending on substituent, can generate lyotropic solutions, soluble polyesters do not. Films cast from solutions of aromatic polyesters are transparent (amorphous) and become turbid (anisotropic) only after annealing above their glass transition temperature. In conclusion, although due to interchain interactions the rigidity of polyesters and polyamides increases versus that predicted by gas phase and dilute solutions of low molar mass compounds, both types of polymers belong to the class of semiflexible polymers. A recent discussion on the persistence length of polyesters, polyamides, polyimides, and even poly(p-phenylene)s, concludes that truly rigid-rod polymers may not result from one-dimensional chains (Tiesler et al., 1994).

8.7 CHEMICAL HETEROGENEITY IN MAIN CHAIN LIQUID CRYSTALLINE COPOLYMERS

As discussed in section 8.6, main chain liquid crystalline polymers and copolymers are synthesized by step polymerizations that are based on reversible or irreversible reactions. A polymerization reaction is reversible depending on its mechanism and on the reaction conditions used. In contrast to chain copolymerizations, step copolymerizations performed in a homogeneous phase at high conversion and with a stoichiometric ratio between comonomers always lead to copolymer compositions that are identical with the comonomer feed. Also, the difference between the reactivities of the various monomers used in step reactions is lower than in the case of monomers used in chain reactions. This implies that the compositional heterogeneity of the copolymers obtained by step reactions is lower than that of copolymers synthesized by chain copolymerizations, where copolymer composition is conversion dependent. However, again in contrast to chain copolymerizations, in step copolymerizations the copolymer's sequence distribution can be either kinetically (in irreversible copolymerizations) or thermodynamically (in reversible copolymerizations) determined. In reversible step copolymerizations, the sequence distribution is thermodynamically controlled and the copolymer microstructure is determined by redistribution reactions. The copolymer sequence distribution (Chen and Lenz, 1985; Lenz et al., 1983) and the configuration of the structural units (Gardner et al., 1988; Irwin et al., 1989; Kricheldorf and

Schwarz, 1987a; Kwolek and Louise, 1986) are both determined by the type of phase (isotropic, liquid crystalline, or crystalline) in which copolymerization or the copolymer reorganization reaction is performed. Microheterogeneous copolymerization reactions complicate the control of the copolymer's microstructure since the concentration of the comonomers in the proximity of the growing chain is determined by the miscibility and/or the association between the growing chain and the monomers. The concept that the growing chain can control its own environment during copolymerization was explained on the basis of the "bootstrap" model, and its implications were recently reviewed (Harwood, 1987). Some of the most recent examples of "bootstrap effects" were observed in the radical copolymerization of macromonomers (Muhlbach and Percec, 1987; Percec and Wang, 1990a, 1992o), in the synthesis of block copolymers from immiscible amorphous segments (Auman et al., 1987a, b), and in the synthesis of ternary copolymers from monomers that can give rise to amorphous and liquid crystalline structural units by reversible copolymerization reactions (Auman and Percec, 1988).

Copolymerization reactions performed in the bulk are frequently used in the preparation of liquid crystalline polyesters, poly(ester imide)s, etc. The starting monomers lead to isotropic melts, nevertheless, above a certain conversion this reversible reaction is performed in the liquid crystalline phase. A heterogeneous composition is generated both by the different reactivities of the monomers and by the polydispersities of the polymers. This heterogeneous composition generates a microphase separated reaction mixture, which at a given temperature contains isotropic, liquid crystalline, and crystalline phases. Each of these phases generates a polymer homologous series of copolymers that have both different compositions, and for the same composition, a different sequence distribution. The sequence distribution is determined by the phase in which the copolymer is generated that is, isotropic, liquid crystalline, or crystalline. This microheterogeneous copolymerization reaction enhances the chemical heterogeneity of the resulting copolymers. Both the polydispersity and the chemical heterogeneity of liquid crystalline copolymers are responsible for the biphasic, or even multiphasic, nature of liquid crystalline copolyesters (D'Allest et al., 1986; Economy, 1990a; Economy et al., 1990b; Laus et al., 1991; Martin and Stupp, 1988; Moore and Stupp, 1988; Ober et al., 1990; Stupp et al., 1988). The chemical heterogeneity of liquid crystalline copolymers is reduced or even eliminated in the case of azeotropic irreversible copolymerizations (Percec and Tsuda, 1990b, 1991r; Sherwood et al., 1994). Excellent discussions on the change in sequence distribution and molecular weight during the annealing of thermotropic polyesters are available (Economy, 1990a; Economy et al., 1990b; MacDonald et al., 1991). The same processes should take place in polyamides, polycarbonates, poly(ether imide)s, polyurethanes, polyanhydrides, and any other polymers containing chemical bonds which can undergo rearrangement reactions during thermal treatment.

8.8 FLEXIBLE AND SEMIFLEXIBLE LIQUID CRYSTALLINE POLYETHERS

Flexible liquid crystalline polyethers are based on flexible mesogenic units which are rod-like mesogenic units based on conformational isomerism. In the gas phase and dilute solutions, these units are as flexible as n-butane (Percec and Zuber, 1991o) (see section 8.4.2). However, in the melt phase, interchain interactions increase their rigidity. The simplest structural units which can be used for the synthesis of flexible liquid crystalline polymers based on conformational isomerism are diphenylethane, phenylbenzyl ether, and methyleneoxy units (FIGURE 8.44). Flexible rod-like mesogens or rod-like mesogens based on conformational isomerism exhibit a number of different conformers which are in dynamic equilibrium. The two most stable conformers are the *anti* and *gauche*. The *anti* conformer has an extended rod-like shape and therefore is expected to display liquid crystallinity. The *gauche* conformer is similar to a "kinked" unit which is occasionally introduced within the structure of main chain liquid crystalline copolymers based on rigid rod-like mesogens to decrease their crystallization ability. Therefore, the insertion of flexible units capable of giving rise to extended and kinked conformers within the main chain of a polymer is expected to provide a liquid crystalline polymer with a dynamic composition. To date, there are no data on the dynamic equilibrium between different conformers of "flexible" rod-like mesogens. However, it is well established that when rod-like or linear conformations are in dynamic

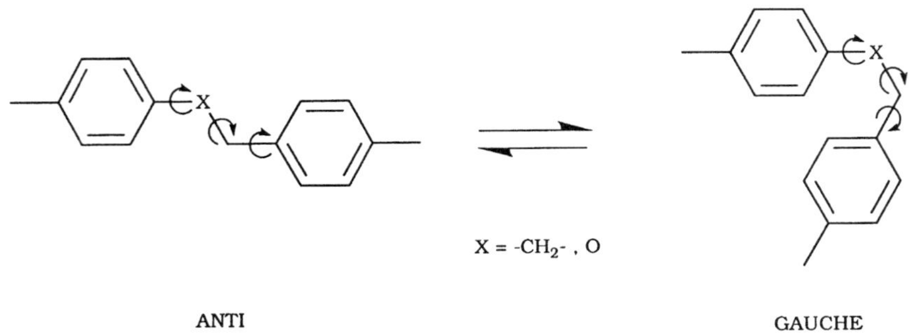

FIGURE 8.44. Anti and gauche conformers of diphenylethane and benzyl ether flexible rod-like mesogens based on conformational isomerism.

equilibrium with random-coil conformations, the rod-like conformation is preferred in the nematic phase (de Gennes and Pinkus, 1977; Flory and Matheson, 1984b; Kim and Pinkus, 1979). Therefore, it is not excluded that in the nematic phase, the preferred conformer may be the *anti*.

Low molar mass liquid crystals based on benzyl ether and diphenylethane units are well known. Table 8.10 summarizes the dependence between the structure of some benzyl ether compounds and the thermodynamic stabilities of their mesophases. A brief inspection of this table shows that all the flexible compounds listed exhibit liquid crystalline mesophases. However, the thermodynamic stability of their mesophase with respect to the crystalline phase is drastically dependent on subtle changes in their structure. For example, compound 1 exhibits a virtual nematic mesophase at −20 °C. Hydrogenation of one of its phenyl rings enhances its isotropization by 68.6 °C. Nevertheless, the nematic phase of compound 2 is only monotropic. The replacement of the 1,4-cyclohexane ring of compound 2 with a bicyclooctane unit generates an enantiotropic mesophase. The difference between the isotropization temperature of compounds 1 and 3 in Table 8.10 is 93 °C. If we replace the cyano group of compound 2 with a pentyl group, the new compound 4 exhibits again a virtual mesophase. The replacement of a methylenic unit of the pentyl group of compound 4 with an oxygen generates compound 6, which displays a monotropic nematic phase. The substitution of the methyleneoxy group of 6 with an ethane unit generates an enantiotropic nematic mesophase (compound 7).

Table 8.11 provides some examples of low molar mass liquid crystals based on diphenylethane, cyclohexylphenylethane, dicyclohexylethane, and dicyclooctylphenylethane (Carr et al., 1983; Carr and Gray, 1985; Kelly and Schad, 1984; Osman, 1982a, b; Osman and Ba, 1983; Praefke et al., 1980; Schadt et al., 1983). The flexible mesogenic units in Table 8.11 follow the same trend as those in Table 8.10. For example, the phase behavior of compounds 1 and 2 in Table 8.10 is almost identical to that of compounds 8 and 9 in Table 8.11. Hydrogenating a phenyl ring from compound 8 increases the thermal stability of the nematic phase of compound 9 by about 75 °C versus that of compound 8. Replacing a phenyl ring from compound 8 with a bicyclooctane ring enhances the thermodynamic stability of the nematic phase by 137 °C.

Let us now consider increasing the molecular weight of compounds 1 or 8 by polymerization. According to the thermodynamic trends discussed in section 8.5.3, the virtual mesophase of the monomeric structural unit should become at higher molecular weights enantiotropic, monotropic, or remain virtual. In the last case, the transition of the virtual mesophase of the polymer should, however, be shifted to higher temperatures.

The synthesis of the first flexible main chain liquid crystalline polyether without spacers is outlined in Figure 8.45 (Percec and Yourd, 1988a). As expected based on the above discussion, the resulting polyethers based on *trans* 1,4-*bis*[(methylsulfonyl)methyl]cyclohexane and methylhydroquinone or phenylhydroquinone display an enantiotropic nematic mesophase. The copolymers containing some amount of *cis* 1,4-*bis*[(methylsulfonyl)methyl]cyclohexane also display a nematic mesophase. DSC traces of the polyethers based on methylhydroquinone and *trans* 1,4-*bis*[(methylsulfonyl)methyl]cyclohexane are identical to those of the corresponding polyesters based on methylhydroquinone and 1,4-cyclohexanedicarboxylic acid (Kricheldorf and Schwarz, 1987a; Kwolek and Louise, 1986;

TABLE 8.11. The dependence between structure and phase transitions of some low molar mass liquid crystals based on disubstituted ethane derivatives (Carr et al., 1983; Carr and Gray, 1985; Kelly and Schad, 1984; Osman and Ba, 1983; Praefke et al., 1980; Schadt et al., 1983).

No.	Structure	Phase transitions (°C)
8	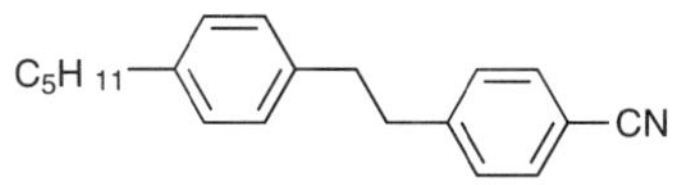	k 62 n [−24] i[a]
9	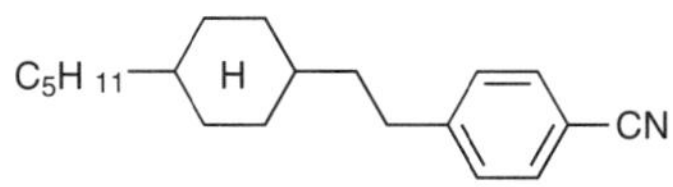	k 30 n 51 i
10	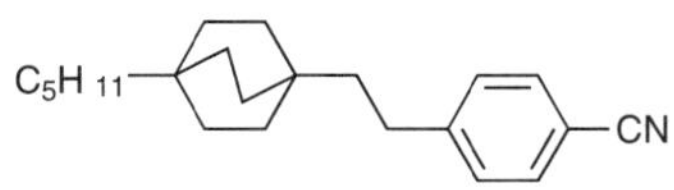	k 76 n 113i
11	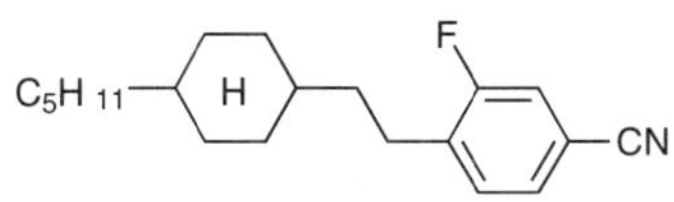	k 45 n 54 i
12	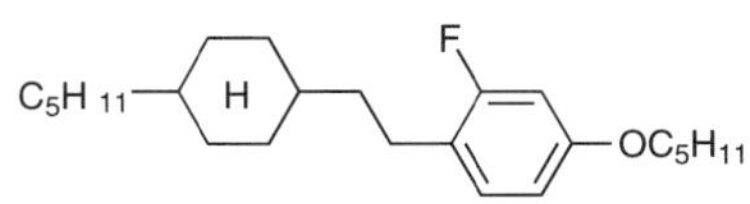	k 26 n (12) i[b]
13	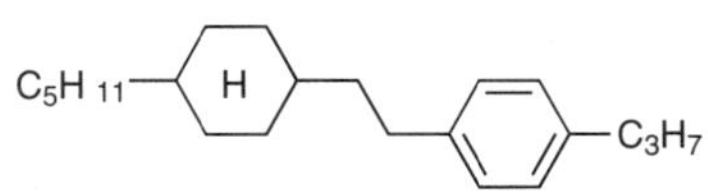	k 4 s_B 18 i
14	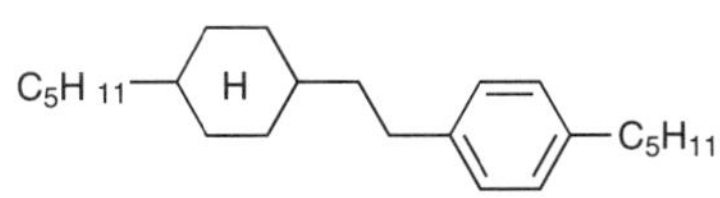	k 46 s_B 109 i

[a] [ ]: Virtual.
[b] ( ): Monotropic.

Kyotari and Manetsuna, 1983), except that the phase transitions of the polyethers are shifted to lower temperatures (Percec and Yourd, 1988a). Based on the thermodynamic principles discussed previously, this is an expected result since the corresponding polyesters are identical to the polyether from FIGURE 8.45, except that the methyleneoxy units from the structure of the polyethers are replaced with ester units. This increases only the overall rigidity of the polyesters versus that of the polyethers.

Polyethers based on substituted hydroquinones and *p*-dibromo- or dichloroxylene also lead to liquid crystalline polyethers (Memeger, 1991).

Memeger (1991) synthesized high molecular weight polyethers based on α,α'-dichloro-p-xylene and methylhydroquinone in dimethylacetamide using K_2CO_3 as base. Fibers with high modulus (233 g/denier) were spun from these polymers. Unfortunately, the polyetherification of substituted hydroquinones with α,α'-dichloro-p-xylene leads both to C and O alkylation reactions. Nematic polyethers with and without flexible spacers were also synthesized by cationic ring opening polymerization of exo-2-methyl-7-oxabicyclo[2.2.1]heptane and 7-oxabicyclo[2.2.1]heptane, as well as by the copolymerization of the latter monomer with tetrahydrofuran and/or ethylene oxide (Kops and

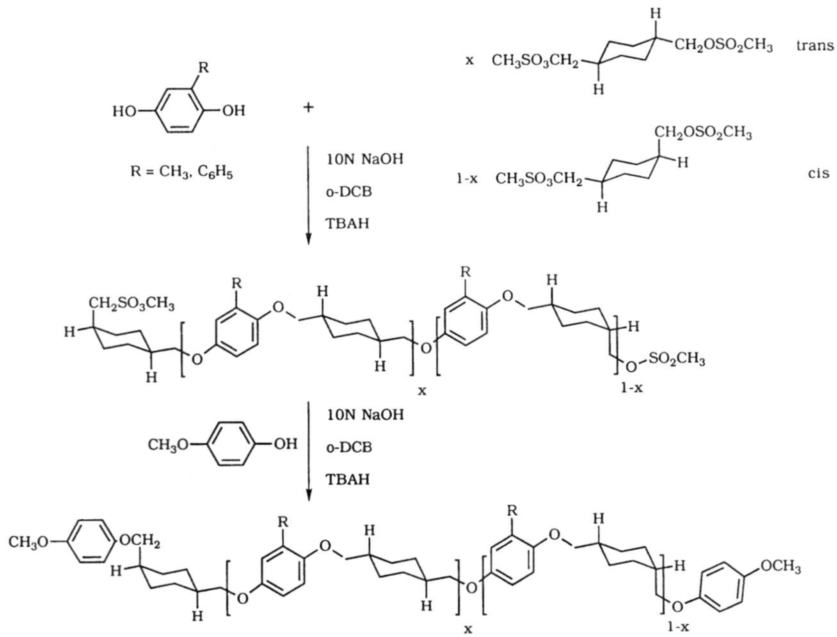

FIGURE 8.45. Synthesis of flexible liquid crystal polyethers based on conformational isomerism by phase transfer catalyzed polyetherification.

Spanggaard, 1986; Sikkema and Hoogland, 1986) (FIGURE 8.46).

A comprehensive series of experiments was performed with polyethers based on 1-(4-hydroxyphenyl)-2-(2-R-hydroxyphenyl)ethane (RBPE) and α,ω-dibromoalkanes containing from four to 20 methylenic units (Percec and Tsuda, 1989n, o, 1990b, d, h, 1991q, r; Percec and Yourd, 1989d, i, 1990f, g; Ungar et al., 1990), particularly those based on 1-(4-hydroxyphenyl)-2-(2-methyl-hydroxyphenyl)ethane (MBPE). All these polyethers are in fact copolyethers, since their structural unit has two constitutional isomers (FIGURE 8.47). With the exception of MBPE-5 and MBPE-9 which exhibit two monotropic nematic mesophases, and MBPE-8 and MBPE-11 which exhibit one monotropic nematic mesophase, all the other MBPE-X

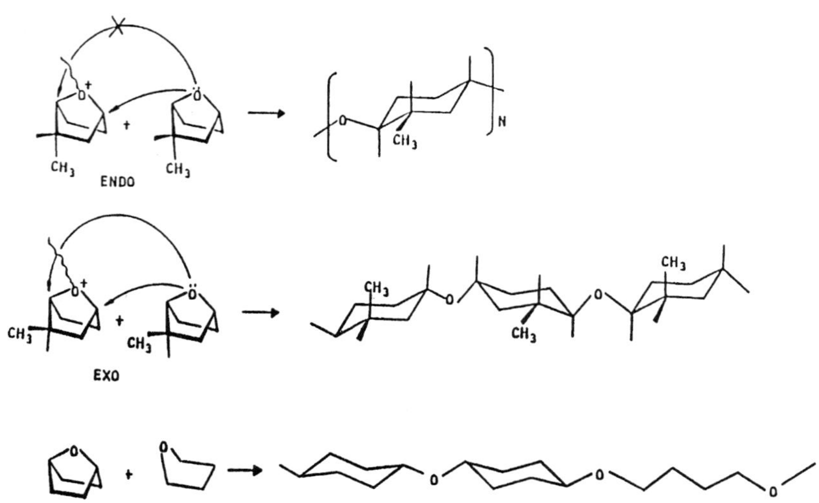

FIGURE 8.46. Synthesis of flexible liquid crystalline polyethers and copolyethers by cationic ring-opening polymerization.

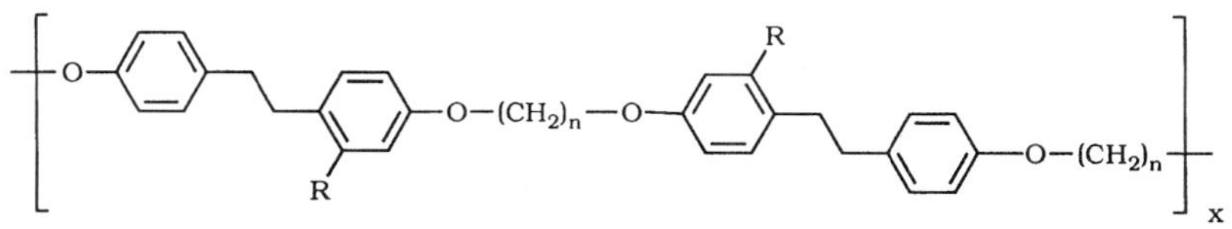

FIGURE 8.47.
Conformationally flexible liquid crystalline polyethers based on 1-(4-hydroxyphenyl)-2-(2-R-4-hydroxyphenyl) ethane (RBPE, where R = Me(M), Br, Cl, F, H, etc.) and α,ω-dibromoalkanes.

polyethers are only crystalline (Percec and Tsuda, 1990d). Nevertheless, again based on the thermodynamic discussion from section 8.5, we can assume that all of them may exhibit virtual mesophases. The virtual transition temperatures and thermodynamic parameters of all MBPE-X polymers were determined by copolymerization experiments of MBPE with two different flexible spacers (Percec and Tsuda, 1989n, o, 1990b, d, h, 1991q, r; Percec and Yourd, 1988a, 1989d, i, 1990f, g; Ungar et al., 1990).

The technique developed and used for the determination of their virtual transition temperatures and thermodynamic parameters is as follows. The structural units of the copolymers resulting from two flexible spacers and one mesogen are isomorphic in their mesophase, but not in their crystalline phases. As a consequence, the crystallization temperature of the copolymer is decreased and exhibits a eutectic point while both the temperature and the enthalpy associated with the mesophase display continuous dependences on composition. Therefore, in the mesophase the two structural units behave as an ideal solution and obey the Schroeder–Van Laar equations (section 8.3) (Achard et al., 1989; Schroeder, 1893; Van Hecke, 1979; Van Laar, 1908). Since the difference between ΔH_1 and ΔH_2 of the two homopolymers is not large and the values of ΔH_1 and ΔH_2 are proportional to their T_1 and T_2 transition temperatures, most of these dependences are linear. Therefore, upon extrapolation to the composition of the two homopolymers, the "uncovered" temperature transitions of the copolymers can be used to determine both virtual transition temperatures and the associated thermodynamic parameters of the parent homopolymers. An example of DSC traces for the copolymer MBPE-5/8 (i.e., based on five and eight methylenic units in the spacer) is presented in FIGURE 8.48. The dependences of transition temperatures and enthalpy changes of MBPE-5/8 versus composition for the results collected from FIGURE 8.48 together with their extrapolations are presented in FIGURE 8.49 (Percec and Yourd, 1990g). The virtual phase transition temperatures and thermodynamic parameters of MBPE-X determined from various pairs of flexible spacers agree very well (Percec and Tsuda, 1990d). FIGURE 8.50 presents the plot of the dependence of the virtual isotropic–nematic transition temperatures as a function of the number of carbons (x) in the flexible spacer and its inverse ($1/x$) (Percec and Tsuda, 1990d). We can observe from FIGURE 8.50 that the virtual mesophase of these polymers does not vanish, even at very long spacer length. This suggests that polyethylene should also exhibit a virtual mesophase, as was indeed theoretically predicted (Ronca and Yoon, 1982). The highest-temperature nematic mesophase of MBPE-X homopolymers and copolymers was confirmed by x-ray experiments (Cheng et al., 1991; Ungar et al., 1990). The entire series of ClBPE-X homopolymers behaves almost identically to the series of MBPE-X polymers and confirms all these data (Zuber, 1991).

The thermodynamic parameters associated with the isotropic–nematic phase transition temperatures of MBPE-X (Percec and Tsuda, 1990d) and ClBPE-X were plotted as functions of the number of carbons in the spacer and the orientational contribution of the mesogen, and conformational contribution per $-CH_2-$ group at the isotropic–nematic transitions were determined (Percec and Tsuda, 1990d; Zuber, 1991). These values are summarized in TABLE 8.12 together with the corresponding data obtained for the polyethers based on 4,4′-dihydroxy-α-methylstilbene (HMS-X) (Percec et al., 1991m) and of the polyesters based on 4,4′-dihydroxy-2,2′-dimethylazoxybenzene and α,ω-alkanedioic acids (DMAB-X) (Blumstein and Blumstein, 1988). The polymers HMS-X and DMAB-X contain rigid rod-like mesogens, while MBPE-X and ClBPE-X contain flexible rod-like mesogens. As we can see from TABLE 8.12 both the enthalpic and entropic contributions per $-CH_2-$ are independent of the nature of the mesogen. They are higher for

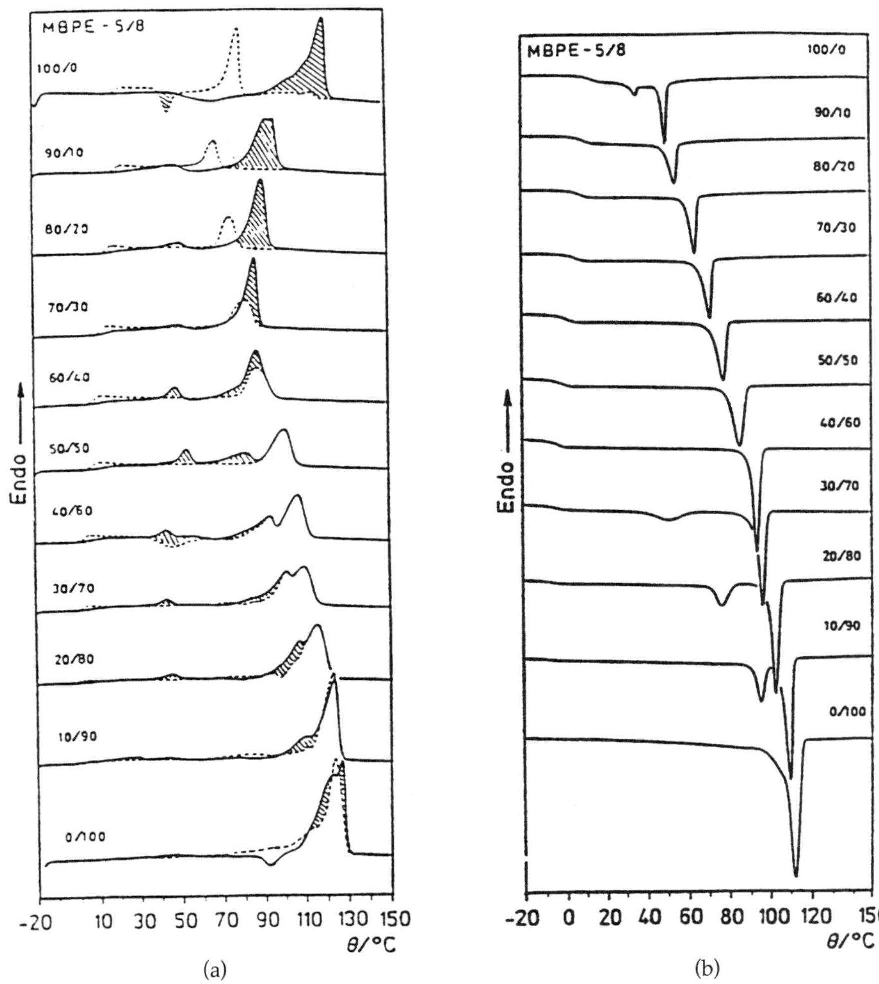

FIGURE 8.48. DSC curves of polyethers based on MBPE and 1,5-dibromopentane (MBPE-5), MBPE, and 1,8-dibromooctane (MBPE-8), and corresponding copolyethers [MBPE-5/8(A/B)]. (a) First (——) and second (- - - -) heating scans; (b) cooling scans.

the polymers based on even spacers and all data are in good agreement. However, the orientational entropic contributions per mesogen are higher in the case of flexible mesogens than in the case of rigid mesogens. This result suggests that at the isotropic–nematic transition, the flexible mesogen undergoes a transition from its *gauche* into its *anti* conformer in addition to the orientational arrangement.

The determination of the virtual phase transition temperatures of polymers also can be done by preparing mixtures with low molar mass liquid crystals, or blends with other virtual liquid crystalline polymers. However, these two techniques are less reliable than the copolymerization-based technique (Percec and Tsuda, 1991q).

It has also been shown that copolymerizations based on two and more than two flexible spacers or mesogenic units can be used to engineer phase transition temperatures and corresponding thermodynamic parameters of copolymers (Percec and Tsuda, 1990b, 1991r). The only requirements for these experiments is that the structural units of these copolymers should be

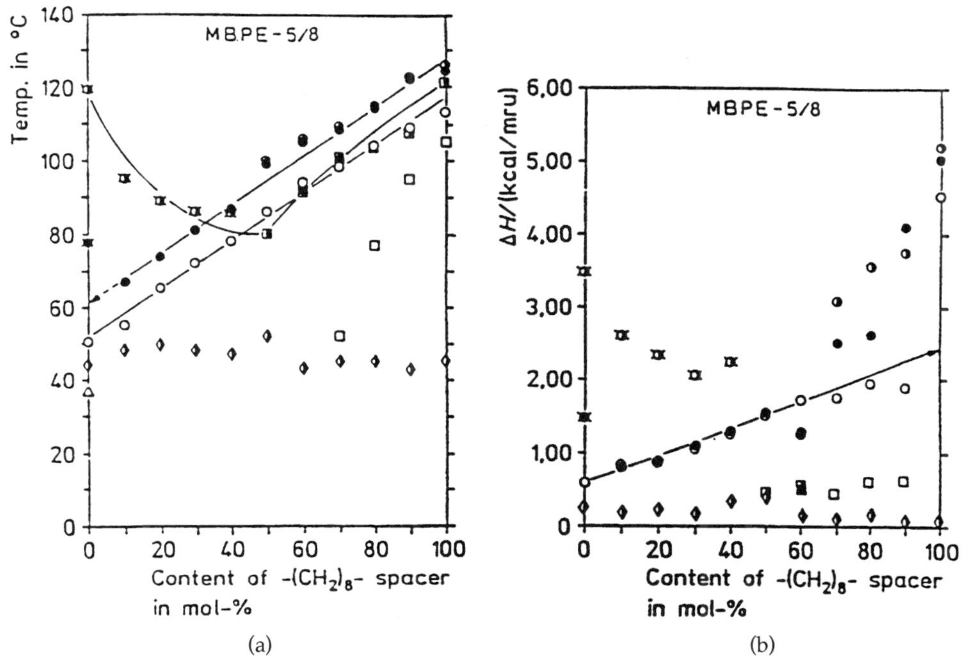

FIGURE 8.49. (a) Thermal transitions of polyethers (MBPE-5 and MBPE-8) and copolyethers [MBPE-5/8(A/B)]. First heating scan [(⌛), T_{ki}; (◑), T_{ni}; (◨), T_{kn}; (◈), T_{kk}]; second heating scan [(⧗), T_{ki}; (●), T_{ni}; (■), T_{kn}]; cooling scan [(⌛), T_{in}; (□), T_{nk}; (△), T_{ns}]. Arrows point to virtual transition temperatures for the homopolymers. (b) Enthalpy changes of polyethers (MBPE-5 and MBPE-8) and copolyethers [MBPE-5/8(A/B)]. First heating scan [(⌛), ΔH_{ki}; (◑), ΔH_{ni}; (◨), ΔH_{kn}; (◈), ΔH_{kk}]; second heating scan [(⧗), ΔH_{ki}; (●), ΔH_{ni}; (■), ΔH_{kn}]; cooling scan [(○), ΔH_{in}; (□), ΔH_{nk}]. Arrows point to enthalpies of virtual transitions for the homopolymers.

isomorphic within the mesophase we want to tailor-make, and the molecular weight of the homopolymers and copolymers should be higher than the molecular weight below which phase transitions are molecular weight dependent. When the thermal transition temperatures and the corresponding enthalpy changes of the homopolymers are not highly dissimilar, both the transition temperatures of the copolymers and their associated enthalpies can be calculated from equations (4) and (5), in which X_n is the mole fraction of the structural unit n from the copolymer and T and ΔH are the transition temperature and the enthalpy change associated with the same phase transition of the parent homopolymer based on an identical structural unit (Percec and Tsuda, 1991r).

MBPE-X and ClBPE-X with X = 5, 7, 9, 11, and 13 exhibit a second uniaxial nematic mesophase, which in certain polymers is monotropic while in others is virtual (Percec and Tsuda, 1990d; Zuber, 1991). This second nematic mesophase was transformed into an enantiotropic one by copolymerization experiments and was subsequently characterized by x-ray experiments (Ungar et al., 1992, 1994b, 1995). This second uniaxial nematic phase was theoretically predicted (Ferrarini et al., 1993; Vasilenko et al., 1984).

The influence of the size of various substituents on the phase transitions of RBPE-X was also investigated (Percec and Zuber, 1992h). It has been demonstrated that phase transition temperatures of polymers depend on the nature of the lateral substituent by following the same trend as that followed by low molar mass liquid crystals. Some data are summarized in TABLE 8.13. A detailed discussion is available (Percec and Zuber, 1992h). When the substituent of RBPE-X polymers is H, that is, BPE, they no longer exhibit a nematic mesophase. For example, BPE-8/10 (i.e., the copolymer containing a 50/50 molar ratio of spacer lengths containing eight and ten carbon atoms) displays an

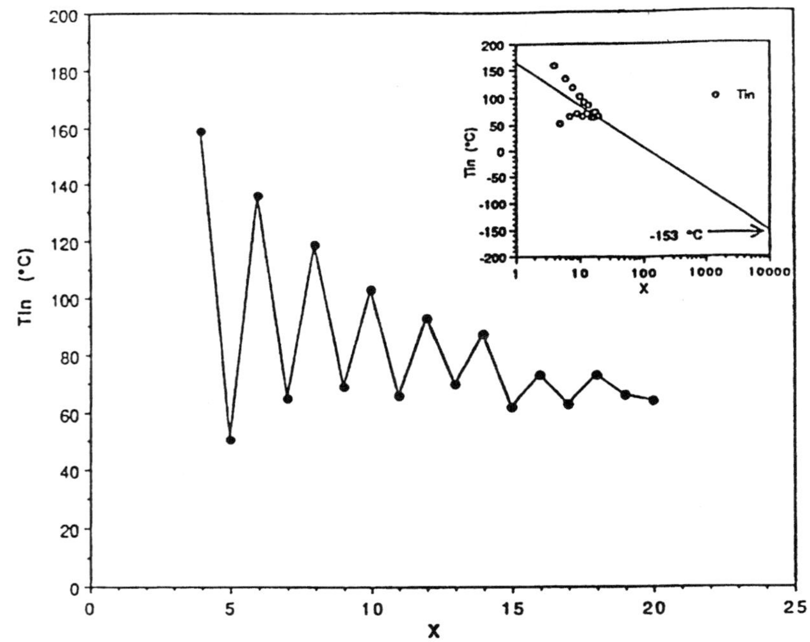

(a)

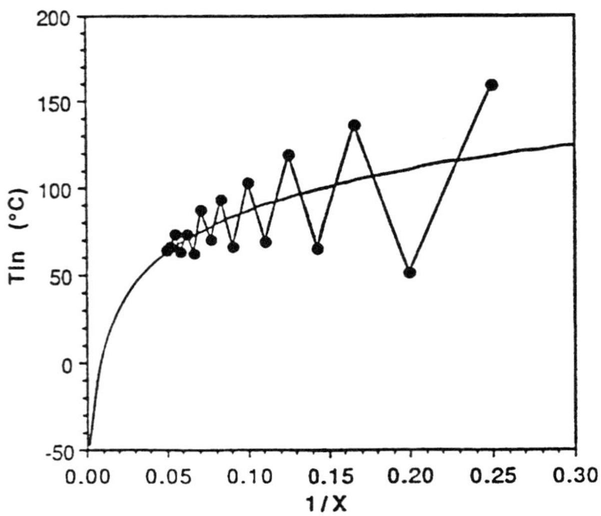

(b)

FIGURE 8.50. (a) Dependence of the virtual isotropic–nematic (T_{in}) phase transition temperature of MBPE-X homopolymers on the number of methylene units in the flexible spacer (X). The graph in the upper right-hand corner presents the logarithmic plot of this dependence. (b) Dependence of the virtual isotropic–nematic (T_{in}) phase transition temperature of MBPE-X homopolymers on the inverse of the number of methylene units in the flexible spacer (1/X). The logarithmic dependence is presented with a continuous line.

enantiotropic s_B mesophase (Ungar et al., 1991b). At the same time, BPE-8/12 (i.e., the copolymer containing a 50/50 molar ratio of spacer lengths containing eight and twelve carbon atoms) exhibits an enantiotropic columnar hexagonal (Φ_h) mesophase (Ungar et al., 1991a). By increasing the length of the mesogenic unit from BPE to 4,4′-*bis*[2-(4-hydroxyphenyl)ethyl]benzene (PEB) (FIGURE 8.51), the resulting homopolyethers

TABLE 8.12. Thermodynamic parameters corresponding to the orientational and conformational order contributions of HMS-X (Perec et al., 1991m), MBPE-X (Perec and Tsuda, 1990d), CIBPE-X (Zuber, 1991), and of polyesters based on 4,4′-dihydroxy-2,2′-dimethylazoxybenzene and α,ω-alkanedioic acids (DMAB-X) (Blumstein and Blumstein, 1988).

	(ΔH_{in}) even (kcal mru^{-1})				(ΔH_{in}) odd (kcal mru^{-1})			
Nature of order	HMS-X	MBPE-X	CIBPE-X	DMAB-X	HMS-X	MBPE-X	CIBPE-X	DMAB-X
Orientational/mole of mesogen	1.45	2.00	1.63	1.12	0.34	0.30	0.20	0.22
Conformational?mole of $-CH_2-$	0.03	0.01	0.03	0.38	0.04	0.03	0.04	0.045
	(ΔH_{in}) even (cal mol^{-1} K^{-1})				(ΔH_{in}) odd (cal mol^{-1} K^{-1})			
Nature of order	HMS-X	MBPE-X	CIBPE-X	DMAB-X	HMS-X	MBPE-X	CIBPE-X	DMAB-X
Orientational/mole of mesogen	2.15	4.02	3.60	1.80	0.45	0.85	0.62	0.34
Conformational?mole of $-CH_2-$	0.15	0.17	0.12	0.21	0.13	0.11	0.07	0.14

TABLE 8.13. Thermal transitions (°C) and corresponding thermodynamic parameters, ΔH_{in} (kcal mru^{-1}) and ΔS_{in} (cal mru^{-1} K^{-1})[a] of 1-(4-methoxyphenyl)-2-(2-R-4-methoxyphenyl) (RBMPE) and 1-(4-hydroxyphenyl)-2-(2-R-hydroxyphenyl)-ethane (RBPE) and of the corresponding RBPE-8 and RBPE-5 polymers. Van der Waal's Radius of R (Å) and electonegativity of R, Ar–R bond length (Å), and breadth of molecule (Å).

R	TMBPE	RBPE	Radius of R	Electronega-tivity of R	Breadth of RBPE	Ar–R bond length	RBPE-8 T_{ni}	RBPE-8 T_{in}	RBPE-8 ΔH_{in}	RBPE-8 ΔS_{in}	RBPE-r T_{in}	RBPE-r ΔH_{in}	RBPE-r ΔS_{in}
H	125–127	200[b]	1.2	2.19	6.69	1.4	155	146	—	—	—	—	—
F	73–74	172–173	1.35	4.0	7.03	1.30	165	152	2.82	6.8	92	0.92	2.5
Cl	62–64	126	1.8	3.0	7.82	1.70	130	118	2.27	5.83	52	0.65	2.0
Br	81–82	114–115	1.95	2.8	8.10	1.85	111	102	2.12	5.3	47	0.59	1.86
CH_3	52–54	148	2.0	2.3	7.76	1.33	132	119	2.33	5.9	38	0.49	1.57
CF_3	69–70	64–65	2.3	3.29	8.15	1.33	76	43	0.93	2.94	—	—	—

[a] Data from Percec and Zuber, 1992H.

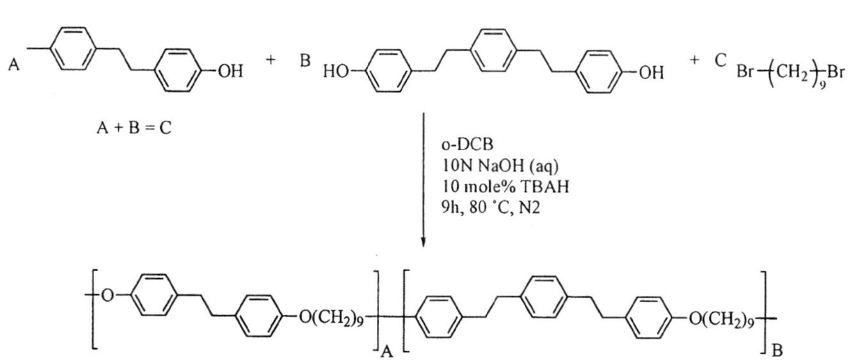

FIGURE 8.51. Synthesis of copolyethers based on 1,2-bis(4-hydroxyphenyl) ethane, 1,4-bis[2-(4-hydroxy-phenyl)ethyl] benzene and 1,9-dibromonane which exhibit a columnar hexagonal mesophase.

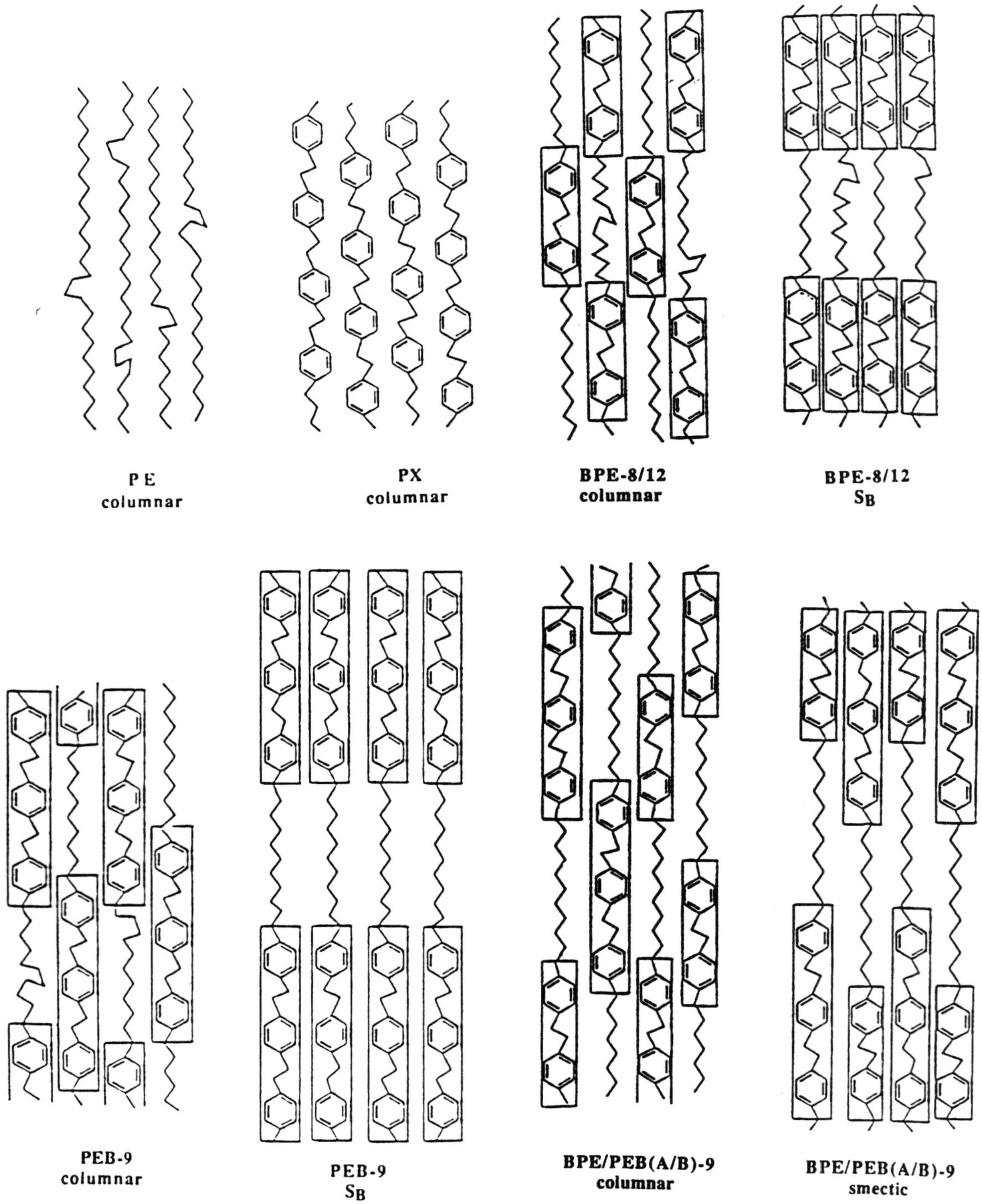

FIGURE 8.52. Schematic representation of columnar and smectic mesophases exhibited by polyethers and copolyethers based on BPE, PEB, and 1,9-dibromononane.

PEB-X and the copolyethers based on PEB, BPE, and various flexible spacers all exhibit columnar mesophases (Percec et al., 1992f, 1992n). Both the phase transition temperatures and the corresponding thermodynamic parameters of the columnar hexagonal mesophases can be tailor-made by similar copolymerization experiments (Percec et al., 1992f, 1992n) as those used to tailor-make nematic mesophases (Percec and Tsuda, 1990b; Ungar et al., 1990). The columnar hexagonal mesophase of BPE and PEB copolymers is not unexpected if we consider them as being "copolymers" of polyethylene and poly(p-xylylene). Since both polyethylene and poly(p-xylylene) display columnar hexagonal mesophases it is expected that RBPE-X with R=H or even F will display the same mesophases (Ungar et al., 1991a). FIGURE 8.52 outlines the columnar hexagonal phases of polyethylene, poly(p-xylylene), and of some of the copolymers based on BPE and PEB (Ungar et al., 1991a). As we can observe from FIGURE 8.52, the columnar hexagonal mesophase is induced by the conformational disorder of the spacer.

So far, with few exceptions, all the experiments with polyethers based on RBPE generate mesophases which are either thermodynamically unstable (virtual) or metastable (monotropic) with respect to the crystalline. Based on the thermodynamic discussion from section 8.5, an increase of the rigidity of the mesogen and a simultaneous increase in its disorder or entropy should decrease the ability for crystallization and increase the ability for the generation of a mesophase. FIGURE 8.53 outlines the development of a series of polyethers based on this concept (Percec and Kawasumi, 1991k). A 4-hydroxyphenyl group from BPE was replaced with a 4-hydroxybiphenyl to increase its rigidity. The ethyl group from BPE was replaced with 1,2-butane group. Thus a new mesogenic group 1-(4-hydroxy-4′-biphenyl)-2-(4-hydroxyphenyl)butane (TPB) results (Percec and Kawasumi, 1991k). TPB has a chiral center and therefore the racemic monomer leads to TPB-X "copolymers" based on four constitutional isomeric structural units (FIGURE 8.53). These polymers are soluble in conventional solvents. The phase transitions of TPB-X (with X=4 to 20) are summarized in FIGURE 8.54 (Percec and Kawasumi, 1991k). The dependences of glass transition, isotropization, and melting transition temperatures versus X are different and as a consequence, they generate polymers exhibiting glassy (TPB-5), noncrystallizable nematic (TPB-4, TPB-6 to TPB-10), crystallizable nematic (TPB-11 to TPB-15), monotropic nematic (TPB-16 and TPB-17) and virtual nematic (TPB-18 to TPB-20) mesophases. This behavior is self-explanatory if we follow the plots from FIGURE 8.54. For example, some polymers are noncrystallizable since their glass transition temperature prohibits their crystallization for kinetic reasons. TPB-5 is only glassy since its glass transition temperature does not allow the formation of the mesophase, again for kinetic reasons. When the melting temperature is higher than the glass transition, these polymers can crystallize, while when melting is at a higher temperature than isotropization, they exhibit a virtual nematic mesophase only.

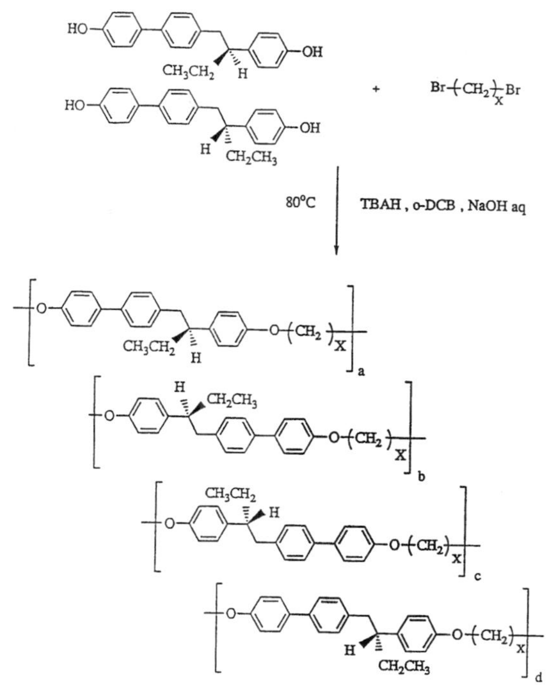

FIGURE 8.53. Liquid crystalline copolyethers based on a construed mesogenic unit containing rigid and flexible groups: 1-(4-Hydroxy-4′-biphenyl)-2-(4-hydroxyphenyl)butane.

TPB-10 was synthesized with a perdeuterated spacer and the conformational and orientational order in the nematic phase was determined by DNMR measurements and conformational calculations (Hardouin et al., 1995; Sherwood et al., 1994). Since this polymer does not undergo redistribution reactions as the polyesters do, its behavior is as

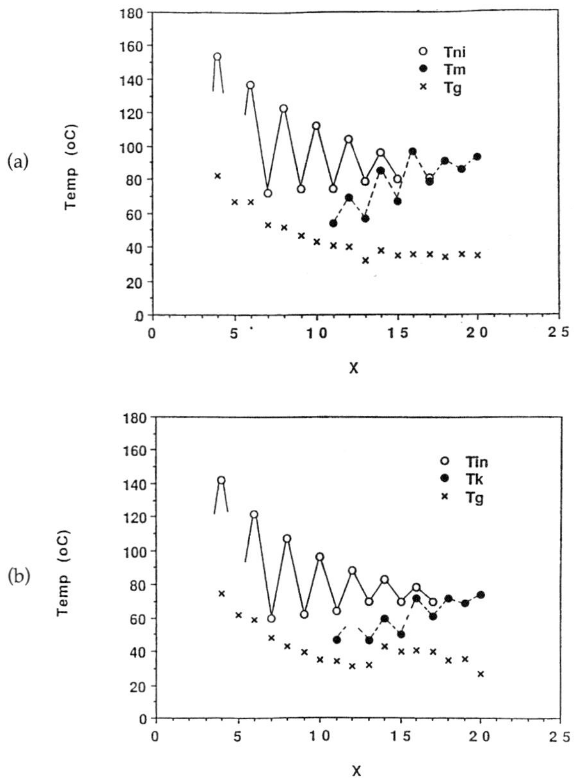

FIGURE 8.54. Dependence of phase transition temperatures of TPB-X polyethers on the number of methylenic units in the flexible spacer (X): (a) data from the second heating scan (T_g data of TPB-11 to TPB-20 are from the first heating scan); (b) data from the first cooling scan.

predicted by theory (Yoon and Flory, 1989; Hardouin et al., 1995). Beginning at the isotropic–nematic transition at 104 °C, the polymer chains in the nematic phase adopt nearly fully extended (alternate *trans*) conformations and a rather high value of the orientational order parameter around 0.85 is attained. As the temperature is lowered to 96 °C, the isotropic peak disappears completely, indicating the absence of any isotropic phase and hairpin conformations. With increasing supercooling, the conformational order shows little change, but the orientational order parameter increases continuously to around 0.95 at 80 °C. These results are therefore in good agreement with the predictions of a recent theory on the nematic phase of a polymer comprising rigid and flexible sequences in alternating succession (Yoon and Flory, 1989). When the 1,2-butane group from TPB is replaced by 1,2-propane, the new mesogen 1-(4-hydroxy-4′-biphenyl)-2-(4-hydroxyphenyl)propane (TPP) has a more cylindrical shape. This new mesogen therefore resembles RBPE with R = H. The polyethers TPP-X with X containing an even number of methylenic units display highly ordered smectic phases while those with an odd number of methylenic units exhibit nematic mesophases (Percec et al., 1994d). A detailed discussion on the molecular engineering of main chain liquid crystalline polyethers exhibiting nematic, smectic, and columnar hexagonal mesophases as well as of their structures is available (Ungar et al., 1995).

8.9 HYPERBRANCHED AND DENDRIMERIC LIQUID CRYSTALLINE POLYMERS

Hyperbranched and dendrimeric macromolecules are a novel class of branched polymers which contain a branching point in each structural unit. Review articles on this topic are available (Hawker et al., 1994; Isberner et al., 1994; Tomalia et al., 1990; Tomalia and Durst, 1993; Tomalia, 1994). Hyperbranched polymers are obtained via step polymerization reactions of AB_2 monomers and therefore exhibit a broad molecular weight distribution. Dendrimers are prepared by using a protection deprotection methodology in a divergent or convergent approach. This strategy yields monodisperse dendrimers.

The first examples of thermotropic hyperbranched polymers with disk-like mesogens (Percec et al., 1992i) and rod-like flexible mesogens based on conformational isomerism (Percec and Kawasumi, 1992l; Percec et al., 1994g) were recently reported. The synthesis of hyperbranched polyethers containing cyclotetraveratrylene units in the branching points was accomplished by the electrophilic polymerization of a 3,4-*bis*(n-alkyloxy)benzyl alcohol with an α,ω-*bis*{[2-(alkoxy)-5-(hydroxymethyl)phenyl]oxy}-alkane (FIGURE 8.55) (Percec et al., 1992i). FIGURE 8.56 outlines the preparation of TPD-b hyperbranched polymers, while their phase behavior is summarized in TABLE 8.14 (Percec and Kawasumi, 1992l).

As observed from the results presented in TABLE 8.14, the phase behavior of these hyperbranched liquid crystalline polymers is strongly dependent on the

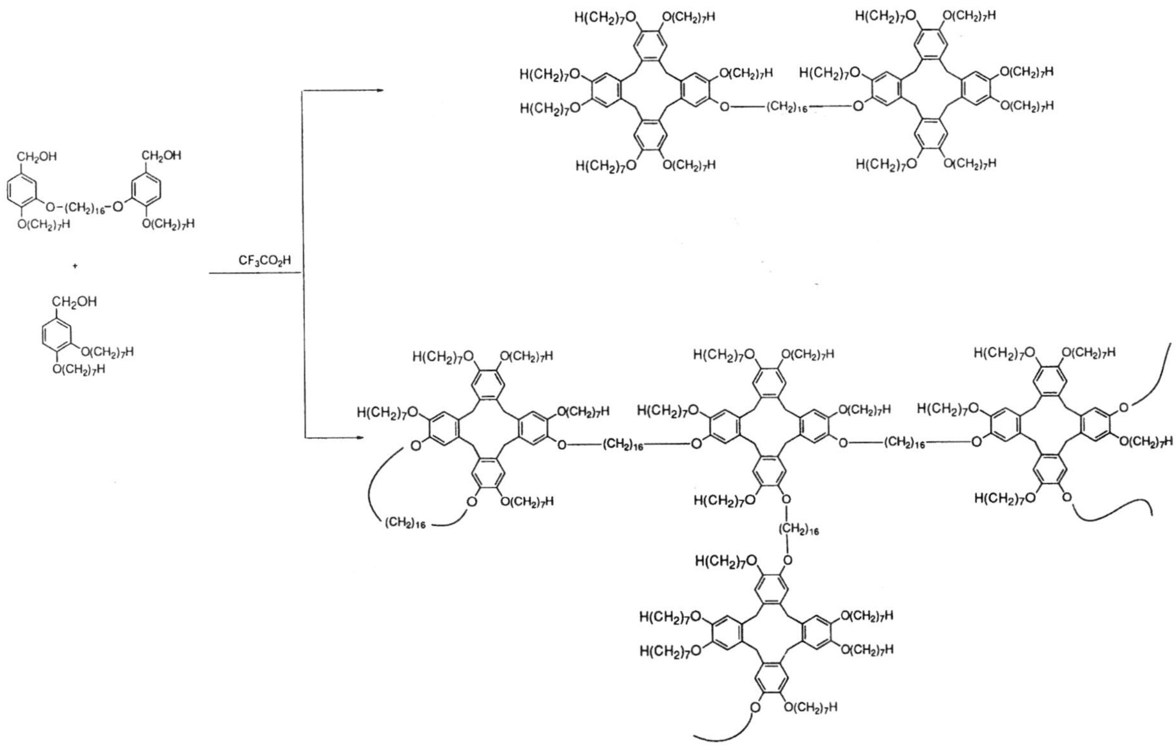

FIGURE 8.55. Synthesis of columnar hexagonal hyperbranched liquid crystalline polyethers containing cyclotetraueratrylene disc-like mesogens.

nature of their chain ends (Percec and Kawasumi, 1992l). Much broader thermotropic mesophases were obtained by using the synthetic procedure from FIGURE 8.57. Of particular interest are the hyperbranched polymers TPT-b-8 since they display a very broad nematic mesophase (TABLE 8.15). FIGURES 8.58–8.60 present the molecular design of four generations of monodisperse dendrons. The structure of the dendrimer based on these dendrons of the fourth generation is outlined in FIGURE 8.61 (Percec, 1995b; Percec and Chu, 1995c). The phase behavior of these dendrons is outlined in TABLE 8.16. An entire new field of liquid crystalline polymers is emerging from hyperbranched and dendrimeric structures. Cholesteric hyperbranched liquid crystals were prepared by attaching chiral tails to hyperbranched nematic polymers (Bauer et al., 1993).

Independent and simultaneous with this work, an aromatic polyamide lyotropic dendrimer was obtained by the polycondensation of 5-amino-isophthaloyl chloride. However, the dendrimer obtained from the polymerization of 3,5-diaminobenzoyl chloride displays only a lyotropic phase (Kim, 1992a, b).

8.10 CYCLIC MAIN CHAIN LIQUID CRYSTALLINE OLIGO- AND POLYETHERS

The membranes of archaebacteria, that is, extremophile microorganisms which proliferate in extreme environments such as high temperature, acid, alkali, or high salt concentration, are formed from macrocyclic oligoether lipids, since these lipids are more stable than the corresponding linear ones (Yamauchi and Kinoshita, 1993). In 1992 it was demonstrated that macrocyclic lipids display a thermotropic mesophase (Menger et al., 1992, 1993), and more recently that they generate highly stable liposomes (Ladika et al., 1994).

Independently, in 1992 it was predicted and demonstrated that certain sizes of low molar mass

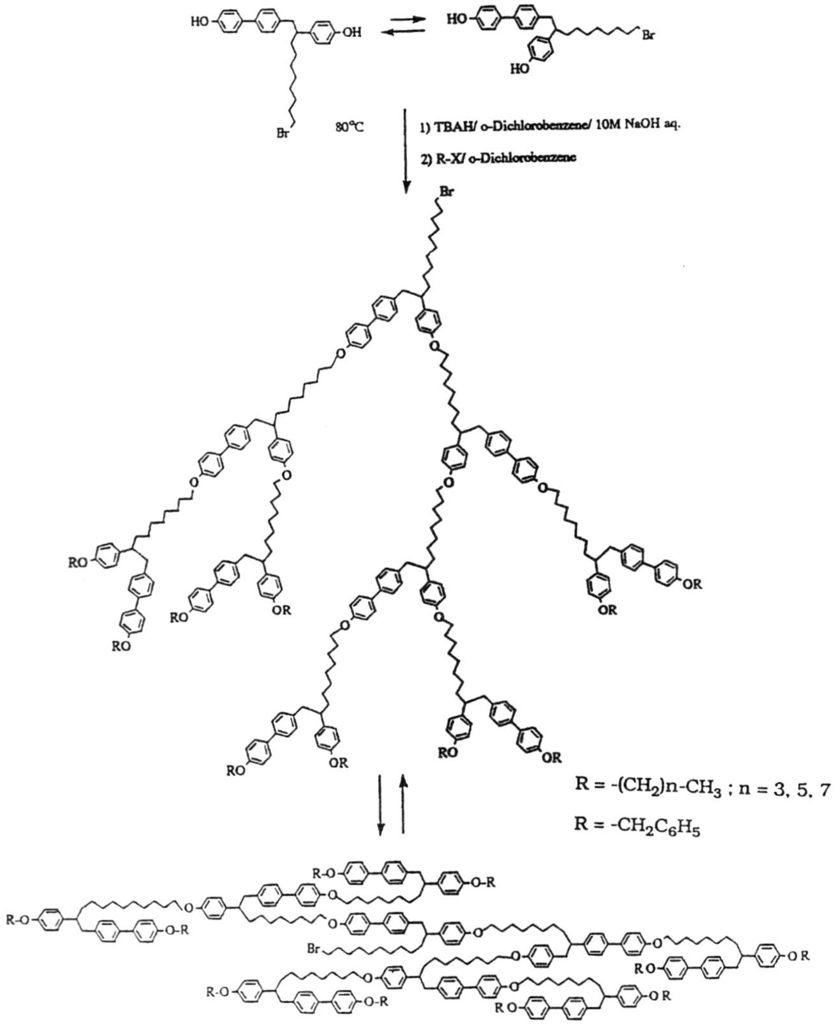

FIGURE 8.56. Hyperbranched liquid crystalline polyethers which display calamitic mesophases.

TABLE 8.14. Characterization of hyperbranched polyethers based on TPD-b with different tail lengths R. Data collected from second heating and first cooling DSC scans (Percec and Kawasumi, 1992l).

Tail R	Yield (%)	M_{nGPC}	M_w/M_{nGPC}	Thermal transitions (°C) and corresponding enthalpy changes (Kcal mru^{-1}) in parentheses: Heating	Cooling
Octyl	80.0	7900	2.6	g 20 n 40 (0.75) i	i 30 (70.0) n 15 g
Hexyl	82.2	4700	3.9	g 24 n 39 (0.15) i	i 30 (0.18) n 18 g
Butyl	70.3	4400	3.5	g 40 i	1 34 g
Benzyl	79.2	4700	3.8	g 48 n 60 (0.10) i	i 53 (0.15) n 43 g

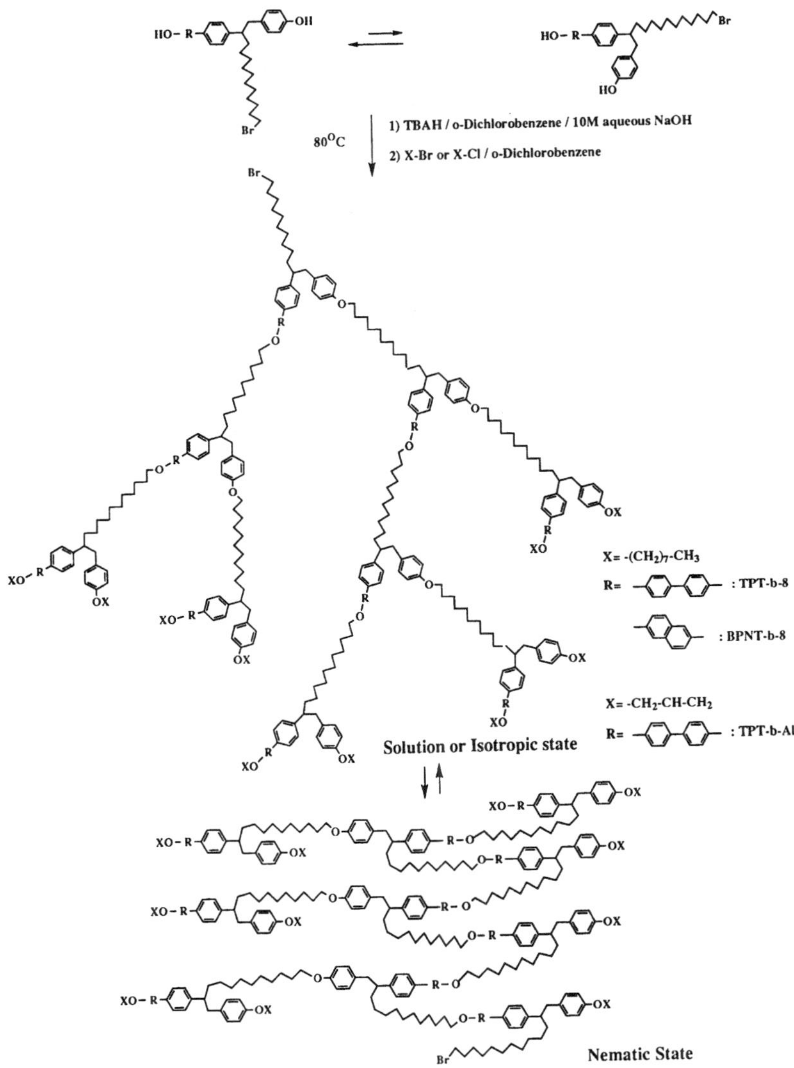

FIGURE 8.57. Hyperbranched liquid crystalline polyethers.

macrocyclics which are homologs of main chain liquid crystalline polymers based on conformationally flexible rod-like mesogens should and do exhibit a higher ability to generate liquid crystalline phases than do their low molar mass linear and both their linear and cyclic high molecular mass compounds (FIGURE 8.62) (Percec et al., 1992m). The trend from FIGURE 8.62 is due to the fact that macrocyclics of a certain size that contain at least a minimum extent of ring strain and a proper combination of spacer and mesogenic lengths, based on which they can generate a liquid crystalline phase, should have a lower entropy and therefore, must display higher transition temperatures (Percec and Keller, 1990e) than the corresponding high molar mass linear and cyclic compounds. FIGURE 8.19 outlines the *anti* and *gauche* conformers of 1-(4-hydroxybiphenyl-4-yl)-2-(4-hydroxyphenyl)butane (TPB) and the macrocyclization of TPB with α,ω-dibromoalkanes by a high dilution phase transfer catalyzed polyetherification. Optimum reaction conditions yield either the regioirregular linear polymer TPB-(l)X(8), or a mixture of up to five sizes of regioirregular

TABLE 8.15. Synthesis and characterization of hyperbranched polymers and comparison of their phase behaviors with that of the linear model polyether (TDP-8)[a].

				Thermal transitions (°C) and corresponding enthalpy changes (kj mru^{-1}) in parentheses	
Polymer	Yield (%)	M_{nGPC}	M_w/M_{nGPC}	Heating	Cooling
TPD-b-8[b]	80.0	7900	2.60	g 19 n 37 (2.41) i	i 25 (−2.17) n 10 g
TPH-b-Bz[c]	94.1	2520	1.69	g 47 i	i 39 5
TPH-b-4	71.9	3140	1.51	g 42 b 48 (0.12) i	i 33 g
TPH-b-6	65.1	3170	1.57	g 29 n 40 (0.20) i	i 33 (−0.08) n 22 g
TPH-b-8	75.7	3220	1.58	g 24 n 41 (1.85) i	i 30(−1.82) n 15 g
BPNT-b-8[d]	71.5	8640	2.49	g 27 n 59 (1.89) i	i 49 (−2.51) n 18 g
TPT-b-8[e]	79.9	11800	2.42	g 50 n 132 (2.96) i	i 125 (−2.88) n 38 g
TPT-b-All	81.4	8333	1.90	k 126 (7.59) n 155 (3.46)	i 149 (−2.90) n 166 (−6.98) k
TPD-8[f]	92.2	27500	2.00	g 19 n 60 (6.61) i	i 47 (−6.61) n 13 g

[a] Yields are after precipitation in methanol. Molecular weights and thermal transitions are after precipitation in acetone. All thermal data are collected from second heating and first cooling DSC scans.
[b] TPD-b-X is the short name for the polymer derived from 10-bromo-1-(4-hydroxy-4′-biphenylyl)-2-(4-hydroxyphenyl)decane (see Percec et al., 1994g).
[c] TPH-b-X is the polymer derived from 8.
[d] BPNT-b-X is the polymer derived from 22.
[e] TPT-b-X is the polymer derived from 30.
[f] TPD-8 is the linear polymer from 1-(4-hydroxy-4′-biphenylyl)-2-(4-hydroxyphenyl)decane and 1,8-dibromooctane (see Percec et al., 1994g).

macrocyclics TPB-(c)X(z) which can be separated (where l stands for linear, X represents the number of methylenic units in the spacer, c stands for cyclic and z is the degree of oligomerization of the macrocyclic compound). Both the regioirregular TPB-(l)X and TPB-(c)X(z) with X = 2 to 20 and z = 1 to 5 were synthesized and their phase behavior was investigated. Questions such as the following were addressed, answered, and explained: What is the shortest spacer X and the minimum ring size z of TPB-(c)X(z) that displays a liquid crystalline phase? What is the lowest size of macrocyclic that displays a liquid crystalline phase whose isotropization temperature is higher than that of its linear high molar mass homolog (Percec et al., 1992m; Percec and Kawasumi, 1993i, 1994j)? What is the probability of transforming kinetically prohibited (Percec and Kawasumi, 1992b, 1993a, i, j, 1994j) and/or kinetically controlled (Percec and Kawasumi, 1993j) mesophases of the linear polymer into enantiotropic mesophases via cyclization? What is the possibility of generating noncrystallizable macrocyclics with a high glass transition temperature and a broad range of temperature of the mesophase (Percec and Kawasumi, 1993a)? What is the dependence of the phase transition temperatures of a certain macrocyclic on its spacer length X and how does it compare with that of the corresponding linear polymer (Percec and Kawasumi, 1993i)?

FIGURE 8.63 outlines the synthesis of macrocyclic and linear polyethers based on TPB and α,ω-dibromoalkanes. The structures of these macrocyclics were confirmed by a combination of 300 MHz 1D and 2D ^{1}H-NMR experiments and by molecular weight determination (Percec et al., 1992m). ^{1}HNMR spectroscopy was used to demonstrate that ring strain is released above the size of pentamer. The phase behavior of individual regioirregular linear monomer, dimer, trimer, tetramer TPB-(l)10 (FIGURE 8.64), and that of the high molecular weight linear polymers with narrow molecular weight distribution TPB-(l)10 is compared with that of the TPB-c(10)z (z = 1 to 5) in FIGURE 8.65 (Percec and Kawasumi, 1993k). Both the linear and cyclic compounds from FIGURE 8.65 exhibit a nematic mesophase. The linear dimer has a higher isotropization temperature than the cyclic one. However, the cyclic trimer displays a much higher ability to form a nematic mesophase than the corresponding linear trimer. At the same time, the cyclic tetramer and pentamer exhibit much higher abilities to form nematic mesophases than even the very high molecular weight linear polymer. *These results demonstrate that at least in the case of the macrocyclic compounds based on*

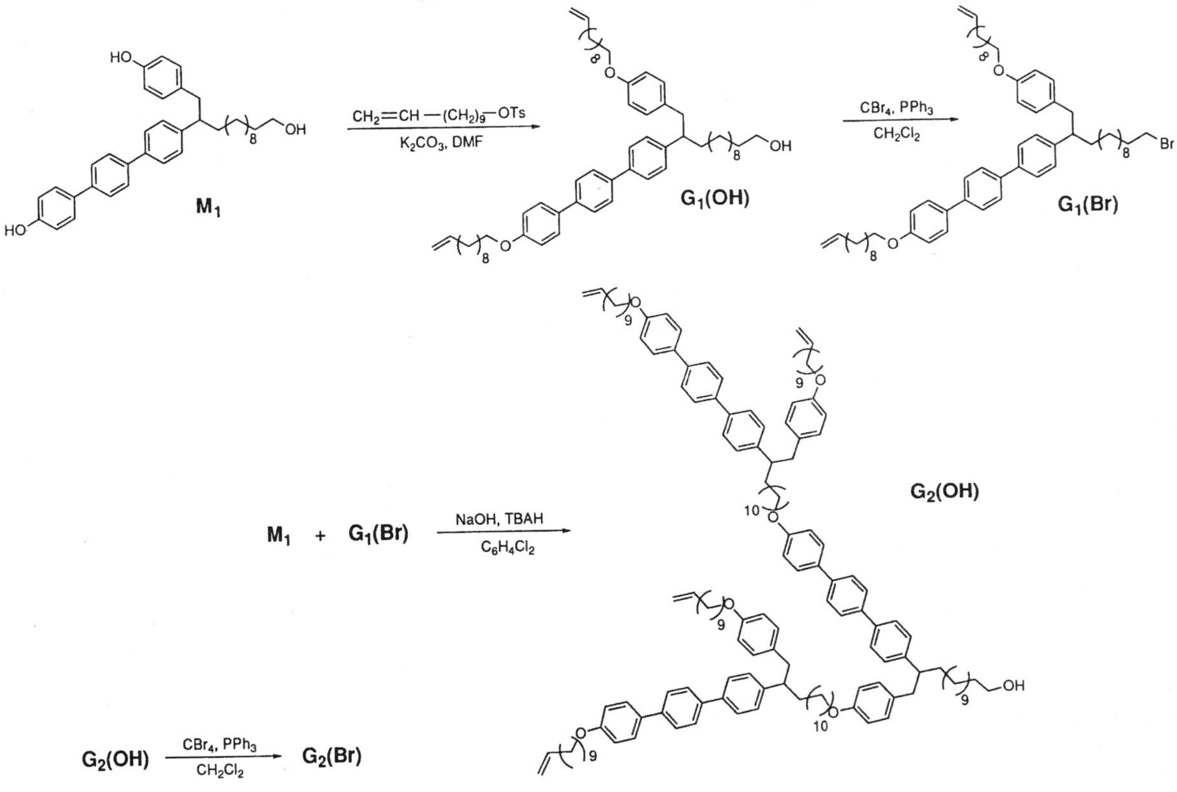

FIGURE 8.58. Synthesis of liquid crystalline monodendrons G_1(OH) to G_2(OH).

TPB and 1,10-dibromodecane, the tetramer and pentamer and not the linear high molar mass polymers exhibit the highest tendency to form nematic mesophases.

FIGURE 8.66(a) compares the isotropization temperatures (T_{ni}) of TPB-(c)10(4) and of TPB-(l)10 as functions of the number of methylenic units in the spacer. All TPB-(c)10(4) have higher T_{ni} values than the corresponding linear polymers. The ΔH_{ni} values of T_{ni} are plotted in FIGURE 8.66(b). ΔH_{ni} values of cyclics are lower than those of the corresponding linear polymers since the difference between the entropies of a macrocyclic in the isotropic and nematic phases is lower than the corresponding parameter of the linear polymer. The odd–even dependences of T_{ni} and ΔH_{ni} of cyclic tetramers and linear compounds are similar.

FIGURE 8.67(a) plots the thermal transitions of TPB-(c)10(3) and FIGURE 8.67(b) shows their ΔH_{ni} values. Both sets of data for TPB-(c)10(3) (FIGURE 8.67) are different from those of TPB-(c)10(4) (FIGURE 8.66). T_{ni} values of cyclic trimers show a very weak odd–even dependence on spacer length. However, this dependence (FIGURE 8.67(a)) follows the reversed trend of the data from FIGURE 8.66(a). T_{ni} values of cyclic trimers increase with an increase in spacer length, while the same data for cyclic tetramers decrease. The cyclic trimers show a higher T_{ni} value for odd spacer length while cyclic tetramers and linear polymers do so for even spacer length. Both the T_{ni} and ΔH_{ni} values from FIGURE 8.67 demonstrate that the rigidity of cyclic trimers increases with an increase in spacer length. Therefore, while cyclic tetramers resemble the behavior of linear polymers, cyclic trimers represent a completely different class of liquid crystal. Finally, cyclic trimers with very short and very long spacers exhibit, in addition to the nematic mesophase, a smectic A (S_A) phase (FIGURE 8.67(a)). Regardless of the spacer length, the layer thickness of the S_A phase for the cyclic trimers is equal to $1.5 \times (L_m + L_s)$, where L_m is the length of the *anti* conformer of TPB and L_s is

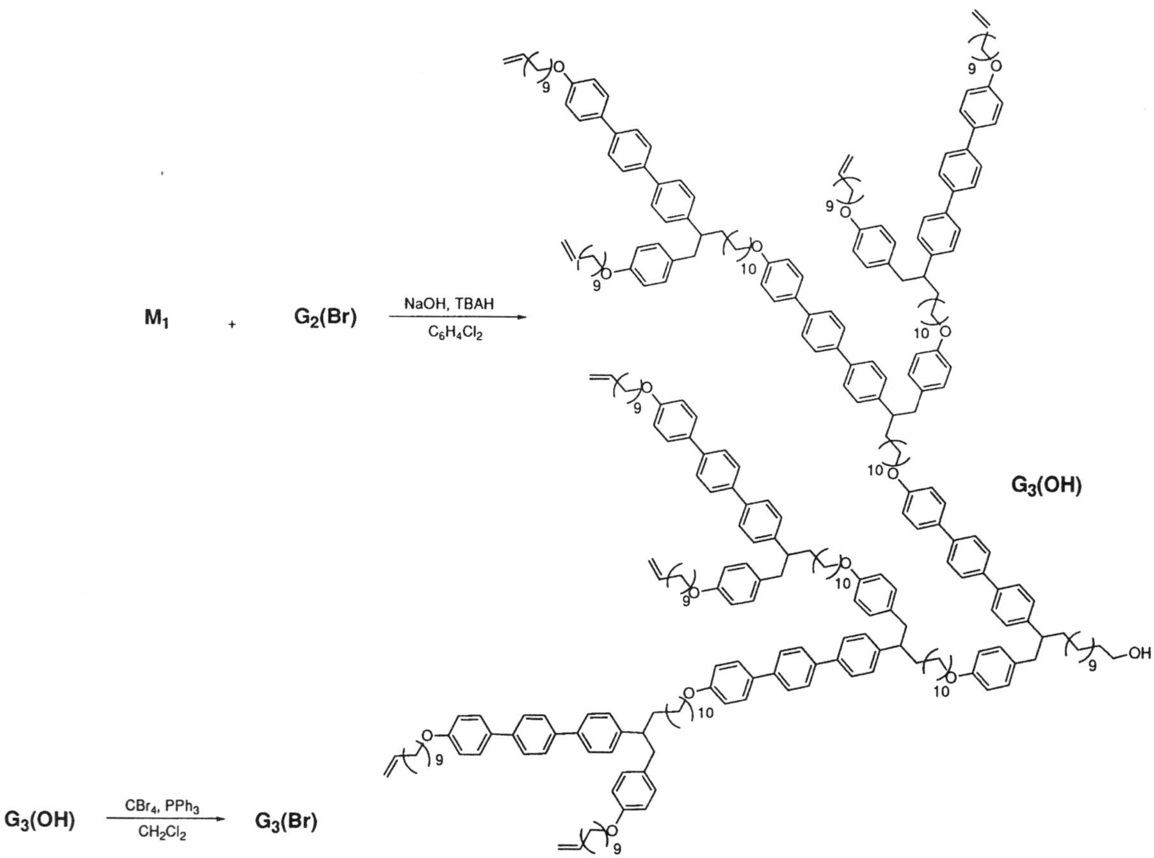

FIGURE 8.59. Synthesis of liquid crystalline monodendron G_3(Br).

equal to the length of the flexible spacer in its fully extended all *trans* conformation (FIGURE 8.68). A trimer conformation like the middle of FIGURE 8.68 requires that $2L_m + L_s = 2 \times (L_s - 2.5) + L_m$, where 2.5 Å is the minimum length of the 180 °C turn-in fold which is generated by two carbon atoms. Under these requirements, the shortest spacer length (L_s) which can accommodate such a structure is equal to or longer than L_m, that is, $L_s \geq L_m$. At shorter spacer lengths, an S_A phase can be found only when one TPB is in its *gauche* conformation and it is part of the fold (middle trimer in FIGURE 8.68) (Percec and Kawasumi, 1993b). Some experiments with macrocyclics based on enantiomerically enriched TPB suggest that intracyclic heterochiral recognition may contribute to the architecture of the larger macrocyclics in the nematic phase (Percec and Kawasumi, 1993l) and that a superhelical shape is not excluded. In addition, the nematic phase of TPB-(c)9(3) seems to be biaxial (Li et al., 1993, 1994).

All these results suggest that in their liquid crystalline phase, these compounds are *supramolecular rigid-rods generated from collapsed macrocyclics* (FIGURE 8.69). This novel class of liquid crystals has recently received interest in other laboratories (Ashton et al., 1994) and opens numerous avenues for the design of new macromolecular and supramolecular architectures based on liquid crystalline macrocyclic building blocks.

8.11 SIDE CHAIN LIQUID CRYSTALLINE POLYMERS

8.11.1 General Considerations

The field of side chain liquid crystalline polymers was recently reviewed (McArdle, 1989). Therefore, we will

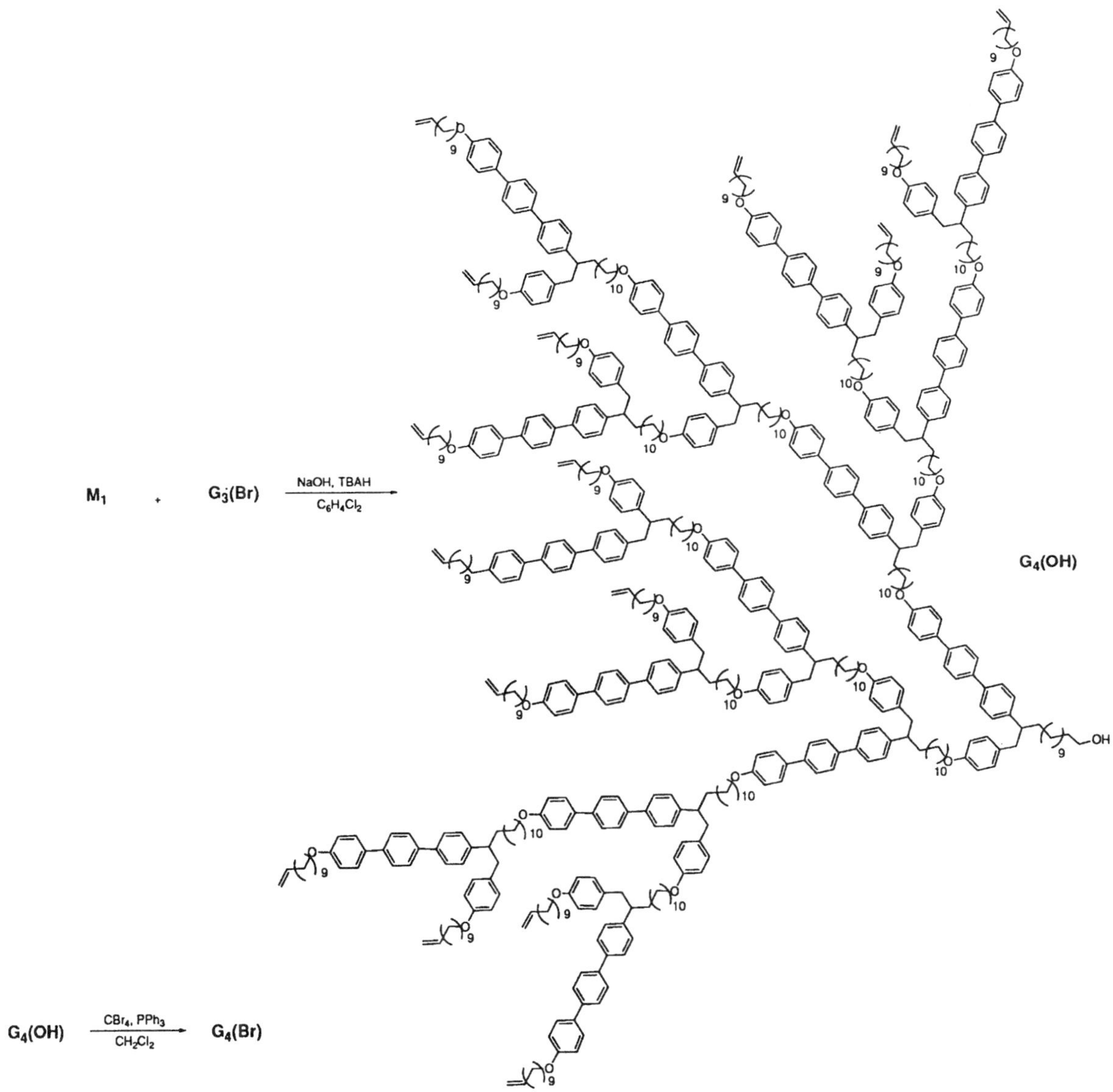

FIGURE 8.60. Synthesis of monodendrons $G_4(OH)$ and $G_4(Br)$.

discuss only recent progress made on the molecular engineering of side chain liquid crystalline polymers. Most of the present discussion will be made on side chain liquid crystalline polymers with mesogenic groups normally attached to the polymeric backbone (Percec and Pugh, 1989a). FIGURE 8.70 outlines the concept of side chain liquid crystalline polymers. It has been theoretically predicted (Warner, 1989, 1993) that the conformation of the polymer backbone should be distorted in the liquid crystalline phase. Both small-angle neutron scattering (SANS) experiments (Davidson et al., 1991; Hardouin et al., 1991; Noël, 1988, 1989; Noirez et al., 1988, 1995; Pepy et al., 1988) and x-ray scattering experiments (Davidson and Levelut, 1992; Kuschel et al., 1990; Mattoussi et al., 1986; Percec et al., 1990c) have shown that the

FIGURE 8.61. Structure of the liquid crystalline dendrimer G_4(D3).

statistical random-coil conformation of the polymer backbone is slightly distorted in the nematic phase and highly distorted in the smectic phase.

Let us now consider very briefly the influence of various parameters (i.e., the nature of the flexible spacer and its length, nature, and flexibility of the polymer backbone and its degree of polymerization) on the phase behavior of a side chain liquid crystalline polymer. According to the thermodynamic schemes from section 8.5, the increase of the degree of polymerization decreases the entropy of the system and therefore, if the monomeric structural unit exhibits a virtual or monotropic mesophase, the resulting polymer should probably exhibit a monotropic or enantiotropic mesophase. Alternatively, if

TABLE 8.16. Phase transitions of G_1(X) to G_4(X) from schemes 34 to 26 (k = crystalline, i = isotropic, n = nematic, g = glassy, X = unknown).

G_1(Br)		k 64 i	i 51 n 40 k
G_2(OH)		X 62 n 92 i	i 85 n 47 X
G_3(OH)		X 62 n 105 i	i 98 n 57 X
G_3(Br)	g 52	X 72 n 104 i	i 97 n 64 X
G_4(OH)	g 62	X 74 n 110 i	i 103 n 59 X

FIGURE 8.62. The theoretical and experimental dependence of the isotropization temperature (Ti) of cyclic and linear main chain liquid crystalline polymers on their degree of polymerization (DP). Both T_i and DP are in arbitrary units.

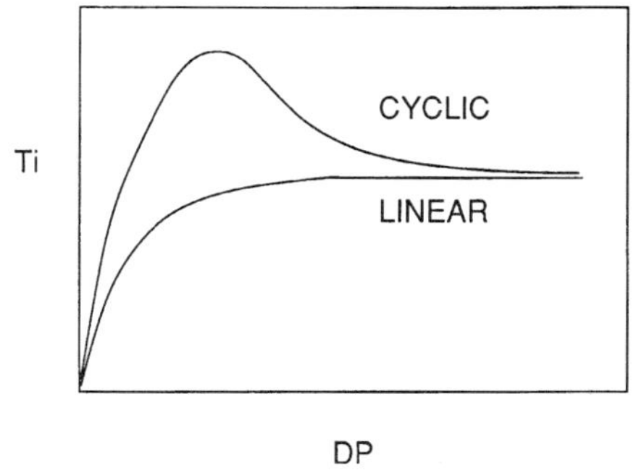

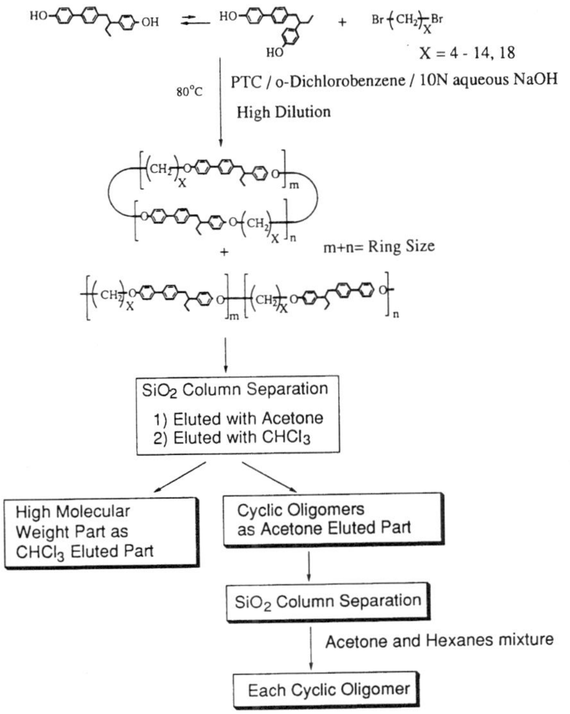

FIGURE 8.63. Conformational isomerism of TPB and its regioirregular cyclization with 1,10-dibromodecane under high-dilution conditions.

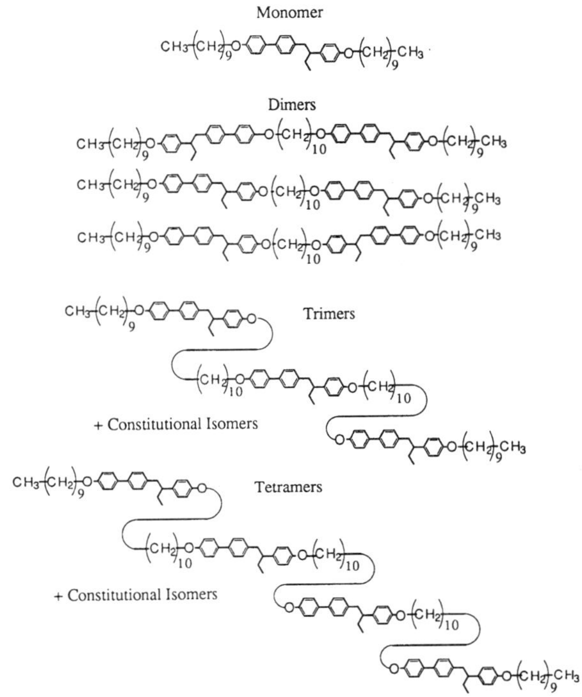

FIGURE 8.64. The structures of the linear regioirregular model compounds of TPB-(l)10.

the monomeric structural unit displays an enantiotropic mesophase, the polymer should display an enantiotropic mesophase over a broader range of temperature. It is also possible that the structural unit of the polymer exhibits more than one virtual mesophase and therefore, at high molecular weights, the polymer will increase the number of its mesophases. All these effects were observed with various systems (Percec and Pugh, 1989a; Percec and Tomazos, 1992a).

The length of the flexible spacer determines the nature of the mesophase. Long spacers favor smectic phases while short spacers favor nematic phases (Craig and Imrie, 1994, 1995; Percec and Tomazos, 1992a). This effect is similar to that observed in low molar mass liquid crystals.

At constant molecular weight, the rigidity of the polymer backbone determines the thermodynamic stability of the mesophase. According to the thermodynamic schemes described in section 8.5, the isotropization temperature of the polymer with the more rigid backbone should be higher. However, experimentally, this situation is reversed. The highest isotropization transition temperature is observed for polymers with more flexible backbones. This conclusion is based on systematic investigations performed with two mesogenic groups which are constitutional isomers, that is, 4-methoxy-4′-hydroxy-α-methylstilbene (4-MHMS) (Percec and Tomazos, 1989b) and 4-hydroxy-4′-methoxy-α-methylstilbene (4′-MHMS) (Percec and Tomazos, 1989h) and polymethacrylate, polyacrylate, polysiloxane, and polyphosphazene backbones (Percec and Tomazos, 1989b, h; Percec et al., 1989m) (FIGURE 8.71). Recently, a large amount of data for the literature on side chain liquid crystalline polymers with different backbones and 4-cyanobiphenyl-based mesogens were compiled and supported the same trend (Craig and Imrie, 1995). This dependence can be explained by assuming that

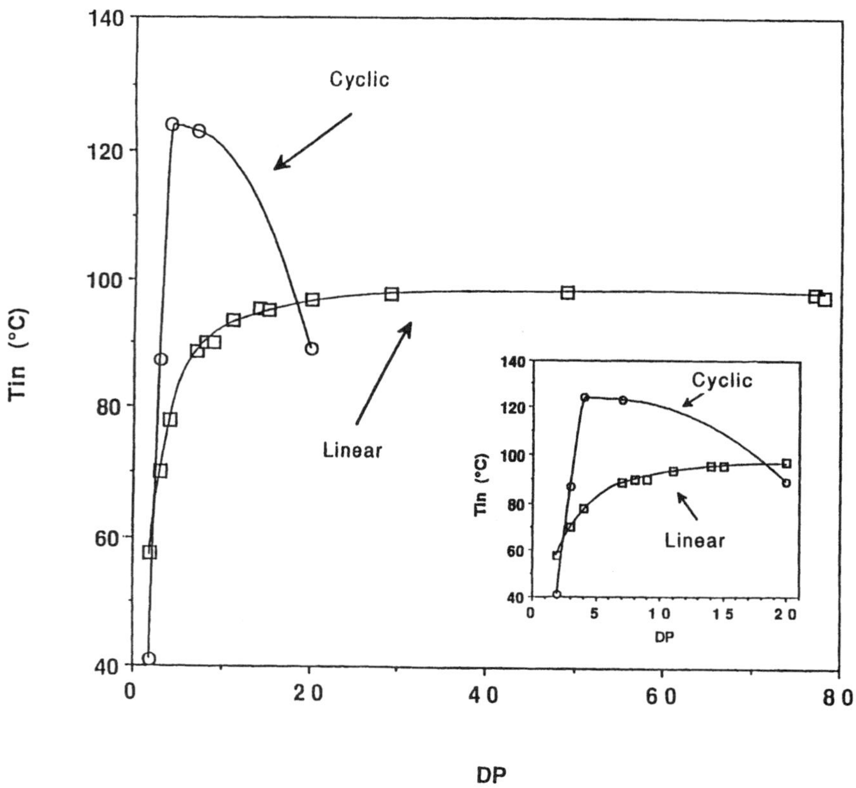

FIGURE 8.65. Dependence of the isotropic–nematic (T_{in}) transition temperatures of regioirregular linear (□) and macrocyclic (○) oligomers and polymers of TPB-(l)10 on their degree of polymerization.

a more flexible backbone uses less energy to be distorted and therefore generates a more decoupled polymer system. In fact, the more flexible backbones do not generate only higher isotropization temperatures, but also a higher ability towards crystallization. However, contrary to all expectations, the entropy change of isotropization is higher for those polymers which are based on more rigid backbones and therefore exhibit lower isotropization temperatures (FIGURE 8.72(a), (b)) (Percec and Tomazos, 1990i). This contradiction between the values of the entropy change and the isotropization temperatures can be accounted for by a different mechanism of distortion for different polymer backbones as outlined in FIGURE 8.73, that is, while a rigid backbone gets more extended and therefore, in the smectic phase can cross the smectic layer, in the case of a flexible backbone it gets squeezed between the smectic layers. The higher configurational entropy of the flexible backbone versus that of the rigid backbone in the smectic phase can account for the difference between the entropy change of isotropization from FIGURE 8.72. At shorter spacer lengths, there is not much difference between the contribution of various backbone flexibilities since probably in order to generate a mesophase, they should be extended. Therefore, the entropy change of isotropization is less dependent on backbone flexibility (FIGURE 8.72).

Based on this discussion, it is quite obvious that copolymers containing structural units with and without mesogenic groups and flexible backbones display a microphase-separated morphology in their smectic phase (FIGURE 8.74) (Kuschel et al., 1990; Percec and Pugh, 1989a; Percec et al., 1990c). Therefore, the highest degree of decoupling is expected for copolymers containing mesogenic and nonmesogenic structural units and highly flexible backbones, that is, microphase-separated systems. In this last case, when the monomeric structural unit of the polymer exhibits a virtual mesophase, the high molecular weight polymer

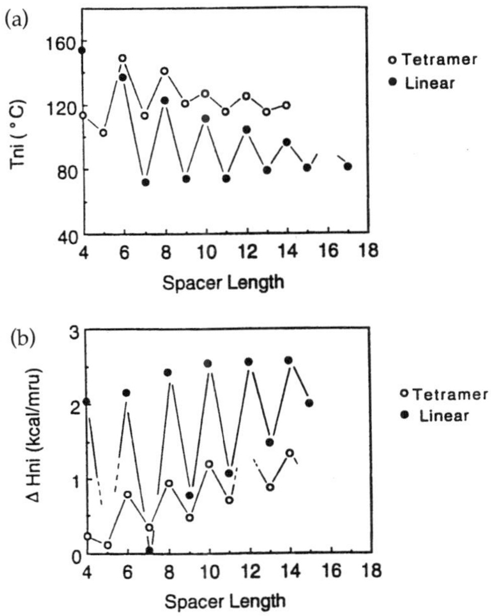

FIGURE 8.66. The dependence of the isotropic–nematic transition temperature (T_{ni}) (a) and of its associated enthalpy changes (ΔH_{ni}) (b) of regioirregular TPB-(c)X(4) (cyclic tetramer) and TPB-(l)X (high molecular weight linear polymer) on spacer length (X).

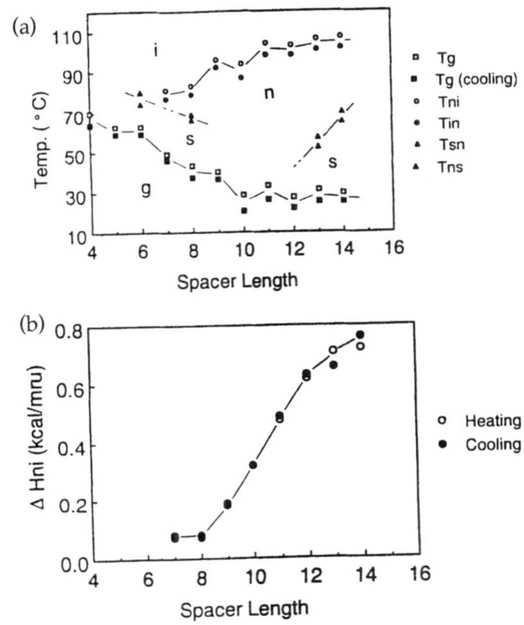

FIGURE 8.67. The dependence of the various transition temperatures (a) and of the enthalpy changes (b) associated with the nematic–isotropic (ΔH_{ni}) transition temperatures of regioirregular TPB-(c)X(3) (cyclic trimer) versus spacer length (X).

might also display only a virtual or a monotropic mesophase. The transformation of a virtual and/or monotropic mesophase of the homopolymers into an enantiotropic mesophase can be most conveniently accomplished by making copolymers based on two monomers which are constitutional isomers, such as monomers based on 4-MHMS and 4′-MHMS (Percec and Tomazos, 1989e, p). Since the structural units of the homopolymers based on 4-MHMS and 4′-MHMS are isomorphic within their liquid crystalline phase but not within their crystalline phase, the crystalline melting transition decreases, while the mesophase exhibits a continuous, almost linear dependence on composition. As a consequence, the virtual or monotropic mesophase of the homopolymer becomes enantiotropic (Percec and Tomazos, 1989e, p). Finally, the molecular weight at which the isotropization temperature becomes independent of molecular weight should be, and indeed is, dependent on the flexibility of the polymer backbone. For example, the isotropization temperature of polysiloxanes (Gray, 1989; Richards et al., 1990) containing mesogenic side groups is molecular weight dependent up to much higher molecular weights than polymethacrylates containing mesogenic side groups (Percec and Pugh, 1989a).

Based on this discussion and on the thermodynamic discussion from section 8.5, we can easily consider that the "polymer effect" can provide, via its molecular weight and backbone flexibility, the same effect. In an oversimplified way it can be considered that it provides an overall change in the entropy of the system. Through this change, it can transform in a reversible way, a virtual mesophase into a monotropic and an enantiotropic one. In addition, the kinetic factors provided by the glass transition and crystallization should always be considered. For example, the formation of a mesophase located in close proximity to a glass transition temperature becomes kinetically controlled or can even can be kinetically prohibited.

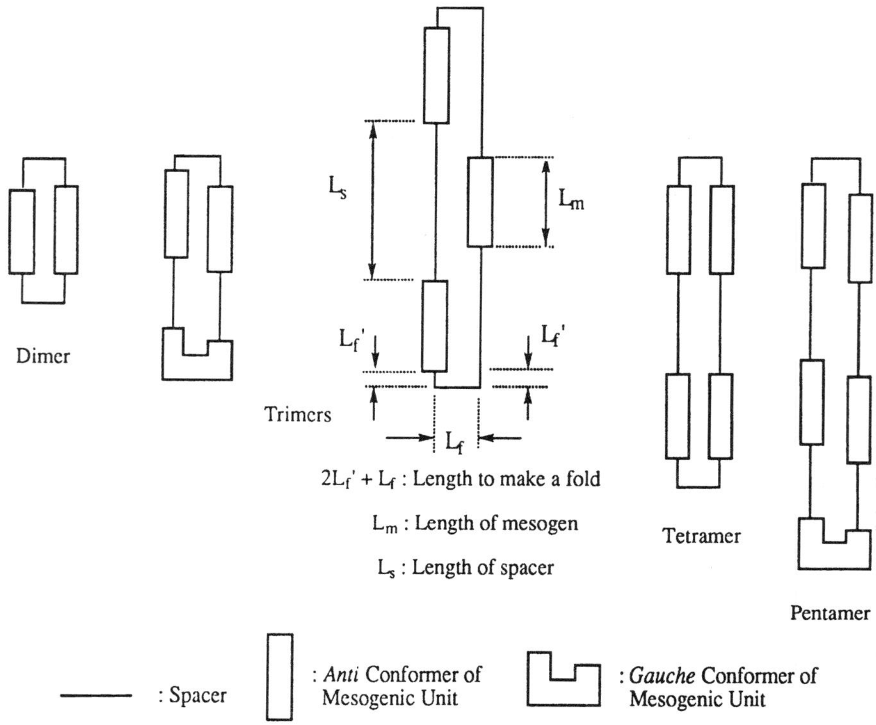

FIGURE 8.68. The ideal architecture of cyclic oligomers (dimer to pentamer) in the crystalline and liquid crystalline phases.

8.11.2 Molecular Engineering of Liquid Crystalline Polymers by Living Polymerization

8.11.2.1 General Considerations

Several polymerization methods were investigated in order to develop living polymerization procedures for the preparation of side chain liquid crystalline polymers with well-defined molecular weights and narrow molecular weight distributions. They include cationic polymerization of mesogenic vinyl ethers (Percec and Tomazos, 1987c; Rodriguez-Parada and Percec, 1986), cationic ring opening polymerization of mesogenic cyclic imino ethers (Rodriguez-Parada and Percec, 1987), group transfer polymerization of mesogenic methacrylates (Kreuder et al., 1986; Percec et al., 1989j; Pugh and Percec, 1985), polymerization of methacrylates with methylaluminum porphyrin catalysts (Kodaira and Mori, 1990) and metathesis polymerization of norbornenes containing mesogenic groups (Komiya et al., 1992a, b; Komiya and Schrock, 1993a, b; Pugh and Schrock, 1992; Pugh et al., 1994). Cationic polymerization has been proved to be the most successful method since the initiators are very simple and it can be used to polymerize under living conditions mesogenic vinyl ethers containing a large variety of functional groups (Heroguez et al., 1990; Jonsson et al., 1991a, b, c; Kostromin et al., 1990a; Papon et al., 1992; Percec et al., 1991e, 1992r; Percec, 1991b, f, p; Percec and Lee, 1991a, g, h; Percec and Johansson, 1993f; Rodenhouse et al., 1990; Rodenhouse and Percec, 1991a, c; Sagane and Lenz, 1988, 1989a, b).

FIGURE 8.75 provides some representative examples of mesogenic vinyl ethers that could be polymerized by a living mechanism and with the following, for example (CF_3SO_3H, $(CH_3)_2S$, CH_2Cl_2, 0 °C) (Cho et al., 1990). As we can observe from FIGURE 8.72, vinyl ethers containing nucleophilic groups such as methoxybiphenyl (Jonsson et al., 1991a), electron-withdrawing groups such as cyanobiphenyl (Jonsson et al., 1991a; Percec et al., 1991e, f, p, 1992r; Percec and Lee,

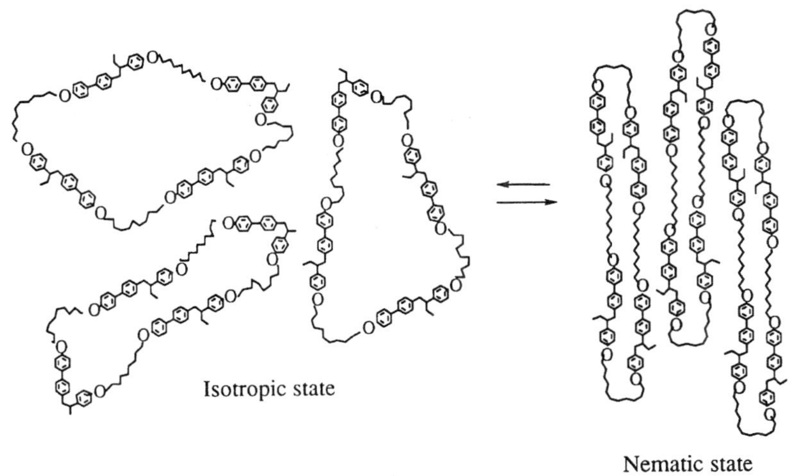

FIGURE 8.69. Schematic representation of the suggested mechanism for the isotropic–nematic transition of TPB-(c)X(4).

FIGURE 8.70. Schematic representation of side chain liquid crystalline polymers showing the necessity of decoupling the mesogenic groups and the polymer backbone through flexible spacers.

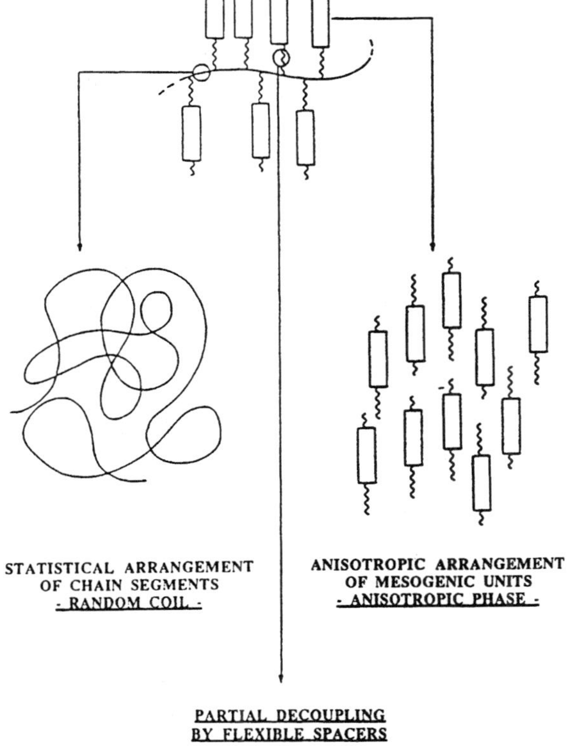

1991a, g, h), nitrobiphenyl and cyanophenylbenzoate (Jonsson et al., 1991a), double bonds like those in 4-alkoxy-α-methylstilbene (Percec et al., 1991p), double bonds and cyano groups like those in 4-cyano-4′-α-cyanostilbene (Percec et al., 1991f), aliphatic aromatic esters (Percec et al., 1991b), acidic protons and

FIGURE 8.71. Side-chain liquid crystalline polymers based on 4-MHMS and 4′-MHMS constitution isomeric mesogens and different backbones.

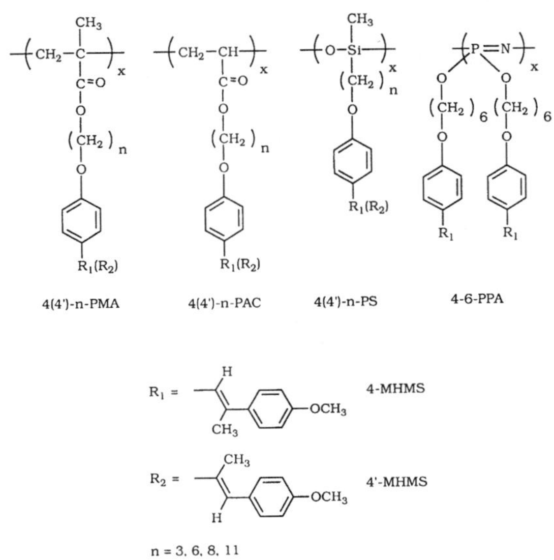

(a)

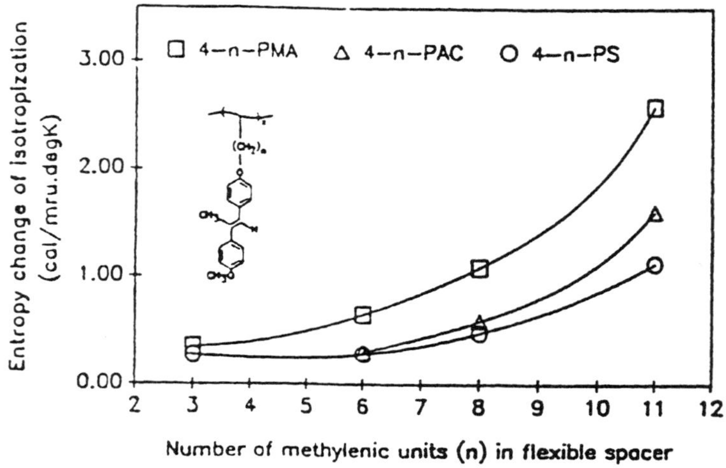

(b)

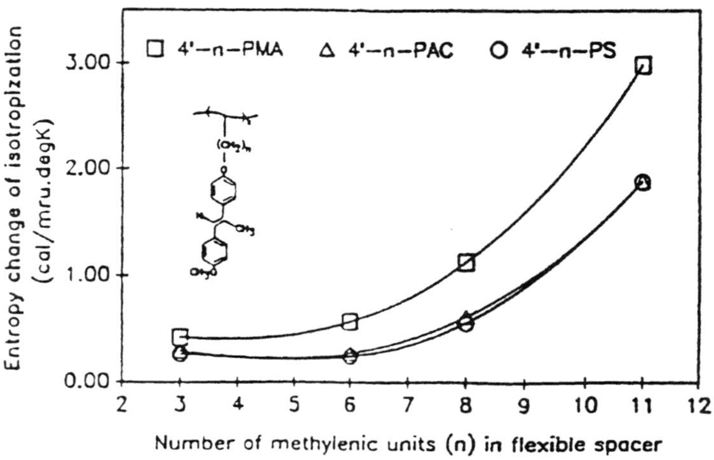

FIGURE 8.72. (a) The dependence of the entropy change of isotropization (ΔS_i) determined from the cooling DSC scans on the nature of the polymer backbone and the number of methylenic units (n) in the flexible spacer for the series of polymers based on 4-MHMS isomer. (b) The dependence of the entropy change of isotropization (ΔS_i) determined from the cooling DSC scans on the nature of the polymer backbone and the number of methylenic units (n) in the flexible spacer for the series of polymers based on 4′-MHMS isomer.

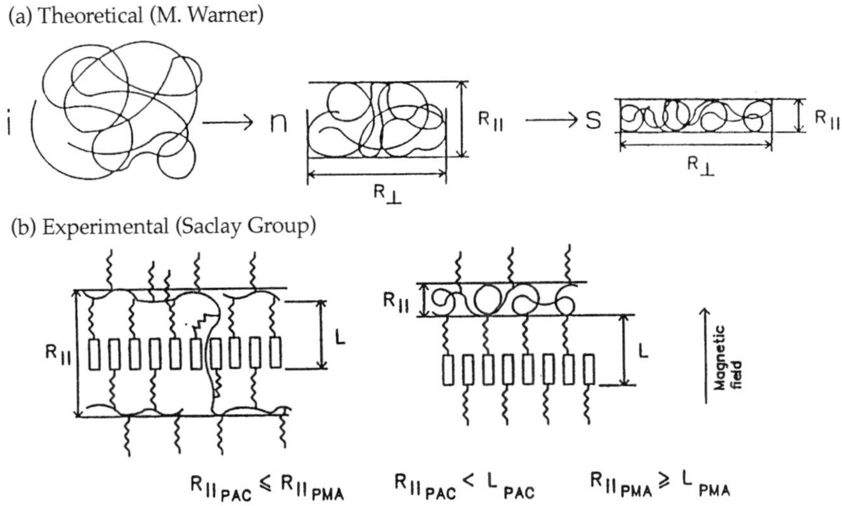

FIGURE 8.73. (a) Schematic representation of the theoretical distortion of the statistical random-coil conformation of the polymer backbone in the nematic and smectic phases; (b) two possible modes of distortion of the random-coil conformation of a rigid (left) and a flexible (right) polymer backbone. $R_{\|}$ refers to the radius of gyration parallel to the magnetic field. The radius of gyration perpendicular to the magnetic field is labeled as $R_{\perp}$.

mesogen(%)	d(Å)	a(Å)	T_i (°C)
100	30.4	1.4	104
82	32.0	3.0	92
56	36.0	7.0	75
33	42.0	13.0	58
21	45.6	16.6	36

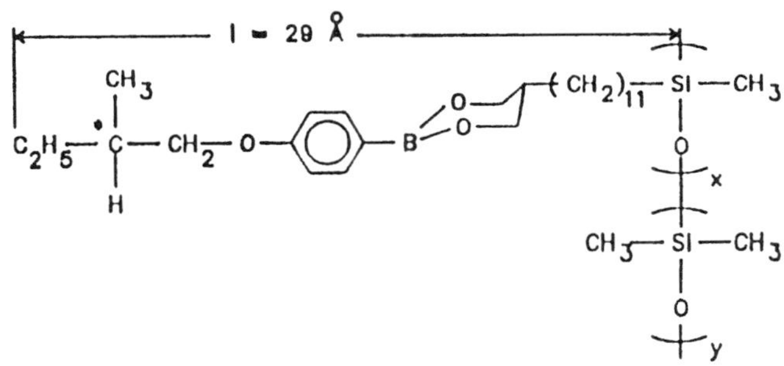

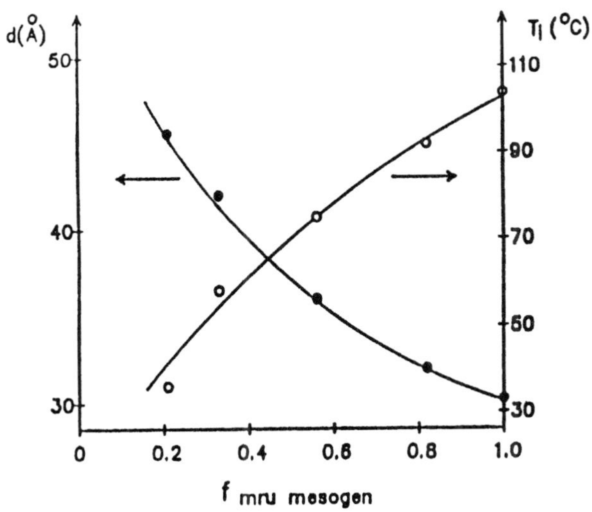

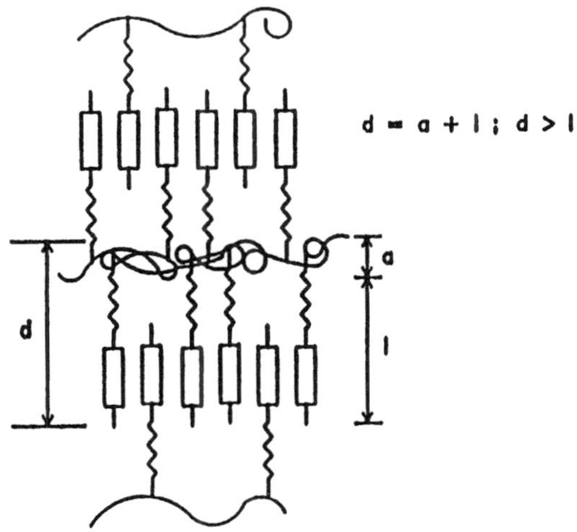

FIGURE 8.74. Microphase-separated morphology of smectic copolymers.

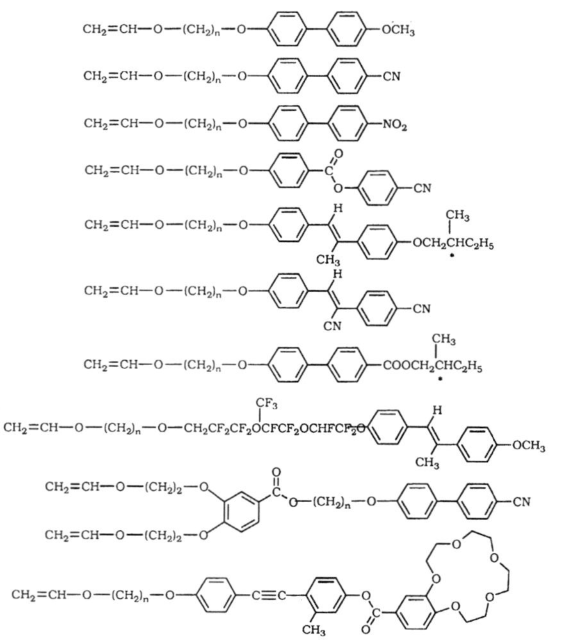

FIGURE 8.75. Examples of functional mesogenic vinyl ethers that could be polymerized by a living cationic mechanism.

perfluorinated groups (Percec et al., 1991b; Jonsson et al., 1991a), oligooxyethylene and aromatic ester groups (Rodenhouse and Percec, 1991a), crown ethers and triple bonds (Percec and Johansson, 1993f; Rodenhouse and Percec, 1991c), all can be polymerized by a living cationic mechanism. In addition, cationic polymerization of any of these monomers can be performed in the melt phase, either in the liquid crystalline or in the isotropic phase by using thermal (Jonsson et al., 1991c) or photocationic initiators (Jonsson et al., 1991b, 1992). When the polymerization is performed in the liquid crystalline phase with aligned films of liquid crystalline monomers, perfectly aligned single-crystal liquid crystalline polymer films are obtained (Jonsson et al., 1991b, 1992).

In the following two sections we will discuss two topics. The first one refers to the influence of molecular weight on the phase transitions of poly{ω-[(4-cyano-4′-biphenylyl)oxy]alkyl vinyl ether}s with alkyl groups containing from four to eleven methylene units. In the second section we will demonstrate the molecular engineering of phase transitions of side chain liquid crystalline polymers by azeotropic living copolymerization experiments.

8.11.2.2 Influence of Molecular Weight on Phase Transitions of Poly{ω-[(4-cyano-4′-biphenylyl)oxy]alkyl Vinyl Ether}s

FIGURE 8.76 outlines the general method used for the synthesis of ω-[(4-cyano-4′-biphenylyl)oxy]alkyl vinyl ethers (6-n) and of the model compound for the polymer with a degree of polymerization of one, that is, ω-[(4-cyano-4′-biphenylyl)oxy]alkyl ethyl ethers (8-n). We will use throughout the discussion the same short notations as in the original publications. The synthesis and characterization of poly(6-n) and (8-n) with n = 2, 3, 4 (Percec and Lee, 1991a), 5, 7 (Percec and Lee, 1992r), 6, 8 (Jonsson et al., 1991a), 9, 10 (Percec and Lee, 1991h), and 11 (Percec and Lee, 1991e) will be briefly discussed. Details are available in the original publications. All polymers have polydispersities of about 1.10. FIGURE 8.77 outlines the polymerization mechanisms and the structures of the resulting polymers which were confirmed by 300 MHz 1D and 2D ^{1}HNMR spectroscopy (Percec et al., 1992g).

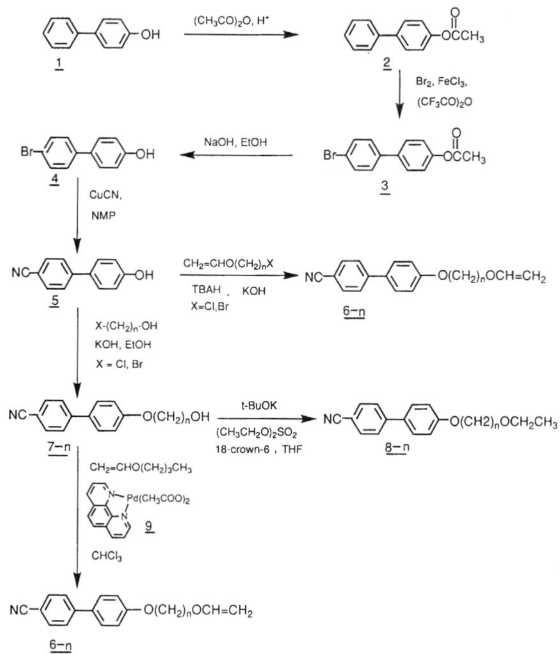

FIGURE 8.76. Synthesis of vinyl ethers 6-n.

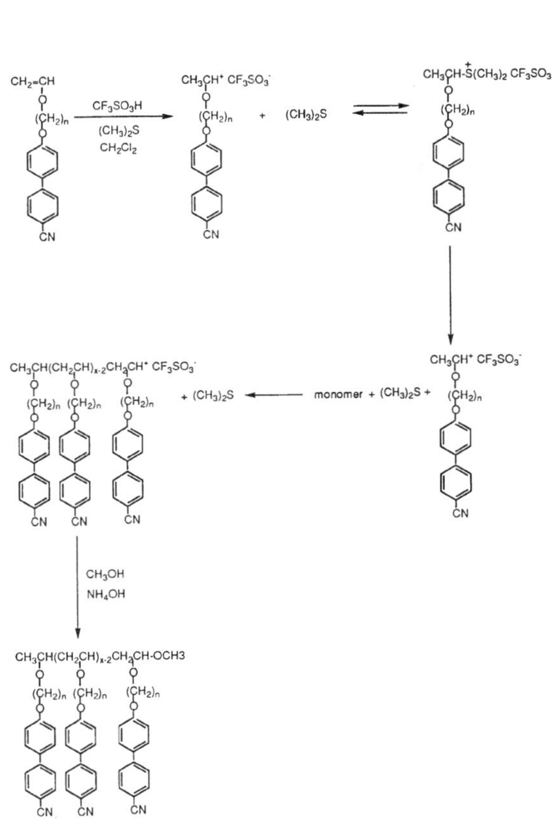

FIGURE 8.77. Living cationic polymerization of mesogenic vinyl ethers.

All data were classified according to their similarities. FIGURE 8.78 presents the dependence of phase transition temperatures of poly(6-n) with n = 3, 4, 7 and 9 as a function of molecular weight. These data were collected from second heating scans. The data for 8-n are not plotted. 8-3 is crystalline, 8-4 and 8-7 exhibit a monotropic nematic mesophase while 8-9 exhibits monotropic nematic and smectic mesophases. As we can observe from FIGURE 8.78, by increasing the molecular weight, all four polymers show a broadening of the thermal stability of their mesophase. The mesophase of 8-7 and 8-9 changes from nematic to s_A by increasing the degree of polymerization from one to about three.

FIGURE 8.79 presents similar data for poly(6-2), poly(6-6), and poly(6-8). In all cases the nature of the mesophase is molecular weight dependent. Poly(6-2) has a nematic mesophase only at degrees of polymerization lower than five. 8-2 is only crystalline. At degrees of polymerization higher than five poly(6-2) is only glassy. This is because its glass transition temperature becomes higher than the isotropization temperature and therefore, the mesophase is kinetically prohibited. 8-6 exhibits an enantiotropic nematic mesophase. At low degrees of polymerization poly(6-6) and poly(6-8) exhibit nematic and s_A mesophases. Due to the difference between the slope of the dependences of the nematic phase transition temperature on molecular weight and of the s_A phase transition temperature on molecular weight, above a certain molecular weight the nematic phase disappears. Both poly(6-6) and poly(6-8) show a second smectic mesophase (s_X, i.e., unassigned). Qualitatively, this behavior is in agreement with the influence of molecular weight on phase transitions predicted by thermodynamics (section 8.5). Quantitative predictions of these phase diagrams require more theoretical research.

Finally, FIGURE 8.80 presents the behavior of poly(6-5), poly(6-10), and poly(6-11). 8-5 shows an monotropic nematic phase, 8-10 shows a monotropic s_A phase, while 8-11 shows an enantiotropic s_A mesophase. Poly(6-5) exhibits above a degree of polymerization of 10 the unusual sequence isotropic–nematic–s_{Ad}–n_{re}–glassy (Percec and Lee, 1992c). This will be discussed in more detail in a subsequent section. At high molecular weights, poly(6-10) and poly(6-11) exhibit s_A and s_X phases.

As a general observation, we can mention that polymers with short (n = 2, 3, 4) and medium length spacers containing an odd number of methylene units (n = 7, 9) do not generate polymorphism at different molecular weights. Polymers with medium length and an even number of methylene units (n = 6, 8), as well as polymers with long length with both even and odd numbers of methylenic units (n = 10, 11), generate a rich polymorphism which is molecular weight dependent. The borderline polymer is poly(6-5), which is the only one displaying n and s_A mesophases over a broad range of molecular weights, and therefore also generates the reentrant nematic mesophase (Percec and Lee, 1992c).

8.11.2.3 Molecular Engineering of Liquid Crystalline Phases by Living Cationic Copolymerization

In order to tailor-make mesophases of side chain liquid crystalline copolymers, we first need to synthesize

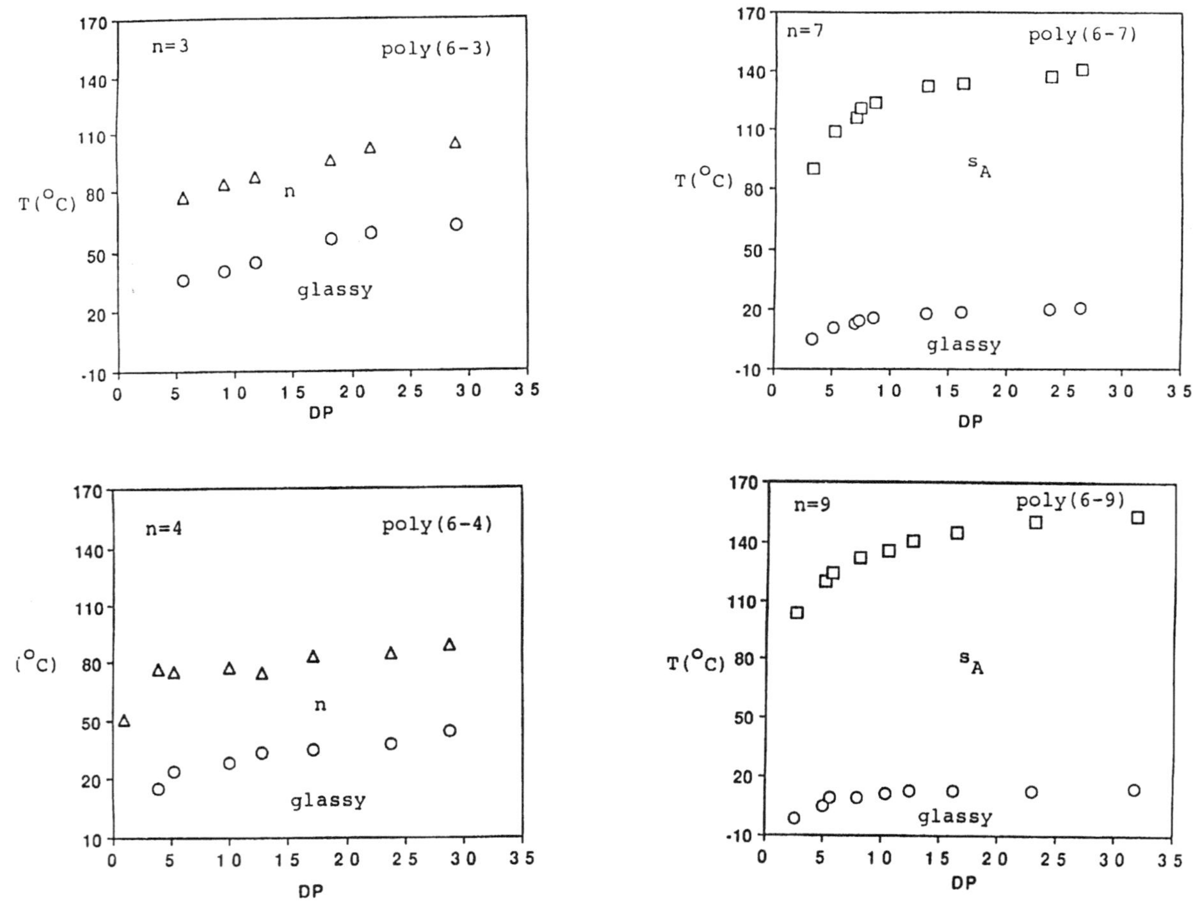

FIGURE 8.78. The influence of molecular weight on the phase behavior of poly(6-3), poly(6-4), poly(6-7), and poly(6-9) (determined from second DSC heating scans).

copolymers with constant molecular weight and controllable composition. Copolymer composition is conversion-dependent in all statistic copolymerizations. The only exception is provided by azeotropic copolymerizations in which the copolymer composition is identical to the monomer feed at any conversion (Tirrell, 1986). This situation is provided by monomers with $r_1 = r_2 = 1$. Since the reactivity of the polymerizable vinyl ether groups is not dependent on spacer length, all 6-n monomers have the same reactivity. Therefore, all 6-n pairs of monomers lead to azeotropic copolymerizations, and when the copolymerization is performed under living conditions it leads to copolymers with controllable molecular weight. The azeotropic copolymerization of various pairs of 6-n monomers is outlined in FIGURE 8.81. We will discuss selected examples of copolymers prepared from monomer pairs which give rise to homopolymers exhibiting nematic and nematic, s_A and s_A, nematic and s_A, and glassy and s_A phases as their highest-temperature mesophases.

FIGURE 8.82 presents the dependence of phase transition temperatures obtained from second DSC heating scans (a), (d), cooling scans (b), (e), and the enthalpy changes associated with the highest-temperature mesophase of copolymers poly[(6-3)-co-(6-5)]X/Y and poly[(6-6)-co-(6-11)]X/Y. The degrees of polymerization of all copolymers are equal toò20 (Percec and Lee, 1991j). Copolymers poly[(6-3)-co-(6-5)]X/Y are based on a monomer pair that gives

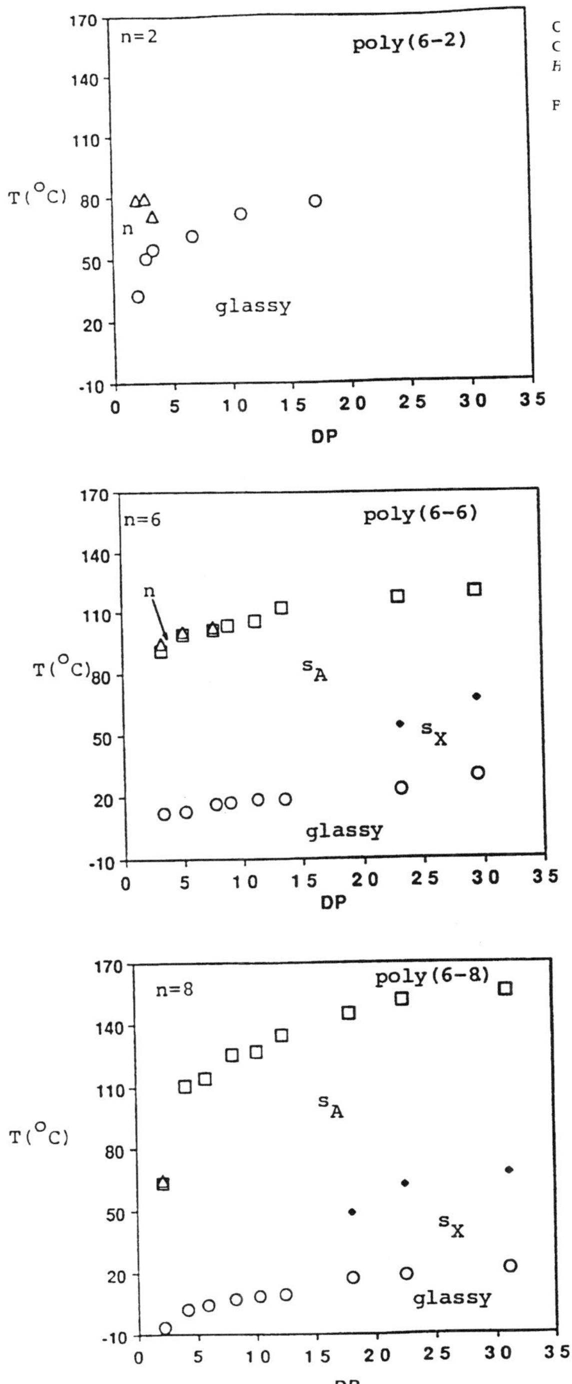

FIGURE 8.79. The influence of molecular weight on the phase behavior of poly(6-2), poly(6-6), and poly(6-8) (determined from second DSC heating scans).

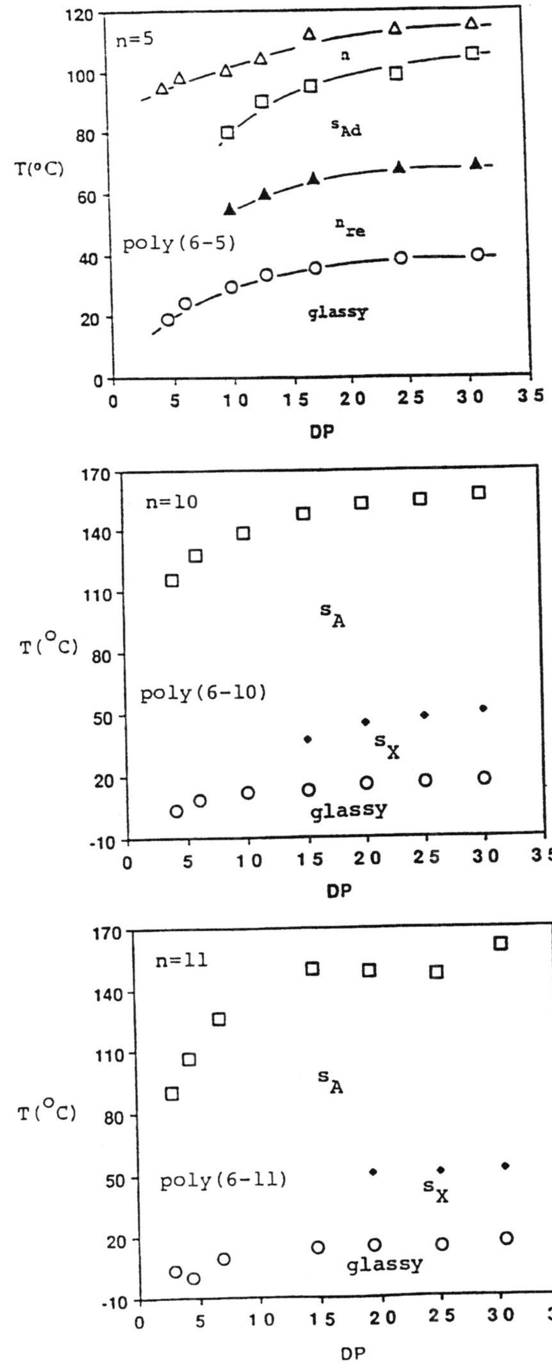

FIGURE 8.80. The influence of molecular weight on the phase behavior of poly(6-5), poly(6-10), and poly(6-11) (determined from second DSC heating scans).

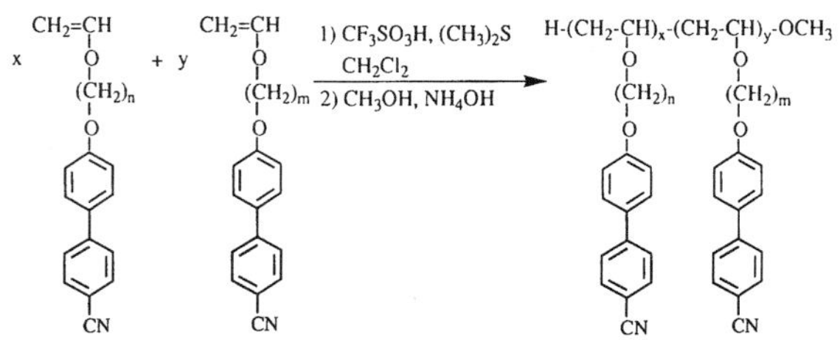

FIGURE 8.81. Living azeotropic copolymerization of mesogenic vinyl ethers.

rise to two homopolymers displaying an enantiotropic nematic mesophase as their highest-temperature mesophase. As we can observe from FIGURE 8.82(a), (b) and (c) the nematic–isotropic transition temperature and its associated enthalpy change show linear dependences with composition. This means that the structural units derived from poly(6-3) and poly(6-5) are isomorphic to their nematic mesophase. However, the same two structural units are isomorphic within the s_A mesophase exhibited by poly(6-5) only over a very narrow range of compositions. The linear dependence of the isotropization temperature is predictable by the Schroeder–Van Laar equations (Achard et al., 1989; Shroeder, 1893; Van Hecke, 1979; Van Laar, 1908). The same discussion is valid for the copolymer system poly[(6-6)-co-(6-11)]X/Y except that the isotropization temperatures of these copolymers exhibit an upward curvature. This upward curvature is also predicted by the Schroeder–Van Laar equations (Achard et al., 1989; Shroeder, 1893; Van Hecke, 1979; Van Laar, 1908) and is due to the more dissimilar enthalpy changes associated with the isotropization temperatures of the two homopolymers.

FIGURE 8.83 presents the phase diagrams of copolymers poly[(6-3)-co-(6-11)]X/Y (Percec and Lee, 1991j) and poly[(6-5)-co-(6-11)]X/Y (Percec and Lee, 1992c). Both sets of copolymers have degrees of polymerization of 20. Both pairs of copolymers are based on monomers which give rise to homopolymers exhibiting nematic and s_A as their highest-temperature mesophases. However, poly(6-5) displays a nematic and an s_A mesophase, while poly(6-3) shows only a nematic mesophase. Both sets of copolymers display continuous dependences of their highest temperature mesophase with a triple point at a certain composition. This triple point generates over a very narrow range of compositions copolymers exhibiting the sequence isotropic–nematic–s_A–n_{re}. Again, the shapes of the dependences of the phase transition temperature on composition obey the Schroeder–Van Laar equations.

FIGURE 8.84 presents two sets of phase diagrams obtained from monomer pairs giving rise to homopolymers that exhibit isotropic and s_A mesophases as their highest-temperature mesophases, that is, poly[(6-2)-co-(6-8)]X/Y with a degree of polymerization of 10 (Percec and Lee, 1991t; Percec et al., 1992r), and poly[(6-2)-co-(6-11)]X/Y with a degree of polymerization of 15 (Percec and Lee, 1991n). Both sets of copolymers display a similar phase diagram. Over a certain range of compositions, the two structural units are isomorphic within the s_A phase, after which follows a triple point. After this triple point the two structural units are isomorphic within a newly generated nematic mesophase. Both copolymers generate within a certain range of compositions on the left-hand side of the triple point the sequence isotropic–nematic–s_A–n_{re} (Percec and Lee, 1992c). Again, the shape of the dependence of the highest-temperature mesophase on composition is predictable by the Schroeder–Van Laar equations. This means that the structural units of all binary copolymers based on an identical mesogenic unit but different spacer lengths behave as an ideal solution. This behavior allows the engineering of mesomorphic phase transition temperatures and of their thermodynamic parameters in a straightforward manner by living azeotropic copolymerizations. The same behavior was demonstrated for monomer pairs, both of which give rise to homopolymers exhibiting a chiral smectic C mesophase (Percec et al., 1991c).

Living homopolymerization and azeotropic copolymerization of monomer pairs which are diastereomers or enantiomers was used to investigate *hetero chiral* molecular recognition in pairs of monomer mixtures and polymer mixtures as well as in copolymers (Percec et al., 1994f; Percec and Oda, 1994h, i, 1995a).

8.11.2.4 Side Chain Liquid Crystalline Polymers Exhibiting a Reentrant Nematic Mesophase

The reentrant nematic phase (n_{re}) was discovered in 1975 in low molar mass liquid crystals (Cladis, 1975). Since then it has received substantial theoretical and experimental interest (Cladis et al., 1977, 1988; Cladis, 1988, 1978; Goodby et al., 1984; Guillon et al., 1978; Hardouin et al., 1983; Hardouin, 1986; Sigaud et al., 1981; Tinh, 1983).

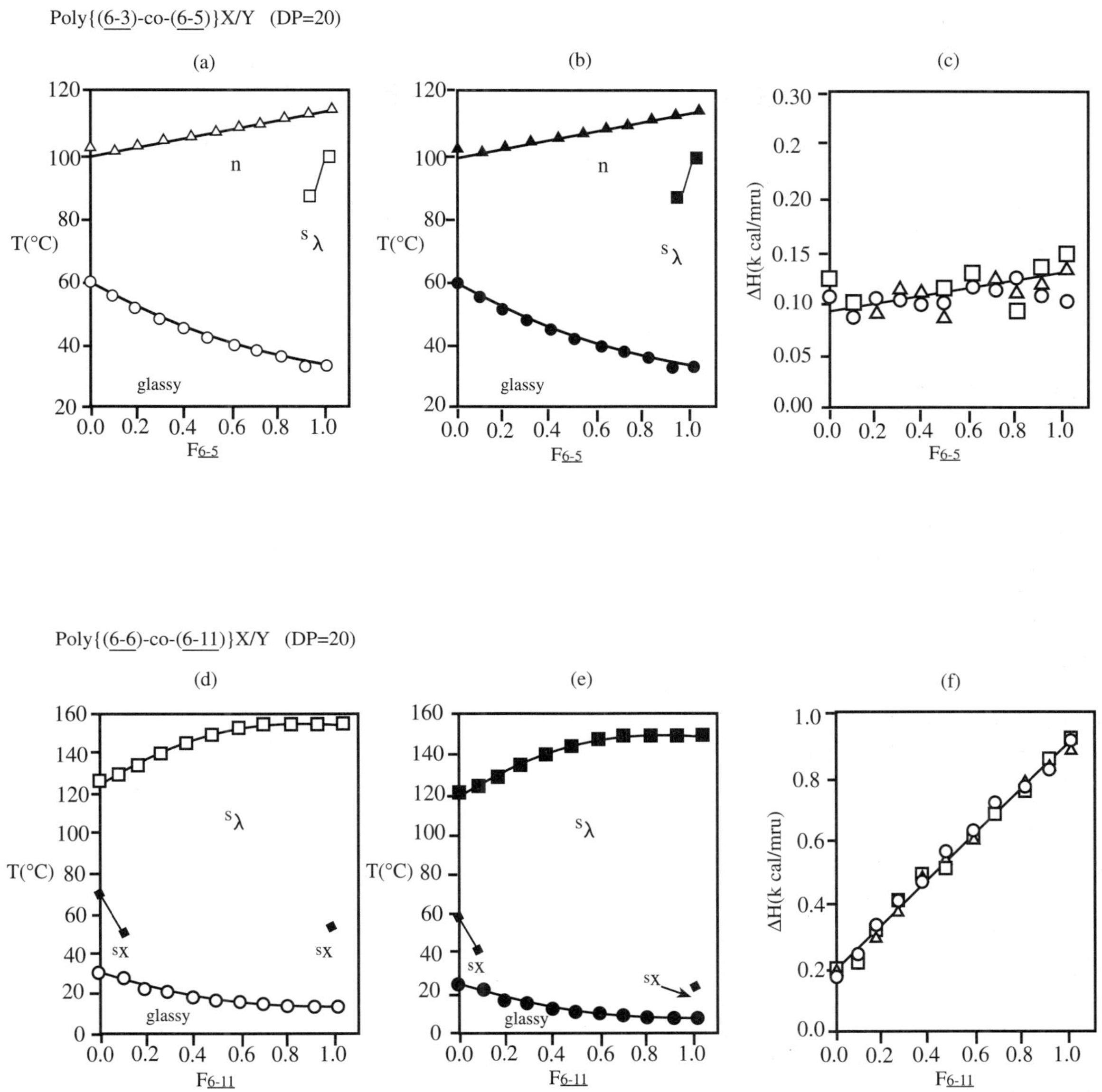

FIGURE 8.82. The behavior of phase transition temperatures obtained from the second heating scan (a), (d), cooling scan (b), (e), and the enthalpy changes associated with their highest-temperature mesophases of copolymers poly[(6-3)-co-(6-5)]X/Y and poly[(6-6)-co-(6-11)]X/Y (all with degrees of polymerization equal to 20).

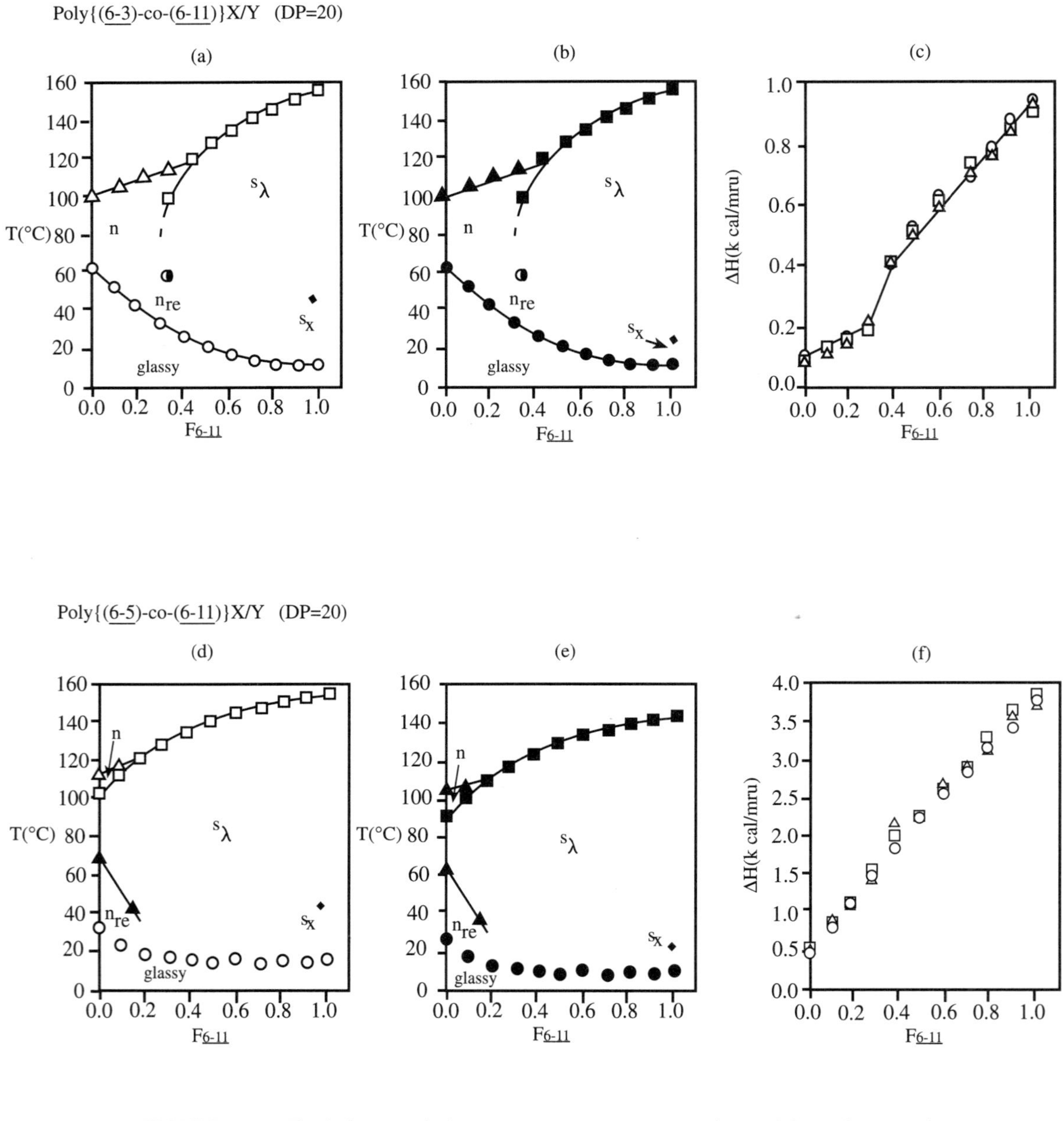

FIGURE 8.83. The behavior of phase transition temperatures obtained from the second heating scan (a), (d), cooling scan (b), (e), and the enthalpy changes associated with their highest-temperature mesophases of copolymers poly[(6-3)-co-(6-11)]X/Y and poly[(6-5)-co-(6-11)]X/Y (all with degrees of polymerization equal to 20).

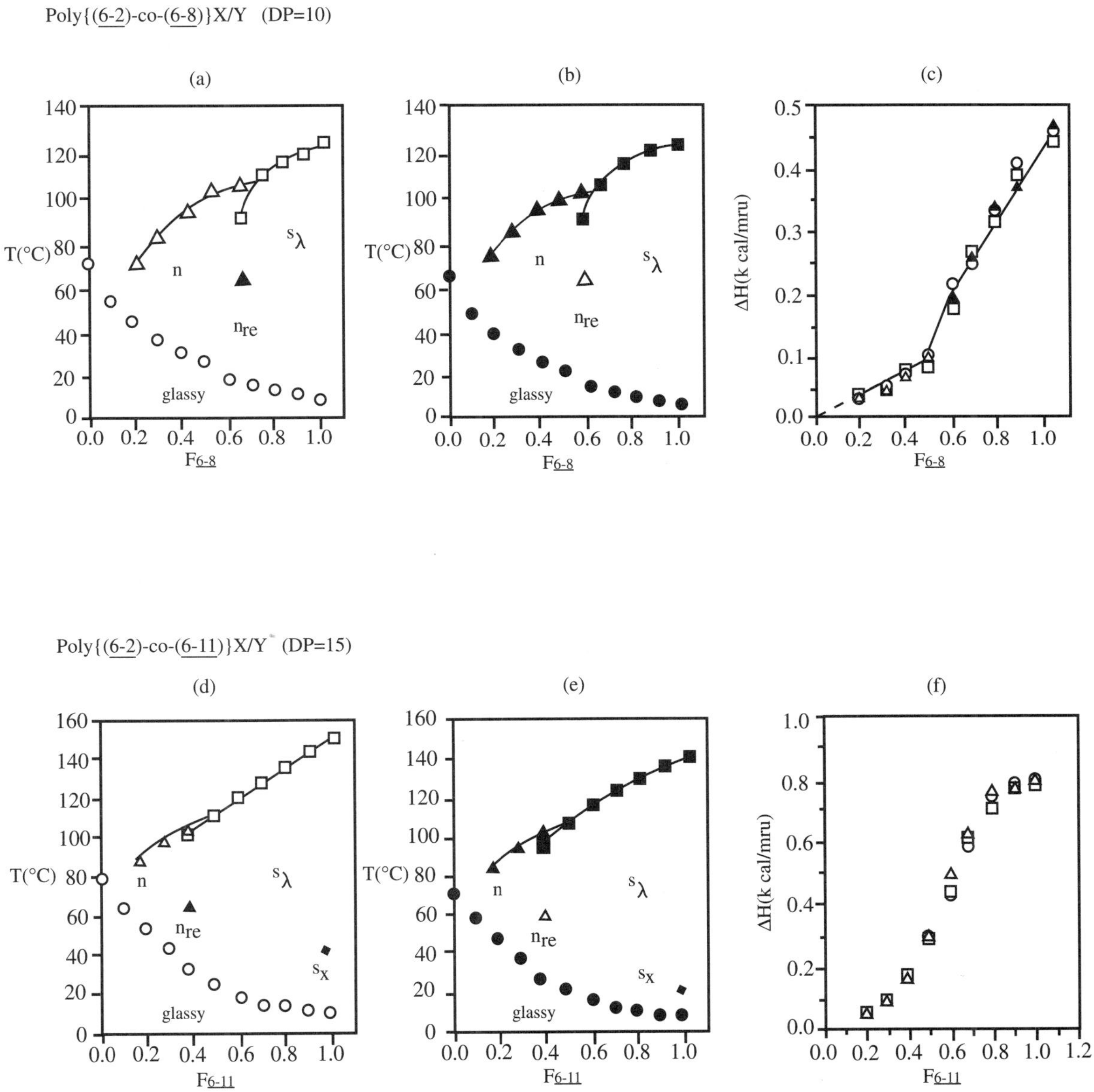

FIGURE 8.84. The behavior of phase transition temperatures obtained from the second DSC heating scan (a), (d), cooling scan (b), (e), and the enthalpy changes associated with their highest-temperature mesophases of copolymers poly[(6-2)-co-(6-8)]X/Y and poly[(6-2)-co-(6-11)]X/Y (with degrees of polymerization equal to 10 and 15, respectively).

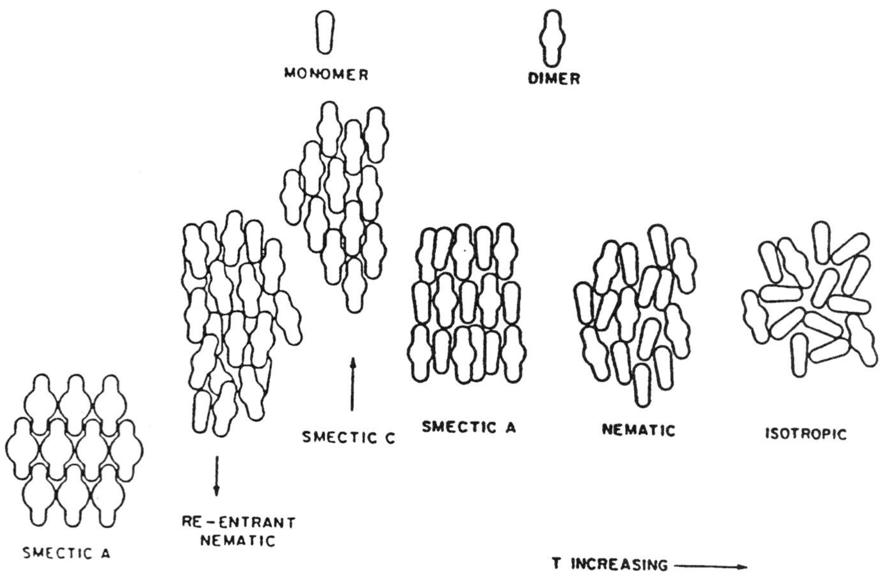

FIGURE 8.85. The mechanism of formation of the reentrant nematic mesophase (Cladis, 1975).

The first side chain liquid crystalline polymers exhibiting an n_{re} phase were reported in 1986 (Gubina et al., 1986; LeBarny et al., 1986). Some other examples of polymers exhibiting the sequence isotropic–nematic–s_{Ad}–n_{re} have been reported since (Gubina et al., 1989; Kostromin et al., 1990b; Legrand et al., 1990; Percec and Lee, 1992c; Shibaev, 1988; Spassky et al., 1989). All these polymers are based on mesogenic units containing a cyano group, five or six atoms in the flexible spacer and a polyacrylate or polyvinyl ether backbone. The replacement of these quite flexible backbones with more rigid ones like polymethacrylate does not allow the formation of the n_{re} phase. As discussed in the previous section, an n_{re} mesophase can be generated by copolymerization of two monomers which lead to homopolymers with nematic or isotropic and s_A as their highest-temperature mesophases, since these copolymers exhibit a triple point on their phase diagrams (Percec and Lee, 1992c). According to experimental results, any polymer that exhibits the sequence isotropic–nematic–s_A should also display an n_{re} phase. The most probable mechanism for the generation of an n_{re} phase is outlined in FIGURE 8.85 (Cladis, 1988). The most stable s_A phase of mesogens containing cyano groups is based on layers containing dimers of mesogens. On cooling, the nematic phase formed directly from the isotropic phase contains both dimeric mesogens and monomeric mesogens, and so does the first s_A phase. In order to go from the less ordered s_A phase to the s_A phase based on dimeric mesogens, an n_{re} phase is required (FIGURE 8.85) (Cladis, 1988).

8.11.2.5 Influence of Tacticity on Phase Transitions

The tacticity of a polymer determines the flexibility and the conformation of its own backbone. Therefore, tacticity should affect the phase behavior of side chain liquid crystalline polymers at least in the same way as backbone flexibility. Liquid crystalline polymethacrylates with atactic, syndiotactic, and isotactic backbones have been synthesized and characterized (Bohnert et al., 1993; Frosini et al., 1981; Hahn et al., 1981; Magagnini, 1981; Nakano et al., 1993; Okamoto et al., 1991). The general trend observed is that the most flexible backbones (i.e., those containing the lowest number of syndiotactic sequences) give the highest isotropization temperatures. Polymers with higher syndiotactic sequences provide intermediary isotropization temperatures between atactic and isotactic ones. Isotactic polymers exhibit the lowest isotropization temperatures. This trend is also dependent on spacer length. Isotactic and syndiotactic polymethacrylates have helical chain conformations while atactic ones have zig-zag (all-*trans*) conformations.

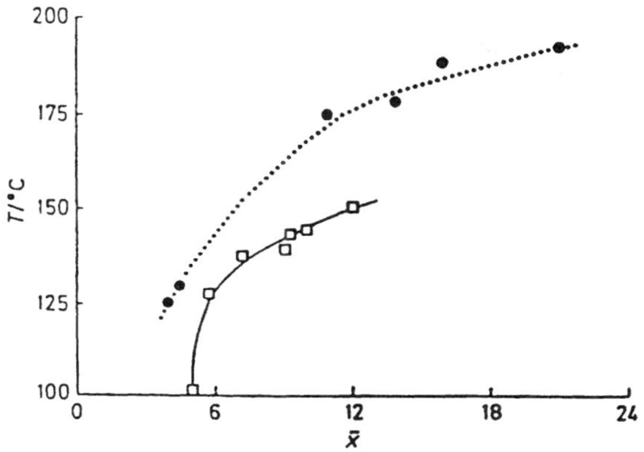

FIGURE 8.86. Isotropization transition temperatures of cyclic (●) and linear (□) polysiloxanes containing cyanophenyl benzoate mesogen connected to the backbone via a pentyloxy group (Richards et al., 1990).

8.11.2.6 Cyclic Polysiloxanes Containing Mesogenic Side Groups

Low molecular weight (Hahn and Percec, 1988; Percec and Hahn, 1989c) and high molecular weight (Richards et al., 1990) cyclic polysiloxanes containing mesogenic side groups were reported. FIGURE 8.86 presents the dependence of the isotropization temperature on the degree of polymerization for linear and cyclic polysiloxanes containing cyanophenyl benzoate groups connected to the backbone by a pentyl spacer (Richards et al., 1990). Isotropization temperatures of cyclic polymers are higher than those of linear polymers. However, more research is required to reach the same level of understanding between structure and properties in this class of polymers as that currently available for cyclic, amorphous, and crystalline polymers (Percec et al., 1989l). A general review of various classes of liquid crystalline polysiloxanes is available (Boileau and Teyssié, 1991).

8.11.2.7 Side Chain Liquid Crystalline Polyacetylenes

Monosubstituted acetylenes (LeMoigne et al., 1989, 1991a, b; Oh et al., 1993a, b; Zhang et al., 1990) and p-substituted phenylacetylenes (LeMoigne et al., 1992) containing various mesogenic side groups were synthesized and polymerized by Ziegler–Natta and metathesis catalysts. Conjugated polyacetylenes exhibiting nematic, cholesteric, and various smectic phases were reported. Additional research is required to establish the dependence between polymer molecular weight and double bond configuration (i.e., *cis*-cisoidal, *cis*-transoidal, *trans*-transoidal, and *trans*-cisoidal) on phase transitions. All these polymers exhibit interesting nonlinear optical properties and electrical conductivity that are anisotropic.

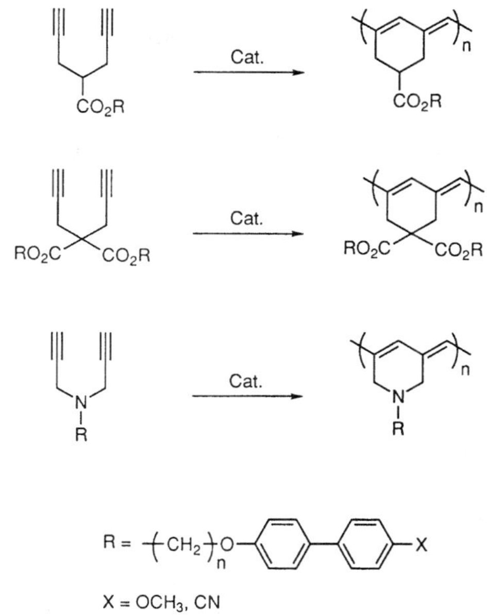

FIGURE 8.87. Side chain liquid crystalline polyacetylenes.

A very interesting class of liquid crystalline polyacetylenes is obtained by the metathesis cyclopolymerization of various dipropargyl derivatives containing mesogenic groups (Choi et al., 1994a, b, c; Jin S.-H. et al., 1991, 1993). A fully conjugated poly(1,6-heptadiyne) backbone containing mesogenic side groups is obtained. Selected examples of monomers and the corresponding polymers are outlined in FIGURE 8.87. Conducting side chain liquid crystalline polymers with nematic and various smectic phases were obtained by this synthetic procedure.

8.12 IONIC LIQUID CRYSTALLINE POLYMERS

Ionic liquid crystalline polymers are of recent research interest but are of great potential from both a synthetic as well as from a materials point of view. We will only mention very briefly a few selected examples from various research trends, since structure–properties relationships in this field are far from being elaborated.

8.12.1 Main Chain Ionene Liquid Crystal Polymers

Quarternization of 4,4′-bipyridyl with α,ω-dibromoalkanes yields thermotropic oligomers bearing viologen mesogenic groups (Paleos et al., 1986; Yu and Samulski, 1984). Other examples are the thermotropic ionenes synthesized by a similar procedure from α,ω-dielectrophilic alkanes and *trans*-1,2-*bis*(4-piridyl)- ethene. The size of the counteranion derived from the electrophilic group of the α,ω-dielectrophilic alkane has a strong influence on the thermotropic properties of the resulting polymer (Yousif et al., 1990). The salts obtained from α,ω-diaminoalkanes and α,ω-alkanedicarboxylates also form thermotropic mesophases (Sellens et al., 1989).

8.12.2 Side Chain Ionic Liquid Crystal Polymers

Alkylation of polyethyleneimine with alkylhalide-containing mesogenic groups yields thermotropic liquid crystalline polyethyleneimines with ammonium halide groups in the main chain. Smectic mesophases are easily obtained (Ujiie and Iimura, 1990, 1991a, b, 1992). Alkylation or protonation of poly(4-vinylpyridine) and of other polymers containing 4-substituted pyridine side groups with ω-halides of mesogenic groups represent other methods for the preparation of ionic side chain polymers (Brandys and Bazuin, 1993).

8.12.3 Liquid Crystalline Polyelectrolyte Complexes

The complexation of poly(acrylic acid) with dodecyltrimethylammoniumchloride and of poly(styrene sulfonate) (Antonietti and Conrad, 1994a) with different alkyltrimethylammonium derivatives (Antonietti et al., 1994b) by common precipitation in water yields liquid crystalline polyelectrolyte–surfactant complexes. These complexes represent a novel class of liquid crystalline materials with very large synthetic capabilities.

8.13 LIQUID CRYSTALLINITY INDUCED VIA INTER- AND INTRACHAIN ELECTRON DONOR–ACCEPTOR (EDA) INTERACTIONS

Both intrachain and interchain electron donor–acceptor (EDA) interactions between repeat units containing one mesogenic group with an electron donor and one with an electron acceptor were used to induce and enhance liquid crystallinity in copolymers and in polymer blends. The positive deviations of the isotropization temperature from ideal behavior were observed in the copolymers obtained from acceptor and donor mesogenic monomers (Imrie et al., 1991; Imrie and Paterson, 1994; Kosaka and Uryu, 1995; Schleeh et al., 1993). A mesophase was induced by blending two amorphous polymers containing mesogenic groups with donor and acceptor mesogenic groups (Imrie and Paterson, 1994). A smectic phase was induced in a blend based on two homopolymers which each exhibit only a nematic mesophase (Kosaka and Uryu, 1994). The use of donor–acceptor interactions to manipulate the phase behavior and to induce new mesophases in mixtures of low molar mass liquid crystals has been more extensively investigated, and the transplant of this concept to the field of polymers is only in its early stages.

8.14 LIQUID CRYSTALLINE POLYMERS CONTAINING CROWN ETHERS AND POLYPODANTS

Mesomorphic host–guest systems of low molecular weight and polymer liquid crystals containing macroheterocyclic ligands and polypodants provide a novel approach to self-assembled systems which combine selective recognition with external regulation (Cram, 1988; Lehn, 1988, 1990, 1993; Pedersen, 1988). Three basic architectures can be considered for liquid crystalline polymers containing crown ethers (FIGURE 8.88): main chain liquid crystalline polymers containing crown ethers in the main chain of the polymer, and side chain liquid crystalline polymers containing crown ethers either in the mesogenic group or in the main chain. Alternatively, the same series of polymers with polypodants instead of crown ethers can be considered.

Main chain polyamides and polyethers containing crown ethers have been reported (Cowie and Wu,

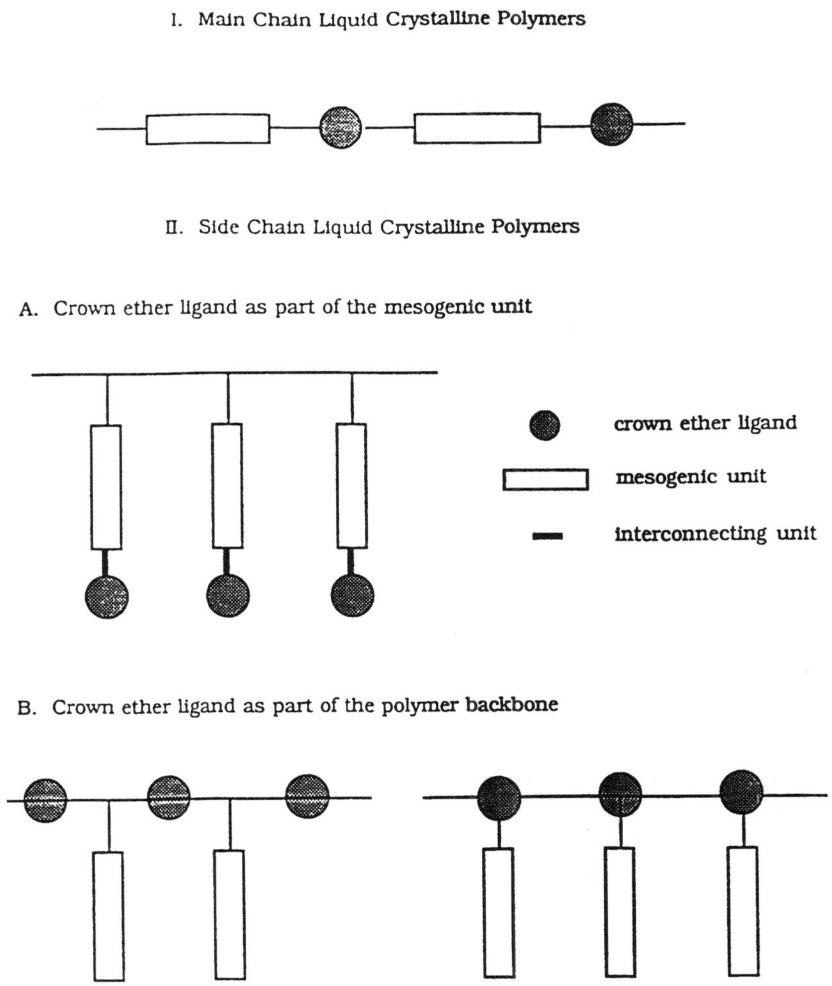

FIGURE 8.88. The molecular architecture of liquid crystalline polymers containing crown ether ligands.

1988; Percec and Rodenhouse, 1989g). A variety of side chain liquid crystalline polymers containing crown ether groups at one end of the mesogenic unit have been designed (Hsiue et al., 1991; Percec and Rodenhouse, 1989k, 1991d; Percec et al., 1992k; Percec and Johansson, 1993f; Rodenhouse and Percec, 1991b, c; Ungar et al., 1991c; Wen et al., 1990). Side chain liquid crystalline polymers containing crown ethers in the main chain were synthesized by living cationic cyclopolymerization and cyclocopolymerization of 1,2-*bis*(2-ethenyloxyethoxy)benzene derivatives containing mesogenic side groups (Rodenhouse et al., 1990; Rodenhouse and Percec, 1991a). Polymers containing crown ethers in the side groups dissolve ion pairs and behave as copolymers containing two different mesogenic groups, that is, complexed and uncomplexed. Their behavior is similar to that of copolymers derived from two different mesogenic groups, therefore, their phase behavior is directed by molecular recognition (Percec et al., 1992k, Percec and Johansson, 1993f). The use of oligooxyethylenic spacers in main chain (Shaffer and Percec, 1987), and side chain (Hsieh et al., 1990b; Percec and Tomazos, 1993g, h; Rodriguez-Parada and Percec, 1986) liquid crystalline polymers leads to liquid crystalline polypodants. Both main chain (Shaffer and Percec, 1987) and side chain (Percec and Tomazos, 1993g, h) liquid crystalline polypodants dissolve large amounts of alkali metal salts, and the resulting liquid crystalline polyelectrolytes are ionic conductors (Hsieh et al., 1990a). A detailed review of molecular, macromolecular, and supramolecular liquid crystals containing macroheterocyclic ligands is available (Percec and Johansson, 1995). A general discussion on reactions and interactions in liquid crystalline media is also available (Percec et al., 1992s).

8.15 ELECTRON DONOR–ACCEPTOR (EDA) COMPLEXES OF LIQUID CRYSTALLINE POLYMERS CONTAINING DISCOTIC MESOGENS

Ringsdorf and coworkers have studied the effects of doping with electron acceptor molecules on the phase behavior of polymers containing disk-like donor mesogens (Bengs et al., 1991; Eigen and DeMaeyer, 1966; Eigen, 1971; Green et al., 1990; Paulus et al., 1991; Ringsdorf et al., 1989; Ringsdorf and Wustefeld, 1990). They demonstrated that charge transfer interactions increase the degree of order within the liquid crystalline phase, and also stabilize the liquid crystalline mesophase. For example, when a discotic main chain polyester was doped with 2,4,7-trinitrofluorenone (TNF), the width of the discotic columnar hexagonal mesophase was increased by $\simeq$50 °C, for molar ratios of polymer/TNF equal to 3/1 and 2/1 (FIGURE 8.89) (Ringsdorf et al., 1989). The intermolecular spacing also decreased considerably, in proportion to the amount of TNF present in the complex. The same authors doped a non-liquid crystalline main chain polyester containing disk-like mesogens, and this

FIGURE 8.89. Main chain liquid crystalline polymers containing discotic mesogens.

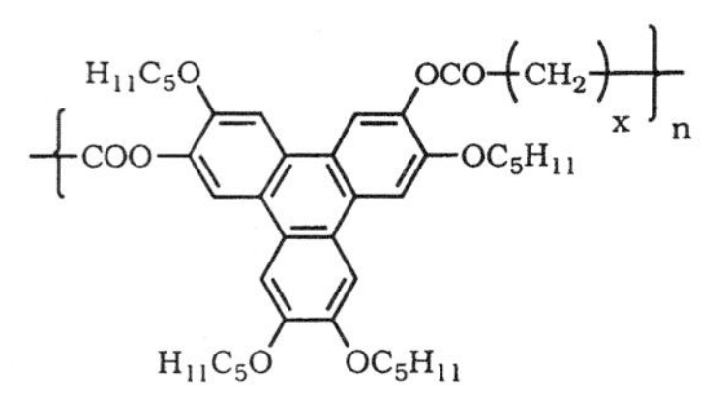

x=10 g 50 °C D_{ho} 195 °C i —TNF (0.25 mole)→ g 50 °C D_{ho} 246 °C i

x=20 g 35 °C i —TNF (0.30 mole)→ g 22 °C D_{ho} 83 °C i

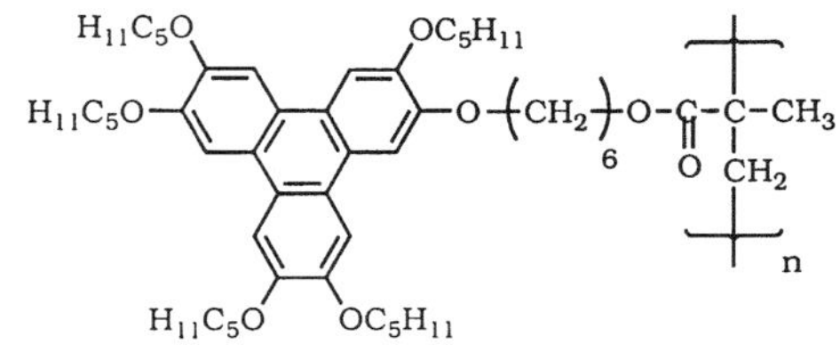

g 30 °C i —TNF (0.25 mole)→ g 50 °C N_c 170 °C i

FIGURE 8.90. Side chain liquid crystalline polymers containing discotic mesogens.

resulted in the induction of a discotic columnar mesophase (FIGURE 8.89). The intercolumnar spacings found were extremely large, and the authors assumed that considerable amounts of TNF were uncomplexed and were located in the intercolumnar spacing. Furthermore, doping of a non-liquid crystalline polymethacrylate containing donor disk-like groups in the side chain resulted in the formation of a nematic columnar mesophase (FIGURE 8.90). Doping also induced compatibility and liquid crystallinity in an incompatible mixture of nonmesogenic electron donor polymers, that is, a polymethacrylate and a polyester. The resulting discotic polymer blend exhibited a nematic columnar phase (FIGURE 8.91) (Ringsdorf et al., 1989). Finally, chirality was induced in a discotic liquid crystalline polymer by doping with a chiral acceptor (Green et al., 1990).

Research on liquid crystalline polymers containing disk-like mesogens has been restricted by their difficult multistep and low yield synthesis. Recently, a very convenient and universal procedure for the preparation of triphenylene-based discotic liquid crystalline polymers was reported (Boden et al., 1995). This development will allow detailed investigations of polymers containing discotic mesogens.

8.16 MOLECULAR RECOGNITION DIRECTED SELF-ASSEMBLY OF SUPRAMOLECULAR LIQUID CRYSTALLINE POLYMERS. H-BONDING AND OTHER INTERACTIONS

The molecular recognition of complementary components leads to systems able to self-assemble, that is,

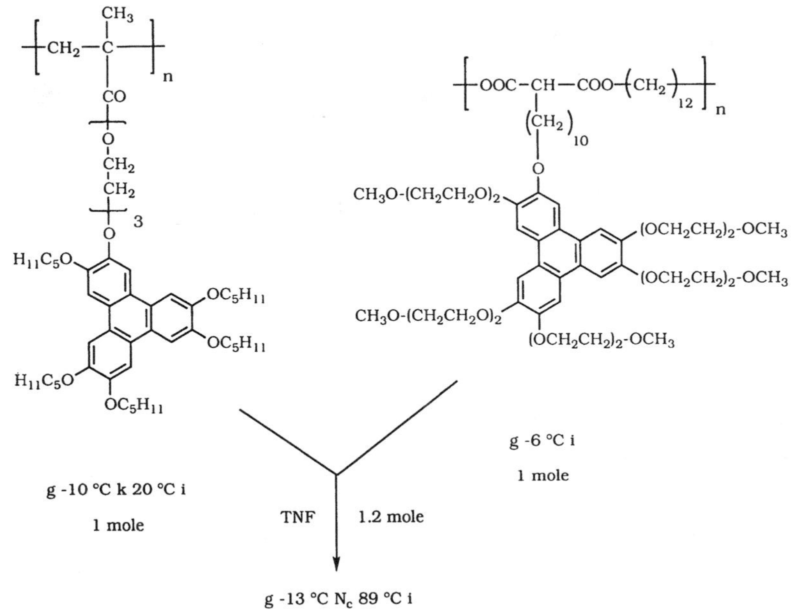

FIGURE 8.91. Induction of liquid crystallinity via electron-donor interactions.

systems capable of generating spontaneously a well-defined supramolecular architecture from their components under a well-defined set of conditions (Lehn, 1988, 1990).

Although self-assembly is a well recognized process in biological systems (Eigen and DeMaeyer, 1966; Eigen, 1971; Klug, 1983), the general concept of self-assembly of synthetic molecules by molecular recognition of complementary components received a revived interest only after it was integrated by Lehn into the new field of supramolecular chemistry (Lehn, 1988, 1990, 1993; Lindsey, 1991; Philp and Stoddart, 1991; Vgtle, 1989, 1991; Weber, 1993; Whitesides et al., 1992).

Several examples in which molecular recognition induces the association of complementary nonmesomorphic components into a low molar mass or polymeric supramolecular liquid crystal are described below.

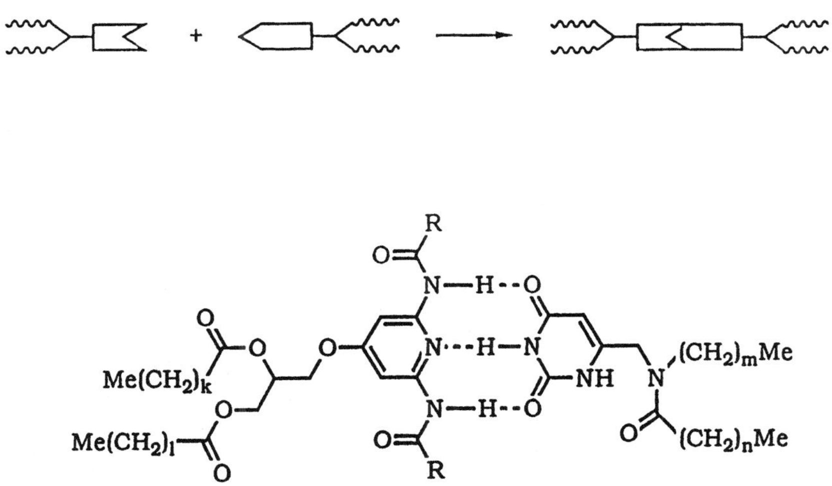

FIGURE 8.92. Formation of discotic mesogens via H-bonding.

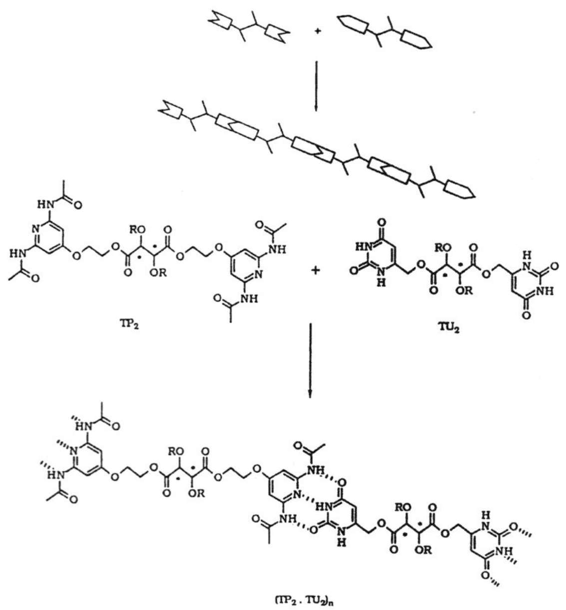

FIGURE 8.93. Formation of columnar polymers via H-bonding.

The principles of formation of a mesogenic supramolecule from two complementary components is outlined in FIGURE 8.92. The particular example used by Lehn and coworkers (Brienne et al., 1989) to generate a supramolecular disk-like mesogenic group that displays a hexagonal columnar mesophase is by formation of an array of three parallel hydrogen bonds between groups of uracil and 2,6-diaminopyridine type (FIGURE 8.93).

The transplant of the same concept to the generation of a supramolecular liquid crystalline polymer is outlined in FIGURE 8.93 (Fouquey et al., 1990). The complementary moieties TU_2 and TP_2 are uracil (U) and 2,6-diacylamino pyridine (P) groups connected through tartaric acid esters (T). The tartaric acid (T) unit provides in addition the opportunity to investigate the effect of chirality on the mesomorphic behavior of the species formed. Thus the components LP_2, LU_2, DP_2, MP_2, and MU_2 are derived from L(+), D(−), and meso (M) tartaric acid, respectively. Although all monomers (LP_2, LU_2, DP_2, MP_2, and MU_2) are only crystalline, the corresponding supramolecular "polymers" obtained through hydrogen bonding ($LP_2 + LU_2$, $DP_2 + LU_2$, and MP_2 and MU_2) exhibit hexagonal columnar mesophases. These hexagonal columnar mesophases are generated from helices that have a cylindrical suprastructure (Forquey et al., 1990).

An additional example of a supramolecular liquid crystalline polymer obtained through the hydrogen

FIGURE 8.94. Examples of dimers and polymers via H-bonding.

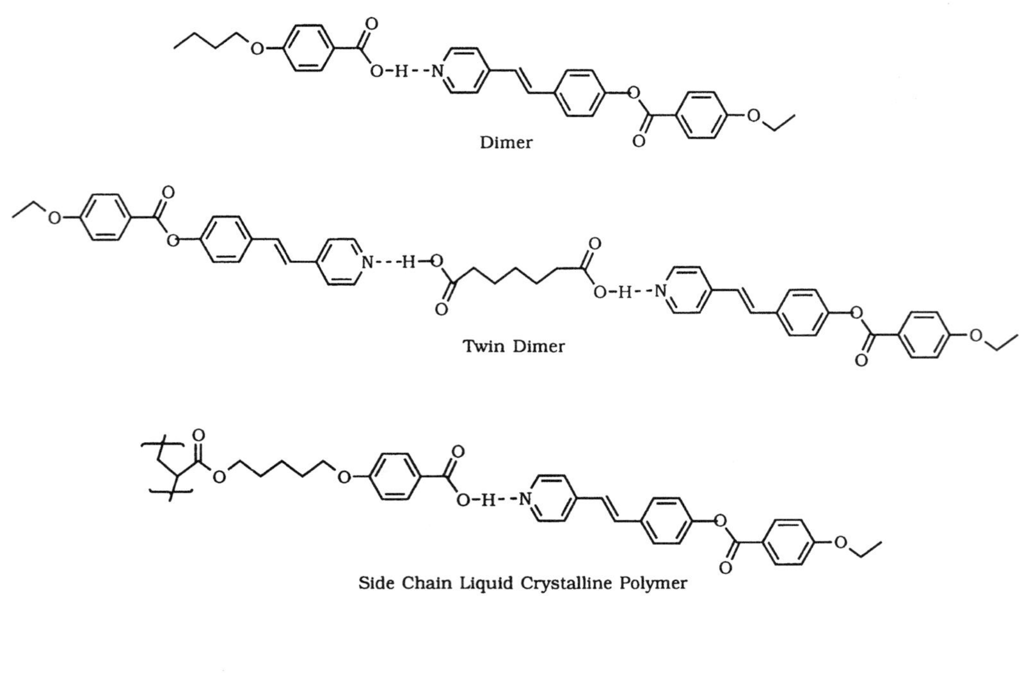

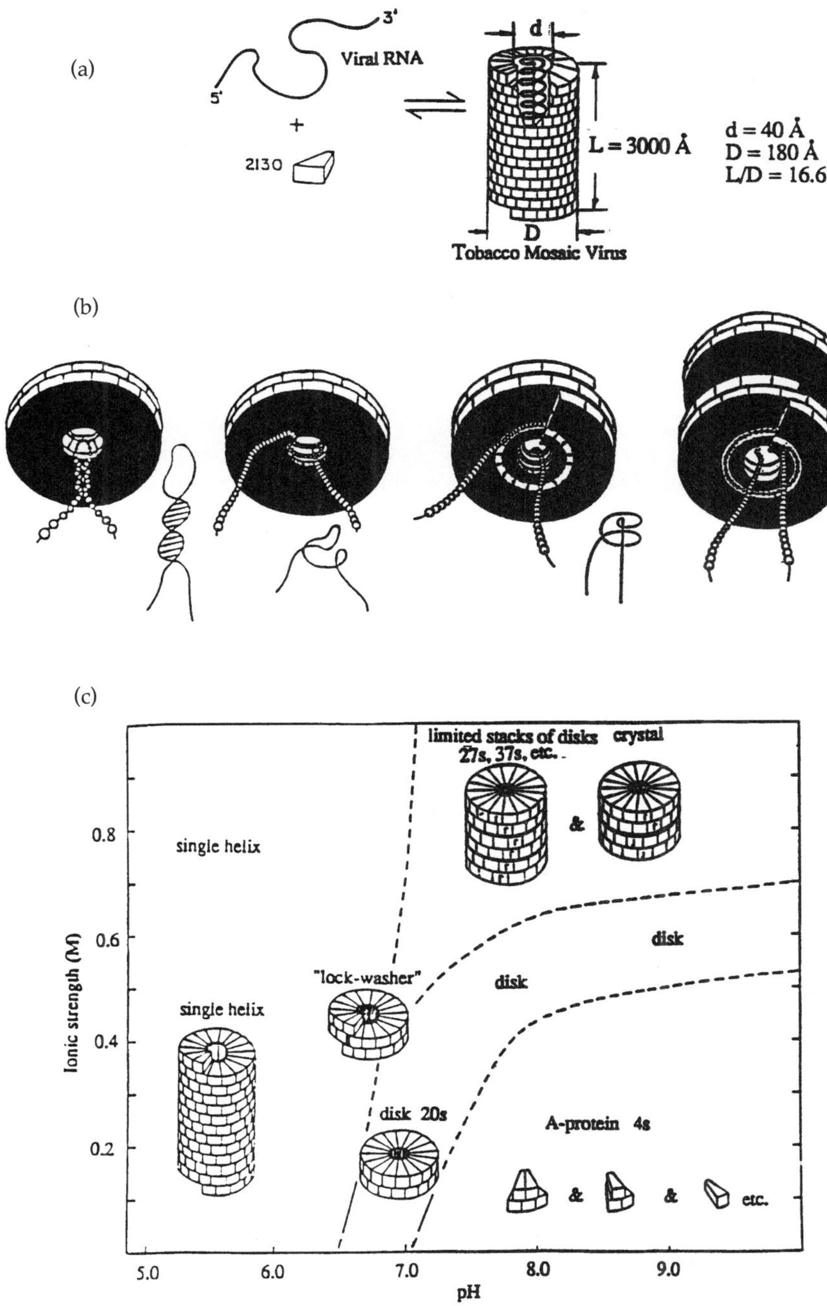

FIGURE 8.95. Self-assembly of tobacco mosaic virus (TMV). The protein subunits define the shape of the helix and the RNA defines the helix length. All information for assembly is contained within the component parts. The structure contains 16 1/3 protein subunits per turn and 2130 identical protein subunits (17,500 daltons each). The virus dimensions are: 3000 Å in length, 180 Å in diameter, a helical pitch of 23 Å, a central hole of 40 Å, with 3 nucleotides bound per protein subunit (Percec et al., 1994c).

bonding of nonmesomorphic monomers was recently reported (Fornasier et al., 1990).

Examples in which a mesophase was generated through dimerization of carboxylic acid derivatives via hydrogen bonding are available in the classic literature on liquid crystals and are extensively reviewed (Eidenschink, 1989; Gray, 1962, 1974, 1979, 1982, 1985). New and interesting examples on the generation of nonsymmetrical liquid crystalline dimers (Kato and Fréchet, 1989a), twin dimers (Kato et al., 1990a, b), side chain liquid crystalline polymers (Kato and Fréchet, 1989b; Kato et al., 1992, 1993; Kumar et al., 1992a, b), main chain liquid crystalline polymers (Alexander et al., 1994; Bladon and Griffin, 1993; Lee, C.-M. et al., 1994), and supramolecular networks (Kato et al., 1994; Wilson, 1994b) by specific hydrogen bonding "reactions" are under active investigation in several laboratories (FIGURE 8.94).

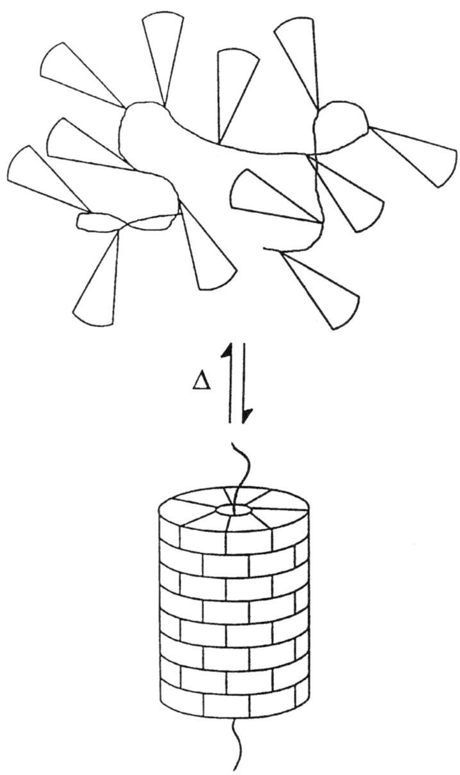

FIGURE 8.96. The self-organization of a randomly coiled flexible polymer containing tapered side groups into a rigid rod-like columnar structure.

Recently, a new approach was reported to molecular recognition directed self-assembly of supramolecular structures displaying liquid crystallinity by a mechanism which resembles that of the self-assembly of tobacco mosaic virus (TMV) (Johansson et al., 1994; Kwon et al., 1994a, 1995; Percec et al., 1991i, 1992d, 1992e, 1993c, 1993d, 1993e, 1994a; Tomazos et al., 1994; Ungar et al., 1994a). The self-assembly mechanism of TMV is outlined in FIGURE 8.95 (Klug, 1983). The synthetic approach can be summarized as follows. A flexible polymer backbone containing tapered side groups self-assembles the side groups into a column that surrounds the polymer backbone (FIGURE 8.96). These polymers exhibit thermotropic hexagonal columnar mesophases. Although the number of chains penetrating through the center of the column is not yet known and requires further research, it seems that this self-assembling system is complementary to those elaborated by Lehn and coworkers (Brienne et al., 1989; Forquey et al., 1990). In the model elaborated by Lehn and coworkers (Brienne et al., 1989; Forquey et al., 1990), the complementary pairs are self-organized through hydrogen bonding-type interactions (endo-recognition), while in the last case (Johansson et al., 1994; Klug, 1983; Kwon et al., 1994a, b, 1995; Percec et al., 1991i, 1992d, e, 1993c, d, e, 1994a; Tomazos et al., 1994; Ungar et al., 1994a), the combination of taper-shaped side groups and the interactions generated by functional groups attached to them are responsible for the generation of a polymeric column. By analogy with the self-assembly of TMV, this self-assembling process can be generated with and without a polymer backbone. Some reviews on this process are available (Percec et al., 1994b, c, e, k).

Acknowledgements
Financial support by the National Science Foundation (DMR-92-06781 and DMR-91-22227) and the Office of Naval Research is gratefully acknowledged.

REFERENCES

Abe, A., Jernigan, R. L., and Flory, P. J. 1966. *J. Am. Chem. Soc.* **88**: 631.

Abe, A., Kimura, N., and Tabata, S. 1991. *Macromolecules* **24**: 6238.

Achard, M. F., Sigaud, G., Keller, P., and Hardouin, F. 1988. *Makromol. Chem.* **189**: 1259.

Achard, M. F., Mauzac, M., Richard, H., Sigaud, G., and Hardouin, F. 1989. *Eur. Polym. J.* **25**: 593.

Alexander, C., Jariwala, C. P., Lee, C.-M., and Griffin, A. C. 1994. *Macromol. Symp.* **77**: 283.

Antonietti, M., and Conrad, J. 1994a. *Angew. Chem. Int. Ed. Engl.* **33**: 1869.

Antonietti, M., Conrad, J., and Thünemann, A. 1994b. *Macromolecules* **27**: 6007.

Antoun, S., Gagnon, D. R., Karasz, F. E., and Lenz, R. W. 1986. *J. Polym. Sci. Polym. Lett. Ed.* **24**: 503.

Arnold, H., and Sackmann, H. Z. 1960a. *Z. Phys. Chem.* **213**: 137.

Arnold, H., and Sackmann, H. Z. 1960b. *Z. Phys. Chem.* **213**: 145.

Ashton, P. R., Joachimi, D., Spencer, N., Stoddart, J. F., Tschierske, C., White, A. P. J., Williams, D. J., and Zab, K. 1994. *Angew. Chem. Int. Ed. Engl.* **33**: 1503.

Auman, B. C., Percec, V., Schneider, H. A., Jishan, W., and Cantow, H. J. 1987a. *Polymer* **28**: 119.

Auman, B. C., Percec, V., Schneider, H. A., and Cantow, H. J. 1987b. *Polymer* **28**: 1407.

Auman, B. C., and Percec, V. 1988. *Polymer* **29**: 938.

Bair, T. I., Morgan, P. W., and Killian, F. L. 1977. *Macromolecules* **10**: 1396.

Ballauff, M. 1986a. *Ber. Bunsen-Ges. Phys. Chem.* **90**: 1053.

Ballauff, M. 1986b. *Makromol. Chem. Rapid Commun.* **7**: 407.

Ballauff, M., and Schmidt, G. F. 1987. *Makromol. Chem. Rapid Commun.* **8**: 93.

Ballauff, M. 1989. *Angew. Chem. Int. Ed. Engl.* **28**: 253.

Bao, Z., Chen, Y., Cai, R., and Yu, L. J. 1993. *Macromolecules* **26**: 5281.

Bartell, L. S., and Kohl, D. A. 1963. *J. Chem. Phys.* **39**: 3097.

Bassett, D. C., ed. 1982. *Developments in Crystalline Polymers*, p. 115. London: Applied Science.

Bauer, S., Fischer, H., and Ringsdorf, H. 1993. *Angew. Chem. Int. Ed. Engl.* **32**: 1598.

Bengs, H., Renkel, R., Ringsdorf, H., Baehr, C., Ebert, M., and Wendorff, J. H. 1991. *Makromol. Chem. Rapid Commun.* **12**: 439.

Benthack-Thoms, H., and Finkelmann, H. 1985. *Makromol. Chem.* **186**: 1895.

Bilibin, A. Y., Tenkovtsev, A. V., and Piraner, O. N. 1989. *Makromol. Chem.* **190**: 3013.

Bladon, P., and Griffin, A. C. 1993. *Macromolecules* **26**: 6604.

Blumstein, A., ed. 1978a. *Liquid Crystalline Order in Polymers*. New York: Academic.

Blumstein, A., ed. 1978b. *Mesomorphic Order in Polymers and Polymerization in Liquid Crystalline Media*. ACS Symp. Series, no. 74. Washington, D.C.: American Chemical Society.

Blumstein, A., Vilasagar, S., Ponrathnam, S., Clough, S. B., Blumstein, R. B., and Maret, G. 1982. *J. Polym. Sci. Polym. Phys. Ed.* **20**: 877.

Blumstein, A., ed. 1985. *Polymeric Liquid Crystals: Proceedings of the 2nd Symposium on Polymeric Liquid Crystals [Washington, D.C., 1983]*. New York: Plenum.

Blumstein, R. B., and Blumstein, A. 1988. *Mol. Cryst. Liq. Cryst.* **165**: 361.

Boden, N., Bushby, R. J., and Cammidge, A. N. 1995. *J. Am. Chem. Soc.* **117**: 924.

Bohnert, R., Finkelmann, H., and Lutz, P. 1993. *Makromol. Chem. Rapid Commun.* **14**: 139.

Boileau, S., and Teyssié, D. 1991. *J. Inorg. Organometal. Polym.* **1**: 247.

Brandys, F. A., and Bazuin, C. G. 1993. *Am. Chem. Soc. Div. Polym. Chem., Polym. Prepr.* **34**: 186.

Brenda, M., Greiner, A., and Heitz, W. 1990. *Makromol. Chem.* **191**: 1083.

Brienne, M. J., Gabard, J., Lehn, J. M., and Stibor, I. 1989. *J. Chem. Soc. Chem. Commun.* **1989**: 1868.

Brown, C. J. 1954. *Acta Crystallogr.* **7**: 97.

Brugging, W., Kampschulte, U., Schmidt, H.-W., and Heitz, W. 1988. *Makromol. Chem.* **189**: 2755.

Campbell, T. W., and MacDonald, R. N. 1959. *J. Org. Chem.* **24**: 1246.

Carr, N., Gray, G. W., and McDonnell, D. G. 1983. *Mol. Cryst. Liq. Cryst.* **97**: 13.

Carr, N., and Gray, G. W. 1985. *Mol. Cryst. Liq. Cryst.* **124**: 27.

Casagrande, C., Veyssie, M., and Finkelmann, H. 1982. *J. Physique Lett.* **43**: 671.

Caseri, W., Sauer, T., and Wegner, G. 1988. *Makromol. Chem. Rapid Commun.* **9**: 651.

Chandrasekhar, S. 1982. *Advances in Liquid Crystals*, vol. 5, ed. G. H. Brown, p. 47. London: Academic.

Chandrasekhar, S., and Ranganath, G. S. 1990. *Rep. Prog. Phys.* **53**: 57.

Chandrasekhar, S. 1992. *Liquid Crystals*. 2d ed. Cambridge: Cambridge University Press.

Chapoy, L. L., ed. 1985. *Recent Advances in Liquid Crystalline Polymers*. Amsterdam: Elsevier.

Chen, G., and Lenz, R. W. 1985. *Polymer* **26**: 1307.

Cheng, S. Z. D., Yandrasits, M. A., and Percec, V. 1991. *Polymer* **32**: 1284.

Chiellini, E., and Lenz, R. W. 1989. *Comprehensive Polymer Science*, vol. 5, ed. G. Allen and J. C. Bevington, p. 701. Oxford: Pergamon.

Chisholm, M. H. 1991. *Angew. Chem. Int. Ed. Engl.* **30**: 673.

Chiu, K. K., Haung, H. H., and Chia, L. H. L. 1972. *J. Chem. Soc. Perkin Trans. II* **1972**: 286.

Cho, C. G., Feit, B. A., and Webster, O. W. 1990. *Macromolecules* **23**: 1918.

Choi, S.-J., Jin, S.-H., Park, J.-W., Cho, H.-N., and Choi, S.-K. 1994a. *Macromolecules* **27**: 309.

Choi, S.-J., Cho, N.-H., and Choi, S.-K. 1994b. *Polym. Bull.* **32**: 11.

Choi, S.-J., Kim, S.-H., Ahn, W., Cho, H.-N., and Choi, S.-K. 1994c. *Macromolecules* **27**: 4879.

Ciferri, A., Krigbaum, W. R., and Meyer, R. B., eds. 1982. *Polymer Liquid Crystals*. New York: Academic.

Ciferri, A., ed. 1991. *Liquid Crystallinity in Polymers: Principles and Fundamental Properties*. New York: VCH.

Cladis, P. E. 1975. *Phys. Rev. Lett* **35**: 48.

Cladis, P. E., Bogardus, R. K., Daniels, W. B., and Taylor, G. N. 1977. *Phys. Rev. Lett.* **39**: 720.

Cladis, P. E., Bogardus, R. K., and Aadsen, D. 1978. *Phys. Rev. A* **18**: 2292.

Cladis, P. E. 1988. *Mol. Cryst. Liq. Cryst.* **165**: 85.

Classen, M. S., Schmidt, H.-W., and Wendorff, J. H. 1990. *Polym. Adv. Technol.* **1**: 143.

Collings, P. J. 1990. *Liquid Crystals: Nature's Delicate Phase of Matter*. Princeton: Princeton University Press.

Cooke, A. W., and Wagener, K. B. 1991. *Macromolecules* **24**: 1404.

Coulter, P., and Windle, A. H. 1989. *Macromolecules* **22**: 1129.

Cowie, J. M. G., and Wu, H. H. 1988. *Br. Polym. J.* **20**: 515.

Craig, A. A., and Imrie, C. T. 1994. *J. Mater. Chem.* **4**: 1705.

Craig, A. A., and Imrie, C. T. 1995. *Macromolecules* **28**: 3617.

Cram, D. J. 1988. *Angew. Chem. Int. Ed. Engl.* **27**: 1009.

D'Allest, J. F., Wu, P. P., Blumstein, A., and Blumstein, R. B. 1986. *Mol. Cryst. Liq. Cryst. Lett.* **3**: 103.

Darsey, J. A., and Rao, B. K. 1981. *Macromolecules* **14**: 1575.

Davidson, P., Noirez, L., Cotton, J. P., and Keller, P. 1991. *Liq. Cryst.* **10**: 111 and references therein.

Davidson, P., and Levelut, A. M. 1992. *Liq. Cryst.* **11**: 469.

Davidson, T., Griffin, A. C., Wilson, L. M., and Windle, A. H. 1995. *Macromolecules* **28**: 354.

Davis, S. J., Johnson, B. F. G., Khan, M. S., and Lewis, J. 1991. *J. Chem. Soc. Chem. Commun.* **1991**: 187.

de Gennes, P. G. 1969. *Phys. Lett.* **28A**: 725.

de Gennes, P. G. 1975. *C. R. Acad. Sci. Paris B* **281**: 101.

de Gennes, P. G., and Pinkus, P. 1977. *Am. Chem. Soc. Div. Polym. Chem., Polym. Prepr.* **18**: 161.

Degiorgio, V., and Corti, M., eds. 1985. *Physics of Amphiphiles: Micelles, Vesicles, and Microemulsions*. Amsterdam: North Holland.

Dembek, A. A., Burch, R. R., and Feiring, A. E. 1993. *J. Am. Chem. Soc.* **115**: 2087.

Demus, D., Diele, S., Grande, S., and Sackmann, H. Z. 1983. *Advances in Liquid Crystals*, vol. 6, ed. G. H. Brown, p. 1. London: Academic.

Demus, D. 1988. *Mol. Cryst. Liq. Cryst.* **165**: 45.

Demus, D. 1989. *Liq. Cryst.* **5**: 75.

Destri, S., Mascherpa, M., and Porzio, W. 1993. *Adv. Mater.* **5**: 43.

Donald, A. M., and Windle, A. H. 1991. *Liquid Crystalline Polymers*. Cambridge: Cambridge University Press.

Duran, R., Ballauff, M., Wenzel, M., and Wegner, G. 1988. *Macromolecules* **21**: 2897.

Economy, J. 1989. *Mol. Cryst. Liq. Cryst.* **169**: 1.

Economy, J. 1990a. *Angew. Chem. Int. Ed. Engl.* **29**: 1256.

Economy, J., Johnson, R. D., Lyerla, J. R., and Muhlebach, A. 1990b. *Liquid Crystalline Polymers*, ed. R. A. Weiss and C. K. Ober, p. 129. ACS Symp. Series, no. 435. Washington, D.C.: American Chemical Society.

Eidenschink, R. 1989. *Angew. Chem. Int. Ed. Engl.* **28**: 1424.

Eigen, M., and DeMaeyer, L. 1966. *Naturwissensch.* **53**: 50.

Eigen, M. 1971. *Naturwissensch.* **58**: 465.

Espinet, P., Esteruelas, M. A., Oro, L. A., Serrano, J. L., and Sola, E. 1992. *Coord. Chem. Rev.* **117**, 215.

Feast, W. J., and Millichamp, I. S. 1983. *Polym. Commun.* **24**: 102.

Feljoo, J. L., Ungar, G., Owen, A. J., Keller, A., and Percec, V. 1988. *Mol. Cryst. Liq. Cryst.* **155**: 487.

Ferrarini, A., Luckhurst, G. R., Nordio, P. L., and Roskilly, S. J. 1993. *Chem. Phys. Lett.* **214**: 409.

Finkelmann, H., Ringsdorf, H., and Wendorff, J. H. 1978a. *Makromol. Chem.* **179**: 273.

Finkelmann, H., Happ M., Portugal, M., and Ringsdorf, H. 1978b. *Makromol. Chem.* **179**: 2541.

Finkelmann, H., Luhmann, B., Rehage, G., and Stevens, H. 1984. *Liquid Crystals and Ordered Fluids*, vol. 4, ed. A. C. Griffin and J. F. Johnson, p. 715. New York: Plenum.

Finkelmann, H. 1987. *Angew. Chem. Int. Ed. Engl.* **26**: 816.

Flory, P. J. 1956. *Proc. R. Soc. London A* **234**: 73.

Flory, P. J., and Ronca, G. 1979. *Mol. Cryst. Liq. Cryst.* **54**: 311.

Flory, P. J. 1984a. *Adv. Polym. Sci.* **59**: 1.

Flory, P. J., and Matheson, R. R. 1984b. *J. Phys. Chem.* **88**: 6606.

Fornasier, R., Tornatore, M., and Chapoy, L. L. 1990. *Liq. Cryst.* **8**: 787.

Fouquey, C., Lehn, J. M., and Levelut, A. M. 1990. *Adv. Mater.* **2**: 254.

Freund, L., Jung, H., Niessner, N., Schmidt, H.-W., and Heitz, W. 1989. *Makromol. Chem.* **190**: 1561.

Frosini, V., Levita, G., Lupinacci, D., and Magagnini, P. L. 1981. *Mol. Cryst. Liq. Cryst.* **66**: 21.

Fyfe, H. B., Miekuz, M., Zargarian, D., Taylor, N. J., and Marder, T. B. 1991. *J. Chem. Soc. Chem. Commun.* **1991**: 188.

Galerne, Y. 1988. *Mol. Cryst. Liq. Cryst.* **165**: 131.

Galli, G., Chiellini, E., Laus, M., Angeloni, A. S., Bignozzi, M. C., and Francescangeli, O. 1994. *Mol. Cryst. Liq. Cryst.* **254**: 429.

Gardner, K. H., Gochanour, C. R., Irwin, R. S., Weeny, S., and Weinberg, M. 1988. *Mol. Cryst. Liq. Cryst.* **155B**: 239.

Gaudiana, R. A., Minns, R. A., Rogers, H. G., Sinta, R., Taylor, L. D., Kalyaranaman, P. S., and McGowan, C. 1987. *J. Polym. Sci. Polym. Chem. Ed.* **25**: 1249.

Gaudiana, R. A., Minns, R. A., Sinta, R., Weeks, N., and Rogers, H. G. 1989. *Prog. Polym. Sci., Japan* **14**: 47.

Giroud-Godquin, A.-M., and Maitlis, P. M. 1991. *Angew. Chem. Int. Ed. Engl.* **30**: 375.

Gleim, W., and Finkelmann, H. 1989. *Side Chain Liquid Crystal Polymers*, ed. C. B. McArdle, p. 287. New York: Chapman and Hall.

Godovsky, Y. K., Mamaeva, I. I., Makarova, N. N., Papkov, V. P., and Kuzmin, N. N. 1985. *Makromol. Chem. Rapid Commun.* **6**: 797.

Goodby, J. W., Leslie, T. M., Cladis, P. E., and Finn, P. L. 1984. *Liquid Crystals and Ordered Fluids*, vol. 4, ed. A. C. Griffin and J. F. Johnson, p. 203. New York: Plenum.

Gordon, M., and Platé, N. A., eds. 1984a. *Liquid Crystal Polymers I*. Adv. Polym. Sci., vol. 59. Berlin: Springer.

Gordon, M., and Platé, N. A., eds. 1984b. *Liquid Crystal Polymers II/III*. Adv. Polym. Sci., vols. 60–61. Berlin: Springer.

Gray, G. W. 1962. *Molecular Structure and the Properties of Liquid Crystals*. London: Academic.

Gray, G. W. 1974. *Liquid Crystals and Plastic Crystals*, ed. G. W. Gray and P. A. Winsor, p. 125. Chichester: Harwood.

Gray, G. W. 1979. *The Molecular Physics of Liquid Crystals*, ed. G. R. Luckhurst and G. W. Gray, p. 14. New York: Academic.

Gray, G. W. 1982. *Polymer Liquid Crystals*, ed. A. Ciferri et al., p. 5. New York: Academic.

Gray, G. W., and Goodby, J. W. 1984. *Smectic Liquid Crystals*. Glasgow: Leonard-Hill.

Gray, G. W. 1985. *Proc. R. Soc. London A* **402**: 1.

Gray, G. W., ed. 1987. *Thermotropic Liquid Crystals*. Chichester: Wiley.

Gray, G. W. 1989. *Side Chain Liquid Crystal Polymers*, ed. C. B. McArdle, p. 106. New York: Chapman and Hall.

Green, M. M., Ringsdorf, H., Wagner, J., and Wustefeld, R. 1990. *Angew. Chem. Int. Ed. Engl.* **29**: 1478.

Greiner, A., and Heitz, W. 1988. *Makromol. Chem. Rapid Commun.* **9**: 581.

Griffin, A. C., and Johnson, J. F., eds. 1984. *Liquid Crystals and Ordered Fluids*. Vol. 4. New York: Plenum.

Gubina, T. I., Kostromin, S. G., Tal'rose, R. V., Shibaev, V. P., and Platé, N. A. 1986. *Vysokomol. Soed. B* **28**: 394.

Gubina, T. I., Kise, S., Kostromin, S. G., Tal'rose, R. V., Shibaev, V. P., and Platé, N. A. 1989. *Liq. Cryst.* **4**: 197.

Guillon, D., Cladis, P. E., and Stamatoff, J. 1978. *Phys. Rev. Lett.* **41**: 1598.

Hagihara, N., Sonogashira, K., and Takahashi, S. 1980. *Adv. Polym. Sci.* **41**: 149.

Hahn, B., Wendorff, J. H., Portugal, M., and Ringsdorf, H. 1981. *Colloid Polym. Sci.* **259**: 875.

Hahn, B., and Percec, V. 1988. *Mol. Cryst. Liq. Cryst.* **157**: 125.

Hardouin, F., Levelut, A. M., Achard, M. F., and Sigaud, G. 1983. *J. Chim. Phys.* **80**: 53.

Hardouin, F. 1986. *Physica A* **140**: 359.

Hardouin, F., Mery, S., Achard, M. F., Noirez, L., and Keller, P. 1991. *J. Physique II* **1**: 511.

Hardouin, F., Sigaud, G., Achard, M. F., Brûlet, A., Cotton, J. P., Yoon, D. Y., Percec, V., and Kawasumi, M. 1995. *Macromolecules* **28**: 5427.

Harris, F. W., and Hsu, S. L. C. 1989. *High Perform. Polym.* **1**: 3.

Harwood, D. J. 1987. *Makromol. Chem. Macromol. Symp.* **10–11**: 331.

Hatke, W., Land, H. T., Schmidt, H.-W., and Heitz, W. 1991. *Makromol. Chem. Rapid Commun.* **12**: 235.

Hawker, J. C., Wooley, K. L., and Fréchet, J. M. J. 1994. *Macromol. Symp.* **77**: 11.

Heitz, T., Rohrbach, P., and Hocker, H. 1989. *Makromol. Chem.* **190**: 3295.

Heitz, T., and Webster, O. W. 1991. *Makromol. Chem.* **192**: 2463.

Heitz, W. 1986. *Chemiker Zeitung* **110**: 385.

Heitz, W., Brugging, W., Freund, L., Gailberger, M., Greiner, A., Jung, H., Kampschulte, U., Niessner, N., Osan, F., Schmidt, H.-W., and Wicker, M. 1988. *Makromol. Chem.* **189**: 119.

Heitz, W. 1989. *Makromol. Chem. Macromol. Symp.* **26**: 1.

Heitz, W., and Niessner, N. 1990a. *Makromol. Chem.* **191**: 225.

Heitz, W., and Schmidt, H.-W. 1990b. *Makromol. Chem. Macromol. Symp.* **38**: 149.

Heitz, W. 1991a. *Makromol. Chem. Macromol. Symp.* **47**: 111.

Heitz, W. 1991b. *Makromol. Chem. Macromol. Symp.* **48–49**: 15.

Heroguez, V., Deffieux, A., and Fontanille, M. 1990. *Makromol. Chem. Macromol. Symp.* **32**: 199.

Heroguez, V., Schappacher, M., Papon, E., and Deffieux, A. 1991. *Polym. Bull.* **25**: 307.

Hohlweg, M., and Schmidt, H.-W. 1989. *Makromol. Chem.* **190**: 1587.

Hopken, J., Pugh, C., Richtering, W., and Möller, M. 1988. *Makromol. Chem.* **189**: 911.

Horhold, H. H., and Opfermann, J. 1970. *Makromol. Chem.* **131**: 105.

Horhold, H. H., Helbig, M., Raabe, D., Opfermann, J., Scherf, U., Stockmann, R., and Weiss, D. 1987. *Z. Chem.* **27**: 126.

Hsiao, B. S., Stein, R. S., Weeks, N., and Gaudiana, R. A. 1991. *Macromolecules* **24**: 1299.

Hsieh, C. J., Hsiue, G. H., and Hsu, C. S. 1990a. *Makromol. Chem.* **191**: 2195.

Hsieh, C. J., Hsu, C. S., Hsiue, G. H., and Percec, V. 1990b. *J. Polym. Sci. Polym. Chem. Ed.* **28**: 425.

Hsiue, G. H., Wen, J. S., and Hsu, C. S. 1991. *Makromol. Chem.* **192**: 2243.

Hsu, C. S., and Percec, V. 1987. *J. Polym. Sci. Polym. Chem. Ed.* **25**: 2909.

Hsu, C. S., and Percec, V. 1989. *J. Polym. Sci. Polym. Chem. Ed.* **27**: 453.

Hudson, S. A., and Maitlis, P. M. 1993. *Chem. Rev.* **93**: 861.

Iimura, K., Koide, N., Tanabe, H., and Takeda, M. 1981. *Makromol. Chem.* **182**: 2569.

Imrie, C. T., Karasz, F. E., and Attard, G. S. 1991. *Liq. Cryst.* **9**: 47.

Imrie, C. T., and Paterson, B. J. A. 1994. *Macromolecules* **27**: 6673.

Irvine, P. A., Dacheng, W., and Flory, P. J. 1984. *J. Chem. Soc. Faraday Trans. I* **80**: 1795.

Irwin, R. S., Weeny, S., Gardner, K. H., Gochanour, C. R., and Weinberg, M. 1989. *Macromolecules* **22**: 1065.

Isberner, J., Moors, R., and Vögtle, F. 1994. *Angew. Chem. Int. Ed. Engl.* **33**: 2413.

Ivanov, P., Pojarlieff, I., and Tyutyulkov, N. 1976. *Tetrahedron Lett.* **1976**: 775.

Jackson, W. J., and Kuhfuss, H. F. 1976. *J. Polym. Sci. Polym. Chem. Ed.* **14**: 2043.

Jackson, W. J. 1989. *Mol. Cryst. Liq. Cryst.* **169**: 23.

Jacobus, J. 1976. *Tetrahedron Lett.* **1976**: 2927.

Jadhav, J. Y., Krigbaum, W. R., and Preston, J. J. 1988. *Macromolecules* **21**: 538.

Jadhav, J. Y., Preston, J. J., and Krigbaum, W. R. 1989. *J. Polym. Sci. Polym. Chem. Ed.* **27**: 1175.

Jahns, E., and Finkelmann, H. 1987. *Colloid Polym. Sci.* **265**: 304.

Jin, J.-I., and Kim, R. 1987. *Polym. J.* **19**: 977.

Jin, J.-I. 1994a. *Mol. Cryst. Liq. Cryst.* **254**: 197.

Jin, J.-I., Kang, C.-S., Lee, I.-H., and Yun, Y.-K. 1994b. *Macromolecules* **27**: 2664.

Jin, S.-H., Kim, S.-H., Cho, H.-N., and Choi, S.-K. 1991. *Macromolecules* **24**: 6050.

Jin, S.-H., Choi, S.-J., Ahn, W., Cho, H.-N., and Choi, S.-K. 1993. *Macromolecules* **26**: 1487.

Johansson, G., Percec, V., Ungar, G., and Abramic, D. 1994. *J. Chem. Soc. Perkin Trans. I* **1994**: 447.

Jonsson, H., Werner, P. E., Gedde, U. W., and Hult, A. 1989. *Macromolecules* **22**: 1683.

Jonsson, H., Gedde, U. W., and Hult, A. 1990a. *Liquid Crystalline Polymers*, ed. R. A. Weiss and C. K. Ober, p. 62. ACS Symp. Series, no. 435. Washington, D.C.: American Chemical Society.

Jonsson, H., Wallgren, E., Hult, A., and Gedde, U. W. 1990b. *Macromolecules* **23**: 1041.

Jonsson, H., Percec, V., and Hult, A. 1991a. *Polym. Bull.* **25**: 115.

Jonsson, H., Andersson, H., Sundell, P. E., Gedde, U. W., and Hult, A. 1991b. *Polym. Bull.* **25**: 641.

Jonsson, H., Sundell, P. E., Percec, V., Gedde, U. W., and Hult, A. 1991c. *Polym. Bull.* **25**: 649.

Jonsson, H., Percec, V., Gedde, U. W., and Hult, A. 1992. *Makromol. Chem. Macromol. Symp.* **54–55**: 83.

Kalinowski, H., and Kessler, H. 1973. *Topics in Stereochemistry*, vol. 7, ed. E. L. Eliel and N. L. Allinger, p. 295. New York: Interscience.

Kato, T., and Fréchet, J. M. J. 1989a. *J. Am. Chem. Soc.* **111**: 8533.

Kato, T., and Fréchet, J. M. J. 1989b. *Macromolecules* **22**: 3819.

Kato, T., Fujishima, A., and Fréchet, J. M. J. 1990a. *Chem. Lett.* **1990**: 919.

Kato, T., Wilson, P. G., Fujishima, A., and Fréchet, J. M. J. 1990b. *Chem. Lett.* **1990**: 2003.

Kato, T., Kihara, H., Uryu, T., Fujishima, A., and Fréchet, J. M. J. 1992. *Macromolecules* **25**: 6836.

Kato, T., Kihara, H., Uryu, T., Ujiie, S., Iimura, K., Fréchet, J. M. J., and Kumar, U. 1993. *Ferroelectrics* **148**: 161.

Kato, T., Kihara, H., Kumar, U., Uryu, T., and Fréchet, J. M. J. 1994. *Angew. Chem. Int. Ed. Engl.* **33**: 1644.

Kawabe, M., Yamaoka, I., and Kimura, M. 1990. *Liquid Crystalline Polymers*, ed. R. A. Weiss and C. K. Ober, p. 115. ACS Symp. Series, no. 435. Washington, D.C.: American Chemical Society.

Kelker, H., and Hatz, R. 1980. *Handbook of Liquid Crystals.* Weiheim: Chemie.

Keller, A., Ungar, G., and Percec, V. 1990. *Advances in Liquid Crystalline Polymers*, ed. R. A. Weiss and C. K. Ober, p. 308. ACS Symp. Series, no. 435. Washington, D.C.: American Chemical Society.

Kelly, S. M., and Schad, H. 1984. *Mol. Cryst. Liq. Cryst.* **110**: 239.

Kern, W., Gruber, W., and Wirth, H. O. 1960. *Makromol. Chem.* **37**: 198.

Kessler, H. 1970. *Angew. Chem. Int. Ed. Engl.* **9**: 219.

Kim, H. Y. 1992a. *J. Am. Chem. Soc.* **114**: 4947.

Kim, H. Y. 1992b. *Adv. Mater.* **4**: 764.

Kim, Y. H., Pincus, P., 1979. *Biopolymers*, **18**: 2315.

Klug, A. 1983. *Angew. Chem. Int. Ed. Engl.* **22**: 565.

Kodaira, T., and Mori, K. 1990. *Makromol. Chem. Rapid Commun.* **11**: 645.

Koide, N., Akiyama, E., and Mihara, T. 1994. *Mol. Cryst. Liq. Cryst.* **254**: 283.

Komiya, Z., Pugh, C., and Schrock, R. R. 1992a. *Macromolecules* **25**: 3609.

Komiya, Z., Pugh, C., and Schrock, R. R. 1992b. *Macromolecules* **25**: 6586.

Komiya, Z., and Schrock, R. R. 1993a. *Macromolecules* **26**: 1387.

Komiya, Z., and Schrock, R. R. 1993b. *Macromolecules* **26**: 1393.

Kops, J., and Spanggaard, H. 1986. *Polym. Bull.* **16**: 507.

Kosaka, Y., and Uryu, T. 1994. *Macromolecules* **27**: 6286.

Kosaka, Y., and Uryu, T. 1995. *Macromolecules* **28**: 870.

Kossmehl, G., and Samandari, M. 1985. *Makromol. Chem.* **186**: 1565.

Kostromin, S. G., Tal'rose, R. V., Shibaev, V. P., and Platé, N. A. 1982. *Makromol. Chem. Rapid Commun.* **3**: 803.

Kostromin, S. G., Cuong, N. D., Garina, E. S., and Shibaev, V. P. 1990a. *Mol. Cryst. Liq. Cryst.* **193**: 177.

Kostromin, S. G., Shibaev, V. P., and Diele, S. 1990b. *Makromol. Chem.* **191**: 2521.

Kretzschmann, H., and Meier, H. 1991. *Tetrahedron Lett.* **32**: 5059.

Kreuder, W., Webster, O. W., and Ringsdorf, H. 1986. *Makromol. Chem. Rapid Commun.* **7**: 5.

Kricheldorf, H. R., and Pakull, R. 1985. *J. Polym. Sci. Polym. Lett. Ed.* **23**: 413.

Kricheldorf, H. R., and Schwarz, G. 1987a. *Makromol. Chem.* **188**: 1281.

Kricheldorf, H. R., and Pakull, R. 1987b. *Polymer* **28**: 1772.

Kricheldorf, H. R., Schwarz, G., and Ruhser, F. 1988a. *J. Polym. Sci. Polym. Chem. Ed.* **26**: 1621.

Kricheldorf, H. R., and Pakull, R. 1988b. *Macromolecules* **21**: 551.

Kricheldorf, H. R., Pakull, R., and Buchner, S. 1988c. *Macromolecules* **21**: 1929.

Kricheldorf, H. R., and Awe, J. 1988d. *Makromol. Chem. Rapid Commun.* **9**: 681.

Kricheldorf, H. R., Doring, V., and Eckhardt, V. 1988e. *Makromol. Chem.* **189**: 1425.

Kricheldorf, H. R., Doring, V., Beuermann, I., and Eckhardt, V. 1988f. *Makromol. Chem.* **189**: 1437.

Kricheldorf, H. R., and Jenssen, J. 1989a, *Eur. Polym. J.* **25**: 1273.

Kricheldorf, H. R., Pakull, R., and Buchner, S. 1989b. *J. Polym. Sci. Polym. Chem. Ed.* **27**: 431.

Kricheldorf, H. R., Beuermann, I., and Schwarz, G. 1989c. *Makromol. Chem. Rapid Commun.* **10**: 211.

Kricheldorf, H. R., and Lubbers, D. 1989d. *Makromol. Chem. Rapid Commun.* **10**: 383.

Kricheldorf, H. R., and Awe, J. 1989e. *Makromol. Chem.* **190**: 2579.

Kricheldorf, H. R., and Awe, J. 1989f. *Makromol. Chem.* **190**: 2597.

Kricheldorf, H. R., and Pakull, R. 1989g. *New Polym. Mater.* **1**: 165.

Kricheldorf, H. R., Schwarz, G., and Nowatzky, W. 1989h. *Polymer* **30**: 936.

Kricheldorf, H. R., and Engelhardt, J. 1990a. *J. Polym. Sci. Polym. Chem. Ed.* **28**: 2335.

Kricheldorf, H. R., and Lubbers, D. 1990b. *Macromolecules* **23**: 2656.

Kricheldorf, H. R., and Huner, R. 1990c. *Makromol. Chem. Rapid Commun.* **11**: 211.

Kricheldorf, H. R., and Lubbers, D. 1990d. *Makromol. Chem. Rapid Commun.* **11**: 303.

Kricheldorf, H. R., and Erxleben, J. 1990e. *Polymer* **31**: 944.

Kricheldorf, H. R. 1994. *Mol. Cryst. Liq. Cryst.* **254**: 87.

Krigbaum, W. R. 1985. *J. Appl. Polym. Sci., Appl. Polym. Symp* **41**: 105.

Krigbaum, W. R., Tanaka, T., Brelsford, G., and Ciferri, A. 1991. *Macromolecules* **24**: 4142.

Kromer, H., Kuhn, R., Pielartzik, H., Siebke, W., Eckhardt, V., and Schmidt, M. 1991. *Macromolecules* **24**: 1950.

Kuhfuss, H. F., and Jackson, W. J. 1973. *US Patent* 3,778,410.

Kuhfuss, H. F., and Jackson, W. J. 1974. *US Patent* 3,804,805.

Kumar, R. S., Clough, S. B., and Blumstein, A. 1988. *Mol. Cryst. Liq. Cryst.* **157**: 387.

Kumar, U., Fréchet, J. M. J., Kato, T., Ujiie, S., and Iimura, K. 1992a. *Angew. Chem. Int. Ed. Engl.* **31**: 1531.

Kumar, U., Kato, T., and Fréchet, J. M. J. 1992b. *J. Am. Chem. Soc.* **114**: 6630.

Kuschel, F., Madicke, A., Diele, S., Utschik, H., Hisgen, B., and Ringsdorf, H. 1990. *Polym. Bull.* **23**: 373.

Kwolek, S. L., Morgan, P. W., Shaefgen, J. R., and Gulrich, L. W. 1977. *Macromolecules* **10**: 1390.

Kwolek, S. L., and Louise, R. R. 1986. *Macromolecules* **19**: 1789.

Kwon, Y. K., Chvalun, S., Schneider, A. I., Blackwell, J., Percec, V., and Heck, J. A. 1994a. *Macromolecules* **27**: 6129.

Kwon, Y. K., Danko, C., Chvalun, S., Blackwell, J., Heck, J. A., and Percec, V. 1994b. *Macromol. Symp.* **87**: 103.

Kwon, Y. K., Danko, C., Chvalun, S., Blackwell, J., Heck, J. A., and Percec, V. 1995. *Macromolecules* **28**: 1552.

Kyotari, M., and Manetsuna, H. 1983. *J. Polym. Sci. Polym. Phys. Ed.* **21**: 379.

Ladika, M., Tisk, T. E., Wu, W. W., and Jons, D. 1994. *J. Am. Chem. Soc.* **116**: 12093.

Lai, C. Y., DeBona, B. T., and Prevorsek, D. C. 1988. *J. Appl. Polym. Sci.* **36**: 819.

Lai, C. Y., DeBona, B. T., and Prevorsek, D. C. 1990. *Liquid Crystalline Polymers*, ed. R. A. Weiss and C. K. Ober, p. 102. ACS Symp. Series, no. 435, Washington, D.C.: American Chemical Society.

Laus, M., Caretti, D., Angeloni, A. S., Galli, G., and Chiellini, E. 1991. *Macromolecules* **24**: 1459.

Le Barny, P., Dubois, J. C., Friedrich, C., and Noël, C. 1986. *Polym. Bull.* **15**: 341.

Lee, C.-M., Jariwala, C. P., and Griffin, A. C. 1994. *Polymer* **35**: 4550.

Lee, K.-S., Lee, W.-K., Lee, S.-M., Kim, H.-O., and Lee, B.-H. 1994. *Mol. Cryst. Liq. Cryst.* **254**: 37.

Legrand, C., Le Borgne, A., Bunel, C., Lacoudre, N., Le Barny, P., Spassky, N., and Vairon, J. P. 1990. *Makromol. Chem.* **191**: 2979.

Lehmann, O. 1922. *Handbuch der biologischen Arbeitsmethoden*, ed. E. Abderhalden, p. 123. Berlin: Urban and Schwarzenberg.

Lehn, J. M. 1988. *Angew. Chem. Int. Ed. Engl.* **27**: 89.

Lehn, J. M. 1990. *Angew. Chem. Int. Ed. Engl.* **29**: 1304.

Lehn, J. M. 1993. *Makromol. Chem. Macromol. Symp.* **69**: 1.

LeMoigne, J., Soldera, A., Guillon, D., and Skoulis, A. 1989. *Liq. Cryst.* **6**: 627.

LeMoigne, J., Hilberer, A., and Strazielle, C. 1991a. *Am. Chem. Soc. Div. Polym. Chem. Polym. Prepr.* **32**: 96.

LeMoigne, J., Hilberer, A., and Kajzar, F. 1991b. *Makromol. Chem.* **192**: 515.

LeMoigne, J., Hilberer, A., and Strazielle, C. 1992. *Macromolecules* **25**: 6705.

Lenz, R. W., Jin, J.-I., and Feichtinger, K. A. 1983. *Polymer* **24**: 327.

Li, J. F., Percec, V., and Rosenblatt, C. 1993. *Phys. Rev. E* **48**: 1.

Li, J. F., Percec, V., Rosenblatt, C., and Lavrentovich, O. D. 1994. *Europhys. Lett.* **25**: 199.

Lindsey, J. S. 1991. *New J. Chem.* **15**: 153.

Luckhurst, G. R., and Gray, G. W., eds. 1979. *The Molecular Physics of Liquid Crystals*, p. 14. New York: Academic.

Luhmann, B., and Finkelmann, H. 1986. *Colloid Polym. Sci.* **264**: 189.

Luhmann, B., and Finkelmann, H. 1987. *Colloid Polym. Sci.* **265**: 506.

MacDonald, W. A., McLenaghan, A. D. W., McLean, G., Richards, R. W., and King, S. M. 1991. *Macromolecules* **24**: 6164.

Magagnini, P. L. 1981. *Makromol. Chem. Suppl.* **4**: 223.

Mahler, W., Guillon, D., and Skoulios, A. 1985. *Mol. Cryst. Liq. Cryst. Lett.* **2**: 111.

Majnusz, J., Catala, J. M., and Lenz, R. W. 1983. *Eur. Polym. J.* **19**: 1043.

Mannschreck, A. 1965. *Tetrahedron Lett.* **1965**: 1341.

Mannschreck, A., Mattheus, A., and Rissmann, G. 1967. *J. Mol. Spectrosc.* **23**: 15.

Martelock, H., Greiner, A., and Heitz, W. 1991. *Makromol. Chem.* **192**: 967.

Martin, P. G., and Stupp, S. I. 1988. *Macromolecules* **21**: 1222.

Mattoussi, H., Ober, R., Veyssie, M., and Finkelmann, H. 1986. *Europhys. Lett.* **2**: 233.

McArdle, C. B., ed. 1989. *Side Chain Liquid Crystal Polymers*. New York: Chapman and Hall.

Memeger, W. 1989. *Macromolecules* **22**: 1577.

Memeger, W. 1991. Experimental Station, DuPont, copy of laboratory notebook. Personal communication, 11 November 1991.

Menger, F. M., Brocchini, S., and Chen, X. Y. 1992. *Angew. Chem. Int. Ed. Engl.* **31**: 1492.

Menger, F. M., Chen, X. Y., Brocchini, S., Hopkins, H. P., and Hamilton, D. 1993. *J. Am. Chem. Soc.* **115**: 6600.

Metzmann, F. H., Ballauff, M., Schulz, R. C., and Wegner, G. 1989. *Makromol. Chem.* **190**: 985.

Meurisse, P., Lauprêtre, F., and Noël, C. 1984. *Mol. Cryst. Liq. Cryst.* **110**: 41.

Moore, J. S., and Stupp, S. I. 1988. *Macromolecules* **21**: 1217.

Morgan, P. W. 1977. *Macromolecules* **10**: 1381.

Mormann, W., and Hissmann, E. 1987. *Tetrahedron Lett.* **28**: 3087.

Mormann, W., and Brahm, M. 1989a. *Makromol. Chem.* **190**: 631.

Mormann, W., and Hohn, E. 1989b. *Makromol. Chem.* **190**: 1919.

Mormann, W., and Baharifar, A. 1990a. *Polym. Bull.* **24**: 413.

Mormann, W., and Brahm, M. 1990b. *Mol. Cryst. Liq. Cryst.* **185**: 163.

Mormann, W., and Benadda, S. 1991a. *Makromol. Chem.* **192**: 2411.

Mormann, W., and Brahm, M. 1991b. *Macromolecules* **24**: 1096.

Mormann, W., Benadda, S., and Brahm, M. 1995. *Macromol. Chem. Phys.* **196**: 543.

Mühlbach, K., and Percec, V. 1987. *J. Polym. Sci. Polym. Chem. Ed.* **25**: 2605.

Nakano, T., Hawegawa, T., and Okamoto, Y. 1993. *Macromolecules* **26**: 5454.

Noël, C 1988. *Makromol. Chem. Macromol. Symp.* **22**: 95.

Noël, C. 1989. *Side Chain Liquid Crystal Polymers*, ed. C. B. McArdle, p. 159. New York: Chapman and Hall.

Noël, C., and Navard, P. 1991. *Prog. Polym. Sci., Japan* **16**: 55.

Noirez, L., Cotton, J. P., Hardouin, F., Keller, P., Moussa, F., Pepy, G., and Strazielle, C. 1988. *Macromolecules* **21**: 2889.

Noirez, L., Keller, P., and Cotton, J. P. 1995. *Liq. Cryst.* **18**: 129.

Noll, A., Siegfried, N., and Heitz, W. 1990. *Makromol. Chem. Rapid Commun.* **11**: 485.

Ober, C. K., McNamee, S., Delvin, A., and Colby, R. H. 1990. *Liquid Crystalline Polymers*, ed. R. A. Weiss and C. K. Ober, p. 220. ACS Symp. Series, no. 435. Washington, D.C.: American Chemical Society.

Oh, S.-H., Ezaki, R., Akagi, K., and Shirakawa, H. 1993a. *J. Polym. Sci. Polym. Chem. Ed.* **31**: 2977.

Oh, S.-H., Akagi, K., Shirakawa, H., and Araya, K. 1993b. *Macromolecules* **26**: 6203.

Okamoto, Y., Asakura, T., and Hatada, K. 1991. *Chem. Lett.* **1991**: 1105.

Onsager, L. 1949. *Ann. NY Acad. Sci.* **51**: 627.

Orthmann, E. A., Enkelmann, V., and Wegner, G. 1983. *Makromol. Chem. Rapid Commun.* **4**: 687.

Orthmann, E. A., and Wegner, G. 1986. *Makromol. Chem. Rapid Commun.* **7**: 243.

Osman, M. A. 1982a. *Mol. Cryst. Liq. Cryst. Lett.* **82**: 47.

Osman, M. A. 1982b. *Mol. Cryst. Liq. Cryst. Lett.* **82**: 295.

Osman, M. A., and Ba, T. H. 1983. *Helv. Chim. Acta* **66**: 1786.

Paleos, C. M., Margomenou-Leonidopoulou, G., and Christias, C. 1986. *Mol. Cryst Liq. Cryst.* **137**: 391.

Panar, M., and Beste, L. 1977. *Macromolecules* **10**: 1401.

Papkov, S. P. 1984. *Adv. Polym. Sci.* **59**: 75.

Papon, E., Deffieux, A., Hardouin, F., and Achard, M. F. 1992. *Liq. Cryst.* **11**: 803.

Paulus, W., Ringsdorf, H., Diele, S., and Palze, G. 1991. *Liq. Cryst.* **9**: 807.

Pedersen, C. J. 1988. *Angew. Chem. Int. Ed. Engl.* **27**: 1021.

Pepy, G., Cotton, J. P., Hardouin, F., Keller, P., Lambert, M., Moussa, F., Noirez, L., Lapp, A., and Strazielle, C. 1988. *Makromol. Chem. Makromol. Symp.* **15**: 251 and references therein.

Percec, V., and Nava, H. 1987a. *J. Polym. Sci. Polym. Chem. Ed.* **25**: 405.

Percec, V., Nava, H., and Jonsson, H. 1987b. *J. Polym. Sci. Polym. Chem. Ed.* **25**: 1943.

Percec, V., and Tomazos, D. 1987c. *Polym. Bull.* **18**: 239.

Percec, V., and Yourd, R. 1988a. *Macromolecules* **21**: 3379.

Percec, V. 1988b. *Makromol. Chem. Macromol. Symp.* **13–14**: 397.

Percec, V. 1988c. *Mol. Cryst. Liq. Cryst.* **155**: 1.

Percec, V., and Pugh, C. 1989a. *Side Chain Liquid Crystalline Polymers*, ed. C. B. McArdle, p. 30. New York: Chapman and Hall.

Percec, V., and Tomazos, D. 1989b. *J. Polym. Sci. Polym. Chem. Ed.* **27**: 999.

Percec, V., and Hahn, B. 1989c. *J. Polym. Sci. Polym. Chem. Ed.* **27**: 2367.

Percec, V., and Yourd, R. 1989d. *Macromolecules* **22**: 524.

Percec, V., and Tomazos, D. 1989e. *Macromolecules* **22**: 1512.

Percec, V., and Hahn, B. 1989f. *Macromolecules* **22**: 1588.

Percec, V., and Rodenhouse, R. 1989g. *Macromolecules* **22**: 2043.

Percec, V., and Tomazos, D. 1989h. *Macromolecules* **22**: 2062.

Percec, V., and Yourd, R. 1989i. *Macromolecules* **22**: 3229.

Percec, V., Tomazos, D., and Pugh, C. 1989j. *Macromolecules* **22**: 3259.

Percec, V., and Rodenhouse, R. 1989k. *Macromolecules* **22**: 4408.

Percec, V., Pugh, C., Nuyken, O., and Pask, S. D. 1989l. *Comprehensive Polymer Science*, vol. 6, ed. G. Allen, p. 281. Oxford: Pergamon.

Percec, V., Tomazos, D., and Willingham, R. A. 1989m. *Polym. Bull.* **22**: 199.

Percec, V., and Tsuda, Y. 1989n. *Polym. Bull.* **22**: 489.

Percec, V., and Tsuda, Y. 1989o. *Polym. Bull.* **22**: 497.

Percec, V., and Tomazos, D. 1989p. *Polymer* **30**: 2124.

Percec, V., and Wang, J. H. 1990a. *J. Polym. Sci. Polym. Chem. Ed.* **28**: 1059.

Percec, V., and Tsuda, Y. 1990b. *Macromolecules* **23**: 5.

Percec, V., Hahn, B., Ebert, M., and Wendorff, J. H. 1990c. *Macromolecules* **23**: 2092.

Percec, V., and Tsuda, Y. 1990d. *Macromolecules* **23**: 3509.

Percec, V., and Keller, A. 1990e. *Macromolecules* **23**: 4347.

Percec, V., and Yourd, R. 1990f. *Makromol. Chem.* **191**: 25.

Percec, V., and Yourd, R. 1990g. *Makromol. Chem.* **191**: 49.

Percec, V., and Tsuda, Y. 1990h. *Polym. Bull.* **24**: 9.

Percec, V., and Tomazos, D. 1990i. *Polymer* **31**: 1658.

Percec, V., and Lee, M. 1991a. *J. Macromol. Sci. A* **28**: 651.

Percec, V., Zheng, Q., and Lee, M. 1991b. *J. Mater. Chem.* **1**: 611.

Percec, V., Zheng, Q., and Lee, M. 1991c. *J. Mater. Chem.* **1**: 1015.

Percec, V., and Rodenhouse, R. 1991d. *J. Polym. Sci. Polym. Chem. Ed.* **29**: 15.

Percec, V., Lee, M., and Jonsson, H. 1991e. *J. Polym. Sci. Polym. Chem. Ed.* **29**: 327.

Percec, V., Gomes, A. D. S., and Lee, M. 1991f. *J. Polym. Sci. Polym. Chem. Ed.* **29**: 1615.

Percec, V., and Lee, M. 1991g. *Macromolecules* **24**: 1017.

Percec, V., and Lee, M. 1991h. *Macromolecules* **24**: 2780.

Percec, V., Heck, J. A., and Ungar, G. 1991i. *Macromolecules* **24**: 4957.

Percec, V., and Lee, M. 1991j. *Macromolecules* **24**: 4963.

Percec, V., and Kawasumi, M. 1991k. *Macromolecules* **24**: 6318.

Percec, V., Asami, K., Tomazos, D., Feijoo, J. L., Ungar, G., and Keller, A. 1991l. *Mol. Cryst. Liq. Cryst.* **205**: 47.

Percec, V., Asami, K., Tomazos, D., Feijoo, J. L., Ungar, G., and Keller, A. 1991m. *Mol. Cryst. Liq. Cryst.* **205**: 67.

Percec, V., and Lee, M. 1991n. *Polym. Bull.* **25**: 131.

Percec, V., and Zuber, M. 1991o. *Polym. Bull.* **25**: 695.

Percec, V., Wang, C. S., and Lee, M. 1991p. *Polym. Bull.* **26**: 15.

Percec, V., and Tsuda, Y. 1991q. *Polymer* **32**: 661.

Percec, V., and Tsuda, Y. 1991r. *Polymer* **32**: 673.

Percec, V., Tomazos, D., and Feiring, A. E. 1991s. *Polymer* **32**: 1897.

Percec, V., and Lee, M. 1991t. *Polymer* **32**: 2862.

Percec, V., and Tomazos, D. 1992a. *Adv. Mater.* **4**: 548.

Percec, V., and Kawasumi, M. 1992b. *Adv. Mater.* **4**: 572.

Percec, V., and Lee, M. 1992c. *J. Mater. Chem.* **2**: 617.

Percec, V., Lee, M., Heck, J. A., Blackwell, H., Ungar, G., and Alvarez-Castillo, A. 1992d. *J. Mater. Chem.* **2**: 931.

Percec, V., Heck, J. A., Lee, M., Ungar, G., and Alvarez-Castillo, A. 1992e. *J. Mater. Chem.* **2**: 1033.

Percec, V., Zuber, M., Cheng, S. Z. D., and Zhang, A. Q. 1992f. *J. Mater. Chem.* **2**: 407.

Percec, V., Lee, M., Rinaldi, P. L., and Litman, V. E. 1992g. *J. Polym. Sci. Polym. Chem. Ed.* **30**: 1213.

Percec, V., and Zuber, M. 1992h. *J. Polym. Sci. Polym. Chem. Ed.* **30**: 997.

Percec, V., Cho, C. G., Pugh, C., and Tomazos, D. 1992i. *Macromolecules* **25**: 1164.

Percec, V., Okita, S., and Weiss, R. A. 1992j. *Macromolecules* **25**: 1816.

Percec, V., Johansson, G., and Rodenhouse, R. 1992k. *Macromolecules* **25**: 2563.

Percec, V., and Kawasumi, M. 1992l. *Macromolecules* **25**: 3843.

Percec, V., Kawasumi, M., Rinaldi, P. L., and Litman, V. E. 1992m. *Macromolecules* **25**: 3851.

Percec, V., Zuber, M., Ungar, G., and Alvarez-Castillo, A. 1992n. *Macromolecules* **30**: 439.

Percec, V., and Wang, J. H. 1992o. *Makromol. Chem. Macromol. Symp.* **54–55**: 561.

Percec, V., and Tomazos, D. 1992p. *Adv. Mater.* **4**: 548.

Percec, V., and Tomazos, D. 1992q. *Comprehensive Polymer Science: First Supplement*, ed. G. Allen, p. 299. Oxford: Pergamon.

Percec, V., Lee, M., and Ackerman, C. 1992r. *Polymer* **33**: 703.

Percec, V., Jonsson, H., and Tomazos, D. 1992s. *Polymerization in Organized Media*, ed. C. M. Paleos, p. 1. New York: Gordon and Breach.

Percec, V., and Kawasumi, M. 1993a. *Chem. Mater.* **5**: 826.

Percec, V., and Kawasumi, M. 1993b. *J. Chem. Soc. Perkin Trans. I* **1993**: 1319.

Percec, V., Johansson, G., Heck, J. A., and Ungar, G. 1993c. *J. Chem. Soc. Perkin Trans. I* **1993**: 1411.

Percec, V., Heck, J. A., Tomazos, D., Falkenberg, F., Blackwell, H., and Ungar, G. 1993d. *J. Chem. Soc. Perkin Trans. I* **1993**: 2799.

Percec, V., Heck, J. A., Tomazos, D., and Ungar, G. 1993e. *J. Chem. Soc. Perkin Trans. I* **1993**: 2381.

Percec, V., and Johansson, G. 1993f. *J. Mater. Chem.* **3**: 83.

Percec, V., and Tomazos, D. 1993g. *J. Mater. Chem.* **3**: 633.

Percec, V., and Tomazos, D. 1993h. *J. Mater. Chem.* **3**: 643.

Percec, V., and Kawasumi, M. 1993i. *J. Mater. Chem.* **3**: 725.

Percec, V., and Kawasumi, M. 1993j. *J. Mater. Chem.* **13**: 83.

Percec, V., and Kawasumi, M. 1993k. *Macromolecules* **26**: 3663.

Percec, V., and Kawasumi, M. 1993l. *Macromolecules* **26**: 3917.

Percec, V., Tomazos, D., Heck, J. A., Blackwell, H., and Ungar, G. 1994a. *J. Chem. Soc. Perkin Trans. II* **1994**: 31.

Percec, V., Heck, J. A., Johansson, G., Tomazos, D., Kawasumi, M., and Ungar, G. 1994b. *J. Macromol. Sci. Pure Appl. Chem. A* **31**: 1031.

Percec, V., Heck, J. A., Johansson, G., Tomazos, D., Kawasumi, M., Chu, P., and Ungar, G. 1994c. *J. Macromol. Sci. Pure Appl. Chem. A* **31**: 1719.

Percec, V., Chu, P., Ungar, G., Cheng, S. Z. D., and Yoon, D. Y. 1994d. *J. Mater. Chem.* **5**: 719.

Percec, V., Heck, J. A., Johansson, G., Tomazos, D., and Ungar, G. 1994e. *Macromol. Symp.* **77**: 237.

Percec, V., Oda, H., Rinaldi, P. L., and Hensley, D. R. 1994f. *Macromolecules* **27**: 12.

Percec, V., Chu, P., and Kawasumi, M. 1994g. *Macromolecules* **27**: 4441.

Percec, V., and Oda, H. 1994h. *Macromolecules* **27**: 4454.

Percec, V., and Oda, H. 1994i. *Macromolecules* **27**: 5821.

Percec, V., and Kawasumi, M. 1994j. *Mol. Cryst. Liq. Cryst.* **238**: 21.

Percec, V., Heck, J. A., Johansson, G., Tomazos, D., Kawasumi, M., Chu, P., and Ungar, G. 1994k. *Mol. Cryst. Liq. Cryst.* **254**: 137.

Percec, V., and Oda, H. 1995a. *J. Macromol. Sci. Pure Appl. Chem.* **A32**: 1531.

Percec, V. 1995b. *Pure Appl. Chem.* **67**: 2031.

Percec, V., Chu, P., Ungar, G., and Zhou, J. 1995c. *J. Am. Chem. Soc.* **117**: 11441.

Percec, V., and Johansson, G. 1996. *Macromolecular Design of Polymeric Materials*, ed. K. Hatada et al. New York: Dekker.

Pershan, P. S. 1979. *J. Physique Coll.* **40**: C3–423.

Petraccone, V., and Pirozzi, B. 1991. *Macromolecules* **24**: 6470.

Petterson, I., and Liljefors, T. J. 1987. *Comput. Chem.* **8**: 1139.

Philp, D., and Stoddart, J. F. 1991. *Synlett* **1991**: 445.

Platé, N. A., and Shibaev, V. P., eds. 1987. *Comb-Shaped Polymers and Liquid Crystals*. New York: Plenum.

Platé, N. A., ed. 1993. *Liquid Crystal Polymers*. New York: Plenum.

Polishchuk, A. P., and Timofeeva, T. V. 1993. *Russ. Chem. Rev.* **64**: 291.

Praefcke, K., Schmidt, D., and Heppke, G. 1980. *Chemiker-Zeitung* **104**: 269.

Pugh, C., and Percec, V. 1985. *Am. Chem. Soc. Div. Polym. Chem., Polym. Prepr.* **26**: 303.

Pugh, C., and Schrock, R. R. 1992. *Macromolecule* **25**: 6593.

Pugh, C., Arehart, S., Liu, H., and Narayanan, R. 1994. *J. Macromol. Sci. Pure Appl. Chem. A* **31**: 1591.

Qian, R., Chen, S., Song, W., and Bi, X. 1994. *Makromol. Chem. Rapid Commun.* **15**: 1.

Rehahn, M., Schlüter, A.-D., Wegner, G., and Feast, W. J. 1989a. *Polymer* **30**: 1054.

Rehahn, M., Schlüter, A.-D., Wegner, G., and Feast, W. J. 1989b. *Polymer* **30**: 1060.

Rehahn, M., Schlüter, A.-D., and Wegner, G. 1990a. *Makromol. Chem.* **191**: 1991.

Rehahn, M., and Schlüter, A.-D. 1990b. *Makromol. Chem. Rapid Commun.* **11**: 375.

Reuvers, A. J. M., Sinnema, A., van Rantwijk, F., Ramijnse, J. D., and van Bekkum, H. 1969. *Tetrahedron* **25**: 4455.

Richards, R. D. C., Hawthorne, W. D., Hill, J. S., White, M. S., Lacey, D., Semlyen, J. A., Gray, G. W., and Kendrick, T. C. 1990. *J. Chem. Soc. Chem. Commun.* **1990**: 95.

Ringsdorf, H., Schmidt, H.-W., and Schneller, A. 1982. *Makromol. Chem. Rapid Commun.* **3**: 745.

Ringsdorf, H., Tschirner, P., Schoenherr, O. H., and Wendorff, J. H. 1987. *Makromol. Chem.* **188**: 1431.

Ringsdorf, H., Schlarb, B., and Venzmer, J. 1988. *Angew. Chem. Int. Ed. Engl.* **27**: 113.

Ringsdorf, H., Wustefeld, R., Zerta, E., Ebert, M., and Wensdorff, J. H. 1989. *Angew. Chem. Int. Ed. Engl.* **28**: 914.

Ringsdorf, H., and Wustefeld, R. 1990. *Phil. Trans. R. Soc. London A* **330**: 95.

Rinne, F. 1933. *Trans. Faraday Soc.* **29**: 1016.

Rodenhouse, R., Percec, V., and Feiring, A. E. 1990. *J. Polym. Sci. Polym. Lett. Ed.* **28**: 345.

Rodenhouse, R., and Percec, V. 1991a. *Adv. Mater.* **3**: 101.

Rodenhouse, R., and Percec, V. 1991b. *Makromol. Chem.* **192**: 1873.

Rodenhouse, R., and Percec, V. 1991c. *Polym. Bull.* **25**: 47.

Rodriguez-Parada, J. M., and Percec, V. 1986. *J. Polym. Sci. Polym. Chem. Ed.* **24**: 1363.

Rodriguez-Parada, J. M., and Percec, V. 1987. *J. Polym. Sci. Polym. Chem. Ed.* **25**: 2269.

Rodriguez-Parada, J. M., Duran, R., and Wegner, G. 1989. *Macromolecules* **22**: 2507.

Rogers, H. G., Gaudiana, R. A., Hollinsed, W. C., Kalyaranaman, P. S., Manello, J. S., McGowan, C., Minns, R. A., and Sahatjian, R. 1985a. *Macromolecules* **18**: 1058.

Rogers, H. G., and Gaudiana, R. A. 1985b. *J. Polym. Sci. Polym. Chem. Ed.* **23**: 2669.

Rogers, H. G., Gaudiana, R. A., Minns, R. A., and Spero, D. M. 1986. *J. Macromol. Sci. A* **23**: 905.

Ronca, G., and Yoon, D. Y. 1982. *J. Chem. Phys.* **76**: 3295.

Roviello, A., and Sirigu, A. 1975. *J. Polym. Sci. Polym. Lett. Ed.* **13**: 455.

Roviello, A., and Sirigu, A. 1979. *Eur. Polym. J.* **15**: 423.

Roviello, A., and Sirigu, A. 1980. *Gazz. Chim. Ital.* **110**: 403.

Sackmann, H. Z., and Demus, D. 1973. *Mol. Cryst. Liq. Cryst.* **21**: 239.

Saeva, F. D., ed. 1979. *Liquid Crystals: The Fourth State of Matter*. New York: Dekker.

Sagane, T., and Lenz, R. W. 1988. *Polym. J.* **20**: 923.

Sagane, T., and Lenz, R. W. 1989a. *Macromolecules* **22**: 3763.

Sagane, T., and Lenz, R. W. 1989b. *Polymer* **30**: 2269.

Sato, M., Nakatsuki, K., and Ohkatsu, Y. 1986. *Makromol. Chem. Rapid Commun.* **7**: 231.

Sato, M., Kurosawa, K., Nakatsuki, K., and Ohkatsu, Y. 1988. *J. Polym. Sci. Polym. Chem. Ed.* **26**: 3077.

Sauer, T., and Wegner, G. 1988. *Mol. Cryst. Liq. Cryst.* **162B**: 97.

Sauer, T. 1993. *Macromolecules* **26**: 2057.

Schadt, M., Petrzilka, M., Gerber, P. R., Villiger, A., and Trickes, G. 1983. *Mol. Cryst. Liq. Cryst.* **94**: 139.

Schleeh, T., Imrie, C. T., Rice, D. M., Karasz, F. E., and Attard, G. S. 1993. *J. Polym. Sci. Polym. Chem. Ed.* **31**: 1859.

Schlüter, A.-D., and Wegner, G. 1993. *Acta Polym.* **44**: 59.

Schmidt, H.-W., and Guo, D. 1988. *Makromol. Chem.* **189**: 2029.

Schmidt, H.-W. 1989. *Makromol. Chem. Macromol. Symp.* **26**: 47.

Schoenherr, O. H., Wendorff, J. H., Ringsdorf, H., and Tschirner, P. 1986. *Makromol. Chem. Rapid Commun.* **7**: 791.

Schroeder, T. 1893. *Z. Phys. Chem.* **11**: 449.

Seddon, J., and Templer, R. 1991. *New Scientist* **130**: 45.

Sellens, R. J., Runcimann, P. J. I., Griffin, A. C., Bryant, E. S., and Hughes, W. E. 1989. *Mol. Cryst. Liq. Cryst.* **166**: 123.

Shaffer, T. D., and Percec, V. 1987. *J. Polym. Sci. Polym. Chem. Ed.* **25**: 2755.

Sherwood, H. M., Sigaud, G., Yoon, D. Y., Wade, C. G., Kawasumi, M., and Percec, V. 1994. *Mol. Cryst. Liq. Cryst.* **254**: 455.

Shibaev, V. P. 1988. *Mol. Cryst. Liq. Cryst.* **155**: 189.

Siegrist, A. E. 1981. *Helv. Chim. Acta* **64**: 662.

Sigaud, G., Tinh, N. H., Hardouin, F., and Gasparoux, H. 1981. *Mol. Cryst. Liq. Cryst.* **69**: 81.

Sigaud, G., Achard, M. F., Hardouin, F., Mauzac, M., Richard, H., and Gasparoux, H. 1987. *Macromolecules* **20**: 578.

Sigaud, G. 1992. *Phase Transitions in Liquid Crystals*, ed. S. Martellucci and A. N. Chester, p. 375. New York: Plenum.

Sikkema, D. J., and Hoogland, P. 1986. *Polymer* **27**: 1443.

Sinta, R., Minns, R. A., Gaudiana, R. A., and Rogers, H. G. 1987. *J. Polym. Sci. Polym. Lett. Ed.* **25**: 11.

Sirlin, C., Bosio, L., and Simon, J. 1987. *J. Chem. Soc. Chem. Commun.* **1987**: 379.

Sirlin, C., Bosio, L., and Simon, J. 1988a. *J. Chem. Soc. Chem. Commun.* **1988**: 236.

Sirlin, C., Bosio, L., and Simon, J. 1988b. *Mol. Cryst. Liq. Cryst.* **155**: 231.

Skoulios, A., and Guillon, D. 1988. *Mol. Cryst. Liq. Cryst.* **165**: 317.

Spassky, N., Lacoudre, N., Le Borgne, A., Vairon, J. P., Le Barny, P., Jun, C. L., Friedrich, C., and Nöel, C. 1989. *Makromol. Chem. Macromol. Symp.* **24**: 271.

Stenhouse, P. J., Valles, E. M., Kantor, S. W., and MacKnight, W. J. 1989. *Macromolecules* **22**: 1467.

Stern, R., Ballauff, M., and Wegner, G. 1989. *Makromol. Chem. Macromol. Symp.* **23**: 373.

Stevens, H., Rehage, G., and Finkelmann, H. 1984. *Macromolecules* **17**: 851.

Stupp, S. I., Moore, J. S., and Martin, P. G. 1988. *Macromolecules* **21**: 1228.

Sung, T. C., Mallon, J. J., Atkins, E. D. T., and Kantor, S. W. 1990. *Liquid Crystalline Polymers*, ed. R. A. Weiss and C. K. Ober, p. 158. ACS Symp. Series, no. 435. Washington, D.C.: American Chemical Society.

Suzuki, M., Lim, J. C., and Saegusa, T. 1990. *Macromolecules* **23**: 1574.

Szasz, G. J., Sheppard, N., and Rank, D. H. 1948. *J. Chem. Phys.* **16**: 704.

Takahashi, S., Morimoto, H., Murata, E., Kataoka, S., Sonogashira, K., and Higara, N. 1982a. *J. Polym. Sci. Polym. Chem. Ed.* **20**: 565.

Takahashi, S., Takai, Y., Morimoto, H., Sonogashira, K., and Hagihara, N. 1982b. *Mol. Cryst. Liq. Cryst.* **82**: 139.

Takahashi, S., Takai, Y., Morimoto, H., and Sonogashira, K. 1984. *J. Chem. Soc. Chem. Commun.* **1984**: 3.

Tanaka, M., and Nakaya, T. 1984. *Makromol. Chem.* **185**: 1915.

Tanaka, M., and Nakaya, T. 1986. *Makromol. Chem.* **187**: 2345.

Tanaka, M., and Nakaya, T. 1987. *J. Macromol. Sci. A* **24**: 777.

Tanaka, M., and Nakaya, T. 1988. *Makromol. Chem.* **189**: 771.

Tashiro, K., Kokayashi, M., and Tadokoro, H. 1977. *Macromolecules* **10**: 413.

Tékély, P., Lauprêtre, F., and Monnerie, L. 1983. *Macromolecules* **16**: 415.

Tiesler, U., Schmitz, L., and Ballauff, M. 1994. *Mol. Cryst. Liq. Cryst.* **254**: 387.

Tinh, N. H. 1983. *J. Chim. Phys.* **80**: 83.

Tinh, N. H., Achard, M. F., Hardouin, F., Mauzac, M., Richard, H., and Sigaud, G. 1990. *Liq. Cryst.* **7**: 385.

Tirrell, D. A. *Encyclopedia of Polymer Science and Engineering*, vol. 4, p. 192. 2d ed. New York: Wiley.

Tomalia, D. A., Naylor, A. M., and Goddard III, W. A. 1990. *Angew. Chem. Int. Ed. Engl.* **29**: 138.

Tomalia, D. A., and Durst, H. D. 1993. *Top. Curr. Chem.* **165**: 193.

Tomalia, D. A. 1994. *Adv. Mater.* **6**: 529.

Tomazos, D., Out, R., Heck, J. A., Johansson, G., Percec, V., and Möller, M. 1994. *Liq. Cryst.* **16**: 509.

Tour, J. M. 1994. *Adv. Mater.* **6**: 190.

Ujiie, S., and Iimura, K. 1990. *Chem. Lett.* **1990**: 995.

Ujiie, S., and Iimura, K. 1991a. *Chem. Lett.* **1991**: 411.

Ujiie, S., and Iimura, K. 1991b. *Chem. Lett.* **1991**: 1969.

Ujiie, S., and Iimura, K. 1992. *Macromolecules* **25**: 3174.

Unanue, A., and Bothorel, P. 1965. *Bull. Soc. Chim. Fr.* **1965**: 2827.

Ungar, G. 1986. *Macromolecules* **19**: 1317.

Ungar, G., Feijoo, J. L., Keller, A., Yourd, R., and Percec, V. 1990. *Macromolecules* **23**: 3411.

Ungar, G., Feijoo, J. L., Percec, V., and Yourd, R. 1991a. *Macromolecules* **24**: 953.

Ungar, G., Feijoo, J. L., Percec, V., and Yourd, R. 1991b. *Macromolecules* **24**: 1168.

Ungar, G., Percec, V., and Rodenhouse, R. 1991c. *Macromolecules* **24**: 1996.

Ungar, G., Percec, V., and Zuber, M. 1992. *Macromolecules* **25**: 75.

Ungar, G. 1993. *Polymer* **34**: 2050.

Ungar, G., Batty, S. V., Percec, V., Heck, J. A., and Johansson, G. 1994a. *Adv. Mater. Opt. Electr.* **4**: 304.

Ungar, G., Percec, V., and Zuber, M. 1994b. *Polym. Bull.* **32**: 325.

Ungar, G., Zhou, J., Percec, V., and Chu, P. 1995. *Macromol. Symp.* **98**: 951.

Van Hecke, G. R. 1979. *J. Phys. Chem.* **83**: 2344.

van Laar, J. J. Z. 1908. *Z. Phys. Chem.* **63**: 216.

Vasilenko, S. V., Khokhlov, A. R., and Shibaev, V. P. 1984. *Macromolecules* **17**: 2270.

Verma, A. L., Murphy, W. F., and Bernstein, H. J. 1974. *J. Chem. Phys.* **60**: 1540.

Viney, C., Russell, T. P., Depero, L. E., and Twieg, R. J. 1989a. *Mol. Cryst. Liq. Cryst.* **168**: 63.

Viney, C., Twieg, R. J., Russell, T. P., and Depero, L. E. 1989b. *Liq. Cryst.* **5**: 1783.

Vogel, W., and Heitz, W. 1990. *Makromol. Chem.* **191**: 829.

Vögtle, F. 1989. *Supramolekulare Chemie.* Stuttgart: Teubner.

Vögtle, F. 1991. *Supramolecular Chemistry.* New York: Wiley.

Voigt-Martin, I.-G., Simon, P., Bauer, S., and Ringsdorf, H. 1995a. *Macromolecules* **28**: 236.

Voigt-Martin, I.-G., Simon, P., Yan, D., Yakimansky, A., Bauer, S., and Ringsdorf, H. 1995b. *Macromolecules* **28**: 243.

Wahlenkamp, T., and Wegner, G. 1994. *Macromol. Chem. Phys.* **195**: 1933.

Warner, M. 1989. *Side Chain Liquid Crystal Polymers*, ed. C. B. McArdle, p. 7. New York: Chapman and Hall.

Warner, M. 1993. *Phil. Trans. R. Soc. London A* **344**: 403.

Weber, E., ed. 1993. *Top. Curr. Chem.* **165**.

Weiss, R. A., and Ober, C. K., eds. 1990. *Liquid Crystalline Polymers.* ACS Symp. Series, no. 435. Washington, D.C.: American Chemical Society.

Wen, J. S., Hsiue, G. H., and Hsu, C. S. 1990. *Makromol. Chem. Rapid Commun.* **11**: 151.

Wenzel, M., Ballauff, M., and Wegner, G. 1987. *Makromol. Chem.* **188**: 2865.

Whitesides, G. M., Mathias, J. P., and Seto, C. T. 1992. *Science* **254**: 1312.

Wilson, L. M., and Griffin, A. C. 1993. *Macromolecules* **26**: 6312.

Wilson, L. M. 1994a. *Liq. Cryst.* **17**: 277.

Wilson, L. M. 1994b. *Macromolecules* **27**: 6683.

Wilson, L. M., and Griffin, A. C. 1994c. *Macromolecules* **27**: 1928.

Wilson, L. M., and Griffin, A. C. 1994d. *Macromolecules* **27**: 4611.

Wilson, L. M. 1995. *Macromolecules* **28**: 325.

Wilteler, H., Lieser, G., Wegner, G., and Schulze, M. 1993. *Makromol. Chem. Rapid Commun.* **14**: 471.

Winsor, P. A. 1968. *Chem. Rev.* **68**: 1.

Woller, P. B., and Garbisch, E. W. 1972. *J. Am. Chem. Soc.* **94**: 5310.

Yamauchi, K., and Kinoshita, M. 1993. *Prog. Polym. Sci., Japan* **18**: 763.

Yoon, D. Y., and Flory, P. J. 1989. *MRS Symp. Proc.* **134**: 11.

Yousif, Y. Z., Jenkins, A. D., Walton, D. R. M., and Al-Rawi, J. M. A. 1990. *Eur. Polym. J.* **26**: 901.

Yu, L. J., and Samulski, E. T. 1984. *Liquid Crystals and Ordered Fluids*, vol. 4, ed. A. C. Griffin and J. F. Johnson, p. 697. New York: Plenum.

Yu, L. J., Bao, Z., and Cai, R. 1993. *Angew. Chem. Int. Ed. Engl.* **32**: 1345.

Zang, N., Wu, R., Li, Q., Pakbaz, K., Yoon, C. O., Wude, F. 1993, *Chem. Mater.* **5**: 1598.

Zentel, R. 1989a. *Angew. Chem. Int. Ed. Engl.* **28**: 1407.

Zentel, R. 1989b. *Comprehensive Polymer Science*, vol. 5, ed. G. Allen and J. C. Bevington, p. 723. Oxford: Pergamon.

Zhou, Q. F., Duan, X. Q., and Liu, Y. L. 1986. *Macromolecules* **19**: 247.

Zuber, M. 1991. *PhD Thesis*, Case Western Reserve University.

CHAPTER 9

Polymer Dispersed Liquid Crystals: Nematic Droplets and Related Systems

GREGORY P. CRAWFORD,
J. WILLIAM DOANE, AND
SLOBODAN ŽUMER

9.1 INTRODUCTION

Dispersions of low molecular weight liquid crystals and polymers belong to a broad class of materials in which the concentration of polymer can be as large as 80% or as small as 2%, depending on the type of polymer used and the application. In their most common form, the concentration of polymer is near 50% and the liquid crystal is separated out in the form of droplets randomly distributed throughout the polymer (Doane, 1990). These materials are called polymer dispersed liquid crystals (PDLCs) and are currently under development for several electrooptical applications such as switchable windows, high-definition projection systems, large flat-panel displays, light shutters for optical signal processing, etc. (Doane, 1991). In the case of low polymer concentrations, the polymer forms a network dispersed in the liquid crystal (Hikmet, 1991). One example involves cholesteric liquid crystals where the dispersed polymer locks-in specific textures useful for bistable color reflective displays; bistability is a much sought after property of liquid crystalline materials, since it is particularly important for low power consumption devices (Yang et al., 1991).

In addition to their potential in electrooptic devices, dispersions have been the principal driving force behind the revitalized interest in confined liquid crystals (Crawford and Doane, 1992d). Independent of the concentration of polymer used, dispersions of liquid crystals and polymers have one common theme: the liquid crystal is constrained by a symmetry breaking, nonplanar confinement imposed by the surrounding polymer. A topic of both applied and basic interest, research centers around the tasks of observing and predicting stable nematic liquid crystal structures under spherical, cylindrical, elliptical, and random-type confinements, and measuring the key surface and material parameters that govern their stabilities. Of even greater importance to the experimentalist, confined liquid crystal systems exhibit a large surface-to-volume ratio enabling integrative methods to be used to probe interactions occurring at the surface, never before assessable with techniques such as magnetic resonance, dielectric spectroscopy, and microcalorimetry. Before we proceed with the physics of confined liquid crystals and the role of nonplanar confinement in future liquid crystal technologies, it is worthwhile to survey the historical milestones that have brought us to our current understanding of these systems.

The idea of confining liquid crystals to a nonplanar geometry is not a spin-off of recent display technology, but began in the early 1900s. In 1904, Lehmann discovered that liquid crystals suspended in an isotropic fluid conform to a specific configuration that depends upon the liquid crystal material and the angle at which the liquid crystal molecules are pinned to the isotropic fluid interface. Studying birefringent textures by optical polarizing microscopy was his principal means of observing ordered structures. Lehmann further discussed the possibility of a volume defect in orientational order, occurring in the center of the droplet being transformed into two point defects as the droplet was squeezed between planar glass plates. He subsequently went on to study the dynamics of these unusual defects in the presence of a magnetic field. Possible applications of his work were not foreseen at that time.

It was not until the late 1960s that interest in liquid crystal droplets began to develop. In 1966, Chandrasekhar addressed the problem of finding the equilibrium shape of a small particle with an emphasis on liquid crystalline materials. In 1969, Dubois-Violette and Parodi applied elastic theory to nematic liquid crystal droplets to predict their stable configurations of the nematic director field. Assuming an anisotropic surface energy, they were able to predict the stability of two configurations, depending on whether the surface induces a perpendicular or parallel orientation of the liquid crystal molecules at the surface boundary. They proposed several configurations that are presented in FIGURE 9.1: perpendicular orientation of the molecules at the surface resulting in a stable *radial structure* with a singularity point in the center of the droplet (FIGURE 9.1(c)); a high magnetic field situation, in which the radial droplet will adopt aligned *axial structure* with a director field configuration within the equatorial plane circular surface disclination line (FIGURE 9.1(d)); parallel orientation of the molecules at the surface resulting in a stable *bipolar structure* with two opposite singularity points (FIGURE 9.1(a)); and an applied magnetic field reorienting the symmetry axis of the bipolar configuration parallel to the field direction (FIGURE 9.1(b)). These theoretical studies were stimulated by the experiments of Meyer in 1969 who earlier observed point disclinations in the center and at opposite poles of nematic droplets dispersed in various isotropic fluids that are similar to those presented in FIGURE 9.2.

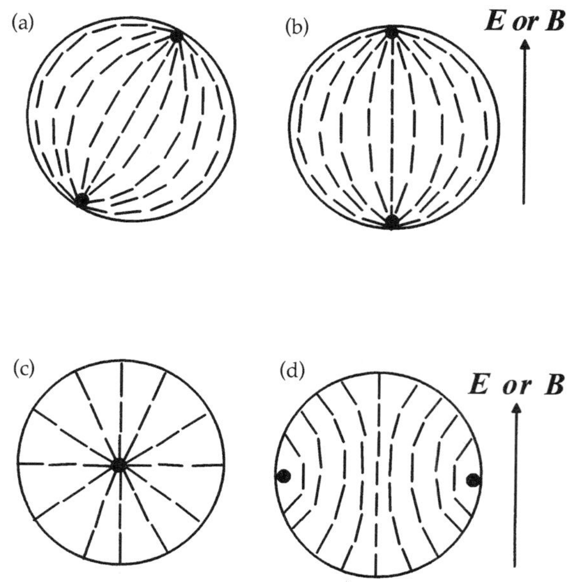

FIGURE 9.1. Stable nematic liquid crystal director fields in spherical droplets: bipolar droplet (a), aligned bipolar droplet (b), radial droplet (c), and aligned axial droplet (d).

In 1973, Meyer extended his work on point disclinations and experimentally prepared circumstances that allowed for their detailed study. In 1973, Candau and coworkers studied cholesteric liquid crystal materials in spherical droplets suspended in an isotropic fluid. The cholesteric droplets adopted a new spiral structure, unlike those reported by Meyer in 1969 for ordinary nematic materials, when the radius of the droplet was much larger than the intrinsic pitch length of the cholesteric material. In 1974, Press and Arrott applied elastic theory to nematic liquid crystal droplets with perpendicular boundary

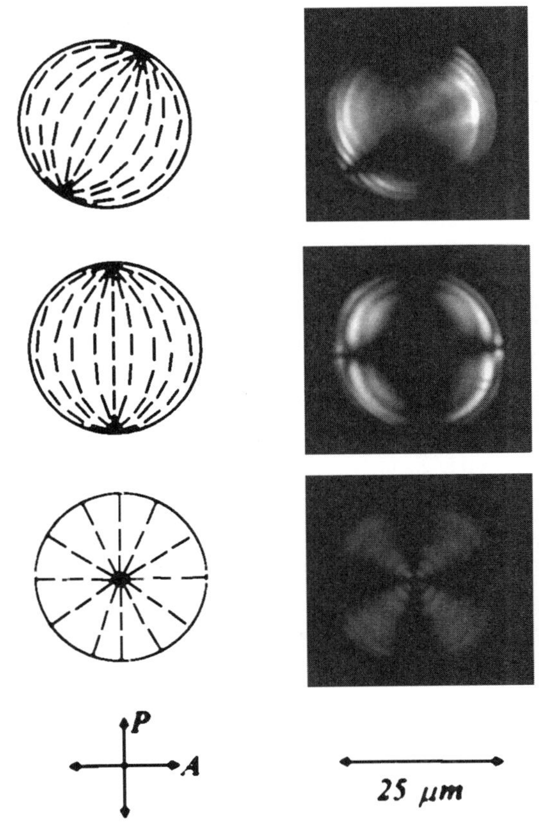

FIGURE 9.2. Microscope textures of nematic liquid crystal droplets suspended in an isotropic fluid illuminated by a white light source between a crossed polarizer (P) and analyzer (A).

conditions. They numerically derived twisted solutions to describe their unusual observation of a twisted star in the center of a radial droplet as viewed between crossed polarizers of an optical microscope.

During this same time period in the early 1970s, a similar problem was being addressed concerning the nature of the ordering of a nematic liquid crystal constrained to cylindrical capillary tubes. It was theoretically predicted by Dzyaloshinskii in 1970 that the nematic director configuration within a cylindrical tube with perpendicular boundary conditions was a pure radial configuration with a line disclination along the cylindrical axis. In 1972, Cladis and Kléman used elastic theory to demonstrate that another configuration is stable for cylinders larger than 0.1 μm in diameter where the line disclination escapes into the third dimension along the cylinder axis; this structure, named the escaped-radial configuration, is continuous and involves no core. Optical observations by Williams and coworkers (1972), Meyer (1973), and Saupe (1973), verified the existence of the escaped-radial configuration and singular point defects along the cylinder axis. They were of the radial and hyperbolic type that alternate along the cylinder axis. Their existence is a consequence of two energetically equivalent nematic director configurations existing where the direction of bend is changed. The escaped-radial configuration and its microscopic textures are presented in FIGURE 9.3.

Cladis continued studies on liquid crystals confined to capillary tubes using a material with a smectic A phase (Cladis, 1974). As bend elastic constants became

FIGURE 9.3. Microscope texture of the escaped-radial configuration (with and without a defect) stable in capillary tubes with homeotropic anchoring. The capillary tubes are oriented between a crossed polarizer (P) and analyzer (A) and are illuminated by a sodium light source.

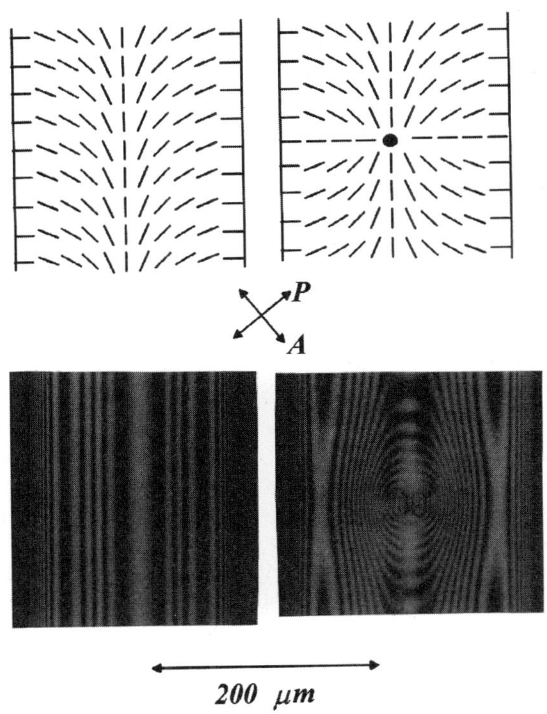

large on approaching the smectic A phase, the escaped-radial configuration transformed into a planar-radial configuration. In 1977, Melzer and Nabarro attempted to study nematic liquid crystals constrained to capillary tubes when concentric parallel ordering is induced at the surface, to verify the stability of the escaped-twisted structure predicted earlier by Cladis and Kléman, in 1972. Due to the complexity of rubbing the inner cavity wall to form concentric grooves they were not able to obtain a uniform director configuration. In 1979, Cladis and coworkers studied defect structures occurring in capillary tubes near the cholesteric–smectic A transition; they proposed that the cholesteric phase grows from the smectic A phase via a spiraling disclination.

The interest in confined liquid crystals shifted back to droplets in the early 1980s when a method of varying the boundary conditions was developed. Volovik and Lavrentovich demonstrated in 1983 that the dynamics of creation, annihilation, and transformation of topological defects (e.g., boojums, hedgehogs, and dislinations) in a closed system follow the predictions of the topological theory. Later the studies on defects were extended by investigating the defect lines that occur in cholesteric liquid crystal droplets (Kurik and Lavrentovich, 1982).

In the 1980s, liquid crystals confined to small volumes also became more than just curious systems: they showed great promise for light modulation devices, based on a concept introduced earlier by Hilsum (1976), in which a microporous filter permeated with nematic liquid crystal of positive dielectric anisotropy is sandwiched between conducting glass substrates (Craighead et al., 1982). The orientation of the liquid crystal confined to the inclusions was varied by an electric field to produce a variation in the refractive index. When the index of refraction of the microporous filter matched that of the liquid crystal, the composite medium was transparent (nonscattering); however, if the refractive indices were mismatched, the composite medium was translucent (scattering). This provided a second demonstration of Hilsum's concept that confined liquid crystals were useful as an electrically controllable light scattering medium. The contrast ratio and the switching times were not impressive and the device was never developed for commercial applications.

In the mid-1980s, two new types of dispersions of liquid crystal and polymers were reported. In 1986,

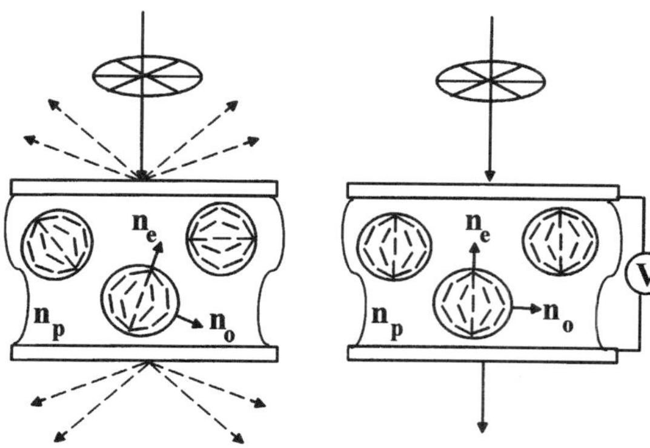

FIGURE 9.4. Operation of a polymer-dispersed liquid crystal (PDLC) display. Scattering (a) and transparent (b) modes are shown.

Doane and coworkers developed phase separation procedures to obtain dispersions of liquid crystal droplets inside a solid polymer binder. These systems were consequently named polymer dispersed liquid crystals (PDLCs). Phase separation techniques gave the control over droplet size and distribution necessary to optimize PDLCs for light shutter applications. Polymer dispersed liquid crystal materials are easily sandwiched between conducting glass plates. The spacing is controlled by micrometer spacers in the 10–25 μm range. The PDLC material strongly scatters light in the zero-field state (FIGURE 9.4(a)). It can be switched to a transparent mode upon application of a voltage (FIGURE 9.4(b)). The transparent condition arises when the electric field is of sufficient strength to align the droplets such that their refractive index matches that of the polymer, subsequently reducing the scattering power of the droplets. Upon removal of the voltage, the droplets return to their original scattering condition. In 1985, Fergason reported an encapsulation procedure to create micrometer-size liquid crystal capsules inside a rigid polymer binder. The material was prepared using a water-based emulsion, referred to as nematic curvilinear aligned phase (NCAP) materials (Drzaic, 1986). The optical shutters based on PDLC and NCAP dispersions possess advantages over other liquid crystal technologies: no preferential substrate treatment is needed; no polarizers are required; and fabrication of large meter sizes is easy.

The success of the PDLC led researchers to test the feasibility of embedding chiral nematic and ferroelectric materials in spherical inclusion of polymer matrices. In 1990, Crooker and Yang first demonstrated that chiral nematics are suitable for applications in colored electrooptic displays. For materials with negative dielectric anisotropy and sufficiently short pitch, they showed that it is possible to induce a texture showing bright selective reflection by applying electric fields. Shortly after, Kitzerow and coworkers (1992d) revealed that the alignments of ferroelectric and antiferroelectric liquid crystals in droplets can be controlled by mechanical shear during the polymerization process, and just recently demonstrated bistability (Molsen and Kitzerow, 1994). They demonstrated helical unwinding, leading to the rotation of the optic axis in the plane perpendicular to the electric field with switching times of $\sim 100\,\mu s$. In 1993, Zyryanov and coworkers also presented electrooptic data for ferroelectric PDLCs. In 1994, Lee and coworkers presented a model for understanding the electrooptic switching in ferroelectric PDLCs. In 1992, Mang and coworkers studied the layer structure of ferroelectric liquid crystals in glass capillary arrays with x-ray diffraction, and more recently in 1994, Aliev and Kelly studied the dynamics, structure, and phase transitions of ferroelectrics confined to porous glass matrices. A recent review of Kitzerow provides details on ferroelectric switching in PDLCs (Kitzerow, 1994); we will not, therefore, treat it further in this chapter.

In 1986, Mariami and coworkers were the first to show that a liquid crystal could be used for an ordering medium of the polymer network. Dispersion materials consisting of low concentrations of polymer for display applications were then reported by Hikmet in 1990, who prepared a light shutter for polarized light by dissolving and then polymerizing suitable monomers in a homogeneous aligned nematic cell. The composite material created a scattering state for one component of the polarization when a voltage was applied to the cell. Yang and coworkers found other uses of ordered polymer networks to stabilize and modify the cholesteric liquid crystal materials. They developed two new types of displays (Yang et al., 1992): (1) haze-free light shutters for unpolarized light and (2) a bistable reflective-mode display. The reflective, bistable mode display is shown in FIGURE 9.5. At high electric fields, the cholesteric texture is untwisted and the cell is transparent (FIGURE 9.5(c)). If the high field is suddenly turned off, the liquid crystal transforms to the planar texture (reflecting state), shown in FIGURE 9.5(a). This texture selectively reflects a certain wavelength of light, depending on the pitch length of the material. When an intermediate electric field is applied to the cell, the liquid crystal transforms into the focal conic texture (FIGURE 9.5(b)). Upon removal of the field, the focal conic texture is normally metastable but can be made stable by a polymer network. Many studies are in progress to understand the role of the fragile networks of the cross-linked polymers that create unusual electrooptic effects not observable in the bulk liquid crystal (Hikmet and Zwerver, 1991; Stannarius et al., 1991; Jákli et al., 1992, 1994; Crawford et al., 1994a, 1995a). Also, many related studies have recently been reported on liquid crystals confined to aerogel (Bellini et al., 1992; Wu et al., 1992; Clark et al., 1993; Maritan et al., 1994) and porous glass materials (Aliev and Breganov, 1989) that are random-type matrices.

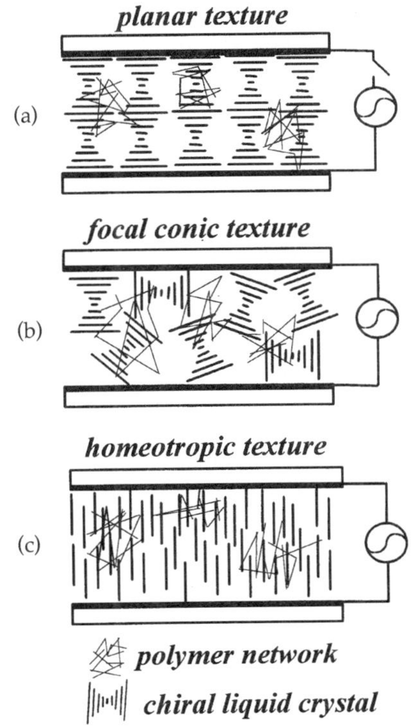

FIGURE 9.5. Operation of a polymer-stabilized, bistable reflective-mode display: reflective mode (a), scattering mode (b), and transparent mode (c).

After the discovery of PDLC materials, a new era burgeoned in the physics of confined liquid crystals, guided by their electrooptic applications. In order to achieve optimal performance of PDLC materials, there are some obvious factors that can be adjusted, such as droplet size and density, choosing liquid crystal materials with large birefringences and dielectric anisotropies, etc. However, there are some aspects of PDLC materials that, if better understood, could improve the electrooptic performance of the display. One of these aspects is the structure of the nematic liquid crystal inside the polymer cavity and its response to the external field. Under micrometer and submicrometer confinement to curved geometries, elastic forces compete with molecular anchoring at the surface (Erdmann et al., 1990) to create nematic director configurations within the droplet. For PDLC device applications, switching from a completely non-scattering to a strongly scattering structure is desirable. In addition, one would like a type of anchoring and director configuration that would require low applied electric field and that would restructure quickly.

These problems prompted researchers to employ a battery of techniques to better understand the structure of liquid crystals under confinement. Optical polarizing microscopy is the most desirable way to study nematic director configurations for droplet sizes greater than $5\,\mu$m (Ondris-Crawford et al., 1991; Xu et al., 1994); however, droplet sizes in light shutters are often much smaller than can be studied with optical microscopy. One of the most useful techniques for configuration characterization in submicrometer cavities is deuterium nuclear magnetic resonance (^{2}H-NMR) in which the nematic director configuration in cavity sizes as small as $0.01\,\mu$m are identifiable (Golemme et al., 1988a, b). Deuterium NMR studies on small PDLC droplets stimulated studies on liquid crystals confined to submicrometer cylindrical cavities of polymeric microporous membranes (Crawford et al., 1992a). Cylindrical cavities have two distinct advantages for studying confinement effects: (1) the cavity walls are easily accessible to chemical treatments for altering anchoring strengths and angles; and (2) the cylindrical symmetry simplifies analysis of the nematic director configuration. Material and surface effects determined from studies on cylindrical cavities can be adopted for other cavity shapes important in electrooptic devices.

The objective of this contribution is to consolidate both the basic and applied perspectives of dispersed liquid crystals in all cavity shapes and sizes, from spherical droplets to more complex geometries. Section 9.2 will review phase separation and emulsification methods used to prepare PDLC and NCAP films, respectively, formation of polymer networks, and related confining matrices. Section 9.3 will address the effects of microconfinement on liquid crystals, including the influence of curvature, surfaces, and external fields. Section 9.4 will be devoted to the various experimental techniques used to investigate structure determination, and section 9.5 will treat ordering and ordering transitions. Section 9.6 is reserved for display applications involving polymer dispersions.

9.2 POLYMER DISPERSIONS AND OTHER CONFINING SYSTEMS

There are three basic methods for preparing dispersions of polymers and liquid crystals: phase separation, encapsulation, and permeation. The first two methods are primarily for application purposes, while the latter method is designated for basic studies on confinement. Phase separation and emulsification methods are fundamentally different, although the end product of liquid crystal droplets randomly dispersed in a solid polymer are essentially the same, with only small differences in droplet morphology. The phase separation technique first involves a mixture of polymer or prepolymer with a liquid crystalline material. During the polymerization process, the liquid crystal material phase separates from the solution, forms droplets, and finally, polymer gelation and solidification occur. The encapsulation material, however, is prepared by dispersing a liquid crystal in a water-borne polymer to form an emulsion. The material is coated on a substrate and allowed to dry; the droplets are permanently locked-in. Liquid crystals have also been permeated into pores of confining matrices for more basic studies. The cavity morphologies of these confining matrices range from random interconnected pores with a distribution of pore sizes, to well-defined cylindrical cavities with a narrow pore size distribution. In this section, we will review only a few specific examples of liquid crystal polymer dispersion systems. More examples of proven materials can be found in the recent review by Kitzerow (1994).

9.2.1 Polymerization-Induced Phase Separation

Polymerization-induced phase separation (PIPS) is useful when the low molecular weight liquid crystal is soluble in a prepolymer, but not in polymer materials (West, 1988). The process begins with a homogeneous mixture of prepolymer and liquid crystal. It is often necessary to mix other substances into the solution for a commercial grade PDLC film, for instance, a catalyst to speed up the reaction or a material to adjust the index of refraction of the polymer to better match that of the liquid crystal. After a homogeneous mixture of all the initial components is formed, polymerization results in a change of the chemical potential of the polymer solvent; phase separation of the liquid crystal results, leading to droplet nucleation and growth. The liquid crystal continues to phase separate until gelation of the polymer limits the growth of the liquid crystal droplets. Purification of the liquid crystal droplets continues; low molecular weight oligomers dissolved within the nematic droplet diffuse into the polymer cavity wall where they react. Purification of the nematic droplets can easily be checked by monitoring the nematic–isotropic transition temperature. As the nematic–isotropic transition temperature becomes narrower and approaches the bulk nematic–isotropic transition temperature value, the liquid crystal within the droplet is close to that of the bulk (West, 1988).

One example of PDLC materials that are formed by the PIPS process is made by mixing an epoxy EPON 828 (Shell Chemical Co.) with the cyanobiphenyl mixture E7 (EM Chemicals) and the curing agent Capcure 3800 (Miller–Stephenson Chemical Co.) in the ratio 1:1:1 by weight percentage (West, 1988). In this system, polymerization proceeds via a condensation reaction in an oven at 70 °C. The phase separation kinetics of this system is well understood. In 1991, Kim and Palffy-Muhoray applied a simple Cahn–Hilliard theory to describe the evolution of homogeneous binary mixtures to PDLC systems. Using light scattering experiments, they concluded that the onset of phase separation occurs well after the start of the polymerization process. In their system, the initial phase separation took place within approximately eight minutes, while the total curing time occurred after approximately two hours.

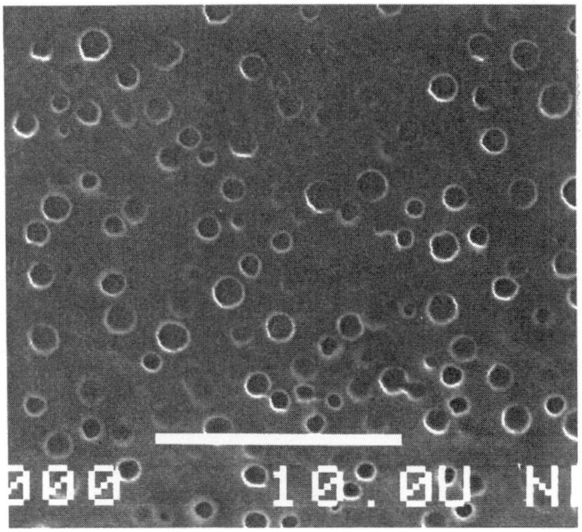

FIGURE 9.6. A scanning electron microscope (SEM) photograph of a polymer-dispersed liquid crystal material prepared by the polymerization-induced phase separation (PIPS) process.

The droplet size and density for materials formed by the PIPS method strongly depend on the cure temperatures. West has shown that the average droplet size for the Epon–Capcure–E7 system has a complicated dependence on cure temperature (West, 1988). The maximum average size occurs at approximately 50 °C, and decreases for higher curing temperatures. Other liquid crystal–epoxy systems exhibit similar dependences. There is a competition between the viscosity of the polymer, rate of diffusion of the liquid crystal material, and solubility of the liquid crystal in the polymer. The types of liquid crystal and polymers used govern the rate of polymerization. A scanning electron microscope photograph is presented in FIGURE 9.6 for an index-matched epoxy system developed at the Liquid Crystal Institute: 9.1 wt.% of Epon 828, 33.8 wt.% of Capcure 3800, 17.1 wt.% of WC 97 or heloxy 97 (Rhone–Poulenc), and the eutectic liquid crystal mixture E7. Usual droplet diameters for display applications are in the range 0.3–3.0 μm in diameter; however, transmission electron microscopy and nuclear magnetic resonance have confirmed droplet diameters down to 0.02 μm in the commercial Bostik brand epoxy and 5CB (4′-pentyl-4-cyanobiphenyl) system (Golemme et al., 1988b). The rate of polymerization, hence the droplet size, can be controlled by cure temperature and relative concentration of the liquid crystal material.

In other phase-separated systems, the polymerization may proceed by a free radical polymerization or through a photo-initiated polymerization, where phase separation is induced by an ultraviolet light. In 1989, Yamagishi and coworkers extensively studied phase separation in the PIPS process using ultraviolet irradiated curing techniques, and discovered a variety of morphologies ranging from conventional spherical cavities to polymer ball morphologies. They reported that all the components of the dispersion, including the photo-initiator, were critical in controlling the morphologies.

The formation energetics, kinetics, and phase behavior of an ultraviolet-cured, polymer-dispersed liquid crystal have been extensively studied using scanning tunneling microscopy and calorimetry (Smith, 1991). He presented several useful results for optimizing PDLC film cure and morphology; for example, the heat of cure and cure time constant are sensitive to cure temperature. The degree of cure was also found to affect both the glass and nematic–isotropic transition temperatures, the amount of liquid crystal contained in the droplets, and droplet morphology. The intensity of the ultraviolet radiation strongly influences the size of the droplets, but evidently has less influence on the amount of liquid crystal contained in the droplets. An analogous curing method using electron beam radiation has also been used (Vaz et al., 1991). The benefits of electron beam curing are that polymerization is fast and no photoinitiator is needed. The disadvantage is that energetic radiation may cause discoloration in the glass substrates. Both the conventional (droplets) and reverse (polymer ball) morphologies were observed, depending on the cure conditions and volume fraction of liquid crystal in the dispersion.

Liquid crystal–polymer dispersion systems that require only small percentages of polymer are also photopolymerized by ultraviolet radiation. These systems are sometimes referred to as gels in the literature (Hikmet, 1992); more recently they have been called polymer networks (Yang et al., 1991; Yang and Doane, 1992; Stannarius et al., 1991). For a recent review see Crawford and Žumer (1996a). One example of low concentration polymer dispersion uses a small amount of the monomer 4,4′-*bis*-acryloylbiphenyl (1–5 wt.%) added into the liquid crystal compound 5CB. The monomer has reactive double bonds at both ends. When the mixture is irradiated with ultraviolet light, the monomer is polymerized to form a cross-linked polymer network. The central part of the monomer is rigid; therefore, the polymer network is also relatively rigid. When the polymerization occurs in the ordered liquid crystal phase, the network retains some of the phase anisotropy. These networks have been formed in many liquid crystal phases: ordinary (Hikmet, 1990) and chiral nematics (Yang et al., 1992), smectic A (Crawford et al., 1994a), smectic B (Mariani et al., 1986), and smectic C* (Pirš et al., 1992, 1995), and various types of monomers have been employed. Fung and coworkers (1994) have explored evacuated polymer networks with SEM. One example is presented in FIGURE 9.7 where the sample was photopolymerized with the electric field switched on.

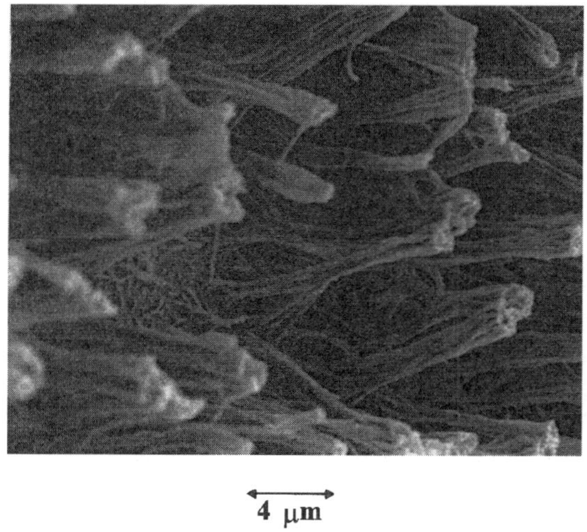

FIGURE 9.7. A scanning electron microscope (SEM) photograph of a polymer network used in polymer-stabilized liquid crystal displays. The sample was cured in a 100 V AC field (500 Hz) and the same is oriented at 45° in this photograph.

9.2.2 Thermally Induced Phase Separation

Preparation of PDLC materials by thermally induced phase separation (TIPS), that is, by controlled cooling of a solution of thermoplastic polymers and liquid crystals was first reported by West in 1988. This method is most useful for thermoplastic materials that melt below their decomposition temperatures. The TIPS process begins by preparing a binary mixture of a liquid crystal and thermoplastic at an elevated temperature to ensure a homogeneous mixture. The mixture is usually sandwiched between two glass substrates with transparent conducting coatings, and cooled into the miscibility gap, resulting in phase separation of the liquid crystal material. The droplet size is predominantly dictated by the cooling rate. Material parameters that govern the droplet size are viscosity and chemical potential.

Typical thermoplastic materials used in PDLC fabrication are polymethylmethacrylate (PMMA) and polyvinylformal (PVF) that are most commonly used in conjunction with cyanobiphenyl liquid crystal materials. Thermoplastic materials can also be engineered from epoxy systems that form extended chain molecules instead of cross-linked systems. A thermoplastic system made up of Epon 828, hexylamine, and E7 in a 1:1:1 weight percentage ratio was studied (West, 1988). In general, because of less complete phase separation, larger concentrations of liquid crystal are required for the TIPS process than for the PIPS process.

9.2.3 Solvent-Induced Phase Separation

Solvent-induced phase separation (SIPS) based on solvent evaporation is possible with thermoplastics that melt above their decomposition temperatures. This method is particularly useful for coating techniques. The liquid crystal and thermoplastics are dissolved in a common solvent forming a homogeneous mixture. A 3:2 weight percentage ratio of the liquid crystal E7 and PMMA in chloroform is a well-characterized system. After a homogeneous solution is achieved, the material is usually deposited on a glass substrate with a transparent ITO coating. The solvent begins to evaporate and eventually progresses through the miscibility gap, resulting in droplet formation and growth until the polymer solidifies. At this point, the droplet morphology is locked in. The average droplet diameter is strongly dependent on the rate of solvent removal. The polymer is plasticized by the presence of dissolved liquid crystal that is found to lower the glass transition temperature. Electrooptic studies on samples prepared from the SIPS process have been reported (Jain et al., 1990).

SIPS can also be used in conjunction with the TIPS process simply by using a thermoplastic that melts below its decomposition temperature (West, 1988). A thermoplastic and liquid crystal are dissolved in a common solvent. The solution is deposited on a glass substrate with an ITO coating. The solvent is quickly removed, ignoring the droplet size and density. By warming the sample to a temperature where the liquid crystal is dissolved back into the polymer, the droplet size and density can then be controlled by the cooling rate.

9.2.4 Encapsulation Process

The process of encapsulation (NCAP) is brought about by emulsifying a liquid crystal in a water-borne polymer. The continuous aqueous medium can contain either a water-soluble polymer or can be a colloidal suspension of a water-insoluble polymer. Drzaic (1986) prepared NCAP films by mixing polyvinyl-alcohol (PVA) and the liquid crystal mixture ZLI 2061 (E. M. Chemicals) in a water solution. The mixture is then blended by a high-speed maceration device. The emulsion is allowed to de-gas, and is deposited onto an ITO polyester substrate and allowed to dry for several hours. After the film is completely dry, it is laminated onto another ITO-coated polyester substrate. The droplet morphology in NCAP systems is usually more prolated than in their PDLC counterparts, and they have a larger droplet diameter distribution than PDLC materials.

9.2.5 Permeation

Permeation simply refers to the filling of liquid crystal into the pores of a prefabricated matrix. These matrices are usually commercially available and generically fall into two distinct categories: those with well-defined geometries and those which are more random in nature.

Nuclepore membranes (Kuzma and Labes, 1983) are polycarbonate filters used principally in critical filtration procedures. The pores are well-defined cylindrical cavities, oriented, on average, perpendicular to the membrane surface and penetrating through its 10 μm thickness (Crawford et al., 1992a). The pores are produced in the polymer film using an irradiation and etching process. The membranes are available in a variety of pore sizes ranging from 0.015 μm to

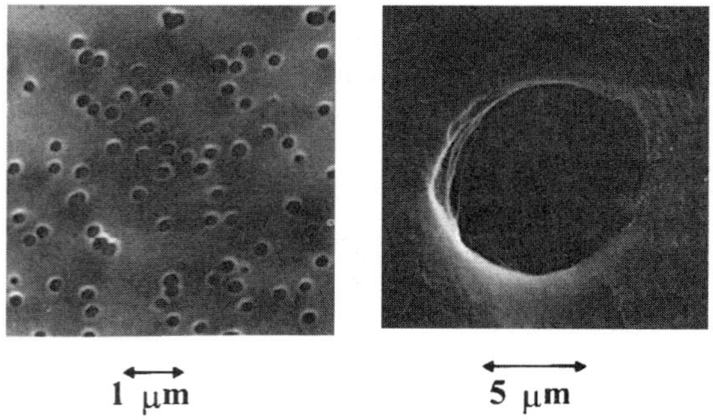

FIGURE 9.8. Scanning electron microscope (SEM) photographs of two Nuclepore membranes.

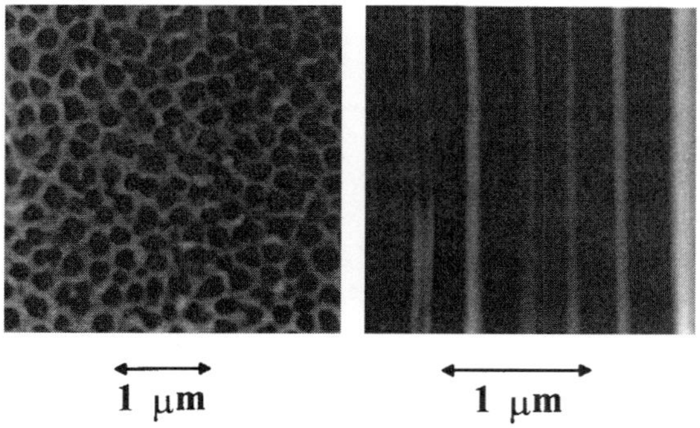

FIGURE 9.9. Scanning electron microscope (SEM) photographs of an Anopore membrane.

micrometer diameters. The pore size distribution is relatively narrow, with 80% of the pores being within 10% of the mean (Chen et al., 1980). The cylindrical nature of the pores and the narrow pore size distribution is shown in the scanning tunneling microscopy photograph presented in FIGURE 9.8. The manufacturers specify the pore density to be $3 \times 10^8\,\mathrm{cm}^{-2}$, and the porosity is approximately 12%. Various measurements have been carried out on Nuclepore membranes to characterize their heat capacity, pore size, and pore density (Chen et al., 1980).

The pores of the membranes are simple to fill using surface tension. The complete filling of the pores is reached after soaking the membrane in an isotropic liquid for one hour. The best way to remove the excess liquid crystal on the outer surfaces of the membranes is to press them between two Whatmann filtration papers. The internal surfaces of the pores are easily accessible to various surface treatments, such as surfactants and polymers, that alter the liquid crystal interactions occurring at the surface (Crawford et al., 1992a, b). A 2% solution of lecithin in hexane is rinsed through the filters. The membrane is then placed in a vacuum oven at 50 °C for several hours to insure complete removal of the solvent. Other surfactants and polymers can be used, as long as they are chemically compatible with the Nuclepore membranes. Uniform homogeneous anchoring can also be achieved by treating the surfaces with polyimide (Ondris-Crawford et al., 1993). The inner corrugations (or grooves) inside the cavities shown in FIGURE 9.8 prove to be the mechanism necessary to enforce unidirectional homogeneous anchoring, analogous to rubbing polymer-treated planar surfaces. After the pores are treated, the filling process is the same as described above.

Anopore membranes (Hoffmann, 1989) are made of aluminum oxide produced by the anodic oxidation of aluminum. They have well-defined cylindrical pores, 0.2 μm in diameter, penetrating straight through the 60 μm thickness. They are commercially available only in the 0.2 μm pore diameter, and exhibit a very narrow pore size distribution. The porosity of Anopore membranes is nearly 40%. The scanning electron microscopy photograph in FIGURE 9.9 reveals that one side of the membrane has asymmetric pores; however, the cross-section photograph shows straight cylindrical channels. Fortunately, the asymmetric nature of the pore only extends 0.1 μm into the cavity, and therefore is negligible compared with the 60 μm length of the cavity (Furneaux, 1989). The inner cavity walls of Anopore membranes can also be treated with many surfactants. Those that have been most successful are the aliphatic acids (C_nH_{2n+1}–COOH), since the acid head group chemically attaches to the cavity wall (Crawford et al., 1993a).

Both Nuclepore and Anopore membrane technologies have relatively well-defined geometries and are ideal for the study of confined liquid crystals. Anopore membranes have a higher porosity and are capable of withstanding a wider variety of surface treatment procedures. Their rigidity also makes them easier to handle. Nuclepore membranes are advantageous because they are available in a variety of pore sizes.

Random porous media like Vycor glass or silica aerogel are becoming popular matrices to study effects

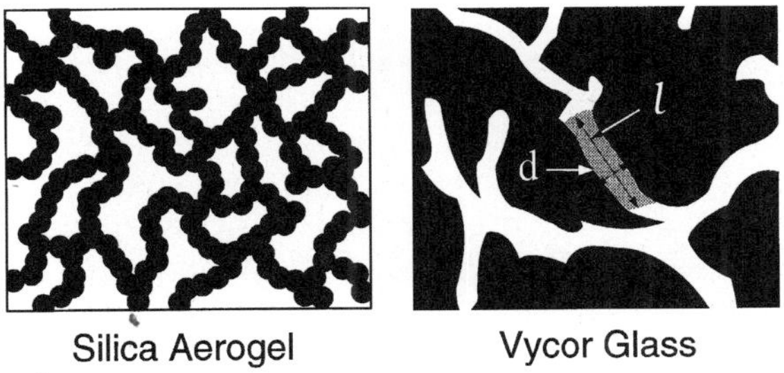

FIGURE 9.10. Model structures of the silica aerogel and Vycor glass matrices used in basic studies of confinement.

of random confinement of liquid crystals. (For details see the relevant chapters of the book edited by Crawford and Žumer, 1996a). Silica aerogel has been well characterized by several groups. One example is engineered from a based catalyzed hydrolysis and condensation of tetramethoxysilane and has a mass density of 0.36 g/cm^{-3} (Schaefer and Keefer, 1986). Small-angle x-ray scattering shows that the mean solid chord length is $\sim$0.0045 μm and the average pore size is $\sim$0.0175 μm. It is useful to envision aerogel matrices as presented in FIGURE 9.10 where structure is to some extent fractal, sometimes referred to as a "jungle gym" or branched polymer structure. This model was proposed by Schaefer and Keefer based on small-angle x-ray scattering experiments (1986). Schaefer (1994) recently reviewed the preparation of these systems. Bellini and coworkers (Bellini et al., 1992) filled aerogels with 8CB under vacuum using capillary action in the nematic phase. Kralj and coworkers (Kralj et al., 1993) have permeated aerogels with 5CB.

Another type of random confining matrix is porous silica glass. Vycor glass, manufactured by Corning, has been extensively characterized to establish the connectivity of the pore network, morphology of the pore volume, and roughness of the pore interface (Levitz et al., 1991). The fabricating process of Vycor glass is well established, and the final product is essentially a pure SiO_2 skeleton. A model network presented in FIGURE 9.10 is based on transmission electron micrograph (TEM) (Levitz et al., 1991). Vycor glass can be roughly envisioned as a three-dimensional interconnected network of pure segments of approximate diameter $d \simeq 0.007\,\mu$m and average length of $l \simeq 0.03\,\mu$m. Vycor glass was filled with liquid crystal (Iannacchione et al., 1993) and other similar glass matrices. After thorough cleaning of the Vycor glass in a hydrogen peroxide solution in an ultrasound bath, the liquid crystal is then capillary-filled into the pores in the isotropic phase. Similar glass matrices have been studied extensively by Aliev and Breganov (1989).

9.3 THEORY OF CONFINED NEMATIC LIQUID CRYSTALS

The statistical mechanics of a complex system is based on a certain knowledge of its basic elements and their mutual interactions. In the case of liquid crystals, the anisotropic molecular shape, described as cylinder-like or disk-like, is particularly crucial for liquid crystallinity. Alone, it can lead to spontaneous long-range orientational ordering (Onsager, 1949). In addition to the shape (stearic interactions), the noncontact anisotropic interactions that depend on molecular properties such as composition, flexibility, chirality, polarity, amphiphilicity, etc., are responsible for the large variety of orientational and positional ordering phenomena in liquid crystal mesophases. This complexity seriously limits all microscopic approached beyond the Maier–Saupe molecular field theory (Maier and Saupe, 1958), to descriptions of particular aspects of liquid crystal ordering. Fortunately, in most situations the macroscopic approach based on an expansion of the free energy in terms of a phenomenological order parameter allows one to develop a consistent description of the liquid crystal phenomena. As will be shown in the following section, this is also true in confined systems.

9.3.1 Phenomenology of Nematic Ordering in Bulk Liquid Crystals

9.3.1.1 Landau Theory

The nematic phases allow us to concentrate only on the orientation order. The general phenomenological approach to nematic ordering requires the introduction of the tensor order parameter $\underline{\mathbf{Q}}$ (de Gennes and Prost, 1993). In the eigenframe defined with nematic director $\mathbf{n}$ and biaxial director $\mathbf{n}_b$, the corresponding degrees of order are measured by the eigenvalues Q (nematic order parameter) and P (biaxial order parameter), respectively. The relevant independent intensive thermodynamic variables are temperature and external magnetic or electric field. According to basic principles, the corresponding thermodynamic potential (usually called the free energy of a system) approaching an equilibrium state tends toward its minimum value. For a homogeneously oriented nematic liquid crystal, one can follow Landau and Lifshitz (1980), expanding the free-energy density f_n in terms of the symmetry-allowed powers of the order parameter $\underline{\mathbf{Q}}$ (Vertogen and de Jeu, 1988):

$$f_n(\mathbf{r}) = f_0(T) + \frac{9}{2}(T - T^*)aQ_{ij}Q_{ji} - \frac{b}{3}Q_{ij}Q_{jk}Q_{ki} + \frac{c_1}{4}(Q_{ij}Q_{ji})^2 + \frac{c_2}{4}Q_{ij}Q_{jk}Q_{kl}Q_{li} \tag{9.1}$$

where $a(T - T^*)$, b, c_1, and c_2 are expansion coefficients. The supercooling limit temperature is denoted by T^*, and the indices i, j, k, and l, represent any of our three coordinate axes. For a uniaxial nematic phase, Q_{ij} can be simply replaced by the scalar Q, the nematic order parameter. The introduction of a homogeneous external magnetic (or similarly electric) field $\mathbf{H}$ adds to the free energy the following anisotropic contribution (de Gennes and Prost, 1993):

$$f_f = -\mu_0 \Delta\chi_0 \mathbf{H}\underline{\mathbf{Q}}\mathbf{H}/2 \tag{9.2}$$

where $\Delta\chi_0$ is the difference between the principal values of the susceptibility tensor in a completely ordered nematic phase.

The resulting free energy is capable of predicting a first-order nematic–isotropic phase transition and its disappearance with increasing external field. Comparisons with experimental data for the order parameter, susceptibility, and specific heat yield the values of the expansion coefficients. To incorporate the effects of an inhomogeneous order parameter, de Gennes and Prost (1993) added to the free-energy density a deformation part, f_d, expanded in terms of symmetry-allowed powers of the order parameter and its space derivatives. Keeping only invariant terms quadratic in the first directive and linear in the second derivative, one finds:

$$\begin{aligned} f_d(r) = {} & L^{(1)}Q_{ij,ij} + L_1^{(2)}Q_{jk,i}Q_{jk,i} + L_2^{(2)}Q_{ij,i}Q_{kj,k} \\ & + L_3^{(2)}Q_{jk,i}Q_{ik,j} + L_5^{(2)}Q_{ik,ij}Q_{jk} + L_6^{(2)}Q_{jk,ii}Q_{jk} \\ & + L_1^{(3)}Q_{ij}Q_{ij,k}Q_{kl,l} + L_2^{(3)}Q_{ij}Q_{ik,j}Q_{kl,l} \\ & + L_3^{(3)}Q_{ij}Q_{ik,k}Q_{jl,l} + L_4^{(3)}Q_{ij}Q_{ik,l}Q_{jk,l} \\ & + L_5^{(3)}Q_{ij}Q_{ik,l}Q_{jl,k} + L_6^{(3)}Q_{ij}Q_{ik,l}Q_{kl,j} + \cdots \end{aligned} \tag{9.3}$$

for the nonhomogeneous part of the free-energy density associated with nonuniform ordering. The expansion coefficients $L_j^{(j)}$ are temperature-independent generalized elastic constants. The notation $Q_{ij,l}$ denotes the partial derivation of Q_{ij} with respect to the lth coordinate. To describe the chiral nematic phase, a term $L_4^{(2)}\varepsilon_{ijk}Q_{il}Q_{jl,k}$ with ε_{ijk} as the Levi–Cevita antisymmetric tensor, should be added to equation (9.3). The total free-energy density $f = f_n + f_f + f_d$ is, in general, not practical to deal with, but it helps to point out relations between expansion coefficients for simplified cases.

9.3.1.2 Frank Elastic Theory

A comparison of the quadratic terms in f_n (equation (9.1)) and f_d (equation (9.3)) enables the introduction of the nematic correlation length $\xi = (a|T - T^*|/L)^{-1/2}$ with L as an average expansion coefficient. The Frank approach applies to cases where the director slowly varies on the scale set by the nematic correlation length, and the order parameter Q is constant. For a uniaxial nematic phase, such a constant order parameter approximation leads to

$$Q_{ij}(\mathbf{r}) = \frac{Q}{2}[3n_i(\mathbf{r})n_j(\mathbf{r}) - \delta_{ij}] \tag{9.4}$$

where nematic structures are simply described by the director field $\mathbf{n}(\mathbf{r}) = (n_1(\mathbf{r}),\ n_2(\mathbf{r}),\ n_3(\mathbf{r}))$ with symmetry $\mathbf{n} = -\mathbf{n}$ (de Gennes and Prost, 1993). Let us now rewrite the free energy in terms of the director field. The external field part of the free energy density transforms to

$$f_f = -\mu_0 \Delta\chi(\mathbf{nH})^2/2 \tag{9.5}$$

with $\Delta\chi = Q\Delta\chi_0$, and the deformation part f_d to the well-known Frank elastic free-energy density (Frank, 1958)

$$\begin{aligned} f_d = {} & \frac{1}{2}\{K_{11}(\nabla\cdot\mathbf{n})^2 + K_{22}(\mathbf{n}\cdot\nabla\times\mathbf{n})^2 \\ & + K_{33}(\mathbf{n}\times\nabla\times\mathbf{n})^2 \\ & - K_{24}\nabla\cdot[\mathbf{n}(\nabla\cdot\mathbf{n}) + \mathbf{n}\times\nabla\times\mathbf{n}]\} \\ & + K_{13}\nabla\cdot[\mathbf{n}(\nabla\cdot\mathbf{n})] \end{aligned} \tag{9.6}$$

where the Frank elastic constants are related in the following way to the L_{ij} constants (Kralj and

Žumer, 1995),

$$K_{11} = \frac{9Q^2}{2}(2L_1^{(2)} + L_2^{(2)} + L_3^{(2)} - L_5^{(2)} - 2L_6^{(2)}) + \frac{9}{4}Q^3(-L_2^{(3)} + 2L_3^{(3)} + L_4^{(3)} + 2L_5^{(3)} - L_6^{(3)}) \quad (9.7a)$$

$$K_{22} = 9Q^2(L_1^{(2)} - L_6^{(2)}) + \frac{9Q^3}{4}L_4^{(3)} \quad (9.7b)$$

$$K_{33} = \frac{9Q^2}{2}(2L_1^{(2)} + L_2^{(2)} + L_3^{(2)} - L_5^{(2)} - 2L_6^{(2)}) + \frac{9Q^3}{4}(2L_2^{(3)} - L_3^{(3)} + L_4^{(3)} - L_5^{(3)} + 2L_6^{(3)}) \quad (9.7c)$$

$$K_{24} = 3QL^{(1)} + \frac{9Q^2}{2}(2L_1^{(2)} + L_3^{(2)} - \tfrac{1}{3}L_5^{(2)} - 2L_6^{(2)}) + \frac{9Q^3}{2}(L_4^{(3)} + 2L_5^{(3)} + L_6^{(3)}) \quad (9.7d)$$

$$K_{13} = 3QL^{(1)} + \frac{3Q^2}{4}L_5^{(2)}. \quad (9.7e)$$

The constants K_{11}, K_{22}, and K_{33} correspond to splay, twist, and bend elastic deformations, respectively. Further, K_{24} and K_{13} are the saddle-splay and mixed splay-bend elastic constants (Frank, 1958; Nehring and Saupe, 1971). The volume integral of divergence terms describing the corresponding deformation in equation (9.6) can be reduced to a surface integral. Therefore, the term "surface elastic" constants is often used. Further, we see that the second derivative of the director field in the K_{24} term in expression (9.6) is only apparent. That is not true for the K_{13} term, which is the only one with a second derivative of the director field. This so-called second-order elasticity causes problems when minimization of the free energy is not performed carefully. There have been several suggestions as to how the inclusion of the splay-bend elasticity also leads to a stable solution (Barbero and Strigazzi, 1989; Pergamenshchik, 1993, 1994; Barbero and Durand, 1993; Faetti, 1994).

In a single elastic constant (K) approximation, where only the constant $L_1^{(2)}$ associated with pure quadratic terms in the expressions 9.7(a–e) are nonzero, we have $K = K_{11} = K_{22} = K_{33} = K_{24} = 9Q^2L_1^{(2)}$, and $K_{13} = 0$.

Taking into account constants associated with other second-order terms in Q, one finds $K_{11} = K_{33} \neq K_{22} \neq K_{24}$ and $K_{13} = 0$, where all nonzero constants are of the same order of magnitude. Including linear and quadratic terms with second derivatives, one also finds $K_{13} \neq 0$ and, further including third-order terms, also $K_{11} \neq K_{33}$. In the chiral nematic case, the twist term has the form $\frac{1}{2}K_{22}(\mathbf{n} \cdot \nabla \times \mathbf{n} - \mathbf{q})^2$ with $2\pi/q$ as the pitch of the helical deformation. Let us recall what a stability consideration can tell us about the elastic constants. If q, K_{24} and K_{13} are zero, all deformations enter quadratically in the free energy. Then a stable uniform solution requires that K_{11}, K_{22}, and K_{33} be positive. The nonzero values of q, K_{24}, or K_{13} that are related to nonquadratic terms in the free energy can, in principle, stabilize a spontaneous deformation. The observations of bulk systems show that such deformations occur only in chiral systems with nonzero $\mathbf{q}$. Therefore, one can show (Ericksen, 1966) that for nonchiral systems, the inequality $0 < K_{24} < 2K_{11}$ or $2K_{22}$ (whichever is smaller) (Polak et al., 1994) must be satisfied. Recent experiments indicate that $K_{24} \sim K_{11}$ (Allender et al., 1991; Crawford et al., 1992a; Lavrentovich and Pergamenshchik, 1994) and that K_{13} is negative and approximately five times smaller than K_{11} (Lavrentovich and Pergamenshchik, 1994).

A rough comparison of the free-energy densities f_f, equation (9.5), and f_d, equation (9.6), enables us to introduce the magnetic coherence length given by the relation $\xi_m = (\mu_0 K/\Delta\chi)^{1/2}/B$, where K is an average of the elastic constants K_{ii}, $\Delta\chi$ is the magnetic susceptibility, and B is the magnitude of the magnetic field. This gives us an estimate above what scales field alignment effects are expected to dominate.

9.3.2 Deformations of Nematic Phase

Let us concentrate on a simple nematic system that cannot exhibit any spontaneous deformation. Thus, all inhomogeneities (nonuniform direct field) are unstable or metastable and relax toward the equilibrium homogeneous state. In real systems, such a situation is difficult to realize because deformations can be stabilized by external constraints (surfaces or fields). It is useful to distinguish between:

- weak inhomogeneities cased by thermal fluctuations that are always present;
- unstable inhomogeneities caused by changes in external conditions that can relax to the stable state by a continuous transformation of the director field;

- large inhomogeneities—singularities (defects) in the director field—created by strong external perturbances that usually cannot relax by a continuous transformation of the director field.

Continuous relaxation of unstable structures and fluctuations both follow nematodynamics, a kind of hydrodynamics (de Gennes and Prost, 1993), which allows for some qualitative estimates to be easily made.

9.3.2.1 Thermal Fluctuations

Fluctuations can be divided into those that vary the director (order director fluctuations—ODF) and those that vary the magnitude of the order parameter (order fluctuations). This division is important because only for ODFs are the associated fluctuations of the free energy proportional to q_f^2 (q_f = wavevector); therefore, they are in a long-wavelength limit energetically "cheap" (amplitude $\sim q_f^{-1}$) and also long lasting. For wavelengths of micrometer size, in a typical liquid crystal with values of $K \sim 10^{-11}$ N m and rotational viscosity $\gamma = 0.02$ N m s^{-1} the relaxation time $\sim q_f^2 K/\gamma$ is already about 10^{-3} s (de Gennes and Prost, 1993). In a large system, this leads to long lasting domains with different director orientations. The allowed fluctuation modes of a nematic confined to a droplet are influenced by particular director fields and by the limited size introducing $(q_f^{-1})_{\min} \sim R$. It is worthwhile to emphasize that once the wavelength approaches the molecular scale, the use of the phenomenological descriptions become questionable.

9.3.2.2 Weak Inhomogeneities

The relaxation of such structures (continuous transformation of the director field) can be straightforward using the predictions of Frank elastic theory and the equations of nematodynamics. Nevertheless, even in a simple bipolar nematic droplet, the calculation of the relaxation after the removal of a strong magnetic field is complicated by the presence of the backflow (Vertogen and de Jeu, 1988; de Gennes and Prost, 1993). Estimates of the relaxation time neglecting backflow usually provide only an order of magnitude (Wu et al., 1989), not much better than a simple estimate $K/(l^2\gamma)$ where l is the characteristic size of the structure. The same is also true for numerical simulations of the evolution of the nonequilibrium nematic structures which neglect the backflow (Kilian and Hess, 1989).

9.3.2.3 Defects

Defects are points or lines in space where the order parameter is singular or undefined. As long as one stays within the constant order approximation, these are singularities of the director field called disclinations. Known for many years from the observations of Lehman (1904) and Fridel (1922), they were first quantitatively treated by Frank (1958). Observation of defects is based on the polarization microscopy described in section 9.4. The line defects are usually classified according to their defect strength $s = \theta/(2\pi)$, where θ is the rotation of the director caused by encircling the defect line. Some simple line defects occurring in the uniaxial nematic phase are shown in FIGURE 9.11. For more details see, for instance, de Gennes and Prost (1993) and Kléman (1983). More complicated are biaxial nematics, where disclinations can occur in all three director fields. In chiral nematics with short pitch where the nematic director, direction of the cholesteric axes, and the normal to both of them, behave as three director fields that can exhibit disclinations (Kléman, 1989).

FIGURE 9.11. Director fields of eight different types of defects with their corresponding strengths, s. In the case of $s = 1$, three structures corresponding to different angles Ω_0 between the positional vector and local director are shown.

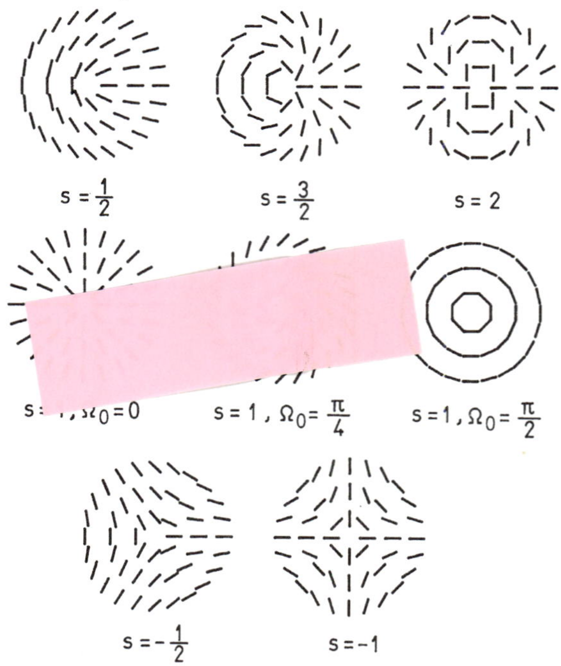

The observation of the instability of the $s = 1$ disclination line in uniaxial nematics (Williams et al., 1972; Meyer, 1973) and other interesting features of defects in uniaxial, biaxial, and chiral nematics are well explained within topological theory using homotopic groups (Toulouse and Kléman, 1976; Volovik and Mineyev, 1977; Mermin, 1979). The defects are in the same homeotropy class if their director fields can be continuously deformed one into the other. For a recent overview, see Kléman (1989). These defects are closely related to the defects in suprafluid helium (Mermin, 1979), and cosmological systems (Pargellis et al., 1991), where the term monopole, instead of point defect, and string, instead of disclination line, are often used.

Going beyond the constant order parameter approximation and solving equations for the tensorial order parameter, one finds that eigenvalues of the tensor $\underline{\mathbf{Q}}$ change continuously. To avoid the elastic energy surpassing the difference between the free energy of the nematic and isotropic phase, the order is depressed (partial melting of the nematic phase) or becomes biaxial. The area where $\mathbf{Q}$ substantially deviates from its bulk value (e.g., defect core) is comparable to the nematic correlation length ξ. Because the structure of defect cores is not relevant for optical properties of the nematic phase, there are only a few studies devoted to cores of point defects (Penzenstadler and Trebin, 1989; Kralj et al., 1991; Berggren et al., 1994) and disclination lines (Schopohl and Sluckin, 1987). More details about a simple hedgehog are given in section 9.5.1.

DYNAMICS OF DEFECTS. In the case of large inhomogeneities with defects that are far from the equilibrium state, local minima in the free energy enable the existence of metastable structures. The relaxation of such systems is very sensitive to external constraints and perturbations and thus difficult to predict theoretically. Studies of the motion of defects in the nematic phases (Cladis et al., 1987; Kilian, 1992; Muzny and Clark, 1992; Pismen and Rubinstein, 1992) have recently been stimulated by cosmological analogies (Pargellis et al., 1991; Yurke et al., 1992).

DRIVEN DYNAMIC PHENOMENA. Among the variety of driven dynamic phenomena, the dissipative structures caused by electric fields in liquid crystals with nonzero conductivity are the most common (de Gennes and Prost, 1993). They are a consequence of the mismatch between the principal directions of the mobility tensor and the director field. This topic, which is very important for planar liquid crystal cells, is not particularly relevant to the PDLC systems and will therefore not be addressed here.

9.3.3 Effects of Confinement on a Nematic Liquid Crystal

A liquid crystal confined to a finite volume in a solid or liquid matrix is affected by interactions and geometry of the confining surface. Unlike in simple solids or liquids, the perturbations of the orientational order induced by the surface penetrates through the entire confined liquid crystal and dramatic effects result. It is instructive to list the confinement induced breakings of the symmetry of the nematic phase:

- *Translational symmetry* is broken by the presence of the boundaries that restrict the liquid crystal to a certain space, and because of the nonuniform positional and orientational ordering caused by anchoring of the liquid crystal molecules on the confining surfaces.
- *Orientational symmetry* is broken on each surface that limits the possible orientations of the liquid crystal molecules. The effect of this orientational anchoring transfers via molecular interactions through the entire confined volume.
- *Rotational symmetry* around the director is lost because of anisotropy in the local molecular motions (fluctuation, rotations) caused by direct interactions on the surface, and indirectly by distortions of the liquid crystalline ordering. This loss of the rotational symmetry is then manifested by induced biaxial ordering and the appearance of flexoelectric polarization (see, for instance, de Gennes and Prost, 1993).
- *Inversion symmetry* is broken in the surface layer if molecular interactions are such that certain parts of the liquid crystal molecules (head, tail, etc.) prefer a certain position relative to the surface, yielding nonzero surface polarization.

9.3.3.1 Geometry and Structure of Surfaces

In a molecular picture, instead of speaking about geometry, the surface is described instead by the *distribution of the atoms or molecules* on the surface of the confining matrix. They can either be elements of

(c)

(b)

(a)

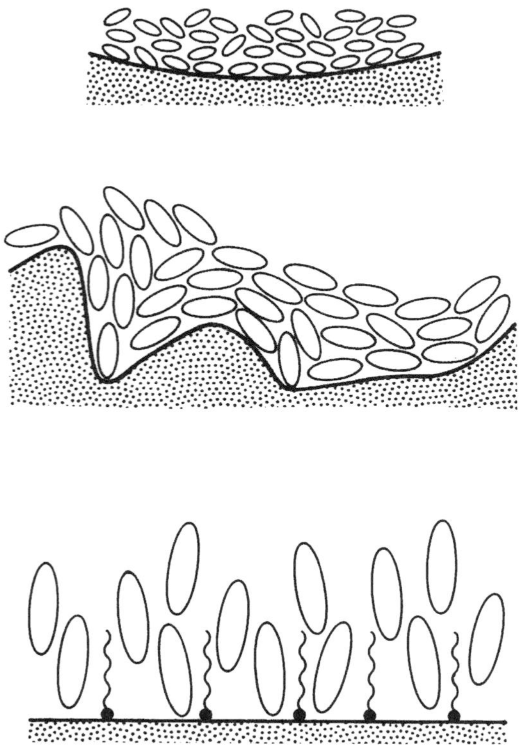

FIGURE 9.12. Schematic presentations of molecular ordering of the liquid crystals near a surface on different scales: the molecular scale, where steric hindrance of surfactant molecules promotes homeotropic anchoring (a); the microscopic scale where the roughness plays a crucial role (b); and the macroscopic scale where the surface is characterized by its curvature (c).

the confining matrix, impurities present in the liquid crystal (Barbero and Durand, 1990; Bechhoefer et al., 1990) or elements of the surface agent (FIGURE 9.12(a)) used to modify the confining surface (Cognard, 1982). In PDLC systems, only the first two situations can be realized, while surface agents can be applied to systems where confining surfaces are accessible during the treatment process. These processes are essential in the preparation of planar liquid crystal cells. In preparing well-defined coverages of planar systems, the Langmuir–Blodgett techniques are particularly useful (Sakuhara et al., 1992). Recently, several surface agents have been used to treat submicron-size cylindrical cavities (Crawford et al., 1991b, 1993b, 1996b).

Further, *anisotropy, orientation, and rigidity* of the interfacial elements are crucial for liquid crystal ordering. For instance, the surface with molecular tails sticking out of it will, depending on their lengths, promote different liquid crystal orientation (Porte, 1976). In case of aluminum oxide, the coverage by aliphatic acids with different chain length causes either parallel or orthogonal liquid crystal orientation (Crawford et al., 1993a, 1996b). If surface molecules are too flexible they cannot strongly influence the ordering of the liquid crystal.

Going beyond the molecular level, speaking about the geometry of the surface becomes justified. It is convenient to distinguish between two scales: a supramolecular one where the surface is characterized by its roughness (FIGURE 9.12(b)), and a macroscopic one (usually much larger than the nematic correlation length) where the surface is characterized by its shape (for instance, the radii of curvature, FIGURE 9.12(c)).

Let us list the main methods for examining the surface on the microscopic and macroscopic levels. Atomic force and electron tunneling microscopy (Sarid, 1991) cover the nanometer to submicrometer range, and electron scanning microscopy (Cosslett, 1987), probes the submicrometer to micrometer range; optical microscopy covers the larger scales. There are several indirect methods: second harmonic generation (Feller et al., 1991), evanescent wave ellipsometry (Chen et al., 1989; Moses and Shen, 1991), and magnetic resonance techniques (Crawford et al., 1991a, 1996b; Vrbančič et al., 1993) which are sensitive to the liquid crystal ordering close to the surface.

9.3.3.2 Surface Interactions

If there is no chemical bonding, the primary surface effects on the liquid crystal molecules are sterical repulsion, van der Waals interactions, quadripolar and dipolar coupling. Depending on the molecular structure of the surface, they yield an ordering "field" in the interfacial region. The induced nematic and positional order can persist far away from the surface. The dipolar part is relevant for establishing the polar order only in the surface layer when molecules are amphiphilic and the surface is either polar or aliphatic (Feller et al., 1991). Simple models describing the effective aligning interactions (Sluckin and Poniewierski, 1986; Tjipto-Margo and Sullivan, 1988;

Teixeira and Telo da Gama, 1991) are within the molecular field approximation used to predict the behavior of the coefficients of the phenomenological Landau–de Gennes description of surface phenomena (Teixeira and Sluckin, 1992a; Teixeira and Sluckin, 1992b) and enable the understanding of orientational and positional anchoring, adsorption, and wetting phenomena. Zannoni's group used a Monte Carlo method to determine directly the ordering of liquid crystal molecules in a spherical cavity (Chiccoli et al., 1990, 1993). A cubic lattice Lebwohl–Lasher model (Lebwohl and Lasher, 1972) with several different surface couplings was used. This approach is particularly powerful in predicting the ordering of very small confined systems where the phenomenological description discussed in this paper breaks down.

9.3.4 Macroscopic Description of Confining Effects

This approach to interfacial phenomena is based on the surface free-energy density (f_s), that is, surface tension. Because of the orientational order, the situation is more complex than it is in isotropic liquids. Particularly interesting are questions such as how wetting of the surface with the nematic phase and anchoring of nematic director field are related to surface-induced orientational order or disorder. For a review of this field, see the review paper by Jerome (1991), and for the discussion of the Landau–de Gennes theory, supplemented by a surface contribution expanded in terms of the surface value of the order parameter $\mathbf{Q}_s$, see Sluckin and Poniewierski (1986). Let us examine two situations where f_s can be simplified: first, the nematic phase far below T_{NI} where anchoring effects are dominant, and second, in the vicinity of T_{NI} where relatively simple ordering phenomena are dominant.

9.3.4.1 Anchoring Energy

At temperatures far enough from T_{NI} where the order parameter varies slowly, dealing with the ordering direction separately from the degree of orientational order is a reasonable simplification. The term "anchoring energy" was introduced to describe that part of surface tension associated with the deviation of the director at the surface from its preferred direction. To emphasize the origin of this effect, a more precise term "orientational anchoring" instead of "anchoring" is often used. This is certainly necessary in phases with partial positional ordering, for example, in a smectic liquid crystal with layers perpendicular to the crystal surface and subjected to a shear force, where the restoring force is associated with the positional anchoring (Cagnon and Durand, 1993). Here we are omitting the word orientational because we focus only on nematic ordering. Even in such a case, there are a large variety of situations, usually classified as homeotropic, tilted, or planar anchoring, that depend on the preferred direction: perpendicular, tilted, or parallel to the surface, respectively. Further, the last two cases can be monostable, bistable, multistable, or degenerate, depending on the nature of the surface (Jerome, 1991). In our theoretical discussion we are going to deal only with homeotropic and monostable (usually called homogeneous) planar anchoring that allows further simplifications. Retaining only the lowest-order terms results in the well-known Rapini–Papoular (1969) form of the anchoring free energy

$$f_s = f_{s0} + \frac{1}{2} W_\theta \sin^2 \theta \qquad (9.8)$$

for homeotropic anchoring and a two-parameter form (Ondris-Crawford et al., 1993)

$$f_s = f_{s0} + \frac{1}{2} (W_\theta \cos^2 \phi + W_\phi \sin^2 \phi) \sin^2 \theta \qquad (9.9)$$

for monostable planar anchoring. The angles are defined in FIGURE 9.13(a) and (b). W_θ represents the polar (zenithal) anchoring strength associated with any variation of the angle θ in the homeotropic case and with out-of-plane variations in the planar case, while W_ϕ stands for the azimuthal anchoring strength associated with in-plane director deviations in the

FIGURE 9.13. Anchoring geometry with the notation defined in the text for homeotropic (a) and unidirectional planar anchoring (b).

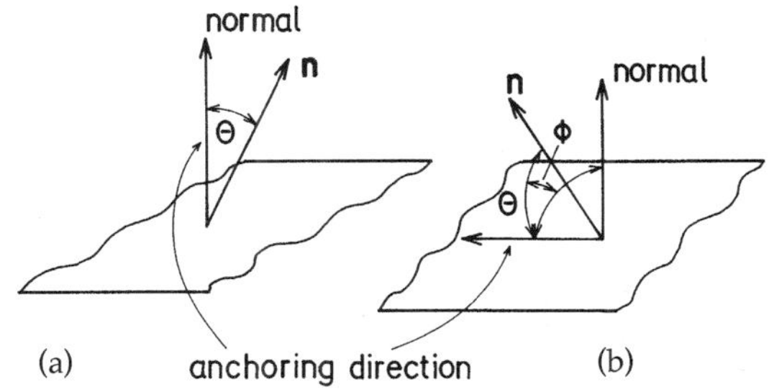

planar case. A rough comparison of the Frank free energy f_d (equation (9.6)) and anchoring free energy f_s (equation (9.8)) enables the introduction of the extrapolation length $d = K/W$ where selections for K and W depend on the details of the geometry (de Gennes and Prost, 1993). This defines the scale where an extended field can cause a substantial deviation of the director from the surface preferred direction.

9.3.4.2 Anchoring Transitions

Anchoring transitions, also called alignment transitions, are discrete transitions between two different anchoring types driven by changing conditions such as temperature (Flatischler et al., 1991), concentration of adsorbed small molecules (Bechhoefer et al., 1990), or surfactant chain length (Crawford et al., 1993a). They have been observed in addition to a continuous change in the tilt angle (Porte, 1976; Patel and Yokoyama, 1993). Theoretical considerations based on a microscopic picture (Teixeira and Sluckin, 1992a, b) and Landau–de Gennes theory (Barbero et al., 1991; Teixeira et al., 1993) are consistent with observations. Nazarenko and Lavrentovich recently reported a temperature-driven anchoring transition that was found to be influenced by ionic impurities in a nematic liquid crystal composed of centrosymmetric molecules (Nazarenko and Lavrentovich, 1994). (For more references, see Jerome, 1991.)

9.3.4.3 Surface-induced Order/Disorder

Let us describe here a simple case with a nondeformed director field where the ordering direction coincides with the surface preferred normal direction. The surface free energy can be expanded only in terms of surface order parameter Q_0, yielding

$$f_s = f_{s0} - GQ_0 + \tfrac{1}{2}UQ_0^2 + \cdots \tag{9.10}$$

or a more intuitive form (Nobili and Durand, 1992):

$$f_s = f_{s0} - G(Q_s - Q_0)^2/2. \tag{9.11}$$

The surface coupling where only the linear term in Q_0 is taken into account was first used by Sheng (1976, 1982) to study the phenomena close to the first-order nematic–isotropic phase transition. Depending on the strength of the coupling G, he predicted possibilities for a prewetting transition (formation of the nematic boundary layer with finite thickness), a complete wetting (divergent nematic layer), and a disappearance of the isotropic–nematic phase transition in a thin layer. The inclusion of the quadratic term does not substantially change this phenomenon in the isotropic phase (Sluckin and Poniewierski, 1986) but can be used to predict disordering effects, for example, surface melting of the nematic phase on rough surfaces (Barbero and Durando, 1991a, b).

More complicated is the degenerate planar anchoring case, where the nematic phase breaks the symmetry of the surface. Depending on the strength of the coupling, the pretransitional ($T > T_{NI}$) ordering close to the surface can be uniaxial with a negative order parameter, weakly biaxial, or strongly biaxial (Sluckin and Poniewierski, 1985; Sheng et al., 1992; L'vov et al., 1993; Kothekar et al., 1994). The symmetry breaking transition in this partially ordered surface layer can be of the first-order type (prewetting transition as in the homeotropic case) or even more interesting, of the second-order type corresponding to the Kosterlitz–Thouless-type transition, as was first pointed out by Sluckin and Poniewierski (1985). The latter is a defect unbinding transition characteristic for two-dimensional systems was xy symmetry (Kosterlitz and Thouless, 1973).

9.3.4.4 Rough and Grooved Surfaces: Effective Anchoring

The enhancement of the azimuthal anchoring strength by using a grooved surface is a well-known method used in nematic liquid crystal cells. As long as grooves characterized by amplitude A and wavelength λ are such that $\lambda^2/A \gg \xi$ (nematic correlation length), the effective anchoring strength is well described by the Berremann–de Gennes model (Berremann, 1972; de Gennes and Prost, 1993):

$$W_\phi = 4KA^2(\pi/\lambda)^3. \tag{9.12}$$

With increasing depth of the grooves, the depression of the nematic order near the surface (surface melting) must be taken into account yielding the upper limit for the azimuthal anchoring to be of the order $L_1^{(2)}Q_{\text{bulk}}/\xi$ (Barbero and Durand, 1991c).

The renormalization of the anchoring strength (i.e., careful interpretation of experiments) must also be done for other anchoring cases as soon as surface roughness (for example, evaporated silica glass surfaces) or director field deformations induced by an external field cause a substantial depression of the order close to the surface (Barbero and Durand, 1991a).

9.3.4.5 Order- and Flexoelectricity

In materials consisting of molecules that have a dipole moment and are not sufficiently symmetric, the inhomogeneous ordering in deformed structures breaks local symmetry of nematic ordering, and therefore results in local polarization. Depending on the main source, inhomogeneous director field, or inhomogeneous degree of order, the phenomenon is either called flexoelectricity (Meyer, 1969) or order electricity (Monkade et al., 1988; Durand, 1990; Osipov and Sluckin, 1993).

In a Landau–de Gennes description, the order electric polarization is proportional to the gradient in the order parameter, and the flexoelectric one is proportional to a combination of the first derivatives of the director field. Both effects lead to deviations from the Rapini–Papoular relation (Barbero and Durand, 1986). Order electricity is important only close to T_{NI} and when deformations are sufficiently large that order is substantially depressed. Flexoelectricity becomes relevant in pure materials where the Debye screening length is too long for effective screening of the flexoelectric polarization.

9.3.4.6 Surface Smectic Order

SMECTIC LAYERING IN THE NEMATIC PHASE. Close to T_{NA}, the effect of pretransitional layering on the Freedericksz transition in a cell with homeotropic anchoring was first observed by Rosenblatt (1984). The free energy associated with the field-induced tilt leads to an effective renormalization of the anchoring energy, introducing an additional term $\frac{1}{2}D_{\text{surf}}\,\xi_s$ where ξ_s is the correlation length describing the width of the smectic layer and the effective elastic constant D_{surf} associated with the molecular tilt in smectic layers. D_{surf} is proportional to the square of the smectic order parameter so that on approaching T_{NA}, the apparent anchoring strength shows a critical increase (Rosenblatt, 1984). Ondris-Crawford has shown that this effect is not observed if confining surfaces are too rough to induce pretransitional layering (Ondris-Crawford et al., 1993).

SMECTIC LAYERING IN THE ISOTROPIC PHASE. Formation of the surface-induced molecular layers in the isotropic phase of liquid crystals was first detected by Horn and coworkers (Horn et al., 1981) measuring the oscillation in the force between the solid surfaces separated by a thin liquid crystalline film. Ocko was the first to show, by observing x-ray reflectivity, that on approaching T_{IA} in the isotropic phase, the smectic ordering proceeds layer by layer (Ocko et al., 1986). Theoretical descriptions based on a lattice model were developed by Pawlowska and coworkers (Pawlowska et al., 1988). For approaches using density functional theory see Somoza and coworkers (1994) and the references therein. Recent NMR (Crawford et al., 1994) and calorimetric (Iannacchione et al., 1994) studies show that there is also partial nematic ordering extending beyond the smectic layers, and that surface-induced smectic layering also occurs in confined systems close to T_{NI} if the material has a smectic phase.

9.3.5 Total Free Energy

In an equilibrium state, a confined system at constant temperature and volume has minimum free energy. Thus, combining the bulk and surface parts of the free energy that we discussed above, one can, in principle, predict all stable and even metastable structures of a confined nematic liquid crystal. Using the tensorial order parameter description to develop a general procedure for the minimization of the total free energy is not practical. In section 9.5, we will discuss a few examples where geometry, surface interaction and size of the system help us to develop a simplified description. We divided the problems according to their characteristic size:

- Suprananometer scale (defined by the nematic correlation length), where the Landau–de Gennes theory is needed to describe ordering. This limitation is realized in very small droplets, in cores of defects, and in surface layers.
- Micrometer scale, where Frank elastic theory including surface elastic terms and weak anchoring is needed to describe the ordering.
- Supra micrometer scale, where Frank elastic theory and strong anchoring is enough to describe the ordering.

Further, we limit our discussion to surfaces that enforce either homeotropic or planar anchoring and materials where flexo- and order electricity can be neglected. The minimization of the free energy is usually performed via solving the corresponding Euler–Lagrange set of differential equations. Depending on the approximation, the description is based on scalar (Kralj et al., 1991), vectorial (Dubois-Violette and Parodi, 1969; Williams et al., 1972; Williams, 1986; Kralj and Žumer, 1992) or tensorial description (Kilian and Hess, 1989; Gartland et al., 1991). It should be emphasized that the vector description of the nematic director field must satisfy the additional $\mathbf{n} = -\mathbf{n}$ symmetry requirement. There are few cases where equations can be solved analytically (Cladis and Kléman, 1972). This is certainly not possible when an external field is present and introduces a nonlinear term into the equations. Usually, generalized numerical relaxation methods are used (Press et al., 1986). The situation where the electric field is present can be more complicated than in the case of the magnetic field. Large values of the dielectric tensor can, in an inhomogeneously ordered system, result in an inhomogeneous internal field in contrast to the small diamagnetism that negligibly affects the magnetic field. A precise treatment, where the back-effect of the nematic ordering on the electric field is taken into account, requires simultaneous solution of the Maxwell and Euler–Lagrange equations (Haas et al., 1991).

9.4 BASIC TECHNIQUES AND FORMALISMS FOR EXAMINING CONFINED LIQUID CRYSTALS

9.4.1 Optical Polarizing Microscopy

The most desirable way to view nematic director in droplets is between the crossed polarizers of an optical microscope. The resulting pattern is an indirect visualization of the nematic director field within the droplet via its overall retardance, which depends on the optical axes of local directors, the optical anisotropy, Δn, the droplet radius, R, and the wavelength, λ, of the incident light.

Certain droplet structures and their corresponding textures are readily identifiable, but others are not. Often, only subtle differences discriminate one nematic director field from another, magnifying the difficulty in characterizing certain director fields. Furthermore, without some sort of analytical tool, it is impossible to deduce the correct nematic director field from the microscope texture. A simple method for simulating microscope textures of nematic droplets has been developed to help characterize the nematic director field (Ondris-Crawford et al., 1991). By first obtaining the nematic director field, either by the minimization of the free energy (see section 9.3.5) or by a trial function that satisfies the appropriate boundary conditions (Xu et al., 1992), the methods model the polarization rotations and phase shifts as light traverses through the polarizer, the droplet, and finally the analyzer.

Before describing the procedure, let us examine the optical properties of the system that helped to select the appropriate approach.

9.4.1.1 Birefringent Droplet

The polarizability of nematic liquid crystals is influenced by the same averaging process as the order parameter $\underline{\mathbf{Q}}$. Therefore the principal axis (optical axis) of the dielectric tensor ε is along the director $\mathbf{n}$ with the corresponding eigenvalue $\varepsilon = \varepsilon_{\text{iso}} + \Delta\varepsilon_0 Q$. The part ε_{iso} is the dielectric constant of the isotropic phase, and $\Delta\varepsilon_0$ is the dielectric anisotropy of a perfectly oriented nematic phase. For simplicity, we limit our discussion to uniaxial cases so that the other two eigenvalues are equal to $\varepsilon_{\text{iso}} - \Delta\varepsilon_0 Q/2$. To emphasize that the dielectric response is in the optical frequency range, it is convenient to use indices of refraction ($=\varepsilon^{1/2}$): n_e, the extraordinary index along the optical axis, and n_o, the ordinary index perpendicular to it, instead of the corresponding dielectric constants. A liquid crystal with an inhomogeneous director field is thus a nonuniform birefringent material (with typical values $n_e \simeq 1.7$ and $n_0 \simeq 1.5$), that strongly influences the propagation of light. It is important to remember from our previous discussion that variations in positional order or in degree of orientational order of a liquid crystal occur only on the molecular scale that is much shorter than the wavelength of visible light. Therefore, their effects on light propagation are negligible in comparison with the effects of inhomogeneities in the director field (direction of the optical axis).

Deformations of the director field in a confined nematic phase (stable or metastable) are always accompanied by thermally driven director fluctuations. Recalling from the above discussion (section 9.3.2.1), one can estimate the relaxation time $\tau \sim 10^{-4}$ s for fluctuations with the wavelength of visible light. This period is very long compared with the inverse frequency of light indicating that all collective dynamic processes in liquid crystals yield only a small line-broadening of the light spectrum.

For instance, broadening using time-resolved spectroscopy (Orsay Liquid Crystal Group, 1969) allows for the determination of correlation times of fluctuation modes, but otherwise has a negligible effect on the distribution of the transmitted and scattered light; therefore, all approaches developed for static distortions can be used. In addition, fluctuations in most cases are little affected by restricted geometry. Therefore, we further simplify the discussion by omitting the effects of fluctuations and focus our attention on the effects of confinement-induced static inhomogeneities of the director field. Nematic droplets are thus inhomogeneously birefringent particles where within a droplet, the optical axis is locally parallel to the nematic director field. A comparison of the characteristic scale of the inhomogeneities l to the wavelength of light λ enables us to discriminate two limiting cases: $l \gg \lambda$, that covers most optical microscopy experiments and allows the use of some aspects of geometrical optics, and $l \sim \lambda$, that is usually present in light scattering experiments, and requires the use of wave optics.

9.4.1.2 Simulated Textures

Following the above discussion, we choose geometrical optics to analyze the microscope textures of supramicron droplets. In performing polarization microscopy experiments, droplets were suspended in the isotropic medium with index of refraction $n_m \sim n_e$ and n_o. This altogether simplifies the description because relatively small variations of the index of refraction enable us to neglect reflections and refractions on droplet interfaces, and even the refraction in the nonuniform interiors of droplets (the limitations of this approach will be examined later). In this "straight ray approximation" (SRA) the only relevant effect is the birefringence-induced phase and polarization modulation of light (see FIGURE 9.14). Following Ondris-Crawford and coworkers (Ondris-Crawford et al., 1991), the distribution of the relative intensity of the transmitted light $I(\mathbf{p})$ can be determined using Jones' matrices (Fowles, 1968):

$$I(\mathbf{p}) = |\mathbf{e}_A \underline{\mathbf{T}}(\mathbf{p}) \mathbf{e}_P|^2. \tag{9.13}$$

The unit vector $\mathbf{e}_A$ describes the polarizer and $\mathbf{e}_P$ describes the analyzer, and $\underline{\mathbf{T}}(\mathbf{p})$ is the transformation matrix describing the modulation of the light polarization when a ray $\mathbf{k}$ traverses through a droplet so that it reaches point $\mathbf{p}$ on the texture (see FIGURE 9.14). Except

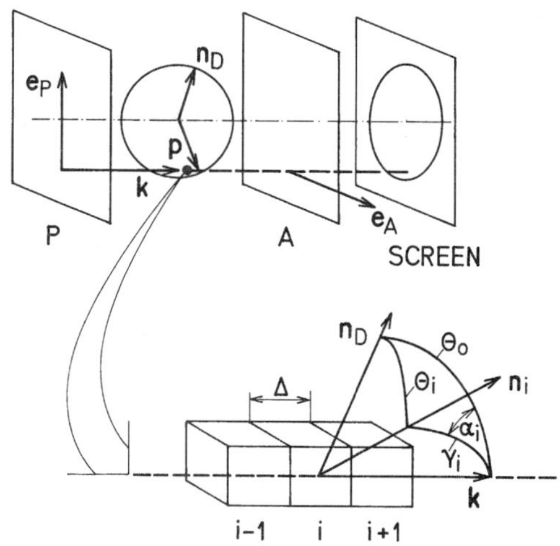

FIGURE 9.14. A schematic defining the notation used in calculating the optical polarizing microscope textures.

for very simple droplet structures (e.g., a radial droplet) the matrix $\underline{\mathbf{T}}(\mathbf{p})$ cannot be calculated analytically; therefore, the path of a ray through the droplet is discretized by dividing it into small intervals where the director field can be assumed to be homogeneous. In this way, the polarization changes in the following steps: change of the polarization vector caused by the rotation of the optical axis on going from one interval to another, and phase shifts within each interval. For N steps, the transformation matrix can be written as the following product:

$$\underline{\mathbf{T}} = \underline{\mathbf{R}}_{N+1} \underline{\mathbf{P}}_N \underline{\mathbf{R}}_N \ldots \underline{\mathbf{P}}_2 \underline{\mathbf{R}}_2 \underline{\mathbf{P}}_1 \underline{\mathbf{R}}_1 \tag{9.14}$$

where $\underline{\mathbf{P}}_i$ induces the phase shift to ordinary and extraordinary components of light in the ith interval, and $\underline{\mathbf{R}}_i$ represent a simple rotation of the optical axis $\mathbf{n}_i$ around $\mathbf{k}$ on going from the $i-1$ interval to the i interval. Introducing $\mathbf{n}_D$ as a direction characterizing a particular structure, we define angle α_i between the reference $\mathbf{n}_D - \mathbf{k}$ plane and an $\mathbf{n}_i - \mathbf{k}$ plane (see FIGURE 9.14), and write the rotation matrix

$$\underline{\mathbf{R}}_i = \begin{bmatrix} \cos(\alpha_i - \alpha_{i-1}) & -\sin(\alpha_i - \alpha_{i-1}) \\ \sin(\alpha_i - \alpha_{i-1}) & \cos(\alpha_i - \alpha_{i-1}) \end{bmatrix}. \tag{9.15}$$

The index i ranging from 1 to N corresponds to the intervals in the droplet. The end points $i = 0$ and $i = N+1$ represent the polarizer and analyzer, respectively. So, α_0 measures the orientation of droplet director $\mathbf{n}_D$ relative to the incoming polarization plane. (If the droplet director is parallel to **k**, α_0 is set to zero.)

The phase shift matrix in the i interval is:

$$\underline{\mathbf{P}}_i = \begin{bmatrix} \exp(in_o 2\pi\Delta/\lambda_0) & 0 \\ 0 & \exp(in_e(\gamma_i)2\pi\Delta/\lambda_0) \end{bmatrix} \quad (9.16)$$

where λ_0 is the wavelength of the incoming light, n_o is the ordinary and

$$n_e(\gamma_i(\mathbf{p})) = \frac{n_o n_e}{\sqrt{n_o^2 \sin^2\gamma_i + n_e^2 \cos^2\gamma_i}} \quad (9.17)$$

is the extraordinary index of refraction corresponding to the angle $\gamma_i = \gamma_i(\mathbf{p})$ between director field $\mathbf{n}_i$ and the ray vector **k**.

To calculate the complete texture, the above procedure of obtaining the transmitted light $I(\mathbf{p})$ is repeated for many points **p** until all needed details of the texture are apparent. To simulate black and white photographs of nonmonochromatic light textures, the laws of color mixing must be used (Ondris-Crawford et al., 1991).

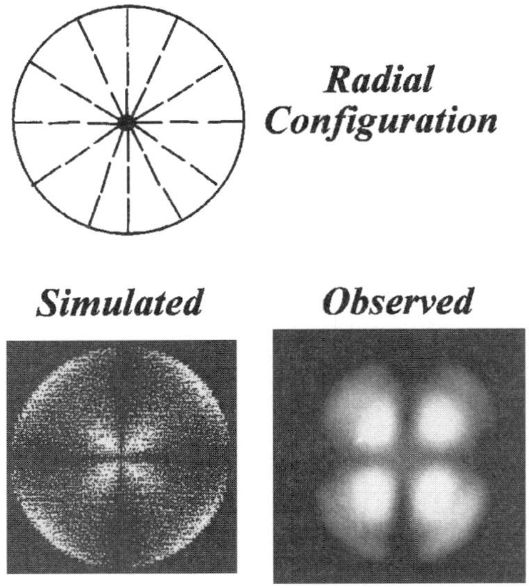

FIGURE 9.15. A model director field for the radial configuration of a droplet with homeotropic anchoring conditions. The observed and simulated white light optical polarizing microscopy textures of a radial droplet with radium 5 μm are in good agreement.

9.4.1.3 Examples of Microscope Textures

DROPLET TEXTURES. The radial configuration is a spherically symmetric radial director field (FIGURE 9.15) that can be stable for homeotropic anchoring at the droplet interface. The angles θ_0 and α_0 are arbitrary, since the radial structure has spherical symmetry. The simulated white light droplet texture presented in FIGURE 9.15 compares nicely to the actual microscope texture for a 5 μm droplet. When an electric field of sufficient magnitude is applied to the radial droplet, it transforms into the axial configuration shown in FIGURE 9.16(a). The difference between the texture of the radial droplet (FIGURE 9.15) and the axial droplet with the axis parallel to the viewing direction (FIGURE 9.16(b)–(d)) is subtle and primarily recognized by the presence of a darker center in the axial case and a bright four-leaf-clover figure near the center of the radial texture. This is more pronounced for the more aligned (high-field) textures presented in FIGURE 9.16(d). The radial configuration is also readily identified because all droplets in the sample have the same texture for all viewing directions. In the case of an axial droplet tilting (changing θ_0), the rotation of the microscope stage alters the texture.

The configuration most commonly observed in PDLC droplets is the bipolar configuration that occurs for parallel anchoring conditions (see FIGURE 9.17(a)–(e)). The droplet is characterized by two point defects known as boojums (Doane, 1991). When this structure is viewed along its axis of symmetry, the resulting texture presented in FIGURE 9.17(e) is invariant under changes of α_0. When $\theta_0 \neq 0$, the texture changes as the microscope stage, α_0, is changed, as shown in FIGURE 9.17(b)–(d). The simulated textures are in very good agreement with the observations for a droplet with a radius of 20 μm. In the case of tangential boundary conditions, the concentric configuration may be stable in droplets of liquid crystals. A few simulations of the concentric configuration are presented in FIGURE 9.18. Drzaic et al. (1992) found that this structure occurred when the bulk elastic constant ratio $K_{11}/K_{33} < 1$. Ding and Yang (1994) performed simulations of bipolar and axial structures using approximate analytical director fields. Xu and coworkers (Xu et al., 1992, 1994) have pushed this approach even further, using trial functions to model their observations of unusual spiral structures in droplets with perpendicular anchoring (FIGURE 9.19).

CAPILLARY TEXTURES. The difficulty in preparing well-defined PDLC droplets has led to the natural

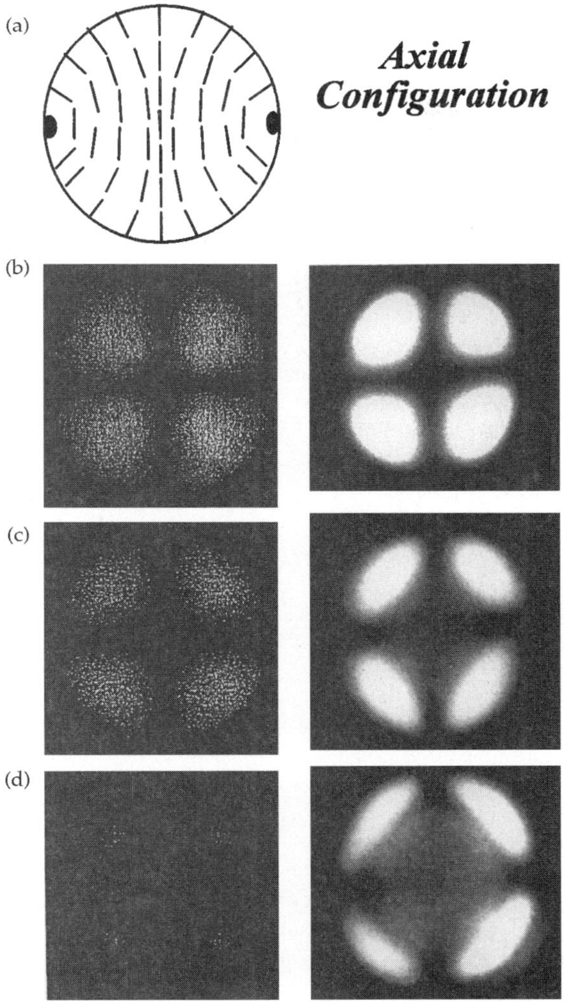

FIGURE 9.16. A model director field for the axial configuration with homeotropic anchoring conditions (a). The simulated (left) are directly compared with the observed (right) white light optical polarizing microscopy textures for no applied electric field (b); a weak applied electric field (c); and a strong applied electric field (d). Droplet radius is 5 μm.

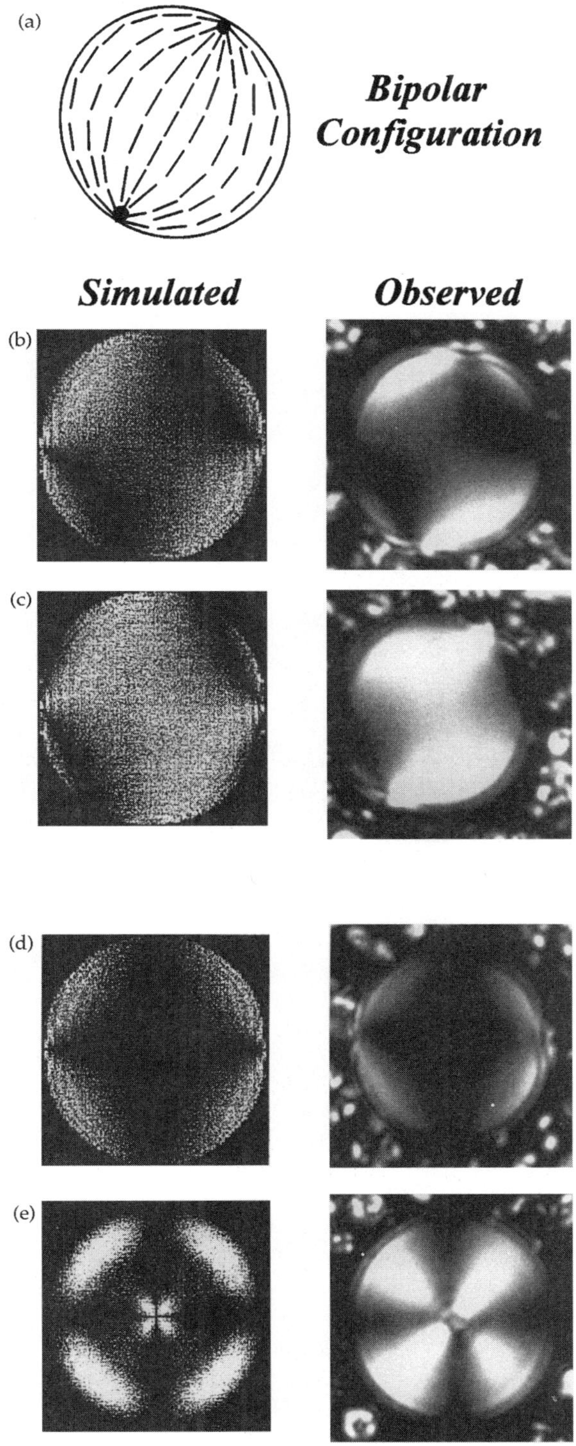

FIGURE 9.17. A model director field for the bipolar configuration with homogeneous anchoring conditions (a). The observed white light textures are all for zero-field situations at various orientations of the director axis n_D in a droplet with radium 20 μm. The observed and simulated optical polarizing microscopy textures are directly compared for $\theta_0 = 90°$, $\alpha = 15°$ (b); $\theta_0 = 90°$, $\alpha_0 = 30°$ (c); $\theta_0 = 90°$, $\alpha_0 = 0°$ (d); and $\theta_0 = 0°$ (e).

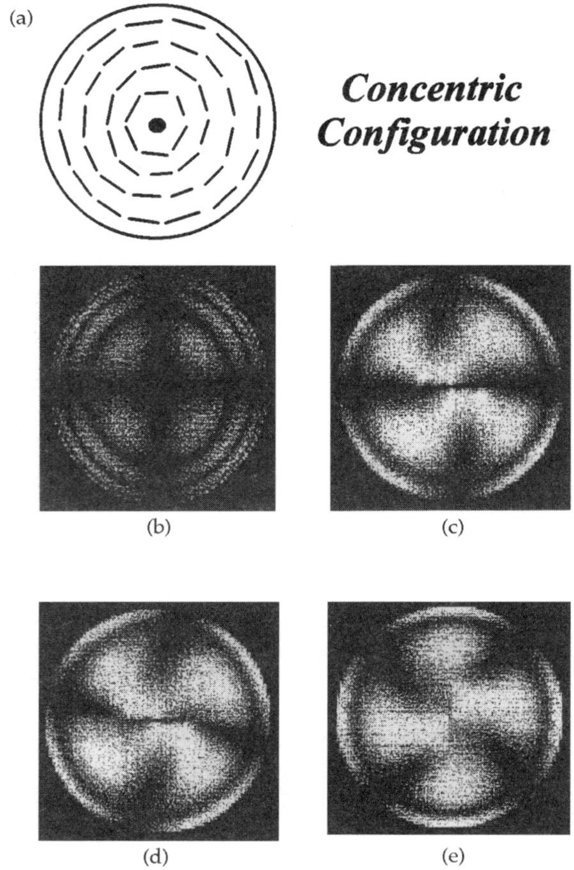

FIGURE 9.18. A model director field for the concentric configuration (a). Simulated microscope textures where the droplet director is oriented at $\theta = \alpha_0 = 0°$ (b); $\theta = 15°$ and $\alpha_0 = 0°$ (c); $\theta = 15°$ and $\alpha_0 = 15°$ (d); and $\theta = 15°$ and $\alpha_0 = 30°$ (e).

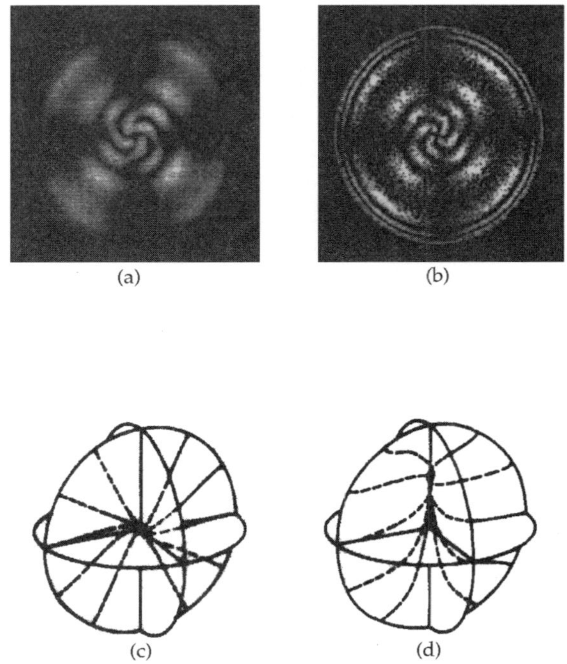

FIGURE 9.19. A texture of a radial nematic droplet between crossed polarizers of an optical microscope (a) alongside its simulation (b). The trial director field (c) that yields the corresponding texture in the simulation is also presented (courtesy of Peter P. Crooker).

extension of the texture simulation technique to liquid crystals confined to capillary tubes ranging in radii from a few micrometers to several hundred micrometers (Crawford et al., 1992c; Scharkowski et al., 1993). Also, in a cylindrical cavity the most important parameters that determine the director configuration are the bend-to-splay bulk elastic constant ratio K_{33}/K_{11}, the saddle-splay surface elastic constant K_{24}, and the surface anchoring strength W_θ. The microscope textures of the escaped-radial configuration are illustrated in FIGURE 9.20, which are stable in supramicron capillary tubes with perpendicular boundary conditions. In tubes with $R \gg K_{11}/W$, the structure is sensitive only to the elastic constant ratio K_{33}/K_{11} (Crawford et al., 1992c; Scharkowski et al., 1993). As K_{33}/K_{11} increases, the amount of bend deformation is reduced for the escaped-radial configuration. By comparing the simulated microscope texture with the observed microscope texture photograph, one can accurately determine K_{33}/K_{11} by matching the number and position of fringes in the textures. This is shown in FIGURE 9.20 for a capillary tube of 100 μm illuminated by a sodium light source oriented with the cylinder axis at 45° between crossed polarizers.

A major accomplishment of this technique in very thin capillaries was the first optical measure of the saddle-splay surface elastic constant K_{24} (Polak et al., 1994). By studying the escaped-radial configuration in small capillary tubes ($R = 5$–$25\ \mu$m) where the elastic and molecular anchoring energies compete favoraby, the simultaneous measurement of K_{24} and W_θ is possible (see section 9.3.4.1). In cavity sizes this small, the microscope textures are sensitive to the

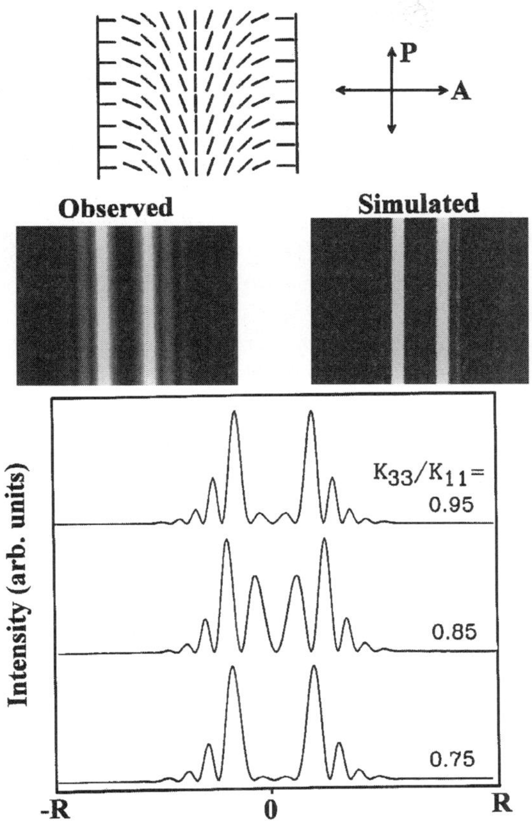

FIGURE 9.20. A model director field for the escape-radial configuration with homeotropic boundary conditions. This structure is stable in large capillary tubes treated with a lecithin aligning layer on the cylinder wall. The observed and simulated textures agree well for $R = 100\,\mu$m and $\lambda = 589$ nm. The only parameter that was varied was K_{33}/K_{11}. Its dramatic effect on the optical interference pattern is shown below. The material was ZLI-2583.

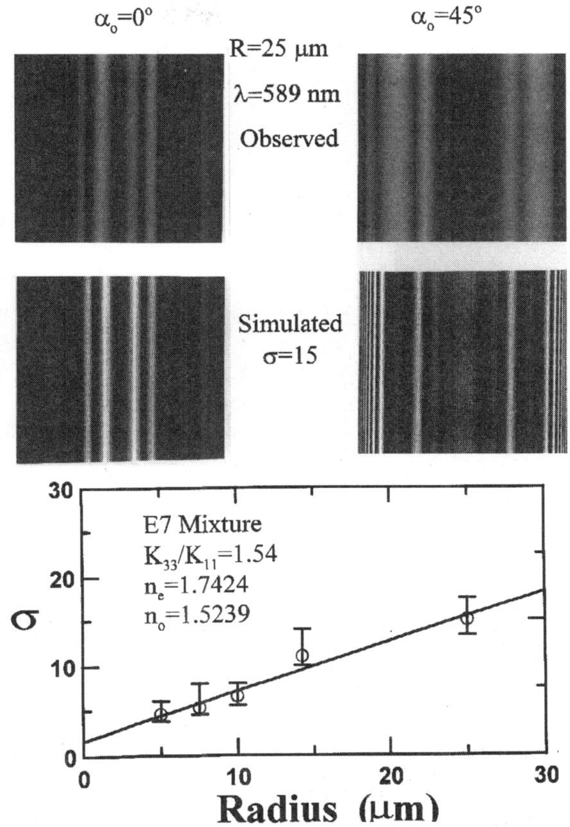

FIGURE 9.21. The optical polarizing microscope textures for the escaped-radial configuration (see FIGURE 9.20) of E7 in a small $R = 25\,\mu$m capillary tube illuminated by a sodium lamp ($\lambda = 589$ nm). The observed and simulated textures are directly compared where the dimensionless parameter $\sigma = RW_\theta/K_{11} + K_{24}/K - 1$ is adjusted for the best fit. The wave diagram shows how the plot σ versus R enables the parameters W_θ/K_{11} and K_{24}/K_{11} to be determined.

dimensionless surface parameter $\sigma = RW_\theta/K_{11} + K_{24}/K_{11} - 1$ and K_{33}/K_{11}. If K_{33}/K_{11} is known, σ is the only fitting parameter. A comparison is possible between theory and experiment as shown in FIGURE 9.21 for a $25\,\mu$m capillary tube oriented with its cylinder axis at 45° between crossed polarizers. Performing the comparisons between simulated and experimental features for a variety of capillary sizes, the value of W_θ/K_{11} and K_{24}/K_{11} can be measured simultaneously by plotting σ versus R and extracting the slope and intercept of the fitted line (see FIGURE 9.21 and Polak et al., 1994).

9.4.2 Nuclear Magnetic Resonance

Nuclear magnetic resonance techniques thus far have been most successful in probing nematic structures well below the resolution of optical microscopy in confined systems such as PDLC droplets (Golemme et al., 1988a), polymer networks (Stannarius et al., 1991), cylindrical cavities (Crawford et al., 1991a, 1996b), silica aerogel matrices (Kralj et al., 1993), and Vycor glass (Iannacchione et al., 1993). These techniques also reveal information about the molecular dynamics occurring near or at the cavity walls of the confining matrix (Vilfan et al., 1988, 1995; Vrbancic et al., 1993).

Deuterium NMR (^{2}H-NMR) is the most extensively used NMR technique to study nematic director fields because the deuterium nucleus has the advantage of carrying the quadrupole moment; therefore, its spectra are dominated by the quadrupole interaction with electric field gradients probing the local liquid crystal ordering. The disadvantage of ^{2}H-NMR is that deuterium's natural abundance is low and one must obtain selectively deuterated compounds or rely on deuterated probe molecules. Most ^{2}H-NMR studies of confined liquid crystals to date have employed selectively deuterated compounds.

The quadrupole splitting frequency for a selectively deuterated compound is given by (Doane, 1979)

$$\delta\nu(\mathbf{r}) = \tfrac{1}{2}\delta\nu_0 |Q(\mathbf{r})(3\cos^2\theta_n(\mathbf{r}) - 1) + P(\mathbf{r})\sin^2\theta_n(\mathbf{r})\cos[2\psi_n(\mathbf{r})]| \tag{9.18}$$

where $\theta_n(\mathbf{r})$ and $\psi_n(\mathbf{r})$ express the orientation of the magnetic field $\mathbf{B}$ in the principal axis frame of the local electric field gradient (EFG) averaged over fast orientational fluctuations; $\delta\nu_0$ is the quadrupole splitting frequency of the perfectly ordered state, $Q(\mathbf{r})$ in the local nematic and $P(\mathbf{r})$ is the biaxial order parameter. $P(\mathbf{r})$ is zero for uniaxial liquid crystals, but can be nonzero for confined nematics if symmetry of orientational fluctuations is broken. As the induced biaxiality is usually small, it has been neglected in all NMR studies to date.

In general, the total NMR spectrum of a droplet is not a simple sum of spectra originating from different parts of the droplet, because of the motional averaging caused by molecular self-diffusion. Therefore, the ^{2}H-NMR spectral distributions are usually calculated from the free induction decay $G(t)$, which is a relaxation function describing the attenuation of nuclear magnetization in the time domain:

$$G(t) = \left\langle \exp\left[\pm i \int_0^t \pi\delta\nu\{\mathbf{r}(t')\}\, dt'\right]\right\rangle \tag{9.19}$$

where the $\mathbf{r}(t')$ describes the diffusion induced motion and $\langle\rangle$ represents the ensemble (space) average. The Fourier transform of (9.19) yields the frequency spectrum

$$I(\nu) = \int_{-\infty}^{\infty} G(t)\exp(-i2\pi\nu)\, dt. \tag{9.20}$$

Lifetime line broadening is usually included by convoluting $I(\delta\nu)$ with a Lorentzian distribution function with the width equal to the longitudinal spin relaxation rate (Abragam, 1961). In a diffusion process characterized by the diffusion constant, the molecule, on average, diffuses a distance $x_0 \sim (D/\delta\nu)^{1/2}$ during the NMR measurement characteristic time $\delta\nu^{-1}$. For the 5CB compound, with $\delta\nu \simeq 50$ kHz and $D \simeq 10^{-11}$ m^2 s^{-1}, this yields $x_0 \sim 0.02\,\mu$m. Therefore, the analysis of nematic structures in cavity sizes $\gtrsim 0.1\,\mu$m is simplified because motional averaging caused by molecular self-diffusion is negligible. The expression in (9.19) is thus a simple integral over volume V of the droplet, yielding

$$I(\nu) = \frac{1}{V}\int \delta(\nu \pm \delta\nu(\mathbf{r})/2)\, dV. \tag{9.21}$$

Before using this expression, one must also consider the influence of the magnetic field on the nematic director when performing an NMR experiment. One way of estimating where the field-aligning effect becomes important is to calculate the magnetic coherence length (see section 9.3.1.2). As an example, we take the constants $\Delta\chi = 2.8 \times 10^{-7}$ and $K = 4 \times 10^{-12}$ J m^{-1}, associated with the liquid crystal compound 5CB and a 4.7 T magnetic field strength; the magnetic coherence length is thus calculated to be $\sim 1.2\,\mu$m. In such a field, a study of a nematic structure where motional averaging and magnetic field disturbances are negligible is limited to cavity radii between 0.1 and 1.0 μm.

The sensitivity of the ^{2}H-NMR technique to the nematic director distribution is nicely illustrated in FIGURE 9.22 for stable nematic director fields that occur in cylindrical cavities. These are escaped-radial (ER) structures with cylindrical symmetry and planar polar (PP) structure when this symmetry is broader. Neither structure has a singular director field: in the ER because of the escape in the third dimension (Cladis and Kléman, 1972); and in the PP because of weak anchoring (for details see section 9.5.2). Studying nematic director fields in cylindrical cavities is advantageous because the cylinder can be oriented in different directions in the magnetic field to remove much of the ambiguity in deciphering the correct director field. FIGURE 9.22 illustrates the spectra of deuterated 5CB in Nuclepore membranes treated with the homeotropic aligning agent lecithin. A structural ER–PP transition is observed as a function of

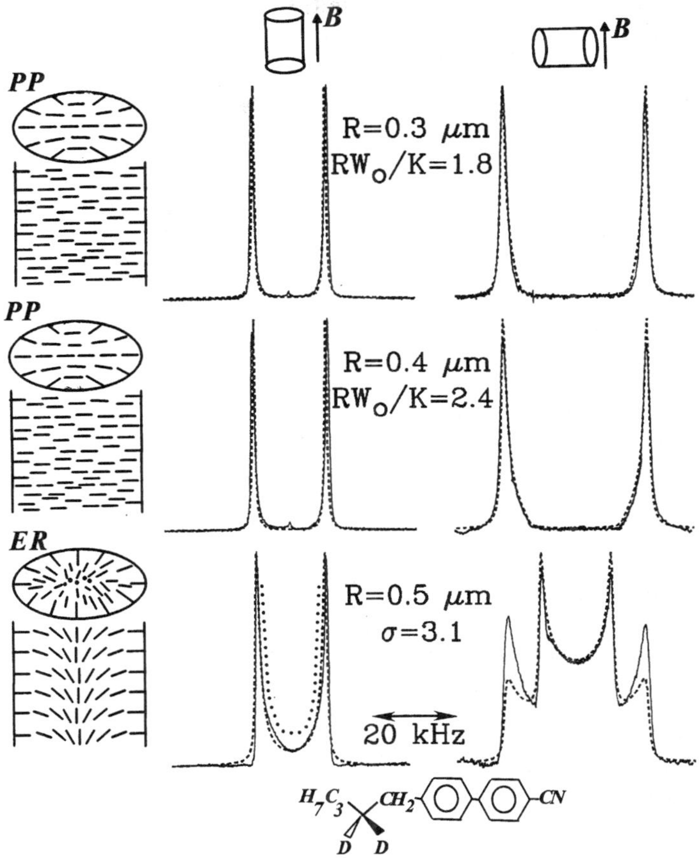

FIGURE 9.22. Deuterium nuclear magnetic resonance spectrum (solid line) directly compared with simulations (dashed line) for the liquid crystal compound 5CB-βd_2 confined in Nuclepore porous membranes, whose inner cavity walls have been treated with lecithin. Parallel ($\theta_B = 0°$) and perpendicular ($\theta_B = 90°$) orientations of the cylinder axis in the magnetic field B are presented. The spectrum reveals a clear transition as a function of radius between the planar–polar (PP) and escaped-radial (ER) director fields. $W_\theta = 3 \times 10^{-5}\,\mathrm{J\,m^{-2}}$ and $K_{24}/K \sim 1.1$ are used in director field calculations (Crawford et al., 1991d).

cavity radius. This manifests itself in a dramatic change in the ^{2}H-NMR spectrum.

Deuterium NMR is also useful when the distance the molecule diffuses during the NMR measurement is comparable to or larger than the cavity radii. This results in a complete motional averaging that, depending on the symmetry of the director field, yields a nonzero line splitting in the spectrum:

$$I(\nu) = \delta\left(\nu \pm \frac{\langle\delta\nu\rangle}{2}\right). \tag{9.22}$$

It was recently demonstrated that liquid crystals confined to various types of porous matrices exhibit a small but measurable quadrupole splitting frequency (<1000 Hz) when the liquid crystal is above the nematic–isotropic transition temperature; this observation is in contrast to the single absorption line observed in bulk isotropic systems, and is attributed to the residual nematic order induced by the cavity walls (Crawford et al., 1991a). Using typical values measured for the averaged quadrupole splitting frequency, $\langle\delta\nu\rangle = 100\,\mathrm{Hz}$, and $D = 10^{-11}\,\mathrm{m^2\,s^{-1}}$ for 5CB, the characteristic length of diffusion is $x_0 \sim (D/\langle\delta\nu\rangle)^{1/2} \sim 0.3\,\mu\mathrm{m}$. If the cavity radius is below this limit, the diffusion-averaged quadrupole splitting frequency can be expressed as:

$$\langle\delta\nu\rangle = \delta\nu_0\langle Q\rangle\left|\frac{3}{2}\langle\cos^2\theta_n(\mathbf{r})\rangle - \frac{1}{2}\right| \tag{9.23}$$

where we assume that averaging over the confining volume of Q and $\cos^2\theta_n$ may be done separately. Combining the experimental $\langle\delta\nu\rangle$ with a model (usually within the Landau–de Gennes framework) the orientational order parameter in a surface layer can be estimated. Surface order parameters in cylindrical cavities of Anopore (Crawford et al., 1991c, 1993a,

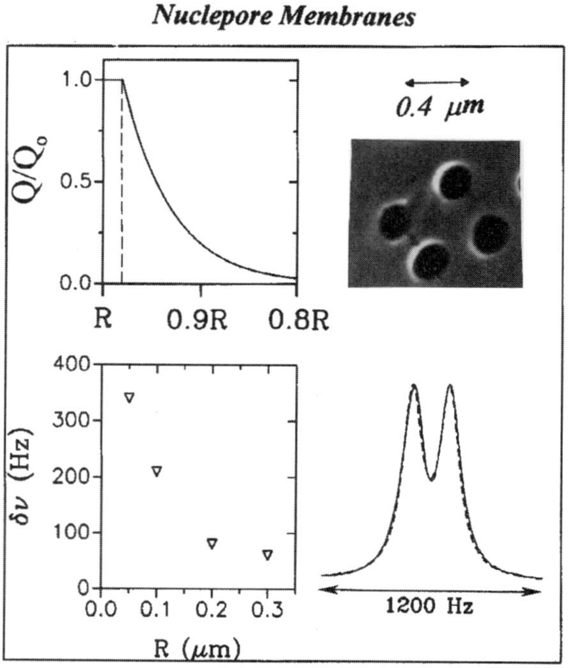

FIGURE 9.23. Deuterium NMR line splitting δv of the liquid crystal 5CB in the isotropic phase confined to Nuclepore membranes (homeotropic anchoring) as a function of pore radius. The spectrum of the 0.1 μm case (solid line) is fitted to the curve (dashed line) corresponding to the order parameter profile presented in the upper left-hand corner.

1996b) and Nuclepore membranes (Crawford et al., 1991a) have been reported to be as small as 0.01 and as large as 0.15 for various surface treatments analyzed. FIGURE 9.23 shows the quadrupole splitting in Nuclepore membranes as measured above the nematic–isotropic transition temperature, which is directly proportional to the surface-induced order parameter. As the radius is reduced, the quadrupole splitting frequency increases because of the increase in the surface-to-volume ratio. There is particular sensitivity of the ^{2}H-NMR spectra to the first molecular layer. By observing an ^{2}H-NMR quadrupole splitting deep in the isotropic phase, it was shown that the orientational order parameter profile must be constant in the first molecular layer with a thickness on the order of the molecular length, and then followed by an exponential decay that behaves according to the Landau–de Gennes theory. In addition, the molecular exchange rate between the liquid crystal and the polymer boundary was deduced from the positional dependence of the diffusion constant needed to fit the line shape of the ^{2}H-NMR spectra. In these systems, the orientational order parameter at the surface was found to be very weak and temperature independent, and the molecular exchange rate of the liquid crystal molecules at the surface with those of the bulk was determined to be $\sim 10^3\ s^{-1}$.

A quadrupole splitting has never been observed in PDLC droplets when probed above the nematic–isotropic transition (Golemme et al., 1988b), only significant broadening of the singlet. This may be attributed to very weak surface coupling or a distribution in droplet sizes smearing out the splitting and resulting in broadening. The latter is probably the dominant reason. Similar is the situation in Vycor glass, where pores have a diameter <10 nm (Iannacchione et al., 1993). Further details of surface order are presented in section 9.5.

The two-dimensional NMR (^{2}D-NMR) of confined liquid crystals was pioneered by the NMR group in Ljubljana (Dolinšek et al., 1991). Simultaneous observation of molecular ordering and dynamics of the surface layer was demonstrated. The spin-lattice, T_1, rotating frame spin-lattice, $T_{1\rho}$, and spin–spin, T_2, relaxation times of deuterium and proton nuclei provide valuable information on the molecular dynamics. It is possible to deduce the average time the liquid crystal molecule resides on the cavity surface. Studies have been performed on both spherical (Vilfan et al., 1988; Žumer et al., 1989; Dolinšek et al., 1991) and cylindrical cavities (Vrbančič et al., 1993). The advantage of the proton spin-lattice relaxation is that one can probe the cross-relaxation between liquid crystal and confining matrix protons. Vilfan and coworkers have reported the cross relaxation rate to be $\sim 10^3\ s^{-1}$, yielding an order of magnitude estimate of 10^{-4}–10^{-3} s for the time the liquid crystal molecule resides on the surface. The difference in the behavior of T_1 and $T_{1\rho}$ or T_2 (see FIGURES 9.24 and 9.25) can be employed to study molecular dynamics in the surface layer (Vilfan et al., 1988).

There have also been limited studies using proton NMR for lineshape studies (Kralj et al., 1989). How the difference in behavior of T_1 and $T_{1\rho}$ can be used to study molecular dynamics in the surface layer is described in detail in the paper by Vilfan and coworkers (Vilfan et al., 1988). The analysis becomes more complicated and has not yet been performed

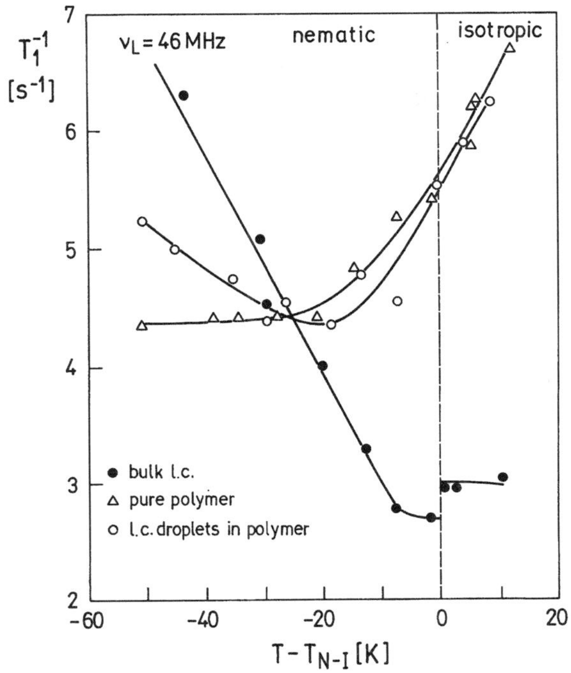

FIGURE 9.24. Temperature dependences of the nuclear magnetic relaxation rate T_1^{-1} of protons in the bulk E7 liquid crystal mixture, in the E7 droplet (average radius 0.12 μm) dispersion in a polymer, and in pure polymer, clearly indicate that relaxation of a droplet dispersion is dominated by the magnetization exchange via polymer.

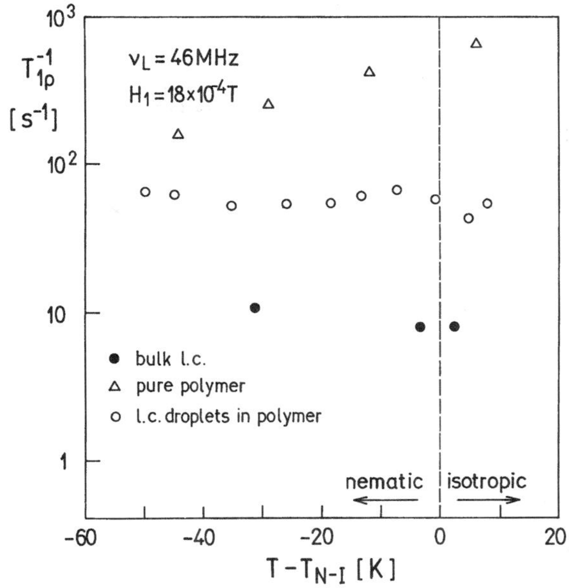

FIGURE 9.25. Temperature dependences of the rotating frame nuclear magnetic relaxation rate $T_{1\rho}^{-1}$ of protons in the bulk E7 liquid crystal mixture, in the E7 droplet (average radius 0.12 μm) dispersion in a polymer, and in pure polymer, clearly show that the magnetization exchange via the polymer is less effective than in the T_1 case. The difference is used to estimate time (~0.001 s) spent by liquid crystal molecules in the surface layer.

for liquid crystal molecules with nonequivalent proton sites.

9.4.3 Other Techniques

Although optical polarizing microscopy and magnetic resonance have been the workhorses in developing an in-depth understanding of nematic structures under confinement, other methods have also been used. Let us briefly mention some of these techniques.

Optical second-harmonic generation is not possible in materials with inversion symmetry, and is therefore not possible in bulk nematic liquid crystals. The large surface-to-volume ratio inherent in PDLC systems, however, breaks the inversion symmetry, and second-harmonic generation occurs also at interfaces between centrosymmetric materials. Li and coworkers (Li et al., 1991) pioneered the second-harmonic generation work on PDLC materials. They demonstrated that the intensity of the second-harmonic generation signal could be modulated by applying a DC field to the fully cured sample that reorients the liquid crystal droplets. They ascribed the second-harmonic generation to the dielectric constant gradient resulting from the inhomogeneity of the PDLC film. Further, there have also been studies on optical field-induced reorientation when PDLC films are subjected to intense laser radiation (Palffy-Muhoray et al., 1990).

Dielectric measurements yield information on the relaxation mechanisms over a broad frequency range, and are extremely important for applications (Kelly and Seekola, 1990; Zhong et al., 1992). Differential scanning calorimetry (Smith, 1993) and heat capacity techniques (Iannacchione et al., 1993) provide an understanding of how the phase transitions of the liquid crystal are modified when under confinement. Recently, PDLC droplets were analyzed by FT-IR microspectroscopy (McFarland et al., 1992), which is useful for determining the amount of liquid crystal present in the polymer. A combination of ultraviolet/visible and infrared absorption spectroscopy has also been shown to be useful in this respect (West and Ondris-Crawford, 1991). Light scattering which is particularly important for the applications of these materials will be detailed in section 9.6.

9.5 ORDERING AND ORDERING TRANSITIONS

Here we are going to illustrate a few examples of structures and transitions in nematic and twisted nematic confined systems.

9.5.1 Surface-Induced Pretransitional Ordering

The phenomenological theory of pretransitional nematic order at a planar surface is introduced in section 9.3.4.3. In microconfined systems, all of these effects are very important because of the relatively high surface-to-volume ratio.

Controlling the interaction strength of the liquid crystal molecules with the surrounding surface has been a topic of great interest to the liquid crystal display community for many years. Controlling the surface interaction in confined systems is relevant to all polymer dispersion technologies. Recently, a way of systematically controlling the liquid crystal–surface interaction in confined systems was developed (Crawford et al., 1993a, b, 1996b) by modifying Anopore membranes with an aliphatic surfactant (C_nH_{2n+1}–COOH). By varying the length of the aliphatic chain, n, the interfacial interaction was varied in a systematic way. The orientational order parameter at the cavity was deduced directly from ^{2}H-NMR experiments. For homeotropic anchoring ($n = 9$ and 15), the value of Q_0 was found to increase as n increased. More interestingly, the $n = 7$ surface was found to support homeotropic anchoring in the nematic phase, but did not yield any evidence of surface-induced ordering in the isotropic phase, indicative of very weak interfacial coupling. For planar anchoring ($n = 5$ and 6), Q_0 was found to decrease with increasing n. The behavior of Q_0 was found to be temperature independent, indicating that the interaction between the long aliphatic chains and the liquid crystal molecules is predominantly sterical and results in homeotropic anchoring. For shorter chain lengths, strong interactions between the Anopore surface and the core of the liquid crystal molecules dominate and enforce planar anchoring (Crawford et al., 1993a, b, 1996b).

The temperature dependence of the quadrupole splitting in these systems was monitored, and the surface-induced orientational order parameter at the cavity wall above the nematic–isotropic transition temperature (T_{NI}) was derived using a phenomenological free-energy density in terms of the orientational order parameter. The data (derived from quadrupole splitting) and corresponding theoretical fits (solid line) appear in FIGURE 9.26. For $n = 15$, the order parameter approaches $Q_0 = 0.1$ near the T_{NI}, which is still less than the bulk nematic order parameter at the

FIGURE 9.26. The surface-induced orientational order parameter Q_0 as a function of reduced temperature for 5CB-αd_2 confined in the cylindrical channels of Anopore membranes treated with C_nH_{2n+1}–COOH (for $n = 5$ and 15). The values of Q_0 were obtained directly from ^{2}H-NMR data. Two very different ordering regimes are evident from the temperature dependence of Q_0. The inset shows the measured ^{2}H-NMR quadrupole splitting $\langle \delta\nu \rangle$ versus $T - T_{NI}$ on a semilogarithmic plot that demonstrates quasi-complete wetting for the $n = 15$ surface and partial wetting for the $n = 5$ surface.

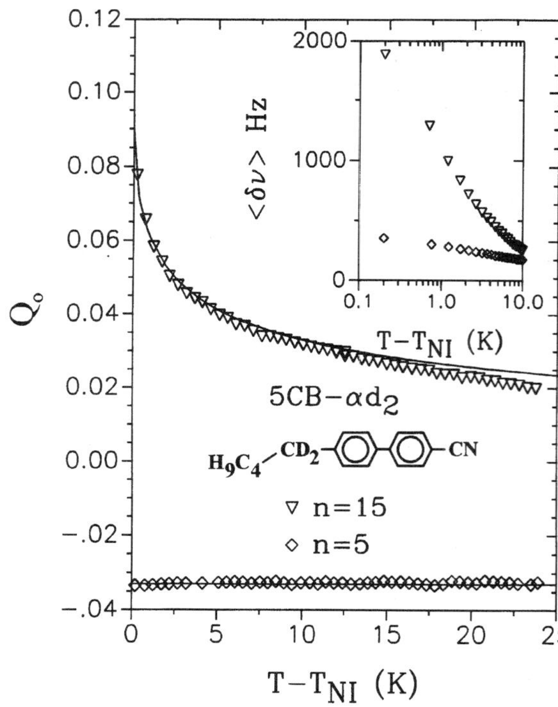

transition, which is $\simeq 0.3$. However, the $n = 5$ sample that promotes parallel anchoring conditions shows a noncritical behavior in Q_0 that is found to have a negative value. A negative Q_0 implies that the liquid crystal molecules are anchored with their long axes parallel to the surface but randomly oriented; therefore, the director is perpendicular to the surface, resulting in a negative Q_0 value (Crawford et al., 1993a, b, 1996b). Deuterium NMR cannot directly distinguish whether Q_0 is positive or negative, but the model can decipher its sign based on the temperature dependence of S_0, that is, temperature dependence of $\langle \delta\nu \rangle$.

The adsorption parameter that characterizes the wetting regime of the confined liquid crystal system can be studied with deuterium NMR (Crawford et al., 1993a, b). It is defined as the integral of the order parameter propagating away from the cavity wall (Sluckin and Poniewierski, 1986). The adsorption parameter $\Gamma = \int_0^\infty Q\,dz$ diverges for complete wetting and remains finite for partial wetting at the nematic–isotropic transition temperature (Chen et al., 1989). This adsorption parameter is easily related to averaged quadrupole splitting frequency by the relation $\Gamma = RQ_B\langle \delta\nu \rangle / 2\delta\nu_B$ where index B stands for bulk. According to theory, Γ exhibits a logarithmic divergence at the nematic–isotropic transition temperature for complete wetting; therefore, plotting $\langle \delta\nu \rangle$ on a semilogarithmic plot, as shown in the inset in FIGURE 9.26 helps characterize the orientational wetting behavior. It appears that Γ exhibits weak divergence corresponding to complete wetting for the $n = 15$ surface; however, Γ definitely remains finite for the $n = 5$ surface.

Another nice example of the power of ^{2}H-NMR is its sensitivity to pretransitional smectic layering (Crawford et al., 1994b). Using a liquid crystal compound that exhibits a smectic A–isotropic transition temperature, the propagation of pretransitional smectic layering is observed (FIGURE 9.27). The ordering of smectic layers manifests itself as discontinuous steps in the quadrupole splitting frequency. These transitions are very sensitive to the surface treatment (C_nH_{2n+1}–COOH) as is evident from FIGURE 9.27.

As pointed out by Sheng (1976, 1982), in a thin film the surface boundary transition is predicted to vanish as the thickness decreases. Even more interesting, the nematic–isotropic coexistence curve is expected to terminate for a critical enclosure size. When this occurs, the isotropic phase is replaced by a paranematic phase where a small but finite amount of order persists, and the first-order nature of the nematic–isotropic transition temperature is replaced by a continuous evolution of order. Kralj and coworkers (Kralj et al., 1991) extended this theory to spherical nematic liquid crystal droplets; this will be summarized below.

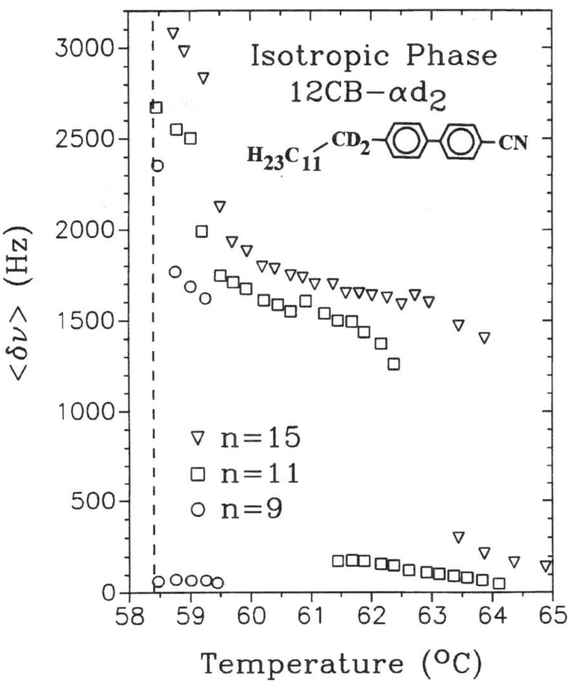

FIGURE 9.27. The averaged quadrupole splitting $\langle \delta\nu \rangle$ versus temperature for 12CB-αd_2 confined in C_nH_{2n+1}–COOH treated Anopore cavities. The data show distinct steps in $\langle \delta\nu \rangle$ as the smectic A–isotropic phase transition temperature ($T_{AI} = 58.4\,°C$) is approached from above. These steps are associated with surface induced smectic ordering transitions that are sensitive to the degree of surface coupling. The surface coupling inducing weaker orientational order also reveals lower layering transition temperatures.

We focus our attention on the effect of large deformation of the director field enforced by surface constraints on nematic ordering. To simplify the calculation, a nematic droplet is assumed with a radial director field dictated by strong homeotropic surface anchoring. Later in our discussion of larger droplets, we will see that material properties needed to stabilize this structure in submicron-size droplets are not available in most systems (Erdmann et al., 1990). The advantage of the radial director field is that one can simply write the free energy in terms of the scalar order parameter. We follow the approach of Kralj and coworkers (Kralj et al., 1991), taking only $L_1^{(2)} \neq 0$ and further assume only linear term $-GQ_0$ in surface coupling given by equation (9.11). Introducing the dimensionless quantities: $\rho = R/r$, $A = \frac{2}{9}(R/\xi_0)^2$, $C = cR^2/(9L_1^{(2)})$, $B = bR^2/(9L_1^{(2)})$, $g = \frac{2}{9}G\sqrt{aL_1^{(2)}T^*2}$, $\tau = (T - T^*)/T^*$, and $\lambda = R/\xi_0$ where $\xi_0 = [2L_1^{(2)}/(aT^*)]^{1/2}$ is the zero temperature coherence length, the free energy can be rewritten as:

$$f(r) - f_0 = 9\frac{L^2}{R^2}\left[\left(\frac{A}{2}\tau + \frac{2}{\rho^2}\right)Q^2 - \frac{BQ^3}{3} + \frac{CQ^4}{4} + \frac{1}{6}\left(\frac{\partial Q}{\partial \rho}\right)^2 - g\lambda Q\delta(1-\rho)\right]. \tag{9.24}$$

The minimization of the functional $\int f(r)\,\mathrm{d}V$ leads to the differential equation:

$$\frac{\partial^2 Q}{\partial \rho^2} + \frac{2}{\rho}\frac{\partial Q}{\partial \rho} - 3\left[\left(A\tau + \frac{4}{\rho^2}\right)Q - BQ^2 + CQ^3\right] = 0 \tag{9.25}$$

that must be solved numerically together with the boundary condition $(\partial Q/\partial \rho)_{\rho=1} = 3g\lambda$. For parameter values corresponding to liquid crystal compound 5CB ($a = 0.1319 \times 10^6\,\mathrm{J\,m^{-3}\,K^{-1}}$, $b = 1.836 \times 10^6\,\mathrm{J\,m^{-3}}$, $c = 4.050 \times 10^6\,\mathrm{J\,m^{-3}}$, $L_1^{(2)} = 1.2 \times 10^{-11}\,\mathrm{J\,m^{-1}}$ and $T^* = 307\,\mathrm{K}$) are taken.

Let us first discuss droplets more than three orders of magnitude larger than ξ_0 with $R = 1\,\mu$m. We expect only a minor influence from the defect, but substantial surface effects, particularly close to T_{NI}. Any nonzero g changes the isotropic phase to a paranematic phase (P) with surface-induced partial nematic order (Sheng 1976, 1982). In FIGURE 9.28, the temperature dependence of the surface order parameter Q_0 is compared with $\langle Q \rangle$, the average droplet order parameters for several coupling strengths g. In FIGURE 9.28 it is evident that for $g < 0.00085$ there is a direct transition between paranematic and nematic phases while for $0.00189 > g > 0.00085$, there are two first-order transitions manifested in two subsequent steps in $\langle Q \rangle$. In the latter case (FIGURE 9.29(a)), the differences in the spatial variation of the order parameter Q (from the surface value Q_0 to zero in the center of the radial defect) clearly show that between nematic (N) and paranematic (P) phases there is, similarly as in the planar systems (Sheng, 1976, 1982), a boundary layer

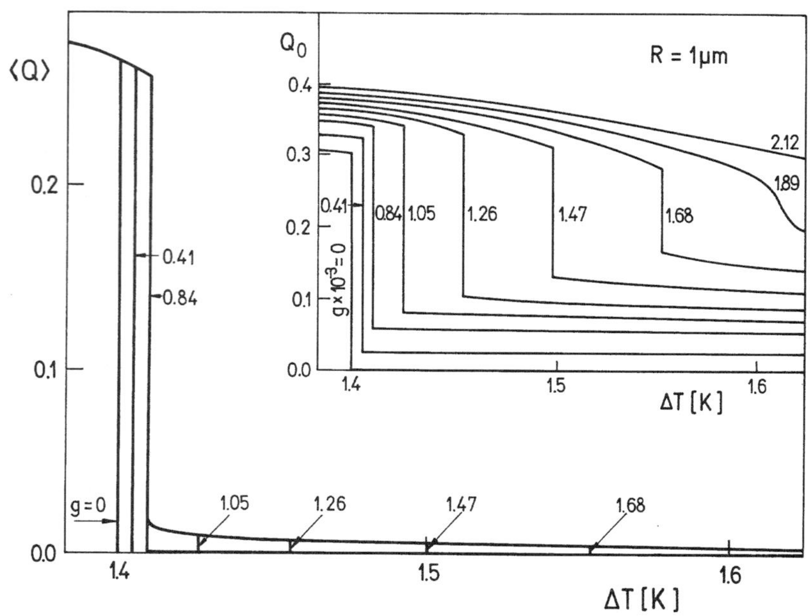

FIGURE 9.28. Surface order parameter Q_0 and average droplet order parameter $\langle Q \rangle$ plotted as functions of the relative temperature ΔT measured from T^* for different surface coupling strengths g and droplet radius 1 mm. The simultaneous jumps in $\langle Q \rangle$ and Q_0 reflect a transition from a paranematic to a nematic phase, and a jump visible only in Q_0 indicates a transition from a paranematic to a boundary-layer nematic phase.

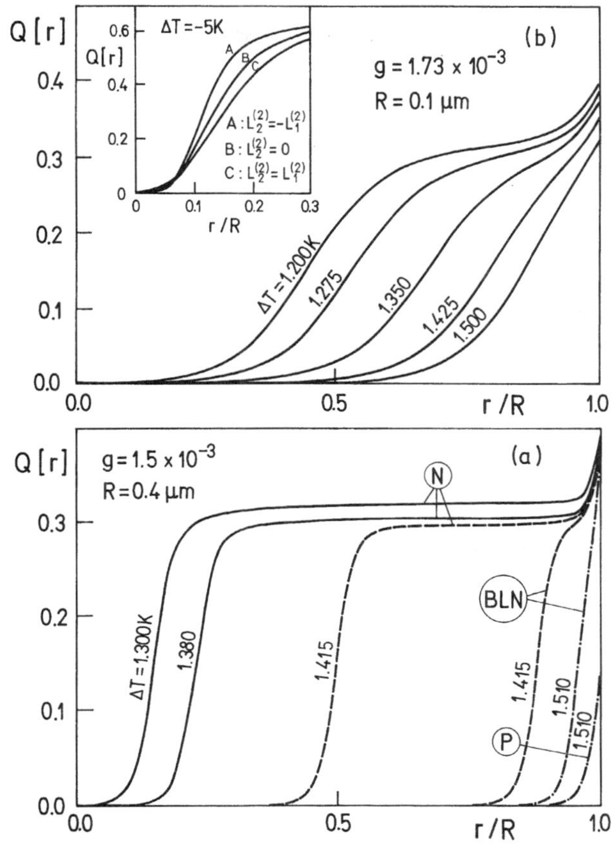

FIGURE 9.29. Spatial variation of the order parameter near T_{NI} within a droplet with the radial director field. Above the critical size (a) profiles for the nematic (N), paranematic (P), and boundary-layer nematic phases (BLN) are shown. Below the critical size and above the critical coupling (b) there is no longer a distinction between phases. The inset shows the dependence of the profile on the expansion coefficient $L_2^{(2)}$ 5 K below T_{NI} in the droplet with radius 0.1 μm.

nematic phase (BLN) with a much thicker ordered surface layer than in the P phase. The additional prewetting (P–BLN) transition with increasing g, that induces stronger surface order, is less pronounced, and at $g = 0.00189$ it becomes continuous.

In smaller droplets, $R = 0.1\,\mu$m, the effect of the confinement is more pronounced. This can be seen in FIGURE 9.30 where the behaviors of Q_0 and $\langle Q \rangle$ are compared. Q_0 exhibits similar behavior as before, but $\langle Q \rangle$ shows only one transition. Above $g = 0.0019$, surface-induced order is strong enough that this transition also vanishes. The continuous evolution of the $Q(r)$ in this regime is shown in FIGURE 9.29(b).

In the inset into FIGURE 9.29(b), the behavior of the order parameter close to the central hedgehog defect is shown deep in a nematic phase ($T_c - T = 5$ K). The radius of the core of the defect is ~10 nm in a droplet with $R = 100$ nm. It does not change much if the free energy term with the coefficient $L_2^{(2)}$ is retained. Cases where $L_2^{(2)}$ varies from $L_1^{(2)}$ to $-L_1^{(2)}$ are illustrated. If inversion symmetry can be broken, defect structure may become more complicated (Schopohl and Sluckin, 1988; Penzenstadler and Trebin, 1989).

A simple presentation of stability regions of nematic (N), paranematic (P), and BLN phases in a radial droplet (a three-dimensional g–R–T phase diagram) is presented in FIGURE 9.31. The phases are separated by "transition surfaces." The N–BLN surface terminates at small radii in a line of critical points R_c ($\simeq 0.22\,\mu$m), where the distinction between two phases vanishes. At small g, the N–BLN surface merges in a line of triple points with the P–BLN surface and the P–N surface. The P–BLN surface terminates at large g in a line of critical points, g_{max}, where the distinction between the P and BLN phases vanish. This phase diagram qualitatively deviates from Sheng's planar system because of the curved-geometry-induced deformations, which suppress the

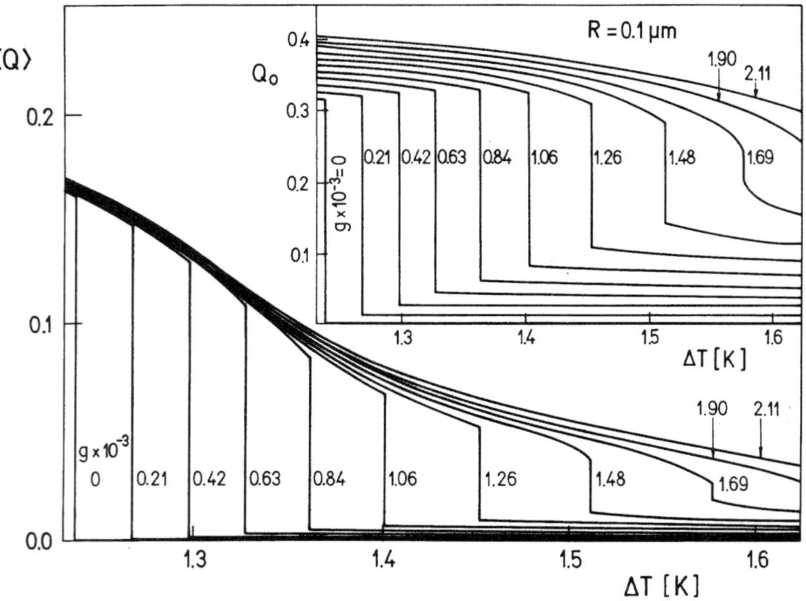

FIGURE 9.30. Surface order parameter Q_0 and average droplet order parameter $\langle Q \rangle$ plotted as functions of the relative temperature ΔT measured from T^* for different surface coupling strengths g, with droplet radius 0.1 μm. The simultaneous jumps in $\langle Q \rangle$ and Q_0 reflect a transition from a paranematic to a nematic phase. There is no longer a boundary nematic phase for this radius, which is well below the critical radius 0.22 μm.

N–P transition at smaller radii (a few tenths of a degree at $R = 0.1\,\mu$m).

Although this example is not particularly realistic (radial structure is unstable in small droplets of common liquid crystals), it describes essential features of confined systems: the competition of the ordering effect of the surface and disordering effect of deformations. It is instructive to illustrate the same phenomena on a completely simplified system confined to a very small volume V where there is no need to distinguish between surface and interior. Introducing an effective scalar order parameter $\langle Q \rangle$, one can write the free energy as:

$$F = V(f_n - G_e\langle Q \rangle + U_e\langle Q \rangle^2/2) \qquad (9.26)$$

where V is the volume of the system, f_n is the free energy of bulk nematic phase given by equation (9.1) with $\langle Q \rangle$ instead of the tensor Q_{ij}. The additional linear and quadratic terms describe the confinement. The $-G_e\langle Q \rangle$ term comes from the surface coupling (equation (9.10)) and describes its ordering effect. Similarly, as the magnetic field at certain strengths in a spin system replaces a discrete transition with a continuous evolution, G_e does that here. The $U_e\langle Q \rangle^2/2$ term comes from possible disordering effects of the surface (see sections 9.3.4.3 and 9.3.4.4) and confinement-induced deformations (terms in the f_d given by equation (9.3)) that are second-order in derivatives. This term simply induces a phase transition temperature depression $\Delta T_c = U_e/a$. A rough estimate yields $G_e = W/R$ and $U_e = L/R^2$, which would in the confined volume be characterized by $R = 20\,\mathrm{nm}$ resulting in transition temperature shift $\simeq 1\,\mathrm{K}$. For more details, see Iannacchione and coworkers (Iannacchione et al., 1993).

It should be mentioned that there are several theoretical studies that predict ordering in related systems: small bipolar droplets with strong anchoring using scalar order parameter (Vilfan et al., 1989), tensorial order parameter (Kilian and Hess, 1989), and Monte Carlo simulations (Chiccoli et al., 1990); cylindrical cavities using tensorial (Gartland et al., 1991) and scalar order parameter (Crawford et al., 1993a); and random pores using effective scalar order parameter (Iannacchione et al., 1993).

Golemme (1988b) provided the first experimental study on the nature of ordering in very small PDLC droplets. Using ^{2}H-NMR, he was able to monitor the quadrupole splitting (see section 9.4.2) as a function of temperature for a variety of droplet diameters (FIGURE 9.32). For droplets 0.035 μm or less in diameter, the isotropic phase shows evidence of paranematic order (broad spectral linewidth). In addition, the isotropic–nematic phase transition was found to be continuous,

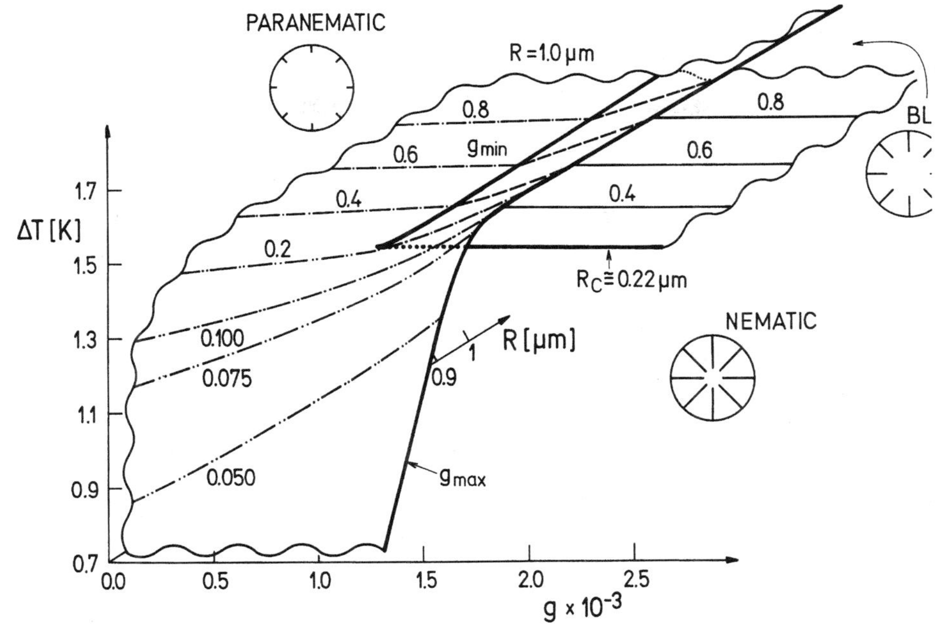

FIGURE 9.31. Three-dimensional phase diagram of a radial nematic structure in a liquid crystal droplet as a function of the surface interaction, droplet radius, and sample temperature. Transitions are first order except along the line of critical radii R_c and along the line of maximum coupling g_{max}, where the transition becomes continuous. There is a line of triple points g_{min}.

as is evident from the continuous evolution of the quadrupole splitting frequency. For droplet diameters 0.35 μm or greater, the quadrupole splitting frequency was found to be discontinuous, in a similar fashion to that of the bulk. This result proved to be the first demonstration of Sheng's prediction that the nematic–isotropic coexistence curve terminates at a critical enclosure size (see FIGURE 9.32). Even more pronounced are effects in very tiny pores ($R \sim 4$ nm) where ^{2}H-NMR and calorimetric studies (Iannacchione et al., 1993) have been performed. The severe confinement yields a strongly depressed nematic order that is intermediate between nematic and isotropic phases, with no evidence of nematic–isotropic phase transition. The average order parameter $\langle Q \rangle$ presented in FIGURE 9.33 was derived from the ^{2}H-NMR quadrupole splitting frequency, which is described (solid line) by a Landau-type formalism based on free energy given by equation (9.26). The evolution of $\langle Q \rangle$ as a function of temperature is continuous through the bulk nematic–isotropic transition temperature. The orientational contribution of the heat capacity of the liquid crystal in the porous network was also obtained and is presented in the inset of FIGURE 9.33. This also shows no signature of the first-order nematic–isotropic phase transition temperature (Iannacchione et al., 1993).

Bellini and coworkers (Bellini et al., 1992) employed light scattering and calorimetry techniques on liquid crystalline materials confined to aerogel matrices, and demonstrated that nematic ordering of the liquid crystal does not occur via the first-order phase transition characteristic of the bulk. Kralj and coworkers (Kralj et al., 1994) found that the nature of the nematic–isotropic phase transition in aerogel matrices also depends strongly on the surface interactions with the matrix. When a lecithin surfactant is applied, the ^{2}H-NMR spectra exhibit a first-order transition, while in the nontreated aerogel surfaces, only a weak discontinuity in the ^{2}H-NMR was found. For more details about liquid crystals in porous media see Crawford and Žumer (1996a).

9.5.2 Micrometer Droplets—Weak Anchoring

Let us now shift our attention to micrometer-size droplets that can be studied optically and which are

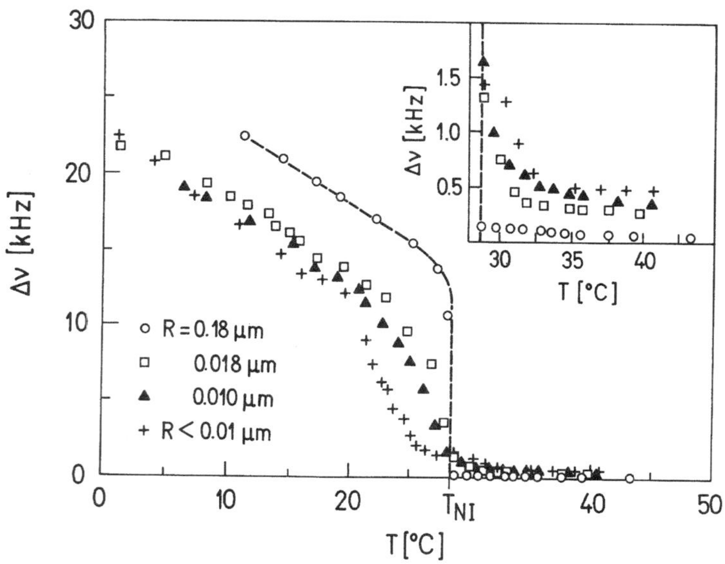

FIGURE 9.32. Deuterium NMR splitting of the deuterated 5CB droplets dispersed in the epoxy-type polymer as a function of temperature for different droplet sizes. With decreasing radius, the average degree of order measured by the splitting in the nematic phase is depressed, but the order in the paranematic phase is increasing (see inset).

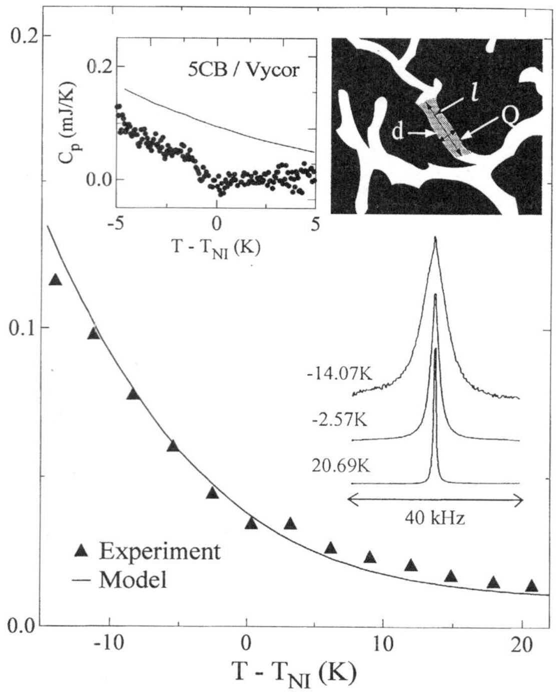

FIGURE 9.33. Average order parameter as a function of temperature for the liquid crystal 5CB confined to the 0.007 μm pores of Vycor glass. The solid line denotes the theoretical fit. The heat capacity against temperature is displayed in the inset.

reasonably well described by Frank elastic energy, but are still small enough to be in the weak anchoring regime (radius R comparable to the extrapolation length $d = K/W_0$). In this regime, director fields of surface defects (points on lines) are relaxed, that is, nonsingular. As an example, see FIGURE 9.34 where the dependence of the director field of the axial droplet structure on the ratio R/d is shown. The transition from the radial (R) to axial (A) director field in nematic liquid crystal droplets with homeotropic anchoring, dispersed in polyurethane, was first studied in detail by Erdmann and coworkers (Erdmann et al., 1990). The radial director field with a central point defect is schematically presented in FIGURE 9.1(c). The first-order R–A transition was either temperature- (FIGURE 9.35(a)) or applied-electric-field- (inset to FIGURE 9.36) induced. Single elastic constant $K_{11} = K_{22} = K_{33} = K$ free energy without surface elastic terms but including Rapini–Papoular surface coupling (equation (9.8)) was used to predict the stability diagram of both structures in terms of the dimensionless variables R/d and d/ξ (see low R/d part of the $K_{24} = 0$ curve in FIGURE 9.35). The inverse coherence length ξ_e^{-1} describes the relative strength of the applied electric field (similar to the magnetic case in section 9.3.1.2). The stability diagram was used to determine the anchoring energy W_0 as a function of temperature (FIGURE 9.34(b)) and to estimate the effective internal

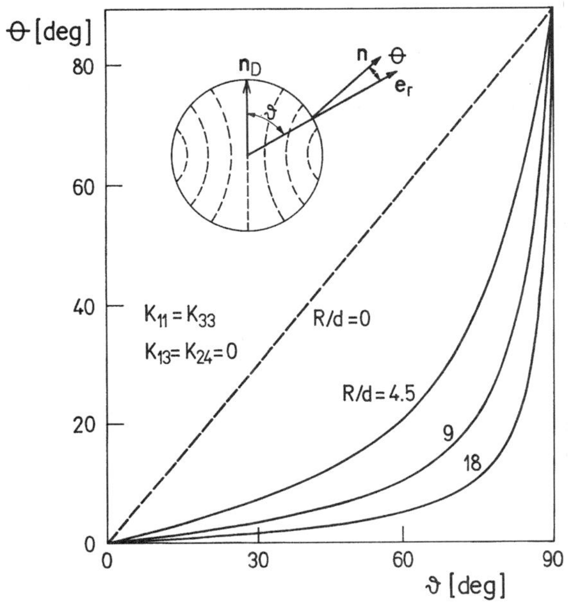

FIGURE 9.34. Illustration of the weak anchoring regime where defects in the director field are relaxed. The angular dependences of the deviations θ of the nematic director field from the surface preferred direction $\mathbf{e}_r$ in droplets with axial structures corresponding to different anchoring strengths R/d are shown.

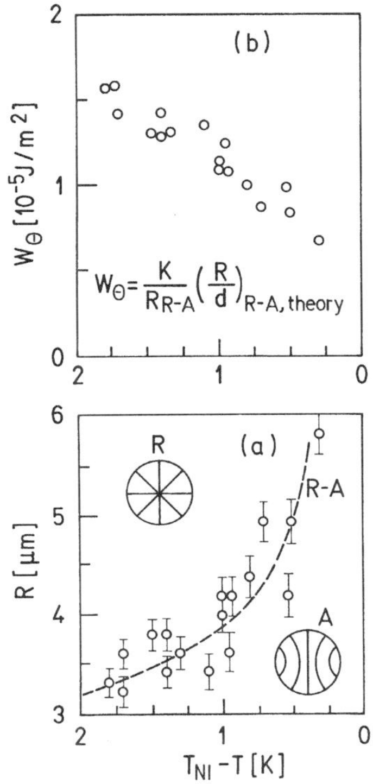

FIGURE 9.35. Experimentally determined radius R versus reduced temperature $T - T_{NI}$ stability diagram of the radial and axial droplet structures (a). The anchoring strength W_θ in droplets as a function of temperature (b).

electric field whose small values (5–7 times smaller than the applied field) were attributed to the dielectric properties of liquid crystals and polymers. A more detailed consideration of the stability diagram needs incorporation of the constant K_{24}. The pronounced effect of the nonzero K_{24} value that broadens the stability range of the radial structure (FIGURE 9.36) can, in principle, be used for the determination of the saddle-splay elastic constant. Two possible methods for simultaneous determination of K_{24} and W_0, based on the radial–axial transition in magnetic fields, are illustrated in FIGURE 9.37. The first is based on observation of the inversion point $((d/\xi)_i, (R/d)_i)$ and the zero-field transition point $((R/d)_0, 0)$ corresponding to the largest d/ξ where the radial structure is still stable. The second method is based on the zero-field transition point and average orientation of the axial director field. In FIGURE 9.37(a), the average orientation of the axial director field is described by the average second Legendre polynomial of the angle between the local director field and the droplet axis. Such information can be extracted from ^{2}H-NMR (see section 9.4.2). It was also shown by Kralj and Žumer that K_{24}/K could be determined from optical microscope simulations (see FIGURE 9.38).

This procedure cannot be used for Erdmann's data because the internal field is unknown. One can use the K_{24}–K transition values from curves b and c in FIGURE 9.37 to re-estimate anchoring energy and internal field, and see that W_0 is four times larger than originally found by Erdmann (FIGURE 9.35), while the internal field is three times smaller than the applied field.

Because of relatively poorly defined PDLC systems, better characterized cylindrical cavities in polycarbonate and alumina membranes filled with nematic liquid crystals (see sections 9.4.1.3 and 9.4.2) were studied. Allender and coworkers (Allender et al., 1991) were the first to determine K_{24} and W_0 using ^{2}H-NMR to study nematic ordering close to structural

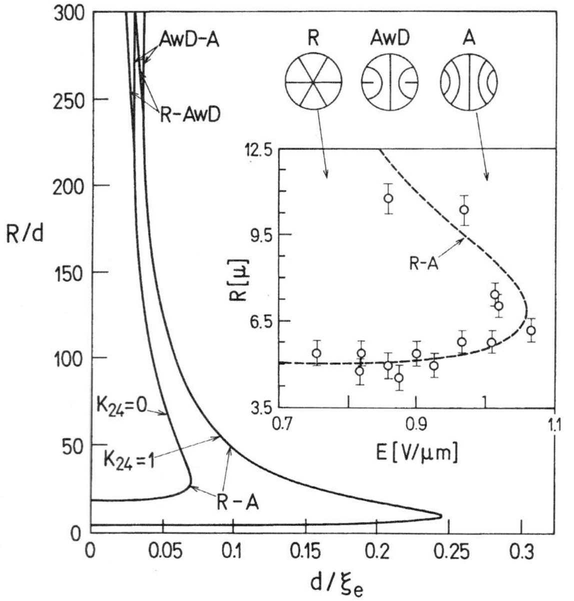

FIGURE 9.36. The calculated stability diagram of droplets with homeotropic anchoring in the external field for two values of the K_{24} constant. The experimental phase diagram is shown as an inset.

transitions in submicrometer cavities with weak surface anchoring. Recently, Polak and coworkers (Polak et al., 1994) succeeded with a similar experiment in supramicrometer cylindrical cavities using polarization optical microscopy, as presented in FIGURE 9.21. Both experiments show that the elastic constant K_{24} is comparable to elastic constants K_{11} and K_{33} (see also Lavrentovich, 1991; Crawford and Žumer 1995).

9.5.3 Large Droplets—Strong Anchoring

The strong anchoring regime defined by the conditions $R/d, \xi_E/d \gg 1$ is characterized by negligible deviations of the director from the preferred direction on the surface (FIGURE 9.34). Therefore, in the strong anchoring limit, W_0 and K_{24} no longer play important roles. In addition, there are two axial structures without a defect (A in FIGURE 9.35) and another with an internal $s = \frac{1}{2}$ ring defect in a diametrical plane (R in FIGURE 9.35). The latter is stable only in the strong anchoring limit in an area between radial and defectless axial structure that finishes in a triple point. Due to the high energy barrier between the two structures in this strong anchoring limit, large hysteresis effects are observed.

In nonchiral nematic phases, particularly interesting are situations where symmetry breaking spontaneous twist deformation appears. The observations of twisted bipolar droplet structures (director field shown in FIGURE 9.39) in nonchiral systems reported by Volovik and Lavrentovich (1983) were theoretically analyzed by Williams (1986). It was shown that for $K_{11} > K_{22} + 0.431K_{33}$, the twisted structure is stable and the symmetry breaking transition between twisted and nontwisted structures is second order. In weakly chiral nematics (large pitch, $R \ll p$) deformations are small and do not differ much from the ones

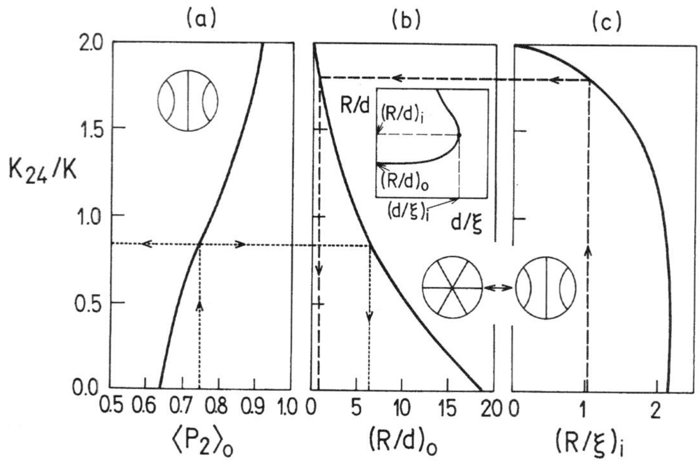

FIGURE 9.37. K_{24}/K_{11} dependences of (a) the average alignment of the axial droplet defined by $\langle P_2 \cos(\Theta)\rangle$; (b) the ratio $(R/d)_0$ at the zero-field A–R coexistence point; (c) at the inversion point $(d/\xi)_i$ of the A–R line in the stability diagram (FIGURE 9.36). The approximation $K_{11} = K_{33} = K$ was used.

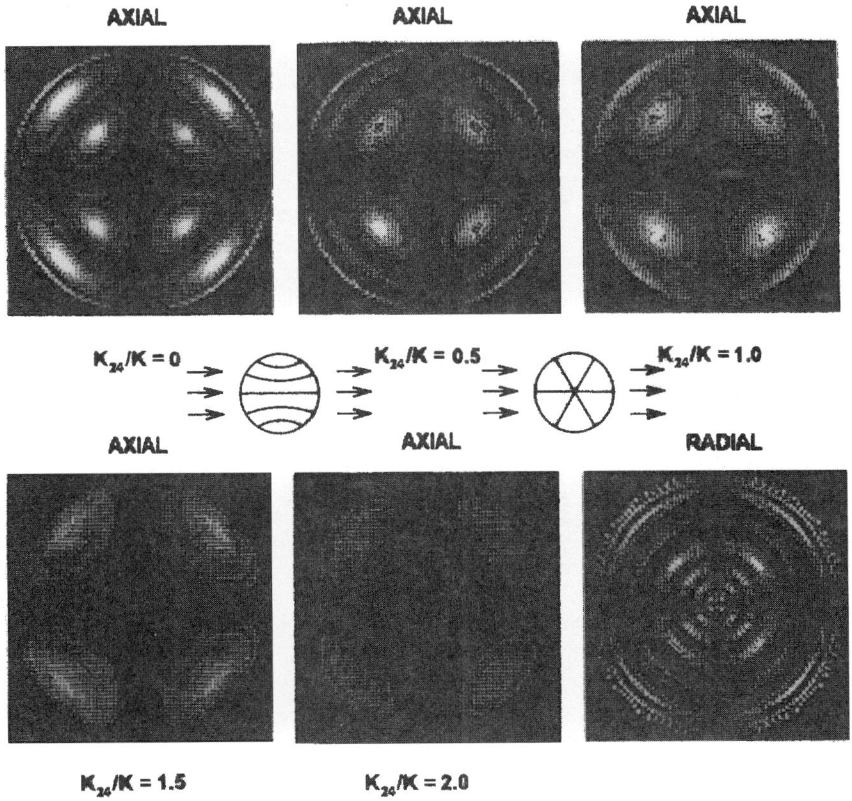

FIGURE 9.38. Radial–axial transition point. Strong dependence of simulated axial textures on the ratio K_{24}/K is illustrated together with a radial texture. The simulation method is described in section 9.4.1.2.

caused by the confinement-induced twist in high K_{11} materials. But there is an important difference: the sense of the twist in chiral materials is intrinsic, while in nonchiral materials, left- or right-hand twists are equally probable.

FIGURE 9.39. Twisted bipolar structure. Director field lines on the surface of the droplet and director field on the diametrical plane. For the latter, the "nail" presentation is used: the length of the nail is proportional to the projection of the director on the paper plane and its head denotes the side which is below the paper plane.

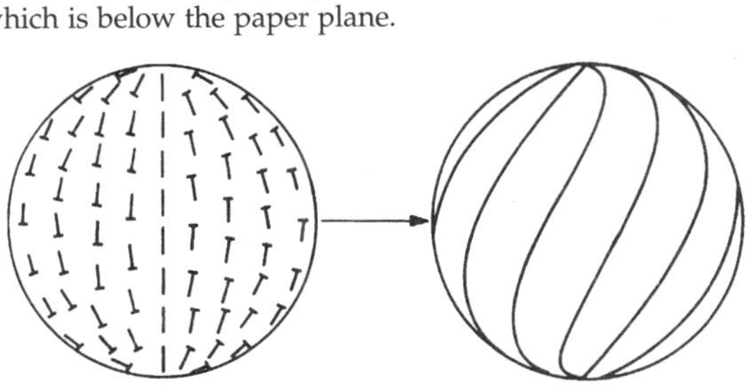

Recently, empirical model structures have been proposed by Crooker's group (Xu et al., 1992, 1994) to describe twisted structures in systems with negative dielectric anisotropy in electric fields subjected to parallel and perpendicular boundary conditions. By comparing droplet observations with texture simulations based on simplified model structures, they pinpointed the stability of very specific director configurations (FIGURE 9.19). For perpendicular anchoring and low fields, the director configuration was found to be essentially radial with a slight azimuthal twist near the hedgehog defect (Xu et al., 1994). At intermediate fields, the defect is displaced along the radius parallel to the field, and at higher fields, the defect moves to the surface and a diametrical line defect along the field direction is observed.

To illustrate ordering in droplets with strong anchoring, we choose highly chiral nematic liquid crystals ($R \gg p$) that are particularly rich in structures. We limit our discussion to cholesteric droplets with planar surface anchoring and negative dielectric anisotropy.

For the description of the highly chiral N* system, it is convenient to use the concept of cholesteric (N*) surfaces that are normal to the helical axes and parallel to the director field. Particularly useful is the fact that the pitch can be assumed constant although the confining surface is far from planar. This explains similarities with textures of smectic liquid crystals (see, for instance, Kléman, 1983). In general, there are three kinds of defect lines in N* phases: parallel to the chiral axis—χ line; in the N* surface parallel to the director field—λ line; and in the N* surface but perpendicular to the director—τ line. In our examples only χ and λ lines are present. Particularly important are χ lines that are perpendicular to the N* surfaces—two-dimensional (2D) nematic director fields. This allows the use of the homotopic theory (Mermin, 1979) for 2D nematics to find the allowed number and strength of defects. In our case, the N* surfaces are spheres or simply connected surfaces with circular border and tangential boundary condition of the director field (Yang and Crooker, 1991; Bezič and Žumer, 1992; Žumer et al., 1992). For instance, topological theory for the director field on a sphere yields a condition relating strengths s_i of surface defects:

$$\sum_i s_i = 2 \tag{9.27}$$

where the sum gives overall defects. For nonclosed surfaces, the sum is equal to unity and must be separated to the part covering internal defects denoted by upper index "int" and the part covering boundary defects denoted by "bound":

$$\sum_i s_i^{\text{int}} + (1/2)\sum_j s_j^{\text{bound}} = 1. \tag{9.28}$$

The factor $\frac{1}{2}$ of a half in front of the boundary part is a result of the fact that once the defect is moved from the interior to the surface, its director field is compressed. The field resembles that of the internal defect with double strength, but it only occupies half the space (Bezič and Žumer, 1993). To select the most stable structure, the free energies must be considered. For a rough estimate one can use the proportionality of the free energy per unit length of the defect line to the square of its defect strength (Bezič and Žumer, 1993). In what follows, we will describe spherical structures and their transitions to planar structures induced by the field.

9.5.3.1 Spherical Structures

The observed spherical structures have either a radial or a diametrical defect line (Robinson et al., 1958; Spencer et al., 1962; Bouligand and Livolant, 1984). Kurik and Lavrentovich (1982) reported a transition between spherical and twisted bipolar structures. The description of spherical structures was first based on geometrical models like the Frank and Price model (Robinson et al., 1958; Bouligand and Livolant, 1984). Bezič and Žumer (1992) have shown that minimization of the free energy within the approximation of constant pitch and nondegenerate N* surfaces yields the director field:

$$n = \cos\Omega e_\phi + \sin\Omega e_\theta \tag{9.29}$$

with

$$\Omega = (s_0 - 1)\phi + \Omega_0 + qr \tag{9.30}$$

which agrees with the geometrical model; r, θ, and ϕ are spherical coordinates; e_r, e_θ, and e_ϕ are the corresponding unit vectors and Ω_0 is an arbitrary initial phase. The $s_0 = 1$ solution $\Omega(s_0 = 1) = \Omega_0 + qr$ corresponds to the spherical structure with diametrical $s = 1$ χ defect line. In FIGURE 9.40(a) it is shown how director field on concentric N* surfaces transforms with change in qr. The field continuously transforms from concentric to bipolar and back via intermediate twisted structures. The $s_0 = 2$ solution $\Omega(s_0 = 2) = \phi + \Omega_0 + qr$ corresponds to the spherical structure with radial $s = 2$ χ defect line. The director field on each spherical cholesteric surface is similar and is described by a set of circles with a common point on the radial defect line (FIGURE 9.40(b)). Going along this line with changing radius r, the field structures rotate relatively according to the value of the product qr. For the propagation of light in the direction parallel to the defect line of the structure, the polarization microscope textures exhibit the spiraling caused by the twisted structure. A comparison of the experimental texture to that simulated using our method developed in section 9.4.1.2 shows qualitative agreement (FIGURE 9.41(a) and (b)) even without matching experimental droplet parameters.

9.5.3.2 Ellipsoidal and Planar Structures

The situation is different when an electric field is applied to chiral nematic droplets with negative dielec-

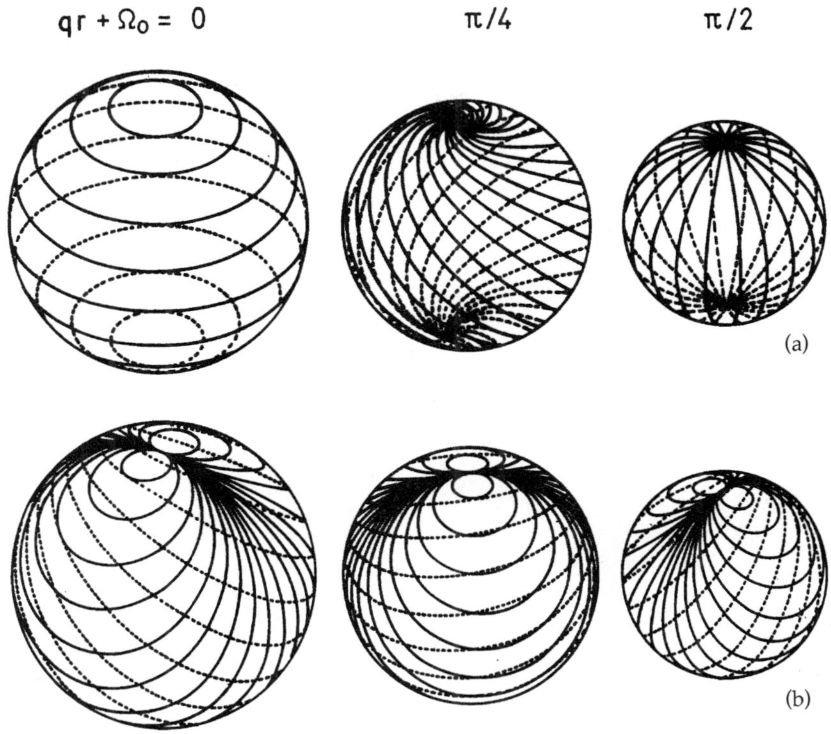

FIGURE 9.40. Spherical structures of chiral nematic droplets. Director field lines on the spherical surfaces correspond to three different rotation angles $qr + \Omega_0$ for structures with diametrical (a) and radial (b) defect lines.

tric anisotropy. There, the spherical N* surfaces become more and more squeezed, and in the limit are completely planar (Yang and Crooker, 1991; Žumer et al., 1992). The squeezed N* surfaces can be approximately described by oblate ellipsoids—closed surfaces, or spherical caps—nonclosed surfaces, limited by the droplet surface. FIGURE 9.42(a)–(d) shows the diametrical cross-sections of these surfaces for zero, low, intermediate, and high applied fields. In the limit of infinitely strong anchoring director fields on spherical caps are either bipolar (see FIGURE 9.43(a) and (b)) (Quigley and Benton, 1977) or monopolar (FIGURE 9.43(c) and (d)) (Bezič and Žumer, 1992; Bajc et al., 1995). These director fields rotate along the symmetry axis as in an unconstrained chiral nematic liquid crystal. In the planar limit, a stack of these 2D nematic fields (FIGURE 9.43(b) and (d)) forms a three-dimensional twisted structure with a spiral (in the monopolar case) or double spiral (in the bipolar case) disclination line on the droplet surface (see FIGURE 9.43(a) and (c)).

The transition from a spherical to a planar structure yields a dramatic change in polarization microscope pattern from a fingerprint to a rather uniform texture (FIGURE 9.41(a) and (b)). The same features are also seen in simulated patterns (FIGURE 9.41(c) and (d)). With the increasingly planar nature of the structure of such droplets, they behave more and more like unconstrained chiral liquid crystals and selectively reflect the light (Yang and Crooker, 1991). The field-induced reflectivity of the chiral nematic droplets with negative dielectric anisotropy is represented in FIGURE 9.44. The use of reflection in such PDLC materials is promising for applications (Kitzerow et al., 1991).

9.6 LIGHT SCATTERING

Confined liquid crystals are interesting scattering objects because of the large variety of mesophases and structures possible with very different scattering properties. Strongly anisotropic molecular polarizability yields large differences between ordinary and extraordinary indices of refraction in ordered phases. Furthermore, the ordered structures strongly depend on temperature, external fields, and confining

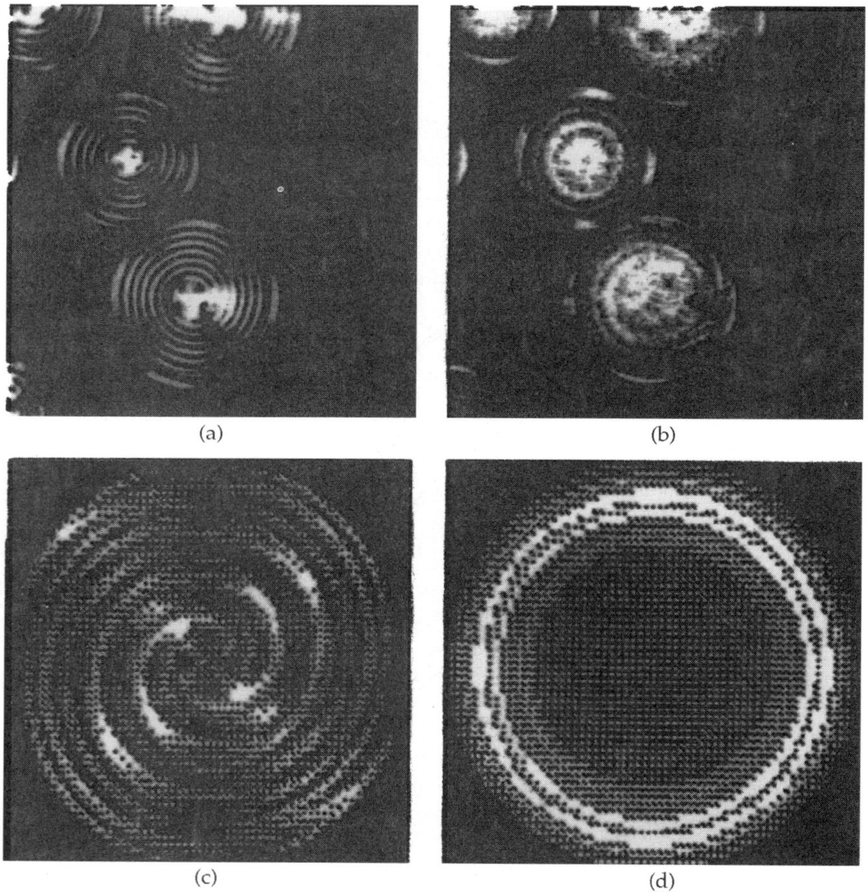

FIGURE 9.41. Textures of chiral droplets. Polarizing microscope textures in zero- (a) and high- (b) field (courtesy of Peter Crooker), and simulated textures for the spherical structure (c) with the defect line along the observation direction and for the planar structure with chiral axis parallel to it. $R/\lambda = 10$ and $Rq = 13$ with $\Delta\,\mathrm{d}n = 0.2$ were used.

geometrical constraints that lead to numerous applications of their optical properties (Doane, 1991; Kitzerow, 1994). For more details see Drzaic (1995).

9.6.1 Scattering From Isotropic Droplets

The effect of small droplets on light is a classical problem extensively studied in the last century. Most studies have been devoted to isotropic objects (van de Hulst, 1957; Bohren and Hoffman, 1983). Besides the exact Mie solution (Mie, 1908) for the scattering on a sphere, and the more recent solution for ellipsoids (Asano and Yamamoto, 1975), several approximations for different droplet sizes (in comparison with the wavelength of light) have been developed (van de Hulst, 1957). The best known are the Rayleigh approximation for very small droplets (van de Hulst, 1957), the Rayleigh–Gans approximation for small weakly

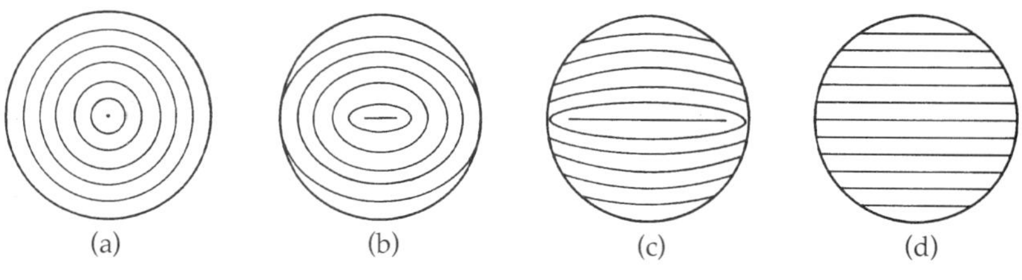

FIGURE 9.42. Chiral nematic droplets. Lines are cross-sections of cholesteric surfaces and paper plane. Restructuring of these surfaces (a) from spherical via ellipsoidal (b), (c) to planar (d) is shown.

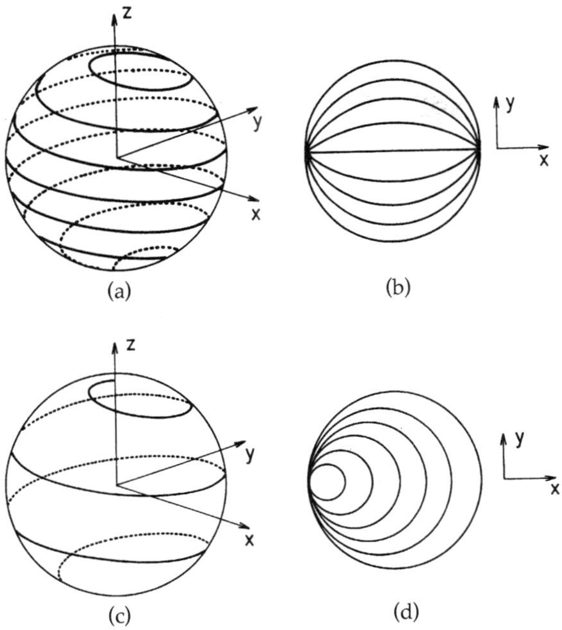

FIGURE 9.43. Model structure of a chiral droplet with negative dielectric anisotropy in a strong electric field. The two spiral-like surface χ disclination lines (a) are formed by a rotation of the planar bipolar director field (b) along the helical axis z. The single spiral-like surface χ disclination line (c) is formed by a rotation of a planar monopolar director field (d) along the helical axis.

refracting droplets (Rayleigh, 1914; Gans, 1925), the anomalous diffraction approach for large weakly reflecting droplets (van de Hulst, 1957), and geometrical optics for very large droplets (van de Hulst, 1957). The step from a single droplet to a collection of droplets introduces interference and multiple scattering (van de Hulst, 1980).

9.6.2 Light Scattering From a Birefringent Droplet

Liquid crystal droplets are also birefringent, to complicate the situation. Although there are many papers devoted to the application of these phenomena (Doane, 1990 and references therein), there are only a few studies devoted to basic aspects of light scattering and propagation in a single nematic droplet (Žumer and Doane, 1986; Montgomery, 1988; Žumer, 1988; Sherman, 1989; Ondris-Crawford et al., 1991; Ding and Yang, 1994) or dispersions of such droplets (Žumer et al., 1989; Kelly et al., 1992; Whitehead et al., 1993).

In dealing with light scattering from nematic droplets, one must start from the basic equations of electromagnetic waves. When a plane wave with electric field

$$\mathbf{E} = \mathbf{E}_0 \exp(i\mathbf{k} \cdot \mathbf{r} + i\omega t) \tag{9.31}$$

where $\mathbf{E}_0$ is the amplitude and $\mathbf{k}$ is the wavevector, scatters on a droplet, in the far-field regime the

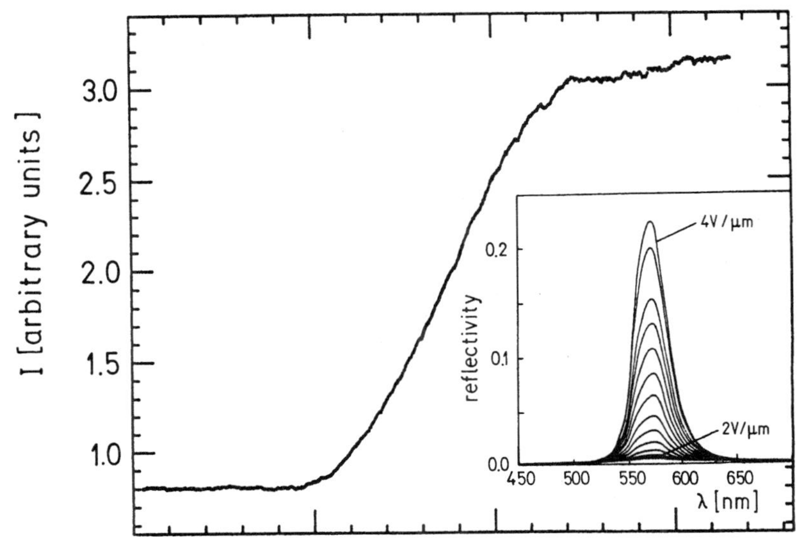

FIGURE 9.44. Reflectivity versus applied field strength of a PDLC made of chiral nematic droplets with the inset showing the wavelength dependence of the reflectivity (from Kitzerow et al., 1992).

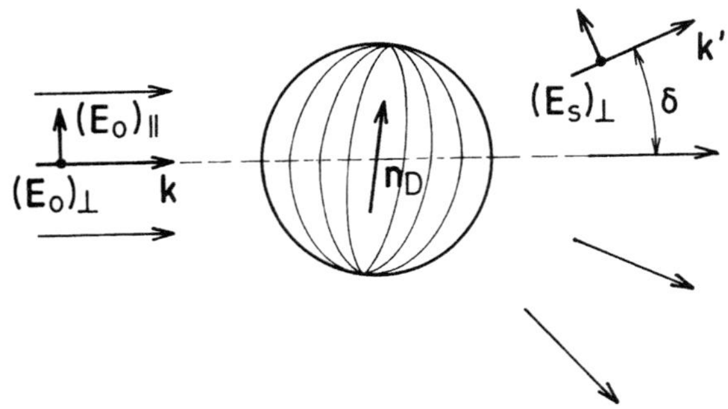

FIGURE 9.45. Light scattering geometry. A separation is shown of the incoming and scattered electric field into the component parallel to the scattering plane and perpendicular to it.

scattered wave can be described by an angularly modulated spherical wave with amplitude (Ishimaru, 1978)

$$\mathbf{E}_s = \underline{\mathbf{S}}\mathbf{E}_0 \exp(ikr)/ikr. \tag{9.32}$$

It is convenient to introduce the van de Hulst scattering matrix $\underline{\mathbf{S}}(\mathbf{k}, \mathbf{k}')$ (van de Hulst, 1957) with $\mathbf{k}'$ (FIGURE 9.45) as the scattering wavevector ($|\mathbf{k}'| = |\mathbf{k}|$), to describe the scattering amplitude $\underline{\mathbf{S}}\mathbf{E}_0/ik$. The distribution of the intensity of the scattered light is usually represented by the differential cross-section:

$$\mathrm{d}\sigma/\mathrm{d}\Omega = |r\mathbf{E}_s/E_0|^2 = (|\underline{\mathbf{S}}\mathbf{e}|/k)^2 \tag{9.33}$$

where $\mathbf{e} = \mathbf{E}_0/\mathbf{E}_0$ is the incoming polarization vector.

To show the relationship between the polarization and intensity of the scattered light, it is useful to separate the cross-section into the part $(\mathrm{d}\sigma/\mathrm{d}\Omega)_{\|}$ with electric field $(\mathbf{E}_s)_{\|}$ parallel to the scattering plane and the part $(\mathrm{d}\sigma/\mathrm{d}\Omega)_{\perp}$ with electric field $(\mathbf{E}_s)_{\perp}$ perpendicular to the scattering plane (FIGURE 9.45). In small-angle light scattering (SALS) (Plaza and Stein, 1959) experiments, the separation of the scattered electric field to $(\mathbf{E}_s)_{\mathrm{VV}}$ and $(\mathbf{E}_s)_{\mathrm{VH}}$ and the corresponding differential cross-sections $(\mathrm{d}\sigma/\mathrm{d}\Omega)_{\mathrm{VV}}$ and $(\mathrm{d}\sigma/\mathrm{d}\Omega)_{\mathrm{VH}}$ are based on the relative polarization of the incident and scattered light, where the index VV stands for the same polarization of the scattering and incident wave, with VH if they are orthogonal.

The total cross-section defined by the integral over the solid angle 4π is

$$\sigma = \int (\mathrm{d}\sigma/\mathrm{d}\Omega)\, \mathrm{d}\Omega \tag{9.34}$$

and more explicitly, using equation (9.33),

$$\sigma = \int (|\underline{\mathbf{S}}\mathbf{e}|k)^2\, \mathrm{d}\Omega \tag{9.35}$$

can also in principle be calculated directly using the optical theorem (van de Hulst, 1957)

$$\sigma = 4\pi\, \mathrm{Re}[\mathbf{e}\underline{\mathbf{S}}(\mathbf{k}, \mathbf{k}' = \mathbf{k})\mathbf{e}]/k^2 \tag{9.36}$$

where Re stands for the real part. The division of the total cross-section into VV and VH parts $\sigma = (\sigma)_{\mathrm{VV}} + (\sigma)_{\mathrm{VH}}$ can be used to describe the change in polarization during the scattering process. The depolarization is usually defined with the ratio $(\sigma)_{\mathrm{VH}}/\sigma$.

Formally, the scattering from a single droplet is reduced to the calculation of the scattering matrix $\underline{\mathbf{S}}$. Knowing the local dielectric tensor of the confined liquid crystal, one can use Maxwell's equation to derive (Stratton, 1941) the following exact expression for the scattering field

$$\mathbf{E}_s = -\frac{1}{4\pi}\int \mathbf{k} \times \mathbf{k}' \times \{[\underline{\varepsilon}_r(\mathbf{r}_n) - \underline{\mathbf{1}}] \cdot \mathbf{E}(\mathbf{k}, \mathbf{r}_n)\} \times \exp(-\mathbf{k}' \cdot \mathbf{r}_n)\, \mathrm{d}V \tag{9.37}$$

where the internal field $\mathbf{E}(\mathbf{r}_n)$ is unknown. Even for simple nematic director fields in spherical liquid crystal droplets, the problem cannot be solved analytically, as it can be for the isotropic droplet (Mie, 1908). Therefore, one is forced to make approximations. Here we are going to discuss approaches that take into account small variations of the index of refraction. It should be emphasized that an estimate using an approximate scattering matrix in the integral of the differential scattering cross-section (equation (9.35)) can yield a reasonably good result for the total cross-section, but will often be completely wrong using the relation (equation (9.36)) based on the optical theorem.

9.6.2.1 Rayleigh Scattering

If droplets are small compared with the wavelength of light, $kR \ll 1$, the phase shifts between scattering contributions from different areas of a droplet are negligible, and one finds

$$\underline{\mathbf{S}} = \mathrm{i}(kR)^3[\langle\underline{\varepsilon} - \underline{\mathbf{1}}\rangle - \mathbf{i}' \otimes (\mathbf{i}'\langle\underline{\varepsilon} - \underline{\mathbf{1}}\rangle)]/3 \tag{9.38}$$

where $\langle\underline{\varepsilon} - \underline{\mathbf{1}}\rangle$ is the dielectric susceptibility averaged over the volume of the droplet, $\mathbf{i}' = \mathbf{k}'/k$, and $\otimes$ stands for diadic product. The differential cross-section obtained after inserting equation (9.38) into equation

(9.33) is the same as for a simple small dipolar antenna with dipole $\langle \underline{\varepsilon} - \underline{1} \rangle \mathbf{E}_0(4\pi R^3/3)$ (Stratton, 1941). There is equivalence between forward $\mathbf{k}' = \mathbf{k}$ and backward $\mathbf{k}' = -\mathbf{k}$ scattering, and the total cross-section exhibits the well known dependence on λ^4 and R^6. The validity condition for the Rayleigh approximation $R < 50\,\text{nm}$ is not met in PDLC materials of practical interest.

9.6.2.2 Rayleigh–Gans Approximation

In the Rayleigh–Gans approximation (RGA), internal phase shifts are no longer neglected (kR is no longer restricted to very small values) but $\varepsilon - 1$ is required to be small enough that the internal field in equation (9.37) can be approximated with the incoming one (Gans, 1925). This approximation is formally equivalent to the well-known Born approximation in quantum mechanics. To be more explicit, the limitations for this approximation are usually written in terms of indices of refraction:

$$|n_{LC}/n_m - 1| \ll 1 \qquad (9.39)$$

where n_m is the index of refraction of the media that surrounds the droplets, and n_{LC} is either n_e or n_o. This inequality is realized in most PDLC materials. The other limitation that must be satisfied requires that the difference in the phase shift of the light induced by inserting a droplet in the optical path is small,

$$2kR|n_{LC}/n_m - 1| \ll 1. \qquad (9.40)$$

In PDLC materials where indices of refraction do not differ by more than 10%, this approach is limited to droplets with diameters smaller than $0.3\,\mu\text{m}$. In our theoretical examples, we use $n_e = 1.70$, $n_o = 1.52$, and $n_m = 1.55$ if not otherwise stated.

Substituting $\mathbf{E}(\mathbf{r}_n)$ with the incoming field $\mathbf{E}_0 \exp(-i\mathbf{k}\cdot\mathbf{r}_n)$, where $\mathbf{r}_n$ denotes the position in the droplet, the scattering matrix defined by equation (9.32) and implicitly given by equation (9.37) is reduced to

$$\underline{\mathbf{S}} = i(kR)^3[\underline{\mathbf{D}} - \mathbf{i}' \otimes (\mathbf{i}'\underline{\mathbf{D}})]/3 \qquad (9.41)$$

where $\underline{\mathbf{D}}$ stands for the spatial Fourier transform of the susceptibility

$$\underline{\mathbf{D}} = 3/(4\pi R^3) \int (\underline{\varepsilon} - \underline{1}) \exp(-i\mathbf{k}_s \cdot \mathbf{r}_n)\, dV \qquad (9.42)$$

which accounts for the internal phase-shift dependence on the scattering vector $\mathbf{k}_S = \mathbf{k}' - \mathbf{k}$. These phase shifts cause the backward and forward scattering intensity to be no longer equal, and also otherwise modify the distribution of the scattered light. The increasing effect that shows up with increasing kR is illustrated in FIGURE 9.46 in the case of the bipolar droplet. The differential cross-section must be calculated numerically (equation (9.33)) using equations (9.41) and (9.42), except for particularly simple radial and aligned director fields.

The expression for the scattering matrix $\underline{\mathbf{S}}$ (equation (9.41)) is unfortunately not accurate enough to allow

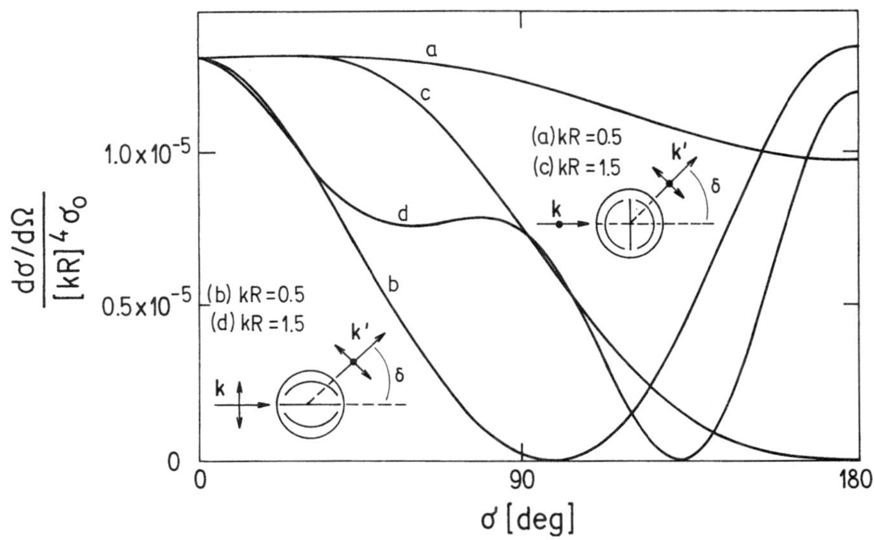

FIGURE 9.46. Differential cross-sections of bipolar droplets in the Rayleigh–Gans regime showing dependences on the scattering angle for two droplet orientations and two different kR values.

the application of the optical theorem (equation (9.36)) in the calculation of the total cross-section. Therefore, the integration of the differential cross-section (equation (9.35)) must be performed:

$$\sigma = \sigma_0(kR)^4 \int [(\underline{\mathbf{D}}\mathbf{e})^2 - (\mathbf{i}' \cdot (\underline{\mathbf{D}}\mathbf{e}))^2]\, d\Omega/(9\pi). \quad (9.43)$$

Here, σ_0 is used for the geometrical cross-section of the droplet. The total cross-sections calculated (Žumer and Doane, 1986) for isotropic and nematic droplets with radial, aligned, and bipolar structures (see FIGURE 9.47) exhibit for $kR < 1$ the same dependence on λ^4 and R^6 as in the Rayleigh limit. On increasing kR above 1, an increasing deviation from the Rayleigh behavior is observed. For the particular case of an aligned

FIGURE 9.47. Dependence of the total cross-sections of different droplet structures on the product kR in the Rayleigh–Gans regime. In the case of a bipolar and aligned droplet the situations with different polarization direction and angles of incidence are illustrated.

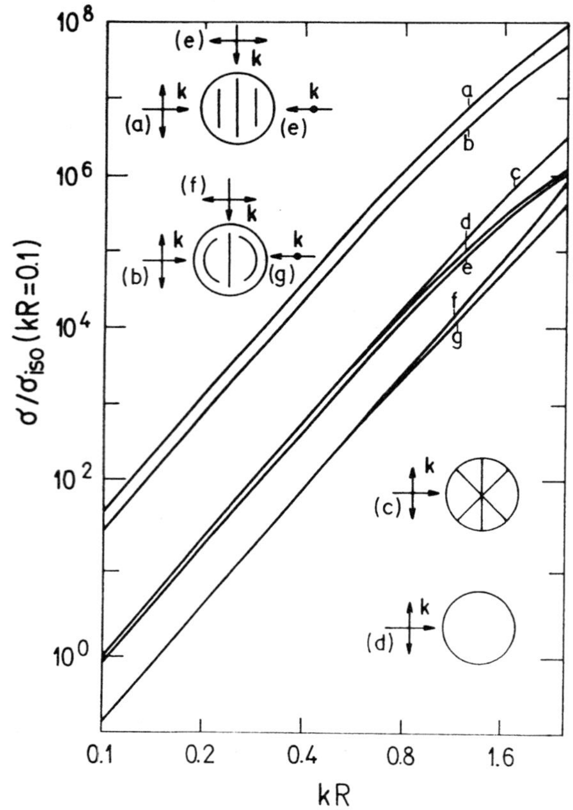

$(\mathbf{n}(\mathbf{r}_m) = \mathbf{n}_D)$ nematic droplet (equivalent to a sphere of the birefringent crystal), σ can even be written analytically in a closed form

$$\sigma = 2\sigma_0(kR)^2\{\cos^2\alpha_0[n_e(\theta)/n_m - 1]^2 + \sin^2\alpha_0[n_0/n_m - 1]^2\} \quad (9.44)$$

with θ as the angle between $\mathbf{n}_D$ and the polarization vector of the extraordinary component, $n_e(\theta)$ given by equation (9.17), and α_0 the angle between incoming polarization vector $\mathbf{e}$ and the principal plane defined by vectors $\mathbf{k}$ and $\mathbf{n}_D$ (FIGURE 9.45). Further, FIGURE 9.47 shows that the radial structure and the aligned structure, where the incident light is polarized orthogonally to the director field, are nearly equivalent to the isotropic droplet in their scattering cross-sections. The scattering from bipolar structures and aligned droplets when light is polarized along their symmetry axes is more than an order of magnitude stronger. The case with the polarization perpendicular to the bipolar axis induces the weakest scattering, because our selection for indices of refraction yields better matching of the effective average dielectric constant $\langle\varepsilon\rangle$ and the corresponding surrounding values, than in the case of a droplet in the isotropic phase, where $n_{iso} = (2n_o^2/3 + n_e^2/3)$.

For the application of PDLC materials in light shutters based on electrically controlled light scattering, the most important feature is the maximum change of total cross-section that can be induced by an external field. As we will see later (section 9.6.3.2), the attenuation coefficient for light propagation through a droplet dispersion is proportional to the total cross-section that directly influences the contrast of a shutter. The contrast is the ratio between the intensity of the direct beam transmitted in the presence of a strong field with $R/\xi_e = \infty$ (ON state), and the rest of it transmitted in zero field (OFF state).

To illustrate this potential, the dependence of the total cross-sections of the bipolar, radial, axial, and aligned droplets (occurring in different external fields) on the incident angle is shown (FIGURE 9.48). The value of the function at $\theta = 0$ (a) determines the transparency of a bipolar or radial PDLC film in the ON state where both structures are completely aligned. The OFF state of a radial droplet is represented by the line (c). Assuming a random distribution of the droplet axes in the OFF state of the bipolar PDLC, the transparency is governed by an appropriate average of the

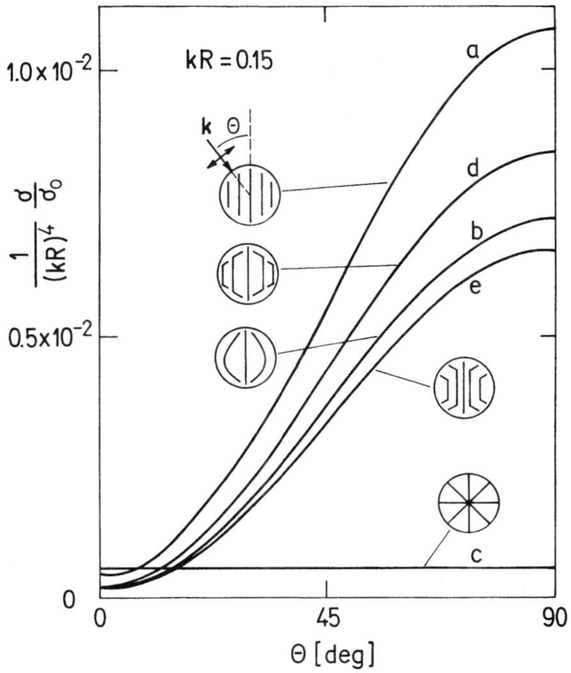

FIGURE 9.48. Angular dependence of total cross-sections of different droplet structures in the Rayleigh–Gans regime.

values on curve (b). Curves (d) and (e) correspond to an intermediate field $R/\xi_e = 5$.

FIGURE 9.49 shows the effect of index (dielectric constant) matching between the liquid crystal and the surrounding matrix. Total cross-sections for bipolar and axial droplets aligned in an intermediate field ($R/\xi_e = 5$) are shown as functions of n_m. For both cases, the minimum scattering occurs at $n_m = 1.55$ for polarization perpendicular to the director field.

9.6.2.3 Anomalous Diffraction Approximation (ADA)

The anomalous diffraction approximation (ADA) approach, similar to Fraunhofer diffraction on a phase grating (Born and Wolf, 1980), was introduced by van de Hulst (1957) for the description of the scattering on an object (here a liquid crystal droplet), when the following conditions are met

$$kR \gg 1 \qquad \text{and} \qquad n_{LC}/n_m - 1 \ll 1. \qquad (9.45)$$

The first limitation enables the use of the ray picture (Born and Wolf, 1980) for the description of light transmission through the object. The second condition allows reflection on external and internal boundaries, and the refraction of the ray crossing through the object to be neglected. This leads to the straight ray approximation (SRA), used for the polarizing microscope

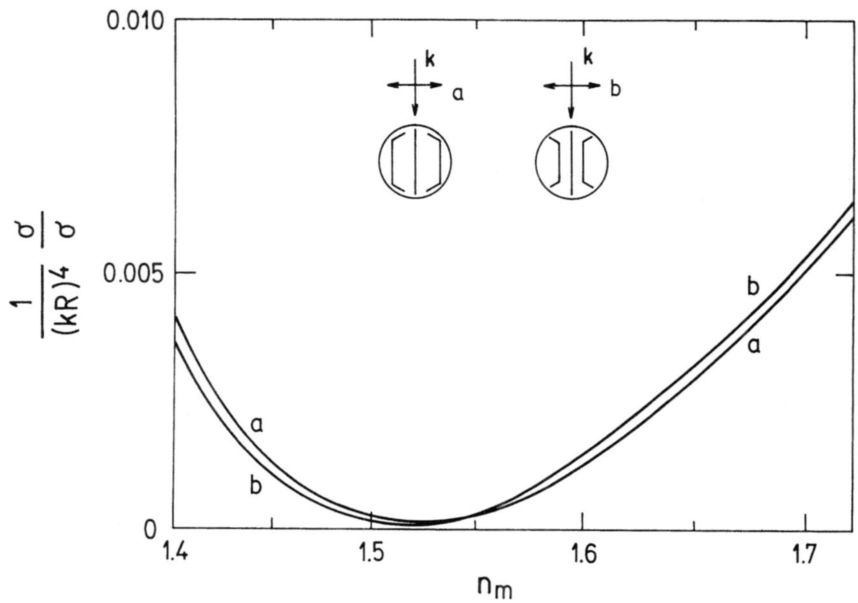

FIGURE 9.49. Total scattering cross-section of partially aligned droplets showing the effect of index matching in the Rayleigh–Gans regime; $n_0 = 1.52$, $n_e = 1.7$, $kR = 0.5$, and $R/\xi e = 5$ are used.

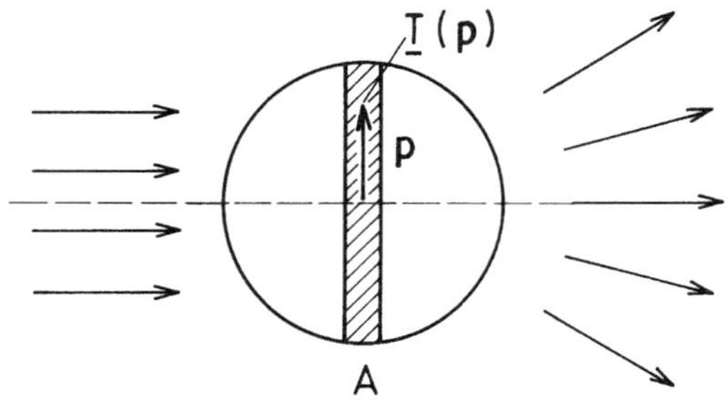

FIGURE 9.50. Anomalous diffraction approximation: schematic presentation of the substitution of the droplet with the phase-modulating screen.

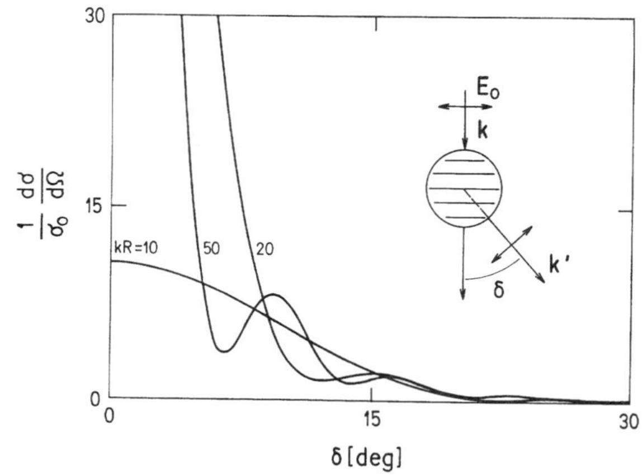

FIGURE 9.51. Angular dependence of the differential cross-section of an aligned droplet in the anomalous diffraction regime.

texture calculation (section 9.4.1.2), where the object does not change either the direction or the amount of light, but introduces phase shifts in the ordinary and extraordinary components of the light. Following methods of Fraunhofer diffraction to calculate the far-field distribution of the scattered light, the three-dimensional object (droplet) is first approximated by a planar screen (A (see FIGURE 9.50)) that modifies the polarization of the transmitted light, that is, induces phase shifts in the ordinary and extraordinary components according to the transformation matrix $\underline{\mathbf{T}}(\mathbf{p})$ developed for texture calculations (equations (9.14)–(9.17)). The scattered field is then decomposed into two parts: (1) light scattered by an opaque screen (equal in size to A), which according to the Babinet principle is equal to the field scattered by a conjugated screen (infinite screen with a hole of size A), but for the π phase shift, and (2) light transmitted through A. One can write the scattered field:

$$\mathbf{E}_s = \frac{k^2}{2\pi}\int [\underline{\mathbf{1}} - \underline{\mathbf{T}}(r'')]\frac{\exp(i\mathbf{k}'\cdot(\mathbf{r}+\mathbf{r}''))}{i\mathbf{k}'\cdot(\mathbf{r}+\mathbf{r}'')}\,\mathrm{d}A\mathbf{E}_0 \qquad (9.46)$$

and taking into account the far-field limit, the scattering matrix is

$$\mathbf{S} = \frac{k^2}{2\pi}\int (\underline{\mathbf{1}} - \underline{\mathbf{T}})\exp(i\mathbf{k}'\cdot\mathbf{r}'')\,\mathrm{d}A. \qquad (9.47)$$

The differential cross-section for the cases of radial and aligned droplets calculated using equations (9.33) and (9.47) show that the majority of the light is scattered in a narrow area within a few degrees of the incoming direction **k** (FIGURE 9.51). The sharpness of the distribution enables us to neglect the correction of the ADA introduced by Meeten (1982). The details in the scattered light distribution depend on kR, polarization, and direction of the symmetry axis of the droplet structure. Pronounced oscillations resemble the Fraunhofer pattern for a diffraction on a hole (for more details, see Žumer, 1988).

The forward scattering is described here by the scattering matrix (equation (9.47)), so that we can use the optical theorem to get the total cross-section (van de Hulst, 1957):

$$\sigma = 2\int [1 - \mathbf{e}(\mathrm{Re}\,\underline{\mathbf{T}})\mathbf{e}]\,\mathrm{d}A. \qquad (9.48)$$

In FIGURE 9.52, the cross-sections of an aligned droplet oriented at different angles θ against the wavevector **k** of the incoming light are shown. At small $kR[n_e(\theta)/n_m - 1]$ ($\lesssim 1$), the cross-section is well described by equation (9.44) obtained within RGA (see $\theta = 0$ curve). For large kR values where $kR[n_e(\theta)/n_m - 1] > \pi$, the cross-section σ starts to oscillate around its asymptotic value, $2\sigma_0$.

9.6.3 Dispersion of Droplets

9.6.3.1 Weakly Scattering Systems

These are systems where multiple scattering (van de Hulst, 1980) can be neglected. Such a situation can be realized in thin films with a single layer of droplets (Whitehead et al., 1993), or in thicker films by diluting the droplets or reducing their scattering. The latter situation is particularly important because it is realized in the ON state of PDLC shutters.

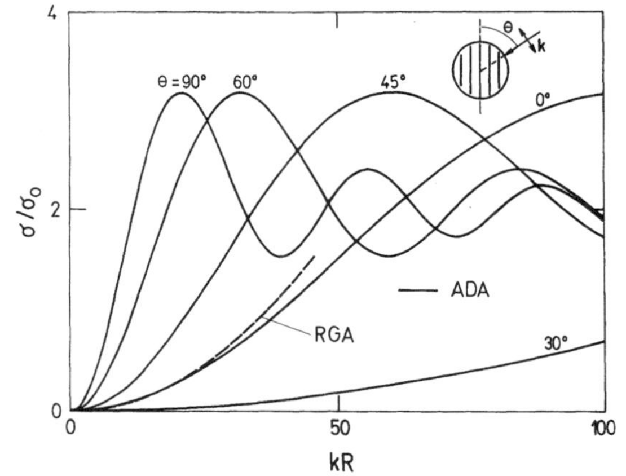

FIGURE 9.52. The kR dependences of the total cross-section of an aligned droplet for different droplet orientations in the anomalous diffraction approximation. Small kR results for the zero-angle case are compared with the results obtained in the Rayleigh–Gans approximation.

Assuming that the coherence length of the light is large compared with interdroplet distances, the scattered far-field $\mathbf{E}_S$ can be written as a sum over all N_D contributing droplets:

$$\mathbf{E}_S = \sum_j \underline{\mathbf{S}}_j \mathbf{E}_0 \exp[ikr + i(\mathbf{k} - \mathbf{k}') \cdot \mathbf{r}_j]/(ikr) \tag{9.49}$$

where $\mathbf{r}_j \ll \mathbf{r}$ describes the position of a droplet in the dispersion and $\mathbf{r}$ is the position of the observer (FIGURE 9.53). As we are interested in the scattering of a collection of droplets, the effective differential cross-section

FIGURE 9.53. Scattering from a collection of droplets: schematic presentation of the scattering geometry.

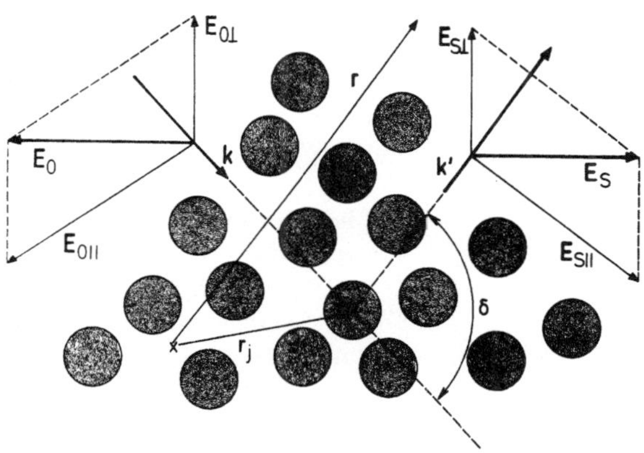

per droplet is introduced:

$$\frac{d\delta_{\text{eff}}}{d\Omega} = \frac{r^2}{N_D}\left(\sum_i \left|\frac{\mathbf{E}_{si}(\mathbf{r})}{E_0}\right|^2 + \sum_{i \neq j} \frac{\mathbf{E}_{si}(\mathbf{r}) \cdot \mathbf{E}_{sj}(\mathbf{r})}{E_0^2}\right) \times \exp[\mathrm{i}(\mathbf{k} - \mathbf{k}') \cdot (\mathbf{r}_i - \mathbf{r}_j)]. \tag{9.50}$$

No short-range correlations between droplet directors and sizes of neighboring droplets have been reported for PDLC materials. Therefore, the summation in equations (9.49) and (9.50) is substituted by averaging over droplet sizes, structures, and positions (Žumer et al., 1989), yielding

$$d\sigma_{\text{eff}}/d\Omega = \langle d\sigma_j/d\Omega \rangle F(\mathbf{k}_S) \tag{9.51}$$

where

$$\langle d\sigma_j/d\Omega \rangle = \langle |\underline{\mathbf{S}}_j \mathbf{e}|^2 \rangle / k^2 \tag{9.52}$$

stands for an average differential cross-section of a droplet and

$$F(\mathbf{k}_S) = 1 + G(\mathbf{k}_S)|\langle \underline{\mathbf{S}}_j \rangle \mathbf{e}|^2 / \langle |\underline{\mathbf{S}}_j \mathbf{e}|^2 \rangle \tag{9.53}$$

for the structure factor with

$$G(\mathbf{k}_S) = \beta \int_V [g(\mathbf{r}') - 1] \exp(-\mathbf{i}\mathbf{k}_S \cdot \mathbf{r}')\, dV' \tag{9.54}$$

as the Fourier transform of the droplet positional pair correlation function $g(\mathbf{r})$. The scattering volume is defined as V, and $\beta = N_D/V$ is the droplet number density. It should be stressed that the structure factor is defined for a large system with many droplets according to the conventional approach for scattering in liquids. The introduction of -1 in equation (9.54) accounts for the singular forward scattering contribution occurring in the limit $V \to \infty$ (Parod, 1982). This separation of the direct beam from the scattered light is useful for $kR \lesssim 1$, when small scattering cross-sections allow for the clear distinction between transmitted and scattered light. The information about the pair correlation function $g(\mathbf{r})$ can be obtained from electronic and optical microscope studies (for details, see Žumer et al., 1989).

To show the effects of interference, let us assume that all droplets have the same radius $R \ll k^{-1}$, and the pair correlation function is of the square-well type ($g(r) = 0$ for $r < 2R$ and 1 for $r > 2R$). The resulting RGA effective differential and total cross-sections are shown in FIGURE 9.54. A system of equal radial droplets (a) and a dispersion of randomly oriented bipolar droplets (b) are represented. The

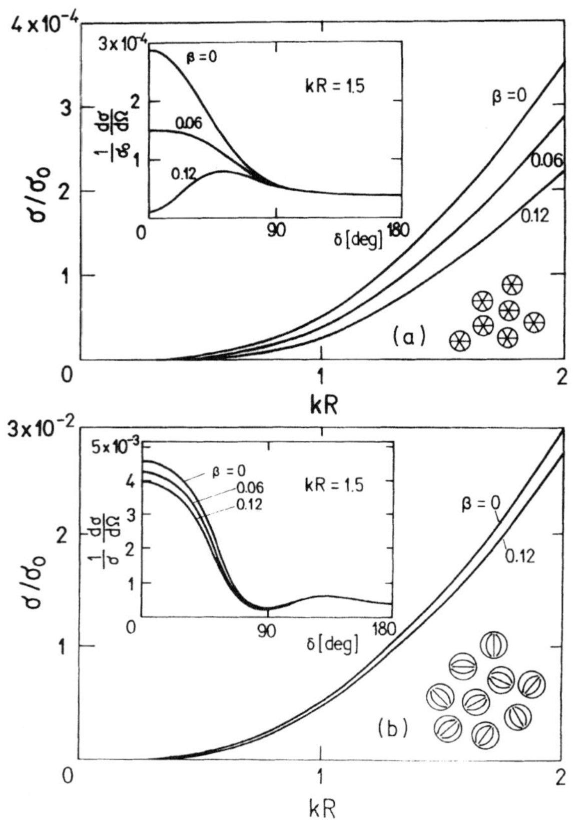

FIGURE 9.54. Scattering from a collection of droplets in the Rayleigh–Gans regime. Dependences of the total cross-sections in radial (a) and bipolar (b) cases are presented for different droplet densities. The insets show the angular dependences of the corresponding differential cross-sections.

interdroplet interference has a minimal effect on the total cross-sections in both cases. The differential cross-sections are significantly affected only in the case of identical (radial) droplets. We can safely conclude that in a more realistic PDLC system, where droplets have different sizes, even further reduction of the interference effects is expected because of additional averaging.

For small collections of scattering droplets, the effect of interdroplet interference must be calculated directly from equation (9.50). If droplets are large, the RGA approach for the calculation of $\underline{\mathbf{S}}$ must be substituted by the ADA. Using the optical theorem (equations (9.36) and (9.48)), this leads to the effective total cross-section:

$$\sigma_{\text{eff}} = 2\left[\sigma_0 - \int \mathbf{e}(\text{Re}\langle\underline{\mathbf{T}}\rangle)\mathbf{e}\,\text{dA}\right] \tag{9.55}$$

and to a conclusion that, in this limit, the interference has no effect on the effective total cross-section.

9.6.3.2 Attenuation of Light

The turbidity of PDLC materials without dissolved dyes is completely determined by light scattering. The decrease in the direct light beam intensity traveling the distance dz in the dispersion is thus given by

$$\text{d}I = -I\sigma_{\text{eff}}\beta\,\text{dz} \tag{9.56}$$

which in a diluted dispersion yields the exponential attenuation

$$I = I_0\,\text{e}^{-\mu z} \qquad \text{with} \qquad \mu = \sigma_{\text{eff}}\beta \tag{9.57}$$

as the attenuation coefficient.

In a real PDLC film, this description is limited to situations in which the contribution of the single or multiple scattered light registered by the detector is small compared with the contribution of the attenuated direct beam. It is thus obvious that the description of the attenuation of light depends on the experimental set-up. We estimate the limitation on the exponential behavior of I/I_0 in the following way

$$I/I_0 \gg \Delta\Omega_0/\Delta\Omega_S \tag{9.58}$$

where $\Delta\Omega_0$ is the collecting solid angle of the detector and $\Delta\Omega_S$ is the solid angle into which the dominant part of the light from a single droplet is scattered (Žumer, 1988).

For small droplets ($kR \lesssim 1$), where backward and forward scattering do not differ much (see FIGURE 9.46), we simply assume that scattering is uniform and that $\Delta\Omega_S$ is 4π. In the case of large droplets ($kR \gg 1$), the scattering is predominantly forward with $\Delta\Omega_S \ll 4\pi$ (see FIGURE 9.51) so that limitations on the use of equation (9.57) are more severe.

9.6.3.3 Experimental Light Scattering Results

The influence of the applied field on the attenuation coefficient of a PDLC with small droplets ($\langle R\rangle = 0.13\,\mu$m) of E7 liquid crystals (BDH Chemicals) in a matrix of epoxy resin (Žumer et al., 1989) is shown in FIGURE 9.55. The volume fraction occupied by liquid crystal droplets C_D was estimated from electron scanning of cross-microscopy of the material to be 0.1.

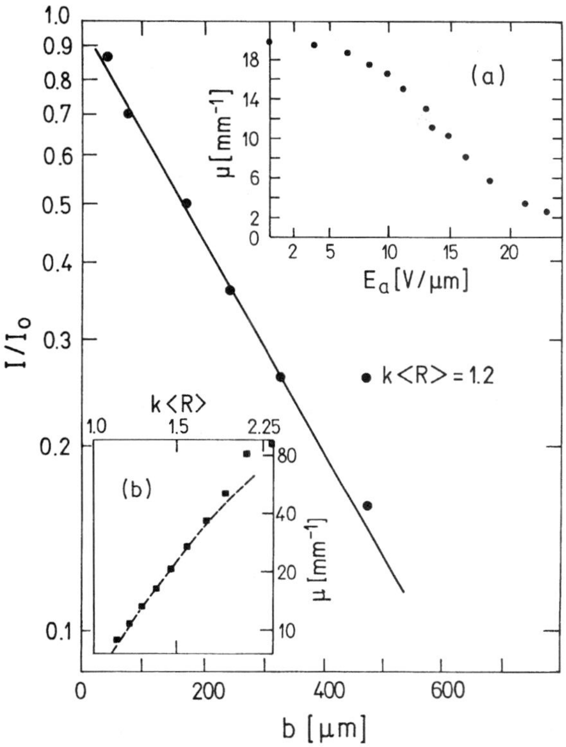

FIGURE 9.55. Attenuation of light traversing a PDLC material in the ON state. Inset (a) shows the dependence of the attenuation coefficient on the applied field and inset (b) shows its wavevector dependence.

The indices of refraction of E7 were $n_o = 1.522$ and $n_e = 1.746$ and the index of the epoxy matrix was estimated to be 1.561. The field-induced transformation from randomly oriented bipolar droplets in zero field to the high-field-induced aligned structures reduces the attenuation coefficient by one order of magnitude (see FIGURE 9.55, inset (a)). This is in good agreement with theoretical estimates based on RGA expressions that yield (Žumer et al., 1989) $\mu(E=0) = 0.019\,\mu\text{m}^{-1}$ and $\mu(\text{strong } E) = 0.0017\,\mu\text{m}^{-1}$. The transmitted intensity shows an exponential decay as a function of PDLC sample thickness b in the ON state. It yields an attenuation coefficient $4 \times 10^{-3}\,\mu\text{m}^{-1}$ at $\lambda = 0.5\,\mu\text{m}$. Inset (b) shows that the coefficient μ follows the theoretical curve rather well, and that in this range it is nearly proportional to the fourth power of k.

The angular dependence of the transmission of a thin film (10 μm) with aligned large droplets

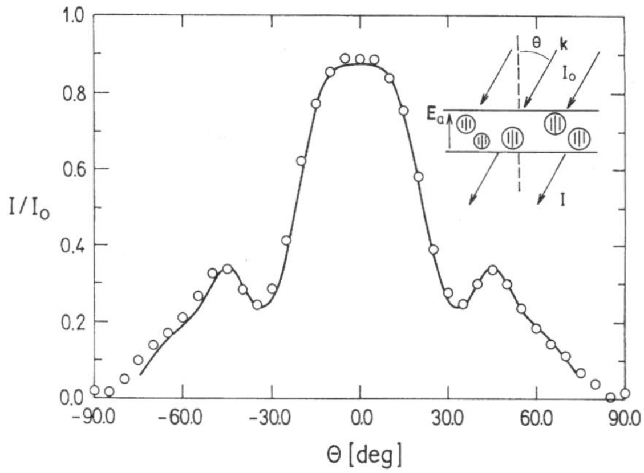

FIGURE 9.56. The angular dependence of a thin PDLC film in the ON state. The solid line corresponds to the prediction of the anomalous diffraction approach for droplets with average radius 2.4 μm.

($R = 2.4(1 \pm 0.15)\,\mu$m) on θ, the angle between the aligned director field and **k** of incoming light (FIGURE 9.56), is reasonably well fitted with ADA predictions for the dependence of the cross-section on θ (for details, see Whitehead et al., 1993).

Studies of the distribution of scattered light (Montgomery, 1988; Stein et al., 1992; Whitehead et al., 1993) showed that the differential cross section for aligned droplets agrees reasonably well with theory based on ADA (FIGURE 9.57).

FIGURE 9.57. The angular dependence of the effective differential cross-section for a distribution of aligned droplets in a PDLC sample. The solid line stands for a fit based on the anomalous diffraction approach (see Whitehead et al., 1993).

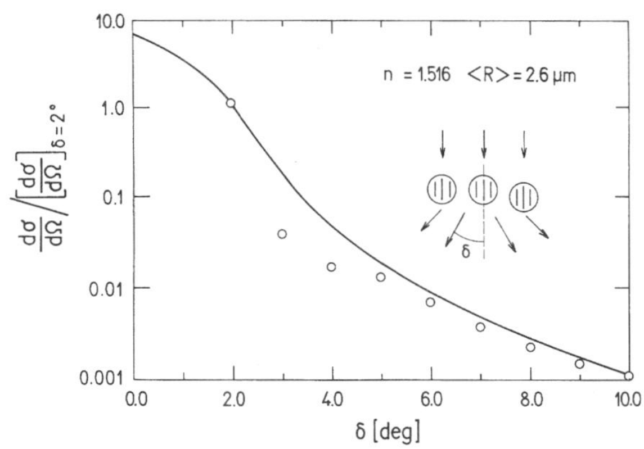

9.6.3.4 Strongly Scattering Systems

With increasing amount of liquid crystal (density of droplets) in the dispersion, and particularly in the extreme case when the polymer (network) is dispersed in the liquid crystal, the light transmission and scattering are progressively more affected by the interference and multiple scattering from neighboring regions (droplets). This area, although important for applications of PDLC materials, has not been investigated much, so we can only summarize some basic aspects here.

INTERDROPLET INTERFERENCE. The effect of the interdroplet interference could be easily visualized by the examination of diffraction gratings. With decreasing distance between adjacent slits of a grating, the wavefront of light passing the opening is progressively less perturbed. In other words, scattering is more and more restricted to the forward direction. This explains why systems of dense but aligned liquid crystal droplets, or dispersed polymer networks in aligned liquid crystals, are very transparent, even if the single droplet/domain scattering is significant because of a substantial refraction index mismatch between the polymer and liquid crystal. This effect is important for scattering-based shutters that use a layer of PDLC material. Nematic droplets depend on the applied-field ordered (transparent) ON state or disordered (scattering) OFF state. A dense dispersion can thus yield a highly transparent ON state that is also relatively insensitive to the direction of the light beam, which also means it is relatively haze-free. Such a reduction of haze is certainly simpler to achieve than using a liquid crystal polymer matrix to achieve index matching for all directions of the light beams.

The situation in the OFF state of a dense system of randomly oriented bipolar and other nonuniform droplets or domains of liquid crystals is equivalent to a phase grating. There, each slit causes a different phase shift, so that even if the separation between slits vanishes there is still substantial light diffraction arising from the variation of the phase going from one slit to another. So one expects that the scattering of randomly oriented droplets is only slightly affected by droplet density, and greatly determined by their sizes and dielectric anisotropies.

MULTIPLE SCATTERING. The propagation of light through a dispersion where the amount of light scattered several times cannot be neglected is complicated, and is relatively well understood only in isotropic systems (van de Hulst, 1980). The preliminary investigation of multiple scattering by nematic droplets using the homogenization method (Friedman, 1992) led to the approximate expansions that are still being evaluated numerically for particular structures. The light absorption in systems with pronounced multiple scattering, where the light path in the medium is much longer than in the predominantly single scattering cases, must also be considered.

SHUTTER CONTRAST. The most important feature of a light shutter is the contrast ratio, which we will define here as the ratio of the power P_0 registered by a small detector when the shutter is in the ON state, and the power reaching the detector when the shutter is in the OFF state, P_c. To estimate P_c for such a shutter, one must integrate scattered light coming into the detector from the whole illuminated area of the shutter (see FIGURE 9.58):

$$P_c = \int j(\mathbf{p})h(\mathbf{m}(\mathbf{p}))\Omega_D(\mathbf{p})\,\mathrm{d}A. \tag{9.59}$$

Here, vector $\mathbf{p}$ describes the point on the shutter, $j(\mathbf{p})$ is the illuminance of the shutter, $\mathbf{m}(\mathbf{p})$ is the unit vector pointing from the point $\mathbf{p}$ towards the detector, $h(\mathbf{m})$ is

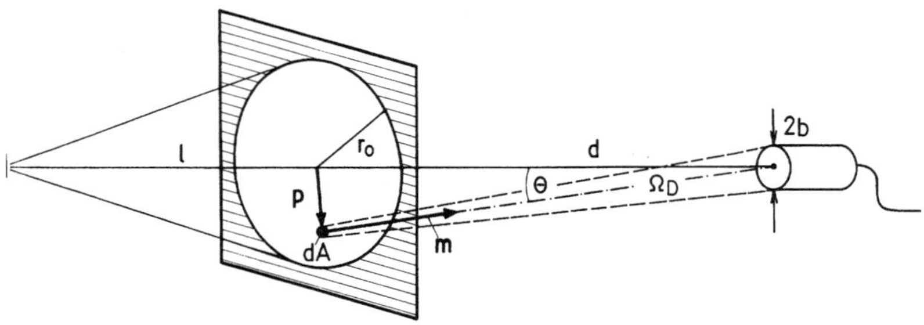

FIGURE 9.58. Schematic presentation used in the calculation of PDLC shutter contrast.

the angular distribution of the light emerging from the shutter, and $\Omega_D(\mathbf{p})$ is the solid angle of the detector as seen from the point $\mathbf{p}$. Let us choose a simplified case where the absorption of light is negligible, the shutter is completely transparent in the ON state, and the PDLC film is thick enough so that in the OFF state, the multiple scattering yields a uniform light distribution in the PDLC material. The refraction of light passing from the high refractive index material ($n = 1.5$ for polymer and glass) to the air modifies the isotropic distribution according to Snell's law of refraction, yielding a simple cylindrically symmetric distribution:

$$h(\theta) = \frac{1}{4\pi} \frac{n}{(n - \sqrt{n^2 - 1})} \frac{\cos\theta}{n\sqrt{n^2 - 1 + \cos^2\theta}} \tag{9.60}$$

characterized by the angle θ between vector $\mathbf{m}$ and the normal to the surface. The first factor in equation (9.60) represents the isotropic distribution, the second represents the limitation induced by total reflection, and the third represents refraction-induced change of angles. A point light source is chosen at a distance l away from the center of a circular shutter with the radius r_0 and a detector with radius R at a distance d behind it (see FIGURE 9.58). Assuming that $l, d \gg r_0, R_D$, the contrast is given by:

$$\text{contrast} = 4n(n - (n^2 - 1)^{1/2})(l/r_0)^2/(1 + l/d)^2. \tag{9.61}$$

This contrast converges to a constant value as the distance d increases. It should be stressed that for a nondivergent beam, realizing the limit $l \to \infty$, the contrast tends to infinity as d increases.

For a shutter with radius of illuminated area $r_0 = 2.2\,\text{cm}$, the distance of the point source $l = 0.5\,\text{m}$, and $d = 2.0\,\text{m}$, one finds a contrast of around 130, that is, not far from the experimental value of approximately 100 (for more details, see Pirš et al., 1990).

9.6.4 Electro-optics

There have been a large number of electrooptics studies of PDLC systems in the last few years to determine the feasibility of PDLC materials in various applications (see section 9.7). Studies of the electro-optic performance parameters, such as response times, contrast ratios, and switching voltage, have been reviewed previously (Doane, 1990, 1991; Kitzerow, 1994); therefore we will present only a short description here. There have been many experimental electrooptic studies to date: affect of droplet shape (Drzaic, 1988; Drzaic and Muller, 1989; Wu et al., 1989); viewing angle (Wu et al., 1987; Doane et al., 1990; West et al., 1992; Drzaic and Gonzales, 1993); materials development (Vaz et al., 1987; Jain et al., 1990; Drzaic, 1991; Fung et al., 1992; Coates et al., 1993); effect of glass transition temperature (West et al., 1992); effect of cell thickness (Akins and West, 1992); effect of droplet size (Nomura et al., 1990; Erdmann et al., 1991); optical power limiting and bistability (Kim and Palffy-Muhoray, 1989); effectiveness in color modulators (Pirš et al., 1990); dual-frequency addressing (Vaz and Montgomery, 1989); magnetic switching (Li et al., 1992); optical switching (Palffy-Muhoray et al., 1990); partial alignment effects (Margerum et al., 1989); materials forming connected microspheres (Nomura et al., 1990); and infrared modulating shutters (McCargar et al., 1992). For a recent review see Drzaic (1995).

One of the most important discoveries relevant to the electrooptic performance of PDLCs is that the shape of the droplet can be a significant factor in the response time for bipolar droplets (Drzaic, 1988; Wu et al., 1989). This feature has been modified by a simple droplet director approach where the viscous torque density competes with the elastic torque in returning the droplet director to its equilibrium orientation upon removal of the field (Wu et al., 1989). Balancing the two torque densities and assuming the inertial term to be negligible, the turn-off time relaxation constant, τ_{OFF}, for a PDLC is estimated as $\tau_{\text{OFF}} = \gamma\alpha^2/K(l^2 - 1)$, where $l = \alpha/\beta$ is the aspect ratio of the droplet with α and β being the lengths of the semimajor and semiminor axes, and K the effective elastic constant. For typical parameters $\gamma = 4 \times 10^{-2}\,\text{kg}\,(\text{ms})^{-1}$, $K = 1 \times 10^{-12}\,\text{J}\,\text{m}^{-1}$, $\alpha = 1.0\,\mu\text{m}$, and $l = 1.1$, the τ_{OFF} value is $\sim$20 ms. This demonstrates that large values of l can significantly reduce the relaxation time. For the spherical droplet case, l is unity and therefore results in an infinite relaxation time, since other relaxation mechanisms were ignored.

The turn-on time, τ_{ON}, of a PDLC film with bipolar droplets is modeled (Wu et al., 1989) by balancing all three torque densities, elastic, electric, and viscous, which results in the following expression:

$$\tau_{\text{ON}}^{-1} = 9\varepsilon_0 \Delta\varepsilon E^2/[\gamma(\rho_p/\rho_{lc} + 2)^2] + K(l^2 - 1)/(\gamma\alpha^2)$$

where ρ_p/ρ_{lc} is the ratio of the resistivities of the polymer and the liquid crystal. This approach shows that the first term usually dominates and that the rise time depends strongly on the resistivity of the components of the dispersion.

Wu and coworkers (Wu et al., 1989) also estimated the switching voltage for the bipolar droplet, calculated to be

$$V \simeq d(\rho_p/\rho_{lc} + 2)[K(l^2 - 1)/(\varepsilon_0 \Delta\varepsilon)]^{1/2}/(3\alpha).$$

Using typical values $K = 5 \times 10^{-12}\,\mathrm{J\,m^{-1}}$, $\Delta\varepsilon = 14$, the cell thickness $d = 25\,\mu\mathrm{m}$, $\rho_p/\rho_{lc} = 10$, $l = 1.1$, and droplet diameters of $1\,\mu\mathrm{m}$ (the diameter in this case is 2α), the switching voltage is $V \simeq 18\,\mathrm{V}$, substantially less than for the radial droplets which are governed by the expression $V \simeq d(\rho_p/\rho_{lc} + 2)[K/(\varepsilon_0 \Delta\varepsilon)]^{1/2}/2\alpha$. Using the same parameter set as above, $V \simeq 60\,\mathrm{V}$ for the radial droplets. Notice that in both cases, the switching voltage can be modified substantially by changing ρ_p/ρ_{lc}.

Recently, Kelly and Palffy-Muhoray (1994) proposed a more general model to describe the optical response of PDLC materials. They introduced a hierarchy of order parameters to describe the orientational order on different length scales, and to express the optical response in terms of these order parameters. The results are essentially the same as those presented earlier by Wu and coworkers (Wu et al., 1989).

It can be very difficult to compare the contrast ratio measurements of PDLC displays from different groups. There is no well-established procedure for measuring the contrast ratio of a PDLC display. The values depend not only on the scattering properties of the films as presented in section 9.6, but also on the details of the experiment, for example, the illumination source, the illumination and viewing directions, the acceptance angle of the detector, etc. Lackner and coworkers (Lackner et al., 1988) have performed an extensive comparison of the various methods and measurements.

9.7 APPLICATIONS

In recent years, applications have been the principal driving force behind liquid crystal dispersions, although polymers were not the first materials dispersed with liquid crystals to make a device. Glass beads (Hilsum, 1976) were dispersed in a liquid crystal to put into practice the concept of index matching in a scattering-mode light shutter device. Hilsum placed micron-size beads in a nematic liquid crystal in which the ordinary refractive nematic index was approximately matched with the index of the beads. In a randomly ordered nematic, brought on by the bead surfaces, the material is scattering. When an electric field is applied to the material, the liquid crystal aligns such that along the direction of view, perpendicular to the cell wall, the refractive indices of the liquid crystal and the beads are matched to remove the scattering condition. Upon removal of the field, the bead surfaces return the nematic liquid crystal in the cell to the scattering condition; hence the making of a light shutter. This same concept was applied later when polymers were introduced. These came in several forms: emulsions (Fergason, 1985); filled porous membranes (Craighead et al., 1982); and phase-separated systems (Doane et al., 1986). Application of PDLCs then focused on index matching for several years to make variable transmission windows, displays, and other light shutters. Currently, however, dispersions of liquid crystals employ a number of different electro-optic phenomena besides index matching, and have extended their application to other kinds of displays, projection light valves, and shuttering devices. In this section we review some of these applications.

9.7.1 Droplet Dispersions

Droplet dispersions, whether they are made from emulsions or by phase-separation techniques, can be coated onto large ITO-coated plastic sheets by roll-to-roll processing methods. This feature has made these materials attractive for privacy windows in which the transparency is electrically controlled. Companies in the United States, such as the Raychem Corporation and the 3M Corporation, have been the most aggressive in getting products into the market place, and there has been considerable development activity in Japan and Europe. FIGURE 9.59 illustrates a PDLC window application taken from a brochure of the 3M Corporation. Their windows are currently under manufacture for indoor privacy windows for applications such as conference room windows, automobile windows (such as sunroofs or in vans), and even outdoor windows on double-pane glass are under development. Cost has been the most troublesome feature in marketing windows. The two largest contributors to cost are the liquid crystal material (requiring about a gram per square foot) and the ITO-coated plastic. Both of these costs are expected to drop

FIGURE 9.59. A PDLC variable transmission window application from a 3M Corporation brochure; courtesy of Dr. Steve Miller.

dramatically under high-volume sales. Off-angle haze that becomes apparent at wide-view angles in PDLCs has not been of much concern, but manufacturers would like to eliminate it if possible. The haze at wide angles is basically a result of one medium of the dispersion, namely the polymer, being optically isotropic while the other medium, the liquid crystal, is birefringent, and the indices can only be matched along one direction of incident light (Wu et al., 1987). This can be corrected by using a birefringent polymer such as a polymer liquid crystal, so that both indices of the liquid crystal droplet can be matched. In this case the material is transparent at all angles of view (Doane et al., 1990). Attempts to produce such a commercial film in this way have not been successful to date.

When PDLCs were first developed, direct-view displays were first considered as an application of these materials. In this application, the display cell is backed by a black, luminescent, or reflective sheet that becomes visible when the PDLC material is in the transparent state. While there were attractive features of these displays, such as their large sizes and flexible plastic substrates, a limiting feature is the loss in contrast because the bulk of the light is forward rather than backward scattered. Limited backward scattering causes the black or colored background to be visible through the PDLC, turning the white opaque state into a gray or pale-colored appearance. Another problem is the off-angle haze that whitens the black background in the transparent state. The net result of these two effects is a loss in contrast. The Raychem Corporation incorporated dyes to help overcome these problems (Drzaic, 1991). They used dichroic dyes in the NCAP droplets with a color that was complementary to the background. FIGURE 9.60 shows one of the company's recent products.

FIGURE 9.60. A direct-view segmental display manufactured by the Raychem Corporation and made from NCAP materials; courtesy of Dr. Paul Drzaic.

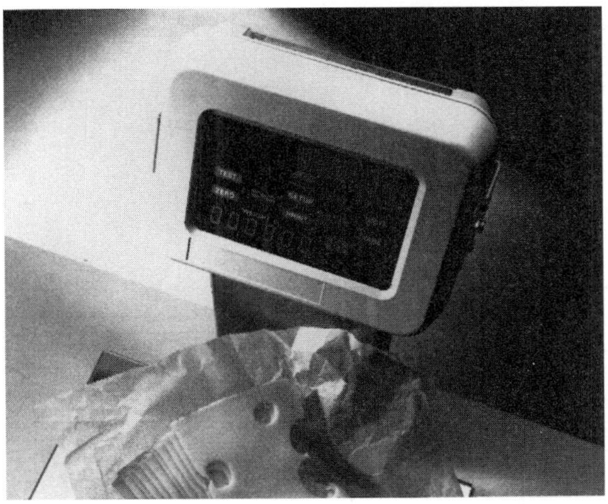

Even isotropic dyes can make a significant difference in the contrast of the display, and they have been heavily studied at Kent State University (West and Ondris-Crawford, 1991). Other methods incorporating cell design, such as the use of gain reflectors (Drzaic et al., 1992), have been used to improve contrast, and polarizers have also been used (Yoshida et al., 1993; West, 1994) to improve the angle of view. The Hughes group has been successful at patterning the droplet size to create unusual effects and has heavily explored lighting and driving techniques (Margerum et al., 1989; Erdmann et al., 1991).

With the improved compatibility of PDLC materials with the active matrix substrate, PDLC materials are becoming attractive for the development of high-resolution paper-like black-on-white displays (Coates, 1993). In spite of the bleed-through that occurs from forward scattering, the brightness of a reflective display can be improved over the TN, STN, or other displays that require polarizers. Good progress in this direction has been made by a joint effort between Merke Ltd. and the University of Stuttgart for MIM-active matrix displays (Nolan et al., 1993). Figure 9.61 shows a reflective direct-view display reported by these two groups. More recently, the groups at Stuttgart and Bell Labs have collaborated to fabricate an impressive 1/2 VGA, 5.4″ diagonal MIM PDLC display to interface with the SONY Magic Link (Huttelmaier et al., 1996). The display encorporates a brightness enhancement film (BEF) and a novel black absorber from the 3M Corporation. Several material issues become important in use of the active matrix (AM): (1) the elimination of the hysteresis in the voltage response curve; (2) the lowering of the drive voltages to a value compatible with αSi and p-Si AM technologies; and (3) increasing the resistivities for desired charge hold times. The PDLC material developed by Merck (Coates et al., 1993) can achieve these characteristics over a limited temperature range for the MIM AM technology. The MIM technology can tolerate higher drive voltages, allowing thicker cell gaps and hence better contrast. Even prior to the Merck group's success, the Raychem group (Reamey et al., 1992) made substantial improvements to the emulsion-based (NCAP) materials, making them AM compatible. Raychem has reported a direct-view AM NCAP system (Jones et al., 1992). In Japan, the group at Asahi Glass Co. (Niiyama et al., 1993) has developed a carefully designed phase-separated PDLC system to eliminate hysteresis and improve resistivity for p-Si AM substrates.

FIGURE 9.61. Photograph of a Merck Ltd. high-resolution direct-view PDLC that is MIM active matrix addressed; courtesy of Dr. David Coates.

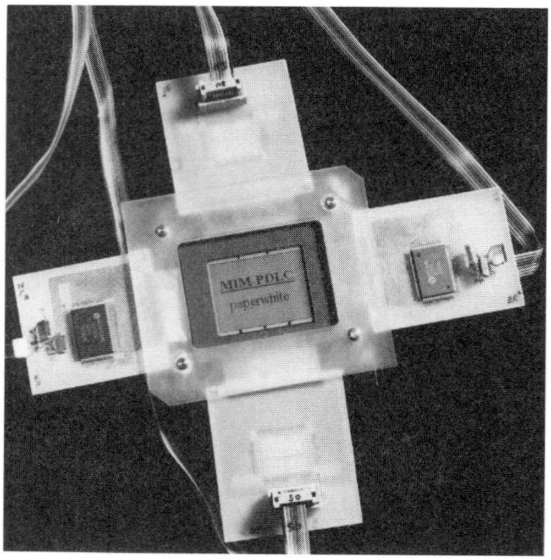

Perhaps most developmental effort on PDLCs has gone into projection light valves. In this application the light is well collimated, avoiding off-angle haze problems, and forward scattering is not an issue with suitable design of the projection system. The attraction of PDLCs is the absence of polarizers, which improves the brightness of the projector by at least a factor of two. The use of brighter lamps is also possible, since the light valve operates on the scattering rather than the absorption principle. The material issues are the same as those described above for the active matrix, since high resolution is required. The first PDLC projection prototype reported was a black and white system that made use of the αSi, by Optical Imaging Systems, and PDLC material, by Kent State University (Yaniv et al., 1989). This was immediately followed by a full-color system using a p-Si technology by Asahi Glass (Kuniga et al., 1990). Since that time, a considerable amount of progress has been made on advanced materials for projection systems, which may be found in papers such as Tomita (1993), Niiyama (1993), and Coates (1993). The Raychem Corporation has described a projection system (Jones and Tomita, 1992), and more recently, the University of Stuttgart, in cooperation with Merck, has reported optimized

PDLC materials for MIM-active matrix substrates (Ginter et al., 1993). An optically addressed PDLC projection system has been reported by NHK (Takizawa et al., 1991), and NCAP materials have even been used for the projection screen itself (Kinugasa et al., 1991).

Applications of droplet dispersions have been explored for liquid crystal phases other than nematics. Crooker and Yang (1990) reported a reflective display from cholesteric droplets in a polymer host. Cholesteric liquid crystal materials with negative dielectric anisotropy and sufficiently short pitch length will Bragg reflect colored light in the presence of an applied electric field. In the absence of the field, the droplet for a texture is colorless and weakly scattered. The droplets need to be large, on the order of several microns in diameter, to give good reflectivities in the powered state and weak scattering (nearly transparent) in the unpowered state.

A challenge has been to develop ferroelectric PDLCs involving dispersed chiral smectic C droplets. Success has been reported by four groups (Molsen et al., 1993; Zyryanov et al., 1993; Lee et al., 1994; Patel et al., 1994). The Kent group has demonstrated a multipixel bistable system that phase shifts rather than scatters light, and therefore requires polarizers. An advantage of this cell over the surface stabilized (SSFLC) cell is the cell thickness: a 4 μm gap rather than the 1.2 μm gap used for SSFLC is much more attractive for uniformity and manufacturability. Another advantage may be that it is less subject to failure by mechanical shock. The cell requires elongated droplets, and one current problem is reproducibility in droplet shape uniformity.

Still another PDLC innovation is the suspension of microsize anisometrically shaped solid particles, such as needles or plate-shaped crystals in the droplets (Young et al., 1994). The Litton group has been successful in prototyping impressively large window shutters using this type of suspended-particle system.

One of the more exciting new applications of PDLCs is in the general area of holography, but it also relates to a different approach to reflective color displays as well as to optical computing and networking. The basic idea is to pattern a PDLC system in a periodic structure that is on the order of the wavelength of light. In fact, the patterning is done by using a coherent light source ad creating standing waves to optically cure the polymer-creating phase separation that mimics the standing-wave pattern.

Different types of patterns have been proposed and prototyped. One that has possibilities for color reflective display is made by irradiating and curing the sample with light normally incident on a monomeric/nematic liquid crystal mixture (Tanaka et al., 1994, 1995). Light reflected back through the sample sets up an interference pattern of planar fringes parallel to the cell walls and stacked along the cell normal. If a suitable multifunctional polymer is used, phase separation occurs such that the dispersion forms sheets of polymers separated by sheets of liquid crystals. After the cell is formed, it will reflect light of a color corresponding to the layer spacing of the sheets. It is an efficient reflector and can be switched off by applying a voltage to cell electrodes that align a positive $\Delta\varepsilon$ material to index match the polymer. The potential for display applications is exciting, since nearly 100% of the light at the selective wavelength can be reflected, yielding high brightness.

Another type of holographic structure has been described (Sutherland et al., 1993), using an acrylate photocurable polymer. The group at the SAIC Corporation and Wright–Patterson Air Force Base created a phase-separated patterning with the layers oriented perpendicular to the cell surface to create a switchable grating, different to the Tanaka structure described above. The SEM photograph in FIGURE 9.62 indicates a droplet structure in the polymer-cured region.

9.7.2 Non-droplet Dispersions

In the past few years there has been an interest in dispersing very low concentration polymer networks in liquids to modify the electrooptical response or stabilize a particular texture. An example is the cholesteric phase that is known to exhibit three textures: a planar or grandjean texture that if tuned to the visible will decompose incident light into its left and right circularly polarized components and totally reflect one of them; a focal–conic texture that can be either weakly or strongly light scattering (white in appearance); and a homeotropic texture that is formed under an applied electric field of sufficient strength to untwist the cholesteric in a cell (positive dielectric anisotropy is required). When certain multifunctional acrylic monomers are dissolved in a liquid

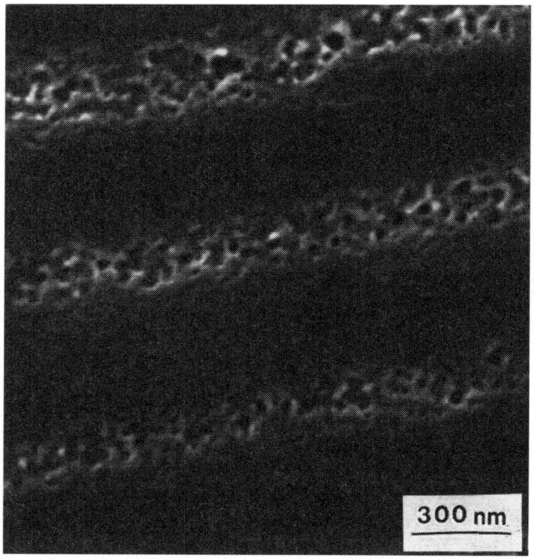

FIGURE 9.62. An SEM photograph of an in-plane view of a PDLC grating. (*Polymer* 36, 1995, fig. 2a; courtesy of Dr. L. V. Natarajan, Science Applications International Corporation.)

crystal, a photocured network will form that mimics the orientational order of the texture in which it is cured (Crawford et al., 1994a).

In the case of the cholesteric, a bistable color reflective display cell with gray-scale memory can be produced simply by polymerizing the network while the cholesteric cell is in the homeotropic texture (Yang et al., 1991). This has the feature of stabilizing the focal–conic texture while maintaining the stability of the planar texture. Since the material can be electrically switched back and forth between the two stable states, a useful bistability is achieved suitable for a ultrahigh-resolution display on a low cost passive matrix. The polymer network creates the domain structure necessary for wide-view angle displays and gray-scale memory (Yang and Doane, 1992). Display cells over various resolutions and with sizes up to 100 dpi on a full-page A4 size have been prototyped (Doane et al., 1992); some of these are now under commercialization by Kent Display Systems. FIGURE 9.63 shows one of their lower-resolution displays manufactured for a sign application. Under certain surface preparations, the use of a polymer network can be avoided in these reflective cells (Yang et al., 1994; Taheri et al., 1996); however, the polymer network gives excellent control of the display performance.

FIGURE 9.63. A reflective display from polymer stabilized cholesteric texture (PSCT) materials manufactured by Kent Display Systems. The display in the photo features a resolution of 10 dpi on a 3 × 5 square inch area; courtesy of Dr. Zvi Yaniv.

Even large quantities of polymers (10–20%) can be used for these display cells (West et al., 1993). Certain types of thermoplastics have been found to give the same bistable and optical performance as the low-concentration multifunctional acrylics. These dispersions are of particular interest for self-sealed systems such as might be used on large-scale coatings on plastic substrates.

Low-concentration network dispersions have been made for a variety of different display cells: ferroelectric smectic C^* cells where the desired structure of the smectic layers is stabilized by the network (Pirš et al., 1992) as opposed to the Clark–Lagerwall surface stabilization technique; modified twisted nematic (TN) cells (Bos et al., 1993); biostable supertwisted nematic (STN) cells (Li et al., 1996); normal and reverse

FIGURE 9.64. Photograph of a high-resolution, reverse-scattering mode PSCT display on a MIM active matrix substrate directly viewed from the front with front lighting conditions; obtained through a joint Kent State University of Stuttgart project.

scattering-mode light shutters for black on paper-like white reflective displays, and projection light valves (Yang et al., 1992). The latter normal and reverse scattering-mode cells described above have been recently incorporated into the MIM AM technology to examine their suitabilities for black on paper-like white displays (Pfeiffer et al., 1994) (see FIGURE 9.64).

In the authors' opinions, we have only scratched the surface of what can be done with polymer networks. For example, the Phillips group has prepared liquid crystal monomers for polymerization and has created most interesting and useful devices from them (Hikmet and Zwerver, 1992). There will be many more innovations to come as the availability of advanced monomers become available. For a recent review of this field see relevant chapters in Crawford and Žumer (1996).

9.8 CONCLUDING REMARKS

During the past decade, we have seen what started from a few simple ideas for light shutters evolve into a fascinating area of physics. These systems are rich in physical phenomena and have given physicists and chemists an opportunity to explore the effects of confinement to a level not previously possible. Surface effects, finite-size effects, and the roles of high curvature and surface elasticity are some of the material issues studied in these systems. Optical and electrooptical phenomena in PDLC systems show potential for new optical devices.

In this chapter, we hope we have given a broad picture of PDLC systems and at the same time provided sufficient depth for a reader to begin a study of these or related systems. The subject is in its infancy; we are only now learning the potential in creating new electrooptical material when liquid crystals are dispersed with polymers.

Acknowledgments

The authors wish to acknowledge support from the NSF Science and Technology Center ALCOM under grant No. DMR89-20147.

REFERENCES

Abragam, A. 1961. *The Principles of Nuclear Magnetic Resonance.* London: Oxford University Press.

Akins, R. B., and West, J. L. 1992. Effect of thickness on PDLC electrooptics. *Liquid Crystal Materials, Devices, and Applications I*, ed. P. Drzaic and U. Efron, p. 280. Proc. SPIE, vol. 1665. Bellingham, Wash.: SPIE.

Aliev, F. M., and Breganov, M. N. 1989. Electric polarization and dynamics of molecular motion of polar liquid crystals in micropores and macropores. *Zh. Eksp. Teor. Fiz.* **95**: 122 (1989. *Sov. Phys.–JETP* **68**: 70).

Aliev, F. M., and Kelly, J. R. 1994. Dynamics, structure, and phase transitions of ferroelectric liquid crystal confined in a porous matrix. *Ferroelectrics* **151**: 263.

Allender, D. W., Crawford, G. P., and Doane, J. W. 1991. Determination of the liquid-crystal surface elastic constant K_{24}. *Phys. Rev. Lett.* **67**: 1442.

Asano, S., and Yamamoto, G. 1975. Light scattering by a spheroidal particle. *Appl. Opt.* **14**: 29.

Bajc, J., Bezič, J., Žumer, S. 1995. Chiral nematic droplets with tangential anchoring and negative dielectric anisotropy in an electric field. *Phys. Rev. E* **51**: 2176.

Barberi, R., and Durand, G. 1990. Order parameter of a nematic liquid crystal on a rough surface. *Phys. Rev. A* **41**: 2207.

Barbero, G., and Durand, G. 1986. On the validity of the Rapini–Papoular surface anchoring energy form in nematic liquid crystals. *J. Physique* **47**: 2129.

Barbero, G., and Strigazzi, A. 1989. Second-order elasticity in nematics: a new anchoring source. *Liq. Cryst.* **5**: 693.

Barbero, G., and Durand, G. 1990. Selective ion adsorption and nonlocal anchoring energy in nematic liquid crystals. *J. Appl. Phys.* **67**: 2678.

Barbero, G., and Durand, G. 1991a. Order parameter spatial variation and anchoring energy for nematic liquid crystals. *J. Appl. Phys.* **69**: 6968.

Barbero, G., and Durand, G. 1991b. Anchoring energy or surface melting in nematic liquid crystals. *Mol. Cryst. Liq. Cryst.* **203**: 33.

Barbero, G., and Durand, G. 1991c. Curvative-induced quasi-melting from rough surfaces in nematic liquid crystals, *J. Physique II* **1**: 651.

Barbero, G., Gabbasova, Z., and Osipov, M. A. 1991d. Surface order transitions in nematic liquid crystals. *J. Physique II* **1**: 691.

Barbero, G., and Durand, G. 1993. Splay-bend curvature and temperature-induced surface transitions in nematic liquid crystals. *Phys. Rev. E* **48**: 1942.

Bechhoefer, J., Duvail, J.-L., Masson, L., Jérôme, B., Hornreich, R. M., and Pieranski, P. 1990. Critical behavior in anchoring transitions of nematic liquid crystals. *Phys. Rev. Lett.* **64**: 1911.

Bellini, T., Clark, N. A., Muzny, C. D., Wu, L., Garland, C. W., Schaefer, D. W., and Oliver, B. J. 1992. Phase behavior of the liquid crystal 8CB in silica aerogel. *Phys. Rev. Lett.* **69**: 788.

Berggren, E., Zannoni, C., Chiccoli, C., Pasini, P., and Semeria, F. 1994. Monte Carlo study of the effect of an applied field on the molecular organization of polymer-dispersed liquid-crystal droplets. *Phys. Rev. E* **49**: 614.

Berreman, D. W. 1972. Solid surface shape and the alignment of an adjacent nematic liquid crystal. *Phys. Rev. Lett.* **28**: 1683.

Bezič, J., and Žumer, S. 1992. Structures of the cholesteric liquid crystal droplets with parallel surface anchoring. *Liq. Cryst.* **11**: 593.

Bezič, J., and Žumer, S. 1993. Chiral nematic liquid crystals in cylindrical cavities: a classification of planar structures and models of nonsingular disclination lines. *Liq. Cryst.* **14**: 1695.

Bodnar, V. G., Lavrentovich, O. D., and Pergamenshchik, V. M. 1992. Threshold of structural hedgehog-ring transition in drops of a nematic in an alternating electric field. *Zh. Eksp. Teor. Fiz.* **101**: 111 (1992. *Sov. Phys.–JETP* **74**: 6067).

Bohren, C. F., and Hoffman, D. R. 1983. *Absorption and Scattering of Light by Small Particles.* New York: Wiley.

Born, M., and Wolf, E. 1980. *Principles of Optics.* 6th ed. Oxford: Pergamon.

Bos, P. J., Rahman, J. A., and Doane, J. W. 1993. A low-threshold-voltage polymer network TN device. *SID Dig. Tech. Papers* **24**: 877.

Bouligand, Y., and Livolant, F. 1984. The organization of cholesteric spherelites. *J. Physique* **45**: 1899.

Cagnon, M., and Durand, G. 1993. Positional anchoring of smectic liquid crystals. *Phys. Rev. Lett.* **70**: 2742.

Candau, S., LeRoy, P., and Debeauvais, F. 1973. Magnetic field effects in nematic and cholesteric droplets suspended in an isotropic liquid. *Mol. Cryst. Liq. Cryst.* **23**: 283.

Chandrasekhar, S. 1966. Surface tension of liquid crystals. *Mol. Cryst.* **2**: 71.

Chen, S. H., and Liang, B. J. 1991. Stability of a hedgehog nematic configuration in a small closed cylindrical cavity. *Appl. Phys. Lett.* **59**: 1173.

Chen, T., Dipirro, M. J., Bhattacharyya, B., and Gasparini, F. M. 1980. Some properties of Nuclepore filters. *Rev. Sci. Instrum.* **51**: 846.

Chen, W., Martinez-Miranda, L. J., Hsiung, H., and Shen, Y. R. 1989. Orientational wetting behavior of a liquid crystal homologous series. *Phys. Rev. Lett.* **62**: 1860.

Chiccoli, C., Pasini, P., Semeria, F., and Zannoni, C. 1990. A computer simulation of nematic droplets with radial boundary conditions. *Phys. Lett.* **150A**: 311.

Chiccoli, C., Pasini, P., Semeria, F., and Zannoni, C. 1993. Are Monte Carlo simulation results affected by sample shape? An investigation based on the Lebwohl–Lasher model. *Phys. Lett.* **176A**: 428.

Cladis, P. E., and Kléman, M. 1972. Nonsingular disclinations of strength $S = +1$ in nematics. *J. Physique* **33**: 591.

Cladis, P. E. 1974. Study of the nematic and smectic A phases of an N-p-cyanobenzylidene-p-n-octyloxyaniline in tubes. *Phil. Mag.* **29**: 641.

Cladis, P. E., White, A. E., and Brinkman, W. F. 1979. Nonsingular defect structures near the smectic A transition. *J. Physique* **40**: 325.

Cladis, P. E., van Saarloos, W., Finn, P. F., and Kortan, A. R. 1987. Dynamics of line defects in nematic liquid crystals. *Phys. Rev. Lett.* **58**: 222.

Clark, N. A., Bellini, T., Malzbender, R. M., Thomas, B. N., Rappaport, A. G., Muzny, C. D., Schaefer, D. W., and Hrubesh, L. 1993. X-ray scattering study of smectic ordering in a silica aerogel. *Phys. Rev. Lett.* **71**: 3505.

Coates, D. 1993a. Normal and reverse mode polymer-dispersed liquid crystal devices. *Displays* **14**: 94.

Coates, D., Greenfield, S., Goulding, M., Brown, E., and Nolan, P. 1993b. Recent developments in materials for TFT/PDLC devices. *Liquid Crystal Materials, Devices, and Applications II*, p. 2. Proc. SPIE, vol. 1911. Bellingham, Wash.: SPIE.

Cognard, J. 1982. Alignment of nematic liquid crystals and their mixtures. *Mol. Cryst. Mol. Liq. Suppl.* **1**: 1.

Cosslett, V. E. 1987. Fifty years of instrumental development of the electron microscope. *Advances in Optical and Electron Microscopy 10*, ed. R. Barer and V. E. Cosslett, p. 215. London: Academic.

Craighead, H. G., Cheng, J., and Hackwood, S. 1982. New display based on electrically induced index matching in an inhomogeneous medium. *Appl. Phys. Lett.* **40**: 22.

Crawford, G. P., Yang, D. K., Žumer, S., Finotello, D., and Doane, J. W. 1991a. Ordering and self-diffusion in the first molecular layer at a liquid crystal–polymer interface. *Phys. Rev. Lett.* **66**: 723.

Crawford, G. P., Vilfan, M., Doane, J. W., and Vilfan, I. 1991b. Escaped radial nematic configuration in submicrometer size cylindrical cavities: DNMR study. *Phys. Rev. A* **43**: 835.

Crawford, G. P., Stannarius, R., and Doane, J. W. 1991c. Surface-induced orientational order in the isotropic phase of a liquid crystal material. *Phys. Rev. A* **44**: 2558.

Crawford, G. P., Allender, D. W., Doane, J. W., Vilfan, M., and Vilfan, I. 1991d. Finite molecular anchoring in the escaped radial nematic configuration: a ^{2}H–NMR study. *Phys. Rev. A* **44**: 2570.

Crawford, G. P., Allender, D. W., and Doane, J. W. 1992a. Surface elastic and molecular anchoring properties of nematic liquid crystals. *Phys. Rev. A* **45**: 8693.

Crawford, G. P., Steele, L. M., Ondris-Crawford, R. J., Iannacchione, G. S., Yeager, C. J., Doane, J. W., and Finotello, D. 1992b. Characterization of the cylindrical cavities of Anopore and Nuclepore membranes. *J. Chem. Phys.* **96**: 7788.

Crawford, G. P., Mitcheltree, J. A., Boyko, E. P., Fritz, W., Žumer, S., and Doane, J. W. 1992c. K_{33}/K_{11} determination in nematic liquid crystals: an optical birefringence approach. *Appl. Phys. Lett.* **60**: 3226.

Crawford, G. P., and Doane, J. W. 1992d. Polymer-dispersed liquid crystals. *Condens. Matt. News* **1**: 5.

Crawford, G. P., Ondris-Crawford, R. J., Žumer, S., and Doane, J. W. 1993a. Anchoring and orientational wetting transitions of confined liquid crystals. *Phys. Rev. Lett.* **70**: 1838.

Crawford, G. P., Ondris-Crawford, R. J., Žumer, S., Keast, S. S., Neubert, M. E., and Doane, J. W. 1993b. Alignment and ordering mechanisms at a liquid crystal–solid interface. *Liq. Cryst.* **14**: 1573.

Crawford, G. P., Polak, R. D., Scharkowski, A., Chien, L.-C., Žumer, S., and Doane, J. W. 1994a. Nematic director-fields captured in polymer networks confined to spherical droplets. *J. Appl. Phys.* **75**: 1968.

Crawford, G. P., Ondris-Crawford, R. J., Doane, J. W., Kranjc, T., and Žumer, S. 1994b. Surface ordering transitions at a liquid-crystal–solid interface above the isotropic–smectic A transition. Unpublished report.

Crawford, G. P., Scharkowski, A., Fung, Y. K., Žumer, S., and Doane, J. W. 1995a. Internal surface orientational order and distribution of a polymer network in a liquid crystal. *Phys. Rev.* **52**: R1273.

Crawford, G. P., and Žumer, S. 1995b. *Int. J. Mod. Phys. B* **9**: 2469.

Crawford, G. P., and Žumer, S. eds., 1996a. *Liquid Crystals in Complex Geometries Formed by Polymer and Porous Networks.* London: Taylor and Francis.

Crawford, G. P., Ondris-Crawford, R. J., Doane, J. W., and Žumer, S. 1996b. Systematic study of orientational writing and anchoring at a liquid-crystal-surfactant interface. *Phys. Rev. E* **53**: 3647.

Crooker, P. P., and Yang, D. K. 1990. Polymer-dispersed chiral liquid crystal color display. *Appl. Phys. Lett.* **57**: 2529.

de Gennes, P. G., and Prost, J. 1993. *The Physics of Liquid Crystals.* 2d ed. Oxford: Oxford University Press.

Ding, J., and Yang, Y. 1994. Small angle scattering from axial droplets. *Mol. Cryst. Liq. Cryst.* **238**: 47.

Doane, J. W. 1979. Magnetic resonance studies of liquid crystal phase transitions. *Magnetic Resonance of Phase Transitions*, ed. F. J. Owens et al., ch. 4. New York: Academic.

Doane, J. W., Vaz, N. S., Wu, B.-G., and Žumer, S. 1986. Field controlled light scattering from nematic microdroplets. *Appl. Phys. Lett.* **48**: 269.

Doane, J. W., Žumer, S., and Golemme, A. 1988. NMR studies of smectic-induced ordering in polymer-dispersed liquid crystals. *Proceedings of the Tenth Symposium on Magnetic Resonance*, ed. R. Blinc, p. 189. Ljubljana: J. Stefan Institute.

Doane, J. W. 1990a. Polymer-dispersed liquid crystal displays. *Liquid Crystals: Applications and Uses*, vol. 1, ed. B. Bahadur, ch. 14 and references therein. Singapore: World Scientific.

Doane, J. W., West, J. L., Whitehead, J. B., and Fredley, D. S. 1990b. Wide angle view PDLC displays. *SID Dig. Tech. Papers* **21**: 224.

Doane, J. W. 1991. Polymer-dispersed liquid crystals: Boojums at work. *MRS Bull.* **16**: 22.

Doane, J. W., Yang, D. K., and Yaniv, Z. 1992. Front-lit flat-panel displays from polymer-dispersed light crystals. *Japan Display '92: Proceedings of the Twelfth International Display Research Conference [Hiroshima]*, p. 73. Playa del Rey, Calif.: Society for Information Display.

Dolinsek, J., Jarh, O., Vilfan, M., Žumer, S., Blinc, R., Doane, J. W., and Crawford, G. P. 1991. Two-dimensional deuterium nuclear magnetic resonance of a polymer-dispersed nematic liquid crystal. *J. Chem. Phys.* **95**: 2154.

Drzaic, P. S. 1986. Polymer-dispersed nematic liquid crystal for large area displays and light valves. *J. Appl. Phys.* **60**: 2142.

Drzaic, P. S. 1988. Reorientation dynamics of polymer-dispersed nematic liquid crystal films. *Liq. Cryst.* **3**: 1543.

Drzaic, P. S., and Muller, A. 1989. Droplet shape and reorientation fields in nematic droplet/polymer films. *Liq. Cryst.* **5**: 1467.

Drzaic, P. S. 1991. Comparison of electrooptic response functions of nematic droplet/polymer films. *Mol. Cryst. Liq. Cryst.* **198**: 61.

Drzaic, P. S., and Gonzales, A. M. 1992a. Phenomenological scattering models for nematic droplet/polymer films: refractive index and droplet correlation effects. *Mol. Cryst. Liq. Cryst.* **222**: 11.

Drzaic, P. S., Gonzales, A. M., Jones, P., and Montoya, W. 1992b. Dichroic-based displays from nematic dispersions. *SID Dig. Tech. Papers* **23**: 571.

Drzaic, P. S., Gonzales, A. M. 1993. Refractive index gradients and light scattering in polymer-dispersed liquid crystal films. *Appl. Phys. Lett.* **62**: 1332.

Drzaic, P. S. 1995. *Liquid Crystal Dispersions.* Singapore: World Scientific.

Dubois-Violette, E., and Parodi, O. 1969. Émulsions nématiques: effets de champ magnétiques et effets piézoélectriques. *J. Physique Coll.* **30**: C4–57.

Durand, G. 1990. Order elasticity in liquid crystals. *Physica A* **163**: 94.

Dzyaloshinskii, E. 1970. Theory of disclinations in liquid crystals. *Zh. Eksp. Teor. Fiz.* **31**: 773 (1970. *Sov. Phys.–JETP* **31**: 773).

Erdmann, J. H., Doane, J. W., Žumer, S., and Chidichimo, G. 1989. Electrooptic response of PDLC light shutters. *Proc. SPIE* **1080**: 32.

Erdmann, J. H., Žumer, S., and Doane, J. W. 1990. Configuration transition in nematic liquid crystals confined to a small spherical cavity. *Phys. Rev. Lett.* **64**: 1907.

Erdmann, J. H., Lackner, A. M., Sherman, E., and Margerum, J. D. 1991. Droplet-size effects and lighting techniques in direct-view PDLC displays. *Proc. SID* **22**: 602.

Ericksen, J. L. 1966. Inequalities in liquid crystal theory. *Phys. Fluids* **9**: 1205.

Faetti, S. 1994. The K_{13} surfacelike elastic constant: analysis of the predictions of different theoretical models. *Mol. Cryst. Liq. Cryst.* **241**: 131.

Feller, M. B., Chen, W., and Shen, Y. R. 1991. Investigation of surface-induced alignment of liquid-crystal molecules by optical second-harmonic generation. *Phys. Rev. A* **43**: 6778.

Fergason, J. L. 1985. Polymer encapsulated nematic liquid crystals for display and light control applications. *SID Dig. Tech. Papers* **16**: 68.

Flatischler, K., Komitov, L., Lagerwall, S. T., Stebler, B., and Strigazzi, A. 1991. Surface-induced alignment transition in a nematic layer in the symmetrical boundary conditions. *Mol. Cryst. Liq. Cryst.* **198**: 119.

Fowles, G. R. 1968. *Introduction to Modern Optics.* New York: Holt, Rinehart and Winston.

Frank, F. C. 1958. Liquid crystals: on the theory of liquid crystals. *Discuss. Faraday Soc.* **25**: 19.

Friedel, G. 1922. *Ann. Physique* **18**: 273.

Friedman, A. 1992. Homogenization approach to light scattering from polymer-dispersed liquid crystals. *Journal of Applied Mathematics* **52**: 46.

Fung, B. M., Heavin, S. D., Lin, Z., Jiang, X. Q., Sluss, J. J., and Batchman, T. E. 1992. Improvement in the electrooptic response of PDLC films. *Proc. SPIE* **1815**: 92.

Fung, Y., Yang, D. K., Ying, S., Chien, L.-C., Žumer, S., and Doane, J. W. 1995. Polymer networks formed in liquid crystals. *Liq. Cryst.* **19**: 797.

Furneaux, R. C., Rigby, W. R., and Davidson, A. P. 1989. The formation of controlled-porosity membranes from anodically oxidized aluminum. *Nature* **337**: 147.

Gans, R. 1925. Strahlungsdiagramme utramikroskopischer Teilchen. *Ann. Physik* **76**: 29.

Gartland, E. C., Palffy-Muhoray, P., and Varga, R. S. 1991. Numerical minimization of the Landau–de Gennes free energy defects in cylindrical capillaries. *Mol. Cryst. Liq. Cryst.* **199**: 429.

Ginter, E., Lüder, E., Kallfass, T., Huttelmaier, S., Dobler, M., Hochbolzer, D., Coates, D., Tillin, M., and Nolan,P. 1993. Optimized PDLCs for active matrix addressed light valves in projection systems. *Eurodisplay '93: Proceedings of the Thirteenth International Display Research Conference [Strasbourg]*, p. 105. Saint Brieuc: Le Club Visu.

Golemme, A., Žumer, S., Doane, J. W., and Neubert, M. E. 1988a. Deuterium NMR of polymer-dispersed liquid crystals. *Phys. Rev. A* **37**: 559.

Golemme, A., Žumer, S., Allender, D. W., and Doane, J. W. 1988b. Continuous nematic–isotropic transition in submicrometer-size liquid-crystal droplets. *Phys. Rev. Lett.* **61**: 1937.

Haas, G., Wohler, H., Fritsch, M. W., Mlynarski, D. A. 1991. Simulation of two-dimensional nematic director structures in inhomogeneous electric field. *Mol. Cryst. Liq. Cryst.* **198**: 15.

Hikmet, R. A. M. 1990. Electrically induced light scattering from anisotropic gels. *J. Appl. Phys.* **68**: 4406.

Hikmet, R. A. M. 1991a. Anisotropic gels and plasticized networks formed by liquid crystal molecules. *Liq. Cryst.* **9**: 405.

Hikmet, R. A. M., and Zwerver, B. H. 1991b. Dielectric relaxation of liquid crystal molecules in anisotropic confinements. *Liq. Cryst.* **10**: 835.

Hikmet, R. A. M. 1992a. Electrically induced light scattering from anisotropic gels with negative dielectric anisotropy. *Mol. Cryst. Liq. Cryst.* **213**: 117.

Hikmet, R. A. M., and Zwerver, B. H. 1992b. Structure of cholesteric gels and their electrically induced light scattering and colour changes. *Liq. Cryst.* **12**: 319.

Hilsum, C. 1976. Electrooptic device. *UK Patent* 1,442,360.

Hoffmann, J. 1989. Inorganic membrane filter for analytic separations. *Am. Lab.* **21**: 70.

Horn, R. G., Israelachvili, J. N., and Perez, E. 1981. Forces due to structure in a thin liquid crystal film. *J. Physique* **42**: 39.

Huang, W., and Tuthill, G. F. 1994. Structure and shape of nematic liquid-crystal microdroplets. *Phys. Rev. E* **49**: 570.

Huttelmaier, S., Lueder, E., Kallfass, T., Baumbach, J., Schkupen, K., LeGrange, J., Newman, R., Wojick, T., and Tong, J. 1996. *Society for Information Display Digest*, vol. XXVII: 622.

Iannacchione, G. S., and Finotello, D. 1992a. Calorimetric studies of phase transitions in confined liquid crystals. *Phys. Rev. Lett.* **69**: 2094.

Iannacchione, G. S., Crawford, G. P., Doane, J. W., and Finotello, D. 1992b. Orientational effects on confined 5CB. *Mol. Cryst. Liq. Cryst.* **222**: 205.

Iannacchione, G. S., Crawford, G. P., Žumer, S., Doane, J. W., and Finotello, D. 1993. Randomly constrained orientational order in porous glass. *Phys. Rev. Lett.* **71**: 2595.

Ishimaru, A. 1978. *Wave Propagation and Scattering in Random Media.* New York: Academic.

Jain, S. C., Rout, D. K., and Chandra, S. 1990. Electrooptic studies on polymer-dispersed liquid crystal films prepared by solvent-induced phase separation. *Mol. Cryst. Liq. Cryst.* **188**: 251.

Jákli, A., Kim, D. R., Chien, L.-C., and Saupe, A. 1992. Effect of a polymer network on the alignment and rotational viscosity of a nematic liquid crystal. *J. Appl. Phys.* **72**: 3161.

Jákli, A., Bata, L., Fodor-Csorba, K., Rosta, L., and Noirez, L. 1994. Structure of polymer networks dispersed in liquid crystals: small angle neutron scattering study. *Liq. Cryst.* **17**: 227.

Jérôme, B. 1991. Surface effects and anchoring in liquid crystals. *Rep. Prog. Phys.* **54**: 391.

Jones, P., Montoya, W., Garza, G., and Engler, S. 1992. Performance of NCAP for high-resolution active-matrix reflective display. *SID Dig. Tech. Papers* **23**: 762.

Kelly, J. R., and Seekola, D. 1990. Dielectric losses in a polymer-dispersed liquid crystal film. *Proc. SPIE* **1257**: 17.

Kelly, J. R., Wu, W., and Palffy-Muhoray, P. 1992. Wavelength dependence of scattering in PDLC films: droplet size effects. *Mol. Cryst. Liq. Cryst.* **223**: 251.

Kelly, J. R., and Palffy-Muhoray, P. 1994. The optical response of polymer-dispersed liquid crystals. *Mol. Cryst. Liq. Cryst.* **243**: 11.

Kilian, A., and Hess, S. 1989. Derivation and application algorithm for the numerical calculation of the local orientation of nematic liquid crystals. *Z. Naturforsch.* **44a**: 693.

Kilian, A. 1992. Disclination movement in nematic liquid crystals—3-D computer simulations and experiment. *Mol. Cryst. Liq. Cryst. Lett.* **8**: 91.

Kilian, A. 1993. Computer simulations of nematic droplets. *Liq. Cryst.* **14**: 1189.

Kim, J. Y., and Palffy-Muhoray, P. 1989. Optical power limiting and bistability in polymer-dispersed liquid crystal films with linear feedback. *J. Appl. Phys.* **66**: 362.

Kim, J. Y., and Palffy-Muhoray, P. 1991. Phase separation kinetics of a liquid crystal–polymer mixture. *Mol. Cryst. Liq. Cryst.* **203**: 93.

Kinugasa, N., Yano, Y., Nagayama, H., Yosida, H., and Watanabe, H. 1991. NCAP liquid crystals for large-area projection screens. *SID Dig. Tech. Papers* **22**: 598.

Kitzerow, H.-S., and Crooker, P. P. 1991. Polymer-dispersed cholesteric liquid crystals: challenge for research and applications. *Ferroelectrics* **122**: 183.

Kitzerow, H.-S., Xu, F., and Crooker, P. P. 1992a. Microscopic observations of axial and radial nematic droplets for a dual-frequency addressable liquid crystal. *Liq. Cryst.* **12**: 1019.

Kitzerow, H.-S., Crooker, P. P., and Heppke, G. 1992b. Chromaticity of polymer-dispersed liquid crystals. *Liq. Cryst.* **12**: 49.

Kitzerow, H.-S., Rand, J., and Crooker, P. P. 1992c. Dynamics of polymer-dispersed cholesteric liquid crystals. *J. Physique II* **2**: 227.

Kitzerow, H.-S., Molsen, H., and Heppke, G. 1992d. Linear electrooptic effects in polymer-dispersed ferroelectric liquid crystal. *Appl. Phys. Lett.* **60**: 3093.

Kitzerow, H.-S., and Crooker, P. P. 1992e. Behavior of polymer-dispersed cholesteric droplets with negative dielectric anisotropy in electric fields. *Liq. Cryst.* **11**: 561.

Kitzerow, H.-S., and Crooker, P. P. 1993. Electric field effects on the droplet structure in polymer-dispersed liquid crystals. *Liq. Cryst.* **13**: 31.

Kitzerow, H.-S. 1994. Polymer-dispersed liquid crystals from the nematic curvilinear aligned phase to ferroelectric films. *Liq. Cryst.* **16**: 1.

Kléman, M. 1983. *Points, Lines, and Walls: In Liquid Crystals, Magnetic Systems, and Various Ordered Media.* New York: Wiley.

Kléman, M. 1989. Defects in liquid crystals. *Rep. Prog. Phys.* **52**: 555.

Köhler, A. E. 1988. Systematische Ableitung nematischer Tropfenstrukturen. *Z. Phys. Chem.* **269**: 196.

Kosterlitz, J. M., and Thouless, D. J. 1973. Ordering, metastability, and phase transitions in two-dimensional systems. *J. Physique Coll.* **C6**: 1181.

Kothekar, N., Allender, D. W., and Hornreich, R. M. 1994. Surface layer phase transitions in nematic liquid crystals. *Phys. Rev. E* **49**: 2150.

Koval'chuk, A. V., Kurik, M. V., Lavrentovich, O. D., and Sergan, V. V. 1988. Structural transformations in nematic droplets located in an external magnetic field. *Zh. Eksp. Teor. Fiz.* **94**: 350 (1988. *Sov. Phys.–JETP* **67**: 1065).

Kralj, S., Vilfan, M., and Žumer, S. 1989. Proton NMR lineshapes in nematic microdroplets. *Liq. Cryst.* **5**: 1489.

Kralj, S., Žumer, S., and Allender, D. W. 1991. Nematic–isotropic phase transition in a liquid crystal droplet. *Phys. Rev. A* **43**: 2943.

Kralj, S., and Žumer, S. 1992. Fréedericksz transitions in supra-nematic droplets. *Phys. Rev. A* **45**: 2461.

Kralj, S., Lahajnar, G., Zidansek, A., Vrbancic-Kopac, N., Vilfan, M., Blinc, R., and Kosec, M. 1993. Deuterium NMR of a 5CB liquid crystal confined in silica aerogel matrix. *Phys. Rev. E.* **48**: 340.

Kralj, S., and Žumer, S. 1995. Saddle-splay elasticity of nematic structures confined to a cylindrical cavity. *Phys. Rev. E* **51**: 366.

Kuniga, M., Yirai, Y., Ooi, Y., Niiyama, S., Asahawa, T., Masumo, K., Kumai, H., Yuki, M., and Gunjima, T. 1990. A full-color projection TV using LC/polymer composite light valves. *SID Dig. Tech. Papers* **21**: 227.

Kurik, M. V., and Lavrentovich, O. D. 1982. Negative-positive monopole transitions in cholesteric liquid crystals. *Zh. Eksp. Teor. Fiz.* **35**: 362 (1982. *JETP Lett.* **35**: 444).

Kurik, M. V., and Lavrentovich, O. D. 1988. Defects in liquid crystals: homotopy theory and experimental studies. *Usp. Fiz. Nauk* **154**: 381 (1988. *Sov. Phys. Usp.* **31**: 196).

Kuzma, M., and Labes, M. M. 1983. Liquid crystals in cylindrical pores: effect on transition temperatures and singularities. *Mol. Cryst. Liq. Cryst.* **100**: 103.

Landau, L. D., and Lifshitz, E. M. 1980. *Statistical Physics.* Part 1. 3d ed. London: Pergamon.

Lavrentovich, O. D., and Terentjev, E. M. 1986. Phase transition altering the symmetry of topological point defects (hedgehogs) in a nematic liquid crystal. *Zh. Eksp. Teor. Fiz.* **91**: 2084 (1986. *Sov. Phys.–JETP* **64**: 1237).

Lavrentovich, O. D. 1988. Flexoelectricity of droplets of a nematic liquid crystal. *Pis'ma Zh. Tekh. Fiz.* **14**: 166 (1988. *Sov. Tech. Phys. Lett.* **14**: 73).

Lavrentovich, O. D., and Sergan, V. V. 1989. The phase transition connected with the origin of the twisted structure in bipolar drops of nematic liquid crystals. *Ukr. Fiz. Zh.* **34**: 1034.

Lavrentovich, O. D. 1991. Hybrid aligned nematic films with horizontally degenerated boundary conditions and saddle-splay elastic term. *Phys. Scr. T* **39**: 394.

Lavrentovich, O. D., and Pergamenshchik, V. M. 1994. Stripe domain phase of a thin nematic film and the K_{13} divergence term. *Phys. Rev. Lett.* **73**: 979.

Lebwohl, P. A., and Lasher, G. 1972. Nematic liquid crystal order: a Monte Carlo calculation. *Phys. Rev. A* **6**: 426.

Lee, K. P., Suh, S.-W., and Lee, S.-D. 1994. Fast linear electro-optical switching properties of polymer-dispersed ferroelectric liquid crystals. *Appl. Phys. Lett.* **64**: 718.

Lehmann, O. 1904. Die Hauptstatze der Lehre von den Flüssigen Kristallen. *Flüssige Kristalle*, by O. Lehmann. Leipzig: Engelmann.

Levitz, P., Ehret, G., Sinha, S. K., and Drake, J. M. 1991. Porous vycor glass: the microstructure as probed by EM, direct energy transfer, small angle scattering, and molecular absorption. *J. Chem. Phys.* **95**: 6151.

Li, J., Hoke, C., Fredley, D. S., Bos, P. J. 1996. *Society for Information Display Digest*, vol. XXVII: 265.

Li, L., Yuan, H. J., and Palffy-Muhoray, P. 1991. Second-harmonic generation by polymer-dispersed liquid crystal films. *Mol. Cryst. Liq. Cryst.* **198**: 239.

Li, Z., Kelly, J. R., Palffy-Muhoray, P., and Rosenblatt, C. 1992. Comparison of magnetic and electric field-induced switching in polymer-dispersed liquid crystal films. *Appl. Phys. Lett.* **60**: 3132.

Liang, B. J., and Chen, S. H. 1991. Discontinuous radial to axial transformation of director configuration in a cylindrical nematic cavity. *Japan. J. Appl. Phys. Lett.* **30**: 1955.

Liang, B. J., and Chen, S. H. 1992. Director-configuration diagram for a closed-cylinder nematic liquid crystal. *J. Appl. Phys.* **71**: 3189.

Lin, H., Palffy-Muhoray, P., and Lee, M. A. 1991. Liquid crystalline cores for optical fibers. *Mol. Cryst. Liq. Cryst.* **204**: 189.

L'vov, Y., Hornreich, R. M., and Allender, D. W. 1993. Theory of symmetry breaking phase transition in a surface layer of nematic liquid crystal. *Phys. Rev. E* **48**: 1115.

Madhusudana, N. V. and Sumathy, K. R. 1983. Nematic droplets with a new structure. *Mol. Cryst. Liq. Cryst.* **92**: 179.

Maier, W., and Saupe, A. 1958. Eine einfache molekulare Theorie des nematische kristallinflüssigen Zustands. *Z. Naturforsch.* **13a**: 564.

Mang, J. T., Sakamoto, K., and Kumar, S. 1992. Smectic layer orientation in confined geometries. *Mol. Cryst. Liq. Cryst.* **223**: 133.

Margerum, J. D., Lackner, A. M., Ramos, E., Lim, K. C., and Smith, W. H. 1989. Effects of off-state alignment in polymer-dispersed liquid crystals. *Liq. Cryst.* **5**: 1477.

Mariani, P., Samori, B., Angeloni, A. S., and Ferruti, P. 1986. Polymerization of bisacrylic monomers within a liquid crystalline smectic B solvent. *Liq. Cryst.* **1**: 327.

Maritan, A., Cieplak, M., Bellini, T., and Banavar, J. R. 1994. Nematic–isotropic transition in porous media. *Phys. Rev. Lett.* **72**: 4113.

McCargar, J. W., Ondris-Crawford, R. J., and West, J. L. 1992. Polymer-dispersed liquid crystal infrared light shutter. *J. Elec. Imaging* **1**: 22.

McFarland, C. A., Koenig, J. L., and West, J. L. 1992. Analysis of polymer-dispersed liquid crystals by infrared spectroscopy. *Appl. Spec.* **47**: 321.

Meeten, G. H. 1982. Small-angle light scattering by spherulites in the anomalous diffraction approximation. *Optica Acta* **29**: 759.

Melzer, D., and Nabarro, F. R. N. 1977. Optical studies of a nematic liquid crystal with circumferential surface orientation in a capillary. *Phil. Mag.* **35**: 901.

Mermin, N. D. 1979. The topological theory of defects in ordered media. *Rev. Mod. Phys.* **51C**: 591.

Meyer, R. B. 1969. Piezoelectric effects in liquid crystals. *Phys. Rev. Lett.* **22**: 918.

Meyer, R. B. 1973. On the existence of even-indexed disclinations in nematic liquid crystals. *Phil. Mag.* **27**: 405.

Mie, G. 1908. *Ann. Physique* **25**: 377.

Molsen, H., Kitzerow, H.-S., and Heppke, G. 1992. Antiferroelectric switching in polymer-dispersed liquid crystal. *Japan. J. Appl. Phys. Lett.* **31**: 1083.

Molsen, H., Bardon, R., and Kitzerow, H.-S. 1993. Electro-optics of polymer-dispersed ferroelectric liquid crystals. *Eurodisplay '93: Proceedings of the Thirteenth International Display Research Conference [Strasbourg]*, p. 113. Saint Brieuc: Le Club Visu.

Molsen, H., and Kitzerow, H.-S. 1994. Bistability in polymer-dispersed ferroelectric liquid crystals. *J. Appl. Phys.* **75**: 710.

Monkade, M., Boix, M., and Durand, G. 1988. Order electricity and oblique nematic orientation on rough solid surfaces. *Europhys. Lett.* **5**: 697.

Montgomery, G. P., and Vaz, N. A. 1987. Contrast ratio of polymer-dispersed liquid crystal films. *Appl. Opt.* **26**: 738.

Montgomery, G. P. 1988. Angle-dependent scattering of polarized light by polymer-dispersed liquid crystal films. *J. Opt. Soc. Am. B* **5**: 774.

Montgomery, G. P., and Vaz, N. A. 1989. Light scattering analysis of the temperature dependent transmittance of a polymer-dispersed liquid crystal film in its isotropic phase. *Phys. Rev. A* **40**: 6580.

Montgomery, G. P., West, J. L., and Tamura-Lis, W. 1991. Light scattering from polymer-dispersed liquid crystal films: droplet size effects. *J. Appl. Phys.* **69**: 1605.

Moses, T., and Shen, Y. R. 1991. Pretransitional surface ordering and disordering of a liquid crystal. *Phys. Rev. Lett.* **67**: 2033.

Muzny, C. D., and Clark, N. A. 1992. Direct observation of the Brownian motion of a liquid-crystal topological defect. *Phys. Rev. Lett.* **68**: 804.

Nazarenko, V. G., and Lavrentovich, O. D. 1994. Anchoring transition in a nematic liquid crystal compound of centrosymmetric molecules. *Phys. Rev. E* **49**: R990.

Nehring, J., and Saupe, A. 1971. On the elastic theory of uniaxial liquid crystals. *J. Chem. Phys.* **54**: 337.

Niiyama, S., Hirai, Y., Ooi, Y., Kunigita, M., Kumai, H., Wakabayashi, T., Iida, S., and Gunjima, T. 1993. Hysteresis and dynamic response effects on the image quality in an LCPC projection display. *SID Dig. Tech. Papers* **24**: 869.

Nobili, M., and Durand, G. 1992. Disorientation-induced disordering at a nematic-liquid-crystal–solid interface. *Phys. Rev. A* **46**: R6174.

Nolan, P., Tellin, M., Coates, D., Ginter, E., Lüder, E., and Kallfass, T. 1993. Reflective mode PDLC displays: paper-white display. *Eurodisplay '93: Proceedings of the Thirteenth International Display Research Conference [Strasbourg]*, p. 397. Saint Brieuc: Le Club Visu.

Nomura, H., Suzuki, S., and Atarashi, Y. 1990a. Electrooptical properties of films consisting of nematic liquid crystals and connected polymer spheres. *J. Appl. Phys.* **68**: 2922.

Nomura, H., Susuki, S., and Atarashi, Y. 1990b. Electrooptic properties of polymer films containing nematic liquid crystal microdroplets. *Japan. J. Appl. Phys.* **29**: 522.

Ocko, B. M., Braslau, A., Pershan, P. S., Als-Nielsen, J., and Deutsch, M. 1986. Quantized layer growth at liquid-crystal surfaces. *Phys. Rev. Lett.* **57**: 94.

Ondris-Crawford, R. J., Boyko, E. P., Wagner, B. G., Erdmann, J. H., Žumer, S., and Doane, J. W. 1991. Microscope textures of nematic droplets in polymer-dispersed liquid crystals. *J. Appl. Phys.* **69**: 6380.

Ondris-Crawford, R. J., Crawford, G. P., Doane, J. W., Žumer, S., Vilfan, M., and Vilfan, I. 1992. Surface molecular anchoring in microconfined liquid crystals near the nematic–smectic A transition. *Phys. Rev. E* **48**: 1998.

Ondris-Crawford, R. J., Crawford, G. P., Žumer, S., and Doane, J. W. 1993. Curvature-induced configuration transition in confined nematic liquid crystals. *Phys. Rev. Lett.* **70**: 194.

Onsager, L. 1949. The effects of shape on the interaction of colloidal particles. *Ann. NY Acad. Sci.* **51**: 627.

Orsay Liquid Crystal Group. 1969. Quasi-elastic Rayleigh scattering in nematic liquid crystals. *Phys. Rev. Lett.* **22**: 1361.

Osipov, M. A., and Sluckin, T. J. 1993. Molecular theory of order electricity. *J. Physique II* **3**: 793.

Palffy-Muhoray, P., and West, J. L. 1988. Nonlinear optical response of polymer-dispersed liquid crystals. *Proc. SPIE* **927**: 226.

Palffy-Muhoray, P., Lee, M. A., and West, J. L. 1990. Optical field-induced scattering in polymer-dispersed liquid crystal films. *Mol. Cryst. Liq. Cryst.* **179**: 445.

Pargellis, A. N., Turok, N., and Yurke, B. 1991. Monopole–antimonopole annihilation in a nematic liquid crystal. *Phys. Rev. Lett.* **67**: 1570.

Parod, G. 1982. General theory in small angle x-ray scattering. *Small Angle X-Ray Scattering*, ed. O. Glatter and O. Krotky, p. 17. New York: Academic.

Patel, J. S., and Yokoyama, H. 1993. Continuous anchoring transition in liquid crystals. *Nature* **362**: 525.

Patel, P., Chu, D., West, J. L., and Kumar, S. 1994. Bistable switching in polymer-dispersed ferroelectric smectic C^* displays. *SID Dig. Tech. Papers* **25**: 845.

Pawlowska, Z., Sluckin, T. J., and Kventsel, G. F. 1988. Systematics of wetting and layering phenomena in smectic materials. *Phys. Rev. A* **38**: 5342.

Penzenstadler, E., and Trebin, H.-R. 1989. Fine structure of point defects and soliton decay in nematic liquid crystals. *J. Physique* **50**: 1027.

Pergamenshchik, V. M. 1993. Phenomenological approach to the problem of the K_{13} surfacelike elastic term in the free energy of a nematic liquid crystal. *Phys. Rev. E* **48**: 1254.

Pergamenshchik, V. M. 1994. *Phys. Rev. E* **49**: 934(Errata).

Pfeiffer, M., Sun, Y., Yang, D. K., Doane, J. W., Sautter, W., Hochholzer, V., Ginter, E., and Lüder, E. 1994. Design of PSCT materials for MIM addressing. *SID Dig. Tech. Papers* **25**: 837.

Pirš, J., Olenik, M., Marin, B., Žumer, S., and Doane, J. W. 1990. Color modulator based on polymer-dispersed liquid crystal shutters. *J. Appl. Phys.* **68**: 3826.

Pirš, J., Blinc, R., Marin, B., Mušerič, S., Pirš, S., Žumar, S., and Doane, J. W. 1992. Volume stabilized ferroelectric liquid crystal display. *14th Int. Liquid Crystal Conf.* Pisa, Italy. Book of Abstracts, p. 278.

Pirš, J., Blinc, R., Marin, B., Pirš, S., and Doane, J. W. 1995. Polymer network volume stabilized ferroelectric liquid crystal display. *Mol. Cryst. Liq. Cryst.* **264**: 155.

Pismen, L. M., and Rubinstein, B. Y. 1992. Motion of interacting point defects in nematics. *Phys. Rev. Lett.* **69**: 96.

Plaza, A., and Stein, R. S. 1959. The scattering of light from polyethylene films at low angles. *J. Polym. Sci.* **40**: 267.

Polak, R. D., Crawford, G. P., Kostival, B. C., Doane, J. W., and Žumer, S. 1994. Optical determination of the saddle-splay elastic constant K_{24} in nematic liquid crystals. *Phys. Rev. E* **49**: R978.

Porte, G. 1976. Tilted alignment of MBBA induced by short-chain surfactants. *J. Physique* **37**: 1245.

Press, M. J., and Arrott, A. S. 1974. Theory and experiment on configurations with cylindrical symmetry in liquid crystal droplets. *Phys. Rev. Lett.* **33**: 403.

Press, W. H., Flammery, B. P., Tenkolsky, S. A., and Vetterling, W. T. 1986. *Numerical Recipes: The Art of Scientific Computing*. Cambridge, Cambridge University Press.

Quigley, J. R., and Benton, W. J. 1977. Some optical properties of dispersed cholesteric liquid crystals. *Mol. Cryst. Liq. Cryst.* **42**: 1053.

Rapini, A., and Papoular, M. 1969. Distorsion d'une lamelle nématique sous champ magnétique conditions d'ancrage aux parois. *J. Physique Coll.* **30**: C4–54.

Rayleigh, Lord. 1914. On the diffraction of light by spheres of small refractive index. *Proc. R. Soc. London A* **90**: 219.

Reamey, R., Malloy, K., and Montoya, W. 1992. Voltage-holding in nematic droplet–polymer films: contributions of matrix polymer, liquid crystal, and morphology. *SID Dig. Tech. Papers* **23**: 769.

Robinson, C., Ward, W. C., and Beevers, R. B. 1958. Liquid crystalline structures in polypeptide solutions. Part 2. *Discuss. Faraday Soc.* **25**: 29.

Rosenblatt, C. 1984. Liquid-crystal Fréedericksz transitions and surface-induced smectic ordering. *Phys. Rev. Lett.* **53**: 791.

Sakuhara, T., Tomita, E., Nakahara, H., and Fukuda, K. 1992. Orientational control of liquid crystallinity by Langmuir–Blodgett films. *Thin Solid Films* **109**: 269.

Sarid, D. 1991. *Scanning Force Microscopy: With Applications to Electric, Magnetic, and Atomic Forces*. New York: Oxford University Press.

Saupe, A. 1973. Disclinations and properties of the director field in nematic and cholesteric liquid crystals. *Mol. Cryst. Liq. Cryst.* **21**: 211.

Schaefer, D. W., and Keefer, K. D. 1986. Structure of random porous materials: silica aerogels. *Phys. Rev. Lett.* **56**: 2199.

Schaefer, D. W. 1994. Structure of mesoporous aerogels. *MRS Bull.* **19**: 49.

Scharkowski, A., Crawford, G. P., Žumer, S., and Doane, J. W. 1993. A method for the determination of the elastic constant ratio K_{33}/K_{11} in nematic liquid crystals. *J. Appl. Phys.* **73**: 7280.

Schopohl, N., and Sluckin, T. J. 1987. Defect core structure in nematic liquid crystal. *Phys. Rev. Lett.* **59**: 2582.

Schopohl, N., and Sluckin, T. J. 1988. Hedgehog structure in nematic and magnetic systems. *J. Physique* **49**: 1097.

Sheng, P. 1976. Phase transition in surface-aligned nematic films. *Phys. Rev. Lett.* **37**: 1059.

Sheng, P. 1982. Boundary-layer phase transition in nematic liquid crystals. *Phys. Rev. A* **26**: 1610.

Sheng, P., Li, B.-Z., Zhou, M., Moses, T., and Shen, Y. R. 1992. Disordered-surface-layer transition in nematic liquid crystals. *Phys. Rev. A* **46**: 946.

Sherman, R. D. 1989. Refraction by a spherical bubble. *Phys. Rev. A* **40**: 1591.

Shimada, E., and Uchida, T. 1992. Control of polymer orientation in polymer-dispersed liquid crystal. *Japan. J. Appl. Phys. Lett.* **31**: 352.

Sluckin, T. J., and Poniewierski, A. 1985. Novel surface phase transition in nematic liquid crystals: wetting and the Kosterlitz–Thouless transition. *Phys. Rev. Lett.* **55**: 2907.

Sluckin, T. J., and Poniewierski, A. 1986. Orientational wetting transitions and related phenomena in nematics. *Fluid Interfacial Phenomena*, ed. C. A. Croxton, p. 215. New York: Wiley.

Smith, G. W., Vaz, N. A., and van Steenkiste, T. H. 1989. The interfacial free energy of nematogen droplets in an isotropic matrix: determination of its temperature dependence from coalescence kinetics. *Mol. Cryst. Liq. Cryst.* **174**: 49.

Smith, G. W. 1991. Cure parameters and phase behavior of an ultraviolet-cured polymer-dispersed liquid crystal. *Mol. Cryst. Liq. Cryst.* **196**: 89.

Smith, G. W. 1993. Mixing and phase separation in liquid crystal matrix systems: determination of the excess-specific heat of mixing. *Phys. Rev. Lett.* **70**: 198.

Somoza, A. M., Mederso, L., and Sullivan, D. E. 1994. Smectic layering transition at the surface of an isotropic liquid. *Phys. Rev. Lett.* **72**: 3674.

Sparavigna, A., Komitov, L., Lavrentovich, O. D., and Strigazzi, A. 1992. Saddle-splay and periodic instability in a hybrid-aligned nematic layer subjected to a normal magnetic field. *J. Physique II* **2**: 1881.

Sparavigna, A., Lavrentovich, O. D., and Strigazzi, A. 1994. Periodic stripe domains and hybrid-alignment regime in nematic liquid crystals: threshold analysis. *Phys. Rev. E* **49**: 1344.

Spencer, M., Fuller, W., Wilkins, M. H. F., and Brown, G. L. 1962. Determination of the helical configuration of ribonucleic acid molecules by x-ray diffraction study of crystalline amino-acid-transfer ribonucleic acid. *Nature* **194**: 1014.

Stannarius, R., Crawford, G. P., Chien, L.-C., and Doane, J. W. 1991. Nematic director orientation in a liquid crystal dispersed polymer: a DNMR approach. *J. Appl. Phys.* **70**: 135.

Stein, R. S., Srinivasarao, M., and Kita, S. 1992. Light scattering studies of liquid crystal dispersions. *Liquid Crystal Materials, Devices, and Applications I*, ed. P. Drzaic and U. Efron, p. 41. Proc. SPIE, vol. 1665. Bellingham, Wash.: SPIE.

Stratton, J. A. 1941. *Electromagnetic Theory*. New York: McGraw-Hill.

Sutherland, R. L., Natarajan, L. V., Tondiglia, V. P., and Bunning, T. J. 1993. Bragg gratings in an acrylate polymer consisting of periodic polymer-dispersed liquid-crystal planes. *Chem. Mater.* **5**: 1533.

Taheri, B., Doane, J. W., Davis, D., and St. John, D. 1996. *Society for Information Display Digest*, vol. XXVII.

Takizawa, K., Kikuchi, H., and Fujikake, H. 1991. Polymer-dispersed liquid-crystal light values for projection displays. *SID Dig. Tech. Papers* **22**: 250.

Tanaka, K., Katoh, K., Tsuru, S., and Sakai, T. 1994. Holographically formed liquid crystal/polymer device for reflective color display. *J. SID* **2**: 37.

Tanaka, K., Kato, K., Date, M., and Sahai, S. 1995. *Society for Information Display Digest*, vol. XXVI: 261.

Teixeira, P. I. C., and Telo da Gama, M. M. 1991. Density-functional theory for the interfacial properties of a dipolar fluid. *J. Phys. Condens. Matt.* **3**: 111.

Teixeira, P. I. C., and Sluckin, T. J. 1992a. Microscopic theory of anchoring transitions at the surfaces of pure liquid crystals and their mixtures. I. The Fowler approximation. *J. Chem. Phys.* **97**: 1498.

Teixeira, P. I. C., and Sluckin, T. J. 1992b. Microscopic theory of anchoring transitions at the surfaces of pure liquid crystals and their mixtures. II. The effect of surface adsorption. *J. Chem. Phys.* **97**: 1510.

Teixeira, P. I. C., Sluckin, T. J., and Sullivan, D. E. 1993. Landau–de Gennes theory of anchoring transitions at a nematic liquid crystal–substrate interface. *Liq. Cryst.* **14**: 1243.

Tjipto-Margo, B., and Sullivan, D. E. 1988. Molecular interactions and interface properties of nematic liquid crystals. *J. Chem. Phys.* **88**: 6620.

Tomita, A. 1993. Status of projection-type polymer-dispersed LCDs. *SID Dig. Tech. Papers* **24**: 865.

Toulouse, J., and Kléman, M. 1976. Principles of a classification of defects in ordered media. *J. Physique Lett.* **37**: 149.

van de Hulst, H. C. 1957. *Light Scattering by Small Particles.* New York: Dover.

van de Hulst, H. C. 1980. *Multiple Scattering.* New York: Academic.

Vaz, N. A., Smith, G. W., and Montgomery, G. P. 1987. A light control film composed of liquid crystal droplets dispersed in a UV curable polymer. *Mol. Cryst. Liq. Cryst.* **146**: 1.

Vaz, N. A., and Montgomery, G. P. 1989. Dual frequency addressing of polymer-dispersed liquid crystal films. *J. Appl. Phys.* **65**: 5043.

Vaz, N. A., Smith, G. W., and Montgomery, G. P. 1991. Polymer-dispersed liquid crystal films formed by electron beam cure. *Mol. Cryst. Liq. Cryst.* **197**: 83.

Vertogen, G., and de Jeu, W. H. 1988. *Thermotropic Liquid Crystals: Fundamentals.* Berlin: Springer.

Vilfan, I., Vilfan, M., and Žumer, S. 1989. Orientational order in bipolar microdroplets close to a phase transition. *Phys. Rev. A* **40**: 4724.

Vilfan, M., Rutar, V., Žumer, S., Lahajnar, G., Blinc, R., Doane, J. W., and Golemme, A. 1988. Proton spin-lattice relaxation in nematic microdroplets. *J. Chem. Phys.* **89**: 597.

Vilfan, M., Lahajnar, G., Zupančič, I., Žumer, S., Blinc, R., Crawford, G. P., and Doane, J. W. 1995. Dynamics of a nematic liquid crystal constrained by a polymer network. *J. Chem. Phys.* **103**: 8726.

Volovik, E. E., and Mineyev, V. P. 1977. Investigation of singularities in superfluid ^{3}He in liquid crystals by the homotopic typology methods. *Zh. Eksp. Teor. Fiz.* **72**: 2256 (1977. *Sov. Phys.–JETP* **49**: 1186).

Volovik, G. E., and Lavrentovich, O. D. 1983. Topological dynamics of defects: Boojums in nematic droplets. *Zh. Eksp. Teor. Fiz.* **85**: 1997 (1983. *Sov. Phys.–JETP* **58**: 1159).

Vrbančič, N., Vilfan, M., Blinc, R., Dolinšek, J., Crawford, G. P., and Doane, J. W. 1993. Deuteron spin relaxation and molecular dynamics of a nematic liquid crystal (5CB) in cylindrical microcavities. *J. Chem. Phys.* **98**: 3540.

West, J. L. 1988. Phase separation of liquid crystals in polymers. *Mol. Cryst. Liq. Cryst.* **157**: 427.

West, J. L., and Ondris-Crawford, R. J. 1991. Characterization of polymer-dispersed liquid crystal shutters by ultraviolet/visible and infrared absorption spectroscopy. *J. Appl. Phys.* **70**: 3785.

West, J. L., Kelly, J. R., Jewell, K., and Ji, Y. 1992a. Effect of polymer matrix glass transition temperature on polymer-dispersed liquid crystal electrooptics. *Appl. Phys. Lett.* **60**: 3238.

West, J. L., Fredley, D. S., and Carrel, J. C. 1992b. Haze-free polymer-dispersed liquid crystals utilizing linear polarizers. *Appl. Phys. Lett.* **61**: 2004.

West, J. L., Akins, R. B., Francl, J., and Doane, J. W. 1993. Cholesteric/polymer–dispersed light shutters. *Appl. Phys. Lett.* **63**: 1471.

West, J. L. 1994. Liquid crystal dispersions. *Technological Applications of Dispersions*, ed. R. B. McKay, ch. 10. New York: Dekker.

Whitehead, J. B., Žumer, S., and Doane, J. W. 1993. Light scattering from a dispersion of aligned nematic droplets. *J. Appl. Phys.* **73**: 1057.

Williams, C. E., Pieranski, P., and Cladis, P. E. 1972. Nonsingular $S = +1$ screw disclination lines in nematics. *Phys. Rev. Lett.* **29**: 90.

Williams, R. D. 1986. Two transitions in tangentially anchored nematic droplets. *J. Phys. A: Math. Gen.* **19**: 3211.

Wu, B.-G., West, J. L., and Doane, J. W. 1987. Angular discrimination of light transmission through polymer-dispersed liquid crystal films. *J. Appl. Phys.* **62**: 3925.

Wu, B.-G., Erdmann, J. H., and Doane, J. W. 1989. Response times and voltages for PDLC light shutters. *Liq. Cryst.* **5**: 1453.

Wu, X., Goldburg, W. I., Liu, M. X., and Xue, J. Z. 1992. Slow dynamics of isotropic-nematic phase transition in silica gels. *Phys. Rev. Lett.* **69**: 470.

Xu, F., Kitzerow, H.-S., and Crooker, P. P. 1992. Electric field effects on nematic droplets with negative dielectric anisotropy. *Phys. Rev. A* **46**: 6535.

Xu, F., Kitzerow, H.-S., and Crooker, P. P. 1994. Director configurations of nematic liquid crystal droplets: negative dielectric anisotropy and parallel surface anchoring. *Phys. Rev. E* **49**: 3061.

Yamagishi, F. G., Miller, L. J., and van Ast, C. I. 1989. Morphological control in polymer-dispersed liquid crystal film matrices. *Proc. SPIE* **1080**: 24.

Yang, D. K., Chien, L.-C., and Doane, J. W. 1991a. Cholesteric liquid crystal/polymer dispersion bistable at zero field. *Proceedings of the Eleventh International Display Research Conference [San Diego, CA]*, p. 49. Playa del Rey, Calif.: Society for Information Display.

Yang, D. K., and Crooker, P. P. 1991b. Field-induced textures of polymer-dispersed chiral liquid crystal microdroplets. *Liq. Cryst.* **9**: 245.

Yang, D. K., and Doane, J. W. 1992a. Cholesteric liquid-crystal polymer-gel dispersions: reflective display application. *Sid Dig. Tech. Papers* **23**: 759.

Yang, D. K., Chien, L.-C., and Doane, J. W. 1992b. Cholesteric liquid crystal/polymer dispersions for haze-free light shutters. *Appl. Phys. Lett.* **60**: 3102.

Yang, D. K., West, J. L., Chien, L.-C., and Doane, J. W. 1994. Control of reflectivity and bistability in displays using cholesteric liquid crystals. *J. Appl. Phys.* **76**: 1331.

Yaniv, Z., Doane, J. W., West, J. L., and Tamura-Lis, W. 1989. Active matrix polymer-dispersed liquid crystal display. *Japan Display '89: Proceedings of the Ninth International Display Research Conference [Kyoto]*, p. 572. Playa del Rey, Calif.: Society for Information Display.

Yoshida, H., Nakamura, K., Tsuda, H., Ohashi, M., Tomita, I., and Okabe, M. 1993. A direct view polymer-dispersed LCD with crossed nichols and a uniaxial film. *Eurodisplay '93: Proceedings of the Thirteenth International Display Research Conference [Strasbourg]*, p. 121. Saint Brieuc: Le Club Visu.

Young, W., Bahadur, B., Lewis, D., and Wan, K. 1994. Performance of polymer-dispersed suspended particle cells. *SID Dig. Tech. Papers* **25**: 611.

Yurke, B., Pargellis, A. N., Chuang, I., and Turok, N. 1992. Coarsening dynamics in nematic liquid crystals. *Physica B* **1–4**: 56.

Zhong, Z. Z., Schulele, D. E., Gordon, W. L., Adamic, K. J., and Akins, R. B. 1992. Dielectric properties of a PMMA/E7 polymer-dispersed liquid crystal. *J. Polym. Sci. Polym. Lett. Ed.* **30**: 1443.

Žumer, S., and Doane, J. W. 1986. Light scattering from a small nematic droplet. *Phys. Rev. A* **34**: 3373.

Žumer, S. 1988a. Light scattering from nematic droplets: anomalous diffraction approach. *Phys. Rev. A* **37**: 4006.

Žumer, S., Vilfan, M., and Vilfan, I. 1988b. Confinement of a liquid crystal to small droplets and its effect on nuclear magnetic resonance. *Liq. Cryst.* **3**: 947.

Žumer, S., Golemme, A., and Doane, J. W. 1989. Light extinction in a dispersion of small nematic droplets. *J. Opt. Soc. Am. A* **6**: 405.

Žumer, S., and Kralj, S. 1992a. Influence of K_{24} on the structure of nematic liquid crystal droplets. *Liq. Cryst.* **12**: 613.

Žumer, S., Kralj, S., and Bezič, J. 1992b. Freedericksz transitions in nematic and cholesteric liquid crystal droplets: determination of K_{24} elastic constant. *Mol. Cryst. Liq. Cryst.* **212**: 163.

Zyryanov, V. Y., Smorgon, S. L., and Shabanov, V. F. 1993. Electrooptics of polymer-dispersed ferroelectrics. *Ferroelectrics* **143**: 271.

Zyryanov, V. Y., Shabanov, V. F., and Smorgon, S. L. 1994. Polymer-dispersed ferroelectric liquid crystal light valves. *SID Dig. Tech. Papers* **25**: 605.

CHAPTER 10

Active Matrix Liquid Crystal Displays

SHUNSUKE KOBAYASHI,
HIROO HORI AND
YASUHARU TANAKA

10.1 INTRODUCTION

Liquid crystal displays (LCDs) have been widely utilized as both direct-view flat-panel displays and projection displays, not only for information equipment such as computers, televisions, and other audio-visual and office equipment, but also for avionics, automobiles, measuring instruments and so forth, by taking advantage of their low-voltage operation, low power consumption, multicolor capabilities, flatness, and lightness in weight.

Historically, early demonstrations of liquid crystal (LC) applications were thermometers using cholesteric LC and alphanumeric displays in the dynamic scattering (DS) mode LCDs developed in the middle and the later 1960s by Westinghouse and RCA in the United States [1,2]. Soon after, the advent of twisted nematic (TN) LCDs [3] began a new era of personal information equipment, including digital watches and calculators using LCDs [4]. The production of these TN-LCDs has not been done without the availability of necessary peripheral components and materials such as CMOS drivers, batteries, transparent electrodes, LC molecular alignment layers, spacers, and even LC materials that exhibit the LC phase at room temperature.

The number of scanning (strobing) row electrodes (N_{ROW}) in the early direct multiplexed (D-MPX or MPX) matrix TN-LCDs was only three in the beginning in the early 1970s, and this continued to increase up to several dozen in 1984; using these LCDs, more advanced instruments, such as portable TVs and pocket computers, were produced in the middle 1980s [4]. However, this increase of the N_{ROW} in the D-MPX TN-LCDs sacrificed their legibility; heavy homogeneous crosstalk appeared, which results in a loss of transparency or whiteness in the nonselect, or off, state. Furthermore, the response speed of TN-LCDs was not high enough. In the middle 1980s, much research and development activity was aimed at resolving the above mentioned difficulties, by exploring the following new technologies that had already been reported in the literature: active matrix (AM) TN-LCDs using thin film transistors (TFTs) or diodes (TFDs) [5]; super-twisted nematic (STN) LCDs [6]; ferroelectric (F) LCDs [7]; superhomeotropic color (SHC) LCDs [8]; and so forth. The LCD modes, which are reported in the literature, are summarized in TABLE 10.1 in appendix A.

Among these, it has been shown that the AM-TN-LCD is one of the most promising technologies for color image quality and response speed; this has been demonstrated through much research by many laboratories. However, large-scale development and production of the AM-TN-LCDs needs a fairly large amount of human resources and investments. Owing to these factors, the real advent of AM-TN-LCD products was delayed until the end of the 1980s.

In 1989, the majority of computer manufacturers made the decision to use AM-TN-LCDs for personal computers—this movement was called down-sizing in the computer business. This development accelerated the R&D activities and the production of AM-TN-LCDs.

Reasonable growth of the LCD technology had been attained prior to the advent of AM-LCDs, in terms of the production of STN-LCDs [6], since 1986. Defect-free STN-LCDs [9] and black and white STN-LCDs [10–14] were developed mostly in Japan from 1986 to 1988. These STN-LCDs have been used for laptop computers, such as the "DynaBook" or "Note Book" types. FLCDs and SHC-LCDs have also been explored as alternatives to the TN and the STN-LCD by many researchers.

Historically, the first idea of active matrix addressing, where the switching transistors or diodes are arranged at each pixel to store the charge, was proposed by Lechner in 1971 [5]. Actual structures and materials for these switching elements will be described in a later section.

This chapter represents an overview of the AM-LCDs by discussing the following: classification of LCDs; the issues that characterize LCD performance; the driving methods of LCDs in relation to AM driving; switching elements for AM-LCDs, requirements for them and their characteristics; the operation and the electrooptic performance of AM-LCDs; a comparison of AM-LCD technologies using different switching devices; the performance and technical issues of the AM-LCD technology, such as image quality, power consumption, resolution and display size, and manufacturing; AM-LCDs using modes other than the TN-LC mode; and finally, future prospects. In principle, almost all the LCD devices can be driven with the active matrix circuit, but we deal mainly with the TN-LCD mode, owing to its current popularity and wide utility.

Regarding LCDs in general and active matrix LCDs, the reader should refer to useful monographs and review papers.

10.2 FORMS OF ELECTRONIC DISPLAY AND THE ISSUES THAT CHARACTERIZE LCD PERFORMANCE

The manner in which the LCDs are used can be classified into the following two groups: (1) direct-view and projection types; and (2) monochrome (black and white) and multicolor LCDs, according to their color capability. These classifications are necessary in developing the discussions in this chapter.

A complete description of the characteristics and performance of an LCD technology may require covering over sixty technical issues. However, in this article we will choose the following major representative six issues:

(1) information content (the number of pixels; each pixel of a multicolor LCD comprises an RGB triad or an RGGB quad, each color dot is called a color subpixel);
(2) display size (diagonal and aspect ratio);
(3) display image quality (legibility and color capability, luminance, luminance contrast ratio, image uniformity, color gamut, color balance, viewing angle, grayscale, image crosstalk, response speed, etc);
(4) electrical driving (direct multiplexing or active matrix; driving voltage and frequency, power consumption including the power consumption of back light, and frequency);
(5) production yield and through-put; and
(6) cost of a unit.

It is necessary to make a technical trade-off between items (1) and (3) (legibility), and generally between the cost and other issues.

FIGURE 10.1. The classification of the driving scheme of matrix LCDs. Legend: TFT, thin film transistor; TFD: thin film diode.

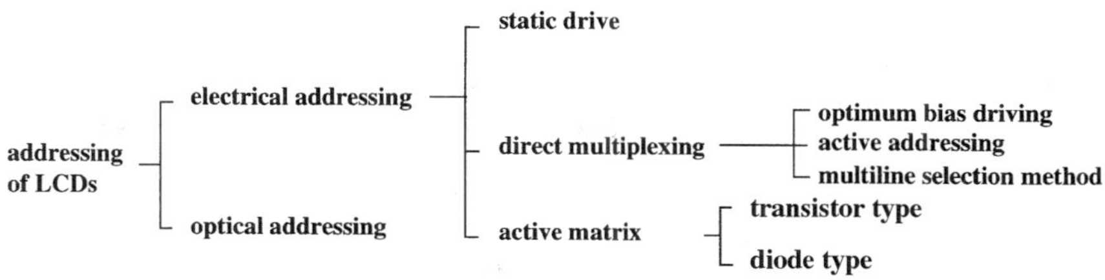

10.3 LCD DRIVING AND THE ADVANTAGES OF ACTIVE MATRIX DRIVING

The driving schemes of LCDs can be classified into static, direct multiplexing (D-MPX), and active matrix driving (FIGURE 10.1). The matrix LCD driven by D-MPX is also called a passive matrix or simple matrix LCD. Detailed description of the various multiplexing schemes are described in this book in chapter 11 by Scheffer.

In nematic (N) LC display devices, such as TN and STN modes, the electrooptic (EO) response is the so-called rms response, which means that the EO response (e.g., optical transmission) is determined by the rms value of the applied voltage, and thus the EO response is temporally accumulative. Furthermore, it takes advantage of the existence of a threshold voltage for D-MPX, whereas the EO characteristic of an FLCD using a chiral smectic LC is a linear function of the applied field.

In FIGURES 10.2 and 10.3, the structure and operational principle of the TN-LCD, together with its optical response for an rms value of the applied voltage, are shown. The voltages V_L and V_H are also defined. The former is slightly higher than the threshold voltage and the latter is lower than the saturation voltage.

In the D-MPX driving of the optimum bias method, as illustrated in FIGURE 10.4, the scanning (strobing) voltage waveforms, which have amplitude $V(Y_i) = aV_0$, where a is a constant, are applied sequentially to the row electrodes. The scanning voltage pulse has a period T and a width t. The ratio t/T is the duty ratio or the duty cycle, and its inverse is the number of row electrodes, N_{ROW}, which is equal to T/t. The waveforms, which are applied to the column electrodes, $V(X_i)$, are called video signals (for TVs) or data signals (for computers), have amplitude $V(X_i) = V_0$.

The select (or "on") and the nonselect (or "off") states can be realized by adjusting the phase relations between two waveforms $V(Y_i)$ and $V(X_i)$ to give $(a+1)V_0$ or $(a-1)V_0$. Furthermore, the luminance level can be controlled by changing V_0.

FIGURE 10.2. The structure and operating principle of the TN-LCD.

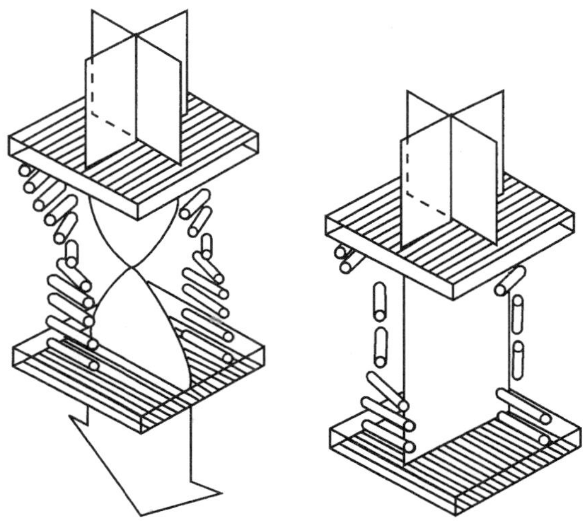

(a) off $(V = 0)$ (b) on $(V > V_{\text{threshold}})$

FIGURE 10.3. The EO performance of the TN-LCD; (A) normally open (white), and (B) normally closed (black).

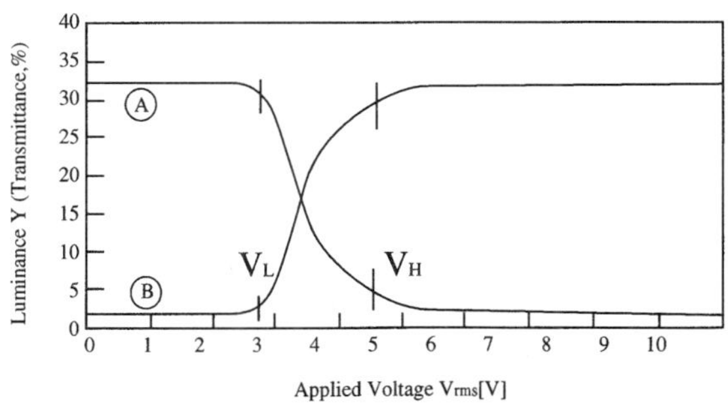

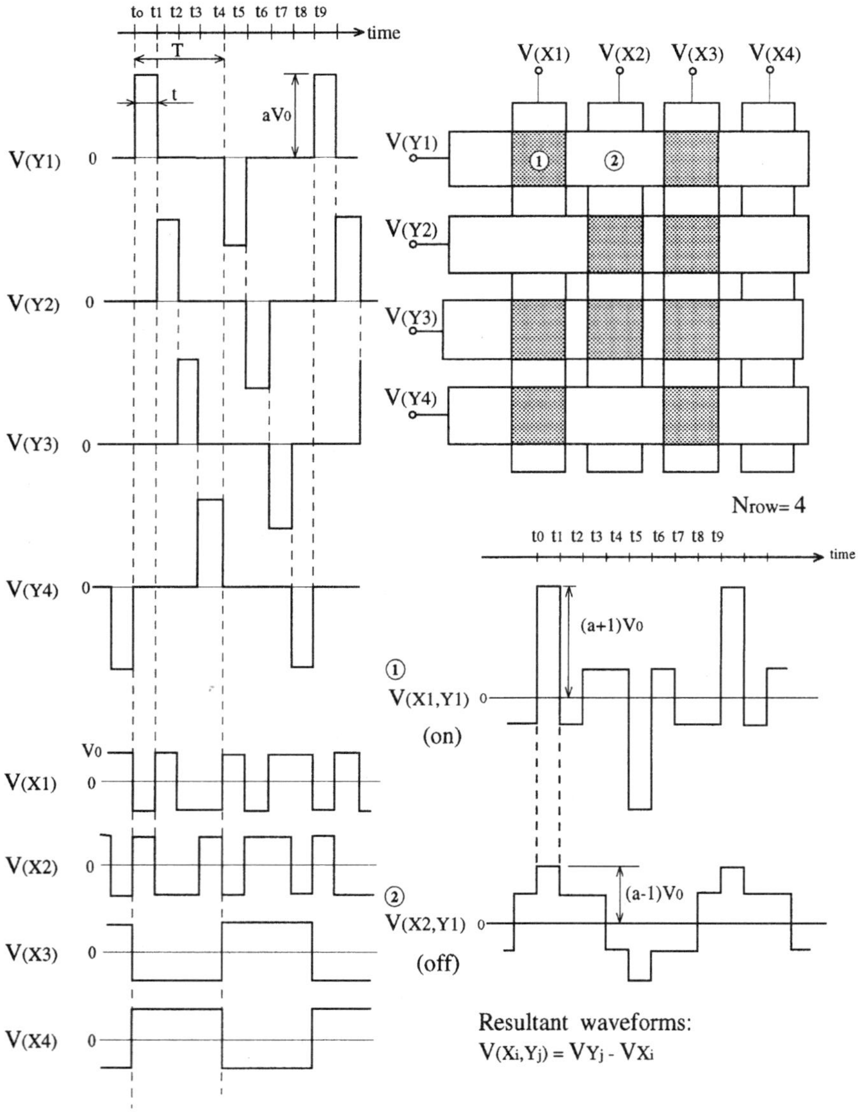

FIGURE 10.4. Typical driving waveforms for direct multiplexing, where shaded pixels are in the on-state in a 4×4 matrix, as an example.

Alt and Pleshko [17] showed that the best contrast condition is attained when N_{ROW} satisfies the equation

$$N_{ROW} = [(M^2 + 1)/(M^2 - 1)]^2 \tag{10.1}$$

where M is the steepness of the rms response, defined as $M = V_H/V_L$ where these two quantities are shown in FIGURE 10.3. The value of the constant a is chosen such that $a\sqrt{N_{ROW}}$ to optimize the performance. In a practical MPX-TN-LCD showing an acceptable legibility, the attainable N_{ROW} is about only 64; this corresponds to $M = 1.11$, which is almost the limit of the multiplexability of the D-MPX-TN-LCD. The actual values of the luminance and the crosstalk of an AM-TN-LCD, which determine their legibility, will be demonstrated and discussed in a later section in comparison with TN-LCD and STN-LCD.

Compared with the MPX-TN-LCD or STN-LCD, an AM-TN-LCD features the following advantages in its characteristics and performance: (1) relative freedom from the limit in N_{ROW}; (2) almost free from crosstalk;

(3) high contrast ratio; (4) fast response speed; (5) grayscale capability; (6) multicolor or full-color capability; and (7) the possibility of large-area display. However, there are still problems in the legibility that will be discussed in a later section.

10.4 SWITCHING ELEMENTS FOR AM-LCDs: STRUCTURE AND FABRICATION

10.4.1 The classification of switching devices

The configurations of AM-LCDs using two-terminal (diode) devices and three-terminal (transistor) devices are shown in FIGURE 10.5 in comparison with D-MPX.

The idea of using amorphous silicon (a-Si) TFT for the AM drive was first proposed by Le Comber in 1979 [18]. Owing to its productivity and performance, the AM-LCD using amorphous a-Si TFT and the TN-LCD medium has taken a major rank in LCD technology. Along with this technology, other approaches have been explored and demonstrated such as polycrystalline silicon (p-Si) TFTs [19], CdSe TFTs [20], MOS transistor arrays on silicon single-crystal wafers [21] or on glass substrates [22], diodes such as zinc oxide (ZnO) [23], metal–insulator–metal (MIM) thin-film diodes on glass substrates using Ta_2O_5 [24,25], a-Si ring diodes [26], Si pn junctions [27] with back-to-back diodes [28], and metal–SiN_x–metal diodes [29]. Regarding diodes, thin-film diodes (TFDs) made of Ta_2O_5 [24,25] and Si pn junctions [27] now dominate over other TFD technologies.

FIGURE 10.5. Simple equivalent circuits of (a) DMPX, (b) AM with TFT, and (c) AM with diode.

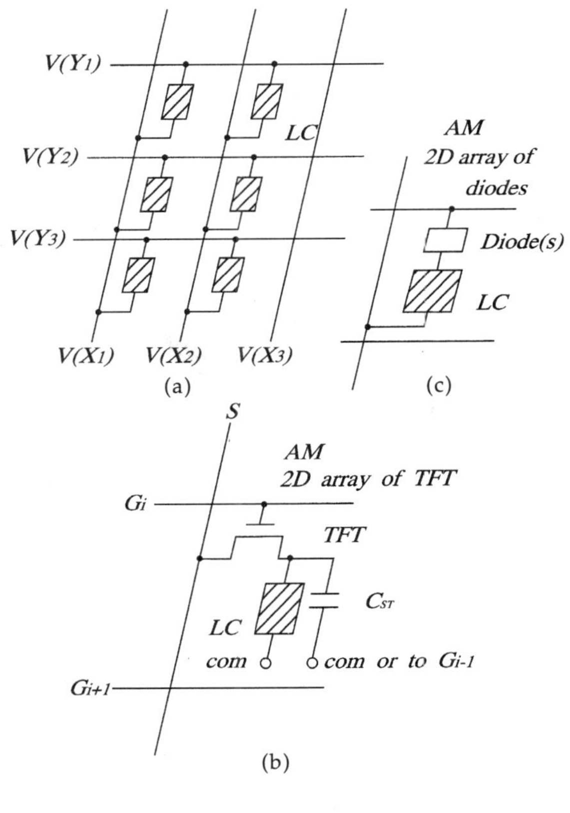

Before entering into detailed discussions of the required electrical characteristics of the switching elements, we will review the fabrication and features of each switching device in the following subsection. Regarding transistor and diode devices, the reader may refer to the monographs cited in appendices B and C on semiconductor devices.

10.4.2 TFTs for AM-LCD Applications

10.4.2.1 a-Si TFTs for AM-LCD Applications

AM-TN-LCDs of the direct-view type using a-Si TFTs with 1.3×10^6 pixels (RGB) with a 13 inch diagonal and 3×10^5 (RGB) pixel with a 21 inch diagonal have been demonstrated, and for their projection use of AM-TN-LCD cells with 2×10^6 pixels has also been demonstrated. These demonstrations are shown in color pages.

FIGURE 10.6 depicts the cross-sectional views of three kinds of TFTs, where (a) and (b) are inverted staggered types most widely used in AM-LCDs, and the (c) is the normally staggered type.

In types (a) and (b), a gate electrode is first formed on a glass substrate, and then source and drain electrodes are deposited, while in type (c) the sequence of the formation of electrodes is reversed. Structure (a) differs from (b) only in the fabrication process: in (a) after deposition of an a-Si layer as a channel, a high-concentration phosphor-doped n^+a-Si layer is deposited for an ohmic contact; after this process the electrodes for source and drain are formed; then the part of the n^+a-Si layer between the two electrodes is removed by etching. In type (b), after depositing the a-Si layer, an insulating layer of SiN_x is deposited, then the part of the insulating layer covering the positions of the two electrodes is removed by etching; after this

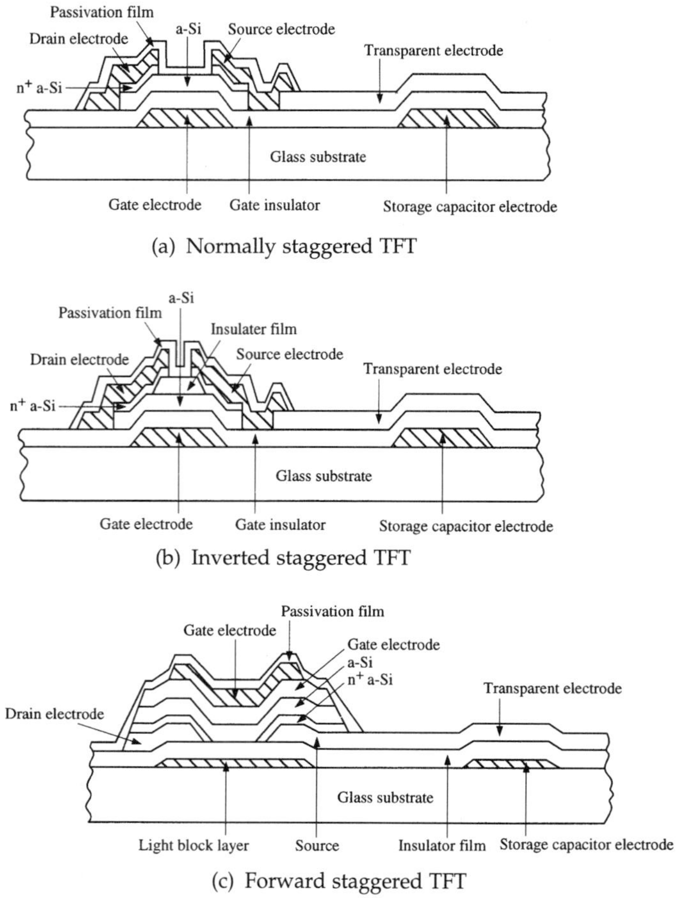

FIGURE 10.6. Two kinds of a-Si TFT: (a) and (b) are the inverted staggered type and (c) is the normally staggered type.

process an n^+a-Si layer and the source and drain electrodes are deposited. The role of the insulating layer is to protect from damage the a-Si layer during etching of the n^+a-Si layer. An example of the fabrication process of the inverted staggered type is shown in FIGURE 10.7.

Commonly, the a-Si film is deposited by glow discharge decomposition (plasma-enhanced chemical vapor deposition = PECVD) of a hydrogen gas diluted monosilane (SiH_4) onto a glass substrate at about 300 °C. The thickness of the a-Si film for TFT application is from 20 to 300 nm. The gate insulation layer is commonly formed using SiO_x or SiN_x, using also CVD, where a double-layer formation is common in order to avoid an electrical short through pin holes; the formation of a Ta_2O_5 film for this purpose is also useful.

Since the resistivity of a-Si is fairly high, it is possible to make the off-resistance of a TFT high without forming a pn junction. This situation makes the manufacturing process simple. The photoconductivity of a-Si must be eliminated by installing a cover or by decreasing the thickness of the a-Si film. The carrier mobility of a-Si is about $1\,cm^2\,V^{-1}\,s^{-1}$ at best, and this value is large enough to switch a TFT, but too small to form an integrated external circuit together with a two-dimensional (2D) TFT array. A typical example of the current–voltage characteristic of an a-Si TFT is shown in FIGURE 10.8. The delicate fabrication process of a 2D array of TFTs and peripheral electrodes is done with photolithography techniques, where the necessary number of photomasks is in the range of five to eight, depending on the process. Obviously, reduction of the number of photomasks is

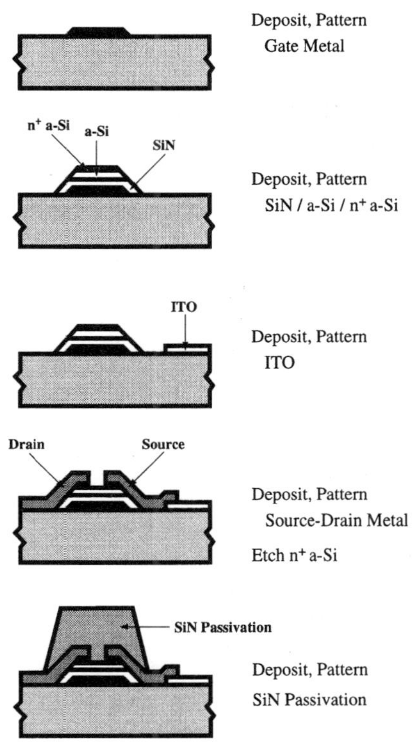

FIGURE 10.7. An example of the fabrication of an a-Si TFT array.

necessary to simplify the process and increase the yield.

In order to increase the production yield and to reduce the cost of the products, it is necessary to reduce the number of process steps including that of photomasks and to remove the origins of the defects. The precise formation of the insulating layer (SiN_x), for example in type (b) in FIGURE 10.6, is done by using the gate electrode as a photomask. This is called the self-alignment technique, which is effective in reducing stray capacitance of the TFT, which will be discussed later.

10.4.2.2 p-Si TFTs for AM-LCD Applications

A p-Si TFT features several advantages, such as a high speed response due to a high carrier mobility, and in turn a suitability for high-resolution display [30].

A p-Si film is usually formed by the thermal recrystallization of a-Si film in a furnace at about 450 °C to 600 °C [31], or with a scanned laser beam

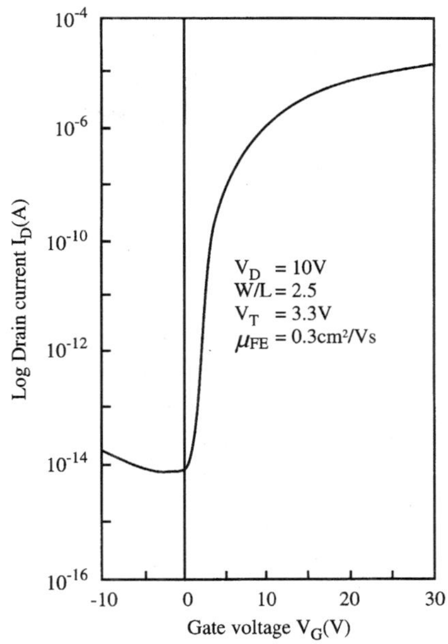

FIGURE 10.8. Typical example of the drain current versus gate voltage characteristic of an a-Si TFT.

[32]. A p-Si film with a thickness from 30 to 300 nm is prepared.

FIGURE 10.9 shows a cross-sectional view of a coplanar-type TFT using p-Si that is commonly utilized to form an AM-LCD. Normally staggered TFTs are also utilized. In the fabrication process of a p-Si TFT, in order to control the type of conductivity (n or p) in a channel, the fabrication of an MOS structure is useful, where impurity ion implantation or thermal oxidation of Si (done at 1000 °C) may be done. This may contribute to increasing the reliability of the p-Si TFT. However, in the p-Si TFT process carried out at 1000 °C there are several drawbacks, such as the

FIGURE 10.9. Coplanar TFT.

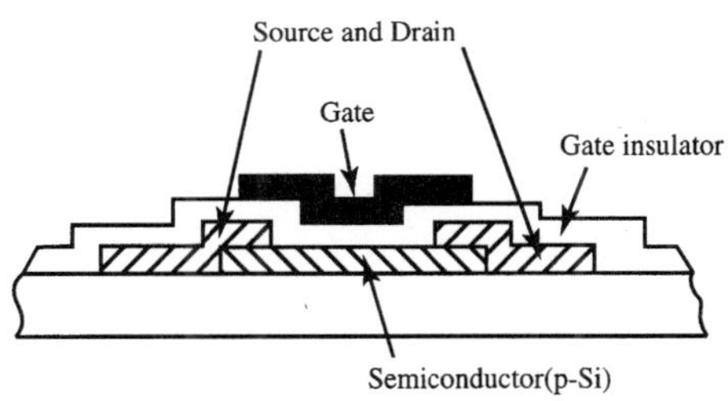

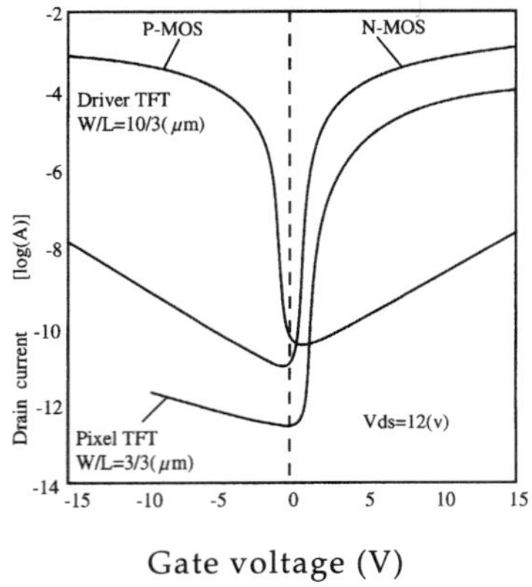

FIGURE 10.10. Characteristics of *n*-channel and *p*-channel TFTs. (H. Ohshima, Fourteenth Intl. Display Research Conf., Digest of Tech. Papers, 6, 1994.)

increase of process steps and the necessity of high-cost quartz as a substrate to accommodate this high temperature. It is necessary to reduce the process temperature for preparing a gate insulator layer to fabricate large-area AM-LCDs using p-Si TFTs; for this purpose, several new technologies that claim the possibility of carrying out the deposition at 400–450 °C have been reported [32]. By using laser annealing, a fairly good p-Si film with a high value of mobility ($350\,cm^2\,V^{-1}\,s^{-1}$) can be prepared [33]. The off-current of p-Si TFT is higher than the a-Si TFT, as shown in FIGURE 10.10. However, it is possible to operate an LCD at TV rate using a pair of serially connected p-Si TFTs in order to overcome its high off-current, by installing an appropriate storage capacitance in each pixel [34].

Taking advantage of the high carrier mobility of p-Si, it is possible to fabricate an AM-LCD cell with a high resolution, installed with an integrated driving circuit. AM-TN-LCD cells using p-Si TFTs are useful as components for projection display instruments having megapixels and viewfinders for video cameras.

10.4.2.3 CdSe TFTs for AM-LCD Applications

The advent of a TFT using CdSe as a semiconductor layer dates back to the early 1960s [20,35], with the pioneering works of Brody and his colleagues. This technology was first applied to a solid state image converting device, and was then applied to driving electroluminescence matrix displays and LCDs [36,37]. The CdSe TFT features a high carrier mobility of $200\,cm^2\,V^{-1}\,cm^{-1}$. Using these TFTs, the fabrication of an AM-TN-LCD was done using photolithography.

10.4.2.4 TFTs Using Organic Materials

It is interesting and necessary to explore a room-temperature manufacturing process of AM-LCDs that uses, in particular, polymer substrates. This technology is useful for fabricating instruments for personal use that are lightweight. One possible way is to use an organic semiconductor for TFT applications, as demonstrated by several groups [38].

General discussion of the current through a TFT as a function of drain or source voltage is given in appendix B.

10.4.3 Diodes for AM-LCD Applications

By introducing an appropriate element with a superlinearity serially to each LC element in a matrix display, it is possible to increase the multiplexability of a matrix TN display. This may seem to be equivalent to an increase of the steepness in its EO characteristics, but diodes act as switching elements. An AM-LCD using diodes can be fabricated in a simple process using two to four masks. Since diodes are two-terminal devices, it is rather difficult to realize a good grayscale capability compared with an AM-LCD using TFTs. For these reasons, the diode-type AM-LCD is something between a D-MPX-LCD and an AM-LCD using TFTs in terms of the quality and cost. Of course, the fabrication of high-quality AM-TN-LCDs using TFTs is also possible by adopting high-level technologies.

The following technologies of AM-LCDs using diodes have been reported.

10.4.3.1 Metal–Insulator–Metal (MIM) Diodes

These are fabricated using films as insulators. The cross-sectional view of the Ta_2O_5 MIM is shown in FIGURE 10.11 together with other main diodes for AM-LCD applications, such as pin and a-SiN_x:H diodes [39,42,44].

The fabrication process of the diode is as follows: first the Ta film electrodes are deposited, then Ta_2O_5 films are formed by anodization of the Ta films. Alternatively, Ta_2O_5 films are formed directly by vacuum deposition; this process is followed by deposition of Cr or Ti films as the second electrodes [39].

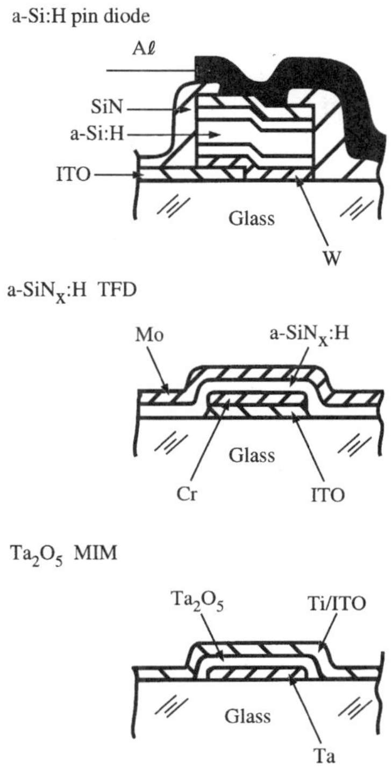

FIGURE 10.11. Schematic cross-sections of the main two-terminal devices. (R. A. Hartman, SID Intl. Symp. Digest Tech. Papers, 7, 1995.)

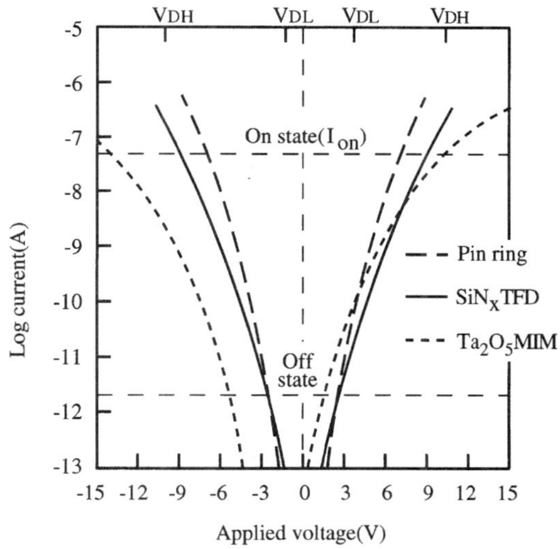

FIGURE 10.12. Current–voltage characteristics of a-Si pin ring, SiN_x TFD, and Ta_2O_5 MIMs. (R. A. Hartman, SID Intl. Symp. Digest Tech. Papers, 7, 1995.)

An example of the current–voltage characteristic of a MIM diode of Ta–Ta_2O_5–Cr is shown in FIGURE 10.12, together with the pin-ring type and SiN_x TFD [40–44]. Theoretical expressions and relevant parameters of diode currents are summarized in appendix C.

To feed sufficient voltage to the diode, the capacitance of the diode, C_0, must be much smaller than that of the LC pad, C_{LC}. This condition can be achieved by fabricating diodes with a very small area, say, $4 \times 4\,\mu m^2$. Small-area MIM diodes can be fabricated by using the edges of the structure [39]. The apparent threshold of the MIM diode is about 2 V and the current required to charge up an LC pad is about 10^{-8} A.

By introducing the MIM diode, it is possible to achieve a high multiplexability, up to 600 lines or more.

10.4.3.2 Diodes Using a-Si

To attain good bidirectionality required for AC operation of the liquid crystals, a simple combination of two or more diodes is used. These techniques are called back-to-back and ring diodes, and their equivalent circuits are shown in FIGURE 10.13(a) and (b), respectively. More multiple diode types are shown in parts (c) to (e). In part (a), each diode is a Schottky-type diode with the structure Ti/a-Si/Ti [43] or Pt/a-Si/Pt [28], and its reverse-biased characteristic is utilized, whereas in (b), each diode is an a-Si pin diode.

The multiplicity of diodes (FIGURE 10.13(c)–(e)) ensures the device has a fairly high value of on/off ratio; however, this multiplicity has the drawback of reducing aperture.

10.4.3.3 TFD Using Mo-a-SiN_x-Cr

This system has been developed extensively by several researchers and several successful demonstrations have been made [42,44].

10.4.3.4 Diodes Using Organic Materials

MIM diodes may also be fabricated using polyimide Langmuir–Blodgett films as insulators, and demonstrations of AM-LCD driving have been made [45].

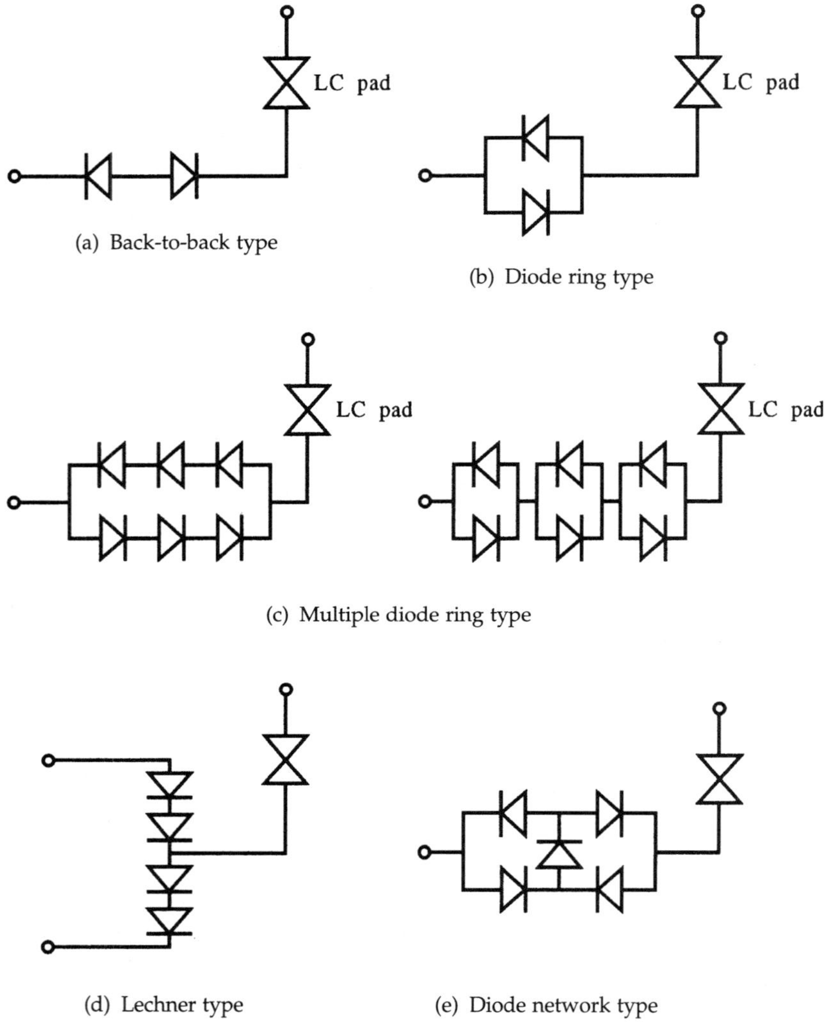

FIGURE 10.13. Equivalent circuits for one pixel using various diodes.

10.5 STRUCTURE OF AM-LCDs

FIGURE 10.14 illustrates a typical bird's eye view of an AM-LCD where a TN medium and TFTs are used. This is the case of a transmissive type and a 2D RGB array color filter is used for multicolor display capability. The actual cross-sectional structure is shown in FIGURE 10.15.

In FIGURE 10.16, a circuit configuration of an AM-LCD using TFTs installed with storage capacitors is shown, together with external driving circuits such as a scanning circuit, column shift register, and sample hold circuit. The TFTs can be replaced by diodes. The scanning voltages are applied to the row electrodes, actually to the gates ($V_{G1}, V_{G2}, \ldots$) sequentially with field time, T. The data ($V_{D1}, V_{D2}, \ldots$) or video signals are applied to the column electrodes through a sample hold circuit.

For TV operation, the frame time is generally 16.7 ms (60 Hz) according to the NTSC system. The electrical requirements for the switching elements for AM-LCD applications will be discussed in the next section.

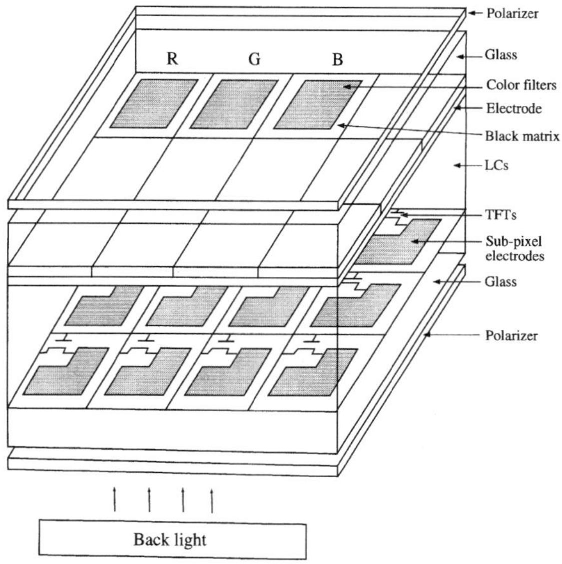

FIGURE 10.14. A bird's eye view of the AM-LCD using TFTs.

10.6 ELECTRICAL REQUIREMENTS FOR SWITCHING ELEMENTS

We will discuss the electrical requirements for the switching elements prior to the discussion of AM-LCD circuits using a TN medium.

The requirements differ according to the application, such as bilevel or grayscale operations.

FIGURE 10.15. A cross-sectional view of a one-color sub-pixel of an AM-LCD using an a-Si TFT.

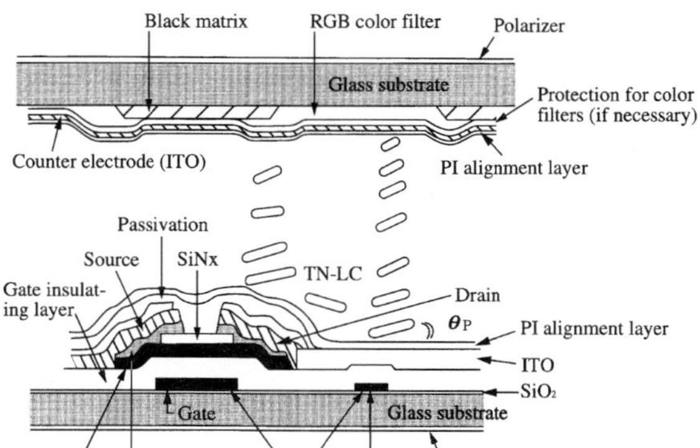

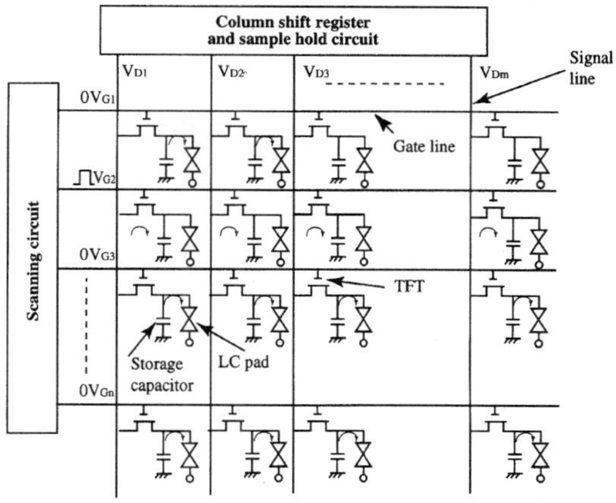

FIGURE 10.16. Driving scheme of an AM-LCD using TFTs.

For bilevel operation, the turn-on voltage V_{ON} might be typically 5 V. The pixel behaves as a capacitor with capacitance C_{LC}, typically 1 pF for an LCD pad of thickness 5 μm and pixel area about $300 \times 300\,\mu m^2$. Thus the charging current, I_{ON}, supplied by the active matrix to charge the pixel by switching from $\pm V_{ON}$ to $\mp V_{ON}$ during $1/N$ of the field time, T, is

$$I_{ON} = 2C_{LC}V_{ON}N/T. \tag{10.2}$$

In the remaining period $(N-1)/N$ of the field time, T, the voltage (or charge) residing on a pixel must be isolated from spurious voltages. However, due to a leakage current, I_{OFF} through the AM element during the off-state, a drop or rise in the voltage, V_{ON}, occurs,

$$I_{OFF} = \Delta V_{ON}C_{LC}N/(N-1)T \simeq \Delta V_{ON}C_{LC}/T \tag{10.3}$$

where ΔV_{ON} is an allowable value.

The AM element must be able to feed enough charging current I_{ON} and have a sufficiently small leakage current, as given by equations (10.2) and (10.3), respectively.

The parameters that specify an AM switching element are I_{ON} and I_{OFF} themselves, and their ratio, R

$$R = I_{ON}/I_{OFF} = 2V_{ON}(N-1)/\Delta V_{ON}. \tag{10.4}$$

It is worth noting that this ratio depends only on N and the ratio $V_{ON}/\Delta V_{ON}$, and is independent of the number of columns and the pixel size.

In practice, both I_{ON} and I_{OFF} depend on temperature, and further, the capacitor is voltage-dependent due to the anisotropy in the dielectric constant of the

LC, so that it is plausible to introduce engineering factors F_{ON} and F_{OFF} quantities, whose magnitudes range from 1 to 10 [15a].

In an AM-LCD, for example, $N = 500$ lines, $V_{ON} = 5\,V$, $\Delta V_{ON} = 0.5\,V$, and the actual levels of currents are $I_{ON} = 3 \times 10^{-7}$ A and $I_{OFF} = 3 \times 10^{-11}$ A (so $R = 1 \times 10^4$). However, in practice the required value of R is 10^5–10^6 by taking account of non-uniformity and other engineering factors.

For grayscale operation, the optical state must be controlled by setting a desired voltage level during the row select time, T/N, and this optical state must be kept at the same level in a divided level of the gray-scale, where the voltage variation, ΔV_{gray}, must be very small compared with ΔV_{ON}. The requirements for the grayscale are more stringent compared with those for saturated bilevel operation.

To keep the error to 3% in the voltage applied to a pixel, the time constant should be 30% of the row selection time. However, since in practice, both C_{LC} and R_{ON} are voltage-dependent, we need a reasonably rigorous simulation.

10.7 BASIC ELECTRICAL OPERATION OF THE AM-LCD AND ITS ELECTRICAL AND ELECTROOPTIC CHARACTERISTICS

10.7.1 AM-LCD with TFTs

FIGURE 10.17 is an equivalent circuit of a pixel with a TFT device, including parasitic capacitances and voltage waveforms. To keep the voltage variation, ΔV_{ON} low, a storage capacitance is usually inserted.

The gate voltage V_G and signal voltage V_D applied to a TFT (here V_D is a simple square wave oscillating around its central value, $V_{SIG(CTR)}$, for simplicity) and the solid line indicates a real pixel electrode voltage V_P.

The deviation ΔV_P appearing in both the upper and lower parts of V_P in FIGURE 10.18(a) is caused by the existence of the parasitic capacitance C_{GS}. Here, we will discuss only the shift ΔV_P. Other effects caused by the stray capacitance C_{DS} will be discussed later in relation to the image quality.

At the instant when V_G terminates, V_P is divided by $C_{LC} + C_S$ (C_S is the storage capacitor) and C_{GS}; this causes level shifts, as shown in FIGURE 10.18:

$$\Delta V_P = C_{GS} V_G / (C_{GS} + C_{LC} + C_S) \qquad (10.5)$$

where C_{LC} is a function of V_P dependent on V_D.

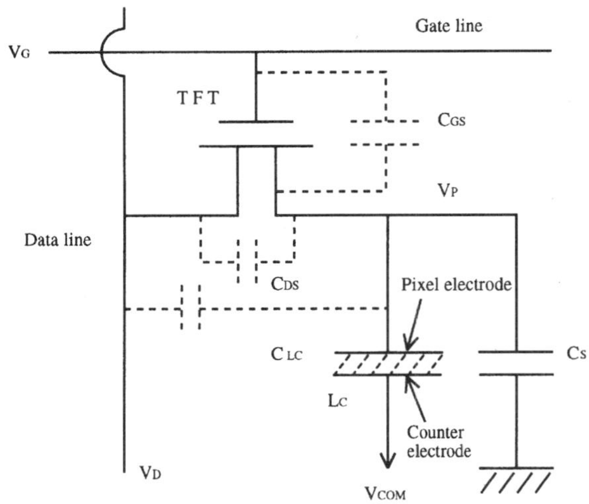

FIGURE 10.17. An equivalent circuit of one pixel of TFT AM-LCD together with parasitic capacitances.

These level shifts cause a DC offset voltage, therefore it is necessary to choose V_{COM} to avoid this. The level shift can be reduced by reducing C_{GS} and inserting a storage capacitance C_S.

The shift in V_P due to C_{DS} is

$$\Delta V_P = C_{DS} V_G / (C_{DS} + C_{LC} + C_S) \qquad (10.6)$$

where C_{LC} is also dependent on V_P. This level shift degrades the display uniformity, as will be discussed in a later section.

To avoid the degradation of LC material, it is necessary to operate an LC medium with an alternating voltage waveform that does not accompany any DC offset. FIGURE 10.18(a) is an example of AC driving where the V_P oscillates around an appropriately chosen V_{COM}. Further, FIGURE 10.18(b) shows a response waveform of TN-LCD. The choice of V_{COM} is such that it minimizes observed flickering. The issue of flickering will be discussed in a later section.

10.7.2 AM-LCD with Diodes

An example of driving waveforms for an AM-LCD with diode switching devices is shown in FIGURE 10.19, where V_S, V_{SIG}, and V_{DH} stand for the scanning, the signal, and the high level of the diode voltages. V_{DH} is shown in FIGURE 10.12, which represents the V–I characteristic of the diode.

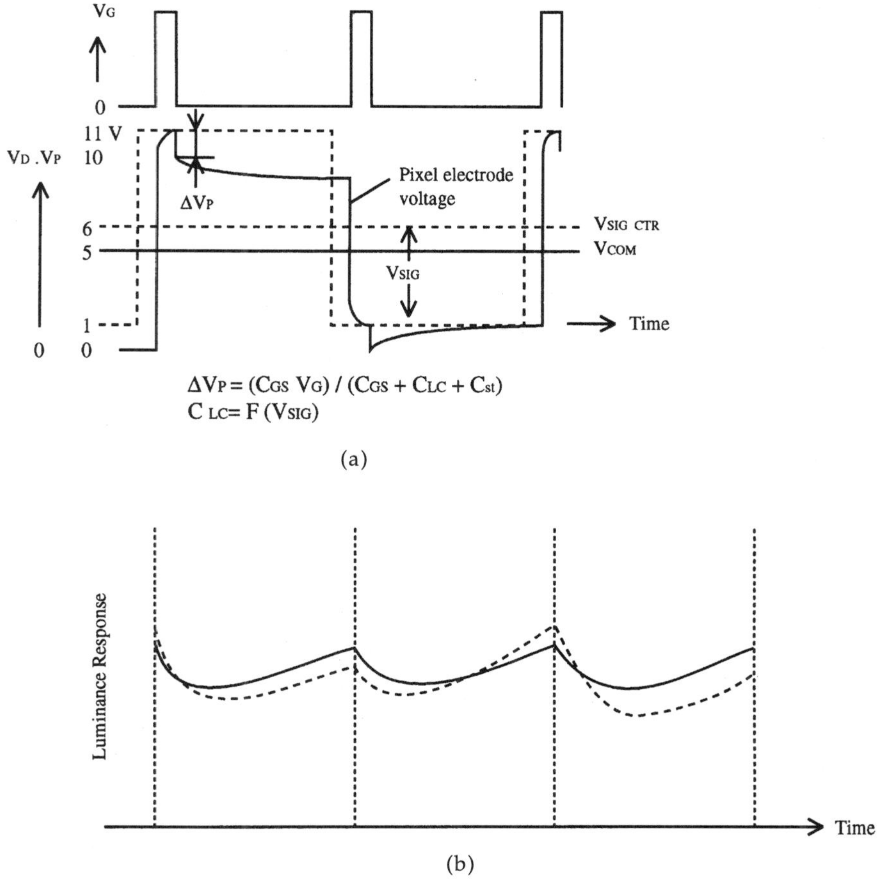

FIGURE 10.18. (a) Waveforms of gate line, data line, and pixel electrode voltage; (b) luminance response of the LC pad. In the response, the solid line represents the response for the optimum V_{COM}, while the dashed line represents the state when V_{COM} is slightly smaller than the optimum value; in the latter case a flicker appears due to the difference in the responses for positive and negative frames.

Generally, the function of the diode as an external element may be thought of as modifying the intrinsic EO characteristic of the LC medium. In this function, the diodes for AM-LCD applications are characterized by the value of V_{DL} (apparent diode threshold voltage, as shown in FIGURE 10.12) and the degree of nonlinearity. Ideally, V_{DL} must be larger than V_{TH} (threshold voltage of LC). Further, a sufficiently large current flow through the diode is necessary to charge up a pixel, together with low leakage current of the diode, the same requirements for a TFT. An example of a diode with a low V_{DL} is the ring diode, while two with high values of V_{DL} are the ZnO varistor and the back-to-back diode. The MIM diodes generally have an insufficient nonlinearity and a low V_{DL}, therefore a large V_S (scanning voltage) is necessary at the sacrifice of a leakage current.

A useful driving waveform for the diode AM-LCD, called the five-level reset drive, is shown in FIGURE 10.20 in comparison with former schemes. This method was discussed and adopted by the Philips group to overcome the nonuniformity of diode characteristics [44].

In order to obtain a uniform display pattern, it is effective to reverse the voltages applied to both the rows and the columns in every sequential field.

A full computer simulation is necessary to obtain details of a full operation of AM-LCD with diodes, as is true for the AM-LCD with TFTs.

10.8 ELECTROOPTICAL PERFORMANCE OF AM-LCDs

The purpose of the fabrication of AM-LCDs is: (1) to increase the contrast ratio and reduce the image

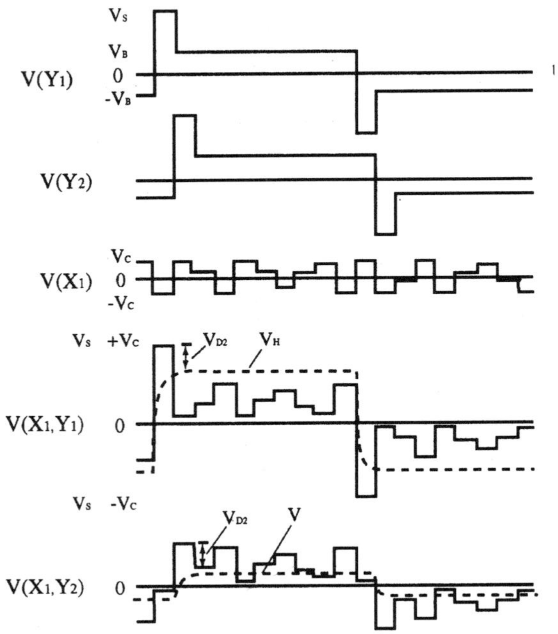

FIGURE 10.19. Voltage waveforms applied to the diode-type matrix LCD. $V(Yn)$ scanning voltage; $V(Xn)$ data voltage; and $V(Xn, Yn)$ pixel voltage, where Xn, Yn stand for the nth column and nth row.

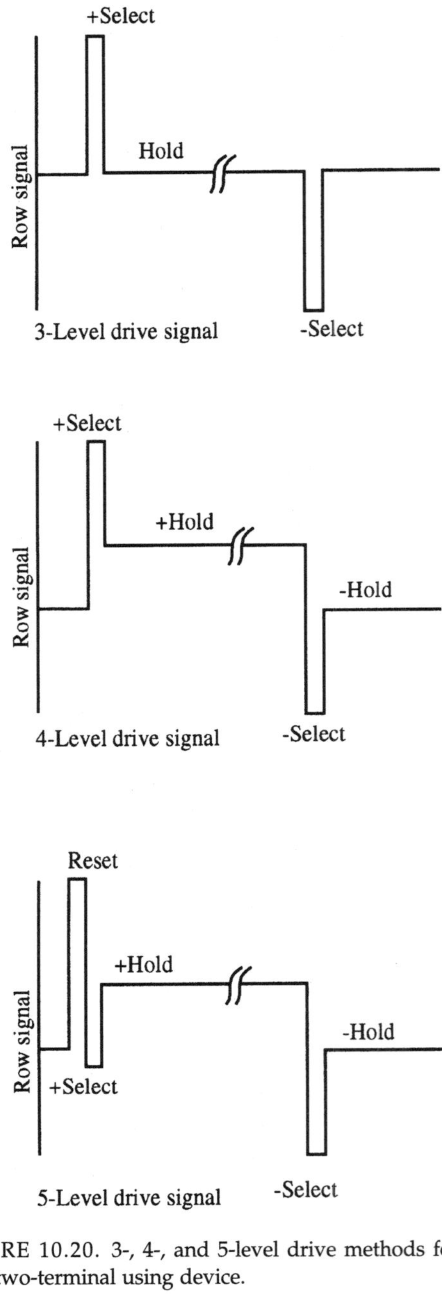

FIGURE 10.20. 3-, 4-, and 5-level drive methods for AM-LCD two-terminal using device.

crosstalk which is heavy in D-MPX TN-LCDs; (2) to increase the information content without degrading grayscale capability, and to enlarge the display area by taking advantage of freedom from the N_{ROW} limit; (3) to obtain a display with a high resolution; and finally (4) to take advantage of fast response speed. These items will be discussed in this section.

10.8.1 Luminance Level

A comparison of the luminance levels of D-MPX-TN, D-MPX-STN, and AM-TN-LCD media is shown in FIGURE 10.21, where in addition to these data, the datum of a back-lit AM-TN-LCD is shown as a reference. Due to the optical loss of a pair of polarizers, the optical transmission falls to typically 40%; the transmittance of the nonselect state in a normally white D-MPX TN-LCD (black and white) is about 16% for $N_{ROW} = 200$ lines, due to crosstalk. In an STN, however (black and white), with $N_{ROW} = 200$ lines, the transmission is 18%, due to less image crosstalk, but accompanies a better black state. In typical AM-TN-LCD TFTs, the image crosstalk is small but overall optical transmission is about 6% due to the limited aperture ratio and multilayered structure, including the color filter for a color LCD, which

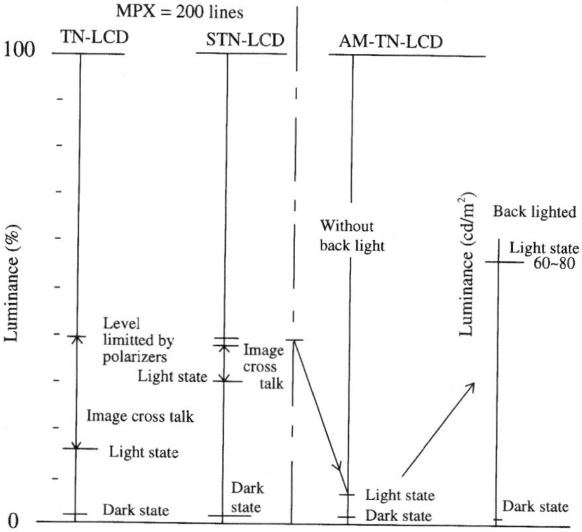

FIGURE 10.21. Luminance levels of multiplexed TN-LCD and STN-LCD with MPX of 200 lines in comparison with AM-TN-LCD with and without back-lighting. In TN and STN-LCDs "image crosstalks" are shown.

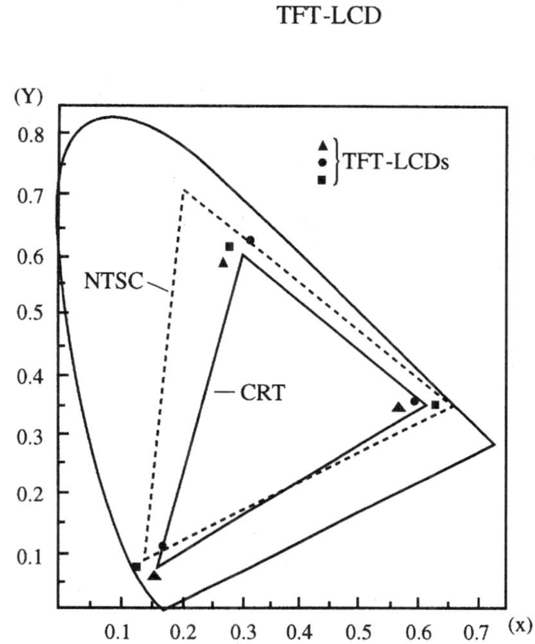

FIGURE 10.22. Examples of color reproductions of some TFT LCDs in comparison with CRT and NTSC specifications.

induces an additional optical loss of about 75%. For this reason to obtain a practical luminance level for a color AM-TN-LCD (at least 60 cd m^{-2} for office equipment), it is necessary to illuminate the cell with a back light; the achieved luminance contrast ratio is about 150:1. Methods to improve the luminance level of AM-LCDs will be discussed in a later section. LCDs without polarizers, such as the PDP [46], the White and Taylor type GH, and the amorphous N* GH [47], are able to exhibit transmissions of 50 to 60%.

10.8.2 Color Gamut

FIGURE 10.22 demonstrates the attained color gamut (CIE 1931 *xy* chromaticity diagram) of an AM-TN-LCD using RGB color filters. A comparison is made with the color gamut of the NTSC standard and that of a typical CRT. FIGURE 10.23 shows a typical example of the spectrum of a cold cathode fluorescent lamp, called a three-wavelength tube, used for backlighting, and the spectra of RGB color filters for color LCDs. Color design has been discussed by the author's group [48].

The methods of color formation in LCDs and some relevant references are shown in appendix D.

10.8.3 Response Times

The rise time of an LCD using a nematic medium is almost inversely proportional to the voltage squared, therefore it is possible to reduce the rise time of an AM-TN-LCD since it is driven with a sufficiently high voltage, without causing image crosstalk. This is one of the advantages of the AM-TN-LCD in comparison with the D-MPX TN-LCD. This is true for a binary operation, but there is still the problem of slow response in grayscale operation. Furthermore, the decay time does not depend on the applied voltage, therefore it is impossible to improve the decay time, even in the AM-TN-LCD.

10.8.4 Resolution

For office equipment use, AM-TN-LCDs with 640 × 480, 1024 × 768, 1152 × 900, and 1280 × 1024 pixels are already commercially available, where the first is the VGA specification and the last is the EWS specification. For video use, products with 200 × 200 to 480 × 720 pixels are also commercially available. An

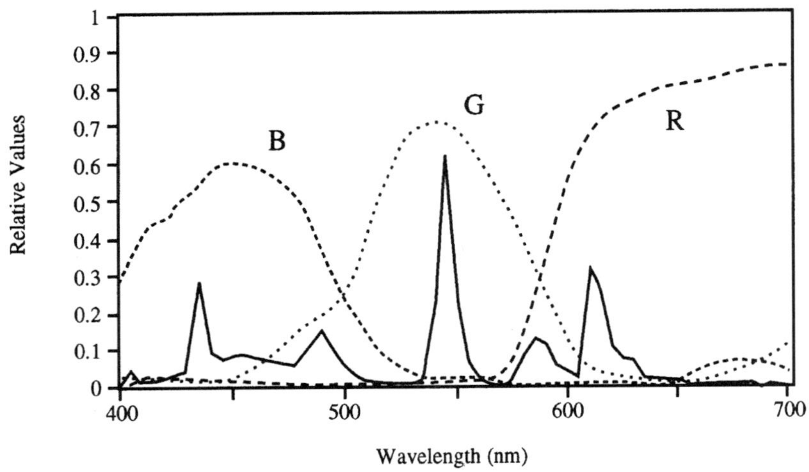

FIGURE 10.23. An example of the spectra of an RGB color filter, and emission spectra of a common "three-wavelength-type" fluorescent lamp used for back-lighting.

example of 21 inch color AM-TN-LCD is shown on the color pages. The projection-type LCDs with mega-pixels corresponding to HDTV specifications are also available. These products are produced using a-Si TFTs. Some projection displays and one-inch-size video camera viewfinders with a very fine pitch of 40 μm are produced using p-Si TFTs (see color pages).

Improvements in the luminance level by aperture enlargement, viewing angle characteristics, and gray-scale capability are discussed in the next section, together with image sticking and nonuniformity of display images.

10.9 TECHNICAL ISSUES AND THEIR RESOLUTIONS IN AM-TN-LCDs

In section 10.8 we summarized the generally achievable characteristics of AM-TN-LCDs. In this section, we will discuss ways to improve the EO performance of AM-TN-LCDs in more detail.

10.9.1 Electrical Properties of AM-LCD Circuits: Influence on Image Quality in a Large-Area AM-TN-LCD

The issues that characterize the image quality of LCDs have already been introduced and discussed in section 10.2. In this section, we will discuss ways of removing several phenomena that degrade image quality, such as image shading, image crosstalk, and flickering, appearing particularly in large-area AM-TN-LCDs.

All these phenomena are caused by electrical properties of the AM circuit, which consist of bus lines, TFTs, and LC pixels and EO characteristics of LCs.

10.9.1.1 Finite *RC* Time Constants for Bus Lines

The *RC* time constants for bus lines, and in particular gate bus lines, has a great influence on the performance of a large-area AM-TN-LCD, which needs long bus lines and a high driving frequency, where much more gate pulse distortion takes place in the end portion of a gate bus line than in the input portion. This gate portion distortion leads to insufficient current to charge the pixel capacitor in one horizon period, resulting in horizontal gradual lightness shading on the LCD screen along the gate bus lines. It may also create a crosstalk phenomenon, due to the delay in switching off the TFTs.

It is necessary to reduce the *RC* time constant of the gate bus lines down to 10% of the writing period. This can be achieved by forming metal film lines with a low resistivity. Useful metals for this purpose are: Mo–Ta and W–Ta alloys [49,50]; Al layer [51]; Al–Cr double layer [52]; and Cu [53]. A signal-line configuration of ITO–Mo–Al triple layer has been used in a large AM-LCD utilizing normal staggered TFTs. From the viewpoint of reducing gate line capacitance, the structure of the auxiliary storage capacitors should be examined, that is, the auxiliary storage capacitors using the former gate line as a capacitor electrode give rise to an undesirably large gate line capacitance [54].

An overlap scanning method, in which the gate pulse width is twice as long as usual and the scan pulse timings overlap in neighboring lines, is an effective way to charge the pixel capacitor in a short write period [54].

10.9.1.2 Voltage Holding Capability—The Problem of Leakage Current in an AM-LCD Circuit

TFT-LCDs use a frame-memory technique, in which charges are held to drive LC pixel elements until they are addressed again.

An insufficient charge (which effectively means voltage across the pixel) holding capability degrades the contrast of the LCD, which sometimes results in a nonuniformity in the display screen due to nonuniform holding capability.

To achieve good charge-holding capability, it is necessary to use an LC material with a high resistivity—typically fluorinated NLCs [55], and appropriate LC molecular aligning layers, such as well developed polyimide materials [56,57], and well developed TFTs.

The holding ratio, HR, is defined as the ratio of the rms voltage for a frame period, divided by the initial value on a pixel electrode, and is given by

$$\mathrm{HR} = \{[1 - \exp(-2T/\tau)](\tau/2T)\}^{1/2} \tag{10.7}$$

where T is the frame period and τ is the RC time constant of the LC pixel and auxiliary storage capacitor. It is estimated that 100% HR is obtained when the resistivity ρ_{LC} exceeds $10^{12}\,\Omega\,\mathrm{cm}$ regardless of the values of the auxiliary capacitor.

FIGURE 10.24 depicts how the pixel electrode voltage drops owing to the increase of the leakage current.

Another origin of the decrease of HR may be the existence of a depolarization field caused by the accumulation of extra charges on the surface of the aligning layer [56]. This effect can be reduced by using a special polyimide with a low electrical polarity [56,57].

10.9.1.3 The Effect of TFT Off-Current, the Origin of Flickering and Vertical Luminance Shading

A TFT that is commonly utilized in AM-LCD has characteristics shown in appendix E (called enhancement type), and it allows the flow of an appreciable amount of off-current in the zero and minus gate voltage regions.

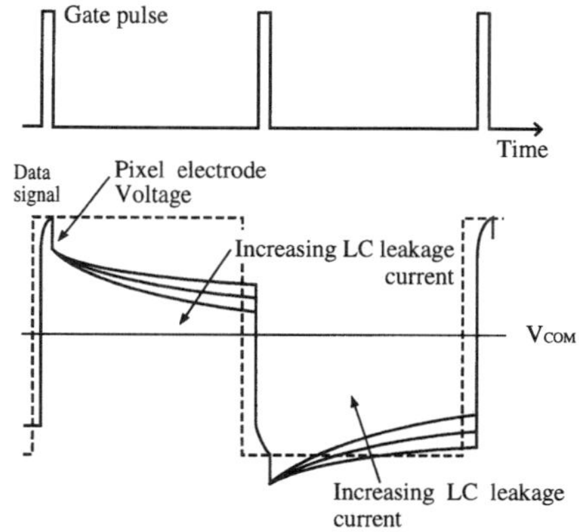

FIGURE 10.24. The effect of liquid crystal leakage current on the pixel electrode voltage.

This causes degradation of image quality through flickering, and vertical luminance shading due to the crosstalk effect.

To avoid chemical degradation of the LC material, an alternating voltage waveform is applied to the LC cell. For this purpose, several modes of polarity inversion driving are adopted. Typical waveforms are shown in FIGURE 10.25 as a time chart of the gate pulses and data signals, where positive and negative periods are drawn alternately. Further, four different periods designated A to D are shown. Detailed voltages across the electrodes and common electrodes are shown in appendix F by giving typical values. In the cases where TFTs of the n-channel depletion type are used, the leak currents behave as follows: (1) the amount of off-current in regions C and D is larger than that in region A, and the order is D > C > B > A; (2) in

FIGURE 10.25. Four periods, A, B, C, and D, having different TFT off-currents (see appendix E) in the timing chart of the gate pulse and data signal. (H. Hori, Proc. SID 32/4, 311, 1991.)

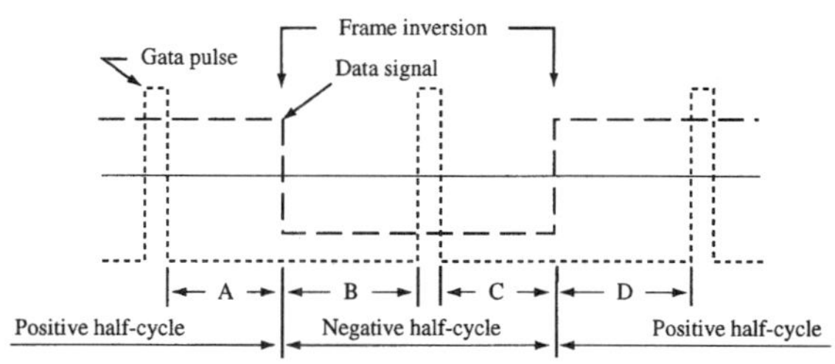

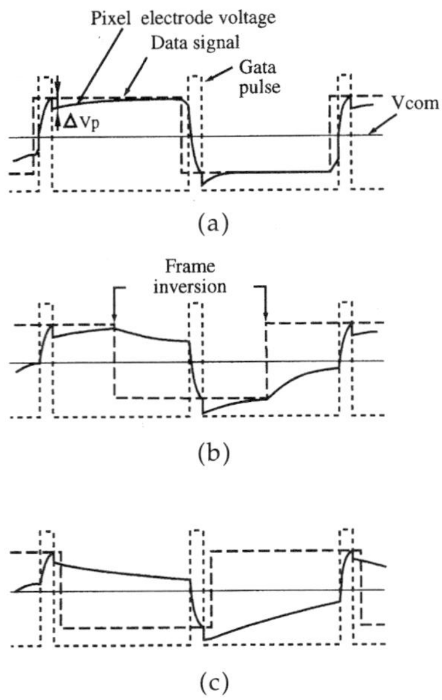

FIGURE 10.26. Pixel electrode voltage waveforms for an AM-LCD using TFTs with large off-currents (a) at top, (b) at middle, and (c) at bottom of the LCD panel, respectively. A large off-current leads to the asymmetrical pixel voltage waveforms in the positive and negative half-cycles, resulting in a flicker and luminance shading from the top to the bottom that is caused from the rms voltage differences in the vertical direction. (H. Hori, Proc. SID 32/4, 311, 1991.)

the negative half-cycle (in the B period), the off-current flows as a leak from the source (pixel) to the drain electrode after frame inversion; and (3) in the positive half-cycle (in the D period), the leak current before the gate pulses are applied is maximum because of the peak drain–source voltage and low gate voltage. Details of current behavior are shown in appendix E and the states of the TFT in regions A to D are given in appendix F.

In this process, the following effects occur: (1) large off-currents lead to asymmetrical pixel voltage waveforms between positive and negative half-cycles, as shown in FIGURE 10.26, resulting in flicker; (2) top-to-bottom luminance shading in the vertical direction, resulting from rms voltage differences (FIGURE 10.26) due to leak current differences in the four periods mentioned above; (3) the source (pixel) voltage V_P drops by ΔV_P owing to the existence of a parasitic capacitance C_{GS} given by equation (5), where the capacitance of the pixel L_C, C_{LC}, changes appreciably according to the on-state and the off-state, resulting in the difference in ΔV_P and making the voltage waveform applied to an LC pixel asymmetric, also causing flicker (these degradation effects are eliminated by adopting a row–line inversion driving [58] and reducing the leak current of the TFTs); (4) another effect of the large off-current is one of the origins of signal crosstalk observed on the same columns: when a white window pattern is displayed in the normally white display mode, the same columns above the window (i.e., the B region in FIGURE 10.27(a)) become

FIGURE 10.27. Signal crosstalk effect due to large TFT off-current. Pixel electrode voltage waveforms at A region (b) and at B region (c), while displaying a window pattern (a) in the normally white mode. The rms voltage of the B region is lower than that of the A region. The B region above the window is brighter than the A region due to the reduced rms voltage caused by the signal crosstalk. On the other hand, the C region under the window is darker than the D region because of the lower source–drain voltage. In this way, an image crosstalk occurs. (H. Hori, Proc. SID 32/4, 311, 1991.)

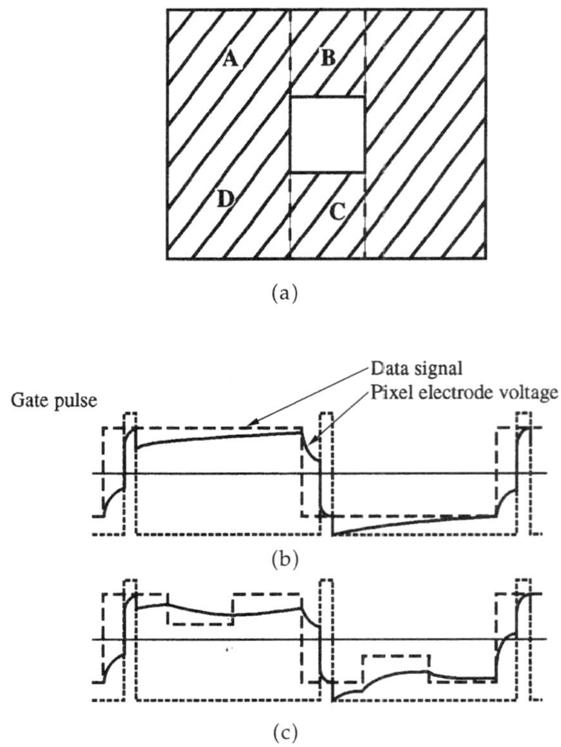

whitish due to the reduced rms voltage caused by higher source–drain voltage across the window region being addressed. FIGURE 10.27(b) and (c) shows the pixel electrode voltage waveform for this case. On the other hand, the C region below the window is darker than the D region because of the lower source–drain voltage [54]. This is a kind of image crosstalk.

10.9.1.4 The Effect of C_{DS} on Grayscale Error and Luminance Shading

The parasitic capacitance between drain and source, C_{DS}, is generally smaller than C_{GS}. However, the effects of C_{DS} cannot be neglected in a high-resolution grayscale display [60]. One of the effects of C_{DS} is an error in grayscale due to voltage crosstalks that are caused in a given pixel by data signals from the other pixels through C_{DS} [59], as shown in FIGURE 10.28(a). Another effect is luminance shading as shown in FIGURE 10.28(b) [59]. The effect of C_{DS} can be eliminated by column–line inversion [61].

The methods for removing the origins of image degradation are summarized as follows. (1) Flickering can be eliminated by decreasing TFT off-currents, adopting a storage capacitor, and reducing C_{GS} (parasitic capacitance between gate and source electrodes) by fabricating small TFTs and adopting self-alignment fabrication methods. (2) The vertical image crosstalk can be reduced by using TFTs with small leak currents, reducing photoconductivity, and improving patterning and driving methods. (3) The horizontal image crosstalk can be eliminated by reducing the time constant of the gate lines. (4) Flickering is also removed by choosing the optimum V_{COM}.

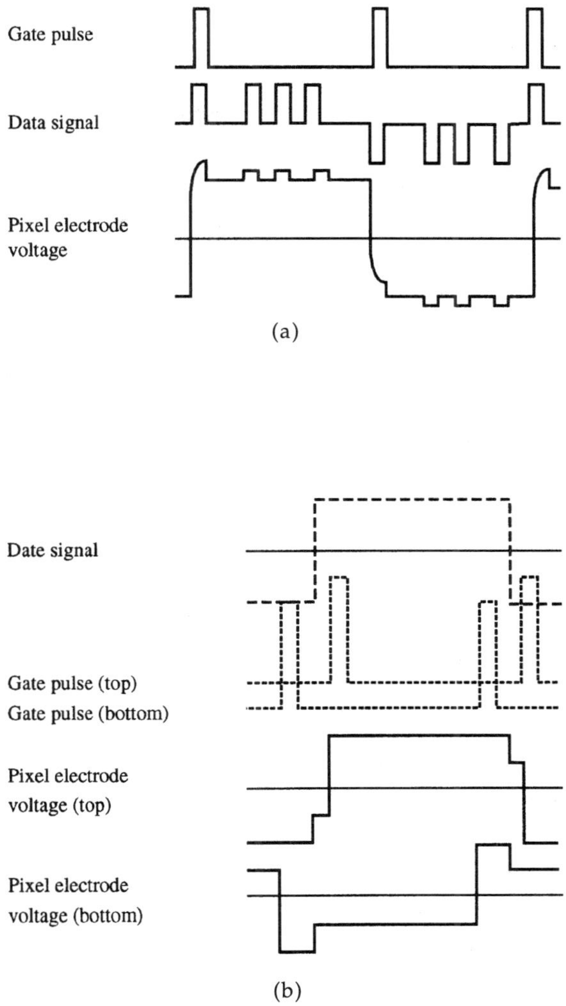

FIGURE 10.28. Effects of drain–source capacitive coupling. (a) Signal crosstalk causing grayscale error, and (b) top–bottom variation in the rms voltages. (H. Hori, Proc. SID 32/4, 311, 1991)

10.9.2 Electrooptical Properties of AM-TN-LCDs: Viewing Angle and Grayscale Capabilities

In this section, we will discuss the electrooptical properties of TN-LC media by focusing on the viewing-angle characteristics expressed by contrast ratio contours and grayscale capability.

10.9.2.1 Contrast Ratio Contours

Conventional TN-LCDs, which are fabricated using the both-side rubbing technique to align NLC molecules, show peculiar viewing-angle characteristics featuring butterfly-like contrast ratio contours (also called isocontrast ratio curves) accompanying a contrast inversion for the vertical direction, as shown in FIGURE 10.29(a), where the contours of an amorphous TN-LCD [69] (b) are shown for comparison. The latter exhibits uniform viewing-angle characteristics. There is a strong need for uniform and wide-viewing angle characteristics and good grayscale capabilities, in particular in large-area AM-TN-LCDs.

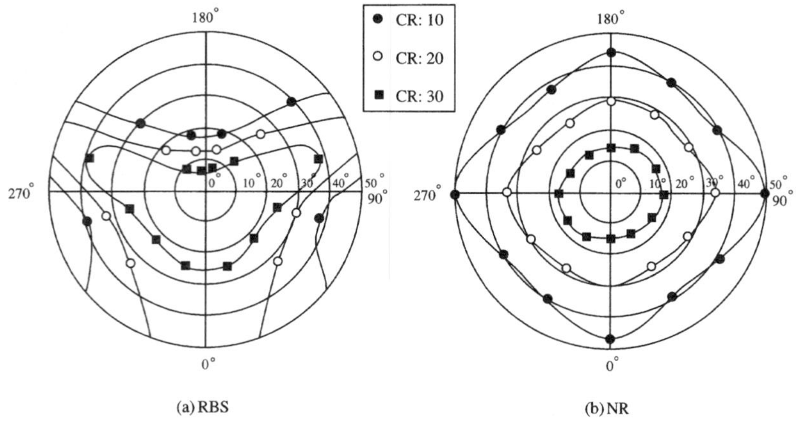

FIGURE 10.29. Luminance contrast ratio contours. (a) conventional TN-LCD fabricated by rubbing both sides; (b) TN-LCD fabricated by nonrubbing, called amorphous TN.

To resolve this problem, the following techniques have been proposed and demonstrated, and some of them have already been adopted in the products of AM-TN-LCDs: (1) Installing two to four control capacitors in each subpixel [62,63]; (2) dividing each subpixel of the RGB color display into two parts—dual subpixel (DSP) by dual rubbing [64–67]; (3) quartered subpixel (QSP) [68]; (4) amorphous TN with several tens of thousands of divisions [69]; and besides TN-LCDs, (5) vertically aligned but bent NLC [70,71]; (6) in-phase-switching mode [91]; and so forth.

The contrast ratio contours of a TN-LCD with DSPs, QSPs, and an a-TN-LCD are shown in FIGURE 10.30(a) and (b). These curves are obtained by a computer simulation, but they agree fairly well with experiments [72]. The QSP configuration is the best for the uniform viewing angle, and experimentally the NLC alignment was carried out using polarized-light irradiated-polymer films (for example, polyvinylcinnamate) [68]. The a-TN-LCD was fabricated by a nonrubbing technique; first a chiral-agent-doped NLC in the isotropic phase is injected into an empty cell whose inner surfaces are covered with nonrubbed polyimide film, and then it is cooled down to room temperature. In an a-TN-LCD, the stabilization of the TN-LC medium using an LC polymer (for example diacrylate) is useful to immobilize reverse-tilt disclination [73].

10.9.2.2 Grayscale Operation

For both TV and computer monitor screens, the necessary value of the grayscale is 8 bit, which equals 256 shades. However, in computer monitors, some graphs or figures are fairly well drawn with a 4 bit (16 shades) or a 5 bit (32 shades) grayscale. The term full color is customarily used for expressing 16.7 million colors ($256 \times 256 \times 256$ for RGB colors).

In a conventional TN-LCD, the grayscale characteristic is very poor and largely dependent on viewing direction. An example is shown in FIGURE 10.31(a), while multidomain (actually amorphous) TNs show a very good grayscale capability, as shown in FIGURE 10.31(b) [69]. As mentioned previously, good grayscale operation can be achieved by controlling the capacitors [62,63]. The approach of QSP is shown to be very effective in achieving a good grayscale capability [68].

10.9.3 Luminance Level and Power Consumption

As discussed in section 10.8.1, the optical transmission of the standard AM-TN-LCD is very low, about 6%. Therefore, the backlighting is necessary. The power consumption of a backlighting system for 10 inch diagonal panels is 2 W to 6 W, depending on the technology. A recent innovation makes it possible to reduce power consumption to 1 W in a practical AM-TN-LCD. Typically, a backlight comprises a cold cathode fluorescent lamp tube that is capable of emitting white light of 1800 cd m^{-2}.

The improvement in optical characteristics of TN-LCD modules makes it possible to reduce its power consumption. This improvement can be attained in the following ways: (1) by improving the color filter's transmission; (2) by increasing the aperture by

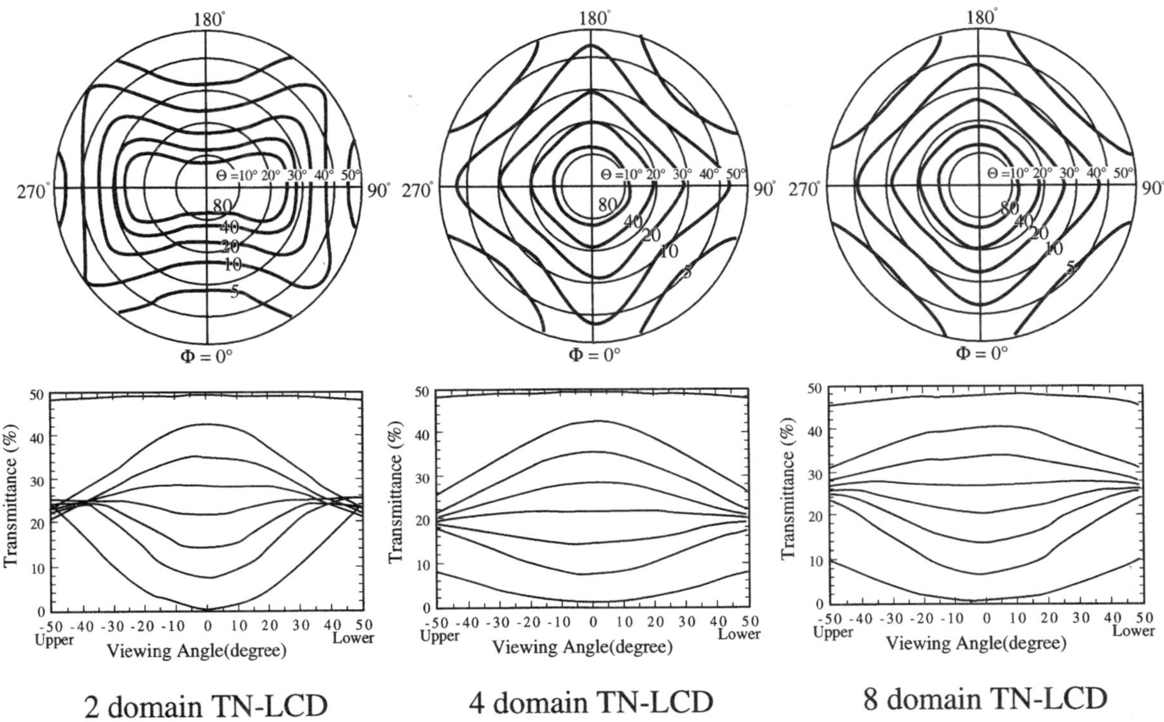

FIGURE 10.30. Isocontrast ratio curves and the polar angle dependence of light transmission for multidomain TN-LCDs with 2, 4, and 8 divided subpixels.

improvement of the fabrication process and design (for example, fabricate small TFTs to reduce the width of electrodes, use self-aligned TFTs, etc); (3) by reducing surface reflection mainly caused by a black matrix normally made of Cr, by adopting other materials such as Cr/CrO or an organic material; (4) by improving backlight efficiency; (5) by decreasing operation voltage by adopting LC materials with lower threshold voltages; (6) by using common electrode voltage reversal drive and by using voltage shift drive; (7) by developing polarizer-free (or at least one-polarizer) technology; and (8) by developing a multicolor LCD without using color filters.

With the increase in resolution, the pixel size decreases to the 100 or 40 μm scale; this trend causes light leakage due to the nonuniform movement of LC molecules in the edge of the pixel. This is actually due to the existence of the electric field across bus lines and pixel electrodes that causes a reverse-tilt disclination, reducing the contrast ratio. 2D analysis of this effect has been carried out by computer simulation by several researchers [74–77]. An analysis of 3D LC director profile also exists [78].

Potentially, all kinds of LCD modes can be driven with a 2D switching array. For this reason, it is worth seeing how many polarizers are necessary to operate individual LCD technology. This matter is shown in TABLE 10.1 in appendix A. Among them, the following polarizer-free technologies have been proposed and actually demonstrated, mainly as reflective LCDs: (1) guest–host-type LCDs using amorphous chiral nematics [47]; (2) cholesteric–nematic phase change mode LCDs [79]; (3) NCAP or polymer-dispersed LCDs [46]; (4) polymer-stabilized Ch-LCDs [80]; (5) TN-LCDs with photopolymer gratings [81]; (6) G-H LCDs with spiral nematic hosts [82], and so forth. All these technologies are expected to be of the reflective type without the need for backlighting. The goal for the reflectivity is 50–70%. An instrument installed with a 6 inch diagonal backlight-free reflective LCD may consume about 0.2 W for operation.

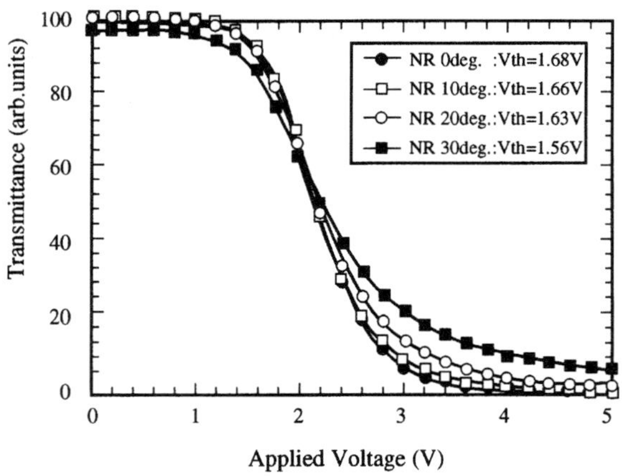

(a) NR

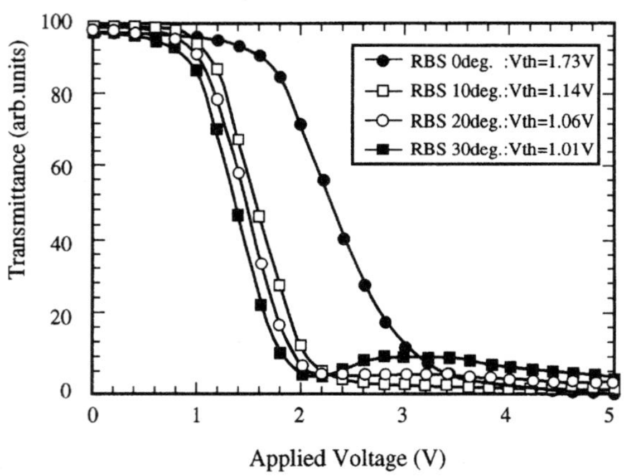

(b) RBS

FIGURE 10.31. Angular dependence of the EO performance of (a) NR and (b) RBS TN cells, respectively. (Y. Toko, T. Sugiyama, K. Katoh, Y. Iimura, and S. Kobayashi, J. Appl. Phys. 74, 1993.)

Regarding the differences in the power consumptions according to the driving schemes, the method of column-line inversion is superior to that of row-line inversion in terms of power consumption.

10.10 ISSUES IN THE PRODUCTION OF AM-LCDs

To fabricate an AM-TN-LCD, about 21 process steps are necessary. We need to develop the technology that will enable fabrication of AM-LCDs that feature in particular large area and/or high resolution. The production yield, Y, is generally expressed by $\exp(-kX)$, where X is the parameter that characterizes the scale of the device, for example, the area of display or the resolution. The overall yield may be expressed by the product of each part, such as TFT-array fabrication, LC cell fabrication, and display module assembly.

In order to increase the production yield, we should do the following: (1) simplify the production procedure; for example, reduce the number of masks from seven or eight to four or five for TFT fabrication (two-terminal devices need two or three masks [83]); (2) devise and fabricate processes that are free from creation defects, for example, lay gate electrodes on the TFT array substrate while the signal electrodes are being deposited on the opposite substrates—this configuration is free from crossover [84]; (3) add appropriate and necessary redundancy, for example, two transistors for one pixel [85], or install the two systems that operate alternately (other ideas are to install dual TFTs [86] and to divide pixels into two parts [87]); (4) repair the defects by, for example, repairing broken parts of electrode lines with MOCVD using a laser, or cutting the short. These repair methods must be further developed.

10.11 AM-LCD USING MODES OTHER THAN THE TN-LCD

As mentioned already, all possible kinds of LCD modes can be operated and driven using a three-terminal or a two-terminal 2D array. In fact, AM-driven polymer-dispersed LCDs [88], and FLCDs are already reported [89], together with the TB mode [90]. In these ways, it may be possible to attain the unique performance discussed in section 10.2.

10.12 FUTURE ASPECTS OF AM-LCDs

FIGURE 10.32 depicts a concise representation of the performance of electronic displays in terms of their information contents and diagonals. LCDs are useful for instruments that require very small sizes for video equipment, to very large sizes.

In terms of the development of AM-LCDs, all areas of information displays shown on FIGURE 10.32 will be filled by LCDs except for huge direct-view displays, for example, for stadium use. The instruments installing these LCDs will be used in a network system.

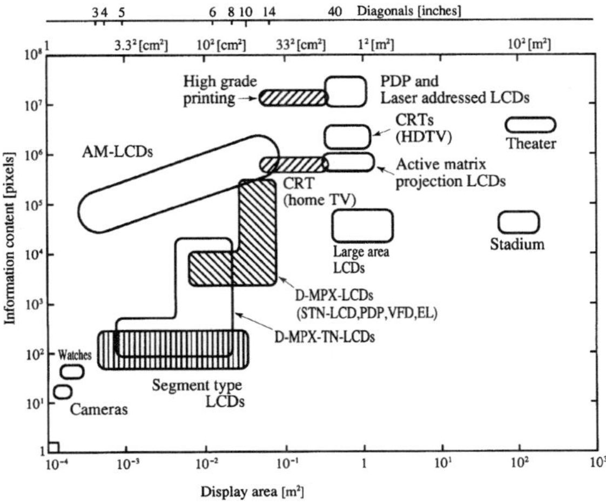

FIGURE 10.32. Information contents and display areas of various electronic displays.

Features of these LCDs will be:

(1) small LCDs having diagonals of several cm and a high resolution of several tens of μm (about 10 μm) driven by p-Si TFTs;
(2) displays for personal (portable) instruments; the display must be of the reflective type with a high reflectivity comparable to that of paper—50 to 70%—without sacrificing legibility by giving a contrast ratio larger than 10:1; multicolors and a fast response speed (this kind of LCD may be called "electronic paper");
(3) personal computers will play the role of televisions in an advanced network society and human factor issues, such as user fatigue, will be largely improved;
(4) most workstations will install LCDs to resolve the spatial issue and human factor issue;
(5) meter-size wall hanging TV screens made of AM-LCDs will be produced by developing manufacturing equipment and by improving the legibility; and
(6) projection displays of both the front and rear projection types for large screens will be developed by improving the luminance levels compared with that of the best OHP currently available.

Another future prospect is materials research. Room temperature AM-LCD processes will be developed, for example, by using organic materials for TFTs and diodes. These technologies may promote the utilization of plastic films for LCD substrates. New materials with high conductivities are necessary for electrodes in large-area direct-view LCDs.

APPENDIX A

Operation principles, nomenclatures, and features of LCDs reported in the literature are listed in TABLE 10.1.

APPENDIX B

For designing practical AM-LCDs using TFTs, a simulation using a software such as SPICE is effectively utilized. In these simulations, the formula for the current (I_{DS}) versus voltage (V_{DS}) characteristics of the MOS-FET is applied to TFTs using a-Si or p-Si.

The formula for below threshold V_{TH} is

$$I_{DS} = (W/L)\mu_{\text{eff}}C_I(V_{GS} - V_{TH} - V_{DS}/2)V_{DS}$$
$$\text{for } V_{DS} < V_{GS} - V_{TH}$$

and above the threshold

$$I_{DS} = (W/2L)\mu_{\text{eff}}C_I(V_{GS} - V_{TH})^2$$
$$\text{for } V_{DS} > (V_{GS} - V_{TH})$$

where W, L, μ_{eff}, C_I, and V_{TH} stand for the width and the length of channel, the effective mobility, the capacitance of the gate insulator per unit area, and the threshold voltage [1].

It is almost possible to have a good fitting to the measured values by choosing relevant parameter appropriately [2,3,4]. No analytical expression for the off-current as shown in appendix E has been obtained yet. However, it is necessary to design and fabricate TFTs having a low leak current by reducing the surface states between semiconductor and insulator and also reducing the photocurrent of the TFTs in terms of the elimination of lights shining on the TFTs.

It is necessary to conduct a simulation of overall characteristics of AM-LCDs by taking account of all the electrooptical effects and detailed equivalent circuits.

The reader may refer to the following literature.

1. Physics of Semiconductor Devices, Sze, S. M. 1969 (New York: Wiley-Interscience).
2. Shur, M. S., Jacunski, M. D., and Hack, M. reference [16] in the text; Physics of Semiconductor Devices,

Table 10.1. Operation modes of LC devices. Legend: E for electric field; I for electric current; T for thermal effect; Y for yes; N for no; MPX for multiplexability

LCs	Modes	Abbreviation	Effects	Principles	Reflective	Transmissive	Projection	Color	MPX	Memory	Number of polarizers
	Tunable Birefringence	TB	E	Birefringence	Y	Y	Y	Y	Y	N	2
	Super Twisted Nematic	STN	E	Birefringence	Y	Y	Y	Y	excellent	N	2
N(N*)	Twisted Nematic	TN	E	Optical rotation	Y	Y	Y	Y	Y	N	2
	Guest Host	GH	E	Dichroism	Y	Y	excellent Y	Y	Y/N	N	1 or 0
	Dynamic Scattering	DS	E	Light Scattering	excellent Y	Y	Y	weak	weak Y	N	0
	Polymer Dispersed	PD	E	Light Scattering	Y	Y	Y	Y	N	N	0
Ch	Optical Bragg Reflection	BR	E or T	Selective reflection	Y	excellent N	N	Y	N	Y	0
	Phase change (unwinding)	PC	E	Light Scattering	excellent Y	Y	Y	Y	Y	N Y	0
SA	Thermally Addressed	TA	T + E	Light Scattering	Y	Y	excellent Y	Y	Y N	Y	0
	Electroclinic	EC	E	Birefringence	Y	Y	excellent Y	Y	N/?	excellent N	2
	Tunable Birefringence	TB	E	Birefringence	Y	Y	Y	Y	Y	Y	2
SC*	Surface Stabilized	SSFLC	E	Birefringence	Y	Y	Y	Y	Y	excellent Y	2
	Guest Host	GH	E	Dichroism	Y	Y	Y	Y	Y	excellent Y	1 or 0
	Unwinding	UW	E	Light Scattering	excellent Y	Y	Y	Y	Y	N	0

Shur, M., 1990 (Englewood Cliffs, NJ: Prentice-Hall).
3. Straub, D., Lueder, E., Kallfass, T., and Schleupen, K. 1994. *Conf. Record 1994 IDRC* **14** (Monterey): 398.
4. Quinn, M. and Milglioto, P. 1994. *Conf. Record 1994 IDRC* **14** (Monterey): 402.

APPENDIX C

The major technologies for AM-LCDs using TFDs are those using Ta_2O_5 and SiN_x as insulators or semi-insulators. For these diodes, the Poole–Frenkel equation is applicable to the current–voltage characteristic:

$$I = AV \exp\left[-\frac{1}{kT}\left(\phi_B - \frac{b}{\sqrt{V}}\right)\right]$$

where ϕ_B is the potential energy for trapped carriers in the insulator layer [1].

As shown in FIGURE 10.12, log I is proportional to $\sqrt{V}$ for the Ta_2O_5 TFD, and to V for the SiN_x TFD.

The actual $I-V$ data of an MIM diode contain both the tunneling and ohmic current parts.

An MIM diode using a polyimide LB film as an insulator is shown to obey the Poole–Frenkel formula.

1. Physics of Semiconductor Devices, Sze, S. M. 1969 (New York: Wiley-Interscience).
2. Reference [44] in the text.
3. Abe, H., Ishizaki, J., Ando, N., Iimura, Y., and Kobayashi, S. 1994. *Trans. Mater. Res. Soc. Japan* **15A** 543.

APPENDIX D

The methods for color formation in LCDs are classified into the following, in terms of components and systems.

Regarding single display component or pixel (or subpixel) of a display, the mechanics of color formations in LCDs are:

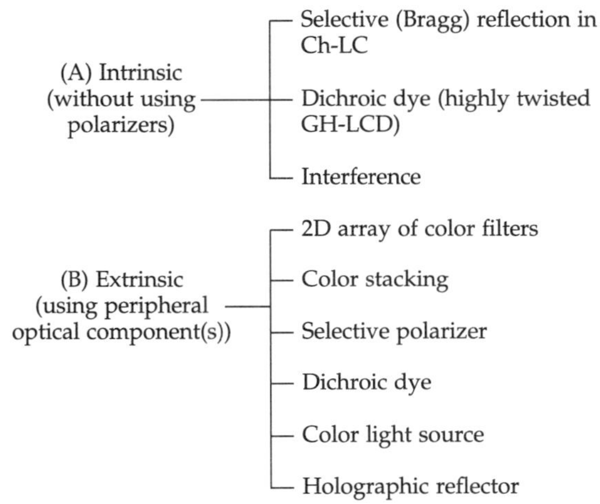

Systems for displaying 2D images:

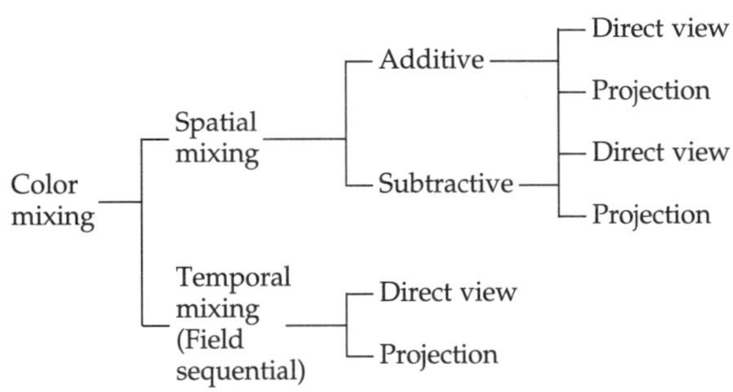

Regarding photometric, colorimetric, and ergonomic characterizations of color LCDs, the reader may refer to, for example, Kobayashi, S. 1983. *Proc. Japan Disp.* **83**, 198.

APPENDIX E

An example of the detailed data of the current–voltage characteristics of a-Si TFTs, including the behavior of leak currents.

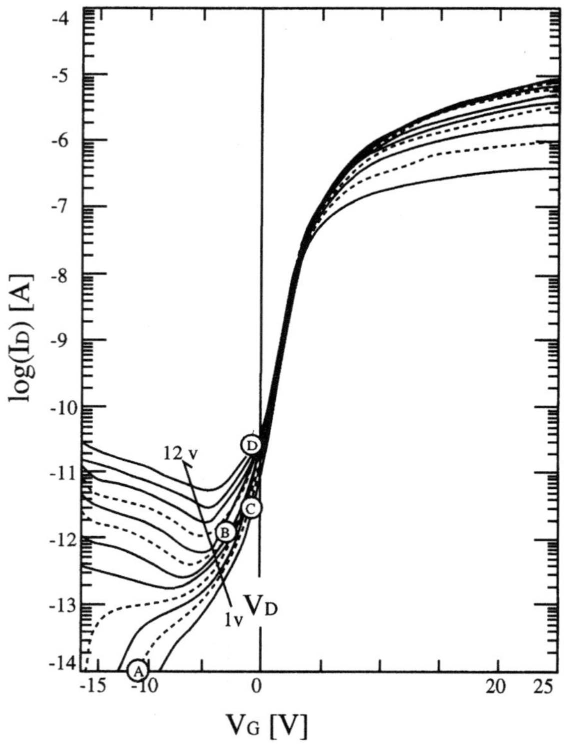

Drain current versus gate voltage of an a-Si TFT, with drain voltage as a parameter. The curves exhibit off-currents (leakage).

APPENDIX F

The states of TFTs in regions A to D, shown in FIGURE 10.25.

V_{SIG} : 13V for on state (postive-cycle)
3V for on state (negative-cycle)
Voltage level shift in pixel electrode ΔV_P=2V

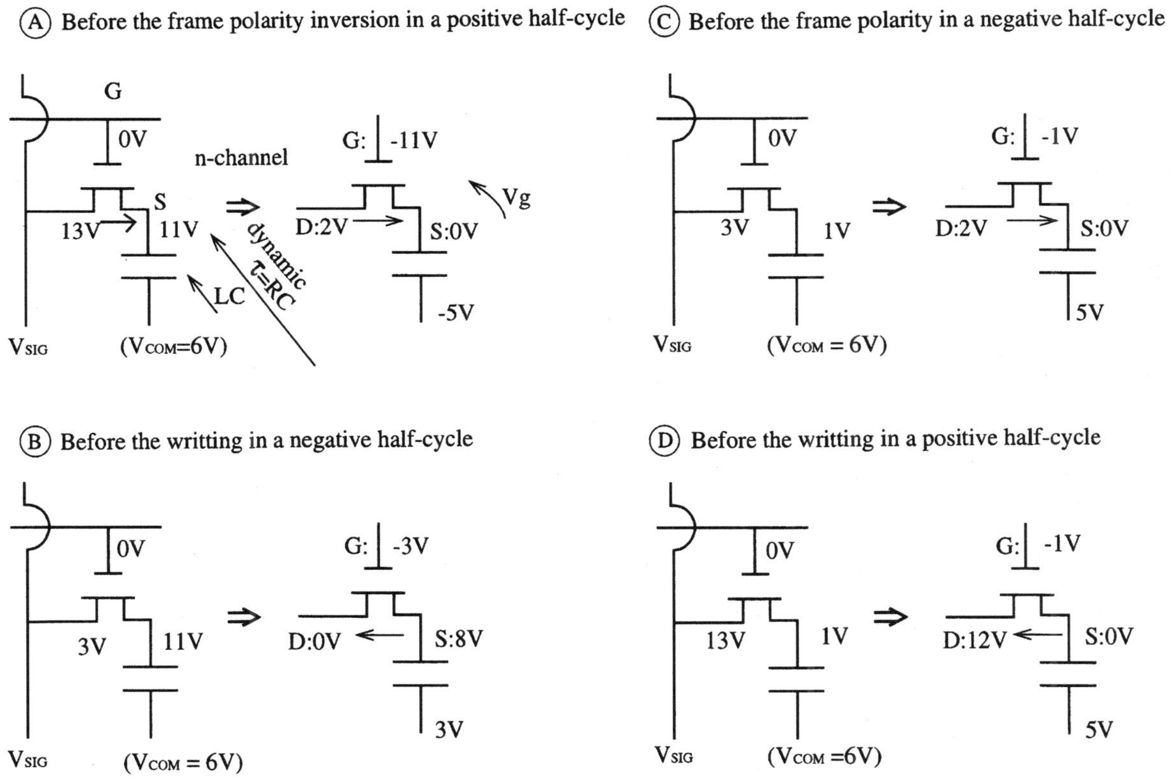

The operation states of TFT in the periods A to D shown in FIGURE 10.25.

REFERENCES

1. Fergason, J. L. 1964. *Sci. Am.* **211**: 77.
2a. Williams, R. 1963. *J. Chem. Phys.* **39**: 384.
2b. Heilmeier, G. H., Barton, L. A., and Zanoni, L. A. 1968. *Appl. Phys. Lett.* **13**: 46.
3. Schadt, M., and Helfrich, W. 1971. *Appl. Phys. Lett.* **18**: 127.
4a. Castellano, J. A. 1988. *Mol. Cryst. Liq. Cryst.* **165**: 389.
4b. Uchida, T., Morozumi, S., and Sasaki, A. 1988. *Mol. Cryst. Liq. Cryst.* **165**: 533.
5a. Lechner, B. J. 1971. *Proc. IEEE* **59**: 1566.
5b. Brody, T. P., Asars, J. A., and Dixon, G. D. 1973. *IEEE Trans. Elec. Dev.* **ED-20**: 995.
5c. Baraff, D. R., and Long, J. R. 1973. *IEEE Trans. Elec. Dev.* **ED-20**: 995.
5d. Streater, R. W. 1980. *Conference Record of 1980 Biennial Display Research Conference*, p. 109. New York: IEEE.
6. Scheffer, T. J., and Nehring, J. 1984. *Appl. Phys. Lett.* **45**: 1021.
7. Clark, N. A., and Lagerwall, S. T. 1980. *Appl. Phys. Lett.* **36**: 899.
8. Yamauchi, S., Aizawa, M., Clerc, J. F., Uchida, T., and Duchene, J. 1989. *SID Dig. Tech. Papers* **20**: 378.
9a. Yoshida, K., Fukuro, H., and Kobayashi, S. 1986. *Japan Display '86: Proceedings of the Sixth International Display Research Conference [Tokyo]*, p. 396. Playa del Rey, Calif.: Society for Information Display.
9b. Fukuro, H., and Kobayashi, S. 1988. *Mol. Cryst. Liq. Cryst.* **163**: 157.
10. Funada, F., Matsuura, M., and Wada, T. 1984. *US Patent* 4,443,065.
11. Nagato, M., Wada, H., and Matsuzawa, K. 1987. *Nikkei Microdevices* **October**: 84.
12. Iwashita, Y., and Wada, H. 1988. *Nikkei Microdevices* **October**: 57.
13. Scheffer, T. J., and Nehring, J. 1990. *Liquid Crystals: Applications and Uses*, vol. 1, ed. B. Bahadur, ch. 10. Singapore: World Scientific.
14. Luo, F. C. 1990. *Liquid Crystals: Applications and Uses*, vol. 1, ed. B. Bahadur, ch. 15. Singapore: World Scientific.
15a. Firester, A. H. 1990. *Sem. Lect. Notes SID* **1**: M–8.
15b. Bruce, R. H. 1991. *Sem. Lect. Notes SID* **2**: F–3.
15c. Howard, W. E. 1992. *Sem. Lect. Notes SID* **2**: F–2.
15d. Morozumi, S. 1993. *Sem. Lect. Notes SID* **1**: M–8.
15e. Wisnieff, R. L. 1994. *Sem. Lect. Notes SID* **1**: M–2.
16a. Howard, W. E. 1994. *Conference Record of the Fourteenth International Display Research Conference [Monterey]*, p. 6. Playa del Rey, Calif.: Society for Information Display. Hereafter cited as *IDRC 14*.
16b. Oshima, H. 1994. *IDRC 14*, p. 26.
16c. Awane, K. 1994. *IDRC 14*, p. 1.
16d. Kobayashi, S., Iimura, Y., and Nishikawa, M. 1994. *IDRC 14*, p. 78.
16e. Knapp, A. G., and Hartman, R. A. 1994. *IDRC 14*, p. 14.
16f. Lüder, E. 1994. *IDRC 14*, p. 30.
16g. Salerno, J. P. 1994. *IDRC 14*, p. 39.
16h. Shur, M. S., Jacunski, M. D., and Hack, M. 1994. *IDRC 14*, p. 45.
16i. Doane, J. W., St. John, W. D., Lu, Z. J., and Yang, D. K. 1994. *IDRC 14*, p. 65.
16j. Okamoto, M., Yamamoto, Y., Fukuoka, N., Hisataka, Y., Yamamoto, T., Tanaka, Y., Hatoh, H., Monma, K., Sunohara, K., Hasegara, R., and Nagata, H. 1994. *IDRC 14*, p. 88.
16k. Hashimoto, T., Katoh, K., Hasebe, H., Takatsu, H., Iwamoto, Y., Iimura, Y., and Kobayashi, S. 1994. *IDRC 14*, p. 484.
17. Alt, P. M., and Pleshko, P. 1974. *IEEE Trans. Elec. Dev.* **ED-21**: 146.
18. Le Comber, P. G. 1979. *Electron. Lett.* **15**: 179.
19. Morozumi, S., Oguchi, K., Yazawa, S., Kodaira, T., Ohshima, H., and Mano, T. 1983. *SID Dig. Tech. Papers* **16**: 156.
20. Brody, T. P., Asars, J. A., and Dixon, G. D. 1973. *IEEE Trans. Elec. Dev.* **ED-20**: 739.
21. Lipton, L. T., Meyer, M. A., and Massetti, D. O. 1975. *SID Dig. Tech. Papers* **6**: 78.
22a. Salerno, J. P., Vu, D. P., Dingle, B. D., Batty, M. W., Ipri, A. C., Stewart, D. L., and Tilton, M. L. 1992. *SID Dig. Tech. Papers* **23**: 63.
22b. Salerno, J. P. 1994. *Proc. SID 14*, p. 39.
23. Castleberry, D. E. 1979. *IEEE Trans. Elec. Dev.* **ED-26**: 1123.
24. Baraff, D. R., Long, J. R., MacLaurin, B. K., Miner, C. J., and Streater, R. W. 1981. *IEEE Trans. Elec. Dev.* **ED-28**: 736.
25. Morozumi, S., Ohta, T., Araki, R., Sonehara, T., Kubota, K., Ono, Y., Nakazawa, T., and Ohara, H. 1983. *Japan Display '83: Proceedings of the Third International Display Research Conference [Kobe]*, p. 404. Los Angeles: Society for Information Display.
26. Togashi, S., Sekiguchi, K., Tanabe, H., Yamamoto, E., Sorimachi, K., Tajima, E., Watanabe, H., and Shimizu, H. 1984. *SID Dig. Tech. Papers* **15**: 324.
27a. Yaniv, Z., Lien, A., den Boer, W., and Catchpole, C. E. 1985. *Conference Record of the Fifth International Display Research Conference [Orlando, Fla.]*, p. 76. Santa Anna, Calif.: Society for Information Display.
27b. Knapp, A. G., and Hartman, R. A. 1994. *Proc. SID 14*, p. 14.
28. Szydlo, N., Chartier, E., Perbet, J. N., Prout, N., Magarino, J., and Hareng, M. 1983. *Japan Display '83: Proceedings of the Third International Display Research Conference [Kobe]*, p. 416. Los Angeles: Society for Information Display.
29. Toyama, M., Harajiri, T., Maeda, T., Kuroda, Y., Tsunoda, Y., Suzuki, M., and Yamazaki, T. 1987. *SID Dig. Tech. Papers* **18**: 155.

30. Awane, K. 1994. *Proc. SID 14*, p. 1.
31. Ohshima, H. 1994. *Proc. SID 14*, p. 26.
32. Ohshima, H. 1994. *Proc. SID 14*, p. 26.
33. Matsueda, Y. 1993. *Proceedings of the Thirteenth International Display Research Conference [Strasbourg]*, p. 601. Saint Brieuc: Le Club Visu.
34. Ueda, T., Sugahara, J., Toeda, H., Akiyama, M., Ikeda, M., and Suzuki, K. 1993. *SID Dig. Tech. Papers* **24**: 739.
35. Weimer, P. K. 1962. *Proc. IRE* **50**: 1462.
36. Brody, T. P., Luo, F. C., Szepesi, Z. P., and Davis, D. H. 1975. *IEEE Trans. Elec. Dev.* **ED-22**: 739.
37. Brody, T. P., Asars, J. T., and Dixon, G. D. 1973. *IEEE Trans. Elec. Dev.* **ED-20**: 995.
38. Tsumura, A., Koezuka, H., and Ando, T. 1986. *Appl. Phys. Lett.* **49**: 1210.
39. Morozumi, S., Ohta, T., Araki, R., Sonehara, T., Kubota, K., Ono, Y., Nakazawa, T., and Ohara, H. 1983. *Japan Display '83: Proceedings of the Third International Display Research Conference [Kobe]*, p. 404. Los Angeles: Society for Information Display.
40. Aota, K., Kato, Y., Hoshini, K., Takahashi, S., Irino, M., Kikuchi, M., Hamano, M., and Togashi, S. 1991. *SID Dig. Tech. Papers* **17**: 219.
41. Nicholas, K. H. 1990. *Eurodisplay '90: Proceedings of the Tenth International Research Display Conference [Amsterdam]*, p. 170. Berlin: VDE.
42. Knapp, A. G., and Hartman, R. A. 1994. *Proc. SID 14*, p. 14.
43. Ast, D. G. 1982. *Proceedings of the Second International Display Research Conference [Philadelphia]*, p. 152. Santa Anna, Calif.: Society for Information Display.
44a. Shannon, J. M. 1995. *Vakuumelektronik und Displays*, p. 291. ITG–Fachbericht, 132. Berlin: VDE.
44b. Hartman, R. A. 1995. *SID Dig. Tech. Papers* **26**: 7.
45. Maeda, H., Otobe, Y., and Kobayashi, S. 1990. *Proc. SID* **31**: 79.
46. Drzaic, P., Wiley, R., McCoy, J., and Guillaume, A. 1990. *SID Dig. Tech. Papers* **21**: 210.
47. Koimai, H., Iimura, Y., Kobayashi, S., Hashimoto, T., Sugiyama, T., and Katoh, K. 1995. *SID Dig. Tech. Papers* **26**: 699.
48. Oh-ishi, H., Takeuchi, M., Kobayashi, S., and Shimomura, T. 1985. *Eurodisplay '84: Proceedings of the Fourth International Display Research Conference [Paris]*, p. 113. Playa del Rey, Calif.: Society for Information Display.
49. Dojho, M., Aoki, T., Suzuki, K., Ikeda, M., Higuchi, T., and Oana, Y. 1988. *SID Dig. Tech. Papers* **19**: 330.
50. Nomoto, T., Sekido, Y., Abiko, I., and Kaminishi, K. 1989. *Japan Display '89: Proceedings of the Ninth International Display Research Conference [Kyoto]*, p. 502. Santa Anna, Calif.: Society for Information Display.
51. Shibata, K., Takahashi, K., Futami, T., Suzuki, K., Ohta, M., Kitajima, M., Hasebe, A., and Mano, H. 1992. *Japan Display '92: Proceedings of the Twelfth International Display Research Conference [Hiroshima]*, p. 479. Santa Anna, Calif.: Society for Information Display.
52. Moriyama, H., Uchida, H., Nishida, S., Nakano, H., Hirai, Y., Kaneko, S., and Tani, C. 1989. *SID Dig. Tech. Papers* **20**: 144.
53. Ikeda, M., Ogawa, M., and Suzuki, K. 1989. *Japan Display '89: Proceedings of the Ninth International Display Research Conference [Kyoto]*, p. 498. Santa Anna, Calif.: Society for Information Display.
54. Hori, H. 1991. *Proc. SID* **32**: 311.
55. Rieger, B., and Kobayashi, K. 1993. *Displ. Imaging* **1**: 191 (in Japanese).
56. Nishikawa, M., Suganuma, T., Tsuda, Y., Bessho, N., Iimura, Y., and Kobayashi, S. 1994. *Japan. J. Appl. Phys. Lett.* **33**: 1113.
57. Kobayashi, S., Iimura, Y., and Nishikawa, M. 1994. *Proc. SID 14*, p. 78.
58. Powell, M. J., Chapman, J. A., Knapp, A. G., French, I. D., Hughes, J. R., Pearson, A. D., Allison, M., Edwards, M. J., Ford, R. A., Hemings, M. C., Hill, O. F., Nicolls, D. H., and Wright, N. K. 1987. *Eurodisplay '87: Proceedings of the Seventh International Display Research Conference [London]*, p. 63. London: Institute of Physics.
59. Nasu, Y., Kawai, S., Kisumi, S., Oki, K., and Hori, K. 1986. *SID Dig. Tech. Papers* **17**: 289.
60. Howard, W. E., Alt, P. M., and Wisnieff, R. L. 1988. *Conference Record of the Eighth International Display Research Conference [San Diego]*, p. 230. Santa Anna, Calif.: Society for Information Display.
61. Takubo, Y., Takeda, M., Tamura, T., Iwai, H., Ukita, K., Bessho, Y., Takahara, H., Komori, K., and Yamashita, I. 1989. *Japan Display '89: Proceedings of the Ninth International Display Research Conference [Kyoto]*, p. 584. Santa Anna, Calif.: Society for Information Display.
62. Sarma, K. S., Franklin, H., Johnson, M., Frost, K., and Bernot, A. 1989. *SID Dig. Tech. Papers* **20**: 148.
63. Sunata, T., Inada, T., Nakagawa, T., Matsushita, Y., Ugai, Y., and Aoki, S. 1991. *Conference Record of the Eleventh International Display Research Conference [San Diego]*, p. 255. Santa Anna, Calif.: Society for Information Display.
64. Yang, K. H. 1991. *Conference Record of the Eleventh International Display Research Conference [San Diego]*, p. 68. Santa Anna, Calif.: Society for Information Display.
65. Koike, Y., Kamata, T., Okamoto, K., Tomita, I., and Okabe, M. 1992. *SID Dig. Tech. Papers* **23**: 798.
66. Takatori, K., Sumiyoshi, K., Hirai, Y., and Kaneko, S. 1992. *Japan Display '92: Proceedings of the Twelfth International Display Research Conference [Hiroshima]*, p. 591. Playa del Rey, Calif.: Society for Information Display.
67. Okamoto, M., Yamamoto, Y., Fukuoka, N., Hisataka, Y., Yamamoto, T., Tanaka, Y., Hatoh, H., Monma, K., Sunohara, K., Hasegara, R., and Nagata, H. 1994. *Proc. SID 14*, p. 88.

68. Hashimoto, T., Sugiyama, T., Katoh, K., Saitoh, T., Suzuki, H., Iimura, Y., and Kobayashi, S. 1995. *SID Dig. Tech. Papers* **26**.
69. Toko, Y., Sugiyama, T., Katoh, K., Iimura, Y., and Kobayashi, S. 1993. *J. Appl. Phys.* **74**: 2021.
70. Bos, P. J., Rahman, J. A., and Doane, J. W. 1993. *SID Dig. Tech. Papers* **24**: 877.
71. Miyashita, T., Kuo, C.-L., and Uchida, T. 1994. *SID Dig. Tech. Papers* **25**: 44.
72a. Sugiyama, T., Toko, Y., Hashimoto, T., Katoh, K., Iimura, Y., and Kobayashi, S. 1993. *Japan. J. Appl. Phys.* **32**: 5621.
72b. Sugiyama, T., Toko, Y., Hashimoto, T., Katoh, K., Iimura, Y., and Kobayashi, S. 1995. *Japan. J. Appl. Phys.* **34**: 2396.
73. Hashimoto, T., Katoh, K., Hasebe, H., Takatsu, H., Iwamoto, Y., Iimura, Y., and Kobayashi, S. 1994. *IDRC 14*, p. 484.
74. Haas, G., Fritsch, M. G., Wohler, H. F., and Mlynski, D. A. 1990. *SID Dig. Tech. Papers* **21**: 102.
75. Lien, A. 1990. *Eurodisplay '90: Proceedings of the Tenth International Display Research Conference [Amsterdam]*, p. 248. Berlin: VDE.
76. Okita, Y., Yamada, T., Nakatani, N., Ohshima, S., Murashita, Y., Arioka, T., and Nakakado, T. 1990. *SID Dig. Tech. Papers* **21**: 411.
77. Sumiyoshi, K., Takatori, K., Takahashi, N., and Hirai, Y. 1993. *Liq. Cryst.* **14**: 1199.
78. Kitamura, M. 1995. *SID Dig. Tech. Papers* **26**: 540.
79. Yabe, Y., Yamada, H., Hoshi, T., Yoshihara, T., Mochizuki, A., and Yoneda, Y. 1993. *J. SID* **1**: 43.
80. Doane, J. W., St. John, W. D., Lu, Z. J., and Yang, D. K. 1994. *Proc. SID 14*, p. 65.
81. Tanaka, K., Katoh, K., Tsuru, S., and Sakai, S. 1993. *Eurodisplay '93: Proceedings of the Thirteenth International Display Research Conference [Strasbourg]*, p. 109. Saint Brieuc: Le Club Visu.
82. Takatsu, H., Umezu, Y., Hasebe, H., Takeuchi, K., Suzuki, K., Iimura, Y., and Kobayashi, S. 1995. *SID Dig. Tech. Papers* **26**: 579.
83. Tanaka, H., Motte, S., Hoshino, M., Takahashi, K., Ohta, M., Sakai, T., and Yamazaki, T. 1987. *SID Dig. Tech. Papers* **18**: 140.
84. Oki, K. 1987. *Eurodisplay '87: Proceedings of the Seventh International Display Research Conference [London]*, p. 55. London: Institute of Physics.
85. Takeda, M., Ogo, S., Tamura, T., Kamiura, H., Noda, H., and Kawaguchi, T. 1986. *Japan Display '86: Proceedings of the Sixth International Display Research Conference [Tokyo]*, p. 204. Playa del Rey, Calif.: Society for Information Display.
86. Tanaka, M., Ogo, S., Tamura, T., Kamiura, H., Noda, H., Kawaguchi, T., Yamashita, J., Ando, J., and Kuroda, H. 1986. *Japan Display '86: Proceedings of the Sixth International Display Research Conference [Tokyo]*, p. 204. Santa Anna, Calif.: Society for Information Display.
87. Nagayasu, T., Oketani, T., Hirobe, T., Kato, H., Mizushima, S., Take, H., Yano, K., Hijikigawa, M., and Washizuka, I. 1988. *Proceedings of the Eighth International Display Research Conference [San Diego]*, p. 56. Santa Anna, Calif.: Society for Information Display.
88. Murai, H., Nakagawa, N., Nakashima, K., Nakanishi, K., Hirose, S., Yuki, M., Shikama, S., and Tanaka, M. 1995. *SID Dig. Tech. Papers* **26**: 329.
89. Tanaka, T., Sakamoto, K., Tada, K., and Ogura, J. 1994. *SID Dig. Tech. Papers* **25**: 430.
90. Koma, N., Baba, Y., and Matsuoka, K. 1995. *SID Dig. Tech. Papers* **26**: 869.
91. Kiefer, R., Weber, B., Windschied, F., and Bauer, G. 1992. *Japan Display '92: Proceedings of the Twelfth International Display Research Conference [Hiroshima]*, p. 547

the on-segments receive a DC-free squarewave with p–p amplitude 2 V. The in-phase/out-of-phase switching at the segment electrode is implemented with dedicated drivers containing exclusive–or gates at each output. The TN display is very compatible with the direct-drive method because a high contrast ratio can be achieved by switching a pixel between zero volts and an arbitrarily high AC voltage.

The 23 segment electrodes of the $3\frac{1}{2}$ digit watch display of FIGURE 11.2 are near the upper limit for the number of segment electrodes in this type of direct-drive display because of electrode routing and interconnect restrictions as well as cost, since every pixel requires its own driver output. To produce displays with more electrodes but fewer interconnections, manufacturers have had to implement matrix addressing methods similar to the time-multiplexing methods used for LED arrays. Even though many LEDs share the same column electrode they can be independently activated in time by cyclically pulsing the matrix rows.

Liquid crystals are non-emissive and respond to voltages differently than LEDs do [3], and so the matrix drive waveforms for an LCD will be different. However, one general addressing principle is common to both LCDs and LEDs, namely that the row signals are independent of the information to be displayed and the column signals carry the pixel information [4]. FIGURE 11.3 shows how the electrodes of the $3\frac{1}{2}$ digit LCD watch display of FIGURE 11.2 can be reconfigured to the electrical equivalent of a four-row, six-column matrix, reducing the drive electrode lead count from 24 to 10.

FIGURE 11.3. The $3\frac{1}{2}$ digit watch display of FIGURE 11.2 configured as a four-row, six-column matrix, reducing the driver lead count from 24 to 10.

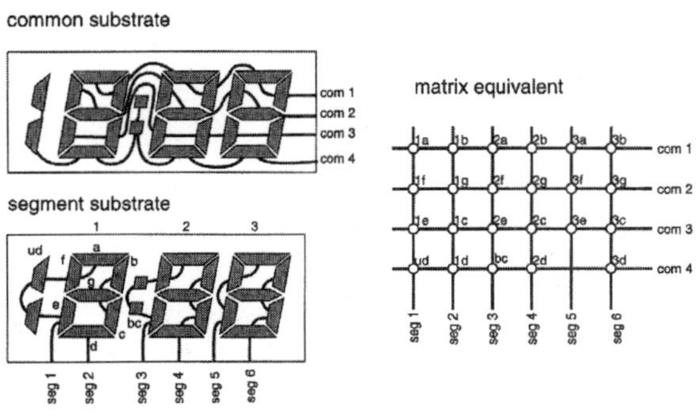

It is important to realize that a drive waveform applied to a column electrode influences the pixel waveforms appearing across every pixel in that column. In order to display a desired pattern of pixels, it is therefore necessary to ascertain the characteristics of the pixel waveform that the electrooptical effect responds to so that this can be exploited by the addressing technique. In the surface-stabilized ferroelectric liquid crystal display, for example, the polarity of the last applied pulse whose amplitude-time product lies above a certain threshold value determines the optical state of the pixel [5]. For the reflective cholesteric display [6] the magnitude of a bipolar selection pulse is used to switch between the highly contrasting planar and focal–conic textures that then remain stored at a lower, nonselecting voltage or even at zero volts. Other electrooptic effects, such as the TN [7], STN [8], and superhomeotropic [9] LCD effects respond, over a wide range of frequencies, only to the root mean square (rms) value of the pixel waveform. The addressing schemes described in the remainder of this chapter will focus on rms-responding electrooptical effects [10].

11.1.3 RMS-Responding Electrooptical Effects

A sufficient condition for rms response is that the characteristic response time of the liquid crystal display, which is governed by viscoelastic properties, be many times longer than the period of the drive waveform. In addition, the interaction between the liquid crystal and the instantaneous applied field must be proportional to E^2, which is the case when the interaction with an electric field occurs purely through induced polarization. (The induced polarization, which is proportional to the electric field, interacts with the electric field itself giving the E^2 dependence.) When these two conditions are met, the optical response is determined by the average value of the voltage squared, or equivalently, the rms voltage.

The assumption of an rms-responding electrooptical effect, as defined above, is all that needs to be known about the electrooptical effect to be able to define the matrix addressing problem. It now becomes only a matter of choosing row and column addressing waveforms that will generate rms voltages appearing across each pixel in the display to produce the desired optical states.

11.2 AMPLITUDE SELECTION METHOD

The method first developed for matrix addressing rms-responding LCDs, known as the amplitude selection method, is still in use today. FIGURE 11.4 shows the amplitude selection method for an N-row matrix. For clarity, only one column is illustrated, although more columns could be included without altering the principle since the columns are independent of one another. The matrix rows, normally held at zero potential, are sequentially "selected" with a momentary voltage level of amplitude S, returning to zero as the next row is selected. The row waveforms are thus independent of the information to be displayed. After the last row has been selected a new frame begins and the process is repeated for as long as power is applied to the display. The selection interval Δt is thus the frame period T divided by the number of rows N, that is, $\Delta t = T/N$. In order to fulfill the rms response condition in a display that is updated at a 60 Hz rate, typical for a CRT interface, the display response time must be 200 ms or longer. For later discussion, it will be convenient to consider the frame period to be subdivided into N addressing intervals, or simply time intervals.

In the amplitude selection method, the drive waveforms applied to the column electrodes carry the information to be presented on the display in the form of polarity modulation. The column waveforms consist of two voltage levels of the same magnitude with a polarity at any particular time that is determined by whether the pixel on the row that is being selected at that time is "on" or "off". When a certain pixel in the column is chosen to be on, that column receives a voltage level $-D$ of opposite polarity to the select voltage S during the time interval when that pixel's row is selected. Similarly, the column electrode receives a voltage level $+D$ of the same polarity as S during the selection interval for an off-pixel. Since the pixel voltage is the difference between the row and column voltages, an on-pixel therefore receives a voltage $S+D$ during the selection interval and an off-pixel receives $S-D$. During the remaining $N-1$ time intervals of the frame period, these pixels receive voltages alternating between $+D$ and $-D$, the exact sequence depending upon the pixel input data for the other pixels in the column (cf. FIGURE 11.4). Of course, the pixel voltage at every time interval contributes to the overall rms pixel voltage, but this does not introduce crosstalk because the pixel voltages corresponding to the unselected pixels in the column all have the same magnitude and only vary in polarity.

Until the early 1970s, it was mistakenly thought that the TN electrooptical response behaved more like a peak detector rather than an rms averager. Following this misconception, it was believed that an off-pixel had to receive a pixel waveform whose magnitude remained below a certain threshold value at all times. Thus the S/D ratio for the amplitude selection method was set at $S/D = 2$ for the so-called 3:1 addressing

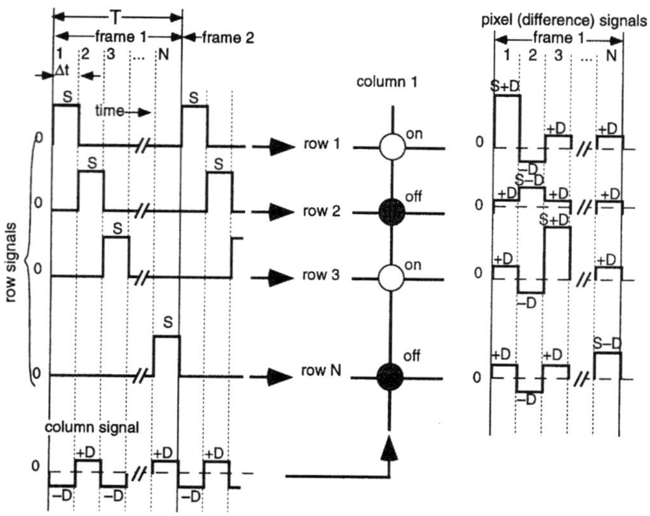

FIGURE 11.4. Illustration of the amplitude selection method on an N-row matrix showing the row addressing waveforms, a column waveform and corresponding pixel difference waveforms.

scheme. With $S/D = 2$, an on-pixel receives $3S/2$ during the selection interval and $|S|/2$ during the remainder of the frame period, and an off-pixel receives $|S|/2$ throughout the entire frame period. By realizing that the LCD was actually an rms averager, Kawakami [11] and Alt and Pleshko [12] independently showed that for matrices with $N > 4$, the efficiency of the amplitude selection method could be greatly improved over the 3:1 scheme resulting in improved display performance. This optimized addressing method is known as Alt and Pleshko addressing.

11.2.1 Optimization of Amplitude Selection

To explain the optimization process, we begin by writing down the expressions for the rms voltage of an on- and off-pixel. Referring to the pixel voltage waveforms on the right-hand side of FIGURE 11.4, an on-pixel receives one time interval of magnitude $S + D$, and $N - 1$ time intervals of magnitude D. The rms voltage of an on-pixel, $\langle \text{on} \rangle$, averaged over one frame period, is obtained by taking the square root of the average value of the squares of the pixel voltage at each addressing interval of the frame period, that is,

$$\langle \text{on} \rangle = \sqrt{\frac{(S+D)^2 + (N-1)D^2}{N}}. \tag{11.1}$$

Again, referring to the pixel waveforms of FIGURE 11.4, an off-pixel receives one time interval of magnitude $S - D$, and $N - 1$ time intervals of magnitude D, so the rms voltage across an off-pixel is

$$\langle \text{off} \rangle \sqrt{\frac{(S-D)^2 + (N-1)D^2}{N}} \tag{11.2}$$

Because the optical response is determined by the rms voltage across the pixel, it is advantageous to make the selection ratio, $\langle \text{on} \rangle / \langle \text{off} \rangle$ as large as possible. $\langle \text{on} \rangle / \langle \text{off} \rangle$ can be expressed in terms of the single variable S/D as:

$$\frac{\langle \text{on} \rangle}{\langle \text{off} \rangle} = \sqrt{\frac{(S/D+1)^2 + N - 1}{(S/D-1)^2 + N - 1}}. \tag{11.3}$$

Optimizing the selection ratio by differentiating equation (11.3) with respect to S/D, and setting the result equal to zero results in:

$$\left.\frac{S}{D}\right|_{\text{opt}} = \sqrt{N} \tag{11.4}$$

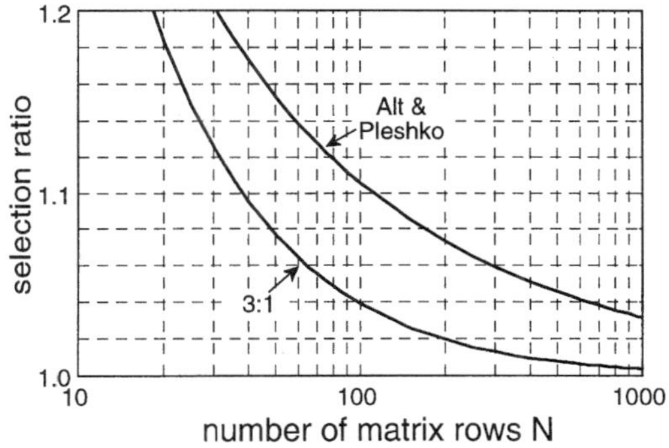

FIGURE 11.5. Selection ratio of the rms voltage of an on-pixel to an off-pixel as a function of the number of matrix rows for the Alt and Pleshko addressing and 3:1 addressing.

yielding

$$\left.\frac{\langle \text{on} \rangle}{\langle \text{off} \rangle}\right|_{\text{opt}} = \sqrt{\frac{\sqrt{N}+1}{\sqrt{N}-1}}. \tag{11.5}$$

For the 3:1 scheme, substituting $S/D = 2$ into equation (11.3) results in:

$$\left.\frac{\langle \text{on} \rangle}{\langle \text{off} \rangle}\right|_{3:1} = \sqrt{\frac{N+8}{N}} \tag{11.6}$$

which is always less than the optimized value for $N > 4$. Note that $\langle \text{off} \rangle$ for a matrix addressed display can never be zero, which explains why the off-segments of a low-level multiplexed watch or calculator display become visible when viewed from below at an oblique angle where the threshold voltage is lowest. Equation (11.5) is often referred to as the "iron law of multiplexing." This relationship, plotted on a semilog scale in FIGURE 11.5, illustrates the asymptotic approach to unity for large N. The results for the poorer performing 3:1 scheme have been included for comparison. According to equation 11.5, for a dual-scan VGA display ($N = 240$) the optimized selection ratio is 1.0667, and for a dual-scan XGA display ($N = 384$) the selection ratio is 1.0524. These 6.7% and 5.2% voltage margins are quite narrow and require a liquid crystal display with a very sharp electrooptical transfer curve in order to provide an adequate contrast ratio. The STN effect and the superhomeotropic effect can provide this sharpness, but the TN effect cannot.

The decrease in selection ratio with increasing number of matrix rows can be visualized from FIGURE 11.6 which depicts the on and off optimized pixel waveforms for one frame of a 10, 20, and 40

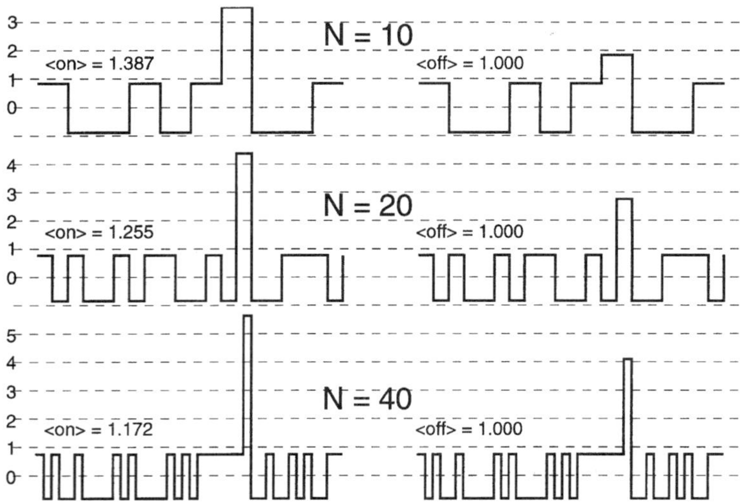

FIGURE 11.6. On and off-pixel waveforms of the optimized amplitude selection technique for LCD matrices of various sizes.

row matrix. For all cases, the voltages have been normalized such that $\langle \text{off} \rangle = 1.0$. For increasing N, the width of the selection interval, the part of the pixel waveform that determines whether the pixel will be on or off, occupies a smaller portion of the frame period and thus has less of an influence in determining the selection ratio, even though the amplitude of the selection pulse is increased. For the 3:1 case where S/D remains fixed at 2 irrespective of N, the selection ratio decreases much more sharply.

In the context of matrix addressing, the terms *row* and *column* refer to those display electrodes that receive the image-independent and image-dependent waveforms, regardless of their orientation in space. To emphasize this point, these electrodes are sometimes referred to as *common* and *segment* electrodes. For a nonsquare matrix, the row electrodes are generally assigned the side having the fewest numbers of electrodes because of the adverse relationship between the selection ratio and the number of addressed rows, indicated by equation (11.5); the number of columns has no influence upon the selection ratio. Thus for a "landscape" format display, the rows would run horizontally and for a "portrait" format display, the rows would run vertically.

11.2.2 Practical Application of Amplitude Selection

The waveforms of FIGURE 11.4 served to illustrate the principle of amplitude selection addressing and optimization of the selection ratio, but they would not be suitable for application to a practical display because of net DC voltages. Depending upon the data pattern, the net DC voltages on the pixel can vary roughly between $+D$ and $-D$. For a liquid crystal material with a 2 V threshold, the net DC voltage could be as large as ± 1.5 V which would cause liquid crystal decomposition due to irreversible electrochemical reactions. This DC problem can be eliminated by reversing the polarity of both the row and column signals during alternate frames, as shown in FIGURE 11.7 (left). This reversal has no effect on the rms value of the pixel waveform.

A disadvantage of reversing the polarities is the doubling of the voltage range requirement of the row drivers from S to $2S$. This increase is significant because a larger-sized silicon chip is needed to accommodate the higher voltages required, and this adds cost and decreases the total number of available outputs per package. However by offsetting (biasing) both row and column voltages by D in the normal frames, and S in the reversed frames (FIGURE 11.7, right), the same DC-free pixel waveform can be obtained with a significantly reduced row voltage of $S + D$. This technique is often referred to as DF switching [13] or the improved Alt and Pleshko technique (IAPT) [10]. However, this technique increases the voltage requirement for the column driver from $2D$ to $S + D$, making it the same as for the row driver. For most display applications this is not viewed as a particular disadvantage because row

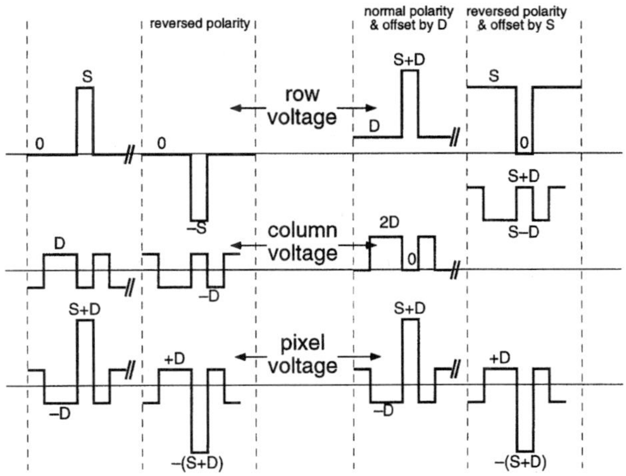

FIGURE 11.7. Reversal of the row and column drive polarities on alternate frames eliminates any net DC component across the pixel (left) but doubles the voltage requirement of the row driver from S to $2S$. However, by additionally offsetting (biasing) both row and column voltages by D in the normal frame, and by S in the reversed frame (right), the identical DC-free pixel waveform can be obtained while significantly reducing the voltage requirement for the row driver to $S + D$.

and column drivers are often used interchangeably. However, for a display with many more columns than rows, one manufacturer [14] has found cost and performance advantages by implementing the unbiased scheme of FIGURE 11.7 (left), trading off a few expensive high-voltage parts for the rows against many inexpensive, low-voltage parts for the columns.

The row and column waveforms in FIGURE 11.7 (right) still do not appear as the actual signals that one would observe on an oscilloscope if one were to probe the driver outputs of an operating display. This is because DF switching is usually performed more frequently than on an alternate frame basis, in order to reduce a certain type of crosstalk (cf. section 11.2.5.1). FIGURE 11.8 shows more representative row, column, and pixel waveforms for a 240-row display where, for clarity, DF switching occurs every 27 addressing intervals. Note that, despite the bizarre appearance of the row and column waveforms, the pixel waveform is still recognizable. To ensure the elimination of average DC it is important that the number of multiplexed rows, N, is not evenly divisible by the number of DF switching intervals. For a 240-row display, the DF switching is usually set at 7 lines.

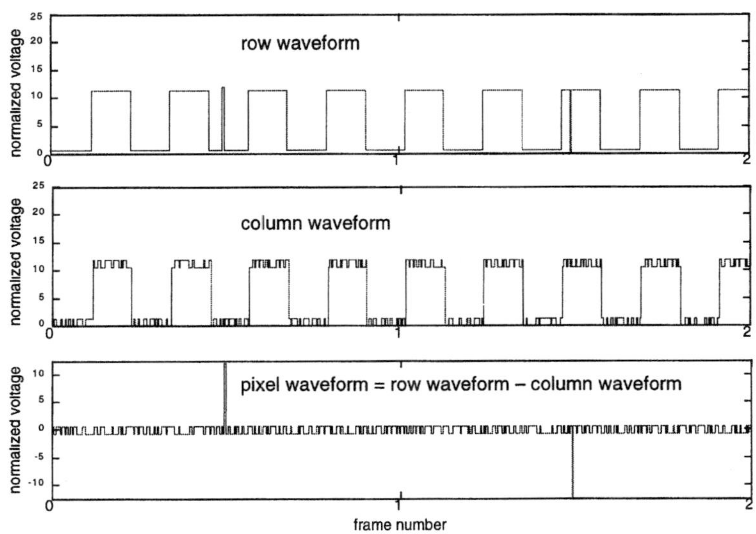

FIGURE 11.8. Row, column, and pixel waveforms for a 240-row display where, to reduce image-dependent crosstalk, DF switching occurs every 27 lines instead of only once per frame, as shown in FIGURE 11.7. These waveforms are more representative of a practical, operating display.

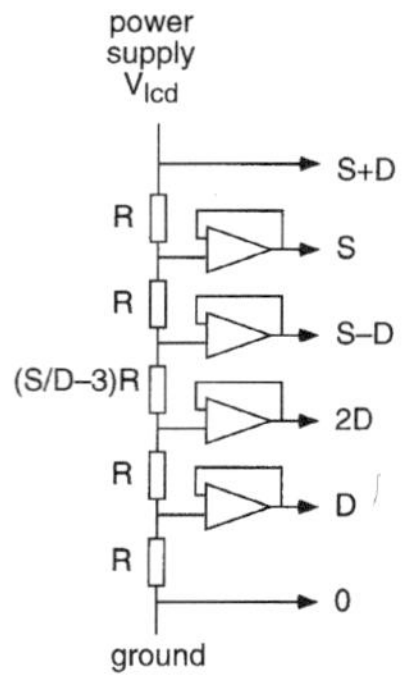

FIGURE 11.9. Voltage ladder network used to generate the six unipolar voltage levels required for standard LCD addressing when $N \geq 9$.

The practical scheme of FIGURE 11.7 (right) requires six unipolar voltage levels, known as bias levels, $S+D$, S, $S-D$, $2D$, D, and 0. These can be conveniently supplied from a single voltage source and a buffered voltage divider network consisting of five resistors in series, and four unity gain buffer amplifiers (FIGURE 11.9). Because the voltage drop across the resistors in each of the upper and lower resistor pairs is D, these resistors have the same relative resistance R. Since the voltage drop across the center resistor is $S-3D$, the center resistor has the relative value $(S/D-3)R$. Very often, the terms bias ratio and V_{lcd} are used when these kinds of circuits are described in a commercial data sheet or applications note. The definitions of these terms are:

$$\text{Bias ratio} = B = \frac{D}{V_{\mathrm{lcd}}} \tag{11.7}$$

where

$$V_{\mathrm{lcd}} = S + D. \tag{11.8}$$

The center resistor of FIGURE 11.9, the bias resistor, therefore has the relative value $(1/B-4)R$. The optimum bias ratio is determined from equation (11.5) to be $B_{\mathrm{optimum}} = 1/(\sqrt{N}+1)$ so in the optimum case the bias resistor has the value $(\sqrt{N}-3)R$.

11.2.2.1 Driver Architectures

As seen from the column waveforms of FIGURE 11.7 (right), the four bias levels 0, $2D$, $S-D$, and $S+D$ need to be applied to the column electrodes at the appropriate times, a task that is performed by the column driver IC whose inner workings are illustrated in FIGURE 11.10. Image data belonging to the next row of pixels to be selected are serially clocked into a shift register at high speed. A loading pulse then initiates the parallel transfer of these data into a latching register, where the data bits set the positions of the state switches that simultaneously apply the appropriate voltages to the column electrodes during the selection interval. During this process, the image data for the next row of pixels gets loaded into the shift register, ready to be latched at the beginning of the next addressing interval. It can be seen from FIGURE 11.10 that although the driver can supply four different voltage levels to the columns, only two output levels are available at any one time; with the DF switch in one position 0 and $2D$ are available, and in the other position $S+D$ and $S-D$ are available.

The column driver IC can also be used as a row driver IC, and its operation is quite similar. In this case,

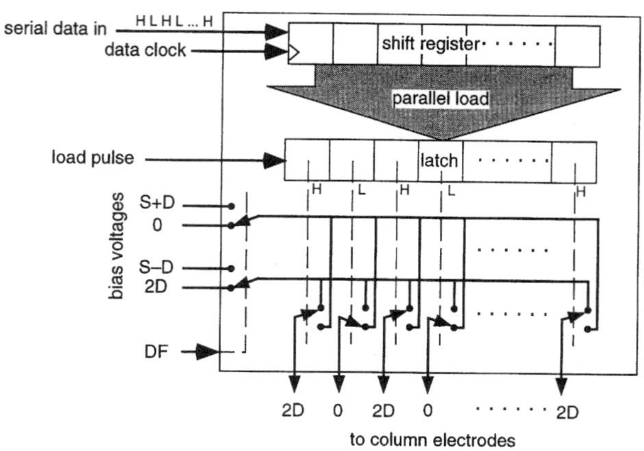

FIGURE 11.10. Functional diagram of the inner workings of a column driver IC. Image data belonging to the next row of pixels to be selected are serially clocked into a shift register at high speed. A loading pulse then initiates the parallel transfer of these data into a latching register where the data bits set the positions of the state switches and all voltages are simultaneously applied to the column electrodes during the selection interval. During this process, the image data for the next row of pixels is being loaded into the shift register, ready to be latched at the beginning of the next addressing interval.

the data bits in the input data packet all have the same state except for the one generating the selection pulse, and the bias levels D and S are substituted for $S - D$ and $2D$ of the previous case. Some manufacturers specify a row driver part that is different from the column driver part. In this case the row driver is designed to supply more current that may be needed because of the larger capacitive load of the rows due to their increased length.

11.2.2.2 Bias Ratio

Although the selection ratio is maximum under the condition of optimum bias, there are several reasons why it may be advantageous to operate an LCD at other than optimum bias conditions. For example to reduce "frame response" (cf. section 11.4), reducing the S/D ratio below the optimum value of $\sqrt{N}$ [15] has been suggested. On the other hand, another suggestion was to increase the S/D ratio above the optimum value to improve display uniformity [16]. In view of these observations, we present here a few useful relationships for nonoptimum operation in terms of the bias ratio B.

First, normalizing all voltages by the rms voltage of an off-pixel, that is, making $\langle \text{off} \rangle = 1.0$ we have, from equations (11.2), (11.7), and (11.8)

$$V_{\text{lcd}} = \sqrt{\frac{N}{1 - 4B + (N+3)B^2}} \tag{11.9}$$

$$D = B\sqrt{\frac{N}{1 - 4B + (N+3)B^2}} \tag{11.10}$$

$$S = (1 - B)\sqrt{\frac{N}{1 - 4B + (N+3)B^2}}. \tag{11.11}$$

These voltages are plotted in FIGURE 11.11 as functions of $1/B$ for the case of a 240-row display with $\langle \text{off} \rangle = 1.0$. Decreasing $1/B$ from the optimum value of 16.49 significantly decreases S and $V_{;cd}$, while only slightly increasing D.

From equations (11.1), (11.2), (11.7), and (11.8) the selection ratio can be expressed in terms of the bias ratio B as:

$$\frac{\langle \text{on} \rangle}{\langle \text{off} \rangle} = \sqrt{\frac{1 + (N-1)B^2}{1 - 4B + (N+3)B^2}} \tag{11.12}$$

and the effective number of multiplexed lines N_{eff} for an optimized system that would have this

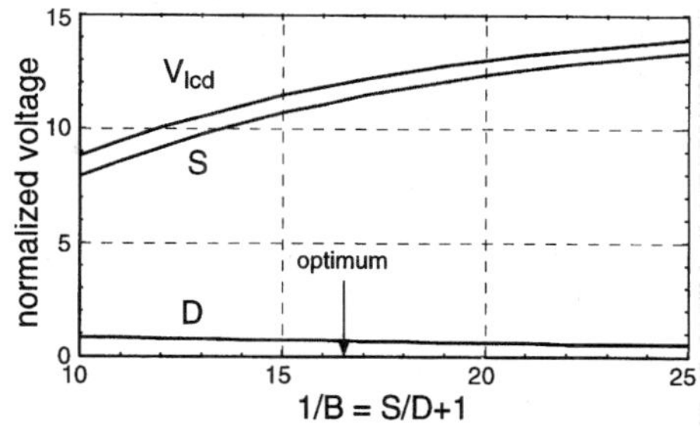

FIGURE 11.11. Dependence of the normalized LCD supply voltage, V_{lcd}, as well as other voltages on the inverse bias ratio $1/B$ of a 240-row display.

selection ratio can be obtained from equations (11.5) and (11.12) as:

$$N_{\text{eff}} = \frac{1}{4B^2}\left[\frac{1 - 2B + (N+1)B^2}{1 - B}\right]^2 \tag{11.13}$$

Equations (11.12) and (11.13) are illustrated in FIGURE 11.12 for a 240-row display. For the optimum inverse bias of 16.49, the selection ratio is maximum at 1.06677; this is a relatively broad maximum and the bias ratio can depart significantly from the optimum value without much effect on the selection ratio. This insensitivity is also reflected in the effective number of multiplexed lines. N_{eff} never exceeds 300 over a range of inverse bias ratios extending from 11 to 26.

11.2.3 Obtaining Gray Shades with Amplitude Selection

Presentation graphics and multimedia applications have demanded the accurate portrayal of grayscale

FIGURE 11.12. Dependence of the selection ratio and effective number of multiplexed lines on the inverse bias ratio $1/B$ for a 240-row display. Because of the broad extrema, the bias ratio can depart significantly from the optimum value without much effect on the selection ratio.

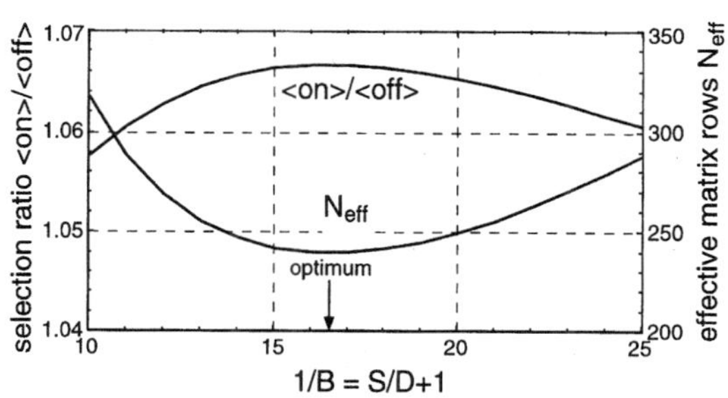

on LCDs. The two methods that are presently used to achieve gray levels are frame modulation [17], sometimes referred to as frame rate control (FRC), and pulse width modulation (PWM) [18]. In frame modulation a shading algorithm turns the pixel on and off during certain frames of a "superframe" that might consist of 16 regular frames. If the pixel is turned on for a certain number of the frames and off for the remaining number, then the pixel rms voltage will have an intermediate rms value lying between the off and on rms values. Ideally, the characteristic response time of the display should be long compared with the superframe period in order to avoid flicker effects, but this would require impractically slow panels. However, the flicker induced in faster responding panels can be significantly reduced by using different shading algorithms on adjacent pixels that have different permutations of the on/off distributions in the superframe. In practice, up to 64 FRC gray levels can be realized on an STN display with a response time of a few hundred ms.

In pulse width modulated (PWM) gray level addressing, a pixel's gray level is determined by adjusting the width of its data pulse during the selection interval. As shown in FIGURE 11.13, the column voltage is held at $-D$, corresponding to an on-pixel, for a certain fraction f of the selection interval and then switched to $+D$ for the remainder of the interval. The pixel voltage amplitude is thus $S+D$ for the fraction f and $S-D$ for the remaining fraction $(1-f)$ of the selection interval. Over the remaining $N-1$ time intervals of the frame

FIGURE 11.13. Example of row, column, and pixel waveforms in the pulse width modulated (PWM) method of generating gray shades.

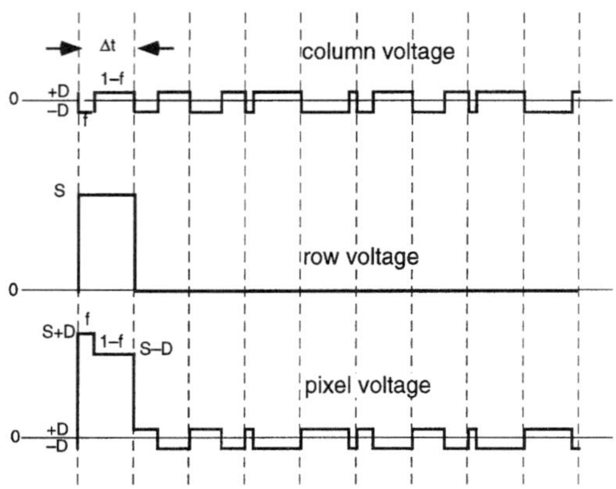

period, the pixel voltage amplitude is either $+D$ or $-D$. Following the same line of reasoning used to derive equations (11.1) and (11.2), the rms voltage appearing across the gray pixel is therefore

$$\langle \text{gray} \rangle = \sqrt{\frac{f(S+D)^2 + (1-f)(S+D)^2 + (N-1)D^2}{N}}$$
$$= \sqrt{f\langle \text{on} \rangle^2 + (1-f)\langle \text{off} \rangle^2} \tag{11.14}$$

which is between the on and off rms voltages $\langle \text{on} \rangle$ and $\langle \text{off} \rangle$.

PWM grayscale generally becomes impractical for more than 16 gray levels because the higher frequencies resulting from the narrower pulses never reach the pixel because of the low-pass, RC filtering action of the panel.

11.2.4 Drivers and Interconnect

The practical implementation of matrix addressing is simplified by the many types of drivers that are currently available in convenient packages. The most common type of chip packaging for high information content displays is the tape automated boding (TAB) flex packages that can be attached directly to the edge of the display with heat-sensitive anisotropic conductive film adhesive [19], resulting in interconnect resistances of only a few ohms. Advanced driver types use narrow, "slim chip" designs that minimize the peripheral area around the display and operate with a V_{lcd} of up to 42 V with 160 outputs per package.

11.2.5 Minimizing Crosstalk Effects

11.2.5.1 Nonideal Frequency Response Crosstalk

Up to this point, we have tacitly assumed that the optical state of the pixel is only determined by the rms value of the voltage difference across its row and column electrodes, regardless of the frequency of the applied signals, that is, an ideal flat response curve. In fact, a number of phenomena contribute to noticeable frequency dependence of the rms response. FIGURE 11.14 shows the rms voltage where the optical transmission of a D-STN display reaches 50% of its maximum transmission, V_{50}, as function of the frequency of an applied sine wave. The decrease in V_{50} for frequencies below 500 Hz is attributed to ionic effects in the liquid crystal [20,21], presumably caused by the

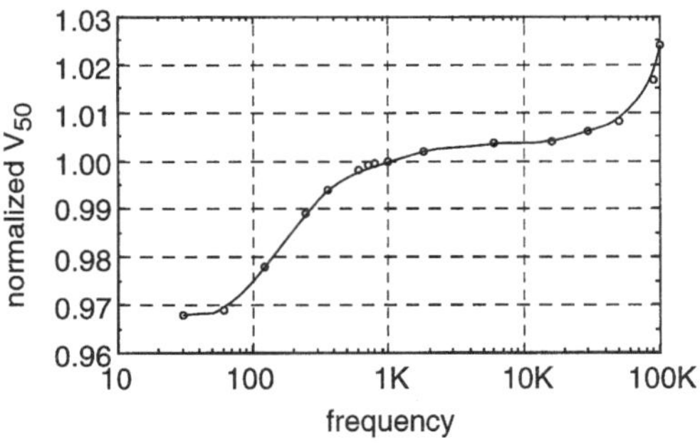

FIGURE 11.14. Frequency dependence of the applied rms voltage required to reach 50% of the maximum of a normally black, D–STN pixel as a function of the applied sine wave frequency. Voltages have been normalized to V_{50} for 1 kHz.

field produced by the separated ionic impurities augmenting the applied electric field at the instant of field reversal. Between 500 and 20 kHz the curve is relatively flat, and above 20 kHz, V_{50} increases sharply due to the to low-pass RC filtering action of the panel caused by the finite-sheet resistance of the transparent electrodes (10–100 Ω/sq.) and the capacitance of the liquid crystal dielectric (≈1000 pf cm^{-2}). Although the relative change in V_{50} is only a few percent over this frequency range, it represents a significant fraction of the available voltage margin for a matrix addressed LCD (cf. section 11.2.1).

The nonideal frequency response illustrated in FIGURE 11.14 introduces an objectionable kind of display crosstalk known as "ghosting" which is particularly noticeable with images containing fine checkerboard patterns. On a white background, a checkerboard image appears to cast dark shadows above and below its width along the columns. This crosstalk can be explained by examining the Fourier spectra of the column signals applied to the display for both a plain background image and a checkerboard image. As shown in FIGURE 11.15, except for the selection pulse, the background pixel waveform (a) is a constant voltage over the frame period and most of its Fourier components occur at low frequencies, with the largest component at 30 Hz. In view of FIGURE 11.14, this would cause the display background to appear brighter than normal because the V_{50} "threshold" is low. Conversely, the column waveform for the checkerboard pattern (b) consists mainly of a high-frequency square wave containing many high-frequency Fourier components where the V_{50} "threshold" is high, making these areas appear darker than normal.

This crosstalk is caused by a combination of the nonideal frequency response and the image-dependent frequency content of the column signals. This crosstalk can be reduced considerably by periodically reversing the polarity of both the row and column signals with the DF signal every few addressing intervals or so [22] to "homogenize" the frequency content of the column waveforms to make them less image-dependent. Curves (c) and (d) of FIGURE 11.15 show the pixel waveforms and corresponding frequency spectra of the background and checkerboard images after their column waveforms have been periodically switched with the DF signal every seven intervals. In both cases, the stronger Fourier components have moved toward the central frequencies, where the rms response shows more ideal character.

11.2.5.2 Column-to-Row Coupling Crosstalk

A second, less severe type of crosstalk is introduced by voltage transitions on waveforms applied to the column electrodes inducing voltage spikes on the row electrodes, through capacitive coupling, resulting in a pixel voltages other than the desired ones. This "coupling" crosstalk, also referred to as horizontal crosstalk [22], is manifested by vertical ghosting of solid images displayed on the screen. One scheme developed to reduce these effects involves a counting method to keep track of the changes in the total number of column voltage transitions at each addressing interval and applying compensating voltage [22]. Another scheme [23] uses negative feedback to the row driver voltage source to correct for the voltage transients detected at a "monitor" row electrode. Still other schemes are now finding their way into commercial displays.

11.2.6 Schemes to Increase the Selection Ratio

11.2.6.1 Restricted Pattern Drive Schemes

If it is known beforehand, for example, that each column will have the same number of pixels turned on or turned off, then it is possible to devise drive schemes where the selection ratio can be increased beyond the iron law limit. These are known as restricted pattern addressing schemes, a classic example of which is the waveform identity [24] addressing method used in an early, high-resolution

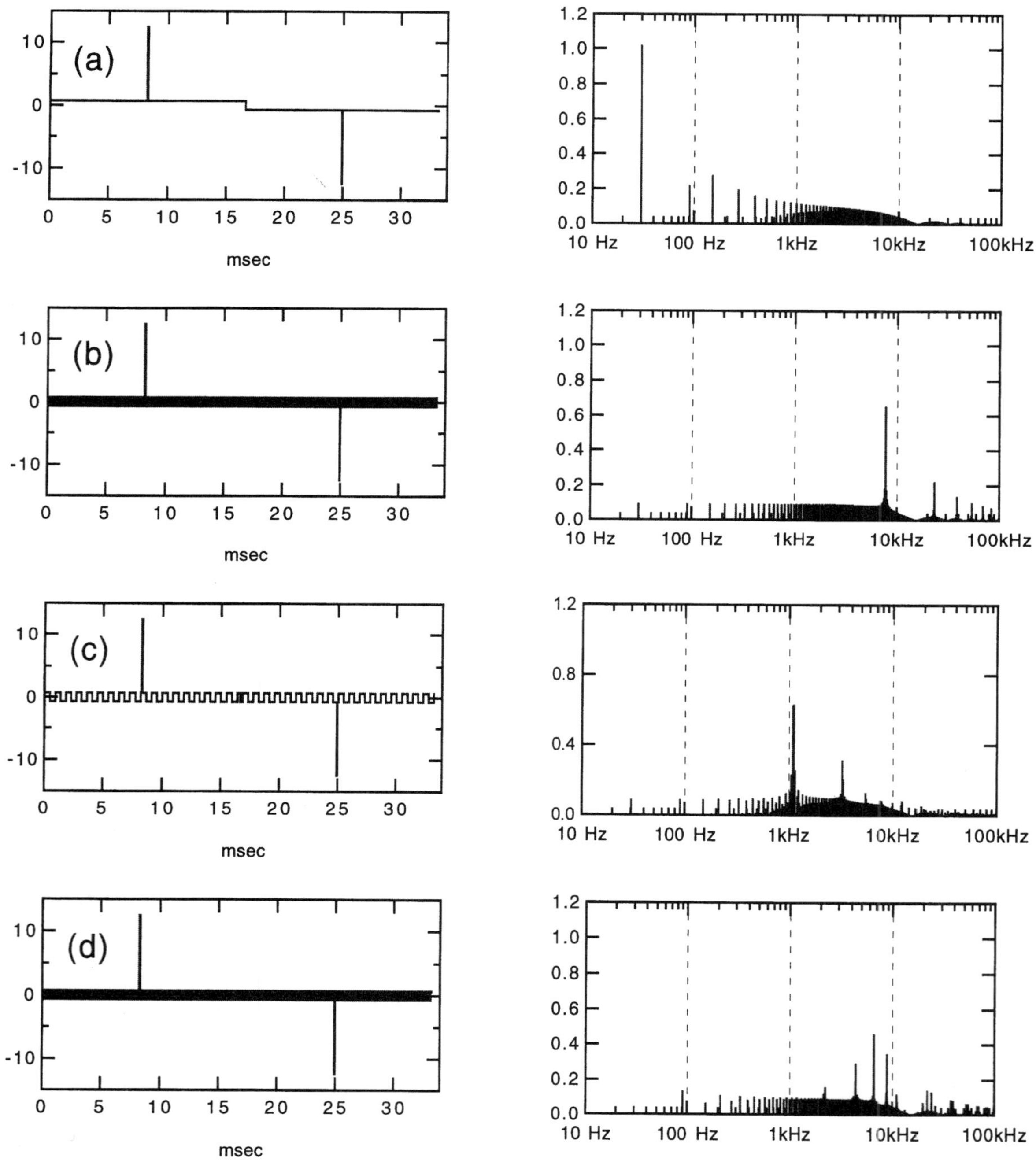

FIGURE 11.15. Pixel waveforms and their corresponding Fourier spectra for different displayed images: (a) blank screen, (b) fine checkerboard, (c) blank screen with seven-line DF, and (d) fine checkerboard with seven-line DF. Frequent DF switching tends to make the frequency content of the column signal less image-dependent.

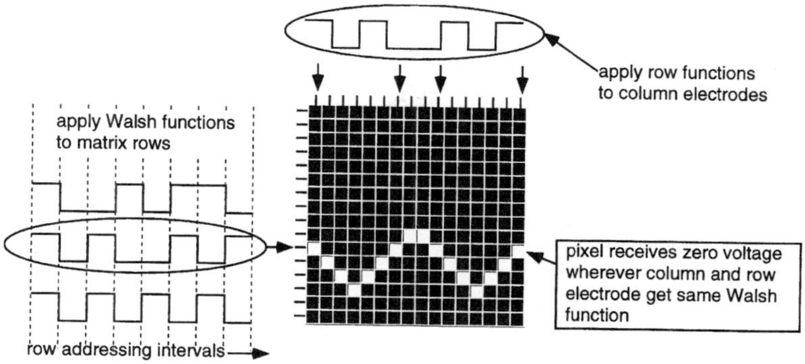

FIGURE 11.16. Illustration of the waveform identity addressing method with a normally white TN display. Example shows circled row waveform being applied to four columns to turn pixels white at the intersections. Waveforms of other white pixels are not shown.

TN LCD oscilloscope (cf. FIGURE 11.16), where one pixel per column is always turned off [25]. Periodic signals corresponding to a normalized, mutually orthogonal set of functions, such as Walsh functions [26], are applied to the matrix rows of this display and selected functions from this set are applied to the matrix columns. When a column receives the same function as a row then the potential difference at the intersection is zero and that pixel is turned off, indicated by the white pixel in the normally white TN display of FIGURE 11.16. Pixels that receive different row and column signals receive a constant rms voltage corresponding to $\sqrt{2}$/times the rms value of the row signals, regardless of which signals are chosen. Other ingenious schemes were developed for multitrace oscilloscopes that even permitted a limited amount of alphanumeric information [27]. In general, these restricted pattern schemes had only limited applications and could not be used in mainstream, high information content notebook computer screens. They enjoyed a brief popularity up to the early 1980s when they became superseded by Alt and Pleshko addressed STN displays, which could achieve both the high contrast and display arbitrary information patterns.

11.2.6.2 Matrix Architecture

The selection ratio of an arbitrary information content display can be increased by breaking the display up into independent sections and driving each section separately. FIGURE 11.17 illustrates this technique by comparing an assortment of LCD matrices having the same resolution, but different selection ratios. Layout (1) in FIGURE 11.17 corresponds to an N-row matrix with a multiplexed duty of $1/N$. By dividing the column electrodes in two along the middle of the panel and driving the column electrodes of the upper and lower panel halves independently, as shown in layout (3), the effective number of multiplexed lines can be reduced to $N/2$, albeit with twice as many column drivers. This is known as the divided electrode scheme and has been used on monochrome TN and STN displays from the very beginning. A variation of this scheme that reduces the number of row drivers by half is shown in layout (4). Both (3) and (4) are in use today with (3) being more popular for TAB interconnects and (4) being used more with elastomer interconnect where the drivers are surface mounted on a multilayered circuit board. The divided electrode scheme increases the selection ratio for VGA displays or higher resolution by about 42% (cf. equation (11.5)).

Up to the early 1990s, color STN displays for notebook computers could not take advantage of the divided electrode schemes (3) or (4) because the 640 × 3 column electrodes required by the RGB color triads reduced the interconnect pitch to approximately 100 μm, which was too fine for the technology at the time. Manufacturers were forced to use the N-row matrix with interdigitated column electrodes as shown in layout (2), because it doubled the pitch. This is why the first STN color displays for notebook computers had relatively poor contrast ratios and slow response times compared with the STN monochrome displays of similar resolution. Later, when more sophisticated interconnect technology became available, color notebook displays were able to take advantage of the higher selection ratio afforded by the divided electrode scheme. Indeed, the performance improvement was so dramatic that marketers coined

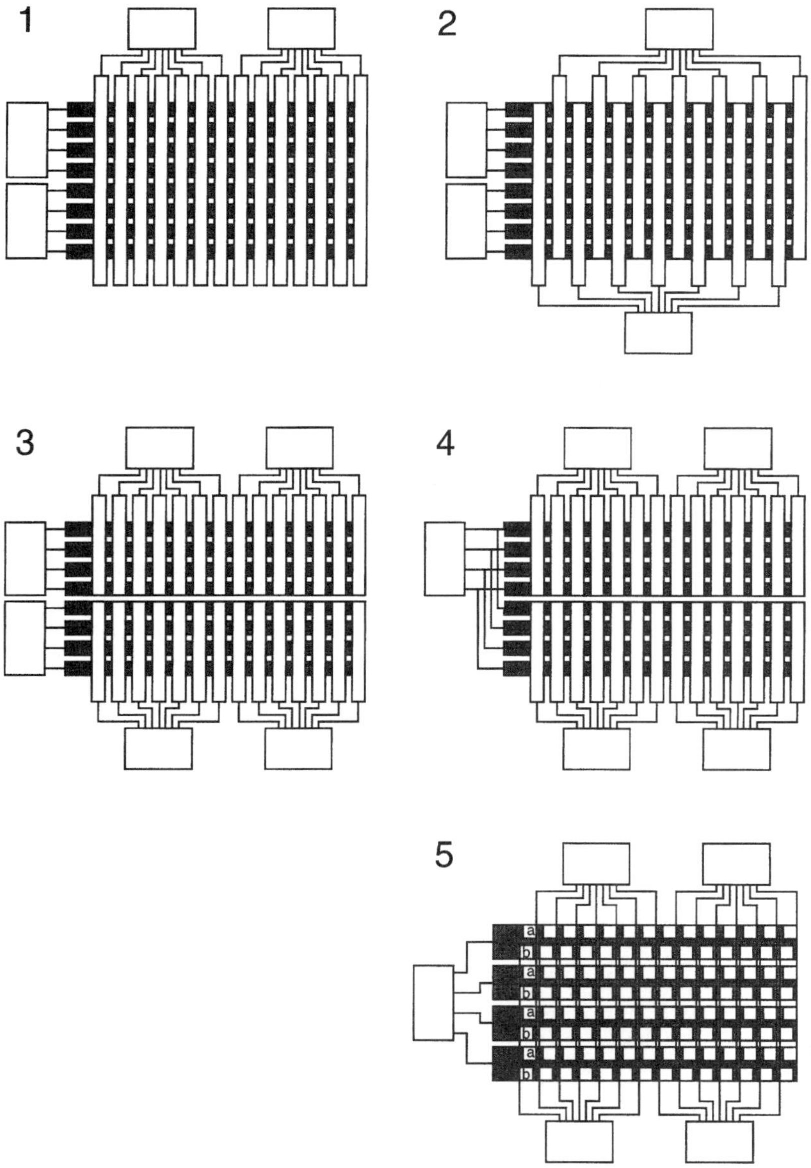

FIGURE 11.17. Different LCD matrix layouts having the same resolution: (1) standard matrix, (2) standard matrix with interdigitated column electrodes, (3) dual-scan matrix, (4) dual-scan matrix with common row driver, and (5) double-matrix structure electrically equivalent to (4).

the new term "dual scan" to describe this old technology.

Finally, layout (5) shows a double-matrix architecture that is electrically equivalent to the dual-scan layout (4). The row electrodes have double width and the column electrodes have an intricate structure composed of a narrow metal electrode with transparent pixel "flags" occupying staggered positions along its length and arranged so that two flags, belonging to different column electrodes, are positioned one above the other and share the same row electrode. This configuration is particularly advantageous for small TV displays, since the picture information arrives in the form of a scanned video stream that is naturally compatible with this architecture and does not require an external frame buffer as the conventional dual-scan layout (4) would. Quad matrices where four pixel flags share an extra wide row have also been used for pocket TVs [28].

11.2.6.3 Plural Line Addressing

Plural line addressing [29] (PLA) is a drive scheme for arbitrary information pattern displays that significantly increases the selection ratio by increasing the width of the selection pulse so that more than one row at a time is selected. Figure 11.18 shows an example of PLA where the width of the standard selection pulse (dotted line) is doubled (solid line) so that two lines are simultaneously selected. With the standard pulse width, the column signal shown produces the pixel pattern on the left. With the double width selection pulses the same column signal produces the pattern on the right, somewhat reminiscent of the first pattern, but with the image edges smeared by an intermediate gray shade. This smearing comes about because the original image states of the simultaneously selected rows get averaged together and can become gray if one of the original image states is black and the other white. The dotted arrows in Figure 11.18 show which states are averaged together. The pixel waveforms for this example are shown on the right of the figure. Each pixel waveform has two selection intervals whose amplitudes are determined by the column signal during these two intervals resulting in three possible states—on, off, and gray. On the basis of this example, the selection ratio for the general case where the selection pulse extends over ℓ addressing intervals, that is, where ℓ matrix rows are simultaneously selected, can be written as

$$\frac{\langle \text{on} \rangle}{\langle \text{off} \rangle} = \sqrt{\frac{\ell(S+D)^2 + (N-\ell)D^2}{\ell(S-D)^2 + (N-\ell)D^2}}. \tag{11.15}$$

The selection ratio is optimized when

$$\left.\frac{S}{D}\right|_{\text{opt}} = \sqrt{N/\ell} \tag{11.16}$$

yielding

$$\left.\frac{\langle \text{on} \rangle}{\langle \text{off} \rangle}\right|_{\text{opt}} = \sqrt{\frac{\sqrt{N/\ell}+1}{\sqrt{N/\ell}-1}}. \tag{11.17}$$

Comparing equations (11.17) and (11.5), PLA can be seen to reduce the effective number of multiplexed rows by a factor of ℓ, which can greatly increase the selection ratio of the display. However, PLA also diminishes the vertical resolution of the display by taking a running average of the pixel states down a column with an averaging window that is ℓ pixels wide. This averaging automatically adds gray levels at the vertical contours of contrasting objects and tends to soften them. For $\ell = 2$, this averaging effect is barely noticeable for "natural" video images that already contain many gray levels, but might be objectionable for computer graphics and alphanumeric images.

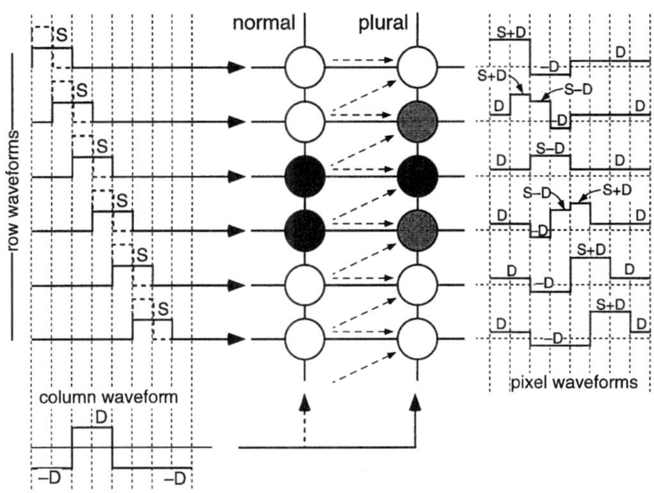

FIGURE 11.18. Example of plural line addressing (PLA) for $\ell = 2$ where two matrix rows are simultaneously selected with double-width selection pulses. This increases the selection ratio by decreasing the effective number of multiplexed lines by a factor of two, but smoothes out the vertical resolution by averaging the pixel states on adjacent rows.

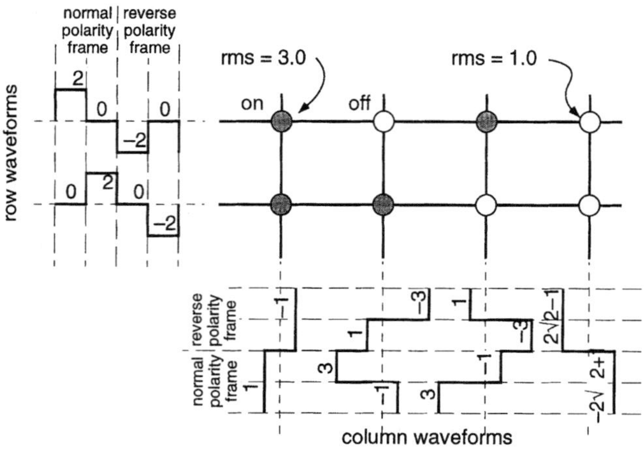

FIGURE 11.19. This two-row matrix does not use the amplitude selection method and has a selection ratio that is 24% higher than that mandated by the "iron law."

11.3 GENERALIZED MATRIX ADDRESSING

In investigating a two-row matrix where more than two voltage levels could be applied to the columns, a violation of the amplitude selection method occurs; Nehring and Kmetz [30] obtained a selection ratio of 3.0, 24% higher than the value of 2.414 mandated by the iron law equation (11.5). A version of this scheme is illustrated in FIGURE 11.19. Note that for this example, the customary usage of the term "row selection" no longer applies because column voltage during the row "selection" is determined by the collective information states of both pixels in the column, not just the information state of the "selected" pixel.

By the late 1970s, it was realized that the electro-optic transfer curve of the TN display could not be made sufficiently sharp to overcome the narrow voltage margins imposed by Alt and Pleshko addressing for high information content displays. The encouraging results obtained for the two-line matrix of FIGURE 11.19 led to a general analysis of rms responding matrix displays to see if alternative addressing schemes could be developed which could overcome the iron law limitation and improve TN LCD performance by increasing the selection ratio. Using mutually orthogonal, but otherwise completely general row functions and column signals formed by linear combinations of the row functions, Nehring and Kmetz were able to derive the ultimate limits for the selection ratio of an rms-responding matrix, regardless of the driving waveforms chosen. For matrices whose number of rows are perfect squares they found no improvement over Alt and Pleshko, and for other sized matrices they found significant improvement only for two-row matrices (cf. FIGURE 11.20). These

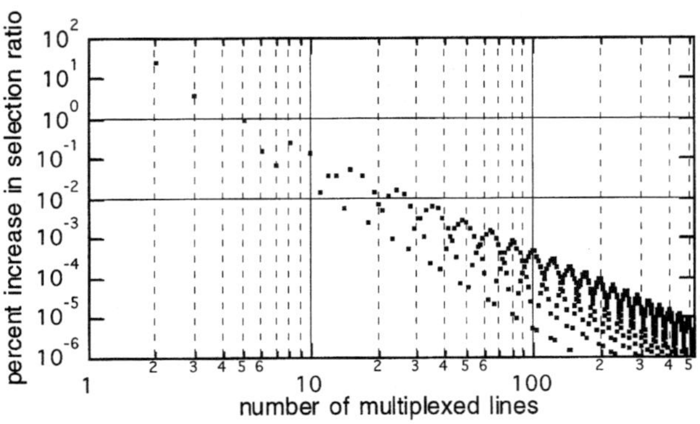

FIGURE 11.20. The improvement of the ultimate limit selection ratio compared with the iron law value (equation (11.5)) for different sized matrices. There is no improvement when N equals a perfect square, and only significant improvement (24%) for a two-row matrix.

results showed that the Alt and Pleshko selection ratio had a wider validity than had previously been recognized, and thus it became known as the iron law.

11.4 MULTIPLE LINE ADDRESSING

To satisfy the rms response condition it is sufficient that the inherent response time of the display be many times longer than the period of the addressing signals. This generally precludes displays having response times short enough to show moving images at video rates. Indeed, if a video-responding, 50 ms panel is operated with Alt and Pleshko addressing at the conventional frame rate of 60 Hz, the breakdown of the rms condition results in a phenomenon known as frame response [15], where the display no longer responds to the rms voltage averaged over a frame period, but strongly reacts to voltage changes occurring within the frame period. As FIGURE 11.21 shows, frame response can significantly reduce the perceived optical transmission and contrast ratio because of the rapid decay of the optical transmission that occurs after each select pulse in the pixel waveform.

FIGURE 11.21. The "on" pixel waveform of a standard addressed LCD should drive the pixel to a constant "white" state, but in a fast LCD the rapid decay in transmitted light intensity after the select pulse results in less perceived brightness.

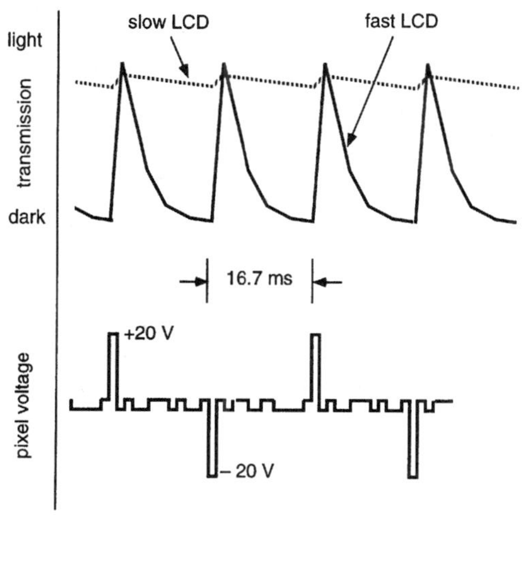

Attempts have been made to decrease frame response by increasing the frame frequency to restore the rms condition, but this increases the power consumption and shifts the Fourier components of the drive signals to higher frequencies where they are strongly attenuated by the low-pass filtering action of the panel (cf. FIGURE 11.14). Another approach has been to decrease the S/D ratio from the optimum value of $\sqrt{N}$ in an attempt to "smooth out" the pixel waveform This affords some improvement in perceived transmission and contrast ratio for panels with response times longer than 100 ms, but is counterproductive for video-rate displays (≤ 50 ms), because to be effective the S/D ratio has to be decreased to the point where the selection ratio is severely compromised.

Because frame response is so clearly related to the pulsed nature of standard LCD addressing waveforms, making the pixel waveforms more uniform should result in more rms-like behavior, even for fast-responding LCDs. Indeed rms response can still be obtained, even if the sufficient condition is not met, if the running rms average of the pixel waveform can be made relatively constant over an averaging window that is narrower than the frame period. The necessary condition for rms response is that the characteristic response of the display be many times longer than the width of this averaging window. Active Addressing™ [31] and the independently developed multi-line selection (MLS) [32] are addressing techniques that generate relatively uniform pixel waveforms and make video-rate operation possible on passive matrix displays. These schemes, generally referred to as Multiple Line Addressing (MLA) schemes, will be discussed next.

11.4.1 Mathematical Analysis

The analysis of generalized matrix addressing is an excellent theoretical framework for developing practical drive waveforms to eliminate frame response from fast-responding LCDs. Consider a liquid crystal panel of N rows and M columns. Let an information matrix **I** represent the desired information to be displayed on the panel, where the elements I_{ij} correspond to the desired state for the pixel on row i and column j. Without loss of generality, we define $I_{ij} = +1$ for a nonselect or "off" pixel and $I_{ij} = -1$ for a select or "on" pixel, assuming for the moment a display

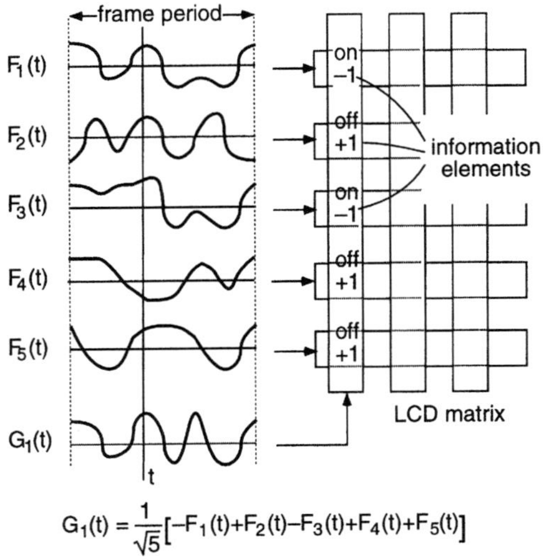

FIGURE 11.22. Example for five-row matrix showing elements of generalized matrix addressing. The $F_i(t)$ are a set of orthonormal row addressing waveforms that are otherwise arbitrary in nature (in practical applications with digital circuitry the waveforms would assume discrete, discontinuous values). Information to be displayed on the matrix is represented by an information element, -1 for an "on" pixel and $+1$ for an "off" pixel. The column signal $G(t)$ is computed from the dot product of row signal amplitudes and corresponding information elements.

without gray shades. Let each matrix row i be driven with periodic row signal $F_i(t)$having a common period T with all the other row signals, and let each matrix column j be driven with a column signal, $G_j(t)$, also periodic over T. The essential features of generalized addressing are illustrated by the five row matrix shown in FIGURE 11.22.

The voltage across a pixel, $U_{ij}(t)$, is the difference between the signals applied to row i and column j,

$$U_{ij}(t) = F_i(t) - G_j(t) \tag{11.18}$$

and the root mean square value of the voltage appearing across this pixel, averaged over the frame period T, is

$$\langle U_{ij} \rangle = \frac{1}{\sqrt{T}} \sqrt{\int_0^T U_{ij}(t)^2 \, dt}. \tag{11.19}$$

In an rms-responding display without frame response the optical state of pixel ij is determined by $\langle U_{ij} \rangle$. Substituting equation (11.18) into equation (11.19) yields

$$\langle U_{ij} = \frac{1}{\sqrt{T}} \times \sqrt{\int_0^T F_i(t)^2 \, dt - 2\int_0^T F_i(t)G_j(t)\, dt + \int_0^T G_j(t)^2 \, dt}. \tag{11.20}$$

Now suppose that the row signals $F_i(t)$ are orthonormal, that is,

$$\frac{1}{\sqrt{T}} \sqrt{\int_0^T F_j(t) F_k(t) \, dt} = \begin{cases} F & \text{if } j = k \\ 0 & \text{if } j \neq k \end{cases}. \tag{11.21}$$

Following Nehring and Kmetz, we consider a column signal $G_j(t)$ that is proportional to the dot product of the information vector and all the row signals at time t:

$$G_j(t) = c \sum_{i=1}^{N} I_{ij} F_i(t) \tag{11.22}$$

where c is the proportionality constant, independent of the information state of the display [33].

In the most general case represented by equation (11.22) the collective information state of the entire column must be known before the column drive signal can be generated. However whenever $F_i(t) = 0$ it is not necessary to know the state of the corresponding pixel. In the limit of standard LCD drive it is possible to generate the column signal "on the fly" since at any particular time it is only necessary to know the information state of the pixel on the selected row since $F(t) = 0$ for all other rows.

Substituting equation (11.22) into equation (11.20), making use of equation (11.21), yields the following expression for the rms voltage across pixel ij:

$$\langle U_{ij} \rangle = F\sqrt{1 - 2cI_{ij} + Nc^2}. \tag{11.23}$$

Note that the rms voltage across pixel ij depends only upon the value of the corresponding information element for that pixel. This is a requirement for the display to be free of crosstalk, that is, where the state of one pixel does not affect the states of other pixels in the column.

As with the amplitude selection addressing method, the selection ratio is defined as the ratio of the select rms voltage $\langle \text{on} \rangle$ ($I_{ij} = -1$) to the nonselect rms voltage $\langle \text{off} \rangle$ ($I_{ij} = +1$). Maximizing this selection ratio with respect to the proportionality constant c

results in

$$c|_{\text{opt}} = \sqrt{N} \tag{11.24}$$

yielding

$$\left.\frac{\langle on\rangle}{\langle off\rangle}\right|_{\text{opt}} = \sqrt{\frac{\sqrt{N}+1}{\sqrt{N}-1}} \tag{11.25}$$

identical with the Alt and Pleshko iron law result of equation (11.5). Thus the iron law selection ratio can therefore be obtained with *any* orthonormal set of row functions if the column voltages are generated by the expression:

$$G_j(t) = \frac{1}{\sqrt{N}}\sum_{i=1}^{N} I_{ij}F_i(t). \tag{11.26}$$

11.4.2 Gray Scale with Multiple Line Addressing

Gray shading with multiple line addressing can be achieved with either the FRC and PWM methods described earlier (cf. section 11.2.3). However, with the fast responding LCDs used with multiple line addressing, only a few flicker-free frame modulated gray shades are possible. PWM can be adapted for multiple line addressing by generating a different $G_j(t)$ value for each gray shade image bit plane and time-weighting the different $G_j(t)$ signals to provide flicker-free gray shades. However, PWM for multiple line addressing is prone to the same high-frequency roll-off problems that are encountered with standard addressing.

A new method to achieve grayscale in passive matrix LCDs is called pulse height modulation (PHM) [34], a subset of which is also referred to as amplitude modulation (AM) [35]. PHM has two implementations, "full-interval" and "split-interval." The split-interval method maintains the correct rms level in the pixel waveform averaged *over each row address time interval* (and consequently over the whole frame period). The full-interval approach maintains the correct rms level in the pixel waveform when averaged *over the whole frame period*. The full interval method has the advantage of not increasing the drive frequencies and will be described in more detail next.

The full-interval PHM gray level method can easily be incorporated into the generalized matrix analysis by extending the definition of information elements to allow them to exist on the continuum between 1 and +1, corresponding to intermediate gray shades. For example, an information element $I_{ij} = 0$ would correspond to a pixel with a 50% gray level.

Starting again from equation (11.20) with this extended definition, equation (11.23) becomes

$$\langle U_{ij}\rangle = F\sqrt{1 - 2cI_{ij} + c^2\sum_{m=1}^{N} I_{mj}^2}\,. \tag{11.27}$$

Allowing the information elements to become continuous variables has introduced crosstalk because now $\langle U_{ij}\rangle$ depends not only on I_{ij} but on all the other information elements of the column as well, through the summation term. However, this crosstalk can be removed by adding an $(N+1)$st "virtual" row, and its associated virtual information element $V_{(N+1)j}$. By summing over all $N+1$ rows, equation (11.27) now becomes:

$$\langle U_{ij}\rangle = F\sqrt{1 - 2cI_{ij} + c^2\sum_{m=1}^{N} I_{mj}^2 + c^2V_{(N+1)j}^2}\,. \tag{11.28}$$

By requiring that the virtual information element $V_{(N+1)j}$ be determined by all the other information elements of the column through the condition:

$$V_{(N+1)j} = \sqrt{N - \sum_{m=1}^{N} I_{mj}^2} \tag{11.29}$$

equation (11.28) reduces to equation (11.23), the desired result with no crosstalk. The grayscale rms voltage at pixel ij now depends only on the grayscale information element for that pixel, I_{ij}, *and not on the state of any other pixel in the column*. Equations (11.24) and (11.25) still apply, so the addition of the $(N+1)$st virtual row does not increase the number of multiplexed rows and the selection ratio remains uncompromised. This becomes an important consideration if many virtual rows are present. Where there are no intermediate gray levels present on the column the virtual information element is identically zero. It achieves its maximum value when the intermediate gray level corresponding to $I_{mj} = 0$ (halfway between "on" and "off") is displayed on all pixels of the column; in this case $V_{(N+1)j} = \sqrt{N} \simeq 15.5$ for a display with 240 multiplexed rows. This falls outside the imposed range of $-1 \leq I_{ij} \leq +1$ specified for an information element, but the virtual pixel is never shown on the display.

To implement the full interval PHM method an additional term must be added to the sum for the column voltage $G_j(t)$ of equation (11.26), corresponding to the product of the virtual information element and the virtual row signal:

$$G_j(t) = \frac{1}{\sqrt{N}} \sum_{i=1}^{N} I_{ij}F_i(t) + \frac{1}{\sqrt{N}} V_{(N+1)j}F_{N+1}(t). \quad (11.30)$$

It is important to realize that no $(N+1)$st row need be physically present on the display; only an orthonormal $(N+1)$st signal, $F_{N+1}(t)$, must be made available from the electronic circuitry for the last term of equation (11.30). The $(N+1)$st row is thus aptly named a virtual or "phantom" row.

11.4.2.1 Example of PHM Full-Interval Grayscale

Further understanding of the application of the full-interval PHM grayscale technique to multiple line addressing can be obtained from the example illustrated in FIGURE 11.23. The display has six real rows, indicated by the horizontal lines numbered 1–6, one virtual row indicated by (7), and one column electrode. The row signals, including the virtual row

FIGURE 11.23. Example of pulse height modulation (PHM) grayscale in a six-row display. Information elements with values between −1 and +1 represent gray-shaded pixels. An "adjustment signal" derived from a seventh, fictitious "virtual" row must be added to the column signal to prevent crosstalk.

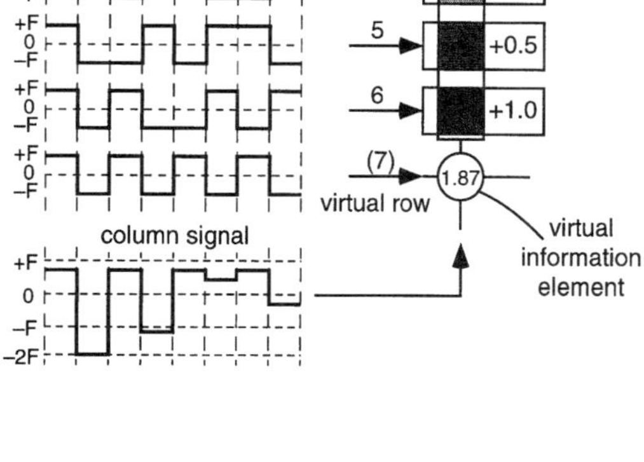

signal, are bilevel for this example, having a value of either $+F$ or $-F$, and form an orthonormal set. Information consisting of "off" and "on" pixels as well as intermediate gray shades are indicated by the information elements, reading top to bottom, −1.0, 0.5, 0, 0, +0.5, and +1.0. In a display where "off" pixels are black and "on" pixels are white, these would represent white, light gray, medium gray, medium gray, dark gray and black pixels, respectively. Using the gray levels of above example, the virtual information element is determined from equation (11.29) to be $V = \sqrt{6 - 2.5} = \sqrt{3.5} \simeq 1.87$. The column signal is then computed from equation (11.30).

11.4.3 Multiple Line Addressing Row Waveforms

It was shown in the preceding section that *any* set of orthonormal row functions can be used to drive an STN display at optimum selection ratio without crosstalk. To effectively reduce or eliminate frame response in a fast-responding display the time intervals when the row function is nonzero, the "selection" intervals, must be distributed over the frame period. This distribution makes the pixel waveform more uniform in time and prevents the optical state of the pixel from significantly decaying before it receives another pulse. In order to be practical, the row signals and column signals must be easy to generate. Also, for good image quality, there can be no artifacts such as row streaking and ghosting.

The choice of orthonormal row functions is the key in determining the complexity of the hardware required to perform the necessary column signal computations. The most practical row functions are either bilevel functions with amplitude $+F$ or $-F$, or trilevel functions with amplitude $+F^*$, zero, and $-F^*$.

11.4.3.1 Bilevel Row Functions

Bilevel row functions are attractive because the selection intervals are *de facto* distributed over the frame period since the rows are selected all of the time with either $+F$ or $-F$. These functions are thus expected to provide the minimum amount frame response. Bilevel functions also have the practical advantage of being ideally suited to the commonly available, two-level integrated circuit STN column drivers (FIGURE 11.10).

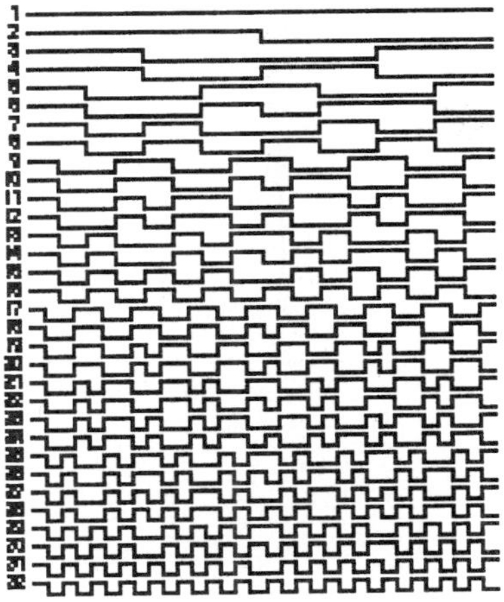

FIGURE 11.24. The complete set of sequency-ordered Walsh functions of order 32 that would be suitable as multiple line addressing row functions for a matrix having between 17 and 32 rows.

Bilevel functions with the minimum number of time intervals per frame period can be represented by Hadamard matrices [36], which are square orthogonal matrices with the elements -1 and $+1$. The rows of the Hadamard matrix correspond to the signals applied to the display rows. The order of any Hadamard matrix is divisible by four, and thus can be represented by $4t$, where t is any positive integer. Walsh functions [26] are a subset of Hadamard matrices, and have an order which is a power of two, that is, there are 2^s time intervals, where s is an integer, such that $2^{s-1} < N \leq 2^s$. FIGURE 11.24 shows the set of sequency-ordered Walsh functions of order 32; a 240-row display would require a set of Walsh functions of order 256.

11.4.3.2 Trilevel Row Functions

Trilevel multiple line addressing functions can be generated from the bilevel Hadamard functions, for example, by forming a supermatrix built up of repeating smaller $L \times K$ Hadamard units down the diagonal as shown in FIGURE 11.25 (left). Since the other elements in the matrix are zero, the expanded function is still orthogonal. The K selection intervals per frame are distributed over the frame period by rearranging the order of the matrix columns as shown in FIGURE 11.25 (right). Each row function simultaneously selects L matrix rows. Addressing matrices

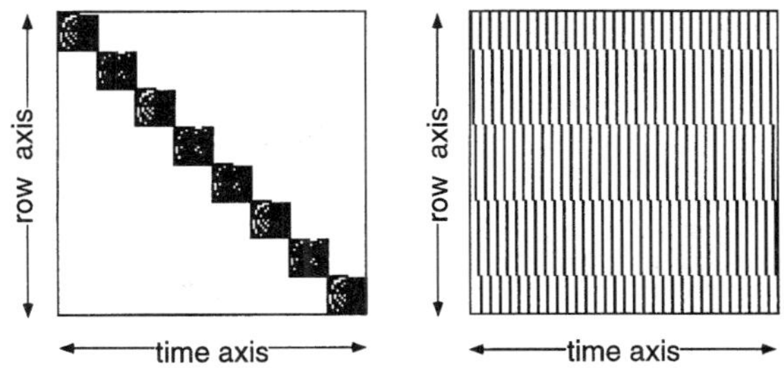

FIGURE 11.25. Example of trilevel multiple line addressing row function matrix built up by repeating smaller $L \times K$ Hadamard units down the diagonal and distributing the K selection intervals over the frame period by rearranging the order of the matrix columns.

composed of a 16×16 Walsh submatrix [37] as well as 4×8 and 7×16 submatrices [38] have been described. Row drivers for trilevel functions are currently under development.

11.4.4 Row and Column Voltage Requirements

We now consider the dependence of the row and column voltages on the number of lines that are selected at a time, L. This information is vital for display driver selection and design. To have a consistent basis for comparison all voltages are normalized with respect to the "off" pixel rms voltage, that is, $\langle \text{off} \rangle = 1.0$. (To convert these normalized voltages to actual voltages they must be multiplied by the threshold voltage of the display, which is typically 1.5–2.3 V for an STN display, depending on the particular device and material parameters.)

To satisfy the normalization requirement, it can be shown from equations (11.23) and (11.24) that the rms voltages of the row signals must be equal to F:

$$F = \sqrt{\frac{\sqrt{N}}{2(\sqrt{N} - 1)}} \tag{11.31}$$

where N is the number of multiplexed display rows. For $N = 256, F \simeq 0.730$.

In view of equation (11.31), if there are fewer selection pulses per frame period, the amplitude of the selection pulses must be higher. If S is the magnitude of the row selection pulses then the following

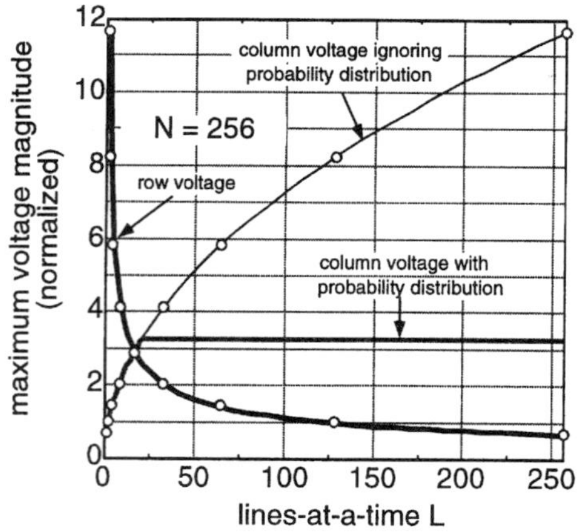

FIGURE 11.26. The dependence of the row and column voltages on the number of lines selected at a time for different multiple line addressing embodiments. The row voltage is equal to the column voltage when $L = \sqrt{N}$.

condition holds:

$$S = \sqrt{J/K}/F \tag{11.32}$$

where J is the number time intervals per frame period, of which K are selection intervals. K is always greater than or equal to the number of lines selected at a time, L. For an efficient system, L and K should be nearly the same because this minimizes the number of dot product computations required to generate the column voltage.

Consider the example for the case where $N = J = 256$ and $L = K$. FIGURE 11.26 shows how the normalized *magnitude* of the row selection pulse S depends upon the number of rows selected at a time, L. A drive scheme where just a few lines are selected at a time requires higher voltage trilevel drivers than schemes where more lines are selected.

For the case where L rows are selected at a time, equation (11.26) is equivalent to:

$$G_j(t) = \frac{S}{\sqrt{N}} \sum_{i=1}^{L} I_{ij} W_i(t) \tag{11.33}$$

where $W_i(t) = \pm 1$ are the values of the embedded orthonormal function. According to equation (11.33) the maximum magnitude of the column signal, $|G_{max}|$, is given by $|G_{max}| = LS/\sqrt{N}$. For the example, with $N = J$ and $L = K$ the expression for $|G_{max}|$ simplifies to $|G_{max}| = \sqrt{LF}$, which is plotted in FIGURE 11.26.

As was demonstrated previously [31], when about half of the row functions are randomly selected and multiplied by -1, the probability of $|G_{max}|$ actually being generated becomes extremely low, obeying a binomial distribution. We have shown [39] that the probability of $|G|$ exceeding a certain critical value G_0 is:

$$P(|G| \geq G_0) = 2\text{Erfc}(G_0/F) \tag{11.34}$$

where Erfc(ξ) is the complementary error function defined by

$$\text{Erfc}(\xi) = \frac{1}{\sqrt{2\pi}} \int_{\xi}^{\infty} \exp\frac{z^2}{2} \text{dz}. \tag{11.35}$$

For $G_0 = 3.25$ with $N = 256$, the probability of a level being generated whose magnitude exceeds G_0 is 8.79×10^{-6}.

Thus, roughly one out of every 114,000 column levels generated exceeds this magnitude. Therefore clamping all voltages at G_0 will have negligible effect on the rms voltage appearing across the pixel. The maximum effective column voltage occurring in practice, taking probability into account, is therefore about 3.25—independent of L. This is illustrated by the flat line in FIGURE 11.26, showing that probability effects start at $L > 16$.

In FIGURE 11.26, the magnitudes of the row selection pulses are larger than the maximum values comprising the column signals for $L \leq 16$. Since the pixel voltage is the voltage difference between the row and column voltages, the row selection pulses are likely to "show through" on the pixel voltage waveform for $L \leq 16$, making it less uniform. This is confirmed by referring to FIGURE 11.27 which shows the computed pixel waveforms for $L = 4$, 32, and 256. For $L = 4$, the four selection pulses ($K = 4$) are readily apparent on the pixel waveform and therefore likely to induce a certain amount of frame response.

11.4.5 Hardware Implementation of Multiple Line Addressing

In the previous sections, we have concentrated on describing multiple line addressing method using examples where the matrix has only a single column. The situation is more complex in a practical display because there are many columns that must be driven simultaneously. Computation of the column voltages according to equation (11.26) may be visualized as a

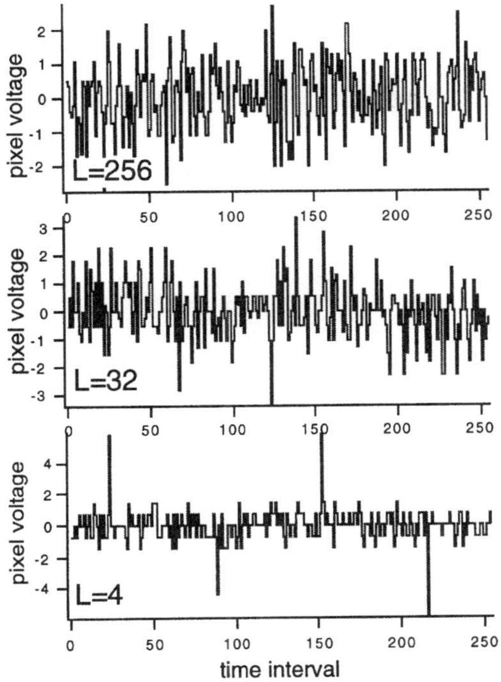

FIGURE 11.27. Computed pixel waveforms for $L = 4$, 32, and 256. The row selection pulses superimposed on the waveform for $L = 4$ are likely to induce a certain amount of frame response. Note changes in vertical scales.

transformation of a bit-map image matrix from the *image domain* to a column voltage matrix in the *time domain* where the matrix multiplication performs the dot product computation, including the product of the virtual pixel with the virtual row if gray levels are present. This operation is illustrated in FIGURE 11.28. In the original bit-map image the columns of the matrix form information vectors composed of information elements where each element represents the pixel data on the respective row and column. In the transformed column voltage matrix, columns of the matrix form column voltage vectors where each element is the column voltage for a column at a particular time t.

The matrix multiplication indicated in FIGURE 11.28 can follow two different progressions. In one progression, column voltages are computed for every column j in the matrix of M columns for one particular time interval k, after which column voltages are computed for every column j for the next time interval $k+1$. In this way, the column voltage matrix is updated *column-sequentially*. In another progression, the column voltages are computed for one particular column for all time intervals after which column voltages for the next column $j+1$ are computed for all time intervals. In this way, the column voltage matrix is updated *time-sequentially*. These two orderings are illustrated in FIGURE 11.29.

Column sequential computation has the advantage that column voltages are computed in an order that allows them to be directly applied to the columns of the display through the display drivers, such as with standard addressing as described in connection with FIGURE 11.10. It has the disadvantage that every computation is performed on a different information vector, which requires a tremendous memory bandwidth when conventional memories are used, since the entire memory must be read out for each time interval k. A bandwidth requirement this large generally requires that the computation be performed within the memory itself, implying a specialized memory device.

Time-sequential computation overcomes this memory bandwidth problem because the image frame buffer memory is read only once per frame; however, it requires that the computed column voltages be stored in a frame buffer so that they may be applied to the column of the display in the proper time sequence.

In either column sequential or time sequential computation, the core operation is the dot product term of equation (11.26) which will be discussed next.

11.4.5.1 Dot Product Computation

The complexity of the dot product computation is not only a function of the method used, but also the row

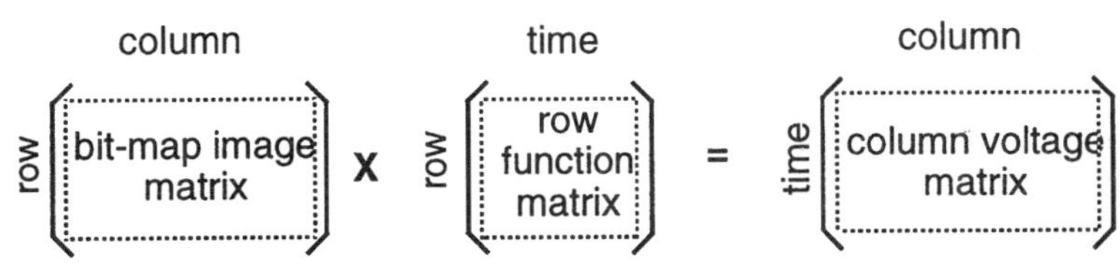

FIGURE 11.28. Illustration of column voltage computation for a display with many columns. The row function matrix transforms the bit-map image from the image domain to the time domain.

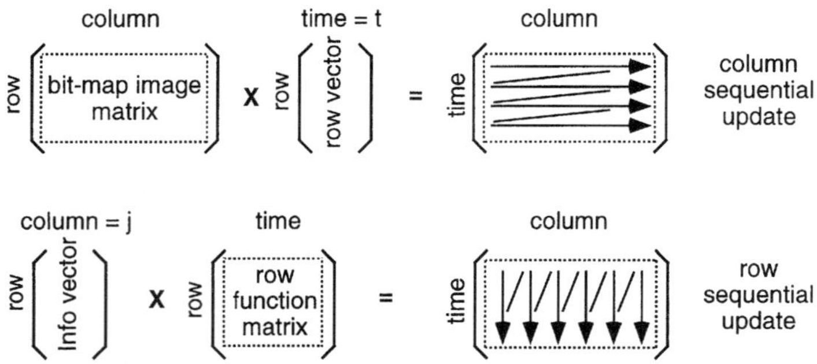

FIGURE 11.29. Comparison between column-sequential and time-sequential update of the column voltage matrix.

functions used. When using trilevel functions, only the information elements corresponding to nonzero (i.e., nonselected) rows need be considered in the dot product computation. In the example of FIGURE 11.25, only L rows are nonzero at any one time; and thus only L information elements are considered for the dot product at time t, indicated by equation (11.33). On the other hand, for the example of FIGURE 11.24, bilevel row signals are used so all the rows are nonzero at all times t, and thus all the information elements must be included in the dot product computation, indicated by equation (11.26). The fewer rows that are selected, the larger the ratio of peak voltage to rms voltage in the pixel waveform, and hence the disposition for frame response (cf. FIGURE 11.24) [40]. This points out the need for careful balancing of frame response and image quality with hardware complexity when selecting row functions.

The multiplication and summation inherent in the dot product can be carried out in many ways. When bilevel and trilevel row functions are used, the multiplication reduces to simply adding, subtracting, or disregarding the value of each information element representing a pixel value, depending on whether the row function corresponding to the pixel at time t is $+F$, $-F$, or zero; this is equivalent to an exclusive–or operation (XOR). The summation can be carried out in a digital pipeline fashion for speed. The XOR operation followed by a summation can be viewed as a correlation, wherein the correlation score for each time interval is proportional to the dot product.

Referring again to FIGURE 11.28, the matrix multiplication with the row function matrix can also be considered as a transform operation transforming the bit-mapped image space to column voltage space. For certain kinds of row functions, fast transform algorithms such as fast Walsh transforms can considerably simplify the number of computations required to generate the column signals.

11.4.5.2 Hardware Example Using Bilevel Row Functions

Many hardware implementations are possible for multiple line addressing, but each method has two main tasks in common. First, it must generate and apply the orthonormal row functions to the row electrodes of the display. Second, it must compute and apply the proper column signals based on the orthonormal row functions and the data to be displayed.

In the example of FIGURE 11.30, the orthonormal row functions are generated from a read-only memory (row function ROM) programmed, for example, with the bilevel Walsh functions. For a 240-row display

FIGURE 11.30. Block diagram of a simple hardware implementation of a display system using multiple line addressing.

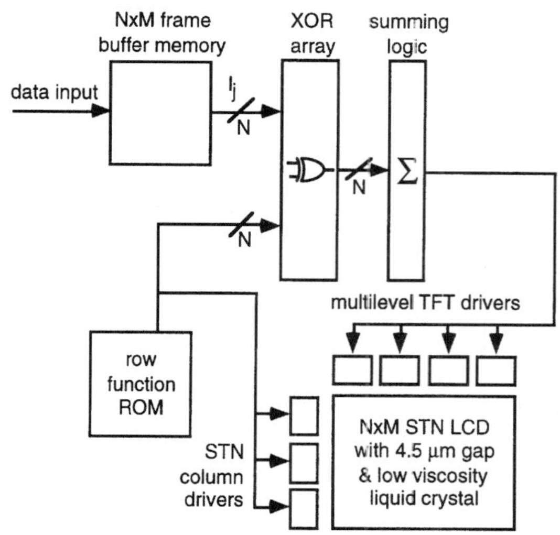

($N = 240$), the row function ROM would contain 240 Walsh functions, each having 256 values corresponding to the 256 time intervals. Using the column sequential update method (FIGURE 11.29), the 240-row function values for time interval k are read from the ROM and latched into the bilevel STN column drivers as explained, in connection with FIGURE 11.10, only this time they are used to drive the row electrodes. The same row function values are also read into a 240-bit-long array of XOR gates where they will participate in the dot product computation of the voltages for all the columns during the time interval k.

The information to be displayed is input to an $N \times M$-bit frame buffer memory, where it is stored as a bit-mapped image. For time interval k, the image is read out, one column at a time, into the other channel of the XOR array where it gets multiplied with the row function values, the XOR function acting as a multiplier. The N outputs of the XOR array are presented to a summing logic network, which completes the dot product. This digital result for each column is sent to multilevel, TFT-type source drivers with digital input, where they are queued by internal latches analogous to the two-level column driver of FIGURE 11.10. When all M column levels have been computed and input to the column drivers for time interval k, the drivers output the queued signals *en masse* to the column electrodes of the display in concert with the row drivers, which output the row signals for time interval k. The whole process is repeated for successive time intervals $k + 1$, until the dot products for all time intervals of the frame period have been processed, at which time a new frame period begins.

11.4.6 Impact of Multiple Line Addressing

Many possible systems would benefit from a low-cost, high-performance display, and multiple line addressing matches the need for hand-held instruments, notably videophones, games, pocket televisions, and perhaps even test equipment. Notebook computers will be served by having a TFT-like image for less cost, provided that the power consumption is in keeping with battery operation. Once multiple line addressing becomes established in high-volume applications, IC manufacturers are likely to embed multiple line addressing algorithms into the display controller circuitry, making it possible to use the frame buffer memory which may be already present in the system, to further reduce cost. Ultimately, the added cost of multiple line addressing will amount to the price differential between the multilevel TFT type drivers and standard STN drivers.

The currently available high-resolution 1024 × 768 STN displays point to the possibility of multiple line addressed HDTV systems. Passive matrix displays are also increasing in size, with one 16.7 inch diagonal LCD notably in full production. Hang-on-the-wall TV may require the manufacturing cost of passive matrix displays in order to be viable in the consumer marketplace.

The many nonideal performance aspects of STN LCDs still need to be improved (i.e., increased viewing angle and elimination of image artifacts) before these LCDs can really be considered a threat to TFT dominance in high-performance systems. Many of these qualities have been improved so remarkably in the last two years that it is tempting to extrapolate and predict very high performance STN LCDs in the near future.

REFERENCES

1. Kahn, F. J., Kendrick, P. N., Leff, J., Livoni, L., Loucks, B. E., and Stepner, D. 1987. A paperless plotter display system using a laser smectic liquid-crystal light valve. *SID Dig. Tech. Papers* **18**: 254.
2. Furuhashi, T., Ikeda, M., Tanaka, T., Kawakami, H., and Suzuki, T. 1993. High-quality TFT–LCD drive using low-voltage driver. *SID Dig. Tech. Papers* **24**: 15.
3. Budin, J.-P. 1993. The basics of flat panel displays addressing. *Sem. Lect. Notes Eurodisplay* **1**: A-1/1.
4. More complex schemes where both the row and column signals are data dependent are under consideration, but their practical applicability remains to be demonstrated. See for example H. Siefker, *Teilbild–Multiplex–Verfahren für bunte hochauflösende Flüssigkristall–Matrixbilder* (Düsseldorf: VDI, 1994).
5. Hughes, J. R., and Raynes, E. P. 1993. A new set of high-speed matrix addressing schemes for ferroelectric liquid-crystal displays. *Liq. Cryst.* **13**: 597.
6. Yang, D. K., Doane, J. W., Yaniv, Z., and Glasser, J. 1994. Cholesteric reflective display: drive scheme and contrast. *Appl. Phys. Lett.* **64**: 1905.
7. Schadt, M., and Helfrich, W. 1971. Voltage-dependent optical activity of a twisted nematic liquid crystal. *Appl. Phys. Lett.* **18**: 127.
8. Scheffer, T. J., Nehring, J., Kaufmann, M., Amstutz, H., Heimgartner, D., and Eglin, P. 1985. 24 × 80 character LCD panel using the supertwisted birefringence effect. *SID Dig. Tech. Papers* **16**: 120.

9. Clerc, J. F., Aizawa, M., Yamauchi, S., and Duchene, J. 1989. Highly multiplexed super homeotropic LCD. *Japan Display '89: Proceedings of the Ninth International Display Research Conference [Kyoto]*, p. 188. Playa del Rey, Calif.: Society for Information Display.
10. A thorough review of RMS–responding liquid crystal addressing schemes through 1988 can be found in T. N. Ruckmongathan, "Some New Addressing Techniques for RMS–Responding Matrix LCDs" (PhD Thesis, Indian Institute of Science, 1988).
11. Kawakami, H. *US Patent* 3,976,362 (1976) Aug. 24.
12. Alt, P. M., and Pleshko, P. 1974. Scanning limitations of liquid crystal displays. *IEEE Trans. Elec. Dev.* **ED-21**: 146.
13. Kawakami, H., Nagae, Y., and Kaneko, E. 1976. Matrix addressing technology of twisted nematic liquid crystal display. *Conference Record of 1976 Biennial Display Research Conference*, p. 50. New York: IEEE.
14. Tottori Sanyo. *Smart Addressing*.
15. Kaneko, Y., Okada, K., Tajima, E., Tsuruma, K., Kurachi, H., Suzuki, A., Ishida, T., Okamoto, M., Togashi, S., and Aoyama, T. 1990. Full-color STN video LCDs. *Eurodisplay '90: Proceedings of the Tenth International Display Research Conference [Amsterdam]*, p. 100. Berlin: VDE.
16. Murata, T. *US Patent* 4,824,211 (1989) April 25.
17. Suzuki, Y., Sekiya, M., Arai, K., and Ohkoshi, A. 1983. A liquid-crystal image display. *SID Dig. Tech. Papers* **14**: 32.
18. Kawakami, E., Hanmura, H., and Kaneko, E. 1980. Brightness uniformity in liquid crystal displays. *SID Dig. Tech. Papers* **11**: 28.
19. Kokogawa, T., Morishita, H., Adachi, K., Otsuki, H., Takasago, H., and Yamazaki, T. 1991. Effective evaluation of anisotropic conductive film. *Proceedings of the Eleventh International Display Research Conference*, California, p. 45. Playa del Ray, Calif.: Society for Information Display.
20. Akatsuka, M., Koh, H., Kuwata, T., Nakagawa, Y., Yuki, M., and Sawada, K. 1991. Material approach for the reduction of crosstalk in simple matrix LCDs. *Proceedings of the Eleventh International Display Research Conference*, California, p. 64. Playa del Rey, Calif.: Society for Information Display.
21. Seiberle, H., and Schadt, M. 1992. Influence of charge carriers and display parameters on the performance of passively and actively addressed LCDs. *SID Dig. Tech. Papers* **23**: 25.
22. Kaneko, Y., Haraguchi, M., Murakami, H., and Yamaguchi, H. 1990. Crosstalk-free driving methods for STN–LCDs. *Proc. SID* **31**: 333.
23. Hirai, H., Kondo, S., Murayama, A., and Hatoh, H. 1992. A new driving method for crosstalk compensation in simple-matrix LCDs. *Japan Display '92: Proceedings of the Twelfth International Display Research Conference [Hiroshima]*, p. 499. Playa del Rey, Calif.: Society for Information Display.
24. Clark, M. G., Glasper, J. L., Shanks, I. A., and Smith, C. J. T. 1985. New applications for waveform identity addressing of LCDs. *Proc. SID* **26**: 147.
25. Of course, a mutually orthogonal function that does not belong to the row set could be applied to the column to turn on *all* the pixels in the column.
26. Beauchamp, K. G. 1984. *Applications of Walsh and Related Functions*. New York: Academic.
27. Ruckmongathan, T. N. 1986. An LCD for multitrace oscilloscopes. *SID Dig. Tech. Papers* **17**: 128.
28. Kaneko, E. 1981. Liquid crystal matrix displays. *Adv. Image Pickup Displ.* **4**: 1.
29. Hirai, H., Kinoshita, Y., Shohara, K., Murayama, A., Hatoh, H., and Matsumoto, S. 1989. Optimization of cell condition and driving method in a VAN LCD for color video display. *Japan Display '89: Proceedings of the Ninth International Display Research Conference [Kyoto]*, p. 184. Playa del Rey, Calif.: Society for Information Display.
30a. Nehring, J., and Kmetz, A. R. 1979. Ultimate limits for matrix addressing of RMS–responding liquid-crystal displays. *IEEE Trans. Elec. Dev.* **ED-26**: 795.
30b. Nehring, J., and Kmetz, A. R. 1980. Ultimate limits for RMS addressing. *The Physics and Chemistry of Liquid Crystal Devices*, ed. G. J. Sprokel, p. 105. New York: Plenum.
31. Scheffer, T. J., and Clifton, B. 1992. Active addressing method for high-contrast video rate STN displays. *SID Dig. Tech. Papers* **23**: 228.
32. Ihara, S., Sugimoto, Y., Nakagawa, Y., Kuwata, T., Koh, H., Hasebe, H., and Ruckmongathan, T. N. 1992. A color STN–LCD with improved contrast, uniformity, and response times. *SID Dig. Tech. Papers* **23**: 232.
33. To derive the ultimate limit of the selection ratio, Nehring and Kmetz made the proportionally constant *c* dependent upon the number of "on" elements in the display. Forfeiting the slight increase in selection ratio by keeping *c* constant greatly simplifies the hardware implementation.
34. Conner, A. R., and Scheffer, T. J. 1992. Pulse-height modulation (PHM) gray shading methods for passive matrix LCDs. *Japan Display '92: Proceedings of the Twelfth International Display Research Conference [Hiroshima]*, p. 69. Playa del Rey, Calif.: Society for Information Display.
35. Ruckmongathan, T. N. 1992. Addressing techniques for RMS–responding LCDs: a review. *Japan Display '92: Proceedings of the Twelfth International Display Research Conference [Hiroshima]*, p. 77. Playa del Rey, Calif.: Society for Information Display.
36. Again, S. S. 1980. *Hadamard Matrices and Their Applications*. Lecture Notes in Mathematics, no. 1168. New York: Springer.

37. Clifton, B., and Prince, D. 1992. Hardware architectures for video rate, active addressed STN displays. *Japan Display '92: Proceedings of the Twelfth International Display Research Conference [Hiroshima]*, p. 503. Playa del Rey, Calif.: Society for Information Display.
38. Ruckmongathan, T. N., Kuwata, T., Ohnishi, T., Ihara, S., Koh, H., and Nakagawa, Y. 1992. A new addressing technique for fast responding STN LCDs. *Japan Display '92: Proceedings of the Twelfth International Display Research Conference [Hiroshima]*, p. 65. Playa del Rey, Calif.: Society for Information Display.
39. Scheffer, T. J., Clifton, B., Prince, D., and Conner, A. R. 1993. Active addressing of STN displays for high-performance video applications. *Displays* **14**: 74.
40. Clifton, B., Prince, D., Leybold, B., Scheffer, T. J., Conner, A. R., and Greenberg, B. 1993. Optimum row functions and algorithms for active addressing. *SID Dig. Tech. Papers* **24**: 89.

CHAPTER 12

Liquid Crystals for Optical Communication Devices

J. S. PATEL

12.1 INTRODUCTION

Optical networks require a variety of active and passive devices to accomplish connectivity [1]. The purpose of these devices is quite simple and stated most simply, it is to allow generation, routing, and detection of photons. While existing optical networks use a single wavelength of light, typically 1.55 μm or 1.3 μm, there is increasing interest in expanding the networks to simultaneously accommodate several wavelengths. This poses additional challenges requiring devices that are sensitive to a particular wavelength, in other words, devices that not only switch an optical signal from one fiber to another, but do so selectively for a particular wavelength. In this chapter we will examine a variety of such devices that may be useful for the optical communications.

An optical device affects light in one of several ways. FIGURE 12.1 shows some of the ways in which light can be affected by passing through a medium. Any one or combination of these effects can be used to produce an optical modulator. The device can be made active if the medium can somehow be changed using external perturbations, or passive when the device function is fixed. An example of the former is a light modulator, while an example of the latter is a beam splitter.

Consider the simple case of absorption in FIGURE 12.1. If the absorption of the medium can be controlled, the passage of light through the medium can be controlled resulting in an optical modulator. Notice that in this case it is not a switch, since the light that is absorbed is lost and cannot be directed to a different destination, as would be required for the strict definition of a switch which requires the signal to be switched from one path to another. This can be achieved, for example, by using an electrically controlled diffraction element. If the grating can be

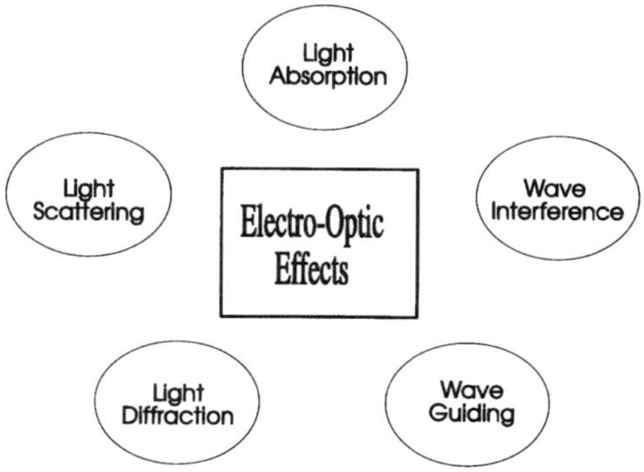

FIGURE 12.1. Various types of electrooptic effects.

switched on at will, then it would be possible to electrically change the direction of light. A device such as this would constitute a switch.

In this chapter, we limit the discussion to those optical devices based on liquid crystals. Some of the earliest liquid crystal devices were light modulators based on the dynamic scattering effect [2]. In this case, the application of an electric field across the liquid crystal causes the liquid crystal to set up convection in the cell, resulting in rapid fluctuation of the local refractive index, which due to index mismatch causes the light to reflect and refract, and undergo multiple scattering. This causes the emerging light to have many different directions, including reflection in the original direction. Such a device is an example of the use of the electrically controlled scattering principle shown in FIGURE 12.1. Polymer-dispersed liquid crystal devices [3], in which tiny phase-separated liquid crystal droplets are dispersed throughout the polymer, also operate by the light scattering principle. In this case, in the absence of an applied electric field, the randomly oriented domains of liquid crystal molecules serve as the light scattering centers, while in the field-on state, the molecules become oriented all along the field, causing index matching between the polymer host and the liquid crystal molecules, thus eliminating light scattering. A more unique liquid crystal device is one in which the optical medium has continuously changing dielectric properties as a function of thickness. One such device is the twisted nematic device [4] in which the light experiences a dielectric tensor which rotates the plane of polarization of the light by 90°. For such devices, properly chosen device parameters are necessary for the light to rotate by exactly 90° as it traverses the device. For devices that are not well engineered, the light may emerge elliptically polarized, and the ellipticity depends on the relative orientation of the plane of the input polarization and the orientation of the long axes of the molecules at the surfaces [5]. There are other devices in which the phase of the light is affected, but not the amplitude. These devices can be used in a variety of ways, as we shall see later.

Liquid crystals are fluids which have orientational order because the long axes of the molecules lie roughly parallel to each other. The local average orientation of the long axis of the molecules is represented by a unit vector called the director [6]. In the simplest geometry of an aligned liquid crystal cell, the directors at both surfaces are parallel to each other. In this geometry, the material behaves essentially as a uniaxial material, whose index can be changed along the director axis by using a small electric field. This ability of a liquid crystal to change its refractive index with very little power makes it an ideal candidate for use in a variety of optical devices. In the next section, we briefly discuss the elementary optical functions that can be performed using these materials.

12.2 ELEMENTARY OPTICAL FUNCTIONS

The way in which liquid crystals affect light usually depends on the input polarization of the light. This is simply because liquid crystals are anisotropic fluids and light propagation through the medium depends on the direction of the electric field of the electromagnetic radiation [7,8]. The polarization state of the light can be represented by Stokes vector, defined by four parameters, and can be graphically represented as a point on the Poincre sphere [9]. Without going into the details of the structure of liquid crystals, the optical properties of these materials can perform the following most elementary functions:

(A) impose a phase shift on plane-polarized light;
(B) change the ellipticity of the light by imposing a relative phase shift between the TE and the TM polarized waves; and
(C) rotate the plane of polarization of light.

These elementary functions are possible by acting upon the real part of the refractive index of liquid crystals. The imaginary part can also be used in electrically controlled light absorption using liquid crystals. Later we will see that these elementary functions can be combined with polarizers, gratings, dielectric mirrors etc., to produce interesting liquid crystal based devices.

12.3 ELEMENTARY LIQUID CRYSTAL STRUCTURES

There are a variety of liquid crystals which basically fall in two categories: the nematics and the smectics [10]. Although lyotropic liquid crystals [11] are interesting and have a variety of fascinating structures, the electrooptic properties of these materials are generally less useful for optical devices and therefore will not be considered here. While nematics are in a class by themselves, various modifications of smectic phases exist, providing a rich range of properties. Depending on the nature of the chemical makeup of these liquid crystals, a wide range of optical and electrical properties can also be obtained [12]. These properties can also be changed dramatically by making the constituent molecules chiral. In fact, nematic liquid crystals made up of chiral molecules are grouped separately in a class of materials called cholesterics. Similarly, the chiral modification of the smectic C phase, denoted as smectic C*, has very different electrical properties compared with the nonchiral one. The smectic C* phase behaves as a ferroelectric phase [13]. The reader is referred to other chapters in this book for a more detailed description of these materials.

An important aspect of using these materials in devices requires an understanding of how the optical properties [14] of these materials change when subjected to an external perturbation, such as an applied electric field. The effect not only depends on the intrinsic structure of these materials but also on the extrinsic structure, produced by controlling the boundary conditions. While nematic liquid crystals are commonly uniaxial by symmetry, the smectics can be optically biaxial. In constructing devices, not only is it important to obtain the proper liquid crystals with appropriate properties, but also to obtain monodomain samples, and uniform cell thickness. The essential aspects of these materials can be summarized as follows.

(A) Nematic liquid crystals allow their optic axes to be reoriented such that it is either along the direction of the applied electric field (for materials with positive dielectric anisotropy), or perpendicular to the direction of the field (for materials with negative dielectric anisotropy).

(B) Similar effects can also be obtained using smectic A liquid crystals at much higher fields, but with some degree of irreversibility because once the layered structure is disturbed, it can not be restored easily.

(C) Smectic C* and the electroclinic effect, as well as the flexoelectric effect in cholesteric liquid crystals, allow the optic axis to move in a plane perpendicular to the direction of the applied electric field. For the ferroelectric effect, the optic axis is typically bistable, with the optic axis choosing one of two orientations. In the case of electroclinic and flexoelectric effects, the rotation of the optic axis could be a variable, depending on the magnitude of the applied electric field.

(D) It is possible to obtain optical devices in which the optical properties change continuously along the thickness of the device. A simple example of this is the twisted nematic structure, which is optically equivalent to a large number of thin birefringent plates on top of each other, with the optic axis in the neighboring plates at a small angle with respect to each other. While such structures can be artificially produced by piling up birefringent plates, in practice, the total amount of twist from the bottom to the top plate is somewhat limited. Using cholesteric liquid crystals on the other hand, the twist can be quite large; from 0 to about $360^\circ\ \mu m^{-1}$ is fairly easy using common liquid crystal materials. Thus using cholesteric liquid crystals, it is possible to have the optic axis undergo a rotation of several thousand degrees using a sample thickness of just $10\ \mu m$. However, in these samples, the structure cannot easily be altered with an applied electric field. In samples with total twist of the order of π, it is, however, possible to easily alter the structure with an applied electric field. The classic example is of a twisted nematic device in which the total optic axis rotation is 90°, but the twist can be destroyed by applying an electric field.

The optical properties outlined in section 12.1, and the structure description given in section 12.2, can be combined to produce useful electrooptic structures which can be incorporated into specific devices designed for a specific operations. This is outlined below with some examples.

The following are simple devices to produce elementary optical function, using either nematic liquid crystals, smectic liquid crystals or both by exploiting various electrooptic effects. Several of these elementary structures can be combined to produce more complex liquid crystal devices, such as those for producing arbitrary polarization transformation [15], which are useful in coherent communications applications [16].

12.3.1 An Electrically Controllable Phase Shifter

One of the simplest liquid crystal devices is one for providing an electrically controlled phase shift of the light falling on the device. The main advantage is that this shift can be accomplished with very small voltages, since liquid crystal structures are easily deformed using external perturbations. However, the birefringent nature of the liquid crystal does not allow for this input polarization of the light, unless the structure is specifically designed for polarization-independent operation. The elementary operation is best described by considering plane-polarized light impinging on the device in a specific way. Consider, for example, a liquid crystal device containing nematic liquid crystals, which has been oriented using surface forces in a way that the directors at the top and the bottom surfaces are parallel to each other, and along the y- axis as shown in FIGURE 12.2. The light propagates along z. By applying an electric field along the z axis, which is typically done using the conductive transparent electrodes that are deposited on the inner surfaces of the cell, the director can be reoriented. Because of the rigid boundary conditions, the molecules in the middle of the cell are much more easily distorted in the middle of the cell compared with those at the surfaces. This produces a nonuniform director distortion across the cell such that the angle of deviation θ away from the xy plane changes as a function of z. However, in the geometry used here, the tilt $\theta(z)$ is confined in the yz plane. The distortion for the original orientation depends on the magnitude of the applied electric field and is shown schematically in FIGURE 12.2. Notice however that the nonlinear variation of $\theta(z)$ is irrelevant for optical purposes since the total phase shift is the integrated value. In the case of a nematic liquid crystal with principal values of the refractive indices $nx = nz$ and ny, the effective index at z in the cell is

$$\frac{1}{n_z(\theta)} = \frac{\cos^2\theta(z)}{n_y^2} + \frac{\sin^2\theta(z)}{n_x^2}. \tag{12.1}$$

FIGURE 12.2. Nematic liquid crystal with parallel boundary conditions at the two surfaces in absence and presence of an applied electric field.

The total phase shift for light polarized along the y axis is

$$\phi_y = \frac{2\pi}{\lambda}\int_0^d n_z(\theta)\,\mathrm{d}z \tag{12.2}$$

which depends on the magnitude of the applied electric field. The phase shift in the perpendicular direction is, however, field-independent and is simply $(2\pi n_x/\lambda)d$ where d is the thickness of the liquid crystal material, and λ is the wavelength of the light.

12.3.2 A Device to Electrically Control the Ellipticity of Light

An ideal polarization control device is one in which the polarization of the light, represented by a point on a Poincre sphere [17], can be changed to any other point on the sphere. Although this is impossible using a single electrooptic element, it be done using multiple electrooptical elements. These devices can be very useful in coherent communication applications, where the polarization state of the light is quite important. This can be done using liquid crystals as has been previously demonstrated [18], but here we limit the discussion to a simpler case, in which the principal axes of the ellipse are fixed, and only the ellipticity changes.

To control the ellipticity, one has to produce a differential phase shift between two optical waves, where the principal directions of vibration are orthogonal to each other for the two waves. This is most easily done using nematic liquid crystals with two different indices along the two orthogonal directions, such as that described above, in which the directors at the top and the bottom surfaces are parallel to each other. A variety of elliptical states are possible by proper choice of the optical thickness and the orientation of the director axis with respect to the input polarization. For example, if the light falling on the liquid crystal cell is at 45° to the rubbing axis, then light excites the two normal modes of the liquid crystal cells equally. However, the propagation of these two modes occurs with different phase velocities, resulting in phase difference when the waves exit. The maximum phase difference is determined by the birefringence of the material, the sample thickness and the wavelength of the light, and is given by $\Delta\phi = 2\pi\,\Delta nd/\lambda$, where the symbols have the meanings defined before, and $\Delta n = n_e - n_o$ is the birefringence. In the presence of the applied field, the director distorts as explained above, and the effective phase difference decreases. Thus the phase difference becomes electrically adjustable since $\Delta\phi(E) = 2\pi\,\Delta n(e)/\lambda$. In this configuration, when the phase difference is π, the output polarization again becomes linearly polarized, but with the output polarization direction orthogonal to the input polarization direction. This can be visualized most easily by noting that the linearly polarized light gets transported to the opposite side of the Poincre sphere for a phase difference of π. Similarly, circularly polarized light can be obtained by introducing a phase difference of $\pi/2$, which transforms linearly polarized light into circularly polarized light since, by transforming the point on the equatorial plane to a point at the pole of the Poincre sphere. A detailed description of how to perform this graphical transformation in a visual way can be found elsewhere [19] and will not be discussed here. FIGURE 12.3 shows the transformation for obtaining circularly and orthogonally linearly polarized light from initial linearly polarized light.

FIGURE 12.3. Polarization transformation shown on a Poincre sphere due to a a quarter-wave and a half-wave plate oriented at 45° to the input polarization axis.

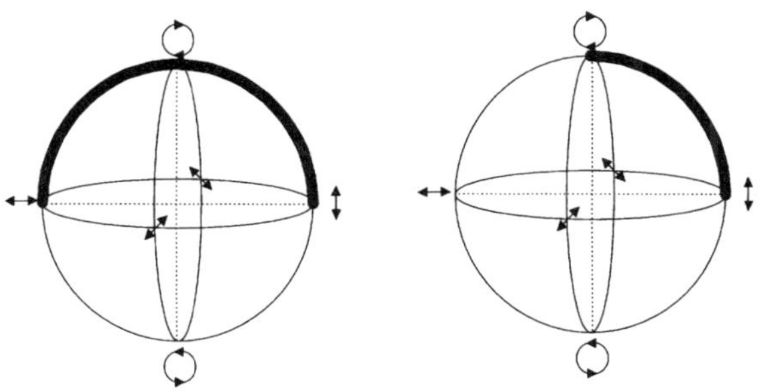

12.3.3 Electrically Controlled Polarization Rotator

Unlike the above two cases which use electrooptic elements with linear polarization as eigenmodes, the twisted structures in general have elliptical modes as eigenmodes. On the Poincre sphere, this means that by using twisted nematic structures, in the most general case, the input polarization can be transformed to a different point on the Poincre sphere, but in a more complex way. The net result, however, is that for a properly designed twisted structure with appropriate thickness and birefringence, the linearly polarized light gets rotated by 90° if the total twist of the nematic structure is 90°. For the case of very large thickness such that $\Delta nd \gg \lambda$, the so-called Mauguin regime [20], the 90° rotation of the director produces a 90° rotation of the polarization and can be represented in the Poincre notation as shown below. Note that in this regime, when linearly polarized light remains linearly polarized as it passes through the twisted nematic structure, it is represented by the dark line.

This rotation can be prevented by destroying the twisted structure by applying an electric field. The light passing through a thick twisted nematic structure is along the equator as shown in FIGURE 12.4. There is, however, no special reason to chose 90° other than convenience, since it easily allows a modulator to be made with polarizers, capable of 0 or 1 transmission states. Thus any arbitrary rotation of the polarized

FIGURE 12.4. Polarization transformation shown on a Poincre sphere due to a 90° twisted nematic structure in the Mauguin regime.

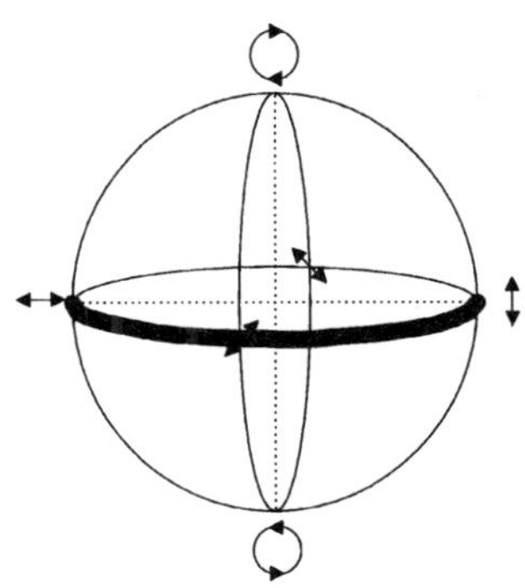

light can be produced by appropriate choice of the total amount of twist in the twisted nematic structure, together with the proper choice of material parameters and thickness. In these types of devices, again most easily constructed with nematic or cholesteric liquid crystals, the directors at the two surfaces make an angle θ with respect to each other, where θ is the desired amount of polarization rotation. In the absence of an applied voltage, such a device would rotate the input polarization by approximately θ. If, however, a voltage is applied to such a structure, and provided that $\theta < 2\pi$, the twist can be eliminated by applying an electric field which eliminates the mechanism for polarization rotation. If the twist is much greater than 2π, the structure can be electrically undone, but upon removal of the field defects and metastable structures are often produced, which are difficult to eliminate. Thus it is well known that for cholesteric liquid crystals, while a sufficiently high electric field destroys the helical structure, upon removal of the field, that structure is quite different from the initial one. This behavior has been used for memory displays. It is also possible to use ferroelectric liquid crystals to produce polarization rotation using a variety of devices [21].

12.3.4 Cross-Connect Devices

Devices with more functional properties often require the combination of one or more of these elementary devices, together with other elements such as polarizers, beam displacers, dielectric mirrors etc. For example, the switching function requires a path selection mechanism that can be used in conjunction with a polarizing beam splitter to choose the direction of the output path of the light, as shown in FIGURE 12.5. Path A is chosen if the light falling on the polarization sorting device is polarized horizontally, while path B is chosen if the light is polarized vertically. A polarizing beam splitter, or some other similar device such as calcite beam displacer, would constitute the polarization sorting device in FIGURE 12.5.

The path chosen by the light beam can thus be controlled by controlling the state of the input polarization falling on the device S. Simple devices would be like those listed above that would produce the desired polarization state, controlled through the application of a voltage across the liquid crystal. While these structures may be fine for some applications, they require modifications to be useful as communication devices. This is because in fiber optic based communications, the input polarization at any stage in the fiber where such a device is to be inserted, is not generally known or well defined. As a result, the devices for communications applications should generally be polarization insensitive. In what follows, we describe such a structure that may be used to produce a $1 \times N$ polarization-independent optical switch.

There are some examples of the use of liquid crystal materials in making optical switches for use in communication systems [22]. Most liquid crystal devices operate by changing the polarization properties of the input light. The operating principles in most of these devices are based on electrically changing the orientation of the molecules to affect the output polarization of the light for a chosen input polarization of the light. Since liquid crystals are inherently polarized, simple liquid crystal devices are undesirable for communication applications because of the random variability of the input polarization of the light.

A common technique is to use the polarization diversity technique, in which the input light is sorted into two beams of well-defined polarizations, and then acted upon by two properly oriented liquid crystal modulators, before combining the two beams back together into a single beam by suitable means. One such device was demonstrated many years ago by Wagner and Chang [23] using conventional twisted nematic liquid crystals and polarization beam splitters. More recently, a similar structure was used by

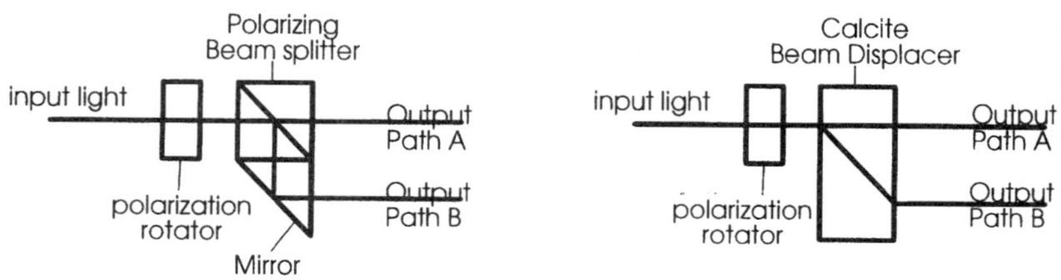

FIGURE 12.5. Beam displacement components which allow beam routing using input polarization for path selection.

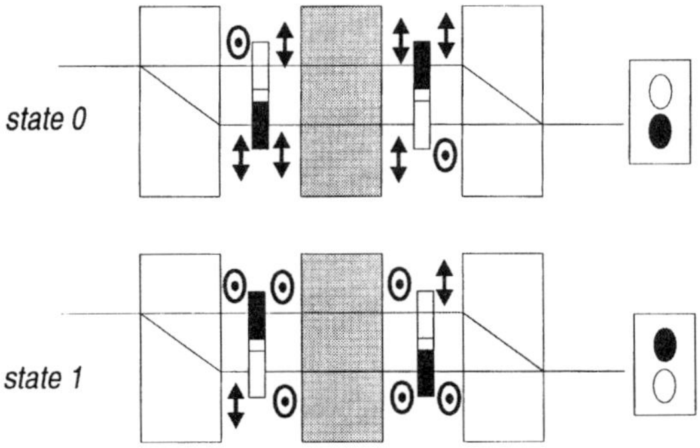

FIGURE 12.6. Configuration of 1×2 optical switch.

Fujii to demonstrate a 1×2 low-crosstalk optical switch [24]. Here we show a different $1 \times N$ polarization-insensitive optical switch, which allows one input beam to have one of N possible output beams determined by the switch configuration. This switch uses birefringent crystals as polarization sorting elements, beam displacers, and beam combiners. Liquid crystals have also been used to produce other $1 \times N$ switching configurations that will not be discussed here in any detail [25].

The basic structure of a 1×2 switch is shown in FIGURE 12.6 illustrating the use of three birefringent crystals and two sets of liquid crystal elements that electrically configure the switch. The switch consists of three birefringent crystals; the first is used to separate the beam into two well-defined orthogonally polarized beams. These beams pass through two liquid crystal modulators. At any one time, one of the modulators is switched on and the other is switched off. This produces two beams with the same polarization state after passing through the liquid crystal device. Let these two polarization states be denoted by A and B. The next birefringent crystal determines the path of the beam, because the path the light follows through the crystal depends on the state of polarization of the two beams. Thus for beam A, the light passes through the birefringent crystal in the plane of FIGURE 12.6, while for beam B (shown at the bottom of the figure) the light follows a different path that is elevated at a height d with respect to the plane of FIGURE 12.6. On emerging from the second birefringent crystal, the light beams once more pass through liquid crystal modulators. The purpose of the modulators is to polarize the two beams orthogonally with respect to each other. Thus once again one of the modulators is on while the other is off. The light beams then enter the third birefringent crystal, whose purpose is to combine the two beams into a single beam. The position of the beam that emerges from the third birefringent crystal depends on the path taken by the light through the middle birefringent crystal. Thus if for the first modulator, the top and the bottom elements are on and off, respectively, and for the second modulator, the top and the bottom elements are chosen to be off and on, respectively, then the beam that emerges from the third birefringent crystal lies in the plane of FIGURE 12.6. If, on the other hand, the switching of the liquid crystal modulating elements is interchanged as shown in the figure, the beam emerges elevated at a distance d from the plane of the figure.

The structure can be extended to a 1×4 switch by adding one more switching elements and another birefringent crystal of twice the original crystal thickness, as shown in FIGURE 12.7. The thickness of the crystal must be twice the original thickness to displace the beam by twice the distance. The operating principle of this device can be understood by following the arguments laid down in the previous paragraph.

FIGURE 12.7. Configuration of 1×4 optical switch.

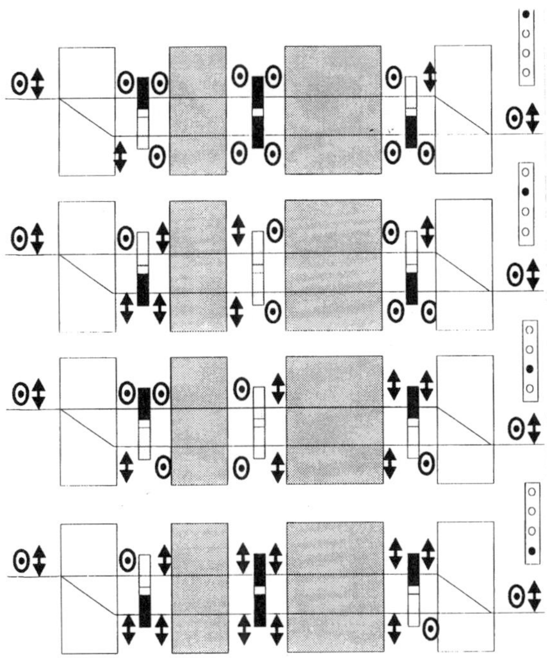

Basically, the insertion of the additional birefringent crystal and the liquid crystal elements allows the beam to follow a different path through the switch, allowing the number of output states to be multiplied by a factor of two for each additional stage. Thus the 1×4 switch can be extended to 1×8 by the addition of one more birefringent crystal of twice the thickness of the original and a liquid crystal element. In this switch configuration the light passes through the same number of liquid crystal elements, which makes the loss through the device uniform for all the states. This is a desirable property for optical switches that should not have path-dependent losses, such as those that may be encountered in cross-bar switches.

Such a device has been constructed by us, using 4 mm thick calcite crystal which allowed beam displacement of 400 μm. The twisted nematic liquid crystal devices were made by conventional techniques. The device was assembled using UV curable epoxy by first attaching the two liquid crystal elements to the middle calcite crystal, taking care to register the two liquid crystal elements properly. This was followed by attaching the other two calcite crystals on either side of this structure with proper orientation. Because of the small beam displacements, collimating GRIN (GRaded INdex) lenses attached to single-mode fibers were used to produce a collimated beam of about 250 μm width that passed through the switch and was collected by similar GRIN lensed fiber when the beam emerged on the other side of the switch. The light source used in these experiments was a DFB (distributed feedback) laser operating at 1.5457 μm. This very high extinction ratio is due to the fact that the light beam has to pass through two modulators and only the light with the correct polarization emerges in the right position. The crosstalk signals are spatially displaced in a different location and are not collected by the output fiber. Thus, consider the case where the first modulator allows through a small amount of light of the wrong polarization (orthogonal to the desired polarization state). Further, assume that light, once more with the wrong polarization, passes through the second liquid crystal stage. This "doubly wrong" polarized light would appear at one of the two output ports as noise. Assuming that we are able to properly align the calcite crystals, which is relatively easy, the off state still lets some light of the wrong polarization through. Thus the transmission at the first stage due to light of the wrong polarization, is about -20 dB, and at the second stage there is another decrease of -20 dB, resulting in an overall transmission of -40 dB for light that is not supposed to pass through. This is close to the observed value of -37 dB, assuming that a single element has a extinction of -20 dB. Light that is of the wrong polarization at only one of the stages follows a different path through the switch, and does not appear at either of the two output ports. Thus, light with the wrong polarization at only one stage, appears on one side or the other of the proper output ports. In addition, the crosstalk signal is either in the plane of FIGURE 12.7 or above it, depending on whether the first or second stage is faulty. Thus in addition to the two desired output ports, the switch has four additional spatial locations where the crosstalk signal appears. This feature may be useful as a switch diagnostic or alignment tool.

FIGURE 12.8 shows the light emerging at one of the output ports as the phase of the two elements was interchanged. The operating voltage was 5 V AC. To produce the out-of-phase train of square-wave voltages, we used two phase-locked programmable generators. We observed that the switch produced an extinction of -37 dB and the switching times, as expected, were on the order of tens of milliseconds. The switch-on and switch-off times were different since one is determined by the magnitude of the square of the applied voltage, while the other is determined by the reciprocal of the square of the thickness of the cell. Note that the cell cannot be too thin because of the required liquid crystal thickness for achieving 90° rotation of the light. The

FIGURE 12.8. Switching performance of the 1×2 switch.

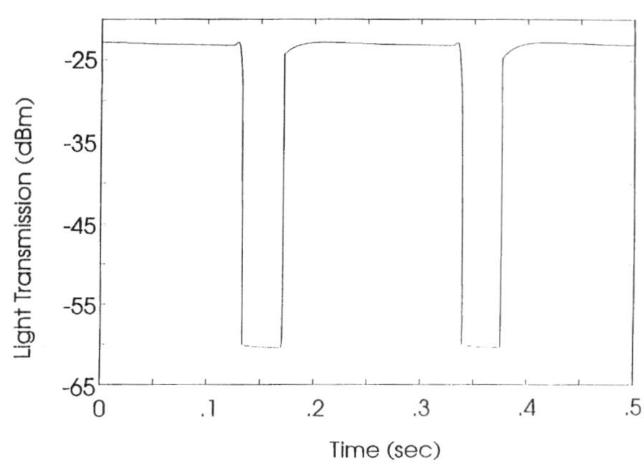

switching speeds can be made symmetric if one uses the ferroelectric liquid crystals in a newly reported structure [26].

The liquid crystal devices used had the conventional twisted nematic structures. It is possible to use other structures, such as electrically controlled birefringence, which can also cause the input polarization of the light to be rotated by 90° as described above. These liquid crystal devices, however, are relatively slow and can only switch on a time scale of milliseconds. A more attractive approach is the use of ferroelectric liquid crystals in these switches, which would allow not only faster switching, but also the ability to latch [27–29]. This is a very desirable property which allows the switch to be reconfigured with the appropriate electrical signal and stay in that state after the electrical power is turned off. Most of these switches are intended to operate at a broad wavelength region, and in general do not allow switching operations such as switching one wavelength from a group of wavelengths. This functionality becomes important for emerging telecommunication architectures based on wavelength-division-multiplexed networks. In these networks, communication signals encoded in terms of optical wavelength are grouped into the same fiber. In the next section, we will discuss several liquid crystal based devices that allow sorting of these wavelength-encoded signals.

12.4 SPECIFIC DEVICES FOR WDM APPLICATIONS

The ability to select a desired wavelength channel from a range of available wavelength channels is of great interest for advanced lightwave systems. This is particularly true for high-density wavelength-division-multiplexed (HD WDM) networks [30], in which components capable of wavelength selections are crucial. While several devices have been proposed to achieve this wavelength selection [31–33], most require significant power to operate or have a complex design. For example, certain acoustooptic tunable filters require about 10 W of power [34], although recent work suggests that these requirements can be relaxed [35]. Liquid crystal devices, constructed in the form of Fabry–Perot etalons, can be used as a wavelength filters in the infrared region, with wide tuning capability and requiring several orders of magnitude less power. Previous work involving liquid crystal and Fabry—Perot structures has been primarily limited to spatial light modulators for use in the visible spectral region [36]. In what follows, we outline a variety of liquid crystal Fabry–Perot devices that may be useful for WDM communications networks [37].

12.4.1 Polarization Sensitive Liquid Crystal Fabry–Perot Structures

The ability of a liquid crystal to change its refractive index with very little power makes it an ideal candidate for use in a Fabry–Perot etalon. Another advantage of using a Fabry–Perot structure is that it is simple and easy to construct, while essentially resembling a liquid crystal cell. For a simple Fabry–Perot etalon, it can be shown that the transmission (T) at a given wavelength λ for radiation incident normal to the surface is given by [38]

$$\lambda = \frac{\tau}{(1-\rho)^2 + 4\rho(\sin^2\delta)} \tag{12.3}$$

where $\delta = \varphi + k_0 dn$. Here τ and ρ are the transmittance and the reflectance, φ is the phase shift experienced upon reflection, d is the thickness of the uniaxial material, n is the refractive index along the director axis, and k_0 is the magnitude of the wavevector outside the layer. It can be seen from equation (12.1) that the width of the transmission peak $\Delta\lambda$ depends essentially on the reflectivity of the surfaces, while the overall transmission is dictated by the absorption losses.

The free spectral range (FSR) (wavelength separation between successive transmission peaks) is determined by the optical thickness nd of the etalon. This range depends on the order in which the interferometer is being operated, and is given by

$$\text{FSR} = \frac{2nd}{m(m+1)} \tag{12.4}$$

where m is the order. The refractive index of liquid crystals is typically in the range 1.5 to 1.7. Thus 1.5 μm light will pass through the etalon when the gap is on the order of 10 μm and the etalon is used in the 20th order. In this wavelength region, the free spectral range for a 10 μm etalon would be about 75 nm. This thickness should ideally be chosen such that the wavelength range of interest is the same as the FSR. The tuning range is estimated to be on the order of 200 nm, which corresponds to a change of about 0.2 in the refractive index. The maximum change in the

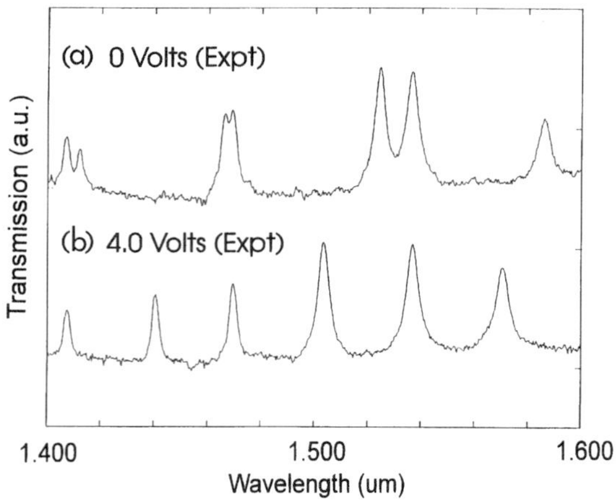

FIGURE 12.9. Measured transmission spectrum of a liquid crystal Fabry–Perot etalon. Curve (a) 0 V and curve (b) 4 V.

index is equal to the birefringence of the uniaxial liquid crystal.

Such devices have been constructed using the commercial nematic liquid crystal E7. The reflectors were made by using a dielectric stack. The electric field was applied to the liquid crystal using transparent indium tin oxide electrodes, which were deposited on the glass plates prior to deposition of the dielectric mirror stacks. A thin alignment layer was deposited on top of the dielectric mirror to promote alignment of the liquid crystal molecules [39].

FIGURE 12.9 shows typical transmission spectra with 0 and 4 V applied across the liquid crystal Fabry–Perot etalon. The increasing linewidth of the transmission peaks with increasing wavelength is an artifact of the mirrors because the reflectivity changes as a function of wavelength. The data suggest that our mirrors were better reflectors at shorter wavelengths than at longer wavelengths. A comparison of the spectra with and without the electrical field clearly shows that while some lines are influenced by an electric field, others are not. This result, which will be explained shortly, can be used to calculate the principal indices of our liquid crystal. Knowing the etalon gap to be about 1 μm, we can calculate the refractive index for the ordinary ray using those transmission peaks that do not move when the field is varied. It was determined to be 1.5 using $m = 25$. The order $m = 25$ was determined from the FSR and the wavelength of the transmitted radiation which was

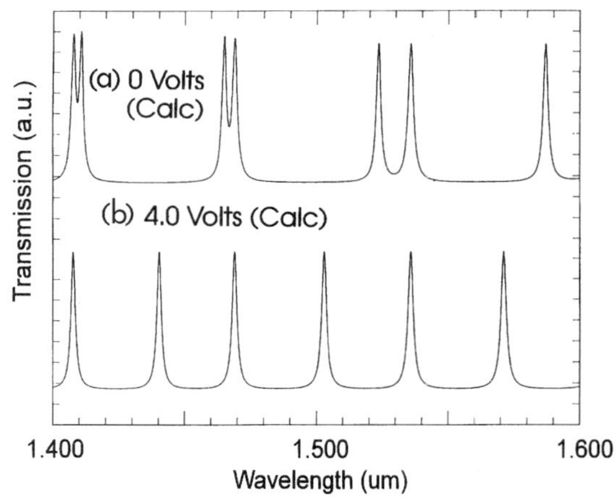

FIGURE 12.10. Calculated transmission spectrum of a liquid crystal Fabry–Perot etalon. Curves (a) and (b) are the calculated transmission spectra of unpolarized light using etalon thickness of 11.32 μm, n_o is 1.5 and n_e is 1.691 for curve (a) and n_o is 1.5 and n_e is 1.5346 for (b). The mirror reflectivity is assumed to be 90% with no absorption loss.

polarized perpendicular to the director. Similarly, the extraordinary index can be determined for the free spectral range corresponding to the lines that do move when the electric field is applied, and was calculated to be 1.7. The difference between the two indices is consistent with previously reported values of birefringence of this material in the infrared region [40].

FIGURE 12.10 is the calculated spectrum assuming no absorption and using parameters which closely match those of the device. The actual values used for the calculation are indicated in the caption. For example, the calculation of the zero-field spectrum required the use of the values $n_0 = 1.5$ and $n_e = 1.7$ whereas $n_0 = 1.5$ and $n_e = 1.53$ were used in calculating the lower spectrum. These values were chosen to obtain a best fit to the observed data. The calculations were done by superimposing the transmission spectra due to both polarizations. This is necessary since the refractive index in the liquid crystal is different in two orthogonal directions. In the sample geometry chosen here, the application of an external electric field alters the refractive index only along one of the axes. This causes the spectral transmission lines corresponding to the ordinary ray to remain unaffected, while others move with the electric field. This can be seen clearly by examining the spectra shown in

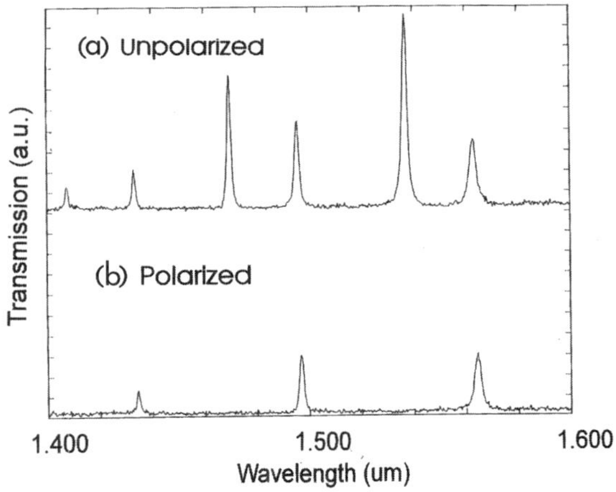

FIGURE 12.11. Measured transmission spectra of a liquid crystal Fabry–Perot etalon (a) without a polarizer and (b) with a polarizer.

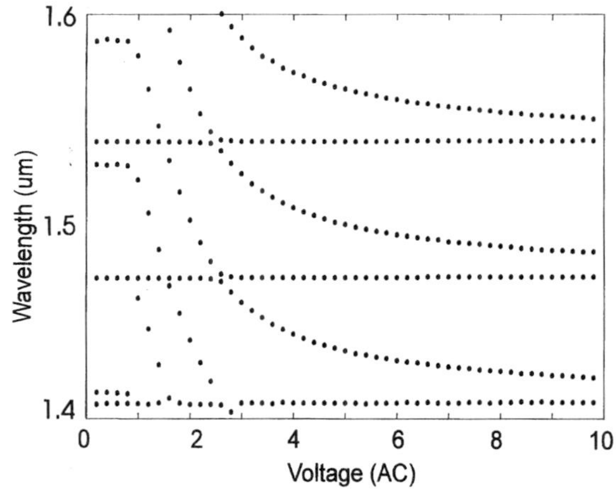

FIGURE 12.12. Positions of the transmission peaks in the unpolarized spectrum as a function of voltage for a 11.32 μm liquid crystal Fabry–Perot etalon.

FIGURE 12.9. The two sets of lines are characterized by orthogonal polarizations.

The two spectra shown in FIGURE 12.11 are obtained with and without an output polarizer oriented parallel to the short axes of the molecules in the device. This clearly illustrates that the two sets of lines have orthogonal polarizations. This is a unique property of a Fabry–Perot etalon in which an anisotropic material is used. In this respect, our device can also be thought of as a narrow band polarizer. This is easy to understand by considering a monochromatic input which is randomly polarized. In this case, the light in the Fabry–Perot etalon is confined to two orthogonal, plane polarization states, which are the modes supported by the uniaxial material. Since the refractive index is different along the two orthogonal directions, this results in different optical thicknesses of the cavity for the two polarizations, although the physical thicknesses are the same. For proper selection of the two indices, one of which can be controlled by an externally controlled electric field, a standing wave can be produced for one mode but not for the other; the output would thus be polarized since only one mode is transmitted.

The wavelengths corresponding to these peaks can be influenced by a small electric field, as illustrated in FIGURE 12.12. In this figure, the wavelengths of all the peaks which appear in an unpolarized spectrum are plotted as functions of field. The figure clearly shows that a wide tunability range is possible with shifts of as much as 172 nm. Note that at sufficiently high fields, when all the molecules align with the field, the output beam is no longer polarized and the six peaks shown in the lower part of FIGURE 12.9 would merge into three peaks since the refractive indices for the two orthogonal polarizations become the same. There exists a threshold voltage below which no wavelength shift is observed, as clearly indicated in FIGURE 12.12. This threshold behavior arises because of a well known effect in liquid crystals called the Fredericksz threshold, which is well understood [41] and will not be discussed here. It exists when the directors at both surfaces are parallel to the surfaces. It is possible to eliminate this threshold and control its characteristics by changing the surface tilt of the molecules. For example, a hybrid aligned sample in which the molecules at one of the surfaces lie parallel to the surface and those at the other lie perpendicular, would have no threshold. For such a structure in the low-voltage regime, the index will change almost linearly with the applied field.

The speed of switching from one wavelength to another is dictated by how fast the molecules can be reoriented. While this speed depends on several factors, such as the dielectric anisotropy of the liquid crystal, the voltage applied, and the thickness of the cell, it is on the order of milliseconds. The advantage of this device is that it is compact, small, and requires practically no power. The power requirement mainly involves the charging and discharging of a capacitor at a rate of about 1 kHz, while the actual power would depend on the size of the device.

12.4.2 Multi-Wavelength Tunable Liquid Crystal Fabry–Perot Filter

In the previous description, the use of unpatterned single-element electrodes restricted the filtering to a single wavelength. However, it is possible to have multiwavelength filtering with the use of multielement patterned electrodes that allow us to apply several independent voltages to the liquid crystal etalon. Multiwavelength filters are a key component in multiwavelength cross-connect switches and in WDM local access networks using subcarrier multiplexing

One approach to multiwavelength filtering is to split the input light into several beams which are filtered separately, and then recombine the beams. This approach, however, requires several splitters and collimators which make the device complex as well as bulky. A better approach is to use a GRIN lens to produce a highly collimated beam so that different spatial portions of the beam are independently filtered. Commercially available assemblies of fiber and GRIN lenses were used in our experiment. These assemblies produce a beam of about 400 μm in diameter for the multimode fiber and about 250 μm for the single-mode fiber. The spatially filtered output beam, which is still collimated, is simply collected into another fiber–lens assembly.

This geometry is very compact and very easy to implement. The liquid crystal etalon filter can be divided into segments simply by defining the electrode pattern. When different voltages are applied to different sections, the local resonance conditions can be controlled independently. In this approach, the region between the electrodes must be minimized to less than a few microns, since these regions act as dead spaces in which active tuning is not possible. There are also limits to how small each of the electrode segments can be made, determined by the amount of fringing fields produced at the edge of the electrodes, and by the distance over which the molecular orientation changes in making a transition from one direction to another [42]. This distance is called the elastic coherence length ζ. This distance depends on external conditions such as the positions of the molecules in the neighboring elements, as well as the magnitude of the fields, and it can be estimated to be about a micron. These two criteria, the fringing fields and the elastic coherence length, limit the element size to being greater than about 10 μm. This suggests that approximately 10 different elements can be incorporated within the beam diameter for a single-mode fiber, if a rectangular geometry of electrodes is chosen. Clearly because of the incommensurability of rectangular and circular geometries, this is not an optimum design. A better design arranges the electrode gaps in a radial manner so that each electrode is pie-shaped. The analysis of the number of elements that can be incorporated within the beam diameter in such an approach is somewhat complicated by the fringing fields and ζ, although structures with more than 10 elements are possible.

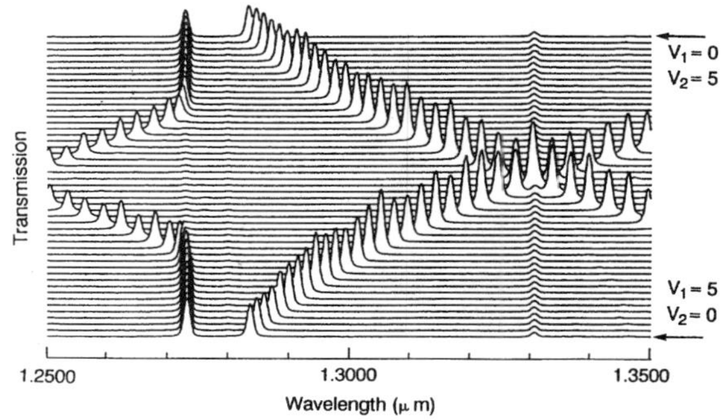

FIGURE 12.13. Top view of the liquid crystal etalon showing segmented electrodes.

FIGURE 12.13 shows the multiwavelength liquid crystal etalon filter. The nematic liquid crystal layer lies between two glass plates that have a highly reflective coating over the transparent electrode layers. The electrode layer of the first glass plate is connected to a common ground. The second glass plate is patterned with arrays of electrodes, where the array element spacing is 100 μm and the gap spacing is 5 μm. The thickness of the liquid crystal layer is about 7 μm. To test the device, the output from an LED source is coupled into a multimode fiber and collimated using a GRIN lens collimator, as shown in FIGURE 12.14. The spot diameter produced by this lens is about 400 μm, hence the beam passes through approximately four elements. The light that emerges from the etalon is coupled back into the fiber and is sent to an optical spectrum analyzer.

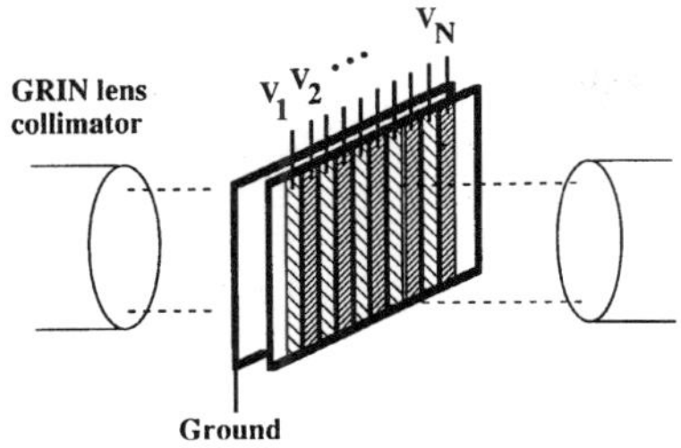

FIGURE 12.14. Transmission spectra showing filtered wavelength dependence on applied voltage. V_1 is increased from 0 to 5 V, V_2 is decreased from 5 to 0 V, and the passband at ~1.331 μm is fixed due to the light passing through the gaps between the electrodes.

More then four independent wavelengths without crosstalk have been demonstrated. Note that a $1/N$ loss is incurred when N channels are to be selected, since we use a space-division-multiplexing technique.

In FIGURE 12.14, the transmission spectra show the two tunable wavelengths with different voltages applied to the two sets of electrodes. The two major peaks are controlled by voltages V_1 and V_2, respectively. The small fixed peak is due to a portion of the beam passing through the etalon where no electrodes are present (i.e., the gaps between the electrodes). By properly selecting the thickness of the Fabry–Perot etalon, the fixed wavelength can easily be made to fall outside the frequency range of interest. The spectra shown in FIGURE 12.14 were obtained by changing V_1 from 0 to 5 V at 10 kHz while simultaneously changing V_2 from 5 to 0 V at 10 kHz. The top-to-bottom symmetry in the curves indicates that the tuning characteristics are essentially identical. Also, the changes in the voltage on one set of electrodes do not change the tuning characteristic for the other. For an applied voltage of 2.5 V, a single transmission peak is observed as expected.

12.4.3 Polarization-Insensitive Fabry–Perot Devices

The Fabry–Perot structures described so far are polarization-sensitive, which makes them less attractive for some applications, such as fiber optic communications in which the state of polarization may be unknown. There are several ways to make the structure polarization-insensitive.

12.4.3.1 Intrinsically Polarization-Insensitive Structure: Twisted Nematic Structure

The polarization-insensitive structure can be realized by using the twisted structure of a nematic liquid crystal such that the molecules at one surface are orthogonal to those at the other surface. In the high-field regime, this configuration can be thought of as a structure composed of two orthogonal birefringent slabs with electrically controllable thickness. At low voltages the device is polarization-sensitive, but it becomes polarization-insensitive at relatively higher voltages. The linewidth of the transmission peak is on the order of 0.5 nm and the tuning range is about 30 nm.

In the absence of an applied electric field, the director **n**, is continuously twisted by 90° in going from one surface to the other. Above the Frederick's threshold, the director in the middle of the FP cavity tends to orient in a direction parallel to the field, and thus coincides with the light propagation direction. The molecules on the surfaces, however, remain undisturbed as long as the anchoring energy is strong, producing two thin birefringent layers at the surfaces. The thickness of the individual birefringent layer depends on the field strength.

For the simplest case where the light is propagating normal to the FP structure, two resonance conditions corresponding to two orthogonally polarized eigenmodes are possible. For the twisted nematic FP structure, these eigenmodes are linearly polarized and are parallel and perpendicular to **n** respectively, when the "wave guiding" conditions are satisfied. These conditions are most easily defined in the absence of the applied field since the director remains perpendicular to the propagation direction. Under these conditions, the wave guiding would occur if $\Delta nd \geq \lambda$ where Δn is the index anisotropy, d is the thickness of the LC, and λ is the wavelength of light. Thus, in the absence of an electric field, if wave guiding occurs, two orthogonally polarized transmission peaks will be observed for a twisted nematic FP structure. In the presence of the field, however, the effective anisotropy becomes smaller and even zero, thereby resulting in the breakdown of the resonance conditions. This leads to mode mixing, and the resultant eigenmodes become a linear combination of the pure polarization modes. A direct consequence of this is that the effective indices for the two modes become equal, thus resulting in a single, polarization-independent index. Therefore, the two

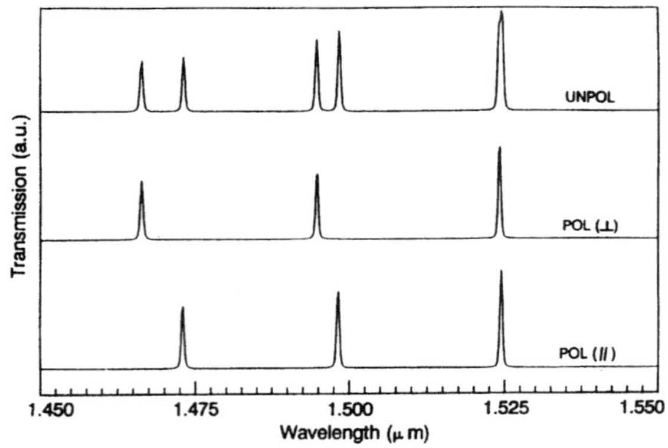

FIGURE 12.15. Measured transmission spectra of a liquid crystal Fabry–Perot etalon in the absence of an applied voltage. The top curve is the unpolarized spectrum, while the bottom two are polarized parallel (||) and perpendicular (⊥) to the buffing axis b.

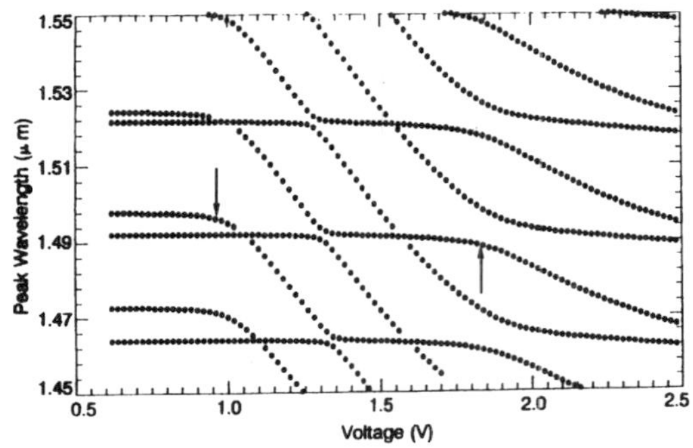

FIGURE 12.16. Position of the transmission peaks in the unpolarized spectrum as a function of the voltage for a 25 μm liquid crystal Fabry–Perot etalon in the low-voltage regime. The two arrows indicate the Frederick's threshold and the onset of mode mixing.

orthogonally polarized transmission bands of the twisted nematic FP, in the absence of the field, would change with the applied electric field in such a way that at high electric fields, the two bands would merge into one. Furthermore, these bands would begin to depolarize with increasing field because of the mode mixing.

FIGURE 12.15 shows typical transmission spectra obtained in the absence of an applied electric field. The top curve shows the unpolarized spectrum, while the bottom two spectra are the polarized spectra with the polarizer parallel (||) and perpendicular (⊥) to the buffing axis b, respectively. The data clearly show the validity of equation (12.1) since the linear combination of the bottom two spectra produces the unpolarized spectrum shown at the top. Furthermore, the complete separation of the two modes indicates that the eigenmodes are polarized parallel and perpendicular to b in the absence of an applied field. This is expected since the wave guiding conditions are satisfied

The transmission wavelength can be changed by a small electric field, as illustrated in FIGURE 12.16. The positions of all the peaks in wavelength that appear in an unpolarized spectrum are plotted as functions of the field strength. At a fixed voltage, for example, at 0.7 V, six peaks are observed. These peaks are divided into two groups with orthogonal polarizations. The three peaks in each group correspond to the higher-order FP modes. At very low voltages, the peak positions do not change significantly until a critical value of the field, the Frederick's threshold, is reached. The threshold is about 1.0 V and corresponds to the onset of molecular distortion in the bulk of the FP cavity. This distortion results in an effective index change for the mode polarized along b. However, the distortion is small and thus the Mauguin criterion for wave guiding is satisfied. This ensures that the modes are decoupled. A further increase in the voltage beyond about 1.75 V results in a significant change in the orientation of the molecules so that the Mauguin conditions are no longer satisfied. Under these circumstances, mode mixing begins to take place and the eigenmodes are no longer polarized either along b or perpendicular to b. Note that the magnitude of the required voltage for mode mixing depends on the wavelength of the light. Around 2.5 V, the spectrum begins to evolve in such a way that produces a pair of transmission peaks which merge into one at higher voltages. These peaks correspond to elliptically polarized modes within the cavity because of the mode coupling, thus an arbitrary linearly polarized light would produce both the peaks. At still higher voltages, the eigenmodes in the cavity again become linearly polarized and lie parallel and orthogonal to b, respectively. The peak transmission for these two modes becomes identical since the optical path length is the same in the high-field regime. This is clearly shown from experimental data depicted in FIGURES 12.17(a) and 12.17(b), which are the transmission spectra of the FP in the high-voltage regime, >2 V). The polarized spectra in FIGURE 12.17(a) is essentially identical to the

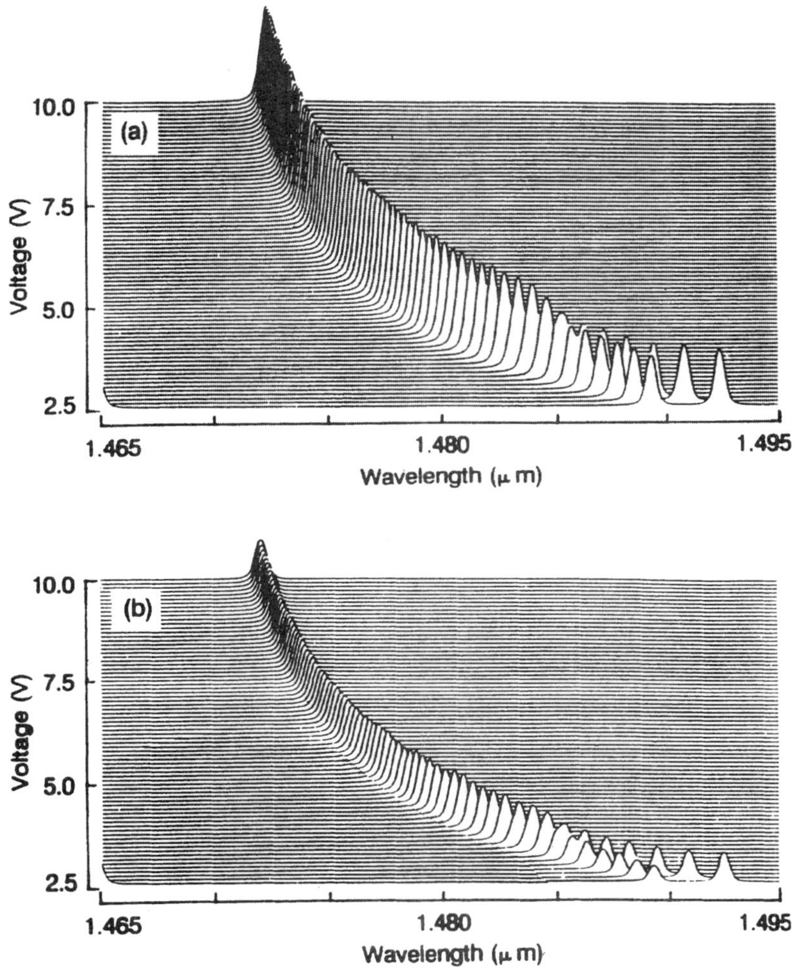

FIGURE 12.17. Measured transmission spectra of a liquid crystal Fabry–Perot etalon in the presence of an applied square wave at 1 kHz; (a) unpolarized and (b) polarized spectra. The magnitude of the rms voltage is indicated to the left of the spectrum.

unpolarized one shown in FIGURE 12.17(b), indicating that the FP is indeed polarization-insensitive in this region. With a linewidth of about 0.5 nm and a tunable range of about 15 nm, the effective finesse is of the order of 30.

12.4.3.2 Polarization-Insensitive Structure: Space Multiplexed Structure

In this type of device, the polarization insensitivity is realized by using two different regions of the same cell with optic axes orthogonal to each other. The input beam of light is divided so that equal amounts pass through the two regions and thus for light of arbitrary input polarization, half of the intensity is transmitted at the resonant wavelength.

In classical approaches, the polarization insensitivity is obtained by splitting the beam into two beams with orthogonal polarizations, and modulating each beam separately. These approaches require optical components which may be bulky and expensive. Furthermore, identical electrooptical elements are required for each channel, which is often difficult when the two elements are physically different devices. The approach that we use in this paper eliminates some of the problems associated with using bulky beamsplitters, and significantly reduces the problem associated with making two identical devices.

In the simple geometry of the liquid crystal Fabry–Perot described in section 12.4.1, the eigenmodes are

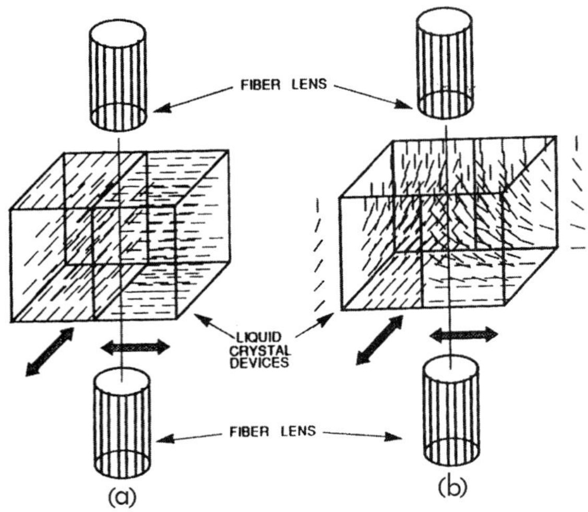

FIGURE 12.18. The structure of dual-orientation liquid crystal Fabry–Perot. The mirrors are not shown in the drawings. (a) Shows the simple geometry that can be used and (b) shows the geometry described here.

linearly polarized and lie parallel and perpendicular to the optic axes. Thus light of arbitrary polarization on entry into such a device will propagate through the device as two waves along these two axes. The amplitudes of the waves will depend on the ellipticity of the input light. Consider now a case where the light is first spatially divided into two parts, and then passed through two identical LCFP elements oriented orthogonal to each other. This is shown in FIGURE 12.18(a), where the input light emerges from a lens and passes through two regions of the liquid crystal cell having optic axes orthogonal to each other.

Since light of arbitrary input polarization can be decomposed into electric field components along the x and y axes, tunability of both components is possible by passing the beam through both regions. The details of operation will be discussed later, but it is easy to show that exactly half of the input incident light will be tunable in this configuration.

Such a device can be made using optical elements such as fiber lens assemblies that allow the beam to be split easily into two halves. By using such an element, the problem is reduced to making a liquid crystal Fabry–Perot etalon with two areas where the optic axes are orthogonal to each other and in close proximity. The output collimated beam is simply collected with another fiber lens assembly, making the geometry very compact and easy to implement. There are two possible geometries for the LC device. In the first case the director **n** and the optic axis in one region lie in the plane of the substrate and parallel to both surfaces as shown in FIGURE 12.18(a). In the second case, shown in FIGURE 12.18(b), the director on one surface lies in the plane of the substrate, but at the other surface lies perpendicular to the surface. In both cases the optics of the structures are essentially identical with the eigenmodes linearly polarized. The magnitude of the maximum refractive index changes induced by the electric field is, however, smaller by about a factor of two for the second case.

The problem of controlling the direction of the optic axis in different regions of the material reduces to controlling the surface properties of the alignment layer. Two different director orientations in the substrate plane can be obtained by first rubbing the whole sample in a chosen direction, covering part of the sample, and then rubbing the sample again in another direction. In our case, these two directions are orthogonal to each other. Two factors determine the width of the director transition between the two regions: (a) the length scale on which the surface structure is modified by rubbing and (b) the length scale on which the liquid crystal responds to the surface forces. The former is determined by the photolithographic resolution of the masking process, typically of the order of a micron. The scale on which the liquid crystal molecules respond to the surface forces is determined by the fundamental limits imposed by the elastic properties of the liquid crystals and the surface anchoring strength [43]. This length is $\zeta = \pi/(2\sqrt{2dK/W})$, where K is the elastic constant, d is the thickness, and W is the surface energy density. Thus for typical values of $K = 10^{-7}$ dyn, and $W = 10^{-4}$ ergs cm^{-2}, ζ is also of the order of a few microns.

The alignment surface modification is required at both surfaces of the liquid crystal cell. Furthermore, in such a device the boundary between the regions with different orientations on one surface has to be aligned exactly with that on the opposing surface, as shown in FIGURE 12.18(a). While this is possible [44], it is nevertheless difficult without the use of special equipment. A simple approach that eliminates this alignment problem is to use a structure such as that shown in FIGURE 12.18(b), explained earlier. Note that optically, these two cases are very similar.

To explain the operation of the device, let the two adjacent regions be labeled "A" and "B" with the optic axes in these two regions along two orthogonal directions called **x** and **y**. When the input polarization of the light is **x**, the optical path length and hence the resonant transmission wavelength of the light passing through the region "A" depends on the magnitude of the applied field. The optical path length through the zone "B," however, remains fixed even though a uniform field is applied across both regions. Similarly, light polarized along **y** experiences a change in the optical path length in region "B," which depends on the applied field but remains unaffected in region "A." For elliptically polarized light the analysis is slightly more complicated since four components of the polarized light, two in each region, must be considered. Since the electrooptic response of the device is the same for the two eigenmodes **x** and **y** in regions A and B, respectively, polarization independence is achieved. The similarity of the electrooptic response is in part due to the close proximity of the two regions, which ensures that the thickness and hence the field and the optical thickness are the same across both the regions.

FIGURE 12.19 shows typical transmission spectra as functions of wavelength for the values of the applied field indicated on the right. There are two resonances, one field-dependent, and another at about 1.5 μm that is independent of the applied electric field. Let this two peaks be named P and Q respectively. Peak Q is due to resonance for polarized light along the axis perpendicular to the director. The polarization state of this peak is the same as that of the input beam because its intensity is due to the sum of contributions from the two regions of the liquid crystal aligned perpendicular to each other. Since the refractive index along the axis perpendicular to the director is not influenced by the field, the resonance conditions are unaltered by changes in the applied field. On the other hand, the refractive index along the director axis changes with the field, and therefore the peak transmission wavelength is influenced by the applied electric field giving rise to peak P. In this case, the output state of the polarization is also identical to the input state of the polarization, for reasons outlined above. The position of the peak Q is determined by the resonance condition $m\lambda = n_s d$, where m is the mode number, λ is the wavelength, n_s is the refractive index along a direction perpendicular to the director, and d is the physical

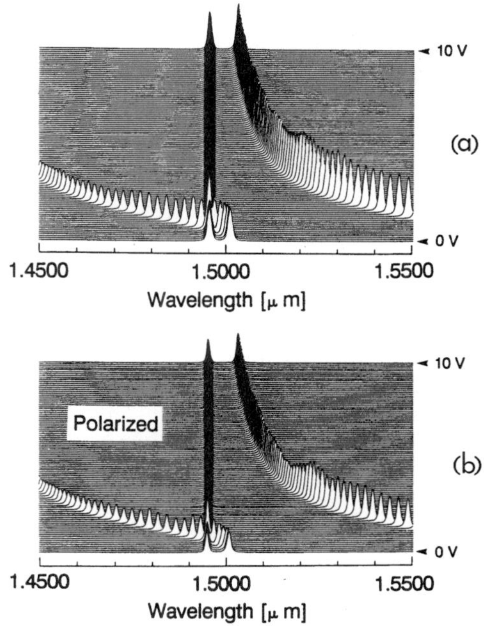

FIGURE 12.19. Measured transmission spectra of a liquid crystal Fabry–Perot etalon 10 μm thick with two regions that are optically identical, at applied voltages between 0 and 10 V. (a) Input light unpolarized; (b) input light polarized along the surface director; (c) input light polarized perpendicular to the surface director.

thickness of the cavity. Thus by changing the thickness of the cavity, the peak position can be changed. The tuning range of peak P is about 100 nm, with a free spectral range of about 70 nm. The width of the transmission peak is about 0.7 nm.

FIGURE 12.19(b) shows data obtained by using polarized light. In this case the polarizer axis coincides with the director axis at the surface, which is collinear with the rubbing direction. Note that the data are essentially identical to those in FIGURE 12.19(a), except for slight intensity differences caused by the selection and attenuation in the polarizers. The data taken with the polarizer axis perpendicular to the rubbing axis are identical to those in FIGURE 12.19(b) and are not shown.

While FIGURE 12.19 shows that polarization-insensitive tuning is possible, it is only because the two regions, A and B, are optically identical. When these two regions are not identical, polarization insensitivity is not obtained, as shown in FIGURE 12.20. In this case,

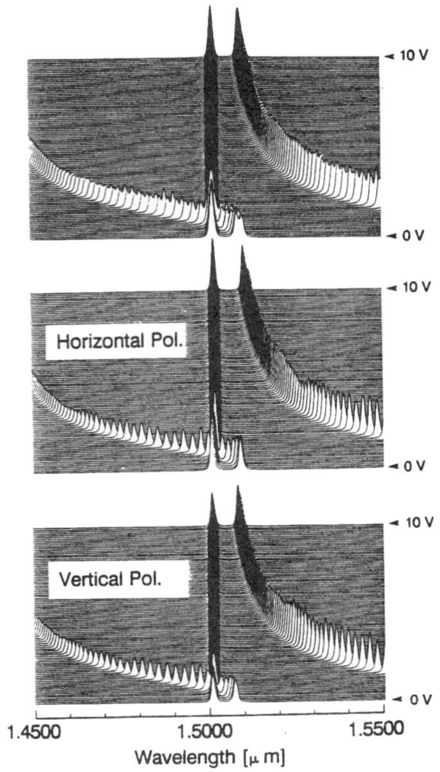

FIGURE 12.20. Measured transmission spectra of a liquid crystal Fabry–Perot etalon 10 μm thick with two regions that are not optically identical, at applied voltages between 0 and 10 V. (a) Input light unpolarized; (b) input light polarized along the surface director; (c) input light polarized perpendicular to the surface director.

the unpolarized spectrum reveals two tunable peaks around 1.5 μm that merge into one peak at high fields. Furthermore, the two peaks are polarized perpendicular to each other, as can be seen by comparing the two polarized spectra with the unpolarized spectrum. This indicates that while the physical thicknesses in the two regions are identical, because the two peaks merge into one, the optical thicknesses are different. If we assume that the director configuration at the homeotropic surface is normal to the surface, then the result can be explained by variations of the tilt angle at the homogeneous surface. This is not surprising since differences in rubbing can cause this angle to change [45]. Although the surface tilt angle depends on the rubbing conditions, this angle saturates, and thus for this type of device the rubbing should be thorough for the two regions to be identical.

12.4.3.3 Polarization-Insensitive Structure: Polarization Diversity Technique

In classical approaches, polarization insensitivity is obtained by splitting the beam into two orthogonal polarizations, and processing each beam separately. In general this requires two electrooptical elements (one for each polarization) and two control voltages to ensure identical electrooptic response. In this section we discuss a compact polarization-insensitive liquid crystal filter using the polarization diversity technique.

FIGURE 12.21 shows the polarization diversity LCFP filter. In this filter the arbitrarily polarized light is split into two beams with orthogonal polarizations using a calcite crystal. The single-mode fiber lens output has a beam diameter of 250 μm, and a 4 mm thick calcite crystal produces two orthogonally polarized beams separated by about 400 μm. The two spatially separated beams are filtered independently, and then the beams are recombined by a second calcite crystal. The tunable filter is a single LCFP device that has spatially distinct regions where the liquid crystal molecular orientations are orthogonal to each other, and parallel to the polarization axes of the two beams emerging from the first calcite crystal. Careful orientation of the axis of the polarization of the beam with that of the long axis of the liquid crystal is necessary, because the LC structure is polarization-sensitive as shown above.

The boundary between the two liquid crystal regions must be abrupt so that the beams go through two uniformly oriented regions. The orientation of the molecules in the two regions can be controlled by surface modification as discussed in the previous section. This is done by first rubbing the whole sample in a chosen direction, covering part of the sample with a photoresist mask, and then rubbing

FIGURE 12.21. Polarization diversity liquid crystal etalon configuration.

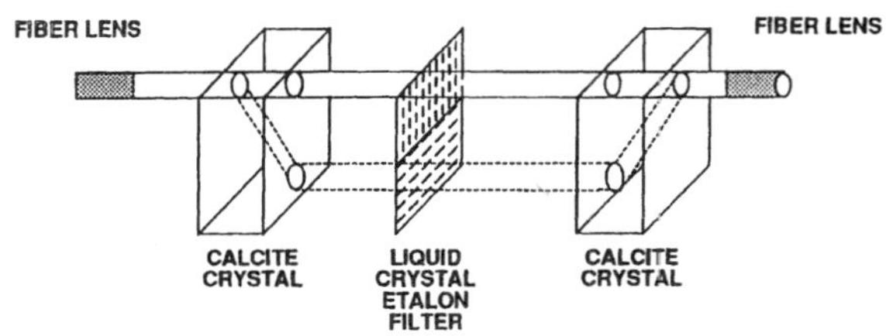

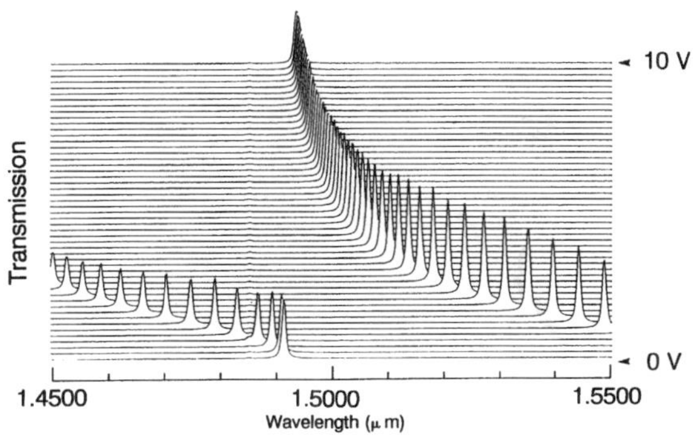

FIGURE 12.22. Transmission spectra showing filtered wavelength dependence on applied voltages.

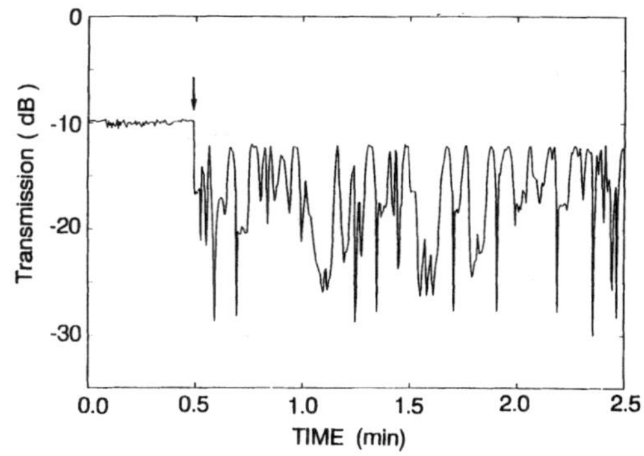

FIGURE 12.23. Demonstration of the polarization independence of the device. The diversity feature is in effect until the time marked by the arrow.

the sample again in an orthogonal direction. The width of the transition region is of the order of a few microns and is determined essentially by the photolithographic resolution of the masking process, and by the fundamental limits imposed by the liquid crystals and the surfaces [46].

The alignment surface modification is required at both the front and the rear surfaces of the cell in order to obtain well-aligned samples. Furthermore, in such a device the boundary between regions with different surface orientations must be oriented exactly parallel with that on the opposing surface. While this is possible, a simpler approach is to control the orientation of the liquid crystal molecules on the opposing surface to be perpendicular to the surface plane, that is, in homeotropic alignment as described above.

FIGURE 12.22 shows typical transmission spectra as functions of wavelength for the values of applied field indicated on the right. The free spectral range is about 70 nm as read from the transmission spectrum obtained at about 1.4 V. The two transmission bands obtained at this voltage correspond to two different FP modes separated by about 70 nm, however, the peaks can be shifted by as much as 100 nm, as can be seen by comparing the transmission spectra at 0 and 10 V.

To verify that the device is indeed polarization-insensitive, we used the output of a DFB laser and adjusted the filter band pass to coincide with the peak wavelength of the laser at 1.534 nm. Using a fiber polarization controller, the polarization was varied randomly, and the power was monitored as a function of time. As shown in FIGURE 12.23, the power variation was negligible for the 30 seconds when the device operated in the diversity mode. To verify that the input polarization of the light was indeed changing, a polarizer was inserted into the path of the input light beam at the time indicated by the arrow in FIGURE 12.23. The offset in the intensity is due to the absorption losses in the polarizer, while the large fluctuations (>15 dB) confirm that the polarization of the input light is indeed changing randomly. From the initial part of the curve we conclude that the diversity feature works well and that the fluctuations in intensity are less then 1 dB.

12.4.4 Frequency Stabilization of a Tunable Liquid Crystal Filter

While liquid crystal Fabry–Perot (LCFP) devices are attractive because relatively small changes in the voltage applied to the liquid crystal within the Fabry–Perot cavity can make significant changes in the optical pass band, this sensitivity also requires voltage control to within a few millivolts, and temperature stabilization to better than a degree to minimize thermally induced changes. It is possible, however, to overcome this problem with a frequency-locking technique. A particularly attractive feature of the liquid crystal Fabry–Perot structure is that the AC control voltage at frequency ω, used to tune the cavity, also produces a small signal modulation at 2ω, which when phase-sensitively detected, can be used for feedback.

Since the resonant wavelength can be controlled by the externally applied voltage, it is possible to use active feedback to compensate for the undesirable resonance shifts caused by temperature fluctuations. The feedback signal must indicate not only that a shift in the resonance has occurred, but also the magnitude and the direction of the shift. Feedback has been used extensively for frequency locking of lasers, typically by comparing the position of the laser line with some standard frequency. The laser is modulated at frequency ω, well below the optical frequency, and the output with respect to the standard frequency is phase-sensitively detected. This allows both the magnitude and the direction of the shift to be determined [47].

Frequency locking generally requires an external reference, but in the approach discussed here, this need is eliminated by using the frequency of the optical signal source as the reference. Because it uses the optical signal source for tracking, any variations in the frequency at the source would also be automatically tracked. Furthermore, a deliberate modulation of the LCFP is rendered unnecessary since the AC voltage that is applied to control the resonance of the LCFP causes a modulation in the transmitted intensity of the light. Although small, this modulation is generally an undesirable feature for most nematic liquid crystal devices. The modulation can be made as small as possible by increasing the frequency of the applied AC voltage and by controlling the shape of the applied waveform. The intensity modulation effect is particularly noticeable in the case of LCFP structures because the sharp resonances amplify the modulation resulting from very small index changes in the cavity.

For nematic liquid crystals, the dielectric torque on the molecules is proportional to the square of the electric field. Consequently, the response should only depend on the rms value of the applied voltage, but this analysis does not include two effects that combine to create the intensity modulation by an AC field. The first is the presence of time-dependent depolarization fields resulting from ion migration; the second is the flexoelectric effect that leads to structural relaxation and the asymmetric distortion of the director field, particularly close to the surfaces [48]. The distortion of the director field in an applied electrical field is caused by opposite signs for the induced flexoelectric moment, which depends on the gradient of the director field. Thus the modulation in the intensity should occur at twice the applied frequency, exactly as observed experimentally. This signal changes phase by 180° when the

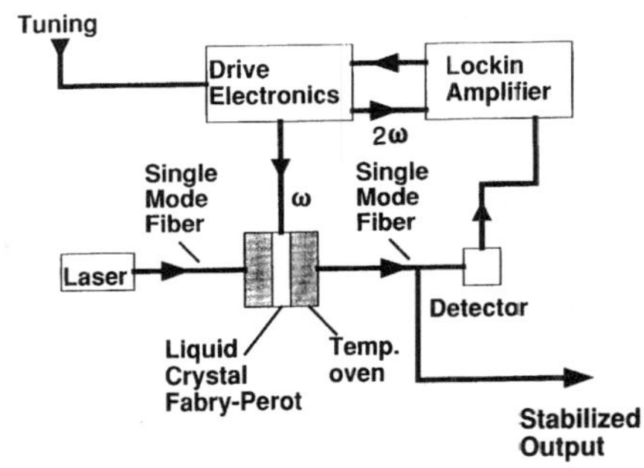

FIGURE 12.24. Block diagram of the feedback control system used to stabilize the LCFP to fluctuations in temperature.

resonance of the LCFP is tuned from one side of the peak of the optical frequency of the signal source to the other, and represents the derivative of the resonance with respect to the applied voltage.

FIGURE 12.24 shows a block diagram of the system that has been used for locking, using a polarization-sensitive liquid crystal Fabry–Perot etalon and a solid state DFB laser operating at 1.5464 μm with a linewidth of 0.05 nm. The output light was collected into a single-mode fiber using a GRIN lens, and detected using a germanium photodetector. The output of the detector was amplified and fed into a voltmeter to measure the average transmitted intensity, and into a lock-in amplifier to detect the amplitude of the modulation at twice the AC voltage frequency. The drive voltage was generated using an arbitrary waveform generator to provide variable waveform shapes and drive frequencies. A second waveform generator was used to generate the second-harmonic frequencies that were used to provide the reference source for the lock-in amplifier.

FIGURE 12.25 shows the transmitted light intensity of a liquid crystal Fabry–Perot etalon as the rms voltage is scanned. The various peaks correspond to different mode numbers. The upper trace shows the signal observed at 2ω, where the derivative-like behavior is clearly evident. The magnitude of the signal depends on the nature of the waveform used to drive the liquid crystal. In particular, sine and triangular-wave drive voltages produced larger modulation signals than square-wave voltages, and the magnitude of the signal decreased with increasing frequency of the drive voltage. Thus the frequency and the shape of the drive voltage can be used to control the amplitude of the modulation.

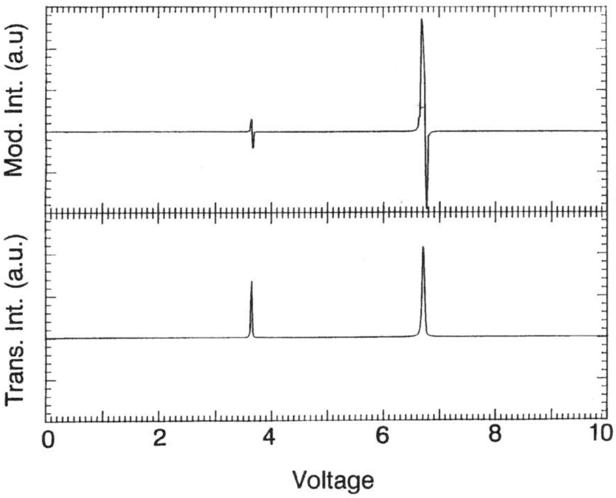

FIGURE 12.25. Response of the LCFP to an applied sine-wave voltage (rms × 1.4) at 1.5 kHz. The lower curve shows the transmission and the upper curve shows the modulation signal at 2ω. The modulation signal is smaller by about three orders of magnitude.

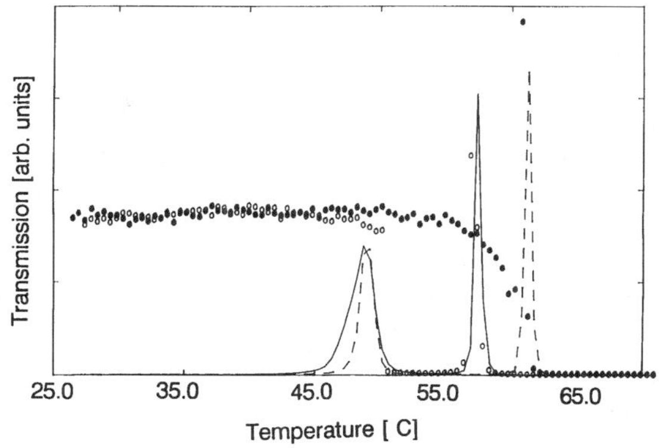

FIGURE 12.26. Response of the LCFP to variations in temperature both without (lines) and with (circles) feedback enabled. The dotted line and the filled circles represent the heating cycle, while the solid line and the open circles represent the cooling cycle.

The performance of the feedback mechanism necessary to prevent frequency drifts due to temperature variations is shown in FIGURE 12.26. In the absence of feedback, the resonance is quickly lost by changing the temperature as seen by the broken (heating) and the continuous (cooling) lines, which show the transmitted light intensity. Two transmission peaks were observed, corresponding to two values of refractive index that satisfy the cavity resonance conditions. Note that the resonance at the higher temperature is narrower because the refractive index of the liquid crystal changes rapidly near the isotropic-to-nematic phase transition at 60.5 °C. On reversing the temperature, there is slight hysteresis for that peak because the transition is first-order. There is also a slight contribution to the hysteresis due to nonadiabatic heating and cooling of the sample.

The open and the filled circles in FIGURE 12.26 represent the transmitted intensity data in the presence of feedback. The LCFP was tuned at about 42 °C, with the locking circuit activated. With locking in effect, the sample was cooled to near room temperature and the data (solid circles) were recorded as the temperature increased from 25 °C. The temperature was raised past the point where the signal was lost, and then data were taken again (open circles) as the temperature was decreased. The intensity remains constant over a large temperature range, including the temperature region near the phase transition temperature, where the index changes occur rapidly.

This shows that the drifts in the pass band of the Fabry–Perot etalon can be controlled by active feedback, extending the operating range of the LCFP by only about a degree; in the absence of tracking, this is extended to several tens of degrees.

12.4.5 Electrically Tunable Ferroelectric Liquid Crystal Fabry–Perot Filter

While nematic liquid crystals allow broad tunability with cavity finesse approaching 400 [49], these devices tend to be slow with typical switching speeds of several tens of milliseconds [50,51]. Because ferroelectric liquid crystals have been shown to have faster switching speeds then nematics, we report their use in Fabry–Perot cavities, in an attempt to improve the speed of the tunable filters.

Ferroelectric liquid crystals consist of chiral molecules arranged in a layered structure [52]. Within each layer, the molecules are tilted with respect to the layer normal at an angle called the tilt angle. The local orientation of the long axes of the molecules, the director, is represented by a unit vector **n**. Although the arguments presented are valid for ferroelectric liquid crystals in general, the discussion is limited primarily to the smectic C* phase. Smectic C* phase materials fall into two categories based on the magnitudes of the tilt angles. One group of materials has a nearly temperature-independent tilt angle of about 45° [53] and the ferroelectric properties allow the molecules to be reoriented within the layers by applying an external electric field through an angle equal to twice the tilt angle. Therefore the slow and the fast axes coincide with the short and the long axes of the molecules, respectively, and are interchanged with respect to the laboratory axes by reversing the direction of the applied electric field. Switching from one orientation to the other can occur in microseconds at large electric fields.

An anisotropic material in a Fabry–Perot structure between two mirrors produces two resonant wavelengths polarized along two orthogonal directions. These two directions are collinear with the principal axes of the anisotropic materials. For randomly polarized light at two resonant wavelengths λ_1 and λ_2, both wavelengths are transmitted but the light will be polarized along one direction for λ_1 and at a perpendicular direction for λ_2. By applying an electric field across a ferroelectric liquid crystal, the slow and the fast axes interchange and thus the polarization directions for the two wavelengths are also interchanged. Alternatively, for polarized light at both the wavelengths, λ_1 is transmitted with one polarity of the applied field, and λ_2 is transmitted when the polarity is reversed.

A ferroelectric liquid crystal Fabry–Perot structure was made using dielectric mirror coatings on top of conductive indium tin oxide coated glass. An alignment layer was deposited on top of the mirrors as previously described, with the rubbing directions aligned orthogonal to each other to maintain symmetric structure with respect to switching between the two states. The ferroelectric liquid crystal material (CS 2004 obtained from Chisso Corporation) was inserted into the sample cell (6.5 μm) in the isotropic phase. An electric field treatment was used to produce monodomain samples [54]. Light from a 1.5 μm LED was coupled into and out of the liquid crystal Fabry–Perot using a GRIN lens and multimode optical fiber. The transmitted light was analyzed using a spectrum analyzer.

Initially, we investigated the transmitted spectrum of the LCFP by applying a DC electric field. Generally, the application of a DC electric field across a liquid crystal for a prolonged period is undesirable because

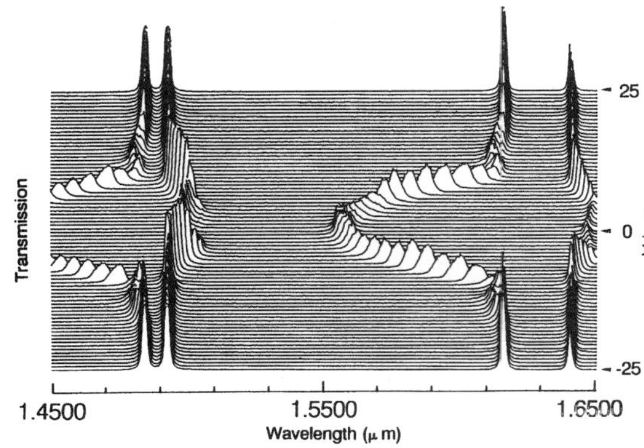

FIGURE 12.27. Gated transmission spectra in the presence of an applied AC square wave. The spectra were taken with short light pulses of 2.86 ms duration that were synchronized with the rising edges of the applied square wave. The numbers on the right refer to the timing of the light pulse with respect to the applied electric field.

of potential electrochemical degradation of the materials, and also because of charge migration within the cell. We observed systematic changes in the spectrum with changes in DC field, but the magnitude of the electric field in the sample was uncertain because of depolarization fields. Therefore all data reported here were obtained by using stroboscopic techniques. An AC electric field across the liquid crystal was synchronized with the pulsed light from the LED using two phase-locked programmable waveform generators, one to drive the liquid crystal and the other to drive the LED.

The data in FIGURE 12.27 is the unpolarized transmission spectrum obtained by varying the amplitude of the square-wave voltage applied to the ferroelectric liquid crystal. The middle trace is obtained in the absence of an applied voltage and corresponds to a structure where the projection of **n** on the cell surface rotates by 90° across the cell thickness. Above this trace, the voltages have positive values because the light source was synchronized with the positive cycle of the AC applied voltage. Similarly, the negative values correspond to the negative cycle of the AC applied voltage. For this set of data, the LED was switched on for 2.86 ms at a frequency of 100 Hz.

The data show about four peaks at each voltage, two with one polarization and two with an orthogonal polarization. Peaks with the same polarization correspond to consecutive mode numbers for the

Fabry–Perot device. At high fields, where the four peaks are most evident, the director is uniformly parallel throughout the sample and the liquid crystal structure is untwisted. The resonant wavelength does not change with a further increase in the field. At low voltages, the twisted structure of the ferroelectric liquid crystal distorts, changing the path length and therefore producing different resonant wavelengths. The decrease in the peak intensity at low fields is caused by spatial variation of the smectic liquid crystal structure in the region being probed. Thus in the low-voltage regime, systematic changes in structural distortion lead to optical path variations, and the ferroelectric liquid crystal acts as an electrically tunable wavelength filter .

The polarization properties of the liquid crystal Fabry–Perot device are shown in FIGURE 12.28. The two traces were obtained as described above, but with the addition of a polarizer. The two traces correspond to switching on the LED during the positive and negative cycles of the applied electric field. As expected, the positions of the transmission peaks for these two traces occur at different wavelengths. The peak positions are interchanged by reorienting the polarizer by 90°. The two peaks at 1.4916 and 1.6388 μm, correspond to eigenmodes polarized along the long axes of the molecules with mode numbers 12 and 13, while the peaks at 1.4836 and 1.6148 μm correspond to the eigenmodes polarized along the short axes with mode numbers 13 and 14, respectively. From a thicker sample we have determined that the two principal indices are 1.38 and 1.48 for the short and the long axes, respectively. This produces a birefringence of 0.10, less than the reported value of 0.15 for this material in the visible region. A smaller value of the birefringence in the infrared region is expected, but the difference is surprisingly large. The wavelength extinction ratio was higher than 100:1, showing that the optic axes indeed reorient by 90°, interchanging the ordinary and the extraordinary axes.

The time dependence of wavelength switching was investigated by driving the liquid crystal at 100 Hz square wave, and recording the spectrum using short light pulses (450 μs) from an LED as the timing was changed with respect to the rising edge of the applied electric field. The results are shown in FIGURE 12.29. The decreased duty cycle of the LED causes the poor

FIGURE 12.28. Transmission polarized spectra of the ferroelectric liquid crystal Fabry–Perot device in the presence of an electric field. Spectra with broken lines are for light polarized parallel to the director axis. The inset shows the orientation of the molecules in the presence of the applied electric field with respect to the laboratory axis.

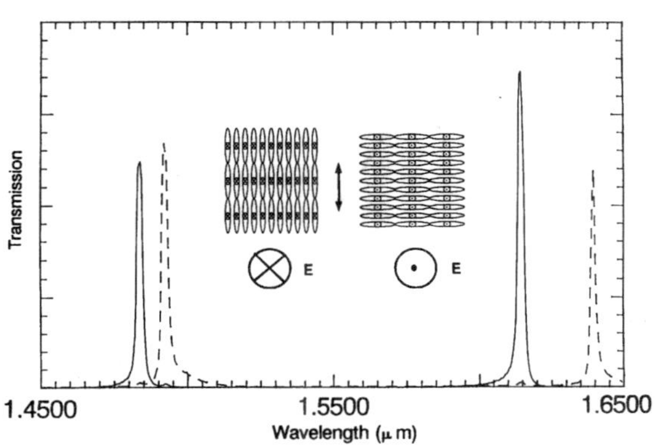

FIGURE 12.29. Transient transmission spectrum of ferroelectric liquid crystal near field reversal of the applied AC electric field. The spectrum is taken near (a) the rising edge of the field, and (b) the falling edge of the field .

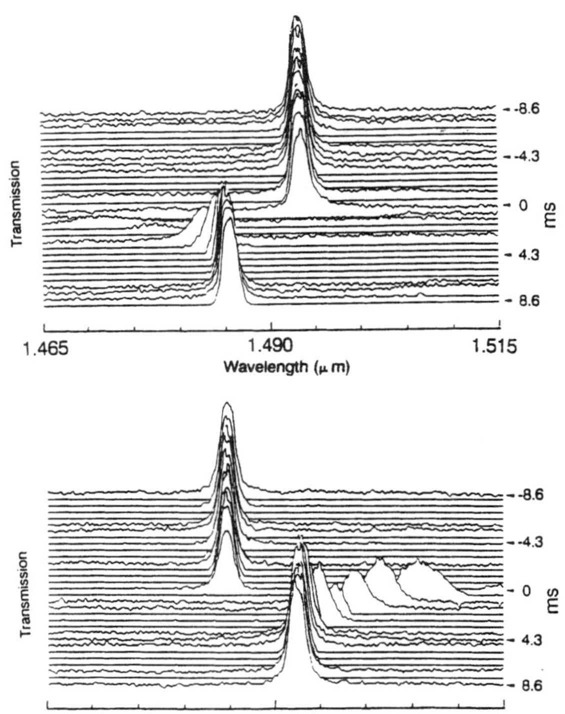

signal-to-noise ratio in these spectra. FIGURE 12.29(a) shows data obtained when the timing of the light pulse was varied with respect to the leading edge of the applied electric field across the liquid crystal; FIGURE 12.29(b) corresponds to the data taken at the falling edge of the field. In both cases, note that switching to a new wavelength occurs in about 2 ms, although the transmission spectrum continues to evolve for 2 ms more. Switching out of resonance occurs faster than 450 μs because the index initially changes rapidly after the electric field is applied. However, the final value of the molecular positions and hence the cavity index is reached asymptotically over a longer time period. Thus "switch-on" is much slower than the "switch-off" process. This asymmetry in switching is expected to be universal, at least for all liquid crystal Fabry–Perot structures.

These results show that ferroelectric smectic liquid crystals with large tilt angles are useful materials for wavelength-tunable filters and binary-wavelength switches. In binary switch applications, switching occurs in milliseconds. The low speed is in part caused by the capacitive voltage drop across the dielectric mirrors. It may be possible to improve the performance with better materials, although the switch-off process will always be faster then the switch-on process because of the dynamics of the index change associated with switching. For tunable wavelength filters, this class of ferroelectric liquid crystal does not yield much higher speeds than the nematic liquid crystal devices. However, higher speeds may be possible when the smectic layers are perpendicular to the light propagation direction. Since bistability has been observed in materials with smaller tilt angles, it may also be possible to make a bistable wavelength filter using this geometry.

12.4.6 Use of Liquid Crystal Fabry–Perot Devices for Making Tunable Lasers

Wavelength tunable lasers at 1.3 and 1.5 μm have a wide range of applications for fiber optic systems. Most lasers that are widely tunable rely on mechanical mode-selective elements such as external gratings [55] or angle-tuned Fabry–Perot devices [56,57]. Electronically tunable semiconductor lasers such as multi-electrode DFB lasers exhibit a limited range of continuous tunability on the order of few nanometers. Other electronically tuned lasers utilizing active filters (e.g., acoustooptic filters) require a large RF electrically tunable fiber laser using a liquid crystal Fabry–Perot etalon filter wavelength-selective element. The advantage of using a Fabry–Perot structure rather than a birefringent filter [58] is the enhanced tuning range and frequency selectivity. With a voltage of only 2.1–3.3 V applied to the filter, the wavelength is tunable over 45 nm (1523–1568 nm), limited primarily by the gain bandwidth of the erbium-doped fiber amplifier. The experiment described here demonstrates the potential for a compact laser that uses IC-compatible voltages for its wavelength control.

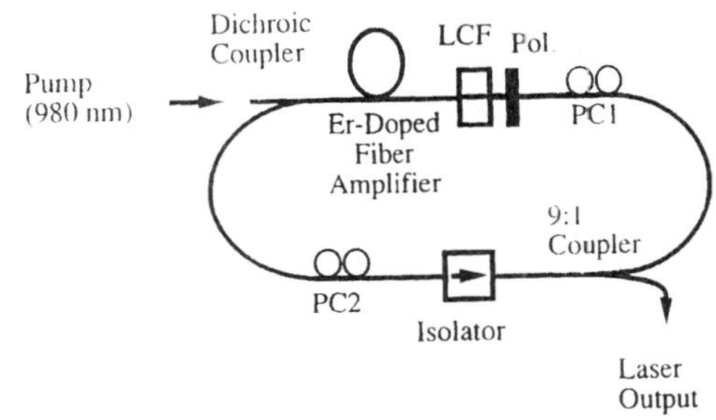

FIGURE 12.30. The configuration of a fiber ring laser. The laser mode travels counterclockwise through the ring. PC1 and PC2 are the polarization control elements.

Figure 12.30 shows the experimental arrangement for the ring laser. The gain in the fiber ring laser is provided by a 2 m long erbium-doped fiber amplifier, pumped by a 980 nm titanium-sapphire laser, which can be replaced by a compact diode laser for similar results. The amplifier fiber has a 5.5 μm diameter germanium–phosphorus–aluminum doped silica core, and a cutoff wavelength of 1042 nm. The erbium-doped silica glass of the above composition provides a wide optical gain bandwidth, and a maximum small signal gain of over 40 dB at 1533 nm.

The wavelength-selective element in the ring laser is the liquid crystal Fabry–Perot filter (LCFP) like those described above. The LCFP is an index-tunable Fabry–Perot etalon, comprised of 10 μm thick nematic liquid crystal layer between two reflectively coated glass plates. The dielectric coating gives a reflectivity of 98.5%, and the filter bandwidth is 0.4 nm with a free-spectral range of over 50 nm. Transparent indium tin oxide electrodes on the glass plates allow the application of voltage across the liquid crystal layer.

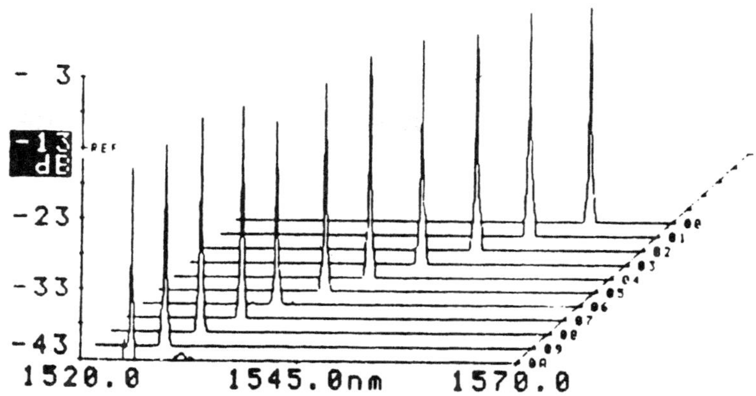

FIGURE 12.31. Ring laser spectra as a function of voltage across the liquid crystal Fabry–Perot filter.

Pigtailed GRIN lens collimators are used to transmit light through the LCFP. A polarizer and a polarization controller (PC1) placed before the LCFP ensure that the light is polarized in the direction that can be tuned. An in-line optical isolator eliminates the clockwise propagating ring mode, which would otherwise cause gain competition. The isolator also suppresses standing-wave laser modes that do not pass through the LCFP. A polarization controller (PC2) is used to align the signal polarization before the isolator. The alignment with PC2 is important for the continuous tunability of the laser, as will be explained later. Approximately 10% of the intracavity light is coupled out for the laser output. The net loss due to the couplers, LCFP, and other fiber-optic components is about 7 dB, and the total fiber length in the ring is approximately 7 m.

Figure 12.31 shows the laser output spectra as the voltage applied to the LCFP is increased from 2.2 V (rms) in 0.1 V steps. The laser tuning range is 45 nm (1523 to 1568 nm) with less than 5 dB variation in the output power. The lasing threshold occurs at 25 mW pump power, and an output power of approximately 1.4 mW is obtained with 50 mW pump power, corresponding to 6% slope efficiency.

Figure 12.32 plots the lasing wavelength as a function of the rms voltage applied to the LCFP. The laser tuning begins when the applied voltage exceeds the threshold value of 1.2 V, and the liquid crystal refractive index decreases with increasing voltage. Because the filter tuning range is much greater than its free spectral range, tunable wavelengths can be obtained between 1558–1523 nm (1.2–1.7 V) and 1568–1523 nm (2.1–3.3 V). A detailed study of the

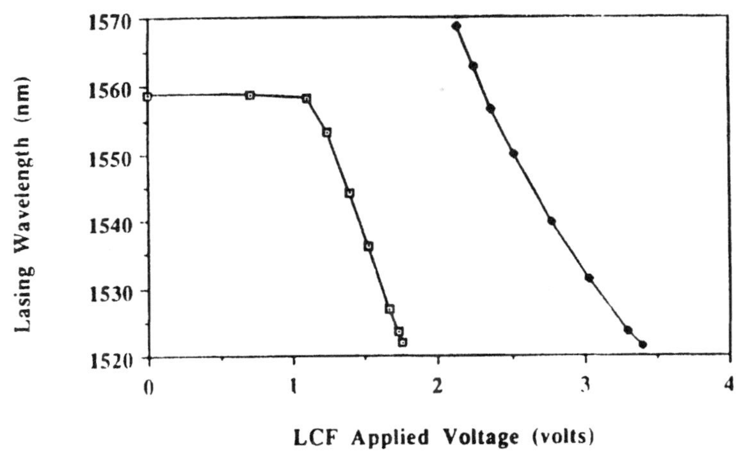

FIGURE 12.32. Laser wavelength versus applied rms voltage. The two curves correspond to the two different Fabry–Perot modes.

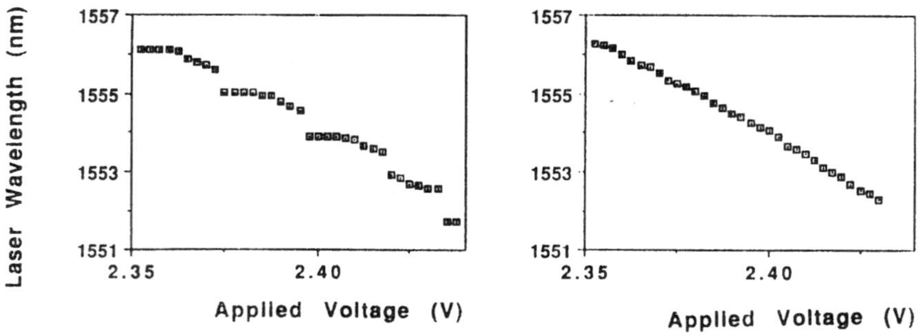

FIGURE 12.33. Fine tuning of laser wavelength as a function of applied voltage in steps of 2.5 mV. The left-hand side shows wavelength pulling and hopping in 1 nm intervals due to birefringence-induced wavelength-dependent loss. The right-hand side shows the result of effectively eliminating the birefringence.

laser tunability reveals wavelength-pulling as well as wavelength-hopping at 1 nm intervals, as shown in FIGURE 12.33(a). Such discrete tuning results from the presence of several wavelength-selection mechanisms in the ring, but the details will not be discussed here. This interesting result occurs because of the use of the polarization-sensitive liquid crystal Fabry–Perot filter.

The ultimate tunability and wavelength-selectivity of the laser are limited by the presence of fixed cavity modes. These modes satisfy the round-trip phase-matching condition and are spaced $c/L_{\text{eff}} = 30$ Mhz where L_{eff} is the optical path length of the ring. A delay shows that the laser operates single-mode with a linewidth of less than 100 kHz. However, a study with a Fabry–Perot spectrometer (FSR = 15 GHz) shows that the lasing mode is unstable and hops between 30 MHz-spaced cavity modes under an envelope of about 10 GHz. Due to this limitation in the frequency stability of the lasing mode, the best wavelength selectivity achieved in the experiment is +0.05 nm. A better wavelength selectivity and more stable operation may be obtained by decreasing the cavity length, thermally and mechanically stabilizing the fibers, or by using a narrower passband filter.

12.4.7 Fabry–Perot Filters in WDM Systems

A tunable optical filter is a key component in high-density wavelength-division-multiplexed (HD WDM) systems [59,60]. Many types of filters have been demonstrated, including fiber Fabry–Perot filters and acoustooptic filters [61]. However, these filters have a limited tuning range and/or require high drive powers or mechanical control.

The use of a liquid crystal Fabry–Perot etalon filter in an HD WDM systems experiment is discussed in this section. We describe the performance of the liquid crystal filter (LCFP) in an 8-channel high-density WDM system, operating at 622 Mb s^{-1}. The experimental setup for the WDM system is depicted in FIGURE 12.34. The sources are eight DFB lasers in the 1539–1556 nm spectral range, tuned to have a 2 nm spacing. The multiplexed signals are passed through a polarization controller and through the LCFP using a GRIN lens collimator. The fiber-to-fiber loss without the LCFP is less than 1 dB and a typical LCFP insertion loss is 5 dB. The LCFP insertion loss can be further reduced by a reflective coating that has a smaller absorptive loss. The LCFP is operated with a 10 kHz square-wave signal. The output of the filter is monitored on an optical spectrum analyzer or transmitted to a receiver followed by bit-error-rate (BER) test set.

FIGURE 12.35 shows the optical spectrum of the lasers before and after the filter, which has been tuned to pass laser No. 6. When all the signal polarizations are aligned, the adjacent channels are 20 dB below the selected channel. The results of BER measurements on laser No. 4 with the filter show negligible

FIGURE 12.34. Experimental setup for 8-channel WDM experiment.

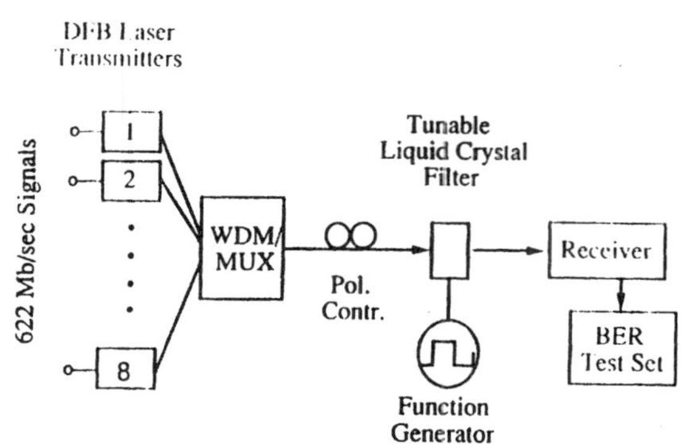

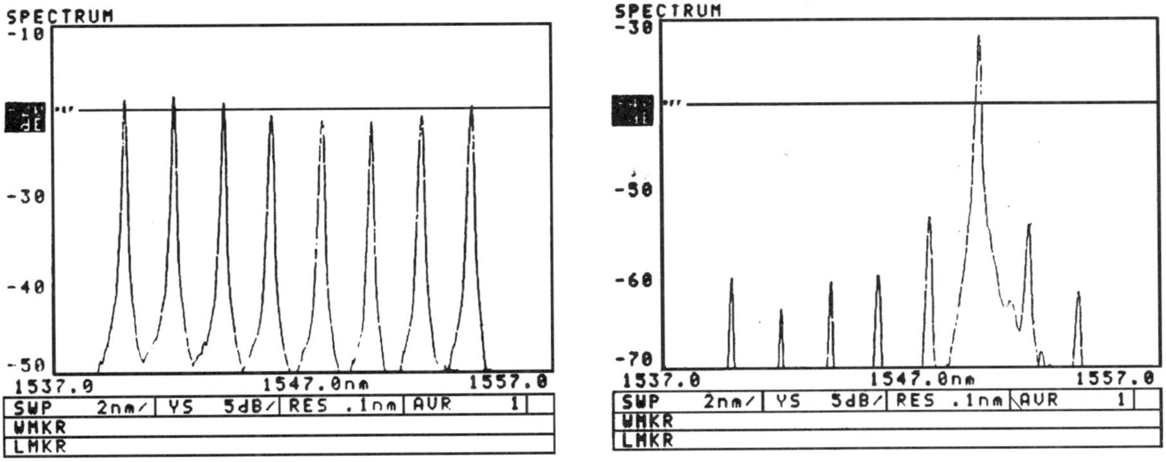

FIGURE 12.35. Optical spectra of the transmitted signal. Left: before filter; right: after filter.

penalty from interchannel crosstalk due to the large optical discrimination of the filter with 2 nm channel spacing. Similar sensitivities have been measured, showing that AC operation of the LCFP has no deleterious effect on the system's sensitivity.

12.4.8 Liquid Crystal and Grating Based Cross-Connect Switches

The simplest form of WDM cross-connect is an add-drop 2X2 switch that can be set in a "bar" or "cross" state for any of the *N* wavelength channels, independently. In general, there are several attributes that are required from such a switch. The crosstalk (i.e., leakage of signal into undesired ports) should be as low as possible. In large networks such leakage may lead to interference, and could be a main source of errors and the main limit of the network size. It has been estimated that 35 dB of extinction might be needed to minimize penalties due to such processes [62]. The switch should be able to tolerate small wavelength variations such that the network will be insensitive to small changes in the operating wavelength of the lasers, but more importantly, it will make it possible to use many sources which tune to the same nominal channel on the network without requiring complicated wavelength stabilization schemes. Ideally, the switch should display a flat response over some wavelength range around the nominal channel wavelengths. Finally, like any other network element, the switch should be polarization-insensitive and have low insertion loss. However, due to the extensive use of fiber amplifiers projected in these networks, some loss in the components can be tolerated. Also, if switching the cross connect is only used for network reconfiguration, high speed is not required.

The switch is constructed from a symmetric grating-based optical system shown in FIGURE 12.36. In this figure the lenses and the liquid crystal based switching element are not shown, to avoid complicating the figure. The optical setup is identical to that used for the shaping of ultrashort optical pulses [63], and it is

FIGURE 12.36. A schematic representation of the principle of operation of the proposed switch.

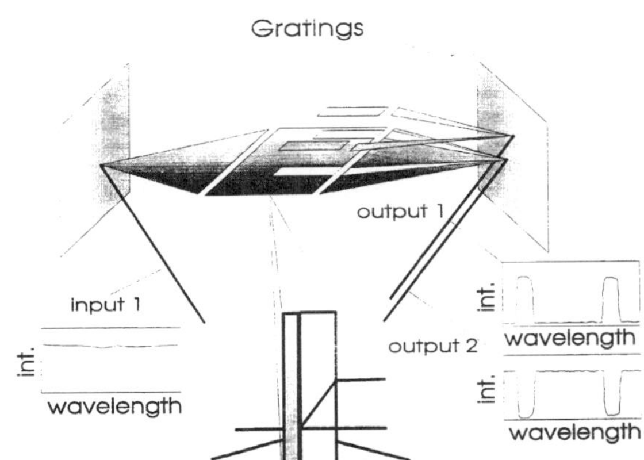

composed of two gratings at the outer focal plane of a telescope. Single-mode optical fibers with appropriate collimators are used on all the ports. FIGURE 12.36 shows schematically the dispersion of the light after striking the grating, and after focusing through the lens, two selected bands of light are deflected using a combination of high-contrast polarization rotators made of liquid crystals and beam displacers made of Wollaston prisms. For clarity, a polarization-dependent version is shown, although the device can be made polarization-independent, as will be discussed later.

The collimated light from the input fiber (input 1), assumed here to be polarized in the vertical plane, is dispersed by the grating. The various wavelength channels are focused by the lens onto separate spots on the liquid crystal spatial light modulators. Without any other perturbations, the channels are recollimated by the output lens and recombined by the output gratings into a single beam that is coupled to output 2. The segmented liquid crystal elements, made by patterning the conductive transparent indium tin oxide electrodes, are designed such that each wavelength channel can be driven independently. The liquid crystal element is of the twisted nematic type, and it either rotates the polarization of the incoming light under the electrode, or leaves it unchanged. The polarization of the light is used to select the direction in which the light beam travels, using polarization-selective deflective elements such as Wollaston prisms or calcite crystals. The high polarization selectivity of these bulk components coupled with the extremely good polarization rotating ability of the liquid crystal element, makes switching possible with very high extinction and very low crosstalk. The selected channels are deflected off the horizontal plane, and are therefore recombined by the grating into another port, labeled output 1 in FIGURE 12.36. Simultaneously, a second input fiber (input 2, not shown in the figure) is polarized in the orthogonal direction, which in the absence of switching would appear at output 1, and those wavelengths that are switched simultaneously by the liquid crystal element would appear at output 2. Thus this switch is a multiwavelength 2X2 cross-connect switch which is independently configurable for any combination of the wavelengths that are multiplexed into the input fibers.

To make a polarization-insensitive device, the technique of polarization diversity is applied. Each input beam is first split into two orthogonal polarization components using a calcite crystal; the beam's component of polarization in the horizontal plane is passed through a 1/2 wave plate that transforms it into a vertically polarized beam. The two parallel beams, now polarized in the same direction, traverse the device together. Because they are initially parallel, they are focused onto the same point at the modulator, and therefore they undergo the same polarization transformation. At the output end, the two beams are combined in a similar fashion into the output port.

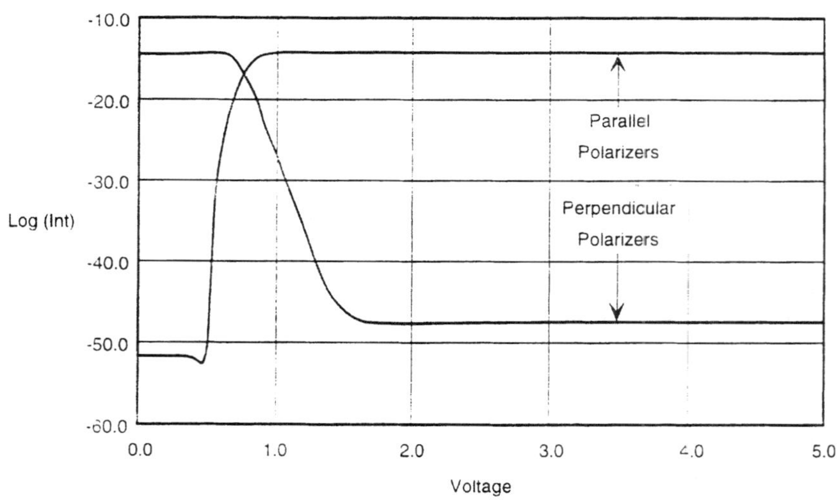

FIGURE 12.37. Extinction of a twisted nematic liquid crystal gate as a function of applied voltage. Polarized input light enter the gate. The two curves show transmission through parallel and perpendicular polarizers after the liquid crystal gate.

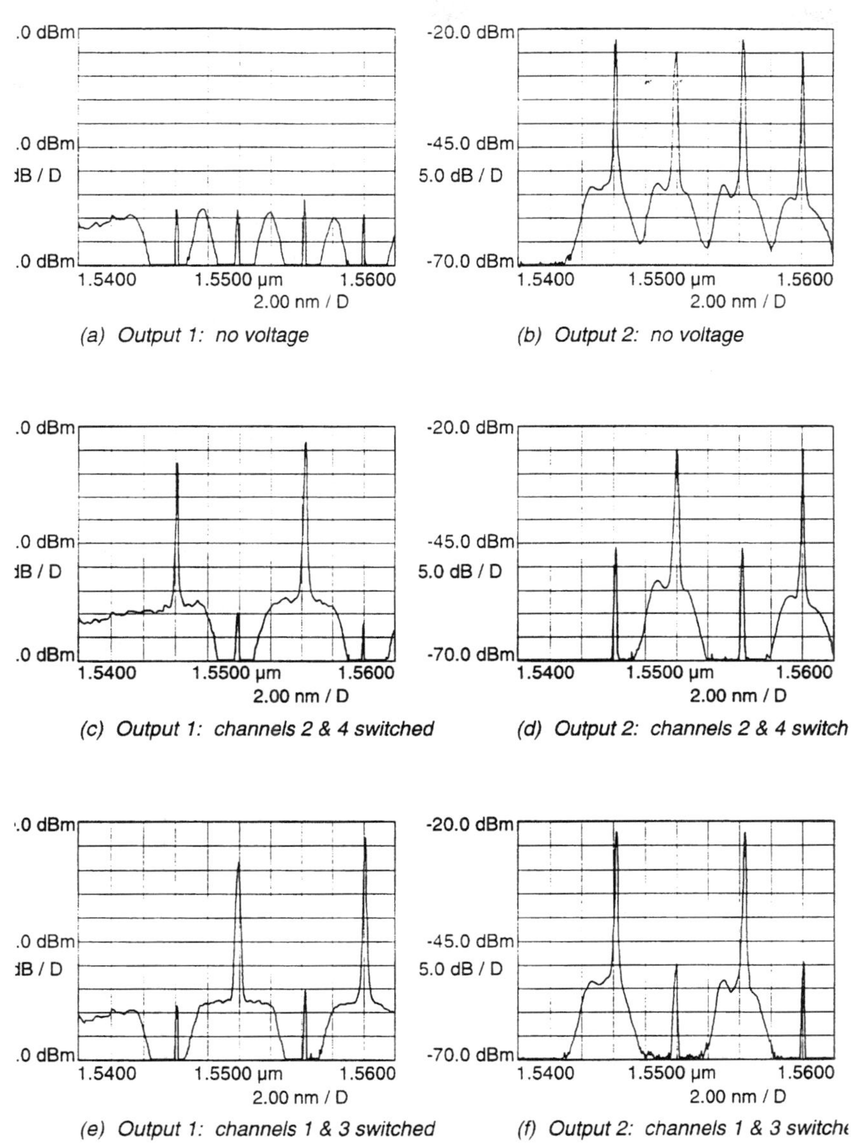

FIGURE 12.38. Demonstration of a 1×2 polarization-sensitive switch. The input light is four wavelength channels at 1546, 1550, 1554, and 1558 nm. A broad ASE signal from an optical amplifier is also visible. Output spectra are shown for three switching permutations.

The polarization conversion is accomplished by the twisted nematic liquid crystal element [64]. In such a structure, in the absence of external electric field, the liquid crystal molecules are arranged in a helical structure, determined by the boundary conditions on the liquid crystal cell. We use cells that induce a 90° rotation to any linearly polarized light introduced into it. When an electric field is applied across the cell, the liquid crystal molecules are aligned parallel to the direction of the applied field, and the polarization rotation is eliminated Since the crosstalk of the switch is related to the performance of the liquid crystal element, we have tested the attainable performance separately. Using the commercial material E7, we constructed a cell with a thickness that was optimized to operate in the 1.5 μm wavelength region. The construction has been described previously [65]. The light transmission through the cell was measured between parallel and crossed polarizers as a function of applied electric field. FIGURE 12.37 shows that polarization extinction better than 30 dB can be obtained.

The main advantage of this design stems from the fact that the frequency response is determined by the structure of the liquid crystal modulator and by the spectral resolution of the instrument, that is, spot size of a monochromatic source at the focal plane. It is important to note that, unlike some grating-based WDM components, the various channels are not coupled to individual wave guides or fibers, but are first transformed into the wavelength domain using gratings and lenses, acted upon by the liquid crystal modulator, and then retransformed back into a single fiber. Thus by acting on individual frequency components, it is possible to attain a flat-top frequency response by designing the widths of each of the liquid crystal switch elements. The switch can accommodate a large number of channels, determined primarily by the spot size in the focal plane. The liquid crystal element should be wider than the spot size, but the number of liquid crystal elements can be arbitrarily large, limited only by the physical size.

A switch designed to accommodate eight wavelength channels with 4 nm separation has been constructed. The first experiment was done using four DFB lasers and a polarization sensitive switch, using four of the eight available channels. The signal was first amplified using an erbium-doped fiber amplifier (EDFA), and then coupled into the input port. The performance of the switch is demonstrated in FIGURE 12.38, which shows the output of the two channels for various switch configurations. The optical loss through the system was about 7 dB. The second experiment was done using all eight channels of a polarization-insensitive switch using the amplified spontaneous emission (ASE) spectrum from an erbium-doped fiber amplifier as the light source. The eight channels, covering a 32 nm wide portion of the spectrum, were coupled to the "out" or "drop" ports. FIGURE 12.39 shows the output of one of the output ports for two different switch configurations, in which alternate channels have been switched. The optical loss through the system was about 10 dB for the polarization insensitive switch.

FIGURE 12.39. Output of one of the ports of a polarization-insensitive switch. The two curves are for two different switch configurations, in which alternate channels have been switched.

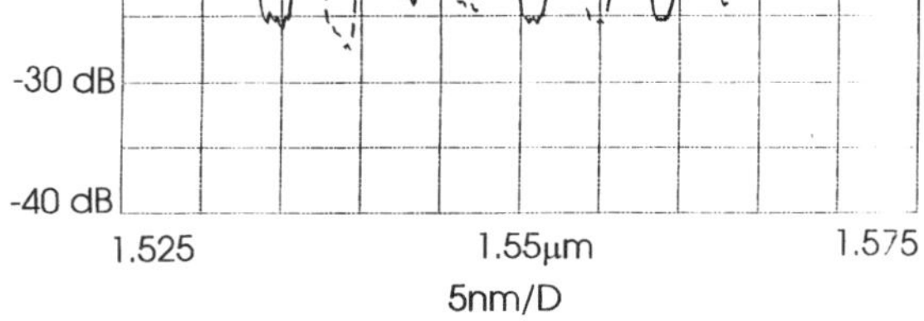

12.5 CONCLUSIONS

In this chapter we have outlined some of the optical properties of liquid crystals and how they can be used in making useful communications devices. It is demonstrated that these materials are extremely versatile in their ability to transform the input polarization of light and, when used in combination with polarization-sensitive optical components, make high-performance modulators or deflectors. By altering the phase properties of light, these materials can also produce a variety of wavelength-selective devices that are important for wavelength-division-multiplexed communications systems. In this chapter, only a small subset of the possible communications devices based on liquid crystals have been highlighted, and no attempt has been made to make this chapter a comprehensive review. Instead, the focus has been on showing the reader the possible types of communications devices that are possible. Undoubtedly, considerable progress in this area is anticipated because of the important role that communication technology plays in our society today.

REFERENCES

1a. Kaiser, P. 1993. *Proc. IEEE* **81**. Special issue on optical interconnection.
1b. *IEEE Commun. Mag.* **32** (1994).
2a. Williams, R. 1963. *J. Chem. Phys.* **39**: 384.

2b. Williams, R., and Heilmeier, G. H. 1966. *J. Chem. Phys.* **44**: 638.
3. See chapter 9.
4. Schadt, M., and Helfrich, W. 1971. *Appl. Phys. Lett.* **18**: 127.
5. Scheffer, T. J., and Nehring, J. 1990. *Liquid Crystals: Applications and Uses*, vol. 1, ed. B. Bahadur, ch. 10. Singapore: World Scientific.
6. de Gennes, P. G., and Prost, J. 1993. *The Physics of Liquid Crystals*. 2d ed. Oxford: Oxford University Press.
7. Berreman, D. W. 1972. *J. Opt. Soc. Am.* **62**: 502.
8. Allia, P., Oldano, C., and Trossi, L. 1991. *Physics of Liquid Crystalline Materials*, ed. I. C. Khoo. Philadelphia: Gordon and Breach, p. 337–359.
9. Born, M., and Wolf, E. 1980. *Principles of Optics*. 6th ed. Oxford: Pergamon.
10. Gray, G. W., and Goodby, J. W. 1984. *Smectic Liquid Crystals: Textures and Structures*. Glasgow: Leonard-Hill.
11a. Charvolin, J., and Tardieu, A. 1978. *Solid State Phys. Suppl.* **14**: 209.
11b. Chapman, D. 1991. *Liquid Crystals: Applications and Uses*, vol. 2, ed. B. Bahadur, p. 185. Singapore: World Scientific.
12a. Cf. Priestley, E. B., Wojotowicz, P. J., and Sheng, P., eds. 1974. *Introduction to Liquid Crystals*. New York: Plenum.
12b. Cf. Blinov, L. M., and Chigrinov, V. G. 1994. *Electrooptic Effects in Liquid Crystal Materials*. New York: Springer.
13. Meyer, R. B., Liebert, L., Strzelecki, L., and Keller, P. 1975. *J. Physique Lett.* **36**: 69.
14. Suresh, K. A. Liquid Crystals in the nineties and beyond, ed. Satyen Kumas, *World Scientific*, Singapore, 1995.
15. Rumbaugh, S. H., Scott, H., Jones, M. D., Michael, D., and Casperson, L. W. 1990. *J. Lightwave Technol.* **8**: 459.
16. Siddiqui, A. S., and Sherief, H. T. 1992. *Int. J. Optoelectron.* **7**: 217.
17. Azzam, R. M. A., and Bashara, N. M. 1977. *Ellipsometry and Polarized Light*. Amsterdam: Elsevier.
18. Reference to arbitrary polarization rotation.
19. Ramachandran, G. N., and Ramaseshan, S. 1961. *Encyclopedia of Physics*, ed. S. Flugge, vol. 25, p. 1. Berlin: Springer.
20. Mauguin, C. 1911. *Bull. Soc. Fr. Minéral. Cristallogr.* **34**: 71.
21. Patel, J. S., and Goodby, J. W. 1987. *Opt. Eng.* **26**: 373.
22. Soref, R. A. 1981. *Opt. Lett.* **6**: 275.
23. Wagner, R. E., and Chang, J. 1980. *Appl. Opt.* **19**: 2921.
24. Fujii, Y. 1993. *IEEE Photon. Technol. Lett.* **5**: 20.
25a. McAdams, L. R., and Goodman, J. W. 1990. *Opt. Lett.* **15**: 1150.
25b. McAdams, L. R., McRuer, R. N., and Goodman, J. W. 1990. *Appl. Opt.* **29**: 1304.
26. Patel, J. S. 1991. *Appl. Phys. Lett.* **60**: 280.
27. Clark, N. A., and Lagerwall, S. T. 1980. *Appl. Phys. Lett.* **36**: 899.
28. Handschy, M. A., and Clark, N. A. 1984. *Ferroelectrics* **59**: 69.
29. Patel, J. S. 1985. *Appl. Phys. Lett.* **47**: 1271.
30. Lin, C., Kobrinski, H., Frankel, A., and Brackett, C. A. 1988. *Electron. Lett.* **B24**: 1215.
31. Harris, S. E., and Wallace, R. W. 1969. *J. Opt. Soc. Am.* **59**: 744.
32. Toba, H., Oda, K., Takato, N., and Nosu, K. 1987. *Electron. Lett.* **23**: 788.
33. Frankel, A., and Lin, C. 1989. *J. Lightwave Technol.* **7**: 615.
34. Stone, J., and Stulz, L. W. 1987. *Electron. Lett.* **23**: 781.
35. Heffner, B. L., Smith, D. A., Baran, J. E., Yi-Yan, A., and Cheung, K. W. 1988. *Electron. Lett.* **24**: 1562.
36. Saunders, F. C., and Parry, G. 1986. *Opt. Quantum Electron.* **18**: 426.
37. Hirabayashi, K., Tsuda, H., and Kurokawa, T. 1993. *J. Lightwave Technol.* **11**: 2033.
38. Levi, L. 1980. *Applied Optics*. Vol. 2. New York: Wiley.
39. Patel, J. S., Leslie, T. M., and Goodby, J. W. 1984. *Ferroelectrics* **59**: 129.
40. Wu, S.-T., Efron, U., and Hess, L. D. 1984. *Appl. Phys. Lett.* **44**: 1033.
41. Strigazzi, A. 1991. *Physics of Liquid Crystalline Materials*, ed. I. C. Khoo. Philadelphia: Gordon and Breach.
42. Kléman, M. 1983. *Points, Lines, and Walls*, p. 54. New York: Wiley.
43. Guyon, E., and Urbach, W. 1969. *Nonemissive Electro-Optic Displays*, ed. A. R. Kmetz and F. K. von Willisen. New York: Plenum, p. 121, 148.
44. Patel, J. S., and Rastani, K. 1991. *Opt. Lett.* **16**: 532.
45. Cf. Mosley, A., Nicholas, B. M., and Gass, P. A. 1987. *Displays* **8**: 17.
46. Patel, J. S. 1991. *Appl. Phys. Lett.* **59**: 1314.
47. Smith, P. W. 1966. *IEEE J. Quantum Electron.* **QE-2**: 666.
48. Lee, S.-D., and Patel, J. S. 1990. *Phys. Rev. Lett.* **56**: 56.
49. Hirabayashi, K., Tsuda, H., and Kurokawa, T. 1991. *IEEE Photon. Technol. Lett.* **3**: 213.
50. McAdams, L. R., McRuer, R. N., and Goodman, J. W. 1991. *Opt. Lett.* **16**: 864.
51. Patel, J. S. 1992. *Liquid Crystal Materials, Devices, and Applications I*, ed. P. Drzaic and U. Efron, p. 244. SPIE, vol. 1665. Bellingham, Wash.: SPIE.
52. Patel, J. S., and Goodby, J. W. 1987. *Opt. Eng.* **26**: 373.
53. Patel, J. S., and Goodby, J. W. 1987. *Mol. Cryst. Liq. Cryst.* **144**: 117.
54. Patel, J. S., and Goodby, J. W. 1986. *J. Appl. Phys.* **59**: 2355.
55. Wyatt, R. 1989. *Electron. Lett.* **25**: 1498.
56. Chen, C. Y., Choy, M. M., Adrejko, M. J., Saifi, M. A., and Lin, C. 1990. *IEEE Photon. Technol. Lett.* **2**: 18.
57. Tsuda, H., Hirobayashi, K., Tohmori, Y., and Kurokawa, T. 1991. *IEEE Photon. Technol. Lett.* **6**: 504.
58. Andrews, J. R. 1990. *IEEE Photon. Technol. Lett.* **2**: 334.
59. Maeda, M. W., Patel, J. S., Lin, C., Spicer, R., and Horrobin, J. 1990. *IEEE Photon. Technol. Lett.* **2**: 820.

60. Maeda, M. W., Chang-Hasnain, C., Lin, C., Patel, J. S., Johnson, H. A., and Walker, J. A. 1991. *IEEE Photon. Technol. Lett.* **3**: 268.
61. Mumai, T., Murata, S., and Mito, I. 1989. *Appl. Phys. Lett.* **54**: 1859.
62. Goldstein, E. L., Eskildsen, L., and Elrefaie, A. F. 1994. *IEEE Photon. Technol. Lett.* **6**: 657.

63a. Heritage, J. P., Weiner, A. M., and Thurston, R. N. 1985. *Opt. Lett.* **10**: 609.

63b. Weiner, A. M., Heritage, J. P., and Kirchner, E. M. 1988. *J. Opt. Soc. Am. B* **5**: 1563.

64. Schadt, M., and Helfrich, W. 1971. Voltage-dependent optical activity of a twisted nematic liquid crystal. *Appl. Phys. Lett.* **18**: 127.
65. Weiner, A. M., Leaird, D. E., Patel, J. S., and Wullert, J. R. 1990. Electronically programmable femtosecond pulse shaping using a multielement phase modulator. *Springer Ser. Chem. Phys.* **53**: 166.

CHAPTER 13

Applications of Liquid Crystals in Image and Signal Processing

Y. OWECHKO

13.1 INTRODUCTION

Over the past twenty years the liquid crystal has evolved from an interesting laboratory curiosity into a technologically and economically important class of material. The general application areas of compact displays and optical information processing depend heavily on liquid crystals because of their attractive combination of properties. As described in detail elsewhere in this book, liquid crystals are fluids which possess partial molecular orientational ordering (at least in a statistical sense) but not the full translational symmetry of crystals. Liquid crystal molecules are highly anisotropic and can be oriented by surface alignment layers or by applied electric fields. The refractive indices along the major and minor molecular axes can differ by as much as $\Delta n = 0.2$, which means that liquid crystal layers as thin as 2–5 μm can be used to modify the phase, polarization, or intensity state of light in periods ranging from 10 μs to 10 ms, depending on the nature of the liquid crystal. The thin layers can be easily and economically formed by encapsulating the liquid crystal material between flat surfaces, which reduces the cost of fabrication and allows flexibility in design.

This unique combination of properties has made compact video and computer displays possible at mass market prices. Liquid crystals are now an enabling technology for such display products as high-resolution projection systems, notebook computers, and pocket-size televisions. They are used in display systems to spatially and temporally modulate light with information. Optical processing systems, for example, computing systems which represent and process information in the form of light, also have a need for devices capable of modulating light in the spatial and temporal domains. It is therefore not surprising that many researchers have adapted commercially available liquid crystal display devices for use in optical processing systems. In addition, liquid crystal based devices have also been custom designed and constructed for optical processing applications (although the capital base available for such work has been much smaller than for displays).

The purpose of this chapter is to discuss applications of liquid crystals in optical processing systems, a market that today is small compared with displays but has the potential for tremendous future growth. The physics and chemistry of liquid crystals will not be discussed in any detail; that is left to other chapters in this book. Instead, I will focus on a limited set of recent image and signal processing applications which utilize liquid crystals. These applications are meant to be representative rather than inclusive with respect to the use of liquid crystals in optical processing. Other important areas related to optical processing which may involve liquid crystals, such as optical interconnects, optical storage, or digital optical computing, are beyond the scope of this chapter.

13.2 OPTICAL PROCESSORS

Progress in computing has occured at an impressive uninterrupted pace for the past 30 years. Computing power is now widely available for prices which are continually dropping. Despite this progress, many important problems are still not economically solvable using present or foreseen conventional computing technologies. In addition, various physical limits will begin to impact on conventional computing in the future. These include such factors as packaging, pinouts, heat dissipation, and interconnection issues. One of the limiting factors for conventional computing is the fact that processors and their interconnections are limited to 2D planes. In many architectures, such as special-purpose neural network integrated circuits for example, most of the area on the chip is dominated by the interconnections rather than by the processing elements. In the future, unconventional processing architectures which use the third dimension to interconnect 2D planes fully populated with processors will begin to supplant conventional architectures, especially in high-end markets. One approach to implementing such 3D connectivity is to use beams of light to interconnect electronic or optical processing units. Many such optical processing systems have already been constructed in the laboratory, and a

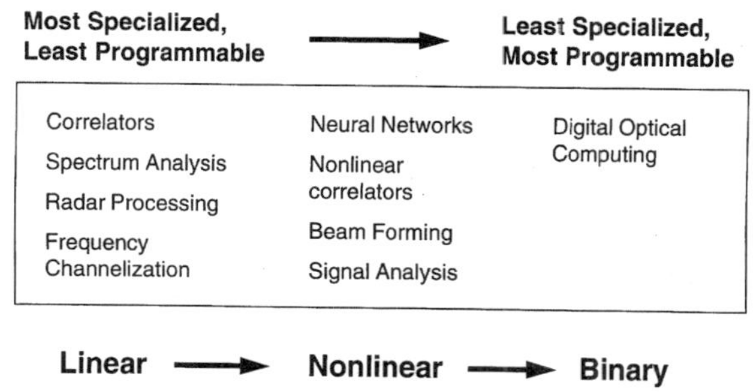

FIGURE 13.1. Some representative applications of optical processing. In general, the programmability increases as the nonlinearity of the processor increases.

limited number have been used successfully, mostly in the military and government markets. The range of applications of optical processing is very broad, as illustrated in FIGURE 13.1. Optical processors have been designed and built for applications which span the gamut from linear to nonlinear problems.

Most optical architectures exploit the parallelism and connectivity of light, made possible by the lack of interaction between light beams in linear media such as free space or glass. The relatively low cost of making many global connections in parallel and the resultant high computational throughput is the primary potential advantage that optics has over electronics. Optical connections using free-space light beams permit the full interconnection of 2D planes of processors using the third dimension. High optical bandwidths are also possible, as has been proven in optical fiber communications. Practical optical processing systems are actually hybrid systems which contain both optical and electronic subsystems. This is due to a variety of reasons. The noninteraction of light beams in linear media that is so useful for connectivity becomes a hindrance when nonlinear processing must be performed, as in neural networks or digital computation. Nonlinearity must be added to the system by including nonlinear materials or electronic circuitry at processing nodes. In addition, most practical optical processors have electronic input (unless raw optical input data are processed directly, as in a smart sensor system or optical image correlator) and a communication link must be maintained with an electronic host computer or output device.

What are the basic features of an optical processing architecture? Due to the wide variety of architectures

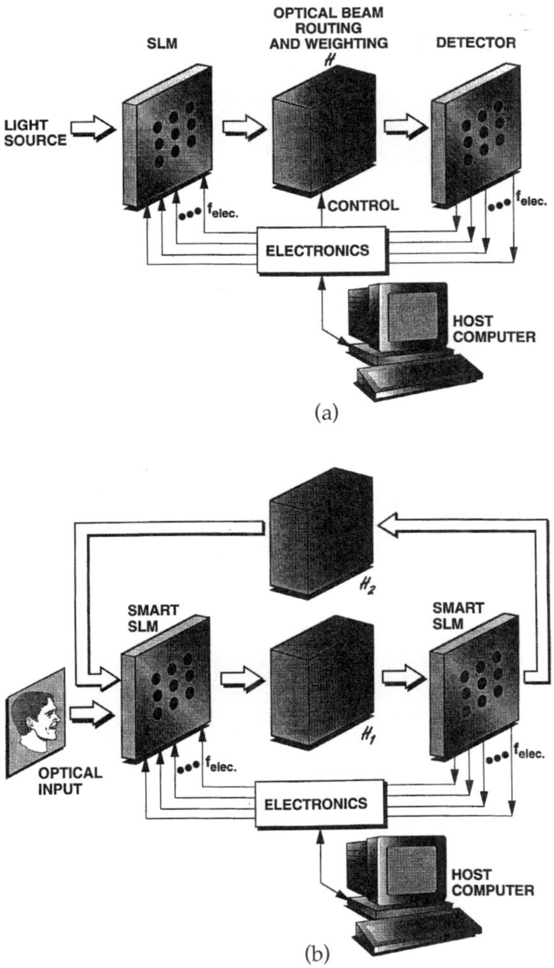

FIGURE 13.2. (a) Generalized optical processing architecture with electronic feedback loop. (b) Generalized optical processing architecture with optical feedback loop.

which have been proposed and constructed in the 30 years since the invention of the laser, any representation of a generic optical processor must necessarily be highly abstracted. Nevertheless, the simplified general schematics of FIGURE 13.2 can be used to illustrate several common features of optical processors. A basic architecture with electronic feedback is shown in FIGURE 13.2(a). Input data in the form of 2D data arrays or images are displayed as pixels of varying transmittance on a spatial light modulator (SLM). SLMs are important components of optical processors. They may be defined as devices which modulate a light beam both spatially and temporally with input data [1]. The original form of the information which is transferred onto the light beam by the SLM may be either optical or electronic. Light beams which are modulated with the information displayed on the SLM are then routed and weighted by passive and active optical components which are lumped together into the optical operator H.

H, which is implemented by lenses, mirrors, holograms, optical fibers, additional SLMs, or other optical components, represents the optical computation performed by the optical interconnection portion of the system. H is often linear, as is the case if passive optical components are used, but it can also be nonlinear. The optical output of H is converted into electronic form by a 1D or 2D detector such as, for example, a CCD (charge-coupled device) array. The output can be read by the host computer for further processing and display, or it can be electronically routed back to the input SLM for further iterative processing. An important point is that once the data are formated on the input SLM in a 2D array, operations can be performed by H in a fully parallel manner. For example, by using a simple lens as the H operator, a fully parallel 2D Fourier transform of a 1000×1000 input array can be theoretically calculated in the time required for light to traverse the distance from the SLM to the detector (<1 ns). (Other more sophisticated operations can be performed as well, as will be described later in this chapter.)

Of course, practical factors limit the actual processing time that can be achieved to values larger than the theoretical minimum. An obvious factor is the optical response time of the SLM, T_{eo}, which is a function of the electrooptic material in the SLM. In addition, the output data cannot be read out until sufficient photocharge is induced in the detector for a required signal-to-noise ratio (SNR). This sets the minimum detector integration time, T_{det}, which depends on the intensity of the light source, the light efficiency of H, and the sensitivity of the detector array. Another basic factor affecting the processing rate is the available electronic input/output (I/O) data bandwidth. In addition to the detector integration time, time must be allowed for the clocking of data into the SLM and out of the detector. In almost all optical processors, I/O of data is performed electronically on $N_{I/O}$ parallel I/O lines with individual bandwidths f_{elec}. Assuming for simplicity the same $N_{I/O}$ and f_{elec} for the input SLM and output detector, the I/O frame time for the processor is given by

$$T_{elec} = \frac{N_{pixels}}{B} \tag{13.1}$$

where $B = N_{I/O} f_{elec}$ is the electronic input/output bandwidth and N_{pixels} is the number of pixels or data items in the input/output arrays. (We assume for simplicity that the input and output arrays are equal in size.)

In most cases, H represents a linear operator which can be thought of as an $N_{pixels} \times N_{pixels}$ matrix which operates on an N_{pixel}-long input vector. This matrix–vector product can be implemented in fully parallel fashion in a single frame time. Thus a practical optical processor achieves a processing rate of

$$R = \frac{N^2_{pixels}}{T} \qquad T = T_{eo} + T_{det} + T_{elec} \tag{13.2}$$

optical operations per second, where an "operation" is normally a multiply-and-add combination. The processing rate is determined by the number of pixels processed per frame by H, the electrooptic response time of the optical modulator, the detector integration time required to achieve the required SNR, and the time required to electronically load or unload data from the SLM. The electronic load time, T_{elec}, depends on the design of the SLM and detector data framing circuitry, the number of parallel I/O lines passing data to and from the SLM and detector, and the electronic clock frequency. In order to achieve a high processing rate, both the detector and the SLM must have fast response times and sufficient electronic bandwidth must be available for a fast interface to the host computer. The processing rate will be limited by the slowest component of T. In this simple analysis, we will assume that the SLM and detector have the same electronic data bandwidth B, and that the processor is I/O limited, for example, both T_{det} and T_{eo} are much less than T_{elec}. Assuming $T = T_{elec}$ and inserting equation (13.1) into equation (13.2) gives

$$R = N_{pixels} B. \tag{13.3}$$

Thus we see that optics achieves a linear increase in computational throughput as the number of pixels or the electronic bandwidth is increased, assuming the optical processor is I/O limited. Increasing B decreases the frame time, which forces the SLM and detector to operate at a faster frame rate. On the other hand, increasing N_{pixels} actually increases the frame time, which relaxes the time response requirements on the SLM and the detector. Of course, this does not come without cost since the SLM and detector must have larger numbers of pixels.

For example, if $N_{pixels} = 10^6$ and $B = 10^8$, then $T_{elec} = 10$ ms, a relatively modest value for T_{det} and T_{eo} to surpass. Assuming T_{elec} is the dominant factor limiting the frame time, this results in $R = 10^{14}$ optical operations per second, an impressive value. The challenge in optical processor design is to harness this large optical throughput to perform useful computations. Proper matching of algorithms and optical architectures is crucial to achieving high performance on practical problems.

The throughput of an optical processor can be further increased by removing the electronic I/O bottleneck in the feedback loop so that the processor is not I/O limited. FIGURE 13.2(b) represents a more advanced architecture which contains a separate optical feedback path. A second optical operator H_2 in an optical feedback loop permits arbitrary global feedback operations. Importantly, recursive operations can be implemented in a fully data parallel manner without limitations due to T_{elec}, except when data must be read from or to the host computer. Such a parallel optical feedback loop requires integration of both optical detection and modulating elements at each pixel. The incorporation of simple processing circuitry in each pixel further increases the power of this architecture. Arrays of such "smart pixels" are an active area of research and several demonstration devices have been constructed [2].

Both digital and analog optical processors have been proposed and constructed. Although in theory any analog computer can be simulated by a digital one [3], in practice analog computers often offer superior computational throughput per unit cost, volume, and power for special-purpose applications. Many analog optical processors have been demonstrated in the laboratory and a number are in use by government agencies. Commercial applications, however, have been relatively few to date. A limitation of this type of analog architecture is the relatively low accuracy and flexibility compared with digital processors. Several groups have studied the accuracy properties of SLMs [4]. Optical processors are best suited to special purpose applications that require very high throughputs with low to moderate accuracy requirements. As is the case for their analog electronic counterparts, analog optical computers are useful primarily for applications in which their limited accuracy and programmability are offset by their superior computational throughput per unit power, volume, and cost. In principle, digital optical computers combine the flexibility and accuracy of programmable general-purpose computers with the high throughput and parallel connectivity of optics. Demonstrations to date (1994), however, have been limited to low integration levels and relatively simple circuits, due to the immature nature of digital optical technology. The most advanced demonstration of digital optical computing so far is the OptiComp DOC II which has emulated a RISC instruction set. It has achieved burst rates of 10^{12} native binary operations per second and is optimized for searching and processing of large databases [5].

No material yet exists that is ideally suited in all respects for modulating light in an optical processor. Selecting such a material inevitably involves compromises between such factors as speed of response and magnitude of the nonlinear optical effect. Liquid crystals offer an attractive combination of properties relative to other materials for many applications, but they do not represent perfect solutions. Due to these material limitations as well as architectural factors, many researchers have added electronic circuitry to the optical processor in order to enhance interactions between light beams, remove electronic bottlenecks, and increase reliability and flexibility. Such circuitry can even be added at the individual pixel level, resulting in so-call "smart pixels," in which light modulation, detection, and simple processing functions are integrated. It is important that a system perspective be maintained when incorporating such innovations, so that the overall performance of the processor is enhanced.

To summarize, virtually all optical processors require SLMs for some or all of the following operations: converting data from electrical to optical formats; incoherent-to-coherent conversion of optical information; dimensionality conversion (1D to 2D and *vice versa*); parallel optical operations on images and data arrays; routing optical beams in the processor in conjunction with passive optical elements (lenses, holograms, etc.); and temporary optical storage. SLMs whose active optical layers utilize liquid crystals, an important class of SLM, are discussed in more detail in the next section.

13.3 SPATIAL LIGHT MODULATOR TECHNOLOGY

Generic schematics of various types of liquid crystal based SLMs are shown in FIGURE 13.3. A widely used class of SLMs is optically addressed SLMs, which operate in the reflective mode. As shown in FIGURE 13.3(a), such SLMs consist of multiple layers. A writing

FIGURE 13.3. (a) Optically addressed liquid crystal spatial light modulator with reflective-mode readout. (b) Electronically addressed liquid crystal spatial light modulator with reflective-mode readout. (c) Electronically addressed liquid crystal spatial light modulator with transmissive-mode readout.

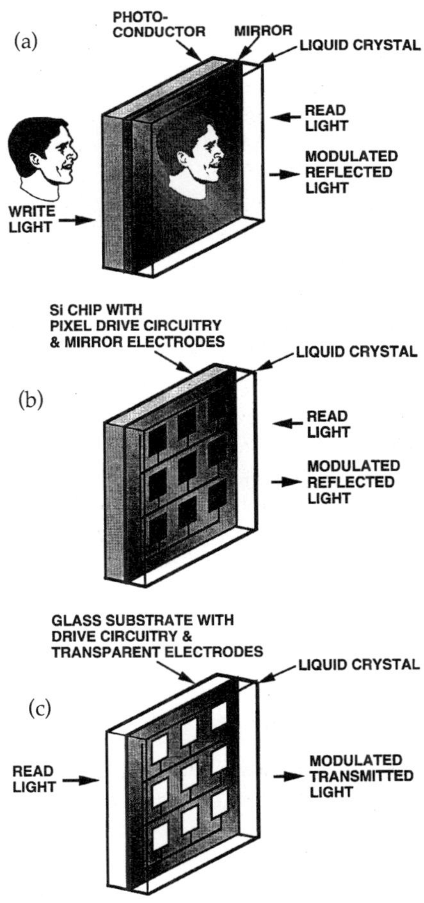

light beam incident on a photoconductor layer modulates the voltage drop across the liquid crystal. As the intensity of the write light increases, a larger fraction of the applied voltage, V_A, drops across the liquid crystal, which changes the polarization state of the reading light beam. A dielectric mirror serves to reflect the read light, as well as optically isolate the photoconductor from the read light. Because of the high lateral resistance of the various layers, the voltage drop across the liquid crystal can be spatially modulated by the write light with high spatial resolution. The read beam normally passes through a polarizing beam splitter which converts differing polarization states into intensity levels. Such an SLM can be thought of as an "optical transistor" since one beam is controlled by another with high isolation between the two beams. Optical gain is easily achieved by using a read beam that is more intense than the write beam. Optically addressed SLMs can convert a weak incoherent image-bearing write beam (such as an image from a real-world naturally illuminated scene) into an intense coherent image-bearing read beam. They can also be used for other optical operations, such as image multiplication, thresholding, phase conjugation, short-term memory, and displays. The liquid crystal light valve (LCLV) invented by Hughes Electronics is perhaps the most commercially advanced example of a liquid crystal reflective SLM [6]. CdS, crystalline silicon, and amorphous hydrogenated silicon (a-Si) have been used as the photoconductor in LCLVs. LCLVs have been successfully used in high-performance military and commercial projection display systems. The latest versions utilize a-Si as the photoconductor for fast response and perpendicularly aligned (homeotropic) nematic liquid crystals for high contrast [7].

SLMs which utilize conventional liquid crystals (e.g., molecules which do not possess permanent dipole moments) in twisted nematic, parallel, or perpendicular alignments have electrooptic response times of 3 to 50 ms, depending on the thickness of the liquid crystal layer. The response times are typically limited by the turn-off time, which is the time required for the molecules to relax back to their zero-field configurations. The viscosity of the liquid crystal limits the relaxation speed. As explained in more detail below, ferroelectric liquid crystals, which possess a permanent dipole moment, can be switched both on and off by an applied field resulting in response times as short as 10 μs. Although more than fast enough for display devices, such response times are still relatively long compared with other electro-optic materials, such as Pockels and Kerr effect crystals and ceramics. It has been pointed out, however, that liquid crystals have other compelling advantages, such as low power dissipation, low voltage operation, and inexpensive device fabrication techniques [8].

Electrically addressed SLMs constitute the second major type of liquid crystal based SLMs. One type of electrically addressed SLM is shown in FIGURE 13.3(b). A liquid crystal layer is deposited directly on a silicon integrated circuit that contains addressing and drive electronics for the individual pixels. Input pins on the chip accept analog or digital input signals which are formated into a 2D array by the on-chip circuitry. Each datum in the array controls the optical properties of a pixel. Two main types of drive circuitry have been developed for electrically addressed liquid crystal SLMs: CCD- and VLSI-based. An example of the former is the CCD LCLV [9,10] in which the photoconductor of the LCLV is replaced with a silicon CCD structure. Analog data are clocked into a linear CCD shift register, and after the shift register is full the contents are shifted into the first row of a 2D CCD array and the entire array is shifted by one row. This process is repeated until the entire CCD array is filled with data, at which time the signal charge is released in parallel. Under the action of an applied field, the charge drifts to the silicon–mirror interface and modulates the voltage drop across the liquid crystal. In this manner, the optical image displayed on the liquid crystal is updated on a frame-by-frame basis. The fact that signal charges in all pixels excite the liquid crystal simultaneously makes this type of SLM suitable for frame-sequential color projector systems in which red, green, and blue components of a scene are projected in sequence. (Row-by-row scanning of the color component images would result in undesirable color artifacts.) The fill-factor (fraction of chip area which modulates the readout light) in such a device can be high, which is necessary for good optical efficiency.

An area of intense research in electrically addressed SLMs is VLSI-based SLMs, in which a liquid crystal layer is deposited directly on the circuitry side of a silicon VLSI chip [11]. An advantage of this approach is that conventional well-developed VLSI processing techniques can be used to form the drive circuits. The liquid crystal is simply poured on as the last step in fabrication. As illustrated in FIGURE 13.3(b), each pixel consists of a storage/drive circuit connected to a metal pad which functions as a combination electrode and mirror. A glass cover slip with a transparent counterelectrode is used to confine the liquid crystal. A readout beam incident on the liquid crystal side is spatially modulated in reflection mode. Row and column addressing is used to load digital data from the host computer. Both SRAM- and DRAM-type circuits have been used for on-chip addressing and driving of the pixels. SRAM SLMs store information at each pixel using several transistors in a flip-flop arrangement. The advantage of SRAM over DRAM is that constant refreshing of the data is not required, however, it requires more transistors per pixel which adversely affects optical uniformity, fill factor, and yield. DRAM SLMs utilize a single transistor and capacitor to store each pixel state. Periodic refreshing of the data is required even if no changes are required in the optical output due to leaking of stored charge from the capacitor. However, the advantages in device uniformity and performance appear to outweigh the extra complexity of periodic refreshing.

Most of the demonstrated VLSI SLMs utilize ferroelectric liquid crystals [12] (FELC). Unlike conventional liquid crystals used in twisted nematic or parallel aligned SLMs and displays, FELCs possess a permanent dipole moment. The molecular orientation and applied electric field therefore couple linearly in these materials. In electrooptic applications, the smectic C^* phase is normally used. Changing the polarity of the applied field flips the molecular director in the plane of the layer. The response time is fast due to the low orientational viscosity for motion in the layer plane and the fact that, due to the permanent dipole moments of FELC molecules, the on and off states are both field-driven, unlike conventional nematic liquid crystals, which must relax back to their off states. The binary nature and fast response times of FELC (tens of microseconds) are good matches to the digital silicon backplanes used in VLSI SLMs. Many optical processing architectures can utilize binary optical inputs. Various multiplexing schemes for achieving grayscale with FELC have also been suggested and demonstrated. An FESLM with 256×256 pixels and a zero-order contrast ratio ranging of 70 : 1 has been recently demonstrated. The pixel pitch is 21.6 μm, T_{eo} is 80 μs at room temperature, and the electronic frame time T_{elec} is 43 μs [13]. Other groups have also demonstrated similar devices [14,15].

The third major group of liquid crystal SLMs is comprised of devices which have been adapted for use in optical processing but were originally designed and manufactured with a different application in mind: compact display of computer and video data. This is an application with a huge market and commensurate large capital investment. From the viewpoint of optical processing, the huge effort invested in designing and manufacturing these devices has resulted in impressive, low-cost technology, but it must be tempered with the fact that the SLM performance requirements for displays and optical processing are not identical. Optically processing typically requires SLMs with higher contrast, higher optical quality, and faster response times than most displays can offer. In addition, displays are usually designed for industry-standard video interfaces, such as NTSC color video or VGA graphics. This is a drawback because the frame time is fixed by the video standard independently of the requirements of the optical processor. Optical processing often requires nonstandard monochrome video interfaces and the ability to read out only a subset of the total number of pixels. Nevertheless, the low cost of liquid crystal displays (LCDs) and miniature computer monitors has made their use attractive in demonstration optical processing systems. In addition, the performance of LCDs continues to rapidly improve, driven by the large computer, consumer, and multimedia markets.

The structure of LCDs is illustrated in FIGURE 13.3(c). The highest-performance LCDs are based on an "active matrix" approach, in which thin film transistor drivers and transparent electrodes are deposited on a glass substrate. A liquid crystal layer is encapsulated between the glass substrate and a transparent counter-electrode, allowing the device to be read out in transmission mode. One thrust in the display industry is to build relatively large versions of these devices for direct-view display panels in portable computers. Another thrust is in smaller versions for projection display systems. The number of pixels is continuing to increase in order to meet the demands of high-resolution computer graphics and high-definition television. For example, the Sharp XG-2000U color LCD projector contains three LCD panels, each of which consists of 479×455 pixels. The LCDs measure approximately 2 inches on each side. The response time is 50 ms and contrast exceeds 200 : 1 for laser readout. The company plans to introduce LCDs with over a million pixels in the near future. Kopin Corporation has recently introduced a 1 square inch active matrix LCD with 640 × 480 pixel resolution, 200 : 1 contrast for laser readout, and 30 ms response time.

LCDs designed for projection display systems can often be used directly in an optical processing experiment with only a few external modifications, such as replacing the glued-on sheet polarizers with high-quality prism polarizers [16]. The performance will fall short of a device custom designed for optical processing for reasons given above. Nevertheless, systems constructed using LCDs can often convincingly demonstrate the power and potential of optical processing. Detailed optical specifications are usually not provided by the manufacturer of the LCD, especially for characteristics important in optical processing but not in the displays arena; for example, the combined phase-amplitude response of the LCD as a function of drive voltage must usually be measured by the optical processor researcher [17,18]. The precision and noise characteristics of SLMs are obviously important to applications in optical processing, yet they are often not specified by manufacturers. Models and methods for measuring SLM noise parameters have been described in the literature [19]. Knowledge of the phase-amplitude response and noise performance is important to many applications of SLMs in optical processing, as will become evident in the next section.

The next section of this chapter deals with a representative set of applications of SLMs in optical processing. The emphasis is on image and signal processing, areas in which the natural parallelism and high throughput can have a large impact. These areas are also potentially less sensitive to the relatively low accuracy of optics compared with digital electronic processors.

13.4 APPLICATION AREAS

The parallelism and high throughput of optical processing architectures are natural matches to many applications in image and signal processing, which deal with input vectors with very large sizes. Described in this section are some representative applications of optical processing in image analysis, restoration, enhancement, and recognition. Applications in signal processing such as spectrum analysis and adaptive beam forming for antenna arrays for radar and communications are also included.

13.4.1 Image Processing

13.4.1.1 Image Analysis

The Fourier transformation and spatial filtering of images using simple lenses was the first application of coherent optical processing [20]. A classic configuration is shown in FIGURE 13.4 where the two-dimensional spatial input pattern to be transformed is located in the front focal plane of lens L_1. In accordance with the physics of light diffraction, when the transparency is illuminated with a coherent plane wave, the light amplitude distribution in the back focal plane of the lens is given by the Fourier transform of the input pattern [21]. Assuming that the light amplitude transmittance of the transparency is given by $t_o(x, y)$, it can be shown (by using the Fresnel approximation for light diffraction, and modeling the lens as a quadratic phase transformation) that the light amplitude spatial distribution in the back focal plane

FIGURE 13.4. (a) Recording of coherent matched filter using off-axis plane-wave reference beam. (b) Coherent spatial filtering optical processor with matched filter $S(u, v)$ in filter plane.

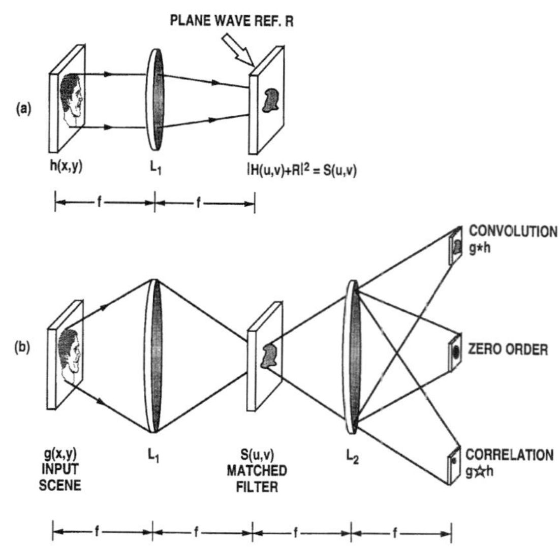

of the lens is given by

$$U(u,v) = \int\int t_o(x,y)\exp\left[-j\frac{2\pi}{\lambda F}(ux+vy)\right]dx\,dy \tag{13.4}$$

where λ is the light wavelength and F is the focal length of the lens. Note that the transformation is an exact Fourier transform with the spatial frequencies in x and y linearly related to their positions in the back focal plane by the relationships $v_x = u/\lambda F$ and $v_y = v/\lambda F$. Since detectors react to light intensity rather than amplitude, the detected quantity is actually proportional to the magnitude squared of the above, for example, the power spectrum of the input pattern. A second lens L_2 can be used to inverse Fourier transform $U(u,v)$ and recover the original image transmittance $t_o(x,y)$ in the output plane. By placing a variable transmittance $H(u,v)$ in the plane of U, optical spatial filtering of $t_o(x,y)$ with the frequency filter $H(u,v)$ can be implemented. The output will be $t_o(x,y)$ convolved with $h(x,y)$, the impulse response of the filter, which is just the Fourier transform of $H(u,v)$. By displaying both $t_o(x,y)$ and $H(u,v)$ on SLMs, programmable high-speed spatial filtering of input images can be performed in real time. The two-lens image filtering system illustrated in FIGURE 13.4 represents one of the earliest applications of coherent optical processing.

The optical Fourier transform illustrates the high computational throughput that can be achieved using optics. By taking advantage of the physics of light diffraction, the computation of the Fourier transform, which would normally require $N^2\log_2 N$ multiplications and additions for an $N \times N$ array if performed digitally using the fast Fourier transform (FFT) algorithm, is theoretically performed in the time required for sufficient light photons to pass through the system to result in a good signal-to-noise ratio or SNR (see above). As an example, for a laser power of $P = 1\,\text{W}$ and a required SNR $= 10^4$, the Fourier transform could in theory be calculated in

$$T = \frac{(\text{SNR})^2 hv}{P} = 4 \times 10^{-11}\,\text{s} \tag{13.5}$$

assuming a Poisson photon probability distribution. Moreover, since optics can handle very large numbers of data elements in parallel; a 2000×2000 Fourier transform would be performed in that amount of time.

Of course, this theoretical performance is an upper bound to what can actually be achieved in practice. Practical optical Fourier transform systems are I/O limited, in other words the computation time is dominated by the time required to input information into the system and by the sensitivity and bandwidth of the output detector (see section 13.2). In addition, the input SLMs invariably introduce noise into the system, which limits the accuracy and dynamic range below what is theoretically possible. The same also holds for detectors. The challenge for optical Fourier transform-based systems, as well as optical processing in general, is to combine architectures and components in such a way that despite imperfect devices, the resultant performance is sufficiently superior to digital electronic processors to be an attractive alternative.

As it turns out, the high throughput of the simple optical Fourier transform by itself has not been sufficient to compete with digital electronics, except for certain high-bandwidth signal processing applications (see section 13.4.2.1). However, the optical Fourier transform can be used as a building block for more sophisticated operations and higher computational complexities. These include image restoration and recognition (described in the next subsection) as well as various other image transforms.

An important transform for analyzing images and signals is the Wigner distribution (WD) of the function $f(x)$:

$$\text{WD}_f(x,\omega) = \int \Phi(x,x_0)\exp(-j\omega x_0)\,dx_0 \tag{13.6}$$

where

$$\Phi(x,x_0) = f\left(x+\frac{x_0}{2}\right)f^*\left(x-\frac{x_0}{2}\right).$$

The WD is given by the Fourier transform of the function Φ. For a signal which has time as a single independent variable, the above integral is one-dimensional. It is straightforward to extend it to two-dimensional images with two independent spatial variables. The dimensionality of Φ, and hence also of WD, is twice that of the original signal. The WD of a 1D signal is two-dimensional, while the WD of a 2D image is four-dimensional. Each pixel in an image has a 2D WD associated with it. (In the following, it is understood that x can represent either 1D time or 2D spatial variables.)

The WD was introduced by Wigner in 1932 as a tool in quantum mechanics to provide a simultaneous representation of a signal in both space and spatial frequency domains [22]. It belongs to a class of bilinear distributions, known as the Cohen class [23,24], which includes the Choi–Williams transform and Rihaczek's complex energy density function. The WD (and other members of the Cohen class) can be interpreted as a local representation of the frequency content of an image or signal. It is not, however, a true or proper distribution which accurately quantifies the amount of signal energy at a particular time and frequency. Although the WD is always real, it can have negative values, unlike a true energy distribution function. In addition, nonzero values can occur for frequencies not present in the signal. This is caused by the bilinear nature of the above integral, which results in cross-terms in the time–frequency plane. The WD does have other desirable properties, however, such as satisfaction of the time and frequency marginal conditions. In other words, integrating the WD over all time results in the power spectrum of the signal. Likewise, integrating over all frequencies results in the magnitude squared of the signal. The WD can be used to extract time–frequency features or signatures of patterns in signals with high time resolution. Unlike the spectrogram, which is normally used for time-dependent spectral analysis, the joint time–frequency resolution of the WD is not limited by a sliding window. Finally, the WD can be used as a starting point to generate other time–frequency distributions, which satisfy positivity and have reduced cross-terms while maintaining good time–frequency resolution [25].

The WD is a good candidate for optical implementation because of its computational complexity and inherent parallel nature, especially for 2D high-resolution images. For images, a 2D Fourier transform of the function $\Phi(x, x_0)$ must be performed for each pixel. Several architectures have been proposed and demonstrated for optical computation of the WD of images. They first generate $\Phi(x, x_0)$ using various optical methods and then use the Fourier transform property of a lens to form the final WD output. Two basic approaches to displaying the four-dimensional WD of a two-dimensional image have been suggested: mechanical scanning and subsampling. In the scanning approach, the optical output consists of the two-dimensional slice of the WD corresponding to the current pixel in the input image. As the input is mechanically scanned in x and y, the output is scanned over the four-dimensional WD space as each image pixel is centered in the input frame.

An architecture which employs the scanning approach is shown in FIGURE 13.5 [26]. An SLM in the input or object plane is used to input a 2D image. A lens and mirror which reflects the input object back through the lens forms the product $\Phi(x, x_0)$ at the object plane between the object and a reversed and shifted version of the object. If the object is centered on the optic axis, the reflection is centered on the object. If the object is shifted, the object and its reflection will be separated by symmetric amounts relative to the optic axis. The separation distance defines the pixel in the input object for which the processor calculates the 2D WD. The Fourier transform lens forms the WD in the

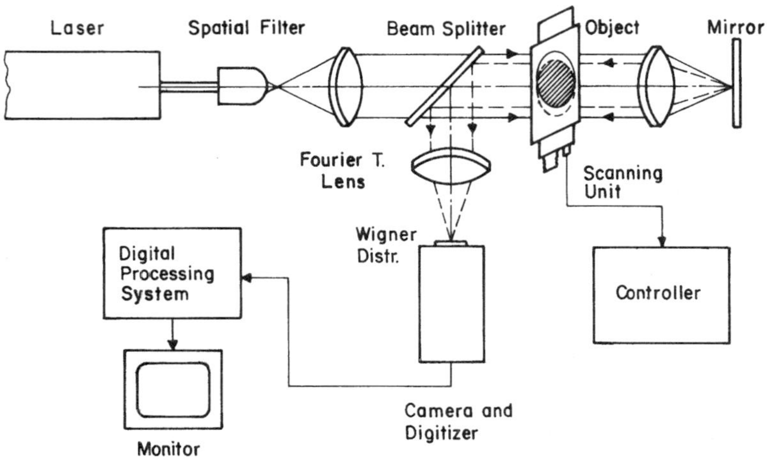

FIGURE 13.5. Scanning optical processor for calculating the Wigner distribution of a 2D image (from reference [26]).

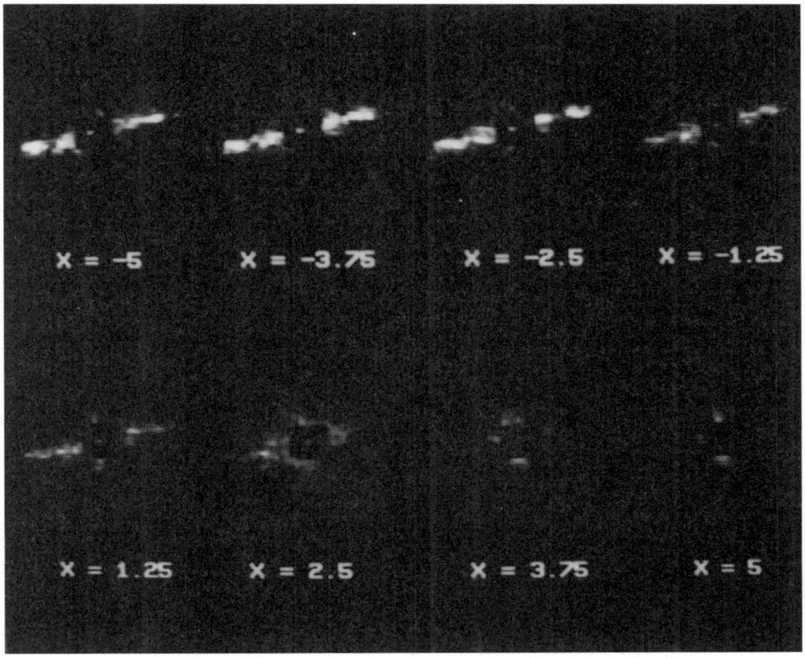

FIGURE 13.6. Example Wigner distributions for texture patterns obtained using the optical processor of FIGURE 13.5 (with kind permission of G. Cristobal, J. Bescos, and J. Santamaria, Institute de Optica CSIC).

detector plane. The output of the detector is digitized and sent to the host computer. As the input image is mechanically scanned in x and y, the processor outputs the full four-dimensional WD for the image. Examples of the output of this hybrid optical–digital processor for some texture pattern inputs are shown in FIGURE 13.6. In reference [26], the WD was used for texture discrimination and classification. Textural features were extracted from the WD and evaluated in terms of quality of separation of texture classes. The WD features were found to be superior to conventional Fourier- and co-occurrence-matrix-based features. A similar architecture can also be used to implement space-variant spatial filtering of images [27].

A disadvantage of the previous method is that mechanical scanning reduces the computational throughput and reduces system reliability. Alternative approaches have been suggested, in which 2D samples of the 4D WD are all output simultaneously without the need for mechanical scanning [28,29]. An architecture which uses a grating to scan the 4D WD is shown in FIGURE 13.7 [30]. A collimated laser beam passes through the object plane where the input image is displayed on an SLM. The beam then passes through a grating which forms a set of beams or orders at various angles. Each order contains the original object. Lens L_1 focuses the orders on a mirror in its back focal plane where a blocking mask selects only the central or zero order. This order retraces its path back through the grating, which now splits the zero order into another set of orders, each of which contain the object image. The orders are imaged onto the object plane, forming a set of shifted versions of the input image. The shifted images multiply the input image, forming $\Phi(x, x_0)$, which is then Fourier transformed by L_2 to form the subsampled 4D WD. The positions of the shifted images determine which x and y samples of the 4D WD are output by the processor. The separation between the shifted images is set by the position z of the grating. An example of the optical output of the processor for an input image consisting of the numeral "4" is shown in FIGURE 13.8. In this case a 1D grating was used, which resulted in sampling of the WD only along the x dimension. Full x and y sampling can be implemented by combining two orthogonal gratings.

The transforms we have considered so far use nonlocalized basis functions (Fourier sines and cosines of different frequencies) for image analysis. The transform of the image is a set of inner-products of the

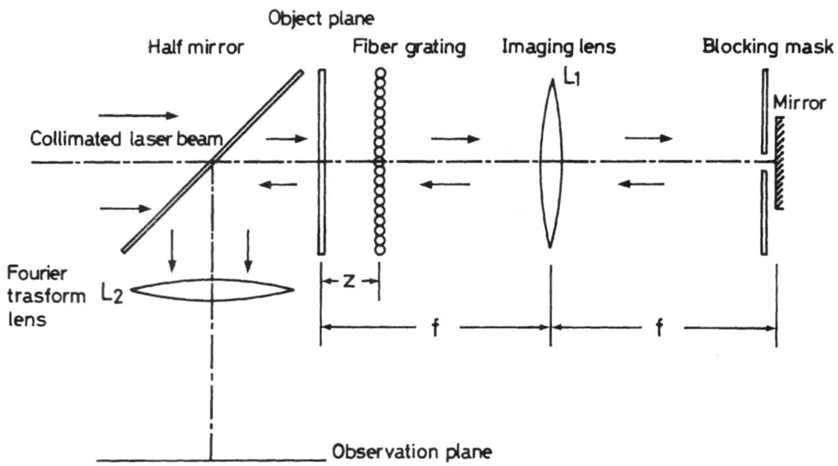

FIGURE 13.7. Grating-based nonscanning optical processor for calculating the Wigner distribution of a 2D image (from reference [30]).

image, or a functional of the image with the basis functions. For example, each value in the WD is the inner-product of $\Phi(x, x_0)$ with a sine wave basis function of a particular orientation, frequency, and phase. The sine wave is localized in frequency but not in position, since it has infinite spatial extent. An area of image processing that has seen an explosion of activity recently is the processing of local features using basis functions, which are localized in both position and frequency. An example of such a localized decomposition is transforms based on wavelet basis sets. Many different wavelet bases have been defined. They all have the required wavelet properties of space and frequency localization and zero mean. Wavelet transforms are linear bandpass filters with impulse responses which satisfy the wavelet properties. The continuous wavelet transform WT of a function $f(x)$ is

FIGURE 13.8. Example Wigner distribution outputs using the optical processor of FIGURE 13.7 for an input consisting of the numeral 4. The illustrated 2D slices of the 4D WD for the indicated spatial positions were all obtained in parallel (from reference [30]).

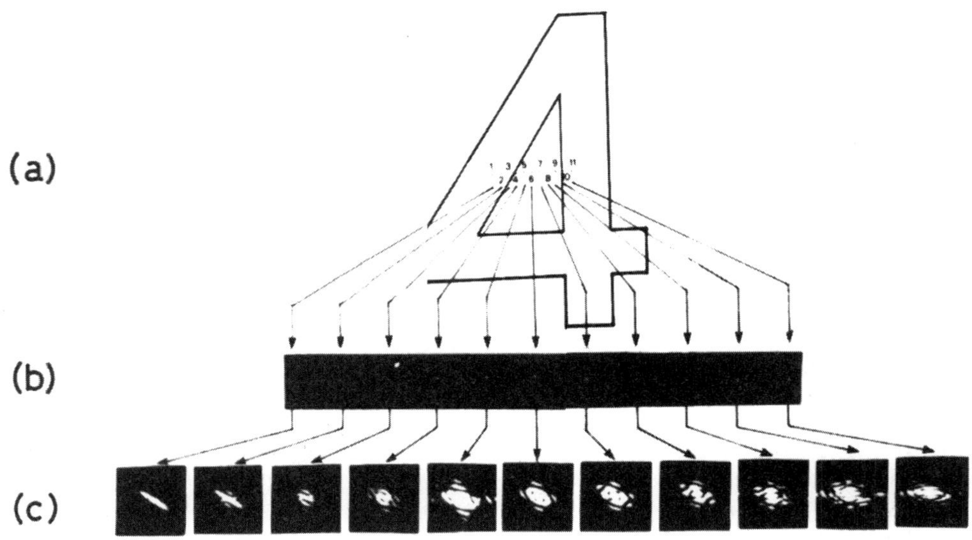

defined as

$$\mathrm{CWT}_f(x,a) = \int f(x_0)\psi_{x,a}(x_0)\,\mathrm{d}x_0 \qquad (13.7)$$

where

$$\psi_{x,a}(x_0) = \frac{1}{\sqrt{a}}\psi\left(\frac{x_0 - x}{a}\right).$$

It is understood in the above equation that x can be a scalar or a vector, depending on the dimensionality of the input function. The wavelet basis functions $\psi_{x,s}$ are scaled and translated versions of the prototype wavelet ψ, which is the impulse response of a bandpass filter. The continuous wavelet transform is the correlation of the input function with the wavelet function at different scales. Like the WD, the continuous WT is two-dimensional for one-dimensional signals and four-dimensional for two-dimensional images.

Discrete versions of the WT (DWT) have been developed with applications in image compression, numerical analysis, and image restoration, among others [31]. Discrete wavelets have been found, which form orthonormal basis sets under integer translations and dyadic scalings [32]. These properties allow the DWT to be calculated with order N operations, where N is the number of samples in the input $f(x)$. The DWT is therefore even faster to compute digitally than the FFT, which makes it practical for applications in real-time image compression. Unlike the WT, however, the DWT is shift-variant. The DWT does not merely shift as the input is shifted, the relative values of the coefficients change as well. This makes the DWT less suitable for applications in image analysis since it is often desired to have the response to a feature be independent of the feature's position, which makes the continuous WT attractive for signal and image analysis.

The continuous WT (CWT) of an image can be computationally intensive for digital electronics since a 2D filtering of an image is required for a range of values of the wavelet scale parameter. Optical systems can implement the 2D filtering with full parallelism using spatial filtering methods. For this reason, several optical CWT architectures have been proposed and demonstrated [33,34]. An optical wavelet processor has been used to analyze the space-variant fractal properties of natural objects [35]. Fractals are complex mathematical patterns that have no characteristic length scales and are characterized by self-similarity across scales [36]. Measurements of the fractal dimension D are relevant to studying chaos in dissipative dynamic systems and to characterizing the geometry of natural objects. D is the exponent in the power law which describes how much fractal "mass" μ is enclosed by an ε-diameter sphere:

$$\mu[B(x,\varepsilon)] = \int_{B(x,\varepsilon)} \rho(x)\,\mathrm{d}x \simeq \varepsilon^D \qquad (13.8)$$

where $\rho(x)$ is the spatial density function for the fractal and $B(x_0, \varepsilon)$ is an ε-diameter sampling sphere centered at x_0. The use of a single quantity D assumes that the fractal dimension is the same over the entire object, that is, D is space-invariant. In general, however, objects may display multifractal properties where the mass μ may scale differently from point to point. A fractal dimension $\alpha(x)$ which varies with position must therefore be defined.

The connection between multifractals and wavelet transforms can be made by examining how the fractal mass changes as the sampling sphere diameter is scaled by a factor λ:

$$\mu[B(x,\lambda\varepsilon)] \simeq \lambda^{\alpha(x)}\mu[B(x,\varepsilon)]. \qquad (13.9)$$

Compare the above with the scaling behavior of the CWT of the fractal object [37,38]:

$$\mathrm{CWT}_f(x+\lambda b, \lambda a) \simeq \lambda^{\alpha(x)-2}\mathrm{CWT}_f(x+b, a). \qquad (13.10)$$

Since the scaling behavior of the CWT mirrors that of the fractal mass, the CWT can be used as a direct measure of the position-dependent fractal dimension of an object. The global fractal dimensions of higher-order correlations of the object (D_q) can also be calculated using the CWT.

As can be seen in equation (13.7), the CWT consists of a set of convolutions of the input with scaled versions of the wavelet function. This is equivalent to filtering the image with bandpass filters of varying scale. In reference [35] a coherent spatial filtering setup similar to that of FIGURE 13.4 was used to implement an optical CWT (OCWT). Slides containing bandpass wavelet filters were mechanically inserted in the filter plane $H(u,v)$. (Future versions would use an SLM instead of slides, which would greatly increase the throughput and flexibility of the system.) Two-dimensional images of the OCWT for a particular scale were detected at the output plane using a video camera. Measurements of the detected intensity of the OCWT at a particular spatial location as a function

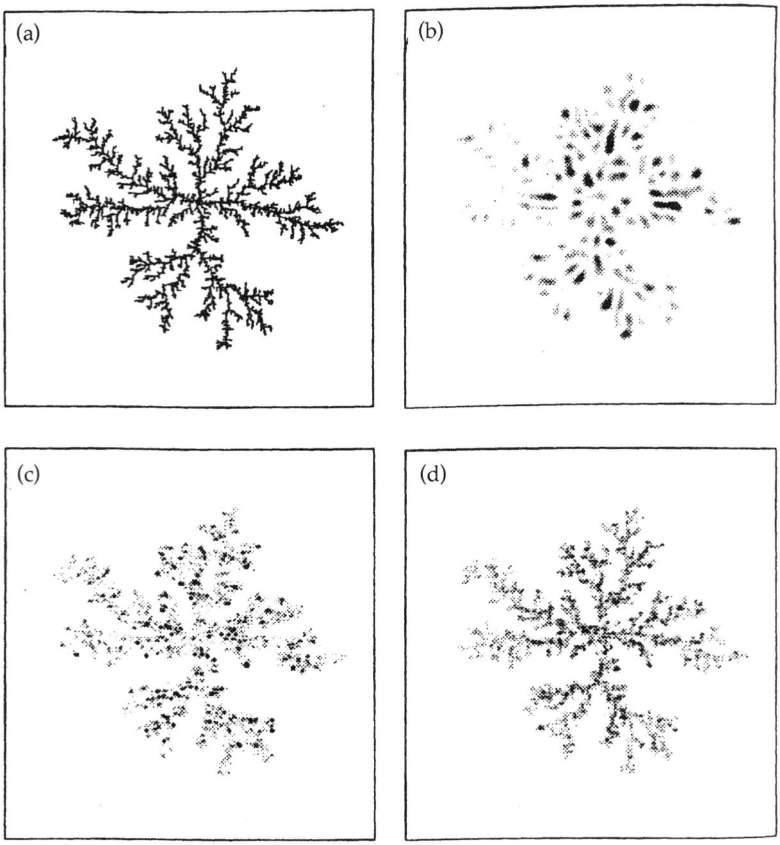

FIGURE 13.9. Optical continuous wavelet transform of a 2D diffusion-limited aggregation (DLA) process for four different scale values (from reference [35]).

of scale were used to measure the position-dependent fractal dimension. The OCWT of a diffusion-limited aggregation (DLA) process is shown in FIGURE 13.9 for several scale values. The DLA model can account for patterns observed in a broad range of physical experiments, including electrodeposition, formation of snowflakes, colloidal aggregation, and dielectric breakdown. The determination of local fractal dimensions of a DLA cluster using the OCWT is shown in FIGURE 13.10.

13.4.1.2 Image Restoration

Restoration of blurred, noisy, or otherwise degraded images is an important subfield of image processing. The restoration of noisy images can be implemented using a Bayesian approach, for example, maximum *a posteriori* (MAP) estimation [39]. In this scheme, prior knowledge about general properties of the class of images, to which the unknown input presumably belongs, is incorporated into a prior probability distribution. Obtaining the MAP estimate corresponds to minimizing a cost function, which includes the prior probability and a measure of the deviation of the current estimate from the original degraded image. The simulated annealing algorithm is guaranteed to find the global minimum of the cost function, but it is a very computationally intensive algorithm [40]. An approximation to simulated annealing, called mean-field annealing, has a good performance and requires fewer calculations [41]. The computationally demanding portion of the algorithm involves iterative spatial filtering of the input image, an operation which we have seen can be efficiently implemented using optics.

Assume the matrix **Y** represents a detected image contaminated by additive Gaussian noise **N**. Let **X** be the original image that we are trying to recover. We

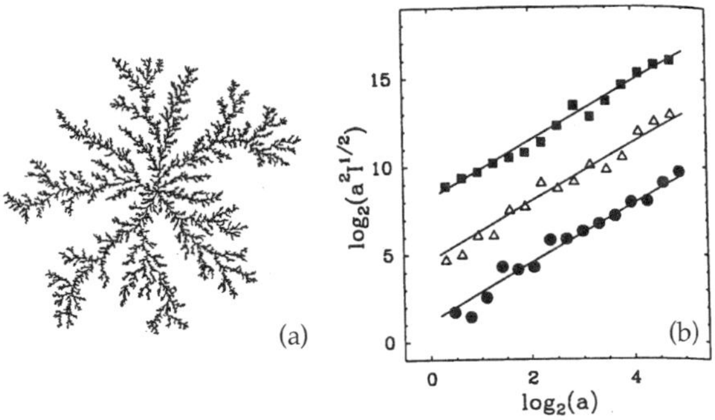

FIGURE 13.10. Determination of the local fractal dimensions of the DLA cluster shown in FIGURE 13.9. The slopes of the three lines are a measure of the fractal dimension for three separate regions (from reference [35]).

then have

$$\mathbf{Y} = \mathbf{X} + \mathbf{N}. \tag{13.11}$$

In forming our MAP estimate of **X**, we wish to find the most probable **X** given the observed image **Y**. In the Bayesian framework, this corresponds to maximizing

$$p(\mathbf{X}|\mathbf{Y}) = \frac{p(\mathbf{Y}|\mathbf{X})p(\mathbf{X})}{p(\mathbf{Y})} \tag{13.12}$$

where $p(\mathbf{X}|\mathbf{Y})$ is the probability of **X** given **Y**, $p(\mathbf{Y}|\mathbf{X})$ is the probability of **Y** given **X**, and $p(\mathbf{X})$ and $p(\mathbf{Y})$ are the probabilities of **X** and **Y**, respectively. (In maximizing $p(\mathbf{X}|\mathbf{Y})$, the denominator $p(\mathbf{Y})$ can be ignored since it is constant.) Since **N** is Gaussian, we can write

$$p(\mathbf{Y}|\mathbf{X}) = (2\pi\sigma^2)^{-m/2}\exp\left[-\frac{1}{2\sigma^2}\sum_{ij}(x_{ij}-y_{ij})^2\right] \tag{13.13}$$

where σ^2 is the variance of **N**, $x_{ij}(y_{ij})$ denotes the elements of the matrix **X** (**Y**), and m is the total number of image pixels. Our prior knowledge of the general properties of **X** is incorporated in $p(\mathbf{X})$. A property that is often chosen is smoothness, for example, adjacent pixels should have similar values, on average. This property can be incorporated in the algorithm by writing $p(\mathbf{X})$ as

$$p(\mathbf{X}) = Z_1^{-1}\exp\left[-\sum_{ijkl}W_{i-k,j-l}(x_{ij}-x_{kl})^2\right] \tag{13.14}$$

where W_{ij} is a local averaging kernel and Z_1^{-1} is a normalizing constant. From equation (13.12) we can obtain an expression for the posterior distribution:

$$p(\mathbf{X}|\mathbf{Y}) = Z_2^{-1}\exp[-E(\mathbf{X},\mathbf{Y})] \tag{13.15}$$

where

$$E(\mathbf{X},\mathbf{Y}) = \sum_{ijkl}W_{i-k,j-l}(x_{ij}-x_{kl})^2 + \frac{1}{2\sigma^2}\sum_{ij}(x_{ij}-y_{ij})^2 \tag{13.16}$$

The MAP estimate $\mathbf{X}_{\mathrm{MAP}}$ is the value of **X** that minimizes the cost function E for a given **Y**. The first term of E emphasizes local continuity, while the second term favors images which are close to the original noisy **Y**.

In practice, the error function E may contain local minima which need to be avoided in order to find the global minimum corresponding to $\mathbf{X}_{\mathrm{MAP}}$. Simulated annealing (SA) is an iterative stochastic algorithm that is analogous to the thermodynamic minimization of the collective energy of a set of molecules as the temperature is lowered. SA is guaranteed to find the global minimum, provided a slow annealing schedule is used. Unfortunately, SA is often too computationally intensive for practical problems. In the mean-field approximation to SA, the number of required operations is reduced by calculating the mean value of **X** with respect to the *a posteriori* distribution $p(\mathbf{X}|\mathbf{Y})_T$ at a low temperature, T. As T approaches zero, the mean $\langle\mathbf{X}|\mathbf{Y}\rangle$ approaches $\mathbf{X}_{\mathrm{MAP}}$. Without going into details here, the mean-field method results in the following iterative equation:

$$\langle x_{ij}|\mathbf{Y}\rangle^{n+1} = \tanh\left[\frac{1}{T_n}\left(\sum_{kl}W_{i-k,j-l}\langle x_{kl}|\mathbf{Y}\rangle^n + \frac{1}{4\sigma^2}y_{ij}\right)\right]. \tag{13.17}$$

The algorithm consists of a convolution operation, addition of a bias, multiplication by the inverse temperature parameter, and a nonlinear threshold. T_n decreases as a function of iteration number n. A schematic diagram of the algorithm is shown in FIGURE 13.11. The spatial filtering properties of optics

FIGURE 13.11. Schematic diagram of mean-field annealing algorithm (from reference [42]).

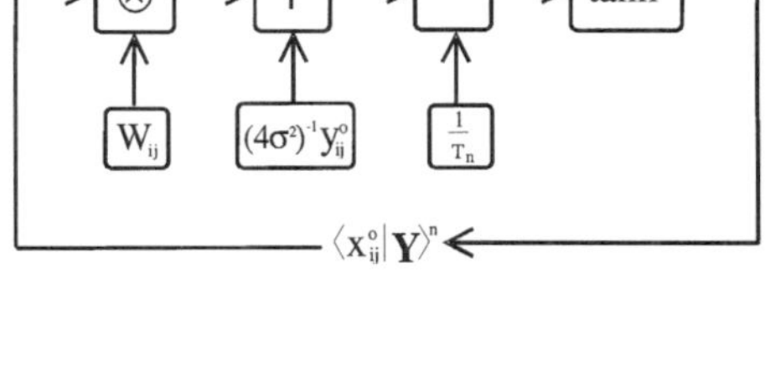

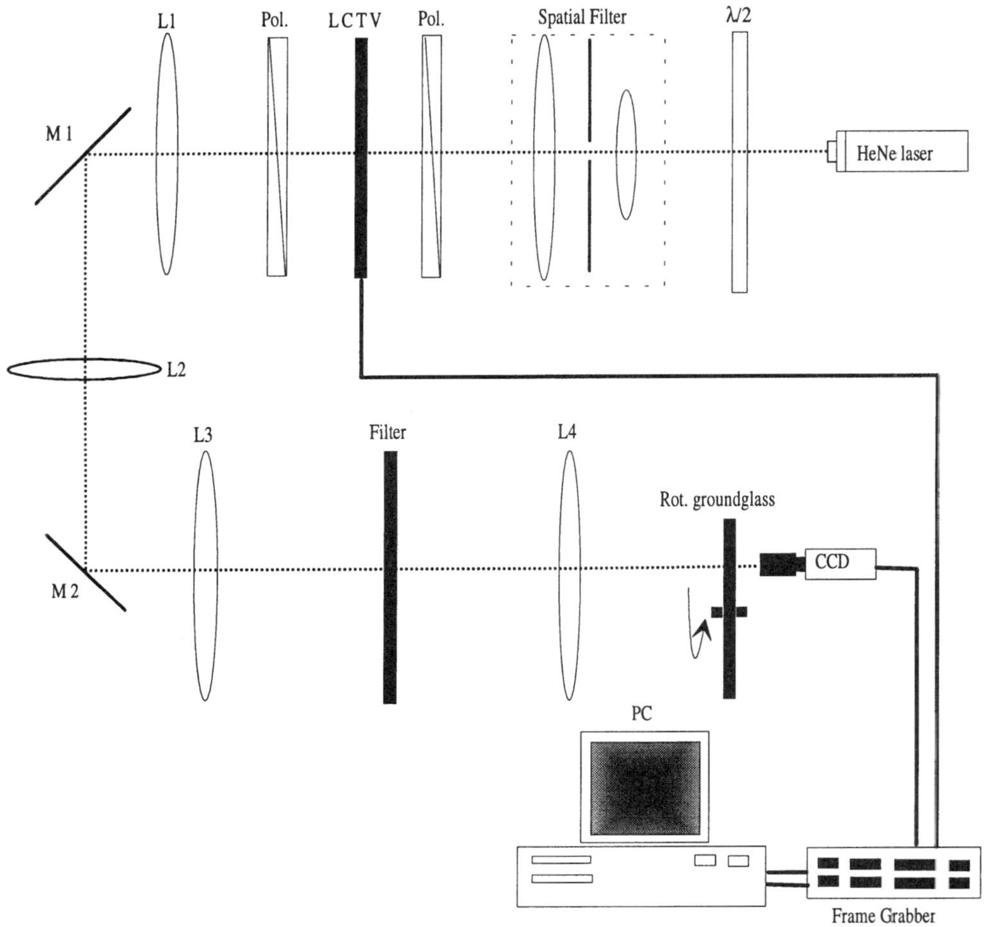

FIGURE 13.12. Optical architecture for implementing the mean-field annealing algorithm (from reference [42]).

suggests that the convolution operation can be implemented with full parallelism using optics. The other nonlinear operations are less computationally intensive and can benefit from the programmability of electronics without a large cost in throughput. A hybrid optoelectronic processor has been constructed which implements the above iterative equation using a closed feedback loop [42]. A diagram of the processor is shown in FIGURE 13.12. The optical portion is essentially a coherent spatial filtering setup, similar to that in FIGURE 13.4, in which the input image **Y** is displayed on a LCTV. The image is optically filtered with the smoothing kernel **W**, the output is then detected by the CCD camera, and the other point-wise operations are implemented electronically using a frame grabber board in the host computer, forming the current estimate $\langle \mathbf{X}|\mathbf{Y}\rangle^n$, which is then redisplayed on the LCTV for the next iteration.

Some experimental results are shown in FIGURE 13.13. The original image **X** is shown in FIGURE 13.13(a). Gaussian noise was then added to the image, forming the input image **Y** with signal-to-noise ratio of approximately 1, as shown in FIGURE 13.13(b). The result of processing **Y** using the opto-electronic processor is shown in FIGURE 13.13(c). The mean-field algorithm was able to extract the unknown image from the noise using only the two constraints of continuity and minimum deviation from the input image. The optically processed result is virtually indistinguishable from the result calculated digitally.

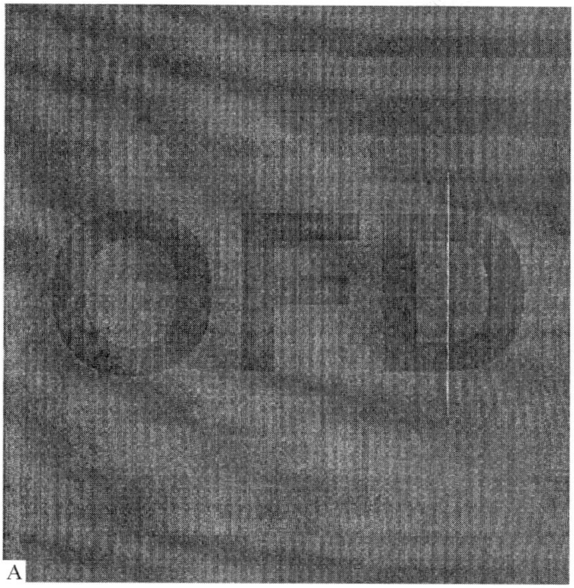

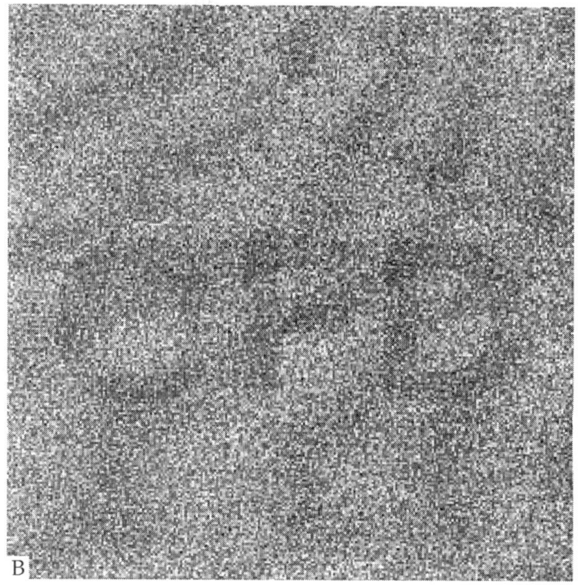

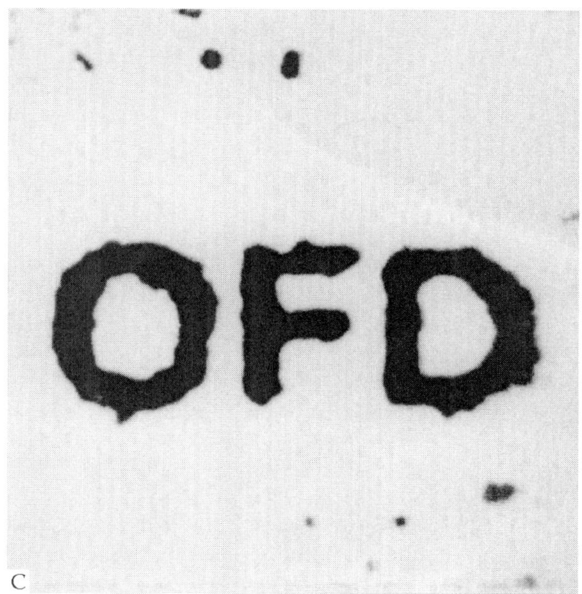

FIGURE 13.13. Sample image restoration results obtained using optical processor of FIGURE 13.12. (a) Original object. (b) Object with added noise. (c) Output of optical processor (from reference [42]).

The optical approach, however, has the potential for very high computational throughput.

13.4.1.3 Image Recognition

Optical architectures have been developed for image recognition and filtering based on the Fourier transform properties of lenses. Most of them are based on the concept of matched filtering or correlation of the input image with the reference or search pattern of interest. The correlation of two optical images with complex amplitudes a and b can be implemented by Fourier transforming the product AB^*, where A and B are the Fourier transforms of a and b and * denotes the complex conjugate. (Convolution results if the product

AB is Fourier transformed.) In the case of 2D spatial images, a matched filter can be formed by recording the sum $A+B$ on photographic film using lenses. Since film responds to the magnitude squared of the incident light amplitude, the amplitude transmittance after development will be proportional to

$$t_{\text{film}} \propto |A+B|^2 = |A|^2 + |B|^2 + AB^* + A^*B. \quad (13.18)$$

If the film is illuminated with the original wavefront A, one of the resultant optical product terms is $|A|^2B$. If the original object scatters light diffusely or is sufficiently complex, then $|A|^2$ will be relatively constant on a large scale and B will be fairly well reconstructed. If this term is then Fourier transformed, the result will be b convolved with the autocorrelation of a. If the autocorrelation of a is strongly peaked, b will be convolved with a near-delta function, reproducing b. Both the phase and amplitude of the original wavefront b will be reconstructed. This process forms the basis for in-line holography, as formulated by Gabor [43]. In Gabor holograms, the reference beam a is derived from the object b. Although this reduces the degree of coherence required of the illumination source, it has the disadvantage of superposition of the other undesired bias and cross-terms in equation (13.18) on top of the desired product term in the reconstruction. This results in poor-quality imaging and low SNR.

A great improvement on in-line holography was made by Leith and Upatnieks in 1962 [44]. They realized that an off-axis plane-wave reference beam A incident on the film at an angle would spatially separate the various product terms in the reconstruction, greatly improving the SNR. The invention of the laser several years earlier mitigated the increased coherence requirements of the off-axis method. Later, Van der Lugt showed that the use of a plane-wave reference allows the off-axis hologram to be used as a matched filter for the stored 2D image [45]. An optical system for spatial filtering based on off-axis holography is illustrated in FIGURE 13.4. The Fourier transform $H(u,v)$ of a 2D image $h(x,y)$ on the input SLM is formed by lens L_1. The magnitude squared of the sum of $H(u,v)$ and an off-axis plane-wave reference is recorded on the film at the hologram plane. Upon illumination of the hologram with the original plane-wave reference, several spatially separated terms are reconstructed, one of which is the original image Fourier transform $H(u,v)$. Lens L_2 inverse Fourier transforms the output, resulting in the original image $h(x,y)$ in the output plane.

The same hologram can be used as a correlator or matched filter for $h(x,y)$ if the roles of the reference and object beams are reversed. If the hologram is illuminated with an input image in the original object beam, the output will consist of three spatially separated terms: an on-axis zero-order term, the convolution of the input with the original image, and the correlation of the input with the original image. Due to the amplitude shift-invariance of the Fourier transform, the correlation spot will appear in the output plane at a position corresponding to the position of the matching image in the input plane. The movement of the correlation spot will match the movements of the input image, which is very useful for tracking applications.

Early optical correlators typically used an SLM for incoherent-to-coherent conversion of the input scene and film to store the holographic filter itself. Real-time updating of the correlation filters requires SLM storage of the filter as well. This is normally difficult because of the large space–bandwidth product of off-axis holographic filters. A disadvantage of the simple Van der Lugt correlator for many practical applications is its lack of rotation and scale invariance. Rotated or scaled versions of recorded images will not be recognized due to the extreme selectivity of matched filters. Rotation and scale invariance have been added to the Van der Lugt correlator by using coordinate transforms [46], wedge-ring filters [47], circular harmonics [48], and by computer-generating composites of simple holograms [49].

It has been shown that the recognition and discrimination capabilities of an optical matched filter are enhanced by removing the amplitude portion of the filter and keeping only the phase [50]. The resultant phase-only filter results in a much narrower correlation peak compared with the original filter, as shown in FIGURE 13.16. The light efficiency is also greatly enhanced. On the other hand, the filter also becomes more sensitive to rotation and scale changes, and to other distortions of the input. The SNR of the filter is only slightly degraded if the phase-only filter is quantized to two levels of phase, equal to 0 and π. This allows rapid programmable updating of the filter function by replacing film with binary phase-only SLMs.

Another approach to programmable optical matched filters for images is the joint-transform optical

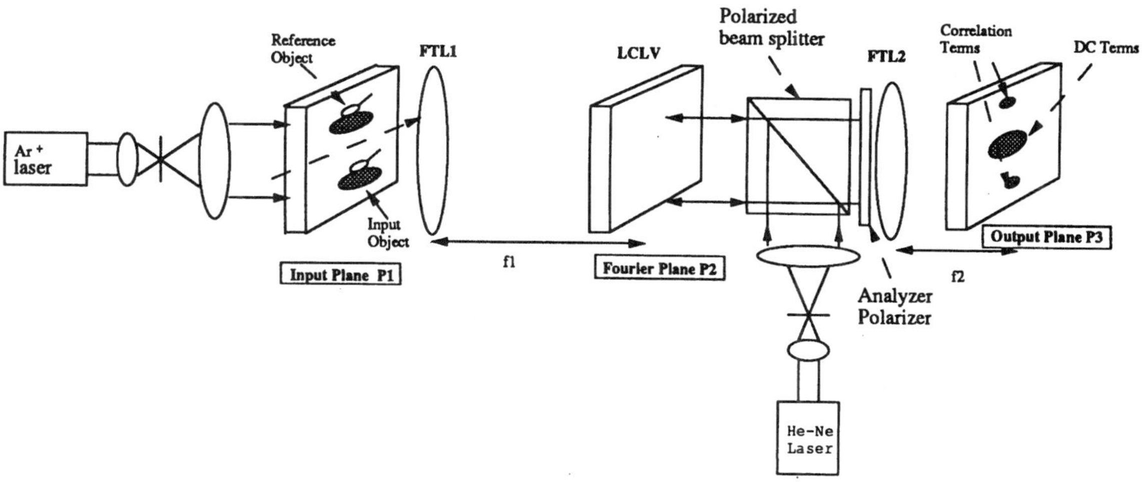

FIGURE 13.14. Optical architecture for joint-transform optical correlator (JTOC) (from reference [52]).

correlator (JTOC), which is illustrated in FIGURE 13.14. In the JTOC, the input object *a* and the reference object *b* are displayed side-by-side (spatially displaced) on an SLM in input plane P1. The sum of the Fourier transforms of *a* and *b*, formed by lens FTL1, is incident on an optically addressed SLM in Fourier plane P2, such as the LCLV. The amplitude transmittance of the SLM is dependent on the magnitude squared of the incident complex optical field. Assuming the SLM is operating linearly, that is, its response is linear in the incident intensity, then the transmittance of the SLM identical to equation (13.18) for the film-based correlator. A major difference is that the reference *b* is not a plane wave, but a reference image which the input is to be correlated against. Also, the SLM is read out using a plane wave, not an input object, so that the output amplitude of the SLM is directly proportional to equation (13.18).

After an inverse Fourier transform by lens FTL2, the output plane P3 consists of three spatially separated images (assuming the SLM responds linearly to intensity). The on-axis image corresponds to the Fourier transform of the first two terms in equation (13.18). The +1 order image corresponds to the Fourier transform of the third term, which gives the convolution of *a* with *b*. The −1 order image, which is the image of interest, corresponds to the fourth term and is the correlation of the input object *a* with the reference *b*. (Obviously, the roles of *a* and *b* as object and reference are arbitrary and can be switched.) Since the output is detected with a square-law detector such as a CCD camera, the final output is actually the magnitude squared of the correlation.

The JTOC has several advantages. Complex matched filters are generated automatically and in real-time, which allows rapid updating of filters. Also, filter alignment requirements are greatly relaxed compared with Van der Lugt type correlators. A disadvantage is that the limited space–bandwidth product of liquid crystal based SLMs compared with holographic recording materials, such as film, limits the complexity of objects that can be recognized. Optical phase information is encoded in the spatial frequency and phase of fringes on the Fourier plane SLM. If the finely spaced fringes cannot be resolved by the SLM, the output orders will not be spatially separated and the JTOC will fail. Another disadvantage is that if multiple similar input objects are present, the output will contain correlations of the input objects with each other, on top of the desired object-reference correlations. These shortcomings can be improved using photorefractive and nonlinear SLMs, as will be discussed below.

A recent experimental example of a JTOC is shown in FIGURE 13.15 [51]. In this case, a photorefractive BSO ($Bi_{12}SiO_{20}$) crystal is used in place of the SLM in plane P2 of FIGURE 13.14. The BSO is capable of film-like fringe resolution using the photorefractive effect, yet like an SLM it is also updatable in real-time. A set of reference images b_i is used so that the input is

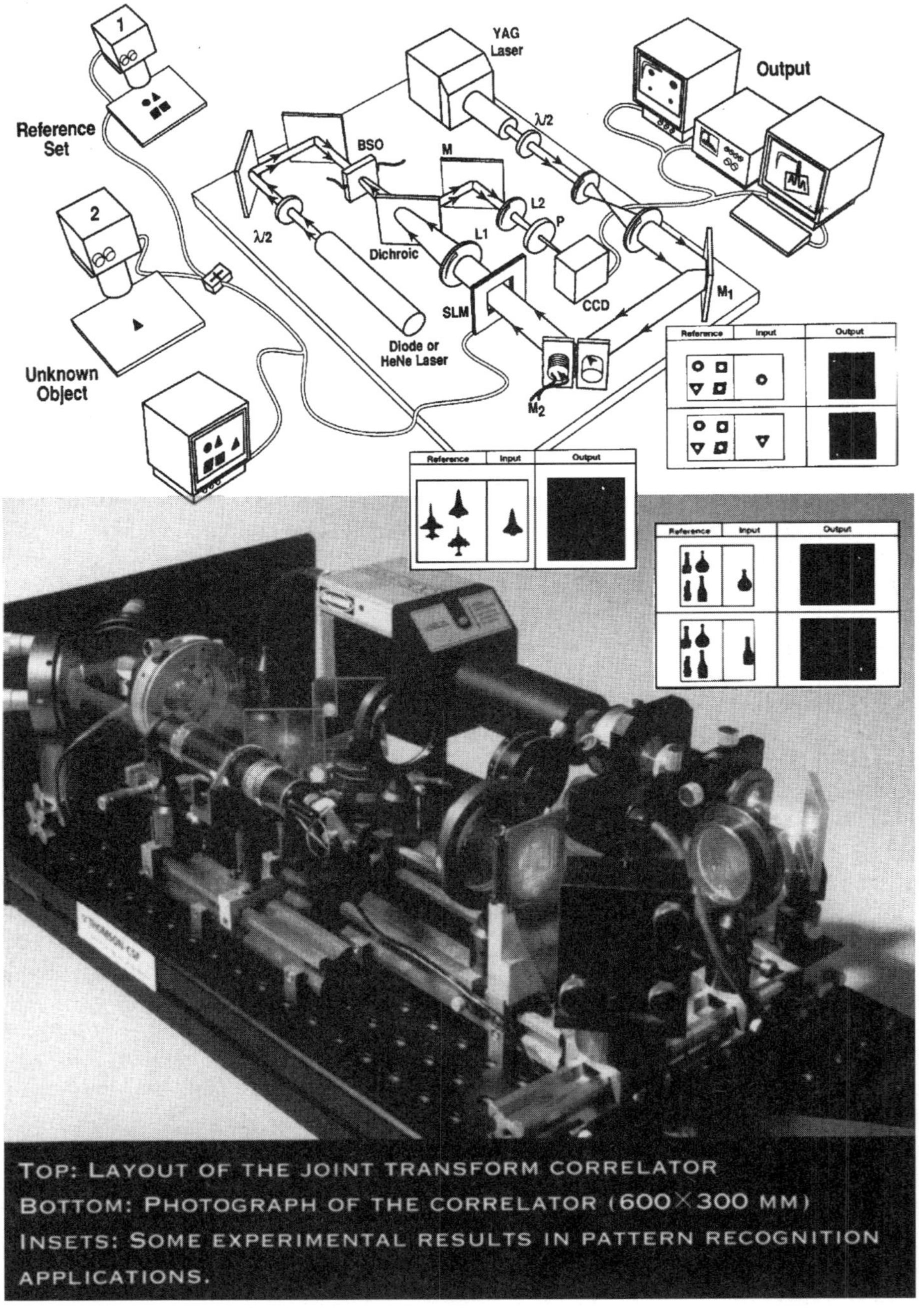

FIGURE 13.15. Experimental JTOC with photorefractive BSO crystal in filter plane (from reference [51]).

correlated against a bank of references in parallel. The BSO is illuminated with a separate coherent beam, which is Fourier transformed to form the set of correlations $a * b_i$. Each correlation is separated in the output plane from the other correlations and the bias and convolution terms. The output plane of correlation spots is detected by a 2D CCD image sensor. Because the reference images are separated in the input plane, correlations of the input with each of the references appear in separate sectors in the output plane. The position of a correlation spot within a sector indicates the position of the recognized object in the input plane. Input scenes can be correlated against multiple reference images simultaneously. (This is also true of Van der Lugt correlators.) Examples of output are also shown in FIGURE 13.15. The system is programmable in real-time since both the input and reference images are stored on an SLM which can be addressed by either a video camera or images stored on a host computer. The trade-off is in the amount of information which can be stored since currently available SLMs have much less storage capacity than film. The system of FIGURE 13.15 was designed for pattern recognition tasks in industrial robotic applications. Video-rate correlation was performed on images containing 10^5 pixels, resulting in an effective processing rate of 250 million multiplications and additions per second. The optical package measured $60 \times 30 \times 30\,\mathrm{cm}^3$.

The performance of the JTOC can also be improved by allowing the output amplitude of the SLM in Fourier plane P2 of FIGURE 13.14 to be a nonlinear function of the incident light intensity $I(x, y)$. Such nonlinear joint-transform optical correlators (NJTOC) can have greatly improved discrimination performance. In addition, the interobject correlations can be greatly reduced [52]. If $g(I)$ represents the nonlinear characteristic of the SLM and $G(\omega)$ is its Fourier transform, then the output of the SLM can be written as a sum of fringes of increasing spatial frequencies [53]. The relative phase between the object and reference Fourier transforms is preserved in the spatial phases of the fringes. The amplitudes of the fringes are complicated functions which depend only on the magnitudes of the object and reference Fourier transforms. Since phase is more important than amplitude in correlators, the NJTOC has good correlation properties.

Varying the nature of the SLM nonlinearity in the Fourier plane P2 produces output correlations with different characteristics. As the nonlinearity increases, high spatial frequencies are emphasized and the NJTOC becomes more sensitive in discrimination between similar inputs. In the limit, if a binary SLM is used in P2, the NJTOC can implement correlation using only the phases of both the object and reference. (This is different from the phase-only correlation filters described earlier which utilize both the phase and amplitude of the object.) However, in order to achieve this, the proper spatial thresholding must be implemented on the binary SLM. The optimum spatially varying threshold level (in terms of the correlation peak signal-to-noise ratio) is given by $V_T(x, y) = |A(x, y)|^2 + |B(x, y)|^2$ [54]. The binary SLM output is 0 for inputs less than V_T and 1 otherwise. The computer simulation results in FIGURE 13.16 compare the binary NJTOC with linear correlation and phase-only filter correlation. The signal-to-noise ratio of the binary NJTOC is the highest. However, this comes at the price of having to determine $V_T(x, y)$ for each reference and input scene.

Experimental demonstrations of the effects of nonlinearities in the Fourier plane were recently described [55]. The setup of FIGURE 13.14 was used with an LCLV SLM in the Fourier plane P2. The nonlinearity of the LCLV was adjusted by varying the frequency of the drive voltage. Plots of the experimental results for the correlation of a military tank in a cluttered background are shown in FIGURE 13.17. As the LCLV nonlinearity was increased, the signal-to-noise ratio (SNR) of the correlation peak improved. The improved performance also has a price, however, since the robustness of the NJTOC to input object distortions is decreased [56]. In other words, the ratio of the correlation peak to its sidelobes decreases more rapidly than for the linear matched filter if the input is slightly different from the original recorded image.

The sensitivity of matched filters to small distortions of the input image limits their usefulness in many pattern recognition tasks, despite the high SNR they can achieve for undistorted inputs. This has motivated a number of researchers to design composite filter correlators which trade-off SNR for robustness [57]. As the name implies, these filters consist of N superimposed subfilters. The superposition may be a weighted linear sum, or a nonlinear function of the subfilters. The weights or parameters in the superposition are determined using various constraints,

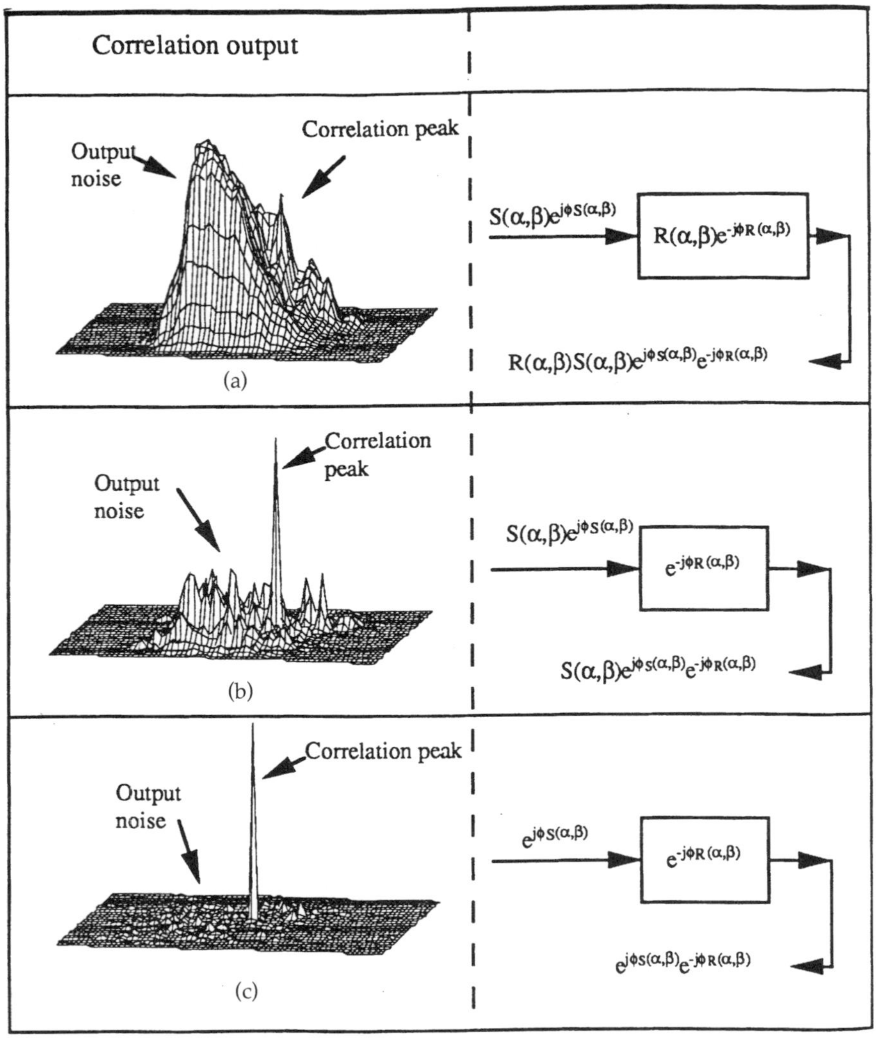

FIGURE 13.16. Demonstration of effects of nonlinearity in filter plane on JTOC performance using computer simulations. (a) Linear correlation. (b) Phase-only filter correlation. (c) Very nonlinear (binary) JTOC. The input image was a military tank in a noisy background (from reference [52]).

depending on the type of composite filter. A training set of *N* images, which represents the span of distortions of the input image that may be expected, is used in the construction of the final composite filter. Except for the linear sum variety, optical composite filters cannot be recorded optically but must be calculated on a digital computer first and then loaded into an electronically addressed SLM in the filter plane. In general, the SLM must be capable of complex modulation of light, although the limitations of the SLM can be incorporated in the constrained optimization calculations of the composite filter.

An early composite filter design was the basic synthetic discriminant function (SDF) approach [58]. SDFs consist of a linear weighted sum of subfilters. The weights are determined by forcing the correlation value at the origin to be large for images from the training set and small for images from the "false set," which the SDF filter is to discriminate against. Although initial results were encouraging in that robustness was improved, the basic SDF has several limitations. The optimization procedure takes into account correlation values at the origin only, thus off-origin values may be larger than at the origin.

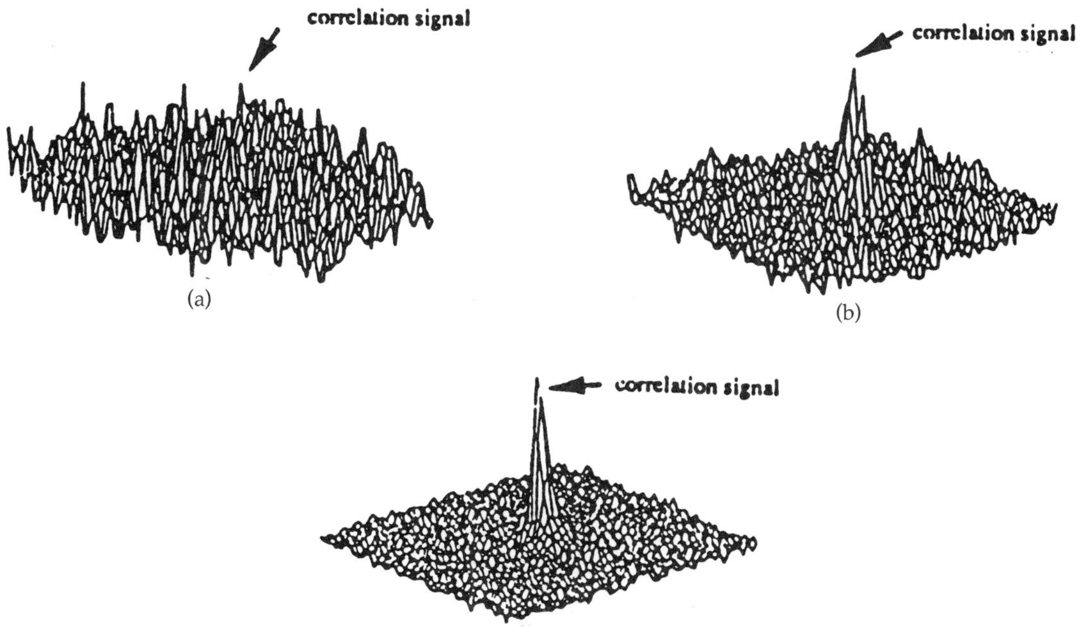

FIGURE 13.17. Experimental results for optical nonlinear JTOC with varying degrees of nonlinearity. The nonlinearity of the LCLV SLM in the filter plane was increased in going from *a* to *b* to *c* (from reference [52]).

In other words, the output is not constrained to be a well-defined correlation peak. Therefore, it is often difficult to unambiguously recognize shifted objects. In addition, the SDF algorithm does not take into account noise in the input scene. The performance rapidly degrades in the presense of noise. Finally, the form of the basic SDF is a linear weighted sum of subfilters, which greatly limits the available degrees-of-freedom in the optimization procedure.

The minimum variance synthetic discriminant function (MVSDF) filter was developed to address the last two limitations of the SDF [59]. It assumes the input can be described by a deterministic portion plus a random portion which represents the noise. The weighted sum constraint on the superposition of subfilters is also removed. The MVSDF is calculated by assuming a statistical model for the noise component and minimizing the variance of the correlation output, resulting in a composite filter which performs better for noisy inputs. (The trickiest part of the computation is inverting the noise covariance matrix.) A complementary composite filter is the minimum average correlation energy (MACE) filter, which emphasizes well-defined correlation peaks [60]. MACE filters are similar to SDFs with the added constraint that the total energy in the correlation plane is minimized, which implies that sidelobe levels are reduced relative to the correlation peak. MACE filters tend to have well-formed correlation peaks, but resistance to the effects of noise is not built into them. MVSDF and MACE filters are complementary in that the former maximizes performance in noise while the latter maximizes the quality of the correlation peak. Recently, efforts have been made to combine the properties of these two filters in a single one, which has good quality correlation peaks and performs well in noise [61].

If the above composite correlation filters are to be implemented optically, a filter plane SLM would have to be capable of implementing arbitrary complex filter values. The phase and amplitude of the SLM transmittance would have to be independently controllable. Liquid crystal based SLMs do not appear to be capable of representing arbitrary complex values since the phase and amplitude responses are linked. However, it may be possible to use two

SLMs in series for complex modulation. A more practical approach would be to incorporate the complex response curve of the SLM in the composite filter algorithm.

13.4.2 Signal Processing

13.4.2.1 Spectrum Analysis

The folded-raster optical processor is an architecture which takes advantage of the large space–bandwidth product (SBWP) of 2D optical systems to calculate the spectrum of high information content 1D signals. The basic idea is to reformat the signal as a series of lines on a 2D SLM and then Fourier transform using a lens, as described in section 13.4.1.1. As described by Thomas [62], this allows signals with large time–bandwidth products (TBWP), for example, those with high information content, to be analyzed with a high degree of frequency resolution. If a 1D signal is represented by N raster-scans across a 2D SLM of width a with center-to-center separation b between scan lines, then the optical Fourier transform of the 2D pattern formed using the setup of FIGURE 13.4 is a pattern consisting of N scan lines rotated 90° relative to the input, as illustrated in FIGURE 13.18. Whereas the scan lines of the input are almost horizontal, the scan lines of the output Fourier transform are almost vertical, with a coarse frequency resolution of $\lambda f/a$ in the horizontal direction and a fine frequency resolution of $\lambda f/b$ in the vertical direction. The operation of the folded-raster processor is best understood if we consider an input consisting of a continuous sine wave. The output then consists of a single spot (ignoring other orders, which are also present but spatially separated). As the frequency is increased, the spot will travel along one of the nearly vertical scan lines until it reaches the end, at which point it will jump to the other end of the next scan line and so on until the maximum bandwidth of $N\lambda f/a$ is reached. The duration T of the signal is given by $T = Na/v$, where v is the scanning rate.

A number of researchers have implemented such a system [63,64]. The system of Anderson and co-workers is shown in FIGURE 13.19. It utilizes an acoustooptic cell (AO cell) to both modulate an optical

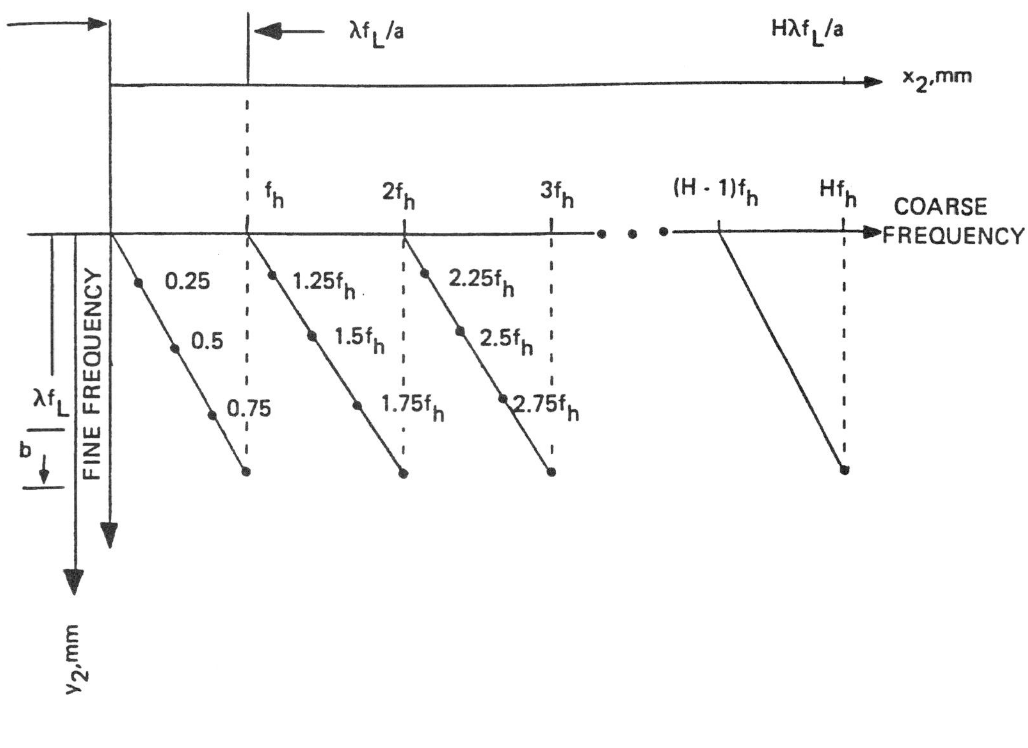

FIGURE 13.18. Frequency plane output of folded-raster optical spectrum analyzer (D. Casasent, in: *Acousto-optic Signal Processing: Theory and Implementation*, edited by N. Berg and J. Lee, New York, 1983, p. 325).

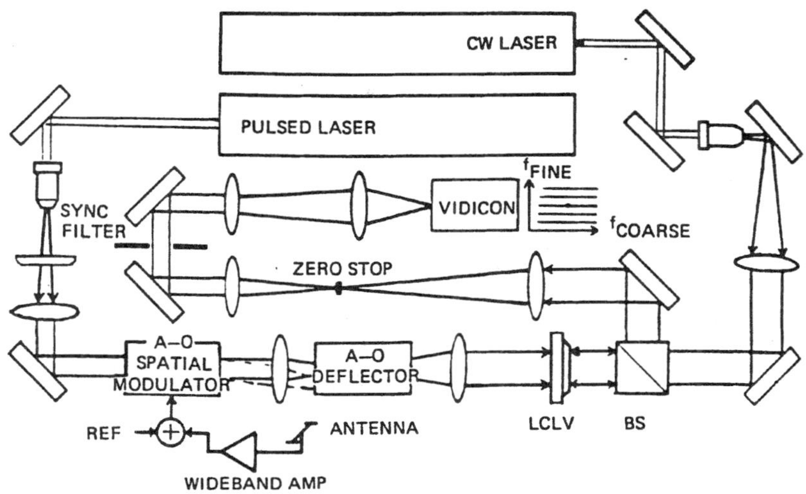

FIGURE 13.19. Optical setup for fold-raster optical spectrum analyzer (from reference [64]).

beam with the input signal and scan it across the input face of a liquid crystal light valve (LCLV). (AO cells are a type of 1D SLM which use sound waves to diffract light.) The LCLV integrated the scan lines and spatially modulated a second laser beam, which was then Fourier transformed and detected on a vidicon camera. The input bandwidth of this system was 150 MHz, limited by the particular AO cells used, but bandwidths up to 1 GHz are feasible using presently available components. A total of 256 raster lines could be recorded on the LCLV, each containing 5 μs of data with 150 MHz bandwidth; thus the total signal duration was $T = 1.28$ ms. However, since the response time of the LCLV was 100 ms, approximately 78 frames of input data were integrated for each output frame. The frequency resolution was 780 Hz for a system TBWP = 192,000. A sample output frame from the system is shown in FIGURE 13.20 in 3D format. It shows the 2D 150 MHz bandwidth folded-raster spectrum of RF signals produced in real-time.

Although the folded-raster spectrum analyzer is suitable for applications requiring fine frequency resolution of long duration signals, it is less successful at detecting transient signals because of the limited frame times of most 2D SLMs. The LCLV, for example, has a minimum response time of a few milliseconds. The SNR for short-duration signals is degraded because of the reduced processing gain. In addition, signals whose durations are less than the frame time cannot be separated in time, due to the time integration of the SLM.

13.4.2.2 Phase Array Antenna Control

An important area of signal processing is the control of phased-array antennas for both receiving and transmitting radar signals. Unlike conventional radars that steer the radar beam by mechanically rotating the antenna, phased-array radars steer the beam electronically by adjusting the phases and amplitudes of the radar waves emitted from a set of antenna elements [65]. The antenna itself is fixed and does not move. The problem of calculating and forming the phase and amplitude distribution necessary to achieve a desired beam pattern in the far field is known as the beam forming problem [66]. It is often desired to make the far-field pattern obey certain constraints, for example, directivity of the beam is usually an important parameter. Multiple beams along specified directions may need to be generated and modified in real-time. In the receive mode, adaptive nulling of jamming signals coming from one or more directions may be required. The ability to independently control the complex weighting factors of the antenna elements permits such flexibility in the antenna gain profile. In general, calculating the factors is an ill-defined inverse problem which is computationally intensive. High-throughput optical processing methods have been proposed both for calculating the complex weighting factors for the antenna elements [67,68] and for driving the elements with high-bandwidth radar signals multiplied by the complex weighting factors [69]. In this section, we will describe an optical architecture which utilizes liquid crystal phase shifters for performing the latter task. Although

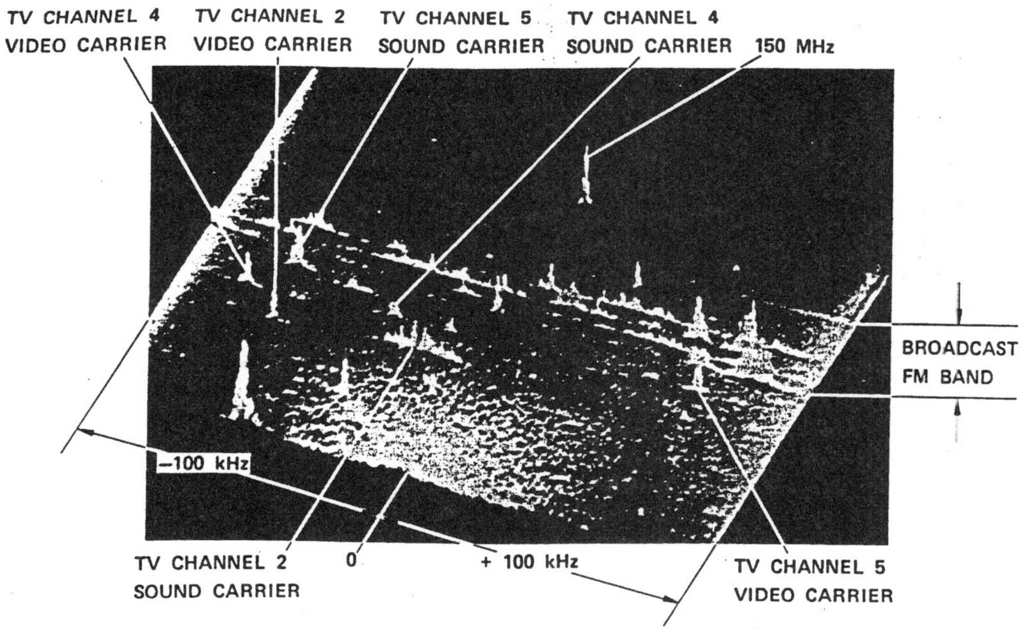

FIGURE 13.20. Experimental optical output of the spectrum analyzer of FIGURE 13.19 displayed in 3D format (from reference [64]).

this technology is discussed here in the context of radar, it is also relevant to applications in communications.

Riza has described an optical beam forming system that uses liquid crystal displays (LCD) to control the phase and amplitude of optical signals which are modulated by the radar signal [70]. A system diagram is shown in FIGURE 13.21. In the transmit mode of the radar, an RF (radio-frequency) beat signal is generated and used to amplitude-modulate an optical beam. LCDs are used to modify the phase and amplitude of the beam in a spatially varying manner. The modified

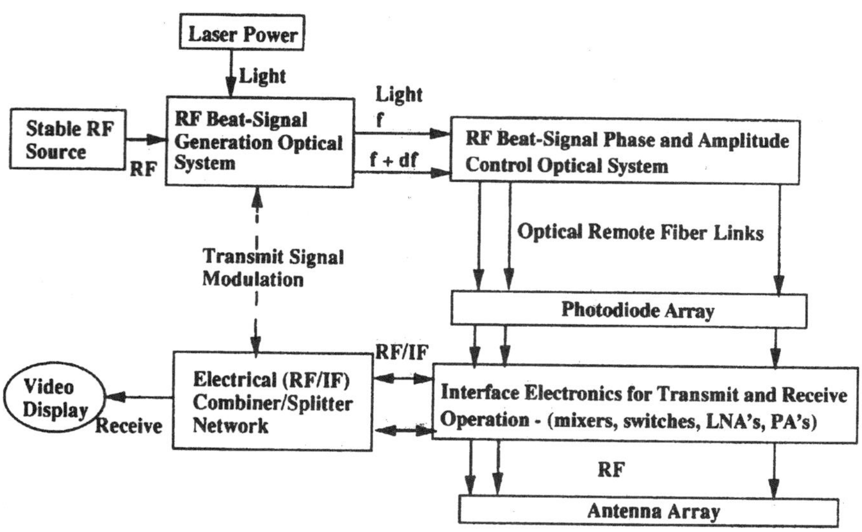

FIGURE 13.21. System diagram for phased-array antenna beam-forming using liquid crystal optical modulators (from reference [70]).

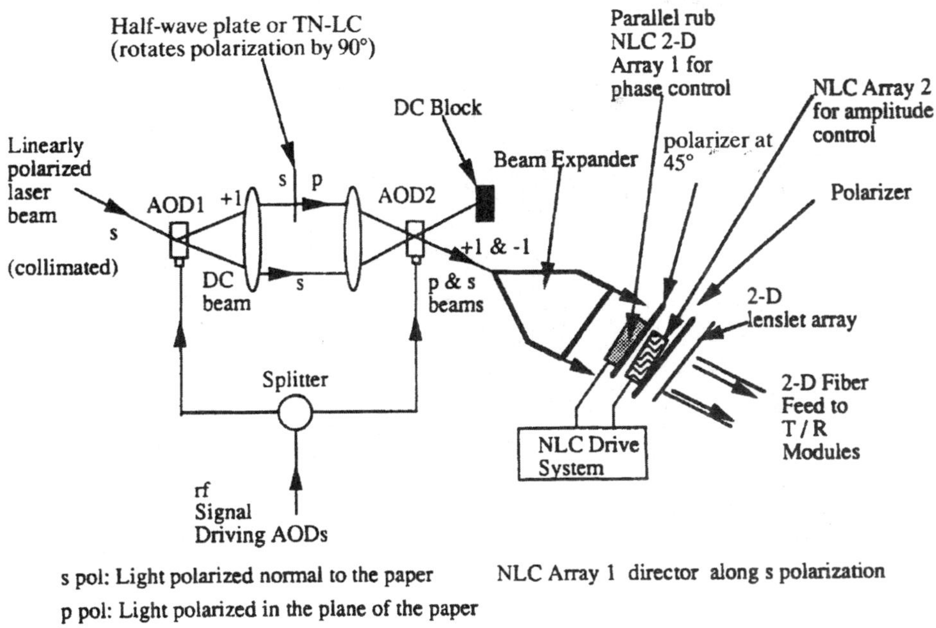

FIGURE 13.22. Optical architecture of RF beat-signal generation and phase-amplitude control subsystems of the beam-forming system of FIGURE 13.21 (from reference [70]).

beam is spatially sampled by a set of optical fibers that acts as a remote link to the antenna array. Each individual fiber transports light weighted by an individual complex factor to a photodiode at the antenna. The photodiodes convert the optical modulation into modulation of the electronic signals that drive the antenna elements. In this way, the optical phase and amplitude variations displayed on the LCDs are carried over to the electronic drive signals. This process can also operate in reverse for beam forming in the receive mode.

A detailed diagram of the RF beat signal generation and phase and amplitude control optical subsystems is shown in FIGURE 13.22. The RF beat signal generator consists of two acousooptic devices (AOD1 and AOD2) arranged as an in-line interferometer. The AODs generate acoustic waves in a transparent crystal in response to transducers which are modulated by RF signals. Light diffracted from the waves undergoes a frequency shift equal to the RF frequency due to the Doppler effect. In this case, AOD1 and AOD2 are driven by the same 70 MHz RF signal. A linearly polarized (s-polarized) laser beam is incident on AOD1 at the Bragg angle for scattering from the acoustic waves. The scattered first-order beam is frequency shifted by 70 MHz. Its polarization is changed to p by a half-wave plate. The s-polarized DC component from AOD1, meanwhile, scatters from AOD2 in such a way that its s-polarized −1 order beam (which is frequency shifted by −70 MHz) is collinear with the +1 order p-polarized beam. Thus the output of the interferometer is two collinear beams with orthogonal polarizations; one is p-polarized and frequency shifted by 70 MHz, while the other is s-polarized and frequency shifted by −70 MHz. These two beams then enter the phase/amplitude control optical subsystem after being expanded and recollimated. The next component is an electronically controlled 2D LCD with a parallel-aligned liquid crystal layer. The p-polarized beam sees the ordinary index of refraction of the liquid crystal, which stays constant. The s-polarized beam undergoes a spatially varying phase shift, which is a function of the individual LCD pixel values. A 45° polarizer converts both beams to the same 45° polarization so that they can interfere. A second electronically controlled 2D LCD between crossed polarizers acts as a spatially varying amplitude modulator. The resultant optical field is then incident on a 2D array of fibers, through which the phase and amplitude

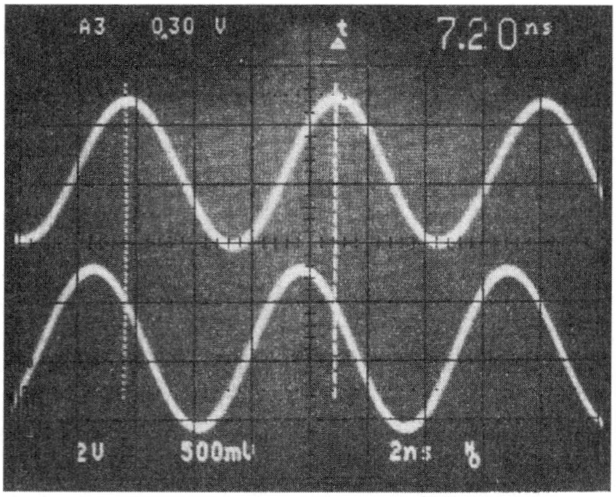

(a)

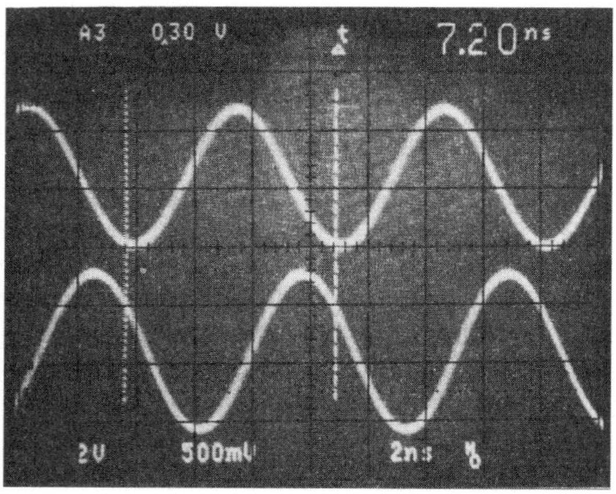

(b)

FIGURE 13.23. Experimental demonstration of controlled phase shifts of 140 MHz RF carrier using the optical system of FIGURE 13.22 (from reference [70]).

modulated light is transported to the photodiode array, which is in turn connected to the antenna elements via the interface electronics.

The optical signal incident on the nth fiber can be written as

$$E(n,t) \propto a_n \exp(-j\omega_c t) + a_n \exp(+j\omega_c t)\exp(-j\phi_n) \qquad (13.19)$$

where the first term results from the +1 order beam from AOD1 while the second term represents the −1 order beam from AOD2. ϕ_n is the relative phase shift induced by LCD 1 while a_n is the amplitude modulation due to LCD 2. The two beams are frequency shifted by ω_c and $-\omega_c$. The output current of the nth square-law-detector photodiode is given by

$$i_n(t) \propto |E(n,t)|^2 \propto A_n + A_n \cos(2\omega_c t - \phi_n) \qquad (13.20)$$

where the first term is a DC or bias level, the second term is the desired 140 MHz RF signal with independent phase and amplitude control for each antenna element, and $A_n = |a_n|^2$. The DC term can be blocked using a high-pass filter. In this way the amplitude, phase, and carrier frequency of signals driving a phased array antenna can be optically controlled. Some experimental results showing phase and amplitude modulated RF signals using this method are shown in FIGURE 13.23.

13.4.3 Neural Networks

Computational problems can be interpreted in terms of many attributes. A fundamental measure is the information entropy or algorithmic complexity. At one extreme are highly structured problems with known solutions, which are best solved using conventional procedural algorithms. On the other hand, many problems (especially those which involve real-world data) have a high degree of randomness or entropy, especially in pattern recognition problems involving noisy data. These are problems at which biological organisms excel compared with procedural operator-programmed computers. Artificial neural network models are useful for solving such high-entropy problems for which no explicit algorithm is known, and the required transformation must be learned from examples. It has been shown that any nonlinear transformation can be represented by a neural network [71]. Although the problem of learning an arbitrary transformation for a completely random problem has been shown by Judd [72] to be NP-complete (therefore probably requiring exponential increases in learning time as the problem size increases), most, if not all, problems in pattern recognition involve at least partially structured data for which the learning time will be a polynomial function of the problem size.

Neural network models of computation consist of many simple processing nodes or "neurons," usually arranged in layers, which communicate with each other via interconnection weights. (This nodes are conventionally called "neurons" although any anthropomorphic similarity to the complexity of real brain

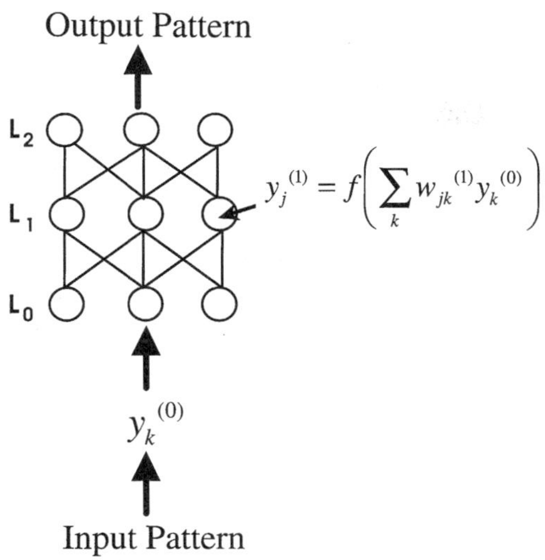

FIGURE 13.24. Structure of feed-forward multilayer neural network. The function $f()$ is a sigmoidal nonlinearity.

neurons is very remote.) A diagram of the logical structure of a multilayer feed-forward neural network is shown in FIGURE 13.24. Patterns which are input to the network through the bottom layer are transformed into top layer output patterns. Each neuron calculates its own activation level by forming a weighted sum of the activity levels of other neurons and passing it through a sigmoidal nonlinearity. This operation is essentially a matrix–vector multiplication followed by a lookup table transformation. Many neural net architectures have been designed and demonstrated for various computing tasks. Techniques such as back-propagation, as well as many others, have been developed to "train" or adjust the interconnection weights to solve problems in pattern recognition, vision, and robotic control. The ultimate capabilities and limitations of neural networks, including problem representation, scaling of network complexity and training time with problem size, and the accuracy required of the interconnection weights, are topics of active research.

Two important parameters which characterize a neural network are the number of neurons, N, and their connectivity, K, where K is the number of synapses or weights connected to a neuron. The computational load rapidly becomes very high as N and K increase, and general-purpose computers are not capable of taking advantage of the inherent parallelism of the neural network paradigm. Thus special-purpose parallel processors (neurocomputers) are often utilized for large neural network problems. Pattern recognition problems involving natural data, especially vision or target recognition, require large numbers of input neurons to handle the large raw input data rate. In addition, large numbers of neurons in the hidden layer are required in order to solve nontrivial problems. As for the connectivity, Abu-Mostafa [73] has shown that it must be large for at least two reasons. First, since the neurons essentially implement K-input threshold functions, one K-input neuron with K associated weights is equivalent to order K^2 two-input neurons with $2K^2$ associated weights. Thus far fewer weights are required if the neurons have high fan-in and fan-out. Second, Abu-Mostafa also showed that in order for a neural network to learn, the connectivity K must exceed the entropy H of the environment where H is the $\log_2$ of the number of input patterns typically generated by the environment. H increases as the randomness of the problem increases. Since neural networks are most useful for partially random problems, we argue that both the number of neurons and their connectivity should be as large as possible in a neurocomputer.

Increasing N and K while maintaining computational parallelism is a daunting task for electronic architectures due to the 2D nature of electronic interconnects. Most of the area on analog or digital electronic neurochips is taken up by the interconnects, while implementing only modest numbers of neurons ($N = O(10^2)$). Optical holographic architectures, on the other hand, can exploit 3D free space interconnects, allowing the input and output planes to be fully populated with highly interconnected neurons ($N = O(10^5)$). Moreover, the weights can be updated in one time step, unlike electronic approaches, which must use some form of time multiplexing when the number of neurons is large [74].

Optical neural networks divide naturally into two classes distinguished by the dimensionality of the weight storage medium. Weights can be stored in 2D or 3D formats. Example 2D weight storage media include film, SLMs, and optical disks while almost all of the 3D formats use photorefractive crystals as the storage medium. SLMs are used as optical input devices in virtually all optical neural networks.

The process of calculating the input to a layer of neurons consists of forming a set of weighted sums. If

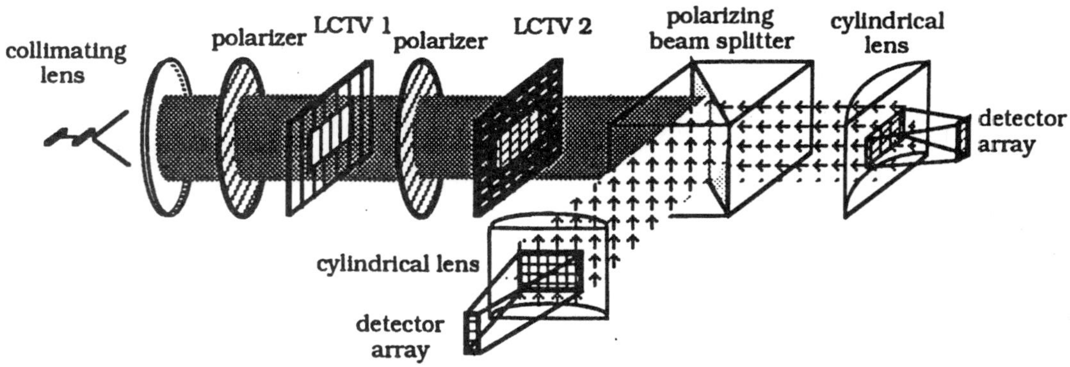

FIGURE 13.25. Architecture for optical neural network based on 2D storage of weights on an SLM (from reference [76]).

the activity values of a layer of neurons is thought of as a vector, then the input to the next layer is formed by calculating a matrix–vector product where the matrix is the set of weights connecting the two layers. A straightforward adaptation of the incoherent optical matrix–vector multiplier [75] can then be used to form a 2D implementation of an optical neural network. The system of FIGURE 13.25 uses line-shaped pixels on an input LCD to represent input neuron values [76,77]. Light from this LCD illuminates a second LCD on which a 2D matrix of weights is encoded using polarization multiplexing to represent bipolar weights. A cylindrical lens then focuses the light onto a 1D detector array, forming an optical matrix–vector product. Both nematic and ferroelectric liquid crystals are used to store the weights. Such systems have been successfully trained using the least-mean-square-error algorithm to perform pattern associations.

A limitation of this type of optical neural network is that only two dimensions are used to store the weights, therefore the size of neural network that can be implemented is limited. Also, the weights need to be loaded and/or updated electronically during the learning phase. For applications in which learning rates are important, architectures which incorporate optical updating of weight values are preferable. One approach with this feature is real-time holographic storage of weights. The three-dimensional storage feature of volume holography also allows very large neural networks to be implemented. Optical neural networks based on holography have been demonstrated by several groups [78–81]. Neural networks with up to 10^8 weights and throughputs of 10^{10} weights per second have been demonstrated in applications such as face and character recognition. Training of large neural networks using supervised learning can be problematic since the amount of required labeled training data (which in the worst case may be proportional to the number of weights [82]) may be too large to be practical. Unsupervised self-organizing optical neural networks which automatically extract features and organize the data may be one approach to dealing with the training set size problem [83,84]. A large unsupervised optical neural network may reduce the dimensionality of the problem space to a level where supervised training with practical training set sizes may be possible.

13.5 SUMMARY

In this chapter, I have attempted to illustrate the range of applications for liquid crystal based devices in optical image and signal processing. Spatial light modulators (SLMs), which are used to impart information onto optical beams using the optical modulation properties of liquid crystals (and also other materials), are key components of optical data processing systems. Two-dimensional optically and electronically addressed SLMs allow the natural parallelism of optics to be exploited for computing. The huge investment that has been made in developing high-performance liquid crystal displays for multimedia, virtual-reality, and high-definition television applications has resulted in technologies which are applicable to optical processing, although modifications need to be made due to the unique requirements of optical processing. Other optical processing components are also benefiting from research being performed for other applications, such as compact lasers for optical storage and high-performance detectors for machine vision.

In order for optical parallel processors to approach their potential computational throughputs, serial bottlenecks, both internally and in the I/O interfaces, must be avoided. Fan-out from the input SLM and fan-in to the output detector should be large and the optical interconnects should be global since locally connected networks (such as mesh and hypercube networks) can be implemented in competing electronic parallel processors. Optics provides a means for practical realization of large cross-bar switch interconnections. These principles result in a large number of optical operations performed per electronic operation, reduce the required electronic bandwidth on the input/output lines, and make it more likely that the optical approach will be superior to a purely electronic solution. The optical systems discussed in this chapter embody these principles. The ongoing progress in liquid crystal based SLMs and other component technologies, driven in part by applications in displays and multimedia, may enable future optical processors to compete with the programmability of digital electronic processors and expand beyond their current position as high-performance processors for a limited number of specialized markets.

REFERENCES

1. See special issue of *Applied Optics* (May 1994) on spatial light modulators and their applications.
2. See special issue of *IEEE Journal of Quantum Electronics* (February 1993) on smart pixels.
3. Vergis, A., Steiglitz, K., and Dickinson, B. 1986. The complexity of analog computation. *Math. Comput. Simul.* **28**: 91.
4. Morelli, M. V., Krile, T. F., and Walkup, J. F. 1994. Temporal intensity noise characteristics and discrete numeric accuracy of analog liquid-crystal–based spatial light modulators. *Appl. Opt.* **33**: 2812.
5. Guilfoyle, P. S., Stone, R. V., and Zeise, F. F. 1994. Digital optical interconnects for photonic computing. *Optoelectronic Interconnects*, vol. 2, ed. R. T. Chen, p. 37. Proc. SPIE, vol. 2153. Bellingham, Wash.: SPIE.
6. Efron, U., Grinberg, J., Braatz, P. O., Little, M. J., Reif, P. G., and Schwartz, R. N. 1985. The silicon liquid-crystal light valve. *J. Appl. Phys.* **57**: 1356.
7. Sterling, R. D., Te Kolste, R. D., Haggerty, J. M., Borah, T. C., and Bleha, W. P. 1990. Video-rate liquid-crystal light valve using an amorphous silicon photoconductor. *SID Dig. Tech. Papers* **21**: 327.
8. Clark, N. A., and Johnson, K. M. 1992. Applications of liquid crystals in optical computing. *Liquid Crystals: Applications and Uses*, vol. 3, ed. B. Bahadur, p. 209. Singapore: World Scientific.
9. Efron, U., Byles, W. R., Goodwin, N. W., Forber, R. A., Sayyah, K., Wu, C. S., and Welkowsky, M. S. 1991. The charge-coupled device-addressed liquid crystal light valve: an update. *Proc. SPIE* **1455**: 237.
10. Efron, U., Byles, W. R., Au, A., Goodwin, N. W., Wu, C. S., Sayyah, K., Wu, S.-T., Forber, R. A., Little, M. J., Esener, S., Marchand, P., Krishnamoorthy, A., and Javidi, B. 1994. Recent progress in optical coprocessor devices. *Euro-American Workshop on Optical Pattern Recognition [La Rochelle, France]*, ed. P. Réfregier and B. Javidi, p. 399. Bellingham, Wash.: SPIE.
11. Johnson, K. M., McKnight, D. J., and Underwood, I. 1993. Smart spatial light modulators using liquid crystals on silicon. *IEEE J. Quantum Electron.* **QE-29**: 699.
12. Dijon, J. 1990. *Liquid Crystals: Applications and Uses*, vol. 1, ed. B. Bahadur, p. 305. Singapore: World Scientific.
13. McKnight, D. J., Johnson, K. M., and Serati, R. A. 1994. 256 × 256 liquid-crystal-on-silicon spatial light modulator. *Appl. Opt.* **33**: 2775.
14. Underwood, I., Vass, D. G., O'Hara, A., Burns, D. C., McOwan, P. W., and Gourlay, J. 1994. Improving the performance of liquid-crystal-over-silicon spatial light modulators: issues and achievements. *Appl. Opt.* **33**: 2768.
15. Handschy, M. A., Cotter, L. K., Cunningham, J. D., Drabik, T., and Gaalema, S. D. 1993. One-transistor DRAM FLC/VLSI SLM. *Spatial Light Modulators and Applications*, p. 14. OSA Technical Digest Series, vol. 6. Washington, D.C.: Optical Society of America.
16. Liu, H.-K., and Chao, T.-H. 1989. Liquid crystal television spatial light modulators. *Appl. Opt.* **28**: 4774.
17. Soutar, C., Monroe, S. E., and Knopp, J. 1993. Complex characterization of the Epson liquid crystal television. *Optical Pattern Recognition*, vol. 4, p. 269. Proc. SPIE, vol. 1959. Bellingham, Wash.: SPIE.
18. Zhang, Z., Lu, G., and Yu, F. T. S. 1994. Simple method for measuring phase modulation in liquid crystal televisions. *Opt. Eng.* **33**: 3018.
19. Morelli, M. V., Krile, T. F., and Walkup, J. F. 1994. Temporal intensity noise characteristics and discrete numeric accuracy of analog liquid-crystal–based spatial light modulators. *Appl. Opt.* **33**: 2812.
20. Abbe, E. 1873. *Archiv. Mikroskopische Anat.* **9**: 413.
21. Goodman, J. W. 1968. *Introduction to Fourier Optics*, p. 86. New York: McGraw-Hill.

22. Wigner, E. 1932. On the quantum correction for thermodynamic equilibrium. *Phys. Rev.* **40**: 749.
23. Cohen, L. 1966. Generalized phase-space distribution functions. *J. Math. Phys.* **7**: 781.
24. Claasen, T. A. C. M., and Mecklenbrauker, W. F. G. 1980. The Wigner distribution: a tool for time-frequency analysis. *Philips J. Res.* **35**: 217.
25. Hlawatsch, F., and Boudreaux-Bartels, G. F. 1992. Linear and quadratic time-frequency signal representations. *IEEE Sig. Proc. Mag.* **April**: 21.
26. Cristobal, G., Bescos, J., and Santamaria, J. 1989. Image analysis through the Wigner distribution function. *Appl. Opt.* **28**: 262.
27. Gonzalo, C., Bescos, J., and Berriel-Valdos, L. R. 1988. Recovery of space-variant filtered images through the Wigner distribution function by a hybrid optical-digital processor. *Proc. SPIE* **963**: 425.
28. Subotic, N., and Saleh, B. E. A. 1984. Generation of the Wigner distribution function of two-dimensional signals by a parallel optical processor. *Opt. Lett.* **9**: 471.
29. Li, Y., Eichmann, G., and Conner, M. 1988. Optical Wigner distribution and ambiguity function for complex signals and images. *Opt. Commun.* **67**: 177.
30. Iwai, T., Gupta, A. K., Asakura, T. 1986. Simultaneous optical production of the sectional Wigner distribution function for a two-dimensional object. *Opt. Commun.* **58**: 15.
31. Rioul, O., and Vetterli, M. 1991. Wavelets and signal processing. *IEEE Sig. Proc. Mag.* **October**: 14.
32. Daubechies, I. 1988. Orthonormal bases of compactly supported wavelets. *Commun. Pure Appl. Math.* **41**: 909.
33. Li, Y., and Zhang, Y. 1992. Coherent optical processing of Gabor and wavelet expansions of one- and two-dimensional signals. *Opt. Eng.* **31**: 1865.
34. Casasent, D. P., Smokelin, J. S., and Ye, A. 1992. Wavelet and Gabor transforms for detection. *Opt. Eng.* **31**: 1893.
35. Arneodo, A., Argoul, F., Muzy, J. F., Pouligny, B., and Freysz, E. 1992. The optical wavelet transform. *Wavelets and Their Applications*, ed. M. B. Ruskai et al., p. 241. Jones and Bartlett Publishers, Boston.
36. Mandelbrot, B. B. 1982. *The Fractal Geometry of Nature.* San Francisco: Freeman.
37. Freysz, E., Pouligny, B., Argoul, F., and Arneodo, A. 1990. *Phys. Rev. Lett.* **64**: 745.
38. Ghez, J. M., and Vaienti, S. 1989. *J. Stat. Phys.* **57**: 415.
39. Whalen, A. D. 1971. *Detection of Signals in Noise.* New York: Academic.
40. Kirkpatrick, S., Gelatt, C. D., and Vecchi, M. P. 1983. Optimization by simulated annealing. *Science* **220**: 671.
41. Peterson, C., and Anderson, J. R. 1987. A mean field theory learning algorithm for neural networks. *Complex Syst.* **1**: 995.
42. Gluckstad, J., and Jorgensen, T. M. 1994. Optoelectronic loop that implements a mean field annealing algorithm for nonlinear noise filtering. *Opt. Eng.* **33**: 1206.
43. Gabor, D. 1948. A new microscopic principle. *Nature* **161**: 777.
44. Leith, E. N., and Upatnieks, J. 1962. Reconstructed wavefronts and communication theory. *J. Opt. Soc. Am.* **52**: 1123.
45. Van der Lugt, A. B. 1964. Signal detection by complex spatial filtering. *IEEE Trans. Inform. Theory* **IT-10**: 2.
46. Casasent, D., and Psaltis, D. 1976. Position, rotation, and scale invariant optical correlation. *Appl. Opt.* **15**: 1795.
47. Casasent, D., and Richards. J. 1988. Optical Hough and Fourier processors for product inspection. *Opt. Eng.* **27**: 258.
48. Hsu, Y. N., and Arsenault, H. H. 1982. Optical pattern recognition using circular harmonic expansion. *Appl. Opt.* **21**: 4016.
49. Ochoa, E., Schils, G. F., and Sweeney, D. W. 1988. Detection of multiple views of an object in the presence of clutter. *Opt. Eng.* **27**: 266.
50. Horner, J. L., and Gianino, P. D. 1984. Phase-only matched filtering. *Appl. Opt.* **23**: 812.
51. Rajbenbach, H. 1991. Demonstration: a photorefractive correlator for robotic applications. *Opt. Photon. News* **December**: 14.
52. Javidi, B. 1994. Nonlinear joint transform correlators. *Real-Time Optical Information Processing*, ed. B. Javidi and J. L. Horner, p. 115. New York: Academic.
53. Javidi, B. 1989. Nonlinear joint power spectrum based optical correlation. *J. Appl. Opt.* **28**: 2358.
54. Javidi, B., Wang, J., and Tang, Q. 1991. Multiple objects binary joint transform correlation using multiple level thresholding crossing. *Appl. Opt.* **30**: 4234.
55. Javidi, B., Tang, Q., Gregory, D. A., and Huson, T. D. 1991. Experiments on nonlinear JTCs using an optically addressed SLM in the Fourier plane. *Appl. Opt.* **30**: 1772.
56. Tang, Q., and Javidi, B. 1992. Sensitivity of the nonlinear joint transform correlator: experimental investigation. *Appl. Opt.* **31**: 4016.
57. Vijaya Kumar, B. V. K. 1992. Tutorial survey of composite filter designs for optical correlators. *Appl. Opt.* **31**: 4773.
58. Hester, C. F., and Casasent, D. 1980. Multivariant technique for multiclass pattern recognition. *Appl. Opt.* **19**: 1758.
59. Vijaya Kumar, B. V. K. 1986. Minimum variance synthetic discrimination functions. *J. Opt. Soc. Am. A* **3**: 1579.
60. Mahalanobis, A., Vijaya Kumar, B. V. K., and Casasent, D. 1987. Minimum average correlation energy filters. *Appl. Opt.* **26**: 3633.
61. Ravichandran, G., and Casasent, D. 1992. Minimum noise and correlation energy (MINACE) optical correlation filter. *Appl. Opt.* **31**: 1823.

62. Thomas, C. E. 1966. Optical spectrum analysis of large space bandwidth signals. *Appl. Opt.* **5**: 1782.
63. Soffer, B. H. 1984. Liquid crystal optical processing. *Solid State Devices 1983: Ten Invited Papers Presented at the Thirteenth European Solid State Device Research Conference (ESSDERC) and the Eighth Symposium on Solid State Device Technology (SSSDT) [Canterbury 1983]*, ed. E. H. Rhoderick, p. 121. Institute of Physics Conference Series, no. 69. Bristol: Institute of Physics.
64. Anderson, J. R., Brown, H., and Markevich, B. 1979. Wideband real-time Fourier analyzer using folded spectrum techniques. *Soc. Photo-Opt. Instru. Eng.* **180**: 128.
65. Knittel, G. H. 1977. Phased-array antennas: an overview. *Radar Technology*, ed. E. Brookner, p. 289. Dedham, Mass.: Artech House.
66. Van Veen, B. D., and Buckley, K. M. 1988. Beamforming: a versatile approach to spatial filtering. *IEEE ASSP Mag.* **April**: 4.
67. Koepf, G. A. 1984. Optical processor for phased array antenna beam forming. *Proc. SPIE* **477**: 75.
68. Anderson, L. P., Boldissar, F., and Chang, D. C. D. 1988. Antenna beamforming using optical processing. *Optoelectronic Signal Processing for Phased-Array Antennas*, vol. 1, p. 228. Proc. SPIE, vol. 886. Bellingham, Wash.: SPIE.
69. Riza, N. A. 1992. Liquid crystal–based optical control system for phased array radars. *IEEE Photon. Technol. Lett.* **4**: 1073.
70. Riza, N. A. 1994. Acoustooptic liquid crystal analog beam formed for phased array antennas. *Appl. Opt.* **33**: 3712.
71. Kreinovich, V. Y. 1991. Arbitrary nonlinearity is sufficient to represent all functions by neural networks: a theorem. *Neural Networks* **4**: 381.
72. Judd, S. 1987. Learning in networks is hard. *IEEE First International Conference on Neural Networks [San Diego]*, vol. 2, ed. M. Caudill and C. Butler, p. 685. Piscataway, N.J.: IEEE Service Center.
73. Abu-Mostafa, Y. 1989. Appendix D: Complexity in neural systems. *Analog VLSI and Neural Systems*, ed. C. Mead, p. 353. Reading, Mass.: Addison-Wesley.
74. Shams, S., and Przytula, K. W. 1991. Implementation of multilayer neural networks on parallel programmable digital computers. *Parallel Algorithms and Architectures for DSP Applications*, ed. M. Bayoumi, p. 225. Deventer: Kluwer.
75. Goodman, J. W., Dias, A. R., and Woody, L. M. 1978. Fully parallel, high-speed incoherent optical method for performing discrete Fourier transforms. *Opt. Lett.* **2**: 1.
76. Robinson, M. G., Johnson, K. M., and Zhang, L. 1990. Network analysis of an optically implemented connectionist architecture. *International Joint Conference on Neural Networks [San Diego]*, vol. 1, p. 603. New York: IEEE Neural Networks Council.
77. Mao, C. C., and Johnson, K. M. 1993. Optoelectronic array that computes error and weight modification for a bipolar neural network. *Appl. Opt.* **32**: 1290.
78. Owechko, Y. 1993. Cascaded-grating holography for artificial neural networks. *Appl. Opt.* **32**: 1380.
79. Owechko, Y. 1991. Optical implementation of back-propagation neural networks using cascaded-grating holography. *Int. J. Optical Comput.* **2**: 201.
80. Li, H.-Y. S., Qiao, Y., and Psaltis, D. 1993. Optical network for real-time face recognition. *Appl. Opt.* **32**: 5026.
81. Hong, J. H., Campbell, S., and Yeh, P. 1990. Optical pattern classifier with perceptron learning. *Appl. Opt.* **29**: 3019.
82. Hertz, J., Krogh, A., and Palmer, R. G. 1991. *Introduction to the Theory of Neural Computation*. Reading, Mass.: Addison-Wesley.
83. Wagner, K., and Slagle, T. M. 1993. Optical competitive learning with VLSI/liquid-crystal winner-take-all modulators. *Appl. Opt.* **32**: 1408.
84. Duvillier, J., Killinger, M., Heggarty, K., Yao, K., and de la Tocnaye, J. L. de Bougrenet. 1994. All-optical implementation of a self-organizing map: a preliminary approach. *Appl. Opt.* **33**: 258.

CHAPTER 14

Liquid Crystals for Nonlinear Optical Studies

E. SANTAMATO and Y. R. SHEN

14.1 INTRODUCTION

The advent of lasers has made light-induced changes in optical properties of a material easily observable, leading to the birth of nonlinear optics. This field has become so important that nowadays it constitutes the main part of laser science. Many analytical probes and optical devices based on harmonic generation, stimulated light scattering, wave mixing, phase conjugation, self-focusing, and others have been developed. Unfortunately, most materials have a very small optical nonlinearity, and powerful lasers are often required to observe nonlinear optical effects. Finding new materials with large optical nonlinearities is currently the most important research topic in optoelectronics [1].

Liquid crystals (LC) are interesting optical materials with unique and extraordinary properties. They consist of highly anisotropic organic molecules with delocalized electron distributions in the chromophore, and possess a number of mesomorphic phases in a readily accessible temperature range [2,3]. Moreover, they can be mixed or doped with other materials to significantly alter their macroscopic properties. In the mesomorphic phases, the orientations of molecules are so strongly correlated that the medium behaves like an anisotropic fluid. The degree of molecular alignment as well as the average molecular orientation can be changed by external perturbations such as mechanical stress, magnetic field, or electric field. Consequently, liquid crystals in the mesophases are highly birefringent materials with large electro- and magnetooptical coefficients. They have therefore attracted a great deal of attention in the past 20 years, and have found many useful applications as sensing, display, and memory devices.

Liquid crystals can also be highly nonlinear [4–8]. The nonlinearity arises from the strong electron delocalization in the molecules and the strong correlation between molecules. In the mesophases, the molecular correlation is extraordinarily strong and the optical nonlinearity is extraordinarily large that even cw laser interaction with these materials can lead to very unusual and interesting nonlinear optical effects. The effects can be so strong that perturbation theories may no longer be valid. Molecular correlation is much weaker in the isotropic phase; optical nonlinearity is then of the same order of magnitude as those of other liquids, but through pretransitional behavior, it can be greatly enhanced near the isotropic–mesomorphic transition.

In the low-intensity limit, the nonlinear optical response of a medium is generally characterized by a series expansion of the material polarization P as a function of the optical field E [8]:

$$P = \chi^{(1)}E + \chi^{(2)}E^2 + \chi^{(3)}E^3 + \cdots. \qquad (14.1)$$

Second-order nonlinear optical effects are dominated by the $\chi^{(2)}$ term, which vanishes in centrosymmetric media under the electric-dipole approximation. They include second-harmonic generation, three-wave mixing, parametric oscillation, and amplification, etc. [9]. The LC molecules are often centroasymmetric and carry conjugate bonds with delocalized electrons. The second-order molecular polarizability can therefore be very large. Unfortunately, in a bulk LC medium, antiparallel pairing of molecules is usually favored, since it lowers the intermolecular Van der Waals interaction energy. Consequently, an overall inversion symmetry normally exists in an LC bulk medium, and the second-order macroscopic nonlinearity vanishes under the electric-dipole approximation. Only in the cholesteric mesophase, where the LC molecules assume a macroscopic helical structure, can a nonzero second-order optical nonlinearity be found, but it is generally too weak for practical applications. Ferroelectric smectic-C LCs also do not have an overall inversion symmetry because of partial alignment of their permanent molecular dipoles. They also possess a nonvanishing, but usually weak, $\chi^{(2)}$ [10]. Only recently, values of $\chi^{(2)}$ comparable to those of inorganic nonlinear optical materials like $LiNbO_3$ have been found by appropriately synthesizing the molecules [11].

At an LC–solid interface, the LC–surface interaction can induce a polar ordering in the LC monolayer on the surface. This breaks the pairing of molecules and hence the inversion symmetry. As a result, $\chi^{(2)}$ of the interfacial region can be dominated by the LC monolayer, making the second-order nonlinear optical effects highly surface-specific [12]. Surface second-harmonic generation (SHG) has been observed in reflection from an LC surface and has been developed into a powerful method to study surface-induced bulk LC alignment [13], which is a problem of great importance in LC device technology. Surface-induced SHG has also been observed in polymer-dispersed LCs that are composed of micron-sized LC droplets dispersed in a polymeric matrix [14].

Third-order nonlinear optical effects characterized by the third-order nonlinear susceptibility $\chi^{(3)}$ can be very large in LCs and have been studied extensively in recent years. The electronic contribution to $\chi^{(3)}$ comes mainly from individual molecules, and is therefore of the same order of magnitude as in other liquids. Related nonlinear optical effects such as third-harmonic generation, stimulated Raman scattering, nondegenerate four-wave mixing, and many others are not expected to be more easily observable than in other liquids, and have little practical importance. Much more important are effects arising from laser-induced correlated molecular reorientation and thermal effects in the mesophases. Both can lead to large changes in the overall refractive index of the medium. In the weak-intensity limit, the induced refractive index change δn is related to the intensity I of a cw laser beam by the simple linear relation

$$\delta n = n_2 I \qquad (14.2)$$

where n_2 is a material parameter characterizing the nonlinear response. The laser-induced δn in turn imposes a phase shift

$$\delta\psi = (2\pi/\lambda)\int_0^L \delta n \, dz$$

on the laser beam as it traverses the medium, where λ is the wavelength, and L is the length of the medium. The phase shift can be measured with interferometric or polarimetric techniques. In simple isotropic media, δn is related to $\chi^{(3)}$ by $\delta n = 2\pi\chi^{(3)}|E|^2/n$, where n is the linear refractive index.

In the mesophases of LCs, the laser-induced molecular reorientation can yield a very large δn because of the large molecular anisotropy [15]. This renders observation of the resulting nonlinear optical effects possible, even with cw lasers. The induced δn

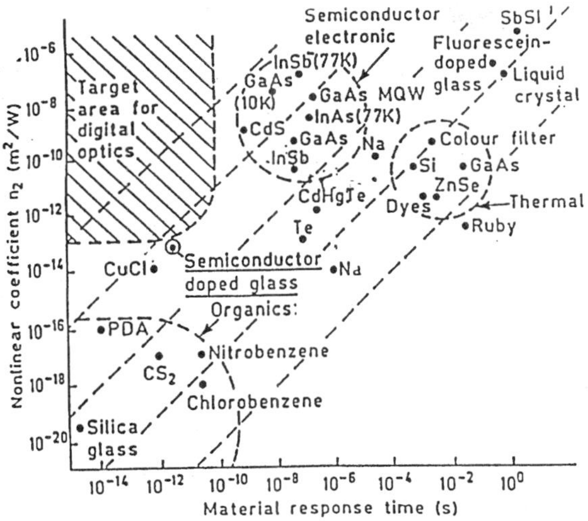

FIGURE 14.1. Nonlinear refractive index n_2 and the material response time τ for various nonlinear optical materials. (After Cotter, D. 1986 *Ultrafast Phenomena V* eds., G.R. Fleming and A.E. Siegman (Berlin: Springer) p. 274.)

can be so large that equation (14.2) is no longer valid, and the whole functional dependence of the refractive index $n(I)$ on the light intensity must be taken into account. Most of the work on nonlinear optical effects in LCs reported in the literature actually concerns effects arising from optical-field-induced refractive index changes. These include polarization ellipse rotation, self-focusing, self-phase modulation, degenerate wave mixing, phase conjugation, optical bistability, and many other effects that cannot be understood in terms of the simple Kerr relation in equation (14.2).

In many applications, one would need not only a large n_2, but also a fast response. Unfortunately, these two requirements generally cannot be met simultaneously; highly nonlinear media usually have a very long response time, as depicted in FIGURE 14.1. In this respect, LCs are not exceptional. They may have a huge n_2, but the response time is on the order of seconds. For example, in LC mesophases, n_2 is typically in the range of 10^{-3} cm^2 W^{-1}, more than 10^8 times the n_2 of a typical organic liquid as CS_2. The only materials having comparable nonlinear responses are photorefractive substances, and they also have very long response times (seconds or more). With their extraordinarily large optical nonlinearities, it is possible to observe in LCs some highly nonlinear optical processes that cannot be imagined in ordinary materials using only cw lasers in the milliwatt range. In studying the dynamics of such nonlinear optical effects, the long response time actually has the advantage of permitting study with very good time resolution.

14.2 ORIGINS OF OPTICAL NONLINEARITIES

We describe here in some detail the physical origins of optical nonlinearities in LCs. We consider first the second-order nonlinearity arising from the electronic responses of molecules to optical fields. Rod-like LC molecules generally have chromophores composed of aromatic rings with delocalized electron distributions that respond easily to external perturbations. If the molecules are asymmetric, then the second-order nonlinear polarizability can be large. Indeed, LC molecules such as cyano-biphenyls and triphenyls have been found to possess nonlinear polarizabilities orders of magnitude larger than those of ordinary molecules [16]. However, in order for an LC bulk to have a large second-order nonlinearity (described by the nonlinear susceptibility $\chi^{(2)}$), the molecules must also be polar-oriented. Unfortunately, this is not usually the case since the molecules like to form antiparallel pairs to minimize the interaction energy, resulting in a medium with inversion symmetry and hence a vanishing $\chi^{(2)}$ in the electric-dipole approximation. An exception can be found for chiral LCs in the ferroelectric smectic-C* phase [10]. The ferroelectric ordering naturally breaks the inversion symmetry and gives a nonvanishing $\chi^{(2)}$. Yet there is still the problem that the polar ordering or broken symmetry is in the direction perpendicular to both the director of molecular alignment and the smectic layer normal, and the resulting nonvanishing $\chi^{(2)}$ is often small. This is because for a rod-like molecule, the nonlinear polarizability $\alpha^{(2)}$ is only large along the long molecular axis, but the polar ordering cannot break the antiparallel pairing of molecules along the molecular alignment in the smectic-C* phase. With properly synthesized LC molecules, however, it is possible to make the components of $\alpha^{(2)}$ perpendicular to the long molecular axis large [11]. This can then yield a $\chi^{(2)}$ for a smectic-C* LC comparable to those of inorganic nonlinear optical crystals. The largest reported to date is $\chi^{(2)} \simeq 5\,\mathrm{pm\,V^{-1}}$. A recent review of nonlinear optical properties of ferroelectric LCs can be found in reference [17].

In another scheme to find large $\chi^{(2)}$ materials, LC side-chain polymers have been studied. The LC molecules hanging on the main polymer chain possess large $\alpha^{(2)}$, but again they do not have polar ordering in the bulk. To achieve polar ordering, one could use an external DC electric field to pole the polymer while it is in the fluid or glass phase, and freeze the resulting polar arrangement of LC molecules by lowering the temperature below the glass transition [18]. LCs as organic $\chi^{(2)}$ materials have been the topic of growing interest in recent years. Many reports on the recent development have appeared in conference proceedings. We refer the readers only to the latest ones [19].

The electronic response certainly also contributes to the third-order nonlinearity of LCs [20], but it is not expected to be larger than those of other organic molecules with delocalized electrons. The main contribution to the third-order nonlinearity comes from laser-induced molecular reorientation, and the effect appears in the form of a refractive index change δn proportional to the laser intensity I. Actually, even with moderate laser intensities, the effect (in the LC mesophases) can be so large that δn is no longer linear in I, but appears as a complex nonlinear function of I. As we shall discuss below, molecular correlation is the reason behind the strong reorientation effect [15]. The large δn not only makes the ordinary third-order nonlinear optical effects such as degenerate four-wave mixing and optical Kerr effect easily observable, but also leads to some very unique, highly nonlinear optical effects in LCs [4–8].

The laser field can affect the LC molecular orientation in two ways: modifying the orientational distribution (as represented by the orientatinal order parameter) and changing the direction of average orientation (specified by the director) [9]. The former is the only mechanism operating if the LC is in the isotropic liquid phase. The latter dominates in the LC mesophases.

For a uniaxial LC system, the orientational order parameter is defined as

$$Q = \int (1/2)(3\cos^2\theta - 1) f(\theta)\sin\theta\, d\theta$$
$$\equiv \langle (3\cos^2\theta - 1)/2 \rangle$$

where θ is the angle between the long molecular axis and the director, $f(\theta)$ describes the normalized orientational distribution, and the angular brackets denote the orientational average. We can write

$$f(\theta) = (1/Z)\exp\left[\left(\sum V_m - 2\mathbf{E}^* \cdot \alpha \cdot \mathbf{E}\right)\big/ k_B T\right]. \quad (14.3)$$

Here, Z is the partition function, α is the linear polarizability tensor of the molecules, V_m is the interaction potential imposed on a molecule by its m-th neighboring molecules, and

$$\mathbf{E} = E\exp(ik\cdot r - i\omega t) \quad (14.4)$$

is the optical field. Symmetry of the system requires $f(\theta) = f(\pi - \theta)$. It is clear from equation (14.3) that $f(\theta)$ can be modified by the applied field, leading to a change in Q.

If we neglect the minor difference between microscopic and macroscopic order parameters, then we can relate Q to the macroscopic optical dielectric tensor ε_{ij}. Let

$$\varepsilon = \varepsilon_\perp \delta_{ij} + \varepsilon_a n_i n_j \quad (14.5)$$

with $\varepsilon_\parallel = \varepsilon_\perp + \varepsilon_a$, and $\varepsilon_\parallel$ and $\varepsilon_\perp$ the components parallel and perpendicular to the director $\mathbf{n}$. The order parameter Q is given by

$$Q = \varepsilon_a / \varepsilon_{a\#} \quad (14.6)$$

where $\varepsilon_{a\#}$ is the maximum of ε_a when molecules are perfectly aligned along $\mathbf{n}$. Since in general, Q is a function of $|E|^2$ (or intensity I) and temperature T, the optical anisotropy ε_a is also a function of I and T.

For the isotropic case of LC, it can be shown that to the first order [21,22],

$$Q \propto \varepsilon_a I/(T - T^*) \quad (14.7)$$

where T^* is the fictitious second-order isotropic–mesomorphic transition temperature, which is lower than the real transition temperature T_c. As $T \to T^*$, the order parameter experiences the pretransitional behavior of a critical divergence. Associated with it is the critical slowing-down behavior of the response of S to the applied field; the response time appears as

$$\tau \propto \nu/(T - T^*) \quad (14.8)$$

where ν is the orientational viscosity.

Both critical divergence and critical slowing down of Q have been observed, as shown in FIGURE 14.2 [21,22]. Because of the critical behavior, the induced

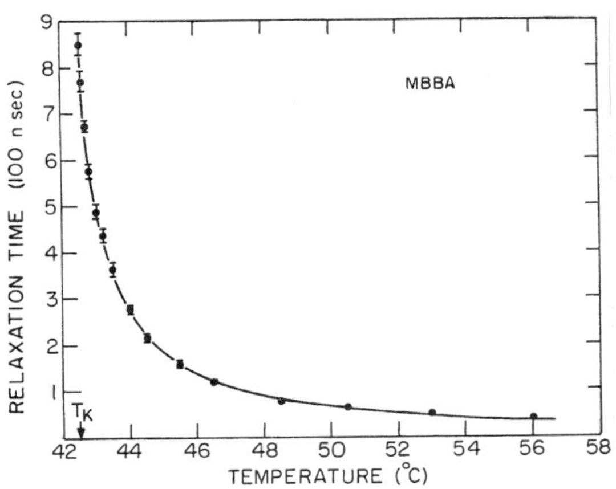

FIGURE 14.2. (a) Nonlinear refractive index n_2 and (b) response time of n_2 as functions of temperature for the liquid crystal MBBA in the isotropic phase, obtained by optical Kerr measurement (△) and by ellipse rotation measurement (◯). The solid curves are theoretical curves proportional to $1/(T - T_c)$, describing the pretransitional behavior (after [22]).

refractive index change (or optical Kerr effect) of an LC in the isotropic phase with $T - T_c \sim 1$ K can be several hundred times larger then that in CS_2 which is a representative liquid possessing a larger optical Kerr constant, but the response time approaching 1 μs is also much longer than that of an ordinary liquid.

Physically, the critical behavior sets in because molecular correlation becomes increasingly strong towards the phase transition. In the mesophases, molecules are highly correlated and the applied field can no longer affect the correlation significantly. It is, however, easy for the field to alter the average direction ($\mathbf{n}$) of orientation of the correlated molecules. This is similar to the case of ferromagnets where the magnetization direction is altered by an applied magnetic field. Director reorientation is the mechanism that leads to the so-called giant optical nonlinearity observed in transparent LC mesophases [15]. Refractive index changes as large as 0.01 to 0.1 can be obtained with a laser intensity of several hundred watts cm^{-2}. Unfortunately, the strong molecular correlation responsible for the giant nonlinearity also leads to a very slow response time, on the order of milliseconds to seconds.

The possibility of reorienting the director of an LC by a DC field was known at the beginning of the century, but the possibility of obtaining a similar effect with optical fields was only recognized in 1979 by Herman and Serinko [15]. Experimental confirmation started to surface a year later [23–26]. The optical-field-induced molecular reorientation can, in principle, occur in all LC mesophases, although because of constraints imposed by the structural requirement, the effect is often more difficult to observe in smectic and cholesteric phases [4].

Molecular reorientation by the applied field works against resistance to deformation of the original molecular alignment imposed on the medium by the boundary conditions through molecular anchoring at the surfaces. To be specific, we consider a thin nematic film between two parallel walls located at planes $z = 0$ and $z = L$. The walls are coated with an appropriate surfactant for either homeotropic or planar alignment. The average local orientation of the LC is generally described by the director $\mathbf{n}$. In the absence of external fields, the equilibrium molecular alignment is uniform with $\mathbf{n}$ independent of $\mathbf{r}$. If the alignment is perturbed, then elastic torques must be generated against the perturbation in order to restore the initial equilibrium. For small deformations, the elastic torques are proportional to the first spatial derivatives of $\mathbf{n}$. They are given, in torques per unit volume, by [1,2]

$$\tau_d = \mathbf{n} \times \mathbf{h} \tag{14.9}$$

where $\mathbf{h} = \mathbf{h}_s + \mathbf{h}_t + \mathbf{h}_B$ with

$$\begin{aligned}
\mathbf{h}_s &= k_{11}\nabla(\nabla \cdot \mathbf{n}) \\
\mathbf{h}_T &= -k_{22}[A\nabla \times \mathbf{n} + \nabla \times (A\mathbf{n})] \\
\mathbf{h}_B &= k_{33}[\mathbf{B} \times (\nabla \times \mathbf{n}) + \nabla \times (\mathbf{n} \times \mathbf{B})] \\
A &= \mathbf{n} \cdot \nabla \times \mathbf{n} \quad \text{and} \quad \mathbf{B} = \mathbf{n} \times (\nabla \times \mathbf{n}).
\end{aligned}$$

In equation (14.9), $\mathbf{h}_S$, $\mathbf{h}_T$, and $\mathbf{h}_B$ are the three torque components originated from the three basic deformations, namely splay, twist, and bend, respectively. They are derived from minimization of the elastic free energy f_d, given by

$$\begin{aligned}
f_d = \int_V (1/2)[k_{11}(\nabla \cdot \mathbf{n})^2 + k_{22}(\mathbf{n} \cdot \nabla \times \mathbf{n})^2 \\
+ k_{33}(\mathbf{n} \times \nabla \times \mathbf{n})^2]\, dV.
\end{aligned} \tag{14.10}$$

The constants k_{11}, k_{22}, and k_{33} are the corresponding elastic constants. The order of magnitude of the elastic constants is u/a, where u is a typical interaction energy between molecules and a is a molecular dimension.

Taking $u \simeq k_B T \simeq 0.026\,\text{eV}$ and $a \simeq 10\,\text{Å}$ we find that $k_{ii} \simeq 3 \times 10^{-7}$ dyn. These elastic constants decrease significantly as T increases, being roughly proportional to the square of the order parameter Q [27]. Although all k_{ii} are of the same order of magnitude, the twist constant k_{22} is usually smaller than the other two. For simplicity in calculations, the so-called one-constant approximation, $k_{11} = k_{22} = k_{33} = K$, is widely used.

If LC molecules are rotating, a viscous torque on the molecules against rotation is expected. The complete analysis of molecular rotation in an LC involves five independent viscosity coefficients [28–31]. For simplicity, a single phenomenological viscous constant γ_1 is often used. In this approximation, the viscous torque is given by

$$\tau_v = -\gamma_1 \mathbf{n} \times (\partial \mathbf{n}/\partial t). \qquad (14.11)$$

Changing a local molecular orientation may also induce fluid motion in the surrounding LC through molecular coupling, known as backflow. Quantitative calculations on backflow are difficult. In many practical cases, its effect on molecular orientation is either neglected or taken into account phenomenologically by a correction on γ_1.

The presence of an optical field yields an optical torque on the molecular director $\mathbf{n}$ [4,5]:

$$\tau_0 = (1/2)(\mathbf{D} \times \mathbf{E}^*) = \varepsilon_a/(2\pi)[(\mathbf{n} \cdot \mathbf{E})(\mathbf{n} \times \mathbf{E}^*)] \quad (14.12)$$

where $\mathbf{D} = \varepsilon \cdot \mathbf{E}$. The balance of the three torques on $\mathbf{n}$ requires

$$\tau_d + \tau_v + \tau_0 = 0 \qquad (14.13)$$

from which the orientational deformation $\mathbf{n}$ is determined, assuming the optical field $\mathbf{E}$ is known.

Appreciable reorientation of $\mathbf{n}$ ($\theta \simeq 1$ rad) can be achieved if the optical torque is comparable to the elastic one. This happens when $\varepsilon_a E^2/4\pi \simeq K/L^2$, where L is the sample thickness. Taking $K \simeq 10^{-6}$ dyn, $\varepsilon_a \simeq 0.1$, and $L \simeq 30\,\mu\text{m}$, we get $E \simeq 10^3\,\text{V cm}^{-1}$, which is equivalent to an optical intensity $I \simeq 1\,\text{kW cm}^{-2}$. As previously noted, the reorientational optical nonlinearity can be so strong that the usual series expansion of equation (14.1) is no longer valid. This is quite common in nematic LCs as we shall see later.

Recently, it was found by Jánossy and coworkers [32,33] that the optical-field-induced molecular reorientation in nematic LCs can be enhanced by more than two orders of magnitude if a small amount (~1%) of antraquinone dye is added into the LC. In some cases, even the sign of the induced refractive index change is reversed [34,35]. It is worth noting that the increase of the nonlinear optical response is not accompanied here by a lowering of the response time. The mechanism producing this remarkable effect is not yet well understood, but all experiments seem to confirm that it should be of molecular origin. The prevailing idea [32,33] is that a small fraction of the dye molecules are excited by the incoming laser beam and the interaction between the dye molecules and the LC molecules is different for dye molecules in the excited state and in the ground state, thereby leading to a laser-excited dye-induced torque on the LC molecules. Being of molecular origin, this torque could be much stronger than the direct optical torque on the LC molecules. We note that this dye-doping effect in LCs seems operative for many dyes soluble in nematics, although the direction and magnitude of the induced torque may depend on the type of dye molecules and the host LC.

Dye molecules in LCs absorbs light and may cause appreciable heating of the LC film, and hence changing its refractive indices. Even in a pure, transparent LC, residual absorption of laser light could increase the sample temperature by a few degrees. Because the refractive indices of LCs often depend strongly on temperature through the order parameter Q, especially near the nematic–isotropic transition, such a small change of temperature alone can yield rather significant changes in the refractive indices. This is known as thermal indexing [36,37]. As we discussed earlier, Q is a function of temperature in the mesophase. For most nematics it can be approximated by

$$Q = (1 - 0.98TV^2/T_c V_c^2)^{0.22} \qquad (14.14)$$

where V and V_c are the specific volumes of the LC at T and T_c, respectively. Close to T_c, the change of Q with T becomes very large; then according to equation (14.5) with $\varepsilon_a = \varepsilon_{a\#}Q$, the optical dielectric tensor and hence the refractive indices ($n_{ii}^2 = \varepsilon_{ii}$) must also have strong temperature dependences as T approaches T_c. Both ordinary and extraordinary refractive indices are affected. This is illustrated in FIGURE 14.3 for a typical LC material 5CB [38].

It can be difficult to find a simple relationship between thermal indexing and laser intensity because of the many experimental parameters entering into the

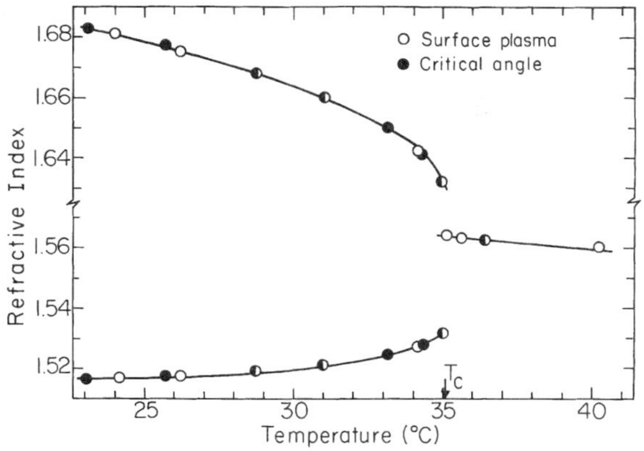

FIGURE 14.3. Ordinary and extraordinary refractive indices of the liquid crystal, 5CB, as functions of temperature, measured by the surface plasmon technique (open circles) and by the critical angle method (solid circle) (after [38]).

heat transport equation, such as the sample cell size and thermal conductivities of the medium and the walls. One can, however, have a rough estimate of the temperature rise δT produced by a beam of intensity I using the equation [39]

$$\delta T \simeq (\alpha I / c_P \sigma)(L/\pi)^2 \tag{14.15}$$

where α is the absorption coefficient, σ is the thermal diffusivity, and c_P is the specific heat per unit volume of the sample. Using typical values such as $c_P \simeq 10^7\,\mathrm{erg\,°C^{-1}\,cm^{-3}}$, $\sigma \simeq 10^{-3}\,\mathrm{cm^2\,s^{-1}}$, $\alpha \simeq 0.1\,\mathrm{cm^{-1}}$, $L \simeq 100\,\mu\mathrm{m}$, we obtain $\delta T \simeq 5\,°\mathrm{C}$, for $I \simeq 5000\,\mathrm{W\,cm^{-2}}$. The resulting refractive index change is on the order of 0.01.

Although this laser-induced thermal indexing generally yields a refractive index change much weaker than the laser-induced molecular reorientation effect in LCs, the former has a significantly faster response time, approximately given by

$$\tau_T \simeq L^2/(\sigma\pi^2). \tag{14.16}$$

With the above values of L and σ, we find $\tau_T \simeq 10^{-2}$ s, which is more than one order of magnitude less than orientational response times. This may render thermal indexing interesting for some applications.

Light absorption in liquid crystals may lead to phenomena other than thermal indexing. Examples are photostimulated chemical reactions [40] or changes in the anisotropy of elastic constants [41] in LCs. These, however, are minor effects and, in general, they are observable only very close to T_c.

14.3 NONLINEAR OPTICAL EFFECTS

Because of the unique characteristics of LCs as nonlinear optical media, nonlinear optical effects in LCs are particularly interesting. They can be divided into two groups: those that commonly occur in all media, and those unique to LCs. We briefly describe here various nonlinear optical processes that have been studied. The first nonlinear optical effect ever observed in LC was the third-harmonic generation (THG) in a cholesteric LC, reported in 1970 [20].

14.3.1 Harmonic Generation

Second-order nonlinear optical processes are forbidden in most LCs because of their possession of inversion symmetry. Only in some specially synthesized ferroelectric LCs in the smectic-C* phase is $\chi^{(2)}$ large enough to be interesting. In these latter media, phase-matched optical second-harmonic generation (SHG) has been observed [10,11]. The possibility of using smectic-C* films as practical harmonic generators is still being investigated [19].

Optical THG is allowed in all media, but $\chi^{(3)}(3\omega)$ for an LC comes from the electronic contribution and is not particularly large. Furthermore, in the normal dispersion region, the collinear phase-matching condition is difficult to satisfy. Consequently, THG in LCs does not appear to be interesting. However, it is known that in a periodic medium, collinear phase matching can be satisfied with the help of a lattice momentum, that is,

$$3k(\omega) = k(3\omega) \pm K \tag{14.17}$$

where $k(\omega)$ is the optical wavevector at frequency ω, and $K = n(2\pi/p)$ is the lattice momentum with n being an integer and p the length of the period. In the language of solid-state physics, the process characterized by equation (14.17) is known as an Umklapp process. A cholesteric LC happens to be an ideal medium to test such an optical Umklapp process. It has a helical structure with a helical pitch or period in the range of optical wavelengths, and is adjustable by temperature or composition. With the laser beam propagating along the helical axis of a properly chosen cholesteric LC, the phase-matching condition

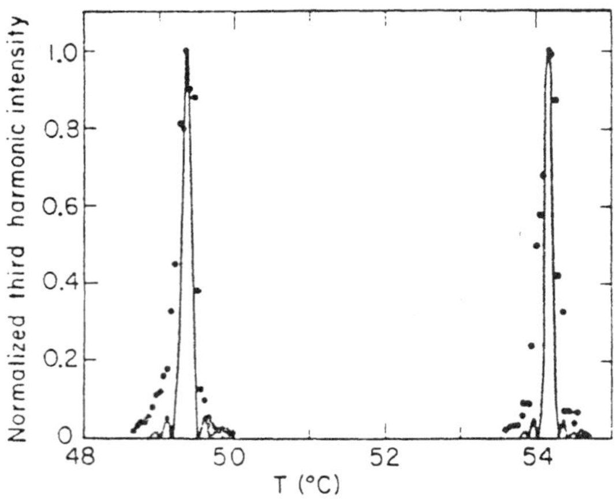

FIGURE 14.4. Normalized third-harmonic output versus temperature near the phase-matching temperatures for the mixture of 1.75 cholesteryl chloride and 1.0 cholesteryl myristate, in a cell 130 μm thick. The peak at low temperature (corresponding to the left helical structure) is generated by a left-circularly polarized fundamental wave, and the one at high temperature (corresponding to right the helical structure) by a right-circularly polarized fundamental wave. The solid line is the theoretical curve and the dots are experimental data points (after [20]).

of equation (14.17) can be easily satisfied by temperature tuning of the helical pitch p. Experimental verification of the prediction is shown in FIGURE 14.4 [20], where the two peaks correspond to phase-matched THG at two temperatures that yield the predicted pitch value with right- and left-hand helical structures, respectively. A number of other THG Umklapp processes with copropagating and counterpropagating fundamental beams have also been measured and found to agree well with predictions [20].

The above discussion can be extended more generally to three- or four-wave mixing in LCs utilizing optical nonlinearities derived from the electronic contribution. These processes are not expected to be very different in LCs than in other condensed media. What is unique for LCs is their giant optical nonlinearities arising from optical-field-induced molecular reorientation, especially in mesophases. In the following, we shall focus our discussion on nonlinear optical effects resulting from such nonlinearities. Because of the slow response of molecular reorientation, the nonlinearities in the form of refractive index changes cannot have a fast time variation. Therefore, for wave mixing, they can only contribute to degenerate and nearly-degenerate cases.

14.3.2 Degenerate Wave Mixing

Degenerate wave mixing in LCs has been studied quite extensively, mostly by Khoo and coworkers [42–54]. It refers to the process in which all waves present have the same frequency. For our purpose, the process can be considered as the combination of two steps: formation of a pattern of refractive index change in the medium as induced by the interference pattern of the incoming waves, and diffraction of the incoming waves by the index pattern. Phase conjugation is often regarded as a special case of degenerate four-wave mixing [9]. In most media, the observation of degenerate wave mixing requires high-power pulsed lasers. This is not so in LCs. In their mesophases, LCs can have very large laser-induced refractive index changes due to molecular reorientation and thermal indexing. Even with moderate cw lasers, degenerate wave mixing with several diffraction orders can be readily observed [55]. An example is shown in FIGURE 14.5. Phase conjugation in LCs was also reported [48] and its use was suggested to reduce speckle noise in a coherent beam [4]. The spatial resolution of phase conjugation using LCs is, however, limited [55]. Because of the strong molecular correlation in the LC mesophases, it is more difficult to induce an orientational variation or order parameter change on a smaller length scale. An elastic energy of the order of K/Λ^2 normally exists against the deformation, where K is the elastic constant and Λ is the characteristic length of the spatial variation.

In many experiments of degenerate wave mixing, laser-induced thermal nonlinearities were exploited. Small amounts of dyes were often added to LCs to enhance thermal indexing. The nonlinearity could become so large that phase conjugation could even be self-started with a laser power in the 100 mW range [49]. The experiment was carried out in the four-wave mixing geometry with two input pump waves crossing in an LC film. If two probe waves were injected into the film in the backward directions of the pump waves, both would experience an exponential gain in the pump-depletionless limit. A gain of exp(50) over a film thickness of 1 mm was observed [52–54]. This indicates that even if the probe waves were not injected but instead originated from noise, the output

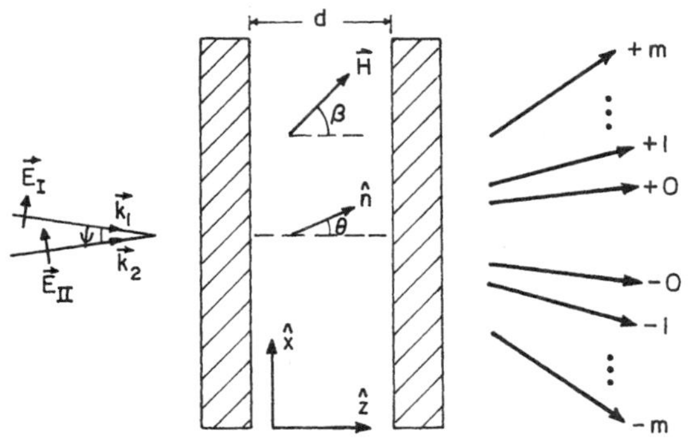

(a)

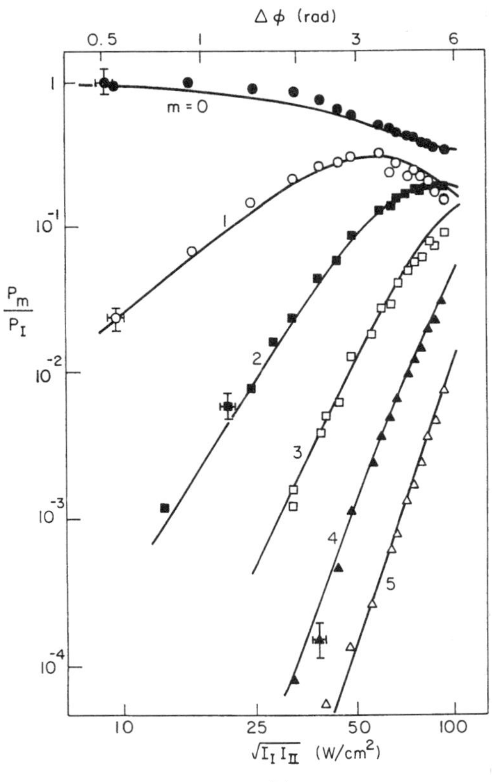

(b)

FIGURE 14.5. (a) Experimental geometry for degenerate wave mixing in a homeotropic nematic film. (b) Diffracted beam powers of various orders from a laser-induced 71 μm grating. Solid lines are theoretical curves (after [55]).

could still be appreciable. Since the backward output radiation is phase conjugated with the forward input pump radiation, this is a case where the phase-conjugated output is generated without an input, that is, phase conjugation is self-started.

Degenerate wave mixing with nanosecond or picosecond laser pulses has also been reported. Detection of the associated transient response can yield information about relaxations of certain excitations in LCs. As an example, we consider the case of a ferroelectric smectic-C* film studied by transient degenerate four-wave mixing [56]. Two interfering picosecond pump pulses were used to create, via laser heating, a transient temperature grating in the film. Since the tilt angle θ of the director $\mathbf{n}$ away from the smectic layer normal is a function of temperature T, the temperature grating induced a molecular orientational grating, and hence a refractive index grating. A time-delayed probe pulse was then used to probe the decay of the grating by measuring its diffraction from the grating as a function of time. The result provided information on how θ, after the initial perturbation by pulsed laser heating, relaxed back to equilibrium. It was found that the relaxation time of θ is in the range of a few microseconds, and shows a critical slowing-down behavior as T approaches the smectic $C^* \rightarrow A$ transition. Measurements on relaxation of laser heating, fluid flow, and director reorientation [57–59], and on relaxation of individual molecular reorientation [60–63] in the mesophases using the pump/probe technique, have also been reported.

Wave mixing can also occur between ordinary and extraordinary waves in mesomorphic LCs. Nonlinear coupling between waves can again come from molecular reorientation and thermal indexing. Actually, depending on the geometry, there can also be a complex interplay between the two. For example, depending on the director tilt angle from the beam wavevector and the proportion of ordinary and extraordinary waves, the local refractive index change may be dominated by either molecular reorientation or thermal indexing, and even the sign of the change can be different. A detailed analysis of the interplay between molecular reorientation and thermal indexing has been reported in [36,37,64].

More detailed theory and experiment on degenerate wave mixing in LCs also exist [8,47]. Recently, wave mixing in polymer-dispersed LCs via thermal indexing nonlinearities has also been reported [51].

14.3.3 Nondegenerate Wave Mixing and Stimulated Light Scattering

In four-wave mixing, it is possible to use two waves of the same frequency to induce an interference index grating in an LC film, and a probe beam with a different frequency to probe the grating [50]. This follows the same idea in holography where the holograph is made, and probed with laser beams at different frequencies. Wave mixing involving waves with different frequencies is generally called nondegenerate wave mixing.

It is also possible to induce a refractive index grating in a medium by two beams with slightly different frequencies, where the grating is moving with a phase term $\exp(i\Delta\mathbf{k}\cdot r - i\Delta\omega t)$ where $\Delta\mathbf{k}$ and $\Delta\omega$ are the differences in the wavevectors and frequencies of the two beams. The probe beam will then yield a diffraction output at a frequency shifted by $\pm\Delta\omega$ from the probe frequency. An interesting case occurs when $\Delta\omega$ is close to the material resonance, as energy exchange between the two beams occurs and part of the energy is deposited into the medium via the material excitation. This is actually a process known as stimulated light scattering: the beam at the lower frequency increases in energy (initially exponentially), while the other loses energy. For material excitation in the form of molecular reorientation, the equation of motion is an orientational diffusion equation, and the resonant frequency is zero. However with a finite orientational response time τ , stimulated scattering can still occur at finite $\Delta\omega$. In fact, the maximum gain appears at $\Delta\omega = 1/\tau$. The process is known as stimulated Rayleigh-wing scattering [9].

In the mesophases of LCs, τ is of the order of 1 s. Stimulated Rayleigh-wing scattering has the Stokes output at practically the same frequency as the pump, but it can appear in a different direction from the pump. With the large reorientation effect, stimulated Rayleigh-wing scattering in LCs has a very high gain and can be easily observed. Together with nondegenerate wave mixing, it has been studied in nematics under both transient and stationary conditions, using pulsed [65,66] and cw lasers [67–70], respectively. Thermal indexing can also contribute to wave mixing and stimulated light scattering (with the laser-induced thermal effect governed by the heat diffusion equation). Usually, thermal indexing dominates in the transient case and molecular reorientation dominates in the steady state. In polymer-dispersed LCs, only transient grating formation due to thermal indexing has been reported [71].

Because liquid crystals are birefringent, stimulated light scattering may also occur between ordinary and extraordinary waves. Consider a beam propagating along a direction perpendicular to the molecular director (z-axis). A spatial grating can be induced by the beam in LC via polarization-dependent induced refractive index change. The period of the grating is $\Lambda \simeq \lambda/(n_e - n_o) \simeq 1\,\mu\text{m}$, which is much shorter than the period $\lambda/\Delta\alpha \simeq 100\,\mu\text{m}$ of a typical grating formed by two beams interfering at an angle $\Delta\alpha$. The coupling between ordinary and extraordinary waves is now provided by a twist deformation of the molecular director. Because of the low value of the elastic constant k_{22} and the absence of backflow for twist deformation, the response time τ is lower than other reorientational effects. Very high ($\simeq 50\%$) conversion efficiencies and very low thresholds ($\simeq 10\,\text{mW}$) for stimulated scattering from ordinary to extraordinary waves have been reported in a thick ($\simeq 1\,\text{mm}$) 5CB nematic film [72–73]. The ordinary–extraordinary wave coupling mechanism is also responsible for the self-oscillation phenomenon observed when an ordinary wave propagates at a small incidence angle into a homeotropically aligned nematic film [74,75]. This effect is related to the interesting phenomenon of laser-induced orientational nonlinear dynamics in nematic LCs, which will be discussed in section 14.3.11.

14.3.4 Intensity-Dependent Phase Retardation

As we have already stated several times, an incoming laser beam can induce a large refractive index change in an LC medium. However, we must also realize that the large refractive index change can also affect the beam propagation in the medium; the immediate effect is the imposition of a phase shift on the beam.

Consider the simplest case of an extraordinary wave obliquely incident at angle α on a homeotropically aligned, transparent nematic film. The laser field in the medium acts to reorient the director $\mathbf{n}$ in the incident plane following equation (14.13). In the limit of a weak laser intensity I, the equation can be linearized in the reorientation angle θ, and reduced to the form [76]

$$-\gamma_1\,\partial\theta/\partial t + K\nabla^2\theta = -(A/c)I \qquad (14.18)$$

where $A = \varepsilon_a n'^3 \sin 2\alpha / 2n_e^2 n_o^2$ and

$$n' = n_o n_e / (n_o^2 \sin^2 \alpha + n_e^2 \cos^2 \alpha)^{1/2}. \quad (14.19)$$

The boundary conditions in the limit of infinitely strong anchoring are $\theta(z=0) = \theta(z=L) = 0$, where L is the film thickness. If the transverse variation of θ can be neglected, then the stationary solution of equation (14.18) is

$$\theta(z) = (A/2cK)Iz(L-z). \quad (14.20)$$

This reorientation leads to a refractive index change

$$\Delta n' = (\partial n'/\partial \alpha)\theta = (A^2/2cK)Iz(L-z) \quad (14.21)$$

seen by the incoming extraordinary wave and, upon traversing the film, an intensity-dependent phase retardation $\Delta\Psi$ on the wave

$$\Delta\Psi = (2\pi/\lambda)\int_o^L \Delta n'(z)\,dz = (A^2/2cK)(\pi L^3/3\lambda)I. \quad (14.22)$$

In a typical nematic film with $L = 100\,\mu$m and $\alpha = 30°$, $\Delta\Psi \approx 2\pi$ can be achieved with I as low as 30 W cm^{-2} or less. Such a phase retardation can easily be measured using an interferometer or polarimeter (see FIGURE 14.6). The above example also shows that LCs could be useful as phase shifters in all-optical devices.

FIGURE 14.6. Average nonlinear refractive index $\langle n_2 \rangle = \delta\varphi/L$ experienced by a laser beam with incidence angle α in a homeotropic nematic film of E7 of thickness L; the laser-induced phase shift $\delta\varphi$ was measured by a Max–Zehnder interferometer (after Santamato, E., Ciaramella, E. and Tamburrini, M. 1994 *Mol. Cryst. Liq. Cryst.* **241** 205; Gordon and Breach Science Publishers).

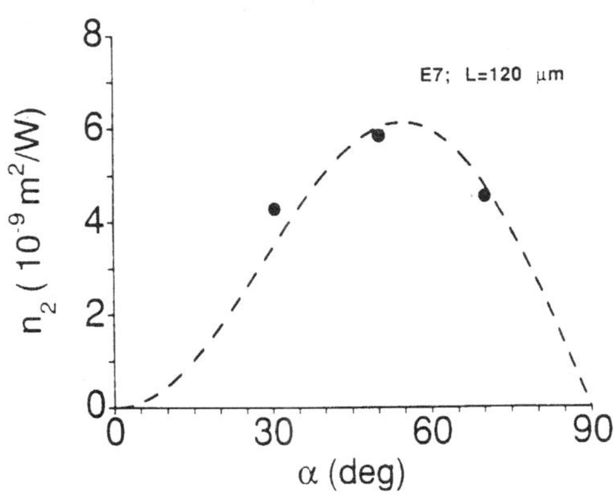

The first demonstrations of self-induced phase retardation were reported in references [23–26].

14.3.5 Self-Phase Modulation and Self-Focusing

The laser intensity profile naturally leads to a director reorientation θ, and hence a phase retardation $\Delta\Psi$ depending on the transverse coordinate ρ. Thus the output wave has its wavefront modulated. This is known as self-phase modulation [9]. If the laser intensity is high, then the linearizing procedure discussed in the last section is no longer valid, and Δn can become a highly nonlinear function of I. Accordingly, self-phase modulation can be very strong, with the maximum $\Delta\Psi$ reaching 100π or more. It gives rise to strong self-focusing and spectacular nonlinear diffraction effects [76].

Consider, for example, that a cylindrically symmetric laser beam suffers a phase modulation $\Delta\Psi$ versus ρ in a bell-shaped form after passing through an LC film. For the sake of discussion, we can assume that $\Delta\Psi(\rho)$ is a Gaussian:

$$\Delta\Psi = \Delta\Psi^0 \exp(-\rho/w^2). \quad (14.23)$$

Because of $\Delta\Psi(\rho)$, the light emerging from the LC sample has a curved wavefront, as depicted in FIGURE 14.7. Using the approximation of geometric optics, we expect the rays to propagate perpendicularly to the wavefront. If $\partial\Delta\Psi/\partial\rho$ is positive, they must all converge towards the beam axis, leading to a self-focused output beam. Self-defocusing would occur if $\partial\Delta\Psi/\partial\rho$ is negative; this may be the case in an absorbing medium. With self-focusing and having a medium act as a thin lens, the focal length of the lens is given by

$$f \simeq nw^2/16\pi L(\Delta n)_{\max} \quad (14.24)$$

where $(\Delta n)_{\max}$ is proportional to the maximum laser intensity. Measurement of f therefore permits the deduction of $(\Delta n)_{\max}$. Self-focusing is actually the basis of the well-known Z-scan method for measuring the nonlinear refractive index, n_2, of a medium [77]. This has recently been applied to LCs [78].

In typical self-focusing experiments in LCs, the sample is placed in an unfocused or weakly focused laser beam, and the effect is observed as a narrowing of the beam diameter in the far field. For higher laser intensities, however, $\Delta\Psi_0$ can be much larger than 2π

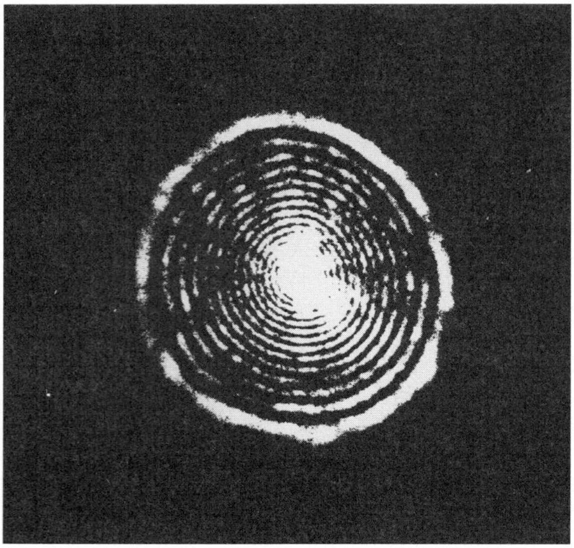

(a)

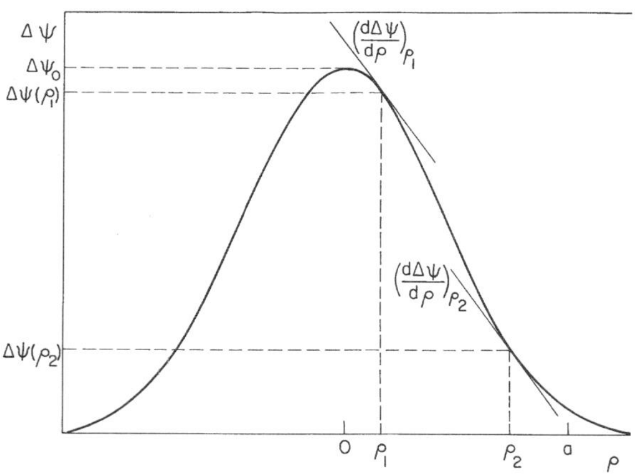

(b)

FIGURE 14.7. (a) Picture of typical far-field diffraction ring pattern due to self-phase modulation in a nematic LC film. (b) Profile of laser-induced phase shift. Light diffracted at ρ_1 and ρ_2 has the same wavevector and interferes in the far field to produce rings (after [76]).

and the thin-lens approximation breaks down. Strong aberration takes place, in the form of a highly divergent output beam with a concentric ring pattern, if the input beam is cylindrically symmetric. An example is shown in FIGURE 14.7(b). This strong self-phase modulation phenomenon can be understood as the result of interference between waves diffracted from different parts of the beam. FIGURE 14.7(b) shows that there are always two points on the bell-shaped curve of $\Delta\Psi(\rho)$ where the slopes $\partial(\Delta\Psi)/\partial\rho$ are the same. Since $\partial(\Delta\Psi)/\partial\rho$ corresponds to the transverse wavevector component, the waves originating from two points with the same $\partial(\Delta\Psi)/\partial\rho$ must propagate in the same direction, and should therefore interfere in the far field. The interference is either constructive or destructive, depending on the phase difference between the two waves, namely, $2n\pi$ for maximum constructive interference and $(2n+1)\pi$ for maximum destructive interference, with n being an integer. It is easy to see that the total number of bright rings in the output pattern is given by $N \sim \Delta\Psi_0/2\pi$ and divergence angle of the outermost ring $\theta_m = (k_\perp)/k = (\partial\Delta\Psi/\partial\rho)_{\max}(\lambda/2\pi)$. Assuming $\Delta\Psi$ takes the Gaussian form in equation (14.23), we find that $\theta_m \sim 2\lambda N/we^{1/2}$. For $N \simeq 100$, $w \simeq 50\,\mu$m, and $\lambda \simeq 0.5\,\mu$m, we obtain $\theta_m \simeq 1$ rad. A more rigorous treatment of the effect, based on diffraction theory, can be found in reference [79].

14.3.6 Optical Bistability

Optical bistability refers to the phenomenon in which an optical system possesses two stable output states for the same input. Which output state the system prefers depends on the history of the input variation, and output versus input shows hysteresis. The effect has potential applications in all-optical memory devices, switching, and others [80].

Optical bistability results from sufficient feedback in the input/output loop of an optical system. It can easily be observed in a nonlinear Fabry–Perot interferometer, that is, a Fabry–Perot (FP) interferometer filled with a nonlinear Kerr medium. This was first proposed by Seidel, Szoke and coworkers [81]. With usual Kerr media, pulsed lasers ($\sim$MW cm^{-2}) are often required to observe the effect. LC in the isotropic phase falls in this category. With its response time adjustable in the nanosecond range, it has been used in the first quantitative investigation of the transient behavior of a nonlinear Fabry–Perot interferometer [82].

Nematic LCs have much stronger nonlinearity, as discussed in section 14.2. Consequently, optical

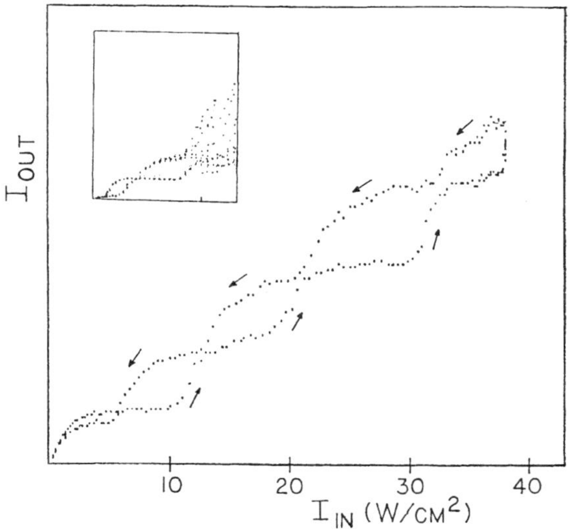

FIGURE 14.8. Multiple bistable loops observed with a Fabry–Perot interferometer filled with an 83 μm homeotropic nematic film of 5CB. Inset shows bistable loops and onset of oscillation with a different magnetic field (after [83]).

bistability is readily observable, even with moderate cw lasers if the FP interferometer is filled with a nematic LC [83–86]. In reference [83], an 83 μm LC film in an FP cavity was used, and optical bistability was observed with a laser intensity of $\sim$10 W cm^{-2}. Actually, with higher laser intensities, multiple bistable loops were also observed. FIGURE 14.8 is an example. In these experiments, molecular reorientation was the dominant mechanism for the required nonlinearity. In other cases, where laser-induced thermal indexing was also important, the interplay between the two mechanisms with different response times could give rise to persistent oscillation of the output from the interferometer [83].

Mirrorless optical bistability can also occur in LCs with special arrangements. Strong feedback in such cases comes internally from within the LC medium, through action/reaction between the LC and the polarized optical field. The situation is rather unique for LCs. The effect has been observed not only in nematic LCs, but also in mixed nematic/cholesteric LCs and in polymer-dispersed LCs. Some examples will be discussed later, in connection with optical Freedericksz transitions.

14.3.7 Optical Switching

Self-switching of a laser beam at an interface between a linear and nonlinear medium can be regarded as a special case of optical bistability [66,87–89]. To observe the effect, an LC film is sandwiched between two prisms. The laser beam incident through one of the prisms is adjusted to have an incident angle only slightly above the critical angle θ_c for linear total internal reflection. As the LC refractive index increases with laser intensity, θ_c increases accordingly. Self-switching of the beam from total reflection to transmission occurs when θ_c increases beyond the incident angle. A hysteresis loop usually shows up in the output/input dependence, as in the case of optical bistability. Self-switching in nematic LCs has been studied by many researchers. The same effect in polymer-dispersed LCs [90] and ferroelectric LCs [91] has also been reported.

14.3.8 Transverse Pattern Formation

A beam with an initially smooth transverse intensity profile can break into a spatially modulated pattern after passing through a nonlinear medium. This phenomenon, generally known as self-organization, which describes the spontaneous formation of regular structures out of chaos in dissipative systems, have recently gained a great deal of attention. It has been observed in a nonlinear Fabry–Perot cavity [92] as well as in systems with a single back-reflecting mirror [93,94]. Usually it happens when the incoming laser intensity is above a threshold value. The final stable pattern is typically hexagonal, but rolls and even pentagons and triangles have been observed. An example of the observed hexagonal pattern using a tilted homeotropically aligned LC film as the nonlinear medium [94] is shown in FIGURE 14.9.

The phenomenon is mainly governed by equation (14.13). For simplicity, we consider the case of weak laser-induced reorientation in a homeotropic nematic film described by the linearized equation (14.18). With the approximation that the self-phase modulation imposed on the wave traversing the medium has the same functional dependence on x and y as θ, we obtain the equation for $\Delta\Psi(x,y,t)$ from equation (14.18) as

$$\begin{aligned} &-\gamma_1 \partial(\Delta\Psi)/\partial t + K\nabla_\perp^2 \Delta\Psi - ((12K)/L^2)\Delta\Psi \\ &\quad = [-(\varepsilon_a n'^3 \sin 2\alpha)/(2n_e^2 n_o^2 c)]I(x,y,t). \end{aligned} \tag{14.25}$$

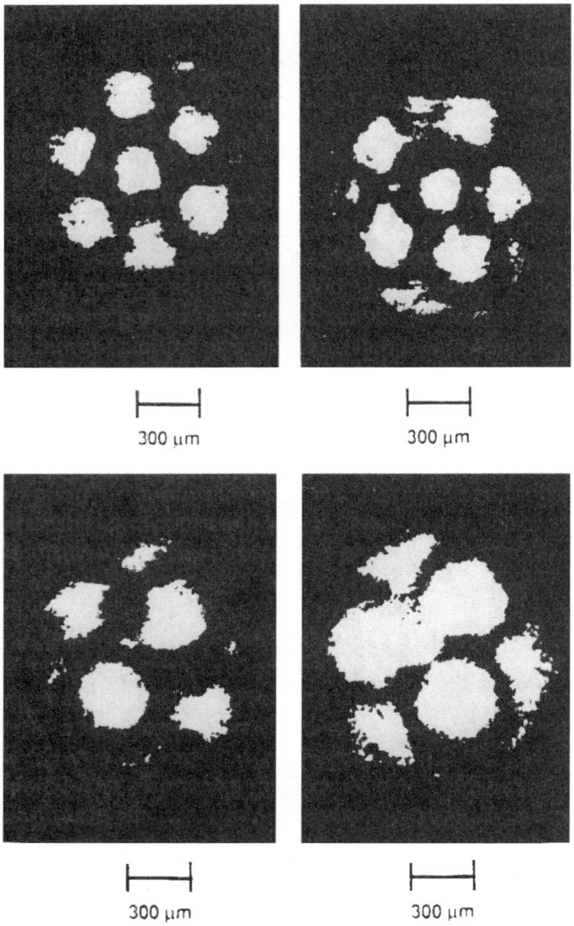

FIGURE 14.9. Near-field patterns of the backward wave transmitted through an LC film for different values of $\eta = w/\Lambda$. (a) D_6 symmetry ($n = 1.75$); (b)D_5 symmetry ($\eta = 1.75$), (c) D_2 symmetry ($\eta = 1.4$); (d) D_3 symmetry ($\eta = 1.35$). D_6 and D_5 patterns have been observed under the same experimental conditions and been ascribed to a sort of multistability. The LC sample is a 50 μm homeotropically aligned nematic film of E7 (after [99]).

We now have

$$I(x,y,t) = I_0(x,y) + I_1(x,y,t) \tag{14.26}$$

with I_0 and I_1 being the intensities of the incident and back-reflected beams, respectively. While $I_0(x,y)$ is known, $I_1(x,y)$ depends on $\Delta\Psi(x,y,t)$ and on the round-trip distance between the exit face of the sample and the back-reflecting mirror. From the Fresnel diffraction formula, we obtain

$$I_1(x,y,t) = \left| 1/(2\lambda d) \iint \mathrm{d}u\,\mathrm{d}v \sqrt{I_0(u,v)} \times \exp\{\mathrm{i}\Delta\Psi + \mathrm{i}\pi[(x-u)^2 + (y-v)^2]/2\lambda d\} \right|^2 \tag{14.27}$$

where d is the mirror-to-sample distance. It is seen that $I_1(x,y,t)$ does depend on the phase factors in the integral. When equation (14.27) is inserted back into equations (14.25) and (14.26), we obtain an integro-differential equation for $\Delta\Psi(x,y,t)$. This was precisely the equation solved numerically by D'Alessandro and Firth [95] exhibiting a transverse hexagonal pattern in the solution.

We can actually understand the phenomenon simply in terms of the well-known Talbot effect [94,96,97], which describes periodic conversion of phase modulation (PM) to amplitude modulation (AM), and *vice versa*, of an optical wave propagating in free space. Consider a wave propagating along z and having a PM with wavenumber q at $z = 0$. Upon propagation over a distance $z_T/4$, where $z_T = 8\pi^2/\lambda q^2$ is the Talbot length, the modulation is totally converted from PM to AM (with the same wavenumber). It is converted back to PM at $z = z_T/2$, and then AM at $z = 3z_T/4$. Any two successive PM or AM are 180° out of phase, so that at $z = z_T$ the initial PM is recovered. Therefore, if $2d = (1/4 + n)z_T(q)$ with $n = 0, 1, 2, \ldots$, an initially PM wave exiting from the sample should return as an AM wave back into the sample. The optical nonlinearity of the medium now converts the AM into PM in the medium, in phase with the original one. The spatial modes of the transverse pattern, having wavenumbers $q_n = (\pi/\sqrt{\lambda d})(1+4n)^{1/2}$ are then selectively enhanced by this feedback and phase-locked together. The phenomenon is a spatial analog to the onset of periodic pulse formation in mode-locked lasers. A more refined analysis shows that hexagonal patterns are preferred. The Talbot model suggests that the hexagon lattice constant Λ should be given by $\Lambda = \Lambda_T = 2\sqrt{\lambda d}$. This is actually the case, as clearly shown in FIGURE 14.10, where the ratio Λ/Λ_T is plotted for different mirror-to-sample distances.

Patterns other than hexagons can be obtained when the overall cylindrical symmetry of the experiment is

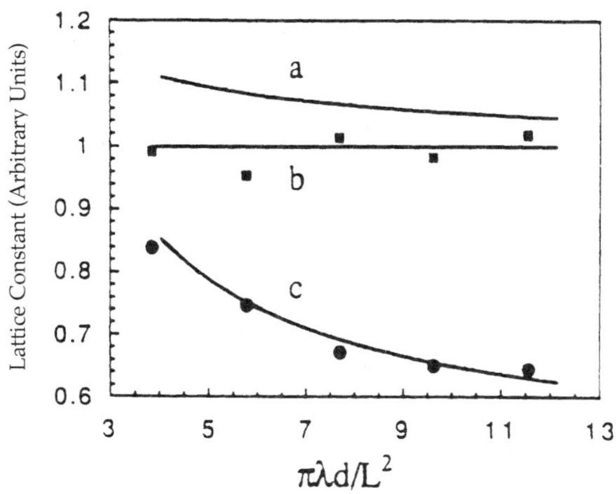

FIGURE 14.10. Lattice constant of hexagonal pattern and threshold intensity for pattern formation as functions of the reduced mirror-to-sample distance $\pi\lambda d/L^2$ for the case of a 50 μm homeotropic nematic film of 5CB (after [94]).

broken. This can be achieved, for example, by introducing a small amount of astigmatism into the incident laser beam, either by changing the incidence angle or by imposing appropriate boundary conditions in the feedback path. Roll-to-hexagon transition and rectangular-shaped patterning have been obtained in this way [97]. Patterns of dihedral symmetry have been predicted for a finite cylindrically symmetric input beam [98] and have actually been observed experimentally [99] when the laser spot at the LC film is reduced to radius $w \sim \Lambda$. In this case, the original cylindrical symmetry can be automatically broken, and according to the general principle of group theory [100], patterns belonging to the dihedral symmetry subgroups $D_1(l = 6)$ should be generated. For decreasing beam radius, successive transitions from hexagonal (D_6) to pentagonal (D_5), to rhombic (D_3), and finally to triangular (D_3) patterns have been observed [99]. Examples are shown in FIGURE 14.9.

14.3.9 Optical Fréedericksz Transitions

The giant optical nonlinearity due to optical-field-induced director reorientation in the mesophases makes LCs unique as nonlinear optical media, and gives rise to some very unusual nonlinear optical effects that do not exist in other media. The optical Fréedericksz transition was the first such effect reported [23–26]. It is analogous to the well-known Fréedericksz transition induced by DC fields [101]. In the optical case, the optical field **E** instead of the DC field tends to align the director **n** parallel to **E**. If initially **n** and **E** are orthogonal, the optical torque on **n** must be zero. However, this is an unstable situation. Fluctuations of **n** away from equilibrium result in a finite optical torque on **n** that tends to amplify the deviation from equilibrium. If the optical torque is smaller than the elastic restoring torque, the fluctuations are damped and the original equilibrium state remains stable. Otherwise, **n** will be reoriented. The transition occurs when the laser intensity I reaches the threshold value I_{th}, and is a clear example of an optical-field-induced structural phase transition.

As an example, consider the case of a linearly polarized laser beam, normally incident on a homeotropically aligned nematic film. The field-induced reorientation is governed by equation (14.18) with $\alpha = \theta$, and the boundary conditions $\theta(z = 0) = \theta(z = L) = 0$. When $I \sim I_{\text{th}}$, we expect $\theta \ll 1$. The stationary solution of equation (14.18) is then given by

$$\theta(z) = 0 \qquad \text{for } I < I_{\text{th}}$$
$$\theta(z) = A\sin(\pi z/L) \qquad \text{for } I \geq I_{\text{th}}$$

with $I_{\text{th}} = \pi^2 n_e^2 cK/\varepsilon_a n_o L^2$ and A being a constant proportional to $(I - I_{\text{th}})^{1/2}$. The field-induced transition exhibited by θ appears to be second-order [102], and I_{th} is inversely proportional to L^2. For $L \sim 100\,\mu$m, I_{th} is typically of the order of $1\,\text{kW cm}^{-2}$. FIGURE 14.11 provides a real example showing $\Delta\Psi_o$ versus I. For I well above I_{th}, $\Delta\Psi_o$ can reach $\sim 100\pi$. This means that the output from the sample should exhibit a multiple-ring pattern in the far field, as discussed in section 14.3.5. (FIGURE 14.7 corresponds to such a situation.) A dynamical analysis of equation (14.18) in the small-θ regime shows that the build-up of the reorientation is exponential in time with a time constant given by [26]

$$\tau_{\text{on}} = (\gamma_1/K)[I_{\text{th}}/(I - I_{\text{th}})](\pi/I)^2 \qquad (14.28)$$

which diverges as $I \to I_{\text{th}}$ from above. This critical slowing-down behavior is the pretransitional characteristic of a second-order phase transition. If I is switched off, the system will return to the undistorted equilibrium state with a relaxation time

$$\tau_{\text{off}} = (\gamma_1/K)(\pi/L)^2 = \tau_{\text{on}}[I_{\text{th}}/(I - I_{\text{th}})]. \qquad (14.29)$$

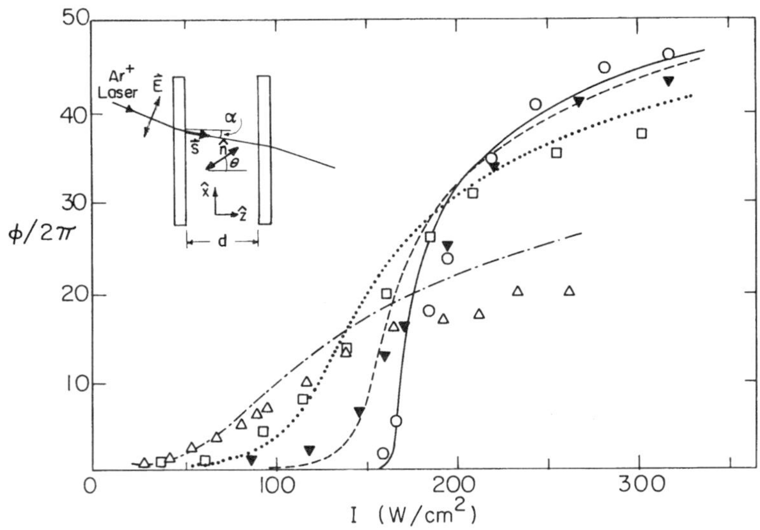

FIGURE 14.11. Experimental points and theoretical curves for laser-induced birefringence at different incidence angles α. Circles and solid curve, $\alpha = 0°$; solid triangles and dashed curve, $\alpha = 3°$; squares and dotted curve, $\alpha = 11°$; open triangles and dot–dashed curve, $\alpha = 30°$. Inset shows experimental geometry (after [26]).

Typical values of τ_{off} are around 1 s for $L \sim 100\,\mu$m. Measurements of τ_{off} can be used to deduce the quantity τ/K [103–105]. An alternative way to measure the ratio γ_1/K is to use self-induced stimulated light scattering, to be described in the next section [106].

In the above analysis, the infinite plane-wave approximation was used. However, if the laser beam is focused into the sample to a diameter comparable to the film thickness, the effect of finite beam size must be considered. The problem has been treated by Tabiryan and coworkers in chapter 5 of their review article [5]. The finite beam increases I_{th} by a factor of approximately $(1 + L^2/w^2)$, where w is a measure of the beam radius in the sample. This can be understood by knowing that only LC molecules in a transverse region $\sim w$ are reoriented, so that, as seen from equation (14.18), the optical torque must now balance an additional elastic torque $\simeq K/w^2$ from transverse molecular interactions in addition to $\sim K/L^2$ in the infinite plane-wave case. We also remark that for the case described here, if the beam is extraordinarily polarized and obliquely incident on the LC film, appreciable laser-induced director reorientation can also be observed, but as expected, there is no threshold intensity for reorientation [107].

The threshold for the optical Fréedericksz transition can be lowered by biasing the LC film with a destabilizing DC electric or magnetic field [98,103]. It is possible to reduce I_{th} to $\sim 10\,\text{W cm}^{-2}$ in this way. Conversely, if a stabilizing DC bias field is used, I_{th} should increase. When the bias field is sufficiently strong, the optical Fréedericksz transition actually becomes first-order [102,108]. The same is true for optical Fréedericksz transitions with circularly and elliptically polarized light. These will be discussed in later sections.

Optical Fréedericksz transitions can, in principle, be observed in other LC films [5], such as homogeneously aligned nematic films, freely suspended smectic-C films [109] etc., although the required experimental conditions may be more difficult to realize, and the theoretical analysis more complicated.

14.3.10 Self-Induced Stimulated Light Scattering

Optical Fréedericksz transitions can also be induced by a circularly [110] or elliptically [111] polarized laser beam. This is certainly different from the DC case and makes optical Fréedericksz transitions unique and more interesting. Actually, as one would expect in general, the laser field can reorient the director in LCs, but the reorientation depends on the polarization of the field. The resulting local birefringence in the medium can in turn modify the polarization as the beam propagates in the medium. It is this interplay between director reorientation and beam polarization that gives rise to the many fascinating nonlinear optical effects observable only in LCs. One of them is the optical Fréedericksz transition and related effects induced by a circularly polarized laser beam [110].

Consider a circularly polarized beam normally incident on a homeotropically aligned nematic film, as described in the last section. It can be shown that the threshold for optical Fréedericksz transitions in this case is higher by a factor of two than for linear polarization [5]. The higher threshold reflects the fact that there is no preferred direction of reorientation with circularly polarized field. Eventually, above the threshold, **n** is forced to reorient, usually in an azimuthal direction determined by some residual asymmetry in the system. The sample then appears to be birefringent to the incoming beam, and the beam polarization must change accordingly as the beam propagates in the film. The initial circular polarization becomes elliptical in general and the director reorientation must be adjusted to the new polarization. Quantitatively, the phenomenon should be described by two coupled equations, one for the LC director reorientation, and the other for the wave propagation in the LC. This description can be found in references [110,112,113].

We can, however, discuss the above problem and obtain the key results using the simple picture of angular momentum conservation [110,112]. If, in traversing the LC film, a circularly polarized input beam is converted into an elliptically polarized output, then a net angular momentum of $(e_{\text{in}} - e_{\text{out}})I/\omega$ per unit time must have been deposited in the film by the beam, where e is the ellipticity of the polarization, with $e_{\text{in}} = \pm 1$ for right and left circular polarizations, respectively. This produces a torque on the director **n** and can set **n** into continuous rotation about the axis along the surface normal. Since the total elastic restoring torque in the LC film must vanish, the total optical torque left in the film must be balanced only by the total viscous torque. We then have

$$(I/\omega)(e_{\text{in}} - e_{\text{out}}) = \gamma_1 \int_0^L (\partial\varphi/\partial t) \sin^2\theta \, dz \qquad (14.30)$$

where $\partial\varphi/\partial t = \Omega$ is the angular velocity of **n** in the azimuthal plane. With $\partial\varphi/\partial t$ taken as constant, and $\int_0^L \sin^2\theta \, dz$ related to ellipticity, Ω can be expressed in terms of I and $(e_{\text{in}} - e_{\text{out}})$ [110]. Experimentally, since the uni-axis of the medium defined by **n** is constantly rotating, the polarization of the output beam must rotate at the same rate, which can be easily measured. The experimental result agrees well with the theoretical prediction [110]. The typical rotation speed is of the order of 0.1 Hz.

The above phenomenon can be regarded as a special type of stimulated light scattering [112]. The rotating elliptical polarization of the beam can be decomposed into two circular polarization components, with one at a frequency 2Ω lower than the other. The process can then be described as an incoming circularly polarized beam of frequency ω coherently scattered by the director rotation in the LC into a Stokes component of opposite circular polarization at frequency $\omega - 2\Omega$. Obviously, photon energy is lost in the medium because of the redshift of the scattered light. This is expected, considering that energy is needed to rotate the director. Unlike ordinary stimulated light scattering, in which the Stokes shift is a characteristic property of the medium only, the process here has a Stokes shift depending on laser intensity. We can therefore call it a self-induced stimulated light scattering process [112].

14.3.11 Laser-Induced Nonlinear Dynamics

Elliptically polarized laser beams can also be used to study optical Fréedericksz transitions in LCs, and the result is yet another set of fascinating observations [111]. Instead of a stationary reorientation of **n** in the case of linear polarization, or a constantly rotating **n** in the case of circular polarization, a variety of complex nonlinear dynamical effects can show up, depending on the beam intensity I and the input beam ellipticity e_{in}.

The main difference between elliptical and circular input polarizations is that in the former case, the azimuthal symmetry of the system is clearly broken at the very beginning so that above the Fréedericksz transition threshold, the direction of molecular reorientation at $z \sim 0$ is well defined, as it must follow the major axis of the elliptical polarization. The birefringence resulting from the molecular reorientation causes the polarization to vary as the beam propagates along z. Accordingly, in exchanging angular momentum with the medium, the beam produces an optical torque on the molecules that also varies along z, sometimes positive and sometimes negative. With increasing laser intensity, this leads first to a damped oscillation of $\mathbf{n}(z)$ and $e(z)$ towards an equilibrium state, and then to a persistent oscillation. If I increases above a certain value, then $\mathbf{n}(z)$ is so reoriented that the

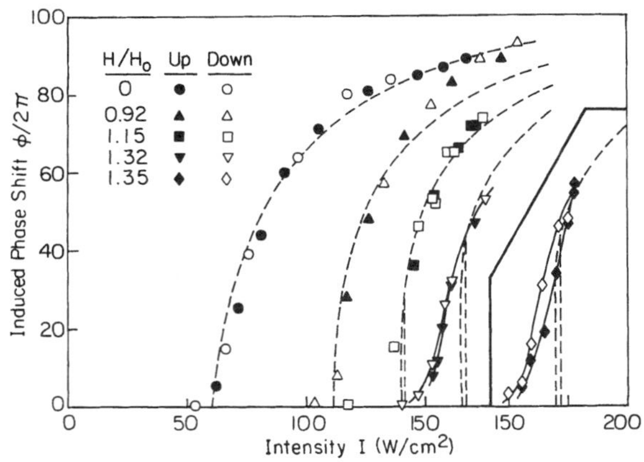

FIGURE 14.12. "Phase diagrams" of various dynamic regimes in the (I, χ) plot, where χ is related to the polarization ellipticity e by $e = \sin(2\chi)$. (a) Experimental observations: squares, circles, and triangles denote distorted-equilibrium, persistent-oscillation, and precession-nutation states, respectively. Open symbols refer to states obtained with increasing laser intensity, and solid symbols to states with decreasing intensity. (b) Theoretical simulation. U, D, O, and P refer to undistorted, distorted-equilibrium, persistent-oscillation, and precession-nutation regimes, respectively. Solid lines describe boundaries between different dynamic regimes. The dashed lines describe boundaries at which P switches back to U, D, or O as the pump intensity increases (after [111]).

optical torque varing along z no longer changes sign, and the angular momentum deposited by the light into the medium is always either positive or negative. This sets $\mathbf{n}(z)$ into a precession about z in conjunction with a nutation. Figure 14.12 describes the "phase diagrams" of different dynamic regimes in the parameter space of I and e predicted by theory and observed by experiment [111]. The semiquantitative agreement between theory and experiment is remarkably good.

Another case that gives similar laser-induced nonlinear dynamic effects has the laser beam incident on a homeotropic nematic film at a small angle (a few degrees), and its polarization perpendicular to the incidence plane [114,115]. This is also a configuration in which the optical Fréedericksz transition can occur. When the laser intensity is above the transition threshold, the initial reorientation of $\mathbf{n}$ makes $\mathbf{E}$ neither parallel nor perpendicular to the plane of incidence. Accordingly, the beam polarization becomes elliptical. The ensuing reorientation of $\mathbf{n}(z)$ and the associated variation of the beam polarization are quite similar to the case with a normally incident elliptically polarized input. Persistent periodic oscillations of $\mathbf{n}$ and e have been observed and studied theoretically. With increasing laser intensities, however, the system may go through a sequence of transitions and enter a deterministic chaotic regime [116]. The route to chaos seems to be a mixing of two well-known scenarios: (a) periodic oscillation $\rightarrow$ quasiperiodic oscillation $\rightarrow$ chaos; and (b) periodic oscillation $\rightarrow$ period-doubling cascade $\rightarrow$ chaos. Because of the presence of both ordinary and extraordinary waves in the medium, and the strong coupling between them via the variation of $\mathbf{n}(z)$ starting almost at the very beginning, the theoretical analysis of the present case is more complicated than the normal-incidence case. While some qualitative understanding could be gained following the ideas learned in the normal incidence case [74,75], a full understanding of the problem has yet to be achieved.

14.3.12 Intrinsic Optical Bistability

We mentioned in section 14.3.9 that for optical Fréedericksz transitions in a homeotropic nematic film, a DC bias field stabilizing the homeotropic alignment is expected to increase the threshold intensity for the transition. When the DC field is sufficiently strong, the transition becomes first-order and is characterized by a hysteresis loop in the medium response versus laser intensity [108]. This is a form of optical bistability. Since it is mirrorless and depends only on the internal feedback in the medium, we call it intrinsic optical bistability. The internal feedback responsible for the first-order structural phase transition in the present case comes from the dependence of the optical torque experienced by $\mathbf{n}$ on the reorientation angle θ. It is easily shown that for $I > I_{\text{th}}$, the optical torque is proportional to $I\varepsilon_a n^3(\theta)\sin\theta\cos\theta$, with $\mathbf{n}(\theta) = n_o/[1 - (\varepsilon_a/\varepsilon_\perp)\sin 2\theta]^{1/2}$. Thus the torque acting on $\mathbf{n}$ increases with θ, and the increase of θ in turn enhances the torque, as long as θ is not close to $\pi/2$. This is a positive feedback with its strength proportional to I. By increasing I_{th} with the DC bias field, the positive feedback at $I \sim I_{\text{th}}$ is also increased. When the feedback is sufficiently strong, the optical Fréedericksz transition naturally becomes first-order (see Figure 14.13). One can, of course, also explain the phenomenon in terms of the usual description of first-order

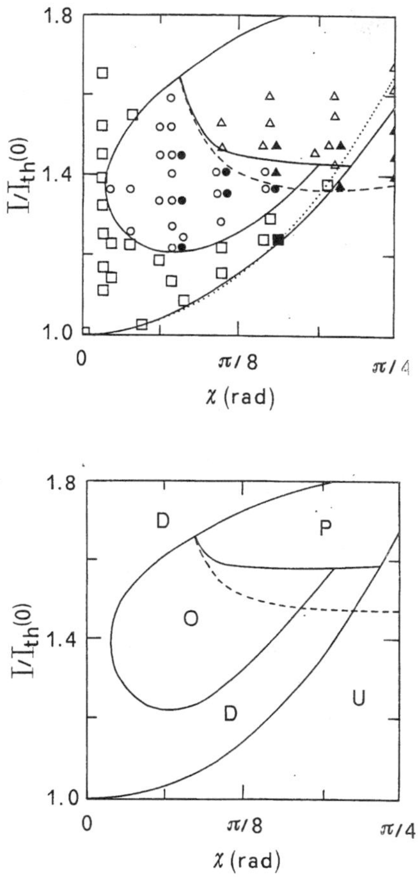

FIGURE 14.13. Phase retardation $\phi/2\pi$ versus laser intensity for fixed magnetic field strengths. Open symbols were measured with increasing laser intensity; solid symbols, with decreasing intensity. Solid lines are guides to the eye. Broken lines are theoretical fits. H_o is the magnetic field at which the second-order optical Fréedericksz transition is expected to become first-order (after [108]).

phase transitions, using the Landau free-energy approach [100].

With circularly or strongly elliptically polarized light, I_{th} is significantly larger than in the linearly polarized case. Even without a bias field, the internal feedback is sufficiently strong to make the transition first-order, as has actually been observed experimentally [110,111].

Another way to obtain intrinsic optical bistability in a homeotropically aligned nematic film is to have two beams of extraordinary polarization obliquely incident on the film in the same incidence plane, but at opposite sides of the surface normal [117]. The intensities of the beams are so adjusted that the optical torques on the molecules from the two beams exactly cancel each other. This fixes the intensity ratio I_1/I_2 of the two beams. However, when the total intensity $I = I_1 + I_2$ reaches a threshold, the optical Fréedericksz transition again must occur. A first-order transition with very large hysteresis can be observed if the angle between the two beams is sufficiently large. The physical reason for the phenomenon is quite similar to that for the DC bias field case. The large I_{th} value makes the positive feedback strong enough for the first-order transition to develop. With the same two-beam geometry, one can use the weak beam to modulate a strong beam, and obtain the so-called "transistor" action [118].

Yet another way to obtain intrinsic optical bistability is to pass a linearly polarized beam through a chiral-doped nematic film with homeotropic anchoring at the walls [119,120]. In this case, the observed hysteresis loops are large, and the jumps between bistable states are appreciable. An example is shown in FIGURE 14.14 using a nematic sample (E7 from BDH) doped with a small amount of a cholesteric (C15 from BDH). The mechanism here is more subtle, and is based on the competition between the chiral dopant, intending to induce an helical alignment in the sample, and the wall anchoring, intending to maintain the homeotropic alignment. In the absence of light, the anchoring force is sufficient to stabilize the homeotropic alignment. The laser beam tends to destabilize the homeotropic

FIGURE 14.14. Large bistable loop observed in the laser-induced phase shift versus the reduced laser intensity I/I_{th} in a chiral-doped nematic film (after [120]).

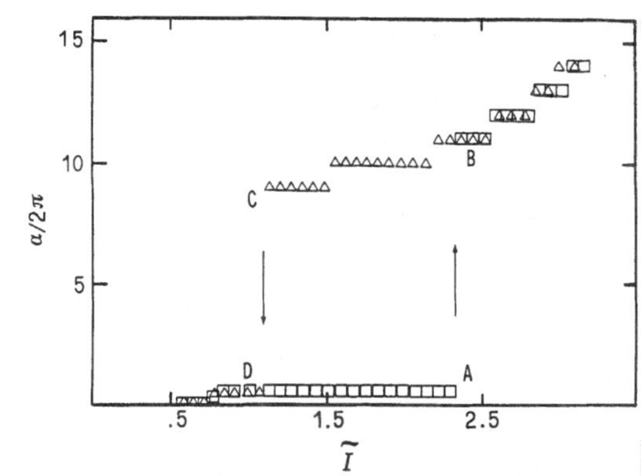

alignment. When the beam intensity increases above the Fréedericksz transition threshold, the molecular director in the film assumes a combined tilt-and-twist configuration. The transition is second-order, exhibiting no hysteresis. If the laser intensity is further increased, the total phase retardation imposed on the beam soon saturates, evidenced by the stationary ring pattern in the far field, due to self-phase modulation. The ring pattern remains "frozen" until a second intensity threshold is reached where more rings suddenly appear. This second transition is clearly first-order. The associated hysteresis loop could be very large. The observed ring pattern remains basically unchanged until I reaches the third threshold intensity [119,120] and more rings appear suddenly. Such intrinsic multistable behavior described by the multiple hysteresis loops is due to interplay between twist and splay-bend distortions in the sample. The number of diffraction rings is proportional to the phase modulation $\Delta\Psi_o$ arising from the splay-bend distortion. In the intensity range where the ring pattern remains unchanged, only the twist distortion is changed by the laser-intensity variation. The splay-bend distortion changes appreciably only at the threshold laser intensities where the first-order transitions occur. The existence of multiple stable reorientation states can be understood in terms of conservation of total angular momentum [119, 120]. The linearly polarized input carries zero angular momentum. For the director **n** to remain constant in time, no angular momentum should be deposited in the medium by light, and hence the transmitted light must be linearly polarized. From the symmetry of the deformed system with respect to $z = L/2$, the output linear polarization must be either parallel or perpendicular to the input polarization, This means that the accumulated phase shift from birefringence due to splay-bend distortion in the film must be $m\pi$, with m being an integer. It can be shown that, following Mauguin's theorem [121], the stable output must correspond to m being odd integers. The observed first-order transitions where the far-field ring number changes abruptly, is a manifestation of a jump between two such consecutive stable output states.

14.4 APPLICATIONS OF NONLINEAR OPTICS

Nonlinear optical effects in LCs may be used to construct nonlinear optical devices such as optical switches, optical memories, phase conjugators, power limiters, and so on. Unfortunately, while the nonlinearities of LCs in their mesophases are large, their response times are very long. The devices could be too slow to be useful for many applications.

Nonlinear optics, however, can be used to study LCs. For example, two-photon-excited luminescence allows the determination of higher-rank orientational order parameters [122]. Laser-induced birefringence permits the study of reorientational dynamics of both individual and correlated molecules [21,22,56–63]. Here, we shall focus on how we can use optical second-harmonic generation (SHG) to probe LC films and surfaces, and hit upon some unexplored areas of research.

Optical SHG originates from the induced nonlinear polarization [9]

$$P^{(2)}(2\omega) = \chi^{(2)} : E(\omega)E(\omega) \tag{14.31}$$

in a medium, where $\chi^{(2)}$ is the nonlinear susceptibility. From symmetry considerations, $\chi^{(2)}$ should vanish under the electric-dipole approximation in a bulk medium with a center of inversion, but at the surface, the inversion symmetry is necessarily broken and $\chi^{(2)}$ could be significant. This symmetry property makes SHG particularly sensitive to surfaces. It so happens that SHG is sufficiently sensitive to detect a submonolayer of molecules in a polar arrangement. The technique can therefore be used as a versatile probe for surface studies [123]. Here, we discuss in particular how we can apply it to study LC monolayers and its orientation on surfaces [16,124], their interaction with surfaces, and surface-induced bulk alignment in LCs [13, 124–126].

Consider an LC monolayer polar-oriented on a substrate, and assume that $\chi^{(2)}$ of the system is dominated by the monolayer. The SHG signal in the reflected direction from such an interface is given by [123]

$$\begin{aligned} S = {} & [(32\pi \sec^2\theta)/(\hbar\omega^3)] \\ & \times |L(2\omega)e_{2\omega} \cdot \chi^{(2)} : e_\omega e_\omega L(\omega)^2|^2 I(\omega)^2 AT \end{aligned} \tag{14.32}$$

where θ is the angle of incidence, e is the polarization unit vector, $L(\omega)$ is the appropriate Fresnel factor at ω, and I, A, and T are the beam intensity, beam cross-section, and laser pulsewidth, respectively. The nonlinear susceptibility $\chi^{(2)}$ of the monolayer can be expressed in terms of the nonlinear molecular polarizability $\alpha^{(2)}$

$$\chi^{(2)} = N\langle\alpha^{(2)}\rangle$$
$$\chi^{(2)}_{ijk} = N\langle G^{\xi\eta\zeta}_{ijk}\rangle\alpha^{(2)}_{\xi\eta\zeta} \tag{14.33}$$

where N is the surface density of molecules, $G^{\xi\eta\zeta}_{ijk}$ denotes the geometric transformation from the molecular coordinates (ξ,η,ζ) to the laboratory coordinates (i,j,k), and the angular brackets refer to an orientational average. Further simplification results when we recognize that many LC molecules are rod-like, with $\alpha^{(2)}$ dominated by a single element $\alpha^{(2)}_{\zeta\zeta\zeta}$ along the long molecular axis ζ.

$$\chi^{(2)}_{ijk} = N\langle(i\cdot\zeta)(j\cdot\zeta)(k\cdot\zeta)\rangle\alpha^{(2)}_{\zeta\zeta\zeta}. \tag{14.34}$$

If the molecules are oriented randomly in the surface plane, then symmetry requires that the only nonvanishing components of $\chi^{(2)}$ are

$$\chi^{(2)}_{zzz} = N\langle\cos^3\theta\rangle\alpha^{(2)}_{\zeta\zeta}$$
$$\chi^{(2)}_{zxx} = \chi^{(2)}_{zyy} = \chi^{(2)}_{xzx} = \chi^{(2)}_{yzy} = (1/2)N\langle\sin^2\theta\cos\theta\rangle\alpha^{(2)}_{\zeta\zeta\zeta} \tag{14.35}$$

with z along the surface normal and θ being the angle between z and ζ. SHG measurements with proper input/output polarization combinations allow us to deduce these nonvanishing $\chi^{(2)}$ elements. From equation (14.35), we can then obtain $\alpha^{(2)}_{\zeta\zeta\zeta}$ and θ, if the orientational distribution is approximated by a δ-function.

The above technique has been used to measure $\alpha^{(2)}_{\zeta\zeta\zeta}$ and θ for a number of LC molecules adsorbed on different substrates. Many LC molecules have large $\alpha^{(2)}_{\zeta\zeta\zeta}$ because of electron delocalization and significant charge transfer in their aromatic chromophores. The angle θ is in the range 70–80°, indicating that the chromophores are lying close to the surface. This orientation is mainly due to Van der Waals interactions between the molecules and the substrates [127].

Knowing $\alpha^{(2)}_{\zeta\zeta\zeta}$, we can use SHG to measure N and monitor *in situ* the adsorption and desorption of LC molecules on a surface [124]. In depositing 8CB molecules on a glass substrate coated with DMOAP for bulk homeotropic alignment, for example, the SHG signal first increases as N^2 and then quickly saturates at one full monolayer. A further increase of N reduces the signal gradually; the signal reduction is appreciable only when N approaches a number equivalent to about ten monolayers. This indicates that after the first monolayer, which is polar-oriented via adsorption, molecules in the additional layers must be in a quadrupole-pair arrangement. The signal decreases with N because the molecules in quadrupole pairs have their heads facing each other (FIGURE 14.14) so that in the homeotropic alignment, the dipole layer and the quadrupole pair layers contribute to $\chi^{(2)}$ with opposite signs. On the other hand, SHG from the air/LC interface is smaller than that from a single monolayer of LC on glass, indicating that no polar ordering of 8CB molecules is present at the air/LC interface. A careful analysis of the experimental results shows that the air/LC interface is dominated by quadrupole pairs [128].

If the substrate surface is anisotropic, the LC molecule–substrate interaction is likely to lead to an anisotropic orientational distribution for the LC monolayer in the surface plane. This can also be measured by SHG. Consider, for example, a rubbed polyimide surface where the polymer chains are presumably stretched and aligned to some extent along the rubbing direction (defined as x) [129]. The polar-oriented LC monolayer adsorbed on the surface now has only a mirror plane along x. Consequently, the independent nonvanishing $\chi^{(2)}$ elements are $\chi^{(2)}_{zzz}$, $\chi^{(2)}_{zyy} = \chi^{(2)}_{yzy}$, $\chi^{(2)}_{zxx} = \chi^{(2)}_{xzx}$, $\chi^{(2)}_{xxx}$, $\chi^{(2)}_{xzz}$, and $\chi^{(2)}_{xyz}$. Again, with $\alpha^{(2)}$ dominated by $\alpha^{(2)}_{\zeta\zeta\zeta}$, χ_{ijk} is related to $\alpha^{(2)}_{\zeta\zeta\zeta}$ by equation (14.34), from which we find

$$\begin{aligned}
\chi^{(2)}_{zzz} &= N\langle\sin^2\theta\rangle\alpha^{(2)}_{\zeta\zeta\zeta}\\
\chi^{(2)}_{zyy} = \chi^{(2)}_{yzy} &= N\langle(\sin\theta\cos^2\theta)\sin^2\phi\rangle\alpha^{(2)}_{\zeta\zeta\zeta}\\
\chi^{(2)}_{zxx} = \chi^{(2)}_{xzx} &= N\langle(\sin\theta\cos^2\theta)\cos^2\phi\rangle\alpha^{(2)}_{\zeta\zeta\zeta}\\
\chi^{(2)}_{xxx} &= N\langle\cos^3\theta\cos^3\phi\rangle\alpha^{(2)}_{\zeta\zeta\zeta}\\
\chi^{(2)}_{xyy} = \chi^{(2)}_{yxy} &= N\langle\cos^3\theta(\sin^2\phi\cos\phi)\rangle\alpha^{(2)}_{\zeta\zeta\zeta}\\
\chi^{(2)}_{xzz} = \chi^{(2)}_{zxz} &= N\langle(\sin^2\theta\cos\theta)\cos\phi\rangle\alpha^{(2)}_{\zeta\zeta\zeta}
\end{aligned} \tag{14.36}$$

where ϕ is the azimuthal angle of ζ from x. We can deduce these nonvanishing $\chi^{(2)}_{ijk}$ for the LC monolayer from SHG measurements with different polarization combinations as functions of the azimuthal angle, ψ, between the plane of incidence and the rubbing direction x. As an example, FIGURE 14.15 shows the results for an 8CB monolayer on a rubbed polyimide surface, together with the theoretical fit that determines the values of $\chi^{(2)}_{ijk}$. They clearly exhibit the mirror symmetry of the orientational distribution the

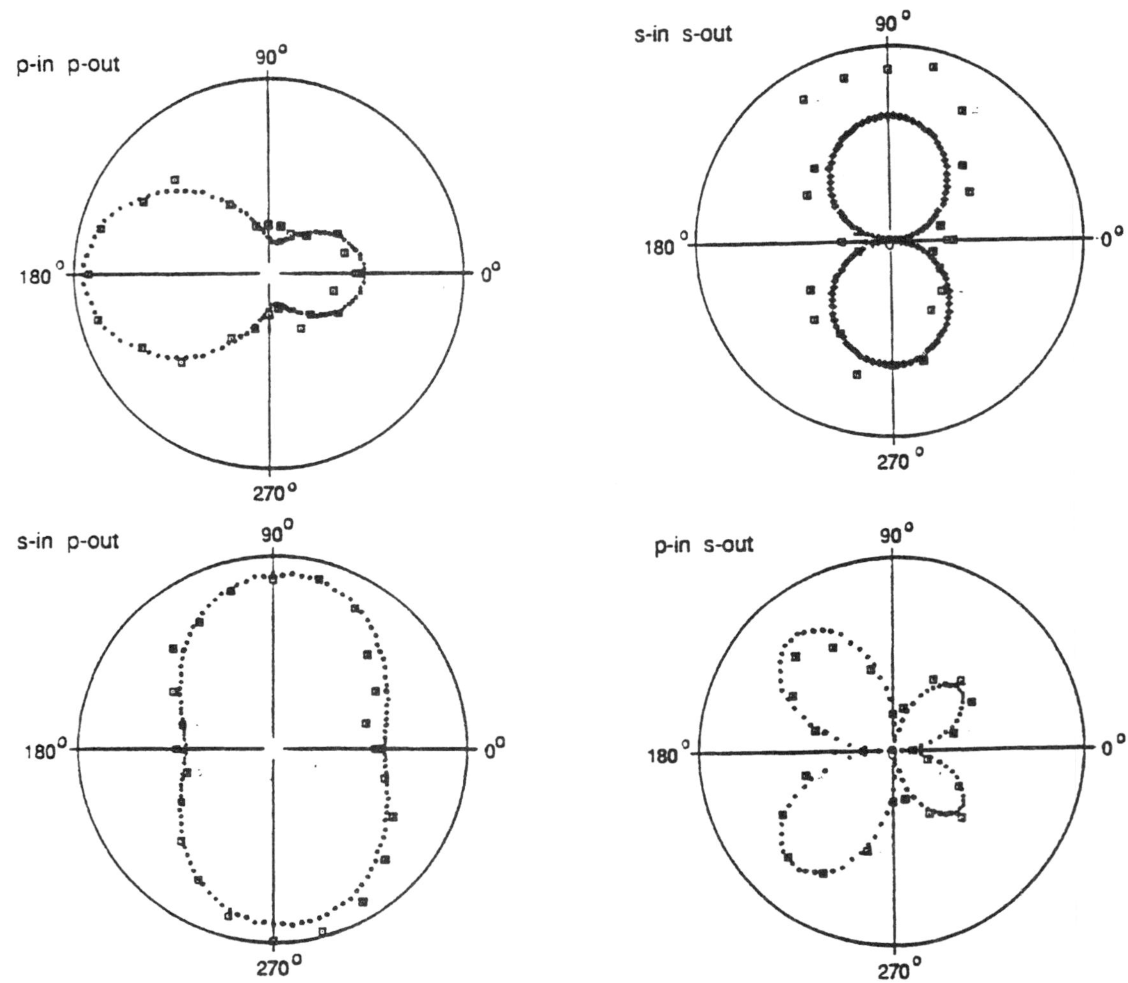

FIGURE 14.15. Square root of the second-harmonic signal (arbitrary units) versus the angle between the rubbing direction and the plane of incidence for four different input/output polarization combinations. p-in and p-out refer to a polarized fundamental input and a polarized second-harmonic output. The sample is an 8CB monolayer on a rubbed polyimide-coated substrate (after [13]).

monolayer is expected to possess if the distribution is induced by the rubbed polymer surface. Moreover, there is a forward–backward asymmetry indicating that the molecules are tilted along the rubbing direction [13].

To find the molecular orientational distribution of the monolayer more quantitatively, we realize that measurement of the six independent nonvanishing $\chi_{ijk}^{(2)}$ should allow us to determine five parameters in the distribution function, in addition to $N\alpha_{\zeta\zeta\zeta}^{(2)}$. We assume the orientational distribution function to be

$$P(\theta, \phi) = f(\theta)g(\phi)$$

with $f(\theta) = \exp[-(\theta - \theta_0)^2/2\sigma^2]$

$$g(\phi) = 1 + d_1 \cos\phi + d_2 \cos 2\phi + d_3 \cos 3\phi.$$

The five parameters to be determined are $\theta_0, \sigma, d_1, d_2$, and d_3. For the data set in FIGURE 14.15, for example, we find $\theta_0 = 16°$, $\sigma = 4°$, $d_1 = 0.18$, $d_2 = 0.55$, and $d_3 = 0.02$. It has been found that θ_0 and σ are rather insensitive to rubbing, and d_3 is almost negligible, but

d_1 and d_2 are appreciable and sensitive to rubbing, with d_1 describing the forward–backward asymmetry and d_2 the x–y anisotropy.

This surface monolayer orientational distribution is of critical importance for surface-induced LC bulk alignment (assuming rubbing is light, so that no grooves are created on the surface). When anchoring of the monolayer on a substrate is strong, as, for example, in the case of cyanobiphenyls on polyimide, the monolayer orientational distribution is stable and hardly affected by the presence of the LC bulk. Knowing this distribution function, we can calculate the surface-order parameter tensor Q_s for the LC surface monolayer. We can then use the Landau–de Gennes free-energy approach to find how the order parameter changes with distance from Q_s to the bulk value Q_b, including the alignment direction. In a recent study [127], it was found that such a calculation for an 8CB film sandwiched between polyimide-coated substrates could predict a homogeneous bulk alignment with a pretilt angle depending on the rubbing strength and the type of polyimide. Experimental measurements of the pretilt angles agreed very well with the predictions.

That the orientation of an LC surface monolayer completely determines the bulk LC alignment is also evidenced in the case of LC films on crystalline surfaces [126]. It was reported that the director orientation of an LC film on a crystal could depend on the atmosphere the film is exposed to [130]. An example is 5OCB on muscovite mica in the presence of ethylene glycol vapor. As the partial pressure of ethylene glycol increases above a critical value, the LC bulk orientation changes suddenly from a direction perpendicular to the mirror plane of the mica surface to a direction parallel to this plane and to the surface. This is known as anchoring transition. It is found, by using SHG to probe the orientation of the surface LC monolayer, that the anchoring transition is actually induced by an orientational transition of the monolayer [126]. Adsorption of a critical amount of ethylene glycol onto the mica substrate after diffusion through the LC film is the cause of the monolayer transition. This is therefore another illustration of how the surface monolayer orientation dictates the LC bulk alignment.

In a surface-stabilized ferroelectric LC cell, it is generally found that the homogeneous bulk alignment direction is not along the rubbing direction but deviates from it. With SHG, we can show conclusively that the surface LC monolayer is still well aligned along the rubbing direction. However, molecules are polar-oriented on the surface, and they can generate a surface DC field. In ferroelectric LCs, this field can give rise to a surface electroclinic effect on the surface–bulk transition layer, forcing the molecules in this region to twist away from the surface monolayer alignment [131].

In the above discussion, the substrate surface is assumed to be smooth. However, hard rubbing of a substrate surface or oblique evaporation of SiO_x on a substrate can also induce homogeneous bulk alignment of LCs. This is because rubbing or oblique evaporation creates a rough surface, with grooves along one direction. Molecules lie preferentially along the grooves in order to minimize the elastic deformation [132]. In such cases, the surface LC monolayer should not be well oriented because of surface roughness on the macroscopic scale. This has also been verified by SHG measurements [13].

To complete our discussion on surface-induced bulk alignment, we should also consider homeotropic LC bulk alignment. Such an alignment is normally induced by coating the substrate with a monolayer of long-chain surfactants. Consider two cases with substrates coated with two different surfactant monolayers, one being DMOAP (n,n'-dimethl-n-octadecyl-3-aminopropyl-trimethoxy-silychloride) which has a long hydrocarbon chain, and the other being MAP (methylaminopropyl-trimethoxy silane) which has essentially the same base as DMOAP but no hydrocarbon chain. The former renders a homeotropic bulk alignment while the latter renders a planar alignment. SHG measurements show that in the two cases, the surface LC monolayers have the same orientation. Therefore, the homeotropic alignment must be the result of interaction of LC molecules with the extended hydrocarbon chains of DMOAP. This can be investigated with the help of surface vibrational spectroscopy using infrared-visible sum-frequency generation (SFG).

FIGURE 14.16 shows the C–H spectra of (a) a loosely packed monolayer of PDA (pentadecanoic acid) (47 Å^2/molecule compared with 22 Å^2/molecule for a fully packed layer); (b) a full monolayer of DMOAP ($\simeq$45 Å^2/molecule); and (c) a 0.7 monolayer of 8CB deposited on the DMOAP monolayer [125]. The first two spectra are very similar because they have nearly the same hydrocarbon chain density. In each

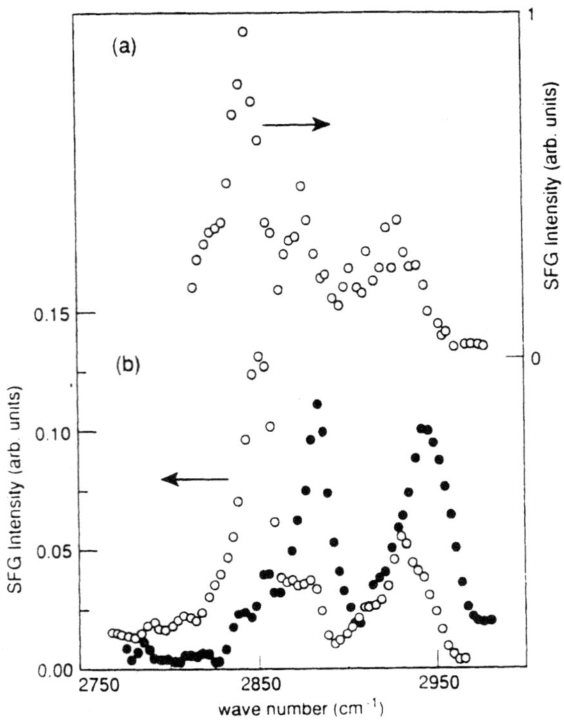

FIGURE 14.16. (a) Spectrum of the PDA monolayer on water surface at a density of 47 Å/molecule. (b) Comparison of the sum-frequency generation spectrum of DMOAP on clean glass (open circles) and the spectrum of the same sample after deposition of 0.7 monolayer of 8CB (solid circles) (after [125]).

spectrum, the peak at 2850 cm^{-1} can be assigned to the CH_3 symmetric stretch mode, and the peaks at 2875 and 2940 cm^{-1} to the asymmetric stretch and Fermi resonance modes, respectively, of the terminal CH_3 group. The CH_2 peak is prominent only because the alkyl chains of the molecules are not straight and upright. They have room to flap around and develop *trans-gauche* defects that break the inversion symmetry and contribute to the CH_2 peak. Deposition of 8CB on the DMOAP layer changes the spectrum drastically, as seen in FIGURE 14.16(b). The CH_2 peak is significantly reduced, and the spectrum now looks more like that for a full PDA monolayer. This indicates that the interaction between 8CB and DMOAP molecules must have straightened the alkyl chains of DMOAP.

From the geometry of the molecular arrangement, it is likely that the chain–chain interaction between 8CB and DMOAP molecules are the most important for the observed conformational change in DMOAP. The same interaction, together with the geometrical restriction, aligns the second 8CB monolayer homeotropically (tails against tails, with the first monolayer to offset the polar arrangement of the first monolayer). This then leads, in the sense of molecular epitaxy, to the homeotropic bulk alignment.

14.5 CONCLUSION

Since the first study of nonlinear optical effects in LCs in the early 1970s, the field of nonlinear optics in LCs has grown significantly, particularly in the last 10 to 15 years. Hundreds of papers have been published on the subject. We have given here only a brief survey, which is not intended to be thorough or complete. As a result, many works must have been neglected, for which we sincerely apologize.

Most of the nonlinear optical effects studied in LCs fall into the class of wave mixing. The optical nonlinearity responsible for these effects derives from either laser-induced thermal indexing, or laser-induced molecular reorientation, or both. Such wave mixing processes are similar to those in other media, but they appear to be particularly strong in LC mesophases because of the intrinsic giant nonlinearities in the mesophases. The response times, on the other hand, are intrinsically very long, making LCs less interesting as practical materials for nonlinear optical devices. There are also other common types of nonlinear optical effects, such as self-phase modulation, self-focusing, stimulated light scattering, and optical bistability, which can easily be observed in LCs. However, most interesting in the study of nonlinear optics in LCs are those unique to LCs. They are so highly nonlinear that they cannot be understood by any perturbation theory, although they can be readily observed with a cw laser. Optical Fréedericksz transitions, self-induced stimulated light scattering, and laser-induced nonlinear dynamics are just few representative examples. These interesting phenomena typically result from strong coupling between the laser field and the LC. While the experiments are often fairly simple, great challenges usually lie in the theoretical interpretation of the experimental results. So far, for highly nonlinear optical phenomena, only a few geometries have been determined. Many others are yet to be explored.

Many LC molecules possess strong second-order nonlinearity because of asymmetry and electron delocalization in their chromophores. If they could only be polar-oriented in the condensed phase, they would constitute ideal materials for second-harmonic and sum-frequency generation. Recently, there has been some success along this line, but for real applications, further advances are needed. Optical SHG and SFG, however, can be used as effective tools to probe LC interfaces, owing to surface specificity of these second-order processes. Such studies are of great importance to LC display technology, considering that most LC devices rely on surface-induced bulk alignment. Indeed, SHG and SFG have permitted the measurements of molecular orientations and distributions in surface LC monolayers, leading to the understanding of how a surface monolayer orientation can completely determine the associated bulk orientation. However, only a few cases have been tested. We still need to explore more systems with different substrates and different LC molecules, particularly those exhibiting ferroelectric or antiferroelectric phases. More generally, we could still expand our study of LC using other well-established nonlinear optical effects.

The field of nonlinear optics in LCs is young and beautiful; more effort and imagination are needed to make it mature and strong.

Acknowledgement

This work was supported by the Director, Office of Energy Research, Office of Basic Energy Sciences, Materials Sciences Division of the U.S. Department of Energy, under Contract No. DE-AC03-76SF00098. One of us (E.S.) acknowledges the support of INFM (Istituto Nazionale di Fisica della Materia) and INO (Istituto Nazionale di Ottica).

REFERENCES

1. Cf. Chemia, D. S., and Zyss, J., eds. 1987. *Nonlinear Optical Properties of Organic Molecules and Crystals*. New York: Academic.
2. de Gennes, P. G., and Prost, J. 1993. *The Physics of Liquid Crystals*. 2d ed. Oxford: Oxford University Press.
3. Chandrasekhar, S. 1992. *Liquid Crystals*. 2d ed. Cambridge: Cambridge University Press.
4. Khoo, I. C., and Shen, Y. R. 1985. *Opt. Eng.* **24**: 579 and references therein.
5. Tabiryan, N. V., Sukhov, A. V., and Zel'dovich, B. Y. 1986. The orientational optical nonlinearity of liquid crystals. *Mol. Cryst. Liq. Cryst.* **136**: 1.
6. Khoo, I. C. 1988. Nonlinear optics of liquid crystals. *Prog. Opt.* **26**: 107.
7. Simoni, F. 1991. Nonlinear optical phenomena in nematics. *Physics of Liquid Crystalline Materials: Based on Lectures Delivered at the Summer School on the Physics of Liquid Crystals [Bra, Italy, 1988]*, ed. I. C. Khoo and F. Simoni, pp. 365–392. Philadelphia: Gordon and Breach.
8. Khoo, I. C., and Wu, S.-T. 1993. *Optics and Nonlinear Optics of Liquid Crystals*. Singapore: World Scientific.
9. Cf. Khoo, I. C., and Shen, Y. R. 1985. *Opt. Eng.* **24**: 579 and references therein.
10. Taguchi, A., Ouchi, Y., Takezoe, H., and Fukuda, A. 1989. *Japan. J. Appl. Phys. Lett.* **28**: 997.
11. Schadt, M. 1993. *Liq. Cryst.* **14**: 73.
12. Shen, Y. R. 1989. *Liq. Cryst.* **5**: 635 and references therein.

13a. Chen, W., Feller, M. B., and Shen, Y. R. 1989. *Phys. Rev. Lett.* **63**: 2665.

13b. Feller, M. B., Chen, W., and Shen, Y. R. 1991. *Phys. Rev. A* **43**: 6778.

13c. Chen, W., Ouchi, Y., Moses, T., Shen, Y. R., and Yang, K. H. 1992. *Phys. Rev. Lett.* **68**: 1547.

13d. Johannsmann, D., Zhou, H., Sonderkaer, P., Wierenga, H., Myrvold, B. O., and Shen, Y. R. 1993. *Phys. Rev. E* **48**: 1889.

13e. Zhuang, X., Marucci, L., and Shen, Y. R. 1994. *Phys. Rev. Lett.* **73**: 1513.

14. Li, L., Yuan, H. J., and Palffy-Muhoray, P. 1991. *Mol. Cryst. Liq. Cryst.* **198**: 239.
15. Herman, R. M., and Serinko, R. J. 1979. *Phys. Rev. A* **19**: 1757.
16. Berkovic, G., Rasing, T., and Shen, Y. R. 1987. *J. Opt. Soc. Am. B* **4**: 945.
17. Drevensec, I., and Blinc, R. 1992. *Condens. Matt. News* **1**: 14.
18. Smith, D. A. McL., and Coles, H. J. 1993. *Liq. Cryst.* **14**: 937.

19a. *Proc. SPIE* **2025** (1993).

19b. *Proc. SPIE* **2042** (1994).

19c. *MRS Symp. Proc.* **328** (1994).

19d. *Proceedings of the International Conference on Optically Nonlinear Organic Materials and Applications* (4th Iketani Conference), ed. K. Sasaki and H. Ito (Hawaii, 1994), published as a volume in the journal [Nonlinear Optics, V. 14, p. 1–326 (1995)].

19e. See articles in the *Journal of Nonlinear Optics* (1993–1994).

20a. Shelton, J. W., and Shen, Y. R. 1970. *Phys. Rev. Lett.* **25**: 23.

20b. Shelton, J. W., and Shen, Y. R. 1971. *Phys. Rev. Lett.* **26**: 538.
20c. Shelton, J. W., and Shen, Y. R. 1972. *Phys. Rev. A.* **5**: 1867.
21. Wong, G. K. L., and Shen, Y. R. 1973. *Phys. Rev. Lett.* **30**: 895.
22. Wong, G. K. L., and Shen, Y. R. 1974. *Phys. Rev. A.* **10**: 1277.
23. Zolot'ko, A. S., Kitaeva, V. F., Kroo, N., Sobolev, N. N., and Csillag, L. 1980. *JETP Lett.* **32**: 158.
24. Zel'dovich, B. Y., and Tabiryan, N. V. 1980. *Sov. J. Quantum Electron.* **10**: 440.
25. Zel'dovich, B. Y., Pilipetskyi, N. F., Sukhov, A. V., and Tabiryan, N. V. 1980. *JETP Lett.* **32**: 263.
26. Durbin, S. D., Arakelian, S. M., and Shen, Y. R. 1981. *Phys. Rev. Lett.* **47**: 1411.
27. Saupe, A. 1960. *Z. Naturforsch.* **A15**: 810.
28. Ericksen, J. L. 1960. *Arch. Ration. Mech. Anal.* **4**: 231.
29. Ericksen, J. L. 1966. *Phys. Fluids* **9**: 1205.
30. Leslie, F. M. 1968. *Arch. Ration. Mech. Anal.* **28**: 265.
31. Parodi, O. 1970. *J. Physique* **31**: 581.
32. Jánossy, I., Lloyd, A. D., and Wherret, B. S. 1990. *Mol. Cryst. Liq. Cryst.* **179**: 1.
33. Jánossy, I., Csillag, L., and Lloyd, A. D. 1991. *Phys. Rev. A* **44**: 8410.
34. Jánossy, I., and Kosa, T. 1992. *Opt. Lett.* **17**: 1183.
35. Khoo, I. C., Hong, L., and Liang, Y. 1993. *IEEE J. Quantum Electron.* **QE-29**: 1444.
36. Khoo, I. C., and Normandin, R. 1985. *IEEE J. Quantum Electron.* **QE-21**: 329.
37. Khoo, I. C., and Normandin, R. 1984. *Opt. Lett.* **9**: 285.
38. Chu, K. C., Chen, C. K., and Shen, Y. R. 1980. *Mol. Cryst. Liq. Cryst.* **59**: 97; Gordon and Breach Science Publishers.
39. Umeton, C., Sgro, A., and Simoni, F. 1987. *J. Opt. Soc. Am. B* **4**: 1938.
40. Odulov, S. G., Reznikov, Y. A., Soskin, M. S., and Khizhnyak, A. I. 1983. *Sov. Phys.–JETP* **58**: 1154.
41. Zel'dovich, B. Y., and Tabiryan, N. V. 1986. *Sov. Phys.–JETP* **63**: 80.
42. Khoo, I. C. 1981. *Appl. Phys. Lett.* **38**: 123.
43. Khoo, I. C., and Zhuang, S. L. 1980. *Appl. Phys. Lett.* **37**: 3.
44. Khoo, I. C. 1982. *Phys. Rev. A* **25**: 1040.
45. Khoo, I. C. 1986. *IEEE J. Quantum Electron.* **QE-22**: 1268.
46. Khoo, I. C. 1986. *Opt. Eng.* **25**: 198.
47. Khoo, I. C., Zhou, P., Lindquist, R. G., and LoPresti, P. 1990. *Phys. Rev. A* **41**: 408.
48. Khoo, I. C., and Zhuang, S. L. 1982. *IEEE J. Quantum Electron.* **QE-18**: 246.
49. Khoo, I. C., Li, H., and Liang, Y. 1993. *Opt. Lett.* **18**: 1490.
50. Khoo, I. C., and Normandin, R. 1985. *Appl. Phys. Lett.* **47**: 350.
51. Simoni, F., Cipparrone, G., Duca, D., and Khoo, I. C. 1991. *Opt. Lett.* **16**: 360.
52. Khoo, I. C., and Liu, T. H. 1986. *IEEE J. Quantum Electron.* **QE-23**: 171.
53. Liu, T. H., and Khoo, I. C. 1986. *IEEE J. Quantum Electron.* **QE-23**: 2020.
54. Khoo, I. C., Yan, P. Y., Finn, G. M., Liu, T. H., and Michael, R. R. 1988. *J. Opt. Soc. Am. B* **5**: 202.
55. Durbin, S. D., Arakelian, S. M., and Shen, Y. R. 1982. *Opt. Lett.* **7**: 145.
56. MacDonald, R., and Grebe, D. 1993. *Phys. Rev. Lett.* **70**: 2746.
57. Hsiung, H., Shi, L. P., and Shen, Y. R. 1984. *Phys. Rev. A* **30**: 1453.
58. Khoo, I. C., Lindquist, R. G., Michael, R. R., Mansfield, R. J., and LoPresti, P. 1991. *J. Appl. Phys.* **69**: 3853.
59. Eichler, H. J., and MacDonald, R. 1991. *Phys. Rev. Lett.* **67**: 2666.
60. Flytzanis, C., and Shen, Y. R. 1974. *Phys. Rev. Lett.* **33**: 14.
61. Lallane, J. R., Martin, B., Pouligny, B., and Kielich, S. 1977. *Mol. Cryst. Liq. Cryst.* **42**: 153.
62a. Lallane, J. R., Buchert, J., Destrade, C., Nguyen, H. T., and Marcerou, J. P. 1989. *Phys. Rev. Lett.* **62**: 3046.
62b. Lallane, J. R., Destrade, C., Nguyen, H. T., and Marcerou, J. P. 1991. *Phys. Rev. A* **14**: 6632.
63. O'Brien, J., Moses, T., Chen, W., Freysz, E., Ouchi, Y., and Shen, Y. R. 1993. *Phys. Rev. E* **47**: R2269.
64. Khoo, I. C. 1983. *Phys. Rev. A* **27**: 2747.
65. Zel'dovich, B. Y., Merzlikin, S. K., Pilipetskyi, N. F., and Sukhov, A. V. 1985. *JETP Lett.* **41**: 514.
66. Khoo, I. C., Michael, R. R., Lindquist, R. G., Zhou, P., and Mansfield, R. J. 1990. *Mol. Cryst. Liq. Cryst.* **179**: 163.
67. Zel'dovich, B. Y., and Tabiryan, N. V. 1979. *JETP Lett.* **30**: 478.
68. Zel'dovich, B. Y., Merzlikin, S. K., Pilipetskyi, N. F., Sukhov, A. V., and Tabiryan, N. V. 1983. *Sov. Phys. Dokl.* **28**: 1038.
69. Zel'dovich, B. Y., Sukhov, A. V., and Tabiryan, N. V. 1987. *Mol. Cryst. Liq. Cryst.* **143**: 7.
70. Khoo, I. C., Michael, R. R., and Yan, P. R. 1986. *IEEE J. Quantum Electron.* **QE-23**: 1344.
71. Simoni, F., Bloisi, F., and Vicari, L. 1992. *Mol. Cryst. Liq. Cryst.* **223**: 169.
72. Khoo, I. C., Wang, W., Simoni, F., Cipparrone, G., and Duca, D. 1991. *J. Opt. Soc. Am. B* **8**: 1433.
73. Goosev, I. V., Kirvoshekov, V. A., and Zel'dovich, B. Y. 1992. *Mol. Cryst. Liq. Cryst.* **222**: 97.
74. Arakelian, S. M. 1987. *Sov. Phys. Usp.* **30**: 1041.
75. Arakelian, S. M., Chilingarian, S. Y., Alaverdian, R. B., Grigorian, G. L., Karaian, A. S., Nersissian, S. T., and Drnoian, V. E. 1989. *J. Physique* **50**: 1393 and references therein.
76. Durbin, S. D., Arakelian, S. M., and Shen, Y. R. 1981. *Opt. Lett.* **6**: 411.
77a. Sheik-Bahae, M., Said, A. A., and Van Stryland, E. W. 1989. *Opt. Lett.* **14**: 955.

77b. Sheik-Babae, M., Said, A. A., and Van Stryland, E. W. 1990. *IEEE J. Quantum Electron.* **QE-26**: 760.

78a. Yuan, H. J., Li, L., and Palffy-Muhoray, P. 1991. *Mol. Cryst. Liq. Cryst.* **199**: 223.

78b. Li, L., Yuan, H. J., Hu, G., and Palffy-Muhoray, P. 1994. *Liq. Cryst.* **16**: 703.

79. Santamato, E., and Shen, Y. R. 1985. *Opt. Lett.* **9**: 564.

80. Cf. Abraham, E., and Smith, S. D. 1982. *Rep. Prog. Phys.* **45**: 815.

81a. Seidel, H. *US Patent* 3,610,731.

81b. Szoke, A., Danean, V., Goldbar, J., and Kurnit, N. A. 1969. *Appl. Phys. Lett.* **15**: 376.

81c. Szoke, A. *US Patent* 3,813,605.

82a. Bischoffberger, T., and Shen, Y. R. 1976. *Appl. Phys. Lett.* **28**: 731.

82b. Bischoffberger, T., and Shen, Y. R. 1978. *Phys. Rev. A* **17**: 335.

83. Cheung, M. M., Durbin, S. D., and Shen, Y. R. 1983. *Opt. Lett.* **8**: 39.

84. Khoo, I. C. 1982. *Appl. Phys. Lett.* **41**: 909.

85. Khoo, I. C., Hou, J. Y., Normandin, R., and So, V. C. Y. 1982. *J. Appl. Phys.* **53**: 7599.

86. Khoo, I. C., Hou, J. Y., Normandin, R., and So, V. C. Y. 1983. *Phys. Rev. A* **27**: 3251.

87. Khoo, I. C. 1982. *Appl. Phys. Lett.* **40**: 645.

88. Khoo, I. C., and Hou, J. Y. 1985. *J. Opt. Soc. Am. B* **7**: 761.

89. Umeton, C., Ciparrone, G., and Simoni, F. 1986. *Opt. Quantum Electron.* **18**: 309.

90. Vicari, L., Bloisi, F., and Simoni, F. 1991. *Mol. Cryst. Liq. Cryst.* **207**: 251.

91. MacDonald, R., Schwartz, J., and Eichler, H. J. 1992. *Int. J. Nonlinear Opt. Phys.* **1**: 1.

92. Kreuzer, M., Balzer, W., and Tschudi, T. 1990. *Appl. Opt.* **29**: 579.

93. MacDonald, R., and Eichler, H. J. 1992. *Opt. Commun.* **89**: 289.

94. Tamburrini, M., Bonavita, M., Wabnitz, S., and Santamato, E. 1993. *Opt. Lett.* **18**: 855.

95. D'Alessandro, G., and Firth, W. J. 1991. *Phys. Rev. Lett.* **66**: 2597.

96. Ciaramella, E., Tamburrini, M., and Santamato, E. 1993. *Appl. Phys. Lett.* **63**: 1604.

97. Ciaramella, E., Tamburrini, M., and Santamato, E. 1994. *Appl. Phys. Lett.* **64**: 3080.

98. Papoff, F., D'Alessandro, G., Oppo, G. L., and Firth, W. J. 1993. *Phys. Rev. A.* **48**: 634.

99. Ciaramella, E., Tamburrini, M., and Santamato, E. 1994. *Phys. Rev. A.* **50**: R10.

100. Golubitsky, M., Stewart, I., and Schaeffer, D. G. 1988. *Singularities and Groups in Bifurcation Theory*, Vol. 2, Series in Applied Mathematical Sciences, vol. 69. New York: Springer.

101. Frederiks, V., and Zolina, V. 1933. *Trans. Faraday Soc.* **29**: 919.

102. Ong, H. L. 1983. *Phys. Rev. A* **28**: 2393.

103. Skarp, K., Lagerwall, S. T., and Stebler, B. 1980. *Mol. Cryst. Liq. Cryst.* **60**: 215.

104. Wu, S.-T., Lackner, A. M., and Efron, U. 1987. *Appl. Opt.* **26**: 3441.

105. Wu, S.-T., and Wu, C. S. 1990. *Phys. Rev. A* **42**: 2219.

106. Marrucci, L., Abbate, G., Ferraiuolo, S., Maddalena, P., and Santamato, E. 1993. *Mol. Cryst. Liq. Cryst.* **14**: 1431.

107. Khoo, I. E., Yan, P. Y., and Liu, T. H. 1987. *J. Opt. Soc. Am. B* **4**: 115.

108. Karn, A. J., Arakelian, S. M., Shen, Y. R., and Ong, H. L. 1986. *Phys. Rev. Lett.* **57**: 448.

109. Karn, A. J., and Shen, Y. R. *Phys. Rev.* **A42**, 2184 (1990).

110. Santamato, E., Daino, B., Romagnoli, M., Settembre, M., and Shen, Y. R. 1986. *Phys. Rev. Lett.* **57**: 2423.

111. Santamato, E., Abbate, G., Maddalena, P., Marrucci, L., and Shen, Y. R. 1990. *Phys. Rev. Lett.* **64**: 1377.

112. Santamato, E., Romagnoli, M., Settembre, M., Daino, B., and Shen, Y. R. 1988. *Phys. Rev. Lett.* **61**: 113.

113. Santamato, E., Abbate, G., and Maddalena, P. 1988. *Phys. Rev. A* **38**: 4323.

114. Zel'dovich, B. Y., Merzlikin, S. K., Pilipetskii, N. F., Sukhov, A. V., and Tabiryan, N. V. 1983. *JETP Lett.* **37**: 676.

115. Zolot'ko, A. S., Kitaeva, V. F., Kroo, N., Sobolev, N. N., Sukhorukov, A. P., Troshkin, V. A., and Czillag, L. 1984. *Sov. Phys. JETP* **60**: 488.

116. Cipparrone, G., Carbone, V., Versace, C., Umeton, C., Bartolino, R., and Simoni, F. 1993. *Phys. Rev. E* **47**: 3741.

117. Santamato, E., Abbate, G., Calaselice, R., Maddalena, P., and Sasso, A. 1988. *Phys. Rev. A* **37**: 1375.

118. Santamato, E., Sasso, A., Bruzzese, R., and Shen, Y. R. 1986. *Opt. Lett.* **11**: 452.

119. Abbate, G., Maddalena, P., Marrucci, L., Saetta, L., Ferraiuolo, A., Ferraiuolo, S., and Santamato, E. 1992. *Mol. Cryst. Liq. Cryst.* **223**: 11.

120. Abbate, G., Ferraiuolo, A., Maddalena, P., Marrucci, L., and Santamato, E. 1993. *Liq. Cryst.* **14**: 1431.

121. Mauguin, C. 1911. *Phys. Z.* **12**: 1011.

122. Durbin, S. D., and Shen, Y. R. 1984. *Phys. Rev. A* **30**: 1419.

123. Cf. Shen, Y. R. 1994. *Surf. Sci.* **299–300**: 551.

124. Mullin, C. S., Guyot-Sionnest, P., and Shen, Y. R. 1989. *Phys. Rev. A* **39**: 3745.

125. Huang, J. Y., Superfine, R., and Shen, Y. R. 1990. *Phys. Rev. A* **42**: 3660.

126a. Jérôme, B., O'Brien, J., Ouchi, Y., Stanners, C., and Shen, Y. R. 1993. *Phys. Rev. Lett.* **71**: 758.

126b. Jérôme, B., and Shen, Y. R. 1993. *Phys. Rev. E* **48**: 4556.

127. Zhuang, X., Marrucci, L., and Shen, Y. R. 1994. *Phys. Rev. Lett.* **73**: 1513.

128. Guyot-Sionnest, P., Hsiung, H., and Shen, Y. R. 1986. *Phys. Rev. Lett.* **57**: 2963.

129a. Geary, J. M., Goodby, J. W., Kmetz, A. R., and Patel, J. S. 1987. *J. Appl. Phys.* **62**: 4100.
129b. Toney, M. F., Russell, T. P., Logan, J. A., Kikuchi, H., Sands, J. M., and Kumar, S. K. 1995. *Nature* **374**: 709.
130a. Jérôme, B. 1991. *Rep. Prog. Phys.* **54**: 391 and references therein.
130b. Pieranskik, P., and Jérôme, B. 1991. *Mol. Cryst. Liq. Cryst.* **199**: 167.
131. Chen, W., Ouchi, Y., Moses, T., Shen, Y. R., and Yang, K. H. 1992. *Phys. Rev. Lett.* **68**: 1547.
132a. Berreman, D. W. 1972. *Phys. Rev. Lett.* **28**: 683.
132b. Berreman, D. W. 1973. *Mol. Cryst. Liq. Cryst.* **23**: 215.

CHAPTER 15

Controlled Textural Bistability in Nematic Liquid Crystals

R. BARBERI AND G. DURAND

15.1 INTRODUCTION

In this chapter, we review the liquid crystal systems that present switchable bistable states, which are useful for potential display application. We do not discuss the bistability of the ferroelectric smectic C* phase, nor that related with the confined systems of polymer dispersed liquid crystals (PDLC). These two systems are described in other chapters. Bistable systems can be classified into two groups: first, the "bulk" bistable, which changes texture without changing the boundary conditions (with infinite anchoring strength, for instance); second, the "surface bistable," for which the surface anchoring changes. In the first class, to switch textures one must find a continuous change of textures which allows the transition from one stable state to another. If this is impossible, one must create (or use already-present) bulk defects, usually "disclination" lines. In the second group, the bistability is controlled by a change of surface orientation, which allows any change of topology.

In this chapter, we first recall why bistability is important for "multiplexed" displays. We explain the two kinds of switching mechanisms: *slow*, with low control fields, using defect motion; or *rapid*, with high control fields, using uniform transitions above high-energy barriers. The "volume" bistable systems are then described, followed by the "surface" bistable systems.

15.2 GENERALITIES

15.2.1 Ordinary Nematic Displays

The electrooptical properties of liquid crystalline phases began to be used for information display in the mid-1960s [1–4]. The first examples of these

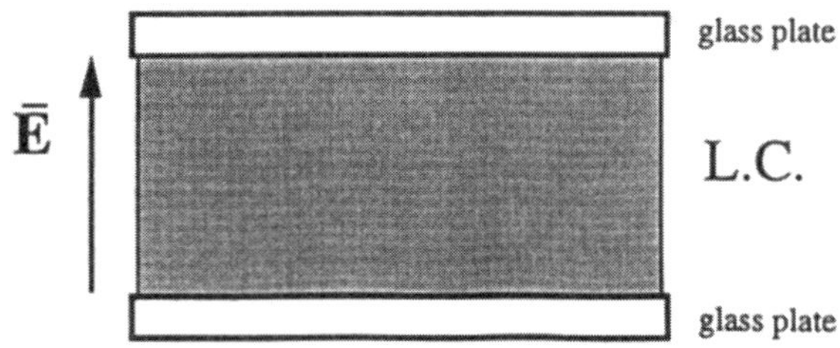

FIGURE 15.1. The most simple geometry of a liquid crystal display. Two conductive glass plates contain the liquid crystal. The electric field **E** is applied between the two boundary plates in the "vertical" direction (not to scale).

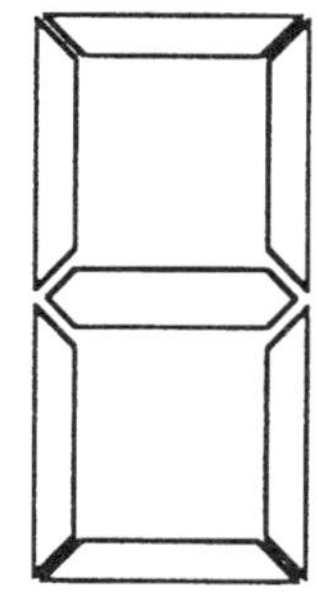

FIGURE 15.2. Arrangement of the electrooptical elements (pixels) of a seven-segment numeric display.

devices were based on the excitation of a liquid crystalline material by an external field. Their simple geometry is shown in FIGURE 15.1 two conductive glass plates, parallel to each other, spaced a few tens of microns apart, containing a liquid crystalline material. In the absence of an external field, the liquid crystal layer is in an equilibrium configuration C_e, corresponding to a minimum of its elastic free energy. The application of an external (generally electric) field between the two boundary plates changes the state of the liquid crystal layer into an excited configuration C_f, and then relaxes back to its starting equilibrium configuration when the field is switched off. These systems are thus "monostable" in the absence of external excitation because they always relax towards the same fundamental unique energy state. The two different configurations, C_e in the absence or C_f in the presence of the external field, are chosen to present different light propagation properties, to obtain, for instance, a "dark" or a "bright" optical state.

15.2.2 Multiplexing

Any actual simple display is composed by several active optical parts that can be independently controlled to pass information to the user. The simplest example of such a device is the well known seven-segment numeric display on which it is possible to show each of the ten digits by switching on or off each independent segment (see FIGURE 15.2).

As each active optical part has no memory of its excited state when the external field is switched off, a digit can be shown only by addressing simultaneously all of its component segments. If one wishes to show more information than a simple digit, more complex addressing of the display-active components is necessary. The simplest addressing scheme is the arrangement of each optical element in a row–column matrix pattern, as shown in FIGURE 15.3. A standard (VGA) large-scale display of this kind is presently composed of 480 rows and 640 columns, with the capability of independently addressing each active small area (pixel). As the total pixel number is larger than 300,000, it is obviously inconceivable to address each pixel directly. The only practical solution is to access, at the same time, all the pixels in a single row with a (hopefully strong) addressing electric signal, and to decide the optical state of each pixel in this row by

FIGURE 15.3. Multiplexable row–column matrix of pixels, with m rows and n columns.

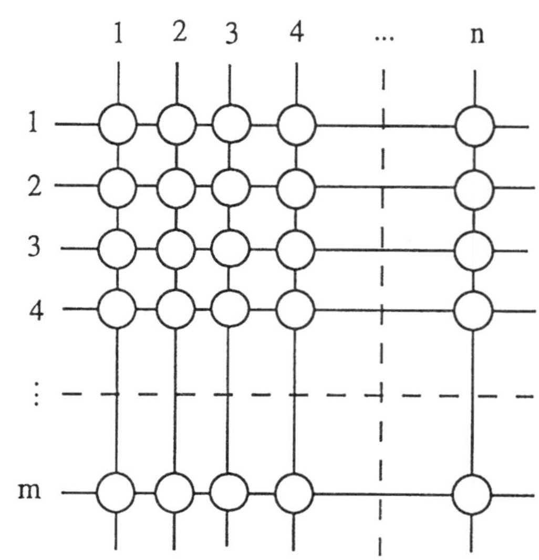

adding (generally weak) electric signals to the addressing signal *in parallel* on all columns. The other rows are sequentially addressed later. This simplified addressing technique is called "multiplexing."

Multiplexing capability is a necessary prerequisite of a matrix display, and depends on the ratio τ_2/τ_1, where τ_1 is the switching time for the transition from C_e to C_f and τ_2 is the switching time for the transition from C_f to C_e. The larger this ratio, the higher the multiplexing capability of monostable displays. Usually, the nematic orientation under electric field is controlled through its dielectric constant anisotropy $\varepsilon_a = \varepsilon_{\|} - \varepsilon_{\perp}$ [5] ($\|$ and $\perp$ for along and across the molecules, respectively). τ_1 depends on the field strength and τ_2 is due to the elastic relaxation of the liquid crystal, and hence is proportional to the square of the cell thickness d. In fact, $\tau_2 \propto \eta d^2/K$, where η is an average viscosity and K is the average curvature elastic constant of the liquid crystal. We have also $\tau_1 \propto \eta \xi_E^2/K$, where $\xi_E = (1/|E|)(4\pi K/\varepsilon_a)^{1/2}$ is the usual electric coherence length [5]. By assuming typical values of common nematics for these parameters [5], for example, $\eta \simeq 0.2$ poise, $K \simeq 10^{-6}$ cgs, $\varepsilon_a \simeq 10$, $\xi_E \simeq 1\,\mu$m, $d \simeq 20\,\mu$m, one obtains $\tau_1 \simeq 2$ ms, $\tau_2 \simeq 40$ ms, that is, $\tau_2/\tau_1 \simeq 20$. This means that only 20 lines could be addressed in such a multiplexing scheme. This is clearly unacceptable for a high-resolution picture.

Given these preliminary statements, it seems that liquid crystal technology does not fit the requirements for large information displays. In fact, this is not the case because several "storage effects," obtained by increasing τ_2 without changing τ_1, are known in liquid crystalline materials, but at present nobody has broken the barrier of large-scale production. A high-definition display demands that $\tau_2/\tau_1 \to \infty$, which is achieved by bistable systems.

15.2.3 Bistable Textures

Let us consider the simpler configuration of a nematic liquid crystal (NLC) in a "vertical geometry" cell, like the one in FIGURE 15.1. The boundary plates are treated so that they impose a strong "planar" anchoring on the nematic, and they are aligned parallel to each other with respect to the anchoring direction. The expected parallel (P) nematic configuration is shown in FIGURE 15.4(a).

As the nematic texture is not distorted at all, it corresponds to the state with the lowest elastic energy.

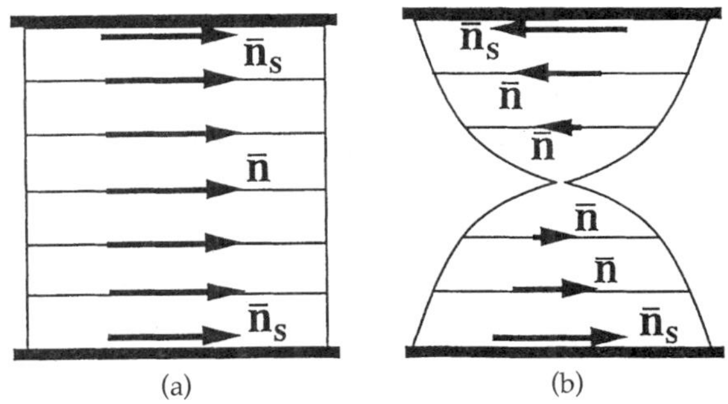

FIGURE 15.4. (a) Vertical section of a planar nematic configuration in an LC cell. The boundary plates impose planar anchoring with $\mathbf{n}_s$ easy axes parallel to each other. $\mathbf{n}$ is the bulk director. The texture does not present any distortion. (b) The same boundary plates can stabilize a higher-energy π-twisted state in the cell.

In fact, this is only one of the possible stable states. The others are twisted by $n\pi$, where n is an integer, and they present higher elastic energies with respect to the P texture. The simplest of these distorted configurations is shown in FIGURE 15.4(b). In this case, with the previous boundary conditions, the nematic exhibits a distortion due to the π-twist of the director over the liquid crystal layer. In principle, this planar cell can exhibit $m/2$ director turns between the two plates, where m is an integer. The associated elastic density energy, F_e, of these stable states increases with m^2. $F_e = (K_{22}/2)(\pi m/d)^2 d$, where K_{22} is the twist elastic constant and d is the cell thickness. After an external excitation, this cell could relax to its unique fundamental state. The system does not exhibit textural bistability. Let us now consider a similar cell with the two glass plates oriented with a relative rotation of 90° (see FIGURE 15.5).

It is evident that, when this cell is filled by a nematic, there are two lowest energy states: "right" (RT) and "left" (LT) $\pi/2$-twisted textures. The RT and LT states are both stable. Also for this cell, there are other more-twisted states with higher energies. The important difference from the first parallel cell is that now, there is a *degeneracy* of the fundamental energy level that corresponds to the two RT and LT energetically equivalent configurations. This is the simplest example of a "bistable" cell. The two states RT and LT can be optically distinguished in polarized light because they show opposite optical activity. Note that,

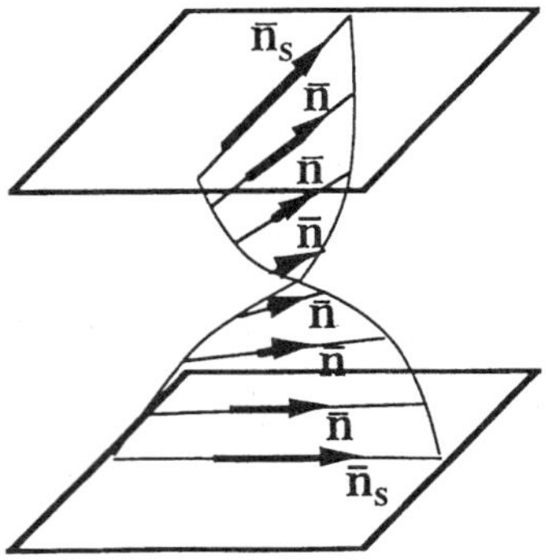

FIGURE 15.5. When the two boundary plates, which impose strong planar anchoring, are oriented with a relative rotation of $\pi/2$ (90°), the nematic texture presents at least a $\pi/2$-twisted deformation. The right and left $\pi/2$-twisted configurations have the same lowest energy.

if the nematic is doped with a small quantity of cholesteric, the degeneracy of the fundamental energy state is lost. One of the two twisted textures is preferred due to the spontaneous twist distortion of the resulting mixture.

More generally, a bistable device must present two energetically degenerate minimum energy states with an intermediate energy barrier, as shown in FIGURE 15.6. The energy barrier must be sufficiently large to prevent thermal activation and consequent spontaneous transitions.

15.2.4 Bistable switching

To make an actual bistable optical device, the second problem to solve, after the definition of a geometry with a degeneracy in its fundamental energy state, is the external control of the transition between the two stable textures. The suitable external field is obviously an electric field, and hence the control mechanisms will depend on its *dielectric*, *flexoelectric*, or *ferroelectric* coupling with the liquid crystal.

FIGURE 15.6. Degenerate minima energy states C_1 and C_2 of a true bistable system. E_b is the energy barrier between C_1 and C_2.

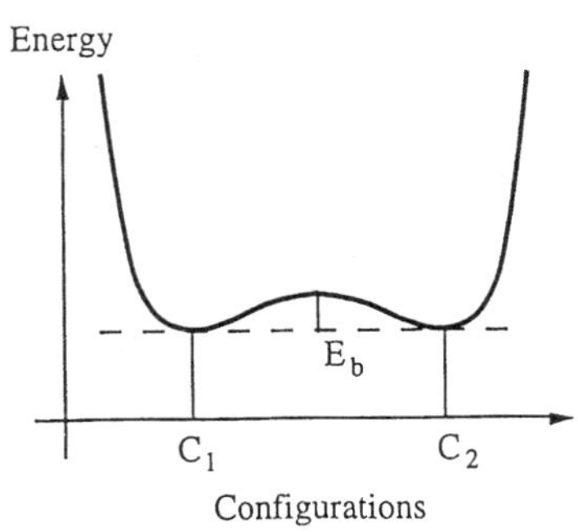

15.2.4.1 Switching by Defect Motion

Slow switching can be achieved with small fields by localized defect motion. Fast and uniform switching will need higher control fields. Let us imagine an ideal bistable device based on a standard planar cell filled with a liquid crystalline material. If the two stable states C_1 and C_2 of its lower energy level are optically distinguishable, one can directly observe at the same time an equally probable distribution of regions in the C_1 and C_2 configurations. These regions will be separated by defect lines, which connect C_1 and C_2, for instance, in the case of the RT and LT textures of FIGURE 15.5; these lines are $\pm\pi$ twist disclinations. It is evident that, if a defect line exists beforehand in the sample, a switching mechanism can be realized by changing with an external perturbation the relative energies of the two stable domains. For instance, using an external electric field, one of the two stable states can be favored with respect to the other by means of dielectric coupling. The free-energy difference (per unit volume) between C_1 and C_2 is then proportional to $K/(\xi_E)^2$ and the elastic energy of the defect line (per unit length) is proportional to K. We consider a circular defect line of radius R, which separates C_1 and C_2. The elastic energy of a segment, Δl, of this defect line is proportional to $K\Delta l$, which must be compared with the corresponding energy difference $(R\Delta l)Kd/(\xi_E)^2$ between the two stable states (d is the sample thickness). One finds that, for $R < (\xi_E)^2/d$, the defect line tends to shrink around the confined region and finally disappears, whereas, for $R > (\xi_E)^2/d$, the confined state is favored. It grows by moving the defect line. The speed of this transition is limited by the time required for the defect line to travel a given distance (the horizontal or vertical pixel size). In principle, this is a *no-threshold* phenomenon. It will be activated by an infinitely small energy difference between C_1 and C_2 under the field. In practice, pinning effects are not excluded, with consequent small apparent activation energies to start to move the existing defect line. In contrast, the *rapid* and *uniform* switching between C_1 and C_2 needs to overcome an actual energy barrier, with the consequent existence of a threshold for the external field intensity.

15.2.4.2 Uniform Switching

The *uniform* switching between the two fundamental states can be obtained in two ways.

(1) There exists an intermediate excited state that allows a "continuous" uniform transition between the two fundamental states, by a simple curvature distortion change of the texture, at constant nematic order.
(2) There is no way to "continuously" change the first stable state into the second one or *vice versa*, that is, the two states are topologically independent. An uniform transition between them implies the creation of a "melted" plane.

Case (1) will be described later (see section 15.3.2.2). Case (2) corresponds, for instance, to the RT and LT textures of the twisted cell previously discussed. Slow switching can be achieved by creating a π-twist disclination line and by moving it in the cell, whereas a fast uniform transition of the cell requires the creation of the melted plane shown in FIGURE 15.7. The barrier energy (per unit surface) is proportional to K/ξ, where ξ is the coherence length of the nematic–isotropic transition ($\xi \sim 100$ Å close to the melting temperature T_c). The application of external electric fields gives a contribution K/ξ_E to the sample energy (per unit area). This means that ξ_E must at least be comparable to ξ to induce the transition between the two stable states. By assuming $K \sim 10^{-6}$ cgs units and $\varepsilon_a \sim 10$, which are usual values for nematics used in displays, the applied electric field E should be on the order of 100 V μm^{-1}: too strong an intensity, and not suitable for practical devices. Note that, if uniform switching could be achieved, the transition time would in principle be very fast (in the microsecond range, the relaxation time of the nematic order parameter), but the required external field intensity is very strong. In contrast, the defect-induced transition works in principle with very weak external fields, but the expected transition is slow.

FIGURE 15.7. Vertical section of the intermediate state that allows a fast uniform transition between the RT and LT textures of the twisted cell. The nematic director in the upper side of the cell is perpendicular to the section plane. In the lower side, it is planar, in the section plane. A melted plane, parallel to the boundary plate, is present in the middle of the cell. Its thickness is $\sim\xi$.

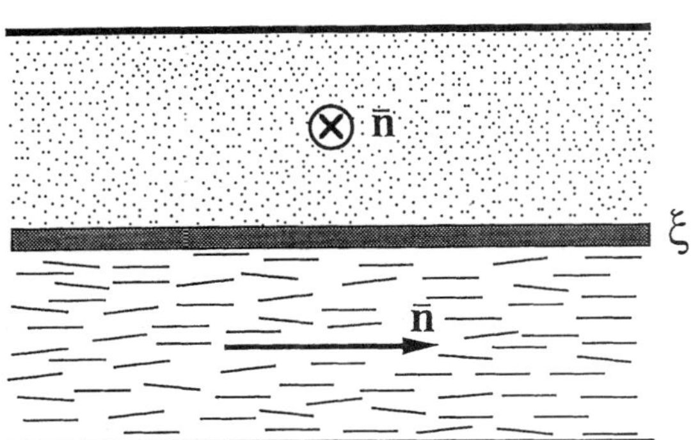

An important point is that, as the simpler cell geometry requires as external excitation an electric field applied perpendicular to the glass plates ("vertical" geometry cell), the practical problem of applying a suitable control field can be very difficult to solve. In fact, there are several examples of liquid crystal bistable optical effects which could work with a combination of electric and magnetic fields, or of "vertical" and "horizontal" electric fields, but for this reason are not very suitable for practical devices.

15.3 BULK BISTABILITY WITH FIXED BOUNDARIES IN LIQUID CRYSTALS

Bistable liquid crystal systems can be divided into two categories: the "ordered–disordered texture" and the "ordered–ordered texture" systems. The ordered–disordered texture devices are based on controlled transitions between a well-defined stable texture and a disordered configuration. The ordered–ordered texture displays are based on transitions between two well defined textures of comparable energies.

15.3.1 Ordered–Disordered Texture Bistable Transitions

A disordered configuration in a liquid crystal sample can be induced practically in two ways: by means of an external electric field or by thermally inducing a transition in the isotropic phase with a fast recrystallization in the nematic phase and the consequent creation of a defect network.

15.3.1.1 Electrooptical Display

The oldest bistable liquid crystal devices are based on the optical storage effect that cholesterics or smectics show when submitted to ordered–disordered texture transitions.

We first consider the monostable electrooptical effect of "dynamic scattering" in nematics. Heilmeier and Goldmacher [3] discovered it in 1968 on a nematic material sandwiched between two parallel transparent conductive plates. The plates induce a planar alignment of the surface "director" (the nematic orientation unit vector **n**); its equilibrium texture, in the absence of external fields, is essentially transparent to visible light for liquid crystal layers of a few tens of microns. This transparent state corresponds to an ordered liquid crystal texture. When a DC or low-frequency AC voltage (usually $\nu < 200\,\text{Hz}$) is applied above a voltage threshold independent of the sample thickness, the nematic becomes turbid and scatters light: it appears white. In practice, when the amplitude of the applied field reaches the threshold, the creation of convective instabilities (Williams domains [6]) occurs, which form a dynamic periodic texture pattern of a size comparable to the cell thickness d, that is, 5–10 times the wavelength of visible light. In fact, by increasing the applied voltage above the threshold, the pattern becomes disordered and mobile, and the dissipative texture induced by the field is disordered. When the external voltage is switched off, the nematic goes back to its starting transparent state. This cell is monostable.

The same researchers discovered at the same time that, if a mixture of nematic and cholesteric is used instead of a pure nematic, the same cell can "remember" the disordered state induced by the low-frequency external field [7]. The external field is responsible for the creation of a dynamic scattering mode in the cell, as in the case of the previous nematic cell. When it is switched off, the cholesteric mixture creates a static defect structure that has a very long relaxation time (tens of hours), probably because the defect lines are attached at local sample "inhomogeneities," on the boundary surfaces. In practice, the surface inhomogeneities hamper the free movement of defect lines in the cell. Moreover, this static scattering state can be erased by a high-frequency electric field, which, by dielectric coupling, realigns the liquid crystal in its ordered state. The typical response time for the order → disorder transition is a few tens of milliseconds, whereas for the inverse disorder → order transition, it is about one second. The observed optical contrast is lower than 10:1.

Note that this "storage" effect does not correspond to a true bistable effect, because only the ordered texture in the cell is actually stable in the absence of external fields. The disordered state is in fact metastable even if its relaxation time is quite long. It has a higher energy than the ordered one. Because the scattering state is obtained by electrohydrodynamic instabilities, which need free electrical charges in the sample, the current consumption of such a cell is a drawback. There are no practical devices in existence using this effect.

A new version of this order → disorder transition has recently been presented using a PDLC with a low polymer content [8]. These effects are reviewed elsewhere and will not be discussed here.

15.3.1.2 Thermoelectrooptic Displays

A real order–disorder device is the thermoelectrooptic *smectic* or *nematic–cholesteric* mixture displays [9,10]. In this case, generally, the liquid crystal is heated above its melting temperature and then rapidly cooled to obtain a metastable disordered scattering state formed by a dense defect pattern [11–13]. Slow cooling or an external high-frequency vertical electric field (dielectric coupling) can instead give an ordered transparent state (planar for nematic–cholesteric mixtures and homeotropic for smectics). The cell, with a typical thickness $d \sim 10\,\mu\text{m}$, is made with two silane-treated conductive glass plates, sometimes coated with infrared-absorbing visible-transmitting film. The liquid crystal can then be locally heated by a focused infrared laser beam or by Joule dissipation from resistive electrodes. Due to the higher density and smaller size of the scattering centers, the smectic cells demonstrate much better optical contrast (30:1) than the nematic–cholesteric mixture cells (10:1). Typical response times of smectic cells (tens of milliseconds) can also be shorter than nematic–cholesteric mixture cells (hundreds of milliseconds), since the transition from ordered–disordered texture is always faster than the opposite transition, from disordered–ordered texture. Note that all these devices can achieve some degree of gray level because the defect density of the disordered scattering state can be slightly modulated.

15.3.1.3 "Filled Nematics"

A more recent (1992) family of order–disorder devices that show erasable optical storage capability using nematic liquid crystals is the so-called "filled nematic" display [14,15]. This is a scattering-type display based on a small quantity (2–3 vol.%) of solid particles (pyrogenic silica) dispersed in a nematic with large optical anisotropy ($\Delta n \simeq 0.2$). The diameter of the silica particles is 7–40 nm. These particles tend to form large agglomerates inside the dispersion via hydrogen bonding. The nematic phase fills the free space between the particle clusters, forming domains of typical size $\simeq$100 nm. The nematic anchoring on the silica particle surface is assumed to be very strong. Cells are made with untreated conductive glass plates with thicknesses ranging between 2 and 20 μm. The cell exhibits two stable states: a static light scattering mode (SS) due to a large amount of randomly aligned domains, and a transparent state (TS) with a homeotropic alignment of the nematic. The SS state can be achieved by local heating of the sample (for example, by a laser), but without the necessity of overcoming the clearing temperature, as for previously analyzed order–disorder displays. The thermal shock caused by the local heating seems to break the silica particle pattern and subsequently induce a random director orientation and thus local scattering. The TS state is obtained by the reorientation of the nematic director in an external vertical electric field. The field rearranges the particle aggregation by locally breaking the hydrogen bond of the aggregate pattern and forming a new aggregate structure compatible with the nematic homeotropic alignment. This electric "erasing," however, requires high applied voltages on the order of 100 V rms at 500 Hz, thus a combination of small thermal shock and vertical AC electric field is preferred to reduce the applied voltage to about 25 V rms at 500 Hz. Full electrical addressing with a high applied voltage can be achieved by using a two-frequency scheme. This uses a nematic with a dielectric anisotropy $\Delta\varepsilon$ sign function of the field frequency. For high frequencies, $\Delta\varepsilon$ is negative. The vertical electric field aligns the nematic director parallel to the glass plates giving a random planar orientation corresponding to the SS state. For low frequencies, $\Delta\varepsilon$ is positive. One obtains the homeotropic orientation of the TS state, with typical switching times of ~10 ms. A positive characteristic of this device is that, due to the low content of dispersed silica particles, no matching between their optical properties and the nematic phase is required to obtain the SS and the TS modes. The typical contrast ratio for filled nematic displays is 30:1 with a resolution of 500 lines mm^{-1}. Gray levels can be obtained, modulating the disorder of the SS state.

15.3.2 Ordered–Ordered Texture Transitions

15.3.2.1 Homeotropic Cholesteric Cell

In 1974, Greubel presented a bistability effect in a cholesteric subjected to homeotropic boundary conditions, in the presence of an external electric field [16]. He used a common nematic doped with a cholesteric. The liquid crystal mixture exhibited a positive dielectric anisotropy $\Delta\varepsilon$, with a pitch $p = q_0/2\pi$ slightly larger than one micron. This mixture filled a planar cell made by two conductive glass plates, treated to give a nematic homeotropic alignment and with thicknesses of about ten microns.

The initial state C of the cell, at zero field, exhibits a cholesteric texture with the helix axis *parallel* to the boundary plates (FIGURE 15.8). Due to the incompatibility between the helicoidal molecular arrangement and the homeotropic boundary conditions, a randomly oriented periodic pattern of disclination lines with a period $p/2$ is present in the cell, close to the plates. In practice, this pattern is more or less randomly oriented. When the applied electric field increases, at a threshold value E^+, the cholesteric

FIGURE 15.8. The homeotropic cholesteric cell. The two plates are treated to give a nematic homeotropic alignment. The C state is represented, with the helix axis parallel to the boundary plates. The vertical arrows represent the projection of the director onto the section plane. D are the disclination lines. p is the cholesteric pitch. In the inset, a schematic hysteresis curve of this display is plotted.

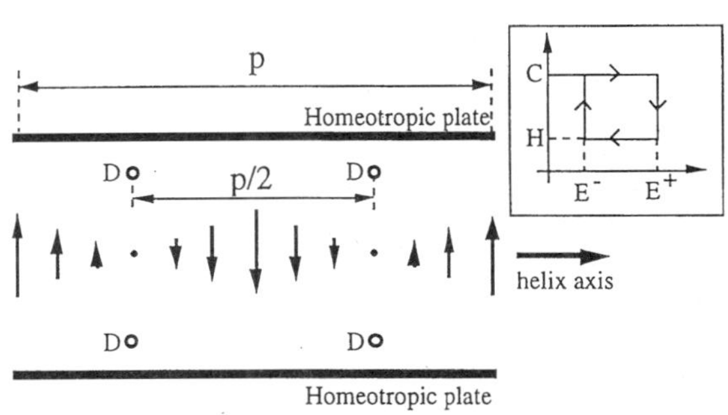

helix unwinds into a homeotropic nematic configuration H. In the absence of disclinations, E^+ would be given by the comparison between the electric contribution of the free energy (per volume unit), proportional to $K/(\xi_E)^2$, and the elastic free energy proportional to Kq_0^2, due to the intrinsic twist deformation: this leads to $q_0\xi_E \sim 1$. In the presence of disclinations, their energy contributions, proportional to Kl where l is the line length, must be added. Integrating the electric and twist free-energy density contributions on the sample thickness d, one obtains $1/(\xi_{E^+})^2 \sim q_0/d + q_0^2$. If the field is now lowered, this homeotropic nematic texture remains stable until the field rises above a second threshold E^-, defined by $1/(\xi_{E^-})^2 \sim q_0^2 - q_0/d$, where the cholesteric defect pattern C reappears. For $q_0 d \sim 1$, ξ_{E^-} is infinite, that is, E^- is zero. Note that the H state is only metastable, and the presence of inhomogeneities in the cell may drastically reduce its lifetime. In the absence of an external field, if d is large enough with respect to p, the transition from H to C is spontaneous and occurs within tens of milliseconds. Experimental data on the transition from C to H have not been reported. We can estimate that it should be slower since it is mediated by defect creation and motion, as for other previously discussed optical effects. In fact, the condition $q_0\xi_{E^+} \sim 1$ ($E^+ \sim 1\,\mathrm{V}\,\mu\mathrm{m}^{-1}$ for usual samples) ensures a very slow transition to the homeotropic state H. To reduce the transition time it is necessary to increase the electric field intensity far above the threshold to overcome the energy of dislocations. Such a cell is not actually bistable in the absence of this field. Bistability is achieved only in the presence of the applied electric field, in the range $E^- < E < E^+$.

15.3.2.2 The Berreman–Heffner Cholesteric Display

The problem of obtaining bistable liquid crystal textures was posed early on, just a few years after the discovery of the first liquid crystal display mechanisms. Analyzing the behavior of a planar twisted cell in a theoretical investigation, Leslie [17] showed that, when the "twist" elastic constant of a nematic is large compared with the "bend" and "splay" elastic constants, the intuitively expected parallel twisted textures are replaced by nonparallel distorted twisted orientations. This means that, in such case, the equilibrium conformation of the nematic is obtained when the bulk director is not parallel to the limiting surfaces, but tilted. Later on, Porte and Jadot, Sheffer, and Kobayashi and coworkers [18–20], using different assumptions, demonstrated that such nonplanar twisted textures could be obtained with common materials in the presence of tilted surface alignment.

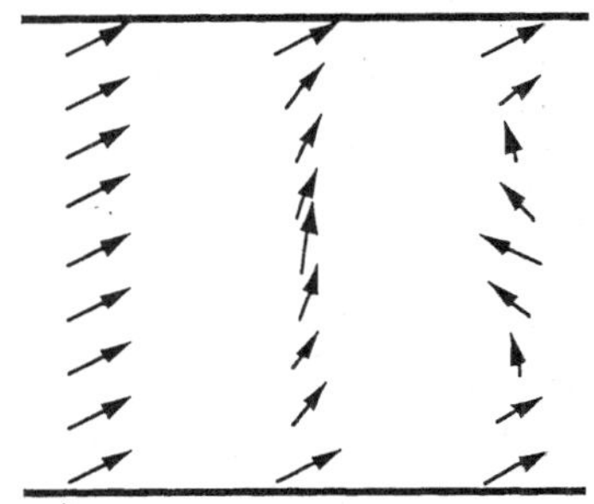

FIGURE 15.9. Geometry of the textures in the Berreman–Heffner display. U is the uniform tilted stable state. T is the 2π-twisted tilted stable state. B is the intermediate quasi-homeotropic twisted texture in the presence of an external field (from [21]).

Starting from these ideas, Berreman and Heffner [21,22] developed a cholesteric bistable cell. They used a mixture of a nematic doped with a small quantity of cholesteric. The cell is made with transparent conducting glass plates coated with a vacuum oblique evaporation of SiO to obtain an oblique surface anchoring $\theta = \theta_0 \simeq 70°$. It is observed between crossed polarizers. The cell can be switched by an external vertical electric field from one state of uniform tilted texture U to another 2π-twisted tilted texture T.

Figure 15.9 shows the geometry of the two stable states U and T, and of the intermediate barrier state B in a weak electric field. Note that the two textures U and T do not have exactly the same energies. The cell thickness is chosen to minimize the elastic energy for the state T. To optically separate the two textures, one uses a sufficiently weak birefringence Δn, that the Mauguin condition [5] $d\Delta n > \lambda$ is not fulfilled. T will transform linearly polarized light into elliptically polarized light, so that it can be optically distinguished from U.

A weak external voltage, comparable to the critical voltage $V_0 = \pi(4\pi K/\varepsilon_a)^{1/2}$ of the electrical Fréedericksz transition of an equivalent parallel nematic cell, puts the liquid crystal in configuration B, intermediate between the U and T states. In B , the nematic bulk director is almost perpendicular to the cell plates, due to alignment with the external electric field. A residual twist remains around the vertical direction. When the field is turned off, the liquid crystal is observed to reorient in a continuous manner towards one of the two states U and T, depending on the field amplitude. The Berreman–Heffner cell does not allow direct coupling between the polarity of the applied electric field and the two final stable textures. One finds that an applied voltage $V_1 \sim 2V_0$ produces the final state U and a slightly stronger field $V_2 \sim 3V_0$ produces the final state T. Since it is observed that a slow decrease of the applied voltage (above threshold) helps the system to relax to the U state, dynamical reasons are invoked by the authors to explain this switching (which is far from clear). At large field, the intermediate state B and the untwisted distorted state U have very comparable energies, since the twist associated with a very small θ is almost negligible. When relaxing the field, θ returns to θ_0. This creates a backflow which could, as suggested by the authors, untwist the B state towards the U state. The final state could then be the untwisted U state, although its energy is slightly higher than that of the T state. Experimentally, the optical switching time τ is on the order of 2–6 ms, multiplied by the square of the cell thickness, in microns (for $d = 25\,\mu$m, $\tau \simeq 200$ ms).

15.3.2.3 Supertwisted Bistable Displays

A development of the Berreman–Heffner cholesteric display is based on the "supertwisted birefringence effect" (SBE) [23]. A planar cell of a few microns thickness ($d \sim 5\,\mu$m) is filled with a mixture of a nematic and a few volume percent of cholesteric. The plate surfaces are treated to obtain oblique anchoring with a pretilt angle $\theta_0 \sim 60^\circ$ with respect to the axis perpendicular to the electrodes. The cell geometry gives an $\simeq 270^\circ$ twist distortion around the same axis. The switching is achieved between the supertwisted state and a quasihomeotropic state, but the system behavior remains very similar to that previously described for the cholesteric cell. This display uses two twisted polarizers (noncrossed) and operates outside the Mauguin regime. Its optical contrast is obtained by the interference of the two optical normal modes that propagate in the birefringent cell. It is sensitive to temperature variations, which change the wavelength of the transmitted light (i.e., the color of the display), and the applied voltage range used to control the bistability depends on the cell thickness. A uniform spacing is therefore essential for correct operation. In practice, the supertwist bistability in the presence of a field is simply used to improve the sharpness of the threshold of the Fréedericksz transition. This allows far better multiplexing of the simplest monostable twisted nematic displays.

Another kind of passive bistable nematic display, which can be considered an improvement of the Berreman–Heffener display, has been recently developed [24] by using a twisted cell with the signs of pretilt angle on the two boundary surfaces mutually opposed. The nematic is doped with a cholesteric to obtain a pitch p of the mixture. If the started texture is twisted of an angle φ, the switching is achieved among a $\varphi + p$ and a $\varphi - p$ metastable distorted bulk states. The life-time of these two states is adjusted by changing p with respect to the cell thickness d. This device is characterized by an optical contrast ratio better than 50 : 1 and a typical response time of a few milliseconds. This short response time for a bulk effect is indicative of small enough cell gap and pitch, in a few μm range.

15.3.2.4 Bulk Texture Bistable Nematic Display

In 1980, Boyd and coworkers presented the first example of a true bistable nematic display working between two ordered and stable nematic textures [25–27]. They used a planar cell with two plates treated to produce a surface-tilted easy axis θ_0. From one plate to the other, the orientation is antiparallel. This cell allows the existence of essentially two ordered textures: the horizontal H or vertical V two-dimensional director orientations (FIGURE 15.10). In the one-constant

FIGURE 15.10. Geometry of the director field in the two stable textures (V–vertical, H–horizontal) of the bulk bistable nematic display by Boyd and coworkers. Disclination lines D are always present between the two stable states. D^- is a $-\pi$ disclination line, D^+ is a $+\pi$ disclination line.The arrows indicate the local director orientation (from [25]).

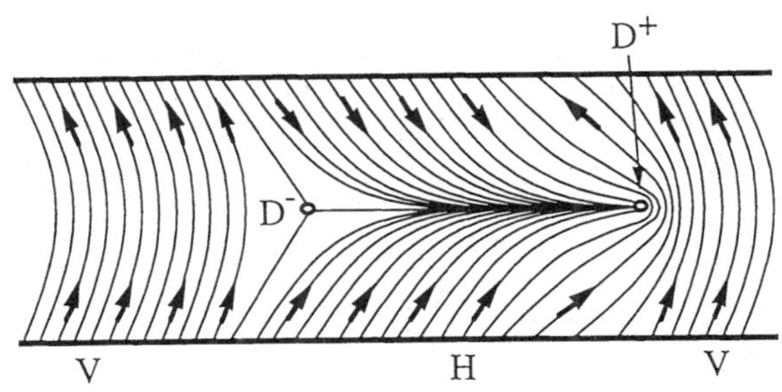

approximation, the free energy (per unit surface) of the V state is $F_V = 2K\theta_0/d$, the free energy of the H state is $F_H = 2K(\pi/2 - \theta_0)/d$, and hence the free-energy difference between the V and H states is $F_V - F_H = (2\pi K/d)(\theta_0 - \pi/4)$, where θ_0 is the surface anchoring angle, measured with respect to the surface normal, and d is the cell thickness. V and H have the same energies for $\theta_0 = \pi/4$.

In FIGURE 15.10, the two textures H and V are shown in contact. According to the relative positions, these textures are matched by $+\pi$ or $-\pi$ disclination lines in the bulk of the sample. The transition between V and H can be achieved by moving these lines, under the influence of a DC horizontal field. This works because these two curved textures present flexoelectric polarizations, proportional to their curvature $\mathbf{P} = e(\mathbf{n}\,\mathrm{div}\,\mathbf{n} + \mathrm{rot}\,\mathbf{n} \times \mathbf{n})$. Here, e is the flexoelectric constant in the quadrupolar approximation and $-e$ is the electric quadrupole density of the material. For a positive value of e, the horizontal component of the flexoelectric polarization is directed towards the left-hand side of the cell for the V texture, and towards the right-hand side for the H texture. The sign of the applied horizontal electric field determines which of the two stable textures is energetically preferred. The domains with the favored texture increase their size, and the unfavored ones disappear. In principle, if the defect lines exist in the sample before the field application, it is possible to move them with a relatively weak electric field. In contrast the creation of a disclination line, starting from a homogeneous texture, requires a strong applied field. The switching time τ, from defect motion, is related to the disclination mobility. Experimental results for a cell width $h \simeq 25\,\mu\mathrm{m}$ give $\tau = h/\mu E \simeq 80\,\mathrm{ms}$, where $E \simeq 3\,\mathrm{V}\,\mu\mathrm{m}^{-1}$ is the applied electric field and $\mu \simeq 10^{-6}\,\mathrm{cm}^2\,\mathrm{V}^{-1}\,\mathrm{s}^{-1}$ is the disclination mobility.

In principle, a fast uniform transition between the two stable state H and V is possible by creating an intermediate state with high energy, as shown in FIGURE 15.11, where a melted plane is present in the center of the cell. Recalling the previous discussion (section 15.2.4), we know that this requires an applied field of about $100\,\mathrm{V}\,\mu\mathrm{m}^{-1}$. This effect has never been observed.

In practice [28], the pixels are arranged to allow a surface pinning of already-present disclinations. The defects are moved with a combination of vertical and transverse AC electric fields, in the range 50–100 V

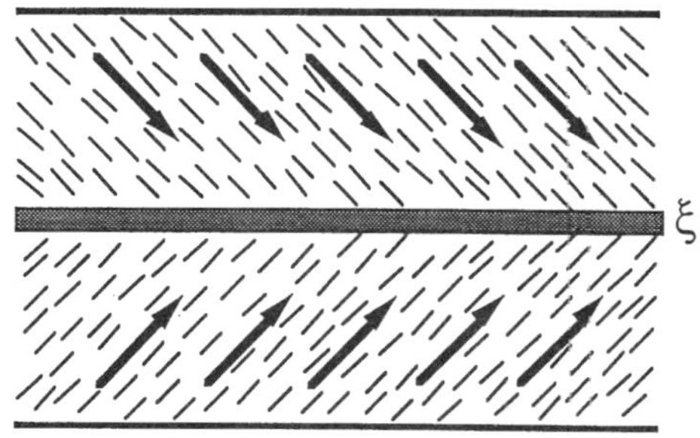

FIGURE 15.11. Vertical section of the intermediate state that would allow a fast uniform transition between the V and H states of the bulk bistable nematic display. The arrows indicate the local director orientation. A melted plane, parallel to the boundary plate, is present in the middle of the cell. Its thickness is $\sim\xi$.

rms, generated by interdigital electrodes and dielectrically coupled with the nematic. The material is chosen to have a positive dielectric anisotropy $\Delta\varepsilon > 0$. To avoid the "horizontal" electric field, a second control mechanism was proposed, based on an electrically driven H $\rightarrow$ V transition and a thermally induced V $\rightarrow$ H transition [29]. In this case, the vertical state V and the horizontal state H are chosen with different intrinsic elastic energies. The H state is selected by imposing $\theta_0 > \pi/4$ on the boundary surfaces, and the V configuration is produced by the dielectric coupling of a vertical AC electric field with the nematic, which induces the disclination line movement. The H texture is created by heating the nematic above its clearing temperature and the consequent spontaneous relaxation to its lower energy state. A system based on a nematic that changes the sign of its dielectric anisotropy with field frequency was also proposed, without relevant experimental results [25].

15.3.2.5 Bistable Boundary Layer Display

Cheng and coworkers presented in 1983 a second bistable nematic device based on two optically distinguishable boundary layer configurations [30]. This device is an improvement on the previously analyzed bistable nematic display. The cell is always made with two conductive glass plates with antiparallel tilted boundary conditions. The surface easy polar angle θ_0 is chosen to be large enough ($\theta_0 > \pi/4$) to favor the horizontal H state of the display discussed last. This is the unique stable state at zero field and bistable behavior is achieved only in the presence of

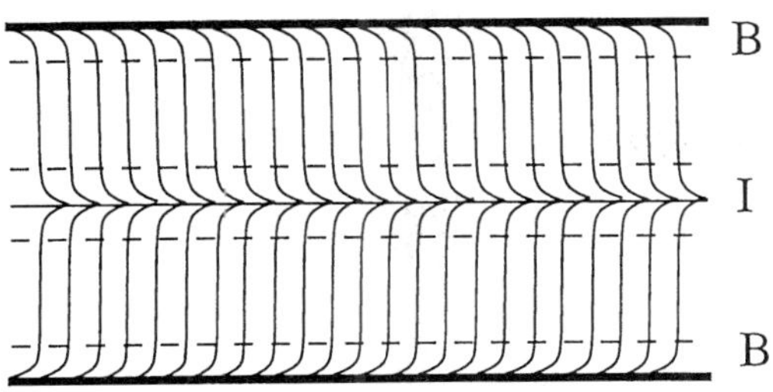

FIGURE 15.12. Vertical section of the director field of the bistable boundary layer display in its symmetric state S. B are the boundary distorted layers close to the plates. I is the inversion layer in the middle of the cell (from [31]).

an external vertical electric field. When an AC external voltage larger than the critical threshold, V_c, is applied to the cell, the H state is transformed into the symmetric state S, shown in FIGURE 15.12. The device contains two boundary layers B close to the plates, and an inversion layer I in the middle of the cell. As the total thickness of the distorted layers (where all the cell distortion is concentrated) in the presence of the external field is $3\xi_E$, where ξ_E is the electric coherence length, the minimum voltage V_c to obtain the S state is given by the relation $3\xi_E = d$, where d is the sample thickness. It results in $V_c = 3(4\pi K/\Delta\varepsilon)^{1/2}$, where K is the average elastic constant and $\Delta\varepsilon$ is the dielectric anisotropy. For most nematics, this means $V_c \simeq 1.5$ V rms.

The configuration with the inversion layer I separated by the boundary layers B is unstable, and its energy can be reduced if the layer I merges with one of the two layers B in the upper or lower sides of the cell (see FIGURE 15.13(a) and (b)), creating the states A^+ (I on the upper plate) and A^- (I on the lower plate).

FIGURE 15.13. (a) Vertical section of the first minima energy state of the boundary layer display with the inversion layer I which merges with the upper boundary layer B. (b) The second minima energy state of the boundary layer display with the inversion layer I that merges with the lower boundary layer B (from [31]).

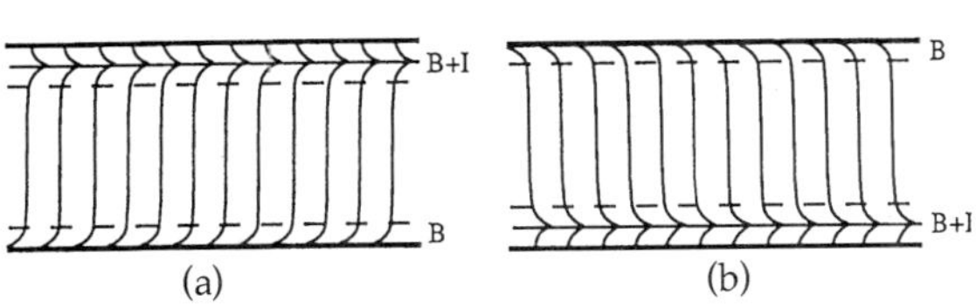

The two resulting asymmetrical states A^+ and A^- are energetically equivalent. They can be switched via the intermediate state S using a vertical DC field [31–33]. The switching voltage is applied before the AC voltage which creates the layer I. The DC pulse has a typical length of a few milliseconds, and the switching AC electric field is applied for a few tens of milliseconds. The switching mechanism is not completely understood. Associated with the bend deformation of layers B, two flexoelectric polarizations exist with opposite vertical components in the sample, but their related energy difference ΔF does not depend on the position of the inversion layer I. Some mechanisms have been proposed for a cell biasing by double-layer fields that arises from the charge of one sign being preferentially adsorbed at the walls. These textures cannot all be optically distinguished in a 2D configuration. To induce a difference of optical properties between A^+ and A^-, the boundary plates of the cell are twisted [30,34]. In this case, as the merged layers BI are thicker and hence more birefringent than the B layer, their higher or lower positions in the cell are equivalent to the rotation of a birefringent retardation plate. Thus a pair of twisted polarizers gives the required optical contrast between A^+ and A^-. No practical device has been built using this mechanism, probably because of its high complexity.

15.4 SURFACE-BISTABILITY-INDUCED NEMATIC DISPLAY (SBIND)

15.4.1 Introduction

The bistability described up to this point has been essentially bulk bistability with fixed boundary conditions. Under applied fields, two liquid crystal textures with comparable energies could exist, depending on the previous history of the system. The choice of one of these bistable textures was made by a suitable volume control field. We present in this section another kind of bistability, induced by the boundary conditions themselves. One can prepare plates that, at any given point on the surface, have *two* easy directions defined for the nematic liquid crystal orientation instead of the usual *one*. These *bistable* plates can be used to induce bistable textures and, in principle, to make bistable displays. The final bulk state of a cell depends on the chosen surface orientation, which can be controlled by a suitably chosen surface field.

We first describe one method of preparation of such bistable plates, from a carefully controlled grazing evaporation of SiO on glass. We discuss the properties of bistable anchoring that exhibits "bifurcation" and the critical behavior at monostable → bistable transitions under external fields. The switching between the two stable surface states implies for the system a chiral symmetry to control the surface bifurcation. Two examples of switching methods are given, where the required chirality comes from the liquid crystal (doped with chiral ions), or from the cell itself (with twisted plates). The performances of one pixel cell controlled with these methods are presented. Such surface-bistability-induced nematic displays (SBIND) are easily multiplexable, and would compare with the well-known surface stabilized ferroelectric liquid crystal displays (SSFLCD), if the surface bistability technique could be industrially mastered.

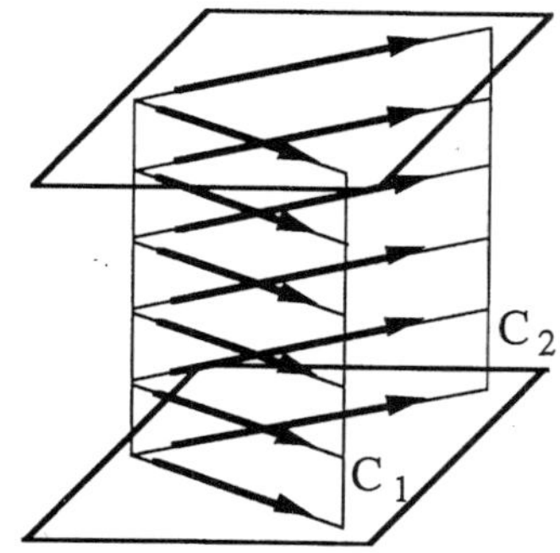

FIGURE 15.14. Ideal bistable nematic sample made with two planar textures C_1 and C_2 corresponding to two hypothetical planar orientations with different azimuths of bistable plates. The arrows represent the uniform director field for C_1 and C_2.

15.4.2 Surface or Bulk Bistability?

Before describing in detail the methods used to achieve nematic surface anchoring bistability, it is important to understand why, in principle, uniform *surface* bistability is intrinsically easier to produce than *bulk* bistability. We recall the discussion of section 15.2.4, where we estimated the height of the energy barrier for uniform switching between two bulk bistable states (FIGURE 15.6); we chose a uniform intermediate state in the bulk of the sample. The barrier energy corresponds to the melting of a layer of thickness ξ, the nematic → isotropic coherence length for the order parameter S. The melting requires $(KS^2/\xi^2)\xi \sim K/\xi$ per square cm. In contrarst, imagine a bulk texture such as the planar one in FIGURE 15.4(a), which will be rotated to another planar orientation of different azimuth, as shown in FIGURE 15.14.

The two possible azimuths correspond to two hypothetical orientations of the bistable plates. From symmetry, the intermediate state in this case would be a planar texture of intermediate azimuth, with two broken anchorings at both surfaces. Calling $(1/2)K/L$ the surface energy as usual, the barrier amplitude is now K/L. When string anchorings are considered, the surface extrapolation length $L \sim 1000$ Å is an order of magnitude larger than ξ. The energy barrier between the two bulk textures controlled by surface bistability is, as expected, an order of magnitude lower than that from the bulk bistability, with "infinitely" strong anchoring and volume melting. Physically, this means that the anisotropic interaction between ordinary orienting plates and nematic liquid crystals is much smaller than the nematic–nematic interaction. Another consequence is that if rapid switching between two bulk bistable states could have been performed, a surface anchoring breaking phenomenon should have been observed. This was not the case since most of the previous bistable systems were switched by *defect motion* and not *uniformly*.

15.4.3 Preparation of Bistable Plates

From the beginning of the century, it has been known that surfaces with low symmetry, like crystal facets, can induce "multistable" anchoring of nematic liquid crystals. Such surfaces are not very appealing for practical applications. We describe here how to produce bistable anchoring on amorphous glass coated with ITO, to be used as transparent electrodes.

The orienting effect of obliquely evaporated oxides, like SiO on glass, is well known for nematics (see FIGURE 15.15 for the usual evaporation geometry). For an angle of incidence α in the 60° region, one finds a single "planar" nematic orientation P, with surface molecules parallel to the glass and normal to the incidence plane of the evaporation S [35]. For grazing incidence ($\alpha \simeq 80°$), one finds a single oblique orientation O of the surface nematic, in the plane S [36]. The nematic tilt is not related to α. When increasing the thickness δ of the evaporated SiO into the

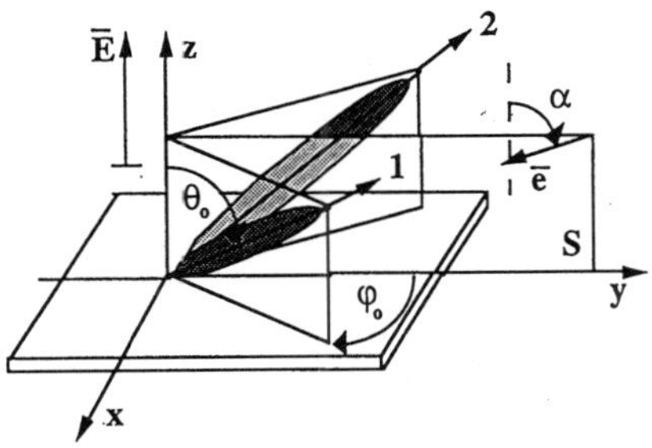

FIGURE 15.15. SiO evaporation geometry. e is the evaporation direction at an angle α with the z axis, in the plane S perpendicular to the xy-plane of the plate. The geometry of the oblique bistable anchoring is also shown, with the two surface orientations 1 and 2 symmetrical with respect to S (from [42]).

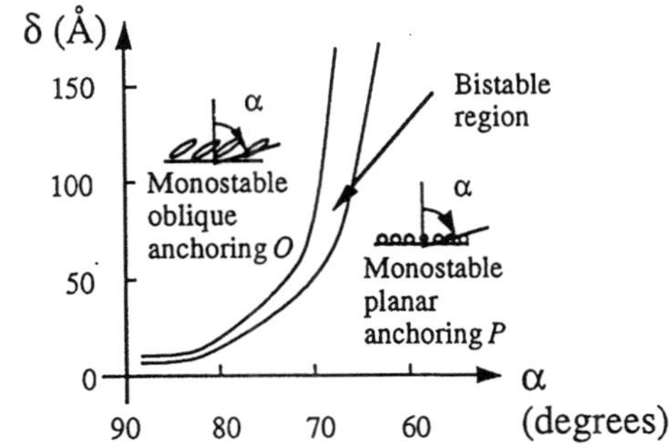

FIGURE 15.17. Phase diagram of the nematic surface orientation (for 5CB) versus evaporation angle α and SiO thickness δ (from [37]).

1000 Å region, the zenithal angle of the tilted state O increases up to $\simeq$55°, at a 35° maximum tilt from the glass. Intermediate evaporation parameters can be found, that is, ranges of δ and α for which the system exhibits bistable anchoring [37,38]. For instance, at grazing incidence ($\alpha \simeq 80°$) for small $\delta < 100$ Å, the anchoring is planar monostable along P. For $100\text{ Å} < \delta < 150$ Å, the anchoring is bistable (FIGURE 15.15), and for $\delta > 150$ Å, the anchoring again becomes monostable, but tilted along O. FIGURE 15.16 shows on a sphere the ends of the surface directors **n** in the bistable region. The two possible nematic easy directions on the surface are symmetrical compared with the plane S of evaporation. Their azimuths rotate continuously from $\varphi_0 = \pm 90°$ (the planar orientation P) to $\varphi_0 = 0$ (the oblique orientation O, tilted by $(90° - \theta_0) \simeq 20°$ from the glass plate). **n** remains in the plane T defined by (O, P). By adjusting the evaporation parameters, one can fix φ_0 to any value between 0° and 90°. The corresponding tilt of **n** is imposed by T.

FIGURE 15.16. The points on the sphere represent the ends of the surface directors for anchorings in the bistable region. The surface orientations remain in the oblique plane T. **P** is the monostable planar anchoring normal to the evaporation plane S. **O** is the monostable oblique anchoring in S (from [39]).

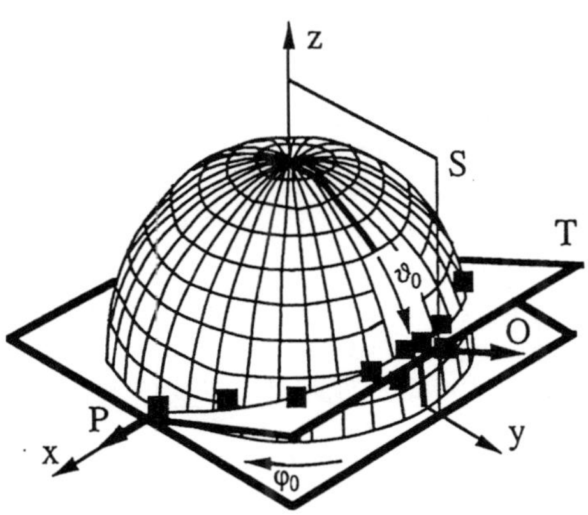

FIGURE 15.17 shows, in the plane (δ, α), the phase diagram of the bistable orientation between the two monostable states P and O. Other methods could in principle be used to create bistable anchoring, but they are still under investigation, and outside the scope of this review.

15.4.4 Bistable Anchoring Properties

The transition from monostable to bistable anchoring is a surface breaking of symmetry for the nematic liquid crystal oriented by such a plate. As previously described, this transition is continuous. One then expects a "critical" behavior of the anchoring at such a transition. An experiment has recently been performed to observe the surface orientation of a nematic above a "bistable" plate under the action of an electric field **E** [39]. The field tends to align the nematic normal to the plate, if the nematic is chosen with a positive dielectric anisotropy ε_a. In the (δ, α) phase diagram, one starts in the planar region, far from the bistable domain; the trajectory of the surface director, when increasing **E**, is the plane (P, N) normal to the glass, as expected. Closer to the bistable transition, one

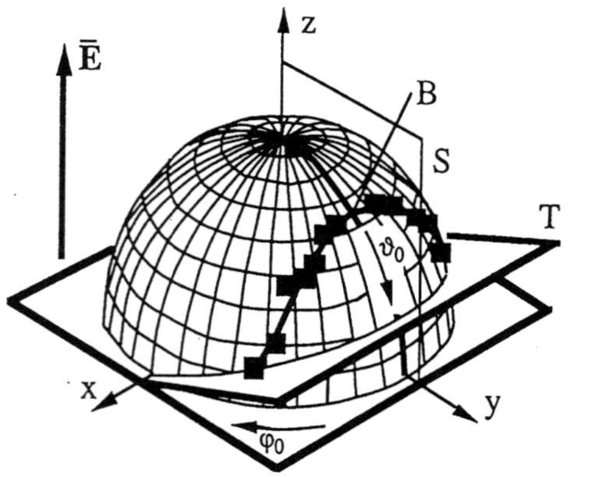

FIGURE 15.18. The points on the sphere represent the ends of the surface director **n** in the bistable region under the action of a vertical electric field **E**. **n** moves from $n_0(\theta_0, \psi_0)$ towards the single state B in the evaporation plane S. B is a bifurcation point for the surface orientation (from [39]).

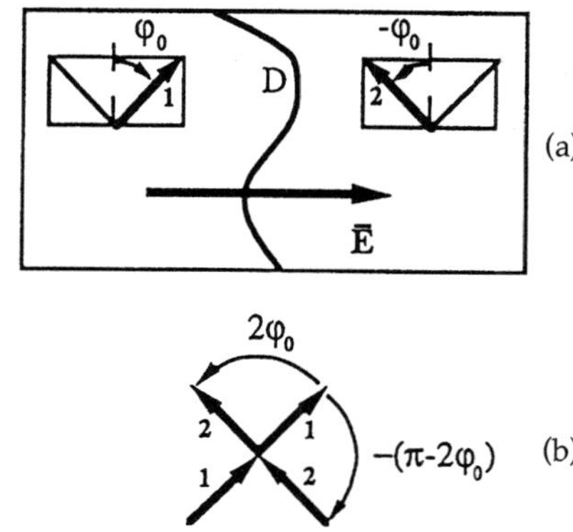

FIGURE 15.19. (a) Upper view of a sample with two surface regions with different anchoring states 1 and 2 separated by a surface wall D. The "horizontal" field **E** creates a surface torque that moves D. This process allows a slow switching between states 1 and 2. (b) Two ways to create surface walls D with different strength $2\varphi_0$ and $-(\pi - 2\varphi_0)$ between states 1 and 2. The wall with strength $2\varphi_0$ follows a line from 1 to 2, inside T, through O; the other one goes from 1 to 2 through P.

observes an oblique trajectory, which becomes tangent to T just at the monostable → bistable transition. The anchoring energy is locally zero; the anchoring presents an infinite susceptibility to the external field.

In the bistable range (FIGURE 15.18), the surface orientation follows z, the field orientation, by first making a *bifurcation* towards a monostable tilted state B in the evaporation plane, and then rotating toward z. The existence of this bifurcation is essential to understanding the surface switching between the two bistable states $\pm\varphi_0$. The surface orientation can be brought to the bifurcation point B with a finite field E. The energy barrier of the bifurcation can be written K/L, with $L \sim 1000\,\text{Å} = 10^{-5}\,\text{cm}$ and $K \sim 10^{-6}$ cgs units. Let us reintroduce the coherence length ξ_E associated with E, defined as usual as: $K/(\xi_E)^2 = \varepsilon_a E^2/4\pi$. The bulk dielectric torque integrated over ξ_E towards the surface is of order $(K/(\xi_E)^2)\xi_E = K/\xi_E$, that is, the condition $\xi_E \sim L$ must be fulfilled. With typical values such as $K \sim 10^{-6}$ cgs units and $\varepsilon_a \sim 10$, one finds $E_B \sim 10\,\text{V}\,\mu\text{m}^{-1}$.

15.4.5 Direct Switching Between Bistable Surface States

We consider only one bistable surface in contact with a nematic material. We suppose that a large fraction of its area is oriented along the surface easy direction 1. We wonder how to switch the system toward the symmetrical state 2. To achieve a *direct* switching, we must use a field in a geometry such that its action can induce a rotation around the normal N to the plate. This leads to the use of "horizontal" fields, parallel to the plates, and along 2–1, for instance, as shown in FIGURE 15.19 and 15.20. As already explained, there are two ways to produce this switching. The problem is overcoming the energy barrier of the surface bifurcation. The lowest-energy route is to localize the high-energy region on a surface wall in the absence of any field. The switching is obtained by moving the surface wall under the action of the field, as later explained. This process is *slow*. A faster way is to switch the whole area *coherently* from state 1 to 2. This process is much more *rapid*, but the energy barrier is much higher,

FIGURE 15.20. Geometry of the surface switching with "horizontal" fields. 1 and 2 are the two stable oblique easy axes. Γ is the surface torque, whose sign depends on the orientation of the applied electric field **E** (or of the velocity field **v**).

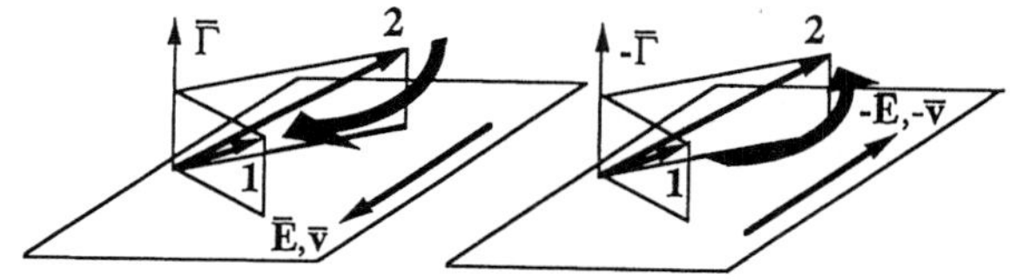

since now it is proportional to the total area, and not the wall area.

15.4.5.1 Surface Walls

By symmetry, the oblique direction O of the tilted plane T in the evaporation plane S is a saddle point for the surface energy as the planar orientation P. In the absence of any field, the director profile of the surface orientation across a surface wall separating two regions of bistable orientations 1 and 2 must follow a line from 1 to 2, going inside T, through O or P. At O or P, the surface energy is maximum along this line. Two kinds of walls then exist (FIGURE 15.19), with core orientations along the bissectrices of the angle $2\varphi_0$ between 1 an 2, that is, perpendicular to each other. The strengths of these "complementary" walls are $\pm 2\varphi_0$ and $\pm(\pi - 2\varphi_0)$. Such pairs of walls originate when a volume disinclination, $\pm\pi$, intersects the boundary surface. The widths δ of surface walls can be estimated with a simplistic equipartition argument. The energy barrier $W_B\,\mathrm{cm}^{-2}$ scales as $(K/L)(\sin 2\varphi_0)^4$ since, for $\varphi_0 \to 0$ or $\pi/2$, $W_B \to 0$ from the merging of the two minima 1 and 2. The integrated curvature energy density across a $\pm 2\varphi_0$ line scales as $(K/\delta^2)(2\varphi_0)^2$. Balancing both energies per unit length of line, one finds $\delta^2 \sim Ld(4\varphi_0^2)/(\sin 2\varphi_0)^4$. For large $\varphi_0 \sim 1$, δ is given by $\delta \sim (Ld)^{1/2}(2\varphi_0) \sim 0.5$ for $d \sim 1\,\mu$m and is *mesoscopic*. For small $\varphi_0 \sim 0.1$, δ is given by $\delta \sim (Ld)^{1/2}/(2\varphi_0) \sim 1.5\,\mu$m and is expected to be *macroscopic*. In this last case, the complementary line $(\pi - 2\varphi_0)$ should exhibit a very small thickness $\delta \sim (Ld)^{1/2}(2\varphi_0) \sim 0.05\,\mu$m.

15.4.5.2 Switching through Surface Wall Motion

We must choose a "field" which will induce a transition from 1 to 2 or, when reversed, from 2 to 1. The general property of this field is to exert alternate surface torques around the surface normal N. We can use two kinds of torque between the surface director and the field: a *dissipative* or an *elastic* one. In principle, there is no threshold for the switching, since one has only to overcome viscous friction to move the wall.

An example of *dissipative* coupling appears when the liquid crystal flows close to the surface [40]. Assume its velocity $\mathbf{v}$ oriented along 2–1 (FIGURE 15.20). The velocity gradient exerts on the wall a space-varying viscous torque, which results in a force tending to move the wall from 1 towards 2. In practice, one observes alternate switching between 1 and 2 when $|\mathbf{v}|$ overcomes a small threshold, probably related to a pinning of the wall on the surface.

An example of *elastic* coupling is given by the application of an electric field $\mathbf{E}$, also oriented along 2–1 [41] (FIGURE 15.20). There is a flexoelectric coupling between $\mathbf{E}$ and the nematic. With e the flexoelectric constant of the material, one can write the surface interaction energy W_E of $\mathbf{E}$ with the nematic as:

$$W_E = e(\mathbf{E}\cdot\mathbf{N})(N\cdot\mathbf{n}) \sim eE\sin\varphi\sin 2\theta/2$$

where N is the normal to the plate and $\mathbf{n}$ the surface director. The surface torque $\delta W_E/\delta\varphi$ comes from the $\sin\varphi$ dependence of W_E. It changes sign with $\mathbf{E}$, as expected. Such a torque can move the surface wall, as previously explained. It can also be used to induce the rapid uniform switching of a whole surface area.

15.4.5.3 Direct Uniform Surface Switching in Electric Field

We now use the flexoelectric coupling previously described with a "horizontal" field parallel to 1–2. The surface anchoring energy versus φ is represented in FIGURE 15.21 with its two minima at $\pm\varphi_0$, the bistable orientations, and an energy barrier K/L. We assume that under the field action, the $\mathbf{E}$-driven bifurcation between 1 and 2 is made at constant zenithal angle θ_0. The field exerts a surface torque $\sim eE$, maximum for the intermediate barrier orientation O. The switching condition is that eE must be larger than the maximum stabilizing surface torque, which appears at $\varphi = \pm\varphi_0/2$ and is $\sim K/L$. The threshold for uniform switching is thus $EL \sim K/e$. With $K \sim 10^{-6}$ cgs units

FIGURE 15.21. Schematic representation of the surface anchoring energy W versus the azimuthal angle φ for the nematic bistable anchoring. The energy barrier between the two stable states is $\sim K/L$.

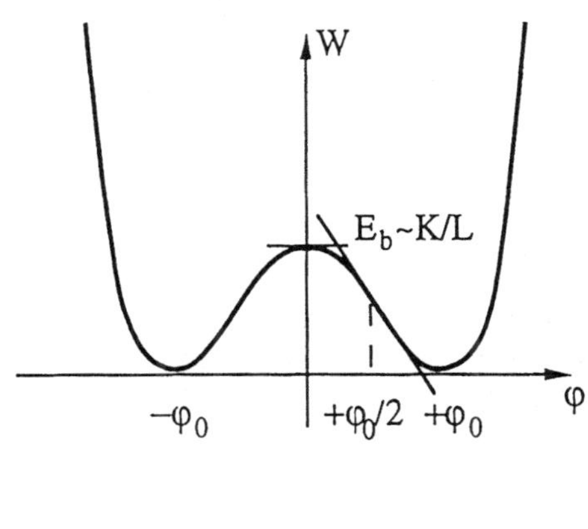

and $e \sim 10^{-4}$, $EL \sim 10^{-2}$ statvolt, ~3 V. The voltage that must be applied across two electrodes at distance d is $V = Ed \sim (K/e)(d/L) \sim 300$ V for $d \sim 10\ \mu$m and $L = 1000$ Å. The experiment confirms these values [41]. In practice, it is difficult to reduce d much below $10\ \mu$m, which keeps V above 100 V, in an inconvenient range for practical applications.

15.4.6 Indirect Uniform Switching of Bistable Surface States

15.4.6.1 Indirect Switching

To use bistable surface switching in practical devices, it is desirable to reduce the threshold voltage by an order of magnitude. In addition, the geometry needed by symmetry for direct switching is awkward for practical purposes: engineers prefer to use "vertical" fields and semitransparent electrodes. Both conditions can be fulfilled if one dismisses the direct switching requirement and uses an *indirect* one. In a "vertical" geometry, the field is applied along the normal N to the plates, separated by a distance $d \sim 1\ \mu$m. The field can now break the surface anchoring by bringing the surface director toward the bifurcation point B. For this purpose, one uses the dielectric coupling with a nematic material of positive dielectric anisotropy. The field is capable of storing potential energy at the surface. It cannot, however, determine with its sign, *which* final state, 1 or 2, will be chosen by the system when its amplitude E is turned off. The choice of the final state is in principle random. In practice, it could depend on a small departure from symmetry, that is, a local variation from one point to another across the surface. In fact, it is because of this random process that we can observe both 1 and 2 domains on bistable plates. There is no way of relating the sign of **E** to one particular state 1 or 2, since the coupling is quadratic in **E**. From the symmetry point of view, this is obvious since **E** is a *polar* vector; for a nonchiral material, there is no way of inducing a rotation ($1 \rightarrow 2$ or $2 \rightarrow 1$), that is, an *axial* vector parallel to a polar vector. The selection of the final state must now be achieved by another mechanism.

15.4.6.2 Final State Selection

To choose a final state, we must break the symmetry of the system to introduce a chiral property. This can be done in two ways: we can use a *chiral material*, or a *chiral geometry*.

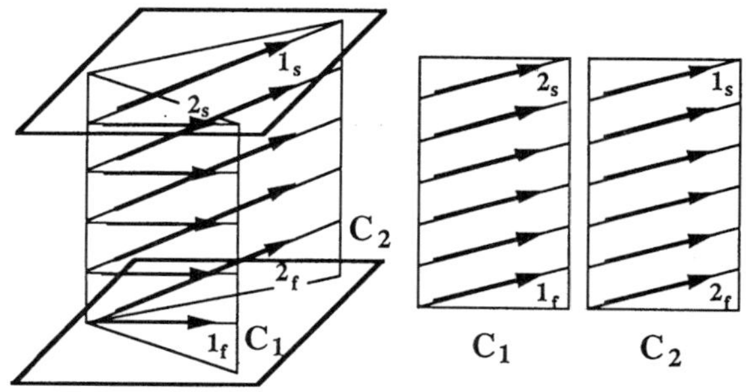

FIGURE 15.22. The two uniform tilted textures of the cell switched by a chiral material. The texture C_1 is created by the stable anchoring direction 1_f on the lower plate that faces the stable anchoring direction 2_s on the upper plate. The texture C_2 is given by the surface easy axis 2_f on the lower plate and the surface easy axis 1_s on the upper plate.

SELECTION FROM A "CHIRAL" SUSCEPTIBILITY MATERIAL [42]. A switchable cell is built with two bistable plates facing each other. 2_f on the first plate is in front of 1_s on the second plate, and *vice versa*. The evaporation directions are antisymmetric. This defines two uniform (constant θ_0) textures for the bulk nematic, as shown in FIGURE 15.22. Imagine first, that under the application of a pulsed electric field, both surface directors **n** have been driven towards their bifurcation orientation B. The system has forgotten which state, 1 or 2, it is coming from. We now turn the field off and wish to force **n** to go towards 1_f on the first plate and towards 2_s on the opposite plate. This is achieved by simply exerting a torque of a given sign to **n**. This torque could originate from a chiral material brought close to each plate by the electric field. We would also like the sign of the chirality to be connected with the sign of the applied field. We would finally demand that, in the absence of any field, the system exhibits no chirality. The way to fulfill this *chiral susceptibility* requirement is to use *chiral ions*. Assume that we can associate a χ^+ chirality (inducing a right-handed screw, for instance) to a positive ion, and a χ^- chirality to a negative ion (or *vice versa*). The ion concentrations are chosen to ensure chiral and electrical neutrality in the absence of any field. When applying a positive **E**, positive ions are dragged towards one plate and negative ions towards the other plate. With this electric dipole is associated a "chiral" dipole. On each plate, the chiral charges induce torques which are mirror images compared

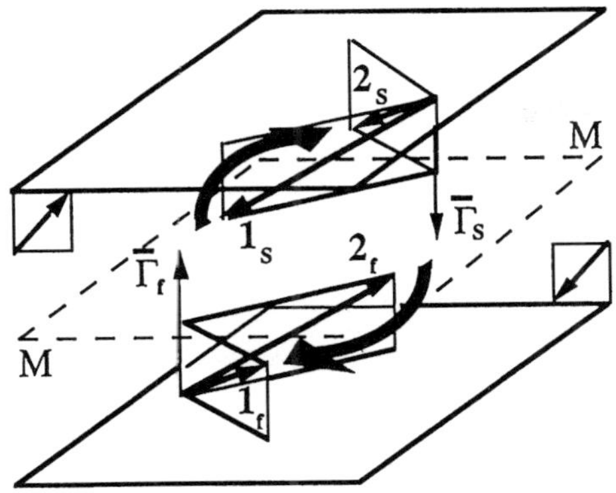

FIGURE 15.23. Geometry of the switch in a bistable cell with uniform textures. The big arrows parallel to the plate represent the effects of the surface torques Γ_f and Γ_s necessary to switch the system from the states 2_f, 1_s to the states 1_f, 2_s. M is the symmetry plane for the two surface torques. The small oblique arrows in the small squares indicate the direction of SiO evaporation on each plate (from [42]).

with a symmetry plane M between the plates (FIGURE 15.23).

Seen from outside, the two rotations have the same sense. As each plate has a surface orientation along B in presence of **E**, both directors will fall, at $E = 0$, into corresponding states with the same projection on M. The bulk texture of the nematic will later reorient itself along this common surface orientation. Applying again a new electric field pulse of opposite polarity would do exactly the opposite: first, it would break the two surfaces again above the bifurcation orientation B, *erasing* the previous state of the system. In the meantime, chiral ions of opposite signs, that is, of opposite chiralities, are dragged by **E** towards the plates. A counter-rotation will be induced from B, towards each opposite surface state. Finally, for $E = 0$, the two surfaces will be oriented with the same projection on M, rotated from the initial projection by the angle 2φ. The bulk orientation will follow to build a new uniform texture, also rotated by 2φ.

What is the required concentration of chiral ions? In principle, at the bifurcation, the surface energy is almost φ-independent. The reorienting torque is very weak so that an infinitely small amount of chiral ions is necessary. In practice, a finite lack of symmetry is expected from imperfection of the bistable treatment. The chiral ion concentration should be large enough to overcome these imperfections. In practice, in a demonstration experiment, we used phenyl lactic acid for the χ^+ anion, and quininium bromide for the χ^- cation, at a few percent molar concentration. For a practical system, the use of electrochemically stable ions would be an important issue. We did find that the lifetimes of our chiral cells, using the "uncarefully" chosen chiral ions, did not exceed one month.

SELECTION FROM A CHIRAL CELL [43]. As explained in the beginning, the two textures we need to control are uniform planar textures. In each state molecules exhibit the same azimuthal angle φ_0 (or $-\varphi_0$), defined by one state (1_f or 2_f) of the first plate in front of the corresponding state (2_s or 1_s) of the second plate. In each texture, the molecules exhibit both identical azimuthal and zenithal angles φ_0, θ_0.

To break the nonchiral symmetry of the system, let us make an "almost" bistable *chiral* plate. Imagine that on a single plate the two states 1 and 2 now have *different* zenithal angles, θ_0 and 90°, for instance. The two surface orientations 1 and 2 and the normal N to the plate now define a helix, that is, they exhibit chiral symmetry. Of course, we are still willing to construct two planar bulk textures with a common azimuthal angle φ_0 (or $-\varphi_0$), and to switch electrically between them. We will again use identical plates, by putting state 1_f of the first opposite state 2_s of the second, and state 2_f of the first opposite state 1_s of the second (FIGURE 15.24). We can again build between 1_s and 2_f, and 1_f and 2_s, two planar textures with azimuths φ_0 and $-\varphi_0$.

The new aspect is that, because of the chiral symmetry, the two textures have opposite *polarities*. The zenithal angle goes continuously from 90° to θ_0 or from θ_0 to 90° from one plate to the other. These opposite splays create opposite *flexoelectric polarizations*, P and -P. Roughly speaking, one can still break the anchoring beyond a bifurcation point. When turning the field off, the final state of the system should now be sensitive to the *field polarity* through its coupling with the bulk texture opposite flexoelectric polarizations. Note that, despite the zenithal angle difference between the two states 1 and 2 of a single plate, the two final textures after switching possess *exactly the same* curvature elastic energy, although on each "almost" bistable plate, the two easy directions have different surface energies.

How can we achieve in practice these "twisted" cells, with such "almost" bistable plates? We have

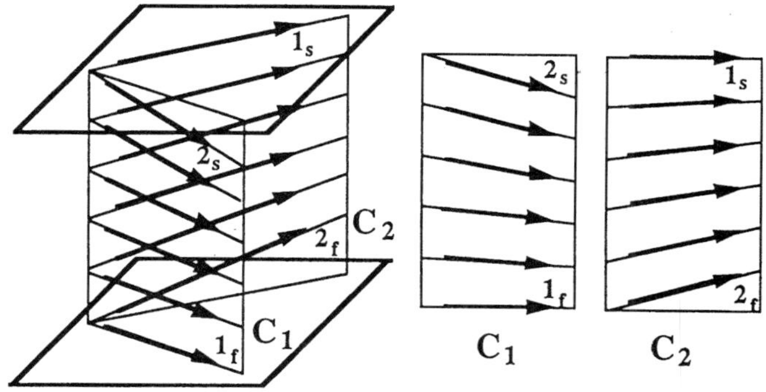

FIGURE 15.24. The two planar textures of the chiral cell, with opposite splay distortions.

used the large susceptibility property of the bistable anchoring close to the monostable → bistable transition. The plates are evaporated in the monostable planar state, close to the bistable transition.

The two plates are twisted towards each other by 45°. One normally expects a single twisted texture between these two monostable planar orientations. In fact, for a sufficiently thin sample, the system prefers to build, at random, one of the two *untwisted* textures, between one planar state on one plate and a metastable state twisted by 45° on the other (FIGURE 15.25). When twisting the anchoring from the planar state, we know that the zenithal angle changes from 90° to θ_0, as required. The condition for this metastable state to appear is that the twist susceptibility K/L is weaker than the surface equivalent twist energy $\sim K/d$, that is, $L \geq d$.

It seems that, turning off the applied voltage, there is no longer a field present in the cell, and that the flexoelectric selection of textures could not work. This is not true: the nematic material always exhibits a finite intrinsic ionic conductivity, at least from the spontaneous ionization of the molecules. These ions are dragged towards the plates after each driving pulse that breaks the bistable anchoring. After the pulse, although the total voltage across the cell is zero, a strong residual field exists in the cell, mainly located close to the plates, with an opposite weak field in the bulk. The surface field sign is related to that of the applied pulse. This remaining field couples with the flexoelectric polarizations of the textures and allows the choice of final state of the system from the sign of the surface-breaking pulse. This flexoelectric model fits with the symmetry of the system. It has not as yet been clearly ascertained. The charge distribution in the cell after the pulse has the same polar symmetry. Its exact contribution should be studied more carefully.

An "after-pulse" selection from the remaining field is demonstrated by an independent experiment. One can apply, after a large breaking pulse of say $V = 20\,\mathrm{V}$ and duration 50 μs, a smaller pulse of the same length but weaker amplitude, $v \simeq 3\,\mathrm{V}$. For $v = 0$, the final state is controlled by the polarity of V. For $v = \pm 3\,\mathrm{V}$, the final state depends only on the polarity of v. This property allows easy multiplexing of this surface-bistability-induced-nematic-display (SBIND), as is further explained in [44]. This after-pulse polarity could also be related to the ionic distribution in the cell.

15.4.7 The Surface Bistability Induced Nematic Display (SBIND)

Two kinds of one-pixel cell have been constructed at the time of writing this review (1993) to demonstrate the potential uses of the surface-induced bistability of nematics for display application. The first one uses an antisymmetric cell and a nematic doped with chiral ions [42], and the second uses a twisted cell with no dopant [43]. Both cells display the same response time behavior and the same range of voltage driving, as explained later. Easy multiplexability has been demonstrated for the twisted cell with flexoelectric coupling [44].

15.4.7.1 The Optical Geometry

Up to now, we have understood that pulse switching, plus some selection mechanism, allowed the commutation between the two planar nematic textures

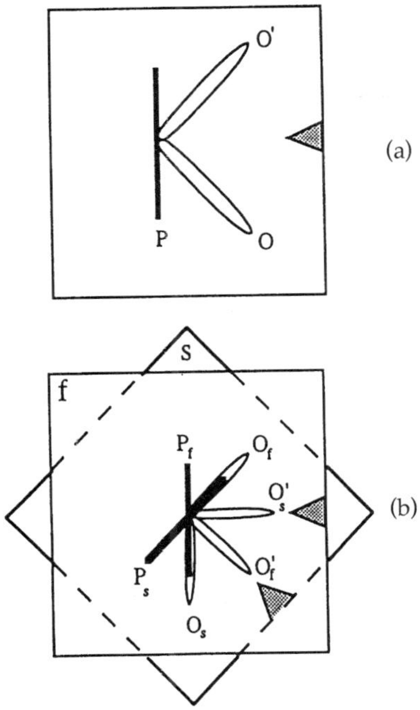

FIGURE 15.25. (a) Top view of the anchoring easy axis for a plate evaporated in the planar region close to the monostable–bistable transition. The triangle indicates the evaporation direction. P is the stable planar anchoring. O and O′ are the two possible metastable oblique anchorings. (b) Top view of the two twisted plates which form a chiral cell. The planar easy axis P_f of the first plate faces the oblique anchoring O_s of the second plate, and O_f faces P_s, creating the two opposite splayed planar textures.

defined by two bistable plates facing each other. We must now couple this bistable cell with light, to produce black or white pixels. We can use, for instance, the standard geometry designed for the ferroelectric smectic C* cell. The cell is placed between crossed polarizers, and the first texture of azimuth $+\varphi_0$ is aligned along one of the two directions of the polarizers, to correspond to a black state. The second texture, of azimuth $-\varphi_0$, is placed on the bissectrix of the polarizers' directions. This implies that $2\varphi_0 = 45^\circ$, obtained in the middle of the bistable phase diagram for the cells with chiral ions, or just by twisting the plates by 45° for flexoelectric coupling. For optimum contrast, an optical delay of $\lambda/2$ is needed, that is, at lowest order $d\Delta n \simeq 0.5\,\mu$m. For a birefringence $\Delta n = 0.2$–0.3, this leads to $d \simeq 2.5$ to $1.5\,\mu$m. In theory, all the linearly polarized light

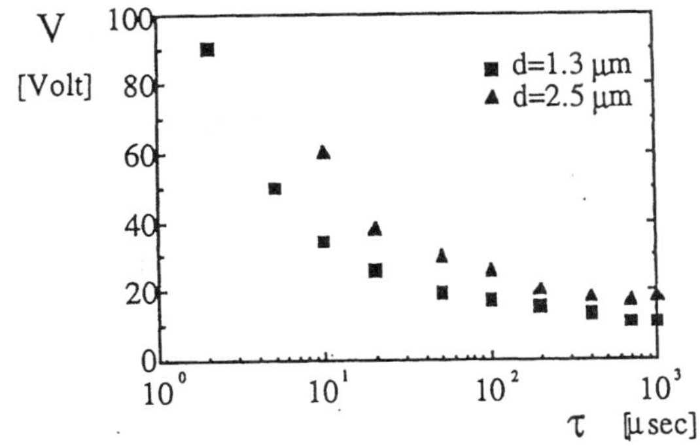

FIGURE 15.26. Voltage threshold versus pulse length for two samples of the cell driven by the electrochiral effect with thicknesses 1.3 and 2.5 μm (from [42]).

along the polarizers will be transformed at the output into linearly polarized light along the analyzer, and the $-\varphi_0$ texture will allow total transparency.

15.4.7.2 The Time–Voltage Diagram

The minimum driving voltage required to break the bistable anchoring is the bifurcation field, $E_B \sim 8\,\text{V}\,\mu\text{m}^{-1}$, for 5CB. In practice, one would like to commute the surface in the shortest possible time τ. This means that we must increase E above E_B, that is, set V above $E_B d$. Experimentally, the two kinds of switching mechanism display similar behavior [42–45], as shown on FIGURE 15.26 and 15.27. The measured amplitude V of a voltage-switching pulse

FIGURE 15.27. Voltage threshold versus pulse length for three samples of the cell driven by the flexoelectric effect with thicknesses 1, 2, and 4 μm (from [43]).

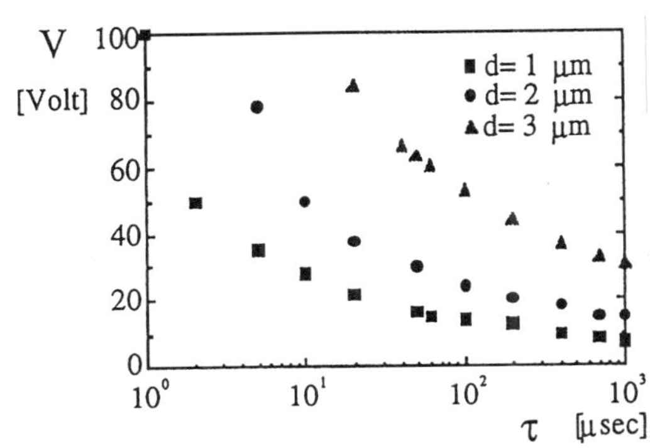

is plotted versus its duration τ. V is indeed found to be proportional to d, since the surface-breaking mechanism is a field effect. For large $\tau \sim 1$ ms, V tends to a limit $V \sim E_B d$. For smaller τ, V increases. One can switch $\tau = 1\,\mu$s for $E = 70\,\mathrm{V}\,\mu\mathrm{m}^{-1}$. For high-definition television where one aims for a writing time of $\tau \sim 30\,\mu$s, E is $\simeq 20\,\mathrm{V}\,\mu\mathrm{m}^{-1}$, which gives a quite acceptable $V \simeq$ 30–40 volts.

The threshold increase for short time can be understood from a simplistic surface friction law [45]: assuming for simplicity that the bifurcation B appears at $\theta = \varphi = 0$, close to the bifurcation, that is, for small θ, the surface torque balance can be written simply as:

$$(K/\xi_E)\theta - (K/L)\theta - \xi_E \eta \theta/\tau = 0.$$

The first term is the surface destabilizing torque from the field E, integrated from the bulk torque density over the thickness ξ_E. The second term is the elastic restoring torque from the surface anchoring. The last term is a friction torque, proportional to the angular velocity $\sim \theta/\tau$, with a surface friction constant $\eta\xi_E$, where η is a standard bulk viscosity. For infinite τ, one finds $\xi_E = L$, which gives the threshold field E for the bifurcation $E = E_B$. For shorter τ, the $E(\tau)$ law becomes simply $E(E - E_B) = (4\pi/\varepsilon_a)\eta/\tau$, which fits the experimental results fairly well. The optical response of the cell has been measured independently. It depends only on the thickness d, and follows the simple orientation–diffusion law: $\tau_D^{-1} = K(\pi^2/d^2)\eta^{-1}$, with $\tau_D = 1$ ms for $d \sim 1\,\mu$m. This apparent discrepancy is simple to understand: the $V(\tau)$ measurements indicate that the *surface* has been commuted in the time $\tau = \tau_S$. After this surface switching, the *bulk* has to reorient to build the new texture during the time τ_D.

In the previous reasoning, we have assumed that the "surface viscosity" is just the bulk viscosity extended over the thickness ξ_E of the field action. In practice, the length extension L' of the surface viscosity $\eta L'$ could be smaller than ξ_E. This could lead to a simpler law of the form:

$$E - E_B \sim \frac{\eta}{\tau} \frac{L'}{(\varepsilon_a K)^{1/2}} 2\pi^{1/2}.$$

The large uncertainty of the experimental results does not allow us to choose with any certainty between the two dynamical behaviors, which are quite close to each other.

15.4.7.3 Multiplexing the SBIND

Let us first show that the response time of the SBIND is well adapted to produce a black and white video display. In a video matrix display, one needs to produce new frames (half a picture) every 1/50 s, that is, a new picture every $T = 40$ ms. If N is the line number, sequential information writing on each line is necessary; with $N \sim 10^3$ lines, one has only $\tau \sim T/N = 40\,\mu$s to write the line information. We see that the SBIND is well adapted to these requirements: $\tau = 40\,\mu$s is obtained on one pixel for $V \sim 30$ V. The optical response time $\tau_D \sim 1$ ms remains small compared with the frame time $T = 40$ ms.

How can we multiplex the flexoelectrically controlled SBIND [44]? To address a line, we must use a large pulse of amplitude $V \simeq 30$ V. After the pulse, it seems that all pixels on this line will be in the same state, depending on the polarity of V. It is not convenient either to use large DC pulses, because of potential nematic decomposition from electrolysis. To prevent these problems, one can simply use an AC square pulse, with one complete period $+V - V$. This AC pulse will break the anchoring exactly like a DC pulse, since the coupling between the field and the surface through the nematic dielectric anisotropy is quadratic in the field. The pixel selection is made from the "column" matrix electrodes, subjected to small voltage pulses $v = \pm$a few volts, *just after* the large pulse V. As previously explained, the final state of the system depends only on the sign of v. The matrix writing process is thus simple: one line "i" is submitted to the large AC pulse $+V - V$, which *erases* all pixels "i" by putting all surface anchoring beyond the bifurcation point. When the large field is released, the polarities of the small column pulses $\pm v$ control the final states of each pixel of the line. v is obviously so much smaller than V that it has no effect on the nonaddressed lines $j \neq i$. The process takes a time τ, after which the line $i+1$ is addressed, and so on. There is no limitation on the picture definition since the textures are bistable. One can stop letting the information in, and the picture remains.

15.4.8 The "Pros" and "Cons" of the SBIND

The experiments previously described have shown that the SBIND is in principle an interesting mechanism to make new nematic liquid crystal bistable displays. It is important now to compare its potential uses with the other interesting bistable displays using ferroelectric smectics, the surface stabilized ferroelectric liquid crystal (SSFLC). We discuss the problems which remain to be solved in order to progress from academic laboratory experiments to eventual real displays. We deal only with the flexoelectrically controlled SBIND.

15.4.8.1 Comparison of SBINDs and SSFLCs

Both systems lead to "fast" displays (fast, at least, for liquid crystal technologies), and their writing speeds are comparable: in the 20–50 μs range for reasonable driving voltages of $V \simeq 20$–30 V. The optical response of the SBIND is that of a nematic cell, and slower than that of the SSFLC. This remains acceptable since the use of thin cells in the micron range is useful to decrease the voltage (from the field effect). The SSFLC uses also thin cells. This point is a common drawback of both systems since it leads to practical difficulties in making large screens. The SBIND seems more flexible in that respect, since it is easier to make a nematic material with small birefringence and large dielectric anisotropy than it is to make an SC* ferroelectric with small birefringence and large polarization. The multiplexing is in principle easier for the SBIND, since it uses a direct small column-selection voltage. The cell filling and orientation is much easier with nematics, which do not exhibit any of the strange defects exhibited by SSFLCs from the mismatching of smectic layers with the boundaries. The SBIND is also less sensitive to mechanical impacts, since nematics can flow as normal fluids, although smectic layers can break irreparably. Nematic materials today are almost optimized in terms of temperature range and properties, although the smectic materials are not fully stabilized; it is difficult, for instance, to increase the spontaneous ferroelectric polarization without increasing the viscosity, that is, the response time. The $2\varphi_0 = 45°$ geometry in the SBIND does not depend on temperature since it is defined for ever by the surface bistable anchoring, although for the SSFLC, the equivalent φ_0 angle depends on temperature and forces the use of complex material adjustments or temperature control to prevent too much variation around 45°. Both effects finally permit simple black and white displays. Their intrinsic bistabilities make the creation of gray scales difficult.

15.4.8.2 The Problems

The main problem with SBINDs is making bistable nematic anchorings with sufficiently good reproducibility. At the present time, the custom-used delicate technology of oblique evaporation is difficult to master. The critical thickness of the evaporated SiO varies, for instance, with the roughness of the ITO-coated glass used. One can expect the situation to improve with careful control of the materials. Another solution is to find different techniques which would lead to bistable anchoring. Up to now, sputtering from gas discharges or ion beams has shown some bistable behavior [46] but not switching between surface states, probably because the anchoring is too strong. Polymer surface treatments could also lead to bistability; some effort should be invested along these lines. A final problem lies in the mechanism itself. The role of ions is not clearly understood in the proposed flexoelectric switching mechanism. The invoked metastable states could be related to the presence of surfaces charges. Further work should clarify these points.

To give an indication of the potential interest of SBIND, let us compare it finally with standard matrix displays using nematic liquid crystals. To solve the difficult problems of the nematic memory time (which is too short compared with the picture time, but too long for the writing time), and of the large multiplexing rate (without contrast decrease), the "active" matrix has been introduced with one transistor (TFT) per pixel. At the cost of making $\sim(10^3)^2 = 10^6$ TFTs without defects, nematic displays have achieved good capabilities in high resolution and response time. This costs a lot of effort and money. So, what is, in principle, a switchable bistable anchoring? It is a "mechanical amplifier": the command field V stores potential energy in the surface anchoring, brought above the bifurcation. The final state is defined by the small control voltage v. The "gain" is V/v. *The bistable surface behaves as a unique transistor*, obtained with a single coating or treatment. Although this single process is delicate, it is in principle simpler than the many processes, with all the mask alignment problems, necessary to make the 10^6 TFTs per screen. The tremendous impact of highly integrated electronics together with Japanese money has solved the 10^6 TFT problem. One might expect that some imagination and a little less money and work could give the SBIND its chance as a new class of liquid crystal display.

15.5 CONCLUSION

Switching between bistable nematic textures is appealing to realize multiplexed matrix displays. For practical systems, one needs to use uniform switching to obtain short response times. This is possible with "volume" bistable systems: the Berreman–Heffner display, for instance, uses the continuous transition between a twisted and a nontwisted texture, when the twist axis is along the molecular axis. Its drawback is that the two states involved are not *really* bistable, and that the switching control depends on the electric field dynamics. The "bulk texture bistable nematic display" really is bistable, but cannot be controlled uniformly. The "boundary layer" display, where the distorted texture interacts with the surface is terribly complex and not very attractive. All the other "volume" bistable effects, presented here for the sake of completeness, present the same difficulties of control and speed.

The more recent "surface" bistable systems now seem more interesting. They can be controlled through uniform switching, at lower fields than the "volume" bistable systems. The reason for this is that it is easier to uniformly break a surface anchoring than it is to melt a bulk layer of nematic. They are intrinsically rapid; they bring attendant new academic questions on anchoring mechanisms. We believe that new progress in the future will lead to a better understanding of surface anchoring, and possibly to a new class of liquid crystal displays.

REFERENCES

1. Fergason, J. L. 1964. *Sci. Am.* **211**: 77.
2. Heilmeier, G. H., and Zanoni, L. A. 1968. *Appl. Phys. Lett.* **13**: 46.
3. Heilmeier, G. H., Barton, L. A., and Zanoni, L. A. 1968. *Appl. Phys. Lett.* **13**: 46.
4. Schadt, M., and Helfrich, W. 1971. *Appl. Phys. Lett.* **18**: 127.
5. de Gennes, P. G., and Prost, J. 1993. *The Physics of Liquid Crystals.* 2d ed. Oxford: Oxford University Press.
6. Williams, R. 1963. *J. Chem. Phys.* **39**: 384.
7. Heilmeier, G. H., and Goldmacher, J. E. 1968. *Appl. Phys. Lett.* **13**: 132.
8. Jákli, A. 1994. *Md. Cryst. Liq. Cryst.* **251**: 289
9. Soref, R. A. 1970. *J. Appl. Phys.* **41**: 3022.
10. Tani, C. 1971. *Appl. Phys. Lett.* **19**: 241.
11. Melchior, H., Kahn, F. J., Maydan, D., and Fraser, D. B. 1972. *Appl. Phys. Lett.* **8**: 392.
12. Kahn, F. J. 1973. *Appl. Phys. Lett.* **22**: 111.
13. Hareng, M., Le Berre, S., and Thirant, L. 1974. *Appl. Phys. Lett.* **25**: 683.
14. Kreuzer, M., Tschudi, T., and Eidenschink, R. 1992. *Mol. Cryst. Liq. Cryst.* **223**: 219.
15. Kreuzer, M., Tschudi, T., de Jeu, W. H., and Eidenschink, R. 1993. *Appl. Phys. Lett.* **62**: 1712.
16. Greubel, W. 1974. *Appl. Phys. Lett.* **25**: 5.
17. Leslie, P. M. 1975. *Pramana Suppl.* **1**: 41.
18. Porte, G., and Jadot, J. P. 1978. *J. Physique* **39**: 213.
19. Scheffer, T. J. 1978. *J. Appl. Phys.* **49**: 5835.
20. Miyaji, A., Yamaguchi, M., Toda, A., Mada, H., and Kobayashi, S. 1977. *IEEE Trans. Elec. Dev.* **ED-24**: 811.
21. Berreman, D. W., and Heffner, W. R. 1980. *Appl. Phys. Lett.* **37**: 109.
22. Berreman, D. W., and Heffner, W. R. 1981. *J. Appl. Phys.* **52**: 3032.
23. Scheffer, T. J., and Nehring, J. 1984. *Appl. Phys. Lett.* **45**: 1021 and van Sprang, H. A., Aartsen, R. G., and de Vaan, A. J. S. M. 1986. *J. Appl. Phys.* **59**: 3087.
24. Tanaka, T., Sato, Y., Inone, A., Momose, Y., Nomura, H., Iino, S., Asia Display '95 (1995) 259.
25. Boyd, G. D., Cheng, J., and Ngo, P. D. T. 1980. *Appl. Phys. Lett.* **36**: 556.
26. Cheng, J., and Boyd, G. D. 1980. *Appl. Phys. Lett.* **37**: 1072.
27. Cheng, J., Thurston, R. N., and Berreman, D. W. 1981. *J. Appl. Phys.* **52**: 2756.
28. Cheng, J. 1981. *J. Appl. Phys.* **52**: 724.
29. Boyd, G. D., Cheng, J., and Thurston, R. N. 1982. *Appl. Phys. Lett.* **40**: 936.
30. Cheng, J., Thurston, R. N., Boyd, G. D., and Meyer, R. B. 1982. *Appl. Phys. Lett.* **40**: 1007.
31. Meyer, R. B., and Thurston, R. N. 1983. *Appl. Phys. Lett.* **43**: 342.
32. Thurston, R. N., Cheng, J., Meyer, R. B., and Boyd, G. D. 1984. *J. Appl. Phys.* **56**: 263.
33. Thurston, R. N. 1984. *J. Appl. Phys.* **55**: 5154.
34. Thurston, R. N., Cheng, J., and Boyd, G. D. 1982. *J. Appl. Phys.* **53**: 4463.
35. Janning, J. L. 1972. *Appl. Phys. Lett.* **21**: 173.
36. Urbach, W., Boix, M., and Guyon, E. 1974. *Appl. Phys. Lett.* **25**: 479.
37. Monkade, M., Boix, M., and Durand, G. 1988. *Europhys. Lett.* **5**: 697.
38. Jérôme, B., Pieranski, P., and Boix, M. 1988. *Europhys. Lett.* **5**: 693.
39. Nobili, M., and Durand, G. 1992. *Phys. Rev. A* **46**: 6174 and Nobili, M., and Durand, G. 1994. *Europhys. Lett.* **25** 527.
40. Barberi, R., and Durand, G. 1991. *Liq. Cryst.* **10**: 289.

41. Barberi, R., Boix, M., and Durand, G. 1989. *Appl. Phys. Lett.* **55**: 2506.
42. Barberi, R., and Durand, G. 1991. *Appl. Phys. Lett.* **58**: 2907.
43. Barberi, R., Giocondo, M., and Durand, G. 1992. *Appl. Phys. Lett.* **60**: 1085.
44. Barberi, R., Giocondo, M., Martinot-Lagarde, P., and Durand, G. 1993. *Appl. Phys. Lett.* **62**: 3270.
45. Gharbi, A., Fekih, F. R., and Durand, G. 1992. *Liq. Cryst.* **12**: 515.
46. Sun, Z., Engels, J., Dozov, I., and Durand, G. 1994. *J. Physique II* **4**: 59.

INDEX